Imaging in Neuroscience

A LABORATORY MANUAL

ALSO FROM COLD SPRING HARBOR LABORATORY PRESS

IMAGING SERIES

Imaging: A Laboratory Manual
Imaging in Neuroscience: A Laboratory Manual
Imaging in Developmental Biology: A Laboratory Manual

RELATED LABORATORY MANUALS

Basic Methods in Microscopy: Protocols and Concepts from Cells: A Laboratory Manual
Drosophila *Neurobiology: A Laboratory Manual*
Live Cell Imaging: A Laboratory Manual, 2nd edition
Single-Molecule Techniques: A Laboratory Manual

HANDBOOKS

Experimental Design for Biologists
Lab Math: A Handbook of Measurements, Calculations, and Other Quantitative Skills for Use at the Bench
Lab Ref, Volume 1: *A Handbook of Recipes, Reagents, and Other Reference Tools for Use at the Bench*
Statistics at the Bench: A Step-by-Step Handbook for Biologists

WEBSITE

Cold Spring Harbor Protocols
http://www.cshprotocols.org/imaging

Imaging in Neuroscience

A LABORATORY MANUAL

EDITORS

Fritjof Helmchen
Brain Research Institute
University of Zurich

Arthur Konnerth
Institute for Neurosciences
Technical University Munich

SERIES EDITOR

Rafael Yuste
Howard Hughes Medical Institute
Columbia University

www.cshprotocols.org/imaging

COLD SPRING HARBOR LABORATORY PRESS
Cold Spring Harbor, New York • www.cshlpress.com

Imaging in Neuroscience: A Laboratory Manual

Publisher	John Inglis
Acquisition Editor	David Crotty
Director of Development, Marketing, & Sales	Jan Argentine
Managing Editor	Michael Zierler
Developmental Editors	Judy Cuddihy, Kevin Griffin, Kaaren Janssen, Virginia Peschke, Maria Smit, Catriona Simpson, Martin Winer, Michael Zierler
Project Managers	Mary Cozza and Inez Sialiano
Permissions Coordinator	Carol Brown
Production Editor	Kathleen Bubbeo
Desktop Editors	Lauren Heller and Susan Schaefer
Production Manager	Denise Weiss
Book Marketing Manager	Jane Carter
Sales Account Managers	Jane Carter and Elizabeth Powers
Cover Designer	Ed Atkeson

Front cover artwork: Multicolor pyramidal neurons in the hippocampus of a Brainbow mouse, in which cells express different fluorescent proteins in a combinatorial fashion. See Chapter 11 of this book and *Nature* **450:** 56–62 (2007) for more details on the Brainbow technique. (Image courtesy of Stéphane Fouquet and Jean Livet, Institut de la Vision, Inserm, France.)

Library of Congress Cataloging-in-Publication Data

Imaging in neuroscience : a laboratory manual / editors, Fritjof Helmchen,
Arthur Konnerth.
p. ; cm. -- (Cold Spring Harbor Laboratory Press' Imaging series)
Includes bibliographical references and index.
ISBN 978-0-87969-937-6 (hardcover : alk. paper) -- ISBN 978-0-87969-938-3
(pbk. : alk. paper)
1. Brain--Imaging--Laboratory manuals. 2. Nervous
system--Imaging--Laboratory manuals. 3. Neurosciences--Laboratory manuals.
I. Helmchen, Fritjof. II. Konnerth, Arthur. III. Series: Cold Spring Harbor
Laboratory Press' Imaging series.
[DNLM: 1. Diagnostic Imaging--methods--Laboratory Manuals. 2. Nervous
System--Laboratory Manuals. WL 25]

QP376.6.I43 2011
612.8'2--dc22

2010047031

Contents

SECTION 8 ▪ BRAIN PATHOLOGY

SECTION 9 ▪ APPENDICES

ACCOMPANYING MOVIES

Movies are freely available online at www.cshprotocols.org/imaging.

CHAPTER 2
Imaging Single Receptors with Quantum Dots

MOVIE 2.1. Diffusion and stabilization of single QD-GlyRs (green). Time-lapse recording (1200 images at 1 Hz; acquisition time, 75 msec). Synapses are labeled with FM4-64 (red).

CHAPTER 68
Imaging Microglia in Brain Slices and Slice Cultures

MOVIE 68.1. Migration of microglia (MG) cells in a rat hippocampal tissue slice following bath application of exogenous ADP (1 mM). The red fluorescence is BodipyTR-ADP.

MOVIE 68.2. Branch extension and subsequent migration of IB4 lectin-labeled MG in rat hippocampal tissue slices following application of exogenous ADP (1 mM).

CHAPTER 86
Two-Photon Imaging of Neuronal Structural Plasticity in Mice during and after Ischemia

MOVIE 86.1. Targeting individual brain arterioles for photoactivation of Rose Bengal. The movie has been recorded as epifluorescence with a charge-coupled device (CDD) camera on a microscope equipped with a 40×, 0.8-NA water-immersion lens over a time period of 120 sec and shows the blood vessel in sextuple time (movie plays in 20 sec) during targeting with 532-nm laser light. While the laser beam position is fixed (bright spot in the middle of the image), the specimen is being moved to target multiple sites within the blood vessel. Approximately 80 sec after start of photoactivation (≈14 sec in the movie), the blood flow appears to be blocked. See Figure 3 for single frames of this movie and further description. A small amount of background light was added to permit viewing of areas not subjected to photoactivation; this procedure was only used for creation of this movie and was never employed during an actual experiment. (Reprinted, with permission, from Sigler et al. 2008, ©Elsevier.)

CHAPTER 87
Two-Photon Imaging of Microglia in the Mouse Cortex In Vivo

MOVIE 87.1. Three-dimensional distribution of enhanced yellow fluorescent protein (eYFP)-expressing neurons (yellow) and enhanced green fluorescent protein (eGFP)-positive microglia (green) in the neocortex of a Thy1-eYFP x Cx3cr1-eGFP mouse implanted with an open skull window. The images are maximum-intensity side projections from a stack of fluorescence images. Individual focal planes were recorded in 2.5-μm steps starting from 800 μm depth below the pia mater to the cortical surface.

MOVIE 87.2. A time-lapse recording showing enhanced green fluorescent protein (eGFP)-expressing microglia (green) and enhanced yellow fluorescent protein (eYFP)-expressing Purkinje cells (red) in the cerebellar cortex of an anesthetized adult Thy1-eYFP x Cx3cr1-eGFP mouse. Images were recorded near a bend in the cerebellar folium, which allowed the molecular and Purkinje cell layers to be visualized within a single image. Microglial processes are highly motile. Neurons appear structurally stable during the same time interval. Elapsed time in minutes is shown in the *upper right* corner. Scale bar, 15 μm.

MOVIE 87.3. A video clip taken in the cerebellar cortex of an anesthetized mouse expressing enhanced green fluorescent protein (eGFP) in microglia. The video shows a microglial process at high magnification. Fine protrusions continually palpate the cellular environment. Elapsed time in minutes is indicated in the *upper right* corner. Scale bar, 5 μm.

MOVIE 87.4. This exemplary video shows phagocytosis of cellular debris by microglia. Fluorescence data were recorded in the cerebellar cortex of an adult Cx3cr1-eGFP mouse expressing enhanced green fluorescent protein (eGFP) in microglia (*left*) following mechanical injury to the cortex and topical application of the red fluorescent dye sulforhodamine 101 (SR101) (*center*). The *right* panel shows an overlay of the eGFP and SR101 signals. The bright red cell is an injured Purkinje neuron that has accumulated SR101. Time after mechanical injury is indicated in the *upper right* corner. Scale bar, 10 μm.

CHAPTER 88

Two-Photon Imaging of Immune Cells in Neural Tissue

MOVIE 88.1.The dynamic interaction between immune cells expressing tdRFP (red) and neural processes expressing eGFP (green) in an experimental autoimmune encephalomyelitis (EAE)-affected mouse. The visualization was performed by dual near-infrared/infrared (NIR/IR) excitation two-photon microscopy in the brain stem of anesthetized mice.

Preface to the Book Series

To train young people to grind lenses... . I cannot see there would be much use...because most students go there to make money out of science or to get a reputation in the learned world. But in lens-grinding and discovering things hidden from our sight, these count for nought.

—Antonie van Leeuwenhoek

Letter to Gottfried Leibniz on 28 September 1715 in response to Leibniz' request that he should open a school to train young people in microscopy

You can observe a lot just by watching. —Yogi Berra

ONE OF THE CENTRAL THEMES OF BIOLOGY IS the constant change and transformation of most biological systems. In fact, this dynamic aspect of biology is one of its most fascinating characteristics, and it draws generation after generation of students absorbed in understanding how an organism develops, how a cell functions, or how the brain works. This series of manuals covers imaging techniques in the life sciences—techniques that try to capture these dynamics. The application of optical and other visualization techniques to study living organisms constitutes a direct methodology to follow the form and the function of cells and tissues by generating two- or three-dimensional images of them and to document their dynamic nature over time. Although it seems natural to use light to study cells or tissues, and microscopists have been doing this with fixed preparations since van Leeuwenhoek's time, the imaging of living preparations has only recently become standard practice. It is not an overstatement to say that imaging technologies have revolutionized research in many areas of biology and medicine. In addition to advances in microscopy, such as differential interference contrast or the early introduction of video technology and digital cameras, the development of methods to culture cells, to keep tissue slices alive, and to maintain living preparations, even awake and behaving, on microscopes has opened new territories to biologists. The synthesis of novel fluorescent tracers, indicator dyes, and nanocrystals and the explosive development of fluorescent protein engineering, optogenetical constructs, and other optical actuators like caged compounds have made possible studies characterizing and manipulating the form and function of cells, tissues, and circuits with unprecedented detail, from the single-molecule level to that of an entire organism. A similar revolution has occurred on the optical design of microscopes. Originally, confocal microscopy became the state-of-the-art imaging approach because of its superb spatial resolution and three-dimensional sectioning capabilities; later, the development of two-photon excitation enabled fluorescence imaging of small structures in the midst of highly scattered living media, such as whole-animal preparations, with increased optical penetration and reduced photodamage. Other nonlinear optical techniques, such as second-harmonic generation and coherent anti-Stokes Raman scattering (CARS), now follow and appear well suited for measurements of voltage and biochemical events at interfaces such as plasma membranes. Finally, an entire generation of novel "superresolution" techniques, such as stimulated emission depletion (STED), photoactivated localization microscopy (PALM), and stochastic optical reconstruction microscopy (STORM), has arisen. These techniques have broken the diffraction limit barrier and have enabled the direct visualization of the dynamics of submicroscopic particles and individual molecules. On the other side of the scale, light-sheet illumination techniques allow the investigator to capture the development of an entire organism, one cell at a time. Finally, in the field of medical imaging, magnetic resonance scanning techniques have provided detailed images of the structure of the living human body and the activity of the brain.

This series of manuals originated in the Cold Spring Harbor Laboratory course on Imaging Structure and Function of the Nervous System, taught continuously since 1991. Since its inception, the course quickly became a "watering hole" for the imaging community and especially for neuroscientists and cellular and developmental neurobiologists, who are traditionally always open to microscopy approaches. The original manual, published in 2000, sprang from the course and focused solely on neuroscience, and its good reception, together with rapid advances in imaging techniques, led to a second edition of the manual in 2005. At the same time, the increased blurring between neuroscience and developmental biology made it necessary to encompass both disciplines, so the original structure of the manual was revised, and many new chapters were added. But even this second edition felt quickly dated in this exploding field. More and more techniques have been developed, requiring another update of the manual, too unwieldy now for a single volume. This is the reasoning behind this new series of manuals, which feature new editors and a significant number of new methods. The material has been split into several volumes, thus allowing a greater depth of coverage. The first book, *Imaging: A Laboratory Manual*, is a background text focused on general microscopy techniques and with some basic theoretical principles, covering techniques that are widely applicable in many fields of biology and also some specialized techniques that have the potential to greatly expand the future horizon of this field. A second manual, *Imaging in Neuroscience: A Laboratory Manual*, keeps the original focus on nervous system imaging from the Cold Spring Harbor Imaging course. A third volume, *Imaging in Developmental Biology: A Laboratory Manual*, now solely deals with developmental biology, covering imaging modalities particularly suited to follow developmental events. There are plans to expand the series into ultrastructural techniques and medical-style imaging, such as functional magnetic resonance imaging (fMRI) or positron emission tomography (PET), so more volumes will hopefully follow these initial three, which cover mostly optical-based approaches.

Like its predecessors, these manuals are not microscopy textbooks. Although the basics are covered, I refer readers interested in a comprehensive treatment of light microscopy to many of the excellent texts published in the last decades. The targeted audience of this series includes students and researchers interested in imaging in neuroscience or developmental or cell biology. Like other CSHL manuals, the aim has been to publish manuals that investigators can have and consult at their setup or bench. Thus, the general philosophy has been to keep the theory to the fundamentals and concentrate instead on passing along the little tidbits of technical knowledge that make a particular technique or an experiment work and that are normally left out of the methods sections of scientific articles.

This series of manuals has only been possible because of the work and effort of many people. First, I thank Sue Hockfield, Terri Grodzicker, Bruce Stillman, and Jim Watson, who conceived and supported the Imaging course over the years and planted the seed blossoming now in these manuals and, more importantly, in the science that has spun out of this field. In addition, the staff at CSHL Press has been exceptional in all respects, with special gratitude to John Inglis, responsible for an excellent team with broad vision, and David Crotty, who generated the ideas and enthusiasm behind this new series. Also, Inez Sialiano, Mary Cozza, Michael Zierler, Kaaren Janssen, Catriona

Simpson, Virginia Peschke, Judy Cuddihy, Martin Winer, Kevin Griffin, Kathleen Bubbeo, Lauren Heller, Susan Schaefer, Jan Argentine, and Denise Weiss worked very hard, providing fuel to the fire to keep these books moving, and edited them with speed, precision, and intelligence. More than anyone, they are the people responsible for their timely publication. Finally, I honor the authors of the chapters in these books, many of them themselves past instructors of the CSH Imaging course and of similar imaging courses at institutions throughout the world. Teaching these courses is a selfless effort that benefits the field as a whole, and these manuals, reflecting the volunteer efforts of hundreds of researchers, who not only have taken the time to write down their technical knowledge but have agreed to generously share it with the rest of the world, are a beautiful example of such community cooperation. As Leibniz foresaw, "lens grinding" is a profession that is indeed meaningful and needs the training of young people.

RAFAEL YUSTE

Preface

DIRECTLY SEEING THE NERVOUS SYSTEM IN ACTION—be it a vesicle releasing transmitter, a neuron integrating synaptic input in its dendrites, or a neuronal population generating patterns of activity—is always a fascinating experience and provides us with a sense of immediate and credible understanding. In recent years, our ability to directly observe neural events at various spatial and temporal scales has enormously expanded because of the development of new imaging technologies as well as novel functional indicators. In this book we try to capture a snapshot of these developments, providing an overview of the currently applied imaging approaches for visualization of neural dynamics. As the field is continually and rapidly developing, our collection of chapters represents, however, an intermediate report at most. It is safe to predict that soon further innovations will permit even more detailed insights into the dynamic organization of the nervous system.

This manual is part of the series of laboratory manuals that has now emerged from the previous laboratory manual *Imaging in Neuroscience and Development*. In the spirit of the whole series the chapters provide short overviews of specific methods and applications as well as step-by-step experimental protocols with many practical tips. *Imaging in Neuroscience* is centered on the original neuroscience focus of the previous editions, and many of the core chapters were maintained and updated. Nonetheless, the volume has undergone substantial changes. Some of the old chapters were merged in revised form into the accompanying *Imaging: A Laboratory Manual* and *Imaging in Developmental Biology: A Laboratory Manual*, which provide details about microscopy techniques and applications in developmental biology, respectively. Most notably, more than 50 entirely new chapters have been included in *Imaging in Neuroscience* in addition to the revised chapters (making up more than half of the book). This large expansion clearly reflects the rapid progress of imaging applications in neuroscience during the past 5 years. For example, in complement to biosensors for reading out neural activity, new classes of "bio-actuators" (e.g., Channelrhodopsin-2) have entered the scene that now enable precise and specific control of neurons with light, thus opening a wide field of applications. Similar great advances have occurred in the area of in vivo imaging of neural activity. Dynamic properties of various cell types (including glia) are now studied in living animals with unprecedented detail, often taking advantage of genetic means for cell identification or labeling. In fascinating work, high-resolution imaging studies have been extended to awake animals, directly linking cellular and network events to behavior. Moreover, the use of animal models of brain diseases has expanded the application of these imaging techniques, thereby helping to identify key cellular processes that occur in brain pathologies.

We have divided the manual into eight sections that are roughly ordered according to the spatial level of neural organization: from molecules to synapses, cells, networks and brain areas. Section 1 starts by introducing molecular tools for labeling receptors and cells and for optical readout and control of neural activity. These tools are applied, for example, to study mechanisms of axonal propagation of excitation and synaptic transmission (Section 2). Section 3 moves to the cellular level, focusing in particular on the integration of synaptic inputs in neuronal dendrites. A further integrative level of neural organization is reached in Sections 4 and 5, where functional properties of neurons and neural circuits are investigated in brain slice preparations (in vitro) and in living animals (in vivo), respectively. Section 6 adds chapters on studies of the dynamic properties of various glial cells, a currently expanding field of investigation. Many new chapters are also found in Section 7, which highlights the emerging approaches of performing high-resolution imaging studies during behavior. Finally, Section 8 provides examples of imaging studies in various animal models of disease. The manual ends with a series of appendices, including a glossary of imaging terms, useful information on spectra, lenses, and filters, and instructions for handling imaging hardware safely.

We enjoy remembering the wonderful time we had discussing the books' concept and content with James Sharpe, Rachel Wong, and Rafael Yuste. Rafa's enthusiasm and drive in putting together this book series were just exceptional. Special thanks go to the entire editorial and production team at CSHL Press for a superb job on finalizing this manual, in particular to David Crotty, Kathleen Bubbeo, Mary Cozza, and Inez Sialiano. We are also thankful to Benjamin Grewe, Anja Gundlfinger, Henry Lütcke, David Margolis, José María Mateos, Morgane Roth, and Marcel van 't Hoff for help with proofreading. Last but not least, we would like to thank our families for their support, patience, and love.

FRITJOF HELMCHEN
ARTHUR KONNERTH

1 Tracking Receptors Using Individual Fluorescent and Nonfluorescent Nanolabels

Laurent Cognet,[1] Brahim Lounis,[1] and Daniel Choquet[2]

[1]Centre de Physique Moléculaire Optique et Hertzienne—CNRS UMR 5798 et Université Bordeaux 1, 33405 Talence, France ; [2]Laboratoire de Physiologie Cellulaire de la Synapse–CNRS UMR 5091 et Université Bordeaux 2, 33077 Bordeaux, France

ABSTRACT

This chapter describes imaging and tracking techniques using nanometer-sized optical labels for the study of the movement of individual or small assemblies of membrane proteins. These labels include fluorescent dyes, luminescent nanocrystals, and absorbing metallic nanoparticles. Different imaging modalities are used to detect and track these labels. Organic dyes and semiconductor quantum dots (QDs) are tracked by standard epifluorescence single-molecule microscopy. Gold nanoparticles are detected by the photothermal heterodyne imaging method (PHI) and tracked using a triangulation scheme.

INTRODUCTION

Experimental results from the last two decades have changed the understanding of the plasma membrane, originally viewed as a homogeneous fluid in which molecules are evenly dispersed (Lommerse 2004). Nowadays the notion has emerged of the membrane being "more a mosaic than a fluid" (Kusumi 2005). Whereas some molecules seem to be randomly distributed, others show a precise compartmentalization that is commonly related to their functional properties.

One of the characteristics of these membrane domains is their submicrometer size, typically close to or below the optical diffraction limit. This makes their study difficult by conventional microscopy. Moreover, these compartments are usually dynamic in size and composition as their component molecules can continuously enter and exit by diffusion. Real-time, high-resolution, live-imaging methods rather than static imaging are thus required to reflect the real behavior of membrane molecules. Fluorescence recovery after photobleaching has been extensively used to study molecule diffusion, but this method yields only average measurements of populations of molecules. In contrast, single-molecule techniques fulfill these requirements, as they provide information about the dynamics of molecules, together with nanometer resolution to study their distribution.

Single-molecule tracking (SMT), with the use of organic dyes and semiconductor QDs, and single-particle tracking (SPT), with the use of gold nanoparticles, allow one to study the diffusion of individual molecules, their sorting into different cellular compartments, and their interactions with other molecules. The main advantage of these approaches is that they eliminate the implicit averaging of conventional optical observations, providing access to heterogeneity and dynamical fluctuations at the molecular level (Saxton and Jacobson 1997; Cognet et al. 2002). Although the advantage of SPT is the possibility to follow the movements of molecules for long periods of time (minutes), the size of the particle may change the normal distribution of the labeled molecules in membrane compartments such as the synapse. In this case, the size of the synaptic cleft could induce steric restrictions. SMT, on the other hand, has the advantage of smaller label size, but it only allows tracking for short time periods (seconds). Both methods are especially suitable for studies in neurons, as the signal-to-noise ratio is optimal in neurites because of their low autofluorescence.

In this chapter, we show the different solutions and methods that we have chosen to address this kind of study. We describe the procedures to detect and automatically localize signals from single molecules as well as the tracking methods suitable for each kind of labeling using small, nanometer-sized optical labels for the study of the movement of individual or small assemblies of membrane proteins. Single dyes are detected by epifluorescence (Moerner and Orrit 1999; Harms et al. 2001; Groc et al. 2007; Triller and Choquet 2008). Gold nanoparticles are detected by the photothermal heterodyne imaging (PHI) method (Berciaud et al. 2004) and tracked using a triangulation scheme (single-nanoparticle photothermal tracking; SNaPT) (Lasne et al. 2006) (see Procedure C, SNaPT Experiments for a brief description of the triangulation scheme). PHI combines a time-modulated heating beam and a nonresonant probe beam, overlapping on the sample. The probe beam produces a frequency-shifted scattered field as it interacts with the time-modulated variations of the refractive index around the absorbing nanoparticle. The scattered field is then detected through its beat note with the probe field, the oscillation of the optical intensity arising from the superposition of the excitation and the scattered fields, which plays the role of a local oscillator as in any heterodyne technique. The complementary SMT, quantum dot tracking (QDT), and SNaPT approaches have been applied to track neurotransmitter receptors or adhesion proteins into confined area of the plasma membrane such as the synaptic cleft. More generally, such approaches allow us to determine the degree of interaction between membrane proteins and cytoplasmic or extracellular stabilizing elements in live cells. They have been applied to a variety of cell culture preparations such as cell lines and primary hippocampal or spinal cord cultures (see Triller and Choquet 2008).

Protocol

Labeling and Imaging Receptors Using Nanolabels

This protocol starts with the attachment of the label to the receptor of interest through a specific high-affinity ligand (a natural ligand, an antibody, or a toxin) that recognizes the extracellular domain of the receptor in live cells. This must be performed at low enough label density to ensure that signals from single molecules or particles are spatially isolated. A set of criteria (see Procedure B below) ensures that single labels are imaged. The use of monovalent labels (toxins or Fab fragments of antibodies) or low coating densities of ligands on particles helps to ensure that labels are bound to single receptors. Imaging of the label, reflecting the behavior or properties of the underlying receptor, is then followed by image analysis. One should keep in mind that it is usually impossible to know whether the label reveals the movement of individual or clustered receptors; indeed, one can be following the movement of a single label bound to an individual receptor, but that receptor can be part of a cluster where the other members are not labeled, and whose presence can thus not be revealed solely from the properties of the movement of the label.

IMAGING SETUP

Single-Molecule Epifluorescence Microscope

This protocol was optimized using an inverted microscope equipped with a 1.4 numerical-aperture (NA) 100× oil-immersion objective (Olympus). For organic dye detection, a sensitive and rapid camera is used (e.g., Cascade back-illuminated electron multiplying charge-coupled device [CCD], Roper Instruments). The fluorescence excitation sources are a CW 488 Semiconductor Laser (Coherent, Inc.), a frequency doubled neodymium-doped yttrium aluminum garnet (Nd:YAG) laser (Coherent, Inc.), or He:Ne lasers (JDS Uniphase). Excitation laser beams enter through the fluorescence epiport of the microscope and illuminate an area of 10 μm^2 of the sample by focusing the beam in the back aperture of the objective. Illumination intensities are of a few kW/cm^2. For each dye, an appropriate set of filters (Chroma Technology or Semrock) is required. The total detection efficiency of a typical experimental SMT setup is in the range of 5%–10%. For QD detection, excitation can be performed with either of the lasers described above, or more simply with a mercury lamp, owing to the large absorption cross section of QDs.

Photothermal Imaging Microscope

A nonresonant probe beam (HeNe, 632.8 nm) and an exciting beam absorbed by the gold nanoparticle (532 nm, frequency-doubled Nd:YAG laser) are overlaid and focused on the sample by means of a high-NA microscope objective (100×, NA = 1.4). The intensity of the exciting beam is modulated at a frequency Ω (typically $\Omega/2\pi$ = 700 kHz) by an acousto-optic modulator. A second microscope objective (80×, NA = 0.8) collects the interfering probe-transmitted and forward-scattered fields. The intensity of the exciting beam sent on the samples is a few hundreds of kW/cm^2, depending on the nanoparticle size. The forward interfering fields are collected on a fast photodiode and fed into a lock-in amplifier to extract the beat signal at a frequency Ω. Integration times of 5 msec are used. Images and tracking are performed by moving the sample over the fixed laser spots by means of a two-dimensional piezoscanner.

MATERIALS

See the end of the chapter for recipes for reagents marked with <R>.

Reagents

Antireceptor antibody
Bovine serum albumen (BSA) (for SMT method)
Cells for labeling (e.g., COS-7 cells, fibroblasts, primary cultured neurons)
Culture medium

For QDT, use conditioned modified Eagle's medium (MEM) without phenol red and with addition of HEPES buffer (10 mM).

HEPES buffer
Labels

For a summary of the available labels and examples of their applications, please see Table 1.

For SMT: Fluorescent labels for single fluorophore detection (SMT) are either semiconductor QDs (~20 nm total diameter; see below) or organic dyes such as cyanine dyes (Cy3 or Cy5, Amersham Corp.) or Alexa dyes (Invitrogen).

For QDT: Qdot 655 goat F(ab')2 anti-rabbit IgG conjugate (H+L), highly cross-adsorbed (Invitrogen).

For SNaPT (Cognet et al. 2003; Lasne et al. 2006): Particles are gold colloids down to a few nanometers in diameter (5-nm gold conjugate goat F(ab')2 anti-mouse IgG; BBInternational).

Phosphate-buffered saline (PBS) <R> (for QD method)

Equipment

Coverslips

For QD, SNaPT, and SMT: Culture cells on glass coverslips. Coverslips must be acid-washed (1 M HCl for 15 min) before use.

Culture chamber, sealed, prewarmed to 37°C
Imaging setup, as described above

When performing SMT, caution should be taken to use low-autofluorescence immersion oils (FF type, Cargille Labs).

EXPERIMENTAL METHOD

Procedure A: Labeling Cells

Prepare labeled cells using SMT (Steps 1–2), QDT (Steps 3–6), or SNaPT (Steps 7–10).

TABLE 1. Labels and areas of application for single-particle tracking and single-molecule detection

Labels	Gold nanoparticles	Quantum dots	Fluorescent proteins	Dyes
Typical applications	Diffusion, stoichiometry	Diffusion, localization	Localization, stoichiometry, interactions	Diffusion, localization, interactions
Size	5–10 nm	10–25 nm	2 × 4 nm	<1 nm
Coupling	Passive absorption, chemical linkage	Chemical linkage	Genetic fusion	Chemical linkage
Excitation	Laser light	Laser light or spectral lamps	Laser light	Laser light
Detection	Fast photodiode	Low-light-level camera	Low-light-level camera	Low-light-level camera
Length of observation	>20 min	Few minutes	Seconds	Seconds
Precision of detection	5–20 nm	10–40 nm	20–50 nm	20–50 nm

SMT

1. Link dyes to the antibody at a mean 1:1 (or lower) ratio by an *N*-hydroxysuccinimide-ester chemical reaction to free amines on the antibody.
2. To label receptors, incubate live cells for a few minutes with the fluorinated antibody (0.1–1 μg/mL).

   Ligands must be sufficiently dilute that the signal arising from a molecule is optically resolved (typically 1 molecule/μm^3, ~1 nM) and also that the signal from a single molecule is above that of all background sources (e.g., noise from the detector, autofluorescence from the sample).

QDT

Receptors are immunolabeled by using QD precoated with the antireceptor antibody.

3. Incubate QDs (0.1 μM) with 1 μg antibody in 10 μL PBS for 30 min.
4. Block nonspecific binding by adding casein to the pre-coated QD 15 min before use (with equal volume of a 10% casein solution; Vector Labs).
5. Incubate neurons 10 min at 37°C in culture medium with precoated QD (final dilution 0.1 nM).
6. Perform four washes in culture medium, 30 sec each.

SNaPT Labeling

7. Incubate cells 3 min at room temperature with 10 μg/mL antireceptor antibody.
8. Rinse twice in culture medium, then perform secondary immunostaining by F(ab)-5nm gold conjugates (a 5-nm gold conjugated with an Fab directed against the specie of the antireceptor antibody used). Incubate 3 min at room temperature, in 0.3% BSA at a concentration of 300 ng/mL.
9. Perform three fast rinses in culture medium.
10. Mount the coverslips in a recording chamber (e.g., a Ludin chamber) with culture medium supplemented with 20 mM HEPES.

Procedure B: Signal Characterization

Follow the guidelines below for the chosen labeling method.

SMT

In SMT experiments, it is important to thoroughly characterize the signal arising from the label-ligand couple immobilized on glass or embedded in a gel or polymers before performing experiments in cells. The signal arising from a single fluorophore should show digital photobleaching: It must be constant over time, and then drop instantaneously to background levels. All molecules should have comparable fluorescence levels. The signal arising from an immobile single molecule should display a diffraction-limited spot.

QDT

For QDs, use blinking between fluorescent and dark states to ascertain the imaging of individual particles. In cells, the signal characteristics should be similar.

SNaPT

In SNaPT experiments, photothermal signals are directly proportional to the nanoparticle's volume, allowing the unambiguous discrimination of individual particles over clusters.

Procedure C: Image Acquisition, Particle Tracking, and Analysis

Perform experiments on labeled cells at 37°C in HEPES buffered culture medium in a sealed chamber to prevent evaporation. Use cells for no more than 30 min and discard them afterward.

Follow the guidelines below for the chosen labeling method.

QDT and SMT Experiments

In fluorescence experiments for QD and SMT, acquire images at fast rates (5–30 Hz). For SMT, a fast shutter in the excitation path prevents sample illumination in between frames to limit dye photobleaching. For image analysis, perform particle or single-dye identification with either custom-made or commercial software (e.g., Metamorph, MDL). Objects are fitted by comparison with reference signals. This allows for subwavelength localization of the labels, typically ~20–40 nm. Trajectories are then reconstructed on image stacks by connecting object positions from one image to the next. Use low labeling densities to facilitate this process.

SNaPT Experiments

In SNaPT experiments, the spatial profile of the PHI signal from an individual gold particle is given by the product of the intensity profiles of the exciting and probe beams (Berciaud et al. 2006). It is well approximated by a Gaussian profile with a constant width and can be defined by three parameters: the central position (x_0, y_0), and the peak signal S_0. By measuring the signal at three well-defined positions around the tentative location of a nanoparticle, both its precise location in space (x, y) and its peak signal S can be unequivocally retrieved.

To do this in practice, first acquire a white light and/or an epifluorescence image, then record a PHI image to check gold-labeling density and specificity. Next, choose a region of interest and the imaging software will take PHI signals at random positions. When a signal above a predefined threshold is obtained, the software takes three data points around this position at the apices of an equilateral triangle and then calculates a first set of particle coordinates and peak signal (x_t, y_t, S_t). If S_t and the three measured signals are above a second predefined threshold, the software repeats this procedure iteratively by recentering the equilateral triangle on (x_t, y_t) for the next three measurements which will give ($x_{t+\Delta t}$, $y_{t+\Delta t}$, $S_{t+\Delta t}$), and thus the track of the moving particle.

DISCUSSION

Example of Application

Figure 1 shows examples of traces recorded with SMT, QDT, and SNaPT used to detect AMPA-subtype glutamate receptor (AMPAR) movements in live cultured hippocampal neurons. AMPARs have been labeled with an antibody directed against the extracellular amino terminus of the glutamate receptor 2 (GluR2) subunit (Chemicon). Short-term observation of receptor diffusion inside synapses was observed by tracking cyanin dye–coupled antibody attached to the receptors (Fig. 1A,B), while receptor diffusion in and out of synapses could be observed by tracking QD-anti-GluR2-labeled receptors (Fig. 1C,D) or 5-nm gold anti-GluR2 labeled receptors (Fig. 1E,F). These experiments reveal that AMPARs diffuse in both the synaptic and extrasynaptic membranes.

Advantages and Limitations

The main advantage of particle- and molecule-tracking approaches is that they eliminate the implicit averaging of conventional optical observations, enabling observation of heterogeneity, dynamic fluctuations, diffusion, reorientation, colocalization, and conformational changes at the molecular level. Because labels can be localized at resolutions well below the diffraction limit, diffusion coefficients can be measured over a wide dynamic range (Saxton and Jacobson 1997; Cognet et al. 2002).

SNaPT, QDT, and SMT have complementary properties. SNaPT and QDT allow for significantly longer observation times than SMT. QDT is the simplest technique to implement; however, the size of the labeling complex may hinder the diffusional properties of the labeled receptors in specific neuronal areas such as the synaptic cleft (Groc et al. 2007). On the other hand, SNaPT presents the advantage of virtually indefinite tracking owing to the absence of bleaching or blinking of gold particles, but it can only be performed on one particle at a time. In contrast, fluorescent dyes and nanoparticles blink and bleach, limiting the observation time, but can be tracked many at a time. Finally, in SMT, fluorescent dyes are short lived, but the small size of the fluorescent ligand is the best suited to reveal receptor dynamics in constrained areas such as synapses (Tardin et al. 2003).

An important limitation of all these techniques is the difficulty in distinguishing the labels moving at the surface of the plasma membrane from those internalized in the cell. This means that rates of receptor endocytosis must be controlled during the experiments. So far, experiments have only been performed in cell culture systems. Future improvements in labeling complex brightness and size will likely allow for the use of these techniques in tissue slices.

RECIPE

CAUTION: See Appendix 6 for proper handling of materials marked with <!>.
Recipes for reagents marked with <R> are included in this list.

Phosphate-Buffered Saline (PBS)

Reagent	Quantity (for 1 L)	Final concentration (10x)
NaH_2PO_4	2.56 g	18.6 mM
Na_2HPO_4	11.94 g	84.1 mM
NaCl	102.2 g	1.75 M

Adjust the pH to 7.4 using NaOH <!> or HCl <!> as necessary.

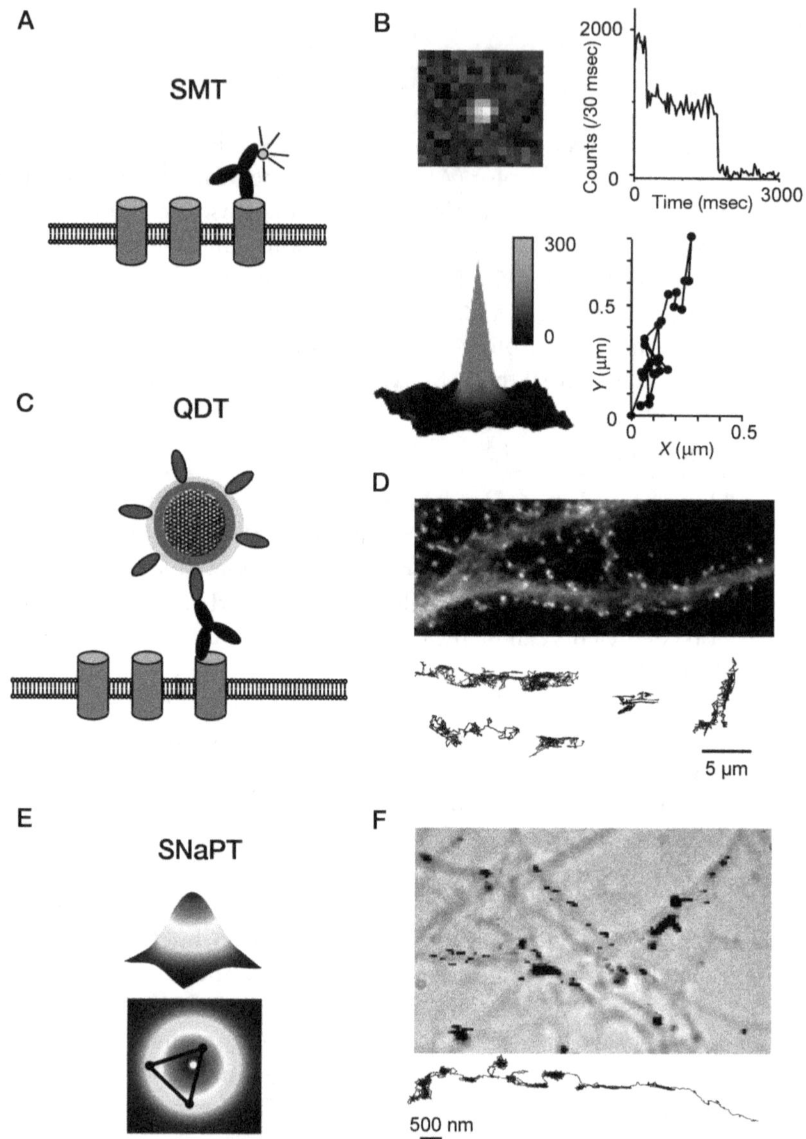

FIGURE 1. Imaging single AMPA receptor movements in neuronal membranes. (*A*) Schematic drawing of the attachment of a single fluorescently labeled antibody to a receptor, as used in single-molecule tracking (SMT) experiments. (*B*) (*Top left*) Image of the fluorescent spot from a single molecule. (*Top right*) Recording over time of the intensity of a fluorescent spot. The intensity displays characteristic sequential discrete photobleaching steps, indicating that the spot initially contained two individual fluorophores. (*Bottom left*) Display of the intensity profile (*x* and *y* are space; *z* is intensity) of a single fluorescent Cy5 molecule. The profile is diffraction limited and matches the point-spread function of the optical system. (*Bottom right*) Example of the trajectory of a single antibody-labeled AMPAR recorded on cultured hippocampal neurons. (*C*) Schematic drawing of the attachment of a quantum dot to receptors via antibodies directed against the extracellular amino terminus of the protein, as used in quantum dot tracking (QDT) experiments. (*D*) Image of cultured hippocampal neurons (20 d in vitro) expressing the synaptic marker Homer-GFP and of sample trajectories on these neurites of QD-tagged surface AMPARs. (*E*) Schematic drawing of the triangulation method used to determine the position of gold particles in SNaPT. (*F*) (*Top*) Superimposed bright-field and photothermal heterodyne imaging (PHI) method images of a 20 × 20-μm portion of live cultured hippocampal neurons with GluR2 receptors labeled with 5-nm gold nanoparticles (AuNPs). The PHI image (black spots and stripes) reveals the presence of pointlike signals that correspond to immobile GluR2-linked AuNPs. Most nanoparticles move during the raster scan of the sample, which produces characteristic stripe signals in the superimposed image of cultured hippocampal neurons (14 d in vitro) and of the photothermal signal. The recording lasted 3 min, during which the bead-bound receptors explored an area of 5 μm^2. (*Bottom*) Trajectory of an individual receptor labeled with a 5-nm AuNP (>5 min, 9158 data points) acquired by SNaPT at video rate on a live neuron.

ACKNOWLEDGMENTS

This work has been supported by grants from the Centre National de la Recherche Scientifique (CNRS), the Council of the Région Aquitaine, and the Agence National pour la Recherche.

REFERENCES

Berciaud S, Cognet L, Blab GA, Lounis B. 2004. Photothermal heterodyne imaging of individual nonfluorescent nanoclusters and nanocrystals. *Phys Rev Lett* **93:** 257402.

Berciaud S, Lasne D, Blab GA, Cognet L, Lounis B. 2006. Photothermal heterodyne imaging of individual metallic nanoparticles: Theory versus experiments. *Phys Rev B* **73:** 045424.

Cognet L, Coussen F, Choquet D, Lounis B. 2002. Fluorescence microscopy of single autofluorescent proteins for cellular biology. *C R Phys* **3:** 645–656.

Cognet L, Tardin C, Boyer D, Choquet D, Tamarat P, Lounis B. 2003. Single metallic nanoparticle imaging for protein detection in cells. *Proc Natl Acad Sci* **100:** 11350–11355.

Groc L, Lafourcade M, Heine M, Renner M, Racine V, Sibarita JB, Lounis B, Choquet D, Cognet L. 2007. Surface trafficking of neurotransmitter receptor: comparison between single-molecule/quantum dot strategies. *J Neurosci* **27:** 12433–12437.

Harms GS, Cognet L, Lommerse PH, Blab GA, Schmidt T. 2001. Autofluorescent proteins in single-molecule research: Applications to live cell imaging microscopy. *Biophys J* **80:** 2396–2408.

Kusumi A, Nakada C, Ritchie K, Murase K, Suzuki K, Murakoshi H, Kasai RS, Kondo J, Fujiwara T. 2005. Paradigm shift of the plasma membrane concept from the two-dimensional continuum fluid to the partitioned fluid: High-speed single-molecule tracking of membrane molecules. *Annu Rev Biophys Biomol Struct* **34:** 351–378.

Lasne D, Blab GA, Berciaud S, Heine M, Groc L, Choquet D, Cognet L, Lounis B. 2006. Single nanoparticle photothermal tracking (SNaPT) of 5-nm gold beads in live cells. *Biophys J* **91:** 4598–4604.

Lommerse PH, Blab GA, Cognet L, Harms GS, Snaar-Jagalska BE, Spaink HP, Schmidt T. 2004. Single-molecule imaging of the H-ras membrane-anchor reveals domains in the cytoplasmic leaflet of the cell membrane. *Biophys J* **86:** 609–616.

Moerner WE, Orrit M. 1999. Illuminating single molecules in condensed matter. *Science* **283:** 1670–1676.

Saxton MJ, Jacobson K. 1997. Single-particle tracking: Applications to membrane dynamics. *Annu Rev Biophys Biomol Struct* **26:** 373–399.

Tardin C, Cognet L, Bats C, Lounis B, Choquet D. 2003. Direct imaging of lateral movements of AMPA receptors inside synapses. *EMBO J* **22:** 4656–4665.

Triller A, Choquet D. 2008. New concepts in synaptic biology derived from single-molecule imaging. *Neuron* **59:** 359–374.

2 Imaging Single Receptors with Quantum Dots

Sabine Lévi,[1] Maxime Dahan,[2] and Antoine Triller[3]

[1]*INSERM UMR-839, IFM, Paris, France;* [2]*Laboratoire Kastler Brossel, CNRS UMR8552, ENS, Paris, France;* [3]*INSERM U497, ENS, Paris, France*

ABSTRACT

This chapter describes a highly sensitive approach for tracking the motion of membrane molecules over extended time periods with single-molecule resolution. This technique uses nanometer-sized quantum dots linked to the extracellular part of the proteins to be followed. Single-fluorophore epifluorescence imaging then reveals the membrane diffusion of the particle of interest.

INTRODUCTION

Quantum dot staining is easy, fast, and nontoxic and has low background fluorescence. Quantum dots (QDs) come in many colors and wide absorption spectra, yet have narrow emission spectra, facilitating multicolor detection. Moreover, QDs, which are ~10–20 nm in diameter, intermediate between latex beads (500 nm) and conventional fluorophores (~1–4 nm), can still access confined cellular domains such as the synaptic cleft. The photostability of QDs significantly extends the recording period duration (see the section, "Using Quantum Dots to Measure GlyR Diffusion"). Another key feature of QDs is their strong fluorescence. With an integration time of 75 msec, the spots are detected with a signal-to-noise ratio of ~50 (an order of magnitude more than that of traditional fluorophores). As a result, the lateral resolution with which individual QD spots can be localized reaches 5–10 nm, well below the 40 nm achieved with Cy3 (Dahan et al. 2003). Furthermore, QDs are brighter than conventional fluorescent molecules such as Cy3 and therefore require less excitation light. They can be excited with a mercury lamp instead of a laser beam, which is required for single Cy3 detection. The laser beam excites only a small region of the field compared to the mercury lamp, which excites the entire field equally and therefore permits imaging of more QDs per field. Consequently, multiple individual fluorescent spots can be observed simultaneously. In addition, the same QD probe can be used both at the optical and the electron microscope levels.

There are two main limitations to QD staining. First, QD blinking makes the analysis of long trajectories more complex and time-consuming, requiring sophisticated software. Second, particular attention should be given to the size of the QDs and the control of stoichiometries.

QD staining, as described in this chapter, has been used successfully in cultured neurons to follow the membrane diffusion of individual glycine receptors (Dahan et al. 2003; Charrier et al. 2006; Calamai et al. 2009); type A GABA receptors (Lévi et al. 2008; Bannai et al. 2009); AMPA receptors (Ehlers et al. 2007; Groc et al. 2007, 2008; Heine et al. 2008; Frischknecht et al. 2009); NMDA receptors (Groc et al. 2006; Michaluk et al. 2009); cannabinoid receptors (Mikasova et al. 2008); and lipid raft markers, glycophosphatidylinositol-anchored green fluorescent protein (GPI–GFP), and cholera toxin (Renner et al. 2009).

Protocol

Labeling Neuronal Membrane Receptors with Quantum Dots

Two methods are presented for labeling neurons with the primary antibody of choice along with a secondary anti-Fab antibody that is either biotinylated or directly coupled to the desired quantum dot. The behavior of QD-labeled molecules can then be followed within the cell using epifluorescence imaging.

IMAGING SETUP

This procedure used an inverted microscope (Olympus, IX70 or 71) equipped with a 60× objective (numerical aperture [NA] = 1.45; Olympus). QD-605 and FM 4-64 were detected using a mercury arc lamp (excitation filter 525DF45) and appropriate emission filters (595DF60 and 695AF55 from Omega Filters). For Cy3 excitation, to achieve single-dye detection, the sample was illuminated with a frequency-doubled YAG crystal laser at 532 nm (~0.5 kW/cm^2). For detection of single QDs, the sample was illuminated with a mercury lamp. Real-time QD and Cy3 recordings were obtained at 13 Hz and 10 Hz, respectively, using a CCD camera (Micromax 512EBFT, Cascade+128, Roper Scientific or ORCA II ER, Hamamatsu Photonics) with up to approximately 1000 consecutive frames under Metaview (Universal Imaging). Single-molecule trajectories were analyzed with custom-written routines (Bonneau et al. 2005) in MatLab (MathWorks). Similar algorithms are freely available (Jaqaman et al. 2008; Sergé et al. 2008).

MATERIALS

CAUTION: See Appendix 6 for proper handling of materials marked with <!>.
See the end of the chapter for recipes for reagents marked with <R>.

Reagents

Antibody, primary
Biotinylated secondary Fab antibody (Jackson ImmunoResearch Laboratories) or
QD F(ab′)2 secondary IgG antibody (Invitrogen)
Casein (1x for labeling Method B)
FM4-64 (*N*-(3-triethylammoniumpropyl)-4-(6-(4-diethylamino)phenyl)hexatrienyl)pyridinium dibromide), 1 μM in 40 mM KCl <!>
MEM medium for quantum dot imaging <R>
Use this medium at 37°C for incubations, washes, and imaging.
Neurons for labeling
QD streptavidin conjugate solution (for labeling Method A) <R>

Equipment

Imaging setup, as described above

EXPERIMENTAL METHOD

1. Label neurons using one of the following two methods:

   Method A:

   i. Incubate the neurons for 5 min with a high dilution (~1–10 µg/mL) of primary antibody to label a small number of molecules.

   ii. Wash the neurons three times without waiting, and then incubate them for 5 min in ~10 µg/mL biotinylated secondary Fab antibody.

   iii. Incubate neurons for 1 min in ~0.5–2 nM QD streptavidin conjugate solution.

   Method B:

   i. Mix 5 nM of the primary antibody with 30 nM QD F(ab′)2 secondary IgG antibody for 30 min at room temperature and then for an additional 15 minutes with 1x casein to block nonspecific binding.

   ii. Incubate neurons for 10 min at 37°C with 0.06 nM of the pre-coupled QD-antibody.

   iii. Rinse off excess antibody.

   If sufficient primary antibody is available, the primary antibody can be bound directly to QD (IgG/QD ratio = 1:1) with the QD antibody conjugation kit (Invitrogen) in a fast (few hours) and specific (coupling of thiols to maleimide groups) manner.

2. Label presynaptic boutons for 30 sec with 1 µM FM4-64 in 40 mM KCl.

   See *Troubleshooting*.

3. Wash the cells, mount the coverslip in a recording chamber, and image the cells for up to 30 min in the imaging medium.

   In our experiments, we perform QD real-time imaging at 13 Hz for approximately 500 consecutive frames. Time-lapse recording is preferred in experiments intended to determine dwell times in compartments (e.g., synaptic) where the molecules might reside for long durations. We often capture a single image of the cell morphology with transmitted light and of the synaptic marker (FM4-64, Venus::gephyrin, or DsRed::Homer1c) with fluorescent light before the QD image sequence.

TROUBLESHOOTING

Problem (Step 2): The diffusion coefficients of the molecule of interest may be changed by neuronal activity stimulation.

Solution: Neuronal activity regulates the neurotransmitter receptor's lateral diffusion properties (Tardin et al. 2003; Groc et al. 2004; Ehlers et al. 2007; Lévi et al. 2008; Bannai et al. 2009). Therefore, the KCl-induced FM4-64 synaptic vesicle loading may change the diffusion coefficients of the molecule of interest. As an alternative, one can use markers that do not require activity stimulation to be integrated into synaptic boutons (e.g., Mitotrackers; Invitrogen). Another option is to lipotransfect neurons with the main components of the inhibitory and excitatory postsynaptic differentiations, such as Gephyrin and Homer1c recombinant proteins tagged with GFP variants.

DISCUSSION

Interpretation of Imaging Data

After imaging, the trajectory is reconstructed from the image sequence of a single QD identified by the intermittency in its fluorescence emission. QDs are classified as synaptic upon co-localization with a synaptic marker, such as glycine receptor (GlyR; see below) (e.g., see Fig. 2). For each QD, we calculate the mean square displacement (MSD), diffusion coefficient (*D*), confinement area, transition between compartments, and dwell time within a compartment. The MSD is determined from

$$\mathrm{MSD}(n\tau)=\frac{1}{N-n}\sum_{i=1}^{N-n}\left[\left(x((i+n)\tau)-x(i\tau)\right)^2+\left(y((i+n)\tau)-y(i\tau)\right)^2\right], \qquad (1)$$

where τ is the frame acquisition time, N is the total number of frames, and n and i are positive integers with n determining the time increment. For simple two-dimensional (2D) Brownian mobility, the MSD as a function of time is linear with a slope of $4D$. If the MSD as a function of time tends to a constant value L, the diffusion is confined in a domain of size L. The diffusion coefficient (D) is determined by a fit on the first four points of the MSD as a function of time, with $\mathrm{MSD}(n\tau) = 4Dn\tau + b$, where b is a constant reflecting the spot localization accuracy. The area in which diffusion is confined can be estimated by fitting the MSD as a function of time to

$$\mathrm{MSD}(n\tau)=\frac{L^2}{3}\left(1-\exp\left(-\frac{12Dn\tau}{L^2}\right)\right)+4D_{\mathrm{mac}}n\tau, \qquad (2)$$

where L^2 is the confined area in which diffusion is restricted and D_{mac} is the diffusion coefficient on a long time scale.

Using Quantum Dots to Measure GlyR Diffusion

GlyR lateral diffusion was studied in cultured neurons using a single-particle tracking (SPT) approach (Dahan et al. 2003). Data obtained with Cy3 fluorophore coupled directly to primary antibody (Fig. 1A) were compared with data obtained using QD-605-streptavidin conjugates (Fig. 1B,C). Cy3- and QD-tagged receptors were detected in extrasynaptic and synaptic regions. Individual Cy3 molecules were identified by their single-step bleaching. Cy3-GlyR could be tracked only for short durations (~2.5 sec in Fig. 1A, white arrow). In contrast, the photostability of QDs allowed QD-GlyR trajectories to be visualized (Fig. 1B,C) for unprecedented durations (20 min) (Fig. 1C; Movie 2.1). Single QDs were identified by the random intermittency of their fluorescence emission (Nirmal et al. 1996). For example, one QD (white arrow in Fig. 1B) temporarily disappeared after 31.5 sec of recording. Long imaging duration enabled the observation of exchanges between extrasynaptic and synaptic domains, in which a GlyR alternated between free and confined diffusion states, respectively (white arrow in Fig. 1B). Silver-intensified and gold-toned QDs were detected using transmission electron microscopy (EM) with the same QD immunolabeling protocol (Fig. 1D). EM analysis provided evidence that QD-GlyR could access the core of the synapse (Fig. 1D). QD trajectories were reconstructed from recordings with custom "Sinema" software written in MatLab (Bonneau et al. 2005). Exploratory maps of trajectories indicated that individual QDs can exchange between extrasynaptic and synaptic compartments (e.g., blue and green trajectories, Fig. 2A). QDs diffused over broad areas of the extrasynaptic plasma membrane, whereas the QD exploratory map is reduced at synapses.

As exemplified (Fig. 2B) for the QD trajectory shown in Figure 2A, the average diffusion coefficient (D) of the QD was lower within the synaptic area (Fig. 1, orange arrows). The plots of mean square displacement function (MSD) versus time were linear and negatively bent in the extrasynaptic and synaptic membrane, respectively (Fig. 2C), indicating a stronger confinement at synapses. This reflects local molecular interactions.

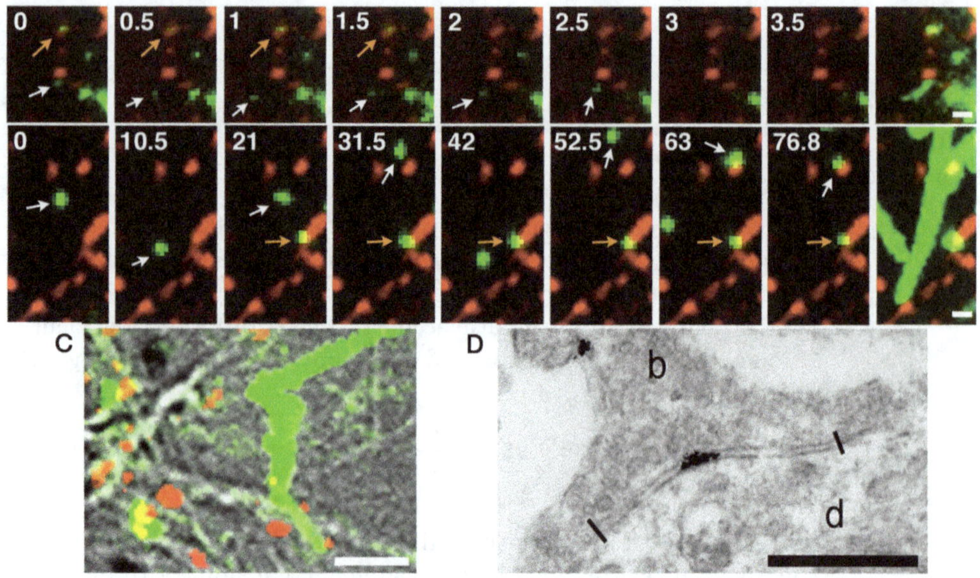

FIGURE 1. Comparison of lateral GlyR motion analyzed with Cy3 and QD probes. GlyRs were detected in cultured neurons with Cy3 (*A*) or QD-streptavidin (*B,C*). (Green) GlyR spots; (red) FM4-64-labeled synaptic boutons. (*A,B*) Images were extracted from a sequence of 35 and 1024 images with an acquisition time of 100 msec for Cy3 and 75 msec for QD. The time after the start of recording (in seconds) is indicated on each image. The last image is a maximum projection of the entire stack of images corresponding to the GlyR trajectory. (*C*) Projection of time-lapse recording (1 Hz, 20 min) of QD-GlyR trajectory (green) overlaid with FM 4-64 staining (red) and bright-field image. (*A–C*) Cy3 and QD diffused rapidly in the extrasynaptic region (white arrows), whereas synaptic GlyRs were stable (orange arrows). Note the short membrane surface explored by Cy3-GlyR compared with QD-GlyR. Long QD imaging duration enabled the observation of a synaptic entry (white arrow in *B*). (*D*) EM detection of QD-GlyR within the synaptic cleft. d, Dendrite; b, synaptic bouton. The edges of the cleft are outlined. Scale bars, 1 µm (*A,B*); 5 µm (*C*); 500 nm (*D*).

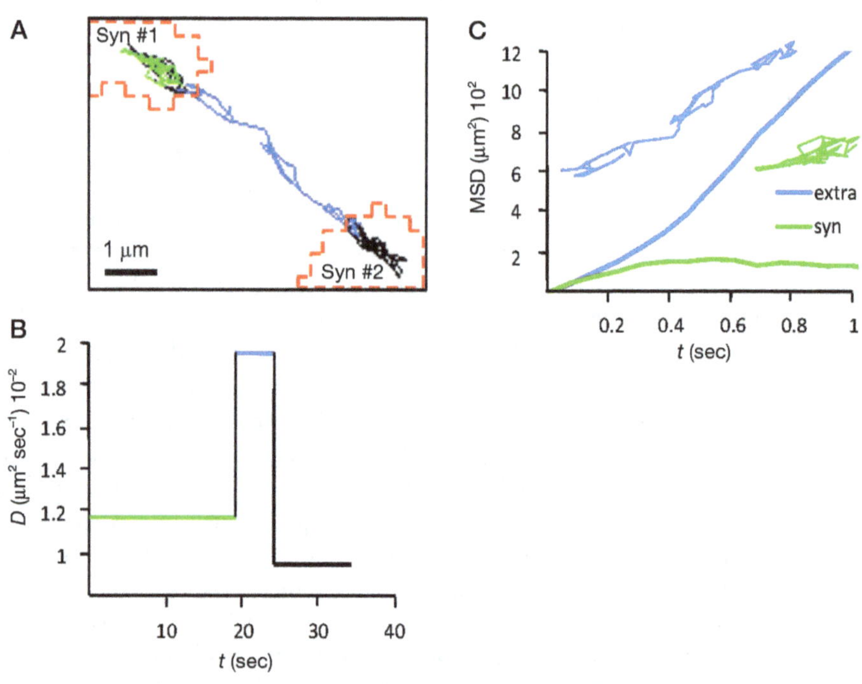

FIGURE 2. QD trajectory analysis. (*A*) Example of surface exploration by an itinerant GlyR-QD visualized on a reconstructed trajectory. QD trajectories in synaptic areas #1 (Syn #1) and #2 (Syn #2) are green and black, respectively, and FM4-64-stained synapses are red. Note the large surface area explored by the GlyR-QD outside (blue) the synaptic areas. Scale bar, 1 µm. (*B*) Average diffusion coefficients of the QD shown in *A* during its extrasynaptic (blue) and synaptic journey (green and blue). Note the changes in the QD diffusion coefficient when exiting synapse #1 or entering synapse #2, since there is one sudden increase and one drop. (*C*) Time-averaged MSD function of individual QD shown in *A* during its exploration of extrasynaptic (blue) and synaptic (green) loci during a recording sequence. The same QD displayed an extrasynaptic linear MSD curve and a synaptic negatively bent MSD curve, characteristic of random walk and confined movement, respectively.

RECIPES

Recipes for reagents marked with <R> are included in this list.

MEM Medium for Quantum Dot Imaging

Prepare MEM Eagle medium (Eagle 1959) without phenol red but containing:

$NaHCO_3$	4 mM
HEPES	10 mM
Glucose	6 g/L
Glutamine	2 mM
Na^+ pyruvate	1 mM
B27 supplement (Invitrogen)	1×

QD Streptavidin Conjugate Solution

QD 605 streptavidin conjugate (Invitrogen)	~0.5–2 nM
Borate buffer (pH 8–8.5)	50 mM
Sucrose	215 mM

The sucrose in the solution increases the osmolarity of the buffer to a physiological level (300 mosmol), whereas the borate buffer minimizes nonspecific QD labeling.

REFERENCES

Bannai H, Lévi S, Schweizer C, Inoue T, Launey T, Racine V, Sibarita JB, Mikoshiba K, Triller A. 2009. Activity-dependent tuning of inhibitory neurotransmission based on $GABA_AR$ diffusion dynamics. *Neuron* **62:** 670–682.

Bonneau S, Dahan M, Cohen LD. 2005. Single quantum dot tracking based on perceptual grouping using minimal paths in a spatiotemporal volume. *IEEE Trans Image Process* **14:** 1384–1395.

Calamai M, Specht CG, Heller J, Alcor D, Machado P, Vannier C, Triller A. 2009. Gephyrin oligomerization controls GlyR mobility and synaptic clustering. *J Neurosci* **29:** 7639–7648.

Charrier C, Ehrensperger MV, Dahan M, Lévi S, Triller A. 2006. Cytoskeleton regulation of glycine receptor number at synapses and diffusion in the plasma membrane. *J Neurosci* **26:** 8502–8511.

Dahan M, Lévi S, Luccardini C, Rostaing P, Riveau B, Triller A. 2003. Diffusion dynamics of glycine receptors revealed by single–quantum dot tracking. *Science* **302:** 442–445.

Eagle H. 1959. Amino acid metabolism in mammalian cell cultures. *Science* **130:** 432–437.

Ehlers MD, Heine M, Groc L, Lee MC, Choquet D. 2007. Diffusional trapping of GluR1 AMPA receptors by input-specific synaptic activity. *Neuron* **54:** 447–460.

Frischknecht R, Heine M, Perrais D, Seidenbecher CI, Choquet D, Gundelfinger ED. 2009. Brain extracellular matrix affects AMPA receptor lateral mobility and short-term synaptic plasticity. *Nat Neurosci* **12:** 897–904.

Groc L, Heine M, Cognet L, Brickley K, Stephenson FA, Lounis B, Choquet D. 2004. Differential activity-dependent regulation of the lateral mobilities of AMPA and NMDA receptors. *Nat Neurosci* **7:** 695–696.

Groc L, Heine M, Cousins SL, Stephenson FA, Lounis B, Cognet L, Choquet D. 2006. NMDA receptor surface mobility depends on NR2A-2B subunits. *Proc Natl Acad Sci* **103:** 18769–18774.

Groc L, Choquet D, Stephenson FA, Verrier D, Manzoni OJ, Chavis P. 2007. NMDA receptor surface trafficking and synaptic subunit composition are developmentally regulated by the extracellular matrix protein Reelin. *J Neurosci* **27:** 10165–10175.

Groc L, Choquet D, Chaouloff F. 2008. The stress hormone corticosterone conditions AMPAR surface trafficking and synaptic potentiation. *Nat Neurosci* **11:** 868–870.

Heine M, Groc L, Frischknecht R, Béïque JC, Lounis B, Rumbaugh G, Huganir RL, Cognet L, Choquet D. 2008. Surface mobility of postsynaptic AMPARs tunes synaptic transmission. *Science* **320:** 201–205.

Jaqaman K, Loerke D, Mettlen M, Kuwata H, Grinstein S, Schmid SL, Danuser G. 2008. Robust single-particle tracking in live-cell time-lapse sequences. *Nat Methods* **5:** 695–702.

Lévi S, Schweizer C, Bannai H, Pascual O, Charrier C, Triller A. 2008. Homeostatic regulation of synaptic GlyR numbers driven by lateral diffusion. *Neuron* **59:** 261–273.

Michaluk P, Mikasova L, Groc L, Frischknecht R, Choquet D, Kaczmarek L. 2009. Matrix metalloproteinase-9 controls NMDA receptor surface diffusion through integrin β1 signaling. *J Neurosci* **29:** 6007–6012.

Mikasova L, Groc L, Choquet D, Manzoni OJ. 2008. Altered surface trafficking of presynaptic cannabinoid type 1 receptor in and out synaptic terminals parallels receptor desensitization. *Proc Natl Acad Sci* **105:** 18596–18601.

Nirmal M, Dabbousi B, Bawendi MG, Macklin JJ, Trautman JK, Harris TD, Brus LE. 1996. Fluorescence intermittency in single cadmium selenide nanocrystals. *Nature* **383:** 802–804.

Renner M, Choquet D, Triller A. 2009. Control of the postsynaptic membrane viscosity. *J Neurosci* **29:** 2926–2937.

Sergé A, Bertaux N, Rigneault H, Marguet D. 2008. Dynamic multiple-target tracing to probe spatiotemporal cartography of cell membranes. *Nat Methods* **5:** 687–694.

Tardin C, Cognet L, Bats C, Lounis B, Choquet D. 2003. Direct imaging of lateral movements of AMPA receptors inside synapses. *EMBO J* **22:** 4656–4665.

MOVIE LEGEND

Movies are freely available online at www.cshprotocols.org/imaging.

MOVIE 2.1. Diffusion and stabilization of single QD-GlyRs (green). Time-lapse recording (1200 images at 1 Hz; acquisition time, 75 msec). Synapses are labeled with FM4-64 (red).

3 Synapto-pHluorins
Genetically Encoded Reporters of Synaptic Transmission

Gero Miesenböck
Department of Physiology, Anatomy and Genetics, University of Oxford, Oxford OX1 3PT, United Kingdom

ABSTRACT

pHluorins are pH-sensitive mutants of green fluorescent protein (GFP). Attached to proteins with defined cellular locations or itineraries, pHluorins report subcellular pH as well as protein transport between compartments of differing pH. Key applications in neurobiology include the optical detection of neurotransmitter release with synapto-pHluorins and their derivatives, as well as measurements of neurotransmitter receptor trafficking.

PHLUORINS

Neurons communicate at chemical synapses by releasing the contents of one or a few transmitter-filled vesicles from a large synaptic vesicle pool. Because it is generally impossible to resolve the several dozen to several hundred vesicles at a typical synapse optically, visualizing chemical synaptic transmission requires that vesicles undergoing exocytosis be distinguished spectrally from resting vesicles. Synapto-pHluorins use pH-sensitive mutants of GFP, termed pHluorins, to afford this distinction (Miesenböck et al. 1998). Synaptic vesicles contain ATP-powered proton pumps that maintain a proton electrochemical gradient across the vesicle membrane; as a consequence, the vesicle interior is acidified to a pH of ~5.7. pHluorins located in these vesicles are in a distinct spectral state—the "off" state (Fig. 1). As a vesicle fuses with the plasma membrane, its interior becomes continuous with the extracellular space and the accumulated protons dissipate, switching the pHluorin "on" (Fig. 1).

Two classes of pHluorins, termed ecliptic and ratiometric, have been developed (Miesenböck et al. 1998). The chromophore of ecliptic pHluorin titrates, with a pK_a of ~7.1 (Sankaranarayanan et al. 2000), between a deprotonated, fluorescent state and a protonated, nonfluorescent state (Miesenböck et al. 1998). The abundance of protons inside synaptic vesicles drives ecliptic pHluorin into the protonated form and quenches ("eclipses") its fluorescence. Exocytic events lead to chromophore deprotonation, which is detected as a sudden increase in fluorescence, excitable in the 450–490-nm band (Fig. 1A). Both the original ecliptic pHluorin (GenBank accession number AF058695) and a newer super-ecliptic variant (GenBank accession number AY533296; this form carries the chromophore substitutions F64L and S65T, in addition to the ecliptic set of mutations) exist as mixtures of two conformers: the ecliptic

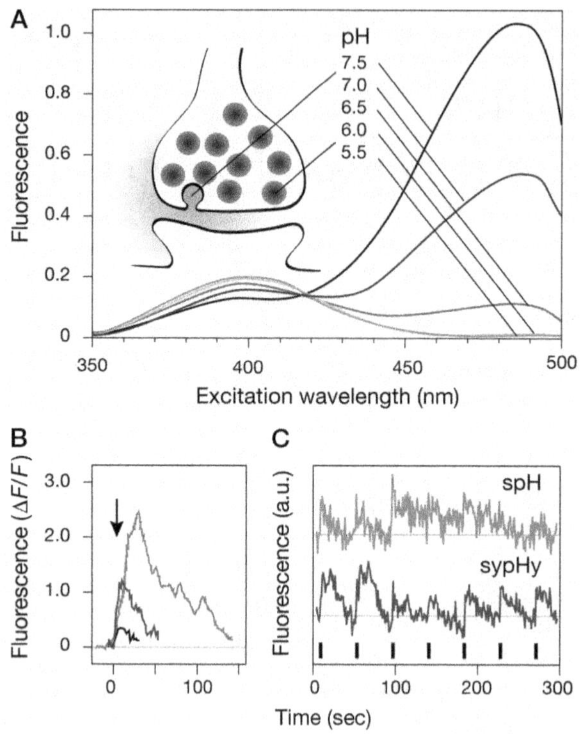

FIGURE 1. Imaging neurotransmission with synapto-pHluorin. (*A*) Fluorescence excitation spectra of super-ecliptic synapto-pHluorin as a function of pH, normalized to fluorescence excited at 491 nm at pH 7.5. In the context of a presynaptic terminal (*inset*), pHluorins located in resting vesicles (internal pH ~5.7) are nonfluorescent when illuminated at ~450–490 nm. Upon exocytosis, the vesicle interior becomes continuous with the extracellular space, pH rises to ~7.4, and the pHluorin complement of the fused vesicle becomes fluorescent. (*B*) Time course of synapto-pHluorin fluorescence of a single presynaptic terminal during 20 (black trace), 100 (dark gray trace), and 300 (light gray trace) action potentials. Action potentials were evoked by electrical field stimulation at 10 Hz; the arrow denotes the onset of the stimulus trains. (Modified, with permission, from Sankaranarayanan and Ryan 2000, ©Nature Publishing Group.) (*C*) Time course of synapto-pHluorin (spH) and sypHy fluorescence of individual presynaptic terminals at hippocampal synapses following seven action potentials at 45-sec intervals (vertical ticks). (Modified, with permission, from Granseth et al. 2006, ©Elsevier.)

pHluorin proper, whose fluorescence is excited at ~475 nm and quenched at pH <6.0, and a mildly pH-sensitive form that is excited at ~395 nm and remains fluorescent at pH <6.0 (Fig. 1A). The effect of the chromophore substitutions in super-ecliptic pHluorin is to populate predominantly the ecliptic species, causing a 5.9-fold increase in its fluorescence, and to redshift its excitation peak from 475 to 491 nm (Fig. 1A). At pH 7.5, super-ecliptic pHluorin emits at 22% the rate of enhanced GFP (EGFP).

Ratiometric pHluorin (GenBank accession number AF058694) possesses a bimodal excitation spectrum, similar to that of wild-type GFP. The excitation maximum lies at 395 nm at neutral pH but gradually shifts to 475 nm as the pH is lowered (Miesenböck et al. 1998; Yuste et al. 2000). Ratiometric comparison of the fluorescence intensities at the two excitation maxima provides a calibrated measure of pH (Miesenböck et al. 1998). Like all ratiometric indices, this measure is insensitive to variations in fluorophore concentration and optical path length. However, because ratiometric pHluorin lacks a nonfluorescent state at pH < 6.0, exocytic events must be distinguished against a background of resting vesicles. This requires that individual secretory vesicles are resolved optically or, if this is impossible, that the fraction of vesicles undergoing exocytosis is sufficiently large to alter the aggregate ratio detectably (Miesenböck et al. 1998).

TARGETING MODULE

The targeting module—the topogenic polypeptide sequence to which the pHluorin module is linked through gene fusion—determines the subcellular location or itinerary of the light emitter and thus the compartment or process to be monitored optically. For imaging neurotransmitter release, super-ecliptic pHluorin has been fused to a range of synaptic vesicle proteins, including the v-SNARE VAMP/synaptobrevin-2 (Miesenböck et al. 1998), the calcium sensor synaptotagmin (Diril et al. 2006), synaptophysin (Granseth et al. 2006), and the vesicular glutamate transporter VGLUT1 (Voglmaier et al. 2006). For imaging the trafficking of postsynaptic transmitter receptors between the plasma membrane and acidified endocytic comparments (pH < 6.0), super-ecliptic pHluorin has been attached to α-amino-3-hydroxy-5-methyl-4-isoxazole propionic acid (AMPA) and *N*-methyl-D-aspartate (NMDA) receptor subunits (Ashby et al. 2004; Kopec et al. 2006).

GENE EXPRESSION

cDNAs encoding the pHluorin and the targeting module are fused via a short, flexible polypeptide linker. The construct is placed under the control of the appropriate constitutive, inducible, or cell type–specific genetic control element and introduced into the cell, tissue, or organism of interest. The characteristics of the expression construct and the optimal DNA transfer method depend on the problem at hand. Approaches range from shotgun methods for transient expression (Miesenböck et al. 1998; Sankaranarayanan and Ryan 2000) to cell type–specific expression in transgenic or "knockin" animals (Ng et al. 2002; Poskanzer et al. 2003; Samuel et al. 2003; Bozza et al. 2004; Li et al. 2005; Shang et al. 2007; Beg et al. 2008; Petzold et al. 2009).

OPTICAL IMAGING

The broad range of cellular processes, cell types, and organisms studied with pHluorins is paralleled by an equally broad range of microscope configurations, light sources, and detectors (Yuste et al. 2000; Burrone et al. 2006; Royle et al. 2008). Options range from wide-field illumination to lasers providing one- or two-photon excitation. The 488-nm argon ion laser line and the excitation band of commercial EGFP filter sets match the super-ecliptic excitation peak for single-photon excitation; efficient two-photon excitation is possible in the 900–940-nm range (Ng et al. 2002; Petzold et al. 2009). Emitted photons in the 500–560-nm band (peak at 508 nm) are separated from excitation light with dielectric mirrors, interference filters, and/or absorptive optics. Laser-scanning microscopy employs photomultiplier tubes (PMTs) in confocal (Sankaranarayanan and Ryan 2000; Sankaranarayanan et al. 2000; Kim et al. 2003; Samuel et al. 2003) or whole-area mode (Ng et al. 2002; Petzold et al. 2009). Instrumentation for excitation ratio imaging of ratiometric pHluorin is discussed elsewhere (Yuste et al. 2000).

EXAMPLES OF APPLICATIONS

Figure 1 illustrates the use of (super-)ecliptic synapto-pHluorin for monitoring synaptic transmission at single boutons. In Figure 1B, transmission from a rat hippocampal neuron in dissociated culture was driven by trains of 20, 100, and 300 electrically evoked action potentials at 10 Hz (Sankaranarayanan and Ryan 2000). Stimulation caused transient increases in fluorescence (due to synaptic vesicle exocytosis) that decayed with time constants ranging from 4 to 50 sec (due to synaptic vesicle endocytosis and re-acidification). Under the assumptions that the synaptic terminal releases at most one vesicle per action potential and that the probability of release is ~0.3, these fluorescence transients represent composites of approximately 6, 30, and 100 single-vesicle signals, respectively. Responses of individual terminals to single action potentials suggest that vesicles add and subtract unitary quanta of fluorescence as they exocytose and recycle (Fig. 1C). The time course of a composite signal elicited by a train of action potentials (Fig. 1B) thus reflects two physiological variables: the timing of individual release events and the distribution of waiting times for the retrieval process (Granseth et al. 2006). The signal-to-noise ratio depends on the choice of targeting module (Fig. 1C), which determines how cleanly pHluorin partitions between synaptic vesicles and the plasma membrane.

Figure 2 demonstrates the detection of synaptic transmission in response to physiological stimuli in vivo. In both cases, synapto-pHluorin was expressed selectively in olfactory receptor neurons of genetically engineered animals—a fruit fly (Ng et al. 2002) and a mouse (Bozza et al. 2004). In the fly (Fig. 2A), fluorescence was measured by two-photon laser-scanning microscopy of a glomerulus in the antennal lobe that receives synaptic input from neurons expressing one particular type of odorant receptor (Ng et al. 2002). Odorants, acting as agonists on this particular receptor, evoke neurotransmitter release, which causes a rapid increase in synapto-pHluorin fluorescence, followed by a slower decay (time constant ~4 sec). In the mouse (Fig. 2B), the olfactory bulbs were imaged through thinned

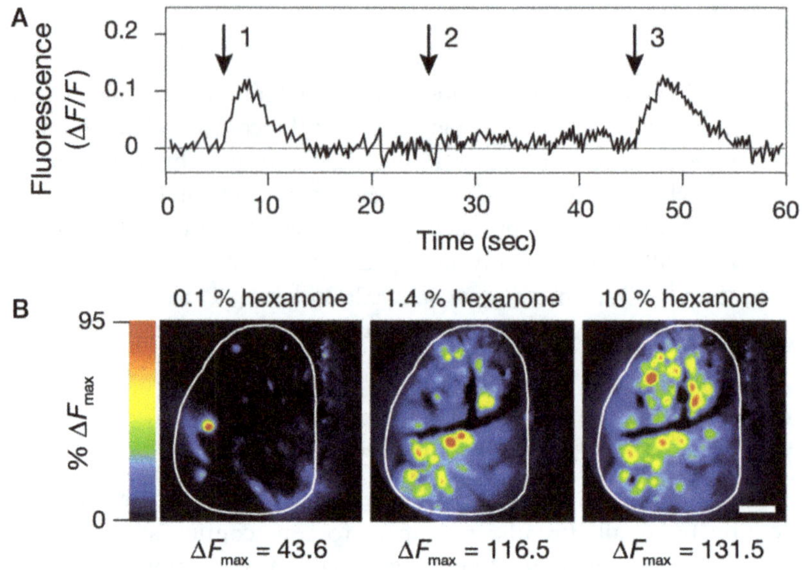

FIGURE 2. (*A*) Responses of olfactory receptor neurons projecting to a single glomerulus in the antennal lobe of a fruit fly. Synapto-pHluorin signals were recorded by two-photon microscopy during the presentation of three different test odors (arrows). Odors 1 and 3 evoke neurotransmitter release, whereas odor 2 is ineffective. (Modified, from Ng et al. 2002, ©Elsevier.) (*B*) Pseudocolored odor-response maps in the olfactory bulb of a knockin mouse expressing synapto-pHluorin from the locus encoding the olfactory marker protein. Responses were elicited by a concentration series of 2-hexanone; images depict peak responses at 1.5–2.5 sec after stimulus onset. Percent dilutions of saturated vapor are given in each panel. Panels are scaled to their maximum ΔF, which is indicated below each panel. Scale bar, 300 µm. (Modified, with permission, from Bozza et al. 2004, ©Elsevier.)

bone by wide-field epifluorescence microscopy; optical signals were detected by a back-illuminated CCD. The pseudocolored response maps depict patterns of glomerular activation (Bozza et al. 2004).

ADVANTAGES AND LIMITATIONS

Synapto-pHluorins possess five unique strengths as indicators of synaptic transmission. First, they are generated biosynthetically, without cofactor requirements. This permits the study of samples that are difficult to stain with exogenous molecules, such as intact tissues and organisms (Ng et al. 2002; Poskanzer et al. 2003; Samuel et al. 2003; Bozza et al. 2004; Shang et al. 2007; Beg et al. 2008). Second, indicator expression can be placed under the control of cell type–specific genetic control elements (Ng et al. 2002; Samuel et al. 2003; Bozza et al. 2004; Li et al. 2005; Shang et al. 2007), ensuring a degree of specificity that synthetic indicator dyes lack (Miesenböck and Kevrekidis 2005). Third, in contrast to "single-shot" methods that involve dye release (Betz and Bewick 1992), the synapto-pHluorin signal is fully regenerative through multiple rounds of vesicle release and recycling. Fourth, the regenerative nature of the synapto-pHluorin signal allows it to provide direct information not only about vesicle release, but also about vesicle recycling (Sankaranarayanan and Ryan 2000; Diril et al. 2006; Granseth et al. 2006; Voglmaier et al. 2006; Wienisch and Klingauf 2006). Finally, expression of synapto-pHluorin does not appear to interfere with neuronal function (Sankaranarayanan and Ryan 2000), which can be a concern when using chelating calcium indicators.

The limitations of synapto-pHluorins are tied to the biology of synaptic vesicles. With an estimated two to three copies per vesicle (R. Sinha and J. Klingauf, pers. comm.), the number of synapto-pHluorins is smaller than the number of lipophilic dye molecules (Betz and Bewick 1992) that can be accommodated, leading to a smaller quantum of fluorescence per vesicle. At hippocampal synapses, ~20% and ~8% of the total synapto-pHluorin and sypHy pools, respectively, are present at the cell surface under unstimulated conditions; these pools elevate background fluorescence and reduce the signal-to-noise ratio (Fig. 1C; Sankaranarayanan and Ryan 2000; Sankaranarayanan et al. 2000; Granseth et al. 2006). If a synapse has a low basal release rate and the vesicular and surface synapto-pHluorin pools do not intermix, prestimulus photobleaching (Samuel et al. 2003) or site-specific proteolysis (Wienisch and Klingauf 2006) can selectively eliminate the fluorescent surface pool. Once a vesicle exocytoses, its synapto-pHluorin complement remains fluorescent for an extended, but variable, period of time that is terminated by vesicle retrieval and acidification (Miesenböck et al. 1998; Sankaranarayanan and Ryan 2000; Sankaranarayanan et al. 2000; Granseth et al. 2006). This period, which may last from a few hundred milliseconds to several dozens of seconds, constrains the temporal resolution of the method. There is another important limitation: There are indications that the pHluorins do not interfere with neuronal function; all the works cited involve an overexpression of synaptic vesicle proteins, which may have profound (and variable, depending on the pHluorin type) effects on the cell biology of the neuron and aspects of neuronal function that have not been assayed. These issues could be tested/ameliorated by titrating expression of pHluorins in a null mutant background for the pHluorin of interest.

Synapto-pHluorin occupies an important position in neurobiology: It is the only probe available to signal the activity of genetically specified synapses in intact neural tissue. Given the key role of synaptic transmission in information processing and storage, knowledge of synaptic dynamics is an integral part of any functional description of neurons and circuits.

REFERENCES

Ashby MC, De La Rue SA, Ralph GS, Uney J, Collingridge GL, Henley JM. 2004. Removal of AMPA receptors (AMPARs) from synapses is preceded by transient endocytosis of extrasynaptic AMPARs. *J Neurosci* **24:** 5172–5176.

Beg AA, Ernstrom GG, Nix P, Davis MW, Jorgensen EM. 2008. Protons act as a transmitter for muscle contraction in *C. elegans*. *Cell* **132:** 149–160.

Betz WJ, Bewick GS. 1992. Optical analysis of synaptic vesicle recycling at the frog neuromuscular junction. *Science* **255:** 200–203.

Bozza T, McGann JP, Mombaerts P, Wachowiak M. 2004. In vivo imaging of neuronal activity by targeted expression of a genetically encoded probe in the mouse. *Neuron* **42:** 9–21.

Burrone J, Li Z, Murthy VN. 2006. Studying vesicle cycling in presynaptic terminals using the genetically encoded probe synaptopHluorin. *Nat Protoc* **1:** 2970–2978.

Diril MK, Wienisch M, Jung N, Klingauf J, Haucke V. 2006. Stonin 2 is an AP-2-dependent endocytic sorting adaptor for synaptotagmin internalization and recycling. *Dev Cell* **10:** 233–244.

Granseth B, Odermatt B, Royle SJ, Lagnado L. 2006. Clathrin-mediated endocytosis is the dominant mechanism of vesicle retrieval at hippocampal synapses. *Neuron* **51:** 773–786.

Kim JH, Udo H, Li HL, Youn TY, Chen M, Kandel ER, Bailey CH. 2003. Presynaptic activation of silent synapses and growth of new synapses contribute to intermediate and long-term facilitation in *Aplysia*. *Neuron* **40:** 151–165.

Kopec CD, Li B, Wei W, Boehm J, Malinow R. 2006. Glutamate receptor exocytosis and spine enlargement during chemically induced long-term potentiation. *J Neurosci* **26:** 2000–2009.

Li Z, Burrone J, Tyler WJ, Hartman KN, Albeanu DF, Murthy VN. 2005. Synaptic vesicle recycling studied in transgenic mice expressing synaptopHluorin. *Proc Natl Acad Sci* **102:** 6131–6136.

Miesenböck G, Kevrekidis IG. 2005. Optical imaging and control of genetically designated neurons in functioning circuits. *Annu Rev Neurosci* **28:** 533–563.

Miesenböck G, De Angelis DA, Rothman JE. 1998. Visualizing secretion and synaptic transmission with pH-sensitive green fluorescent proteins. *Nature* **394:** 192–195.

Ng M, Roorda RD, Lima SQ, Zemelman BV, Morcillo P, Miesenböck G. 2002. Transmission of olfactory information between three populations of neurons in the antennal lobe of the fly. *Neuron* **36:** 463–474.

Petzold GC, Hagiwara A, Murthy VN. 2009. Serotonergic modulation of odor input to the mammalian olfactory bulb. *Nat Neurosci* **12:** 784–791.

Poskanzer KE, Marek KW, Sweeney ST, Davis GW. 2003. Synaptotagmin I is necessary for compensatory synaptic vesicle endocytosis in vivo. *Nature* **426:** 559–563.

Royle SJ, Granseth B, Odermatt B, Derevier A, Lagnado L. 2008. Imaging pHluorin-based probes at hippocampal synapses. *Methods Mol Biol* **457:** 293–303.

Samuel AD, Silva RA, Murthy VN. 2003. Synaptic activity of the AFD neuron in *Caenorhabditis elegans* correlates with thermotactic memory. *J Neurosci* **23:** 373–376.

Sankaranarayanan S, Ryan TA. 2000. Real-time measurements of vesicle-SNARE recycling in synapses of the central nervous system. *Nat Cell Biol* **2:** 197–204.

Sankaranarayanan S, De Angelis D, Rothman JE, Ryan TA. 2000. The use of pHluorins for optical measurements of presynaptic activity. *Biophys J* **79:** 2199–2208.

Shang Y, Claridge-Chang A, Sjulson L, Pypaert M, Miesenböck G. 2007. Excitatory local circuits and their implications for olfactory processing in the fly antennal lobe. *Cell* **128:** 601–612.

Voglmaier SM, Kam K, Yang H, Fortin DL, Hua Z, Nicoll RA, Edwards RH. 2006. Distinct endocytic pathways control the rate and extent of synaptic vesicle protein recycling. *Neuron* **51:** 71–84.

Wienisch M, Klingauf J. 2006. Vesicular proteins exocytosed and subsequently retrieved by compensatory endocytosis are nonidentical. *Nat Neurosci* **9:** 1019–1027.

Yuste R, Miller RB, Holthoff K, Zhang S, Miesenböck G. 2000. Synapto-pHluorins: Chimeras between pH-sensitive mutants of green fluorescent protein and synaptic vesicle membrane proteins as reporters of neurotransmitter release. *Methods Enzymol* **327:** 522–546.

4 Spatially Resolved Flash Photolysis via Chemical Two-Photon Uncaging

Diana L. Pettit[1] and George J. Augustine[2,3]

[1]*Dominick Purpura Department of Neuroscience, Albert Einstein College of Medicine, Bronx, New York 10641;* [2]*Center for Functional Connectomics, Korea Institute of Science and Technology, Seoul, South Korea 136-791;* [3]*Program in Neuroscience and Behavioral Disorders, Duke–NUS Graduate Medical School, Singapore 169547, Singapore*

ABSTRACT

Here we describe "chemical two-photon uncaging," an approach that provides high spatial resolution during flash photolysis experiments. Chemical two-photon uncaging relies on attaching two light-sensitive cages to a single molecule, so that absorption of two UV photons is required to photoactivate the molecule. Because generation of the photoactivated molecule is proportional to the square of light intensity, the volume occupied by uncaged molecules is much smaller than with conventional flash photolysis. Thus, chemical two-photon uncaging provides high spatial resolution with relatively simple and inexpensive UV light sources.

INTRODUCTION

Photolysis of caged compounds has long been used to study the kinetics and spatial range of rate-limiting steps in biological processes (McCray and Trentham 1989; Parker and Yao 1991; Adams and Tsien 1993). However, restricting the spatial extent of photolysis is difficult when caged compounds are present throughout thick specimens, such as brain slices, because light on its way to and from the focal point will photolyze the caged compound in untargeted tissue (Fig. 1A).

Such out-of-focus photolysis can be eliminated using "optical" two-photon excitation, in which two long-wavelength photons are needed to provide sufficient energy to photolyze a single caging group (Göppert-Mayer 1931; Denk 1994; Denk et al. 1995). Whereas such an approach has been very useful for mapping the distribution of glutamate receptors on neurons in brain slices (Matsuzaki et al. 2001; Smith et al. 2003; Tanaka et al. 2008; see Chapter 25), this approach typically yields little uncaging because, until very recently (Zayat et al. 2011), the caging groups used in most caged compounds were not efficiently photolyzed by long-wavelength photons (Denk 1994). For a broader discussion of caged compound photochemistry, see Hess et al. (2011), Ellis-Davies (2011), and Zayat et al. (2011).

An alternative approach, described in this chapter, relies on attaching two caging groups to a single molecule, a concept proposed by Adams and Tsien (1993). The resulting requirement for two

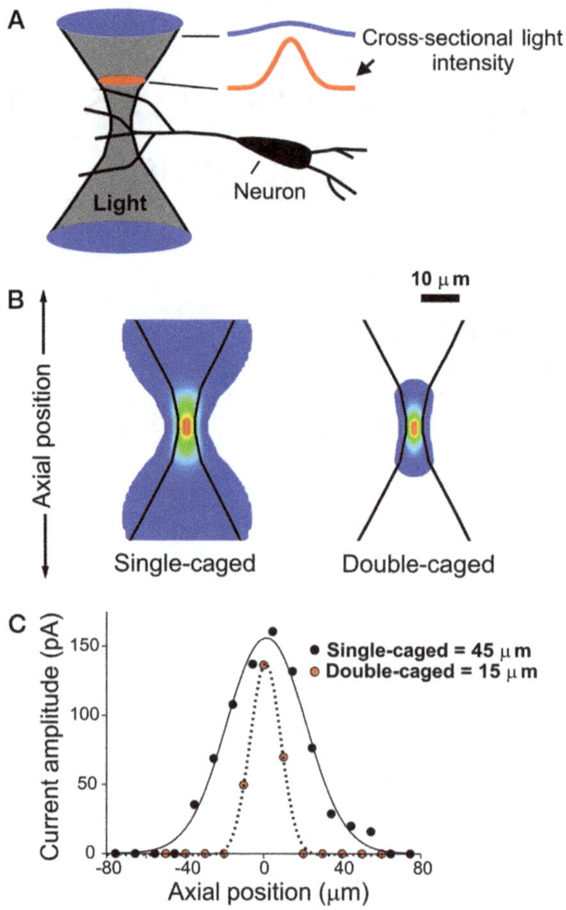

FIGURE 1. (*A*) (*Left*) Schematic side view of a UV-light beam focused onto a neuron (black). At different horizontal planes (red, blue), the spot contains a constant total amount of light (*right*), although the light density increases as the beam comes into focus. (*B*) Side view of expected production of free glutamate following photolysis of single-caged and double-caged glutamate. (Red) Highest glutamate concentration; (blue) lowest glutamate concentration. (*C*) Relationship between axial position and the peak amplitude of currents evoked by single- and double-caged glutamate, obtained while varying the distance between the neuronal cell body and the focal plane of the UV light beam. The half-width of the Gaussian functions fit to these data is 45 μm for responses to single-caged glutamate and 15 μm for responses to double-caged glutamate. (Modified, with permission of Elsevier, from Pettit et al. 1997.)

photons produces the free, uncaged molecule in proportion to the square of the light intensity and, thereby, reduces the volume where uncaged molecules are (Fig. 1B). Because of its reliance on two chemical cages, this phenomenon is called "chemical" two-photon uncaging (Pettit et al. 1997).

Protocol

Chemical Two-Photon Uncaging

This protocol describes chemical two-photon uncaging, a simple and economical alternative to eliminate out-of-focus uncaging during flash photolysis experiments. In this example described here, double-caged glutamate is used to visualize the distribution of glutamate receptors.

IMAGING SETUP

Microscope

The experimental setup must be capable of simultaneous whole-cell voltage-clamp recordings and imaging of the specimen (Wang and Augustine 1995). Mounting the microscope on a translation stage in combination with a fixed microscope stage provides the most stable recording configuration. Alternatively, mobile microscope stages can be used as long as movement is sufficiently vibration-free to maintain whole-cell recordings. Neurons are visualized using a 40x water-immersion objective (Olympus) that is carefully chosen to have maximal UV transmission and minimal chromatic aberration in the UV–visible light range (e.g., 350–700 nm), to minimize differences between the focal point for UV photolysis and the visible excitation light that we typically use for fluorescent imaging of the specimen.

Light Sources

With currently available cages, chemical two-photon uncaging requires a UV light source. For this purpose, we typically use a continuous-emission argon laser (Coherent Innova 305A). We have found that 0.2–2 mJ of UV light is sufficient to obtain responses with the α-carboxy-2-nitrobenzyl (CNB) double-caged glutamate, corresponding to an energy density of 0.01–0.1 $\mu J/\mu m^2$ per 10-msec-long flash. Thus, two-photon photolysis can also be achieved with an inexpensive arc lamp, a capacitor-discharge flash lamp, or an arc lamp connected to a pulse generator (Rapp and Guth 1988; Denk 1997). Continuous light sources require an electronic shutter to control the duration of illumination. Pulsed lasers with high repetition rates can also be used. Diode-pumped solid-state lasers with a pulse length of <65 nsec and a repetition rate of 150 kHz provide near-continuous emission and allow for short uncaging durations without a shutter (DPSS Lasers, Santa Clara, CA).

MATERIALS

CAUTION: See Appendix 6 for proper handling of materials marked with <!>.

Reagents

Brain tissue for analysis

We typically use brain tissue from 13–18-d-old mice or rats; the relatively young age makes it easier to obtain whole-cell patch-clamp recordings, although the age of the animal is of little consequence for the uncaging process itself.

Double-caged glutamate compound

Chemical two-photon uncaging requires a molecule that has two sites where inactivating cages can be attached. It is critical that each cage is sufficient to inactivate independently, so that the single-caged intermediate has no biological activity. For mapping glutamate receptor distribution, we use a double-caged glutamate compound with cages at both the α- and γ-carboxyl groups (Wieboldt et al. 1994; Invitrogen). Double-caged glutamate is hydrophobic, so it must be dissolved in dimethylsulfoxide (DMSO) <!> to a final concentration of 0.1% before adding it to the physiological saline solution.

Physiological saline, bicarbonate-buffered

Equipment

Our experiments are performed on a setup typical for brain slice electrophysiological recording.

Micromanipulators (assembled from Newport Corporation optical bench components)
Patch-clamp amplifier (Axon Instruments/Molecular Devices)
Recirculating pump
Recording chamber
Upright microscope (Nikon) with water-immersion objectives (see Imaging Setup)
UV light source (see Imaging Setup)

EXPERIMENTAL METHOD

1. Prepare slices from brain tissue using standard techniques (Pettit et al. 1997).
2. Place the slices in a recording chamber and continuously perfuse with physiological saline containing the double-caged glutamate (100–200 μM). Use a recirculating pump at a rate of 3–4 mL/min.

   The solution is stable for several hours.

   See *Troubleshooting.*
3. Focus the output of a continuous emission argon laser onto the specimen, through the objective, to form a spot 3 μm in diameter.
4. Use conventional whole-cell patch-clamp recording methods (see, e.g., Pettit and Augustine 2000) to record the electrical currents that result from photolysis of the double-caged glutamate.

   See *Troubleshooting.*

TROUBLESHOOTING

Problem (Step 2): The caging compound exhibits insolubility and toxicity.
Solution: Double-caged glutamate is poorly water-soluble and must be dissolved in DMSO or alcohol. These reagents can have deleterious effects on cell health and should be kept to an end concentration of ≤0.3 % if possible.

Problem (Step 4): There is photodamage to the cells.
Solution: At the energy levels required for photolysis, it should be possible to deliver hundreds of uncaging pulses with little or no signs of UV photodamage. Phototoxic effects may simply require much higher light levels than those required for photolysis. For example, Mendez and Penner (1998) showed that a nonselective Ca^{2+}-permeable cation current is induced by UV irradiation of whole cells on the order of 1 $\mu J/\mu m^2$, whereas photolysis requires light levels on the order of 0.01–0.1 $\mu J/\mu m^2$. However, repeated imaging of the cell with visible light can cause substantial damage. As a result, limiting any fluorescent imaging until the end of the experiment is recommended.

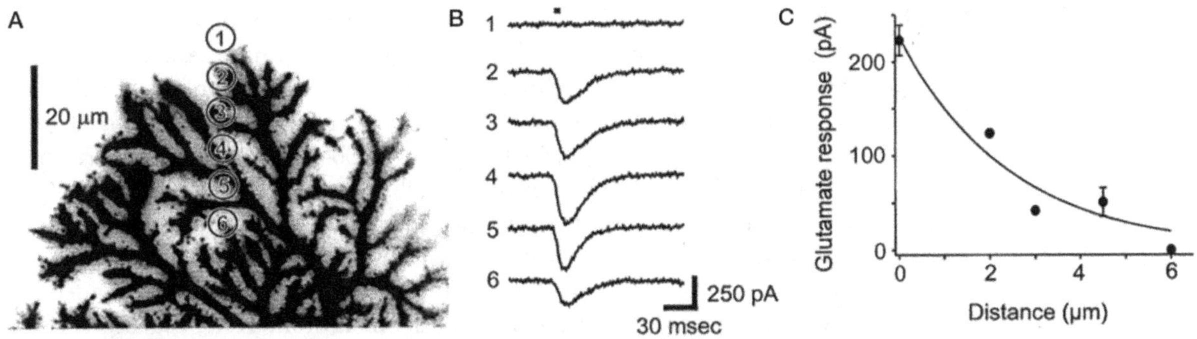

FIGURE 2. Localized activation of glutamate receptors on cerebellar Purkinje cells by chemical two-photon uncaging. (*A*) Glutamate was uncaged by moving a UV-light spot to the locations indicated by numbered circles. This Purkinje cell was filled with Oregon Green 488 BAPTA-1; fluorescence appears dark because the contrast of this image is reversed. (*B*) Currents elicited in voltage-clamped Purkinje cell by uncaging glutamate (small bar at *top*) at the locations shown in the image on the *left*. (*C*) Relationship between the amplitude of the glutamate-induced current and the position of the UV-light spot beyond the end of a Purkinje cell dendrite. The curve is an exponential function with a length constant of 2.4 µm. (Reprinted from Wang et al. 2000.)

EXAMPLE OF APPLICATION

Adding a second caging group to glutamate should improve axial resolution in focal photolysis experiments. We tested this prediction by measuring electrical currents elicited while varying the axial distance between the cell and the focal point of the UV light beam. The relationship between the peak amplitude of glutamate-induced currents and axial position was described by Gaussian functions whose half-widths were used as a measure of axial resolution (Fig. 1C). For hippocampal pyramidal neurons, the mean half-width of this function was 43.5 µm ± 1.3 µm (S.E.M., $n = 21$) for single-caged glutamate and 18.7 ± 1.1 µm (S.E.M., $n = 18$) for double-caged glutamate. This 57% improvement in axial resolution over single-caged glutamate is in excellent agreement with the theoretical predictions (Pettit et al. 1997). Very similar results were obtained in cerebellar Purkinje neurons (Wang et al. 2000).

The spatially restricted production of glutamate from the double-caged compound also yields high lateral resolution. This was shown in experiments on patch clamped cerebellar Purkinje cells (Fig. 2). By filling a Purkinje cell with a fluorescent dye (Oregon Green 488 BAPTA-1), the structure of its dendrites could be resolved clearly and the UV uncaging beam could be positioned relative to the dendrites (Fig. 2A). Glutamate-induced electrical currents were seen only when the light beam was directly over a dendrite; when the uncaging spot was moved 6 µm away from a dendrite there was no response (Fig. 2B). By plotting the relationship between the amplitude of the glutamate-induced current and the position of the uncaging spot, it was determined that the length constant for glutamate production was <3 µm (Fig. 2C).

This high spatial resolution has permitted measurements of the spatial distribution of glutamate receptors on hippocampal pyramidal neuron dendrites (Pettit and Augustine 2000) and visualization of the spread of long-term depression from active to inactive synapses in Purkinje cells (Wang et al. 2000).

DISCUSSION

Chemical two-photon uncaging is useful for a wide range of applications, including both mapping of receptor location (e.g., Pettit et al. 1997; Pettit and Augustine 2000) and localized photostimulation of neurons via activation of excitatory glutamate receptors (Callaway and Katz 1993; Pettit et al. 1999; Dodt et al. 2002).

This method has several advantages over optical two-photon uncaging. Perhaps the biggest practical advantage is that chemical two-photon uncaging works more easily than optical two-photon uncaging, which is limited by the poor two-photon cross section of most existing caging groups (Denk 1994). Chemical two-photon uncaging is also more economical; effective light sources include an argon laser or even an arc lamp, which cost ~$10,000–$20,000 (for comparison, a titanium-sapphire laser capable of the long excitation wavelengths required for optical two-photon uncaging costs ~$150,000). In principle, the short-wavelength light used for chemical two-photon uncaging also offers higher lateral resolution than the long-wavelength light used for optical two-photon uncaging.

A disadvantage of chemical two-photon uncaging is the need for two independently effective caging sites, which may not be available on some agonists of interest. Because the effective spatial resolution of the method is limited by diffusion of singly uncaged glutamate from the site of photolysis, high spatial resolution also requires a high rate of photolysis. Thus, high spatial resolution is most readily achieved with a laser. Finally, because UV light is scattered more than longer wavelength light, the depth of penetration is shorter for chemical two-photon uncaging than for optical two-photon uncaging.

In summary, chemical two-photon uncaging is an effective and inexpensive procedure that offers high spatial resolution for local photolysis of caged compounds. This technique can be used in a variety of experimental preparations, including brain slices, cultured neurons, and perhaps whole brains in vivo. Given the higher lateral resolution, but higher degree of scattering, associated with UV light, chemical two-photon uncaging is the method of choice for two-dimensional preparations such as cultured neurons.

ACKNOWLEDGMENTS

We thank S. Wang and K. Gee for helping to develop chemical two-photon uncaging. This work was supported by National Institutes of Health (NIH) grants and funds provided by Duke–NUS Graduate Medical School.

REFERENCES

Adams SR, Tsien RY. 1993. Controlling cell chemistry with caged compounds. *Annu Rev Physiol* **55:** 755–784.

Callaway EM, Katz LC. 1993. Photostimulation using caged glutamate reveals functional circuitry in living brain slices. *Proc Natl Acad Sci* **88:** 7661–7665.

Denk W. 1994. Two-photon scanning photochemical microscopy: Mapping ligand-gated ion channel distributions. *Proc Natl Acad Sci* **91:** 6629–6633.

Denk W. 1997. Pulsing mercury arc lamps for uncaging and fast imaging. *J Neuro Methods* **72:** 39–42.

Denk W, Piston DW, Webb WW. 1995. Two-photon molecular excitation in laser scanning microscopy. In *Handbook of biological confocal microscopy*, 2nd ed. (ed. Pawley JB), pp. 445–458. Plenum, New York.

Dodt HU, Eder M, Schierloh A, Zieglgansberger W. 2002. Infrared-guided laser stimulation of neurons in brain slices. *Sci STKE* **120:** PL2. doi: 10.1126/stke.2002.120pl2.

Ellis-Davies GCR. 2011. Nitrobenzyl-based caged neurotransmitters. In *Imaging: A laboratory manual* (ed. Yuste R), pp. 889–896. Cold Spring Harbor Laboratory Press, Cold Spring Harbor, NY.

Göppert-Mayer M. 1931. Über Elementarakte mit zwei Quantensprüngen. *Ann Phys* **9:** 273.

Hess GP, Lewis RW, Chen Y. 2011. Caged neurotransmitters and other caged compounds: Design and application. In *Imaging: A laboratory manual* (ed. Yuste R), pp. 869–888. Cold Spring Harbor Laboratory Press, Cold Spring Harbor, NY.

Matsuzaki M, Ellis-Davies GC, Nemoto T, Miyashita Y, Iino M, Kasai H. 2001. Dendritic spine geometry is critical for AMPA receptor expression in hippocampal CA1 pyramidal neurons. *Nat Neurosci* **4:** 1086–1092.

McCray JA, Trentham DR. 1989. Properties and uses of photoreactive caged compounds. *Annu Rev Biophys Biophys Chem* **18:** 239–270.

Mendez F, Penner R. 1998. Near-visible ultraviolet light induces a novel ubiquitous calcium-permeable cation current in mammalian cell lines. *J Physio.* **507:** 365–377.

Parker I, Yao Y. 1991. Regenerative release of calcium from functionally discrete subcellular stores by inositol trisphosphate. *Proc R Soc London Ser B Biol Sci* **246:** 269–274.

Pettit DL, Augustine GJ. 2000. Distribution of functional glutamate and GABA receptors on hippocampal pyramidal cells and interneurons. *J Neurophysiol* **84:** 28–38.

Pettit DL, Wang SS-H, Gee KR, Augustine GJ. 1997. Chemical two-photon uncaging: A novel approach to optical mapping of glutamate receptors. *Neuron* **19:** 465–469.

Pettit DL, Helms MC, Lee P, Augustine GJ, Hall WC. 1999. Local excitatory circuits in the intermediate gray layer of the superior colliculus. *J Neurophysiol* **81:** 1424–1427.

Rapp G, Guth K. 1988. A low cost high intensity flash device for photolysis experiments. *Pflugers Arch* **411:** 200–203.

Smith MA, Ellis-Davies GC, Magee JC. 2003. Mechanism of the distance-dependent scaling of Schaffer collateral synapses in rat CA1 pyramidal neurons. *J Physiol* **548:** 245–258.

Tanaka J, Horiike Y, Matsuzaki M, Miyazaki T, Ellis-Davies GC, Kasai H. 2008. Protein synthesis and neurotrophin-dependent structural plasticity of single dendritic spines. *Science* **319:** 1683–1687.

Wang SS-H, Augustine GJ. 1995. Confocal imaging and local photolysis of caged compounds: Dual probes of synaptic function. *Neuron* **15:** 755–760.

Wang SS-H, Khiroug L, Augustine GJ. 2000. Quantification of spread of cerebellar long-term depression with chemical two-photon uncaging of glutamate. *Proc Natl Acad Sci* **97:** 8635–8640.

Wieboldt R, Gee KR, Niu L, Ramesh D, Carpenter BK, Hess GP. 1994. Photolabile precursors of glutamate: Synthesis, photochemical properties, and activation of glutamate receptors on a microsecond time scale. *Proc Natl Acad Sci* **91:** 8752–8756.

Zayat L, Baraldo LM, Etchenique R. 2011. Uncaging with visible light: Inorganic caged compounds. In *Imaging: A laboratory manual* (ed. Yuste R), pp. 897–902. Cold Spring Harbor Laboratory Press, Cold Spring Harbor, NY.

5 Engineering Light-Regulated Ion Channels

Doris L. Fortin, Timothy W. Dunn, and Richard H. Kramer
Department of Molecular and Cell Biology, University of California Berkeley, Berkeley, California 94720-3200

ABSTRACT

Optical regulation provides unrivaled spatial and temporal precision for the control of protein function. We describe a strategy to engineer photosensitivity onto ion channels based on small photoisomerizable molecules. Photosensitivity is targeted to the ion channel of interest by combining the photoisomerizable moiety with a specific agonist or antagonist. Inclusion of a cysteine-reactive maleimide group enables covalent attachment of the photoswitch to a genetically engineered cysteine on the surface of the ion channel. Exposure to different wavelengths of light triggers changes in the properties of the photoswitch that enable control of protein function.

INTRODUCTION

Optical control of ion channel function offers several advantages over traditional pharmacological regulation. Light can be turned on and off rapidly, allowing control of channel activity with exceptional temporal precision. Because light can be projected precisely in space, ion channel function can also be controlled at specific locations within a cell or tissue. In addition, the ability of light to penetrate tissue enables control of ion channel function from afar, minimizing invasiveness and potential damage to the structure under study.

Several strategies have been developed to control the function of ion channels and receptors with light, including the use of caged molecules (Ellis-Davies 2007) and the development of chimeric proteins that combine the protein of interest with a naturally light-sensitive protein (Airan et al. 2009). This chapter focuses on a third strategy: the use of small photoswitchable molecules to photosensitize genetically engineered or native ion channels (Fig. 1). These molecules contain distinct chemical moieties responsible for regulation of protein function (the ligand), photoisomerization (the photoswitch), and, if applicable, covalent attachment (the reactive moiety). Light absorption induces changes in the properties of the photoisomerizable moiety that are harnessed to trigger biophysical, and ultimately cellular, events. The modular nature of photoswitch molecules allows flexibility in the design of each functional group, yielding a combinatorial tool kit for optical regulation of ion channel function.

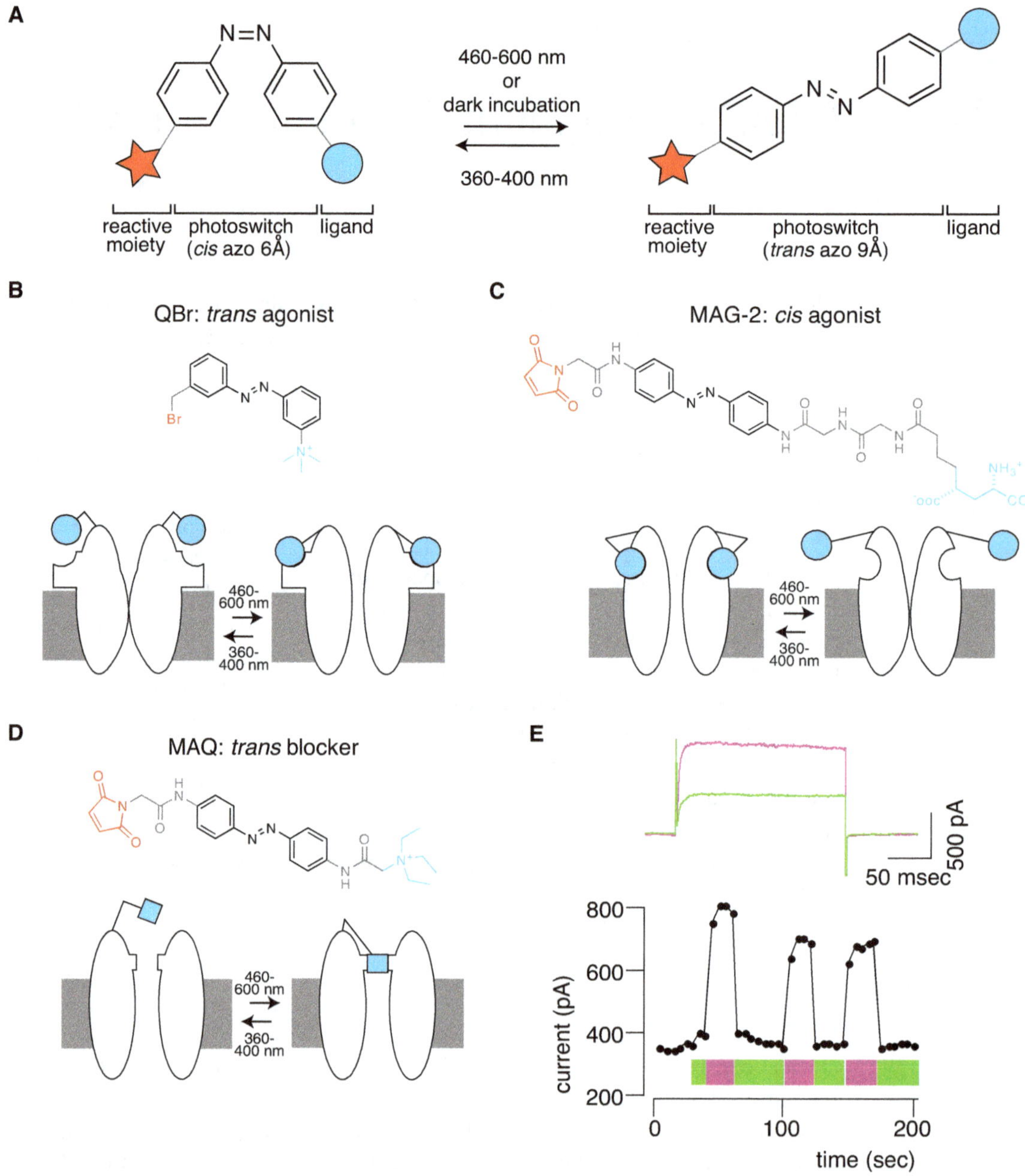

FIGURE 1. Strategies for engineering light-regulated ion channels with azobenzene-containing molecules. *(A)* A photoswitchable tethered ligand (PTL) consists of a reactive moiety (red), a ligand (blue), and the photoswitch (here azobenzene [azo], black). Azobenzene undergoes *trans* to *cis* isomerization under 360–400-nm light. Visible light (>460 nm) or prolonged time in darkness returns the molecule to the *trans* state. The end-to-end distance between the para positions on the *trans* and *cis* azobenzenes is ~9 Å and 6 Å, respectively. *(B)* Tethered QBr is a *trans*-acting agonist of the nAChR. The receptor is activated when the agonist reaches its binding site under 460–600-nm light, whereas illumination with 360–400-nm light retracts the agonist, abrogating ion conduction through the channel. *(C)* MAG-2 is a *cis*-acting iGluR6 agonist that tethers to a genetically engineered cysteine. The agonist reached its binding site only in the *cis* configuration (i.e., under 360–400-nm light). *(D)* MAQ reacts with an engineered cysteine on the surface of a Shaker K^+ channel and acts as a photoisomerizable tethered blocker, blocking ion conduction in the *trans* but not in the *cis* configuration. *(E)* *(Top)* Tethering MAQ to a genetically engineered cysteine (E422C) in a modified Shaker channel enables photoregulation of ion conduction. Under 500-nm light (green), MAQ is in the *trans* form; the QA group reaches the pore of the channel and blocks ion flow. Photoisomerization to the *cis* configuration with 380-nm light (violet) retracts the blocker, allowing ion flow. *(Bottom)* MAQ-treated Shaker channels can be regulated repetitively and persistently with light.

PHOTOSWITCHABLE TETHERED LIGANDS (PTLs)

The Photoswitch

Azobenzene is ideal for use as a photoswitch molecule because it can be photoisomerized repetitively with a high quantum yield and no detectable photobleaching (Renner and Moroder 2006). The *cis* and *trans* azobenzene isomers differ in length by as much as 3 Å (Fig. 1A); this can be exploited to use azobenzene as a mechanical lever that maneuvers a functional group between two different positions depending on the wavelength of illumination. The *trans* (i.e., extended) azobenzene isomer is predominant after prolonged incubation in darkness because it is more thermodynamically stable than the *cis* isomer. Exposure to near-ultraviolet (UV) light (~360–400 nm) triggers photoisomerization to the *cis* state. The reverse *cis*-to-*trans* conversion occurs spontaneously in darkness but can be accelerated by exposure to visible light (~460–600 nm). Photoconversion between isomers occurs within picoseconds (Lednev et al. 1996; Nägele et al. 1997; Ihalainen et al. 2007).

Complete photoconversion to either *cis* or *trans* azobenzene cannot be achieved because the two isomers have overlapping absorption spectra. The relative amounts of *cis* and *trans* isomers at a given wavelength depend on the quantum yield for the isomerization processes and the lifetime of the *cis* isomer, both of which vary with chemical substitution to the azobenzene core, light intensity, temperature, and the environment surrounding the molecule (Yager and Barrett 2006). Typically, exposure to near-UV or visible light yields populations of azobenzene molecules containing up to ~85% *cis* or ~90% *trans* isomer, respectively (James et al. 2001). Near full conversion (>99%) to the *trans* state can be reached only after prolonged incubation in darkness (Renner and Moroder 2006).

The Reactive Moiety

Current strategies for photosensitizing ion channels with photoswitch-containing molecules rely on cysteine–maleimide chemistry. Surface-exposed endogenous cysteine(s) are first replaced with nonreactive amino acids using site-directed mutagenesis. Individual cysteine residues are then introduced on the surface of the protein and assayed for reactivity with the photoswitch-containing molecule. Maleimide reacts rapidly and irreversibly with cysteines, although the local environment of cysteines can influence their reactivity. The recent introduction of chemical approaches that target distinct genetically encoded protein tags (Marks and Nolan 2006) provides potential alternatives for orthogonal labeling if removing endogenous cysteines has a negative impact on ion channel function.

The Ligand

The ligand chosen in the design of the PTL defines which protein will become photosensitized. To be effective, there must be a "handle" on the ligand, such that it can be extended and conjugated to the azobenzene moiety without destroying the ability of the ligand to bind and activate the target protein. The success of affinity labeling as a technique to characterize enzyme function suggests that ligands can be modified with reactive groups and still maintain at least some biological activity (Silman and Karlin 1969; Wold 1977).

When designing PTLs, it can help to introduce flexible linkers between the different functional moieties to generate molecules slightly longer than the optimal distance between the ligand binding and attachment sites (Krishnamurthy et al. 2007). Flexible linkers impart conformational flexibility that might be necessary for tethering but can also compromise the effectiveness of the ligand by increasing the entropic cost of binding (Chung et al. 2009). The presence of flexible linkers can also introduce uncertainty by making the average length of the molecule different from its extended length (Kramer and Karpen 1998; Green et al. 2001). However, the increasing availability of computational tools, including molecular simulation and docking software, combined with the growing number of solved three-dimensional (3D) protein structures should aid in the rational design of PTL/target protein interactions.

TARGETED PROTEINS

Nicotinic Acetylcholine Receptor

Erlanger and colleagues were the first to use azobenzene-containing molecules to regulate the activity of the nicotinic acetylcholine receptor (nAChR) (Bartels et al. 1971). They synthesized 3-(α-bromo-methyl)-3′-[α-(trimethylammonium)methyl]azobenzene bromide (QBr), a photoisomerizable agonist that attaches covalently to a previously reduced natively encoded cysteine on the surface of the nAChR. Once tethered, QBr activates the nAChR only in its *trans* configuration (Fig. 1B). The ability to toggle rapidly between activating and nonactivating QBr using light minimized nAChR desensitization allowing the kinetics of receptor activation to be studied with great temporal accuracy (Lester et al. 1980).

Ionotropic Glutamate Receptor

If available, protein structure information can guide the development of PTLs and engineering of target ion channels. For example, a light-activated ionotropic glutamate receptor (LiGluR) was designed based on the X-ray structure of the ligand-binding domain of the ionotropic glutamate receptor-6 (iGluR6) bound to the agonist (2*S*,4*R*)-4-methyl glutamate (Mayer 2005). Binding of an agonist causes the ligand-binding domain to engulf the ligand, except for a small cavity through which a molecular tether that includes an azobenzene moiety and a maleimide reactive group can be threaded (Volgraf et al. 2006). A cysteine was engineered into iGluR6 such that maleimide–azobenzene–glutamate-2 (MAG-2), a PTL for this receptor, reaches the ligand-binding domain and activates the receptor only in its *cis* configuration (Fig. 1C; Volgraf et al. 2006). Activation of the resulting LiGluR with near-UV light depolarizes neurons and induces action potential firing (Szobota et al. 2007). A shorter PTL, MAG0, activates iGluR in its *cis* and *trans* configurations, depending on the position of the cysteine attachment site. Molecular dynamics simulations revealed that the site of attachment for MAG0 influences the occupancy of the ligand-binding domain and its closure after binding of the PTL, thus determining which photoisomer is "active" (Numano et al. 2009).

Voltage-Gated Potassium Channel

The engineering of the synthetic photoisomerizable azobenzene-regulated K^+ (SPARK) channel (Banghart et al. 2004) was also guided by the abundant biophysical and structural information available on voltage-gated K^+ channels. In particular, a 4 Å difference in the length of tethered tetraethylammonium pore blockers (11 Å vs. 15 Å) attached to an engineered cysteine (E422C) made the distinction between ineffective or complete blockage of the channel (Blaustein et al. 2000). Based on this, the PTL maleimide–azobenzene–quaternary ammonium (MAQ) was synthesized to photoregulate the activity of a Shaker K^+ channel. Once MAQ is tethered to its target cysteine (E422C), the blocking quaternary ammonium can reach the pore of the channel only in its extended (~17 Å) *trans* configuration but not in its short (~10 Å) *cis* form (Fig. 1D,E). Additional mutations were engineered into the channel to abolish fast inactivation (Hoshi et al. 1990) and shift the voltage sensitivity of activation (Lopez et al. 1991), generating a channel with a high resting K^+ conductance that hyperpolarizes and silences neurons illuminated with near-UV light. Illumination with visible light induces photoisomerization to the *trans* configuration and closes the chemical gate, restoring normal membrane potential and action potential firing (Banghart et al. 2004). Mutation of another amino acid in SPARK (V443Q) generates a nonspecific depolarizing cation channel (D-SPARK) by altering the relative permeability of the channel for Na^+ ions (Heginbotham and MacKinnon 1992). Exposure to near-UV light triggers action potential firing in MAQ-treated neurons expressing D-SPARK that can then be halted by illumination with visible light (Chambers et al. 2006).

MODIFICATIONS TO EXISTING LIGHT-GATED CHANNELS

Tuning the Photoswitch

Cis azobenzene reverts spontaneously to the *trans* configuration in the dark, a process known as thermal relaxation. Thus, continuous illumination is required to maintain the *cis* configuration on modified proteins. In the case of LiGluR, the protein environment surrounding the PTL stabilizes the *cis* state (Gorostiza et al. 2007). Because this might not be the case for other proteins, it can be desirable to design azobenzene-based molecules with enhanced thermal stability. Modification of the nature and position of substituents on the azobenzene core can decrease the rate of thermal relaxation from milliseconds to days (Forber et al. 1985; Pozhidaeva et al. 2004; Chi et al. 2006; Sadovski et al. 2009).

Existing PTLs undergo maximal photoisomerization to the *cis* configuration with exposure to near-UV light. Because short-wavelength light is scattered by living tissue, it can be advantageous to shift the action spectra of PTLs from near-UV toward longer wavelengths by modifying the azobenzene core to shift its spectral sensitivity (Chi et al. 2006; Sadovski et al. 2009). Similarly, it can be useful to modify current PTLs for use in deep tissue by improving their two-photon absorption cross-section. Although the two-photon absorption cross-sections of several azobenzene-containing molecules have been investigated (De Boni et al. 2005), these photoswitches have not been incorporated into the design of PTLs. Also, nonlinear optics have not yet been used to regulate ion channels modified with azobenzene-containing PTLs, but this is an important future goal.

Engineering Ion Channel Targets

The introduction of previously characterized trafficking motifs in light-regulated ion channels can enable better control of biological function. For example, SPARK channels localize to the cell body and axon when expressed heterologously in neurons. This localization is adequate for the photocontrol of action potential firing (Banghart et al. 2004). However, depending on the study, it might be more efficacious to target SPARK to dendrites, using a dendritic localization motif from Kv4.2 (Rivera et al. 2003) or a myosin-binding domain (Lewis et al. 2009), to silence inputs to the SPARK-expressing neuron. The cytoplasmic loop linking domains II and III of voltage-gated Na^+ channels (Garrido et al. 2003; Lemaillet et al. 2003) can be used to target proteins to the axon initial segment. Targeting light-regulated depolarizing channels such as D-SPARK and LiGluR to the axon initial segment will mimic the way spikes are initiated naturally in vivo and could be more likely to induce normal downstream events, such as synaptic plasticity.

In addition to providing a means to photoregulate biological functions (e.g., neuronal firing), the use of PTLs also enables functional studies of target proteins in cells and tissues. This can be particularly useful in cases in which few or no specific pharmacological regulators of the protein of interest exist (e.g., between different members of a K^+ channel subfamily). Traditional gene knockouts have provided a wealth of information about ion channel function but sometimes result in unintended lethality or induce compensation from related channels. Replacement of a native ion channel with a cysteine-containing variant in combination with a PTL provides a "chemical knockout" strategy. As a proof-of-concept for this method, we are using the previously characterized MAQ to photosensitize various K^+ channels containing a cysteine at a position analogous to E422C in Shaker. Because of the divergence of loop sequences between different K^+ channels, cysteine scanning might be required to identify the optimal tethering position. The chemical knockout strategy is not limited to K^+ channels; the modular nature of PTLs allows flexibility in the design of each functional group. The choice of ligand and the introduction of the cysteine attachment site will determine which ion channel/receptor is targeted.

Photochromic Ligands

The use of PTLs involves engineering target proteins to contain a cysteine attachment site followed by heterologous expression of the protein in cells. However, if genetic manipulation is not available or not desirable, photosensitive chemicals that target native proteins can provide an alternative one-component system for photosensitization. We have described photoswitchable molecules that confer photosensitivity to endogenously expressed K^+ channels in genetically unadulterated cells or tissues. These molecules contain a pore-blocking quaternary ammonium, a photoisomerizable azobenzene and an electrophilic group for attachment to the channel. In particular, acryl–azobenzene–quaternary ammonium (AAQ) efficiently photosensitizes K^+ channels expressed natively in neurons and enables control of neuronal excitability in dissociated cultures, slices, and semi-intact neuronal preparations (Fortin et al. 2008). Further characterization suggested that AAQ acts as an open channel blocker and thus at the internal tetraethylammonium-binding site of K^+ channels (Banghart et al. 2009). Replacement of the acryl moiety with a series of nonreactive aliphatic tails results in nontethered photoswitch-containing molecules whose potency for K^+ channel blocking generally correlates with tail length and hydrophobicity (Banghart et al. 2009). Similar nontethered agonists have been generated for the control of the nAChR and iGluRs (Deal et al. 1969; Bartels et al. 1971; Volgraf et al. 2007). Unlike photolysis of caged molecules, which consumes the starting material and cannot be reversed, nontethered azobenzene-containing ligands can be toggled repeatedly between *cis* and *trans* configurations to turn ion channel function on and off without genetic modification.

PRACTICAL CONSIDERATIONS FOR USING PTL-MODIFIED ION CHANNELS

Light

We have successfully regulated the activity of PTL-modified ion channels using lasers and other light sources such as mercury and xenon arc lamps equipped with band-pass filters. Our typical light output is in the low milliwatts per square centimeter (mW/cm^2) range or below (measured with a Newport light power meter and normalized to the illuminated area at the sample plane). For azobenzene-based PTLs, the intensity of light influences the relative amount of *cis* and *trans* photoisomers generated at a given wavelength and the rate at which photoequilibrium is reached.

Cysteine–Maleimide Reactivity

Maleimides react specifically and rapidly with cysteine residues; for example, *N*-ethyl maleimide reacts within seconds at neutral pH (Gregory 1955). Cysteine residues are deprotonated and thus most reactive at slightly alkaline pH, which can, however, accelerate the nonspecific hydrolysis of maleimide. For efficient labeling of cysteine-containing ion channels with PTLs, we typically apply the PTL in pH 7.4–8 saline solution. The redox state of cysteine residues will also affect the efficiency and extent of maleimide–cysteine labeling. When photoregulating K^+ channels expressed in HEK293 cells, pretreatment with a reducing agent (1 mM dithiothreitol or 1 mM *tris*(2-carboxyethyl)phosphine) is essential for efficient tethering, presumably because the high levels of cystine in Dulbecco's minimum essential-based growth media oxidize surface-exposed cysteine residues. If ion channel function requires disulfide bonds, follow the reducing treatment with a "resting" incubation in normal saline solution to allow native disulfide pairs to reform before PTL treatment.

CONCLUSION

The method described here for engineering light-regulated ion channels and receptors is based on exogenously applied synthetic photoisomerizable molecules. The versatility of the photoswitch approach has already been demonstrated with the development of PTLs and photochromic ligands for different classes of proteins, including an ionotropic glutamate receptor (Volgraf et al. 2006) and voltage-gated K^+ channels (Banghart et al. 2004; Fortin et al. 2008). Based on the increasing availability of structural data and extensive pharmacological knowledge of various ion channels and receptors, extensions of this approach to other proteins are likely forthcoming.

ACKNOWLEDGMENTS

This work was supported by the NIH Eye Institute (R01EY018957) and the Nanomedicine Development Center for the Optical Control of Biological Function (PN2EY018241). We thank the UC Berkeley College of Chemistry for use of computing resources.

REFERENCES

Airan RD, Thompson KR, Fenno LE, Bernstein H, Deisseroth K. 2009. Temporally precise in vivo control of intracellular signalling. *Nature* **458:** 1025–1029.

Banghart M, Borges K, Isacoff E, Trauner D, Kramer RH. 2004. Light-activated ion channels for remote control of neuronal firing. *Nat Neurosci* **7:** 1381–1386.

Banghart MR, Mourot A, Fortin DL, Yao JZ, Kramer RH, Trauner D. 2009. Photochromic blockers of voltage-gated potassium channels. *Angew Chem Int Ed Engl* **48:** 9097–9101.

Bartels E, Wassermann NH, Erlanger BF. 1971. Photochromic activators of the acetylcholine receptor. *Proc Natl Acad Sci* **68:** 1820–1823.

Blaustein RO, Cole PA, Williams C, Miller C. 2000. Tethered blockers as molecular "tape measures" for a voltage-gated K^+ channel. *Nat Struct Mol Biol* **7:** 309–311.

Chambers JJ, Banghart MR, Trauner D, Kramer RH. 2006. Light-induced depolarization of neurons using a modified Shaker K^+ channel and a molecular photoswitch. *J Neurophysiol* **96:** 2792–2796.

Chi L, Sadovski O, Woolley GA. 2006. A blue-green absorbing cross-linker for rapid photoswitching of peptide helix content. *Bioconjug Chem* **17:** 670–676.

Chung S, Parker JB, Bianchet M, Amzel LM, Stivers JT. 2009. Impact of linker strain and flexibility in the design of a fragment-based inhibitor. *Nat Chem Biol* **5:** 407–413.

Deal WJ, Erlanger BF, Nachmansohn D. 1969. Photoregulation of biological activity by photochromic reagents, III. Photoregulation of bioelectricity by acetylcholine receptor inhibitors. *Proc Natl Acad Sci* **64:** 1230–1234.

De Boni L, Misoguti L, Zílio SC, Mendonça CR. 2005. Degenerate two-photon absorption spectra in azoaromatic compounds. *Chemphyschem* **6:** 1121–1125.

Ellis-Davies GCR. 2007. Caged compounds: Photorelease technology for control of cellular chemistry and physiology. *Nat Methods* **4:** 619–628.

Forber CL, Kelusky EC, Bunce NJ, Zerner MC. 1985. Electronic spectra of *cis*- and *trans*-azobenzenes: Consequences of ortho substitution. *J Am Chem Soc* **107:** 5884–5890.

Fortin DL, Banghart MR, Dunn TW, Borges K, Wagenaar DA, Gaudry Q, Karakossian MH, Otis TS, Kristan WB, Trauner D, et al. 2008. Photochemical control of endogenous ion channels and cellular excitability. *Nat Methods* **5:** 331–338.

Garrido JJ, Giraud P, Carlier E, Fernandes F, Moussif A, Fache M-P, Debanne D, Dargent B. 2003. A targeting motif involved in sodium channel clustering at the axonal initial segment. *Science* **300:** 2091–2094.

Gorostiza P, Volgraf M, Numano R, Szobota S, Trauner D, Isacoff EY. 2007. Mechanisms of photoswitch conjugation and light activation of an ionotropic glutamate receptor. *Proc Natl Acad Sci* **104:** 10865–10870.

Green NS, Reisler E, Houk KN. 2001. Quantitative evaluation of the lengths of homobifunctional protein cross-linking reagents used as molecular rulers. *Protein Sci* **10:** 1293–1304.

Gregory JD. 1955. The stability of *N*-ethylmaleimide and its reaction with sulfhydryl groups. *J Am Chem Soc* **77:** 3922–3923.

Heginbotham L, MacKinnon R. 1992. The aromatic binding site for tetraethylammonium ion on potassium channels. *Neuron* **8:** 483–491.

Hoshi T, Zagotta WN, Aldrich RW. 1990. Biophysical and molecular mechanisms of Shaker potassium channel inactivation. *Science* **250:** 533–538.

Ihalainen JA, Bredenbeck J, Pfister R, Helbing J, Chi L, van Stokkum IHM, Woolley GA, Hamm P. 2007. Folding and unfolding of a photoswitchable peptide from picoseconds to microseconds. *Proc Natl Acad Sci* **104:** 5383–5388.

James DA, Burns DC, Woolley GA. 2001. Kinetic characterization of ribonuclease S mutants containing photoisomerizable phenylazophenylalanine residues. *Protein Eng* **14:** 983–991.

Kramer RH, Karpen JW. 1998. Spanning binding sites on allosteric proteins with polymer-linked ligand dimers. *Nature* **395:** 710–713.

Krishnamurthy VM, Semetey V, Bracher PJ, Shen N, Whitesides GM. 2007. Dependence of effective molarity on linker length for an intramolecular protein–ligand system. *J Am Chem Soc* **129:** 1312–1320.

Lednev IK, Ye T-Q, Hester RE, Moore JN. 1996. Femtosecond time-resolved UV-visible absorption spectroscopy of *trans*-azobenzene in solution. *J Phys Chem* **100:** 13338–13341.

Lemaillet G, Walker B, Lambert S. 2003. Identification of a conserved ankyrin-binding motif in the family of sodium channel α subunits. *J Biol Chem* **278:** 27333–27339.

Lester HA, Krouse ME, Nass MM, Wassermann NH, Erlanger BF. 1980. A covalently bound photoisomerizable agonist: Comparison with reversibly bound agonists at *Electrophorus* electroplaques. *J Gen Physiol* **75:** 207–232.

Lewis TL Jr, Mao T, Svoboda K, Arnold DB. 2009. Myosin-dependent targeting of transmembrane proteins to neuronal dendrites. *Nat Neurosci* **12:** 568–576.

Lopez GA, Ian YN, Ian LY. 1991. Hydrophobic substitution mutations in the S4 sequence alter voltage-dependent gating in Shaker K^+ channels. *Neuron* **7:** 327–336.

Marks KM, Nolan GP. 2006. Chemical labeling strategies for cell biology. *Nat Methods* **3:** 591–596.

Mayer ML. 2005. Crystal structures of the GluR5 and GluR6 ligand binding cores: Molecular mechanisms underlying kainate receptor selectivity. *Neuron* **45:** 539–552.

Nägele T, Hoche R, Zinth W, Wachtveitl J. 1997. Femtosecond photoisomerization of *cis*-azobenzene. *Chem Phys Lett* **272:** 489–495.

Numano R, Szobota S, Lau AY, Gorostiza P, Volgraf M, Roux B, Trauner D, Isacoff EY. 2009. Nanosculpting reversed wavelength sensitivity into a photoswitchable iGluR. *Proc Natl Acad Sci* **106:** 6814–6819.

Pozhidaeva N, Cormier M-E, Chaudhari A, Woolley GA. 2004. Reversible photocontrol of peptide helix content: Adjusting thermal stability of the *cis* state. *Bioconjug Chem* **15:** 1297–1303.

Renner C, Moroder L. 2006. Azobenzene as conformational switch in model peptides. *Chembiochem* **7:** 868–878.

Rivera JF, Ahmad S, Quick MW, Liman ER, Arnold DB. 2003. An evolutionarily conserved dileucine motif in Shal K^+ channels mediates dendritic targeting. *Nat Neurosci* **6:** 243–250.

Sadovski O, Beharry AA, Zhang F, Woolley GA. 2009. Spectral tuning of azobenzene photoswitches for biological applications. *Angew Chem Int Ed Engl* **48:** 1484–1486.

Silman I, Karlin A. 1969. Acetylcholine receptor: Covalent attachment of depolarizing groups at the active site. *Science* **164:** 1420–1421.

Szobota S, Gorostiza P, Del Bene F, Wyart C, Fortin DL, Kolstad KD, Tulyathan O, Volgraf M, Numano R, Aaron HL, et al. 2007. Remote control of neuronal activity with a light-gated glutamate receptor. *Neuron* **54:** 535–545.

Volgraf M, Gorostiza P, Numano R, Kramer RH, Isacoff EY, Trauner D. 2006. Allosteric control of an ionotropic glutamate receptor with an optical switch. *Nat Chem Biol* **2:** 47–52.

Volgraf M, Gorostiza P, Szobota S, Helix MR, Isacoff EY, Trauner D. 2007. Reversibly caged glutamate: A photochromic agonist of ionotropic glutamate receptors. *J Am Chem Soc* **129:** 260–261.

Wold F. 1977. Affinity labeling—An overview. *Methods Enzymol* **46:** 3–14.

Yager KG, Barrett CJ. 2006. Novel photo-switching using azobenzene functional materials. *J Photochem Photobiol A Chem* **182:** 250–261.

6 Microbial Opsins

A Family of Single-Component Tools for Optical Control of Neural Activity

Ofer Yizhar,[1] Lief Fenno,[1] Feng Zhang,[1] Peter Hegemann,[2] and Karl Deisseroth[1,3]

[1]Department of Bioengineering, Stanford University, Stanford, California 94305; [2]Institute of Biology, Experimental Biophysics, Humboldt University, D-10115 Berlin, Germany; [3]Department of Psychiatry and Behavioral Sciences, Stanford University, Stanford, California 94305

ABSTRACT

Members of the microbial opsin gene family have emerged recently as crucial tools for optogenetics, a new neuroscience technology. "Optogenetics" can be defined as the integration of optics and genetics to control well-defined events (such as action potentials) within specified cells (such as a targeted class of projection neurons) in living tissues (such as the brains of freely behaving mammals). In this chapter, we focus on the diversity of the microbial opsin genes and the structure–function properties of their corresponding proteins.

INTRODUCTION

Light is the ultimate source of energy, and a major source of environmental information, for almost all living things. For this reason, the efficient capture and use of photons is a widespread and exquisitely evolved process that is represented across all kingdoms of life. Microorganisms have evolved particularly diverse light-dependent systems to generate energy, avoid maladaptive situations, and identify environments suitable for nutrition and symbiosis, in many cases using different members of the seven-transmembrane-segment opsin gene family. For example, using microbial opsin genes (which include bacteriorhodopsins, halorhodopsins, sensory rhodopsins, and channelrhodopsins) (Fig. 1), prokaryotes can transduce light to shift proton gradients, modulate chloride balance, or switch flagellar motor direction, whereas motile algae transduce light to change flagellar beating to direct locomotion toward environments optimally illuminated for their photosynthetic requirements.

The central operating principle of these elegant molecular machines for ion flux is their unitary, simple nature: They combine the two tasks of light sensation and ion flux into a single pro-

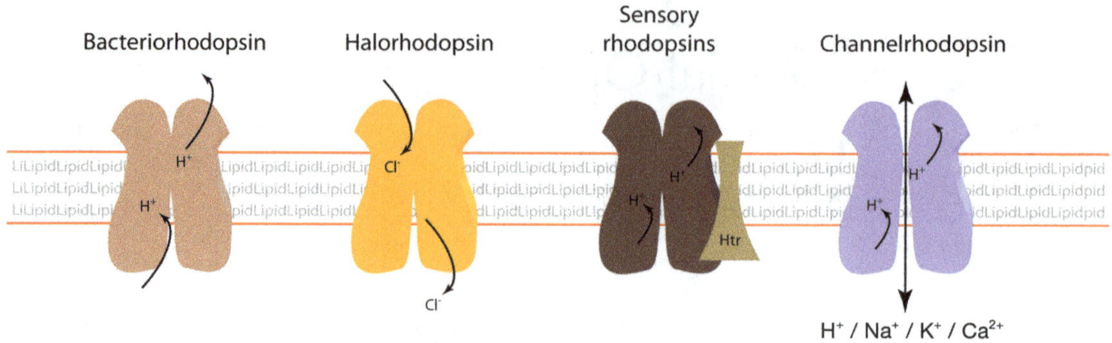

Opsin family	Ion transported	Mode of transport	Peak activation WL (nm)
Bacteriorhodopsin	H^+	Pump	568
Halorhodopsin	Cl^-	Pump	580
Sensory rhodopsin I / II	-	-	587 / 487
Channelrhodopsins (ChR1/ChR2/VChR1/VChR2)	$H^+ > Na^+ > K^+ > Ca^{2+}$	Channel	500/470/540/470

FIGURE 1. Overview of type I microbial rhodopsins. Bacteriorhodopsin and proteorhodopsins pump protons from the cytoplasm to the extracellular medium, whereas halorhodopsins pump chloride into the cytoplasm; all three change the membrane voltage in the same direction. In sensory rhodopsins, the internal proton movement is converted to conformational changes that drive signaling cascades. Algal channelrhodopsins flux cations across the membrane in both directions but always along the electrochemical gradient of the transported ions. In sensory and channelrhodopsins, proton translocation within the protein is required for efficient photocycle progression, but these protons are not necessarily exchanged between the intra- and extracellular space.

tein (with a bound small organic cofactor) encoded by a single gene (Oesterhelt and Stoeckenius 1971). Because of this unitary operating principle, the diverse and efficient microbial opsins have (in retrospect) always had potential for application to neuroscience. However, this potential remained latent for several decades after discovery; many technical challenges needed to be overcome for this possibility to be realized. It has since been shown that expressing microbial opsin genes in neurons can enable millisecond-precision optical control of genetically targeted neuronal populations in mammals (Boyden et al. 2005; Li et al. 2005; Ishizuka et al. 2006; Adamantidis et al. 2007; Zhang et al. 2007, 2008).

This approach, which we termed "optogenetics" (Deisseroth et al. 2006), is fundamentally distinct from earlier designs for optical control (Kramer et al. 2005; Miesenböck and Kevrekidis 2005). It capitalizes on the portability of microbial opsins in efficiently coding for light sensation and effector function within a single gene, and the endogenous presence of the essential organic cofactor retinal in all vertebrate tissues tested, including brain. Thus, the single-component nature of the system is preserved, and no exogenous chemicals need to be added. Furthermore, neurons transduced with opsins are responsive to visible-spectrum light with biologically well-tolerated wavelengths and intensities. Also, the ability to deliver temporally precise light pulses to deep mammalian tissues in vivo with novel, engineered optical devices provides a great degree of experimental versatility. Together, these properties have triggered a surge of interest not only into immediate neuroscience applications, but also into using genomics and molecular engineering to expand the repertoire of tools and generate novel classes of functionality for neuroscience. Neuroscience applications will be reviewed in subsequent chapters (Chapters 79 and 80); here we focus on the fundamentals of microbial opsin gene relationships and the structure–function properties of the ancient and powerful class of proteins encoded by these genes.

COMMON PRINCIPLES OF OPERATION

All opsins require retinal, a vitamin A–related organic cofactor that serves as the antenna for photons, to function (Fig. 2A); when retinal is bound, the functional opsin proteins are termed rhodopsins. After retinal has diffused into the binding pocket of the seven-transmembrane-helix oval, it is covalently attached to a conserved lysine residue of helix 7 by formation of a protonated retinal Schiff base ($RSBH^+$) (Fig. 2A). The ionic environment of the $RSBH^+$, defined by the residues of the binding pocket, dictates the spectral and kinetic characteristics of each individual opsin. Upon

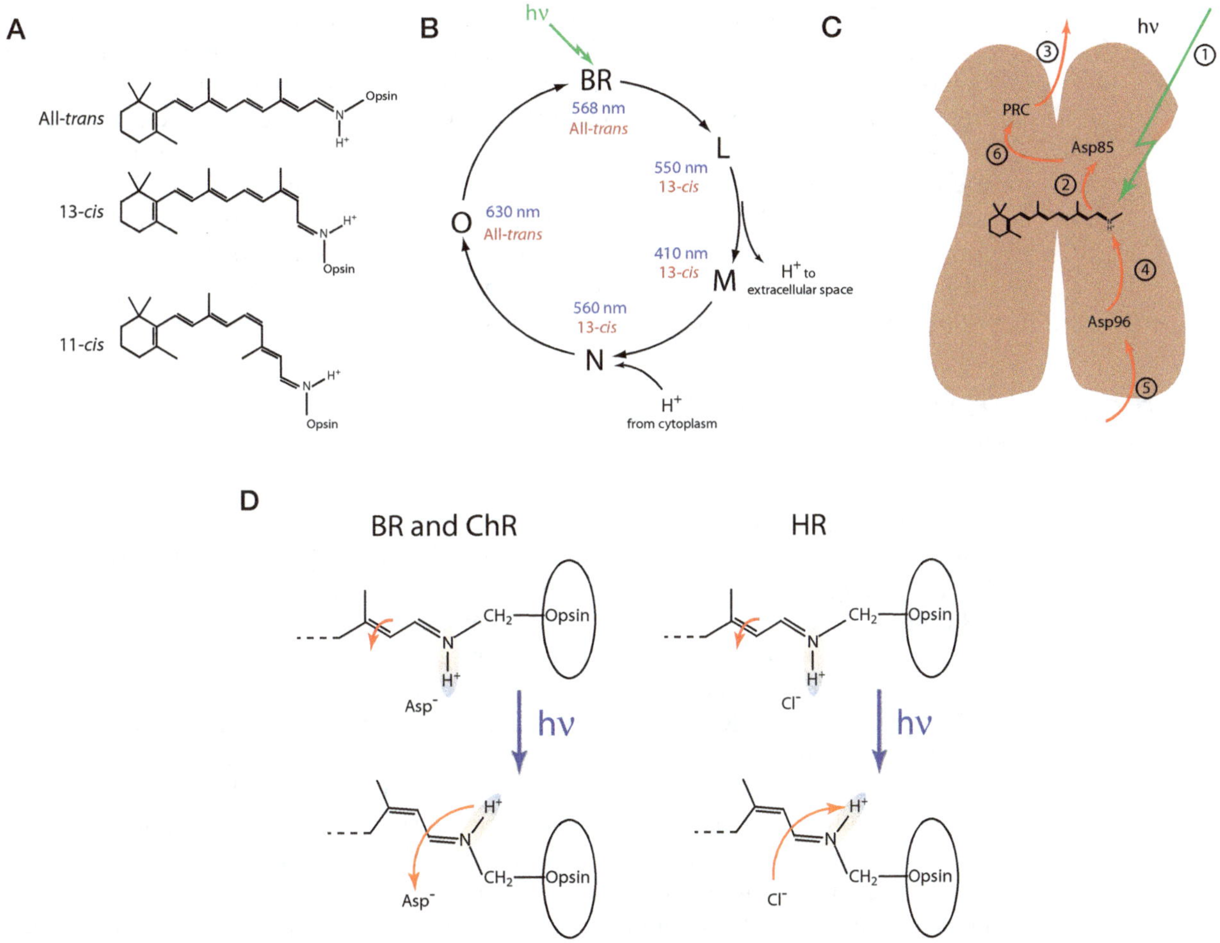

FIGURE 2. Photoreaction mechanisms. (*A*) Distinct configurations of the retinal Schiff base (RSB). (*Top*) Retinal in the all-*trans* state, as found in the dark-adapted state of microbial rhodopsins and in the light-activated forms of type II rhodopsins of higher eukaryotes. (*Middle*) Upon absorbing a photon, protonated microbial RSB converts to the 13-*cis* retinal configuration. (*Bottom*) 11-*cis* retinal is found only in type II rhodopsins, where it binds to the opsin in the dark state before isomerizing to the all-*trans* position after photon absorption. (*B*) The photocycle of bacteriorhodopsin (BR) is initiated from the dark-adapted state; photon absorption activates a sequence of photochemical reactions and structural changes represented by the indicated photointermediates. Also shown are the configuration of the RSB at each step (red) and the wavelength at which each intermediate absorbs light maximally (blue). (*C*) Summary of proton transport reactions during the BR photocycle. Photon absorption (1) initiates the conformational switch in the RSB, leading to transfer of a proton to Asp-85 (2), release of a proton from the proton release complex (PRC) (3), reprotonation of the RSB by Asp-96 (4), uptake of a proton from the cytoplasm to reprotonate Asp-96 (5), and the reprotonation of the PRC from Asp-85 (6). (*D*) Light-induced switching of dipole orientation in response to photon absorption in BR, ChR, and HR. The configuration switch triggers the transfer of the RSB proton to Asp-85 or Glu-123 in BR or ChR2, respectively. In HR, dipole switching facilitates the transfer of a Cl^- ion from the cavity formed between the RSB and Thr-143, to a Cl^- binding site cytoplasmic to the RSB, enabling the key transport steps of these pumps.

absorption of a photon, retinal isomerizes and triggers a wave of conformational changes within the opsin partner. The photoisomerized retinal is the trigger for all subsequent structural rearrangements and all activities these proteins perform.

Opsin genes are divided into two distinct superfamilies: microbial-type opsins (type I) and animal-type opsins (type II). Although both families encode seven-transmembrane structures, sequence homology between the two is practically nonexistent; homology within families, however, is high (25%–80% residue similarity) (Man et al. 2003). Type I opsin genes are found in prokaryotes, algae, and fungi, where they control diverse functions such as phototaxis, energy storage, development, and retinal biosynthesis (Spudich 2006). Type II opsin genes are present only in higher eukaryotes and are mainly responsible for vision, but also play roles in circadian rhythm and pigment regulation (Sakmar 2002; Shichida and Yamashita 2003).

Type II opsins encode G-protein-coupled receptors (GPCRs). In the dark, these proteins bind retinal in the 11-*cis* configuration (Fig. 2A, bottom). Upon illumination, retinal isomerizes to the all-*trans* configuration (Fig. 2A, top) and initiates the reactions that underlie the visual phototransduction second messenger cascade. After photoisomerization, the retinal–protein linkage is hydrolyzed; free all-*trans* retinal then diffuses out of the protein and is replaced by a fresh 11-*cis* retinal molecule for another round of signaling (Hofmann et al. 2009). In contrast, type I opsins encode proteins that use retinal in the all-*trans* configuration, which isomerizes upon photon absorption to the 13-*cis* configuration (Fig. 2A, middle). Unlike the situation with type II rhodopsins, the activated retinal molecule in type I rhodopsins does not dissociate from its opsin protein, but thermally reverts to the all-*trans* state while maintaining a covalent bond to its protein partner (Haupts et al. 1997). Type I opsins encode distinct categories of protein, discussed in more detail below.

LIGHT-GATED ION PUMPS: BACTERIORHODOPSINS, PROTEORHODOPSINS, HALORHODOPSINS

Bacteriorhodopsin (BR) was described by Oesterhelt and Stoeckenius (1971) as the first single-component transmembrane protein known to be capable of actively translocating ions across the cell membrane in the light; it was also recently shown to function as an optogenetic tool (Gradinaru et al. 2010). BR is highly expressed in haloarchaeal membranes under low oxygen conditions to maintain a proton gradient across the membrane when respiration ceases (Oesterhelt and Stoeckenius 1971) and to drive ATP synthesis and maintain cellular energetics (Racker and Stoeckenius 1974; Michel and Oesterhelt 1976). Early work on BR noted that the protein cycles between states that absorb light at 568 nm and 400 nm (Oesterhelt and Hess 1973). Further research into the mechanism of this reaction led to detailed descriptions of a more complete photocycle, in which BR shuttles among several functional states called photointermediates (Fig. 2B; Stoeckenius et al. 1977). Each photointermediate has a distinct spectral signature and represents a separate reaction in the proton translocation process. The photocycle is essentially comprised of a series of proton transfer reactions, each with a unique role in the process of routing protons from the intracellular to the extracellular space (Luecke et al. 1999).

Photon absorption by BR initiates the isomerization of bound retinal from the all-*trans* to the 13-*cis* configuration (Fig. 2B), thereby triggering a series of proton-transfer reactions that constitute the proton translocation mechanism of BR (Fig. 2C). The conformational shift and changing dipole of the $RSBH^+$ raises its pK, resulting in release of the RSB proton and generating the M intermediate, named after the active state of vertebrate rhodopsin, which maximally absorbs light at 410 nm; the released proton is accepted by the nearby Asp-85, which is part of the counterion complex. A distinct proton is then released from a group of amino acids near the extracellular side of the opsin, termed the proton release complex (PRC), out to the extracellular medium. At this point, the RSB captures a proton from Asp-96, and the protein enters the N intermediate, which absorbs at 560 nm. Asp-96 is then reprotonated by resident cytoplasmic protons, which is the key step in modulating cellular proton concentration. During this Asp-96 proton uptake, the Schiff base reisomerizes back to all-*trans* (defining the O state), which absorbs at 630 nm. In the final step, Asp-85 transfers its proton to the PRC, thereby completing one full photocycle and returning BR to its original state. The net outcome is that, upon absorption of a single photon, BR releases one proton to the extracellular milieu and absorbs a second proton from the cytoplasm in a sequence of steps that are spatially disconnected. As discussed below, amino acids involved in the proton translocation reactions are commonly conserved across other type I opsins with homology to BR and appear to function as a conserved pathway for proton translocation.

Light-driven proton pumps are not restricted to archaea but are also found in eubacteria. Marine proteobacteria represent a diverse and widespread class of eubacteria found from the Arctic to the Antarctic Oceans, including equatorial and temperate waters at many depths (Sabehi et al. 2007; Atamna-Ismaeel et al. 2008). Their associated opsin genes encode proteorhodopsins (PRs), with proton pump photocycles similar to that of BR (Váró et al. 2003). However, although PRs can be used to drive ATP production, some proteobacteria do not grow faster in response to increased light, and others have photocycles too slow to contribute significantly to cellular energy production, suggesting other functions for PR (Spudich 2006). The spectra of marine PRs are tuned to ocean depth and latitude, and PRs share a high degree of sequence similarity across species, suggesting that genomic techniques might contribute to our understanding of opsin spectral tuning (Man et al. 2003). Interestingly, absorption variances between blue and green wavelengths can depend on a single amino acid residue (Béjà et al. 2001; Man et al. 2003), but attempts to transfer mutations conferring spectral tuning from PR to other microbial opsins have met with limited success (Yoshitsugu et al. 2009). Nevertheless, the utility of single-amino-acid modifications in shifting absorption spectra has been exploited in vitro to screen randomly mutagenized PR libraries, yielding both red- and blue-shifted mutants (Kim et al. 2008).

Although BR and PR can certainly generate outward (i.e., hyperpolarizing) currents that could in principle be useful for neuronal inhibition, these pumps can produce local juxtamembranous (especially extracellular) pH changes (Chow et al. 2010; Gradinaru et al. 2010). In contrast, halorhodopsin (HR) is a distinct class of outward-current-generating archaeal opsin that uses chloride ions (Matsuno-Yagi and Mukohata 1977). HR is an electrogenic ion pump that controls gradients across the cell membrane by transporting chloride ions from the extracellular medium into the cell (Schobert and Lanyi 1982; Bamberg et al. 1984). Its primary photocycle, although qualitatively similar to that of BR, does not show $RSBH^+$ deprotonation (Oesterhelt et al. 1985; Essen 2002) because of a single amino acid change at the acceptor Asp (Fig. 3A): After the light-induced retinal isomerization and $RSBH^+$ dipole switch, the proton cannot be released because the acceptor Asp has been replaced by Thr. Instead, a Cl^- ion is shuffled from the external side of the $RSBH^+$ chromophore to the internal side, which is then funneled in subsequent steps to bulk intracellular phase; the external Cl^- binding site is refilled by an extracellular Cl^- ion (Fig. 2D; Kolbe et al. 2000).

Halorhodopsin was the first archaeal protein to be successfully applied in neuroscience, when an experimental screen (Zhang et al. 2007) revealed that the best-known HR (from *Halobacterium salinarum*) failed to maintain stable photocurrents when expressed heterologously, but the HR from the less halophilic *Natronomonas pharaonis* (Scharf and Engelhard 1994) expressed and functioned in metazoan neurons even in vivo and could block cellular activity and action potentials by hyperpolarizing neurons with electrogenic inward Cl^- currents. Inactivation is modest, allowing stable, step-like currents over many tens of minutes in response to steady yellow light (Zhang et al. 2007). Because of the stoichiometry of only one transported Cl^- per photocycle, robust expression is required. This is best achieved by addition of mammalian membrane trafficking signals (Gradinaru et al. 2008), which could become a common theme when adapting prokaryotic membrane proteins for expression in mammalian cells. Under these conditions, nanoampere-scale currents and hyperpolarizations of >100 mV can be achieved, with responsivity to yellow or even red light pulses (Gradinaru et al. 2010).

Following the discovery of BR and HR, two more retinal-binding proteins were identified from *H. salinarum*. Sensory rhodopsin I (SR-I) and phoborhodopsin (now termed SR-II) were identified initially as behaviorally relevant photosensors (Hildebrand and Dencher 1975; Takahashi et al. 1985). SRs use a photocycle similar to that of BR (Spudich and Bogomolni 1984; Spudich 1998), except that light-initiated conformational changes and internal proton movement within the opsin are used to

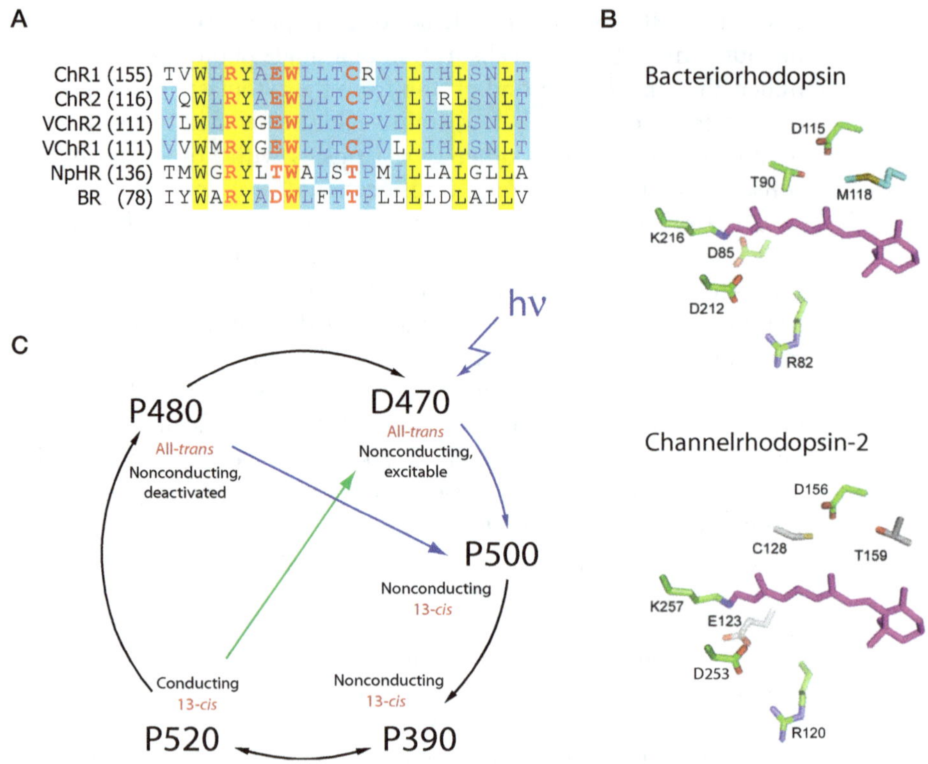

FIGURE 3. Structural and functional homologies of BR and ChR. (*A*) Sequence homology between the algal ChRs and BR within the third transmembrane helix. Identical and conserved residues are highlighted in yellow and blue, respectively; amino acids interacting with the chromophore are shown in red. In BR, R82 functions as a connector between counterion and the PRC; D85 is the counterion to which the RSB proton is transferred; W86 provides hydrophobicity for retinal binding; and the hydrogen bond between T90 on helix 3 and D115 on helix 4 (not shown) stabilizes the structure of the binding pocket and enables both residues to control rapid reprotonation and reisomerization of retinal to the all-*trans* state following a light pulse. (NpHR) *Natronomonas pharaonis* HR. (*B*) Anticipated homology in the retinal binding pocket between BR and ChR2. The BR pocket is shown based on structure 1KGB (Facciotti et al. 2001) with key amino acids that are involved in the proton transfer reaction. The ChR2 pocket was modeled after the BR structure, based on sequence homology in the relevant positions. (*C*) Model for the photocycle of ChR2. The RSB configuration and conductive state of each photocycle component are shown in red and black, respectively. The D470 dark state is converted by a light-induced isomerization of retinal via the early intermediate P500 and the transient intermediate P390 to the conducting state P520. The recovery of the D470 dark state proceeds either thermally via the nonconducting P480 intermediate or photochemically via possible short-lived intermediates (green arrow). The late or desensitized P480 state can also be activated (blue arrow) to yield the early intermediate P500.

activate a closely associated transducer molecule, Htr (Büldt et al. 1998; Chen and Spudich 2002). The activated transducer initiates a phosphorylation cascade that controls the directionality of the flagellar motor and facilitates phototaxis to either the green–yellow range of wavelengths (SR-I, peak absorption 587 nm) or away from the blue range (SR-II, peak absorption 487 nm) (Spudich and Bogomolni 1984; Spudich 2006). Because prokaryotic kinase cascades are fundamentally different from eukaryotic second messengers, opportunities to translate SR function to other systems are more complicated than with the ion conductance-regulating microbial opsins such as HR, although SR proton pumping is possible (Haupts et al. 1996).

LIGHT-GATED ION CHANNELS: CHANNELRHODOPSINS

Continuing the theme of single-component microbial transmembrane ion conductance regulators, channelrhodopsin-1 (ChR1) was identified as a light-gated ion channel in *Chlamydomonas rein-*

hardtii, a green unicellular alga from temperate freshwater environments. Although it was originally identified as proton-selective (Nagel et al. 2002), a broader cation conductance for ChR1, including Na^+ and K^+, was later described (Lin et al. 2009; Tsunoda and Hegemann 2009). Channelrhodopsin-2 (ChR2) was later discovered in the same organism: ChR2 also conducts protons, but its Na^+ and K^+ conductances are more than twofold greater than those of ChR1 (Nagel et al. 2003; Tsunoda and Hegemann 2009). ChR2 was found to express robustly enough in mammalian neurons that precisely timed light-driven action potentials were achievable (Boyden et al. 2005).

Both ChR1 and ChR2 bear some homology to BR and other type I rhodopsins; this homology is strong in the retinal-binding pocket and proton-conducting network (Hegemann et al. 2001; Nagel et al. 2002, 2003; Sineshchekov et al. 2002; Suzuki et al. 2003), suggesting related mechanisms of ionic phototransduction (Fig. 3B). Indeed, the photocycle of ChR2 is similar to that of BR but with different spectral characteristics, likely resulting from variations in the ionic environment near the RSB, larger conformational changes within the protein (Fig. 3C; Ritter et al. 2008; Radu et al. 2009), and possibly higher water content within the proton network. In ChR2, a dark-adapted state absorbing at 470 nm (D470) converts rapidly upon illumination to the conducting state (P520) via the short-lived photointermediates P500 and P390. Illumination of the open channel at this step with green light terminates the photocurrent (Bamann et al. 2008; Berndt et al. 2009) by photochemically shifting the channel back to a closed state, which could be the dark-adapted state (D470) or the light-adapted state (P480) (K. Stehfest and P. Hegemann, unpubl.), effectively resetting the photocycle.

This photocycle shortcut pathway could be relevant only at very high light intensities with wild-type ChR2, but acquires a special utility with certain molecularly engineered slow mutants (e.g., step-function opsin genes or SFOs) (Berndt et al. 2009). Indeed, as with the light-activated ion pumps, molecular engineering and genomic strategies have begun to bear fruit for enhancing function of the light-activated ion channels for particular purposes. The long history of structure–function mutagenesis with regard to photocycle kinetics and absorption spectra in BR, along with the high degree of similarity between BR and ChRs in the chromophore region, has been capitalized on to modify the ChR photocycle and extend the array of tools available to control neuronal activity. For example, SFO mutations at Cys-128 in ChR2 (homologous to Thr-90 in BR) (Perálvarez-Marin et al. 2004) gives rise to a several-hundred-fold extension of the open state (P520) lifetime and resulting long-acting (step function-like) ion fluxes in ChR2 in response to single pulses of light. This property leads to useful bi-stable behavior in the form of sustained subthreshold activation of neurons lasting up to minutes (Berndt et al. 2009; Stehfest et al. 2010). A histidine-to-arginine mutation in the residue homologous to the proton donor Asp-96 in BR (i.e., H134R in ChR2) increases photocurrents, but also slows the deactivation of ChR2 photocurrents, impairing temporal resolution (Nagel et al. 2005; Gradinaru et al. 2007). Chimeric proteins containing fragments of ChR1 and ChR2 reduce inactivation and alter the overall expression in mammalian cells toward higher or lower expression, depending on the individual helices shuffled (Lin et al. 2009; Wang et al. 2009). Finally, modification of the proton acceptor (homologous to Asp-85 in BR) in the Glu-123 position (i.e., ChETA mutations [E123T and E123A]) (Gunaydin et al. 2010) leads to much more rapid channel closing and enables ultrafast optical control of spiking at least up to 200 Hz, with reduced inactivation and improved overall performance.

Genomic strategies have also been productive. For example, novel ChRs have been identified by genomic search strategies in the colonial alga *Volvox carteri*, expanding the family of known ChRs to four members. *Volvox* ChR2 absorbs similarly to ChR2 with a maximum near 470 nm, but the *Volvox* ChR1 (VChR1) absorption is strikingly red-shifted to a maximum at 540 nm (Fig. 1) and allows light-driven spike firing at 589 nm, which has profound advantages for optogenetic applications (Zhang et al. 2008; Kianianmomeni et al. 2009). Generally, light at longer wavelengths is less deleterious for biological tissue. Moreover, tools with diverse spectral tuning offer the potential of addressing multiple distinct neuronal populations that are anatomically intermixed, with different colors of light. Identification of natural opsins in this way, along with rational engineering for novel properties, will likely continue to be productive.

QUANTITATIVE CONSIDERATIONS: OPSIN BIOPHYSICAL PROPERTIES

Many factors control the efficacy of microbial opsins expressed in heterologous systems, not the least of which are cell biological properties ranging from effective transcription, translation, and folding, to proper membrane trafficking and targeting. However, biophysical properties will in many cases be similarly important limiting factors. The most basic of these is the efficiency of light absorption, expressed by the extinction coefficient (ε_{max}). For rhodopsins, ε_{max} is typically between 50,000 and 70,000 M^{-1} cm^{-1}, less than that of chlorophyll (100,000 M^{-1} cm^{-1}) but more than that of most other biological chromophores such as flavins (12,000 M^{-1} cm^{-1}). Another important parameter is quantum efficiency (Φ, the fraction of absorbed photons that are efficacious in driving the relevant conformational change), which varies between 0.3 and 0.7 depending on the opsin species.

Also worthy of detailed consideration is the turnover time of the photocycle, a critical figure of merit both for native function and for neuroscience applications. For the most active transporters (HR and BR), the photocycle turnover time is ~10–20 msec. Turnover is slower for the blue PR (80–100 msec) (Wang et al. 2003), limiting the development of blue PRs for neuroscience applications. However, even these values come from measurements at zero membrane voltage; the turnover for the fast pumps at a more physiologically negative voltage slows dramatically from 10–20 msec to 100–400 msec (Geibel et al. 2001). Thus, the more negative membrane voltage becomes, the slower the photocycle becomes, and more light is needed to achieve a given increment of hyperpolarization. Interestingly, this also means that by hyperpolarizing the membrane, the pump slows down its own photocycle, an effect that will be enhanced by changes in the electrochemical gradient of the transported ion. In general, however, pump direction is not actually inverted by physiologically achievable membrane potentials or ion gradients. Although such an inversion has been reported for PR, the driving force against the normal pump direction arose from extreme negative voltage and a strong pH gradient; under these conditions, the pump can become leaky as protons are forced onto an internal acceptor (Lörinczi et al. 2009). Generally, pump inversion per se is not expected, especially in physiological settings.

For ChRs the kinetic issues are different. Ion transport (i.e., current) is coupled to occupancy of the conducting state. This, in turn, is determined by light intensity and wavelength, extinction coefficient, and quantum efficiency, as well as by a new factor, the lifetime of the resulting conducting state, dictated by the kinetics of the P520–P480 transition (Fig. 3C). For temporally precise depolarization with a single flash, wild-type ChR2 is adequate. In contrast, because of its extended conducting-state lifetime, the H134R mutation of ChR2 noted above delivers approximately twofold higher photocurrents, although temporal precision is somewhat reduced as a result of moderately slower deactivation (Nagel et al. 2005; Gradinaru et al. 2007; Lin et al. 2009). The SFO variants of ChR2, with the greatly slowed deactivation described above (e.g., C128X) (Berndt et al. 2009), achieve chronic depolarization with even brief light delivery. Moreover, in principle, these variants can achieve the same maximal current magnitudes as wild-type ChR but at much lower light levels (Schoenenberger et al. 2009). For a 10-msec lifetime open state (Ritter et al. 2008), ChR2 would have to be excited ~100 times per second for a half-saturating stable current. Assuming an absorption cross-section of 2×10^{-20} m^2, this would require 5×10^{21} photons/m^2-sec (i.e., 5×10^{12} photon/mm-msec or 2 mW/mm^2) chronically. However, as a result of the extended-lifetime conducting states, C128T needs 100 times less light and the C128S mutant more than 1000 times less chronic light delivery for the same activity (Berndt et al. 2009).

Accelerating steps in the photocycle can also present certain advantages. For high-frequency neuronal firing, both fast-on and fast-off kinetics are important, requiring fast formation and decay of the P520 photocycle intermediate. Only the ChETA mutations (e.g., ChR2-E123T or E123A) can provide this unique functionality at present. Total current per flash is slightly smaller than for wild-type ChR2 because of the shorter open time, but as long as good expression is achieved, this can be compensated for by slightly more intense or longer flashes (e.g., 2 msec instead of 1 msec) (Gunaydin et al. 2010). Inactivation must also be considered, which is fast but weak in ChR1 and slow but strong in ChR2. Inactivation is more pronounced at high voltage (i.e., weakly negative or positive) and approaches 70% of peak current at pH 7.5. If inactivation of the current becomes problematic, ChR-E123T (Gunaydin et al. 2010), the ChIEF variant (Lin et al. 2009), or ChR2-E90Q (Ritter et al. 2008) might be suitable.

In addition to conducting-state lifetime, the unitary conductance of a channelrhodopsin will also control efficacy, and estimates for this important value range from 30 fS to ~1 pS for ChR2 (Feldbauer et al. 2009; Lin et al. 2009). In the living alga *Chlamydomonas*, half-maximal current is carried by 1×10^6 ions; if there were 10,000 ChRs per eyespot with quantum efficiency of 0.67 (as with BR), this would correspond to 100 ions per ChR (Harz et al. 1992) and a unitary conductance of 0.3 pS. More recent estimates of opsin number (~100,000 per eyespot) (Sineshchekov et al. 2002; Berthold et al. 2008) put the ions per ChR at 10, and conductance therefore at 30 fS. Recently, this value has been supported by application of noise analysis to purified ChR2 (40 fS) (Feldbauer et al. 2009), although nonstationary fluctuation analysis had resulted in somewhat higher values (>100 fS) (Lin et al. 2009). Despite the uncertainties inherent in all of these methods, the resulting estimates are still reasonably close and together provide useful boundary conditions.

The size of the unitary conductance is also relevant to the prospect of achieving two-photon (2P)-mediated control. Conventional 2P methods pose challenges in this regard, as small excitation volumes with raster-scanning 2P laser systems will typically not recruit sufficient numbers of small-conductance, rapidly deactivating molecules to, for example, drive spiking. Two main approaches are being developed to address this issue, and 2P optogenetic control can now be achieved. First, control of light delivery can be adapted to individual imaged cells to provide spatiotemporally suitable illumination patterns that will recruit sufficient conductance within a single cell and within a temporal window set by current deactivation properties (Rickgauer and Tank 2009). Second, engineered opsins (e.g., SFOs and related mutants in which the deactivation kinetics are altered) can facilitate recruitment of threshold levels of conductance within a given temporal window. In this way, 2P-based recruitment of sparsely distributed cell populations within living tissue can become practical, pointing to the value of continuing to probe the structure–function relationships of the microbial opsins.

The action spectrum is also important, and developing absorbance at longer wavelengths than ChR2 (λ_{max} = 470 nm) is particularly useful for depth penetration, safety, and combinatorial experiments. ChR1 is slightly red-shifted, but does not express well and is pH-dependent. VChR1 (Zhang et al. 2008) absorbs at markedly red-shifted wavelengths (λ_{max} = 540 nm), and spiking can be driven even at 589 nm in hippocampal neurons, but expression levels are significantly lower than that of ChR2 in most host cells. The slower decay of the open state (Zhang et al. 2008) partially compensates, but improvement of VChR1 expression will be very important, and a strongly expressing channelrhodopsin that absorbs beyond 550 nm will outperform and be preferable to ChR2 in the future. Indeed, as with VChR1, the channelrhodopsin variants have diverse useful properties but also corresponding limitations. For example, the SFO mutants have reduced dynamic range. Moreover, because of SFO hypersensitivity and the long lifetime of intermediates, in vitro experiments should be maintained at low light to avoid unwanted activation by ambient light. When imaging is required, illumination should be restricted to nonactivating wavelengths (>560 nm). Complete deactivation of the SFO prior to experiments with a long pulse of 590-nm light will assure that any baseline activity is eliminated and will enable full activation of the SFO with 470-nm light at the desired time (Schoenenberger et al. 2009; Stehfest et al. 2010).

SUMMARY

In this chapter, we have focused on the diversity of the microbial opsin genes and the structure–function properties of the corresponding proteins. Methods for applying microbial opsins to neuroscience are covered in Chapters 79 and 80. Although much has been learned in applying the use of microbial opsins to neuroscience, technological challenges lie ahead, including development of variants with further red-shifted absorption or altered ion selectivity, as well as more complex activation schemes with multiple optogenetic proteins. Biochemical tools are also now available for specific GPCR pathways (Airan et al. 2009), and increasingly complex avenues of control and spectral separation will aid in the design of more refined experiments that do justice to the combinatorial complexity of the mammalian brain. Further rapid expansion of the optogenetic toolkit now will be a natural consequence not only of ever-increasing genomic and metagenomic resources, but also of molecular engineering spurred by increased understanding of the structure–function relationships of these ancient and powerful molecular machines.

REFERENCES

Adamantidis AR, Zhang F, Aravanis AM, Deisseroth K, de Lecea L. 2007. Neural substrates of awakening probed with optogenetic control of hypocretin neurons. *Nature* **450:** 420–424.

Airan RD, Thompson KR, Fenno LE, Bernstein H, Deisseroth K. 2009. Temporally precise in vivo control of intracellular signalling. *Nature* **458:** 1025–1029.

Atamna-Ismaeel N, Sabehi G, Sharon I, Witzel K-P, Labrenz M, Jürgens K, Barkay T, Stomp M, Huisman J, Béjà O. 2008. Widespread distribution of proteorhodopsins in freshwater and brackish ecosystems. *ISME J* **2:** 656–662.

Bamann C, Kirsch T, Nagel G, Bamberg E. 2008. Spectral characteristics of the photocycle of channelrhodopsin-2 and its implication for channel function. *J Mol Biol* **375:** 686–694.

Bamberg E, Hegemann P, Oesterhelt D. 1984. Reconstitution of halorhodopsin in black lipid membranes. *Prog Clin Biol Res* **164:** 73–79.

Béjà O, Spudich EN, Spudich JL, Leclerc M, DeLong EF. 2001. Proteorhodopsin phototrophy in the ocean. *Nature* **411:** 786–789.

Berndt A, Yizhar O, Gunaydin LA, Hegemann P, Deisseroth K. 2009. Bi-stable neural state switches. *Nat Neurosci* **12:** 229–234.

Berthold P, Tsunoda SP, Ernst OP, Mages W, Gradmann D, Hegemann P. 2008. Channelrhodopsin-1 initiates phototaxis and photophobic responses in *Chlamydomonas* by immediate light-induced depolarization. *Plant Cell* **20:** 1665–1677.

Boyden ES, Zhang F, Bamberg E, Nagel G, Deisseroth K. 2005. Millisecond-timescale, genetically targeted optical control of neural activity. *Nat Neurosci* **8:** 1263–1268.

Büldt G, Heberle J, Dencher NA, Sass HJ. 1998. Structure, dynamics, and function of bacteriorhodopsin. *J Protein Chem* **17:** 536–538.

Chen X, Spudich JL. 2002. Demonstration of 2:2 stoichiometry in the functional SRI–HtrI signaling complex in *Halobacterium* membranes by gene fusion analysis. *Biochemistry* **41:** 3891–3896.

Chow BY, Han X, Dobry AS, Qian X, Chuong AS, Li M, Henninger MA, Belfort GM, Lin Y, Monahan PE, et al. 2010. High-performance genetically targetable optical neural silencing by light-driven proton pumps. *Nature* **463:** 98–102.

Deisseroth K, Feng G, Majewska AK, Miesenböck G, Ting A, Schnitzer MJ. 2006. Next-generation optical technologies for illuminating genetically targeted brain circuits. *J Neurosci* **26:** 10380–10386.

Essen L-O. 2002. Halorhodopsin: Light-driven ion pumping made simple? *Curr Opin Struct Biol* **12:** 516–522.

Facciotti MT, Rouhani S, Burkard FT, Betancourt FM, Downing KH, Rose RB, McDermott G, Glaeser RM. 2001. Structure of an early intermediate in the M-state phase of the bacteriorhodopsin photocycle. *Biophys J* **81:** 3442–3455.

Feldbauer K, Zimmermann D, Pintschovius V, Spitz J, Bamann C, Bamberg E. 2009. Channelrhodopsin-2 is a leaky proton pump. *Proc Natl Acad Sci* **106:** 12317–12322.

Geibel S, Friedrich T, Ormos P, Wood PG, Nagel G, Bamberg E. 2001. The voltage-dependent proton pumping in bacteriorhodopsin is characterized by optoelectric behavior. *Biophys J* **81:** 2059–2068.

Gradinaru V, Thompson KR, Zhang F, Mogri M, Kay K, Schneider MB, Deisseroth K. 2007. Targeting and readout strategies for fast optical neural control in vitro and in vivo. *J Neurosci* **27:** 14231–14238.

Gradinaru V, Thompson KR, Deisseroth K. 2008. eNpHR: A *Natronomonas* halorhodopsin enhanced for optogenetic applications. *Brain Cell Biol* **36:** 129–139.

Gradinaru V, Zhang F, Ramakrishnan C, Mattis J, Prakash R, Diester I, Goshen I, Thompson KR, Deisseroth K. 2010. Molecular and cellular principles for diversifying and extending optogenetics. *Cell* **141:** 1–12.

Gunaydin L, Yizhar O, Berndt A, Sohal VS, Deisseroth K, Hegemann P. 2010. Ultrafast optogenetic excitation. *Nat Neurosci* **13:** 387–392.

Harz H, Nonnengässer C, Hegemann P. 1992. The photoreceptor current of the green alga *Chlamydomonas*. *Philos Trans R Soc Lond B Biol Sci* **338:** 39–52.

Haupts U, Bamberg E, Oesterhelt D. 1996. Different modes of proton translocation by sensory rhodopsin I. *EMBO J* **15:** 1834–1841.

Haupts U, Tittor J, Bamberg E, Oesterhelt D. 1997. General concept for ion translocation by halobacterial retinal proteins: The isomerization/switch/transfer (IST) model. *Biochemistry* **36:** 2–7.

Hegemann P, Fuhrmann M, Kateriya S. 2001. Algal sensory photoreceptors. *J Phycol* **37:** 668–676.

Hildebrand E, Dencher N. 1975. Two photosystems controlling behavioural responses of *Halobacterium halobium*. *Nature* **257:** 46–48.

Hofmann KP, Scheerer P, Hildebrand PW, Choe H-W, Park JH, Heck M, Ernst OP. 2009. A G protein-coupled receptor at work: The rhodopsin model. *Trends Biochem Sci* **34:** 540–552.

Ishizuka T, Kakuda M, Araki R, Yawo H. 2006. Kinetic evaluation of photosensitivity in genetically engineered neurons expressing green algae light-gated channels. *Neurosci Res* **54:** 85–94.

Kianianmomeni A, Stehfest K, Nematollahi G, Hegemann P, Hallmann A. 2009. Channelrhodopsins of *Volvox carteri* are photochromic proteins that are specifically expressed in somatic cells under control of light, temperature, and the sex inducer. *Plant Physiol* **151:** 347–366.

Kim SY, Waschuk SA, Brown LS, Jung KH. 2008. Screening and characterization of proteorhodopsin color-tuning mutations in *Escherichia coli* with endogenous retinal synthesis. *Biochim Biophys Acta* **1777:** 504–513.

Kolbe M, Besir H, Essen L-O, Oesterhelt D. 2000. Structure of the light-driven chloride pump halorhodopsin at 1.8 Å resolution. *Science* **288:** 1390–1396.

Kramer RH, Chambers JJ, Trauner D. 2005. Photochemical tools for remote control of ion channels in excitable cells. *Nat Chem Biol* **1:** 360–365.

Li X, Gutierrez DV, Hanson MG, Han J, Mark MD, Chiel H, Hegemann P, Landmesser LT, Herlitze S. 2005. Fast noninvasive activation and inhibition of neural and network activity by vertebrate rhodopsin and green algae channelrhodopsin. *Proc Natl Acad Sci* **102:** 17816–17821.

Lin JY, Lin MZ, Steinbach P, Tsien RY. 2009. Characterization of engineered channelrhodopsin variants with improved properties and kinetics. *Biophys J* **96:** 1803–1814.

Lörinczi E, Verhoefen M-K, Wachtveitl J, Woerner AC, Glaubitz C, Engelhard M, Bamberg E, Friedrich T. 2009. Voltage- and pH-dependent changes in vectoriality of photocurrents mediated by wild-type and mutant proteorhodopsins upon expression in *Xenopus* oocytes. *J Mol Biol* **393:** 320–341.

Luecke H, Schobert B, Richter H-T, Cartailler J-P, Lanyi JK. 1999. Structure of bacteriorhodopsin at 1.55 Å resolution. *J Mol Biol* **291:** 899–911.

Man D, Wang W, Sabehi G, Aravind L, Post AF, Massana R, Spudich EN, Spudich JL, Béjà O. 2003. Diversification and spectral tuning in marine proteorhodopsins. *EMBO J* **22:** 1725–1731.

Matsuno-Yagi A, Mukohata Y. 1977. Two possible roles of bacteriorhodopsin; a comparative study of strains of *Halobacterium halobium* differing in pigmentation. *Biochem Biophys Res Commun* **78:** 237–243.

Michel H, Oesterhelt D. 1976. Light-induced changes of the pH gradient and the membrane potential in *H. halobium*. *FEBS Lett* **65:** 175–178.

Miesenböck G, Kevrekidis IG. 2005. Optical imaging and control of genetically designated neurons in functioning circuits. *Annu Rev Neurosci* **28:** 533–563.

Nagel G, Ollig D, Fuhrmann M, Kateriya S, Musti AM, Bamberg E, Hegemann P. 2002. Channelrhodopsin-1: A light-gated proton channel in green algae. *Science* **296:** 2395–2398.

Nagel G, Szellas T, Huhn W, Kateriya S, Adeishvili N, Berthold P, Ollig D, Hegemann P, Bamberg E. 2003. Channelrhodopsin-2, a directly light-gated cation-selective membrane channel. *Proc Natl Acad Sci* **100:** 13940–13945.

Nagel G, Brauner M, Liewald JF, Adeishvili N, Bamberg E, Gottschalk A. 2005. Light activation of channelrhodopsin-2 in excitable cells of *Caenorhabditis elegans* triggers rapid behavioral responses. *Curr Biol* **15:** 2279–2284.

Oesterhelt D, Hess B. 1973. Reversible photolysis of the purple complex in the purple membrane of *Halobacterium halobium*. *Eur J Biochem* **37:** 316–326.

Oesterhelt D, Stoeckenius W. 1971. Rhodopsin-like protein from the purple membrane of *Halobacterium halobium*. *Nat New Biol* **233:** 149–152.

Oesterhelt D, Hegemann P, Tittor J. 1985. The photocycle of the chloride pump halorhodopsin. II: Quantum yields and a kinetic model. *EMBO J* **4:** 2351–2356.

Perálvarez-Marin A, Márquez M, Bourdelande J-L, Querol E, Padrós E. 2004. Thr-90 plays a vital role in the structure and function of bacteriorhodopsin. *J Biol Chem* **279:** 16403–16409.

Racker E, Stoeckenius W. 1974. Reconstitution of purple membrane vesicles catalyzing light-driven proton uptake and adenosine triphosphate formation. *J Biol Chem* **249:** 662–663.

Radu I, Bamann C, Nack M, Nagel G, Bamberg E, Heberle J. 2009. Conformational changes of channelrhodopsin-2. *J Am Chem Soc* **131:** 7313–7319.

Rickgauer JP, Tank DW. 2009. Two-photon excitation of channelrhodopsin-2 at saturation. *Proc Natl Acad Sci* **106:** 15025–15030.

Ritter E, Stehfest K, Berndt A, Hegemann P, Bartl FJ. 2008. Monitoring light-induced structural changes of channelrhodopsin-2 by UV-visible and Fourier transform infrared spectroscopy. *J Biol Chem* **283:** 35033–35041.

Sabehi G, Kirkup BC, Rozenberg M, Stambler N, Polz MF, Béjà O. 2007. Adaptation and spectral tuning in divergent marine proteorhodopsins from the eastern Mediterranean and the Sargasso Seas. *ISME J* **1:** 48–55.

Sakmar TP. 2002. Structure of rhodopsin and the superfamily of seven-helical receptors: The same and not the same. *Curr Opin Cell Biol* **14:** 189–195.

Scharf B, Engelhard M. 1994. Blue halorhodopsin from *Natronobacterium pharaonis*: Wavelength regulation by anions. *Biochemistry* **33:** 6387–6393.

Schobert B, Lanyi JK. 1982. Halorhodopsin is a light-driven chloride pump. *J Biol Chem* **257:** 10306–10313.

Schoenenberger P, Gerosa D, Oertner TG. 2009. Temporal control of immediate early gene induction by light. *PLoS One* **4:** e8185. doi: 10.1371/journal.pone.0008185.

Shichida Y, Yamashita T. 2003. Diversity of visual pigments from the viewpoint of G protein activation—Comparison with other G protein-coupled receptors. *Photochem Photobiol Sci* **2:** 1237–1246.

Sineshchekov OA, Jung K-H, Spudich JL. 2002. Two rhodopsins mediate phototaxis to low- and high-intensity light in *Chlamydomonas reinhardtii*. *Proc Natl Acad Sci* **99:** 8689–8694.

Spudich JL. 1998. Variations on a molecular switch: Transport and sensory signalling by archaeal rhodopsins. *Mol Microbiol* **28:** 1051–1058.

Spudich JL. 2006. The multitalented microbial sensory rhodopsins. *Trends Microbiol* **14:** 480–487.

Spudich JL, Bogomolni RA. 1984. Mechanism of colour discrimination by a bacterial sensory rhodopsin. *Nature* **312:** 509–513.

Stehfest K, Ritter E, Berndt A, Bartl F, Hegemann P. 2010. The branched photocycle of the slow cycling Channelrhodopsin-2-C128T. *J Mol Biol* **398:** 690–702.

Stoeckenius W, Lozier RH, Niederberger W. 1977. Photoreactions of bacteriorhodopsin. *Biophys Struct Mech* **3:** 65–68.

Suzuki T, Yamasaki K, Fujita S, Oda K, Iseki M, Yoshida K, Watanabe M, Daiyasu H, Toh H, Asamizu E, et al. 2003. Archaeal-type rhodopsins in *Chlamydomonas*: Model structure and intracellular localization. *Biochem Biophys Res Commun* **301:** 711–717.

Takahashi T, Mochizuki Y, Kamo N, Kobatake Y. 1985. Evidence that the long-lifetime photointermediate of s-rhodopsin is a receptor for negative phototaxis in *Halobacterium halobium*. *Biochem Biophys Res Commun* **127:** 99–105.

Tsunoda SP, Hegemann P. 2009. Glu 87 of channelrhodopsin-1 causes pH-dependent color tuning and fast photocurrent inactivation. *Photochem Photobiol* **85:** 564–569.

Váró G, Brown LS, Lakatos M, Lanyi JK. 2003. Characterization of the photochemical reaction cycle of proteorhodopsin. *Biophys J* **84:** 1202–1207.

Wang W-W, Sineshchekov OA, Spudich EN, Spudich JL. 2003. Spectroscopic and photochemical characterization of a deep ocean proteorhodopsin. *J Biol Chem* **278:** 33985–33991.

Wang H, Sugiyama Y, Hikima T, Sugano E, Tomita H, Takahashi T, Ishizuka T, Yawo H. 2009. Molecular determinants differentiating photocurrent properties of two channelrhodopsins from *Chlamydomonas*. *J Biol Chem* **284:** 5685–5696.

Yoshitsugu M, Yamada J, Kandori H. 2009. Color-changing mutation in the E–F loop of proteorhodopsin. *Biochemistry* **48:** 4324–4330.

Zhang F, Wang L-P, Brauner M, Liewald JF, Kay K, Watzke N, Wood PG, Bamberg E, Nagel G, Gottschalk A, et al. 2007. Multimodal fast optical interrogation of neural circuitry. *Nature* **446:** 633–639.

Zhang F, Prigge M, Beyrière F, Tsunoda SP, Mattis J, Yizhar O, Hegemann P, Deisseroth K. 2008. Red-shifted optogenetic excitation: A tool for fast neural control derived from *Volvox carteri*. *Nat Neurosci* **11:** 631–633.

7

Measuring Membrane Voltage with Fluorescent Proteins

Jordan Patti[1] and Ehud Y. Isacoff[1–3]

[1]Department of Molecular and Cell Biology, University of California, Berkeley, California 94720; [2]Helen Wills Neuroscience Institute, University of California, Berkeley, California 94720; [3]Physical Biosciences and Material Science Divisions, Lawrence Berkeley National Laboratory, Berkeley, California 94720

ABSTRACT

Measuring signal transduction in large numbers of cells with high spatial and temporal resolution is fundamental to studying information processing in the nervous system. DNA-encoded sensors have an advantage in that they can be introduced into an organism noninvasively and targeted to specific brain regions, cell types, or subcellular compartments. The last decade has seen the development of a variety of chimeric proteins that report transmembrane voltage. The prototype sensor, FlaSh, is a green fluorescent protein fused to a voltage-sensitive K^+ channel, where voltage-dependent rearrangements in the channel induce changes in the protein's fluorescence. Subsequent sensors have refined this basic design using a monomeric voltage-sensing phosphatase domain from *Ciona intestinalis* and pairs of fluorescent proteins to produce a larger fluorescent signal.

OPTICAL APPROACHES TO STUDYING MEMBRANE VOLTAGE

Fluorescent Dyes as Sensors

Fluorescent indicator dyes, and calcium indicators in particular, have revolutionized our understanding of cellular physiology. The many available membrane voltage dyes can be roughly divided into two categories: faster (millisecond-range) dyes with a small response, and slower dyes with a large response (Apell and Bersch 1987; Loew et al. 1992; Sjulson and Miesenböck 2007). Typically, the slower dyes depend on partitioning of the dye across the membrane, and the faster dyes use a redistribution of intramolecular charge. Although calcium-indicator fluorescence can change manyfold on binding calcium, in practice, fast voltage probes often show fluorescence changes of $<10\%$ per 100 mV, the size of the largest physiological voltage signal in neurons (Homma et al. 2009).

Recently, fast dyes that are excited in the far-red have been introduced, and hybrid systems consisting of a fixed fluorophore and a small mobile quencher or a Förster resonance energy transfer (FRET) partner have seen more use (González and Tsien 1997; Salama et al. 2005; Wuskell et al. 2006; Bradley et al. 2009). However, apart from the signal size, a general drawback for using organic dyes for neuroscience applications is that they cannot be targeted directly to a specific cell population. In the absence of targeting, optical measurements in neural tissue usually cannot distinguish whether a signal originates from activity in neurons or glia, nor can it determine which types of neurons are involved. Moreover, signals that occur in small numbers of cells are often lost in the noise.

Protein-Based Sensors

Because protein-based sensors are DNA-encoded, they can be placed under the control of cell-specific promoters, introduced in vivo or in vitro using gene transfer techniques, and even targeted to specific subcellular compartments. In general, these sensors are chimeras between a signal transduction protein fragment (detector) and a fluorescent protein (reporter). Whether based on a single fluorescent protein or a FRET pair, they typically depend on a movement or change in orientation to alter fluorescence. Voltage reporters depend on a protein with a membrane voltage-induced conformational change, such as a voltage-gated K^+ or Na^+ channel.

Protein-based sensors also have been developed for a variety of biologically important molecules, such as Ca^{2+} and cAMP (Nagai et al. 2001; Miyawaki 2003; Knöpfel et al. 2006). FlaSh, a protein-based voltage reporter, was one of the first protein-based sensors (Siegel and Isacoff 1997); since its creation much effort has been devoted to producing better reporters. Here we describe FlaSh as the prototypical voltage sensor and review subsequently developed voltage reporters (see Table 1).

FLUORESCENT SHAKER (FlaSh) K^+ CHANNEL: A SENSOR TO MEASURE CHANGES IN TRANSMEMBRANE POTENTIAL

For FlaSh, a modified green fluorescent protein (GFP) was fused in-frame at a site just after the sixth transmembrane segment (Fig. 1A) of the voltage-activated Shaker K^+ channel (Tempel et al. 1987; Chalfie et al. 1994; Guerrero et al. 2002). To prevent FlaSh from loading target cells with additional potassium current, a point mutation was engineered into the pore of the channel to prevent ion conduction but preserve the channel's gating rearrangements. The voltage-dependent structural rearrangements in the channel are transmitted to GFP, resulting in a fluorescence change. Correlating closely with the voltage-dependent rearrangements of Shaker, the dynamic range of FlaSh is approximately –50 mV to –30 mV, following the sigmoidal relationship between gating charge and voltage and corresponding to suprathreshold voltages for action potential initiation. This voltage range can be shifted in a predictable manner in the negative direction by mutating the S4 voltage sensor (Guerrero et al. 2002).

The speed of the fluorescence change depends on the version of GFP used, ranging from a time constant of hundreds of milliseconds in the case of yellow fluorescent protein (YFP), which tracks the process of slow inactivation, to ~10 msec in the case of enhanced GFP (eGFP) tracking activation at room temperature (Guerrero et al. 2002). Although the temporal amplification does make apparent events that would otherwise be difficult to detect, these rates are too slow to distinguish multiple rapid events such as neuronal spiking, even at modest frequencies. For a train of brief pulses, the integral of the fluorescence response is constant at frequencies of ≤20 Hz.

IMPROVED VOLTAGE-SENSING FLUORESCENT PROTEINS

Several variants fusing a voltage sensor to a fluorescent protein have been designed to improve on the characteristics of FlaSh. The voltage-sensing fluorescent protein 1 (VSFP1) introduced a trun-

TABLE 1. Properties of protein-based fluorescent voltage sensors

Properties[a]	First generation[b]		
	FlaSh[e]	SPARC[f]	VSFP1[g]
Sensor	Shaker K^+	$Na_v1.4$	$K_v2.1$ (S1–S4)
Fluorescent protein(s)	GFP	eGFP	CFP/YFP
Insertion point	Post-S6	2nd domain, post-S6	Post-S4
Dynamic range	−50 mV to −30 mV	−80 mV to −20 mV	(not reported)
Reported change (100 mV)[c]	5%	0.50%	2%
Activation time constant[d]	10–100 msec	1–7 msec	0.8 msec
Excitation wavelength	450/50 nm	470/44 nm	432 nm (CFP)
Emission wavelength(s)	510/50 nm	525/50 nm	>530 nm (YFP)

	Second generation[b]			
	VSFP2(.1)(.3)[h]	VSFP3.1[i]	Mermaid[j]	VSFP2.4[k]
Sensor[a]	Ci-VSP	Ci-VSP	Ci-VSP	Ci-VSP
Fluorescent protein(s)	Cerulean/Citrine	Cerulean	mUKG/mKOκ	mCitrine/mKate2
Insertion point	Post-S4	Post-S4	Post-S4	Post-S4
Dynamic range	−100 mV to 0 mV	−100 mV to 0 mV	−100 mV to 20 mV	−100 mV to 0 mV
Reported change (100 mV)[c]	4%–13% (ratio)	1%–2%	13%–40% (ratio)	12% (ratio)
Activation time constant[d]	30–100 msec	2–4 msec	10–60 msec	30–60 msec
Excitation wavelength	440/20 nm (Cerulean)	440/20 nm	450/20 nm (mUKG)	490/6 nm (mCitrine)
Emission wavelength(s)	480/40 nm, >505 nm	480/40 nm	507/65 nm, 590/60	520/35, 625/25

[a]Data for FlaSh and SPARC were collected in oocytes. The remaining data were collected in various mammalian cell lines.

[b]First-generation sensors use multimeric proteins as their voltage transducers and target the plasma membrane poorly in mammalian cells. Second-generation sensors rely on the Ci-VSP VSD and show improved localization.

[c]Reported voltage changes are steady-state values and typically correspond to the slow fluorescence response. Ratios are given for constructs with two fluorescent proteins.

[d]Because the response is nonlinear, activation time constants vary depending on the voltage range. Some can also include multiple variants, and are thus best used to characterize constructs as either fast or slow.

[e]Siegel and Isacoff 1997; Guerrero et al. 2002.

[f]Ataka and Pieribone 2002.

[g]Sakai et al. 2001; Knöpfel et al. 2003.

[h]Dimitrov et al. 2007; Lundby et al. 2008.

[i]Lundby et al. 2008.

[j]Tsutsui et al. 2008.

[k]Mutoh et al. 2009.

cated (i.e., consisting of S1–S4 of the pore region, with S5 and S6 removed) voltage-sensing domain from the potassium voltage-gated channel, Shab-related subfamily, member 1 ($K_v2.1$) fused with a cyan fluorescent protein (CFP)–YFP FRET pair connected to each other by a short linker (Fig. 1C; Sakai et al. 2001; Knöpfel et al. 2003). The placement of the two fluorescent proteins adjacent to each other in the amino acid sequence separated by a short (and presumably inert) linker is distinct from other protein-based sensors that rely on FRET, such as calcium sensors like Cameleon or TN-XL in which the moving part (i.e., the calmodulin or troponin calcium-binding sensor domain) links the CFP donor to the YFP acceptor (Miyawaki et al. 1997; Mank et al. 2006). However, this short passive linker donor–acceptor design does create an effective, detectable fluorescent change and has been used in subsequent VSFPs. It has been hypothesized that the FRET change results from a change in the angle between the two fluorophores (Fig. 2C,D; Sakai et al. 2001).

SPARC (sodium channel protein-based activity reporting construct) (Fig. 1B), another of the first generation of protein-based voltage sensors, uses a complete $Na_v1.4$ sodium channel as its sensor, with GFP inserted in the intracellular loop between the second and third domains (Ataka and Pieribone 2002). The voltage-dependent fluorescent change is small but rapid, corresponding to the gating-charge movement characteristic of sodium channels.

The two biggest obstacles to using these proteins in neuroscience are the small fluorescence change (a few percent or less for a 100-mV step) and, more importantly, their poor membrane tar-

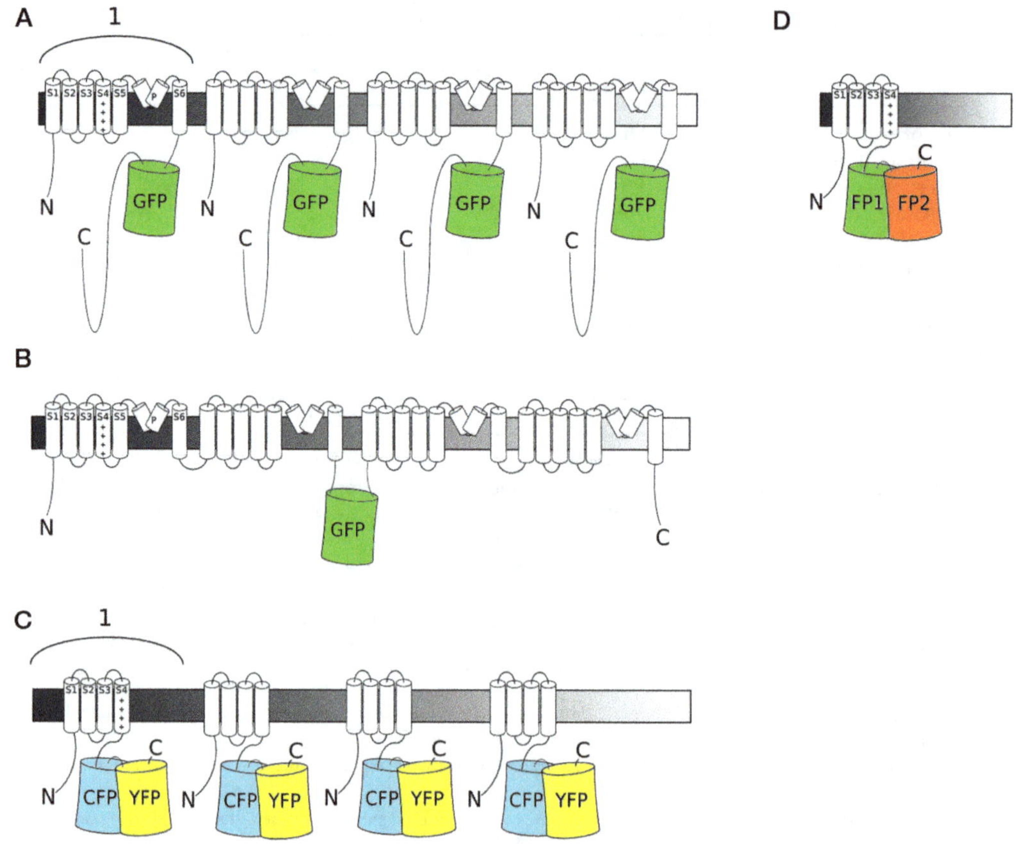

FIGURE 1. Diagram of protein-based voltage indicators. (*A*) FlaSh combines GFP with the complete Shaker K^+ channel. One GFP is found on each of the four subunits. This Shaker-based sensor has relatively slow on/off kinetics. (*B*) SPARC incorporates a single GFP between the second and third domains of the $Na_v1.4$ sodium channel, producing a fast sensor. (*C*) VSFP1 truncates $K_v2.1$ and places a CFP/YFP FRET pair after S4. The quaternary structure is unclear. (*D*) The VSFP2 family and Mermaid use the monomeric Ci-VSP voltage-sensing domain with a pair of fluorescent proteins. This results in better cell membrane localization than the channel-based sensors.

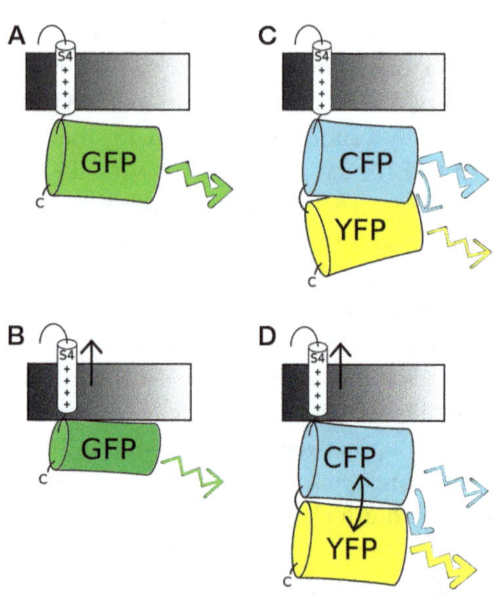

FIGURE 2. Possible mechanisms for voltage-dependent fluorescence changes. The (*A*) resting and (*B*) depolarized membrane of a single-FP sensor. The force exerted by the motion of the voltage-sensing domain destabilizes the fluorescent protein and causes a decrease in fluorescence. The (*C*) resting and (*D*) depolarized membrane of a FRET-based sensor. (*D*) When the cell depolarizes, the angle between the chromophore dipoles changes, increasing FRET efficiency. This decreases the observed donor emission and increases acceptor emission. For clarity, only the charged S4 helix of the voltage-sensing domain is shown.

geting in mammalian cells. These chimeras suffer from poor localization: The majority of the proteins never make it to the plasma membrane, staying instead within the cells, where they contribute to the fluorescence background (Baker et al. 2007). The fact that most of the initial experiments on protein-based voltage indicators were performed in oocytes delayed the recognition of this problem. In retrospect, this is not surprising, considering that the fluorescent proteins are linked so closely to the membrane in a region crowded by four channel subunits and (in the cases of the potassium channel-based Shaker and $K_v2.1$ constructs) multiple fluorescent proteins (Fig. 1).

The discovery of the *C. intestinalis* voltage-sensing phosphatase (Ci-VSP), in which a voltage-sensing domain (VSD) is connected to a cytoplasmic enzyme, enabled the construction of a new class of voltage-sensing proteins (Fig. 1D; Murata et al. 2005). Replacing ion channel VSDs with that of Ci-VSP has two advantages. First, Ci-VSP already couples its VSD to a cytoplasmic protein that is sensitive to its voltage-sensing motion. Second, because Ci-VSP is a monomer, fluorescent protein fusions do not face potential steric clashes between multiple fluorescent proteins in the crowded juxtamembranous space of a channel. Indeed, Ci-VSPs localize more efficiently to the cell membrane in mammalian cells (Dimitrov et al. 2007).

First among this second generation of proteins is the VSFP2 family (Dimitrov et al. 2007; Lundby et al. 2008). Several variants of Ci-VSP-based sensors have been developed that, like VSFP1, use a pair of adjacent fluorescent proteins with varying lengths of linker that connect the VSD to the phosphatase of Ci-VSP (Mutoh et al. 2009). The sensors incorporate the charge neutralization of the first S4 arginine (R217Q) to shift the voltage range negatively into the physiological range (Dimitrov et al. 2007; Kohout et al. 2008). The majority of the voltage-dependent fluorescence change occurs between −80 mV and 0 mV, with a sigmoidal relationship between fluorescence and voltage. VSFP2 has been shown to respond to physiological neuronal membrane signals (Dimitrov et al. 2007). Finally, in a detour from paired fluorescent protein designs, VSFP3.1 consists of the Ci-VSP VSD coupled to a single CFP. Like the SPARC construct, this provides a faster response at the cost of a reduced fluorescent change (Lundby et al. 2008).

The voltage sensor Mermaid borrows the VSFP design of adjacent fluorescent proteins and currently produces the largest reported fluorescence change (Tsutsui et al. 2008). Mermaid replaces the classical CFP–YFP donor–acceptor pair with mUKG and mKOκ (monomeric umi-kinoko green and kusabira orange, respectively), new green and orange GFP-like proteins that have excellent spectral overlap for FRET. Large fluorescence changes can be observed by examining the emission ratio of the two fluorophores, in some cases >10% per 100 mV. Like VSFP2, the voltage-dependent fluorescence change occurs between −80 mV and 0 mV, with a sigmoidal relationship between fluorescence and voltage. Although the magnitude of fluorescence response has improved greatly since the first generation of sensors, like those earlier proteins these still require hundreds of milliseconds to reach their maximum response to a voltage change (Baker et al. 2008; Mutoh et al. 2009). Nevertheless, it is possible with Mermaid to optically detect single slow cardiac action potentials in cultured cardiac myocytes and even low-frequency neuronal action potentials.

MECHANISM OF FLUORESCENCE CHANGE

Most of the protein-based optical voltage sensors have two components to their voltage response: a fast, small change followed by a slower, larger change (Lundby et al. 2008; Villalba-Galea et al. 2009). The fast change closely corresponds to the gating charge movement of the voltage sensor, but it is typically <1% ($\Delta F/F$). Not present in constructs like SPARC, but prevalent in Ci-VSP constructs and FlaSh, the slow component can be an order of magnitude larger. It has been proposed that this derives from the relaxation of the voltage-sensing domain with prolonged depolarization (Villalba-Galea et al. 2009). However, for both components of the response, the mechanism by which the enclosed chromophore of the fluorescent protein is influenced by structural rearrangements remains unclear. It is possible that the movement of the voltage-sensing domain destabilizes the fluorescent protein, decreasing its fluorescence perhaps by increasing chromophore exposure to solvent (Fig. 2). In Mermaid and VSFPs, the FRET effect could result from changes in the angle between the two chromophores (Sakai et al. 2001).

EXPERIMENTAL SETUP AND APPLICATIONS

The improved Ci-VSP-based optical voltage sensors are optimally used with a microscope configured for emission ratio fluorescence imaging. Excitation is performed at a single wavelength appropriate to the FRET donor. A first dichroic mirror separates the emission and excitation light. A second dichroic mirror in the light path then separates the emissions of the donor and acceptor. Two detectors simultaneously record the changes at the respective wavelengths. With a camera as the detector, commercial beamsplitters can allow the collection of both wavelengths on a single charge-coupled device. For faster recording, photomultiplier tubes provide an efficient detector. For most reported values, the ratio of the two wavelengths is used, with corrections made for background fluorescence (Tsutsui et al. 2008; Mutoh et al. 2009).

Currently, the only physiological results from these fluorescent voltage sensors have come from their developers. Investigators have successfully detected action potentials when it is possible to average across multiple sweeps, such as replayed voltage traces (Dimitrov et al. 2007). A single application of voltage-sensing fluorescent proteins in natively excitable cells has been presented (Tsutsui et al. 2008). The Mermaid sensor was used in two types of cultured cells: rat cardiomyocytes and cortical neurons. In both cases, a single sweep was sufficient to observe membrane activity. The Mermaid construct was transfected, and after a few days of expression, imaging was conducted using separate charge-coupled cameras for each fluorescent protein wavelength. The spontaneous cardiomyocyte depolarization cycle was clearly visible with Mermaid (~22% ratio change) (Fig. 3A,B). In cortical neurons, field stimulation triggered firing in cells in which synaptic transmission was blocked pharmacologically. Fluorescence responses to both a single induced action potential (~1.5%) and a pulse train (~5%) were detected. As might be expected, the magnitude of the fluorescence ratio change was larger for the longer, larger events (Fig. 3C,D).

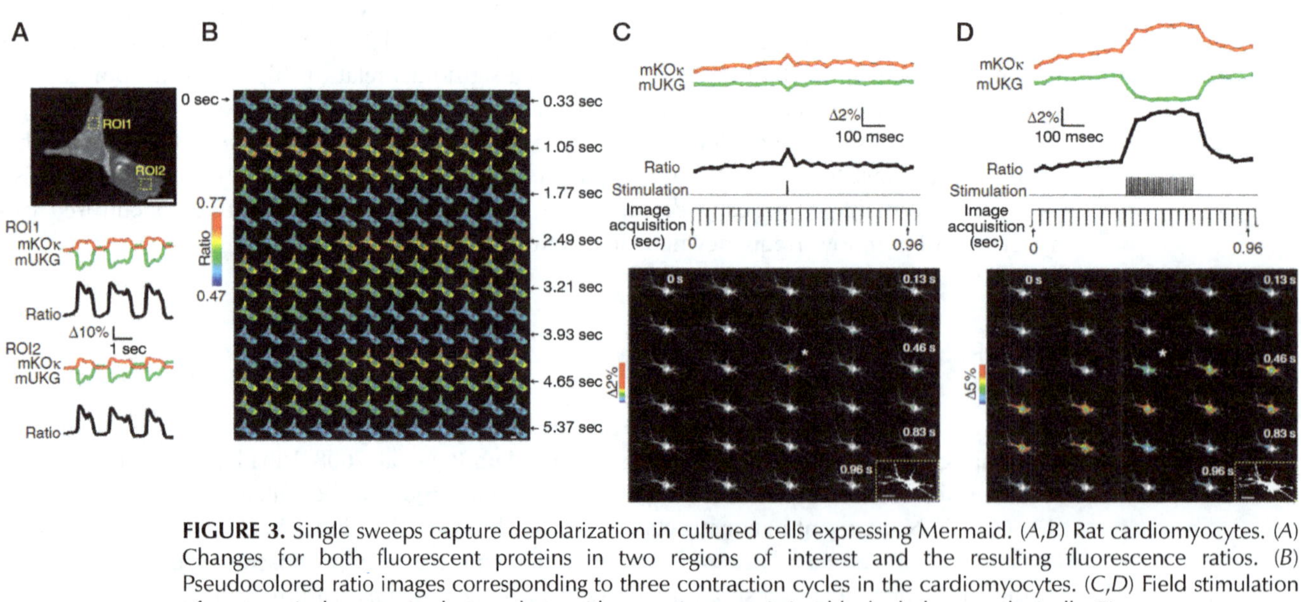

FIGURE 3. Single sweeps capture depolarization in cultured cells expressing Mermaid. (*A,B*) Rat cardiomyocytes. (*A*) Changes for both fluorescent proteins in two regions of interest and the resulting fluorescence ratios. (*B*) Pseudocolored ratio images corresponding to three contraction cycles in the cardiomyocytes. (*C,D*) Field stimulation of a rat cortical neuron in dense culture with synaptic transmission blocked pharmacologically. Time courses present fluorescence changes in each protein, as well as the ratio change. Pseudocolored ratio images show Mermaid response. (*C*) Response to a single 1-msec pulse. (*D*) Thirty pulses at 100 Hz elicit a larger response. Scale bars, 20 µm. (Reprinted, with permission of Macmillan Publishers Ltd., from Tsutsui et al. 2008.)

DISCUSSION AND CONCLUSION

Since the introduction of FlaSh, improvements to voltage-sensing fluorescent protein sensors have been somewhat slow. In comparison, methods for genetically targeting and optically stimulating neurons (a complementary set of techniques) have developed much more rapidly (Zhang et al. 2007; Gorostiza and Isacoff 2008; Miesenböck 2009). Protein-based voltage sensors still have yet to be adopted widely as neuroscience tools. Concerns such as capacitive load from VSFP VSDs remain secondary to another limitation shared with chemical voltage dyes: The fast fluorescence response is small relative to the noise (Sjulson and Miesenböck 2007; Akemann et al. 2009); protein-based sensors do not yet approach the signal-to-noise ratio seen with calcium indicator dyes. Additionally, although genetically encoded subcellular targeting is promising, organic dyes still localize to the plasma membranes of homogeneous cultured cells more uniformly than do even second-generation VSFPs. In more heterogeneous systems or in experiments long enough for dye internalization to be a problem, cell-specific targeting of genetically encoded sensors is more advantageous.

Improvements to these sensors have focused largely on modifying the voltage-sensing domain and the linker that connects it to the fluorescent protein. There has been some success in tuning these sensors through point mutations, primarily by using already well-characterized mutations. Further advancement will require a more detailed model of how GFP fluorescence is affected by conformational change in the voltage-sensing domain. Difficulties crystallizing membrane proteins make it unlikely that this will be clarified by analysis of crystal structures (e.g., as in the case of the mechanism for the GCaMP calcium sensor) (Wang et al. 2008; Akerboom et al. 2009). Still, a better understanding of the mechanism might allow for more rational design approaches, including tuning the fluorescent protein to better represent the state of the voltage-sensing domain, an aspect of the sensor that has not yet been examined systematically.

With the goal of capturing high-speed events, one reasonable focus might be to increase the size of the fast fluorescence response known to correspond to the gating charge movement. This could be explored by placing greater emphasis on smaller voltage changes that fall within the dynamic range of the voltage sensor, instead of on large fluorescence changes that might correspond to a slower, secondary movement of the voltage-sensing domain. Perhaps mutations that destabilize the fluorescent protein would yield a better sensor. Alternatively, new classes of fluorescent proteins could provide new components for these sensors (Shu et al. 2009). Also, fusion proteins are not the only approach to genetically targeted fluorescent voltage reporting. A hybrid system combining GFP with the small, mobile quencher dipicrylamine has also been described (Chanda et al. 2005; Sjulson and Miesenböck 2008). Although under some conditions this system can detect single action potentials in cultured neurons, a purely genetically encoded sensor would be superior for longer studies, as well as for investigating locations that dipicrylamine cannot reach easily.

In vivo systems benefit the most from DNA-encoded sensors, particularly heterogeneous ones such as the brain where many different cell types are present. By resolving targeting, the signal-to-noise ratio for a phenomenon present in a subpopulation of cells can be greatly increased. The signaling events that govern development, sensory transduction, learning, and memory require simultaneous observation of events occurring in multiple cells, and thus optical techniques are preferable to direct electrical measurement. Although work remains to optimize protein-based voltage reporters as neuroscience tools, fusing VSDs to fluorescent proteins produces sensitive and accurate reporters of membrane voltage; new developments will doubtless provide the detection levels and speed necessary to capture fast neuronal events.

ACKNOWLEDGMENTS

This work was supported by the National Institutes of Health (U24NS57631).

REFERENCES

Akemann W, Lundby A, Mutoh H, Knöpfel T. 2009. Effect of voltage sensitive fluorescent proteins on neuronal excitability. *Biophys J* **96:** 3959–3976.

Akerboom J, Vélez Rivera JD, Rodríguez Guilbe MM, Alfaro Malavé EC, Hernandez HH, Tian L, Hires SA, Marvin JS, Looger LL, Schreiter ER. 2009. Crystal structures of the GCaMP calcium sensor reveal the mechanism of fluorescence signal change and aid rational design. *J Biol Chem* **284:** 6455–6464.

Apell HJ, Bersch B. 1987. Oxonol VI as an optical indicator for membrane potentials in lipid vesicles. *Biochim Biophys Acta* **903:** 480–494.

Ataka K, Pieribone VA. 2002. A genetically targetable fluorescent probe of channel gating with rapid kinetics. *Biophys J* **82:** 509–516.

Baker BJ, Lee H, Pieribone VA, Cohen LB, Isacoff EY, Knöpfel T, Kosmidis EK. 2007. Three fluorescent protein voltage sensors exhibit low plasma membrane expression in mammalian cells. *J Neurosci Methods* **161:** 32–38.

Baker BJ, Mutoh H, Dimitrov D, Akemann W, Perron A, Iwamoto Y, Jin L, Cohen LB, Isacoff EY, Pieribone VA, et al. 2008. Genetically encoded fluorescent sensors of membrane potential. *Brain Cell Biol* **36:** 53–67.

Bradley J, Luo R, Otis TS, DiGregorio DA. 2009. Submillisecond optical reporting of membrane potential in situ using a neuronal tracer dye. *J Neurosci* **29:** 9197–9209.

Chalfie M, Tu Y, Euskirchen G, Ward WW, Prasher DC. 1994. Green fluorescent protein as a marker for gene expression. *Science* **263:** 802–805.

Chanda B, Blunck R, Faria LC, Schweizer FE, Mody I, Bezanilla F. 2005. A hybrid approach to measuring electrical activity in genetically specified neurons. *Nat Neurosci* **8:** 1619–1626.

Dimitrov D, He Y, Mutoh H, Baker BJ, Cohen L, Akemann W, Knöpfel T. 2007. Engineering and characterization of an enhanced fluorescent protein voltage sensor. *PLoS ONE* **2:** e440. doi: 10.1371/journal.pone.0000440.

González JE, Tsien RY. 1997. Improved indicators of cell membrane potential that use fluorescence resonance energy transfer. *Chem Biol* **4:** 269–277.

Gorostiza P, Isacoff EY. 2008. Optical switches for remote and noninvasive control of cell signaling. *Science* **322:** 395–399.

Guerrero G, Siegel MS, Roska B, Loots E, Isacoff EY. 2002. Tuning FlaSh: Redesign of the dynamics, voltage range, and color of the genetically encoded optical sensor of membrane potential. *Biophys J* **83:** 3607–3618.

Homma R, Baker BJ, Jin L, Garaschuk O, Konnerth A, Cohen LB, Bleau CX, Canepari M, Djurisic M, Zecevic D. 2009. Wide-field and two-photon imaging of brain activity with voltage- and calcium-sensitive dyes. *Methods Mol Biol* **489:** 43–79.

Knöpfel T, Tomita K, Shimazaki R, Sakai R. 2003. Optical recordings of membrane potential using genetically targeted voltage-sensitive fluorescent proteins. *Methods* **30:** 42–48.

Knöpfel T, Díez-García J, Akemann W. 2006. Optical probing of neuronal circuit dynamics: Genetically encoded versus classical fluorescent sensors. *Trends Neurosci* **29:** 160–166.

Kohout SC, Ulbrich MH, Bell SC, Isacoff EY. 2008. Subunit organization and functional transitions in Ci-VSP. *Nat Struct Mol Biol* **15:** 106–108.

Loew LM, Cohen LB, Dix J, Fluhler EN, Montana V, Salama G, Wu J-y. 1992. A naphthyl analog of the aminostyryl pyridinium class of potentiometric membrane dyes shows consistent sensitivity in a variety of tissue, cell, and model membrane preparations. *J Membr Biol* **130:** 1–10.

Lundby A, Mutoh H, Dimitrov D, Akemann W, Knöpfel T. 2008. Engineering of a genetically encodable fluorescent voltage sensor exploiting fast Ci-VSP voltage-sensing movements. *PLoS ONE* **3:** e2514. doi: 10.1371/journal.pone.0002514.

Mank M, Reiff DF, Heim N, Friedrich MW, Borst A, Griesbeck O. 2006. A FRET-based calcium biosensor with fast signal kinetics and high fluorescence change. *Biophys J* **90:** 1790–1796.

Miesenböck G. 2009. The optogenetic catechism. *Science* **326:** 395–399.

Miyawaki A. 2003. Fluorescence imaging of physiological activity in complex systems using GFP-based probes. *Curr Opin Neurobiol* **13:** 591–596.

Miyawaki A, Llopis J, Heim R, McCaffery JM, Adams JA, Ikura M, Tsien RY. 1997. Fluorescent indicators for Ca^{2+} based on green fluorescent proteins and calmodulin. *Nature* **388:** 882–887.

Murata Y, Iwasaki H, Sasaki M, Inaba K, Okamura Y. 2005. Phosphoinositide phosphatase activity coupled to an intrinsic voltage sensor. *Nature* **435:** 1239–1243.

Mutoh H, Perron A, Dimitrov D, Iwamoto Y, Akemann W, Chudakov DM, Knöpfel T. 2009. Spectrally-resolved response properties of the three most advanced FRET based fluorescent protein voltage probes. *PLoS ONE* **4:** e4555. doi: 10.1371/journal.pone.0004555.

Nagai T, Sawano A, Park ES, Miyawaki A. 2001. Circularly permuted green fluorescent proteins engineered to sense Ca^{2+}. *Proc Natl Acad Sci* **98:** 3197–3202.

Sakai R, Repunte-Canonigo V, Raj CD, Knöpfel T. 2001. Design and characterization of a DNA-encoded, voltage-sensitive fluorescent protein. *Eur J Neurosci* **13:** 2314–2318.

Salama G, Choi B-R, Azour G, Lavasani M, Tumbev V, Salzberg BM, Patrick MJ, Ernst LA, Waggoner AS. 2005. Properties of new, long-wavelength, voltage-sensitive dyes in the heart. *J Membr Biol* **208:** 125–140.

Shu X, Royant A, Lin MZ, Aguilera TA, Lev-Ram V, Steinbach PA, Tsien RY. 2009. Mammalian expression of infrared fluorescent proteins engineered from a bacterial phytochrome. *Science* **324:** 804–807.

Siegel MS, Isacoff EY. 1997. A genetically encoded optical probe of membrane voltage. *Neuron* **19:** 735–741.

Sjulson L, Miesenböck G. 2007. Optical recording of action potentials and other discrete physiological events: A perspective from signal detection theory. *Physiology* **22:** 47–55.

Sjulson L, Miesenböck G. 2008. Rational optimization and imaging in vivo of a genetically encoded optical voltage reporter. *J Neurosci* **28:** 5582–5593.

Tempel BL, Papazian DM, Schwarz TL, Jan YN, Jan LY. 1987. Sequence of a probable potassium channel component encoded at Shaker locus of *Drosophila*. *Science* **237:** 770–775.

Tsutsui H, Karasawa S, Okamura Y, Miyawaki A. 2008. Improving membrane voltage measurements using FRET with new fluorescent proteins. *Nat Methods* **5:** 683–685.

Villalba-Galea CA, Sandtner W, Dimitrov D, Mutoh H, Knöpfel T, Bezanilla F. 2009. Charge movement of a voltage-sensitive fluorescent protein. *Biophys J* **96:** L19–L21.

Wang Q, Shui B, Kotlikoff MI, Sondermann H. 2008. Structural basis for calcium sensing by GCaMP2. *Structure* **16:** 1817–1827.

Wuskell JP, Boudreau D, Wei M-d, Jin L, Engl R, Chebolu R, Bullen A, Hoffacker KD, Kerimo J, Cohen LB, et al. 2006. Synthesis, spectra, delivery and potentiometric responses of new styryl dyes with extended spectral ranges. *J Neurosci Methods* **151:** 200–215.

Zhang F, Aravanis AM, Adamantidis A, de Lecea L, Deisseroth K. 2007. Circuit-breakers: Optical technologies for probing neural signals and systems. *Nat Rev Neurosci* **8:** 577–581.

8 Imaging Neuronal Activity with Genetically Encoded Calcium Indicators

Lin Tian, S. Andrew Hires, and Loren L. Looger
Howard Hughes Medical Institute, Janelia Farm Research Campus, Ashburn, Virginia 20147

ABSTRACT

Based on chimeric fluorescent proteins, genetically encoded calcium indicators (GECIs) can be used to monitor calcium transients in living cells and organisms. Because they are encoded by DNA, GECIs can be delivered to the intact brain noninvasively and targeted to defined populations of neurons and specific subcellular compartments for long-term, repeated measurements in vivo. GECIs have improved iteratively and are becoming useful for imaging neural activity in vivo. Here we summarize extrinsic and intrinsic factors that influence a GECI's performance and provide guidelines for selecting the appropriate GECI for a given application. We also review recent progress in GECI design, optimization, and standardized testing protocols.

INTRODUCTION

Calcium ions are universal second messengers regulating essential cellular signaling events in a wide range of tissues and organisms. In neurons, action potentials (APs) trigger large and rapid calcium influx through voltage-gated channels. Similarly, activation of neurotransmitter receptors during synaptic transmission causes calcium transients in dendritic spines. With appropriate indicators, imaging intracellular calcium dynamics can be used to measure neuronal spiking and synaptic activity across populations of neurons in vitro and in vivo.

There are many fluorescence imaging techniques for visualizing calcium dynamics in living cells and animals. Bulk or intracellular loading of synthetic calcium indicators has been used widely (Yuste et al. 1992; Svoboda et al. 1997; Fetcho et al. 1998; Stosiek et al. 2003). However, small-molecule indicators have many drawbacks that prevent their applicability for a wide variety of questions in neuroscience. They cannot easily be targeted to specific cell types, populations, or subcellular locations (but see Tour et al. 2007). Because loading is highly invasive and can be damaging to tissue, they also are incompatible with repeated, chronic in vivo measurements.

Genetically encoded calcium indicators (GECIs) based on chimeric fluorescent proteins address many of these limitations and provide an alternative to small-molecule calcium dyes. GECIs can be targeted to specific cell types, populations, or subcellular compartments, allowing imaging of structures such as synapses, axons, and dendrites (for review, see Palmer and Tsien 2006). The delivery of GECIs into intact tissues and organisms (e.g., by using transgenesis or viral constructs [see Chapters 57 and 58]) is minimally invasive and therefore compatible with long-term, repeated in vivo measurements (Mank et al. 2008), although calcium buffering and cellular perturbation can occur.

In general, GECIs consist of a calcium-binding domain (e.g., calmodulin or troponin C), fused to one or two fluorescent proteins (FPs) (Fig. 1A–C). In single-FP GECIs, the fluorescence intensity of a circularly permuted or split FP is modulated by calcium binding–dependent changes in the chromophore environment (Baird et al. 1999; Nagai et al. 2001; Nakai et al. 2001). In two-FP GECIs, calcium binding allosterically modulates the relative donor–acceptor emission spectra through a distance- and orientation-dependent change in fluorescence resonance energy transfer (FRET) (Miyawaki et al. 1997; Heim and Griesbeck 2004; Palmer et al. 2006). In many cases, a conformational actuator (e.g., the M13 peptide) is included in the fusion protein to enhance the conformational change and fluorescence modulation.

Through iterative cycles of optimization, GECIs have been improved to the point that they are useful for in vivo neuronal imaging.

- The calmodulin-based FRET sensor D3cpVenus (D3cpV; an improved cameleon variant) (Palmer et al. 2006) has been used to detect single APs in organotypic mouse brain slices and in vivo in layer 2/3 somatosensory cortical neurons (Wallace et al. 2008).
- Recently, another cameleon variant, YC3.60 (Nagai et al. 2004), allowed detection in vivo of single APs with a sensitivity comparable to that of D3cpV, but with faster kinetics and minimal saturation up to 10 APs (Lütcke et al. 2010).
- The troponin C–based FRET sensor TN-XXL has been used for chronic in vivo activity imaging in mouse and fly (Mank et al. 2008).
- The single-FP sensor G-CaMP3 showed high expression and sensitivity with no apparent cytotoxicity or behavioral phenotype in *Caenorhabditis elegans* and *Drosophila* (Tian et al. 2009). G-CaMP3 allowed detection of single APs in pyramidal neurons in acute mouse brain slices. Long-term imaging (months) with G-CaMP3 in the motor cortex of behaving mice showed large fluorescence changes (detecting two to three APs) with linearity up to 20 APs (Tian et al. 2009).

Characterizing several GECIs in the context of a specific application is important, because they have different strengths and limitations (Fig. 1D). To help guide GECI selection and experimental design, we have analyzed factors affecting GECI performance, and we review recent progress in GECI optimization.

PROPERTIES THAT INFLUENCE THE PERFORMANCE OF GECIs

A uniform standard for comparing performance of calcium indicators is the signal-to-noise ratio (SNR). SNR is defined as the ratio of the fluorescence signal change ($\Delta F = F_{obs} - F_0$, where F_{obs} is sensor fluorescence at peak [Ca^{2+}] and F_0 is sensor fluorescence at baseline [Ca^{2+}]) to the shot noise on the baseline fluorescence ($F_0 N^{-1/2}$, where N is the number of photons detected) (Yasuda et al. 2004). An equivalent calculation for FRET sensors takes the difference in donor/acceptor ratio divided by the combined shot noise on the baseline fluorescence of both FPs.

To optimize SNR, the appropriate combination of intrinsic GECI parameters should be matched to the extrinsic factors of the system studied. Extrinsic factors include the size, time course and frequency of calcium transients, and basal [Ca^{2+}] of the target cell type in the organism studied. Intrinsic GECI parameters include the sensor affinity, kinetics, dynamic range, brightness, expression level, fluorescence properties, and independence from endogenous interference.

Underlying Calcium Dynamics

Neurons typically maintain extremely low cytoplasmic free [Ca^{2+}] (30–200 nM) at rest (Hille 1992). When stimulated, calcium flows into the cytoplasm through calcium channels (voltage-gated calcium

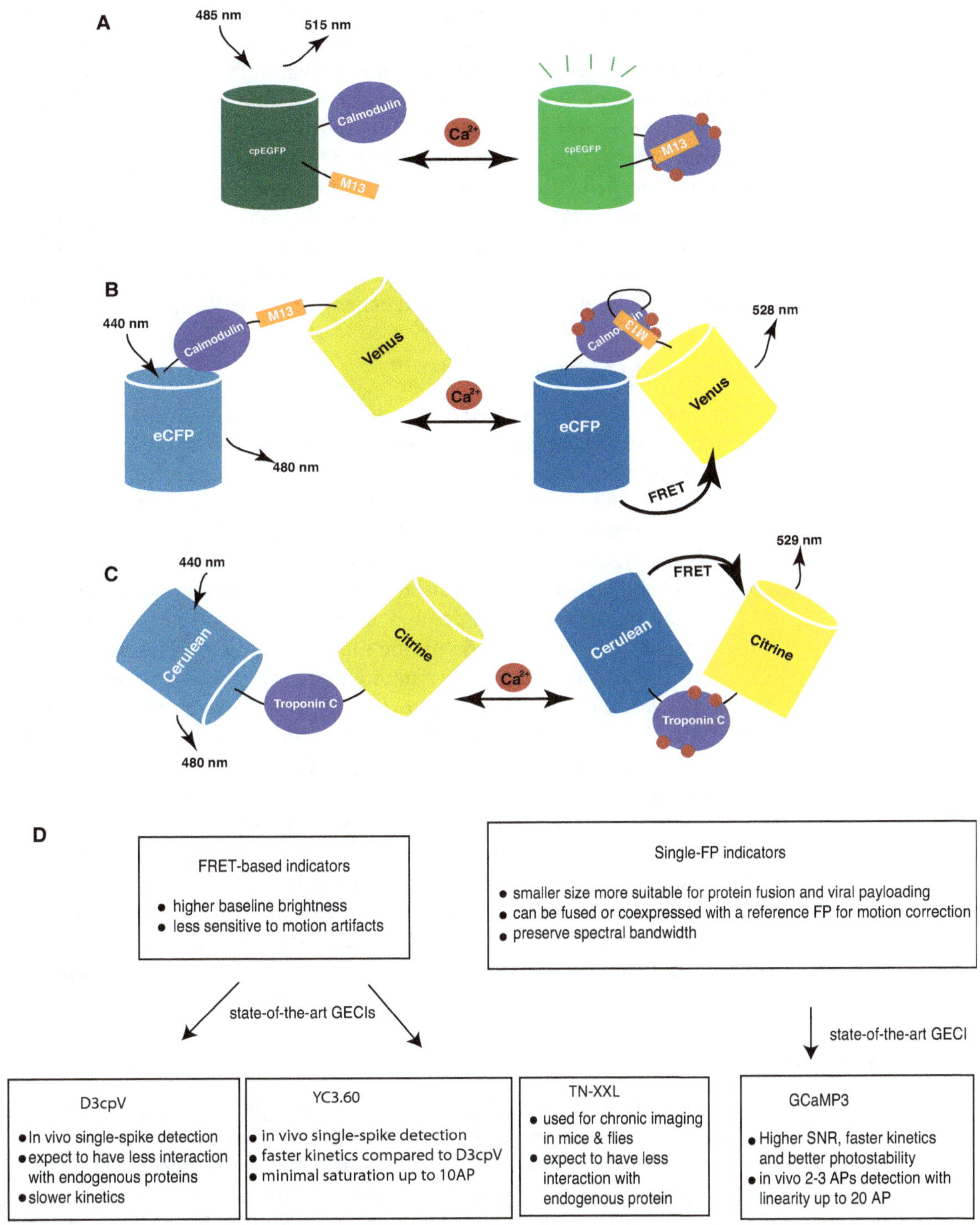

FIGURE 1. Schematic representation of the three major GECI classes. GECIs are based either on florescence intensity changes of split or circularly permuted single FPs (*A*) or changes in FRET efficiency between two FPs (*B,C*). (*A*) Schematic of the G-CaMP-type sensing mechanism. Upon calcium binding, conformational changes in the calmodulin/M13 complex induce fluorescence changes in the circularly permuted enhanced GFP (cpEGFP). (*B*) The cameleon family of FRET-based GECIs. A calcium-dependent increase in FRET between a CFP and YFP FRET pair is coupled to the binding of calmodulin to the M13 peptide. (*C*) Troponin C–based FRET GECIs. Binding of calcium to troponin C induces conformational changes and an increase in FRET between CFP and YFP. (*D*) Practical considerations for choosing the most appropriate GECI.

channels or calcium-permeable, receptor-operated channels) or is released from calcium-loaded organelles, rapidly raising the local [Ca^{2+}] (Hille 1992). Calcium levels then return to baseline via extrusion from the cell and reloading of intracellular buffers and stores. The spatiotemporal evolution of calcium transients is shaped by the localization, mobility, affinity, and kinetics of these processes, including buffering by endogenous proteins and any exogenous GECI present (Helmchen et al. 1996).

In neurons, APs occur over a wide range of frequencies. When sparse, each induced calcium transient is independent, fast (<2 msec rise time), and modest (~250 nM in pyramidal neurons) (Helmchen et al. 1996). When closely spaced, the calcium transients accumulate (Helmchen et al. 1996). Depending on the application, users can choose to optimize any of the following: detection of sparse APs, dynamic range of burst detection, or temporal resolution of individual APs in spike trains. This guides GECI selection as discussed below; for a detailed treatment, see Hires et al. (2008a).

Calcium Affinity, Kinetics, and Dynamic Range

The fluorescence response of a GECI to calcium transients is critically dependent on the affinity, kinetics, and dynamic range of the sensor. Sensors with a linear relationship between fluorescence response and target stimulus are preferred for quantitative measurement of calcium transients (Yasuda et al. 2004). Nonlinearity in sensor or cellular calcium dynamics can prevent precise spike quantification (e.g., firing rate and the number of APs) during action potential bursts (Sasaki et al. 2008).

To obtain large signal changes in response to small calcium transients (such as those triggered by sparse AP firing), a high-affinity sensor is preferred. However, such a sensor will tend to buffer more Ca^{2+} and can influence the kinetics and amplitude of transients during chronic expression (Mank and Griesbeck 2008). It will also saturate more easily, decreasing the linearity at high calcium concentrations, such as those resulting from long or closely spaced spike trains. A combination of high affinity and fast kinetics is optimal for a sensor to resolve incremental signal increases with additional APs. However, it is challenging to maintain both in GECI engineering because affinity is inversely dependent on dissociation rate (Mank et al. 2006). A low affinity provides a baseline F_0 (fluorescence at resting [Ca^{2+}]) closer to F_{min} (fluorescence at zero [Ca^{2+}]), and therefore a greater effective dynamic range (the maximum signal change at saturating [Ca^{2+}]). Thus, such a sensor is useful for quantitative measurement of larger calcium transients (e.g., those through synaptic neurotransmitter receptors or from high-frequency spike trains).

It is important to calibrate and compare the calcium affinity and dynamic range of each GECI under the same experimental conditions because those parameters can be influenced by external factors such as total ionic strength (Linse et al. 1991), pH, temperature, and presence of competing ions (e.g., Mg^{++}) (Ogawa and Tanokura 1984; Hires et al. 2008a).

GECI Expression Level

GECIs, and other kinds of calcium monitors, bind calcium and thus inherently add to calcium buffering. Higher [GECI] improves SNR by increasing the number of photons detected. However, very highly expressed GECIs can become dominant calcium buffers and spread transients over longer times and lower amplitudes, thus changing the calcium signaling in cells (Helmchen et al. 1996). Early work on Yellow Cameleon (YC) showed that HeLa cells expressing YC3.1 at <20 mM had poor SNR, whereas cells with [YC3.1] above 300 mM showed significant calcium buffering (Miyawaki et al. 1999). Similarly, in G-CaMP2 imaging in acute mouse cortical slices, cells with [G-CaMP2] < 5 μM required unacceptably high laser intensity, whereas cells with [G-CaMP2] > 20 μM showed attenuated responses for short AP trains (Hires et al. 2008a).

To estimate [GECI], the fluorescence of the purified sensor can be compared with that of the expressed sensor under identical imaging conditions (Tour et al. 2007). For FRET-based sensors, the emission of the directly excited acceptor chromophore is used because this is theoretically insensitive to calcium levels (Miyawaki and Tsien 2000). For intensity-modulated single-FP sensors, [GECI] estimates can vary with resting calcium levels in different systems. Estimates can be made by calibration with a second calcium dye (Tay et al. 2007), application of ionomycin or calcium chelators

(McCombs and Palmer 2008), or knowledge of the basal spike rates and calcium levels. With all methods, misfolded, constitutively fluorescent, or nonfluorescent GECI forms can contribute to errors in estimates of [GECI] and calcium buffering.

In addition to buffering Ca^{2+}, GECI components (e.g., calmodulin, M13) can be sequestered by endogenous proteins, resulting in high background and nonfunctional indicators (Hasan et al. 2004; Heim and Griesbeck 2004). GECIs constructed with components mutated relative to endogenous calcium signaling molecules (Palmer et al. 2006) or with calcium-binding proteins not typically found in target cells (Heim and Griesbeck 2004) can reduce such interference.

Both altered calcium handling and endogenous interference can perturb cell-signaling pathways and lead to behavioral phenotypes. For example, *C. elegans* expressing G-CaMP3 in chemosensory neurons showed local search turning behavior similar to that of wild-type animals. However, decreased turning was observed in animals expressing YC3.60, YC2.12, G-CaMP1, and G-CaMP2 (Tian et al. 2009).

GECI Fluorescence Properties

Intrinsic factors influencing GECI performance include brightness (extinction coefficient × quantum yield), pH sensitivity, folding stability, and photobleaching of the sensor.

Substituting FPs with enhanced fluorescence properties has proven to be a reliable method of improving sensor brightness and signal change. In the cameleon family, when yellow fluorescent protein (YFP) was substituted by Venus (Nagai et al. 2002) or Citrine (Griesbeck et al. 2001), the resulting sensors showed lower pH sensitivity and improved folding efficiency. The maximum FRET change was improved further to ~500% in vitro by using a circularly permuted Venus as the FRET acceptor (Nagai et al. 2004). In the troponin family, the sensor's brightness and FRET change have been optimized by substitution both of cyan fluorescent protein (CFP) with its brighter variant Cerulean (Heim et al. 2007) and of YFP with circularly permuted Citrine (Mank et al. 2006). In the G-CaMP family, incorporating a subset of the "superfolder green fluorescent protein (GFP)" mutations (Pédelacq et al. 2006) improved brightness at physiological temperatures (Tallini et al. 2006).

State-of-the-art GECIs include G-CaMP3, D3cpV, YC3.60, and TN-XXL. In vitro dynamic ranges are 560%, 400%, and 200% FRET increases for D3cpV, YC3.60, and TN-XXL (Mank and Griesbeck 2008), respectively, and a 1200% fluorescence intensity increase for G-CaMP3 (Tian et al. 2009). In neurons, the smaller fluorescence changes produced by the FRET indicators resulted in lower SNR than G-CaMP3 (Tian et al. 2009), although they are brighter at submicromolar [Ca^{2+}].

Improved photostability increases the duration that activity can be continuously monitored. G-CaMP3 showed greater photostability than the FRET sensors D3cpV and TN-XXL (Tian et al. 2009), presumably because the dark basal state is less susceptible to bleaching. For both intensity and FRET imaging, GECIs whose signals increase on Ca^{2+} binding are preferable, because this reduces misinterpretation of potential artifacts such as sensor bleaching, degradation, or hardware errors. The comparison of GECI performance in acute slice is shown in Table 1.

TABLE 1. Properties of state-of-the-art GECIs in acute mouse brain slices

Indicator	$\Delta R/R$, $\Delta F/F$[a]		Kinetics[a]		Photostability[b]
	1 AP	40 APs	Rise time (msec)	Decay time (msec)	
D3cpV	5% ± 3%	40% ± 20%	100 ± 30	9500 ± 3400	CFP: 59%; YFP: 84%
TN-XXL	4% ± 2%	50% ± 10%	80 ± 20	1600 ± 600	CFP: 36%; YFP: 70%
G-CaMP3	14.3% ± 3%	500% ± 200%	100 ± 30	650 ± 200	GFP: 109%

From Tian et al. 2009.

GECI, genetically encoded calcium indicator; AP, action potential.

[a]Mean ± S.D.

[b]Fluorescence after 30-min imaging.

GECI OPTIMIZATION

Practical Improvement

Protein engineering efforts have significantly improved the properties of several genetically encoded indicator classes. Overall sensor brightness and fluorescence changes have been improved by substituting superior FPs (Griesbeck et al. 2001), linker truncation (Deuschle et al. 2005; Hires et al. 2008b), and FP circular permutation (Nagai et al. 2004) and by site-directed and random mutagenesis (Tallini et al. 2006). Computational protein design of "bump-hole" calmodulin/M13 pairs (Palmer et al. 2006) can decrease interference with endogenous proteins (Palmer et al. 2006). Alternatively, troponin C, a skeletal calcium sensor not normally expressed in neurons, has been used in place of calmodulin (Heim and Griesbeck 2004). Affinity and specificity of binding domains for Ca^{2+} have also been tuned by rational design (Mank et al. 2006, 2008).

From G-CaMP2 to G-CaMP3

We recently reported the improvement of the single-FP GECI G-CaMP2 (Nakai et al. 2001; Tallini et al. 2006). Our aims for maximizing the SNR of the G-CaMP scaffold were threefold: increased fluorescence change to small and rapid calcium transients, larger dynamic range, and increased baseline brightness.

To optimize G-CaMP, we used semirational library screening guided by the G-CaMP2 protein structure (Wang et al. 2008; Akerboom et al. 2009). To increase baseline brightness, we targeted positions that improve protein folding and stability. To increase the dynamic range, we mutagenized sites near the enhanced GFP chromophore. To increase calcium affinity and allow better detection of sparse APs, we mutated the Ca^{2+}-binding EF hands and the interface between the M13 peptide and calmodulin.

Relative to G-CaMP2, the improved variant, named G-CaMP3, is three times brighter, possesses greater protein stability, and has three times the dynamic range and a higher affinity (1.3×) for calcium (Fig. 2A; Tian et al. 2009). G-CaMP3 detected single AP-triggered calcium transients in pyramidal cell dendrites, with SNR, response linearity, and photostability significantly better than those of G-CaMP2 (Fig. 2B). In *C. elegans*, chemosensory neurons, and the *Drosophila* antennal lobe, sensory stimulation-evoked fluorescence responses were significantly enhanced with the new indicator (4×–6×), and no apparent cytotoxicity was detected (Fig. 2B). In somatosensory and motor cortical neurons in the awake, behaving mouse brain, the sensor detected three to 10 AP trains with good response linearity and photostability. It also allowed long-term repeated imaging of targeted neuronal populations in the motor cortex of behaving mice (Fig. 2C; Tian et al. 2009). G-CaMP3 plasmid DNA in a eukaryotic expression vector is now available from AddGene. AAV-G-CaMP3 constructs and live virus are available from the University of Pennsylvania Vector Core (http://www.med.upenn.edu/gtp/vectorcore/); fly and worm stocks are available on request.

GECI Expression

A GECI's expression level influences its performance in cells and organisms. Low GECI expression produces poor SNR and difficulty in visualizing positive cells. High expression increases baseline brightness and SNR, but can alter calcium homeostasis and reduce signal change. Prenatal expression (induced by in utero electroporation of G-CaMP3, D3cpV, or TN-XXL driven by the strong *CAG* promoter) altered subcellular sensor distribution and neuronal excitability in some instances (Tian et al. 2009). Postnatal expression of G-CaMP3 from the weaker synapsin-1 promoter via viral-mediated gene transfer alleviated many of these effects (Tian et al. 2009). For each application, GECI expression timing and magnitude should be optimized to balance signal and cytotoxicity by testing multiple promoters, regulatory sequences, and transduction methods (e.g., screening different viral serotypes).

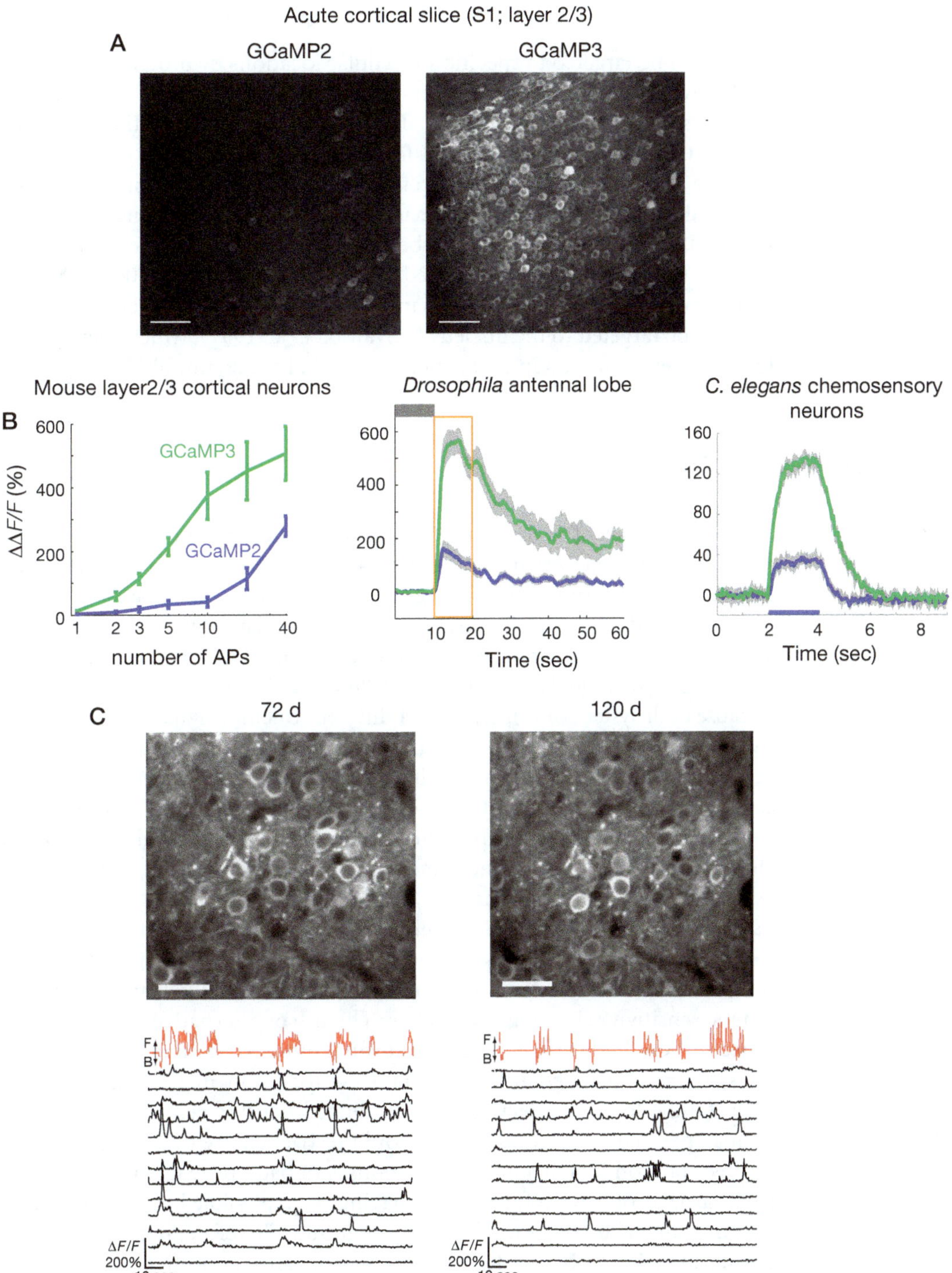

FIGURE 2. Imaging neural activity with improved G-CaMPs (Tian et al. 2009). (*A*) The improved baseline fluorescence of G-CaMP3 compared with that of G-CaMP2. Both indicators were virally delivered to layer 2/3 cortical neurons, and images were taken 12 d post-viral injection. Scale bar, 50 µm. (*B*) The average fluorescence change of G-CaMP3 is greater than that of G-CaMP2 in response to trains of APs given at 83 Hz in layer 2/3 cortical neurons (*left*). Sensory stimulation-evoked fluorescence responses of G-CaMP3 were significantly enhanced in the *Drosophila* antennal lobe (*middle*) and in *C. elegans* chemosensory neurons (*right*). (*C*) G-CaMP3 is suitable for long-term imaging of behaviorally correlated activity in neuronal populations over extended periods of time. Repeated imaging of the same neuronal population (L2/3 neurons of the primary motor cortex) at 72 (*left*) and 120 d (*right*) postinfection showed remarkably constant (*top*) G-CaMP3 expression (scale bar, 30 µm) and (*bottom*) signal change ($\Delta F/F$ of individual cells; black lines). (Red line) Relative treadmill movement (F, forward; B, backward). Scale bars, 20 µm. (Adapted, with permission, from Tian et al. 2009.)

Subcellular Targeting

GECIs can be targeted to specific subcellular locations by protein fusions or signaling peptides. This could reveal underlying Ca^{2+} dynamics in particular organelles (e.g., the pre- and postsynapse) and might also improve the SNR of AP detection without altering the sensor itself. For example, a low-affinity GECI might be less useful for probing small calcium transients in the cytosol but highly effective at measuring large calcium transients in the endoplasmic reticulum (Miyawaki et al. 1999). For neurons exhibiting small somatic calcium changes during spiking, a high-affinity GECI is required. Alternatively, GECIs can be targeted to presynaptic terminals, where Ca^{2+} flux is high, to increase the signal change associated with APs (Dreosti et al. 2009). The rise of Ca^{2+} could be even larger in the vicinity of calcium channel microdomains or synaptic vesicles (Roberts 1994). GECIs can also be targeted to the nucleus (Miyawaki et al. 1997), which might provide an efficient solution for facile segmentation in image processing (L. Looger, unpubl.).

GECI TESTING STANDARDIZATION

The performance of available GECIs has been compared in a wide variety of systems, both in vitro (including purified protein, cultured cells, and hippocampal neurons [Palmer et al. 2006; Tian et al. 2009], and mouse brain slice [Pologruto et al. 2004; Mao et al. 2008; Tian et al. 2009]), and in living animals including worm chemosensory neurons (Tian et al. 2009), *Drosophila* antennal lobe (Tian et al. 2009) and larval neuromuscular junction (Reiff et al. 2005; Hendel et al. 2008), and rodent cortical layer 2/3 pyramidal cells (Mank et al. 2008; Wallace et al. 2008; Tian et al. 2009; Lütcke et al. 2010). Because each system or cell type has a different calcium signaling tool kit, a GECI with high sensitivity in one setting might not be the best fit when conditions change. GECI performance in vitro correlates poorly with that in more intact preparations (Tian et al. 2009). Therefore, we sought to establish a standardized, multistep screening paradigm for GECI developers.

For developmental efficiency, the initial testing system should allow screening for brightness, calcium sensitivity, dynamic range, and kinetics with high throughput. It should also capture as many aspects of intact preparations (e.g., calcium buffering and cytopathicity) as possible. Our method uses agonized endogenous G-protein coupled receptors in cultured mammalian cells to create synthetic Ca^{2+} transients lasting tens of seconds (Tian et al. 2009). Incorporating neurotransmitter receptors into mammalian cells mimics the fast calcium transient in neurons, providing better kinetic sensitivity (L. Looger, unpubl.). The use of alternative cell types could obscure the dynamic aspects of Ca^{2+} signaling unique to neuronal morphology and physiology; emphasis must be placed on systematic analyses in neurons and eventually in intact animals. The recent development of microfluidic cell sorter devices could facilitate library screens of GECI with single-cell resolution (A. Palmer, pers. comm.), although this could restrict choice of cell types and depolarizing stimuli.

Two-photon imaging of GECI responses to back-propagating APs in acute brain slices captures many aspects of in vivo functional imaging (e.g., possible long-term expression and developmental effects) while allowing a moderate throughput. Therefore, it establishes a good standard for comparing a small number of high-quality sensors after an initial screen. Defined numbers of APs can be triggered at specific intervals by intermittent short depolarizing pulses (Pologruto et al. 2004). Alternatively, neurons can be depolarized by continuous current injection (Hendel et al. 2008) from a patch pipette. The calcium transients associated with back-propagating APs have been well characterized. Because the GECI is introduced prenatally by in utero electroporation, potential interference with development should be captured (Mao et al. 2008; Tian et al. 2009).

Owing to its greater imaging depth, motion artifacts, and hemodynamics, in vivo imaging has more challenging SNR requirements. Therefore, it is necessary to confirm the performance of the best candidates in in vivo preparations. Also, to better help end-users choose the most appropriate GECI, the in vivo performance of state-of-the-art GECIs should be compared under identical conditions (e.g., expression cassette, expression time and level, cell type, stimuli).

USING GECIs TO MEASURE NEURAL ACTIVITY

Quantitative neural activity measurements in vivo require calibration of the GECI response to electrophysiological measurement of spike trains in an example set of neurons. In the mammalian brain, calibration via cell-attached recording of spontaneous APs is preferred. Further efforts (e.g., whole-cell patch clamp or synaptic input mapping in slice) should be made to verify that GECI expression does not significantly alter the neurons' physiological properties or network interactions (Tian et al. 2009).

Analytical techniques developed for quantitative imaging with small-molecule calcium dyes can be adapted to GECI imaging, provided the response parameters of the GECI (ΔF/AP, linearity) are determined. Possibilities include reconstruction of firing rate changes via deconvolution (Yaksi and Friedrich 2006), machine learning algorithms (Sasaki et al. 2008), or Monte Carlo spike detection (Vogelstein et al. 2009).

THE FUTURE OF GECI DESIGN

The primary focus of current GECI engineering efforts is to further increase the SNR of existing GECIs for reliable detection of low-firing rates in neurons. Engineering high-affinity sensors for probing small stimuli while preserving fast kinetics is challenging. To balance high SNR with minimal cytotoxicity, expression cassettes that are inducible and reversible or that hold [GECI] steady for long periods of time would facilitate signal calibration and further reduce toxicity concerns in chronic imaging. Careful experiments are required to determine the effect of long-term GECI expression on cells and circuits. Finally, novel GECI scaffolds could be engineered for even greater performance and broader application in neural activity imaging.

ACKNOWLEDGMENTS

We thank Daniel Huber, Jasper Akerboom, Vivek Jayaraman, and Karel Svoboda (all HHMI Janelia Farm) for valuable discussions.

REFERENCES

Akerboom J, Rivera JD, Guilbe MM, Malavé EC, Hernandez HH, Tian L, Hires SA, Marvin JS, Looger LL, Schreiter ER. 2009. Crystal structures of the GCaMP calcium sensor reveal the mechanism of fluorescence signal change and aid rational design. *J Biol Chem* **284:** 6455–6464.

Baird GS, Zacharias DA, Tsien RY. 1999. Circular permutation and receptor insertion within green fluorescent proteins. *Proc Natl Acad Sci* **96:** 11241–11246.

Deuschle K, Okumoto S, Fehr M, Looger LL, Kozhukh L, Frommer WB. 2005. Construction and optimization of a family of genetically encoded metabolite sensors by semirational protein engineering. *Protein Sci* **14:** 2304–2314.

Dreosti E, Odermatt B, Dorostkar MM, Lagnado L. 2009. A genetically encoded reporter of synaptic activity in vivo. *Nat Methods* **6:** 883–889.

Fetcho JR, Cox KJ, O'Malley DM. 1998. Monitoring activity in neuronal populations with single-cell resolution in a behaving vertebrate. *Histochem J* **30:** 153–167.

Griesbeck O, Baird GS, Campbell RE, Zacharias DA, Tsien RY. 2001. Reducing the environmental sensitivity of yellow fluorescent protein. Mechanism and applications. *J Biol Chem* **276:** 29188–29194.

Hasan MT, Friedrich RW, Euler T, Larkum ME, Giese G, Both M, Duebel J, Waters J, Bujard H, Griesbeck O, et al. 2004. Functional fluorescent Ca^{2+} indicator proteins in transgenic mice under TET control. *PLoS Biol* **2:** e163. doi: 10.1371/journal.pbio.0020163.

Heim N, Griesbeck O. 2004. Genetically encoded indicators of cellular calcium dynamics based on troponin C and green fluorescent protein. *J Biol Chem* **279:** 14280–14286.

Heim N, Garaschuk O, Friedrich MW, Mank M, Milos RI, Kovalchuk Y, Konnerth A, Griesbeck O. 2007. Improved calcium imaging in transgenic mice expressing a troponin C–based biosensor. *Nat Methods* **4:** 127–129.

Helmchen F, Imoto K, Sakmann B. 1996. Ca^{2+} buffering and action potential-evoked Ca^{2+} signaling in dendrites of pyramidal neurons. *Biophys J* **70:** 1069–1081.

Hendel T, Mank M, Schnell B, Griesbeck O, Borst A, Reiff DF. 2008. Fluorescence changes of genetic calcium indicators and OGB-1 correlated with neural activity and calcium in vivo and in vitro. *J Neurosci* **28:** 7399–7411.

Hille B. 1992. Pumping ions. *Science* **255:** 742. doi: 10.1126/science.255.5045.742.

Hires SA, Tian L, Looger LL. 2008a. Reporting neural activity with genetically encoded calcium indicators. *Brain Cell Biol* **36:** 69–86.

Hires SA, Zhu Y, Tsien RY. 2008b. Optical measurement of synaptic glutamate spillover and reuptake by linker optimized glutamate-sensitive fluorescent reporters. *Proc Natl Acad Sci* **105:** 4411–4416.

Linse S, Helmersson A, Forsén S. 1991. Calcium binding to calmodulin and its globular domains. *J Biol Chem* **266:** 8050–8054.

Lütcke H, Murayama M, Hahn T, Margolis DJ, Astori S, Zum Alten Borgloh SM, Göbel W, Yang Y, Tang W, Kügler S, et al. 2010. Optical recording of neuronal activity with a genetically-encoded calcium indicator in anesthetized and freely moving mice. *Front Neural Circuits* **4:** 9. doi: 10.3389/fncir.2010.00009.

Mank M, Griesbeck O. 2008. Genetically encoded calcium indicators. *Chem Rev* **108:** 1550–1564.

Mank M, Reiff DF, Heim N, Friedrich MW, Borst A, Griesbeck O. 2006. A FRET-based calcium biosensor with fast signal kinetics and high fluorescence change. *Biophys J* **90:** 1790–1796.

Mank M, Santos AF, Direnberger S, Mrsic-Flogel TD, Hofer SB, Stein V, Hendel T, Reiff DF, Levelt C, Borst A, et al. 2008. A genetically encoded calcium indicator for chronic in vivo two-photon imaging. *Nat Methods* **5:** 805–811.

Mao T, O'Connor DH, Scheuss V, Nakai J, Svoboda K. 2008. Characterization and subcellular targeting of GCaMP-type genetically-encoded calcium indicators. *PLoS One* **3:** e1796. doi: 10.1371/journal.pone.0001796.

McCombs JE, Palmer AE. 2008. Measuring calcium dynamics in living cells with genetically encodable calcium indicators. *Methods* **46:** 152–159.

Miyawaki A, Tsien RY. 2000. Monitoring protein conformations and interactions by fluorescence resonance energy transfer between mutants of green fluorescent protein. *Methods Enzymol* **327:** 472–500.

Miyawaki A, Llopis J, Heim R, McCaffery JM, Adams JA, Ikura M, Tsien RY. 1997. Fluorescent indicators for Ca^{2+} based on green fluorescent proteins and calmodulin. *Nature* **388:** 882–887.

Miyawaki A, Griesbeck O, Heim R, Tsien RY. 1999. Dynamic and quantitative Ca^{2+} measurements using improved cameleons. *Proc Natl Acad Sci* **96:** 2135–2140.

Nagai T, Sawano A, Park ES, Miyawaki A. 2001. Circularly permuted green fluorescent proteins engineered to sense Ca^{2+}. *Proc Natl Acad Sci* **98:** 3197–3202.

Nagai T, Ibata K, Park ES, Kubota M, Mikoshiba K, Miyawaki A. 2002. A variant of yellow fluorescent protein with fast and efficient maturation for cell-biological applications. *Nat Biotechnol* **20:** 87–90.

Nagai T, Yamada S, Tominaga T, Ichikawa M, Miyawaki A. 2004. Expanded dynamic range of fluorescent indicators for Ca^{2+} by circularly permuted yellow fluorescent proteins. *Proc Natl Acad Sci* **101:** 10554–10559.

Nakai J, Ohkura M, Imoto K. 2001. A high signal-to-noise Ca^{2+} probe composed of a single green fluorescent protein. *Nat Biotechnol* **19:** 137–141.

Ogawa Y, Tanokura M. 1984. Calcium binding to calmodulin: Effects of ionic strength, Mg^{2+}, pH and temperature. *J Biochem* **95:** 19–28.

Palmer AE, Tsien RY. 2006. Measuring calcium signaling using genetically targetable fluorescent indicators. *Nat Protoc* **1:** 1057–1065.

Palmer AE, Giacomello M, Kortemme T, Hires SA, Lev-Ram V, Baker D, Tsien RY. 2006. Ca^{2+} indicators based on computationally redesigned calmodulin-peptide pairs. *Chem Biol* **13:** 521–530.

Pédelacq JD, Cabantous S, Tran T, Terwilliger TC, Waldo GS. 2006. Engineering and characterization of a superfolder green fluorescent protein. *Nat Biotechnol* **24:** 79–88.

Pologruto TA, Yasuda R, Svoboda K. 2004. Monitoring neural activity and [Ca^{2+}] with genetically encoded Ca^{2+} indicators. *J Neurosci* **24:** 9572–9579.

Reiff DF, Ihring A, Guerrero G, Isacoff EY, Joesch M, Nakai J, Borst A. 2005. In vivo performance of genetically encoded indicators of neural activity in flies. *J Neurosci* **25:** 4766–4778.

Roberts WM. 1994. Localization of calcium signals by a mobile calcium buffer in frog saccular hair cells. *J Neurosci* **14:** 3246–3262.

Sasaki T, Takahashi N, Matsuki N, Ikegaya Y. 2008. Fast and accurate detection of action potentials from somatic calcium fluctuations. *J Neurophysiol* **100:** 1668–1676.

Stosiek C, Garaschuk O, Holthoff K, Konnerth A. 2003. In vivo two-photon calcium imaging of neuronal networks. *Proc Natl Acad Sci* **100:** 7319–7324.

Svoboda K, Denk W, Kleinfeld D, Tank DW. 1997. In vivo dendritic calcium dynamics in neocortical pyramidal neurons. *Nature* **385:** 161–165.

Tallini YN, Ohkura M, Choi BR, Ji G, Imoto K, Doran R, Lee J, Plan P, Wilson J, Xin HB, et al. 2006. Imaging cellular signals in the heart in vivo: Cardiac expression of the high-signal Ca^{2+} indicator GCaMP2. *Proc Natl Acad Sci* **103:** 4753–4758.

Tay LH, Griesbeck O, Yue DT. 2007. Live-cell transforms between Ca^{2+} transients and FRET responses for a troponin-C-based Ca^{2+} sensor. *Biophys J* **93:** 4031–4040.

Tian L, Hires SA, Mao T, Huber D, Chiappe ME, Chalasani SH, Petreanu L, Akerboom J, McKinney SA, Schreiter ER, et al. 2009. Imaging neural activity in worms, flies and mice with improved GCaMP calcium indicators. *Nat Methods* **6:** 875–881.

Tour O, Adams SR, Kerr RA, Meijer RM, Sejnowski TJ, Tsien RW, Tsien RY. 2007. Calcium Green FlAsH as a genetically targeted small-molecule calcium indicator. *Nat Chem Biol* **3:** 423–431.

Vogelstein JT, Watson BO, Packer AM, Yuste R, Jedynak B, Paninski L. 2009. Spike inference from calcium imaging using sequential Monte Carlo methods. *Biophys J* **97:** 636–655.

Wallace DJ, Meyer zum Alten Borgloh S, Astori S, Yang Y, Bausen M, Kügler S, Palmer AE, Tsien RY, Sprengel R, Kerr JN, et al. 2008. Single-spike detection in vitro and in vivo with a genetic Ca^{2+} sensor. *Nat Methods* **5:** 797–804.

Wang Q, Shui B, Kotlikoff MI, Sondermann H. 2008. Structural basis for calcium sensing by GCaMP2. *Structure* **16:** 1817–1827.

Yaksi E, Friedrich RW. 2006. Reconstruction of firing rate changes across neuronal populations by temporally deconvolved Ca^{2+} imaging. *Nat Methods* **3:** 377–383.

Yasuda R, Nimchinsky EA, Scheuss V, Pologruto TA, Oertner TG, Sabatini BL, Svoboda K. 2004. Imaging calcium concentration dynamics in small neuronal compartments. *Sci STKE* **2004:** pl5. doi: 10.1126/stke.2192004pl5.

Yuste R, Peinado A, Katz LC. 1992. Neuronal domains in developing neocortex. *Science* **257:** 665–669.

9 Imaging Synaptic Inhibition with a Genetically Encoded Chloride Indicator

Ken Berglund,[1] Thomas Kuner,[2] Guoping Feng,[1] and George J. Augustine[3,4]

[1]*Department of Neurobiology, Duke University Medical Center, Durham, North Carolina 27710;* [2]*Institute of Anatomy and Cell Biology, University of Heidelberg, 69120 Heidelberg, Germany;* [3]*Program in Neuroscience and Behavioral Disorders, Duke–National University of Singapore Graduate Medical School, Singapore 169547;* [4]*Center for Functional Connectomics, Korea Institute of Science and Technology, Seoul, South Korea 136-791*

ABSTRACT

Although several techniques are available to image excitatory processes in the brain, synaptic inhibition has remained largely invisible. Most synaptic inhibition in the brain arises from transmembrane fluxes of chloride ions (Cl^-), so imaging intracellular Cl^- concentration ($[Cl^-]_i$) is, in principle, a natural way to visualize the spatiotemporal dynamics of inhibition. This chapter discusses the suitability of Clomeleon, a genetically encoded indicator of Cl^-, as a tool for monitoring synaptic inhibition.

INTRODUCTION

Clomeleon is a fusion protein consisting of the cyan and yellow fluorescent proteins (CFP and YFP, respectively), joined by a flexible 24-amino acid linker (Fig. 1A; Kuner and Augustine 2000). Because of the close spatial proximity of the two fluorophores, fluorescence resonance energy transfer (FRET) occurs. Thus, exciting the CFP donor causes the YFP acceptor to be excited and to emit yellow fluorescence. Binding Cl^- to YFP quenches the yellow fluorophore (Jayaraman et al. 2000) and decreases the degree of FRET. Thus, measuring the ratio of emission of YFP relative to that of CFP (while exciting CFP) provides an absolute measure of the internal Cl^- concentration ($[Cl^-]_i$). Imaging of the fluorescence lifetime of the CFP donor is another way to use Clomeleon to monitor changes in $[Cl^-]_i$ (Jose et al. 2007).

Clomeleon imaging has found a wide range of applications both in vitro and in vivo (Berglund et al. 2008). Such applications are by no means limited to the brain; any cellular system that permits expression of exogenous DNAs or RNAs, or introduction of a recombinant protein that has been expressed in some other way, is appropriate for Clomeleon-based Cl^- imaging (Lorenzen et al. 2004). The specific goals of the work described here are to document the ability to target Clomeleon expression to neurons in the mouse brain and to use Clomeleon to measure $[Cl^-]_i$ in these neurons.

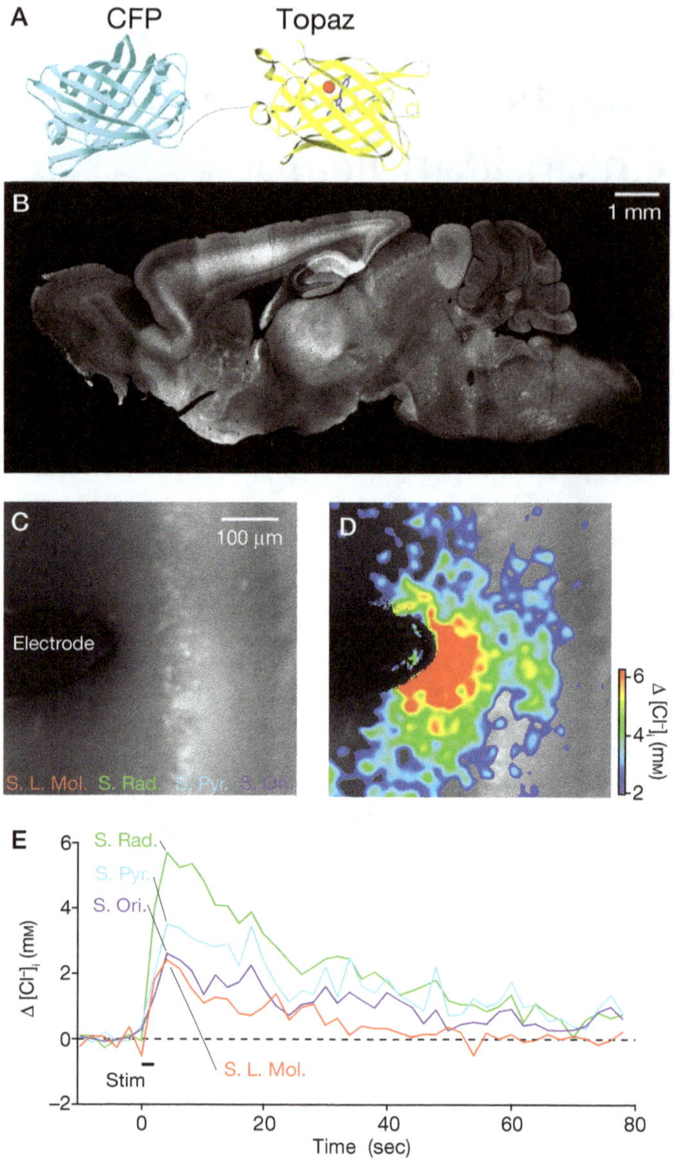

FIGURE 1. (*A*) Structure of Clomeleon. CFP (cyan) and Topaz (yellow) are connected by a short linker (gray). Shown inside Topaz are the halide binding site (red) and the nearby chromophore (blue). (*B*) Expression of Clomeleon in the brain of an adult *thy1*::*Clomeleon* mouse. The image shows YFP fluorescence in a paramedial sagittal section of fixed brain from a CLM11 mouse. (*C*) Image of YFP Clomeleon fluorescence (grayscale) expressed in CA1 pyramidal cells from a CLM1 hippocampal slice. (*D*) Cl^- changes (pseudocolor scale) of the same areas shown in *C* in response to a train of electrical stimuli delivered through the stimulating electrode (*left side* of image). The Cl^- image was acquired 4 sec after the stimulus. S. L. Mol., *stratum lacunosum moleculare*; S. Rad., *stratum radiatum*; S. Pyr., *stratum pyramidale*; S. Ori., *stratum oriens*. (*E*) Time course of changes in $[Cl^-]_i$ calculated from each layer. Duration of the 2-sec stimulus train (Stim) is indicated by a bar.

The studies described here use lines of transgenic mice that express Clomeleon under the control of a neuron-specific promoter, *thy1* (Feng et al. 2000). The transgene has been used to produce several lines of transgenic mice, each of which expresses Clomeleon in specific subsets of neurons (Berglund et al. 2006). These mosaic patterns of expression are heritable and are therefore found in all individuals of a given line. One of the lines expresses Clomeleon within CA1 pyramidal cells in the hippocampus, as well as in neurons of many other brain regions (Fig. 1B). We are in the process of making these mice commercially available via the Jackson Labs (www.jax.org).

Protocol

Chloride Imaging with Clomeleon

The following are the procedures that we use when imaging neuronal $[Cl^-]_i$ in brain slices prepared from Clomeleon transgenic mice. With only minor adjustments, the same procedures should be suitable for imaging from cultured cells as well.

IMAGING SETUP

Optical Imaging

An upright epifluorescence microscope (e.g., Nikon's Eclipse E600FN or equivalent) equipped with a mercury lamp, a 515-nm dichroic mirror, a filter wheel, and excitation (440 ± 10 nm) and emission filters (485 ± 15 nm for CFP; 530 ± 15 nm for YFP) is used to image Clomeleon fluorescence. Fluorescence excitation is produced by 200-msec-long light pulses (0.5 Hz) and fluorescence emission is collected alternately at each wavelength with a cooled CCD camera with on-chip multiplication gain control (Cascade 512B, Photometrics).

Two-Photon Imaging Setup

Clomeleon fluorescence also can be imaged with a two-photon microscope (Ultima IV, Prairie Technologies) using 860-nm laser light (Chameleon Ultra II, Coherent) and a 40x water-immersion objective (LUMPLFL40xW/IR/0.80, Olympus) with total output power <50 mW. In this case, a fluorescence filter cube with an emission beam splitter (71007, Chroma) is used for fluorescence emission.

MATERIALS

CAUTION: See Appendix 6 for proper handling of materials marked with <!>.

Reagents

Artificial cerebrospinal fluid (ACSF)
Isoflurane <!>
Mice, Clomeleon transgenic, 2–3-wk-old

Equipment

Dissection tools
Microscope, epifluorescence or two-photon (as described in Imaging Setup)
Vibratome

SLICE PREPARATION

Hippocampal slices are prepared from transgenic mice using conventional methods (see, e.g., Berglund et al. 2006).

1. Anesthetize the mice with isoflurane.
2. Decapitate the animals. Immerse the skulls in chilled ACSF.
3. Working quickly, remove the brains. Keep the brains in chilled ACSF.
4. Using a vibratome, cut 250-µm-thick sagittal sections of the hippocampus. Transfer sections to an incubation chamber containing ACSF.
5. Incubate the slices for 30 min at 37°C before imaging.

CALIBRATION PROCEDURE

The Clomeleon fluorescence emission ratio can be converted into $[Cl^-]_i$ (Berglund et al. 2009). Briefly, brain slices are incubated in situ at pH 7.1 in the presence of the OH^-/Cl^- antiporter tributyltin (Inglefield and Schwartz-Bloom 1997; Berglund et al. 2006). Because pH-dependent exchangers such as the Cl^-/HCO_3^- exchanger (Krapf et al. 1988) can change $[Cl^-]$, the K^+/H^+ ionophore nigericin is also added to clamp the intracellular pH to the value of the extracellular pH. Under these conditions, tributyltin equilibrates $[Cl^-]_i$ with $[Cl^-]_o$. The slices are then incubated with solutions of F^- or gluconate to determine the values when Clomeleon is Cl^--bound (R_{min}) or Cl^--free (R_{max}). To calculate the effective Cl^- dissociation constant of Clomeleon (K_d'), slices are incubated with two known concentrations of Cl^- (50 and 134 mM). The $[Cl^-]_i$ is then calculated using the equation $[Cl^-]_i = K_d' \times [(R_{max} - R)/(R - R_{min})]$.

EXAMPLE OF APPLICATION

To determine if chronically expressed Clomeleon could report $[Cl^-]_i$ changes associated with synaptic inhibition, changes in Clomeleon signals were measured in response to electrical stimulation (800 µA for 1 sec at 20 Hz) via a concentric bipolar metal electrode placed in the *stratum lacunosum moleculare* of hippocampal slices from transgenic mice. All measurements were made in the presence of 3 mM kynurenic acid (a glutamate antagonist) to prevent activation of excitatory synapses (Fig. 1C). A train of stimuli produced a transient increase in $[Cl^-]_i$ of ~3 mM (Fig. 1D), which returned to baseline within 30–50 sec (Fig. 1E). The Cl^- transient was almost completely blocked by 10 µM of the $GABA_A$ receptor antagonist gabazine, indicating that the transient was the result of activation of $GABA_A$ receptors. These results indicate that Clomeleon can report changes in neuronal $[Cl^-]_i$ associated with synaptic inhibition. By measuring $[Cl^-]_i$ over many (e.g., >20) cells, Clomeleon can detect $[Cl^-]_i$ transients of ~200 µM caused by only five inhibitory postsynaptic potentials (Berglund et al. 2006). In addition, Clomeleon allows detection of gradients in $[Cl^-]_i$ associated with synaptic activity (Fig. 1E).

Not all FRET-based indicator proteins are suitable for two-photon microscopy. However, two-photon excitation spectra of cerebellar granule cells in slices prepared from Clomeleon transgenic mice (Fig. 2A) showed that emission of both CFP and YFP was brightest when excited by 860-nm light (Fig. 2B). The YFP/CFP emission ratio was also relatively high at 860 nm (Fig. 2B). To determine the sensitivity of Clomeleon to Cl^-, the Clomeleon emission ratio was measured at different excitation wavelengths and at different $[Cl^-]$ values (Fig. 2C). Half-maximal quenching of Clomeleon occurred at 120 mM $[Cl^-]$, irrespective of excitation wavelength. Thus, excitation at 860 nm is optimal for Clomeleon-based measurements of $[Cl^-]_i$. These results indicate that chronically expressed Clomeleon is suitable for two-photon imaging of $[Cl^-]_i$ (Duebel et al. 2006). This permits in vivo measurement of $[Cl^-]_i$ in the brains of Clomeleon transgenic mice (Berglund et al. 2009) and is being used to image tonic inhibition of granule cells by ambient GABA in vivo, as well as in slices in vitro (Lee et al. 2010).

ADVANTAGES AND LIMITATIONS

Clomeleon offers two distinct advantages over existing Cl^- indicator dyes. The first is that Clomeleon can be targeted to specific cells via genetic targeting. Using specific promoters, stereotaxic delivery of viruses, and/or Cre recombinase techniques, Clomeleon expression already has been targeted to a variety of neuronal types (Berglund et al. 2008), and it is very likely that such applications will expand in the future. Second, Clomeleon is ratiometric and, therefore, permits determination of absolute $[Cl^-]$ within the cells of interest. In addition, unlike other FRET-based indicator proteins, Clomeleon does not require a conformational change between the units of the FRET pair because quenching of YFP by Cl^- is sufficient to induce a change in the YFP/CFP emission ratio. Thus, Clomeleon permits efficient two-photon excitation by Ti:sapphire lasers, enabling studies in visible light-sensitive retinas (Duebel et al. 2006) and in the mouse brain in vivo (Berglund et al. 2008).

There are three significant limitations when using Clomeleon to image $[Cl^-]_i$.

- Clomeleon is sensitive to protons, with lower pH increasing the affinity of Clomeleon for Cl^- (Kuner and Augustine 2000). In practice, interference by protons is minimal when $[Cl^-]_i$ is low, as is the case under most physiological conditions. In an experimental paradigm in which higher $[Cl^-]_i$ and/or significant pH changes are expected, the internal pH of the cells under investigation should be measured by conventional, organic pH indicator dyes so that the known relationship between pH and Clomeleon affinity (Kuner and Augustine 2000) can be used to calibrate Clomeleon responses accurately (Pond et al. 2006; Berglund et al. 2009).

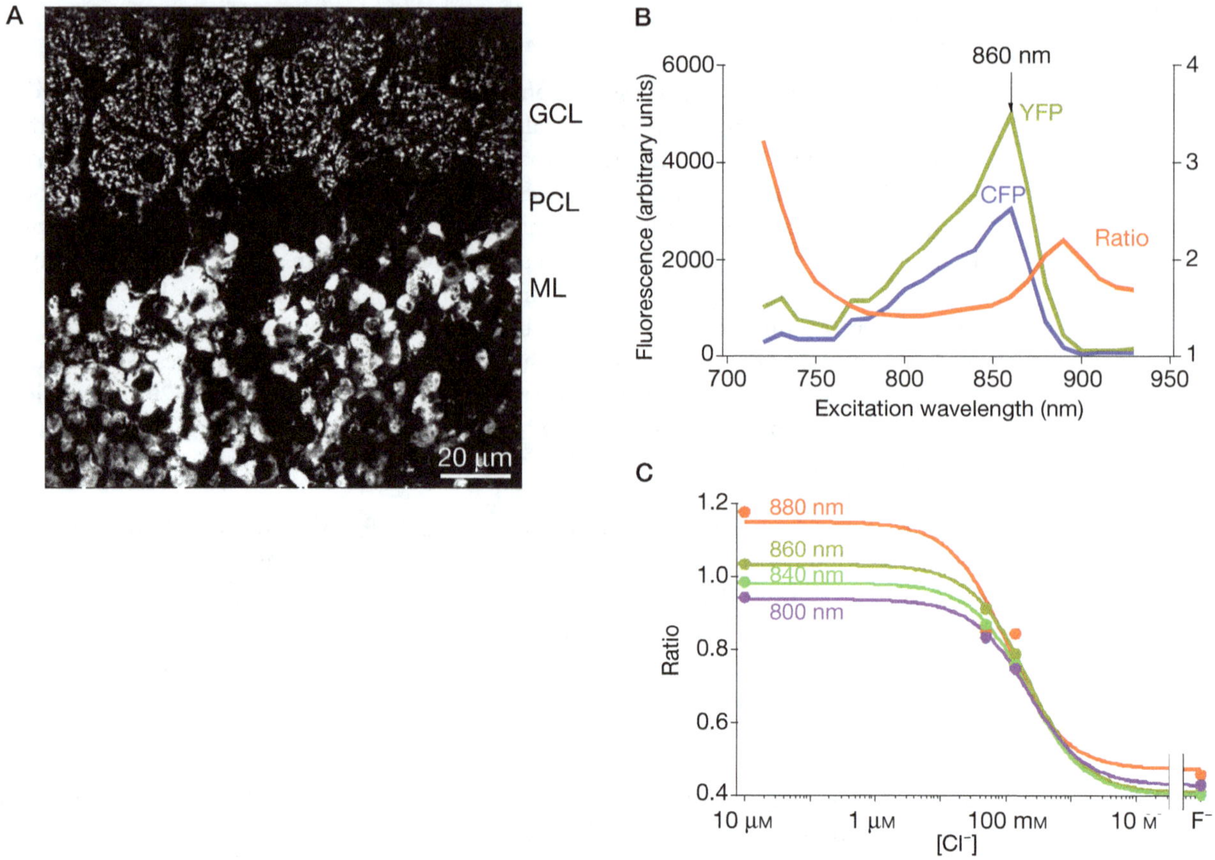

FIGURE 2. *(A)* A two-photon image of YFP fluorescence in the cerebellar cortex, showing expression of Clomeleon exclusively in granule cells. Both the molecular layer (ML) and granule cell layer (GCL) are fluorescent, indicating the presence of Clomeleon in granule cell bodies/ dendrites and parallel fibers. Note that Purkinje cell bodies do not express Clomeleon, as indicated by the absence of YFP fluorescence in the Purkinje cell layer (PCL). *(B)* Two-photon excitation spectrum of Clomeleon. Both the CFP and YFP fluorescences of Clomeleon were brightest when excited by 860-nm light. The YFP/CFP emission ratio (Ratio) was also relatively high at 860 nm. *(C)* Cl^- titration curves for Clomeleon measured at various excitation wavelengths. $[Cl^-]_i$ was clamped at $[Cl^-]_o$ using a Cl^-/H^+ antiporter/ ionophore cocktail, as described in the text. Using an 860-nm excitation light, the dynamic range of Clomeleon was nearly maximal.

- Intense illumination of Clomeleon causes differential bleaching of CFP and YFP, leading to a decrease in the Clomeleon emission ratio. This artifact can incorrectly indicate an increase in $[Cl^-]_i$. In practice, this problem can be avoided by minimizing the intensity and duration of Clomeleon excitation and by using highly sensitive charge-coupled device (CCD) cameras or photomultiplier tubes.
- The affinity of Clomeleon for Cl^- is not optimal, being ~90 mM (Kuner and Augustine 2000), whereas $[Cl^-]_i$ in mature neurons is on the order of 5 mM (Berglund et al. 2006). This means that the sensitivity of Clomeleon for detecting synaptic inhibition is limited, particularly in neurons in which the membrane potential is not voltage clamped and the changes in $[Cl^-]_i$ associated with synaptic inhibition are $<<$1 mM (Berglund et al. 2006). This limitation can be overcome by signal averaging over many cells and/or trials (Berglund et al. 2006), although the ultimate solution is to reengineer Clomeleon to improve its Cl^- binding properties.

Despite these limitations, the utility of Clomeleon for measuring $[Cl^-]_i$ makes this indicator a valuable tool for analysis of the spatiotemporal dynamics of inhibitory networks in living brain tissue.

ACKNOWLEDGMENTS

KB was a Research Fellow of the Japan Society for the Promotion of Science. This work was partially supported by Deutsche Forschungsgemeinschaft grant KU1983/1-2 (FOR 577) awarded to TK, National Institutes of Health grants awarded to GJA and GF, and an Alfred P. Sloan Research Fellowship, a Klingenstein Fellowship Award in the Neurosciences, and a Beckman Young Investigator Award to GF.

REFERENCES

Berglund K, Schleich W, Krieger P, Loo LS, Wang D, Cant NB, Feng G, Augustine GJ, Kuner T. 2006. Imaging synaptic inhibition in transgenic mice expressing the chloride indicator, Clomeleon. *Brain Cell Biol* **35:** 207–228.

Berglund K, Schleich W, Wang H, Feng G, Hall WC, Kuner T, Augustine GJ. 2008. Imaging synaptic inhibition throughout the brain via genetically targeted Clomeleon. *Brain Cell Biol* **36:** 101–118.

Berglund K, Kuner T, Augustine GJ. 2009. Clomeleon, a genetically encoded chloride indicator. In *Physiology and pathology of chloride transporters and channels in the nervous system* (ed. Alvarez-Leefmans FJ, Delpire E), pp. 125–140. Academic, London.

Duebel J, Haverkamp S, Schleich W, Feng G, Augustine GJ, Kuner T, Euler T. 2006. Two-photon imaging reveals somatodendritic chloride gradient in retinal ON-type bipolar cells expressing the biosensor Clomeleon. *Neuron* **49:** 81–94.

Feng G, Mellor RH, Bernstein M, Keller-Peck C, Nguyen QT, Wallace M, Nerbonne JM, Lichtman JW, Sanes JR. 2000. Imaging neuronal subsets in transgenic mice expressing multiple spectral variants of GFP. *Neuron* **28:** 41–51.

Inglefield JR, Schwartz-Bloom RD. 1997. Confocal imaging of intracellular chloride in living brain slices: Measurement of $GABA_A$ receptor activity. *J Neurosci Methods* **75:** 127–135.

Jayaraman S, Haggie P, Wachter RM, Remington SJ, Verkman AS. 2000. Mechanism and cellular applications of a green fluorescent protein–based halide sensor. *J Biol Chem* **275:** 6047–6050.

Jose M, Nair DK, Reissner C, Hartig R, Zuschratter W. 2007. Photophysics of Clomeleon by FLIM: Discriminating excited state reactions along neuronal development. *Biophys J* **92:** 2237–2254.

Krapf R, Berry CA, Verkman AS. 1988. Estimation of intracellular chloride activity in isolated perfused rabbit proximal convoluted tubules using a fluorescent indicator. *Biophys J* **53:** 955–962.

Kuner T, Augustine GJ. 2000. A genetically encoded ratiometric indicator for chloride: Capturing chloride transients in cultured hippocampal neurons. *Neuron* **27:** 447–459.

Lee S, Yoon BE, Berglund K, Oh SJ, Park H, Shin HS, Augustine GJ, Lee CJ. 2010. Channel-mediated tonic GABA release from glia. *Science* **330:** 790–796.

Lorenzen I, Aberle T, Plieth C. 2004. Salt stress-induced chloride flux: A study using transgenic *Arabidopsis* expressing a fluorescent anion probe. *Plant J* **38:** 539–544.

Pond BB, Berglund K, Kuner T, Feng G, Augustine GJ, Schwartz-Bloom RD. 2006. The chloride transporter Na^+-K^+-Cl^- cotransporter isoform-1 contributes to intracellular chloride increases after in vitro ischemia. *J Neurosci* **26:** 1396–1406.

10 Recombinant Fluorescent Rabies Virus Vectors for Tracing Neurons and Synaptic Connections

Nadin Hagendorf and Karl-Klaus Conzelmann

Max von Pettenkofer Institute and Gene Center, Ludwig-Maximilians-University Munich, 81377 Munich, Germany

ABSTRACT

Recombinant rabies virus (RV) vectors expressing fluorescent proteins allow staining of neurons from many mammalian species and enable the study of neuron morphology. Because viral spread occurs only between neurons that have synaptic connections, these vectors also permit transsynaptic tracing. A recently established system for restriction of transsynaptic tracing to a single transsynaptic jump, dubbed monosynaptic tracing, uses glycoprotein gene–defective, pseudotyped RV. This allows infection of defined cells and transient complementation with the glycoprotein in situ to support a single step of transsynaptic crossing to presynaptic cells. The present protocol describes the production of RV vectors, including the recovery of recombinant RV from complementary DNA (cDNA) and virus pseudotyping in vitro. This allows retrograde staining of neurons projecting to the inoculation site.

INTRODUCTION

Rabies Virus

RV is a highly neurotropic, enveloped, negative-strand RNA virus of the Rhabdoviridae family with a very broad host range including virtually all mammals and birds. The major determinant for neurotropism and host range is the viral transmembrane glycoprotein (G), which mediates attachment to target cells, endocytosis, transport of endocytosed vesicles, and virus–vesicle membrane fusion to release the viral nucleocapsid into the cytoplasm of target cells (Finke and Conzelmann 2005; Klingen et al. 2008). Notably, the typical RV retrograde spread from an infected neuron to presynaptic neurons seems to occur exclusively via actual synaptic connections (Ugolini 1995, 2008; Wickersham et al. 2007). This feature is unique among viruses and makes RV a true transsynaptic tracer (Callaway 2008). Transsynaptic transmission requires the G protein, as is evident from recombinant RV lacking the G gene (Street–Alabama–Dufferin [SAD] ΔG) (Mebatsion et al. 1996), which is not able to spread in vitro or in vivo (Etessami et al. 2000). However, the mechanisms behind the specificity of transsynaptic transmission are not yet fully understood. An increased density of high-affinity receptors for the RV G at presynaptic terminals may contribute. So far, p75 neurotrophin receptor (p75NTR) and neuronal cell adhesion molecule (NCAM)have been described as RV attachment and/or receptor proteins, along with gangliosides, phospholipids, and the nicotinic acetylcholine receptor (nAChR; Lafon 2005). Receptor usage of G proteins from different RV strains may differ, and for transsynaptic tracing experiments, the use of G proteins from virulent (neuroinvasive) RV strains (e.g., challenge virus standard [CVS]) is suggested.

Fortunately, nonneuronal cells and cell lines can also be infected in vitro by RV, including hamster kidney cell lines (baby hamster kidney [BHK]) and BSR, a BHK-derived cell line that is used for virus stock production. In these nonpolarized cells, budding of free RV particles into the culture supernatant appears to occur all over the cell membrane. When injected into the brain or incubated with brain slice cultures, such supernatant virus predominantly infects neurons, although glia, astrocytes, and oligodendrocytes are also infected by some RV strains.

Complementation, Pseudotyping, and Envelope Switching

Each of the five genes encoded in the 12-kb RV RNA in the order 3′-N-P-M-G-L-5′ is essential for unrestricted virus propagation and spread (Fig. 1A). Deletion of the N (nucleoprotein), P (phosphoprotein), or L (RNA polymerase) genes leads to failure of RNA synthesis; the lack of G results in

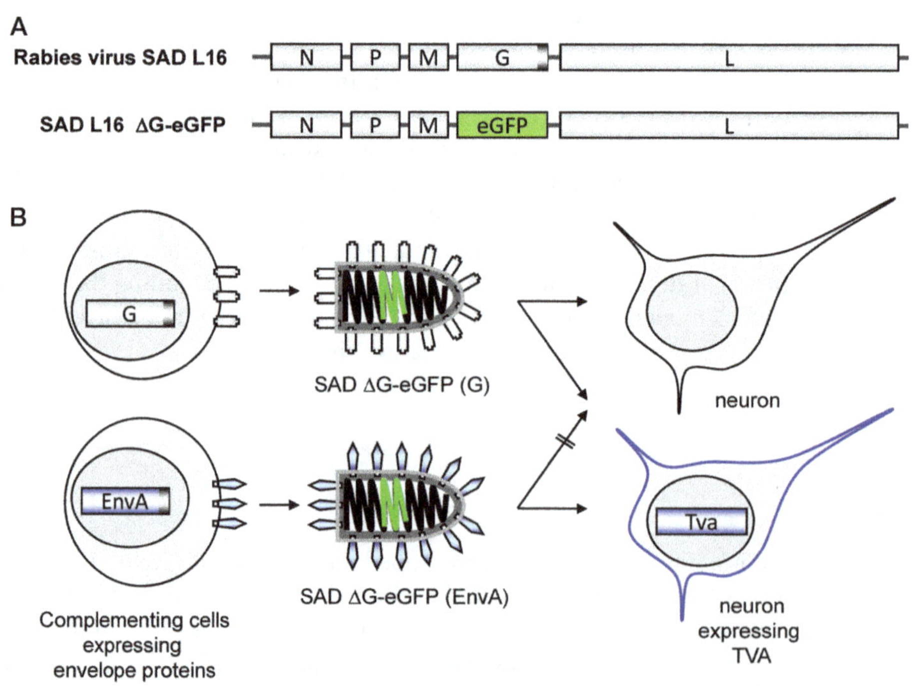

FIGURE 1. *(A)* Organization of the recombinant rabies virus (RV) RNA genome. The 12-kb RNA of RV (SAD L16) encodes five proteins—nucleoprotein (N), phosphoprotein (P), matrix protein (M), glycoprotein (G), and the RNA polymerase (L). N, P, L, and RNA are associated with the viral ribonucleoprotein (RNP); M forms a layer connecting the RNP with the lipid bilayer membrane; and G is a trimeric type I transmembrane spike protein mediating attachment of the virus to cell receptors and membrane fusion. SAD ΔG-eGFP lacks a G gene and encodes eGFP (enhanced green fluorescent protein) instead. *(B)* Pseudotyping of G gene–deficient virus. To yield virus particles able to infect cells, the G deficiency of SAD ΔG-eGFP must be complemented in cells expressing an analogous glycoprotein (i.e., a protein that binds to a receptor and mediates membrane fusion). For effective incorporation of heterologous proteins, such as the retroviral EnvA, their carboxy-terminal domains should be exchanged with that of G. As EnvA has a highly specific avian receptor (TVA), EnvA-pseudotyped RV infects only mammalian cells engineered to express TVA.

the formation of spikeless particles, at least in nonpolarized cells; and in the absence of M (matrix protein), cell-free particles are not produced. Recovery and propagation of such gene-deleted single-round vectors, therefore, requires complementation with the respective gene product in *trans*, by transient expression or in cell lines expressing the protein (Fig. 1B). G deficiency can be complemented also by analogous envelope proteins of heterologous viruses (e.g., envelope [env] proteins of retroviruses, such as EnvA of avian sarcoma and leucosis virus [ASLV], or HIV-1 Env [gp160]). For effective incorporation of the heterologous spike proteins into the RV envelope, their carboxy-terminal cytoplasmic domains have to be replaced with that of the RV G (e.g., $EnvA^{RGct}$). The resulting pseudotype viruses infect cells according to the receptor specificity of the heterologous env protein (Mebatsion and Conzelmann 1996). The receptor for EnvA (or $EnvA^{RGct}$) is TVA, a small avian protein not present in mammalian cells (Bates et al. 1993). Infection of TVA-expressing mammalian neurons by EnvA pseudotype virions is, therefore, highly specific with little background (Federspiel et al. 1994). This combination is, therefore, ideal for monosynaptic tracing experiments such as those pioneered by Callaway and colleagues (Wickersham et al. 2007; Marshel et al. 2010): Neurons expressing TVA and G by standard transgene approaches are selectively infected by RV(EnvA) pseudotype viruses such as SAD ΔG-eGFP(EnvA). The G gene deficiency is complemented in the infected cells by the G protein, allowing for a single transsynaptic crossing by the newly formed SAD ΔG(G) pseudotype viruses. If the TVA-positive target (postsynaptic) cells express a fluorescent protein such as dsRed or mCherry in addition to TVA and G, they can be easily identified by fluorescence microscopy. After infection with SAD ΔG-eGFP (enhanced green fluorescent protein), they appear yellow in the merge (red and green), whereas the presynaptic cells only stain green from SAD ΔG-eGFP gene expression (Fig. 2).

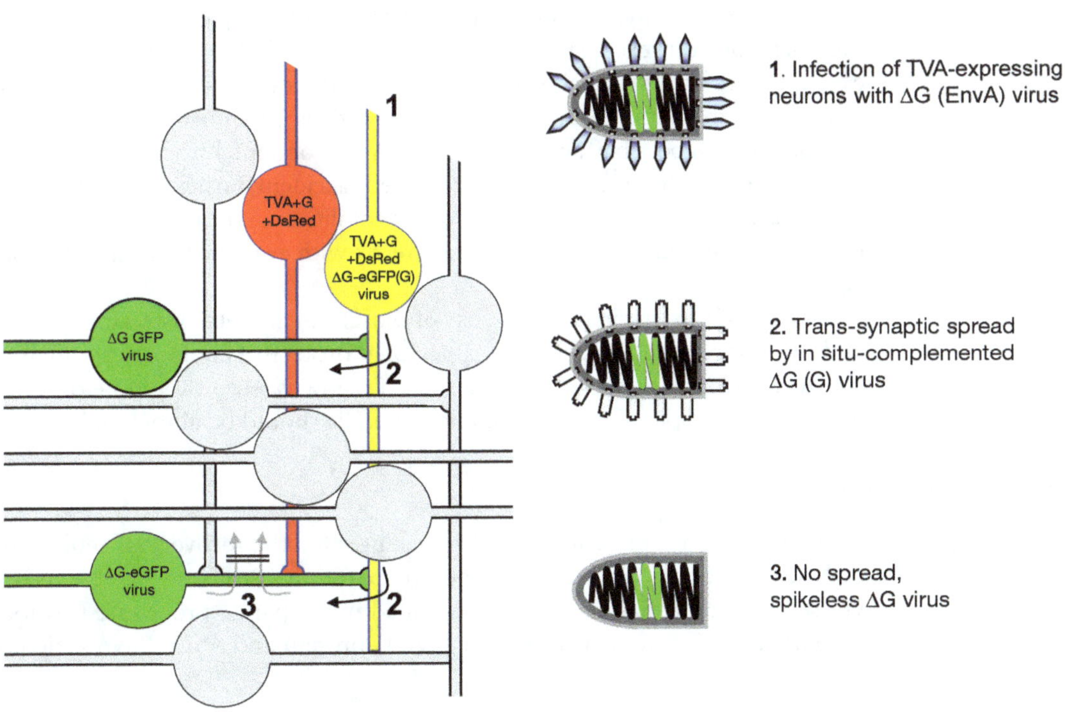

FIGURE 2. Principle of monosynaptic tracing with pseudotyped RV. Neurons are engineered by standard techniques (plasmid transfection, electroporation, gene gun, or transgenics) to express the EnvA receptor TVA, the RV glycoprotein G, and a fluorescent marker (dsRed), thereby allowing identification of potential target cells and infected cells. (1) Successful infection of target cells by EnvA-pseudotyped RV expressing eGFP is observed by yellow (merge) fluorescence; uninfected cells remain red. (2) The G-deficient RV is complemented in situ by the expressed G protein, allowing transsynaptic spread to presynaptic neurons exclusively. (3) The lack of G in infected (green) presynaptic cells prevents spread to secondary order neurons.

Genetic Engineering of RV

The RNA of negative-strand RNA viruses such as RV is not infectious per se, and recovery of the virus from cDNA-derived RNA in cells is a highly inefficient process. The system available involves simultaneous expression of an RNA that corresponds to viral antigenome RNA (cRNA), mostly by T7 RNA polymerase, and three helper proteins (N, P, and L) to promote assembly of a transcriptionally active nucleocapsid in which the RNA is encapsidated by N protein and recognized by the coexpressed viral polymerase as a template (Schnell et al. 1994). In contrast to cotranscriptional encapsidation of virus RNA as it occurs during replication of nucleocapsids, such de novo encapsidation of preformed long RNAs by N is extremely ineffective (Conzelmann 2004). Initiation of an infectious cycle is observed in only one of approximately 100,000 cells expressing cRNA and helper proteins. Because of this low recovery rate, premade amplified virus stocks have to be used in most experiments.

Biosafety

Rabies is a preventable viral illness. All users of RV are advised to read the safety notes and to follow them carefully.

The use and handling of any live virus is hazardous and requires precautions. Because nondeficient RV may cause lethal infection and effective, inexpensive, and completely innocuous vaccines are available, immunization is considered the best solution to prevent any would-be hazard. Gene-deleted RVs such as pseudotyped SAD ΔG are single-round agents categorized in biosafety level 1 and do not pose a threat per se when handled according to good microbiological practice. The risk of recombination of virus RNA and G mRNA in complementing cells is probably close to 0 because of the tight viral ribonucleoprotein (RNP) and the lack of homologous sequences. Recombination of viral RNA with foreign/cellular RNA (or DNA) has not been observed so far for any of the Mononegavirales (Chare et al. 2003). DNA recombination of viral and complementing G cDNA during rescue experiments, however, has not been excluded. It is, therefore, not advisable to use vaccinia virus for T7 polymerase expression (as used in initial protocols for these studies) because this virus is known to cause high-level plasmid recombination. Also, if the full-length virus is handled in a laboratory, accidental contamination of stocks is possible, and G-deficient viruses obtained from other laboratories may be contaminated as well. In the event of an (a recognized) accident, postexposure prophylaxis is still possible, but prophylactic vaccination is advisable to prevent illness in the event of any such accident. We, therefore, recommend that all personnel working directly with even the G-deleted versions be vaccinated against rabies.

In addition to vaccination, the use of biosafety cabinets, disposable gloves, and protective glasses and the other usual biosafety level 2 personal protective equipment is advised. All materials in contact with the virus (tips, tubes, plates, gloves, gowns, etc.) should be decontaminated by treatment with 70% isopropanol, and disposable materials should be autoclaved.

Experimental Protocols

In this chapter, we present detailed procedures for the recovery of replication-competent RV from cDNA, virus titration, and virus stock preparation (Protocol A). The recovery of G gene–deleted viruses is more demanding and is described in a separate procedure (Protocol B). Pseudotyping of G-deficient RV with the neurotropic CVS G protein and ASLV EnvA is described in Protocol C.

Protocol A

Recovery of Replication-Competent RV from cDNA

This protocol describes the recovery of replication-competent viruses such as SAD L16 or SAD L16-eGFP (for a schematic overview, see Fig. 3). It is suggested that at least three parallel transfection experiments are performed to increase the success rate (in three 3.5-cm^2 dishes or in three wells in a six-well plate). The entire protocol takes 10 d, and successful rescue can be obtained after 6 d. Please note that appropriate biosafety measures are needed.

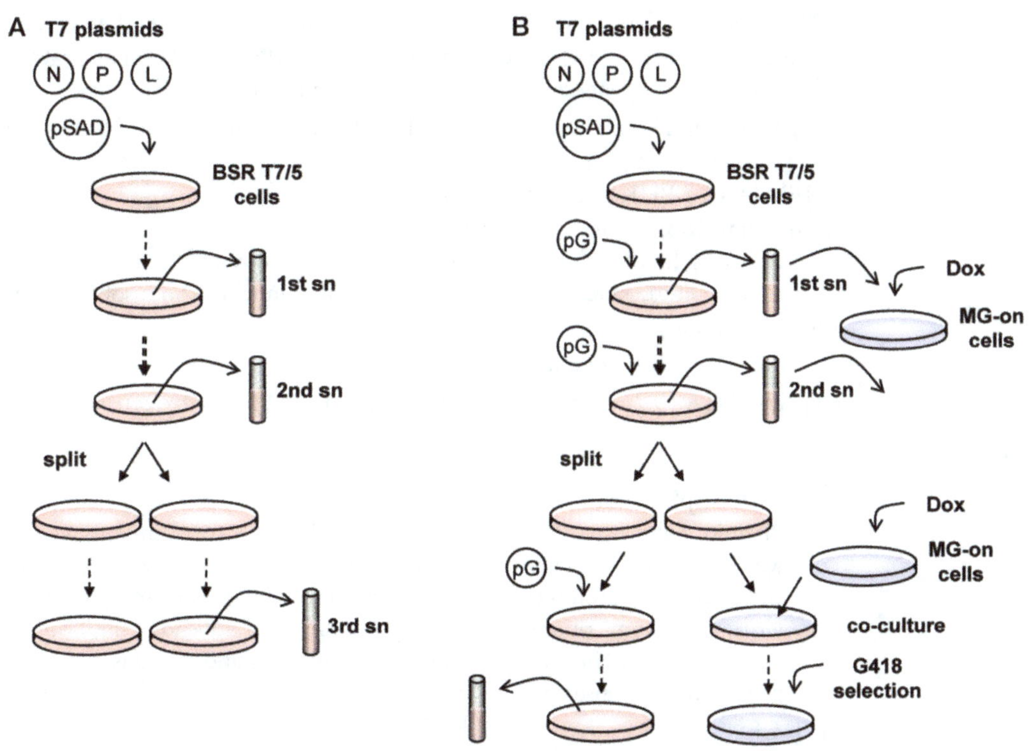

FIGURE 3. Schematic overview showing rescue of nondeficient (*A*) and G gene–deficient RV (*B*) from transfected T7 RNA polymerase-controlled plasmids. (*A*) Recombinant viruses can be recovered from cell culture supernatants (sn) directly after transfection of the plasmids into BSR T7/5 cells, which express T7 RNA polymerase or, in case of poor recovery rates, after splitting and longer-term incubation of transfected/infected cells. (*B*) Recovery of G-deficient virus requires continuous complementation by transfected plasmids or G-expressing cell lines. For details, see Protocol A.

MATERIALS

CAUTION: See Appendix 6 for proper handling of materials marked with <!>.
See the end of the chapter for recipes for reagents marked with <R>.

Reagents

Acetone <!> (80% in H_2O) (Roth)
Store and use at 4°C.

BSR T7/5 cells
BSR cell line (derived from BHK cells) constitutively expressing the T7 RNA polymerase and carrying a neomycin resistance gene (Buchholz et al. 1999).

Centocor (fluorescein isothiocyanate [FITC]-conjugated, monoclonal IgG against RV N; Centocor Inc., Malvern, PA).
Dissolve in 5 mL of H_2O, and store in 200-μL aliquots at –20°C. Prepare working solution by diluting 200 μL of Centocor in 40 mL of phosphate-buffered saline (PBS). Store working solution in the dark at 4°C for up to 4 wk.

Dulbecco's modified Eagle's medium without supplements (DMEM; Invitrogen)

Geneticin (G418; 200 mg/mL dissolved in sterile ddH_2O) (PAA Laboratories GmbH)
Store in aliquots at –80°C. Add to cell culture medium to give a final concentration of 1 mg/mL.

Glasgow minimal essential medium without supplements (GMEM; Invitrogen) (for BHK and BSR cell lines)

GMEM4+ (for BHK and BSR cell lines) <R>

Mammalian transfection kit (Stratagene)
Aliquot solution #1 under sterile conditions, and store in 100-μL aliquots at –20°C. Store solution #2 in 1000-μL aliquots at –20°C. For use, thaw aliquots of both solutions on ice.

PBS (Invitrogen)
For cell culture without $CaCl_2$ and $MgCl_2$.

Trypsin-EDTA (0.25% trypsin <!>/1-mM Na-EDTA) (Invitrogen)

Plasmids

pSAD L16 or pSAD L16-eGFP
Plasmid comprising the full-length cDNA of SAD L16 (GenBank accession number M31046), which is a molecular clone of the vaccine strain SAD B19 (SAD B19). The viral cDNA is flanked by an upstream T7 promoter and three G residues and downstream by the 84-bp HDV (hepatitis delta virus) antigenome ribozyme for exact processing of the 3′ end of the T7 polymerase-derived antigenomic RV RNA. The vector is a modified pBluescript and contains an ampicillin resistance cassette. pSAD L16 eGFP also contains an eGFP gene.

pTIT-GFP
Plasmid for T7-promoter-dependent expression of eGFP. Used as control for T7 expression and for transfection efficiency.

pTIT-N, pTIT-P, and pTIT-L
Expression plasmids (pBluescript-based; Amp^{res}) for the N, P, and L proteins of RV SAD L16, respectively. The genes are under the control of the T7 promoter and the internal ribosomal entry site (IRES) element of encephalomyocarditis virus (EMCV) to facilitate 5′ cap–independent translation. A downstream T7 terminator promotes transcription termination and reinitiation. These helper proteins are needed for encapsidation of the T7 polymerase–made cRNA transcripts by N (N + P) and expression of genes from the newly generated nucleocapsid by the viral polymerase (P + L).

Equipment

Cell culture dishes, 3.5-cm^2
Cell culture flasks, 25-cm^3
Falcon tubes, 15-mL

Microcentrifuge tubes, 1.5-mL
Plates, six-well and 96-well

EXPERIMENTAL METHOD

Cell Culture

1. Cultivate BSR T7/5 cells under standard conditions in GMEM4+ medium in an incubator set at 37°C with 5% CO_2 and 100% humidity. Twice a week, detach adherent cells by trypsinization, and split 1/6 to new cell culture flasks. Every second passage, add Geneticin (G418) to the culture medium to a final concentration of 1 mg/mL to select for T7 polymerase expression.

Transfection and Virus Recovery

2. Approximately 18 h before plasmid transfection, seed 3 x 10^5 BSR T7/5 cells cultivated in GMEM4+ medium without G418 into a 3.5-cm^2 cell culture dish or in a well of a six-well plate. Seed at least one additional dish or well for a GFP-transfection control.
3. Day 0: When cells have reached 80%–90% confluence, wash them once with DMEM without supplements, and incubate in 1 mL of fresh DMEM without supplements per 3.5-cm^2 cell culture dish for 1 h.
4. Prepare the DNA-transfection mix.
   i. Thaw solutions #1 and #2 from the mammalian transfection kit on ice.
   ii. For each reaction, mix 5 µg of pTIT-N, 2.5 µg of pTIT-P, 2.5 µg of pTIT-L, and 10 µg of pSAD L16 with H_2O to give a final volume of 90 µL. Place on ice.
   iii. For the GFP transfection control, mix 10 µg of pTIT-GFP with H_2O to give a final volume of 90 µL. Place on ice.
   iv. Add 10 µL of solution #1 to the DNA dilution on ice, and mix by gently pipetting up and down three to four times on ice.
   v. Add 100 µL of solution #2, and mix by gently pipetting up and down once.
   vi. Incubate the DNA-transfection mix for 10–20 min at room temperature to allow formation of $CaPO_4$/DNA precipitates. Again, gently pipette up and down two to three times.
5. Distribute the 200 µL of DNA-transfection mix dropwise over the cells without removing the 1 mL of DMEM without supplements. To reach an optimal size and distribution of the precipitate in the medium, do not pipette the medium up and down; instead, carefully move the dishes horizontally back and forth in a figure-eight shape. Incubate the culture without moving for 3.5 h at 37°C with 5% CO_2 and 100% humidity.
6. Wash the transfected cells once with GMEM4+ medium, and incubate in 2 mL of GMEM4+ medium for 72 h at 37°C with 5% CO_2 and 100% humidity.
7. Check the control cells (transfected with pTIT-GFP) 24 h after transfection by fluorescence microscopy for the number of green cells to determine transfection efficiency. At transfection efficiencies of <20%, successful virus rescue is unlikely.
8. To assay the presence of newly generated virus in the supernatants of transfected cells, transfer early (day 1–3) and late (day 4–6) supernatants to fresh cells (passages #1A and #1B, respectively).
9. For passage #1A, at 72 h posttransfection, seed 3 x 10^5 fresh BSR T7/5 cells into a 3.5-cm^2 cell culture dish in a final volume of 1 mL.
10. Remove the whole supernatant from the transfected cell monolayer. Pellet the cell debris by centrifuging at 1000*g* at 4°C. Mix the cleared supernatant with the freshly seeded cells, which are still in suspension (passage #1A).

11. Add 2 mL of fresh GMEM4+ medium onto the transfected cell monolayer.
12. Incubate the passage #1A cells and the transfected cells for a further 72 h at 37°C with 5% CO_2 and 100% humidity.
13. At 144 h (6 d) posttransfection, collect the supernatant from passage #1A cells in a 15-mL Falcon tube. Pellet the cell debris by centrifuging 1000 *g* at 4°C. Transfer the cleared supernatant into a 1.5-mL microcentrifuge tube, and store at –80°C.
14. After removal of the supernatant, fix the passage #1A cells, and stain against the RV N protein by direct immunofluorescence (see Steps 20–21), or check GFP fluorescence in the case of SAD L16 GFP.

    If foci are detected (i.e., if virus rescue is successful), the supernatant from passage #1A can be titrated and can be used to prepare virus stocks (see Steps 29–35).
15. Also at 6 d, after the first transfection, prepare passage #1B by seeding 3 x 10^5 fresh BSR T7/5 cells into a 3.5-cm^2 cell culture dish in a final volume of 1 mL. As for passage #1A, collect the whole supernatant from the transfected cell monolayer (72–144 h supernatant), and mix it with the freshly seeded cells, which are still in suspension (passage #1B). Incubate passage #1B for a further 72 h at 37°C.
16. Also at 6 d, in case viral foci are not detectable in cell monolayers from passage #1A, detach the transfected cells from the 3.5-cm^2 cell culture dishes with trypsin-EDTA, and mix them with fresh GMEM4+. Transfer one-eighth of the transfected cells into a new 3.5-cm^2 cell culture dish, and mix them with GMEM4+ to give a final volume of 2 mL. Transfer the remaining seven-eighths of the transfected cells into a 25-cm^3 cell culture flask, and mix them with GMEM4+ to give a final volume of 8 mL. Incubate all transfected cells for a further 72 h at 37°C with 5% CO_2 and 100% humidity.
17. At 9 d posttransfection (and 72 h post-passage #1B), transfer the supernatant from passage #1B cells into tubes, and store them at –80°C.
18. After removing the supernatant, fix the cells from passage #1B with 80% acetone, and stain against the RV N protein by direct immunofluorescence (see Steps 20–21), or check GFP expression.

    If foci are detected (i.e., if virus rescue is successful), the supernatant from passage #1B can be titrated and can be used to prepare virus stocks (see Steps 29–35).
19. Also at 9 d posttransfection, discard the supernatant from the split transfected cells from the 3.5-cm^2 cell culture dish (Step 16). Fix the cells with 80% acetone, and stain against the RV N protein by direct immunofluorescence (see Steps 20–21).

    If foci are detected, the 8-mL supernatant from the split transfected cells in the cell culture flask can be titrated and can be used to prepare virus stocks (see Steps 29–35).

    If no foci are detected in cells from supernatant passages or from split cells, the attempt to generate recombinant RV from cDNA has failed.

Fixation of Cells and Direct Immunofluorescence

For detection of foci of SAD L16 (and virus titration; see below), fix cells, and stain with an FITC-coupled antibody against the N protein.

20. Wash cells once with PBS and once with ice-cold 80% acetone. Incubate cells with 2 mL of ice-cold 80% acetone per 3.5-cm^2 dish for 20 min at 4°C. Remove acetone, and let the dish air-dry (~30–60 min).
21. Prepare a working solution of FITC-conjugated RV N antibody by diluting 200 µL of Centocor in 40 mL of PBS. Incubate the cells in 1 mL of diluted Centocor per 3.5-cm^2 dish for 2 h at 37°C with 5% CO_2 and 100% humidity. Wash the stained cells twice with PBS. Cover the cells with 1 mL of PBS per 3.5-cm^2 dish, and inspect FITC fluorescence by ultraviolet microscopy.

Virus Titration

22. Detach BSR T7/5 cells from a 25-cm^3 cell culture flask (3×10^6 cells) by trypsinization. Resuspend the cells in 25 mL of fresh GMEM4+ medium. Seed 100 µL of the cell suspension (1.2×10^4 cells) into each concavity of a 96-well plate, and incubate for 2 h at 37°C with 5% CO_2 and 100% humidity so that the cells can attach to the bottom of the wells.

    For every virus sample, you need 2 x 8 wells.

23. Thaw the virus-containing supernatants on ice.
24. Prepare seven serial 10-fold dilutions of the virus supernatants (10^{-1}–10^{-7}) in GMEM without any supplements. To do this, prepare seven tubes each containing 900 µL of GMEM without any supplements. Transfer 100 µL of the virus supernatant into the first tube, and mix gently by vortexing. Transfer 100 µL of the first dilution to the second tube, mix gently by vortexing, and repeat this procedure until you have prepared seven dilutions.
25. Transfer 100 µL of every dilution onto the seeded BSR T7/5 cells in the 96-well plate to give a final volume of 200 µL per well.

    Remember to apply 100 µL of the undiluted virus supernatant (dilution 10^0) as well. Every dilution series should be plated twice.

26. Incubate the plate for 48 h at 37°C with 5% CO_2 and 100% humidity.
27. Fix the cells with 80% acetone, and stain them against the RV N protein by direct immunofluorescence (see Steps 20–21).
28. Determine virus titers by counting the foci in the well of the dilution in which the fewest foci are detectable and in the next lower dilution (with 10-fold more foci); count each in duplicate. The virus titer is the mean of foci counted in four wells multiplied by the dilution (Fig. 4).

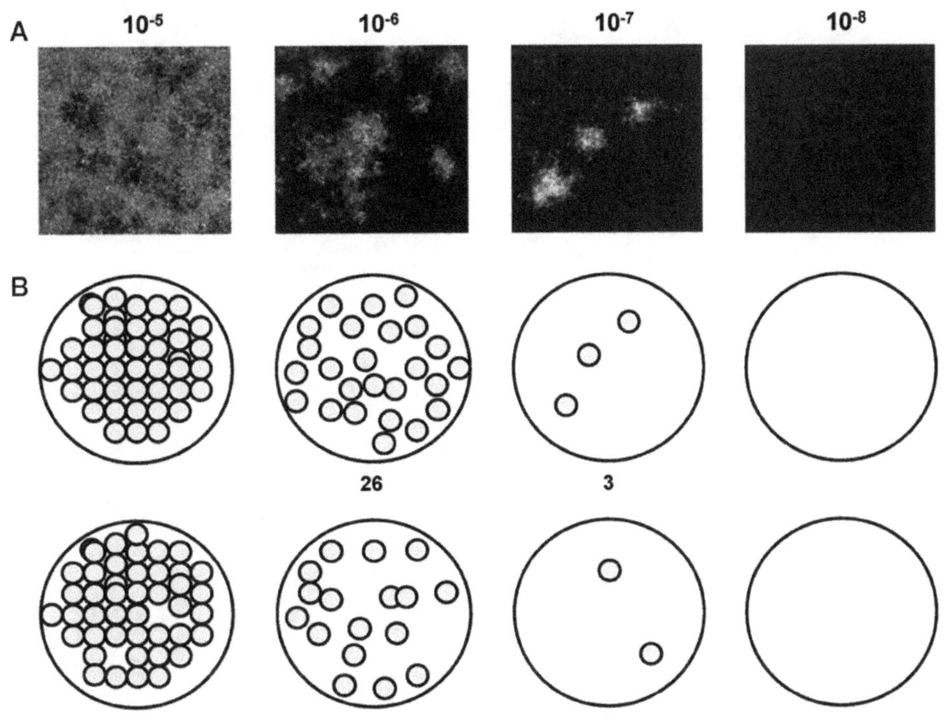

Virus titer = $(26 \times 10^6 + 3 \times 10^7 + 17 \times 10^6 + 2 \times 10^7) / 4 = 2.3 \times 10^7$ focus forming units (ffu)/mL

FIGURE 4. Titration of RV. (*A*) Typical images of RV foci after incubation of cells in a 96-well plate with a serial 10-fold dilution of virus-containing cell culture supernatant. RV N protein was stained with an FITC-conjugated antibody (Centocor). For details, see Protocol A. In *B*, a simple means of calculating virus titers is illustrated.

Virus Stock Preparation

29. Seed 7.5 x 10^5 BSR T7/5 cells into 25-cm^3 cell culture flasks in a final volume of 8 mL, and infect them in suspension with the supernatants from positive rescue experiments (i.e., supernatants of passage #1A or #1B) at a multiplicity of infection (MOI) of 0.01–0.1 (i.e., one focus-forming unit, as determined by titration, per 100 or 10 cells, respectively.)
30. Incubate the infected cells for 48 h at 37°C with 5% CO_2 and 100% humidity.
31. Transfer the supernatant into Falcon tubes, and centrifuge at 1000 *g* for 10 min at 4°C to remove cellular material (harvest #1).
32. Prepare 500-µL aliquots of harvest #1, and store them at –80°C.
33. Incubate the infected cells in 8 mL of fresh GMEM4+ for another 48 h at 37°C with 5% CO_2 and 100% humidity.
34. Transfer the supernatant into tubes, and centrifuge at 1800 rpm for 10 min at 4°C to remove cells (harvest #2).
35. Prepare 500-µL aliquots of harvest #2, and store them at –80°C.

Protocol B

Recovery of G-Gene Deleted Virus

This procedure describes the necessary steps only; for all other experimental details, see Protocol A. A schematic overview is presented in Figure 3.

MATERIALS

CAUTION: See Appendix 6 for proper handling of materials marked with <!>.

Reagents

Doxycyclin <!> (Sigma-Aldrich)

Hygromycin <!> (200 mg/mL in dissolved in sterile ddH_2O) (PAA Laboratories GmbH)

Store in aliquots at –80°C. Add to cell culture medium to give a final concentration of 1 mg/mL.

MG139-on cells

BSR cell line for doxycyclin-inducible expression of the SAD L16 G (and M) protein; hygromycin resistant.

For other reagents, see Protocol A.

Plasmids

pCAGGS-G

Chicken β-actin promoter-driven expression plasmid (Amp^{res}) for the G protein of RV SAD L16.

pSAD ΔG eGFP2 (or SAD ΔG mCherry)

Plasmids comprising a G gene–deleted cDNA of SAD L16 and encoding eGFP or mCherry, respectively, with unique restriction sites for introduction of additional genes upstream and downstream of mCherry or eGFP.

For other plasmids, see Protocol A.

Equipment

See Protocol A.

EXPERIMENTAL METHOD

1. Transfect BSR T7/5 cells at 80%–90% confluence using reagents from the mammalian transfection kit (see Protocol A) and 5 µg of pTIT-N, 2.5 µg of pTIT-P, 2.5 µg of pTIT-L, and 10 µg of pSAD ΔG eGFP or pSAD ΔG mCherry.
2. Transfect the cells 1 and 2 d after the first transfection using reagents from the mammalian transfection kit and 5 µg of pCAGGS-G.
3. On day 3, process the transfected cell culture by (i) supernatant transfer and (ii) splitting of transfected cells.
    i. Transfer the cleared supernatant to 3×10^5 freshly seeded MG139-on cells in a 3.5-cm^2 dish or well of a six-well plate (passage #1A). Incubate for 1 d, then add 10 mg of doxycyclin

per mL of medium to induce expression of G. Incubate for another day, and check for foci expressing GFP. If foci are present, incubate until the entire culture is infected; if not, dispose.

ii. Split the transfected cells 1:1 into (a) a new 3.5-cm^2 dish and (b) a 25-cm^3 flask containing freshly split (1:3) MG139-on cells in 7 mL of GMEM+4 medium.

a. At 24 and 48 h after the split, transfect the cells in the 3.5-cm^2 dish with 5 µg of pCAGGS-G. At 72 h after the split, transfer cleared supernatants to fresh MG139-on cells, and proceed with doxycyclin induction and analysis as described in Step 3i.

b. At 24 h after seeding, add doxycyclin (10 mg/mL) to the flask containing the BSR T7/5 and MG139-on cell mixture to induce expression of G from MG139-on cells. At 24 or 48 h postinduction, check for the presence of green fluorescent foci. If present, add hygromycin to 1 mg/mL to remove residual BSR T7/5 cells, and incubate until the entire monolayer is infected.

4. Harvest the supernatants from positive well-infected cultures. Clear the supernatants by centrifuging at 1000*g* at 4°C, and transfer them into 1.5-mL microfuge tubes. Store at –80°C. Determine titers by counting single green cells (see Protocol A, Step 28) or by counting infectious foci on MG139-on cells; the latter is a more reliable method.

Protocol C

Pseudotyping

G-deleted fluorescent RV pseudotyped with RV G proteins, SAD ΔG eGFP (RV CVS-G), can be used as single-round vectors for efficient retrograde labeling of neurons (Wickersham et al. 2007). For these experiments, as well as for monosynaptic tracing, which involves pseudotyping in situ, the use of the CVS strain G is recommended because of its high tropism for neurons. Pseudotype virus stocks generated by transfection of pCAGGS-G (or in MG139-on cells) contain the G protein of the vaccine strain SAD L16, which is broader in its tropism, and infects astrocytes, glia, and oligodendrocytes (Fig. 5). We also describe a procedure for pseudotyping with ASLV Env A, which uses a cell line expressing a version of the EnvA protein that is incorporated efficiently into the RV envelope ($\text{EnvARG}^{\text{RGct}}$).

MATERIALS

See the end of the chapter for recipes for reagents marked with <R>.

Reagents

$\text{BHK-EnvA}^{\text{RGCD}}$

BHK cell line constitutively expressing a modified EnvA spike protein from ASLV.

DMEM (Invitrogen) (for 293T-TVA800 cells)

DMEM3+ (for 293T-TVA800 cells) <R>

293T-TVA800 cells

HEK293T cell line constitutively expressing the ASLV receptor TVA.

For other reagents, see Protocol A.

Plasmids

pCAGGS-CVS-G

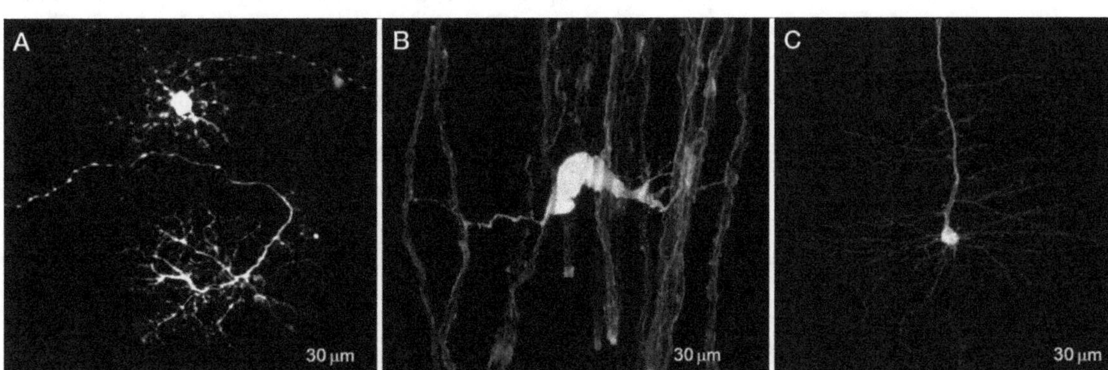

FIGURE 5. Examples of infection of diverse cells types by SAD ΔG-eGFP (G). (*A*) Horizontal cells in an explanted mouse retina. (Courtesy of Botond Roska, Friedrich Miescher Institute, Basel.) (*B*) Oligodendrocyte after injection of SAD ΔG-eGFP (G) into the spinal cord. (*C*) Layer V neuron in neocortex. (*B,C,* Courtesy of Martin Kerschensteiner, Ludwig-Maximilians-University, Munich and Thomas Misgeld, Technical University, Munich.)

Expression plasmid (Amp^{res}) for the G protein of the virulent RV strain CVS.

Equipment

See Protocol A.

EXPERIMENTAL PROCEDURE

Pseudotyping with RV CVS-G

1. To generate SAD ΔG eGFP (CVS-G), transfect all six wells of a BSR T7/5-seeded six-well plate (each corresponding to a 3.5-cm^2 dish) with pCAGGs-CVS-G by setting up six individual transfection mixes.
2. At 48 h after transfection, trypsinize the cells, and combine the cells from two wells into one 25-cm^3 cell culture flask, yielding three flasks of transfected cells. Mix the cell suspensions in the flasks with SAD ΔG-eGFP (G) stocks at an MOI of 0.1.
3. At 48 h after infection, check GFP expression by live microscopy. At this time point, the entire cell monolayer should express GFP; if not, incubate further for an additional day. Harvest the clarified supernatant. Prepare 500-µL aliquots of harvest #1, and store at –80°C. Add fresh medium to the cells, and repeat the harvest (harvest #2) 3 d later.

   Note that the infectious titer of SAD ΔG eGFP (CVS-G) is about 10-fold higher in neuronal cell lines than in BSR T7/5 cells (i.e., neurons are better infected). Titers of SAD ΔG eGFP (SAD-G) are equivalent in neuronal cell lines and in BSR T7/5.

Pseudotyping with ASLV EnvA

4. To generate SAD ΔG eGFP (EnvA), resuspend a confluent culture of BHK-EnvA^{RGCD} cells in 12 mL of GMEM4+. Seed in three 75-cm^3 cell culture flasks, and infect with SAD ΔG-eGFP (G) stock virus at an MOI of 0.1.
5. After 6 h, when the cells are attached, remove the entire medium, and wash at least four times with GMEM4+ to remove free input (G-complemented) virus. Add GMEM4+ to give a final volume of 8 mL. Alternatively, the cells can be trypsinized and can be reseeded in fresh medium.
6. At 48 h after infection, check GFP expression by live microscopy. At this time point, the entire cell monolayer should express GFP; if not, incubate further for an additional day. Harvest the virus in the supernatant, and store harvest #1 in 500-µL aliquots at –80°C. Add fresh medium to the cells, and harvest again after 48 h.

   Note that SAD ΔG-eGFP (EnvA) must be titrated (72 h incubation of titer plates) both on 293T-TVA800 cells and on BSR T7/5 cells to determine the level of contamination by input (G-containing) virus.

EXAMPLES OF APPLICATION

The recombinant G-deleted virus and RV pseudotyped with RV G can be used directly for retrograde staining of all neurons projecting to the inoculation site, both in vitro, as described by Wickersham et al. (2007), and in vivo. By using pseudotype viruses produced in MG139-on cells and, therefore, by carrying the SAD L16 G, which has a less pronounced neurotropism (Finke and Conzelmann 2005), both neurons and glial cells like astrocytes or oligodendrocytes can be infected and identified by fluorescence. The examples in Figure 5 show horizontal cells infected by adding virus to isolated retinas (Fig. 5A) and identified by GFP fluorescence. In the experiment shown in Figure 5B, virus was injected into the spinal cord of a mouse, and explants were analyzed 3 d after injection. Figure 5C shows a layer V neuron after injection into the cortex.

For identification of direct synaptic connections to the cells initially infected by any SAD ΔG virus, it is necessary that the missing G gene is delivered in those cells using standard genetic tools, including cDNA transfection, injection, or delivery by additional viral vectors. The current method of choice for monosynaptic (only first-order) connections is the use of SAD ΔG (EnvA) pseudotype viruses because this allows unequivocal identification of the starting (postsynaptic) neuron. The latter cell is defined by expression of (1) the receptor TVA to allow specific entry of EnvA-pseudotyped virus; (2) RV CVS-G to allow formation of viruses able to cross the synapse efficiently; and (3) a fluorescent protein to identify the cells. Although few examples are published to date, monosynaptic tracing is being used to study several sensory circuits, including those of the retina, and connections within the spinal cord and cortex.

DISCUSSION

RV vectors allow proficient gene expression in probably all types of neurons and yet cause limited neuronal damage so they do not destroy the connections required for viral spreading to and within the central nervous system. Nevertheless, neurons infected with full-length virus or SAD ΔG viruses will continuously produce and will accumulate virus RNA and proteins, which ultimately (within a couple of days) will lead to damage. Attempts to engineer vectors with reduced RNA and protein synthesis for a prolonged time window are underway. Future developments are predicted to include the expression of fluorescent proteins targeted to organelles and synaptic structures as well as the expression of functional proteins such as channelrhodopsins and genetically encoded calcium indicators. Although expression of multiple and diverse proteins, including DNA *trans*-acting factors such as recombinases (e.g., Cre) and transcription factors (e.g., Tet repressors), is straightforward with RV vectors, the use of the respective *cis*-acting DNA sequences (e.g., lox P and Tet operator) is so far limited to DNA viruses and host cell DNA.

The current poor recovery rate may be sufficient to isolate the desired recombinant virus mutants from transfected cell cultures, but it limits the application of RV for neuronal tracing to work with premade amplified virus stocks. Transfection or injection of neurons with RV cDNA or RNA for standard RV tracing experiments is, therefore, not yet feasible. Similarly, conditional activation of RV cDNA in transgenic animals (e.g., by Cre-mediated recombination) is predicted to be too inefficient for standard tracing experiments.

RECIPES

CAUTION: See Appendix 6 for proper handling of materials marked with <!>.
Recipes for reagents marked with <R> are included in this list.

DMEM3+

DMEM supplemented with 10% fetal calf serum (Invitrogen), 2-mM L-glutamine, and 0.4% penicillin/streptomycin <!> (Invitrogen).

GMEM4+

GMEM supplemented with 10% newborn calf serum (Invitrogen), 2% tryptose phosphate broth, 2% MEM (modified Eagle's medium) amino acids (Invitrogen), and 0.4% penicillin/streptomycin <!> (Invitrogen).

ACKNOWLEDGMENTS

Work presented here was funded by the DFG through SFB 455 (virus functions and immune modulation) and SFB 870 (assembly and function of neuronal circuits in sensory processing).

REFERENCES

Bates P, Young JA, Varmus HE. 1993. A receptor for subgroup A Rous sarcoma virus is related to the low density lipoprotein receptor. *Cell* **74:** 1043–1051.

Buchholz UJ, Finke S, Conzelmann KK. 1999. Generation of bovine respiratory syncytial virus (BRSV) from cDNA: BRSV NS2 is not essential for virus replication in tissue culture, and the human RSV leader region acts as a functional BRSV genome promoter. *J Virol* **73:** 251–259.

Callaway EM. 2008. Transneuronal circuit tracing with neurotropic viruses. *Curr Opin Neurobiol* **18:** 617–623.

Chare ER, Gould EA, Holmes EC. 2003. Phylogenetic analysis reveals a low rate of homologous recombination in negative-sense RNA viruses. *J Gen Virol* **84:** 2691–2703.

Conzelmann KK. 2004. Reverse genetics of mononegavirales. *Curr Top Microbiol Immunol* **283:** 1–41.

Etessami R, Conzelmann KK, Fadai-Ghotbi B, Natelson B, Tsiang H, Ceccaldi PE. 2000. Spread and pathogenic characteristics of a G-deficient rabies virus recombinant: An in vitro and in vivo study. *J Gen Virol* **81:** 2147–2153.

Federspiel MJ, Bates P, Young JA, Varmus HE, Hughes SH. 1994. A system for tissue-specific gene targeting: Transgenic mice susceptible to subgroup A avian leukosis virus–based retroviral vectors. *Proc Natl Acad Sci* **91:** 11241–11245.

Finke S, Conzelmann KK. 2005. Replication strategies of rabies virus. *Virus Res* **111:** 120–131.

Klingen Y, Conzelmann KK, Finke S. 2008. Double-labeled rabies virus: Live tracking of enveloped virus transport. *J Virol* **82:** 237–245.

Lafon M. 2005. Rabies virus receptors. *J Neurovirol* **11:** 82–87.

Marshel JH, Mori T, Nielsen KJ, Callaway EM. 2010. Targetting single neuronal networks for gene expression and cell labeling in vivo. *Neuron* **67:** 562–574.

Mebatsion T, Conzelmann KK. 1996. Specific infection of CD4+ target cells by recombinant rabies virus pseudotypes carrying the HIV-1 envelope spike protein. *Proc Natl Acad Sci* **93:** 11366–11370.

Mebatsion T, Konig M, Conzelmann KK. 1996. Budding of rabies virus particles in the absence of the spike glycoprotein. *Cell* **84:** 941–951.

Schnell MJ, Mebatsion T, Conzelmann KK. 1994. Infectious rabies viruses from cloned cDNA. *EMBO J* **13:** 4195–4203.

Ugolini G. 1995. Specificity of rabies virus as a transneuronal tracer of motor networks: Transfer from hypoglossal motoneurons to connected second-order and higher order central nervous system cell groups. *J Comp Neurol* **356:** 457–480.

Ugolini G. 2008. Use of rabies virus as a transneuronal tracer of neuronal connections: Implications for the understanding of rabies pathogenesis. *Dev Biol* **131:** 493–506.

Wickersham IR, Lyon DC, Barnard RJ, Mori T, Finke S, Conzelmann KK, Young JA, Callaway EM. 2007. Monosynaptic restriction of transsynaptic tracing from single, genetically targeted neurons. *Neuron* **53:** 639–647.

11 Generating and Imaging Multicolor Brainbow Mice

Tamily A. Weissman,[1] Joshua R. Sanes,[1] Jeff W. Lichtman,[1] and Jean Livet[2–4]

[1]*Department of Molecular and Cellular Biology, Center for Brain Science, Harvard University, Cambridge, Massachusetts 02138;* [2]*Institut National de la Santé et de la Recherche Médicale, U968, Paris, F-75012 France;* [3]*Université Pierre et Marie Curie-Paris 06, UMR_S 968, Institut de la Vision, Paris, F-75012 France;* [4]*Centre National de la Recherche Scientifique, UMR_7210, Paris, F-75012 France*

ABSTRACT

Visualizing the precise morphology of closely juxtaposed cells and their interactions can be highly informative, particularly when studying the complex organization of neuronal and glial networks in the nervous system. To this end, one can use optical approaches to image distinct markers that are differentially distributed among the cells of interest, such as fluorescent proteins of various colors (XFPs). The Brainbow strategies use Cre/*lox* recombination to stochastically express two to four XFPs in a cellular population from a single promoter. Integration of multiple *Brainbow* transgene copies results in combinatorial expression of these XFPs, creating a wide range of hues. In the nervous system, the multicolor labeling thus generated can be used to distinguish adjacent neuronal or glial cells and to verify the identity of neuronal processes while tracing circuitry. Here, we describe the generation of *Brainbow* transgenes and mice and describe their use to image and digitally reconstruct nerve cells and their interactions in fixed samples. This method also holds potential for studies in other tissues and model organisms as well as live imaging in vivo.

INTRODUCTION

Overview of the Brainbow Approach

The complex circuitry of neuronal processes poses serious challenges to visualizing and understanding the behaviors of individual neurons and their connectivity. For brain studies, one would ideally want to view the precise morphology of multiple (or all) neighboring neurons in a brain circuit. To achieve this goal, a method is needed to highlight neurons while, at the same time, distinguishing them from their neighbors. The Brainbow approach is designed to label cells with distinct fluorescent proteins (FPs), such as cyan fluorescent protein (CFP), yellow fluorescent protein (YFP), or red fluo-

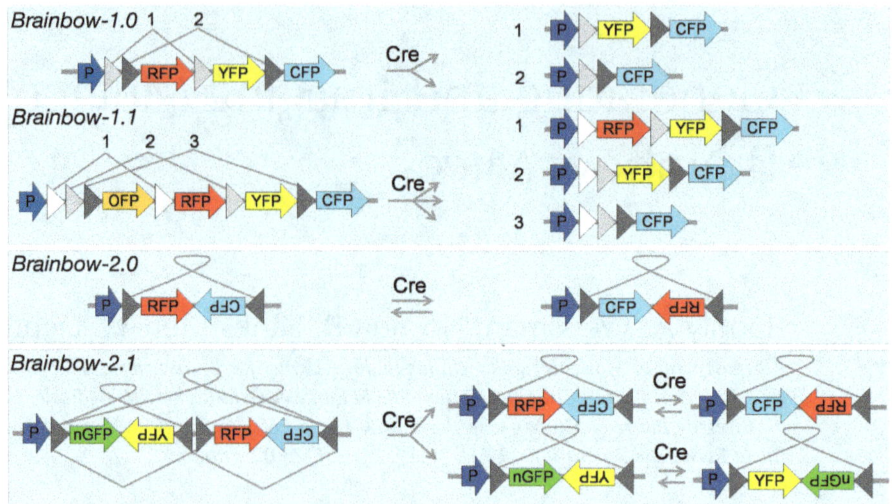

FIGURE 1. Principle of Brainbow strategies. In *Brainbow-1.0* and *Brainbow-1.1*, pairs of incompatible (heterospecific) *lox* variants are interleaved, creating two or three mutually exclusive excision possibilities. Each of these excisions triggers the expression of a distinct XFP. In *Brainbow-2.0*, *loxP* sites are positioned in opposite orientation, defining an invertible DNA segment in which two XFP genes are placed head to head. This construct can invert as long as Cre is active. When it stabilizes, the XFP gene ending in a sense orientation is expressed. In *Brainbow-2.1*, two invertible units are positioned in tandem, offering additional excision and inversion options, and yielding a total of four expression possibilities.

rescent protein (RFP), using a single transgene (Livet et al. 2007; Lichtman et al. 2008) (Fig. 1). Based on Cre/*lox*-mediated DNA excision or inversion (Branda and Dymecki 2004), *Brainbow* transgenes trigger the stochastic expression of two to four XFPs in a cellular population such that each cell randomly adopts one color. If only one copy of the *Brainbow* construct is present in cells, the recombination choice leads to mutually exclusive expression of XFPs. For example, cell A will express one XFP, whereas cell B will express a different XFP. However, when multiple *Brainbow* transgene copies coexist in cells, each copy makes its own stochastic choice, yielding combinatorial expression of XFPs and creating a wide range of additional colors (Fig. 2). In other words, cell A may express one or a combination of XFPs, whereas cell B may express a different combination or ratio of XFPs.

Importantly, because the choice is made in the genome, the entire cell, including all processes, can adopt a single color. These hues allow one to distinguish cells in complex tissues and to visualize their contacts with other cells. In the nervous system, the multicolor labeling provides a method to verify the identity of neuronal processes while tracing them in three-dimensional (3D) image data sets. Individual axons can be distinguished from their neighbors and traced over large distances (Fig. 3). Here, we provide a description of the steps required to (1) generate *Brainbow* transgenes and mice, (2) acquire multichannel images from Brainbow samples, and (3) use those images to digitally trace cellular contours and connections. We also discuss the advantages and limits of this technique.

Brainbow Strategies

Several types of *Brainbow* transgenes are available (Table 1; Livet et al. 2007), using Cre-mediated DNA excision (*Brainbow-1.0* and *Brainbow-1.1*), inversion (*Brainbow-2.0*), or both excision and inversion (*Brainbow-2.1*) (Fig. 1). An important parameter that may guide the user's choice of transgene is the number of desired recombination outcomes, which determines how many different XFPs can be expressed in the mice. A second parameter is the source of Cre available to trigger recombination. *Brainbow-1.0* and *Brainbow-1.1* transgenes lead to dead end products following Cre-mediated DNA excision and are, therefore, largely compatible with any form of Cre, whereas the inversion-based *Brainbow-2.0* and *Brainbow-2.1* transgenes undergo repeated reaction cycles that might lead to

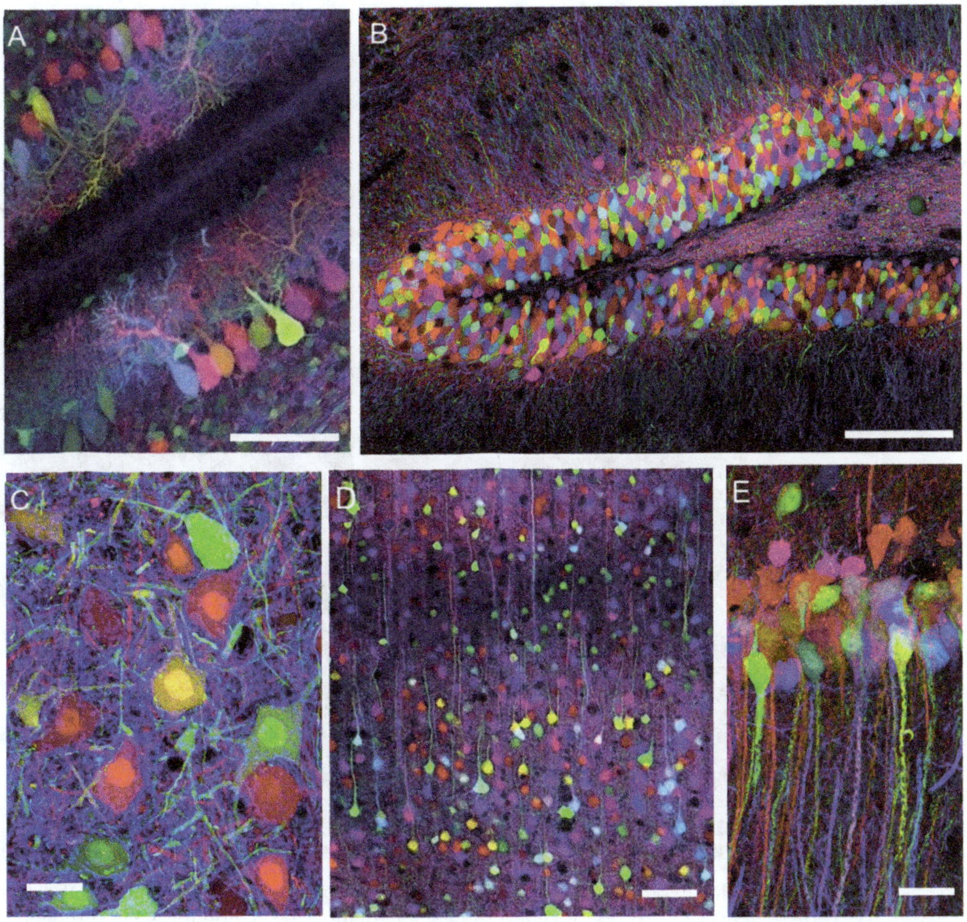

FIGURE 2. Mosaic expression of fluorescent proteins in Brainbow mice. (*A,B,D,E*) *Thy1.2-Brainbow-1.0* line L; (*C*) line H. Combinatorial expression of FPs is observed throughout the brain. (*A*) Purkinje neurons of cerebellum; (*B*) dentate gyrus of the hippocampus; (*C*) brain stem; (*D*) cortex (layers 3–5); (*E*) hippocampus CA1. Scale bars, 125 µm (*A*); 150 µm (*B*); 40 µm (*C*); 100 µm (*D*); 40 µm (*E*).

chromosome breaks; they, therefore, appear to require transient Cre activity to stabilize their orientation (Table 1). Several strategies are available to limit Cre activity in time (Lam and Rajewsky 1998; Silver and Livingston 2001). Here, we use the CreER/tamoxifen system in which administration of a synthetic estrogen ligand specifically activates a modified form of Cre (Metzger et al. 1995).

Colors

Depending on the purpose of the experiment, it may be useful to express either a large or a restricted number of XFP combinations or, in some instances, only one XFP per cell. In Brainbow mice, the number of those color combinations may be influenced by several parameters—in particular, how many *Brainbow* transgene copies are integrated into the genome and can recombine independently. In mice generated by embryonic stem cell technology, a single copy of the *Brainbow* construct should be present, yielding only exclusive XFP expression. Conversely, in mice produced by pronuclear injection, multiple tandem transgene copies are likely to integrate (Palmiter and Brinster 1986), enabling a greater range of color combinations. In such an array, transgene copy number may be modulated by intercopy recombination (Garrick et al. 1998). One way to do this is to use a site-specific recombinase such as Flp; to this end, an *FRT* site has been inserted in several *Brainbow* constructs. In mice bearing tandem *Brainbow* transgene copies, some amount of Cre-mediated

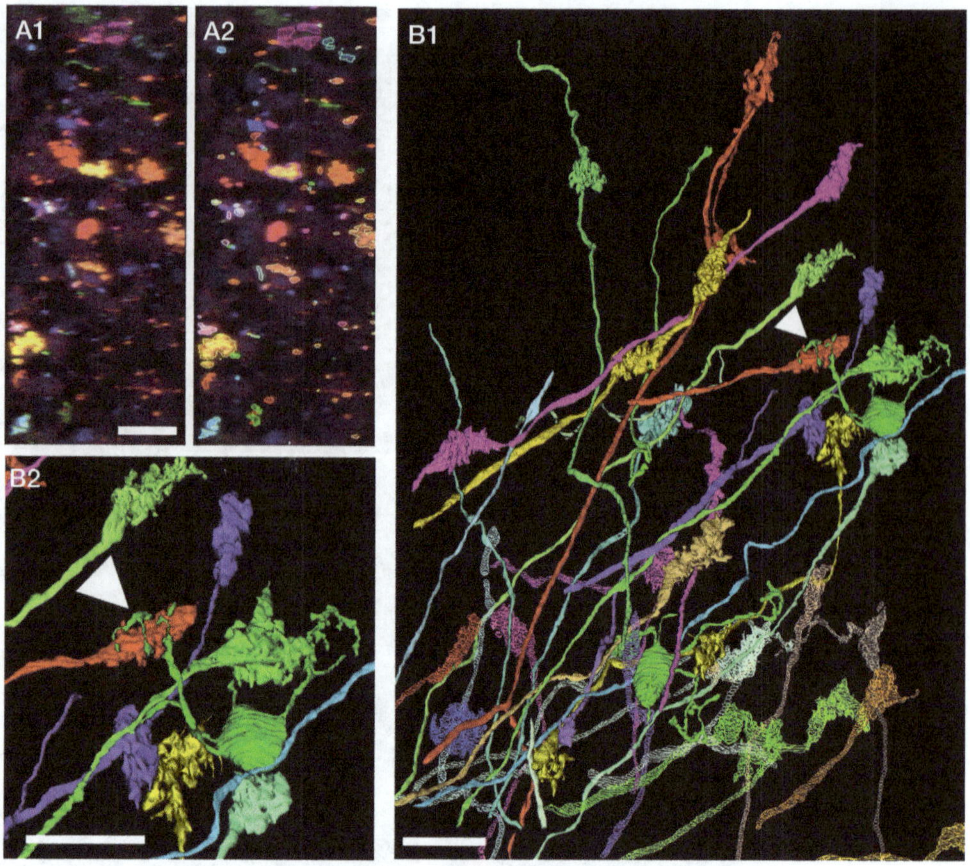

FIGURE 3. Circuit tracing using Brainbow multicolor labeling. (*A*) The confocal data set of the inner granular layer of the cerebellum was resampled in the *y–z* orientation to provide this view orthogonal to most axons (*A1*). In this orientation, each cellular component was traced with Reconstruct (*A2*). Small spots are axons; enlarged, bright objects are mossy fiber rosettes. (*B*) Three-dimensional reconstruction of cerebellar mossy fiber axons and granule cells. A mossy fiber contact with a granule cell is visible (indicated by the white arrowhead and enlarged in *B2*). *B1* also shows two different rendering options in Reconstruct: Boissonant surface (*upper left*) and individual traces (*lower right*). Scale bars, 20 µm (*A*); 10 µm (*B*).

intercopy recombination (favored by extended recombination times, high levels of Cre, and short target transgenes) may also occur and could reduce both the intensity of the labeling and the number of color combinations generated. The length of the *Brainbow* transgene as well as the particular Cre driver line used are, therefore, important parameters to consider when planning an experiment.

TABLE 1. Brainbow strategies

Brainbow strategy	Principle used for stochastic expression	Total number of transgene configurations (prior to and after recombination)	*Brainbow* lines available at Jackson Laboratories	Summary of outcomes in available constructs (before Cre → after Cre)
Brainbow-1.0	Excision	3	H, L[a]	R → Y/C
Brainbow-1.1	Excision	4	M[b]	O → R/Y/C
Brainbow-2.0	Inversion	2	—	R → (C↔R)
Brainbow-2.1	Inversion and excision	4	R[c]	G → (C↔R)/(G↔Y)

The nomenclature for construct and mouse lines follows the one used in Livet et al. (2007): R, red fluorescent protein (RFP); Y, yellow fluorescent protein (YFP); C, cyan fluorescent protein (CFP); O, orange fluorescent protein (OFP); G, green fluorescent protein (GFP).

[a]Line H expresses membrane-targeted YFP and CFP, whereas line L expresses three cytoplasmic XFPs.

[b]Expression in this line is mostly restricted to astrocytes and hippocampal mossy fibers; OFP is undetected.

[c]Expression is in scattered neurons and astrocytes.

Color-Assisted Circuit Tracing

There is a growing interest among neuroscientists to develop tools for the automatic segmentation of 3D data sets. At present, the tracing of multicolor images still relies on human input aided by semiautomated tools (see Experimental Method). Brainbow image data sets can be segmented in this manner, allowing axonal processes to be traced over distances of at least several hundred micrometers (Fig. 3). Such efforts, coupled with the visualization of the neuron's contact with neighboring cells, can provide important information about circuitry, both within and among brain regions. For example, in several instances it is possible to visualize a neuron and immediately determine whether its multiple synaptic inputs arise from distinct (differently colored) cells. This is an advantage over serial reconstruction methods with unlabeled tissues that require tracing nerve processes back to their origin or at least a branch point. There are several current options for digital reconstruction of Brainbow data sets, and new programs are under development.

GENERATION OF *BRAINBOW* TRANSGENES

Several *Brainbow* cassettes expressing XFPs from a *Thy1.2* or *CMV* promoter are available through Addgene.com, which is a nonprofit plasmid sharing resource. Alternatively, one may assemble a *Brainbow* transgene from basic elements (see Fig. 1).

Promoter

In principle, any promoter can drive *Brainbow* construct expression. We advise the choice of a strong promoter such as *Thy1.2* (Caroni 1997; Feng et al. 2000), which is able to express readily detectable levels of GFP in transgenic mice.

lox Sites

Recombination of *Brainbow-1* constructs relies on incompatible *lox* variants. We advise using *lox2272* (Lee and Saito 1998) and *loxN* (Livet et al. 2007), each of which possesses distinct substitutions at positions 2 and 6 in the spacer of the *lox* sequence. In our experience with expression in mammalian cells, these sites constitute, along with the canonical *loxP*, a set of three efficient and mutually incompatible *lox* variants.

XFPs

A wide palette of FPs is presently available for transgene design (Shaner et al. 2007; Day and Davidson 2009), but few have been tested in transgenic mice. Enhanced green fluorescent protein (eGFP) and enhanced yellow fluorescent protein (eYFP) have been the most successful variants by far. Cyan proteins (eCFP and Cerulean) and red proteins (DsRed2, tdimer2(12), and dTomato) can give good signals but do not match green and yellow proteins in terms of photostability and longevity in fixed tissues. To facilitate multichannel imaging, one should select XFPs that are spectrally well separated. Cyan (eCFP or Cerulean), yellow (eYFP), and red (DsRed2, d/tdTomato, and mCherry) FPs constitute a first set of three easily distinguishable XFPs. Green (eGFP) or orange (Kusabira, mOrange2), when added to this palette, may require linear unmixing techniques for good optical separation. New proteins in the blue, red, and far-red range may allow the creation of new and improved transgenes (Ai et al. 2007, Shaner et al. 2008; Shcherbo et al. 2009).

Subcellular Localization Signals

Cytoplasmic XFPs provide a good rendering of the volume of neuronal cell bodies and large neurites (Feng et al. 2000). *Brainbow* line L, which expresses three such proteins, is, therefore, recommended for most applications. However, a more restricted repertoire of structures can be highlighted when fusing XFPs to specific localization tags. In particular, membrane-targeted XFPs, such as those used in *Brainbow* line H, are well suited to highlight fine axonal processes (De Paola et al. 2003) and axons of specific neuronal populations such as cerebellar and hippocampal mossy fibers. However, compared to cytoplasmic proteins, membrane-tagged XFPs often make cell bodies harder to visualize, especially in densely labeled samples. Other useful tags used in mice include nuclear and mitochondrial localization sequences (Misgeld et al. 2007).

Elements for Optimized Expression

Open reading frames in *Brainbow* constructs should have a translation initiation consensus sequence (Kozak 1987), a codon usage optimized sequence (Haas et al. 1996), and an efficient polyadenylation signal (Kakoki et al. 2004).

FRT Site

One *FRT* site, the target sequence for Flp recombinase, can be inserted in a *Brainbow* construct for transgene copy number reduction in tandem arrays (Garrick et al. 1998). Crossing Brainbow animals with germline-expressing Flp mice (Rodriguez et al. 2000) triggers recombination between *FRT* sites located on distinct *Brainbow* copies, thus, reducing the transgene array. With large transgenes, this process may sometimes take several generations.

BRAINBOW TRANSGENIC MICE

Several *Thy1-Brainbow* lines (H, L, M, and R) can be obtained from Jackson Laboratories (JAX stock numbers 007901, 007910, 007911, and 007921, respectively; see Table 1). Line L, which expresses three cytoplasmic proteins in a combinatorial manner, should offer the most flexible usage. It shows strong XFP expression, although not as high as in some *Thy1-XFP* lines selected earlier, such as YFP-16, YFP-H, or GFP-M (Feng et al. 2000). Lines H, M, and R, which express mixed cytoplasmic and membrane-targeted XFPs, will likely have more restricted applications.

CRE MICE

Recombination can be triggered with various Cre driver lines, which are available from several locations including Jackson Laboratories. We use *CAGGS-CreER* mice (Guo et al. 2002) and tamoxifen (Metzger et al. 1995) to create a widespread pulse of recombinase activity in Brainbow mice, which robustly triggers expression of XFPs from *Brainbow* transgenes. *Brainbow-1* mice, which can tolerate tonic Cre expression, have also been used with retina- and motor neuron–specific Cre lines.

Protocol

Generation and Imaging of Brainbow Mice

We discuss how to generate *Brainbow* transgenes and mice, how to induce recombination in Brainbow mice with tamoxifen, and how to fix tissue samples from these animals. We also describe how to acquire multichannel images from Brainbow samples and use those images to digitally trace cellular contours and connections.

IMAGING SETUP

Imaging System

Virtually any fluorescence microscope can be used to image Brainbow samples, provided that each XFP is imaged using its specific excitation and emission parameters. The conditions described here are used for imaging *Brainbow-1.0* samples, which express three very distinct XFPs (cyan, yellow, and red). The samples are fixed and imaged with oil-immersion objectives, in our case, 20x (0.8 numerical aperture [NA]), 40x (1.3 NA), or 60x (1.45 NA). On an epifluorescence microscope, separation of XFP signals can be accomplished with standard CFP, YFP, and tetramethyl rhodamine isothiocyanate (TRITC) filter cubes. For confocal imaging, which provides Brainbow images superior to epifluorescence, one should use a setup equipped with three detectors and laser lines for exciting each XFP (e.g., an Olympus FV1000 equipped with four photomultiplier tubes and a 440/515/568/633-nm beam splitter [for CFP, YFP, RFP, and an additional far-red channel]). In this system, CFP is excited using a 440-nm laser diode and collected with a 510-nm dichroic mirror and 465–505-nm barrier filter. YFP is excited using the 515-nm argon laser line and collected using a 560-nm dichroic mirror and 525–560-nm barrier filter. For RFP, we use either a 561-nm diode laser or a 568-nm krypton laser and a 585–620-nm barrier filter. When XFPs with significant excitation/emission spectra overlap are used (e.g., CFP/GFP/YFP or YFP/OFP/RFP), their fluorescent signals can be separated using linear unmixing, for instance with the ImageJ Spectral Unmixing plug-in.

Reconstruct Software

There are many current options for digital reconstruction software. The tool that we discuss here is called Reconstruct (http://synapse-web.org/; Fiala 2005). This software allows one to load, manipulate, and digitally segment red–green–blue (RGB) multiTIFF data stacks (called series). The Reconstruct program provides the user with several basic advantages, including its ability to reconstruct the full-volume view of each segmented object. (In other words, reconstructed neurons appear as surface-rendered objects displaying all the details of their traced contours; see Fig. 3.) However, Reconstruct does not currently allow users to export the data into vector format, a helpful tool for users who are interested in coding interactions among cells. Other programs allow for tracing such skeletons (see, for instance, Imaris, Bitplane Scientific Software). Furthermore, the processing of large image data sets with many objects identified in Reconstruct requires a powerful computer for efficient processing. (See the Reconstruct user manual for specific system requirements.) Currently, the Reconstruct software is not compatible with Macintosh systems.

MATERIALS

CAUTION: See Appendix 6 for proper handling of materials marked with <!>.
See the end of the chapter for recipes for reagents marked with <R>.

Reagents

Brainbow constructs (available from Addgene.com or assembled from basic elements)
Corn oil (Sigma-Aldrich C-8267)
CreER mice (several lines are maintained at the Jackson Laboratory)
Paraformaldehyde (PFA) <!>
Phosphate-buffered saline (PBS) <R>
Sodium pentobarbitone <!>
Tamoxifen <!> (Sigma-Aldrich T-5648)
Thy1-Brainbow mice (Jackson Laboratory)
Vectashield (Vector Laboratories, Inc.)

Equipment

Coverslips
Dissection tools
Glass slides
Microcentrifuge tubes, 1.5-mL
Nail polish
Needle, 30.5-gauge
Perfusion apparatus (e.g., optional peristaltic pump)
Syringe, 1-mL, disposable
Vibrating microtome <!> (e.g., Leica VT1000)

EXPERIMENTAL METHOD

Production of *Brainbow* Constructs

1. Order *Brainbow* constructs from Addgene.com, or assemble and amplify your own using standard molecular biology methods (Fig. 1). Gene synthesis is an increasingly attractive option for generating novel *Brainbow* cassettes. See the previous section for a list of the elements that can be used.

   When designing Brainbow-2 plasmids, avoid placing similar XFPs head to head because such inverted repeats may be unstable in bacteria. Plasmid amplification in *Escherichia coli* has, at times, yielded truncated Brainbow cassettes. The integrity of Brainbow plasmids should, therefore, be checked at all stages of their preparation.

   See *Troubleshooting.*

Generation of Brainbow Mice

2. Several lines of Thy1-Brainbow mice are available from Jackson Laboratories (Table 1). Alternatively, new mouse lines can be generated by pronuclear injection or homologous recombination in embryonic stem cells, yielding two possible modes of XFP expression depending on the number of active transgene copies (see previous discussion of colors).

Tamoxifen-Induced Recombination in Brainbow Mice

3. Mate Brainbow mice with CreER mice.

   See *Troubleshooting.*

4. Prepare tamoxifen solution in sterile conditions. Weigh 10–50 mg of tamoxifen in a 1.5-mL microcentrifuge tube. Add 1 mL of corn oil, and vortex. Heat for 10 min at 56°C, and sonicate for 10 sec. Repeat until the solution clears. Dilute to a working concentration of 0.1–5 mg/mL. The solution appears stable for at least 1 wk at 4°C.

   See *Troubleshooting.*

5. Inject pups at P0–P1 by dorsal subcutaneous injection with 20–30 µL of tamoxifen solution using a 30.5-gauge needle mounted on a disposable 1-mL syringe. Alternatively, for increased precision, use a gas-tight Hamilton syringe.

   The dose dependence of CreER to tamoxifen provides a convenient way to control the sparseness of the labeling (Buffelli et al. 2003).

   See *Troubleshooting.*

Fixed Tissue Preparation

6. Deeply anesthetize mice with sodium pentobarbitone (or suitable alternate) before performing intracardiac perfusion with 2%–4% PFA.
7. Dissect the desired tissues for analysis. (Brain and other tissues can be postfixed in 2% PFA for 4–10 h at 4°C.) Rinse in PBS for at least 1 h.
8. Section tissue using a vibrating microtome at a thickness ranging from 60 to 200 µm in cold PBS (4°C). Immediately mount tissue sections in Vectashield or other fluorescence-preserving medium on glass slides. Cover with a coverslip, and store at –20°C before imaging.

   XFP fluorescence is well maintained with tissue fixed with PFA.

   The fluorescent signal will decrease significantly in sections kept floating in PBS, but it remains strong for ~2 mo when sections are mounted and stored at –20°C. (Multiple thaw/refreeze cycles should be avoided.) Skeletal muscle and other thin structures may be whole mounted directly onto slides.

   In some cases, immunohistochemical techniques may be combined with Brainbow labeling (e.g., to label other intracellular proteins). Permeabilization appears to decrease the XFP signal within cells; thus, the concentration of Triton X-100 should be kept as low as possible (e.g., 0.02%). XFPs can also be detected with antibodies either to enhance signal or to adjust the emitted wavelengths for detection. However, many XFPs such as the GFP derivatives eYFP and Cerulean are very similar in structure and cannot be distinguished with currently available commercial antisera.

Confocal Image Acquisition and Processing

9. Acquire confocal images sequentially for each channel to obtain optimal separation of XFP signals. Keep laser power minimal to avoid photobleaching. (See Imaging Setup section above for advice on confocal settings.)

   See *Troubleshooting.*

10. Process data for image display. For each channel, generate a maximal projection of confocal image stacks using MetaMorph or ImageJ. Perform uniform adjustment of levels, contrast, brightness, and gamma. Assign colors to each channel, and overlay them to generate the final rendering.

Tracing and 3D Reconstruction

11. Format the image data before loading into Reconstruct:
    1. To begin analyzing a data set in Reconstruct, first format the data as a series of RGB TIFFS. Within each TIFF, overlay the individual channels (from each XFP) into one RGB image.
    2. To optimize the rendering of neuronal processes, adjust RGB brightness levels on each optical plane using the Photoshop batch process mode.

iii. (*Optional*) Resample the image stack. When neurites run in a preferred direction of the sample (e.g., along axonal tracts), the confocal image data set may be resampled (using MATLAB, ImageJ, MetaMorph, or other software) in an orientation that will allow the neurites to be viewed and traced in cross section. In any one image section, an axon will then appear as a small dot or circle of fluorescence, which aids its distinction from neighboring cells (see Fig. 3A).

12. Trace the series in Reconstruct. In this program, a palette of tracing tools allows the user to outline the extent of individual cells or processes on serial sections through the data stack (Fig. 3A). Tracing can be performed manually (by mouse or wand) or can be aided by Reconstruct's automated wildfire tool that selects a region of similar hue and brightness on adjacent sections.
13. View the traced series. At any time, Reconstruct's rendering tool can build a 3D view of the traced series (Fig. 3B). Each reconstructed neuron (which should be named differently by the user during the tracing process) is considered a separate object, and can be loaded individually or with other objects using a number of different rendering options (e.g., see Fig. 3B1). The 3D view may be exported as a basic image file or as a VRML (virtual reality modeling language) file, which can then be imported into other programs to generate animations.

TROUBLESHOOTING

Problem (Step 1): DNA amplification in *E. coli* yields a truncated plasmid.
Solution: Part of the construct may have been excised in bacteria. Restart from fresh plasmid stock, use recombination-defective bacteria, and grow them for a shorter time. Check plasmid at all stages of the preparation.

Problem (Steps 3–5): Tamoxifen and Cre exhibit toxicity.
Solution: High doses of tamoxifen can be toxic for mice pups, resulting in poor growth and vascular problems. Use lower tamoxifen doses, repeating the injection if strong recombination is desired. In addition, the CD1 outbred mouse strain is more resistant to tamoxifen toxicity than the more widely used C57B6 inbred strain. Thus, outcrossing Brainbow or Cre lines can be useful. Cre may also be toxic if expressed at high levels (Naiche and Papaioannou 2007; Schmidt-Supprian and Rajewsky 2007), an effect that should be controlled in mice expressing only Cre and may be reduced by using a transiently active Cre (Heidmann and Lehner 2001).

Problem (Step 9): There are imaging difficulties.
Solution: Although visible with epifluorescence, CFPs and RFPs typically tend to bleach quickly compared to eYFP. Image fresh samples, and store them at –20°C in glycerol-based medium. Use objectives with a high-NA/magnification ratio and alternative modes of imaging (e.g., spinning-disk confocal). Minimize epifluorescence illumination while searching a field of interest. Acquire images close to the surface of the sample. Amplify fluorescent signal using antibodies (e.g., an anti-dsRed rabbit polyclonal can be combined with a monoclonal or chicken polyclonal anti-GFP to reveal RFP and CFP/YFP, respectively).

DISCUSSION

Because they can be genetically encoded, spectrally distinct FPs (XFPs) represent ideal labels for in situ and live imaging studies. Several approaches have emerged for combining two or more of these XFPs in the same animal to distinguish adjacent cells and highlight their interactions. A first set of approaches involves generating several transgenes, each encoding a different color, and coexpressing them in the same animal through various strategies—for example, generation of chimeric animals (Hadjantonakis et al. 2002), interbreeding (Feng et al. 2000; Kasthuri and Lichtman 2003; Walsh and Lichtman 2003), or viral injection (Boldogkoi et al. 2009). A second set of approaches swaps XFP genes expressed from a single locus using Cre- or Flp-mediated excision (Muzumdar et al. 2007) or even interchromosomal recombination (Zong et al. 2005).

Among these methods, Brainbow strategies offer several advantages. First, using the divergent recombination pathways proposed by *Brainbow* transgenes, up to four genes can be stochastically expressed in a cellular population from a single construct. Furthermore, combinatorial expression from multiple transgene copies recombining independently vastly increases the number of expression possibilities. Importantly, it is possible to control whether mutually exclusive or combinatorial expression is obtained by manipulating copy number. In Brainbow mice, cells of a same type, although a priori genetically identical, have a large chance of being labeled with distinct colors, which allows for visualizing their interactions.

This multicolor labeling can be visualized in fixed tissues, as presented here, and also in live conditions with confocal or epifluorescence microscopy (see Livet et al. 2007). However, those techniques are best suited for surface imaging. Multiphoton microscopy allows for increased penetration as well as lower phototoxicity. But for optimal multiphoton excitation of CFP, YFP, and RFP, it appears necessary to either tune the wavelength of the pulsed laser source or simultaneously produce at least two spectrally distinct pulsed excitations. In part because of these technical issues, the potential of the Brainbow approach for volume live imaging in vitro or in vivo has, thus far, been largely unexplored. New setups and Brainbow mice should solve these difficulties, with the possible help of spectral unmixing strategies (Ducros et al. 2009).

Current *Brainbow* transgenes could potentially be improved in several aspects, in particular, regarding the photochemical characteristics and expression level of their fluorochromes. Adequate promoters and Cre drivers should allow for expanding *Brainbow* transgene expression to neuronal types and stages not targeted with the current *Thy1.2* promoter. The method appears broadly applicable to other cell types and models—see, in particular, Pan et al. (2010) for studies demonstrating Brainbow in zebrafish. Importantly, it could, in principle, be used to express genetic markers of any type.

Perhaps the main drawback of the Brainbow method is that, like any optical-based method, it is currently limited in resolution by the diffraction of light. *Brainbow* transgenes allow, to some extent, this limit to be bypassed when attributing unambiguous labels to two adjacent cells. However, when several subdiffraction processes overlap in the same focal volume, their individual spectral signatures become difficult to extract (Helmstaedter et al. 2008). One way to overcome this problem is to analyze samples presenting intermediate sparse labeling of neurons. In the future, restricting the combinatorial labeling to synapses, or combining the Brainbow approach with other approaches such as electron microscopy, array tomography, or superresolution imaging could provide promising methods to overcome the resolution issue (e.g., Kasthuri and Lichtman 2007; Micheva and Smith 2007; Huang et al. 2009; Mishchenko 2010).

RECIPE

CAUTION: See Appendix 6 for proper handling of materials marked with <!>.
Recipes for reagents marked with <R> are included in this list.

Phosphate-Buffered Saline (PBS)

Reagent	Quantity (for 1L)	Final concentration (10×)
NaH_2PO_4	2.56 g	18.6 mM
Na_2HPO_4	11.94 g	84.1 mM
NaCl	102.2 g	1.75 M

Adjust the pH to 7.4 using NaOH <!> or HCl <!> as necessary.

ACKNOWLEDGMENTS

We thank Ryan W. Draft, Ju Lu, Hyuno Kang, and Robyn Bennis for help in developing Brainbow mice with Katie Matho, Robert M. Gould for a critical reading of the manuscript, and Stéphanie Fouquet for help with imaging. Our research is supported by grants from the National Institutes of Health (JRS and JWL), and by Inserm Avenir, Ville de Paris, Ecole des Neurosciences de Paris, ANR, FROm, and ELA (JL).

REFERENCES

Ai HW, Shaner NC, Cheng Z, Tsien RY, Campbell RE. 2007. Exploration of new chromophore structures leads to the identification of improved blue fluorescent proteins. *Biochemistry* **46:** 5904–5910.

Boldogkoi Z, Balint K, Awatramani GB, Balya D, Busskamp V, Viney TJ, Lagali PS, Duebel J, Pásti E, Tombácz D, et al. 2009. Genetically timed, activity-sensor and rainbow transsynaptic viral tools. *Nat Methods* **6:** 127–130.

Branda CS, Dymecki SM. 2004. Talking about a revolution: The impact of site-specific recombinases on genetic analyses in mice. *Dev Cell* **6:** 7–28.

Buffelli M, Burgess RW, Feng G, Lobe CG, Lichtman JW, Sanes JR. 2003. Genetic evidence that relative synaptic efficacy biases the outcome of synaptic competition. *Nature* **424:** 430–434.

Caroni P. 1997. Overexpression of growth-associated proteins in the neurons of adult transgenic mice. *J Neurosci Methods* **71:** 3–9.

Day RN, Davidson MW. 2009. The fluorescent protein palette: Tools for cellular imaging. *Chem Soc Rev* **38:** 2887–2921.

De Paola V, Arber S, Caroni P. 2003. AMPA receptors regulate dynamic equilibrium of presynaptic terminals in mature hippocampal networks. *Nat Neurosci* **6:** 491–500.

Ducros M, Moreaux L, Bradley J, Tiret P, Griesbeck O, Charpak S. 2009. Spectral unmixing: Analysis of performance in the olfactory bulb in vivo. *PLoS One* **4:** e4418. doi: 10.1371/journal.pone.0004418.

Feng G, Mellor RH, Bernstein M, Keller-Peck C, Nguyen QT, Wallace M, Nerbonne JM, Lichtman JW, Sanes JR. 2000. Imaging neuronal subsets in transgenic mice expressing multiple spectral variants of GFP. *Neuron* **28:** 41–51.

Fiala JC. 2005. Reconstruct: A free editor for serial section microscopy. *J Microsc* **218:** 52–61.

Garrick D, Fiering S, Martin DI, Whitelaw E. 1998. Repeat-induced gene silencing in mammals. *Nature Genet* **18:** 56–59.

Guo C, Yang W, Lobe CG. 2002. A Cre recombinase transgene with mosaic, widespread tamoxifen-inducible action. *Genesis* **3:** 8–18.

Haas J, Park EC, Seed B. 1996. Codon usage limitation in the expression of HIV-1 envelope glycoprotein. *Curr Biol* **6:** 315–324.

Hadjantonakis AK, Macmaster S, Nagy A. 2002. Embryonic stem cells and mice expressing different GFP variants for multiple non-invasive reporter usage within a single animal. *BMC Biotechnol* **2:** 11.

Heidmann D, Lehner CF. 2001. Reduction of Cre recombinase toxicity in proliferating *Drosophila* cells by estrogen-dependent activity regulation. *Dev Genes Evol* **211:** 458–465.

Helmstaedter M, Briggman KL, Denk W. 2008. 3D structural imaging of the brain with photons and electrons. *Curr Opin Neurobiol* **18:** 633–641.

Huang B, Bates M, Zhuang X. 2009. Super-resolution fluorescence microscopy. *Annu Rev Biochem* **78:** 993–1016.

Kakoki M, Tsai YS, Kim HS, Hatada S, Ciavatta DJ, Takahashi N, Arnold LW, Maeda N, Smithies O. 2004. Altering the expression in mice of genes by modifying their 39 regions. *Dev Cell* **6:** 597–606.

Kasthuri N, Lichtman JW. 2003. The role of neuronal identity in synaptic competition. *Nature* **424:** 426–430.

Kasthuri N, Lichtman JW. 2007. The rise of the 'projectome'. *Nat Methods* **4:** 307–308.

Kozak M. 1987. An analysis of 5′-noncoding sequences from 699 vertebrate messenger RNAs. *Nucleic Acids Res* **15:** 8125–8148.

Lam KP, Rajewsky K. 1998. Rapid elimination of mature autoreactive B cells demonstrated by Cre-induced change in B cell antigen receptor specificity in vivo. *Proc Natl Acad Sci* **95:** 13171–13175.

Lee G, Saito I. 1998. Role of nucleotide sequences of *loxP* spacer region in Cre-mediated recombination. *Gene* **216:** 55–65.

Lichtman JW, Livet J, Sanes JR. 2008. A technicolour approach to the connectome. *Nat Rev Neurosci* **9:** 417–422.

Livet J, Weissman TA, Kang H, Draft RW, Lu J, Bennis RA, Sanes JR, Lichtman JW. 2007. Transgenic strategies for combinatorial expression of fluorescent proteins in the nervous system. *Nature* **450:** 56–62.

Metzger D, Clifford J, Chiba H, Chambon P. 1995. Conditional site-specific recombination in mammalian cells using a ligand-dependent chimeric Cre recombinase. *Proc Natl Acad Sci* **92:** 6991–6995.

Micheva KD, Smith SJ. 2007. Array tomography: A new tool for imaging the molecular architecture and ultrastructure of neural circuits. *Neuron* **55:** 25–36.

Misgeld T, Kerschensteiner M, Bareyre FM, Burgess RW, Lichtman JW. 2007. Imaging axonal transport of mitochondria in vivo. *Nat Methods* **4:** 559–561.

Mishchenko Y. 2010. On optical detection of densely labeled synapses in neuropil and mapping connectivity with combinatorially multiplexed fluorescent synaptic markers. *PLoS One* **5:** e8853. doi: 10.1371/journal.pone.0008853.

Muzumdar MD, Tasic B, Miyamichi K, Li L, Luo L. 2007. A global double-fluorescent Cre reporter mouse. *Genesis* **45:** 593–605.

Naiche LA, Papaioannou VE. 2007. Cre activity causes widespread apoptosis and lethal anemia during embryonic development. *Genesis* **45:** 768–775.

Palmiter RD, Brinster RL. 1986. Germ-line transformation of mice. *Annu Rev Genet* **20:** 465–499.

Pan YA, Livet J, Sanes JR, Lichtman JW, Schier AF. 2011. Multicolor brainbow imaging in zebrafish. In *Imaging in developmental biology: A laboratory manual* (ed. Wong R, Sharpe J), pp. 191–198. Cold Spring Harbor Laboratory Press, Cold Spring Harbor, NY.

Rodríguez CI, Buchholz F, Galloway J, Sequerra R, Kasper J, Ayala R, Stewart AF, Dymecki SM. 2000. High-efficiency deleter mice show that FLPe is an alternative to Cre-*loxP*. *Nat Genet* **25:** 139–140.

Shaner NC, Patterson GH, Davidson MW. 2007. Advances in fluorescent protein technology. *J Cell Sci* **120:** 4247–4260.

Shaner NC, Lin MZ, McKeown MR, Steinbach PA, Hazelwood KL, Davidson MW, Tsien RY. 2008. Improving the photostability of bright monomeric orange and red fluorescent proteins. *Nat Methods* **5:** 545–551.

Shcherbo D, Murphy CS, Ermakova GV, Solovieva EA, Chepurnykh TV, Shcheglov AS, Verkhusha VV, Pletnev VZ, Hazelwood KL, Roche PM, et al. 2009. Far-red fluorescent tags for protein imaging in living tissues. *Biochem J* **418:** 567–574.

Schmidt-Supprian M, Rajewsky K. 2007. Vagaries of conditional gene targeting. *Nat Immunol* **8:** 665–668.

Silver DP, Livingston DM. 2001. Self-excising retroviral vectors encoding the Cre recombinase overcome Cre-mediated cellular toxicity. *Mol Cell* **8:** 233–243.

Walsh MK, Lichtman JW. 2003. In vivo time-lapse imaging of synaptic takeover associated with naturally occurring synapse elimination. *Neuron* **37:** 67–73.

Zong H, Espinosa JS, Su HH, Muzumdar MD, Luo L. 2005. Mosaic analysis with double markers in mice. *Cell* **121:** 479–492.

12 The Use of BAC Transgenic Mice to Label Genetically Defined Cell Populations and the GENSAT Database of Engineered Mouse Strains

Eric F. Schmidt,[1] Laura Kus,[2] Shiaoching Gong,[2] and Nathaniel Heintz[1,2]

[1]*Laboratory of Molecular Biology, Howard Hughes Medical Institute, New York, New York 10065;* [2]*GENSAT Project, The Rockefeller University, New York, New York 10065*

ABSTRACT

The brain is a complex tissue comprising hundreds of distinct cell types, each of which has unique circuitry and plays a discrete role in nervous system function. Large-scale studies mapping gene-expression patterns throughout the nervous system have revealed that many genes are exclusively expressed in specific cell populations. The GENSAT (Gene Expression Nervous System Atlas) Project created a library of engineered mice utilizing bacterial artificial chromosomes (BACs) to drive the expression of enhanced green fluorescent protein (eGFP) in genetically defined cell populations. BACs contain large segments of genomic DNA and retain most of the transcriptional regulatory elements directing the expression of a given gene, resulting in more faithful reproduction of endogenous expression patterns. BAC transgenic mice offer a robust solution to the challenging task of stably and reproducibly accessing specific cell types from a heterogeneous tissue such as the brain. A significant advantage of utilizing eGFP as a reporter is the fact that it can fill entire cells, including neuronal dendrites and axons as well as glial processes, making GENSAT reporter mice a powerful tool for neuroimaging studies. This chapter provides a primer on the generation of BAC transgenic mice and advantages for their use in labeling genetically defined cell types. It also provides an overview of searching the GENSAT database and ordering engineered mouse lines.

INTRODUCTION

The cellular diversity of the central nervous system (CNS) has been well documented since the classic silver impregnation studies by Ramon y Cajal and his students (Ramon y Cajal et al. 1899). In the century that followed, new tools for studying the nervous system were invented, including histochemical stains, antibody labeling, tract tracing, electrical recording, and in situ hybridization. These techniques allowed neuroanatomists to define specific cell populations based not only on size, shape, and position, but also on chemical signature, electrophysiological properties, and axonal projections. However, to date, relatively few tools exist that allow an investigator to reproducibly target and label molecularly distinct cell populations in vivo. This is particularly true for heterogeneous tissues composed of morphologically similar cells for which there are limited molecular markers, such as the brain.

Although it is difficult to discern cell type based solely on in situ hybridization studies, early efforts to study gene expression in the brain by in situ hybridization clearly showed that a significant fraction of the genome is expressed in the brain, and that many transcripts could be localized to distinct brain regions using this methodology (Bloch et al. 1986; Kato 1990; Kuhar et al. 1993). Recently, with the ability to perform in situ hybridization studies on a genome-wide scale, additional significant insights into CNS development, organization, and function have been obtained through the analysis of gene expression (Visel et al. 2004; Magdaleno et al. 2006; Lein et al. 2007). The GENSAT Project was established for three main purposes: (1) to use reporter gene technology in transgenic animals to provide high-resolution anatomical data on the cell types expressing genes that are regionally expressed or developmentally regulated in the brain; (2) to identify BAC vectors that can be used to efficiently target gene expression to specific CNS cell types; and (3) to provide genetically modified transgenic mice to the research community to stimulate neuroscience research. Accordingly, the GENSAT project has analyzed more than 10,000 transgenic lines (in some cases, multiple lines from the same BAC) and has made all anatomical data, BAC transgenic vectors, and archived mouse lines available to the community (Gong et al. 2003).

The ability to reproducibly access genetically defined cell types has broad implications for neuroscience research. Thus, the use of BAC drivers has become a powerful tool for reaching previously inaccessible cell populations for functional studies, such as differential responses to drugs (Bateup et al. 2008), translational profiling (Doyle et al. 2008; Heiman et al. 2008), and developmental analysis (Osheroff and Hatten 2009). Cell type–specific labeling allows (1) multiple components of neural circuits to be teased apart (O'Connor et al. 2009); (2) the functions of a single cell type to be examined in the context of multiple circuits (Groh et al. 2010); and (3) characterization of biochemical (Selimi et al. 2009) and epigenetic mechanisms (Kriaucionis and Heintz 2009) that play key roles in the control of neuronal function. Furthermore, the existence of a large collection of BAC transgenic mice expressing eGFP in identified cell types has enabled electrophysiological studies aimed at understanding the unique physiological properties of even closely related CNS neurons (Gertler et al. 2008; Valjent et al. 2009). In this chapter, we discuss the use of BAC transgenic mice for labeling CNS cell types and provide a primer on finding and ordering mouse lines generated by the GENSAT Project.

BAC TRANSGENIC REPORTER MOUSE STRAINS

Generation of BAC Mice

The regulatory elements that control expression of mammalian genes are scattered over large regions of the genome; in most cases, the sequences controlling expression of a gene of interest have not been identified. Thus, the transcribed unit of the average mammalian gene is ~30 kilobases (kb) in length, and the *cis* regulatory elements that direct transcription are both within introns and in the 5′ and 3′ intergenic regions that separate the gene from its neighbors. Because most traditional transgenic vectors can only accommodate 10–50 kb of DNA, a small portion (~10 kb) of the 5′ untranslated promoter region is normally used to create transgenic mice, most often leading to unpredictable expression patterns caused by position effects that are both irreproducible and inaccurate. BAC vectors can stably carry very long stretches (100–300 kb) of foreign DNA (Giraldo and Montoliu 2001; Heintz 2001) and are commonly used to archive genomic DNA for large-scale sequencing studies (McPherson et al. 2001; Gregory et al. 2002). Libraries of BACs exist for many species, and almost all genes can be mapped onto a specified BAC. Because of their large size, most of the transcriptional regulatory elements are preserved on the BAC. As such, BACs have been shown to restore function when the chromosomal gene is mutated suggesting that enough information is contained on the DNA segment to direct physiological expression patterns (Antoch et al. 1997; Yang et al. 1999).

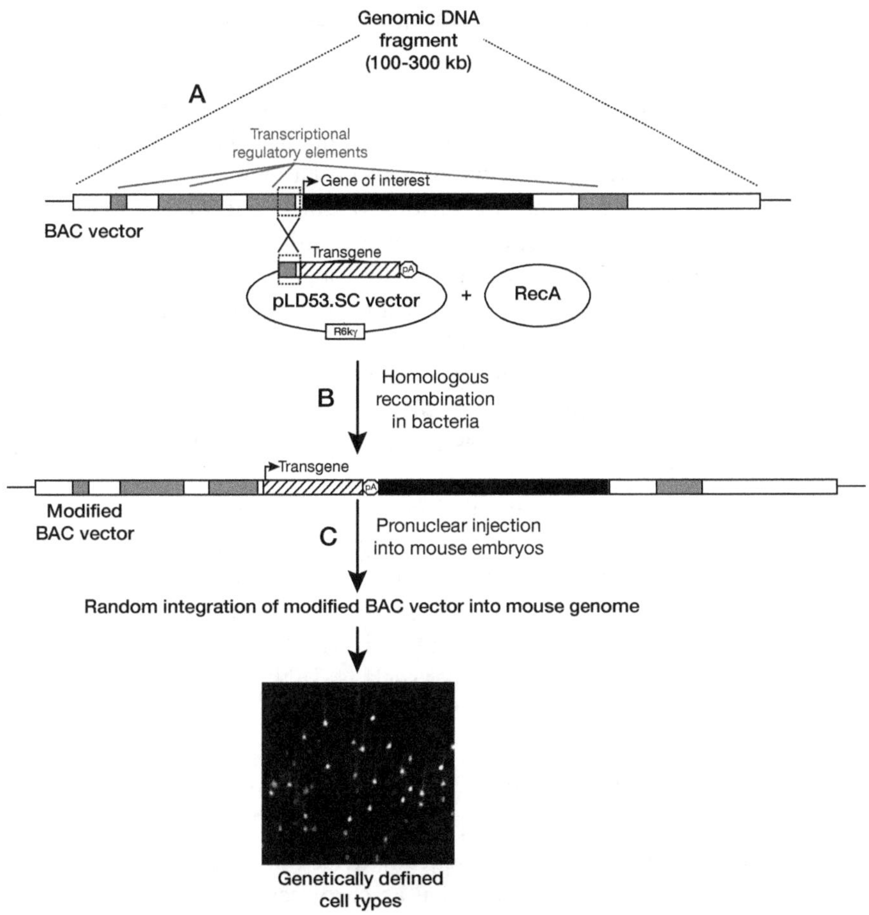

FIGURE 1. Diagram illustrating the generation of BAC transgenic mice. (*A*) BAC vectors consist of large segments (100–300 kb) of genomic DNA and include most of the upstream and downstream transcriptional regulatory elements (gray areas) that direct the expression of a gene of interest (black area). (*B*) A shuttle vector with a *Pir*-dependent R6K γ origin of replication (e.g., pLD53.SC2 for eGFP) containing a <1 kb homology arm corresponding to the sequence just 5′ of the ATG start codon of the gene of interest (dotted boxes), and the coding region for a transgene is co-integrated into the BAC vector by homologous recombination in bacterial hosts. This process is dependent on re-introducing *RecA* into the recombination-deficient bacteria. The transgene contains a polyadenylation (pA) tail to terminate transcription. (*C*) The modified BAC vector is then purified, linearized, and injected into the pronucleus of mouse embryos where it is randomly integrated into the genome to create transgenic founders. The transgene is now expressed in genetically defined cell types that express the endogenous gene.

Over the last decade or so, protocols have been developed and have been optimized for easy and efficient engineering of BACs for generating reporter lines to mimic the expression of almost any gene of interest (Yang et al. 1997; Gong et al. 2002). The coding region of the gene is replaced by a transgene, such as eGFP, just upstream of the start codon by homologous recombination in bacteria using a shuttle vector (Fig. 1). This process is dependent on re-introducing the *RecA* gene into the replication-deficient host bacteria strain (Yang et al. 1997), as well as using the R6Kγ origin of replication for the shuttle vector to prevent false positive co-integrated colonies (Gong et al. 2002). In most cases, the transgene contains a 3′ polyadenylation signal to prevent transcriptional interference from the remaining endogenous gene. Transgenic mice are generated by pronuclear injection of purified modified BACs into mouse zygotes in which the BAC DNA is randomly incorporated into the mouse genome. Founder mice (F_0) inheriting the transgene are born at rates comparable to traditional transgenic methods. Because the chromosomal location and copy number of the integrated BAC DNA varies between each founder from a given injection, it is important to maintain them as separate strains and to examine transgene expression in each case.

Advantages of BACs to Label Genetically Defined Cell Types

Transgenic expression of BACs recapitulates endogenous gene expression following integration into the genome. Because the transgene is stable and heritable, expression patterns are reproducible between individual mice and across generations of the same founder line. This obviates the need to use other less reproducible and more labor-intensive technologies (such as viral-mediated gene transfer, transfection, bioballistics, or microinjection) to label specific groups of cells. With BAC mice, the same cell population can be manipulated and can be studied in repeated experiments using a number of different assays. In addition, large DNA constructs (>100 kb) are less prone to position effects caused by the integration site, which can greatly influence expression when small promoters are used (Giraldo and Montoliu 2001). The reproducible nature of BACs is particularly advantageous when attempting repeated labeling of genetically defined cell populations with different genetic tools, such as eGFP and Cre. Thus, the same BAC vector can be modified numerous times so that a variety of transgenes can be expressed in the same cell populations in multiple transgenic strains (Fig. 2).

Another strategy to label genetically defined cells is to place the transgene directly into the genomic locus using the knockin approach. Although this has many advantages, a normal copy of the gene is disrupted. With BAC transgenics, both endogenous alleles of the gene are preserved, preventing any adverse gene-dosage effects on mouse development and physiology. Additionally, BAC modifications are faster, less expensive, and easier than gene targeting, making them more attractive for high-throughput resources such as GENSAT.

APPLICATION OF BAC REPORTER MICE TO NEUROIMAGING STUDIES

The labeling of unique genetically determined cell populations with a fluorescent marker such as eGFP has unlimited potential as a neuroimaging tool. eGFP reporter strains allow for direct imaging of specific cell populations in a variety of applications as shown in Figure 3. Among the most common uses are gene-expression mapping (Bishop et al. 2006; Head et al. 2007) and electrophysiological recording studies (Gertler et al. 2008; Partridge et al. 2009; Groh et al. 2010). Both are facilitated by the ability to identify prelabeled neurons of a known genotype and can be combined with indicator dyes or optogenetic techniques to measure cell-specific responses within neural circuits or to specifically control subpopulations. In addition, eGFP is soluble and is, thus, able to fill cellular processes and, if the concentrations are high-enough, neuronal dendrites and axons. This allows projections to be visualized and tracked either directly via epifluorescence or through antibody staining (see Fig. 2C). Other studies have taken advantage of the stability of the eGFP reporter for lineage tracing studies (Salero and Hatten 2007; Placantonakis et al. 2009) and to study cell migration.

In addition to fluorescent reporter mice, a growing number of BAC transgenic mouse lines that drive the expression of recombinases, such as Cre, are available from GENSAT and the Allen Institute for Brain Science (Gong et al. 2007; Madisen et al. 2010). These strains can be combined with a number of existing Cre-inducible reporter mouse strains and/or virus technologies to give the experimenter more spatial and temporal control over transgene expression or to perform genetic manipulations in a cell type–specific manner.

Things to Consider When Choosing a Mouse Line

There are a number of considerations to take into account when selecting a line to use for imaging studies. It is important to confirm that the gene is expressed in the cell population of interest at the time the experiment is to be performed. For example, many genes are developmentally regulated, and their expression may turn off too early or turn on too late depending on the timing of experiments. Similarly, the ontology of cell types expressing a given gene may change at different stages of development or be induced in novel cell types following an experimental manipulation, such as CNS injury

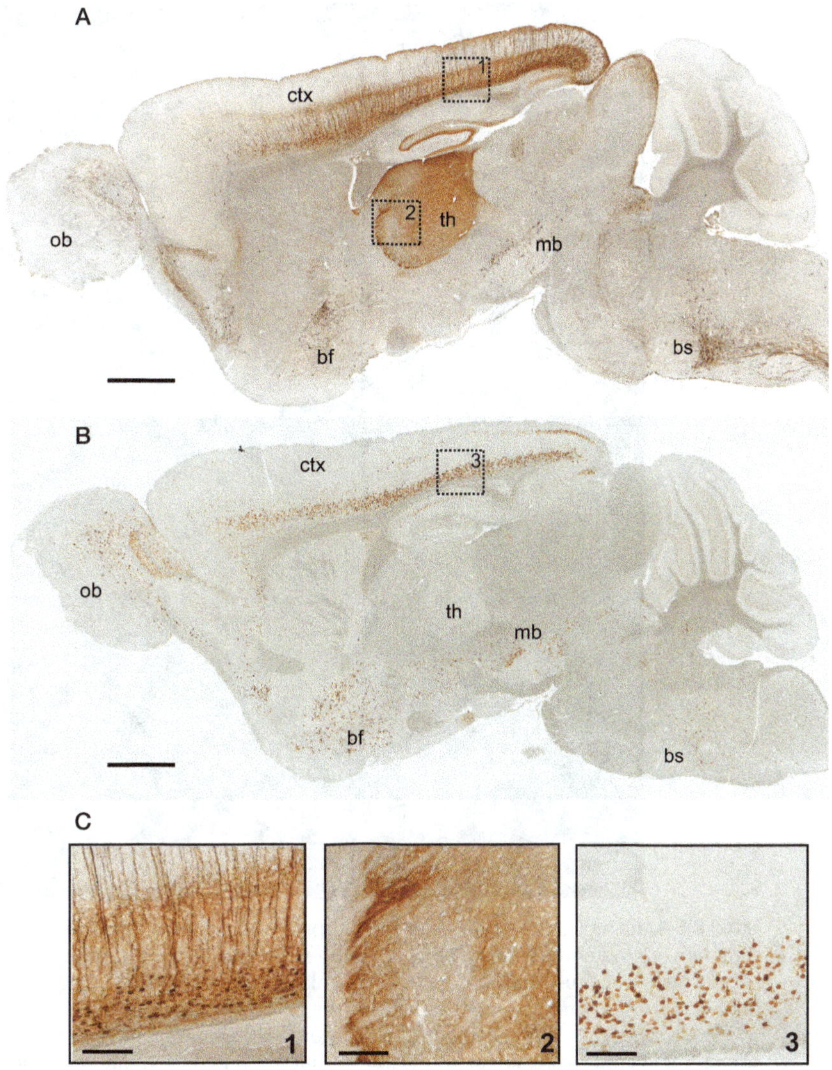

FIGURE 2. Reproducible labeling of neural circuits using the same BAC driver to express different transgenes. (*A*) Micrograph of an eGFP-stained sagittal section through the brain of a GENSAT Ntsr1-eGFP BAC transgenic mouse. eGFP fills the dendrites, soma, and axons of Ntsr1-expressing, layer-6 pyramidal cells in the neocortex (ctx), with other cell populations seen in the olfactory bulb (ob), basal forebrain (bf), midbrain (mb), and brain stem (bs). (*B*) Micrograph of an eGFP-stained sagittal section of an Ntsr1-bacTRAP mouse. In this case, the Ntsr1 BAC is driving the expression of eGFP fused to a ribosomal subunit, therefore, restricting eGFP labeling to the cell body. The distribution of eGFP expression is very similar to that in *A*. (*C*) Higher-magnification images of boxes from *A* and *B* show the cell bodies and dendrites in the cortex from the Ntsr1-eGFP line (*1*), axonal fibers entering the thalamus (th) (*2*), and cell body labeling from an Ntsr1-bacTRAP mouse (*3*). Scale bars, 1 mm (*A,B*); 200 µm (*C*).

or pharmacological treatments. The GENSAT database provides expression information for three developmental time points—embryonic, neonatal, and adult—for every strain available (Gong et al. 2003). The developmental and functional history of a gene is especially important when considering Cre lines because transgene expression is cumulative when crossed to constitutive reporter strains.

The level of fluorescence can vary between BAC eGFP reporter lines. This is most influenced by transgene copy number. Direct eGFP fluorescence may be difficult to visualize when only one copy is incorporated into the genome. Fluorescence intensity also varies with cellular size and shape. eGFP diffuses throughout the cytoplasm, thus, fluorescence from a cell with large complex processes (such as a corticospinal projecting pyramidal cell) may appear relatively dimmer than a cell that possesses the same level of expression but lacks complex cellular arborization.

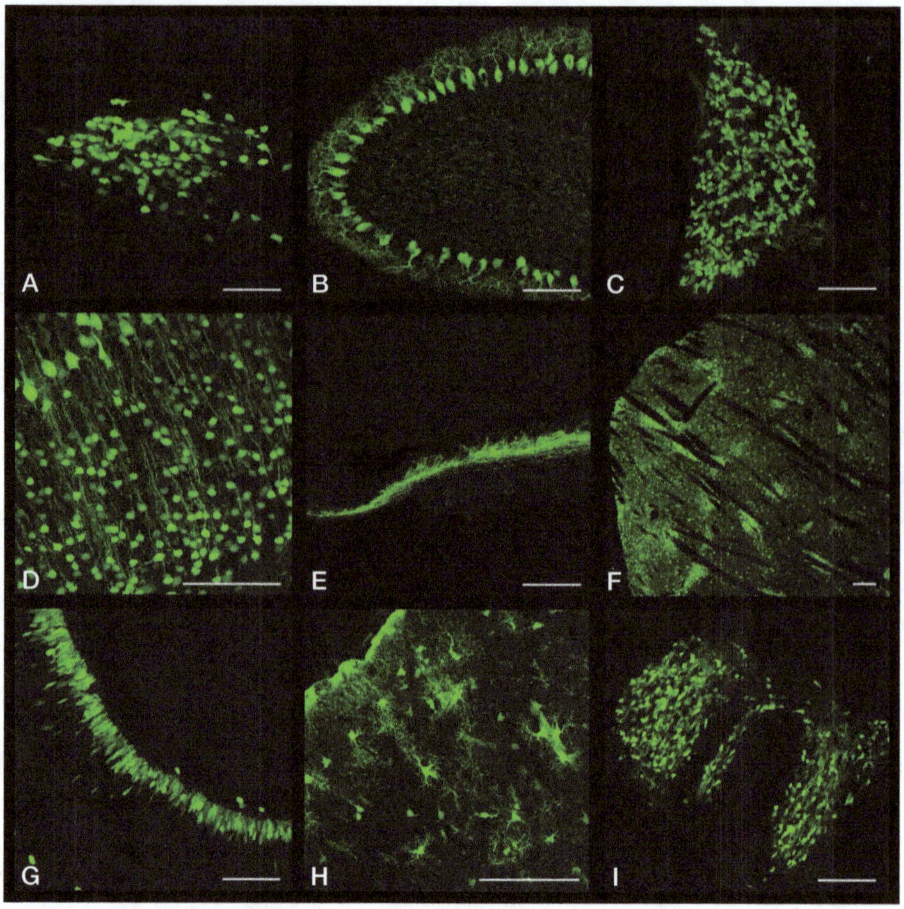

FIGURE 3. Confocal micrographs of direct epifluorescence obtained from tissue sections of the central and peripheral nervous system of nine eGFP BAC transgenic mice. (*A*) Cdh9-expressing neurons in the paraventricular nucleus of the hypothalamus of a P7 mouse. (*B*) Camk2a-expressing Purkinje neurons in the cerebellum of a P7 mouse. (*C*) Gal-expressing primary sensory neurons in the dorsal root ganglion of an E15.5 mouse. (*D*) Crym-expressing pyramidal neurons in the cerebral cortex of a P7 mouse. (*E*) Fluorescent fibers in the median eminence of a P7 Avp-eGFP mouse. (*F*) Gad2-expressing neurons in the caudate putamen of an adult mouse. (*G*) Slc18a2-expressing neurons in the retina of an E15.5 mouse. (*H*) Glud1-expressing astrocytes in the cerebral cortex of a P7 mouse. (*I*) Fxyd7-expressing neurons in the olfactory tubercle of a P7 mouse. Scale bars, 100 µm.

Although the stability of eGFP protein may differ from the endogenous gene product, the relative levels of expression will be the same; for example, if there is strong expression of the gene in cerebellum and weak expression in thalamus, the same will be true for eGFP expression (Heintz 2001; Gong et al. 2003). This is particularly important in applications in which eGFP fluorescence is to be used as the readout of endogenous gene transcription or developmental studies because the half-life of eGFP may be longer than that of the native protein.

THE GENSAT DATABASE

The goal of the GENSAT Project is to use BAC transgenic reporter mice to provide detailed expression maps of thousands of genes expressed in the CNS and to make the transgenic mouse lines publicly available to the scientific community (Gong et al. 2003). Although the majority of mouse lines have used eGFP as a reporter, GENSAT is providing a growing number of engineered Cre lines as well (Gong et al. 2007). A database of all mapped genes is available on the project's website at http://www.gensat.org. GENSAT expression data can also be accessed through the National Center for Biotechnology (NCBI) website at http://www.ncbi.nlm.nih.gov/projects/gensat.

There are a number of search options on the GENSAT website. The "Search" link in the main menu bar allows for a general search for genes and mouse lines using a number of key words. For example, the search terms "calcium channel" and "striatum" will return all of the genes for calcium channels with expression annotated in the striatum. Alternatively, specific genes can be found by entering a gene name or symbol in the "Search for Genes" link in the menu bar or using the drop-down menu to browse an alphabetized list of all the genes in the GENSAT database. The "Search Annotations" link provides a more comprehensive search option to find labeled cell populations. Here, a categorical search can be performed on annotations of each image in the database using a series of drop-down menus. The user can filter search results based on certain criteria including brain structure, fine structure (classically defined cell type), expression level, pattern of expression, or mouse age. All search results will return a list of genes with links to images for eGFP and Cre mouse lines as well as in situ data from the brain gene-expression map (BGEM; Magdaleno et al. 2006). Clicking the "Cre Lines" tab in the main menu bar will also allow the user to browse images and annotations for all available Cre lines.

The NCBI website also provides a number of search options. A general search can be performed by entering a gene name or symbol. In addition, a series of drop-down options allows for a more refined search by specifying age, type of data (bright field or confocal), plane of section, brain structure, cell type, or expression level. Data are returned as either a list of genes or an overview indicating the number of genes and images satisfying the search criteria with links to images.

Ordering Mouse Lines

Almost all eGFP and Cre mouse strains generated by the GENSAT Project are available through the Mutant Mouse Regional Resource Center (MMRRC). A link for purchasing each strain exists at the top of the image page when viewing on the GENSAT website. Alternatively, a list of all GENSAT strains deposited at the MMRRC is available at the MMRRC site (http://www.mmrrc.org). Most strains are cryopreserved and must be resuscitated before distribution.

REFERENCES

Antoch MP, Song EJ, Chang AM, Vitaterna MH, Zhao Y, Wilsbacher LD, Sangoram AM, King DP, Pinto LH, Takahashi JS. 1997. Functional identification of the mouse circadian clock gene by transgenic BAC rescue. *Cell* **89:** 655–667.

Bateup HS, Svenningsson P, Kuroiwa M, Gong S, Nishi A, Heintz N, Greengard P. 2008. Cell type–specific regulation of DARPP-32 phosphorylation by psychostimulant and antipsychotic drugs. *Nat Neurosci* **11:** 932–939.

Bishop GA, Tian JB, Stanke JJ, Fischer AJ, King JS. 2006. Evidence for the presence of the type 2 corticotropin releasing factor receptor in the rodent cerebellum. *J Neurosci Res* **84:** 1255–1269.

Bloch B, Popovici T, Le Guellec D, Normand E, Chouham S, Guitteny AF, Bohlen P. 1986. In situ hybridization histochemistry for the analysis of gene expression in the endocrine and central nervous system tissues: A 3-year experience. *J Neurosci Res* **16:** 183–200.

Doyle JP, Dougherty JD, Heiman M, Schmidt EF, Stevens TR, Ma G, Bupp S, Shrestha P, Shah RD, Doughty ML, et al. 2008. Application of a translational profiling approach for the comparative analysis of CNS cell types. *Cell* **135:** 749–762.

Gertler TS, Chan CS, Surmeier DJ. 2008. Dichotomous anatomical properties of adult striatal medium spiny neurons. *J Neurosci* **28:** 10814–10824.

Giraldo P, Montoliu L. 2001. Size matters: Use of YACs, BACs and PACs in transgenic animals. *Transgenic Res* **10:** 83–103.

Gong S, Yang XW, Li C, Heintz N. 2002. Highly efficient modification of bacterial artificial chromosomes (BACs) using novel shuttle vectors containing the R6Kγ origin of replication. *Genome Res* **12:** 1992–1998.

Gong S, Zheng C, Doughty ML, Losos K, Didkovsky N, Schambra UB, Nowak NJ, Joyner A, Leblanc G, Hatten ME, et al. 2003. A gene expression atlas of the central nervous system based on bacterial artificial chromosomes. *Nature* **425:** 917–925.

Gong S, Doughty M, Harbaugh CR, Cummins A, Hatten ME, Heintz N, Gerfen CR. 2007. Targeting Cre recombinase to specific neuron populations with bacterial artificial chromosome constructs. *J Neurosci* **27:** 9817–9823.

Gregory SG, Sekhon M, Schein J, Zhao S, Osoegawa K, Scott CE, Evans RS, Burridge PW, Cox TV, Fox CA, et al. 2002. A physical map of the mouse genome. *Nature* **418:** 743–750.

Groh A, Meyer HS, Schmidt EF, Heintz N, Sakmann B, Krieger P. 2010. Cell-type specific properties of pyramidal neurons in neocortex underlying a layout that is modifiable depending on the cortical area. *Cereb Cortex* **20:** 826–836.

Head K, Gong S, Joseph S, Wang C, Burkhardt T, Rossi MR, LaDuca J, Matsui S, Vaughan M, Hicks DG, et al. 2007. Defining the expression pattern of the *LGI1* gene in BAC transgenic mice. *Mamm Genome* **18:** 328–337.

Heiman M, Schaefer A, Gong S, Peterson JD, Day M, Ramsey KE, Suarez-Farinas M, Schwarz C, Stephan DA, Surmeier DJ, et al. 2008. A translational profiling approach for the molecular characterization of CNS cell types. *Cell* **135:** 738–748.

Heintz N. 2001. BAC to the future: The use of BAC transgenic mice for neuroscience research. *Nat Rev Neurosci* **2:** 861–870.

Kato K. 1990. A collection of cDNA clones with specific expression patterns in mouse brain. *Eur J Neurosci* **2:** 704–711.

Kriaucionis S, Heintz N. 2009. The nuclear DNA base 5-hydroxymethylcytosine is present in Purkinje neurons and the brain. *Science* **324:** 929–930.

Kuhar SG, Feng L, Vidan S, Ross ME, Hatten ME, Heintz N. 1993. Changing patterns of gene expression define four stages of cerebellar granule neuron differentiation. *Development* **117:** 97–104.

Lein ES, Hawrylycz MJ, Ao N, Ayres M, Bensinger A, Bernard A, Boe AF, Boguski MS, Brockway KS, Byrnes EJ, et al. 2007. Genome-wide atlas of gene expression in the adult mouse brain. *Nature* **445:** 168–176.

Madisen L, Zwingman TA, Sunkin SM, Oh SW, Zariwala HA, Gu H, Ng LL, Palmiter, RD Hawrylycz, MJ Jones, et al. 2010. A robust and high-throughput Cre reporting and characterization system for the whole mouse brain. *Nat Neurosci* **13:** 133–140.

Magdaleno S, Jensen P, Brumwell CL, Seal A, Lehman K, Asbury A, Cheung T, Cornelius T, Batten DM, Eden C, et al. 2006. BGEM: An in situ hybridization database of gene expression in the embryonic and adult mouse nervous system. *PLoS Biol* **4:** e86. doi:10.1371/journal.pbio.0040086.

McPherson JD, Marra M, Hillier L, Waterston RH, Chinwalla A, Wallis J, Sekhon M, Wylie K, Mardis ER, Wilson RK, et al. 2001. A physical map of the human genome. *Nature* **409:** 934–941.

O'Connor DH, Huber D, Svoboda K. 2009. Reverse engineering the mouse brain. *Nature* **461:** 923–929.

Osheroff H, Hatten ME. 2009. Gene expression profiling of preplate neurons destined for the subplate: Genes involved in transcription, axon extension, neurotransmitter regulation, steroid hormone signaling, and neuronal survival. *Cereb Cortex* **19:** i126–i134.

Partridge JG, Janssen MJ, Chou DY, Abe K, Zukowska Z, Vicini S. 2009. Excitatory and inhibitory synapses in neuropeptide Y–expressing striatal interneurons. *J Neurophysiol* **102:** 3038–3045.

Placantonakis DG, Tomishima MJ, Lafaille F, Desbordes SC, Jia F, Socci ND, Viale A, Lee H, Harrison N, Tabar V, et al. 2009. BAC transgenesis in human embryonic stem cells as a novel tool to define the human neural lineage. *Stem Cells* **27:** 521–532.

Ramon y Cajal S. 1899. *Histology of the nervous system of man and vertebrates.* Editorial Moya, Madrid.

Salero E, Hatten ME. 2007. Differentiation of ES cells into cerebellar neurons. *Proc Natl Acad Sci* **104:** 2997–3002.

Selimi F, Cristea IM, Heller E, Chait BT, Heintz N. 2009. Proteomic studies of a single CNS synapse type: The parallel fiber/Purkinje cell synapse. *PLoS Biol* **7:** e1000083. doi:10.1371/journal.pbio.1000083.

Valjent E, Bertran-Gonzalez J, Herve D, Fisone G, Girault JA. 2009. Looking BAC at striatal signaling: Cell-specific analysis in new transgenic mice. *Trends Neurosci* **32:** 538–547.

Visel A, Thaller C, Eichele G. 2004. GenePaint.org: An atlas of gene expression patterns in the mouse embryo. *Nucleic Acids Res* **32:** D552–D556.

Yang XW, Model P, Heintz N. 1997. Homologous recombination based modification in *Escherichia coli* and germline transmission in transgenic mice of a bacterial artificial chromosome. *Nat Biotechnol* **15:** 859–865.

Yang XW, Wynder C, Doughty ML, Heintz N. 1999. BAC-mediated gene-dosage analysis reveals a role for Zipro1 (Ru49/Zfp38) in progenitor cell proliferation in cerebellum and skin. *Nat Genet* **22:** 327–335.

13 Presynaptic Calcium Measurements Using Bulk Loading of Acetoxymethyl Indicators

Stephan D. Brenowitz[1] and Wade G. Regehr[2]

[1]*National Institute on Deafness and Other Communication Disorders, NIH, Bethesda, Maryland 20892;* [2]*Department of Neurobiology, Harvard Medical School, Boston, Massachusetts 02130*

ABSTRACT

We describe a method for labeling presynaptic terminals in mammalian brain slices by focal application of calcium indicators conjugated with acetoxymethyl esters. Fluorescence signals from populations of fibers labeled in this manner can be measured using a photodiode or photomultiplier tube. Considerations for selecting an appropriate indicator and determining the optimum loading conditions are discussed.

INTRODUCTION

Calcium ions entering presynaptic terminals regulate neurotransmitter release as well as multiple types of both short- and long-term synaptic plasticity (Zucker and Regehr 2002). To directly examine the role of calcium in synaptic transmission in the mammalian brain, it is necessary to measure presynaptic calcium levels and determine the time course of presynaptic calcium entry. However, the small size of presynaptic boutons in the mammalian brain (~1 μm) makes measurement of presynaptic calcium technically challenging. Classic studies measuring presynaptic calcium used the squid giant synapse and neuromuscular junction, which are large presynaptic structures that enabled direct calcium measurements with microspectrophotonic absorption and charge-coupled device (CCD) imaging, respectively. Because these structures are many orders of magnitudes larger than typical mammalian presynaptic boutons, different approaches have been used to measure presynaptic calcium in the mammalian central nervous system.

One approach that has proven useful in measuring presynaptic calcium signals in small presynaptic structures is the use of membrane-permeant indicators to label populations of presynaptic boutons (Regehr and Tank 1991). Rather than attempting to measure calcium levels in individual terminals, the strategy is to make an aggregate measurement from many terminals. These methods have been used for studies of presynaptic calcium in presynaptic terminals of cerebellar granule cells (Regehr and Atluri 1995) and hippocampal mossy fibers (Regehr et al. 1994). Many other synapses are also well suited to this method. The loading technique described in this chapter is not difficult and can provide calcium measurements with signal-to-noise ratios much greater than can be obtained with single-bouton measurements using two-photon microscopy. Also, problems associated with presynaptic rundown that can occur with whole-cell recordings of presynaptic neurons can be avoided with bulk loading because the cell contents are not dialyzed with pipette solution.

Membrane-permeant calcium indicators have been developed by conjugating the negatively charged dye molecule to acetoxymethyl (AM) esters (Tsien 1981). The AM-conjugated dye is uncharged and passes across the cell membrane. Once inside cells, the AM group is removed by endogenous esterases, releasing the ion-sensitive acid form of the indicator. AM-conjugated calcium indicators are available with a range of optical properties and calcium affinities, providing flexibility for measuring calcium in different concentration ranges and in combination with other fluorophores (Zhao et al. 1997).

Protocol

Loading Cerebellar Granule Cell Axons in Brain Slices with AM-Conjugated Calcium Indicators

This protocol describes a simple and reliable technique for measuring calcium transients in presynaptic boutons of parallel fibers in transverse cerebellar brain slices. A solution of membrane-permeant AM-conjugated calcium indicator is focally applied to the brain slice and allowed to equilibrate throughout the parallel fiber tract. Fibers are then stimulated with an extracellular electrode and fluorescence transients are measured from a location several hundred micrometers from the loading site.

IMAGING SETUP

The methods described here require an upright microscope equipped with epifluorescence optics and appropriate filter sets for the selected indicator. Illumination is provided by either a 150-W xenon bulb or a 100-W halogen bulb powered by a low-noise power supply. Fluorescence is measured with either a photodiode or a photomultiplier tube. The collected light is focused onto the surface of the detector by a lens placed in the image plane.

To optimize the signal-to-noise ratio, it is important to control both the area of fluorescence excitation and the region in which fluorescence emission is collected from the specimen. This can be performed with apertures placed in the excitation and emission light paths. Movement artifacts caused by opening the shutter can be minimized by using a remote light source coupled to the microscope by a light guide.

To monitor the placement of loading and suction pipettes, it is convenient to use a low-power air objective with a long working distance. For fluorescence measurements, it is desirable to use a high-power water-immersion objective with a high numerical aperture.

MATERIALS

CAUTION: See Appendix 6 for proper handling of materials marked with <!>.
See the end of the chapter for recipes for reagents marked with <R>.

Reagents

AM-conjugated calcium indicator (e.g., Fura-2 from Invitrogen)

See the Discussion section below for considerations regarding the selection of calcium indicators. AM-conjugated calcium indicators are available from Invitrogen in 50-µg ampoules, which is a convenient size for preparing a loading solution for short-term use.

Artificial cerebrospinal fluid (ACSF) <R>
Dimethylsulfoxide (DMSO; anhydrous) <!>
Fast Green (1% stock solution) <R>
Pluronic F-127

Equipment

Aquarium pump

Fluorescence excitation light source

Should have aperture in a conjugate specimen plane to enable adjustment of the field of illumination.

Microcentrifuge tube filters (e.g., Costar Spin-X; pore size 0.22 µm)

Micromanipulators (two; positioned on opposite sides of the specimen)

Microscope (upright)

Objectives, 5X air-immersion (numerical aperture [NA] 0.15) and 60X water-immersion [NA 0.90]

Photodiode or photomultiplier tube

Should have aperture in a conjugate image plane to control the exposure area.

Pipettes (loading and suction)

Tubing and valves

These are used to connect loading and suction pipettes to the aquarium pump and vacuum lines, respectively.

EXPERIMENTAL METHOD

Preparation of AM-Conjugated Calcium Indicator Loading Solution

1. Dissolve 50 µg of an AM-conjugated calcium indicator (e.g., Fura-2) in 20 µL of a solution of 25% pluronic F-127/75% DMSO.
2. Mix by vortexing for 30 sec and sonicate in an ice-filled bath for 2 min.
3. Add 400 µL of ACSF and 20 µL of 1% Fast Green to the vial.

   Fast Green is included in the labeling solution to allow visualization during loading.
4. Mix by vortexing for 30 sec, sonicate for 2 min, and filter using a microcentrifuge tube filter with 0.22-µm pore size. Use a freshly prepared solution and keep on ice to prevent premature deesterification.

Loading Granule Cell Presynaptic Terminals in Rodent Transverse Cerebellar Brain Slices

5. Place a brain slice in the recording chamber of the microscope and perfuse the slice with oxygenated artificial cerebrospinal fluid (Swanson and Contractor 2004).
6. Position a suction pipette with a 20–30 µm tip diameter above the surface of the slice.
7. Fill a loading pipette (tip diameter of 20–30 µm) with labeling solution (containing the AM-conjugated calcium indicator) and position near the suction pipette.

   An aquarium pump or picospritzer can be used to supply pressure for the loading pipette. To allow regulation of pressure in a very low range, we use branched tubing to connect the pressure source and pipette, with a valve at the end of one arm of the tubing to allow pressure to escape. When the valve is fully open, little or no pressure reaches the loading pipette, and partially closing the valve provides gentle pressure for loading.
8. Apply positive pressure to the loading pipette and suction to the other pipette. Using transmitted light to visualize the Fast Green present in the loading pipette, adjust the flow rate to maintain a rapidly flowing, well-confined stream of dye that is collected entirely by the suction pipette (Fig. 1).

   See *Troubleshooting*.
9. Lower the pipettes together near the surface of the slice.
10. Gently lower the loading pipette to the slice surface and place into the slice until the flow of solution from the pipette is reduced or stops entirely.

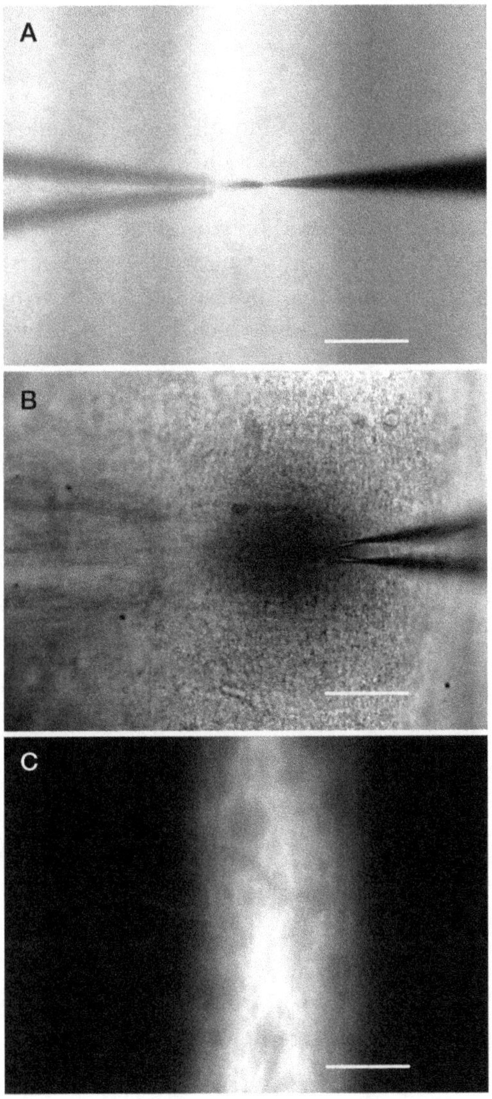

FIGURE 1. Procedure for loading granule cells in cerebellar brain slices with AM-conjugated calcium indicator. (*A*) Low-power (4x), transmitted-light image of the loading pipette (*right*) and the suction pipette (*left*) as they approach the brain slice. The plume of Fast Green expelled from the loading pipette is captured by the suction pipette. The transverse cerebellar slice (out of focus) is visible below. Scale bar, 200 µm. (*B*) Configuration of pipettes during loading with AM-conjugated calcium indicator. In this transmitted-light image, the tip of the loading pipette (on *right*) is placed ~15 µm below the surface of the slice and a small cloud of Fast Green is visible. To obtain focal labeling with the AM dye, the loading solution is collected by the suction pipette (on *left*, out of focus), which is located ~50 µm above the slice. Scale bar, 25 µm. (*C*) Labeled parallel fiber band. This epifluorescence image was taken shortly after loading ~200 µm from the loading site. Scale bar, 25 µm.

11. Gradually increase the pressure on the loading pipette until the flow resumes.

    Typical loading times range from 2 to 20 min depending on the preparation, the indicator used, and the desired degree of loading (see Discussion). The brain slice should be continuously superfused with oxygenated ACSF during loading to minimize undesirable background labeling. The total volume of loading solution applied to the brain slice varies depending on the loading time, pressure, and pipette tip diameter, but it is typically 1–5 µL.

12. After loading, leave for 1–2 h to allow the indicator to equilibrate within the cell.

Imaging Calcium Transients in Parallel Fiber Presynaptic Terminals

13. Position the stimulus electrode in the center of the labeled parallel fiber tract near the loading site.
14. Image parallel fibers 600–800 µm from the loading site in a 150-µm-diameter region.

    Near the loading site, both neurons and glia are indiscriminately labeled, and dye may enter intracellular organelles that do not respond to intracellular calcium changes. However, more distant sites are not exposed to the AM indicator and the only dye reaching these sites is the membrane impermeant acid form.

    See *Troubleshooting*.

TROUBLESHOOTING

Problem (Step 8): The loading pipette clogs.
Solution: This can be corrected by further sonication and filtration of the loading solution. If the problem persists, a loading pipette with a larger tip can help.

Problem (Step 14): No fluorescence change is observed on stimulation.
Solution: It is often necessary to reposition the stimulus electrode and adjust the position of the labeled fibers by moving the slice to optimize the fluorescence transient. Stimulation of a large proportion of labeled fibers will maximize the relative change in fluorescence on stimulation.

Problem (Step 14): The stimulus-evoked calcium transient is small or decays slowly.
Solution: This is a symptom of excessive loading. To reduce the level of calcium indicator in the labeled fibers, shorten the loading time or increase the dilution of calcium indicator in the loading solution.

Problem (Step 14): Indicators are not retained in presynaptic boutons at physiological temperatures.
Solution: For experiments conducted at room temperature (20°C–24°C), the acid form of the indicator is stable inside cells for hours. At elevated temperatures, the indicator is removed from neurons, limiting the time available for fluorescence measurements (Beierlein et al. 2004). It may therefore be necessary to use calcium indicators conjugated to high-molecular-weight dextrans (see Chapter 14).

DISCUSSION

Selecting the Appropriate Indicator

Although the same basic procedure is used for all experiments of this type, the choice of the indicator and the degree of loading depend on the specifics of the experiment. To illustrate this we consider the use of high- and low-affinity indicators for (1) measurement of the time course of presynaptic calcium transients and (2) detection of changes in the amplitude of presynaptic calcium influx.

Measurement of the Time Course of Presynaptic Calcium Transients

Calcium transients on the tens-of-milliseconds to tens-of-seconds timescale have been implicated in several use-dependent forms of plasticity that include facilitation, post-tetanic potentiation, and calcium-dependent recovery from depression. Accurate determination of the time course of calcium transients on these timescales relies on the correct choice of indicator (Regehr and Atluri 1995; Feller et al. 1996). A comparison of the time course of fluorescence transients measured with two different indicators is shown in Figure 2A. Magnesium Green, which has a dissociation constant (K_d) for calcium of 7 μM, is effective at detecting large calcium transients (Zhao et al. 1996) and provides an accurate measure of the time course of the calcium transient. In contrast, the high-affinity indicator Fura-2 (K_d ~ 200 nM) is sufficiently sensitive to slow the transient as a consequence of saturation of the indicator (Regehr and Atluri 1995).

Detection of Changes in the Amplitude of Presynaptic Calcium Influx

Such measurements are required in determining the relationship between calcium influx and transmitter release, as well as in studies of the contribution of presynaptic calcium-channel modulation to alterations in synaptic strength (Sabatini and Regehr 1995). Consider the measurement of fluorescence changes ($\Delta F/F$) for stimulus trains in which fibers are activated twice in rapid succession (Fig. 2B). With a low-affinity indicator, such as Magnesium Green, the second stimulus produces the same increment in fluorescence, but with Fura-2 the fluorescence transient produced by the second stimulus is smaller. The difference between the responses of these two indicators is a reflection of their different affinities. For low-affinity calcium-sensitive indicators, such as BTC (benzothiazole coumarin), Magnesium Green, MagFura-5, Furaptra, Fura-FF, and others, there is a roughly linear

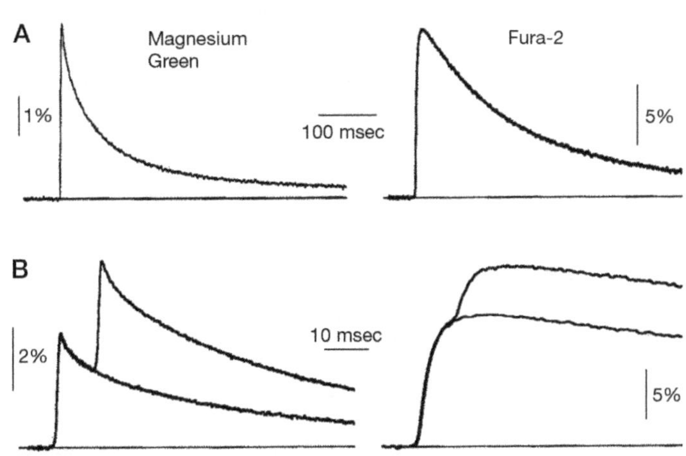

FIGURE 2. Detection of presynaptic calcium transients and changes in calcium influx in cerebellar parallel fibers. Magnesium Green (K_d for calcium = 7 μM) and Fura-2 (K_d for calcium = 200 nM) were used to measure the time course of $\Delta F/F$ signals. Fluorescence changes were evoked by one pulse (*A*) and by two pulses separated by 10 msec (*B*). $\Delta F/F$ signals for Fura-2 are inverted for clarity. Experiments were conducted in cerebellar brain slices from young rats at room temperature.

relationship between calcium entry and the resulting $\Delta F/F$ signals, and it is a simple matter to detect changes in calcium entry. Although some low-affinity indicators such as Furaptra and Magnesium Green are sensitive to magnesium as well as calcium, other low-affinity indicators are insensitive to magnesium (e.g., Fura-FF and Fluo-4FF). For high-affinity calcium indicators, such as Fura-2, Calcium Orange, and Calcium Crimson, there is a distortion of the calcium transient owing to saturation of the indicator. Despite the saturation of high-affinity indicators such as Fura-2, these indicators are useful for the studies of presynaptic calcium. Their sensitivity makes them well suited to detecting small elevations of calcium, and they also have the advantage that they are relatively insensitive to magnesium. Furthermore, their degree of saturation provides a way of quantifying calcium levels reached in presynaptic terminals and, with care, can be used to quantify changes in calcium entry (for a detailed description, see Regehr and Atluri 1995; Sabatini and Regehr 1995).

Determining the Appropriate Loading Conditions

Because the introduction of an indicator into a cell always alters the calcium dynamics to some extent, with a loading method such as described here, it is important to determine how much indicator has been introduced into the presynaptic terminal and what effect this might have on the parameter being measured. With high-affinity indicators, such as Fura-2, it is straightforward to use the degree of saturation as a measure of the extent of calcium buffering. For example, compare the measurements of fluorescence transients produced by one and two stimuli in Figure 3A. For a fiber tract that has a low concentration of Fura-2 (Fig. 3A, left), there is a smaller incremental increase in fluorescence for the second stimulus than for the first, indicating that the calcium transients are sufficiently large to begin to saturate the response of the indicator. In a fiber tract that has a higher concentration of Fura-2 (Fig. 3A, right), the size of the calcium transient is reduced by the increased buffer capacity, and there is less saturation of the indicator. Thus, for a high-affinity indicator, the degree of saturation is sensitive to the amount of indicator introduced into the presynaptic terminal. Changes in the degree of saturation can be used to quantify the amount of buffer introduced into the presynaptic terminal relative to the endogenous buffer capacity, using an approach similar to that described previously (Sabatini and Regehr 1995).

For low-affinity indicators, the extent of loading and the degree to which calcium signaling is distorted can be examined by measuring the time course of the calcium transient, as in Figure 3B. Lightly labeled fiber tracts provide an accurate measure of the calcium transient, whereas transients in heavily labeled tracts are significantly slowed. The degree of slowing reflects the extent to which the buffer capacity of the presynaptic terminal has been increased by the indicator (Neher and Augustine 1992; Tank et al. 1995). Because the loading technique described here can be used to measure fluorescence in a population of presynaptic terminals, an advantage of this approach is that low intracellular concentrations of indicator can provide a fluorescence signal with minimal perturbation of the endogenous calcium transient and yet retain excellent signal-to-noise characteristics.

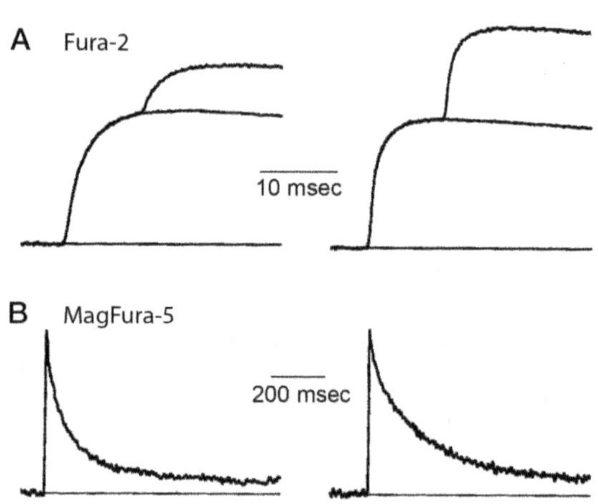

FIGURE 3. Assessing the degree of indicator loading. The effects of loading time are shown for Fura-2 (*A*) and MagFura-5 (*B*) fluorescence transients. Presynaptic terminals were loaded for 5 min (*left*) and 30 min (*right*). Experiments were conducted at room temperature in cerebellar parallel fibers from young rats. Peak changes in $\Delta F/F$ produced by the first stimulus are in *A*, 4.85% (*left*) and 5.9% (*right*), and the single stimulus in *B*, 0.96% (*left*) and 0.97% (*right*).

CONCLUSION

The loading procedure described here has proven to be a useful tool in the study of calcium's role in synaptic transmission in the mammalian brain (Regehr et al. 1994; Wu and Saggau 1994; Mintz et al. 1995; Atluri and Regehr 1996; Dittman and Regehr 1996; Sabatini and Regehr 1996, 1997, 1998; Chen and Regehr 1997). Provided that care is taken in the choice of indicator and loading conditions, it is possible to measure the time course of presynaptic calcium transients, to quantify changes in calcium influx, and to measure the time course of presynaptic calcium entry.

RECIPES

CAUTION: See Appendix 6 for proper handling of materials marked with <!>.
Recipes for reagents marked with <R> are included in this list.

Artificial Cerebrospinal Fluid (ACSF)

Reagent	Final concentration
NaCl	125 mM
$NaHCO_3$	26 mM
NaH_2PO_4	1.25 mM
KCl	2.5 mM
$MgCl_2$	1.0 mM
$CaCl_2$ <!>	2.0 mM
Glucose	25 mM

Bubble 95% O_2 <!> and 5% CO_2 <!> through the solution.

Fast Green (1% stock solution)

Dissolve 100 mg of Fast Green <!> in 10 mL of H_2O. Vortex for 30 sec and sonicate for 2 min.

ACKNOWLEDGMENTS

This work was supported by the Intramural Program of the NIDCD (SDB) and by NIH R01DA024090 and R37NS032405 (WGR).

REFERENCES

Atluri PP, Regehr WG. 1996. Determinants of the time course of facilitation at the granule cell to Purkinje cell synapse. *J Neurosci* **16:** 5661–5671.

Beierlein M, Gee KR, Martin VV, Regehr WG. 2004. Presynaptic calcium measurements at physiological temperatures using a new class of dextran-conjugated indicators. *J Neurophysiol* **92:** 591–599.

Chen C, Regehr WG. 1997. The mechanism of cAMP-mediated enhancement at a cerebellar synapse. *J Neurosci* **17:** 8687–8694.

Dittman JS, Regehr WG. 1996. Contributions of calcium-dependent and calcium-independent mechanisms to presynaptic inhibition at a cerebellar synapse. *J Neurosci* **16:** 1623–1633.

Feller MB, Delaney KR, Tank DW. 1996. Presynaptic calcium dynamics at the frog retinotectal synapse. *J Neurophysiol* **76:** 381–400.

Mintz IM, Sabatini BL, Regehr WG. 1995. Calcium control of transmitter release at a cerebellar synapse. *Neuron* **15:** 675–688.

Neher E, Augustine GJ. 1992. Calcium gradients and buffers in bovine chromaffin cells. *J Physiol* **450:** 273–301.

Regehr WG, Atluri PP. 1995. Calcium transients in cerebellar granule cell presynaptic terminals. *Biophys J* **68:** 2156–2170.

Regehr WG, Tank DW. 1991. Selective fura-2 loading of presynaptic terminals and nerve cell processes by local perfusion in mammalian brain slice. *J Neurosci Methods* **37:** 111–119.

Regehr WG, Delaney KR, Tank DW. 1994. The role of presynaptic calcium in short-term enhancement at the hippocampal mossy fiber synapse. *J Neurosci* **14:** 523–537.

Sabatini BL, Regehr WG. 1995. Detecting changes in calcium influx which contribute to synaptic modulation in mammalian brain slice. *Neuropharmacology* **34:** 1453–1467.

Sabatini BL, Regehr WG. 1996. Timing of neurotransmission at fast synapses in the mammalian brain. *Nature* **384:** 170–172.

Sabatini BL, Regehr WG. 1997. Control of neurotransmitter release by presynaptic waveform at the granule cell to Purkinje cell synapse. *J Neurosci* **17:** 3425–3435.

Sabatini BL, Regehr WG. 1998. Optical measurement of presynaptic calcium currents. *Biophys J* **74:** 1549–1563.

Swanson GT, Contractor A. 2004. Recording in the cerebellar slice. *Curr Protoc Neurosci* **2004:** 6.18.1–6.18.14.

Tank DW, Regehr WG, Delaney KR. 1995. A quantitative analysis of presynaptic calcium dynamics that contribute to short-term enhancement. *J Neurosci* **15:** 7940–7952.

Tsien RY. 1981. A non-disruptive technique for loading calcium buffers and indicators into cells. *Nature* **290:** 527–528.

Wu LG, Saggau P. 1994. Adenosine inhibits evoked synaptic transmission primarily by reducing presynaptic calcium influx in area CA1 of hippocampus. *Neuron* **12:** 1139–1148.

Zhao M, Hollingworth S, Baylor SM. 1996. Properties of tri- and tetracarboxylate Ca^{2+} indicators in frog skeletal muscle fibers. *Biophys J* **70:** 896–916.

Zhao M, Hollingworth S, Baylor SM. 1997. AM-loading of fluorescent Ca^{2+} indicators into intact single fibers of frog muscle. *Biophys J* **72:** 2736–2747.

Zucker RS, Regehr WG. 2002. Short-term synaptic plasticity. *Annu Rev Physiol* **64:** 355–405.

14 Imaging Presynaptic Calcium Transients Using Dextran-Conjugated Indicators

Stephan D. Brenowitz[1] and Wade G. Regehr[2]

[1]*National Institute on Deafness and Other Communication Disorders, NIH, Bethesda, Maryland 20892;* [2]*Department of Neurobiology, Harvard Medical School, Boston, Massachusetts 02130*

ABSTRACT

Dextran-conjugated calcium indicators are stably retained within neurons. As a result they are well suited to measuring presynaptic calcium at physiological temperatures. In addition, dextran indicators can be used to label neurons and their presynaptic boutons in vivo. This has allowed measurements of calcium in the presynaptic boutons of projection fibers that cannot be stably loaded with other types of indicators. In this chapter, we describe in vivo loading and imaging of presynaptic calcium in long-range axonal projections. We also discuss important considerations for using dextran-conjugated indicators to measure presynaptic calcium.

INTRODUCTION

Calcium ions entering a nerve terminal evoke neurotransmitter release and regulate several forms of synaptic plasticity (Zucker and Regehr 2002). Many critical determinants of presynaptic function are temperature dependent, such as calcium channel gating kinetics, calcium extrusion mechanisms, and endogenous buffering properties. To understand the roles of calcium in presynaptic function it is therefore important to perform calcium measurements at physiological temperature.

Unfortunately, calcium indicators in their widely used free-acid form have limitations with regard to measuring presynaptic calcium at physiological temperatures. Low-molecular-weight, acid-based indicators are extruded from neurons in a temperature-sensitive manner (Roe et al. 1990). For indicators that are loaded into a cell with a pipette, the continual replenishment of indicator from the pipette can offset the indicator that escapes the cell. For the cell-permeant acetoxymethyl (AM)-conjugated indicators, the acid form is released following cleavage of the AM group by intracellular esterases (Tsien 1981). Following AM loading, there is a limited amount of indicator within the cell and the temperature-dependent extrusion of the indicator can be particularly pronounced for fine neuronal processes (Beierlein et al. 2004). This greatly limits the experimental time period during which calcium measurements can be made. To address this limitation,

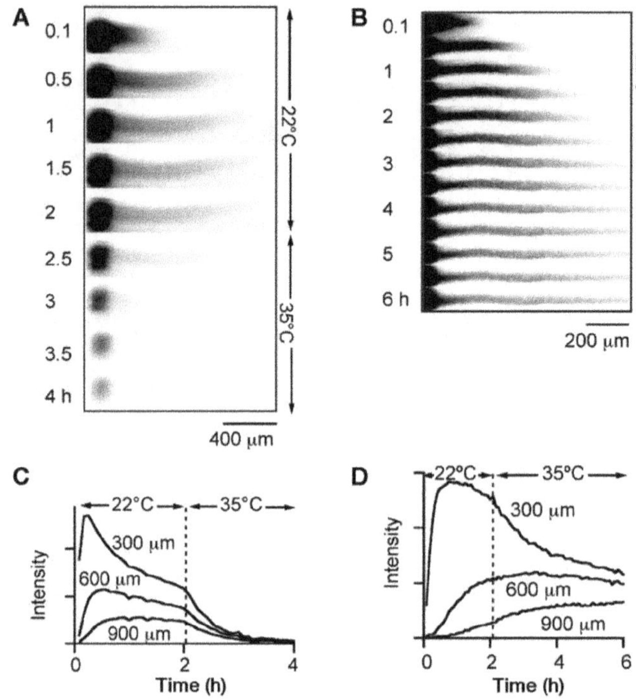

FIGURE 1. Effects of temperature on loading with AM- and dextran-conjugated indicators. (*A*,*B*) Low-magnification images were taken of the parallel fiber band in a transverse cerebellar slice after focal loading of (*A*) MagFura-5 AM or (*B*) Texas Red dextran (MW 10,000) for 5 min at room temperature. Image acquisition times relative to the start of loading and incubation temperatures are indicated. (*C*) Fluorescence intensities as a function of time are shown for locations 300, 600, and 900 μm from the loading site for a slice loaded with MagFura-5 AM. Although some dye extrusion is evident at room temperature, the rate of dye loss greatly accelerates when the temperature is shifted to 35°C. (*D*) Fluorescence intensities are plotted as in C for a slice loaded with Texas Red dextran (MW 10,000). The data show that dextran-conjugated indicators are not extruded from parallel fibers at elevated temperatures. (Modified, with permission, from Beierlein et al. 2004.)

dextran-conjugated calcium indicators have been developed; these molecules are stably retained in neurons at physiological temperature for prolonged periods of time and have properties favorable for the measurement of presynaptic calcium (Fig. 1) (Beierlein et al. 2004; Martin et al. 2004).

Here we describe an approach for measuring presynaptic calcium using calcium-sensitive indicators that have been conjugated to dextrans. As mentioned, these high-molecular-weight polysaccharides are retained within neurons for prolonged periods at physiological temperature (Nance and Burns 1990; Kreitzer et al. 2000; Beierlein et al. 2004). This property makes dextran-conjugated indicators well suited for measuring presynaptic calcium levels at physiological temperatures. As with any calcium indicator, dextran-conjugated indicators compete with endogenous calcium buffers and can perturb intracellular calcium transients when the added buffer capacity of the indicator is significant relative to the endogenous calcium buffering capacity of the neuron (see Chapter 36). Because of their high molecular weights (3–10 kDa), however, dextran-conjugated indicators are less mobile and may therefore have less effect on localized calcium microdomains than the low-molecular-weight acid forms of indicators.

Because of their stable retention within neurons, dextran dyes can be injected into the brain of an intact animal with a stereotaxic instrument to label specific populations of neurons and their associated axons and presynaptic boutons. It is not known how hydrophilic dextran-conjugated dyes are taken up by neurons, although mechanisms such as pinocytosis or endocytosis and membrane damage near the injection site may all play roles. The application of some compounds, such as the nonionic detergent Triton X-100, may enhance loading by locally permeabilizing membranes (Helenius and Simons 1975; van de Ven et al. 2009). Following stereotaxic injection, the dye is taken up nonspecifcially by neurons in the vicinity of the injection site, and because of its long-term intra-

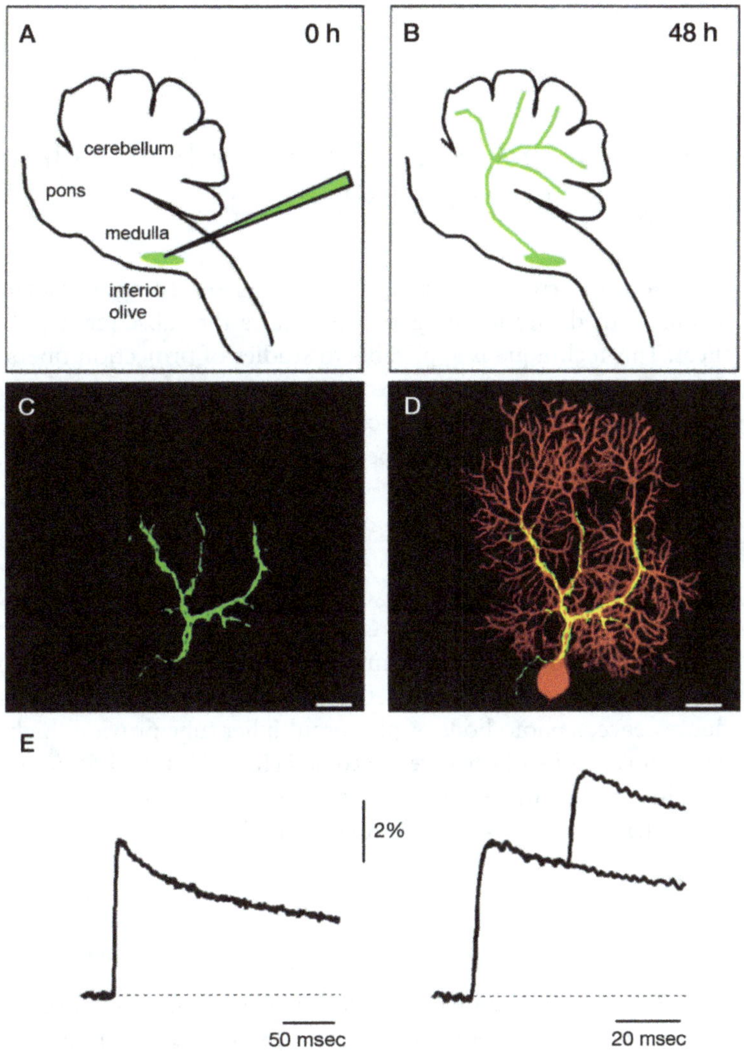

FIGURE 2. In vivo loading of climbing fibers with dextran-conjugated indicators. This diagram illustrates the in vivo loading method. (*A*) Dextran-conjugated indicator is injected into the inferior olive of a juvenile mouse. (*B*) Dye is taken up by neurons near the injection site and labels fibers that project to the cerebellar cortex. (*C*) Confocal image of a climbing fiber labeled with Calcium Green dextran from a sagittal cerebellar slice. Slices were prepared 2–3 d after stereotaxic injection. (*D*) Composite image of the labeled climbing fiber and the postsynaptic Purkinje cell loaded with Alexa Fluor 568 hydrazide. Scale bar, 20 µm. (*E*) Stimulus-evoked fluorescent transient from a Calcium Green dextran–labeled climbing fiber in response to one (*left*) and two (*right*) stimuli. (Modified, with permission of Elsevier, from Kreitzer et al. 2000.)

cellular stability, the dextran indicator can diffuse throughout neuronal processes over periods from hours to days. Using this method, axonal projections extending several millimeters from the injection site can be labeled. This approach permits calcium imaging in presynaptic terminals of projection fibers present in brain slices subsequently prepared from these animals. This method has been applied to measurements of presynaptic calcium in cerebellar climbing fibers that originate in the inferior olive (Fig. 2) (Kreitzer et al. 2000; Kreitzer and Regehr 2001). In vivo injection of dextran-conjugated calcium indicators has also been used in studies of presynaptic calcium in several systems including *Xenopus* tectum (Edwards and Cline 1999), spinal cord of lamprey (McPherson et al. 1997) and chick (O'Donovan et al. 1993), zebrafish hindbrain (Takahashi et al. 2002), and the olfactory bulb in frog (Delaney et al. 2001) and mouse (Wachowiak and Cohen 2001). The in vivo loading technique described here allows calcium measurements in presynaptic terminals that are not suitable for loading with other techniques and is widely applicable to synapses throughout the brain.

Protocol

Presynaptic Imaging of Projection Fibers by In Vivo Injection of Dextran-Conjugated Calcium Indicators

This protocol describes a technique for in vivo loading of the climbing fiber projection to the cerebellum with dextran-conjugated indicators for subsequent presynaptic calcium imaging in brain slices. This technique is applicable to studies of projection fibers in many species from which brain slices can be prepared. The dextran indicator is injected into the inferior olive using a stereotaxic device. After a period of 1–3 d, cerebellar slices are prepared and presynaptic calcium transients are measured at physiological temperature in labeled climbing fibers.

IMAGING SETUP

For presynaptic calcium measurements in brain slices, an upright microscope with standard epifluorescence optics is used. Water-immersion objectives with high numerical aperture (NA) are important for efficient visualization and detection of fluorescence (e.g., Olympus 60x, 0.9 NA). For detection of fluorescence, a photodiode or photomultiplier tube provides high temporal resolution and a high signal-to-noise ratio. Alternatively, a cooled charge-coupled device (CCD) camera may also be used. The signal-to-noise ratio of the fluorescence signal can be optimized by positioning an aperture in the excitation light path to restrict illumination to the region of the labeled fiber of interest.

MATERIALS

CAUTION: See Appendix 6 for proper handling of materials marked with <!>.
See the end of the chapter for recipes for reagents marked with <R>.

Reagents

Dextran-conjugated calcium indicator

See the Discussion section below for considerations regarding the selection of calcium indicators.

Fast Green (1% stock solution) <R>

Mineral oil

Triton X-100 <!>

Equipment

Fluorescence excitation light source

Should have aperture in an image plane conjugate with the specimen to enable adjustment of the field of illumination. The timing of illumination is restricted with a computer-controlled shutter.

Injection apparatus (use a 1-µL Hamilton syringe or an oocyte injector [e.g., Drummond Scientific Nanoject])

Injection pipette

Micropipettes with a tip diameter of 20–50 µm should be attached to the Hamilton syringe or oocyte injection apparatus.

Micromanipulator (for positioning the stimulus electrode)

Objectives, 5x air (NA 0.15) and 60x water-immersion (NA 0.90)

Photodiode or photomultiplier tube

It should have its aperture in a conjugate image plane to restrict light detection from the region of interest.

Stereotaxic instrument (with a fine micromanipulator for positioning the injection apparatus)
Stimulus isolation unit (e.g., World Precision Instruments A365)
Upright microscope

EXPERIMENTAL METHOD

Preparation of Loading Solution

1. For indicators with sufficient fluorescence for visualization at resting levels of calcium (e.g., Calcium Green dextran), combine 5 μL of a 20% stock solution of dextran-conjugated indicator and 5 μL of H_2O.

   To prepare the stock solution of dextran-conjugated indicators, add 250 μl H_2O to a 5-mg vial of the indicator. Vortex for 30 sec and sonicate for 1 min. Several dextran-conjugated calcium indicators have extremely low resting fluorescence (e.g., Fluo-4 dextran and Rhod dextran) and must be combined with a calcium-insensitive dextran-conjugated fluorophore to visualize labeled fibers. In this case, combine 5 μL of a 20% stock solution of the calcium indicator with 5 μL of a 20% solution of a calcium-insensitive dye. For green calcium indicators, dextran conjugates of Alexa-568, Alexa-594, or Texas Red can be used. For red calcium indicators, use Alexa-488 dextran.

In Vivo Injection of Dextran Indicators

2. Prepare an injection pipette by breaking the tip of a micropipette so that it has an open tip diameter of 20–50 μm.
3. Place an anesthetized animal in the stereotaxic instrument.
4. Attach the injection pipette to the injection apparatus (Hamilton syringe or oocyte injector) and attach the injection apparatus to the stereotaxic instrument.
5. Backfill the injector with mineral oil and fill the tip with 1–3 μL of injection solution.
6. Advance the injection pipette into the brain and position at the desired loading site.
7. Deliver a volume of 0.2–0.3 μL of dextran injection solution.
8. Leave the syringe in place for 5–10 min after injection to prevent leakage of dye from the electrode track.
9. Slowly withdraw the pipette and allow the animal to recover from surgery.

Preparation of Brain Slices

10. After allowing sufficient time for diffusion of dextrans from the injection site to presynaptic terminals of projection fibers, prepare brain slices from the experimental animal.

    The time required for indicator diffusion must be empirically determined. For labeling of cerebellar climbing fibers by injections into the inferior olive of a juvenile mouse, 1–3 d is sufficient.

Presynaptic Imaging

11. Identify a labeled climbing fiber and restrict the area of fluorescence illumination to the region of interest.

    See *Troubleshooting*.

12. Stimulate the climbing fibers using a glass electrode placed near the ascending axon in the granule cell layer. Gate the fluorescence illumination and measure the stimulus-evoked change in fluorescence.

    See *Troubleshooting*.

TROUBLESHOOTING

Problem (Step 11): No dye is present in climbing fibers of interest.
Solution: First verify that the injection has been made into the correct site. This can be performed by adding Fast Green (final concentration 0.1%) to the injection solution. Perform an injection and immediately section the brain to determine the location of the loading site. This method can provide useful feedback for determining the correct stereotaxic coordinates for in vivo injections. If the dye has been injected into the correct site but no labeled fibers are visible, try allowing more time for the dextrans to diffuse. If fibers are still not visible after longer postoperative survival times, add Triton X-100 (final concentration 0.1%) to the loading solution.

Problem (Step 12): Stimulus-evoked calcium transients have small amplitudes or slow decay times.
Solution: This is a symptom of overloading. Because of the variable uptake of dextran dyes into neurons, there will often be a wide range of indicator concentrations in labeled fibers from a single animal. Often the basal fluorescence of the calcium indicator, or the fluorescence of a colabeled fluorophore, can be used to select fibers that are appropriately labeled (enough indicator to provide a reasonable signal-to-noise ratio recording without significantly perturbing calcium signaling). However, if signals are consistently small or slow, increase the dilution of the dextran-conjugated calcium indicator in the loading solution.

DISCUSSION

One of the primary advantages of using dextran-conjugated indicators is their stability inside cells at physiological temperatures. The in vivo injection approach described in this chapter has shown that the dextrans are stable for a period of days, allowing presynaptic imaging of projection fibers that are not suitable for loading by other methods.

A detailed analysis of the temperature dependence of indicator stability for AM- and dextran-conjugated indicators has been performed by Beierlein et al. (2004). These experiments, performed by loading parallel fiber bands in transverse cerebellar slices, provide quantitative information about indicator diffusion and extrusion that cannot be obtained with in vivo injections (Fig. 1). After focal application of the low-affinity calcium indicator MagFura-5 AM to parallel fibers at room temperature, steady-state indicator concentrations were reached in <1 h at distances of 600–900 μm from the loading site, and indicator levels were stable for at least 4 h at room temperature. However, when the temperature was shifted to 35°C, the indicator was rapidly extruded and about half of the indicator was lost after 20 min. Such rapid extrusion severely limits the time during which presynaptic calcium measurements can be made at elevated temperatures in brain slices. Moreover, in vivo loading of projection fibers with AM indicators is impractical because of the time required for diffusion. It should be pointed out that extrusion appears to be most rapid in small fibers such as the parallel fibers, and acid forms of indicators may be extruded more slowly from larger structures such as cell bodies, perhaps because of their lower surface area to volume ratio.

In sharp contrast to the rapid loss of indicator in AM-loaded fibers at elevated temperature, indicator levels in dextran-loaded parallel fibers were stable for several hours of incubation at 35°C with no evidence of extrusion (Beierlein et al. 2004). Other studies in which dextrans were injected in vivo also establish that the dextran indicators are retained in neurons for days (Fritzsch et al. 1990; O'Donovan et al. 1993; Rogers et al. 2006). Although diffusion of the dextran dyes along axons proceeds more slowly at room temperature than in fibers loaded with AM dyes, at a temperature of 35°C diffusion of the dextrans was more rapid. These findings illustrate that dextran-conjugated indicators are stably retained in parallel fiber axons, a property that makes them extremely useful for presynaptic calcium measurements at physiological temperatures. No detrimental effects of dextran loading are apparent. Synaptic responses evoked by dextran-labeled fibers are identical to responses from unlabeled fibers (Kreitzer et al. 2000), and in parallel fibers calcium transients have similar kinetics with long-term dextran loading compared to short-term loading with AM indicators (Beierlein et al. 2004).

Besides their improved stability at physiological temperatures, dextran-conjugated indicators are not prone to compartmentalization within internal organelles, in contrast to the more commonly used acid-based indicators that with time often enter intracellular compartments. Indicator compartmentalization is problematic for quantification of fluorescence signals because dye that has entered an intracellular compartment is not responsive to changes in cytosolic calcium, thus increasing the fluorescence background (Steinberg et al. 1987; Blatter and Wier 1990; Roe et al. 1990; Zhao et al. 1997). A further advantage of dextran dyes is that their affinity for calcium is not altered by the intracellular environment (Luby-Phelps et al. 1986). This stands in contrast to indicators in the acid form that can interact with proteins in the cytoplasm, thereby decreasing the calcium affinity of the indicator by 3–5-fold (Konishi et al. 1988).

Properties of Dextran-Conjugated Indicators

In comparison with AM-conjugated indicators, there are fewer calcium indicators available in dextran-conjugated form. However, several are well suited to presynaptic calcium measurements. Dextran-conjugated indicators that emit in the green (Fluo-4, Calcium Green, Oregon Green 488 BAPTA-1) and red (Rhod-2 and X-Rhod-1) portion of the spectrum are commercially available from Invitrogen. Also available is Fura-2 dextran, which is excited by ultraviolet wavelengths and allows for ratiometric measurements.

One of the primary considerations in selecting an indicator is its affinity for calcium, and dextran indicators are available with dissociation constants that range from 240 nM to 5 µM. For both Fluo-4 dextran and Rhod-2 dextran, high- and low-affinity versions are available. The low-affinity dextran conjugates of both Fluo-4 and Rhod-2 are suitable for linear measurements of presynaptic calcium transients in response to action potential trains and report rapid fluorescence decay times (Kreitzer et al. 2000; Beierlein et al. 2004). In the case of the dextran-conjugated calcium indicators, the calcium affinity can vary among different lots of the same dye, so it is important to obtain calibration data for each batch of indicator.

The red dextran indicators are particularly advantageous in tissues that show high levels of autofluorescence because there is generally lower tissue autofluorescence in the 550–600-nm wavelength range in which they are excited. Also, the availability of both red and green indicators allows for the possibility of simultaneous recording from different populations of labeled cells. Moreover, red dextrans can be used to make fluorescence measurements in neurons expressing green fluorescent protein because their excitation and emission wavelengths are sufficiently well separated that the two fluorophores can be independently imaged.

One property of many of the dextran-conjugated calcium indicators is that they have very low levels of fluorescence when they are not bound to calcium. For these indicators, particularly those with low affinity, the fluorescent signal at resting levels of calcium can be difficult to detect, posing difficulties for experimental identification of labeled fibers. For this reason, it is often desirable to label with two dextran-conjugated fluorophores, one calcium-sensitive and the other calcium-insensitive and of a different color. There are a variety of Alexa dextran conjugates available that are suitable for this purpose such as Alexa-488 dextran for use with red calcium indicators and Alexa-594 dextran or Texas Red dextran for use with green calcium indicators.

CONCLUSION

Dextran-conjugated indicators allow presynaptic calcium measurements to be made at physiological temperatures and in the presynaptic boutons of projection neurons. Different color indicators are available with a range of calcium affinities. In vivo injection of dextran-conjugated dyes provides a powerful technique that can be used to measure calcium in presynaptic terminals of projection neurons.

RECIPE

CAUTION: See Appendix 6 for proper handling of materials marked with <!>.
Recipes for reagents marked with <R> are included in this list.

Fast Green (1% Stock Solution)

Dissolve 100 mg of Fast Green <!> in 10 mL of H_2O. Vortex for 30 sec and sonicate for 2 min.

ACKNOWLEDGMENTS

This work was supported by the Intramural Program of the NIDCD (SDB) and by NIH R01DA024090 and R37NS032405 (WGR).

REFERENCES

Beierlein M, Gee KR, Martin V, Regehr WG. 2004. Presynaptic calcium measurements at physiological temperatures using a new class of dextran-conjugated indicators. *J Neurophysiol* **92:** 591–599.

Blatter L, Wier WG. 1990. Intracellular diffusion, binding, and compartmentalization of the fluorescent calcium indicators indo-1 and fura-2. *Biophys J* **58:** 1491–1499.

Delaney K, Davison I, Denk W. 2001. Odour-evoked [Ca^{2+}] transients in mitral cell dendrites of frog olfactory glomeruli. *Eur J Neurosci* **13:** 1658–1672.

Edwards JA, Cline HT. 1999. Light-induced calcium influx into retinal axons is regulated by presynaptic nicotinic acetylcholine receptor activity in vivo. *J Neurophysiol* **81:** 895–907.

Fritzsch B, Sonntag R, Dubuc R, Ohta Y, Grillner S. 1990. Organization of the six motor nuclei innervating the ocular muscles in lamprey. *J Comp Neurol* **294:** 491–506.

Helenius A, Simons K. 1975. Solubilization of membranes by detergents. *Biochim Biophys Acta* **415:** 29–79.

Konishi M, Olson A, Hollingworth S, Baylor SM. 1988. Myoplasmic binding of fura-2 investigated by steady-state fluorescence and absorbance measurements. *Biophys J* **54:** 1089–1104.

Kreitzer AC, Regehr WG. 2001. Retrograde inhibition of presynaptic calcium influx by endogenous cannabinoids at excitatory synapses onto Purkinje cells. *Neuron* **29:** 717–727.

Kreitzer AC, Gee KR, Archer EA, Regehr WG. 2000. Monitoring presynaptic calcium dynamics in projection fibers by in vivo loading of a novel calcium indicator. *Neuron* **27:** 25–32.

Luby-Phelps K, Taylor DL, Lanni F. 1986. Probing the structure of cytoplasm. *J Cell Biol* **102:** 2015–2022.

Martin VV, Beierlein M, Morgan JL, Rothe A, Gee KR. 2004. Novel fluo-4 analogs for fluorescent calcium measurements. *Cell Calcium* **36:** 509–514.

McPherson DR, McClellan AD, O'Donovan MJ. 1997. Optical imaging of neuronal activity in tissue labeled by retrograde transport of Calcium Green dextran. *Brain Res Brain Res Protoc* **1:** 157–164.

Nance DM, Burns J. 1990. Fluorescent dextrans as sensitive anterograde neuroanatomical tracers: Applications and pitfalls. *Brain Res Bull* **25(1):** 139–145.

O'Donovan MJ, Ho S, Sholomenko G, Yee W. 1993. Real-time imaging of neurons retrogradely and anterogradely labelled with calcium-sensitive dyes. *J Neurosci Methods* **46:** 91–106.

Roe MW, Lemasters JJ, Herman B. 1990. Assessment of Fura-2 for measurements of cytosolic free calcium. *Cell Calcium* **11:** 63–73.

Rogers RC, Nasse JS, Hermann GE. 2006. Live-cell imaging methods for the study of vagal afferents within the nucleus of the solitary tract. *J Neurosci Methods* **150:** 47–58.

Steinberg SF, Bilezikian JP, Al-Awqati Q. 1987. Fura-2 fluorescence is localized to mitochondria in endothelial cells. *Am J Physiol* **253:** C744–C747.

Takahashi M, Narushima M, Oda Y. 2002. In vivo imaging of functional inhibitory networks on the mauthner cell of larval zebrafish. *J Neurosci* **22:** 3929–3938.

Tsien RY. 1981. A non-disruptive technique for loading calcium buffers and indicators into cells. *Nature* **290:** 527–528.

van de Ven AL, Adler-Storthz K, Richards-Kortum R. 2009. Delivery of optical contrast agents using Triton-X100, part 1: Reversible permeabilization of live cells for intracellular labeling. *J Biomed Opt* **14:** 021012.

Wachowiak M, Cohen LB. 2001. Representation of odorants by receptor neuron input to the mouse olfactory bulb. *Neuron* **32:** 723–735.

Zhao M, Hollingworth S, Baylor SM. 1997. AM-loading of fluorescent Ca^{2+} indicators into intact single fibers of frog muscle. *Biophys J* **72:** 2736–2747.

Zucker RS, Regehr WG. 2002. Short-term synaptic plasticity. *Annu Rev Physiol* **64:** 355–405.

15 Confocal Spot Detection of Presynaptic Ca^{2+} Domains

David DiGregorio

Unit of Dynamic Neuronal Imaging, Department of Neuroscience, Institut Pasteur 25, 75724 Paris Cedex 15, France

ABSTRACT

Advances in rapid imaging of fluorescence probes are crucial to the study of neuronal signaling and information processing within the nervous system. Fluorescence indicators of cellular signaling (e.g., Ca^{2+} or voltage) can report changes on the millisecond timescale. Such signals can be localized to submicrometer structures such as dendritic spines, or within diffusion-delimited regions such as Ca^{2+} microdomains (Soler-Llavina and Sabatini 2006). There are significant methodological challenges, however, involved in making measurements at such high temporal and spatial resolution. I describe here the principle and implementation of a confocal spot detection method for rapid and localized detection of fluorescence changes, which we have used for the detection of action potential-induced presynaptic Ca^{2+} domains.

INTRODUCTION

It is thought that the time course and efficacy of neurotransmitter release is set by the transient nature of localized intracellular $[Ca^{2+}]$ ($[Ca^{2+}]_i$) changes mediated by action potential (AP) invasion of the nerve terminal (Katz 1969; Barrett and Stevens 1972; Yamada and Zucker 1992). Ca^{2+} channels have been shown to be localized to active zones at the neuromuscular junction (Robitaille et al. 1990) and in the central nervous system (Bucurenciu et al. 2008). Theoretical studies have suggested that AP activation of clustered Ca^{2+} channels can give rise to "domains" of $[Ca^{2+}]_i$ that are localized to within tens to hundreds of nanometers (nanodomains and microdomains, respectively) and reach concentrations in the tens to hundreds of micromolar range (Neher 1998; Oheim et al. 2006). Moreover the rapid dissipation of such $[Ca^{2+}]_i$ gradients may be sufficiently fast to terminate release mediated by a rapid, low-affinity Ca^{2+} sensor. Indeed it has recently been shown that rapid (<0.5 msec) Ca^{2+} transients are necessary to produce submillisecond neurotransmitter release at mammalian nerve terminals (Bollmann and Sakmann 2005). Nevertheless, the study of presynaptic $[Ca^{2+}]_i$ gradients produced by single APs within individual boutons is hampered by the slow temporal resolution of point scanning imaging methods. Using total internal reflection fluorescence imaging, Ca^{2+} domains have also been detected in response to long step depolarizations used to capture "standing" $[Ca^{2+}]_i$ gradients in cultured cells (Becherer et al. 2003; Zenisek et al. 2003; Beaumont et al. 2005). These methods, however, are not amenable to synapses formed close to a coverslip, nor are they sufficiently fast to characterize the transient $[Ca^{2+}]_i$ gradients produced by APs. The confocal spot method described below can be applied to the study of synapses formed in culture and in brain tissue (in situ or in vivo) and can be used to study Ca^{2+} gradients on the millisecond timescale.

Limits of Current Scanning Confocal Microscopes

Traditional fluorescence imaging devices, such as scanning one-photon or two-photon confocal microscopes, provide submicrometer resolution and are based on galvanometer-driven mirrors. Although there are many advantages to using these approaches for studying fast neuronal signals, there are also two significant drawbacks. First, the inertial mass of these scanning devices limits the speed of movement of the confocal detection spot. Image rates of up to 20 Hz for several-micrometer areas are possible but are still far too slow for the submillisecond changes (>2 kHz) expected for presynaptic Ca^{2+} domains. Fluorescence detection rates can be accelerated if scans are restricted to a single dimension (0.5–2 kHz depending on whether single axis scanning or arbitrary line scans are used). Resonant-based galvanometer scanning mirrors can improve rates (between 5 and 10 kHz) but at the cost of inflexible pattern generation because they must be driven with sine wave commands at their resonant frequency. The second major limitation of scanning methods is that a particular pixel location only contributes photons for a small fraction of the line scan period, resulting in a reduced signal-to-noise ratio (SNR; see below).

Single Photon Confocal Spot Detection

The confocal spot method is based on classical diffraction-limited confocal microscopy, producing an optical resolution of ~250 nm in x–y and ~1000 nm in the z dimension (when using 488-nm light). Rather than scanning the confocal spot (which limits the temporal resolution to 1–2 msec), the spot is positioned in a fixed location within the nerve terminal and fluorescence transients are recorded in response to single APs. The spatial dependence of $[Ca^{2+}]$ gradients is characterized by displacing the detection spot and then repeating AP stimulation. Traditional scanned confocal images are used as a reference map between the location in the image plane and the galvanometer command voltage. Static voltages are then delivered to the galvanometers to point them at a particular location for the duration of the fluorescence recording.

HARDWARE OVERVIEW

The critical hardware for implementing confocal spot detection is summarized below and is based on that described previously (Escobar et al. 1994; DiGregorio and Vergara 1997; Escobar 1997; DiGregorio et al. 1999; Bradley et al. 2009). Figure 1 diagrams the hardware layout.

- An acousto-optical tunable filter (AOTF) is used to select the wavelength and modulate the amplitude and duration of illumination by an air-cooled krypton-argon laser (laser lines 488, 568, 647 nm).
- Single-mode fiber-optic coupling of the laser beam directs light into the scan head and acts as a spatial filter to improve laser beam quality, which is especially important for use with diode lasers.
- High-numerical-aperture (NA) dipping objectives (NA 0.9–1.1, Nikon, Olympus, or Zeiss) or oil/glycerol objectives are used for inverted setups (NA 1.2–1.4).
- A galvanometer-mirror-based scan head and associated software and hardware are used to accurately park the confocal spot within the field of view with an effective galvanometer jitter of <30 nm) (we use a modified Ultima Scan Head, Prairie Technologies).

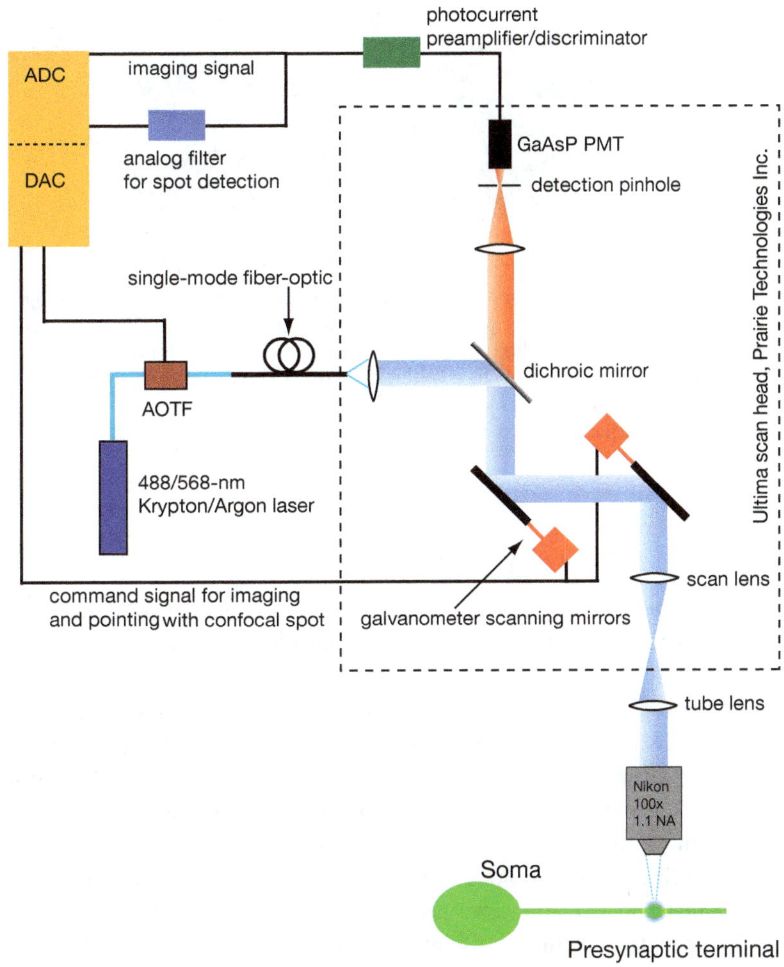

FIGURE 1. Schematic diagram of a typical galvanometer-based confocal spot detection system. Laser light is coupled into the scan head using a single-mode optical fiber. The output of the fiber is then collimated and aligned to reflect off a dichroic mirror and two computer controlled (via a digital to analog conversion [DAC] device, National Instruments) galvanometer mirrors. The excitation light then passes through a scan and tube lens to backfill the back aperture of the objective. Fluorescence emission is detected back through the same galvanometers ("descanned") and, because of its longer wavelength, passes through the dichroic and is finally focused onto a gallium arsenide phosphide-based PMT. Single photocurrent events are detected and converted, by the discriminator, to stereotypic voltage pulses that are then acquired by an analog-to-digital conversion (ADC) card. For confocal spot detection (i.e., fixed galvanometer positions), the photon voltage pulses are electrically filtered before acquisition (generally simultaneous with electrophysiological recordings). AOTF, acousto-optical tunable filter.

- A high quantum efficiency and low noise detector, such as a gallium arsenide phosphide-based photocathode photomultiplier tube (GaAsP PMT; Hamamatsu) with a quantum efficiency of ~45%, is used.
- The current output of GaAsP PMT is passed through a discriminator (Hamamatsu). The use of a discriminator eliminates the hardware noise so that the background fluctuations are determined by photon or "shot" noise (Saleh and Teich 1991).

OPTIMAL CONDITIONS FOR Ca^{2+} DETECTION WITH A CONFOCAL SPOT

In brief, the detection spot should be small, the Ca^{2+} indicator should be bright and sensitive, and the indicator should have a rapid response time. The remainder of this section details each of these parameters.

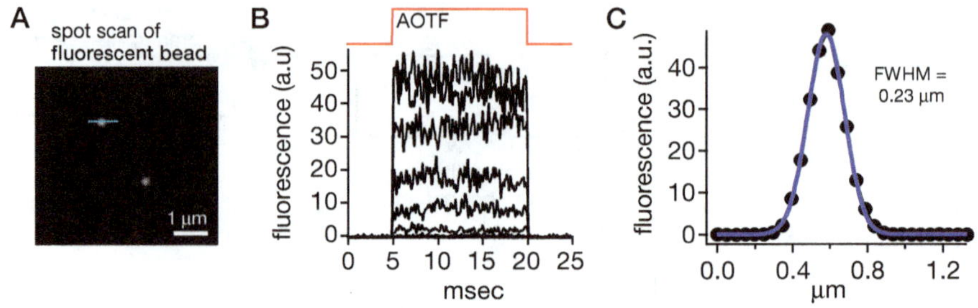

FIGURE 2. Confocal spot scan of a fluorescent bead. (*A*) confocal image of a 110-nm green fluorescent bead. (*B*) Fluorescence responses when the confocal spot is placed at different locations in one direction across the bead (48-nm steps, shown as the dashed blue line in *A*). The laser illumination was controlled with an acousto-optical tunable filter (AOTF) in response to a voltage command (red) changing from 0 to 0.3 µW. (*C*) Plot of the average fluorescence intensity of each trace plotted as a function of spot location. The fluorescence intensity was measured by averaging from 5 to 20 msec.

Characterization of the Confocal Volume (Point Spread Function, PSF)

To detect $[Ca^{2+}]_i$ gradients within neuronal structures, the detection spot must be smaller than the neuronal structure and preferably should approach or be smaller than the size of the domain itself. With a well-characterized PSF and Ca^{2+} indicator behavior, numerical simulations make it possible to predict the spatial and temporal profile of the underlying Ca^{2+} domain. Unlike wide-field microscopy, where out-of-focus photons have an indeterminate origin, the confocal spot is well defined, making such predictions surprisingly robust (DiGregorio et al. 1999).

The confocal PSF is commonly characterized using a scanning confocal image of a point source of light (approximated by a 110-nm fluorescent bead; yellow-green fluospheres, Invitrogen). Under ideal conditions, this would also predict the size of the spot used for spot detection. However, potential uncertainties in galvanometer mirror positions during the ramplike voltage commands used for scanned images may result in differences in the effective PSF for parked beams. It is therefore important to characterize the PSF by scanning a fluorescent bead using a series of "parked" confocal spots, as is performed for Ca^{2+} domain experiments. Such characterization is also an opportunity to estimate the pointing accuracy and reliability in response to fixed command voltages (i.e., jitter in pointing accuracy).

The smallest PSF can be attained by overfilling the back pupil of the highest usable NA objective and using a pinhole size that corresponds to the distance between the first Airy minimum (Wilson 1990). In Figure 2A, a 1.1-NA 100x dipping objective (Nikon) was used with a 60-µm pinhole in a conjugate image plane. A traditional scanning confocal image is used as a reference map to identify the location within the field of view where the confocal beam can be parked (Fig. 2A). For improved accuracy, high zoom factors are generally used, corresponding to images <10 x 10 µm. The confocal spot is then "centered" on the bead by optimizing the fluorescence intensity response to a 2-msec light pulse of very low laser intensity while moving the spot in 50-nm steps in the vicinity of the initial point. Once the confocal spot is "centered" on the fluorescent bead, a "spot" scan is performed by displacing the spot only in the *x* or the *y* direction and eliciting light pulses at each location along a line traversing the bead (50-nm step sizes, Fig. 2B). The amplitude of the fluorescence signal is then plotted as a function of spot location and fit to a Gaussian function to estimate the PSF width (full width at half maximum: 0.23 µm; Fig. 2C). The same procedure can be performed in the *z* direction, although the *z* profile is more simply estimated from traditional scanned images at different focal planes. For numerical simulations, the point spread function can be simulated directly or approximated by a 3D Gaussian function based on fits of the *x*, *y*, and *z* intensity profiles (not shown in Fig. 2).

Choosing a Ca^{2+} Indicator with the Appropriate Brightness and Sensitivity

Because the confocal volume is so small, the number of emitted photons can have a dramatic influence on the ability to detect Ca^{2+}-dependent fluorescence transients. When choosing a Ca^{2+} indicator, both its brightness and its sensitivity to a given $[Ca^{2+}]_i$ change ($\Delta[Ca^{2+}]_i$). Both properties determine the ability to detect the peak change in fluorescence signal (ΔF_{peak}) relative to the noise of the signal prior to the stimulus (background noise or standard deviation (σ_b). The SNR can then be defined as

$$\mathrm{SNR} = \Delta F_{peak} / \sigma_b.$$

The standard deviation is calculated from the background variance (σ_b^2), which in practice is estimated from the mean background fluorescence ($\bar{F}_b$) over a time window from n number of time points (i) before the onset of the Ca^{2+} transient:

$$\sigma_b = \sqrt{\sigma_b^2} = \sqrt{\frac{1}{n-1}\sum_i^n (F_i - \bar{F}_b)^2}\,.$$

The relative brightness of an indicator (amount of emitted fluorescence) is proportional to the product of the extinction coefficient (ε, ability to absorb a photon for a given excitation energy and concentration of dye molecules, $M{*}cm^{-1}$) and the quantum yield (ϕ, probability of emitting a photon following the absorption of one). Such values are usually quoted in product catalogues and serve as a good rule-of-thumb for the relative brightness of dyes. However, ultimately, the brightness can be limited by the bleaching rate of the indicator. A simple relation for relative brightness of a fluorescence molecule ($\bar{B}_F$), or an indicator, at a given $[Ca^{2+}]_i$ is

$$\bar{B}_F \approx \frac{\varepsilon \cdot \phi}{\rho_{bleach}},$$

where ρ_{bleach} is the probability that a dye molecule will be bleached.

An indicator's brightness plays two important roles for confocal spot detection. (1) Qualitatively, the brightness must be high enough to image small structures and to correctly estimate resting fluorescence. The latter is important for estimating $[Ca^{2+}]_i$. (2) Quantitatively, the brightness of an indicator contributes to the background fluorescence noise, which in turn influences the SNR.

The brightness of a fluorescence Ca^{2+} indicator can be considered quantitatively as the number of photons per unit time. The typical photon count rate for 100 µM OGB-5N using spot detection (~0.25 fL volume) is low and can range from 1 to 10 MHz. Under ideal detection conditions (i.e., when the noise of a fluorescence signal is dominated by photons or shot noise rather than instrument noise) the temporal distribution of photon flux will conform to a Poisson process (Saleh and Teich 1991). Thus, the variance in photon number (σ_b^2) is equal to the number of photons (N_b) and the SNR of a change in fluorescence is

$$\mathrm{SNR} = \frac{\Delta F_{peak}}{\sigma_b} \approx \frac{\Delta N_{peak}}{\sqrt{N_b}}.$$

This relation supports what is intuitive—that a brighter indicator will improve SNR. Indeed, under ideal conditions SNR improves by $1/\sqrt{N_b}$. For a more thorough consideration of SNR under less than ideal conditions one should consider other noise and fluorescence sources in addition to the Ca^{2+} indicator (e.g., autofluorescence and detector noise [Saleh 1991]).

Another way to improve the SNR of a given Ca^{2+} dependent fluorescence transient is to increase the signal or ΔF_{peak} without changing background fluorescence noise. This can be attained by using Ca^{2+} indicators with a high sensitivity (i.e., large relative changes in fluorescence for a given $[Ca^{2+}]_i$ change). The parameter most commonly referred to as describing this sensitivity is the ratio (R) of the maximum to minimum fluorescence in low and high $[Ca^{2+}]_i$, respectively. The most commonly used organic Ca^{2+} indicators show $R > 10$ and in some cases up to 100. Thus more sensitive indicators will show improved SNR because ΔF_{peak} is increased for each Ca^{2+} transient under ideal conditions (i.e., when all of the fluorescence and associated noise comes from the indicator and the $[Ca^{2+}]_i$ changes R occur within the linear range of the indicator). This dominant influence of R on SNR means that it is generally better to use an indicator with double the sensitivity rather than double the brightness. But if the brightness differs by an order of magnitude, as is the case for Ca^{2+} indicators based on the Oregon Green fluorophor in comparison with the Fluo-based indicators (see the Molecular Probes Handbook 2010), the influence of a lower sensitivity on SNR can be compensated by dye brightness (i.e., by increasing N_b).

Rapidly Responding Indicator

To measure the rapid $[Ca^{2+}]_i$ dynamics produced by diffusion and buffering, the Ca^{2+} indicator must be able to both bind and release Ca^{2+} rapidly. Nevertheless, because no Ca^{2+} indicator is fast enough to report the most rapid temporal aspects of Ca^{2+} domains (nanodomains dissipate within tens of microseconds), one must consider the time evolution of Ca^{2+} binding when interpreting indicator fluorescence transients. This response is not uniquely determined by the equilibrium behavior of the dye (e.g., K_d), but requires explicit knowledge of the on and off rate constants (k_{on} and k_{off}). To gain an understanding of the influence of rate constants on the response time of Ca^{2+} indicators, it is useful to consider the normalized fluorescence change of a Ca^{2+} indicator in response to an extremely brief $[Ca^{2+}]_i$ change or a quasi-δ-function (Escobar et al. 1997). See the Box for additional details.

Given that the time course and spatial dependence of the confocal spot-detected Ca^{2+} transients are dictated by nonequilibrium binding and dye diffusion, we generally rely on numerical simulations (including diffusion of all mobile species and Ca^{2+} buffering) to predict the underlying time course and cellular determinants of the spatiotemporal evolution of Ca^{2+} domains (DiGregorio et al. 1999, 2007). To perform simulations that provide useful insights, a detailed characterization of Ca^{2+} dye properties is quintessential. The characterization of any dye should first be performed in a cuvette (at physiological pH and osmolarity) to estimate the binding affinity (K_d) and R (Escobar et al. 1997; Faas et al. 2005). Ideally one should make the same characterization inside neurons, because R tends to be lower inside cells (Bollmann and Sakmann 2005). Using flash photolysis (Escobar et al. 1997; Faas et al. 2005) or temperature jump (Kao and Tsien 1988) methods, the kinetics of calcium binding by the

Influence of Ca^{2+} Indicator Rate Constants on the Speed of the Ca^{2+}-Dependent Fluorescence Response

The normalized fluorescence change of a Ca^{2+} indicator in response to an extremely brief $[Ca^{2+}]_i$ change, or a quasi-δ-function (Escobar et al. 1997), can be described by the following equation:

$$\overline{\Delta F}(t) = k_{on} \bullet [Ca^{2+}]_{peak} \bullet \int [Ca^{2+}](t) \bullet e^{-k_{off}*t}.$$

This equation can be derived using the assumptions that the concentration of total indicator is much greater than that of the fraction bound to Ca^{2+}. Given the large concentrations of indicator used, >100 μM, and an estimated free $[Ca^{2+}]_i$ change averaged over the confocal spot of ~1 μM, such assumptions are applicable to confocal spot detection. The k_{on}, the peak $[Ca^{2+}]_i$, and the time integral of the free $[Ca^{2+}]_i$ change determine the peak of the fluorescence response, whereas the time course is solely influenced by the k_{off}. The value of $1/k_{off}$ is equal to the exponential time constant ($\tau_{response}$) and describes the intrinsic filter properties of the Ca^{2+} indicator. For longer duration $[Ca^{2+}]_i$ changes, both k_{on} and k_{off} play a role in the time course and the resulting time course can be estimated from the convolution of the above impulse response and the time course of the $[Ca^{2+}]_i$.

One can roughly estimate the speed of a dye from its equilibrium binding affinity (K_d). Because the on-rates of the commonly used Ca^{2+} indicators have been shown to be diffusion limited (2.4 x 10^8 $M^{-1}sec^{-1}$ for Fluo-3 [Escobar et al. 1997] and OGB-5N [Faas et al. 2005]), then the equilibrium binding affinity ($K_d = k_{off}/k_{on}$) can provide a good proxy for k_{off}, or the speed of a Ca^{2+} indicator. As examples, the $\tau_{response}$ of the high affinity indicator Fluo-3 is ~2.5 msec (k_{off} 175 sec^{-1} [Escobar et al. 1997]), whereas for the low affinity indicator OGB-5N the $\tau_{response}$ is 0.115 msec (k_{off} 8670 sec^{-1} [Faas et al. 2005]).

indicators can be characterized. Finally, because the Ca^{2+} indicator is diffusible it can contribute to diffusional blurring of underlying $[Ca^{2+}]_i$ gradients. Its diffusion coefficient can be estimated using fluorescence correlation spectroscopy or fluorescence recovery after photobleaching.

POTENTIAL SHORTCOMINGS OF THE CONFOCAL SPOT DETECTION METHOD

- The spatial distribution of the underlying Ca^{2+} domain must be invariant from trial to trial, because multiple trials are required for its characterization. This problem can be tested and minimized by signal averaging over three to ten trials at each spot location and comparing intratrial versus spot location variability (DiGregorio et al. 1999). Drifts in the preparation must also be minimized for reliable characterization of the spatiotemporal profile.
- The very small detection spot and the rapid changes in $[Ca^{2+}]_i$ can make it difficult to locate a Ca^{2+} domain change, particularly in large structures such as somata where the increased volume dilutes the $[Ca^{2+}]_i$ to low levels. For small presynaptic boutons this is generally not a problem because the distinct bouton anatomy is a good indicator of the presence of a Ca^{2+} channel cluster. High frequency trains of APs and line scans can facilitate finding hot spots.
- In nerve terminals at the surface of brain slices, fluorescence transients can be detected using 100 μM of the low affinity calcium indicator (OGB-5N). At such concentrations, resting photon fluxes are in the range of 1–10 MHz and allow for the detection of >10% relative changes. The use of high indicator concentrations, however, can lead to perturbation of the free $[Ca^{2+}]_i$ thereby distorting the time course and amplitude of the free $[Ca^{2+}]_i$ change. This can be tested by either numerical simulations or comparing the peak amplitude and decay of spot measurements using lower indicator concentrations.

IMPROVEMENTS IN SNR WHEN USING CONFOCAL SPOT DETECTION

In addition to improved temporal resolution, the SNR for the same laser intensity tends to be much better than rapid line scans of the same structures (Fig. 3). The improved SNR results, in large part, from an increased number of collected photons per duty cycle. In traditional line scans, little time is spent in a particular location relative to the acquisition rates (dwell time 1–5 μsec/pixel vs. 1–2 msec/line). As an example, consider a 10-μm line scan at 1 msec/line. From a particular location (e.g., 0.33 μm, as defined by the point spread function), a line scan would collect photons for only 1/30 of the time as would be the case for confocal spot detection for a 1-msec epoch. This would result in a SNR of $\sqrt{30}$ or 5.5 times less. This is illustrated in a simple experiment, shown in Figure 3. Here the same laser intensity was used for a line scan of a fluorescent bead and a confocal spot measurement of the same bead. The SNR is clearly larger for the spot measurement (48 and 9 in spot detection and line scan, respectively). When the intensity of the laser is reduced by ~30-fold (similar to the fraction of the line scan that is spatially averaged, 0.33-μm/10-μm line length), then the confocal spot fluorescence amplitude decreases, resulting in a SNR more similar to the line scan (SNR 15). Other factors such as bandwidth may also contribute the difference in SNR of the line scan versus the confocal spot detection. Moreover, improved photon count comes at a cost of increased fluorophore excitation over time and can produce more bleaching. However in practice, small reductions in intensity can minimize bleaching while still preserving superior SNR.

Application of the Confocal Spot Detection Method to the Study of Presynaptic $[Ca^{2+}]$ Microdomains

An estimate of the spatial and temporal profile of AP-induced $[Ca^{2+}]_i$ gradients can be studied by recording stimulated Ca^{2+} transients at various locations within a nerve terminal (see Fig. 4A,B). In the case of an AP, the time course of the measured Ca^{2+} transient is stereotypic. When the confocal

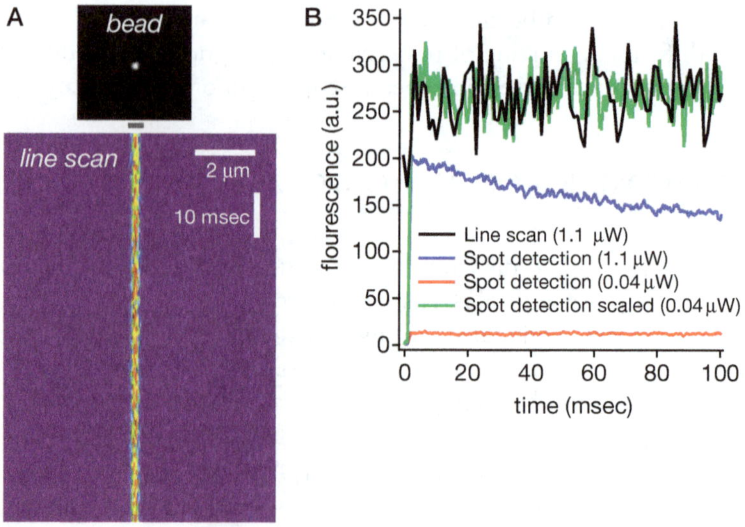

FIGURE 3. Signal-to-noise ratio of confocal spot-detected fluorescence and line scan-detected fluorescence. (*A*) An *x–y* scanned (*top*) and *x*-line scan image (*bottom*) of a single 110-nm green fluorescent bead. The line scan rate was 1.15 msec/line. (*B*) Fluorescence as a function of time for line scans (black; each point is the average for one scan over spatial pixels indicated by gray bar in *A*), and confocal spot detected fluorescence at the same (blue) and lower (red) laser intensities. The confocal spot fluorescence traces were acquired at 100 kHz and boxcar filtered offline using a 100-point running average. The green trace is the low-intensity confocal spot fluorescence scaled to the mean intensity of the line scan.

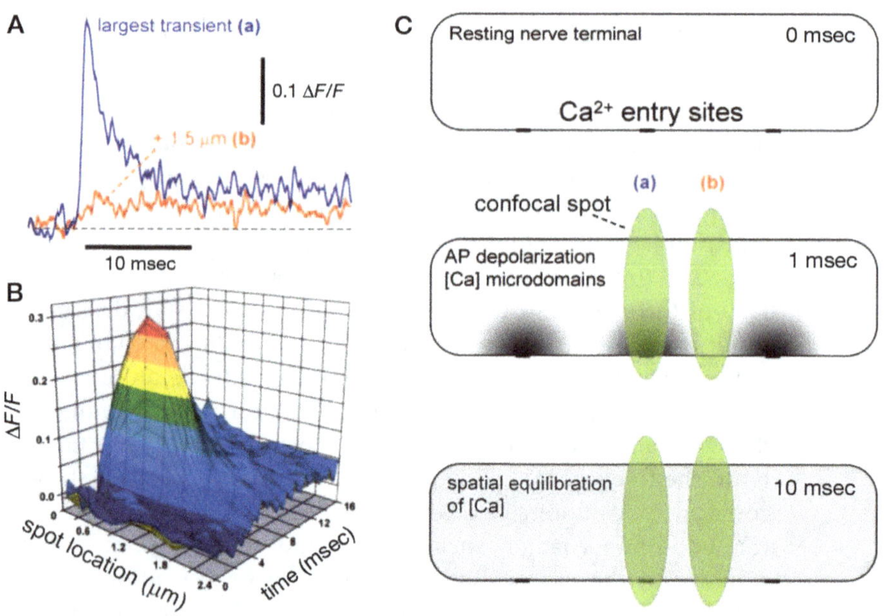

FIGURE 4. Principle of confocal spot detection method. (*A*) Ca^{2+} transients were evoked with single action potentials at two different locations within a nerve terminal of cultured motor neuron. (Modified, with permission, from DiGregorio et al. 1999.) (*B*) All nine transients of the spot scan from *A*, plotted according to spot location within the nerve terminal. (*C*) Schematic diagram illustrating the relationship between spot location and the detection of Ca^{2+} domains. Early in time (~1 msec), $[Ca^{2+}]$ gradients are restricted primarily to sites of calcium entry, resulting in spot detected transients that tend to be large, rapidly rising, and rapidly decaying. Spot locations at a distance have smaller amplitudes, are slowly rising, and have slower decay times. Late in time (10 msec), all the gradients have largely dissipated and fluorescence transients at all spot locations within the terminal have the same amplitude and decay time.

spot is "centered" on a calcium entry site (Fig. 4Aa) the rise time of the calcium transient is maximal, the amplitude large, and the decay fast—owing to diffusion away from the entry site and our recording spot. If the recording spot is displaced from the site of entry, then, within a certain distance, one can see evidence of diffusion of Ca^{2+} and dye molecules into the spot with a delay, and much reduced amplitude (Fig. 4Ab). These position-dependent changes in kinetics and changes in amplitude suggest that a Ca^{2+} entry site is within the spot or extremely close to (<0.5 μm) the location of fast-rising transients. The amplitude of Ca^{2+} transients are then a rough estimate of relative density of Ca^{2+} sources/channels at the Ca^{2+} entry site (DiGregorio et al. 1999).

SUMMARY

- The spatial and temporal profile of action potential-induced Ca^{2+} domains can be characterized using laser spot-delimited volumes that are smaller than neuronal compartments, such as nerve terminals.
- Good SNR can be attained using such a method if the Ca^{2+} indicators are bright and sensitive.
- To estimate the submillisecond rise and decay of action potential-induced Ca^{2+} domains, the Ca^{2+} indicator must have a rapid response time.
- Numerical simulations of buffered Ca^{2+} diffusion, where Ca^{2+} indicator and PSF parameters have been well characterized, provide a quantitative estimation of the spatial temporal profile of Ca^{2+} domains.

ACKNOWLEDGMENTS

I thank Jack Waters, Therese Abrahamsson, and Tom Otis for comments and Jeremy Dittman for helpful discussion and comments on this chapter.

REFERENCES

Barrett EF, Stevens CF. 1972. The kinetics of transmitter release at the frog neuromuscular junction. *J Physiol (Lond)* **227:** 691–708.

Beaumont V, Llobet A, Lagnado L. 2005. Expansion of calcium microdomains regulates fast exocytosis at a ribbon synapse. *Proc Natl Acad Sci* **102:** 10700–10705.

Becherer U, Moser T, Stuhmer W, Oheim M. 2003. Calcium regulates exocytosis at the level of single vesicles. *Nat Neurosci* **6:** 846–853.

Bollmann JH, Sakmann B. 2005. Control of synaptic strength and timing by the release-site Ca^{2+} signal. *Nat Neurosci* **8:** 426–434.

Bradley J, Luo R, Otis TS, DiGregorio DA. 2009. Submillisecond optical reporting of membrane potential in situ using a neuronal tracer dye. *J Neurosci* **29:** 9197–9209.

Bucurenciu I, Kulik A, Schwaller B, Frotscher M, Jonas P. 2008. Nanodomain coupling between Ca^{2+} channels and Ca^{2+} sensors promotes fast and efficient transmitter release at a cortical GABAergic synapse. *Neuron* **57:** 536–545.

DiGregorio DA, Vergara JL. 1997. Localized detection of action potential-induced presynaptic calcium transients at a *Xenopus* neuromuscular junction. *J Physiol* **505.3:** 585–592.

DiGregorio DA, Peskoff A, Vergara JL. 1999. Measurement of action potential–induced presynaptic calcium domains at a cultured neuromuscular junction. *J Neurosci* **19:** 7846–7859.

DiGregorio DA, Rothman JS, Nielsen TA, Silver RA. 2007. Desensitization properties of AMPA receptors at the cerebellar mossy fiber granule cell synapse. *J Neurosci* **27:** 8344–8357.

Escobar AL, Monck JR, Fernandez JM, Vergara JL. 1994. Localization of the site of Ca^{2+} release at the level of a single sarcomere in skeletal muscle fibres. *Nature* **367:** 739–741.

Escobar AL, Velez P, Kim AM, Cifuentes F, Fill M, Vergara JL. 1997. Kinetic properties of DM-nitrophen and calcium indicators: rapid transient response to flash photolysis. *Pfugers Arch Eur J Physiol* **434:** 615–631.

Faas GC, Karacs K, Vergara JL, Mody I. 2005. Kinetic properties of DM-nitrophen binding to calcium and magnesium. *Biophys J* **88:** 4421–4433.

Kao JP, Tsien RY. 1988. Ca^{2+} binding kinetics of fura-2 and azo-1 from temperature-jump relaxation measurements. *Biophys J* **53:** 635–639.

Katz B. 1969. *The release of neural transmitter substances*. Liverpool University Press, Liverpool.

Molecular probes: The handbook. 2010. Invitrogen, Carlsbad, CA.

Neher E. 1998. Vesicle pools and Ca^{2+} microdomains: New tools for understanding their roles in neurotransmitter release. *Neuron* **20:** 389–399.

Oheim M, Kirchhoff F, Stuhmer W. 2006. Calcium microdomains in regulated exocytosis. *Cell Calcium* **40:** 423–439.

Robitaille R, Adler EM, Charlton MP. 1990. Strategic location of calcium channels at transmitter release sites of frog neuromuscular synapses. *Neuron* **5:** 773–779.

Saleh BEA, Teich MC. 1991. *Fundamentals of photonics*. Academic, New York.

Soler-Llavina GJ, Sabatini BL. 2006. Synapse-specific plasticity and compartmentalized signaling in cerebellar stellate cells. *Nat Neurosci* **9:** 798–806.

Wilson T. 1990. *Confocal microscopy*. Academic, New York.

Yamada WM, Zucker RS. 1992. Time course of transmitter release calculated from simulations of a calcium diffusion model. *Biophys J* **61:** 671–682.

Zenisek D, Davila V, Wan L, Almers W. 2003. Imaging calcium entry sites and ribbon structures in two presynaptic cells. *J Neurosci* **23:** 2538–2548.

16 Ca^{2+} Uncaging in Nerve Terminals

Olexiy Kochubey and Ralf Schneggenburger

Laboratory of Synaptic Mechanisms, Brain-Mind Institute, École Polytechnique Fédérale de Lausanne, 1015 Lausanne, Switzerland

ABSTRACT

Ca^{2+} uncaging can be used to create a spatially homogenous elevation of the intracellular free Ca^{2+} concentration, $[Ca^{2+}]_i$, in cells. When applied to nerve terminals or secretory cells, this technique allows one to elicit transmitter release with a $[Ca^{2+}]_i$ signal of measurable amplitude, and therefore to directly relate the rate of transmitter release to the measured $[Ca^{2+}]_i$. When combined with patch-clamp measurements, Ca^{2+} uncaging is done by introducing a Ca^{2+}-loaded photolyzable Ca^{2+} chelator (like DM-Nitrophen) into the cell via the whole-cell patch-pipette. A brief light pulse from a flash lamp or a pulsed laser is used to photolyze the DM-Nitrophen. The resulting increase in $[Ca^{2+}]_i$ is measured with ratiometric fluorescent indicators of suitable Ca^{2+} affinity, such as Fura-2, Fura-4F, Fura-2FF, or Fura-6F, depending on the post-flash $[Ca^{2+}]_i$ values. To quantitatively measure $[Ca^{2+}]_i$, an accurate calibration of the fluorescent indicator in the presence of the photolyzable Ca^{2+} chelator is necessary, which will be described here. Ca^{2+} uncaging in nerve terminals has proven useful for investigating Ca^{2+}-dependent functions like transmitter release, short-term plasticity, and exocytosis–endocytosis coupling in the presynaptic compartment of neurons.

INTRODUCTION

In presynaptic terminals and many secretory cells, an action potential leads to the opening of voltage-gated Ca^{2+} channels, and the ensuing elevation of $[Ca^{2+}]_i$ causes transmitter release. The $[Ca^{2+}]_i$ signal relevant for transmitter release is a microdomain signal in the immediate vicinity of open Ca^{2+} channels, which cannot, in most cases, be imaged directly with accurate information in both time and space (see also Chapter 15 and Neher 2011). Therefore, if one wishes to determine the relationship between the rate of transmitter release and the underlying $[Ca^{2+}]_i$ signal, transmitter release needs to be evoked with a spatially homogenous $[Ca^{2+}]_i$ signal to ensure that the transmitter release sites and the Ca^{2+} indicator in the cell's cytoplasm "see" the same $[Ca^{2+}]_i$ elevation. This can be achieved using Ca^{2+}-uncaging methods. A Ca^{2+}-loaded photolyzable Ca^{2+} chelator, like DM-

Nitrophen, is introduced into the cell under study through a whole-cell patch pipette (see Chapter 35), and then photolyzed by a brief UV light flash. Because the photolyzable Ca^{2+} chelator can be assumed to be distributed homogenously in the cytosol, the resulting elevation of $[Ca^{2+}]_i$ is spatially homogenous, given a homogenous illumination by the flash light. This technique does not cause spatial gradients that are associated with the brief opening of voltage-gated Ca^{2+} channels.

In this chapter, we first outline some components of the optical equipment that we use for Ca^{2+} uncaging. We then provide a protocol to perform a three-point calibration procedure, which is necessary to correctly calculate the $[Ca^{2+}]_i$ from the measured fluorescence ratio of a Ca^{2+} indicator dye in flash-photolysis experiments (see also Helmchen 2011). Finally, we summarize an example of a presynaptic Ca^{2+} uncaging experiment in the large nerve terminals of the calyx of Held synapse.

Protocol

A Three-Point Calibration Procedure for Ca^{2+} Uncaging Experiments

Imaging Setup for Ca^{2+} Uncaging

We use an upright microscope (ZEISS Axioskop FS2, or Olympus BX51WI) equipped with a 60× water-immersion objective (Olympus, LUMPlanFL, NA 0.9) and a Fura-2 filter set (409-nm beam splitter, 440 LP emission filter; AHF Analysentechnik). A xenon short-arc flash-lamp system (SP 20, Rapp Optoelektronik; half-width of flash ≈ 0.45 msec) is used for Ca^{2+} uncaging, with the output light limited to wavelengths <400 nm by an SP395 UV filter. A monochromator (TILL Photonics; see also Uhl 2011) is used for excitation of Fura-2FF. Both light sources are coupled via quartz light guides into a two-port epifluorescence condenser (TILL Photonics), with a 40%/60% beam splitter, or alternatively a sapphire window (transmission, 85%; reflection, 15%) to combine the two light sources in the microscope. We use a slow-scan charge-coupled device (CCD) camera (640 × 480 pixels; TILL Photonics) for imaging the fluorescence of Fura-2FF. On-chip binning (8 × 8 or 8 × 15 pixels) is used to achieve short exposure times, making it possible to measure the $[Ca^{2+}]_i$ signal at high time resolution (~140 frames/sec).

MATERIALS

CAUTION: See Appendix 6 for proper handling of materials marked with <!>.
See the end of the chapter for recipes for reagents marked with <R>.

Reagents

Brain-stem slices from 8–10-d-old rats (or your neural tissue of interest)
Calibration and uncaging solutions (see Table 1)
DM-Nitrophen, tetrasodium salt: 1-(2-nitro-4,5-dimethoxyphenyl)-1,2-diaminoethane-*N,N,N′,N′*-tetraacetic acid, $4Na^+$; Merck) <R>

A titration procedure to estimate the purity of the DM-Nitrophen is included with the recipe.

Alternative photolyzable Ca^{2+} chelators that are often used are NP-EGTA and DMNPE-4. The photochemical properties, Ca^{2+} and Mg^{2+} binding, and the relative advantages of each of the photolyzable Ca^{2+} chelators are discussed by Ellis-Davies (2003, 2011).

EGTA and DPTA (1,3-diamino-2-hydroxypropane-*N,N,N′,N′*,-tetraacetic acid) <!>

These are Ca^{2+} chelators for making solutions with defined and buffered $[Ca^{2+}]$.

Fura-2FF pentapotassium salt (TefLabs)
Patch-pipette solutions (see Table 1)

Equipment

Glass cuvettes, 50-μm path length (VitroCom)
Imaging equipment (see the section Imaging Setup for Ca^{2+} Uncaging)
Whole-cell patch pipettes
Whole-cell recording equipment

TABLE 1. Solutions for Ca^{2+} uncaging and calibrations

Solution	Purpose	Composition[a]	
1	2× stock solution of basal components for Solutions 2–5	125 mM	Cs-gluconate
		20 mM	TEA-Cl <!>
		20 mM	HEPES
		5 mM	Na_2-ATP
		0.3 mM	Na_2-GTP
		(final concentrations after 2× dilution)	
		(pH 7.2 with CsOH <!>)	
2	Determination of R_{min}	10 mM	EGTA
		0.5 mM	$MgCl_2$
		1 mM	DMN[b]
		0.1 mM	Fura-2FF
		in 1× Solution 1	
3	Determination of $R_{10\mu M}$	10 mM	DPTA <!>
		2.5 mM	$CaCl_2$ <!>
		0.5 mM	$MgCl_2$
		1 mM	DMN[b]
		0.1 mM	Fura-2FF (free $[Ca^{2+}]$ = 10.3 μM)
		in 1× Solution 1	
4	Determination of R_{max}	10 mM	$CaCl_2$ <!>
		0.5 mM	$MgCl_2$
		1 mM	DMN[b]
		0.1 mM	Fura-2FF
		in 1× Solution 1	
5	Intracellular solution for Ca^{2+} uncaging	0.8 mM	$CaCl_2$ <!>
		0.5 mM	$MgCl_2$
		1 mM	DMN[b]
		0.1 mM	Fura-2FF
		in 1× Solution 1	
6	Intracellular solution for in-cell calibration of $R_{10\mu M}$	80 mM	Cs-DPTA <!>
		10 mM	$CaCl_2$ <!>
		20 mM	HEPES
		0.1 mM	Fura-2FF
		1 mM	DMN[b]
		(pH 7.2 with CsOH <!>; free $[Ca^{2+}]$ = 10. 3 μM)	
7	Extracellular solution for in-cell calibration of $R_{10\mu M}$	145 mM	*N*-methyl-D-glucamine
		10 mM	HEPES
		2.5 mM	KCl
		10 mM	NaCl
		1 mM	DPTA <!>
		0.14 mM	$CaCl_2$ <!>
		1 mM	$MgCl_2$
		(pH 7.4 with HCl <!>; free $[Ca^{2+}]$ = 10 μM)	

CAUTION: See Appendix 6 for proper handling of materials marked with <!>.

[a]Solutions 2–5 are based on the 2× stock solution prepared as Solution 1, and additionally contain the components listed in the third column. DMN, DM-Nitrophen. Adjust to the final volume with sterile ddH_2O.

[b]Take the purity of the DM-Nitrophen stock solution into account when making the desired final solutions (see Recipes).

EXPERIMENTAL METHOD

Theory: Ratiometric Measurements of $[Ca^{2+}]_i$ in Ca^{2+} Uncaging Experiments

During experiments, the fluorescent ratio R of fura-2FF ($R = F_{350}/F_{380}$) is measured before and after flashes, where F_{350} and F_{380} are the background-corrected fluorescence values with excitation at 350 nm and 380 nm, respectively. The intracellular free Ca^{2+} concentration, $[Ca^{2+}]_i$, is calculated from the measured ratio R according to (Grynkiewicz et al. 1985)

$$\left[Ca^{2+}\right]_i = K_{eff} \cdot \frac{R - R_{min}}{R_{max} - R}. \qquad (1)$$

By rearranging,

$$K_{\text{eff}} = \left[Ca^{2+}\right]_i \cdot \frac{R_{\max} - R}{R - R_{\min}}. \tag{2}$$

The calibration constants R_{max}, R_{min}, and K_{eff} must be determined by a three-point calibration, which is described below.

In Vitro Calibration

1. Using Table 1 as a guide, prepare buffered Ca^{2+} Solutions 2, 3, and 4. For Solutions 2 and 3, use the Ca^{2+} buffer EGTA and the low-affinity Ca^{2+} buffer DPTA (K_d for Ca^{2+} = 80 µM at pH 7.2; see Heinemann et al. 1994), respectively. Solutions 2, 3, and 4 are used to obtain the calibration constants R_{min}, R_{max}, and K_{eff}, respectively. To calculate the necessary concentrations of $CaCl_2$ to be added, a chelator program that calculates the free $[Ca^{2+}]$ in the presence of various buffers including the photolyzable Ca^{2+} chelator must be used. We use the "Frecon" insert to Igor Pro (available from the authors upon request).
2. Measure the calibration constants in vitro, using small glass cuvettes (50-µm path length). Measure the fluorescence of the solutions in the cuvettes placed below the microscope objective (Helmchen 2011). The minimum calibration comprises measurements at three different free $[Ca^{2+}]$, to determine the three calibration constants R_{min}, R_{max}, and K_{eff} (see Equations 1 and 2).
   i. Measure R_{min} in a strongly Ca^{2+} buffered solution without added Ca^{2+} (Solution 2; nominally Ca^{2+}-free).
   ii. Measure R_{max} in a solution with excess $[Ca^{2+}]$ (Solution 4).
   iii. To obtain K_{eff}, measure the spectrum of Fura-2FF at a buffered free $[Ca^{2+}]$ close to the expected half-maximal binding (K_d) of the Ca^{2+} indicator (Solution 3; free $[Ca^{2+}]$ of 10.3 µM).

      Fluorescence spectra for these three solutions are shown in Figure 1A (black symbols), and the ratio R can be extracted from these spectra. In addition, we usually measure the spectra at two additional free $[Ca^{2+}]$ which bracket the value of 10.3 µM $[Ca^{2+}]$ (here, 2.0 and 20.1 µM free $[Ca^{2+}]$; see Fig. 1A, gray spectra). We obtain such in vitro spectra, including two intermediate $[Ca^{2+}]$ around the point close to half-maximal Ca^{2+} binding of the indicator dye (Fig. 1A), as a consistency check and to guard against pipetting errors in Solution 3 ($R_{10\mu M}$). All fluorescence spectra are normalized to the isosbestic point of Fura-2FF (usually at a wavelength of 353–358 nm).
3. Determine the fraction of the indicator dye Fura-2FF bound to Ca^{2+}, according to the ratio, $\Delta F_{380,XCa}/\Delta F_{380,max}$ (see Fig. 1A). Plot this quantity against the free $[Ca^{2+}]$ of each calibration solution (Fig. 1B).
4. Fit with a Michaelis–Menten equation (BF = $[Ca^{2+}]/([Ca^{2+}] + K_d)$) to determine the K_d value of Ca^{2+} binding to Fura-2FF (Fig. 1B; line; K_d = 10.7 ± 2.0 µM $[Ca^{2+}]$; n = 3).

   A K_d value of ~10 µM for Ca^{2+} binding to Fura-2FF has been found reproducibly in measurements made in our laboratory (Felmy et al. 2003; Lou et al. 2005; Wölfel et al. 2007).

In-Cell Calibration

The properties of fluorescent dyes are likely different in cytoplasm as compared with pure aqueous solutions. Therefore, obtain the final calibration in in-cell measurements (Fig. 1C–E). As an example, we perform the following in our laboratory.

5. Make whole-cell voltage-clamp (holding potential = –80 mV) recordings of calyces of Held in brain-stem slices from 8–10-d-old rats. Use Ca^{2+}-calibration solutions as the (intracellular) pipette solutions, and measure the fluorescence ratio R every 5 sec (Fig. 1C). In this way, perform an in-cell measurement of R_{min}.

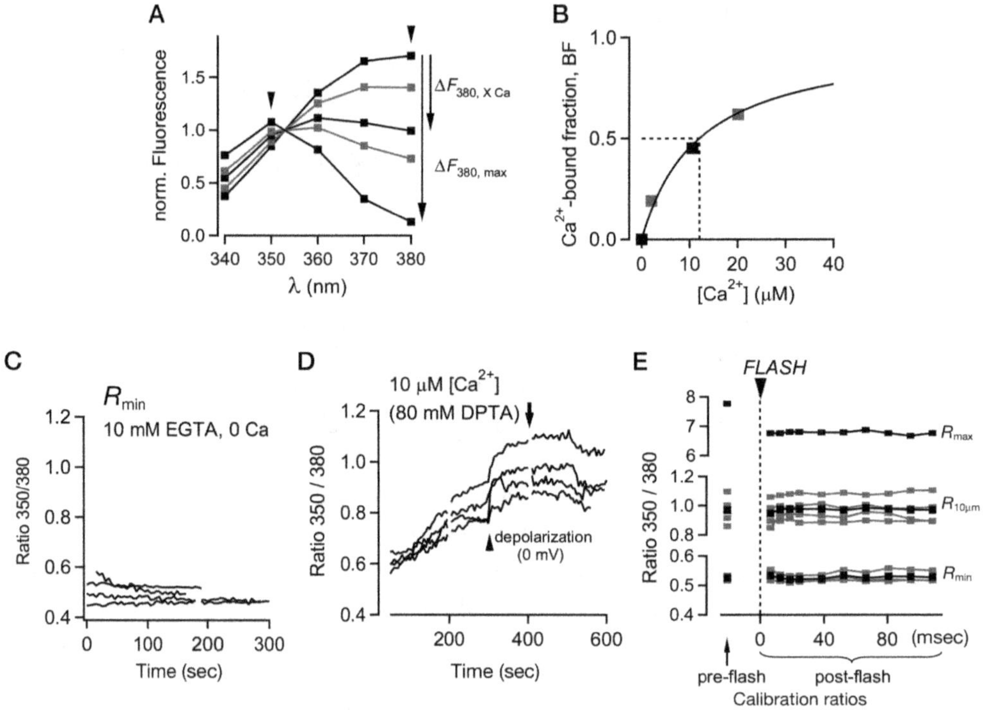

FIGURE 1. Calibration procedures for Ca^{2+} imaging in combination with Ca^{2+} uncaging. (*A*) Spectra of Fura-2FF, measured in thin glass cuvettes at five different [Ca^{2+}] ranging from nominal Ca^{2+}-free solution (highest traces at 380 nm; Solution 2) to 10 mM [Ca^{2+}] (lowest traces at 380 nm; Solution 4). The intermediate [Ca^{2+}] were 2.0, 10.3, and 20.1 μM (see Solution 3 in Table 1 for 10 μM [Ca^{2+}]). Arrowheads indicate the wavelengths of 350 and 380 nm relevant for determining the fluorescence ratio. (*B*) Ca^{2+}-bound fraction of Fura-2FF (BF) calculated from the measurements in *A*. The fit with a Michaelis–Menten equation (line) indicates half-maximal binding at 12 μM [Ca^{2+}] in this example. (*C*) *In-cell* calibration measurements of R_{min} in single calyx of Held nerve terminals, using a nominally free Ca^{2+}, EGTA-buffered intracellular solution (Solution 2; see Table 1). (*D*) *In-cell* calibration measurements with a strongly buffered intracellular solution (10 μM free [Ca^{2+}], 80 mM DPTA; Solution 6) and an extracellular solution with reduced Na^+ concentration and 10 μM [Ca^{2+}] (Solution 7). Note the slow additional increase of the fluorescence ratio when the membrane potential was depolarized from –80 mV to 0 mV. The value for $R_{10\mu M}$ was taken when the fluorescence ratio had equilibrated (arrow). (*E*) Determination of the change in the calibration ratios R_{min} (*lower* panel), $R_{10\mu M}$ (*middle* panel), and R_{max} (*upper* panel) after photolyzing DM-Nitrophen with a maximal flash. Black traces in the *middle* and *lower* panels were obtained by averaging several individual fluorescence ratio traces obtained in separate recordings (gray symbols).

The in-cell determination of R_{min}, using the same solution as used for the cuvette measurements (Solution 2; see Table 1), gave a value of R_{min} of 0.49 ± 0.03 (Fig. 1C), a value similar to that obtained in the cuvette measurement.

6. Perform an in-cell calibration with a free [Ca^{2+}] (10.3 μM) close to the K_d of Fura-2FF (Fig. 1D). Use a strongly Ca^{2+}-buffered intracellular solution without ATP (e.g., Solution 6; Table 1) to suppress the intracellular Ca^{2+} extrusion mechanisms in the nerve terminals. In addition, we use an extracellular solution with low [Na^+] to further suppress Ca^{2+} extrusion via Na^+/Ca^{2+} exchangers (Solution 7; Table 1).

   In four recordings shown in Fig. 1D during the in-cell calibration measurements, the fluorescence ratio *R* slowly increased during whole-cell recording. After 300 sec of whole-cell recording, the holding potential was set to 0 mV for 200 sec. This resulted in a further, gradual increase in the fluorescence ratio, likely caused by equilibration of intracellular and extracellular Ca^{2+} concentrations by Ca^{2+} flux through open Ca^{2+} channels (note that the extracellular [Ca^{2+}] was set to 10 μM; Solution 7, Table 1). The calibration ratio for this intermediate [Ca^{2+}] of 10 μM ($R_{10\mu M}$) was taken as the value reached after 100 sec of depolarization onset (Fig. 1D, arrow). This value ($R_{10\mu M}$ = 0.993 ± 0.11; mean ± S.D.; n = 4 cells) was similar to the corresponding value obtained in vitro at 10 μM free [Ca^{2+}] (R = 1.00 ± 0.1; n = 3) using Solution 3. With less drastic suppression of the Ca^{2+} extrusion mechanisms, we found significantly lower values for $R_{10\mu M}$, which are likely caused by insufficient control of the [Ca^{2+}]$_i$ inside the nerve terminal (not shown). Thus, care must be taken to design appropriate experimental conditions for the in-cell determination of fluorescence calibration constants, especially for calibrating dyes at elevated [Ca^{2+}]$_i$. For determining the calibration ratio R_{max}, we rely on in vitro measurements because it is impractical to load nerve terminals with saturating [Ca^{2+}] in the micromolar range. Also, the expected decrease of F_{380} at saturating [Ca^{2+}] (see Fig. 1A) will make in-cell calibrations prone to errors (e.g., by incorrectly estimating the background fluorescence).

Determination of Post-Flash Calibration Constants

Because the fluorescent properties of DM-Nitrophen change after photolysis (Zucker 1992), the calibration constants used for calculating [Ca^{2+}]$_i$ following a flash must be corrected accordingly.

7. Make whole-cell recordings for 300–400 sec (see Fig. 1D, arrow). Then expose the cells to flashes of maximal intensity.

   In Fig. 1E, corresponding flash responses for calibration measurements are shown. In the case of R_{min} (Solution 2; Table 1), the fluorescence ratio is unchanged before and after the flash. In the case of $R_{10\mu M}$ (Fig. 1E; middle panel; Solutions 6 and 7), there was no apparent change in fluorescence ratio. However, assuming a photolysis efficiency of 75%, one expects that a flash elevates [Ca^{2+}]$_i$ from 10 μM to 12.3 μM, using the "Frecon" program. The fact that the measured ratio was unchanged is caused by the change in fluorescent properties of the photoproducts of DM-Nitrophen (Zucker 1992). Figure 1E (upper panel) shows the change of fluorescence ratio R in an in vitro determination of R_{max} (determined with Solution 4).

Calculate K_{eff}

8. Insert the measured values of R_{min}, $R_{10\mu M}$, and R_{max} into Equation 2 to calculate K_{eff}.

   Two sets of calibration constants are obtained for the calculation of [Ca^{2+}]$_i$ values *before* and *following* a flash of 75% photolysis efficiency, respectively (post-flash values are given in brackets): R_{min} = 0.53 (0.53), K_{eff} = 150.4 μM (158.2 μM), R_{max} = 7.77 (6.79). Linearly interpolated post-flash values can be used for experiments in which the flash intensity, and therefore photolysis efficiency, is reduced by neutral density filters.

DISCUSSION

Using Ca^{2+}-Uncaging to Examine Transmitter Release at the Calyx of Held Synapse

Figure 2 shows an example of a Ca^{2+}-uncaging experiment at the calyx of Held synapse, a glutamatergic synapse in the auditory brainstem. Owing to the large size of its presynaptic nerve terminal, which surrounds more than half of the postsynaptic nerve cell body (see Fig. 2), the nerve terminal is directly accessible to presynaptic whole-cell recordings (Forsythe 1994; Borst et al. 1995), which in turn has made it possible to perform presynaptic Ca^{2+}-uncaging and Ca^{2+}-imaging experiments (Bollmann et al. 2000; Schneggenburger and Neher 2000).

In the example illustrated in Figure 2, a calyx of Held in a brainstem slice from a P14 mouse was investigated. Figure 2A shows the appearance of the MNTB (medial nucleus of trapezoid body) principal cell and its presynaptic calyx of Held under infrared (IR) trans-illumination; Figure 2B shows the fluorescence image of the presynaptic nerve terminal after loading with the Fura-2FF-containing intracellular solution (Solution 5; Table 1). Before and after the flash, Fura-2FF was excited at 350- and 380-nm wavelengths (see Fig. 2C, top, for the illumination protocol). The background-corrected cellular fluorescence values were analyzed using the six brightest binned pixels selected in the area of the calyx (Fig. 2B), and $[Ca^{2+}]_i$ was calculated from the fluorescence ratio R using Equation 1. In this example, a full flash elevated $[Ca^{2+}]_i$ to ~8.5 μM. Note that the $[Ca^{2+}]_i$ elevation was similar at the different locations within the calyx terminal, indicating spatial homogeneity in the *x–y* plane (see also Wölfel et al. 2007). The elevation of $[Ca^{2+}]_i$ in the nerve terminal induced strong

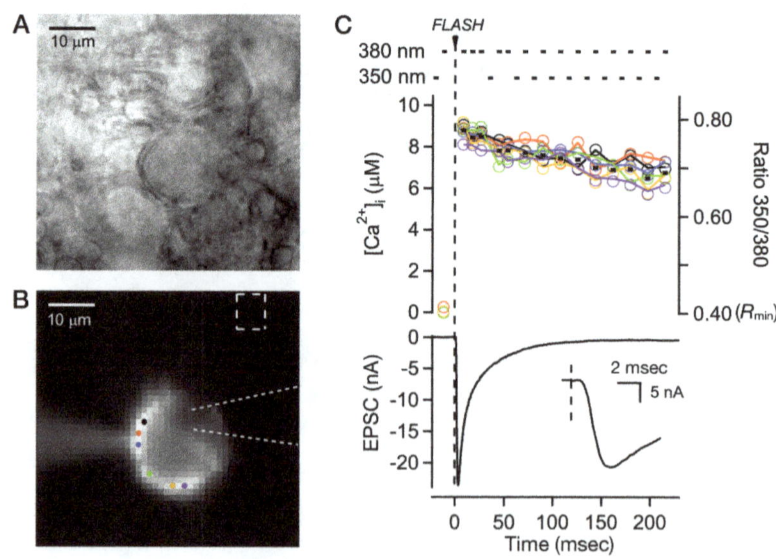

FIGURE 2. Example of presynaptic Ca^{2+} uncaging at the calyx of Held synapse. (*A*) Transmission image of a calyx of Held synapse in a slice preparation of the mouse MNTB under infrared illumination. (*B*) Fluorescence image of the nerve terminal shown in *A* during whole-cell recording, taken at 380-nm excitation (8 × 8 binning). The binned pixels indicated by circles were used to obtain the $[Ca^{2+}]_i$ traces shown in C. The dotted box shows the region used to measure the background fluorescence. (*C*) (*Top*) Fluorescence excitation protocol used for this Ca^{2+} uncaging experiment. Only a single ratio image pair is acquired before triggering the flash, to keep undesired photolysis of DM-Nitrophen low before the flash. (*Middle*) Fluorescence ratio changes (lines, *right* y-axis) and the corresponding change in $[Ca^{2+}]_i$ (open circles, *left* y-axis) for the six binned pixels indicated in *A*, before and after flash-photolysis of Ca^{2+}-loaded DM-Nitrophen. The $[Ca^{2+}]_i$ averaged over all pixels is shown by the filled black symbols. Note that in this set of calibrations R_{min} was 0.40. (*Bottom*) The postsynaptic EPSC evoked by the presynaptic $[Ca^{2+}]_i$ jump; the *inset* shows the rising phase of the flash-evoked EPSC at higher time resolution.

transmitter release, as measured by the simultaneous recording of the excitatory postsynaptic current, EPSC (Fig. 2C, bottom). The EPSC can be deconvolved with the mEPSC waveform (Neher and Sakaba 2001) to yield the time course of quantal transmitter release following Ca^{2+} uncaging. Thus, a quantitative relation between the rate of transmitter release and the presynaptic $[Ca^{2+}]_i$ in the presynaptic nerve terminal can be achieved (Schneggenburger and Neher 2000).

SUMMARY

By stimulating nerve terminals with Ca^{2+} uncaging, spatial gradients of $[Ca^{2+}]_i$, which develop during the opening of voltage-gated Ca^{2+} channels, are avoided. By studying transmitter release in response to a spatially homogeneous $[Ca^{2+}]_i$ signal of measurable amplitude, valuable insights into the Ca^{2+} dependence of transmitter release and into exocytosis–endocytosis coupling can be gained (Bollmann et al. 2000; Schneggenburger and Neher 2000; Felmy et al. 2003; Lou et al. 2005; Wölfel et al. 2007; Sun et al. 2007; Hosoi et al. 2009).

RECIPES

CAUTION: See Appendix 6 for proper handling of materials marked with <!>.
See the end of the chapter for recipes for reagents marked with <R>.

DM-Nitrophen

Dissolve DM-Nitrophen (nominally 100 mM) in sterile ddH_2O. Protect the DM-Nitrophen solution from ambient light whenever possible. Aliquot and store at –80°C.

Titration to Estimate Purity of DM-Nitrophen

To assess the content of unphotolyzed DM-Nitrophen in the stock solution, titrate with $CaCl_2$ <!> as follows.

1. Prepare a solution containing (nominally) 1 mM DM-Nitrophen and 100 µM of the high-affinity Ca^{2+} indicator, Fura-2.
2. Incrementally add $CaCl_2$ in steps of 0.1 mM, and measure the resulting Fura-2 ratios in a glass cuvette (50-µm path length).
3. When the free $[Ca^{2+}]_i$ exceeds ~150 nM (based on a suitable calibration of Fura-2), we assume that all DM-Nitrophen is bound to Ca^{2+}, and therefore the effective DM-Nitrophen concentration can be determined from the total added $[Ca^{2+}]$.

We usually find a purity of ~60%–70% for DM-Nitrophen.

ACKNOWLEDGMENTS

We thank Erwin Neher for fruitful discussions on the use of photolyzable Ca^{2+} chelators.

REFERENCES

Bollmann J, Sakmann B, Borst J. 2000. Calcium sensitivity of glutamate release in a calyx-type terminal. *Science* **289:** 953–957.

Borst JG, Helmchen F, Sakmann B. 1995. Pre- and postsynaptic whole-cell recordings in the medial nucleus of the trapezoid body of the rat. *J Physiol* **489:** 825–840.

Ellis-Davies GCR. 2003. Development and application of caged calcium. *Methods Enzymol* **360:** 226–238.

Ellis-Davies GCR. 2011. Nitrobenzyl-based caged neurotransmitters. In *Imaging: A laboratory manual* (ed. Yuste R), pp. 889–896. Cold Spring Harbor Laboratory Press, Cold Spring Harbor, NY.

Felmy F, Neher E, Schneggenburger R. 2003. Probing the intracellular calcium sensitivity of transmitter release during synaptic facilitation. *Neuron* **37:** 801–811.

Forsythe ID. 1994. Direct patch recording from identified presynaptic terminals mediating glutamatergic EPSCs in the rat CNS, in vitro. *J Physiol* **479:** 381–387.

Grynkiewicz G, Poenie M, Tsien R. 1985. A new generation of Ca^{2+} indicators with greatly improved fluorescence properties. *J Biol Chem* **260:** 3440–3450.

Heinemann C, Chow RH, Neher E, Zucker RS. 1994. Kinetics of the secretory response in bovine chromaffin cells following flash photolysis of caged Ca^{2+}. *Biophys J* **67:** 2546–2557.

Helmchen F. 2011. Calibration of fluorescent calcium indicators. In *Imaging: A laboratory manual* (ed. Yuste R), pp. 403–416. Cold Spring Harbor Laboratory Press, Cold Spring Harbor, NY.

Hosoi N, Holt M, Sakaba T. 2009. Calcium dependence of exo- and endocytotic coupling at a glutamatergic synapse. *Neuron* **63:** 216–229.

Lou X, Scheuss V, Schneggenburger R. 2005. Allosteric modulation of the presynaptic Ca^{2+} sensor for vesicle fusion. *Nature* **435:** 497–501.

Neher E. 2011. Quantitative aspects of calcium fluorimetry. In *Imaging: A laboratory manual* (ed. Yuste R), pp. 417–426. Cold Spring Harbor Laboratory Press, Cold Spring Harbor, NY.

Neher E, Sakaba T. 2001. Combining deconvolution and noise analysis for the estimation of transmitter release rates at the calyx of Held. *J Neurosci* **21:** 444–461.

Schneggenburger R, Neher E. 2000. Intracellular calcium dependence of transmitter release rates at a fast central synapse. *Nature* **406:** 889–893.

Sun J, Pang ZP, Qin D, Fahim AT, Adachi R, Sudhof TC. 2007. A dual-Ca^{2+}-sensor model for neurotransmitter release in a central synapse. *Nature* **450:** 676–682.

Uhl R. 2011. Arc lamps and monochromators for fluorescence microscopy. In *Imaging: A laboratory manual* (ed. Yuste R), pp. 117–122. Cold Spring Harbor Laboratory Press, Cold Spring Harbor, NY.

Wölfel M, Lou X, Schneggenburger R. 2007. A mechanism intrinsic to the vesicle fusion machinery determines fast and slow transmitter release at a large CNS synapse. *J Neurosci* **27:** 3198–3210.

Zucker RS. 1992. Effects of photolabile calcium chelators on fluorescent calcium indicators. *Cell Calcium* **13:** 29–40.

17 Multiple Light Scattering Changes Associated with Secretion from Peptidergic Nerve Terminals

Brian M. Salzberg, Martin Muschol,[1] Paul Kosterin, and Ana Lia Obaid

Departments of Neuroscience and Physiology, University of Pennsylvania School of Medicine, Philadelphia, Pennsylvania 19104-6074

ABSTRACT

Intrinsic optical changes, viz., light scattering signals, occur in mammalian nerve terminals during and immediately following the arrival of the action potential. In the neurohypophysis (posterior pituitary gland), the action potential is coupled to calcium-mediated secretion of the neuropeptides oxytocin and vasopressin, and it is this excitation–secretion coupling that is intimately related to the extremely rapid changes in light scattering that are detailed here. These optical signals provide a millisecond–time-resolved monitor of events in the terminals that follow the arrival of the action potential and the entry of calcium.

INTRODUCTION

In this chapter we describe the use of light scattering measurements to directly monitor events associated with the secretion of neuropeptides. In the neurosecretory terminals of the vertebrate neurohypophysis, the arrival of the action potential is coupled to the secretion of the neuropeptides, oxytocin and vasopressin. This excitation–secretion (E-S) coupling is mediated by an increase in intracellular calcium concentration (Douglas 1963, 1978), and, in mammals, the secretory event is accompanied by extremely rapid changes in light scattering, measured as transparency (Salzberg et al. 1985), that are relatively large ($\Delta I/I \sim 3 \times 10^{-4}$, where I is the fractional change in intensity) compared with other intrinsic signals from neurons. These intrinsic optical signals provide a millisecond–time-resolved monitor of events in the terminals that follow the arrival of the action potential and the entry of calcium, and they are intimately related to the release of neuropeptides. Whereas the identification of the optical signals described in this chapter with particular steps in the sequence of events that couples excitation to secretion is not yet certain, recent experiments using dynamic high bandwidth atomic force microscopy (Kim et al. 2007) have begun to clarify this important question.

[1]Present address: Department of Physics, University of South Florida, Tampa, Florida 33620.

Secretion of peptides in the neurohypophysis results from bursts or trains of action potentials invading the terminals (Poulain and Wakerley 1982) rather than from the occurrence of individual impulses. The top trace in Figure 1B illustrates the changes in transmitted light intensity at 675 nm, when the terminals of an unstained CD-1 mouse (or any mouse) neurohypophysis (illustrated in Fig. 1A) are stimulated at 16 Hz for 400 msec. The lower red trace shows the same measurement, at a reduced gain, after the same preparation was stained for 25 min in a 100 µg/mL solution of the potentiometric merocyanine-oxazolone dye, NK2367 (Salzberg et al. 1977). The intrinsic signal, at the top, consists of at least three separable components (Salzberg et al. 1985; Obaid et al. 1989). The rapid upstroke, which is an increase in large-angle light scattering, referred to as the E wave (for excitation), marks the arrival of the propagated action potential and action currents in the terminals, and probably has voltage- and current-dependent components (Cohen et al. 1972a,b), whereas the large, long-lasting decrease in scattered-light intensity (an increase in transparency), the S wave (for secretion), is considerably more interesting. (The S wave is followed, in turn, by a variable, long-lasting recovery, the R wave, which seems to reflect changes in the *pars intermedia*.) This S wave has been

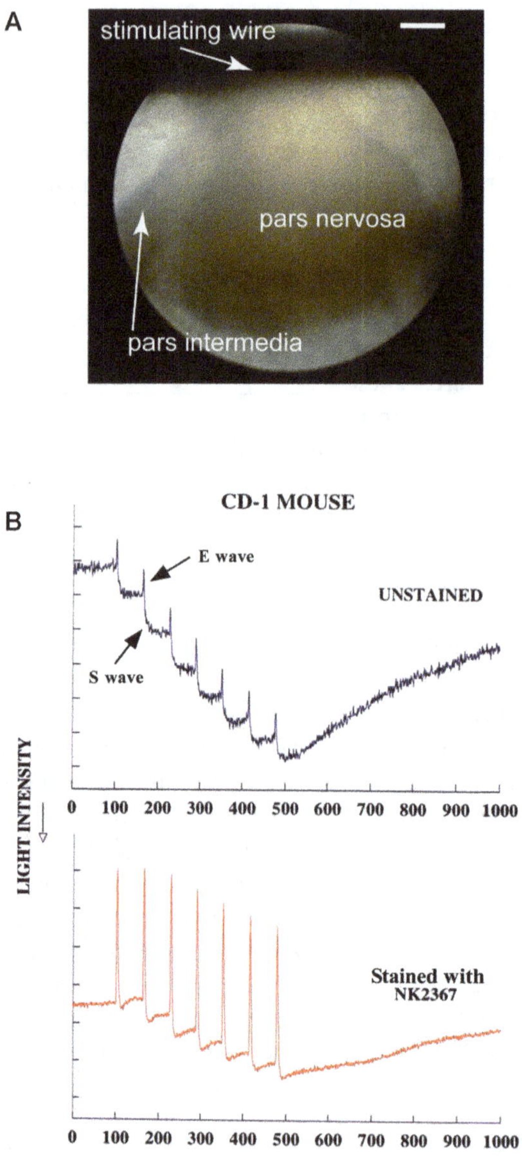

FIGURE 1. (*A*) A dissection microscope image of an isolated mouse neurohypophysis with surrounding *pars intermedia* and with the infundibular stalk contacting stimulation leads. Scale bar, 200 µm. (*B*) The optical response to trains of stimuli (500 µsec at 16 Hz for 410 msec) of unstained and stained nerve terminals of the CD-1 mouse neurohypophysis. (*Top* trace) The intrinsic optical signal recorded at 675 ± 35 nm from the nerve terminals of an unstained CD-1 mouse neurohypophysis during a stimulus train. The rapid upstrokes (E waves) reflect the arrival of the action potentials (and action *currents*) in the terminals, and the sustained downward deflections (increases in transparency; S waves) represent decreases in large-angle light scattering that are closely coupled to secretion of arginine vasopressin and oxytocin. The R waves are too slow to be seen in this figure. (*Bottom* trace) After staining the CD-1 mouse terminals with the potentiometric merocyanine-oxazolone dye NK2367 (100 µg/mL), extrinsic absorption signals that monitor the changes in membrane potential are superimposed on the light scattering signals. All of the traces were recorded without averaging using a single photodiode (PV-444a from EG&G) mounted behind a set of adjustable slits in the image plane of a compound microscope (Model UEM, Carl Zeiss, Inc.). 10x/0.4 numerical aperture (NA) objective (Wild Heerbrugg); ac coupling time constant 1.7 sec; rise time of the light measuring system (10%–90%) 330 µsec; temperature 24°C–26°C.

studied for a number of years (Salzberg et al. 1985, 1986; Obaid et al. 1989; Parsons et al. 1992; Salzberg and Obaid 1997) and, although it is intimately related to peptide release from these terminals, and it monitors some event that depends on Ca^{2+} entry, it may or may not reflect vesicle fusion, because it is absent in the frog, whose vesicles appear to be indistinguishable from those of the mouse. This rapid-onset, but long-lasting change in the intrinsic optical properties of the mammalian terminals shares many of the characteristics of the behavior of fast neurosecretory systems. These include dependence on stimulation frequency (Salzberg et al. 1985), paired-pulse facilitation and exhaustion with high-frequency stimulation, dependence on extracellular Ca^{2+} (Salzberg et al. 1985), sensitivity to Ca^{2+}-channel blockers (Obaid et al. 1989) and to aminoglycoside antibiotics (Parsons et al. 1992), and sensitivity to other interventions known to influence secretion, such as D_2O substitution for water (Salzberg et al. 1985). Indeed, changes in the amplitude of the S wave are covariant with the effects on secretion of vasopressin and oxytocin produced by a wide variety of agents, and all of the manipulations that alter the light scattering signal produce comparable effects on arginine vasopressin release, as determined by radioimmunoassay (Gainer et al. 1986).

EFFECTS OF Ca^{2+} ON THE INTRINSIC OPTICAL SIGNAL COMPONENTS

Consider, for example, the effects of $[Ca^{2+}]_o$, illustrated in Figure 2. As has been known since the pioneering work of Douglas (Douglas 1963; Douglas and Poisner 1964), calcium ion profoundly influences neurosecretory activity, and the extracellular concentration of Ca^{2+} would be expected to have large effects on the size of an intrinsic signal related to secretion from the terminals of the neurohypophysis. The records shown in A and B are the averages of 16 sweeps, in 2.2 mM Ca^{2+} and 10 mM Ca^{2+}, respectively. The effects of Ca^{2+} are clear, and the block of the S wave by Cd^{++} (not shown) is dramatic. Calcium, however, has other effects, including influence on threshold for action potential. Some of the results of threshold variation can be eliminated by always using supramaximal stimulation, but

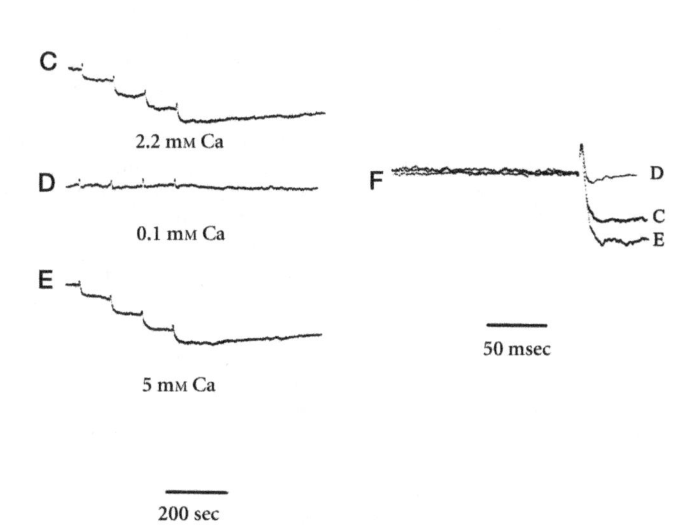

FIGURE 2. The effects of $[Ca^{2+}]_o$ on the amplitude of the S wave of the light scattering signal in the mouse neurohypophysis, when account is taken of the number of active terminals. (*A*) Light scattering changes at 675 ± 26 nm accompanying stimulation at 10 Hz of an unstained mouse neurohypophysis in normal mouse Ringer's solution (2.2 mM Ca^{2+}). (*B*) Light scattering changes in the same preparation 40 min after exposure to a Ringer's solution containing 10 mM Ca^{2+}. Traces *A* and *B* are each the average of 16 sweeps. (*C*) Light scattering changes accompanying stimulation at 10 Hz in a different preparation, in normal mouse Ringer's solution. (*D*) Light scattering changes in the same preparation as C recorded 50 min after a reduction in $[Ca^{2+}]_o$ to 0.1 mM by Mg^{++} substitution. (*E*) Light scattering changes in the same preparation as C recorded 30 min after an increase in $[Ca^{2+}]_o$ to 5 mM. Records *C–E* were each recorded without averaging. (*F*) Averages of 16 trials recorded under the same conditions as traces *C–E*. These records were obtained immediately after the corresponding single trains, and are shown here normalized to the height of the E wave (see text). The normalization factor for each trace is given in parentheses: trace *C* (1.0); trace *D* (0.72); trace *E* (1.16). Objective 10x, 0.4 NA; ac coupling time constant 3 sec; rise time of the light measuring system (10%–90%), 1.1 msec. Temperature 24°C–26°C.

notice that there remains some variation in the size of the E waves that coincide in time with the terminal action potential. This very small light scattering signal is related to the action potential, or the action current, or probably, to both (Cohen et al. 1972a,b), and is therefore roughly proportional to the number of terminals activated at a given time. On this admittedly crude assumption, the intrinsic signal is normalized to the size of the E wave to compensate for changes in the invisibility of the tissue. This is shown most clearly in traces C–F of Figure 2, where the effect of Ca^{2+} on the light scattering signal, per active terminal, is illustrated. In Figure 2, the records labeled C (2.2 mM Ca^{2+}), D (0.1 mM Ca^{2+}), and E (5.0 mM Ca^{2+}) are here single sweeps but, between each trial, 16 trials were averaged to improve the normalization to the initial upstroke, and these traces are shown on the right in Figure 2F. Here, it is evident that $[Ca^{2+}]_o$ has a powerful effect on the amplitude of the S wave, per active terminal. This is in contrast to the observation that micromolar concentrations of 4-aminopyridine enhance secretion by increasing the extent to which the action potential invades the highly ramified terminal arborization of the magnocellular neurons (Obaid and Salzberg 1996; data not shown).

POSSIBLE MECHANISMS UNDERLYING SECRETION-COUPLED LIGHT SCATTERING

There is now preliminary evidence linking the light scattering changes, E wave and S wave, to particular steps in the sequence of events that couples excitation to secretion in the mammalian neurohypophysis. Dynamic high bandwidth atomic force microscopy (HBAFM) (Kim et al. 2007) has determined that the E wave reflects the occurrence of a "mechanical spike" that probably results from the inward movement of H_2O along with Na^+ ions during the action potential upstroke, followed by an opposite movement during K^+ efflux. As for the S wave, the fusion of secretory vesicles during exocytosis should result in the loss of relatively high refractive index particles and, thereby, reduce the refractive index gradients in the terminals. The extrusion and dissolution of the dense core granules contained in the vesicles should also result in a long-lasting volume *decrease*. Such a volume decrease has also been detected by dynamic HBAFM (Kim et al. 2007) (Fig. 3), and there is the strong possibility that this mechanical event is associated with the S wave. However, the very early onset of the S wave suggests that the optical signal could also arise as a result of some calcium-dependent

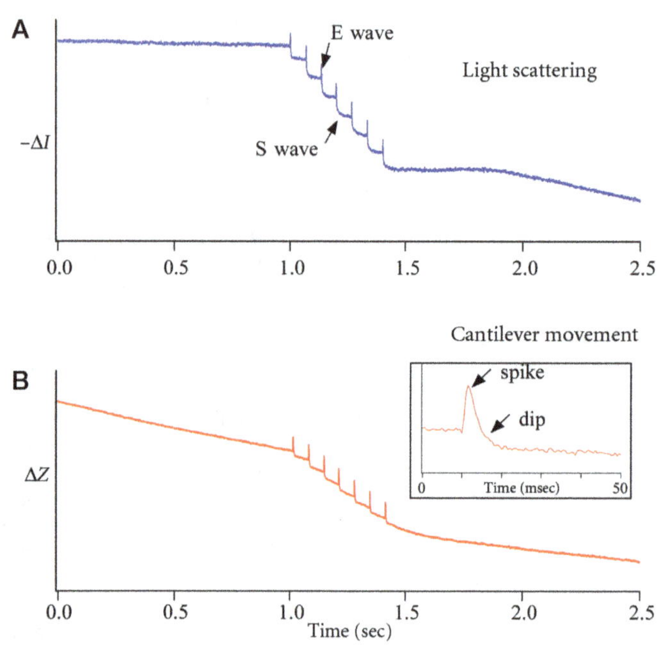

FIGURE 3. Comparison of the light scattering signals with mechanical changes in the mouse neurohypophysis. (*A*) Light scattering signals recorded as changes in transmission (–Δ*I*) at 675 ± 25 nm of unstained nerve terminals of the CD-1 mouse neurohypophysis in response to a train of stimuli (seven 500-µsec stimuli delivered at 15 Hz). An increase in transmitted light intensity is plotted downward. (*B*) Mechanical events triggered by action potentials, monitored with a modified high bandwidth atomic force microscope (HBAFM). The *inset* shows the first mechanical spike and dip from this train on an expanded time base. All traces were low-pass filtered at 1 kHz with an 8-pole Bessel filter and sampled at 2 kHz. (Modified, with permission, from Kim et al. 2007, ©Elsevier.)

process before the fusion of the vesicles and the release of their contents. For example, it is possible that the light scattering changes reflect a phase transition in the contents of the vesicles, or some rapid change in intracellular calcium stores, following calcium entry, but before secretion (Salzberg and Obaid 1997). Further experiments are required to finally resolve this question. In any event, these intrinsic optical signals are quite easily measured in a tissue like the mammalian neurohypophysis, where the relative homogeneity of the preparation and the enormous proliferation of magnocellular neuron terminals undoubtedly contribute to the generation of these relatively large optical changes. Magnocellular neurons, located in the hypothalamus, project their axons as bundles of fibers through the median eminence and infundibular stalk to terminate in the neurohypophysis, where neuropeptides and proteins are secreted into the local circulation. In the rat, for example, some 18,000 magnocellular neurons give rise to 40,000,000 terminals in the neurohypophysis, and nearly 99% of the excitable membrane belongs to terminals or secretory swellings (Nordmann 1977). At the same time, the precise control of the timing of the action potential in the terminals, the absence of any postsynaptic structure to further confound the interpretation of the optical signals, and the absence of any barrier to the collection of the secretory products for biochemical assay make the neurohypophysis a particularly favorable system for understanding the intrinsic optical changes and how they are related to secretion of neuropeptides.

Protocol

Measuring Intrinsic Optical Signals from Mammalian Nerve Terminals

Light scattering procedures are designed to measure intrinsic optical signals (light scattering changes) from mammalian nerve terminals. In practice, these signals are remarkably simple to record from any of the mammalian neurohypophyses (posterior pituitary; *pars nervosa*) that have been studied. To date this approach has been used successfully in mouse, rat, and guinea pig. Instrumentation requirements and a protocol for preparation of the neurohypophysis are given below.

IMAGING SETUP

The only requirements (Salzberg et al. 1977, 1985) are a light source that is stable to a few parts in 10^5, a vibration-isolation table, an imaging system such as a microscope, and a simple low-noise photodetector in the image plane of the objective. The latter can be a single silicon PIN photodiode, an avalanche photodiode, an array of photodiodes, a photomultiplier (although this is not recommended as the incident light intensities need to be quite high), or even a scientific charge-coupled device (CCD) camera, having at least 14 bits of gray-scale resolution. (Because photodiode detectors are readily ac coupled, 8-bit resolution of the time-varying component of the signal is generally adequate when photodiodes are used.)

MATERIALS

See the end of the chapter for recipes for reagents marked with <R>.

Reagents

Mice, CD-1 females, 30–40-d-old
Mouse Ringer's solution, oxygenated <R>

Equipment

Dissecting microscope
Dissection dish, Sylgard-lined
Dissection instruments (including iridectomy scissors, fine [Dumont #5] forceps, and pins)
Guillotine
Imaging microscope and associated instruments (see Imaging Setup)
Pt-Ir electrodes, Teflon coated (see Step 5)
Stimulus isolator

EXPERIMENTAL METHOD

Isolation of the Neurointermediate Lobe (Comprising Neurohypophysis and *Pars Intermedia*)

1. Kill female CD-1 mice, 30–40 d old, by guillotine and exsanguinate. Pin the head to the bottom of a Sylgard-lined dissection dish and remove the skin from the skull. Then open the skull along the dorsal midline and remove the dorsal cranium bilaterally.

2. Under a low-power dissecting microscope, cut the optic and olfactory nerves and reflect the brain caudally before removing it. During this procedure, the infundibular stalk is automatically ruptured, leaving an infundibular stump and the entire pituitary gland in the base of the skull, held in place by thin connective tissue.
3. Circulate oxygenated mouse Ringer's solution over the pituitary. Use iridectomy scissors and fine (Dumont # 5 INOX) forceps to remove the gland.
4. Once the pituitary has been isolated, separate the anterior pituitary (*pars anterior*) from the neurointermediate lobe (neurohypophysis, or *pars nervosa*, plus *pars intermedia*). The *pars intermedia* consists of a delicate lacework of tissue surrounding the neurohypophysis and provides a convenient site for pinning to the bottom of an experimental chamber.

Experimental Setup

5. Pin the neurohypophysis to the thin Sylgard bottom of a chamber so that the infundibular stalk lies clasped between a pair of Pt-Ir electrodes that are coated with Teflon, except where they contact the infundibulum.
6. Deliver brief shocks (100–200 V), lasting between 300 and 500 μsec, through a stimulus isolator. The resulting changes in the transparency of the tissue are recorded by the photodetector, which is positioned in the image plane of the optical system behind an adjustable diaphragm, or mask, so that only light passing through the preparation is monitored. With adequate oxygenation, the resulting intrinsic optical signals can easily be recorded for several hours.

CONCLUSION

In the case of mammalian peptidergic neurosecretory terminals, the challenge lies not in detecting the changes in their optical properties, but in divining their meaning. The E wave apparently reflects the occurrence of a very rapid, action potential-coupled volume increase in the terminals of the neurohypophysis (a "mechanical spike"). The S wave of the intrinsic optical signal, in turn, has its onset within a few milliseconds of the arrival of the action potential, suggesting that it could arise as the result of some calcium-dependent process before the fusion of secretory vesicles and the release of their contents, or, alternatively, it could reflect the actual extrusion and dissolution of the dense core granules. The first category includes the possibility that the S wave described here may reflect a phase transition of the contents of the secretory granules (e.g., as the result of intragranular pH changes). Another possibility is that it monitors the contents or status of intraterminal calcium stores and that these reservoirs, located within a few hundred nanometers of the plasma membrane, play a direct role in exocytosis in mammalian neurosecretory terminals. However, in light of the recently discovered experimental correlation between mechanical events and intrinsic optical changes in the neurohypophysis (Kim et al. 2007), the possibility that the S wave monitors a volume decrease related to secretory granule extrusion must be considered very seriously. In any event, the weak dependence on wavelength of the signal reflecting secretion (the S wave), contrasted with the strong wavelength dependence of the extrinsic signals provided by calcium-indicator dyes, permits one to monitor simultaneously, in an AM-dye-loaded preparation, the Ca^{2+} transients in the nerve terminals and the time course of events intimately associated with the release of neuropeptides. The inherently fast responses of the two optical measurements should certainly improve our ability to resolve early events in the coupling of excitation to secretion.

RECIPE

CAUTION: See Appendix 6 for proper handling of materials marked with <!>.
Recipes for reagents marked with <R> are included in this list.

Mouse Ringer's Solution

NaCl	154 mM
KCl	5.6 mM
$MgCl_2$	1 mM
$CaCl_2$ <!>	2.2 mM
Glucose	10 mM
HEPES	20 mM

Adjust the pH to 7.4 with NaOH <!>.
Oxygenate the solution before use.

ACKNOWLEDGMENTS

This work was supported by U.S. Public Health Service grants NS-16824 and NS-40966 (BMS) and NS-35561 (ALO).

REFERENCES

Cohen LB, Keynes RD, Landowne D. 1972a. Changes in light scattering that accompany the action potential in squid giant axons: Potential-dependent components. *J Physiol* **224:** 701–725.

Cohen LB, Keynes RD, Landowne D. 1972b. Changes in light scattering that accompany the action potential in squid giant axons: Current-dependent components. *J Physiol* **224:** 727–752.

Douglas WW. 1963. A possible mechanism of neurosecretion-release of vasopressin by depolarization and its dependence on calcium. *Nature* **197:** 81–82.

Douglas WW. 1978. Stimulus-secretion coupling: Variations on the theme of calcium activated exocytosis involving cellular and extracellular sources of calcium. *Ciba Found Symp* **54:** 61–90.

Douglas WW, Poisner AM. 1964. Stimulus secretion coupling in a neurosecretory organ and the role of calcium in the release of vasopressin from the neurohypophysis. *J Physiol* **172:** 1–18.

Gainer H, Wolfe SA Jr, Obaid AL, Salzberg BM. 1986. Action potentials and frequency-dependent secretion in the mouse neurohypophysis. *Neuroendocrinology* **43:** 557–563.

Kim G-H, Kosterin K, Obaid AL, Salzberg BM. 2007. A mechanical spike acompanies the action potential in mammalian nerve terminals. *Biophys J* **92:** 3122–3129.

Nordmann JJ. 1977. Ultrastructural morphometry of the rat neurohypophysis. *J Anat* **123:** 213–218.

Obaid AL, Salzberg BM. 1996. Micromolar 4-aminopyridine enhances invasion of a vertebrate neurosecretory terminal arborization: Optical recording of action potential propagation using an ultrafast photodiode-MOSFET camera and a photodiode array. *J Gen Physiol* **107:** 353–368.

Obaid AL, Flores R, Salzberg BM. 1989. Calcium channels that are required for secretion from intact nerve terminals of vertebrates are sensitive to ω-conotoxin and relatively insensitive to dihydropyridines. Optical studies with and without voltage-sensitive dyes. *J Gen Physiol* **93:** 715–729.

Parsons TD, Obaid AL, Salzberg BM. 1992. Aminoglycoside antibiotics block voltage-dependent calcium channels in vertebrate nerve terminals. *J Gen Physiol* **99:** 491–504.

Poulain DA, Wakerley JB. 1982. Electrophysiology of hypothalamic magnocellular neurons secreting oxytocin and vasopressin. *Neuroscience* **7:** 771–808.

Salzberg BM, Obaid AL. 1997. Triggered calcium release from intraterminal stores may play a direct role in neuropeptide secretion: Evidence from light scattering in mammalian nerve terminals. *Biophys J* **72:** A227.

Salzberg BM, Grinvald A, Cohen LB, Davila HV, Ross WN. 1977. Optical recording of neuronal activity in an invertebrate central nervous system: Simultaneous monitoring of several neurons. *J Neurophysiol* **40:** 1281–1291.

Salzberg BM, Obaid AL, Gainer H. 1985. Large and rapid changes in light scattering accompany secretion by nerve terminals in the mammalian neurohypophysis. *J Gen Physiol* **86:** 395–411.

Salzberg BM, Obaid AL, Gainer H. 1986. Optical studies of excitation and secretion at vertebrate nerve terminals. In *Optical methods in cell physiology* (ed. DeWeer P, Salzberg BM), pp. 133–164. Society of General Physiologists and Wiley-Interscience, New York.

18 Imaging Synaptic Vesicle Recycling Using FM Dyes

Peer Hoopmann,[1] Silvio O. Rizzoli,[1] and William J. Betz[2]

[1]*European Neuroscience Institute Göttingen, 37077 Göttingen, Germany;* [2]*Department of Physiology and Biophysics/C-240, University of Colorado Medical School, Denver, Colorado 80262*

ABSTRACT

The synaptic vesicle is the essential organelle of the synapse. The vesicles fuse to the plasma membrane (exocytosis) to release neurotransmitters and are then retrieved into the nerve terminal (endocytosis). Afterward, they are redirected to the vesicle cluster for further rounds of fusion (vesicle recycling). Many approaches for studying synaptic vesicle recycling have been devised, one of which, the styryl (FM) dye, is well suited for this purpose and is discussed here. The FM dyes have a unique set of properties that allows them to selectively label recycling vesicles: They reversibly stain, but do not permeate, membranes, so can specifically label membrane-bound organelles, and their quantum yield is drastically higher when bound to membranes than when in aqueous solution. Here, we review the main characteristics of FM dyes and their use in studying vesicle recycling. We also discuss recent FM dye monitoring techniques that have been used to investigate the kinetics of synaptic vesicle fusion in an attempt to distinguish between the alternative scenarios of full-collapse and transient (kiss-and-run) fusion.

INTRODUCTION

Release of neurotransmitters from nerve terminals requires the fusion of synaptic vesicles with the presynaptic plasma membrane (exocytosis). Next, the vesicles are internalized (endocytosis) and directed to the pool of vesicles inside the nerve terminal (vesicle recycling). As a consequence of this vesicle cycle, the inner leaflet of the vesicle membrane is temporarily exposed to the outside space, which provides an opportunity for membrane-staining compounds to specifically label recycling vesicles. Because the endocytic process also results in the uptake of a volume of fluid, even fluid-phase tracers can be used. Indeed, various markers have been used, such as sulforhodamine (Lichtman et al. 1985), fluorescently labeled antibodies (Kraszewski et al. 1996; Westphal et al. 2008), styryl dyes (i.e., FM dyes; Betz and Bewick 1992), and fluorescent dextrans (Holt et al. 2003). Among these markers, FM dyes have proven to be valuable in a number of different preparations, including neuromuscular junctions (NMJ) from the frog (Betz and Bewick 1992), snake (Teng et al. 1999), lizard (Lindgren et al. 1997), and larval *Drosophila* (Ramaswami et al. 1994); hippocampal neurons in culture (Ryan et al. 1993) and acute slice (Zakharenko et al. 2001); goldfish retinal bipolar cells (Zenisek et al. 2000); *Caenorhabditis elegans* neurons (Kay et al. 1999); rat calyx of Held (de Lange et al. 2003); inner and outer hair cells (Griesinger et al. 2002; Kaneko et al. 2006); and a variety of nonneuronal preparations (for a review, see Cochilla et al. 1999).

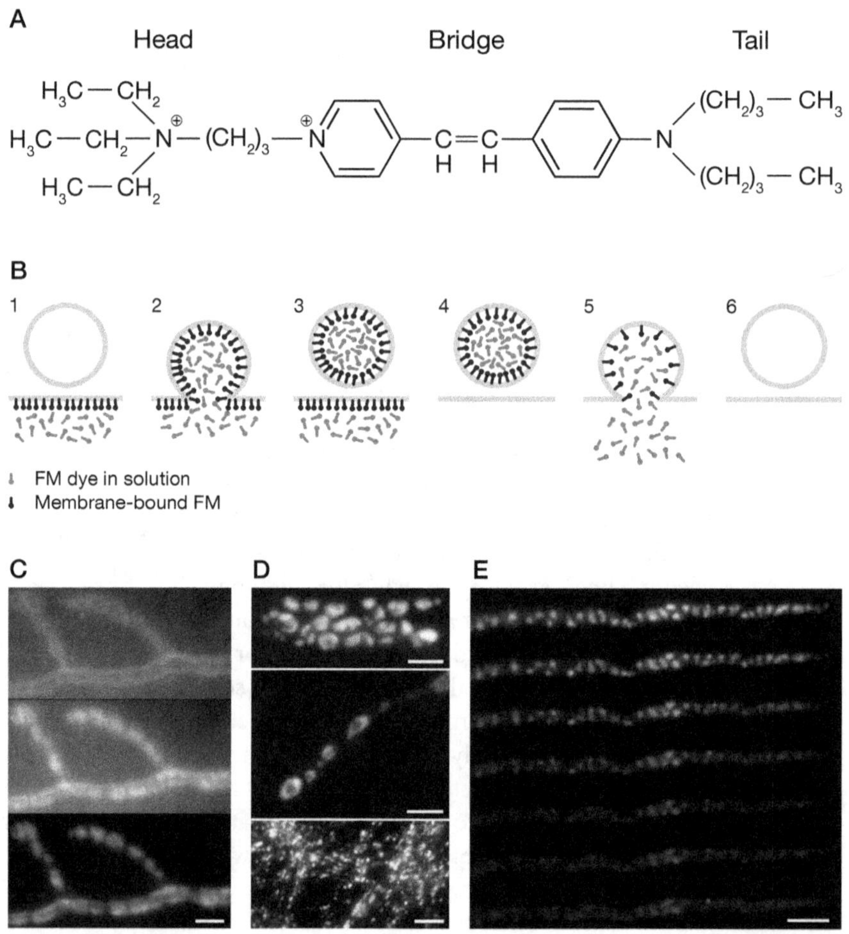

FIGURE 1. (*A*) Chemical structure of FM1-43. The molecule is composed of a positively charged head group, which prevents the dye from permeating membranes and a lipophilic tail, which ensures dye partitioning into membranes. These components are connected by a double bond bridge, the number of double bonds determining the spectral characteristics. (*B*) FM dye staining and destaining. (*1*) The dye is added to the solution bathing the preparation. It does not have access to the intracellular space. (*2*) The preparation is stimulated. Vesicles exocytose and come in contact with the dye. (*3*) Vesicles endocytose and take up dye. (*4*) The dye is washed from the extracellular solution; it departitions from the plasma membrane but remains trapped in internalized vesicles (staining). (*5*) The stained preparation is stimulated in the absence of the dye. Vesicles fuse with the membrane and release dye (destaining). (*6*) Destained vesicles are internalized; a new experiment of staining and destaining can be performed. (*C*) Frog motor nerve terminal shown during staining. Images were acquired and processed identically, and are, therefore, directly comparable. (*Top*) The preparation had been bathed in FM dye-containing solution for 1 min. The surface membrane of the nerve terminal is faintly stained. (*Middle*) After a 1-min tetanus (30 Hz), followed by 15 min of rest in dye-containing solution. The terminal is brightly stained, but background fluorescence is also high. (*Bottom*) After dye washout, the background staining is greatly reduced, and clusters of vesicles can be observed as distinct fluorescent spots. (*D*) FM-labeled nerve terminals in different preparations: (*Top*) *Drosophila* larval NMJ. Scale bar, 3 µm. (*Middle*) Lizard (Anolis) NMJ. Scale bar, 2 µm. (*Bottom*) Cultured hippocampal neurons (fluorescent spots represent single presynaptic boutons). Scale bar, 5 µm. (Photograph courtesy of Dr. V. Murthy.) (*E*) Nerve terminal destaining. A stained preparation was stimulated at 2 Hz for ~40 min in the presence of 10-µM curare to prevent muscle movement. The images were taken approximately every 5 min. The nerve terminal loses most of its initial fluorescence. Scale bar, 5 µm. (*E*, Reprinted, with permission, from Betz and Henkel 1994, ©The Rockefeller University Press.)

FM Dyes: Properties and Mechanism of Action

Several FM dyes have been synthesized (Betz et al. 1996), the most widely used being FM1-43 (chemical structure shown in Fig. 1A). Because of its amphipathic nature, the FM dye reversibly partitions into lipid membranes without permeating them, so that when dye is present in the bathing solution, some of it becomes trapped in endocytic organelles. Because the quantum yield of FM dye is several 100-fold higher when in membranes versus aqueous solution, these dyes are exceptionally well suited for labeling membrane-bound compartments.

All FM dyes share the same general structure: a positively charged hydrophilic head group and a lipophilic tail (usually formed by two aliphatic hydrocarbon chains) connected by a double bond bridge (Fig. 1A). The length of the tail determines the dye's hydrophobicity and its ability to be washed out of membranes (e.g., the time constants for dye washout are ~1 sec for the short-tailed FM2-10, and 3–6 sec for FM1-43; see also Fig. 3A). The length of the double bond bridge connecting the two aromatic rings (i.e., the extent of the conjugated double bond system) determines the dye's spectral properties. Dyes with one double bond (FM1-43, FM2-10) fluoresce at shorter wavelengths (green) than ones with two or three double bonds (FM4-64, red).

A typical FM dye experiment at a synapse is illustrated in Figure 1B. Preparations are bathed in a solution containing the dye, which labels the extracellular membranes (Fig. 1B1). Exocytosis is then induced (e.g., by electrical stimulation), and inner leaflets of vesicles come in contact with the dye-containing solution, thus becoming labeled (Fig. 1B2). After a fraction of the vesicles are internalized through compensatory endocytosis (Fig. 1B3), the dye is washed from the solution (Fig. 1B4), leaving label only within those organelles that formed while the dye was present. Exocytosis can then be monitored by stimulating these preparations in the absence of dye. The fluorescent molecules wash out of the vesicles into the bath, which also dramatically lowers quantum yield (Fig. 1B5) so that the decrease in fluorescence gives a direct measurement of vesicle release. After internalization of destained vesicles (Fig. 1B6), the preparation is ready for another round of staining and destaining.

Protocol A

Staining and Destaining Synaptic Vesicles with FM Dyes

The general procedure for staining synaptic vesicles presented in this protocol is outlined in Figure 1B. Nerve terminals are stimulated (electrically or by depolarization with high K^+) in the presence of dye, next their vesicles are allowed to recycle, and finally dye is washed from the chamber (Fig. 1C). In NMJ preparations, movements of the muscle must be inhibited if imaging during stimulation is desired (e.g., by application of curare, a potent acetylcholine receptor inhibitor). Dissection and experiments should be performed under the optimal physiological conditions for the experimental preparation. If stimulation is performed by high-K^+ depolarization, the bathing solution should be adjusted by replacing KCl for NaCl to reach the desired final K^+ concentration (typically 60 mM).

IMAGING SETUP

An upright epifluorescence microscope equipped with a 40x–60x water-immersion objective is optimal. When working with cell cultures, the advantages of high-resolution oil-immersion objectives (e.g., 100x/1.4 numerical aperture [NA]) can be exploited by using an inverted setup. For FM1-43, maximal fluorescence is at 465-nm excitation/560-nm emission (Henkel et al. 1996); a conventional fluorescein filter set (435-nm excitation) can also be used. Because the major concern in FM imaging is phototoxicity (which appears usually before any clear signs of photobleaching and alters the ability of the preparations to recycle vesicles), keep the illumination to a minimum by including neutral-density filters. For example, when using a 40x/0.75-NA water-immersion objective with a 100-W mercury lamp, 5%–15% transmission is sufficient. On confocal setups, an argon 488-nm laser line is typically used for excitation, but special care has to be taken to avoid phototoxicity. Images are typically acquired by charge-coupled device (CCD) cameras or photomultiplier tubes, which allows for fast acquisition rates and software-based quantitative image analysis.

MATERIALS

Reagents

FM dye(s)

Dye concentrations are usually 2–10 μM for FM1-43 or FM4-64, and 25–40 μM for FM2-10. The dyes readily dissolve in water, and stocks of 1–5 mM concentration should be prepared and can be stored at 4°C.

Neuronal tissue or cell sample of interest

Equipment

Microscope and associated instruments (see Imaging Setup)

Neuronal stimulator (see Step 3)

EXPERIMENTAL METHOD

Sample Preparation

1. Prepare neuronal tissue or cell sample. For neuromuscular junctions, clean the muscle preparation by removing surface connective tissue, being careful not to damage muscle fibers or nerve cells (FM dyes intensely stain intracellular membranes of injured cells, creating unwanted background fluorescence). For cultured hippocampal neurons, cultures with minimal astrocytic background are preferable.

Staining

2. The dye-containing solution is washed onto the preparation. Incubate for ~1 min, to allow the dye to diffuse into the synaptic clefts.

   The dye stains the surface membranes of muscle cells (in NMJ) and nerve terminals (typical image shown in top panel of Fig. 1C, frog NMJ).

3. Stimulate the preparation in the presence of the dye. This will cause the preparation to fill with dye as vesicles are retrieved after ongoing exocytosis.

   The increased fluorescence intensity is proportional to the number of exocytosed vesicles. Electrical tetanic stimulation of the nerve (30 Hz for 1–5 min) triggers recycling of 30%–80% of all synaptic vesicles in frog NMJ, ensuring adequate staining. High-K^+ (60-mM) treatment of the preparation results in similar levels of dye uptake.

4. Allow the preparation to recover while still in the presence of the dye, to ensure complete recycling of the vesicles.

   The length of the recovery period depends on the length of the stimulus—for example, 1–3 min tetani require 10–20 min of rest. Figure 1C (middle panel) shows the same nerve terminal after stimulation for 1 min (30 Hz), followed by 15 min of recovery. The terminal took up dye, and it appears much brighter than in the top panel. However, as the dye is still present in the chamber, background fluorescence is also high.

5. Wash the dye from all surface membranes, for 30–60 min, at 4°C.

   Washing at low temperatures prevents dye release by spontaneous exocytosis. Dye-scavenging compounds, such as ADVASEP-7 (see Discussion), can be used to shorten washing times.

6. Image the preparation.

   As can be seen in Figure 1C (bottom panel), fluorescent spots are now visible within the nerve terminal, and background fluorescence is low. The spots (usually 1–3 µm in diameter) represent clusters of synaptic vesicles containing a few hundred to several thousand vesicles. Most synapses investigated to date stain with FM dyes in a similar manner (see several examples in Fig. 1D), although the synaptic morphology may vary substantially. Note, for example, the ring-like appearance of the *Drosophila* larval NMJ boutons, which is caused by their vesicles being concentrated at the bouton periphery.

Destaining

7. Stimulate labeled preparations in the absence of the dye. On synaptic vesicle exocytosis, the dye will be released and will diffuse into the aqueous solution.

   Figure 1E shows images from a time-lapse sequence from an experiment in which a frog NMJ nerve terminal was stimulated for 40 min at 2 Hz, losing most of its FM dye. Tetanic stimulation causes faster dye release than low-frequency stimulation (stimulation at 30 Hz for 2–3 min releases similar amounts of dye as 2-Hz stimulation for tens of minutes). During destaining, minimize the intensity of the excitation light to avoid phototoxicity without compromising image quality. A good rule of thumb is to adjust both the intensity with neutral-density filters and the exposure time such that the dynamic range of the initial (predestaining) images is ~8 bits (meaning that the brightest pixel is ~250–300 counts brighter than the dimmest one). Additionally, the number of images acquired should be kept as low as possible.

Protocol B

FM Dye Photoconversion

FM dyes can also be used as endocytic markers in electron microscopy (EM) through a procedure termed photoconversion (or photooxidation). Fluorescent dye molecules generate free radicals (reactive oxygen species) when subjected to strong illumination. These short-lived radicals readily oxidize any molecules found in the immediate vicinity of the fluorophore. When photoconversion of FM dyes is performed while the preparation is bathing in diaminobenzidine (DAB), a dark brown precipitate forms after the DAB is oxidized. Thus, illumination turns FM-labeled organelles into dark electron-dense ones (Sandell and Masland 1988; Henkel et al. 1996). The technique results in a substantial increase in the resolution of FM dye labeling studies (with the obvious caveat that it is restricted to fixed preparations).

The general protocol is illustrated in Figure 2. In brief, fluorescently labeled preparations are loaded with DAB and illuminated (Fig. 2A, top; nerve terminals fluoresce brightly), which induces DAB precipitation (Fig. 2A, bottom), causing the individual labeled vesicles to be easily identifiable by EM (Fig. 2B).

MATERIALS

CAUTION: See Appendix 6 for proper handling of materials marked with <!>.
See the end of the chapter for recipes for reagents marked with <R>.

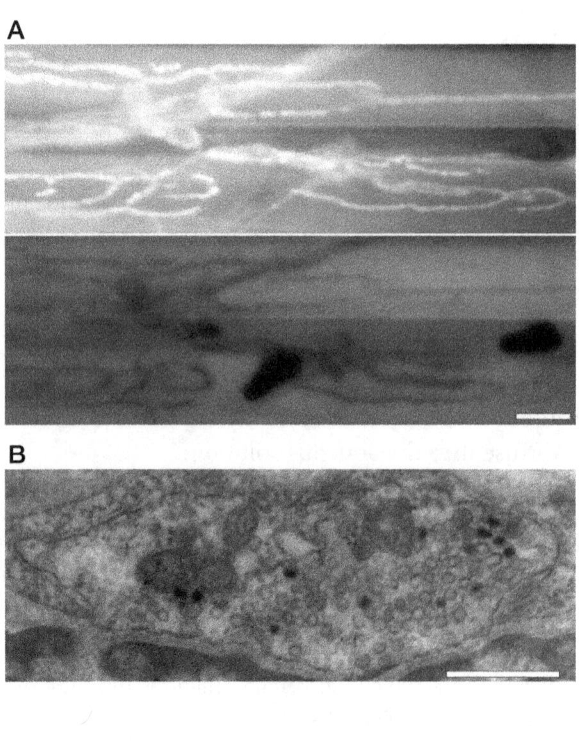

FIGURE 2. FM dye photoconversion. (*A*) Preparations are stained, chemically fixed, and soaked with diaminobenzidine (DAB). The dye within nerve terminals is brightly visible under the fluorescence microscope (*top*). The images are taken with fluorescein optics; both FM fluorescence and glutaraldehyde autofluorescence are visible. When the photoconversion is complete, the DAB precipitate accumulates, and nerve terminals are seen as dark stripes on the muscle fibers (*bottom*), which still show glutaraldehyde autofluorescence. The image was taken after 35 min of illumination. Note also darkening of erythrocytes (almost invisible in the *top*, but clearly shown in the *bottom* panel), which indicates adequate photoconversion. Scale bar, 100 µm. (*B*) Typical image of a mildly stimulated photoconverted nerve terminal in cross section. Labeled vesicles appear electron dense (dark). Scale bar, 500 nm.

Reagents

Ammonium chloride <!>, 100 mM in phosphate-buffered saline (PBS)
DAB <!>, 1.5 mg/mL in PBS
Epon resin, 50% in propylene oxide
Epon resin, 100%
Ethanol <!> for dehydration series (50%, 70%, 90%, 95%, and 100%)
FM1-43 dye

The concentration is usually 2–10 μM. The dye readily dissolves in water. Prepare a stock at 1–5-mM concentration, and store it at 4°C.

Glutaraldehyde <!>, cold 2%–3% in 0.1-M PBS (pH 7.2)
Neuronal tissue or cell sample of interest
OsO_4 <!>, 2% in PBS
PBS <R>
Propylene oxide <!>

Equipment

Electron microscope
Heat block or incubator set at 60°C
Light source for photoconversion of FM dye (e.g., 100-W mercury lamp)
Ultramicrotome
Vials for EM preparation of samples

EXPERIMENTAL METHOD

Sample Staining and Fixation

1. Dissect and stain the preparation with FM1-43, as in Protocol A.
2. Fix the preparation with cold 2%–3% glutaraldehyde in 0.1-M PBS (pH 7.2) on ice, for 60–90 min.
3. Dissect the region of interest from the preparation (eliminating unwanted material, such as muscle ends, skin, and bone in neuromuscular junctions). Wash the dissected tissue in PBS for 30–60 min at 4°C.
4. Incubate the tissue in 100 mM ammonium chloride in PBS for 10–20 min to quench the non-reacted glutaraldehyde. Wash with PBS for 20–30 min.

Photoconversion

5. Incubate the preparation in 1.5-mg/mL DAB in PBS for 30–60 min at 4°C.
6. Illuminate the preparation to photoconvert the FM dye and to oxidize the DAB.

   The illumination provided by 100-W mercury lamps is ideal when using a 20x/0.5-NA objective. We use low magnification objectives because they provide a larger illumination field and, therefore, more photoconverted synapses. The nerve terminals appear bright initially, losing fluorescence completely after 20–30 min; formation of the DAB precipitate follows, and the nerve terminals turn brown–black (Fig. 2A).

7. Remove the preparation, wash it in PBS for 5–10 min, and cut out the photoconverted spot (because it will not be possible to identify it after processing for EM).

Processing for EM

8. Postfix samples in 2% OsO_4 in PBS for 60 min.

9. Dehydrate through an ascending series of ethanol solutions (50%, 70%, 90%, 95% for 10 min each and 100% ethanol for 30 min), followed by a 30-min incubation in propylene oxide.
10. Infiltrate overnight with a 50% dilution of Epon resin in propylene oxide.
11. Replace the resin solution with 100% Epon, and allow the propylene oxide to evaporate for 6–8 h in open vials at room temperature.
12. Embed the preparations in Epon, and cure the resin for 24–48 h at 60°C.
13. Slice the samples into 50–80-nm sections using an ultramicrotome.
14. Keep poststaining of the thin sections to a minimum to ensure that the DAB precipitate is clearly seen.

    We typically image the samples without poststaining. If, however, this step is necessary to visualize sample morphology, we recommend a 1-min incubation with 2% uranyl acetate dissolved in 50% ethanol. A typical image of a photoconverted nerve terminal in cross section is shown in Figure 2B.

DISCUSSION

Slice Preparations

FM studies in preparations such as acute brain slices are difficult because damaged membranes soak up a large amount of FM dye, which is then hard to wash out. This problem can be alleviated by quenching the fluorescence of the FM dye that is located outside of the membrane-bound organelles. Pyle et al. (1999) applied sulforhodamine to FM1-43-labeled hippocampal slices and obtained a rapid decrease in background fluorescence. This technique requires that the sulforhodamine be present throughout the imaging experiments because the FM dye is not washed from the slice. Another approach to improve the image quality has been to increase the amount of dye that is washed out of the preparation by including the oligosaccharide ADVASEP-7 during the washing steps. ADVASEP-7, which is a derivative of β-cyclodextrin, has a higher affinity for FM dyes than do biological membranes (Kay et al. 1999). The addition of ADVASEP-7 to washing solutions greatly reduces background staining, making it possible to image systems such as rat brain and *C. elegans* slices. Zakharenko et al. (2001) used this technique in acute hippocampal slices and showed an increase in presynaptic activity (increased rates of FM destaining) during long-term potentiation (the first such description of this phenomenon in a noncultured preparation).

Using FM Dyes to Image the Fine Dynamics of Exocytosis

Presynaptic boutons of hippocampal neurons in culture are the most frequently used preparation for studying vesicle recycling in central nervous system synapses. At the frog NMJ (see Fig. 1C,E), most vesicles recycle through a slow endocytic pathway (Richards et al. 2000), which relies on endosome-like intermediates and clathrin-dependent endocytic vesicle formation (Heuser and Reese 1973). Such a slow pathway may represent a significant bottleneck for vesicle recycling under conditions of high activity, especially for the hippocampal synapses, which typically contain just 100–200 vesicles. Thus, an alternative mechanism has been proposed, known as the kiss-and-run pathway (Fesce et al. 1994), in which vesicles transiently fuse with the plasma membrane through a relatively small pore, and endocytosis occurs simply by closure of the pore and retraction of the vesicle. A kiss-and-run mechanism might be able to recycle vesicles much more quickly than could the classical pathway. However, in hippocampal synapses, there was little evidence supporting fast recycling until investigators began using FM dyes. Some examples follow.

In full fusion, the vesicle membrane completely collapses into the plasma membrane so that FM dye departitions into the bathing solution, leaving behind a dye-free membrane to be endocytosed. As kiss-and-run is thought to only involve a transient fusion of the membranes, FM dye release on fusion should only be partial. A number of experimental schemes has been devised to exploit this concept.

One experiment focused on the release of different FM dyes from vesicles during destaining (Klingauf et al. 1998; Pyle et al. 2000; Fig. 3A). The destaining behavior of dyes with different hydrophobicities (the hydrophilic FM2-10 and the lipophilic FM1-43) was compared. Vesicles collapsing into the plasma membrane are expected to release both of the dyes completely and with similar kinetics. However, if the process of endocytosis is faster than the departitioning of the dye from the membrane, proportionally more FM2-10 should be released, which was indeed the observed result. Furthermore, the more lipophilic FM1-84 was released even more slowly than FM1-43 (Klingauf et al. 1998), suggesting that hippocampal boutons rely on fast vesicle recovery. One should note, however, that this differential release of dyes is not necessarily a general phenotype for all preparations (see the frog NMJ, Richards et al. 2000).

Aravanis et al. (2003) labeled a single vesicle per hippocampal bouton with FM1-43. Stimulating at low frequency, they mainly observed a partial loss of fluorescence per fusion event, indicating that the vesicle retained much of the dye after fusion. They also observed that individual vesicles fused repeatedly during the experiment (Fig. 3B). Complete loss of fluorescence intensity, indicative of full collapse fusion, was only seen rarely after a single action potential (but compare Chen et al. 2008, in which only evidence for full collapse was obtained).

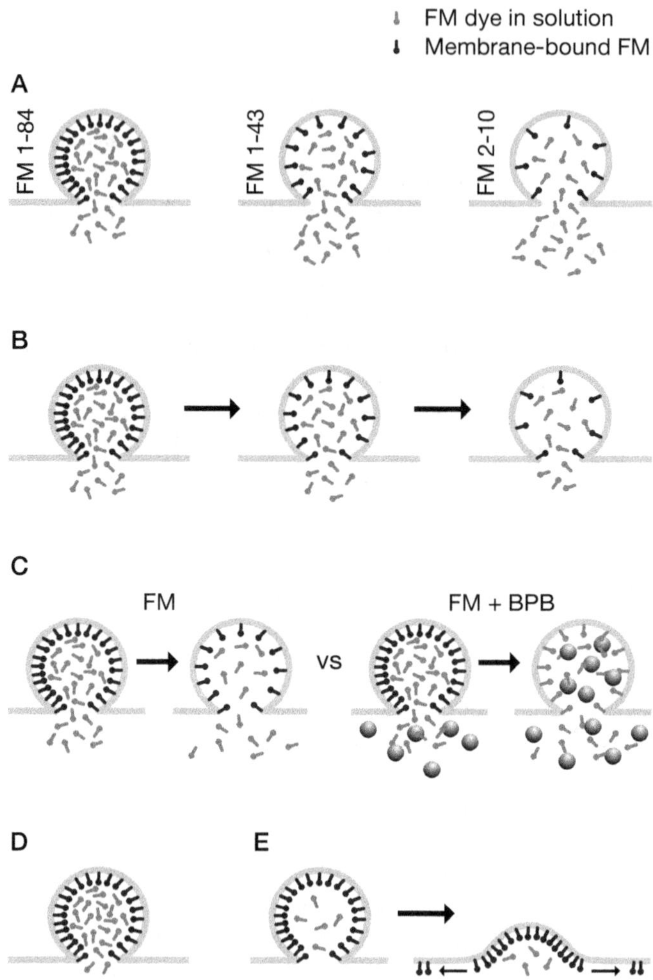

FIGURE 3. Approaches to imaging exocytosis with FM dyes. (*A*) FM dye release kinetics depend on dye hydrophobicity. FM dyes with lower hydrophobicity (FM1-84 > FM1-43 > FM2-10) departition faster from the membrane on exocytosis. (*B*) Partial FM dye release in kiss-and-run. Exocytosis through a transient kiss-and-run fusion pore releases only a fraction of the dye, leaving partially filled vesicles to release the remaining dye in successive rounds of fusion. (*C*) Probing partial FM dye release by quenching. If partial dye release occurs as a consequence of kiss-and-run, vesicles retain some dye after exocytosis (*left*). If the hydrophilic quencher bromophenol blue (BPB) is included in the bathing solution (*right*), it quenches any dye remaining in the vesicles (including any membrane-bound dye molecules), leading to a difference in fluorescence (when compared with the situation without BPB). For full collapse with complete loss of dye, application of BPB would not lead to a difference in fluorescence. (*D*) Complete dye retention. As shown by experiments combining electrophysiology and imaging, a fraction of vesicles might release a neurotransmitter, without the ability to lose or to take up FM dye. These results were interpreted in favor of exocytosis through a restrictive fusion pore. (*E*) Direct imaging of exocytosing vesicles by total internal reflection microscopy revealed little FM dye release into the fluid, but rather a lateral diffusion in the plasma membrane (with the dye later departitioning from the membrane at a distance from the point of exocytosis), indicative of full collapse fusion.

Another elegant way of testing kiss-and-run fusion by use of FM dyes was devised by Harata et al. (2006). They labeled hippocampal cultures with FM1-43 and destained in either the presence or the absence of BPB (a small hydrophilic quencher; Fig. 3C). In this experiment, one expects to find different destaining behaviors depending on the fusion mechanism at work: For full collapse, all the FM dye will escape the vesicle and, hence, the presence of BPB will not affect the fluorescence signal. For kiss-and-run, some amount of dye is retained in the vesicle so that the presence of BPB causes a decrease in the fluorescence signal. The results suggested that at least a fraction of the vesicles retain a fraction of their FM dye contents during exocytosis.

Stevens and Williams (2000) combined imaging of FM2-10 and FM1-43 with electrophysiological recordings, stimulating vesicle release by hyperosmolarity. During hyperosmolarity-induced release, a proportion of the vesicles could have released glutamate, but neither lost nor took up dye (Fig. 3D), possibly because the vesicle connected to the plasma membrane through an exceedingly small (or transient) pore. An estimated 20% of all vesicles also have a similar behavior during normal release.

Interestingly, in the only preparation in which the fusion of the vesicle with the membrane can be directly followed, a different result was obtained: By imaging single vesicles of retinal bipolar cells of the goldfish with total internal reflection microscopy, Zenisek et al. (2002) did not detect a partial release of dye. On vesicle fusion, the dye spread laterally into the membrane within a few milliseconds, indicative of full fusion or of a large fusion pore that does not interfere with dye movements in the plane of the membrane (Fig. 3E).

CONCLUSION

Three main characteristics have made FM dyes extremely useful for the study of vesicle recycling: (1) They reversibly stain membranes so that both exo- and endocytosis can be studied. (2) They do not permeate through membranes and, thus, can specifically label membrane-bound organelles, such as synaptic vesicles. (3) Their quantum yield is dramatically higher when bound to membranes than when in aqueous solution, making them the dye of choice for specific membrane labeling with low background staining.

FM dye imaging was developed using the frog NMJ, and subsequently adopted for investigations of most other synaptic systems. It continues to be a major technique for studies of synaptic vesicle recycling. FM dye experiments have provided substantial insight into the mechanisms of vesicle recycling (e.g., in the controversy over full collapse and kiss-and-run fusion). Because of their unique properties and membrane-staining characteristics, FM dyes remain a valuable tool to the imaging community, for studies of both neuronal and nonneuronal preparations.

RECIPE

CAUTION: See Appendix 6 for proper handling of materials marked with <!>.
Recipes for reagents marked with <R> are included in this list.

Phosphate-Buffered Saline (PBS)

Reagent	Quantity (for 1 L)	Final concentration (10x)
NaH_2PO_4	2.56 g	18.6 mM
Na_2HPO_4	11.94 g	84.1 mM
NaCl	87.7 g	1.5 M

Adjust the pH to 7.4 using NaOH <!> or HCl <!> as necessary.

REFERENCES

Aravanis AM, Pyle JL, Tsien RW. 2003. Single synaptic vesicles fusing transiently and successively without loss of identity. *Nature* **423:** 643–647.

Betz WJ, Bewick GS. 1992. Optical analysis of synaptic vesicle recycling at the frog neuromuscular junction. *Science* **255:** 200–203.

Betz WJ, Henkel AW. 1994. Okadaic acid disrupts clusters of synaptic vesicles in frog motor nerve terminals. *J Cell Biol* **124:** 843–854.

Betz WJ, Mao F, Smith CB. 1996. Imaging exocytosis and endocytosis. *Curr Opin Neurobiol* **6:** 365–371.

Chen X, Barg S, Almers W. 2008. Release of the styryl dyes from single synaptic vesicles in hippocampal neurons. *J Neurosci* **28:** 1894–1903.

Cochilla AJ, Angleson JK, Betz WJ. 1999. Monitoring secretory membrane with FM1-43 fluorescence. *Annu Rev Neurosci* **22:** 1–10.

de Lange RP, de Roos AD, Borst JG. 2003. Two modes of vesicle recycling in the rat calyx of Held. *J Neurosci* **23:** 10164–10173.

Fesce R, Grohovaz F, Valtorta F, Meldolesi J. 1994. Neurotransmitter release: Fusion or 'kiss-and-run'? *Trends Cell Biol* **4:** 1–4.

Griesinger CB, Richards CD, Ashmore JF. 2002. FM1-43 reveals membrane recycling in adult inner hair cells of the mammalian cochlea. *J Neurosci* **22:** 3939–3952.

Harata NC, Choi S, Pyle JL, Aravanis AM, Tsien RW. 2006. Frequency-dependent kinetics and prevalence of kiss-and-run and reuse at hippocampal synapses studied with novel quenching methods. *Neuron* **49:** 243–256.

Henkel AW, Lübke J, Betz WJ. 1996. FM1-43 dye ultrastructural localization in and release from frog motor nerve terminals. *Proc Natl Acad Sci* **93:** 1918–1923.

Heuser JE, Reese TS. 1973. Evidence for recycling of synaptic vesicle membrane during transmitter release at the frog neuromuscular junction. *J Cell Biol* **57:** 315–344.

Holt M, Cooke A, Wu MM, Lagnado L. 2003. Bulk membrane retrieval in the synaptic terminal of retinal bipolar cells. *J Neurosci* **23:** 1329–1339.

Kaneko T, Harasztosi C, Mack AF, Gummer AW. 2006. Membrane traffic in outer hair cells of the adult mammalian cochlea. *Eur J Neurosci* **23:** 2712–2722.

Kay AR, Alfonso A, Alford S, Cline HT, Holgado AM, Sakmann B, Snitsarev VA, Stricker TP, Takahashi M, Wu LG. 1999. Imaging synaptic activity in intact brain and slices with FM1-43 in *C. elegans*, lamprey, and rat. *Neuron* **24:** 809–817.

Klingauf J, Kavalali ET, Tsien RW. 1998. Kinetics and regulation of fast endocytosis at hippocampal synapses. *Nature* **394:** 581–585.

Kraszewski K, Daniell L, Mundigl O, De Camilli P. 1996. Mobility of synaptic vesicles in nerve endings monitored by recovery from photobleaching of synaptic vesicle-associated fluorescence. *J Neurosci* **16:** 5905–5913.

Lichtman JW, Wilkinson RS, Rich MM. 1985. Multiple innervation of tonic endplates revealed by activity-dependent uptake of fluorescent probes. *Nature* **314:** 357–359.

Lindgren CA, Emery DG, Haydon PG. 1997. Intracellular acidification reversibly reduces endocytosis at the neuromuscular junction. *J Neurosci* **17:** 3074–3084.

Pyle JL, Kavalali ET, Choi S, Tsien RW. 1999. Visualization of synaptic activity in hippocampal slices with FM1-43 enabled by fluorescence quenching. *Neuron* **24:** 803–808.

Pyle JL, Kavalali ET, Piedras-Renteria ES, Tsien RW. 2000. Rapid reuse of readily releasable pool vesicles at hippocampal synapses. *Neuron* **28:** 221–231.

Ramaswami M, Krishnan KS, Kelly RB. 1994. Intermediates in synaptic vesicle recycling revealed by optical imaging of Drosophila neuromuscular junctions. *Neuron* **13:** 363–375.

Richards DA, Guatimosim C, Betz WJ. 2000. Two endocytic recycling routes selectively fill two vesicle pools in frog motor nerve terminals. *Neuron* **27:** 551–559.

Ryan TA, Reuter H, Wendland B, Schweizer FE, Tsien RW, Smith SJ. 1993. The kinetics of synaptic vesicle recycling measured at single presynaptic boutons. *Neuron* **11:** 713–724.

Sandell JH, Masland RH. 1988. Photoconversion of some fluorescent markers to a diaminobenzidine product. *J Histochem Cytochem* **36:** 555–559.

Stevens CF, Williams JH. 2000. "Kiss and run" exocytosis at hippocampal synapses. *Proc Natl Acad Sci* **97:** 12828–12833.

Teng H, Cole JC, Roberts RL, Wilkinson RS. 1999. Endocytic active zones: Hot spots for endocytosis in vertebrate neuromuscular terminals. *J Neurosci* **19:** 4855–4866.

Westphal V, Rizzoli SO, Lauterbach MA, Kamin D, Jahn R, Hell SW. 2008. Video-rate far-field optical nanoscopy dissects synaptic vesicle movement. *Science* **320:** 246–249.

Zakharenko SS, Zablow L, Siegelbaum SA. 2001. Visualization of changes in presynaptic function during long-term synaptic plasticity. *Nat Neurosci* **4:** 711–717.

Zenisek D, Steyer JA, Almers W. 2000. Transport, capture and exocytosis of single synaptic vesicles at active zones. *Nature* **406:** 849–854.

Zenisek D, Steyer JA, Feldman ME, Almers W. 2002. A membrane marker leaves synaptic vesicles in milliseconds after exocytosis in retinal bipolar cells. *Neuron* **35:** 1085–1097.

19 Imaging Exocytosis with Total Internal Reflection Fluorescence Microscopy

Christina Joselevitch,[1] David Zenisek,[1] and David Perrais[2]

[1]*Yale University School of Medicine, Department of Cellular and Molecular Physiology, New Haven, Connecticut 06520;* [2]*Laboratoire de Physiologie Cellulaire de la Synapse, CNRS UMR 5091 and Université Bordeaux 2, Institut François Magendie, 33077 Bordeaux, France*

ABSTRACT

Although electrophysiological techniques such as membrane capacitance measurements, electrochemical detection, and postsynaptic recordings are powerful methods for studying exocytosis, information concerning any steps prior to vesicle fusion must be inferred indirectly. To investigate the steps leading up to vesicle fusion, it is useful to image synaptic vesicles and dense core granules before and including exocytosis. One technique that can be used to visualize these events with good signal-to-noise ratio is total internal reflection fluorescence microscopy (TIRFM). This technique allows selective imaging of fluorescent molecules that are closest to a high-refractive-index substance such as glass. In this chapter, we show how TIRFM can be applied to the study of exocytosis of single synaptic vesicles stained with FM1-43 in living goldfish retinal bipolar neurons, and exocytosis of single dense core granules stained with neuropeptide Y–enhanced green fluorescent protein (NPY-eGFP) in living bovine chromaffin cells. Details on the transfection of chromaffin cells are also provided.

INTRODUCTION

Principle of Total Internal Fluorescence Microscopy

Light directed from a material of high refractive index (n_1) to one of lower refractive index (n_2) is totally reflected at angles equal to or greater than the critical angle α_c, where

$$\alpha_c = \sin^{-1}(n_2/n_1). \tag{1}$$

At these angles that lead to total reflection of light at the interface, an electromagnetic "evanescent field" is formed in the lower-refractive-index medium. The intensity of this field decays exponentially with distance from the interface with length constants that can be much shorter than the wave-

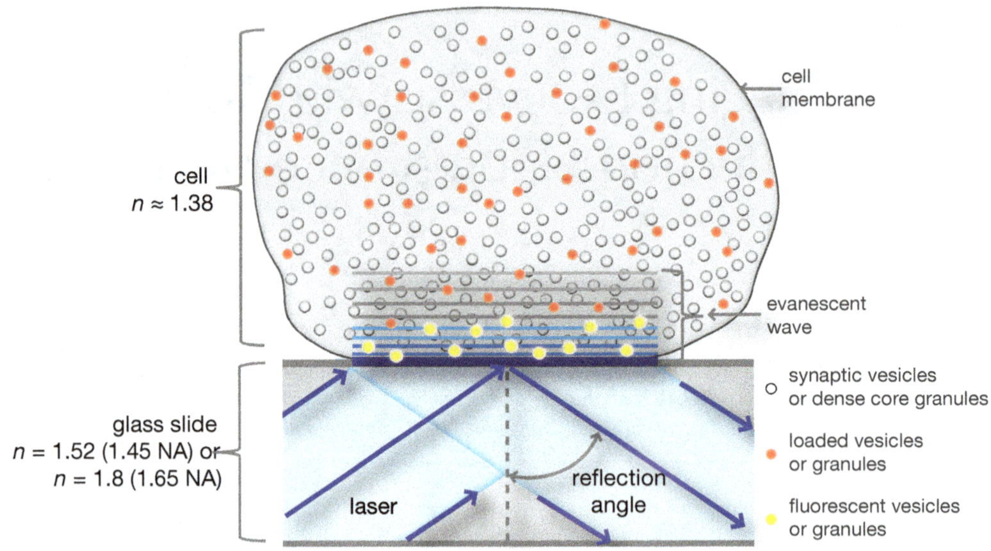

FIGURE 1. Total internal reflection fluorescence microscopy of secretory vesicles. Light directed at an angle beyond the critical angle from a high-refractive-index coverslip toward a lower-refractive-index solution is totally internally reflected. Within the lower-refractive-index cell, an exponentially decaying "evanescent field" is used to excite fluorophores near the interface without exciting fluorophores deeper within the cell.

length of illumination light. The length constant (d) of the exponential decay of this evanescent wave is governed by the function

$$d = (\lambda/4\pi)(n_1^2 \sin^2 \alpha - n_2^2)^{-1/2}, \tag{2}$$

where α is the angle at which the excitation beam hits the interface, and λ is the wavelength of the light.

TIRFM (also known as "evanescent field fluorescence microscopy") uses the thin evanescent wave to selectively excite fluorescent molecules in the portion of the lower-refractive-index medium nearest the interface of the two media (Hellen and Axelrod 1987; Axelrod 2001; Yildiz and Vale 2011). In cell biological or neurobiological studies, the low-refractive-index medium is usually the cytoplasm of a cell adherent to a higher-refractive-index coverslip (see Fig. 1). When a cell with fluorescently labeled secretory vesicles is firmly attached to a coverslip, the dye molecules will be excited only in those vesicles that lie within the thin evanescent field, thereby providing a high signal-to-noise ratio for vesicles within this region. In addition, the exponential drop in light intensity with distance allows small movements relative to the coverslip to be monitored by tracking object fluorescence. Using this method, single vesicles near (i.e., less than ~100 nm) the plasma membrane can be monitored as they travel to the membrane and undergo exocytosis.

Considerations for Optical Design

Several different microscope setups can be used for TIRFM (for review, see Axelrod 2001). A common setup uses a prism or hemicylinder that is in optical contact with the coverslip via a layer of immersion oil, in order to couple the laser beam to the interface. An alternative approach, first pioneered by Stout and Axelrod (1989), uses a high-numerical-aperture (high-NA) oil-immersion objective lens to guide the excitation light in an epi-illumination configuration onto the interface.

This "prismless" or "objective-type" approach is preferable for our own experiments because it offers easy access to the cell for recording electrodes and pipettes. In addition, it provides better image quality and efficient light collection owing to the high NA of the objective lens.

Selection of Microscope Objective

The NA of an objective lens is equal to

$$NA = n_1 \sin \alpha_{max}, \tag{3}$$

where α_{max} is the maximum angle of incident light rays that can be achieved with the objective. Therefore, substituting Equation 3 into Equation 2 gives the equation for the shortest possible exponential length constant as a function of the objective:

$$d_{min} = \lambda/4\pi(NA^2 - n_2^2)^{-1/2}. \tag{4}$$

To achieve total internal reflection, the objective must have an NA higher than the refractive index of the cell, typically ~1.37. With the growing popularity of TIRFM, Olympus, Zeiss, and Nikon now sell objectives with 1.45–1.49 NA, suitable for TIRFM. These objectives enable total internal reflection using standard glass coverslips and standard immersion oils and are the most convenient objectives to use for TIRFM. For these objectives, the theoretical value of d_{min} is 66–82 nm, with λ = 488 nm. Although these objectives are suitable for most TIRFM applications, an Olympus 100X/1.65 NA objective (Caldwell 1997) is preferred for imaging vesicles in bipolar neurons, which are densely packed within the nerve terminal. The higher NA objective allows shallower ($d_{min} \approx 45$ nm) evanescent fields and, hence, better signal-to-noise ratios. However, this objective requires special immersion fluid and high-refractive-index (n = 1.78) glass coverslips that render its use difficult.

IMAGING SETUP

We use this imaging setup to monitor exocytosis in both living goldfish retinal bipolar neurons and living bovine chromaffin cells (Protocols A and B, respectively).

Illumination

The setup is based on a commercial inverted microscope (IX70; Olympus) equipped with a side illumination port and side-facing filter cube (both available as options from Olympus), which allows for illumination through the side of the microscope. Upon shutter opening, light from the 488-nm laser (Coherent Sapphire 488) is reflected by a mirror and focused to the back focal plane by a focusing lens mounted on the side illumination port of the microscope. To achieve total internal reflection, the mirror is mounted on a translation stage, which can be moved to direct the beam to locations off-axis in the back focal plane of the high-NA objective lens (Apo 100× O HR; Olympus, NA 1.65).

Light Sources

For exciting FM1-43 and eGFP fluorescence, the setup is equipped with a 488-nm solid-state laser with an output power of 25 mW (Coherent Sapphire 488). To excite the rhodamine-based RIBEYE-binding peptides used to determine the localization of synaptic ribbons in retinal bipolar cells, a second, 561-nm solid-state laser line is used (Melles Griot 85 YCA-010), with a maximum output power of 10 mW.

Data Capture and Analysis

The side-mounted filter cube contains a dichroic mirror (Semrock FF498/581-Di01) and an emission filter (Semrock NF01/405-488-568). An EM-CCD camera (Roper Scientific Cascade 512B) serves as the imaging device. Images are collected and analyzed using a PC running MetaMorph 7.1 (Molecular Devices). For bipolar cell movies, the capture frame rate is 33 Hz, and for chromaffin cell movies, 20 Hz.

For experiments that include simultaneous patch-clamp recordings, the camera shutter is used as the trigger for the shutter controllers (Uniblitz VMM-D1 and VMM-D3), shutters (Uniblitz), and patch-clamp amplifier (HEKA EPC-9) via the trigger input and D/A outputs of the amplifier.

Protocol A

Loading and TIRF Imaging of Dissociated Bipolar Cells

TIRFM is used to follow exocytosis in goldfish retinal neurons pre-loaded with the styryl dye FM1-43.

MATERIALS

CAUTION: See Appendix 6 for proper handling of materials marked with <!>.
See the end of the chapter for recipes for reagents marked with <R>.

Reagents

ADVASEP-7 (Sigma-Aldrich A3723) (optional; see Step 3)
FM1-43 (Invitrogen T35356)
High-K^+ Ringer's solution <R>
Low-Ca^{2+} Ringer's solution <R>
Retinal tissue, digested
Standard Ringer's solution <R>
Trolox <!> (Sigma-Aldrich 56510) (optional; see Step 2)

Equipment

Analysis software (MatLab R2008a)
Coverslips, glass (having a high refractive index)
Micromanipulator (Sutter Instrument MP-285)
Patch-clamp equipment and software (HEKA Pulse 8)
Patch-clamp pipettes
TIRF microscopy setup (see "Imaging Setup")

EXPERIMENTAL METHOD

1. Triturate digested pieces of retina, prepared as described elsewhere (Zenisek et al. 2002), on washed and rinsed high-refractive-index coverslips. After allowing the cells to settle, search the coverslip for bipolar cells with tightly adherent synaptic terminals.

   Bipolar cells have distinctive morphology, containing a pear-shaped cell body with a single axon terminating in a single bulbous synaptic terminal.

   See *Troubleshooting.*

2. Stain bipolar cell vesicles with 5 μM FM1-43 in high-K^+ Ringer's containing 1 mM Trolox via a local perfusion pipette aimed directly at the axon terminal for 10 sec at room temperature.

   The use of Trolox is optional but improves cell viability.

   FM1-43 reversibly inserts into the outer leaflet of an exposed plasma membrane and is taken up into vesicles via endocytosis. The dye is fluorescent in lipid environments, but nonfluorescent in aqueous solution.

3. Wash cells thoroughly with dye-free low-Ca^{2+} Ringer's containing 1 mM ADVASEP-7 using bath exchange for 5 min followed by dye-free low-Ca^{2+} Ringer's alone for another 30 min to remove fluorescent dye from the bipolar cell plasma membrane and from the coverslip.

   The use of ADVASEP is optional, but may improve dye removal. FM1-43 tends to adhere to glass.

4. Replace low-Ca^{2+} Ringer's with standard Ringer's and patch a bipolar cell. Elicit exocytosis by applying depolarizing voltage pulses with a whole-cell patch pipette.

   See *Troubleshooting.*

5. Image cells simultaneously using TIRF illumination and the 488-nm laser. To record the position of synaptic ribbons, take pictures of the axon terminal with the 561-nm laser.

   i. Run the CCD camera in focus mode to focus on the footprint of the cell. There should be only one focal plane in which objects are in focus.

   ii. Use camera-control software to open the shutter and start recording TIRFM images.

   See *Troubleshooting.*

TROUBLESHOOTING

Problem (Step 1): Cells do not load.
Solution: Make sure that the cells you are loading are firmly attached to the glass. If cells are floating, the evanescent field will not be able to excite the molecules of FM dye. To check for well-adhered cells, tap the microscope lightly and look for movement of the adherent terminal.

Problem (Step 4): Cells die soon after patching.
Solution: Check the osmolarity of the intracellular solution. We have obtained our best results when the osmolarity of the internal solution was ~10 mOsm lower than that of the Ringer's solution. Also, avoid exposing cells to the laser before patching, and use neutral density filters in front of laser lines in order to decrease the risk of phototoxicity.

Problem (Step 5): No release events are seen.
Solution: Minimize phototoxicity by using the methods outlined in the Solution above. Co-puffing Trolox with the FM dye may help keep cells healthy and viable longer. To reduce background, it may be helpful to include 1 mM ADVASEP-7 in washing solution for 5 min after loading, before switching to washing solution without ADVASEP-7. Wash the imaging/recording chambers with distilled and deionized water only; soaps and detergents may leave residues behind that are toxic to cells. With use, coverslips tend to adsorb more and more FM dye, leading to increased background fluorescence. If this is the case, replace the glass in your chamber with a new coverslip.

Protocol B

Transfection and TIRF Imaging of Cultured Bovine Chromaffin Cells

Cultured chromaffin cells are transfected with neuropeptide Y–enhanced green fluorescent protein (NPY-eGFP). After allowing for growth and production of the fluorescent protein, exocytosis is imaged using TIRFM.

MATERIALS

CAUTION: See Appendix 6 for proper handling of materials marked with <!>.
See the end of the chapter for recipes for reagents marked with <R>.

Reagents

Dissociated bovine chromaffin cells
Culture medium <R>
Phosphate-buffered saline <R>
Plasmid DNA (NPY-eGFP)
Poly-D-lysine hydrobromide (Sigma-Aldrich P0296)
Standard Ringer's solution <R>
Standard Ringer's solution containing 5 mM $CaCl_2$ <!> and 10 mM glucose

Equipment

Analysis software (MatLab R2008a)
Coverslips, glass (having a high refractive index)
Electroporation cuvettes, 0.4-mm gap size (BTX)
Electrosquare porator (BTX T820)
Micromanipulator (Sutter Instrument MP-285)
Patch-clamp software (HEKA Pulse 8)
TIRF microscopy setup (see Imaging Setup)
Tissue culture dishes, plastic

EXPERIMENTAL METHOD

1. Prepare dissociated bovine chromaffin cell cultures as described previously (Parsons et al. 1995), and plate them on plastic culture dishes in culture medium equilibrated at 37°C and 5% CO_2 <!>.
2. After 24 h, resuspend the cells (the weakly adhering chromaffin cells are readily washed off while fibroblasts remain adherent to the culture dish) in phosphate-buffered saline.
3. Place the cell suspension (10^6 cells/mL) in an electroporation cuvette (0.4-mm gap size) with 20 µg of plasmid DNA (NPY-eGFP) (Lang et al. 1997) and electroporate with one voltage pulse (20 msec, 220 V).
4. Plate the cells onto coverslips coated with 0.1 mg/mL poly-D-lysine and that are in pre-equilibrated culture medium.

5. Renew the medium after 1 d. Image the cells 2–4 d after transfection in standard Ringer's containing 5 mM $CaCl_2$ and 10 mM glucose.
6. View the cells under epifluorescence illumination (either by using a conventional epifluorescence illumination or by moving the focusing lens of the TIRFM setup to obtain propagation of the laser light into the cell).

   The transfection process is highly inefficient (<1% of cells treated are transfected). This step is necessary to speed up the detection of transfected cells.
7. Image transfected cells using TIRF illumination:
   i. Run the CCD camera in focus mode to focus on the footprint of the cell. There should be only one focal plane in which objects are in focus.
   ii. Use camera-control software to open the shutter and start recording TIRFM images.

   See *Troubleshooting.*

TROUBLESHOOTING

Problem (Step 7): No release events are seen.

Solution: Minimize phototoxicity by using the methods outlined in the Troubleshooting for Protocol A. Co-puffing Trolox with the FM dye may help keep cells healthy and viable longer. To reduce background, it may be helpful to include 1 mM ADVASEP-7 in washing solution for 5 min after loading, before switching to washing solution without ADVASEP-7. Wash the imaging/recording chambers with distilled and deionized water only; soaps and detergents may leave residues behind that are toxic to cells. With use, coverslips tend to adsorb more and more FM dye, leading to increased background fluorescence. If this is the case, replace the glass in your chamber with a new coverslip.

DISCUSSION

Examples of Imaging Exocytosis Using TIRFM

Figure 2 shows an FM1-43-labeled bipolar cell, imaged with evanescent field microscopy. Single fluorescent synaptic vesicles appear as diffraction-limited solitary spots with TIRFM. Vesicles appear only within the footprint where the cell adheres to the coverslip, consistent with selective excitation of the cell surface by TIRFM. In sequences of images, the fluorescence of most vesicles fluctuates rapidly. Vertical motion of vesicles can be tracked by following the fluorescence intensity over time (Zenisek et al. 2000). When recording at 30 Hz, a step depolarization applied through a patch-clamp electrode elicits a transient increase in fluorescent intensity, followed by the lateral diffusion of fluorescent signal indicating exocytosis of a vesicle (Fig. 2C–I). This rise in fluorescence results from the released dye coming closer to the coverslip and changing orientation, thus being more strongly excited by the evanescent wave (Zenisek et al. 2002).

Figure 3 shows a chromaffin cell, transfected with NPY-eGFP and imaged with TIRFM. Single granules are visible as fluorescent spots. Unlike most synaptic vesicles in bipolar cells, granules remained fairly stationary during 5-sec illumination periods. A step depolarization applied through a patch-clamp electrode caused a granule to release its fluorescent content, recorded at 20 Hz. The fluorescence rose rapidly, owing to the released NPY-eGFP moving closer to the coverslip and to the dequenching of eGFP fluorescence following deacidification of the granule interior. The free NPY-eGFP diffused in the cleft between the cell and the coverslip (Perrais et al. 2004).

Advantages and Limitations of TIRFM

If fluorescent objects near the cell membrane are to be imaged, TIRFM offers several advantages over confocal or conventional epifluorescence microscopy: (1) TIRFM eliminates most background fluorescence because there is very little out-of-focus light; (2) TIRFM can provide improved *z*-discrimination of fluorescent objects; (3) TIRFM has a high collection efficiency of the emitted fluorescence, because no pinhole is needed; and (4) real-time imaging is readily achievable, because the whole-object plane is imaged at once, without scanning. Thus, with the development of digital cameras with high speeds and good collection efficiencies, continuous image acquisition at a frequency >50 Hz is achievable, allowing release and diffusion kinetics of fluorescent molecules to be measured from single secretory vesicles (Zenisek et al. 2002). It is noteworthy, however, that the util-

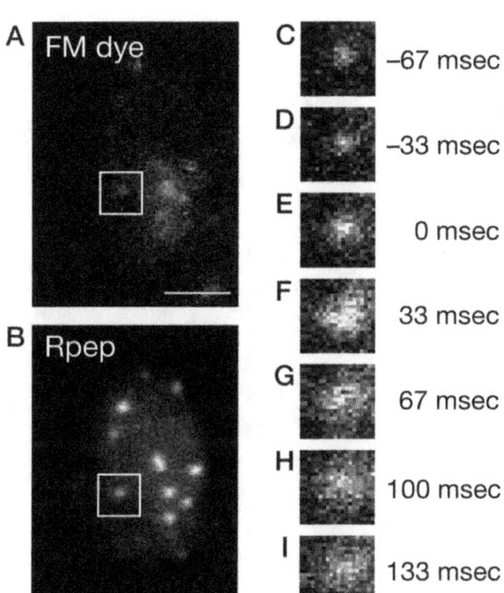

FIGURE 2. TIRFM imaging of bipolar cell vesicle exocytosis. *(A)* Image of a bipolar cell labeled with FM1-43 using TIRFM. The "footprint" of the terminal fluoresces faintly where it adheres to a glass coverslip. *(B)* TIRFM image of ribbon binding peptide. Bright spots indicate the location of synaptic ribbons. Scale bars, 1 µm. *(C–I)* Sequence of images of vesicle indicated by box in *A*. Images taken at the indicated times relative to the beginning of a 500-msec step depolarization. During the depolarization, the vesicle undergoes exocytosis. Note that the fluorescence of the vesicle increases and subsequently spreads over the surface of the terminal. Scale bar, 1 µm. For clarity, a piece of brightly fluorescent debris is masked out in *A*.

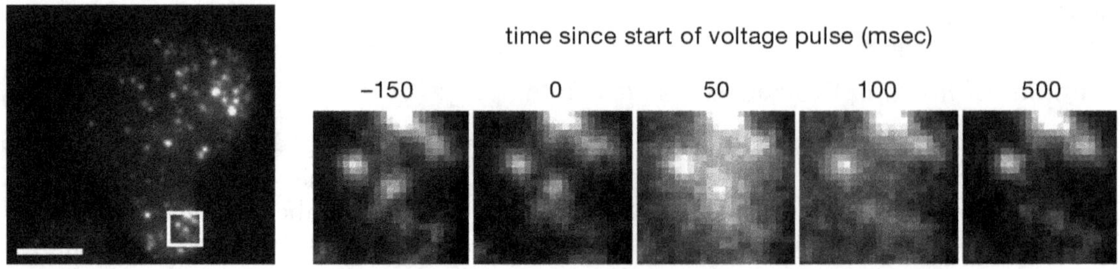

FIGURE 3. TIRFM imaging of a chromaffin cell transfected with NPY-eGFP and stimulated to exocytose. (*Left*) Image of a whole cell. Fluorescent spots correspond to single dense core granules stained with NPY-eGFP. (*Right*) Sequence of images, corresponding to the square in the *left* image, at times indicated relative to the start of a 500-msec voltage pulse. The granule in the center releases NPY-eGFP that disappears completely in <500 msec. Scale bar, 1 μm.

ity of TIRFM is limited to applications where the objects of interest reside within 200 nm of a flat, highly refractive substrate, usually a coverslip. In cell biology, this limitation restricts the practical uses of TIRFM to events at or near the plasma membrane in cells that adhere to glass.

RECIPES

CAUTION: See Appendix 6 for proper handling of materials marked with <!>.
Recipes for reagents marked with <R> are included in this list.

Culture Medium

DMEM	45%
Ham's F-12	45%
Fetal calf serum	10%
Penicillin <!>/streptomycin <!>	1 unit/mL

High-K^+ Ringer's Solution

NaCl	97.5 mM
KCl	25 mM
$MgCl_2$	1 mM
$CaCl_2$ <!>	2.5 mM
HEPES	10 mM

Final solution is pH 7.4 and 265 mOsm.

Internal Solution

CH_3O_3SCs	100 mM
TEA-CL <!>	10 mM
$MgCl_2$	4 mM
HEPES	10 mM
EGTA	0.5 mM
ATP-Mg	10 mM
GTP-Li	1 mM
Glutathione	1 mM
RIBEYE-binding peptide (rhodamine and EQTVPVDLSVARDR-COOH, MW = 1997.75)	5 mM

Final solution is pH 7.2 and 250 mOsm.

Low-Ca^{2+} Ringer's Solution

NaCl	120 mM
KCl	2.5 mM
$MgCl_2$	1 mM
$CaCl_2$ <!>	0.5 mM
HEPES	10 mM
Glucose	10 mM
Glutathione	2 mM
EGTA	0.75 mM

Final solution is pH 7.4 and 265 mOsm.

Phosphate-Buffered Saline

NaCl	155 mM
KH_2PO_4	1 mM
NaH_2PO_4 (pH 7.3)	6.5 mM

Filter-sterilize the solution.

Standard Ringer's Solution

NaCl	120 mM
KCl	2.5 mM
$MgCl_2$	1 mM
$CaCl_2$ <!>	2.5 mM
HEPES	10 mM
Glucose	10 mM
Glutathione	2 mM

Final solution is pH 7.4 and 265 mOsm.

ACKNOWLEDGMENTS

CJ and DZ are supported by National Institutes of Health grant RO1-EY014990-01. DP was supported by a fellowship from the Human Frontier Science Program Organization.

REFERENCES

Axelrod D. 2001. Total internal reflection fluorescence microscopy in cell biology. *Traffic* **2:** 764–774.

Caldwell JB. 1997. Ultra high NA microscope objective. *Opt Photonics News* **8:** 44–46.

Hellen EH, Axelrod D. 1987. Fluorescence emission at dielectric and metal-film interfaces. *J Opt Soc Am B Opt Phys* **4:** 337–350.

Lang T, Wacker I, Steyer J, Kaether C, Wunderlich I, Soldati T, Gerdes HH, Almers W. 1997. Ca^{2+}-triggered peptide secretion in single cells imaged with green fluorescent protein and evanescent-wave microscopy. *Neuron* **18:** 857–863.

Parsons TD, Coorssen JR, Horstmann H, Almers W. 1995. Docked granules, the exocytic burst, and the need for ATP hydrolysis in endocrine cells. *Neuron* **15:** 1085–1096.

Perrais D, Kleppe IC, Taraska JW, Almers W. 2004. Recapture after exocytosis causes differential retention of protein in granules of bovine chromaffin cells. *J Physiol* **560:** 413–428.

Stout AL, Axelrod D. 1989. Evanescent field excitation of fluorescence by epi-illumination microscope. *Appl Opt* **28:** 5237–5242.

Yildiz A, Vale RD. 2011. Total internal reflection fluorescence microscopy. In *Imaging: A laboratory manual* (ed. Yuste R), pp. 597–608. Cold Spring Harbor Laboratory Press, Cold Spring Harbor, NY.

Zenisek D, Steyer JA, Almers W. 2000. Transport, capture and exocytosis of single synaptic vesicles at active zones. *Nature* **406:** 849–854.

Zenisek D, Steyer JA, Feldman ME, Almers W. 2002. A membrane marker leaves synaptic vesicles in milliseconds after exocytosis in retinal bipolar cells. *Neuron* **35:** 1085–1097.

20 Interferometric Detection of Action Potentials

Arthur LaPorta[1] and David Kleinfeld[2]

[1]*Department of Physics, Institute for Physical Science and Technology, Biophysics Program, University of Maryland, College Park, Maryland 20742;* [2]*Department of Physics, University of California, La Jolla, California 92093-0374*

ABSTRACT

Individual action potentials in axons can be detected in vitro using intrinsic optical signals. The principles of this detection technique are described here. The technique has been demonstrated using neuronal axons dissected from lobster. It would also be applicable to axons and active dendrites of cultured neurons with diameters somewhat larger than an optical wavelength. Using index-matching techniques, it may be possible to extend the technique to smaller structures.

INTRODUCTION

Early work showed that light scattering from a network of small neuronal processes is modulated by the propagation of action potentials (Cohen et al. 1968, 1972a,b). Studies of the scattering pattern obtained from a single axon are consistent with a change in the refractive index of the cell membrane. This change is proportional to the voltage across the membrane (Stepnoski et al. 1991) and presumably arises from alignment of dipoles in the membrane. Later work detected displacement of nerves during action potentials (Fang-Yen et al. 2004; Akkin et al. 2007; McCluskey et al. 2007). The magnitude of the change was determined for isolated axons from *Aplysia* neurons, for which the relative change in refractive index was $\Delta n/n = 50 \times 10^{-6}$ per mV. This corresponds to an equivalent optical phase shift of 0.1 to 0.2 mrad per spike. These relatively small changes could be detected, in single trials, by measuring changes in light scattering with dark-field illumination (Stepnoski et al. 1991).

PRINCIPLES OF INTERFEROMETRIC DETECTION

The goal of this procedure is to directly confirm the refractive index change by constructing a two-beam interferometer in which a laser beam that passes through an axon interferes with a beam that bypasses the axon. A change in the relative optical phase of the two beams is a measure of the change in total optical density of the preparation. In order to avoid randomly fluctuating phase shifts, the two beams must be transported by the same optical components. To accomplish this, the apparatus shown in Figure 1A is used. The apparatus uses polarization optics to create two distinct beams that propagate through the same optical system.

A polarization-preserving single-mode optical fiber that transports light from a 5-mW HeNe laser acts as a point source for a microscope objective, producing a collimated Gaussian beam whose linear polarization angle is chosen to be at 45° with respect to the polarization optics that follow. The beam passes through an inclined quartz birefringent plate, which causes a displacement of the S and P polarization components with respect to each other. The beams then pass through a Wollaston prism, which introduces a relative angular deviation of the two polarization components, causing them to cross at a point beyond the rear face of the prism. The combined effect of the birefringent plate and the Wollaston prism is similar to that of a Nomarski prism with an unusually large divergence angle. The S and P polarized beams are incident on the rear aperture of a 20× water-immersion microscope objective. The beam diameters are chosen to match the rear aperture of the objective, and the crossing point is placed at the rear focal plane of the objective. The microscope objective brings the collimated beams to diffraction-limited Gaussian focal spots in the image plane. The angular deviation of the two beams causes the focal spots to occur at distinct positions in the sample plane. The Wollaston angle was chosen to make this separation ~5 μm; the choice of separation distance depends on the application. Because the two beams cross at the rear focal plane of the objective, their central rays are parallel. The specimen chamber is positioned so that the two beams are normally incident, with their waist at the bottom reflective surface. As a result, the two beams are retro-reflected back into the microscope. The two beams are recombined by the Wollaston and birefringent plate and are coupled out of the apparatus by a 50% reflector.

The beam that emerges from the apparatus consists of a vertical polarization component that has passed through the axon and a horizontal polarization component that has bypassed the axon. The optical phase shift of the resting axon changes the polarization state of the output beam from linear to an arbitrary elliptical state. A change in the optical density or path length of the axon during an action potential would result in a perturbation to the elliptically polarized state.

To detect the small phase shifts expected in response to an action potential, the measurement must employ a differencing scheme to cancel technical fluctuations in the laser light intensity. Here, the output beam is passed through an EOM, which is used to adjust the relative phase shift of the S and P polarization components for the resting neuron. There are two ways the EOM and polarization analyzer may be used (Fig. 1D). The nominal phase shift between the two polarizations may be set to 90°. In this case, the output beam has circular polarization (solid blue curve in Fig. 1B), and the analyzer, which is oriented to measure diagonal components I_u and I_v, sees a balanced signal. This corresponds to 90° in Fig. 1C. A deviation in the phase causes a deformation of the circular state (dashed blue curve in Fig. 1B), which imbalances I_u and I_v. As a result, the difference signal, $I_u - I_v$, is a measure of the phase in which the laser intensity cancels.

A more effective detection method (called the "dark fringe method") is to set the nominal phase shift between the two polarizations to 180° so that the output state is linearly polarized along v and the intensity I_u is 0 (solid red curve in Fig. 1B). An additional phase shift results in a small deviation from linear polarization that causes I_u to become nonzero. This signal cannot be used directly because the intensity $I_u \propto 1 + \cos(180° + \phi)$ is second order in the phase ϕ and would have a vanishing signal-to-noise ratio. However a first-order signal can be obtained by superimposing a sinusoidal modulation of frequency f_s on the EOM phase shift. For the resting axon, the modulation in the neighborhood of 180° produces no first-order signal in I_u because the slope of the I_u versus phase curve is 0 (see line b in Fig. 1C). However, a positive phase shift in the axon would bias the net phase toward higher values (line c in Fig. 1C), where the slope of the curve is positive. In this case, the phase modulation produces a signal at frequency f_s in I_u with amplitude proportional to the axon phase shift. Similarly, a negative axon phase would bias the phase toward negative values (line a in Fig. 1C) and produce a modulation with negative amplitude in I_u. The amplitude of the signal at frequency f_s (and therefore the axon phase retardation) can be measured using a lock-in amplifier referenced to the EOM modulation signal as shown in Figure 1D. The primary advantage of this technique is

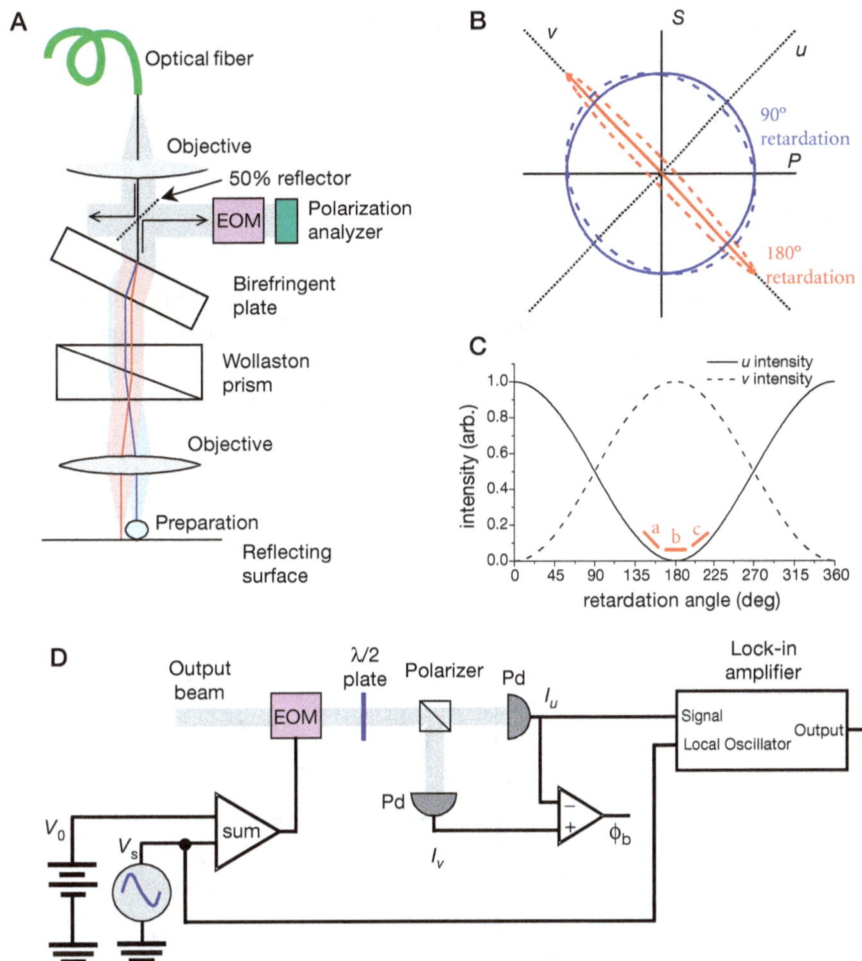

FIGURE 1. Schematic of the interferometer and the polarization-based detection system for changes in optical path length. The interferometric microscope needs to be compact and stiff to avoid uncontrolled fluctuations in the relative phase of the two optical paths, a combination that can be achieved with Microbench components (LINOS Photonics). (*A*) Essential optics in the interferometer. The broad gray bar represents the total extent of the beam, and the black line represents the central ray. After separation by polarization components, red and blue shading indicates S and P polarization components of the light, respectively. Laser light is introduced via a polarization-preserving optical fiber at the top; this light propagates down to the preparation plane through a microscope objective, polarization components, and a second water-immersion objective. The preparation plane is reflective at the excitation wavelength, and the light is retro-reflected and ultimately coupled out of the microscope using a 50% reflector. The output beam is passed through an electro-optic modulator (EOM) and is detected with a polarization analyzer shown in greater detail in *D*. The principles of operation are described in the text. (*B*) The electric field as a function of optical phase shift, where the *S* and *P* axes correspond to the polarization directions of the two rays (red and blue) in *A*, and the *u* and *v* axes correspond to the two channels of the polarization analyzer. The projection of the curves on the *u* and *v* axes gives the amplitude of the component that is transmitted to the detectors. (*C*) The intensity measured by the two channels of the analyzer as a function of optical phase shift. The short red lines are explained in the text. (*D*) Detailed schematic of the phase-sensitive detection scheme. The EOM introduces a phase shift of the S polarized field with respect to the P polarized field. A DC voltage, denoted V_0, is applied to correct for the phase shift of the resting axon. The λ/2 plate rotates the polarization state so that the analyzer detects the intensity (I) along two diagonal directions, *u* and *v*. For balanced detection, the intensities I_u and I_v are nominally equal (see text) and are subtracted to give a phase signal that is denoted ϕ_b. For dark fringe detection, V_0 is set so that all light goes to the I_v detector. A phase modulation, denoted V_s with frequency f_s, is superimposed on the EOM drive signal. A deviation of the optical phase from the dark fringe (ϕ_d) causes a modulation at frequency f_s to appear in the photocurrent I_u (see text), which may be detected with a lock-in amplifier (LIA). The modulation signal V_s is connected to the reference input, and the photocurrent I_u is connected to the signal input of the LIA, which is configured to measure the in-phase first harmonic component of the signal. The optical phase ϕ_d is obtained from the output of the LIA, which is equal to $(1/2)I_t(\phi_d)$, where I_t is the total intensity.

that the photodetectors have much less noise near the carrier frequency (chosen to be >100 kHz) than near DC (0 Hz). In this configuration, both fluctuations in laser intensity and $1/f$ noise in the photodetectors are avoided, and the principal sources of noise are shot noise and fluctuations in the preparation itself.

SHORT EXAMPLE OF APPLICATION

Figure 2 shows the preparation chamber for the walking leg nerve dissected from lobster and the optical phase shift recorded from a nerve. The nerve was microdissected so that only a small number of axonal fibers remained in the bundle. The second of the above schemes for phase detection was used. Action potentials were elicited by stimulating the nerve with a voltage pulse set just above

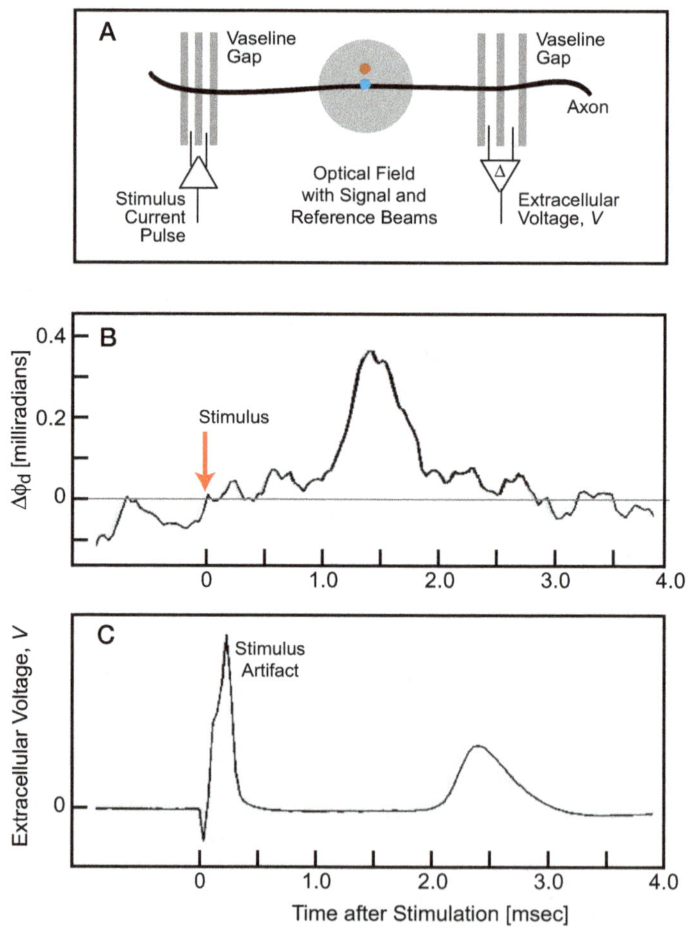

FIGURE 2. Results for interferometric measurements through a walking leg nerve from lobster, *Homarus americanus*. The nerve was maintained in artificial seawater. The nerve was microdissected down to a few, but more than one, axonal fibers. (*A*) The chamber that holds the preparation is a multicompartment structure that positions the nerve across a dielectric mirror that has a maximal reflection at 630 nm, close to the HeNe emission line. One beam (P polarization, blue) passes through the nerve, and the second (S polarization, red) passes ~5 μm away from the nerve. The nerve further rests on two platinum wires at one end, which serve as the voltage stimulation source, and two at the other end, which serve as extracellular recording electrodes. Thin walls with slits that are filled with Vaseline provide electrical isolation between the chambers. (*B*) The optical signal, denoted by the change in output phase ϕ_d between the two beams, in response to a just-supra-threshold stimulus. Shown is the stimulus-triggered average of 100 samples. (C) The concomitantly measured and averaged electrical signal. The time at which the optical signal is approximately halfway between the application of the stimulus and the detection of the electrical signal at the downstream electrodes. This is consistent with uniform propagation of the action potential along the fibers.

threshold; this minimized the number of axons excited. The extracellular signals from the action potential(s) were detected using electrodes placed beyond the imaging point. An optical phase shift of ~0.4 mrad was measured after averaging 100 pulses. This signal may contain contributions from more than one axon.

ADVANTAGES AND LIMITATIONS

The advantage of this technique is that a direct measure of change in optical density is made at a specific point. The measurement is performed in such a way that noise from the measurement process itself (e.g., fluctuations in laser intensity and detector noise) is minimized. The main limitation is that it requires an optically thin, in vitro sample; any distortions in the beams degrade the sensitivity of the polarization analyzer. The final signal-to-noise ratio seems to be dominated by fluctuation in the specimen itself. Although these fluctuations are difficult to quantify, the main causes are probably Brownian motion of the physical structures and biochemical fluctuations in the axon cytoplasm.

REFERENCES

Akkin T, Joo C, de Boer JF. 2007. Depth-resolved measurement of transient structural changes during action potential propagation. *Biophys J* **94:** 1347–1353.

Cohen LB, Keynes RD, Hille B. 1968. Light scattering and birefringence changes during nerve activity. *Nature* **218:** 438–441.

Cohen LB, Keynes RD, Landowne D. 1972a. Changes in axon light scattering that accompany the action potential: Current-dependent components. *J Physiol* **224:** 727–752.

Cohen LB, Keynes RD, Landowne D. 1972b. Changes in light scattering that accompany the action potential in squid giant axons: Potential-dependent components. *J Physiol* **224:** 701–725.

Fang-Yen C, Chu MC, Seung HS, Dasari RR, Feld MS. 2004. Noncontact measurement of nerve displacement during action potential with a dual-beam low-coherence interferometer. *Opt Lett* **29:** 2028–2030.

McCluskey MD, Sable JJ, Foust AJ, Gratton G, Rector DM. 2007. Recording invertebrate nerve activation with modulated light changes. *Appl Optics* **46:** 1866–1871.

Stepnoski RA, LaPorta A, Raccuia-Behling F, Blonder GE, Slusher RE, Kleinfeld D. 1991. Noninvasive detection of changes in membrane potential in cultured neurons by light scattering. *Proc Natl Acad Sci* **88:** 9382–9386.

21 Imaging Sodium in Axons and Dendrites

William Ross,[1] Ilya Fleidervish,[2] and Nechama Lasser-Ross[1]

[1]*Department of Physiology, New York Medical College, Valhalla, New York 10595;* [2]*Department of Physiology, Faculty of Health Sciences, Ben Gurion University, Beer-Sheva 84105, Israel*

ABSTRACT

This chapter describes a one-photon method to detect the time course and the spatial distribution of intracellular sodium ion concentrations $[Na^+]_i$, which result from action potentials and synaptic activity in different regions of neurons in brain slices. For two-photon sodium imaging, see Chapter 30. The one-photon technique has been applied to several different cell types and can likely be applied to other neurons in brain slices, particularly those that are relatively flat. Imaging of $[Na^+]_i$ changes is much less common than imaging of $[Ca^{2+}]_i$ changes, in part because typical signals are much smaller and more difficult to detect. However, with careful experiments, accurate and interesting results can be obtained.

INTRODUCTION

In principle, imaging $[Na^+]_i$ changes in neurons is very similar to measuring $[Ca^{2+}]_i$ changes in cells. A cell is loaded with an indicator that responds to Na^+, the cell is stimulated, and the spatial and temporal characteristics of the fluorescence changes are recorded with an appropriate detector or camera. However, there are several differences between $[Na^+]_i$ and $[Ca^{2+}]_i$ that may make $[Na^+]_i$ measurements more difficult but interesting in other ways. First, resting $[Na^+]_i$ is >10,000× higher than resting $[Ca^{2+}]_i$. Therefore, the $[Na^+]_i$ changes take place against a much higher background. Second, Na^+ is not buffered in cytoplasm. As a consequence, Na^+ diffuses much faster than Ca^{2+}; concentration gradients dissipate rapidly; and high-speed imaging is required if the measurements are to accurately reflect the dynamics in the cell. Third, much less effort has been put forth to synthesize Na^+ indicators. Therefore, the available Na^+ indicators are probably not as optimized as the current generation of Ca^{2+} indicators. Nevertheless, interesting measurements have been made in a number of preparations.

Protocol

Loading Cells and Detecting $[Na^+]_i$ Changes

In this protocol, a cell is loaded with an indicator that responds to Na^+, the cell is stimulated, and the spatial and temporal characteristics of the fluorescence changes are recorded with an appropriate detector or camera.

MATERIALS

See the end of the chapter for recipes for reagents marked with <R>.

Reagents

Fluorescent indicator dye, sodium sensitive

The most commonly used indicator is sodium-binding benzofuran isophthalate (SBFI; Molecular Probes). Prepare a concentrated stock solution of SBFI (10–20 mM) in distilled H_2O. This indicator comes in both a cell-permeant form and an impermeant form. The impermeant form is used in experiments in which cells are loaded from patch pipettes. Other available indicators, such as CoroNa Green (Molecular Probes), bleach more rapidly and have smaller fluorescence changes but may be more appropriate when longer wavelength excitation is required (e.g., when combined with ultraviolet uncaging or when using laser excitation).

Inert fluorescent compound that is excited at longer wavelengths (e.g., Alexa Fluor 488) (optional; see Step 1)

Internal pipette solution <R>

Equipment

Camera and software

A high-speed charge-coupled device (CCD) camera and appropriate software are used to correlate electrophysiological and optical recordings. The Quantix 57 or Evolve 512 (Photometrics), both with frame-transfer CCD chips, and the NeuroCCD-SMQ (RedShirt Imaging) are recommended. These low-noise cameras have high quantum efficiency in the appropriate wavelength range and have sufficient spatial and temporal resolutions for interesting measurements. Our laboratory custom wrote the software for the Photometrics cameras (an early version is described in Lasser-Ross et al. 1991). For the NeuroCCD-SMQ, we use Neuroplex (RedShirt Imaging).

Electrical recording device

Most conventional bridge amplifiers (e.g., Axoclamp 900A, Multiclamp 700B, Axon Instruments; or equivalent) and analog-to-digital converters can be used to record the membrane current and voltage. It is important that the software be designed to synchronize the optical and electrical data streams.

Filter set

Fluorescence changes can be detected with the same filter set often used to detect fluorescence changes of the fura series of calcium indicators. However, to maximize the fluorescence signal, we recommend a wider excitation filter than usual (FW 25 nm; Omega Optical or Semrock), a dichroic mirror, centered at 420 nm, and a glass post filter (GG455; Edmund Scientific). However, the increased excitation energy can lead to more photodynamic damage. If this is a problem, then some effort in choosing the optimal excitation bandwidth is recommended.

Microscope

We use an Olympus BX50WI equipped with an epifluorescence port, a computer-controlled shutter, a 75-W xenon arc lamp, and 40× (NA 0.8) and/or 60× (NA 0.9) water-immersion lenses (Olympus). Equivalent microscopes from other manufacturers could be used. The usual requirements for a slice patching rig are appropriate. The microscope should be stable, have good transmission characteristics, and have differential interference contrast (DIC) to improve cell visibility when patching.

EXPERIMENTAL METHOD

1. Add an aliquot of the SBFI stock to the internal pipette solution. Add H_2O to achieve a final concentration of ~1–2-mM SBFI with proper osmotic balance.

   For some experiments, it is useful to add ~5 μM of an inert fluorescent compound that is excited at longer wavelengths (e.g., Alexa Fluor 488). No special procedures are required other than those that are normally used to prepare internal pipette solutions. Use the fluorescence of this dye to position and to focus the cell, avoiding photodynamic damage from exciting the high concentration of SBFI. Then, for the $[Na^+]_i$ measurements, switch the excitation filter to the appropriate wavelength.

2. Use standard patch-pipette procedures (see Chapter 35) to load the chosen neuron with SBFI. Allow the pipette solution to diffuse throughout the cell, until all parts of the dendrites are visible (typically 15 min for nearby dendrites or axon, longer for more distal processes).

3. Stimulate the neuron either with the patch pipette used to load the cell or with an extracellular electrode. Record the $[Na^+]_i$ changes as changes from the baseline fluorescence levels recorded before stimulation.

   The fluorescence changes of SBFI are typically only a few percent following electrical or synaptic stimulation. Therefore, it is important to arrange the imaging system to minimize noise and to maximize the detection of emitted fluorescence. Recording a trial without stimulation and subtracting the drift in this trial from one with stimulation can usually correct for changes in fluorescence caused by an indicator or tissue bleaching.

DISCUSSION

Calibration of SBFI Fluorescence Changes

Changes in SBFI fluorescence are small and are usually proportional to $[Na^+]_i$ changes. SBFI binds Na^+ with a 1:1 stoichiometry (Minta and Tsien 1989). However, the binding constant is difficult to determine. In the original papers describing the synthesis and use of SBFI, Tsien and colleagues found that the properties of SBFI in the intracellular environment of lymphocytes and in the cuvette were quite different (Harootunian et al. 1989). The cuvette K_D was ~17 mM. Other groups have calibrated SBFI in astrocytes (Rose and Ransom 1996). Rose et al. (1999) estimated that the apparent K_D was ~26 mM in pyramidal neurons in brain slices. In addition, all measurements are a spatial average of $[Na^+]_i$ changes in a volume in which the magnitude varies from point to point. Therefore, in most experiments, we find it more useful to present the results in terms of the fractional fluorescence change $\Delta F/F$, without calibration. By doing this, information can be obtained about the time course, the spatial distribution, and the relative amplitude of the changes in $[Na^+]_i$ in different parts of the neuron. When comparing amplitudes, it is important to correct for background fluorescence in the slice and to consider only the indicator fluorescence. An approximate way to make this correction is to measure the fluorescence of the slice at a corresponding position away from the filled cell and to subtract this value from the total fluorescence of the slice at the position of the filled neuron. High time resolution is important because Na^+ diffuses rapidly in cytoplasm and localized $[Na^+]_i$ elevations dissipate rapidly (e.g., Fleidervish et al. 2010).

Examples of Application

This technique has been applied to many preparations, including cerebellar Purkinje neurons (Lasser-Ross and Ross 1992; Callaway and Ross 1997), hippocampal pyramidal neurons (Jaffe et al. 1992), presynaptic terminals (Regehr 1997), and neocortical pyramidal cell axons (Fleidervish et al. 2010). The following examples illustrate the ability to detect signals from thin axons and dendrites, and the heterogeneity of $[Na^+]_i$ changes in different parts of a neuron.

Figure 1 shows two measurements of $[Na^+]_i$ changes in guinea pig cerebellar Purkinje cells in response to synaptic and electrical stimulation. Figure 1A compares the $[Na^+]_i$ changes from climbing fiber (CF) activation to the $[Na^+]_i$ changes due to a train of antidromic action potentials. The spike signals are only in the soma, corresponding to the known failure of Na^+ spikes to invade the dendrites of these cells (Stuart and Hausser 1994). Figure 1B shows the spatial distribution and the time course of the $[Na^+]_i$ changes in response to CF activation. The spatial distribution corresponds closely to the known distribution of these synapses along the main dendrites of these cells, with weaker delayed changes over the fine dendrites. The delay probably reflects Na^+ diffusion from the sites of synaptic contact on the thick dendrites. There is also a strong signal in the soma caused by voltage-gated Na^+ entry.

Figure 2 shows $[Na^+]_i$ changes at different locations along an axon of a neocortical L5 pyramidal neuron in response to a train of five intrasomatically evoked action potentials. The largest increases are in the axon initial segment (AIS) close to the soma. These signals recover rapidly because of Na^+ diffusion into the soma and under the myelin. There is a localized signal (green arrow and trace) at ~100 µm from the soma, which corresponds to Na^+ entry into the first node of Ranvier. In the intervening myelinated region, there is either no $[Na^+]_i$ increase or a slow-rising increase resulting from Na^+ diffusion from the AIS.

Limitations of the Method

Because resting $[Na^+]_i$ in neurons is in the millimolar range, high concentrations of SBFI can be used to improve detection efficiency. SBFI is nontoxic, and we have not found any pharmacological effects of the indicator. SBFI responds to changes in $[K^+]_i$. However, changes in $[K^+]_i$ are not significant in most experiments.

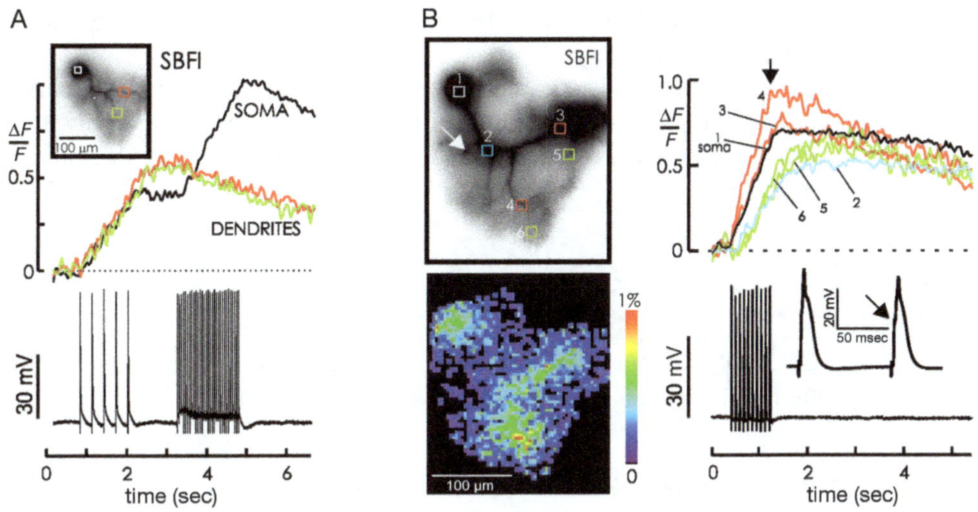

FIGURE 1. (*A*) $[Na^+]_i$ changes in response to a train of five climbing fiber (CF) stimuli followed by a train of 40 antidromic spikes. The SBFI fluorescence changes in response to the spikes were only detected in the soma. Optical data are an average of 20 sweeps. (*B*) Spatial distribution and time course of $[Na^+]_i$ changes in response to a train of 10 CF responses. The fluorescence image in the upper left shows a guinea pig Purkinje cell filled with SBFI from a patch electrode on the dendrites (white arrow). The labeled boxes show the regions from which the individual traces on the *right* were measured. The pseudocolor image (*lower left*) shows the spatial distribution of the $[Na^+]_i$ changes measured at the time of the vertical black arrow over the traces. The *inset* at the *lower right* shows the first two dendritically recorded CF electrical responses. Note the large Ca^{2+} spike at the peak of the response and the attenuated Na^+ spike (small arrow) on the rising edge of the potential change. Optical data are an average of 10 sweeps. (Adapted, with permission of the American Physiological Society, from Callaway and Ross 1997.)

We note two additional limitations. First, the signals are small. The change in fluorescence was only ~1%–10% in the examples illustrated by the figures. Signal averaging was needed to reduce the noise, and trains of stimuli were used to increase the signal. Therefore, this kind of measurement is very difficult to make if the responses cannot be repetitively activated. There is not much room for improvement in the current generation of cameras, because they have almost 100% quantum efficiency, low noise, and higher frame rates than needed for these experiments. Therefore, better sig-

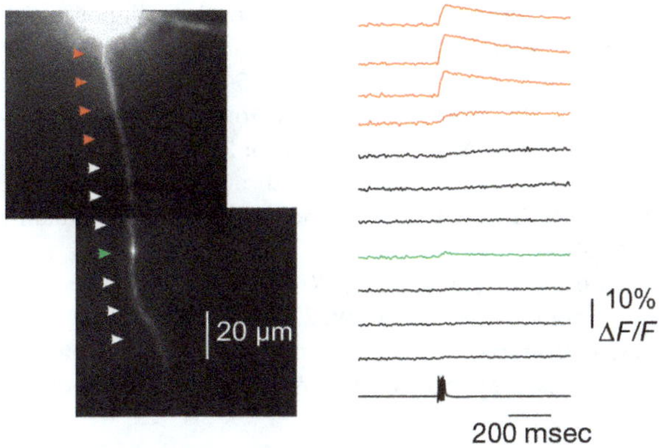

FIGURE 2. $[Na^+]_i$ changes at different locations along an axon of a neocortical L5 pyramidal neuron in response to a train of five intrasomatically evoked action potentials. Five sweeps were averaged. The largest increases were in the axon initial segment (AIS) close to the soma (red arrows and traces). There was a localized signal (green arrow and trace) at ~100 µm from the soma, which corresponds to Na^+ entry into the first node of Ranvier. The bright fluorescence at this location indicates that it is close to a branch point. In the intervening myelinated regions (white arrows and black traces) there was either no $[Na^+]_i$ increase or a slow-rising increase resulting from Na^+ diffusion from the AIS. (Modified from Fleidervish et al. 2010.)

nals are more likely to come from the synthesis of new optimally tuned and more fluorescent indicators in the future.

Second, photodynamic damage limits the duration of the experiments. This probably results from the high concentration of SBFI needed to generate signals with a reasonable signal-to-noise ratio. Therefore, sweep durations should be kept as short as possible, especially when signal averaging is used.

RECIPE

CAUTION: See Appendix 6 for proper handling of materials marked with <!>.
Recipes for reagents marked with <R> are included in this list.

Internal Pipette Solution

Reagent	Final concentration
K-gluconate <!>	140 mM
NaCl	4 mM
Mg-ATP	4 mM
Na-GTP	0.3 mM
HEPES buffer	10 mM

Adjust the pH to 7.2–7.4 with KOH <!>.

REFERENCES

Callaway JC, Ross WN. 1997. Spatial distribution of synaptically activated sodium concentration changes in cerebellar Purkinje cells. *J Neurophysiol* **77:** 145–152.

Fleidervish IA, Lasser-Ross N, Gutnick MJ, Ross WN. 2010. Na^+ imaging reveals little difference in action potential–evoked Na^+ influx between axon and soma. *Nat Neurosci* **13:** 852–860.

Harootunian AT, Kao JPY, Eckert BK, Tsien RY. 1989. Fluorescence ratio imaging of cytosolic free Na^+ in individual fibroblasts and lymphocytes. *J Biol Chem* **264:** 19458–19467.

Jaffe DB, Johnston D, Lasser-Ross N, Lisman JE, Miyakawa H, and Ross WN. 1992. The spread of sodium spikes determines the pattern of dendritic Ca^{2+} entry into hippocampal neurons. *Nature* **357:** 244–246.

Lasser-Ross N, Ross WN. 1992. Imaging voltage and synaptically activated sodium transients in cerebellar Purkinje cells. *Proc R Soc Lond B Biol Sci* **247:** 35–39.

Lasser-Ross N, Miyakawa H, Lev-Ram V, Young SR, Ross WN. 1991. High time resolution fluorescence imaging with a CCD camera. *J Neurosci Methods* **36:** 253–261.

Minta A, Tsien RY. 1989. Fluorescent indicators for cytosolic sodium. *J Biol Chem* **264:** 19449–19457.

Regehr W. 1997. Interplay between sodium and calcium dynamics in granule cell presynaptic terminals. *Biophys J* **73:** 2476–2488.

Rose CR, KovalchukY, Eilers J, and Konnerth A. 1999. Two-photon Na^+ imaging in spines and fine dendrites of central neurons. *Pflugers Arch* **439:** 201–207.

Rose CR, Ransom BR. 1996. Intracellular sodium homeostasis in rat hippocampal astrocytes. *J Physiol* **491:** 291–305.

Stuart GS, Hausser M. 1994. Initiation and spread of sodium action potentials in cerebellar Purkinje cells. *Neuron* **13:** 703–712.

22 Generation and Screening of Mice with Transgenic Neuronal Labeling Controlled by *Thy1* Regulatory Elements

Petar Marinković, Leanne Godinho, and Thomas Misgeld

Chair of Biomolecular Sensors, Center for Integrated Protein Sciences (Munich) at the Institute of Neuroscience, Technische Universität München, Munich 80802, Germany

ABSTRACT

During the past decade, major progress has been made using in vivo imaging in mice to study mammalian nervous system development, plasticity, and disease. This progress has depended in part on the wide availability of two-photon microscopy, which is capable of penetrating deep into scattering tissue. Equally important, however, is the generation of suitable transgenic mouse models, which provide a "Golgi staining"-like labeling of neurons that is sparse and bright enough for in vivo imaging. Particularly prominent among such transgenic mice are the so-called *Thy1*-XFP mice (in which XFP stands for any fluorescent protein) that are used in numerous studies, especially to visualize spine plasticity in the cortex and remodeling in peripheral synapses. New generations of *Thy1*-XFP mice are now being generated at a high rate, and these have allowed previously difficult experiments to become feasible. Moreover, with easy access to core facilities or commercial providers of pronuclear injections, generating simple *Thy1* transgenic mice is now a possibility even for small laboratories.

In this chapter, we discuss the *Thy1* regulatory elements used to generate transgenic lines with neuronal labeling. We provide a brief overview of currently available *Thy1* transgenic mice, including lines labeling neuronal organelles or reporting neuronal function. We provide protocols for inserting new fluorescent indicators into the *Thy1* vector for pronuclear injection and for screening the offspring of transgenic founders using cryosections and/or acute nerve–muscle explants.

INTRODUCTION

Thanks to Golgi's *reazione nera*, the value of high-contrast labeling of stochastic subsets of nerve cells has been apparent since the dawn of modern neuroscience (Ramon y Cajal 1904). However, translating these early achievements into a method to observe nerve cells in vivo over time has required another century of technological development. With the advent of two-photon microscopy and major advances in fluorescent protein development, such studies have now become feasible. Although many different approaches to such studies have been explored (Young and Feng 2004), most recent chronic in vivo imaging studies of nerve cells in the central nervous system (CNS) and peripheral nervous system (PNS) have taken advantage of transgenic mice that express high levels of fluorescent proteins in subsets of nerve cells driven by murine *Thy1* regulatory elements (Lichtman and Sanes 2003; Misgeld and Kerschensteiner 2006; Pan and Gan 2008; Holtmaat and Svoboda 2009; Wilt et al. 2009). A collection of such transgenic mice (*Thy1*-XFP mice) was first described by Feng et al. (2000), and since then numerous studies have taken advantage of these original transgenic lines. In addition, many new variants of such mice have been engineered (see *Thy1* Lines below). The central role that *Thy1*-XFP mice have come to play in some areas of in vivo imaging, and the plethora of lines now available, motivate our attempt in this chapter to provide background information about the *Thy1* promoter itself, the available *Thy1* lines, and the generation of new ones. Moreover, as core facilities that generate transgenic founder mice for a reasonable fee are now available at most major research institutions, generating new *Thy1* transgenic animals is an option even for relatively small laboratories. We, therefore, provide protocols for *Thy1* vector generation and screening of *Thy1* XFP mice.

The *Thy1* Expression Vector

The *Thy1* expression vector (Fig. 1) is derived from a gene that encodes a small immunoglobin superfamily member, *Thy1* (GeneID 21838; Ingraham et al. 1986), which is glycosylphosphatidylinositol (GPI) anchored to the plasma membrane. Endogenously, murine *Thy1* is widely expressed on numerous cell types, including immune cells and neurons (Morris 1992). As one of the first discovered surface markers of specific immune cell subsets, effective antibodies became available early. In addition to immunological studies, this allowed detailed exploration of the expression of *Thy1* in the nervous system (Morris 1985). Studies of transcriptional regulation followed suit, revealing the strong and broad expression in nerve cells controlled by *cis*-acting elements within the locus. This

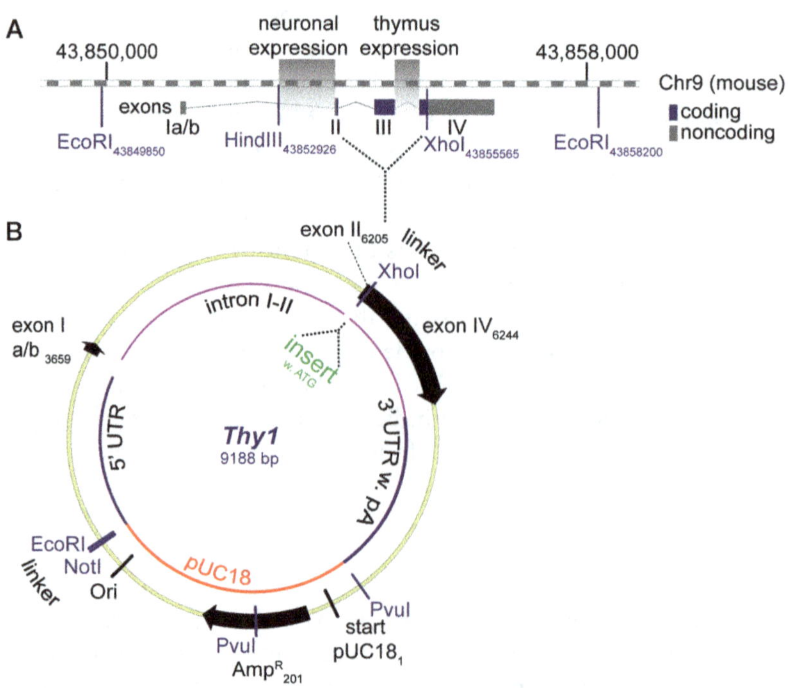

FIGURE 1. The murine *Thy1* locus and the *Thy1* cloning vector. (*A*) Schematic map of the endogenous *Thy1* locus. (*B*) *Thy1* cloning vector. The critical restriction sites for cloning are indicated. Important features within the plasmid are annotated. (Modified from Ingraham et al. 1986, Vidal et al. 1990, and Caroni 1997.)

allowed tailoring expression toward neuronal specificity, while largely suppressing immune cell expression (Vidal et al. 1990). Together, these features made the *Thy1* gene the source of well-characterized pan-neuronal regulatory elements and an obvious choice for the creation of neuron-specific transgenic mice (Caroni 1997). Originally, *Thy1*-derived regulatory elements were used to express disease- and remodeling-associated genes (Aigner et al. 1995; Andra et al. 1996). Because the *Thy1* gene is up-regulated only postnatally in most neuronal populations, the *Thy1*-driven gene expression in these studies allowed circumventing undesirable developmental effects, without resorting to complex conditional gene-expression strategies. In these studies, the strong variation in expression levels among different classes of neurons, but also, more importantly, among individual neurons of the same population, was already noted (Caroni 1997). Although this variation represents a potential problem for studies that use *Thy1* regulatory elements to establish pan-neuronal expression of transgenes to test for their biological effects, it has proven an unexpected blessing for studies that aim at revealing neuronal morphology using in vivo imaging.

Stochastic Subset Labeling

In 2000, Feng and colleagues took advantage of the strong but variable neuronal expression driven by *Thy1* regulatory elements to label neurons in vivo (Feng et al. 2000). They described what have become known as *Thy1*-XFP transgenic mice, a set of 25 transgenic lines with individual expression patterns of green fluorescent protein (GFP), yellow fluorescent protein (YFP), cyan fluorescent protein (CFP), or DsRed. Among these originally described lines were some in which the majority of neurons endogenously expressed the *Thy1* gene, as well as others that showed a striking Golgi-like labeling pattern (Fig. 2). These *Thy1*-XFP lines of the first generation are still widely in use; they can be easily obtained from The Jackson Laboratory. They have been featured in numerous seminal in vivo imaging studies of developmental, physiological, and pathological neuronal plasticity (e.g., Grutzendler et al. 2002; Trachtenberg et al. 2002; Gan et al. 2003; Walsh and Lichtman 2003; Xu et al. 2009; Yang et al. 2009).

One of the most useful features of *Thy1*-driven expression of fluorescent proteins has proven to be its baffling sensitivity to chromosomal position effects. This results in a wide variety of expression patterns in different founders that originate from pronuclear injection of the same *Thy1* construct (Fig. 3). Expression varies with respect to both the neuronal subpopulations that are labeled as well as the percentage of cells within a population that expresses the transgene. Although few general rules have held up without exception, certain neuron populations still tend to show expression in most *Thy1* transgenic lines. These neuronal populations typically give rise to long-distance projections (e.g., motor neurons, large-diameter dorsal root ganglion neurons, retinal ganglion cells, or cerebellar mossy fibers), but some cells with local projections (e.g., dentate gyrus granule cells in the hippocampus) are also typically labeled. In many lines, subpopulations of excitatory cortical neurons (e.g., pyramidal cells in the cortex or hippocampal region CA1) are labeled to a variable degree. Less commonly, local interneurons, Purkinje cells, climbing fibers, or small-diameter sensory neurons are labeled. However, many unexpected labeling patterns can be found, even in nonneuronal cells. For example, lines have been reported with expression in astrocytes or terminal Schwann cells (Livet et al. 2007). As a general rule, transgene expression becomes detectable during the first postnatal week in most *Thy1* lines in accordance with the developmental profile of endogenous *Thy1* expression (Morris 1985). However, *Thy1*-XFP lines show a wide range of variation with regard to expression onset as well. For example, *Thy1*-YFP[16] (Feng et al. 2000) shows strong expression in motor neurons as early as embryonic day 13 (Lichtman and Sanes 2003).

A rather surprising feature of *Thy1*-driven expression is the fact that within a seemingly homogenous population of neurons (say, the motor neurons of a given muscle's motor pool) individual neurons can show strong labeling, while in adjacent cells, such labeling is undetectable. This does not represent a slight difference in the labeling intensity, such that the detection sensitivity creates an arbitrary cutoff above which some brighter neurons would be scored as labeled, whereas dimmer labeling would be scored as absent. Indeed, in some of the so-called "subset" *Thy1* lines, labeling of the few

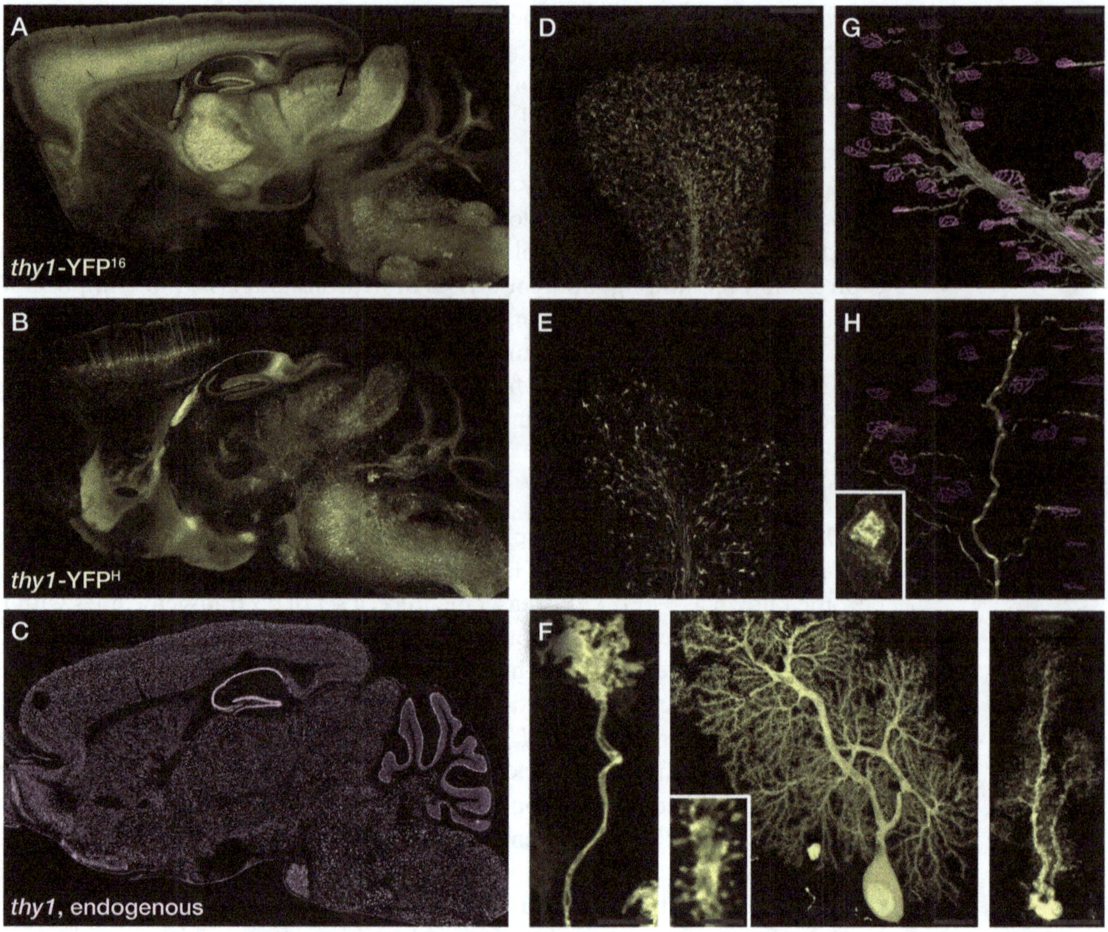

FIGURE 2. Two *Thy1* lines with typical sites of expression. (*A,D,G*) Line *Thy1*-YFP[16], a typical full line. (*B,E,F,H*) Line *Thy1*-YFP[H], a subset line. (*C*) In situ hybridization of endogenous expression pattern (modified from Allen Brain Atlas; contrast inverted). Confocal analysis of parasagittal sections (*A,B*) shows that even a full line only labels subsets of most neuronal populations, albeit in this case, at a significantly higher frequency than in the subset line. (*D,E,F*) Higher power views confirm this for the cerebellum. In-line *Thy1*-YFP[16] mossy fibers and parallel fibers are densely labeled, whereas line *Thy1*-YFP[H] shows sparse subset labeling in these axon populations (see isolated mossy fiber rosette in *F, left*). In addition, however, line *Thy1*-YFP[H] very occasionally also harbors labeled Purkinje cells (*F, middle; inset* shows spiny dendrites at full resolution) and even nonneuronal Bergmann glia (*F, right*). In the neuromuscular system, full lines, such as *Thy1*-YFP[16], show complete overlap between preterminal axon labeling (yellow) and bungarotoxin-labeled postsynaptic sites (magenta label; *G*). In subset lines, such as line *Thy1*-YFP[H], motor units can be traced instead (*H*; inset shows a labeled nonneuronal cell). Lines were originally described in Feng et al. (2000). Scale bars: in *A*, 1 mm (for *A,B*); in *D*, 100 µm (for *D,E*); in *F* (*left*), 5 µm; in *F* (*center, right*), 50 µm (*inset*, 2.5 µm); in *G*, 50 µm (for *G,H*). Images acquired on an FV1000 confocal system (Olympus) with 4x/0.13-NA air (*A,B*), 20x/0.85-NA oil (*D,E*), and 60x/1.42-NA oil (remaining panels) objectives. (*C*, Modified, with permission, from the Allen Mouse Brain Atlas 2010 [http://mouse.brain-map.org/viewImage.do?imageId=68400987]; Lein et al. 2007.)

expressing cells is brighter than the brightest cells in any of the "full" lines. These labeling patterns are characteristic for a given *Thy1*-XFP line and are heritable, and numerous distinct lines that differ in labeling density, intensity, and expressed fluorophore are now available. With regard to subset labeling and the related nomenclature, it should be noted that the original *Thy1*-XFP lines were categorized as full or subset mostly based on the percentage of motor neurons labeled in a limited number of motor pools because the completeness of motor neuron labeling is easily scored by counterstaining synaptic sites in the periphery (Fig. 2). This nomenclature does not imply that a full line necessarily shows complete labeling of any other neuronal population, or even of all motor neurons in all motor pools. Still, a general tendency exists that lines with subset labeling in one population of neurons also show subset labeling of other neuronal populations. However, all *Thy1* lines lack labeling of

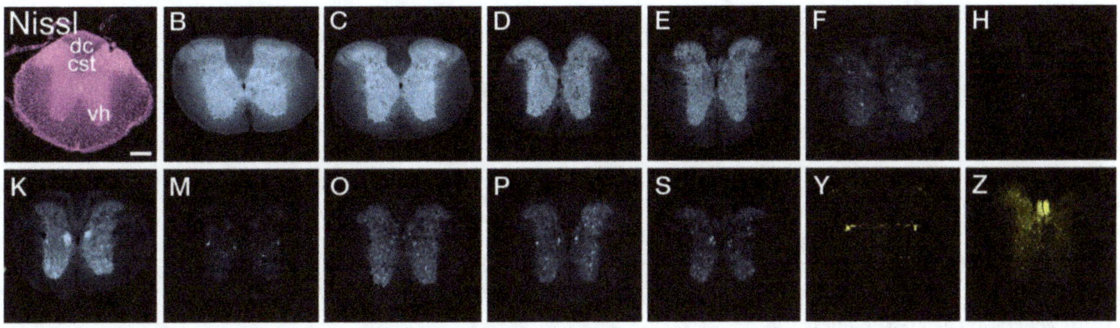

FIGURE 3. Spectrum of expression patterns found in a typical screen of founders. This figure shows the full spectrum of expression patterns found in the spinal cord of the MitoMice (*Thy1*-mitoCFP, cyan; Misgeld et al. 2007). Labeling density ranges from almost complete (e.g., lines *B,C*) to extremely sparse (line *H*). Subset labeling is not restricted to *Thy1*-transgenic lines. Lines *X* and *Y* were generated using neuron specific enolase–derived regulatory elements to express mitoYFP (yellow). (*Nse* lines courtesy of R.W. Burgess, The Jackson Laboratory; images were taken in collaboration with M. Kerschensteiner, Ludwig-Maximilians University, Munich.) First image shows Nissl-like fluorescent stain (magenta) of a matched spinal cord section (cst, corticospinal tract; dc, dorsal column; vh, ventral horn). Scale bar, 250 µm. Images acquired on an FV1000 confocal system (Olympus) with a 4x/0.13 air objective.

some neurons or neuronal subsets; hence, *Thy1*-driven expression should never uncritically be accepted as pan-neuronal. One unfortunate feature of *Thy1*-driven subset expression is its developmental dynamics. In many subset lines, the density of labeling increases as the animal ages, which can be an impediment in studies aimed at longitudinal reidentification of the same neurons. Moreover, at least based on anecdotal observations, the density of subset labeling in individual animals of a given line seems to be influenced by its genetic background (and even gender). As a consequence, different substrains of the original *Thy1*-XFP lines now likely exist, and in planning studies based on these animals, such variation should be taken into consideration.

The exact reasons for the subset expression found in *Thy1*-XFP lines remain unknown. Generally, this variability is ascribed to positional effects relating to insertion site and copy number of the transgene. Moreover, effects akin to position effect variegation could explain the subset character of some lines (Feng et al. 2000). No study to date, however, has systematically explored this question or taken advantage of subset *Thy1* lines to define new neuronal cell classes based on the integration loci of the *Thy1* transgenes.

Thy1 LINES

Overview of Available *Thy1* Lines

Since the description of the first *Thy1*-XFP lines by Feng et al. (2000), numerous additional sets of *Thy1*-driven transgenic lines that express a wide variety of conventional and functionalized fluorescent proteins have been reported. Initially, efforts aimed at widening the spectrum of available labeling patterns and colors (second generation). Additionally, attempts were initiated to systematically control subset labeling using the cre-loxP recombination system (third generation). In parallel, with the advent of well-characterized mammalian organelle targeting sequences and the proliferation of fluorescent protein-based functional reporter constructs, numerous *Thy1* lines were generated that permit in vivo assaying of organelle distribution and transport, intraneuronal ion concentrations, neuronal spiking activity, and synaptic release. An overview of these lines, and the corresponding references, is provided in Table 1. Many of these lines are now available either through The Jackson Laboratory (www.jax.org) or directly from the laboratories that generated them. Considerable challenges for laboratories that generate *Thy1* lines are the financial and infrastructural requirements for line maintenance. Moreover, expression patterns in many of the reported lines have not been systematically documented, at least not in centralized public domain databases. Therefore, centralized

TABLE 1. *Thy1* lines described in the literature

Reference	Name	Purpose	Fluorophores	Available at the Jackson Laboratory?
First-generation *Thy1*-XFP lines				
Feng et al. (2000)	*Thy1*-XFP	Assessment of neuronal morphology	CFP, GFP, RFP, YFP	Y
Second-generation *Thy1*-XFP lines				
De Paola et al. (2003)	*Thy1*-membXFP	Neuronal plasma membrane labeling	GFP	
Hirrlinger et al. (2005)	*Thy1*-RCFP	Assessment of neuronal morphology	Coral FPs	
Third-generation *Thy1*-XFP lines				
Buffelli et al. (2003)	*Thy1*-STOP-YFP	Subset labeling and neuronal Cre reporter	YFP	Y
Livet et al. (2007)	Brainbow	Multicolor labeling of neurons	CFP, GFP, RFP, YFP	Y
Organellar *Thy1*-XFP lines				
De Paola et al. (2003)	*Thy1*-synaptophysin-XFP	Synapse localization and SV trafficking	GFP, RFP	
Umemori et al. (2004)	*Thy1*-synaptophysin-YFP	Synapse localization and SV trafficking	YFP	Y
Südhof 2008[a]	*Thy1*-syntaxin1a-YFP	Synapse localization and SV trafficking	YFP	Y
Han et al. (2005)	*Thy1*-synaptotagmin-YFP	Synapse localization and SV trafficking	CFP	Y
Südhof 2008[b]	*Thy1*-VAMP2-CFP	Synapse localization and SV trafficking	CFP	Y
Misgeld et al. (2007)	MitoMouse (*Thy1*-mitoCFP)	In vivo assessment of mitochondrial transport	CFP, YFP	Y
Functionalized *Thy1*-XFP lines				
Berglund et al. (2006)	*Thy1*-clomeleon	Ratiometric measurement of chloride levels	CFP/YFP	
Heim et al. (2007)	*Thy1*-CerTN-L15	FRET measurement of calcium levels	Cerulean/citrine	
Arenkiel et al. (2007)	*Thy1*-ChR2-YFP	Optogenetic control of neuronal firing	YFP	Y
Young et al. (2008)	SLICK	Cre-mediated recombination in labeled neurons	YFP	Y
Li et al. (2005)	*Thy1*-synapto-pHluorin	Synaptic transmitter release	YFP	
Tabares et al. (2007)	*Thy1*-synapto-pHluorin	Synaptic transmitter release	YFP	
Wyatt et al. (2008)	*Thy1*-synapto-pHluorin	Synaptic transmitter release	YFP	

CFP, cyan fluorescent protein; GFP, green fluorescent protein; YFP, yellow fluorescent protein; RFP, red fluorescent protein variants (e.g., DsRed1, Tdimer2); FPs, fluorescent proteins; FRET, fluorescence resonance energy transfer.

[a]MGD Direct Data Submission [MGI Ref ID J:132968].

[b]MGD Direct Data Submission [MGI Ref ID J:131425].

deposition of frozen embryos or sperm, in addition to public domain documentation of expression patterns, should become a future priority.

Generating *Thy1* Lines

The generation of new *Thy1* transgenic lines generally consists of five steps: (1) engineering and characterization of the desired fluorescent reporter protein, (2) cloning of the reporter protein into the *Thy1* vector, (3) linearization and purification of the new *Thy1* construct, (4) pronuclear injection to generate founders, and (5) screening of founder progeny to establish transgenic lines.

Here, we provide protocols for steps 2 and 3 above (Protocol A) and step 5 above (Protocols B and C). Step 1 above (generation and characterization of desired reporter proteins) depends critically on the exact reporter being studied. Therefore, a general protocol cannot easily be provided. Often, fluorescent reporters are first characterized in simple cell lines, such as human embryonic kidney (HEK) cells. However, in our experience, a difficult problem remains that the performance in transgenic mice is often not easily predictable on this basis. For example, many fluorescent proteins that appear unproblematic in cell culture, show aggregation when expressed at high levels in vivo. This represents a particular problem with fusion proteins, in which in vivo functionality can often not be proven. One approach is to cross the transgenic animal into a mutant strain that carries a deletion of the gene under study. Phenotypic rescue would suggest intact function of the fusion protein, provided expression is properly targeted. Such a stringent assay, however, is rarely possible and requires considerable breeding efforts. Methods that allow transient overexpression in mice (such as in utero electroporation) can provide valuable hints with regard to in vivo performance of a new reporter. Given these challenges, plus the cost and effort involved in making transgenic animals, it often is advisable to err on the side of caution and use transgenic fluorescent proteins that have been tested in vivo already, rather than use the product of the latest improvement in fluorescent protein engineering.

After the desired fluorescent protein reporter sequence is cloned into the *Thy1* vector (Protocol A), it must be sent to a facility specializing in pronuclear injection (Step 4). (Because few imaging laboratories maintain such a facility themselves, we do not describe the pronuclear injection step.) After successful pronuclear injection, the transgenic facility will generally transfer a set of polymerase chain reaction (PCR)-confirmed founders (i.e., animals that, on tail biopsy, were found to contain the *Thy1* vector sequences). To screen for functional expression, we cross founders with commercially obtained inbred mice (typically C57Bl/6). Animals of the F_1 generation are screened at 3 wk of age, when *Thy1*-driven expression has typically reached high levels. In most cases, more than half of the PCR-confirmed founders have protein-expressing offspring. To not jeopardize lines during screening, it is important to always keep sufficient numbers of PCR-confirmed breeders until the screening process for a given line is complete.

For screening, we typically prepare acute nerve–muscle explants (Protocol C) in parallel to transcardial perfusion (Protocol B) because fixation can weaken fluorescence and create misleading autofluorescence. In addition, this allows the first functional tests of the expressed biosensors. Positive lines are then established by further crossing into the in-bred strain, and the lines are maintained by PCR-based selection of positive animals. Once the desired level of backcrossing to a defined genetic background is reached, animals can be homozygosed for the transgene by brother–sister inbreeding.

Because screening, for cost and space reasons, is typically performed in the F_1 generation, changes in expression patterns are possible in future generations because of multiple insertions of the transgene that segregate. Similarly, changes in genetic background can affect expression patterns. Therefore, new lines should be regularly checked for stability of the desired expression patterns, and care should be taken to indicate generation numbers when first describing new lines.

Fluorescent proteins can modulate protein function or even exert toxic effects when fused to functional domains (Han et al. 2005) or when expressed at very high levels (Bridge et al. 2009). Therefore, careful functional studies of fusion proteins before and after transgenic mouse generation are necessary. Generally, it is advisable to maintain lines that show lower levels of fluorescent reporter protein in addition to highly expressing lines.

Protocol A

Generation of *Thy1* Constructs for Pronuclear Injection

In this protocol, the sequence for the desired fluorescent reporter protein is cloned into the XhoI restriction site of the *Thy1* vector. This usually involves blunt-end cloning, because the traditional *Thy1* vector does not carry an intact multiple cloning site. Following successful cloning, the DNA is prepared for pronuclear injection by linearizing it using EcoRI and PvuI restriction enzymes. The purified linearized DNA must then be sent to a facility specializing in pronuclear injection to generate transgenic founder mice.

MATERIALS

CAUTION: See Appendix 6 for proper handling of materials marked with <!>.
See the end of the chapter for recipes for reagents marked with <R>.

Reagents

Agarose gels (1%)
Calf intestinal phosphatase (CIP)
Competent cells for DNA transformation (e.g., DH5α cells)
Deoxyribonucleotide triphosphates (dNTPs)
DNA for fluorescent reporter protein to be cloned into the *Thy1* vector
DNA polymerase I (large Klenow fragment)
Ethylenediaminetetraacetic acid (EDTA) <!>, 0.5 M
Gel extraction kit (e.g., QIAquick Gel Extraction Kit, Qiagen)
HEK 293 cells
Luria Bertani (LB) broth and agar plates containing ampicillin <!>
Microinjection buffer <R>
Plasmid DNA purification kit (mini- and midiprep kits)
Restriction enzymes and buffers

EcoRI, PvuI, and XhoI are required for the *Thy1* vector. Other enzymes may be required to digest the DNA to be cloned and to assess successful cloning. Bovine serum albumin may be required for some digestions.

Sequencing and PCR reagents as necessary (see Steps 12 and 14)
Sterile, double-distilled H_2O
T4 DNA ligase and buffer
Thy1 vector (Addgene, Inc., Cambridge, USA)

Generate 10–20 µg of the *Thy1* vector before cloning. A midiprep should yield sufficient amounts of DNA.

Tris-acetate-EDTA (TAE) buffer, with and without 2-µg/mL ethidium bromide <!>

Equipment

Gel electrophoresis equipment
Heating block, variable-temperature
Incubator at 37°C

Scalpel blade
Sequencing and PCR equipment as necessary (see Steps 12 and 14)
Spectrophotometer to determine DNA concentration
UV light box <!>

EXPERIMENTAL METHOD

Cloning of Fluorescent Reporter Protein Sequence

Because enzyme activities can vary from batch to batch, cross-check the amounts here with the manufacturers' current recommendations.

1. Linearize 10 µg of the *Thy1* vector with 60 U of XhoI overnight (at least 4 h) at 37°C.
2. Stop the restriction digest by heat inactivating the enzyme for 20 min at 65°C.
3. Dephosphorylate the vector by adding 5 U of CIP, and incubate for 1 h at 37°C.
4. Blunt the DNA by adding 10 U of DNA polymerase I (large Klenow fragment) and dNTPs at a final concentration of 33 µM (each dNTP) and then by incubating the sample at room temperature for 15 min.
5. Stop the enzymatic reaction by adding EDTA at a final concentration of 10 mM, and then by heat inactivating the sample for 20 min at 75°C.
6. Run the entire volume of linearized dephosphorylated blunted DNA on a 1% agarose gel, and purify the fragment using a commercial gel extraction kit (e.g., QIAquick Gel Extraction Kit). Elute the DNA in sterile double-distilled H_2O.
7. Determine the concentration of the eluted *Thy1* vector DNA using a spectrophotometer.
8. Prepare the DNA for the fluorescent reporter protein to be cloned into the *Thy1* vector by digesting it with the appropriate restriction enzyme(s), and then blunting, gel purifying, and quantifying it as described in Steps 4–7.
9. Calculate the amount of insert to be used in the ligation reaction according to the following formula:

   $$\text{amount insert required [ng]} = 3 * \text{amount of vector [ng]} * (\text{insert size [bp]} / \text{vector size [bp]}).$$

   This corresponds to a ratio of 1:3 (vector:insert). Fifty nanograms of *Thy1* vector are recommended for a ligation reaction.
10. Perform the ligation reaction in a 10 µL volume by combining the insert and vector with 400 U of T4 DNA ligase and 1 µL of 10x ligase buffer. Incubate at 16°C for at least 12 h.

    It is important to include a ligation negative control in which the reaction mix does not include the insert.
11. Transform half of the ligation reaction (5 µL) into chemically competent bacterial cells, and grow on LB agar plates with ampicillin overnight at 37°C.
12. Pick colonies, and extract DNA with commercially available kits (miniprep kits should suffice). Perform a restriction digest or a PCR reaction to verify successful insertion of the fragment of interest into the *Thy1* vector in the correct orientation.
13. Transfect DNA from putative clones into HEK cells to examine and to verify expression of the construct.

    Thy1 regulatory elements are not very efficient in HEK cells. As a result, the number of expressing cells tends to be low compared with expression driven by regulatory sequences optimized for expression in cell lines (such as cytomegalovirus [CMV]). Therefore, *Thy1* vector–transfected cell cultures need to be inspected carefully to detect expressing cells.

14. Sequence putative clones to further confirm successful insertion of the fragment of interest into the *Thy1* vector in the correct orientation.
15. Proceed with one clone that has been verified with respect to its sequence and that yielded the expected expression pattern in HEK cells. Prepare a midiprep of plasmid DNA from this clone.

Preparation of DNA for Pronuclear Injection

In the subsequent steps, we describe one way to purify DNA for pronuclear injection. For good results, it is paramount to adjust this protocol to the recommendations of the specific facility that will carry out the pronuclear injections. See *Troubleshooting*.

16. To prepare DNA for pronuclear injection, begin by digesting 10 µg of the DNA from Step 15 with 20 U of EcoRI and 20 U of PvuI overnight (at least 4 h) at 37°C. This should yield three fragments (see Fig. 1) provided there are no EcoR1 or PvuI sites in the cloned fluorescent reporter: 896 base pairs (bp), 1467 bp, and a fragment >6825 bp, which should contain the cloned reporter gene.
17. Run the entire volume of the DNA, except for a small portion (e.g., 5 µL), on a 1% agarose gel *without* ethidium bromide. To verify that the *Thy1* DNA has been digested appropriately, run the reserved small volume of DNA in an outlying lane of the same gel.

    It is important not to expose the DNA that will be used for pronuclear injection to UV light.
18. Cut away the entire outlying lane, and stain it for 30 min with TAE containing 2 µg/mL ethidium bromide.
19. Visualize the ethidium bromide–stained agarose lane using UV light, and mark the position of the *Thy1* fragment containing the cloned reporter gene using a scalpel blade. (This fragment should be >6825 bp.)
20. Accurately align the ethidium bromide–stained agarose lane with the rest of the unstained agarose gel. Using the scalpel markings in the stained gel as a reference, cut out the band containing the *Thy1* vector and cloned reporter gene from the unstained gel.
21. Purify the fragment using a commercial gel extraction kit (e.g., QIAquick Gel Extraction Kit). Elute the DNA in microinjection buffer.
22. Measure the concentration of the eluted DNA.
23. Provide the DNA, along with details about its concentration, to a facility where pronuclear injections are performed.

    Gel photographs that document the purity and integrity of the DNA submitted for pronuclear injection are generally required. The facilities may also require primer sequences for identifying founder animals by tail-clip PCR.

TROUBLESHOOTING

Problem: The inserted reporter contains EcoRI and/or PvuI restriction sites, which are the same sites used for linearizing the *Thy1* vector.
Solution: The *Thy1* vector provides a limited choice in restriction sites for linearization. Try partial restriction and gel isolation of the correct linearized full-length product. Alternatively, perform silent mutagenesis in the reporter sequence to remove undesired cut sites.

Problem: Founders are not obtained.
Solution: DNA of insufficient quality is a major cause of failures to produce founders. Therefore, extreme care should be taken to follow the respective core facilities' recommendations for DNA amount and quality. Avoiding ethidium bromide and UV exposure are critical factors.

Protocol B

Generation of Tissue Sections for Screening *Thy1* Lines

In this protocol, transcardial perfusion is performed on 3-wk-old F_1 mice that have been produced by crossing *Thy1* transgenic founders and commercially obtained inbred mice. Cryosections are generated, and *Thy1*-driven expression is detected by histological characterization.

RELATED INFORMATION

A neuromuscular explant can be obtained in parallel for use in Protocol C.

IMAGING SETUP

We use an Olympus FV1000 confocal microscope equipped with the following objectives: 4×/0.13-NA air, 20×/0.85-NA oil, and 60×/1.42-NA oil. The microscope has standard filters for fluorescent proteins. However, any conventional wide-field epifluorescence microscope equipped with narrow-band fluorescence filters will do.

MATERIALS

CAUTION: See Appendix 6 for proper handling of materials marked with <!>.
See the end of the chapter for recipes for reagents marked with <R>.

Reagents

Ethanol (70%, v/v) <!> in spray bottle
Glycerol-based mounting medium (e.g., VECTASHIELD, Vector Laboratories, Inc.)
Isoflurane <!> or any other lethal anesthetic <!> used in accordance with local animal protection legislation
Isopentane (2-methylbutane) <!>
Liquid nitrogen <!>
Nail polish
Paraformaldehyde (PFA; 4%) <!>, prepared in 1x phosphate-buffered saline (PBS)
PBS (1x) <R>
Sucrose (20%), prepared in a solution of 1x PBS and 0.01% sodium azide <!>
Tissue-freezing medium (e.g., OCT <!>, Tissue-Tek)

Equipment

Coverslips (#1.5 thickness)
Cryostat with disposable blades <!>
Embedding molds for tissue freezing
Eye protection (goggles)
Forceps (e.g., Dumostar, Dumont #3, Fine Science Tools, 11293-00)

Glass slides, gelatine coated
Hemostat, straight
Hypodermic needles
Metal beaker that fits tissue molds
Paper tissue
Razor blade
Reaction tube, 10-mL

Fill the 10-mL reaction tube with the 20% sucrose solution prepared in 1x PBS and 0.01% sodium azide.

Reaction tubes, 50-mL

Three 50-mL reaction tubes are required. Fill one each with 1x PBS, 4% PFA, and distilled water.

Rotary perfusion pump
Russian organ-holding forceps (e.g., Fine Science Tools, 11026-15)
Scissors, large medical (e.g., Fine Science Tools, 14108-09)
Small spring scissors
Staining jar containing 1x PBS
Styrofoam board
Styrofoam box that fits metal beaker
Tissue culture dish (10 cm) containing 1x PBS
Tubing for pump with hypodermic needle attached to one end
Wide-field or confocal fluorescence microscope

EXPERIMENTAL METHOD

1. Set up the perfusion system by filling a line with 1x PBS (from a 50-mL reaction tube; avoid air bubbles), and turn off the pump.
2. Lethally anesthetize a 3-wk-old mouse by isoflurane overdose. Spray the mouse with 70% ethanol to prevent fur hair from being released during opening of the thorax and to sterilize.
3. Pin down the animal on a styrofoam board with three hypodermic needles (one through each forelimb, one at the tail base).
4. Remove the skin, and cut off the pectoral muscles, making incisions close to the muscle insertions on the sternum.
5. Using large scissors, open the abdominal wall, and make incisions parallel to the rib cage all the way to the vertebral column. Then cut the diaphragm open just below the xiphoid cartilage, and detach the diaphragm from the rib cage by cutting along its costal insertions.
6. Use forceps to hold the rib cage by the xiphoid process, and then cut the right and left ribs off the vertebral column, as close as possible to their insertions. The two cuts should converge above the heart, level with the manubrium sterni. Make sure not to cut major blood vessels. Better visibility is achieved by gently lifting the preparation while holding onto the xiphoid process with forceps.
7. The anterior thoracic wall contains the triangularis sterni muscle, and this can be cut off for the explant described in Protocol C. If the explant is not desired, fold the anterior thoracic wall back, and pin it down with a hypodermic needle—to reveal the heart.
8. Hold the heart with organ-holding forceps. Carefully insert the hypodermic needle at the end of perfusion pump tubing into the left ventricle. Make sure the needle does not slide out of heart when released.

   The left ventricle comprises the cardiac apex and can be recognized by its lighter color (compared with the dark red of the right ventricle).

9. Open the right atrium with spring scissors.

   Dark red venous blood should flow out.

10. Start the perfusion pump. Perfuse the animal with 1x PBS (run at ~5 mL/min) for 1 min.

    The liver should turn from dark red to clay-like brown.

11. Turn off the pump, and carefully move the tubing end from 1x PBS to 4% PFA. Avoid sucking in air bubbles by making sure that the pump wheel has come to a complete halt before removing the tubing from the PBS.

12. Restart the pump; perfuse at the same speed with 25 mL of 4% PFA.

    Once PFA reaches the animal's circulation, muscle contractions set in. These contractions lead to curling of the tail and adduction of the hind limbs, which indicate efficient perfusion.

13. At the end of the perfusion (~5 min, i.e., 25 mL), remove the needle from the heart, and flush the perfusion system with water.

14. Cut the skin off of the animal. If the triangularis sterni was not used for explant, remove the anterior thoracic wall now.

15. Isolate a tissue block consisting of the vertebral column and the head. Open the spinal cavity with a transverse scissor cut between two vertebrae, and open the cranium with a rostral cut from the nasal cavity.

    This ensures access of the PFA to CNS tissue during postfixation.

16. Transfer the tissue blocks to reaction tubes containing PFA. Make sure the tissues are submerged. Post-fix the tissues at 4°C.

    i. Post fix the anterior thorax wall with the triangularis sterni muscle for only 1 h.

    ii. Post fix the CNS tissue for 24–48 h.

17. After the postfixation period, rinse the tissues twice with 1x PBS.

18. In a 10-cm tissue culture dish filled with 1x PBS, isolate tissues of interest (e.g., brain, spinal cord, and eyes).

    i. Split the brain along the medial longitudinal fissure with a razor blade. Further divide one hemisphere by making two frontal cuts at midcortical level and between the forebrain and the cerebellum.

    ii. Isolate small 5-mm blocks (e.g., the cervical swelling) from the spinal cord. Ideally, include some attached dorsal root ganglia.

    iii. Puncture the eyes in front of the pupil with a hypodermic needle.

19. Transfer the tissue to a 10-mL reaction tube containing 20% sucrose for cryoprotection. Store at 4°C until the tissue sinks; this typically takes 24 h.

20. After the tissue is cryoprotected, prepare for freezing under a fume hood. Cover the bottom of a metal beaker with a 0.5-cm deep layer of isopentane. Fill a styrofoam container with liquid nitrogen so that the bottom is covered to several centimeters. (*Important:* Wear protective goggles!)

21. Blot tissue blocks dry on paper tissue. Place them in tissue molds: The complete brain hemisphere on the sagittal cut side, the frontal cut pieces on their frontal side, the spinal cord as a standing cylinder, and the eye on its side.

22. Cover the tissues with tissue-freezing medium, avoiding air bubbles. Make sure that tissues do not shift.

23. Transfer the metal beaker with isopentane into liquid nitrogen, and wait until the isopentane becomes turbid and viscous.

24. Using the hemostat, transfer the tissue mold into the isopentane, making sure that the tissues do not shift and that no isopentane leaks into the tissue-freezing medium.

25. Wait until the tissue-freezing medium almost completely hardens, and then, when only the central part of the block is still liquid, transfer it to −20°C storage.
26. After complete freezing, attach the frozen tissue block onto the cryostat chuck. Cut 10–60-µm sections, and take them up on gelatine-coated glass slides.

    The thickness depends on whether a confocal microscope is used for documentation.
27. Allow the sections to air-dry for 10 min at room temperature. Rinse them in a staining jar with 1x PBS to remove the tissue-freezing medium.
28. Add a drop of mounting medium to each coverslip. Cover each slide by slowly lowering the tissue-carrying side down onto a coverslip (carefully avoiding air bubbles).
29. Seal with nail polish. Store at –20°C.
30. Document expression patterns using fluorescence microscopy.

    Typically, overview images taken at 4x magnification of the sagittal section of the brain and the spinal cord, and higher-magnification views (e.g., 20x) of the cortex, hippocampus, cerebellum, retina, and dorsal root ganglion, provide a good basis for classifying a *Thy1* founder. Fully resolved imaging (e.g., using a 60x oil lens at Nyquist sampling) can be used to confirm subcellular localization of a fluorescent tag.

TROUBLESHOOTING

Problem: Low levels of expression are obscured in fixed tissue.

Solution: This may be caused by high background (e.g., for CFP) or by the instability of the fixed fluorescent reporters (e.g., during PFA fixation of some red fluorescent proteins). When using new fluorescent proteins, test different fixatives. To reduce autofluorescence buildup, do not expose tissues to fixatives for excessive periods of time. Inspect preparations soon after generation, and store them at −20°C.

Protocol C

Imaging of *Thy1* Lines Using an Acute Neuromuscular Explant

In this protocol, acute neuromuscular explants are obtained from 3-wk-old F_1 mice that have been produced by crossing *Thy1* transgenic founders and commercially obtained inbred mice. *Thy1*-driven expression is detected by fluorescence microscopy.

RELATED INFORMATION

The neuromuscular explant can be obtained in parallel to perfusion of a mouse to produce cryosections for histological characterization (see Protocol B). This protocol is adapted from methods presented in Misgeld et al. (2007) and Kerschensteiner et al. (2008).

IMAGING SETUP

We use an Olympus BX51WI microscope equipped with a 4x/0.13-NA air objective and the following long-working distance dipping-cone objectives: 20x/0.5-NA water and 100x/1.0-NA water. The microscope has standard filters for the most common fluorescent proteins.

MATERIALS

CAUTION: See Appendix 6 for proper handling of materials marked with <!>.
See the end of the chapter for recipes for reagents marked with <R>.

Reagents

α-bungarotoxin (BTX) <!> tagged with a fluorescent dye that can be separated from neuronally expressed fluorescent protein (50 μg/mL) (optional; see Step 7)
Store the 50 μg/mL stock solution at –20°C. Before use, dilute it 1:20 in 1x Ringer's solution.

Ringer's solution (1x) <R>

Equipment

Dissection microscope with cold-light source
Forceps (e.g., Dumostar, Dumont #3, Fine Science Tools, 11293-00)
Forceps, beveled, with fine shanks (e.g., Dumostar, Dumont #5, Fine Science Tools, 11295-00)
Minutien pins (10, shortened to <4 mm; e.g., Fine Science Tools, 26002-20) to fix the explant in the dish
Scissors, large medical (Fine Science Tools, 14108-09)
Scissors, small angled spring (Fine Science Tools, 15033-09)
Tissue culture dish, 3.5-cm, filled with Sylgard polymer (e.g., Sylgard 184 Silicone Elastomer Kit, Dow Corning)
Tissue culture dish, 10-cm
Upright fluorescence microscope with suitable dipping-cone objectives

EXPERIMENTAL METHOD

Perform Steps 3–5 under a dissection microscope.

1. Perform Steps 1–7 of Protocol B.
2. Make a transverse cut just below the manubrium sterni to release the anterior thoracic wall. Transfer the liberated anterior thoracic wall to a 10-cm dish containing cooled (~4°C) 95% O_2/ 5% CO_2 -bubbled Ringer's solution.
3. Using small spring scissors and fine forceps, remove the remnants of intrathoracic organs (inside) and pectoral muscles (outside).
4. Cut off ribs that do not insert to the sternum. This is best performed using a small spring scissors to cut the tissue in between the ribs.
5. Immobilize the triangularis sterni explant in a Sylgard-coated 3.5-cm dish using minutien pins.
6. Transfer the dish to the imaging setup, and screen for expression.
7. If a detailed characterization of labeling frequency within the triangularis motor pool is desired, stain the explant with BTX as follows.
    i. Replace the Ringer's solution with fluorescently tagged BTX (2.5 µg/mL) in Ringer's solution.
    ii. Stain for 15–30 min.
    iii. Rinse twice with fresh Ringer's solution.
    iv. Determine the percentage of synapses with presynaptic labeling.

RECIPES

CAUTION: See Appendix 6 for proper handling of materials marked with <!>. Recipes for reagents marked with <R> are included in this list.

Microinjection Buffer

Reagent	Final concentration
Tris <!>	5 mM
EDTA <!>	0.1 mM

Adjust the pH to 7.6, filter sterilize, and store at 4°C.

Phosphate-Buffered Saline (PBS, 10x)

Reagent	Quantity (for 1 L)	Final concentration (10x)
$NaH_2PO_4 \cdot H_2O$	2.6 g	18.8 mM
$Na_2HPO_4 \cdot H_2O$	14.4 g	80.9 mM
NaCl	87.5 g	1.5 M

Ringer's Solution (10x)

Reagent	Quantity (for 1 L)	Final concentration
$NaHCO_3$	21.84 g	260 mM
$NaH_2PO_4 \cdot H_2O$	1.72 g	12.5 mM
KCl <!>	1.86 g	25 mM
NaCl	73.05 g	1.25 M
H_2O	to 1L	

Store at 4°C and use within 1 wk.

Ringer's Solution (1x)

Reagent	Quantity (for 1 L)	Final concentration
$CaCl_2$ <!>, 1 M	2 mL	2 mM
H_2O	900 mL	
$MgCl_2$, 1 M	1 mL	1 mM
Ringer's solution (10x) <R>	100 mL	125-mM NaCl; 2.5-mM KCl; 1.25-mM NaH_2PO_4; 26-mM $NaHCO_3$
Glucose	3.60 g	20 mM

$CaCl_2$ and $MgCl_2$ are prepared monthly as 1 M stock solutions. On the day of the experiment, mix the reagents in the order listed. Bubble with 95% O_2 <!>/5% CO_2 <!> for 30 min before use. The 1x Ringer's solution should have a pH of 7.8 without 95% O_2/5% CO_2 and a pH of 7.3 after 30 min of bubbling with 95% O_2/5% CO_2.

ACKNOWLEDGMENTS

We thank Martin Kerschensteiner, Ludwig-Maximilians University Munich, Rob Burgess, Jackson Laboratory, and Jeff Lichtman, Harvard University, for collaboration on obtaining the pictures shown in Figure 3. We also acknowledge the kind permission of the Allen Brain Institute to use Figure 2C. Work in our laboratory is supported by the Alexander-von-Humboldt foundation, the TUM Institute for Advanced Study, and the Bundesministerium für Bildung und Forschung (BMBF).

REFERENCES

Aigner L, Arber S, Kapfhammer JP, Laux T, Schneider C, Botteri F, Brenner HR, Caroni P. 1995. Overexpression of the neural growth–associated protein GAP-43 induces nerve sprouting in the adult nervous system of transgenic mice. *Cell* **83:** 269–278.

Andra K, Abramowski D, Duke M, Probst A, Wiederhold KH, Burki K, Goedert M, Sommer B, Staufenbiel M. 1996. Expression of APP in transgenic mice: A comparison of neuron-specific promoters. *Neurobiol Aging* **17:** 183–190.

Arenkiel BR, Peca J, Davison IG, Feliciano C, Deisseroth K, Augustine GJ, Ehlers MD, Feng G. 2007. In vivo light-induced activation of neural circuitry in transgenic mice expressing channelrhodopsin-2. *Neuron* **54:** 205–218.

Berglund K, Schleich W, Krieger P, Loo LS, Wang D, Cant NB, Feng G, Augustine GJ, Kuner T. 2006. Imaging synaptic inhibition in transgenic mice expressing the chloride indicator, Clomeleon. *Brain Cell Biol* **35:** 207–228.

Bridge KE, Berg N, Adalbert R, Babetto E, Dias T, Spillantini MG, Ribchester RR, Coleman MP. 2009. Late onset distal axonal swelling in YFP-H transgenic mice. *Neurobiol Aging* **30:** 309–321.

Buffelli M, Burgess RW, Feng G, Lobe CG, Lichtman JW, Sanes JR. 2003. Genetic evidence that relative synaptic efficacy biases the outcome of synaptic competition. *Nature* **424:** 430–434.

Caroni P. 1997. Overexpression of growth-associated proteins in the neurons of adult transgenic mice. *J Neurosci Methods* **71:** 3–9.

De Paola V, Arber S, Caroni P. 2003. AMPA receptors regulate dynamic equilibrium of presynaptic terminals in mature hippocampal networks. *Nat Neurosci* **6:** 491–500.

Feng G, Mellor RH, Bernstein M, Keller-Peck C, Nguyen QT, Wallace M, Nerbonne JM, Lichtman JW, Sanes JR. 2000. Imaging neuronal subsets in transgenic mice expressing multiple spectral variants of GFP. *Neuron* **28:** 41–51.

Gan WB, Kwon E, Feng G, Sanes JR, Lichtman JW. 2003. Synaptic dynamism measured over minutes to months: Age-dependent decline in an autonomic ganglion. *Nat Neurosci* **6:** 956–960.

Grutzendler J, Kasthuri N, Gan WB. 2002. Long-term dendritic spine stability in the adult cortex. *Nature* **420:** 812–816.

Han W, Rhee JS, Maximov A, Lin W, Hammer RE, Rosenmund C, Sudhof TC. 2005. C-terminal ECFP fusion impairs synaptotagmin 1 function: Crowding out synaptotagmin 1. *J Biol Chem* **280:** 5089–5100.

Heim N, Garaschuk O, Friedrich MW, Mank M, Milos RI, Kovalchuk Y, Konnerth A, Griesbeck O. 2007. Improved calcium imaging in transgenic mice expressing a troponin C–based biosensor. *Nat Methods* **4:** 127–129.

Hirrlinger PG, Scheller A, Braun C, Quintela-Schneider M, Fuss B, Hirrlinger J, Kirchhoff F. 2005. Expression of reef coral fluorescent proteins in the central nervous system of transgenic mice. *Mol Cell Neurosci* **30:** 291–303.

Holtmaat A, Svoboda K. 2009. Experience-dependent structural synaptic plasticity in the mammalian brain. *Nat Rev Neurosci* **10:** 647–658.

Ingraham HA, Lawless GM, Evans GA. 1986. The mouse Thy-1.2 glycoprotein gene: Complete sequence and identification of an unusual promoter. *J Immunol* **136:** 1482–1489.

Kerschensteiner M, Reuter MS, Lichtman JW, Misgeld T. 2008. Ex vivo imaging of motor axon dynamics in murine triangularis sterni explants. *Nat Protoc* **3:** 1645–1653.

Lein ES, Hawrylycz MJ, Ao N, Ayres M, Bensinger A, Bernard A, Boe AF, Boguski MS, Brockway KS, Byrnes EJ, et al. 2007. Genome-wide atlas of gene expression in the adult mouse brain. *Nature* **445:** 168–176.

Li Z, Burrone J, Tyler WJ, Hartman KN, Albeanu DF, Murthy VN. 2005. Synaptic vesicle recycling studied in transgenic mice expressing synaptopHluorin. *Proc Natl Acad Sci* **102:** 6131–6136.

Lichtman JW, Sanes JR. 2003. Watching the neuromuscular junction. *J Neurocytol* **32:** 767–775.

Livet J, Weissman TA, Kang H, Draft RW, Lu J, Bennis RA, Sanes JR, Lichtman JW. 2007. Transgenic strategies for combinatorial expression of fluorescent proteins in the nervous system. *Nature* **450:** 56–62.

Misgeld T, Kerschensteiner M. 2006. In vivo imaging of the diseases nervous system. *Nat Rev Neurosci* **7:** 449–463.

Misgeld T, Kerschensteiner M, Bareyre FM, Burgess RW, Lichtman JW. 2007. Imaging axonal transport of mitochondria in vivo. *Nat Methods* **4:** 559–561.

Morris R. 1985. *Thy-1* in developing nervous tissue. *Dev Neurosci* **7:** 133–160.

Morris R. 1992. *Thy-1*, the enigmatic extrovert on the neuronal surface. *Bioessays* **14:** 715–722.

Pan F, Gan WB. 2008. Two-photon imaging of dendritic spine development in the mouse cortex. *Dev Neurobiol* **68:** 771–778.

Ramon y Cajal S. 1904. *Texture of the nervous system of man and the vertebrates*. Springer, New York.

Tabares L, Ruiz R, Linares-Clemente P, Gaffield MA, Alvarez de Toledo G, Fernandez-Chacon R, Betz WJ. 2007. Monitoring synaptic function at the neuromuscular junction of a mouse expressing synaptopHluorin. *J Neurosci* **27:** 5422–5430.

Trachtenberg JT, Chen BE, Knott GW, Feng G, Sanes JR, Welker E, Svoboda K. 2002. Long-term in vivo imaging of experience-dependent synaptic plasticity in adult cortex. *Nature* **420:** 788–794.

Umemori H, Linhoff MW, Ornitz DM, Sanes JR. 2004. FGF22 and its close relatives are presynaptic organizing molecules in the mammalian brain. *Cell* **118:** 257–270.

Vidal M, Morris R, Grosveld F, Spanopoulou E. 1990. Tissue-specific control elements of the *Thy-1* gene. *EMBO J* **9:** 833–840.

Walsh MK, Lichtman JW. 2003. In vivo time-lapse imaging of synaptic takeover associated with naturally occurring synapse elimination. *Neuron* **37:** 67–73.

Wilt BA, Burns LD, Wei Ho ET, Ghosh KK, Mukamel EA, Schnitzer MJ. 2009. Advances in light microscopy for neuroscience. *Annu Rev Neurosci* **32:** 435–506.

Wyatt RM, Balice-Gordon RJ. 2008. Heterogeneity in synaptic vesicle release at neuromuscular synapses of mice expressing synaptopHluorin. *J Neurosci* **28:** 325–335.

Xu T, Yu X, Perlik AJ, Tobin WF, Zweig JA, Tennant K, Jones T, Zuo Y. 2009. Rapid formation and selective stabilization of synapses for enduring motor memories. *Nature* **462:** 915–919.

Yang G, Pan F, Gan WB. 2009. Stably maintained dendritic spines are associated with lifelong memories. *Nature* **462:** 920–924.

Young P, Feng G. 2004. Labeling neurons in vivo for morphological and functional studies. *Curr Opin Neurobiol* **14:** 642–646.

Young P, Qiu L, Wang D, Zhao S, Gross J, Feng G. 2008. Single-neuron labelling with inducible Cre-mediated knockout in transgenic mice. *Nat Neurosci* **11:** 721–728.

WWW RESOURCE

http://mouse.brain-map.org Allen Mouse Brain Atlas ©2010. Allen Institute for Brain Science, Seattle, WA.

23 Correlated Light and Electron Microscopy of Green Fluorescent Protein–Labeled Dendrites and Axons

Graham W. Knott

Bio EM Facility, Centre of Electron Microscopy, Ecole Polytechnique Fédérale de Lausanne, Lausanne, Switzerland

ABSTRACT

This chapter describes how green fluorescent protein (GFP)-labeled axons and dendrites, in live neural tissue, and imaged with two-photon microscopy, can be prepared for analysis in the electron microscpe. It explains how animals, or cultured slices, are fixed, labeled, and serially thin-sectioned so that the ultrastructure of specific parts of neurites can be analyzed in three dimensions.

INTRODUCTION

The ability to observe axons and dendrites with transmission electron microscopy (EM) after they have been previously imaged live with laser-scanning microscopy is a useful technique to study their synaptic connectivity. The method described in this chapter allows for a complete three-dimensional (3D) reconstruction to be made of the imaged structures from serial electron micrographs.

This technique has been used to analyze dendrites and axons of layer 5 neurons in the somatosensory cortex of adult mice that were imaged repeatedly live using two-photon microscopy (Trachtenberg et al. 2002; Holtmaat et al. 2005, 2006; Knott et al. 2006). It was also used to study the synapses on dendrites of neurons that were imaged in cultured slices (Zito et al. 2004; Nikonenko et al. 2008), including dendritic spines that were probed with caged glutamate (Zito et al. 2009).

Protocol

Imaging GFP-Labeled Neurons Using Light and Electron Microscopy

This protocol provides a detailed method by which neurons that were imaged in a live brain or slice culture can be reimaged using EM. First, brain tissue expressing GFP is chemically fixed. Then, an immunocytochemistry process is used to render the fluorescent protein electron dense so that it can first be located using light microscopy and then serial sectioned for EM. Patterns of blood vessels observed in the live brain are used to locate the previously imaged neurons.

IMAGING SETUP

Light Microscopy Equipment

The imaging of the resin-embedded sections is performed with any light microscope and digital camera so that the sections can be aligned onto one another. It can also be achieved with a camera lucida drawing tube attached to a light microscope. We use a light microscope that has a motorized stage and is computer controlled with Neurolucida software (MBF Bioscience).

EM Equipment

The EM is a 120-kV transmission EM with a digital camera. We use a CM12, Philips (now FEI Company) equipped with a Morada digital camera (Olympus).

MATERIALS

CAUTION: See Appendix 6 for proper handling of materials marked with <!>.
See the end of the chapter for recipes for reagents marked with <R>.

Reagents

Agarose, 5% in 0.01 M phosphate-buffered saline (PBS), pH 7.4
Antibodies
  Primary antibody: Anti-GFP antibody (cat. no. AB3080, Millipore)
  Secondary antibody: Biotin-SP-AffiniPure Goat Anti-Rabbit IgG, F(ab′)2: Jackson ImmunoResearch Laboratories, USA; cat. no. 111-065-047
Avidin-biotin complex (ABC; Vectastain, Vector Laboratories; cat. no. PK-6100)
Brain tissue for imaging (live mouse or brain slice culture expressing GFP in neurons)
Bovine serum albumin (BSA)-c (Aurion, The Netherlands; cat. no. 900.022)
Cyanoacrylate glue <!>
3,3′-diaminobenzidine tetrahydrochloride (DAB) <!>
Dimethylsulfoxide (DMSO) <!>
Double-distilled water
Durcupan resin <!>
Ethanol <!> (50%, 70%, 90%, 95%, 100%)
Glutaraldehyde <!> (25% solution)

Glycerol
Hydrogen peroxide <!>
Lead citrate <R>
Liquid nitrogen <!>
Osmium tetroxide <!>
Paraformaldehyde powder <!>
PBS (pH 7.4) <R>
Sodium cacodylate buffer <!>, 0.1 M (pH 7.4)
Sodium pentobarbitone <!> (Sigma-Aldrich Chemie GmbH, Switzerland; Cat. No. P3761)
Toluidine blue (optional; see Step 5)
Tris-buffered saline (TBS) (0.1 M Tris <!>, 0.9% NaCl, pH 8)
Tris-buffered saline (TBS) (0.05 M Tris, 0.9% NaCl, pH 7.6)
Uranyl acetate <!>

Equipment

Culture dishes, 24-well
Deionizer (One Point Ionizer, Haug, Switzerland) (Optional; see Step 45)
Diamond knife
Dissecting forceps
Dissection board (polystyrene block)
Glass histology slides
Glass knives
Glass scintillation vials, 20-mL
Grid staining plate
Grid staining plate, clear plastic (Electron Microscopy Sciences [EMS], USA; cat. no. 71560-10)
Grid storage box
Hot plate (maintained at 60°C)
Image manipulation software

For stacking images, we use Photoshop (Adobe), although free software packages such as Gimp and ImageJ can do the same. (Fiji is the version of ImageJ that has been bundled with the tools that best serve this function).

Mold-separating agent (Glorex, Switzerland; cat. no. 62407445)
Needle, 23-gauge
Oven, 65°C
Paint brush, size 6
Perfusion pump, tubing, and needle
Razor blades
Reverse-action tweezers
Rodent heart clamp: Marcel Blanc et Cie, Switzerland; cat. no. BH020R
Rotary mixer (Taab Laboratory Supplies, UK)
Shaker table
Single-slot grids, carbon-coated with Formvar support film; slot dimensions, 2 × 1-mm
Surgical scissors (Aichele Medico AG, Switzerland; cat. no. 14110-15, 14088-10)
Ultramicrotome
Vibratome <!> (Leica)
Wooden cocktail sticks

EXPERIMENTAL METHOD

If using brain slice cultures, begin procedure at Fixation of Imaged Slices (Step 10).

Fixation and Immunocytochemistry

1. Once the animal is deeply anesthetized with sodium pentobarbitone, pin the forelimbs to a dissection board to get access to the underside and chest cavity.
2. With scissors and forceps, quickly open the chest cavity, exposing the heart and lungs, and insert the perfusion needle into the left ventricle. Once inside, clamp in place, and immediately cut the right auricle.

   Do not take >1 min; otherwise the ultrastructure will be compromised. As soon as the chest cavity is opened, the lungs cannot function and the tissues will become hypoxic. It is important that fixation takes place as soon as possible and the brain is not left in a hypoxic state for too long.
3. Start the perfusion pump. Ensure that the fixative is perfused at a rate not >30% of total body volume per minute. For example, a 30-g animal should not be perfused faster than 10 mL/min of fixative. Perfuse with a total volume of 300 mL of 0.1 M phosphate buffer (pH 7.4) containing 4% paraformaldehyde and 0.2% glutaraldehyde.

   It is important that the perfusion rate is not so high that it would cause vessels to rupture. There is no need to wash out the animal with a buffer solution before introducing the fixative.
4. Leave the animal at room temperature for 2 h. Do not leave for >3 h, as this will reduce the immunoreactivity of the tissues, owing to epitope masking.
5. Using a fine-gauge needle (e.g., 23-gauge), puncture the surface of the brain rostral and caudal to the region that was imaged with laser-scanning microscopy. These holes will be visible when the brain is sectioned, giving a rough indication of the location of the imaged neuron.

   The needle can also be dipped into 2% toluidine blue before inserting it into the brain to make the holes visible. This helps to orient the brain, by eye, in the agarose before it is sectioned in the Vibratome.
6. Remove the brain from the skull, and keep in PBS (0.01 M, pH 7.4).
7. Embed the brain in 5% agarose (in 0.01 M PBS), and allow the agar to solidify in the refrigerator.
8. Mount the agarose-embedded brain in the vibratome so that the imaged region is uppermost and so that sections will be cut from this region in the same plane as the focal place used to image the brain in vivo.
9. Using the vibratome, cut 60-µm-thick sections in PBS (0.01 M, pH 7.4), and, using the paint brush, place into the 24-well culture dishes.

Fixation of Imaged Slices

Imaged cultured slices are normally cut at a thickness of 300–400 µm but then flattened during the culture process. Therefore, it is usually not necessary to resection them, and they can be placed immediately into the same fixative as above for ~2 h at 4°C.

10. Wash sections (or slices) in PBS (0.01 M, pH 7.4) three times, 5 min each wash, agitating continuously, but gently, on the shaker table.
11. Cryoprotect the sections in 2% glycerol and 20% DMSO in PBS (0.01 M, pH 7.4), and agitate continuously for 10 min.
12. Freeze each section, one by one, in liquid nitrogen, collecting them from the solution with the wooden cocktail sticks. Keep them submerged in the liquid nitrogen until completely frozen, or until the fizzing stops, and then place them back into the same well containing the cryoprotectant.

13. Repeat Step 12.
14. Wash sections in PBS (0.01 M, pH 7.4) for 5 min.
15. Block sections in a solution of 0.3% peroxide in PBS (0.01 M, 0.9% NaCl), and agitate for 10 min.

    As bubbles may emerge from the sections, this may cause them to float, therefore ensure that they are completely immersed.

16. Wash sections in PBS (0.01 M, pH 7.4, 0.05% BSA-c) for three changes of 5 min each change, agitating continuously.
17. Incubate sections in primary antibody (1:500) with PBS/BSA-c (PBS 0.01 M + 0.05% BSA-c, pH 7.4) overnight at 4°C, agitating continuously.
18. Wash in PBS/BSA-c for three changes of 5 min each.
19. Incubate sections in secondary antibody in PBS/BSA-c for 90 min.
20. Wash sections in TBS (0.1 M, pH 8, 0.9% NaCl) for three changes of 5 min each change.
21. Incubate sections in ABC solution (100 μL solution A + 100 L solution B in 10 mL TBS (0.1 M, pH 8.0) for 90 min, agitating continuously.
22. Wash sections in TBS (0.05 M, pH 7.6, 0.9% NaCl) for three changes of 5 min each change.
23. Incubate sections in TBS (0.05 M, pH 7.6, 0.9% NaCl) with 0.04% DAB and 0.015% H_2O_2 for 20 min. During this step, the sections can be monitored with a dissecting microscope to ensure that the sections do not become stained too dark with the DAB.
24. Wash sections in TBS (0.05 M, pH 7.6, 0.9% NaCl) for three changes of 5 min each change.

Resin Embedding for EM

25. Transfer sections to a glass scintillation vial containing sodium cacodylate buffer (0.1 M, pH 7.4), and wash three times.
26. Fix, and stain sections with 1.0%(w/v) osmium tetroxide in sodium cacodylate buffer (0.1 M, pH 7.4) for 40 min. Gently agitate at all times to ensure that the sections are always as flat as possible while they are in this solution.
27. Wash sections for 5 min in double-distilled water.
28. Stain sections for 15 min in 1% (w/v) uranyl acetate in water
29. Wash sections for 5 min in double-distilled water.
30. Dehydrate sections in graded alcohol series, 2 min each change (1 × 50%, 1 × 70%, 1 × 90%, 1 × 95%, 2 × 100%).
31. Replace half the ethanol in the rinse above with Durcupan resin, and agitate gently. The sections are fragile at this point, so take care that the agitation is not too strong that the sections break. Place on the shaker table to agitate for 30 min.
32. Replace half the Durcupan/ethanol mix with Durcupan, and place on rotary mixer for 30 min.
33. Replace all the Durcupan/ethanol mix with Durcupan, and place on rotary mixer for a minimum of 6 h.
34. Replace with fresh Durcupan, and place on rotary mixer for 1 h.
35. Remove sections from the resin, using the wooden cocktail sticks, and place onto a glass slide coated with mold-separating agent. Once all sections are arranged onto the slice, cover with another coated slide, and place in a 65°C oven for a minimum of 24 h.

Finding the Imaged Neuron in the Sections

36. Using a low-power objective on a transmitted light microscope, photograph each resin-embedded section.

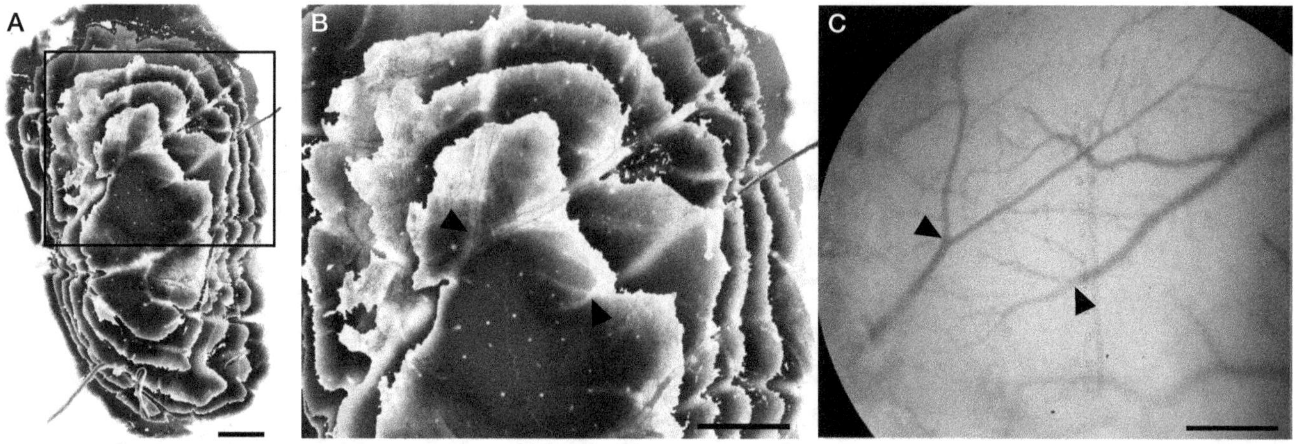

FIGURE 1. Locating the region of the imaged cell in the resin sections. (*A*) Stack of resin-embedded sections with the *box* indicating the region shown in *B*. The sections were aligned onto one another so that the uppermost sections reveal the pattern of the large blood vessels on the surface of the brain (*B*). This arrangement of vessels around where the cell was imaged can also be seen in the image taken through the eyepiece of the two-photon microscope using direct illumination (*C*). The same vessels in *B* and *C* are indicated with *arrowheads*. Scale bars, 1 mm (*A*), 0.5 mm (*B,C*).

37. Align the images on top of one another so that all the blood vessels are aligned, and the smallest sections cut from the surface of the brain are arranged uppermost (Fig. 1A,B).

    For this, we use Photoshop software, placing the image of each section into a separate layer. This alignment is made easier by using the needle holes made in the brain as fiducial markers. The image will show the blood vessels on the surface of the brain. This image can be compared with an image of the surface of the cortex taken during the live imaging experiments (Fig. 1C).

38. Align the live image of cortical surface with that of the image of the resin-embedded section stack.
39. Locate the blood vessels in the live image, next to which the imaged neuron was located, and find the same vessels in the resin sections (Fig. 1).
40. Using a high-power objective on the light microscope, locate the labeled cell (dendrites and axons) that was imaged in vivo (Fig. 2A) in the resin-embedded section (Fig. 2B).

    See *Troubleshooting*.

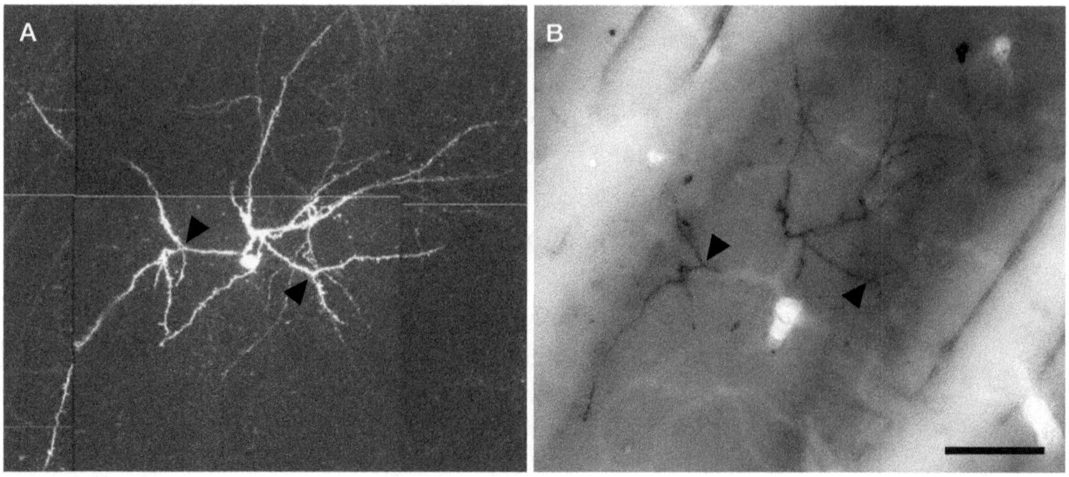

FIGURE 2. Identifying the imaged cell. (*A*) A tiled low-resolution image of the cell will show the pattern of its dendritic arborizations. (*B*) The same pattern can also be seen in the resin-embedded section. Here, a projection of a series of images taken at different depths through the material shows the pattern of labeled dendrites. The same dendrites in both images are indicated with *arrowheads*. Scale bar, 100 μm.

Serial Sectioning and Imaging in the EM

41. Using a new razor blade, separate the sections from the glass slides, and cut a small (~2 × 2-mm) square containing the required imaged neurites.
42. Stick this piece to a blank resin block with cyanoacrylate glue, and leave to dry.
43. Using new glass knives and an ultramicrotome, trim this block carefully until a trapezoid block face remains that contains the required structures within. This block face is typically 400-µm wide and 50–100-µm high. The labeled neurite is within 20 µm of the surface of this face so that only a short series of sections (less than 500 sections) needs to be cut.
44. Take a picture of the finished block to show the position of the neurites within its face. Indicate the depth of each neurite from the surface. This diagram will help find the labeled structures in the serial sections.
45. Serially section the block face using a diamond knife.

    To achieve good ribbons of sections, the static electricity must be kept low. This can be achieved by using an antistatic device (deionizer), with the probe pointed toward the surface of the water onto which the sections float. Adjust the intensity of the probe to a level in which each section sticks well to the previous one without any separations between them. Make sure this is turned off when the sections are collected.

    See *Troubleshooting*.
46. Separate the ribbon of sections into lengths that will each fit onto the single-slot grids. We use slot grids with a hole dimension of 1 × 2 mm, with a support film of Formvar and a thin carbon layer. Between 10 and 20 sections should fit onto one grid (Fig. 3). The grids are held, with the reverse action forceps, in the bath of the diamond knife, and the ribbons of sections collected onto them (Fig. 3).

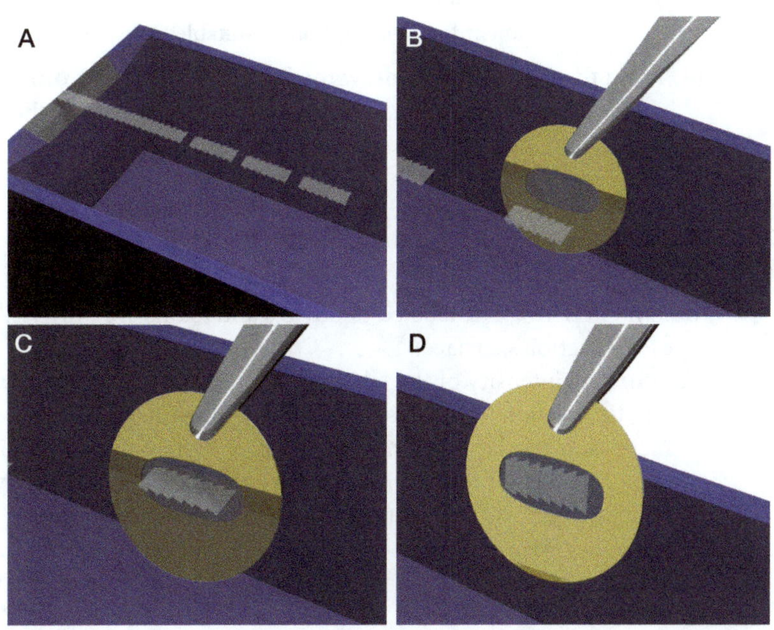

FIGURE 3. Collecting the serial thin sections from the water bath. (*A*) Ribbon of sections, cut with a diamond knife, floating on the surface of the water, with short lengths separated from the main ribbon using a fine hair or eyelash. Each short length will fit entirely onto a single-slot grid. Care is taken to ensure that the order of the groups is maintained. (*B*) Each length of sections is then manipulated, with the use of an eyelash, toward a slot grid held in the water at a height that allows the meniscus to almost cover the entire window. (*C*) The strip of sections is then brought close to the window but not quite touching. (*D*) As soon as the grid is lifted clear of the water, the sections will be left attached to the grid window.

47. Dry the grids on a hot plate maintained at 60°C. This will also ensure that no dust falls onto the grids during the drying process.

48. Load the grids into a grid-staining plate.

    Use of a grid-staining plate ensures that grids are stained quickly, without the risk of dirt accumulating onto the sections that often occurs when staining grids individually on the surface of drops of solutions.

49. Stain the grids with 1% uranyl acetate for 10 min and lead citrate for an additional 10 min.

    To avoid the contamination of the grids with dust and crystals of stain, first flood the plate with distilled water. Then, throughout the staining process, when the solutions are changed, do not allow the meniscus to descend low enough that the surface of the sections is exposed to air. For each change of solution, reduce the volume using a pipette until almost all the solution is removed, leaving enough to just cover the grids without exposing their surface. This will ensure that no drying of the staining solutions can occur onto the grids, and no dust particles on the surface of the solution touches the sections.

50. Remove the final water rinse from the grids, dry, and then place back in the grid box.

Imaging the Dendrites in the EM

51. Insert a grid into the EM, and using low-power magnification (3000x), scan all the sections, one after the other, until DAB labeling can be seen. Use the location of this labeling and the image of labeled dendrites in the block face (taken in Step 44) to determine which neurite has been found.

    See *Troubleshooting.*

52. Photograph the same region on each section at a higher magnification (2–3 nm per pixel), placing the labeled structure at the center of each image. Often the labeling inside the dendrites is not continuous with the attached spines, so be careful to take images that include all the spines.

53. When the images have been taken, arrange them into a stack with those taken on one section stitched together as one.

    See Equipment for a description of suitable software programs for image manipulation.

54. From the aligned stack, draw the labeled structures from each image to reconstruct the labeled structure. For this, we use Neurolucida software with the confocal module included (Fig. 4).

TROUBLESHOOTING

Problem (Step 40): The labeled dendrites and axons are only visible within the first few micrometers of the section's surface.

Solution: The intensity of the labeling is dependent on many parameters. Without labeling throughout the section thickness, recognizing the imaged neurites in the sections is more difficult. Increasing the labeling intensity can be achieved by exposing the tissue to the antibodies for longer periods of time; however, this may decrease the quality of the ultrastructure. Reducing the fixation time and concentration of glutaraldehyde to 0.05% (v/v) will enable more labeling but again, decrease quality, as will more cycles of freeze-thawing. If possible, it is best to try to maximize the levels of GFP in the imaged material before trying to boost the intensity of DAB stain.

Problem (Step 40): It is not possible to identify the correct imaged neurites in the resin sections.

Solution: Ensure that an image of the surface of the brain is taken before the animal is perfused so that the pattern of blood vessels can clearly be seen. Also, ensure that the location of the imaged cell is known in relation to this vessel pattern. This method will allow you to locate an exact region in the resin-embedded sections in which you should then see the correct cell.

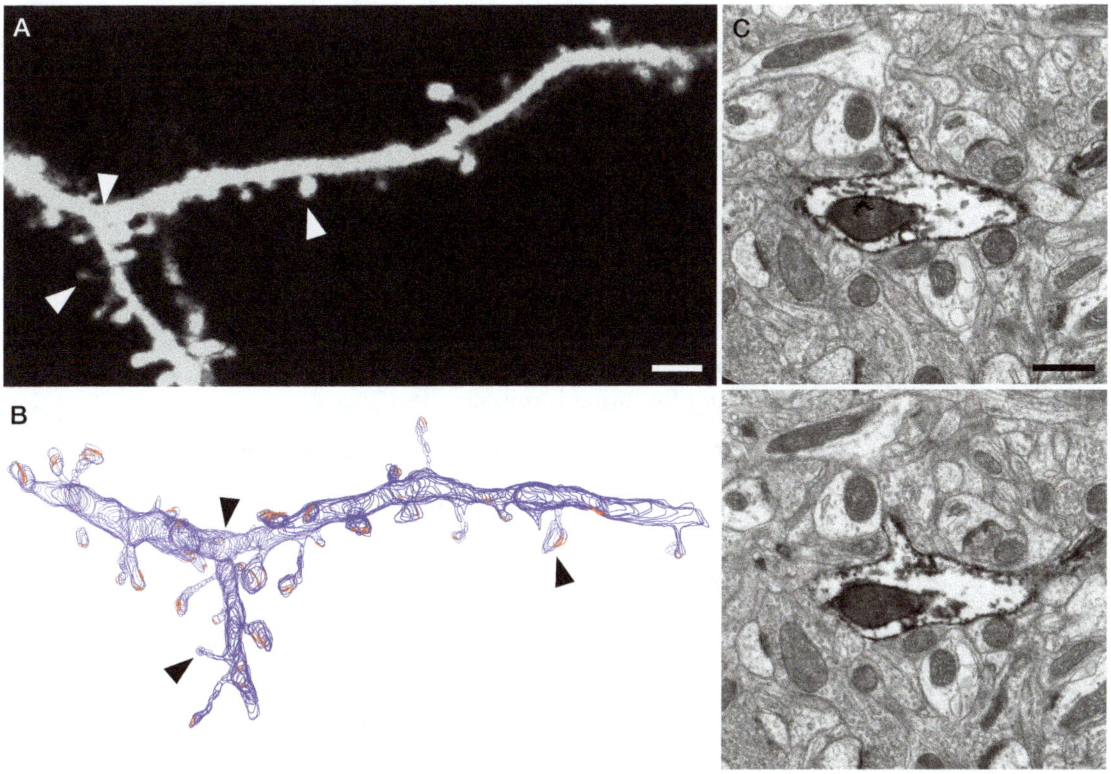

FIGURE 4. Serial section electron microscopy of the imaged dendrite. (*A*) Two-photon image of a dendrite in vivo with the corresponding 3D reconstruction (*B*) made from serial electron micrographs (*C*). The reconstruction can be used to find which spines in the EM images correspond to particular spines seen in the two-photon image. The white arrows in *A* and black arrows in *B* indicate the same spines. Red areas in *B* show the synaptic connections seen in the EM images. Scale bars, 2 μm (*A*), 0.5 μm (*C*).

Problem (Step 45): It is not possible to cut long ribbons of serial sections.

Solution: Serial sectioning depends on having a good quality clean diamond knife with a small trapezoid block that has straight sides. The block must be trimmed in the final stages with a glass knife. Serial sections often separate from one another because of electrostatic forces that are typically greater in conditions of low humidity. Increasing the humidity (above 60%) or using an antistatic probe will reduce the static and enable the sections to stick together more easily.

Problem (Step 51): It is difficult to see the DAB-labeled structures in the serial thin sections.

Solution: Too much contrast in the membranes can result in dark sections in which it is difficult to see the labeled structures. Reducing the uranyl acetate and lead-citrate staining of the sections will improve this. Otherwise, reduce the time the sections are left in the osmium stain.

DISCUSSION

This protocol relies on being able to see the DAB labeling in the resin sections but also on preserving the ultrastructure enough so that the integrity of axons, dendrites, and synapses is maintained. For this reason, no detergent is used, and the penetration of the reagents relies on the disruption to the membranes that is caused by the freeze-thawing process. This will reveal the large dendrites and axons, but other thinner processes, including the dendritic spines, may not contain DAB. Although this makes it difficult to follow these structures, they can be identified with the 3D reconstruction, and it gives the advantage that more of the internal structures, such as postsynaptic densities and the spine apparatus, can be seen.

RECIPE

CAUTION: See Appendix 6 for proper handling of materials marked with <!>.
Recipes for reagents marked with <R> are included in this list.

Lead Citrate

1. Dissolve 1.33 g of lead nitrate (!) (Sigma-Aldrich Chemie GmbH, Switzerland; cat. no. 15334) in 30 ml of distilled water.
2. Add 1.76 g of sodium citrate (Calbiochem, cat. no. 567446) and shake well, until solution is milky.
3. Add 8 mL of 1 M sodium hydroxide <!> and shake well.
4. Make up the volume to 50 mL with distilled water.

Phosphate-Buffered Saline (PBS)

Reagent	Quantity (for 1 L)	Final concentration (10x)
NaH_2PO_4	2.56 g	18.6 mM
Na_2HPO_4	11.94 g	84.1 mM
NaCl	102.2 g	1.75 M

Adjust the pH to 7.4 using NaOH <!> or HCl <!> as necessary.

REFERENCES

Holtmaat AJ, Trachtenberg JT, Wilbrecht L, Shepherd GM, Zhang X, Knott GW, Svoboda K. 2005. Transient and persistent dendritic spines in the neocortex in vivo. *Neuron* **45:** 279–291.

Holtmaat A, Wilbrecht L, Knott GW, Welker E, Svoboda K. 2006. Experience-dependent and cell-type-specific spine growth in the neocortex. *Nature* **441:** 979–983.

Knott GW, Holtmaat A, Wilbrecht L, Welker E, Svoboda K. 2006. Spine growth precedes synapse formation in the adult neocortex in vivo. *Nat Neurosci* **9:** 1117–1124.

Nikonenko I, Boda B, Steen S, Knott G, Welker E, Muller D. 2008. PSD-95 promotes synaptogenesis and multiinnervated spine formation through nitric oxide signaling. *J Cell Biol* **183:** 1115–1127.

Trachtenberg JT, Chen BE, Knott GW, Feng G, Sanes JR, Welker E, Svoboda K. 2002. Long-term in vivo imaging of experience-dependent synaptic plasticity in adult cortex. *Nature* **420:** 788–794.

Zito K, Knott G, Shepherd GM, Shenolikar S, Svoboda K. 2004. Induction of spine growth and synapse formation by regulation of the spine actin cytoskeleton. *Neuron* **44:** 321–334.

Zito K, Scheuss V, Knott G, Hill T, Svoboda K. 2009. Rapid functional maturation of nascent dendritic spines. *Neuron* **61:** 247–258.

24 Stimulated Emission Depletion (STED) Imaging of Dendritic Spines

Katrin I. Willig[1] and U. Valentin Nägerl[2]

[1]*Max Planck Institute for Biophysical Chemistry, 37077 Göttingen, Germany;* [2]*Inserm U862/University of Bordeaux, 33076 Bordeaux, France*

ABSTRACT

The confluence of innovations in transgenic labeling and light microscopy techniques in recent years has greatly advanced our understanding of dynamic cell biological events underlying neuronal function and plasticity. Increasingly, it has become possible to perform fundamental experiments inside the relevant subcellular compartments of a neuron embedded in three-dimensional living tissues. Overcoming the limiting role of diffraction in far-field light microscopy, nanoscopy is advancing our ability to see and manipulate cellular events well below the diffraction barrier of ~200 nm. The first concrete and implemented concept of nanoscopy was STED (stimulated emission depletion) microscopy. In this chapter, we give an example of the power that STED microscopy holds for neuroscience research: live-cell time-lapse imaging of the dynamic morphology of dendritic spines of pyramidal neurons. In addition, we describe the basic elements needed to assemble a custom-built STED microscope capable of live cell imaging and how to use it for physiology experiments.

INTRODUCTION

Fluorescence microscopy is one of the most powerful and widely used imaging techniques in biological research, because it can visualize dynamic processes inside living cells with exquisite sensitivity and specificity. On the downside, it has lacked the power to resolve organelles and cellular structures that are smaller than about one-half the wavelength of the light that is used, because of the diffraction of the focused light. However, thanks to a number of microscopy innovations in recent years, the classic limit of light microscopy, as defined by Ernst Abbé more than 100 years ago, has been resoundingly shattered. While not abolishing diffraction per se, the new nanoscopy techniques cleverly circumnavigate its limiting effects on spatial resolution. As a result, it is now possible to resolve details at the nanoscale in biological specimen without forgoing the customary benefits of fluorescence microscopy (live-cell imaging, biomolecular specificity, etc.).

So far, the lack of spatial resolution in far-field light microscopy has posed a problem for fluorescence imaging of the morphology of synapses, in particular of dendritic spines, which are tiny specializations of the postsynaptic membrane that mediate excitatory synaptic transmission in the central nervous system. Spines can go up to 3–4 μm in length as measured by electron microscopy (Harris and Kater 1994) and can thus be readily identified by confocal laser-scanning fluorescence microscopy. However, measuring changes in spine substructure, such as changes in spine shape and size, has been difficult because they are usually smaller than what conventional optical systems can reliably resolve. This is unfortunate given the critical role that the size and shape of synaptic structures, such as the width of spine necks, is thought to play for synaptic function and plasticity.

In this chapter, we will give a brief outline of the principles of stimulated emission depletion (STED) microscopy and its application to a neurobiological context. STED microscopy is a variation of part of the RESOLFT (reversible saturable optical fluorescence transitions) concept (Hell 2003, 2009). Although the STED principle was originally conceived some time ago (Hell and Wichmann 1994), its practical implementation continues to be refined at a rapid clip. The first application of STED microscopy to a neurobiologically relevant topic was the study of distributions of immunohistochemically labeled proteins in fixed tissue preparations (Kittel et al. 2006; Sieber et al. 2006; Willig et al. 2006); since then, its scope has been extended in several important ways. STED was recently used to monitor the movements of fluorescently labeled vesicles inside nerve terminals (Westphal et al. 2008) and to image activity-driven changes in the morphology of dendritic spines (Nägerl et al. 2008), using genetically encoded fluorescence (yellow fluorescent protein [YFP]) as a label. As a consequence, it is now possible to image and capture the dynamics of dendritic spines of YFP-positive CA1 pyramidal neurons in living hippocampal slices with a resolution of better than 100 nm. Fine details, such as the shape of the spine head or the width of the spine neck, can be faithfully resolved and measured, even well below the surface of the tissue. As such, STED is suited for studying activity-driven structural changes of synaptic morphology, revealing a diversity of shapes and dynamics that cannot be detected by established techniques. STED has been complemented recently by powerful methods based on the sequential stochastic switching of individual photoactivatable fluorophores in wide-field illumination (PALM) (Betzig et al. 2006; Hess et al. 2006; Rust et al. 2006; Hell 2009).

Protocol

STED Imaging of Dendritic Spines in Living Hippocampal Slices

This protocol provides a method for live-cell time-lapse imaging of dendritic spines of pyramidal neurons. Imaging is performed in an organotypic hippocampal slice culture system, with YFP used as a volume marker for the synaptic structures.

IMAGING SETUP

Principle of STED Microscopy

A typical STED microscope uses a coaligned pair of focused laser beams, one for excitation and one for STED (Fig. 1A). The STED beam is a doughnut-shaped beam for deexciting potentially excited molecules by stimulated emission in the outer rim of the excitation spot. The resolution is given by $\Delta r = \lambda/(2\,\mathrm{NA}\sqrt{1+I/I_s})$ with NA denoting the numerical aperture of the lens, λ the wavelength, and I the intensity at the crest of the doughnut-shaped STED beam. I_s is the STED intensity required to cut down the fluorescence probability of the molecules by one-half (Harke et al. 2008). The fluorescent volume (or effectively excited spot) is dramatically reduced in size by the STED beam, thereby effectively increasing the spatial resolution of the STED microscope. Therefore, STED microscopy is not fundamentally limited by diffraction, unlike confocal or two-photon microscopy (Hell and Wichmann 1994).

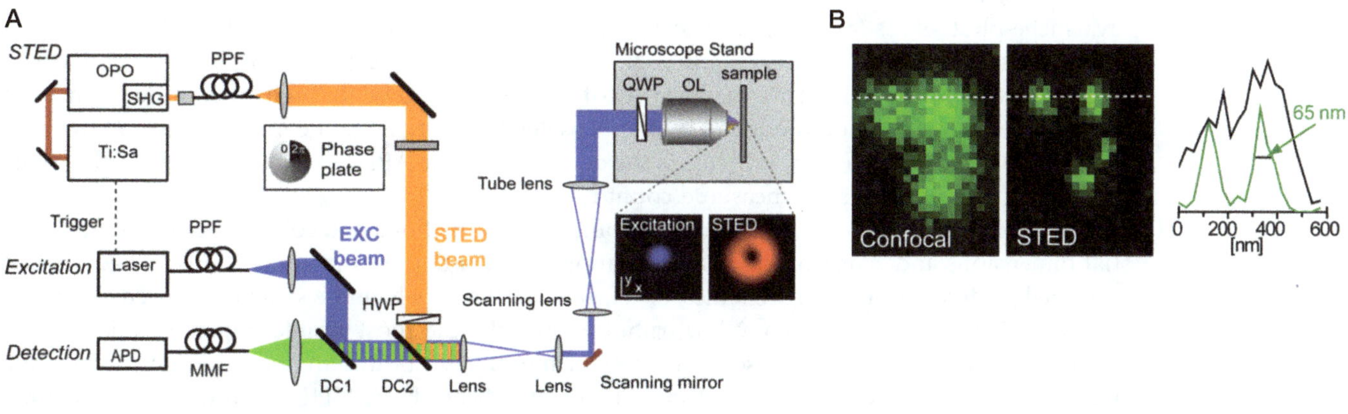

FIGURE 1. (*A*) Optical beam path of a STED microscope. APD, avalanche photodiode; DC, dichroic mirror; EXC, excitation; HWP, half-wave plate; MMF, multimode optical fiber; OL, objective lens; OPO, optical parametric oscillator; PPF, polarization-preserving optical fiber; QWP, quarter-wave plate; SHG, second-harmonic generation; Ti:Sa, Ti:sapphire laser. The dotted line indicates the electronic trigger to synchronize both laser sources. (*B*) Breaking the diffraction barrier. Yellow-green fluorescent beads (40-nm; Invitrogen) show the resolution improvement of STED (*middle* panel) versus confocal microscopy (*left* panel). These beads are used to check the spatial resolution of the microscope because they can be imaged with the same settings as the YFP-labeled neurons. Plots (*right* panel) show the intensity distribution across the dotted lines over the bead images. Green line, STED; black line, confocal microscopy. The full width at half-maximum (FWHM) reduces to 65 nm in the STED mode, making it possible to discriminate the beads as distinct fluorescent structures.

STED Microscopy Setup

Figure 1A shows the STED microscopy setup we used for imaging living neurons. For excitation, a pulsed laser diode (80-MHz pulses of 70-psec duration at a wavelength of 490 nm; PicoTA; PicoQuant, Berlin, Germany) was used. The STED light was provided by a Ti:sapphire laser (MaiTai; Spectra-Physics, Mountain View, CA) coupled to an optical parametric oscillator (OPO; APE, Berlin, Germany) running at 598 nm central wavelength. Originally of 200-fsec duration, the OPO pulses are stretched by dispersion using a 100-m-long polarization-preserving fiber (OZ Optics, Ottawa, Canada) to ~300 psec. The expanded STED beam passes through a polymeric phase plate imprinting a helical phase delay of $\exp(i\phi)$, with $0 < \phi < 2\pi$ (vortex phase plate; RPC Photonics, Rochester, NY) on the wave front, which produces an intensity distribution shaped like a doughnut in the focal plane of the lens (Fig. 1A). The excitation and STED laser pulses need to be precisely synchronized in time; this can be achieved by externally triggering the laser diode and tuning the delay by an electronic delay generator.

A high-NA objective lens (e.g., PL APO, 100×, oil, NA 1.4; Leica, Mannheim, Germany) should be used to focus the incident light into the sample. The average focal power of the STED beam, which reduces the fluorescence by one-half, is on the order of 1 mW for YFP at 80-MHz repetition rate (measured without phase plate). By increasing the STED power above that level, the resolution can be continuously improved. For excitation, an average focal power of 1–10 µW is sufficient; however, it can be adjusted depending on the brightness and photostability of the sample.

The epifluorescence is separated from the excitation and STED by two dichroic mirrors. After passing through a band-pass filter for cleanup (HQ535/50, Chroma Technology) the signal photons are piped onto a photodetector, which can be a multimode optical fiber (MMF) connected to an avalanche photodiode (APD, SPCM-AQR-13; PerkinElmer, Fremont, CA). When the diameter of the fiber is ~80% of the back-projected Airy disk, it serves as a confocal pinhole, which conveniently blocks out light from out-of-focus planes and ambient light. In principle, a pinhole is not needed, as the STED resolution improvement is accomplished by the interaction of the laser beams in the focal plane with the fluorophore.

The setup can be readily built around a commercial microscope base (upright or inverted), such as the BX61 from Olympus or the DMI 6000 from Leica, which are standard research microscopes with motorized control of the wide-field epifluorescence and the *z* focus for convenience. We use an avalanche photodiode for detection because its quantum yield is very high (50%–60% at 535 nm) and the dark counts are negligible, although its signal response becomes significantly nonlinear at count rates <2 MHz. To improve the linearity of the signal response for higher count rates, the dead time of the avalanche photodiode can be compensated by multiplying the measured count rate with a correction factor of $1/(1 - t_d * C_R)$; t_d is the dead time of the APD, which is typically in the range of 10–100 nsec, and C_R is the measured count rate.

After bringing the excitation and STED point-spread functions into coincidence in all three spatial dimensions and synchronizing the excitation and STED pulses in time, STED images can be acquired by different scanning schemes: (1) stage scanning, in which the sample is moved by a piezo stage (P-733; Physik Instrumente, Karlsruhe) is optically the most simple implementation as the laser beams stay fixed; (2) beam scanning, in which both laser beams are moved in unison over the sample by galvo-driven mirrors (Yanus IV Digital Scan Head; TILL Photonics, Gräfelfing, Germany); or (3) a combination of the two, in which the beam is rapidly moved along one axis by means of a resonant mirror (15-kHz, SC-30; EOPC, Glendale, NY), while the sample is moved more slowly along the perpendicular axis by the piezo stage (Westphal et al. 2008). Although any scheme can be used in principle, the preferred choice will depend on practical issues of how well it integrates into the setup and which type of additional applications are intended (choice of software control, need for mounting micromanipulators, etc.).

Image Acquisition and Analysis

Given that the resolution is greatly improved in STED imaging (see Fig. 1B) over conventional imaging modes, pixel and image sizes should be set accordingly. To comply with the Nyquist sampling theorem, images should be acquired with a pixel size of less than one-half of the measured STED resolution (i.e., about 30 nm or less). The size of the field of view will depend on which type of scanning mode is used as well as on practical issues such as frame rate or hard disk space. Whereas piezo stages usually do not come with large travel ranges, beam scanning certainly allows for a larger field of view.

Several types of software packages can be used for data acquisition and instrument control of a STED setup, much like for confocal or two-photon microscopes for which a range of commercial or custom-made solutions exist. However, STED microscopy requires special attention to the adjustment of the excitation beam as well as the STED beam and conventional software packages can be somewhat clumsy for this purpose. The commercially available software package "Imspector" is a useful solution to this issue, as it facilitates rapid PSF measurements (http://www.max-planck-innovation.de/de/industrie/technologieangebote/software). The choice will therefore depend on the expertise and programming support available in the laboratory. Likewise, there are no special restrictions on the type of software to analyze the images (e.g., ImageJ, MATLAB, Imaris). STED images can be further enhanced by deconvolution algorithms as is done for confocal images, but unlike other nanoscopy techniques image processing is in principle not needed to obtain superresolution images.

MATERIALS

See the end of the chapter for recipes for reagents marked with <R>.

Reagents

Artificial cerebrospinal fluid (ACSF) <R>
Chicken plasma (Sigma-Aldrich)
Horse serum (Invitrogen/Gibco)
Mice (e.g., YFP transgenic, postnatal day 5–7)

> Hippocampal slices from YFP transgenic mice (prepared on postnatal day 5–7) are suitable for this protocol, as several distinct populations of neurons, including the CA1 pyramidal neurons, are strongly but sparsely labeled (Feng et al. 2000).

Other standard culture media (see Gähwiler 1981 for details)
Poly-L-lysine (Sigma-Aldrich)
Thrombin (VWR/Merck)

Equipment

Coverslips, thin glass
Heated chamber, 35°C
Imaging setup, as described above
Roller incubator, 35°C (optional, see Step 1)

EXPERIMENTAL METHOD

1. Prepare hippocampal slices (300-µm-thick) from YFP transgenic mice (see Materials) and embed them in a plasma clot on thin glass coverslips (see Gähwiler 1981). Slices can be readily incubated for several weeks in a roller incubator at 35°C.

   > Slices can be readily incubated for several weeks in a roller incubator at 35°C, according to the Gähwiler method (Gähwiler 1981).

2. For the imaging experiments, transfer cultures to a heated chamber (35°C) and perfuse continuously with carbogenated (95% O_2, 5% CO_2) ACSF.
3. Record confocal images by scanning the excitation beam through the sample. To obtain STED microscopy images, also switch on the STED beam.

   See *Troubleshooting*.

TROUBLESHOOTING

Problem (Step 3): Phototoxicity and bleaching are observed.

Solution: Both phototoxicity and bleaching are potential concerns, as the fluorophore is not only subjected to the blue excitation light, but also to the red STED laser beam, which is three orders of magnitude more intense. However, we found it is possible to work with laser intensities that allow for repeated acquisition of superresolved image stacks without any noticeable damage to neuronal morphology or excessive loss of signal owing to bleaching. This is likely to be explained by the following facts.

1. The STED beam intensity (I_{peak} = 400 MW/cm^2, time-averaged focal power 24 mW) is still about two to three orders of magnitude below the intensity typically used in two-photon microscopy (200 GW/cm^2).
2. The wavelength of the STED beam is at the red edge of the emission spectrum of the dye, producing little interaction with the dye beyond stimulated emission, which returns the fluorophore back to the nonreactive ground state. Moreover, water absorbs much less at 598 nm (the wavelength of the STED beam) than in the near-infrared range (by ~5–20-fold).
3. Bleaching is not really an issue for STED imaging of volume-labeled structures, as the fluorescent protein (YFP) is highly expressed inside the cytosol of the neurons, where it can diffuse freely and thus replenish itself rapidly.

Problem (Step 3): Mechanical or sample drift is observed.

Solution: There are several sources of drift, which impact measurement quality over the course of the experiment. Some of them need to be suppressed a priori; others can be corrected for offline.

Mechanical drift of the optical components can greatly compromise the performance of the STED system, degrading signal strength and spatial resolution. Therefore, proper alignment of the excitation and STED laser beams should be routinely checked before the experiment (and for confirmation afterward, too). This can be readily performed by using 80-nm reflective gold beads to measure and optimize the focal intensity distributions of the laser beams (i.e., their point-spread functions). In addition to a solid and judicious design of the optical system (using high-quality mechano-optical components, minimizing mechanical lengths and beam paths, etc.), it is important to make sure that the temperature is kept relatively stable in the laboratory (varying by <1°C), as thermal drift can otherwise be a problem. An interesting new design of a STED microscope effectively deals with this problem by combining the optical paths used for the excitation and STED beams. However, this design relies on a phase plate, which selectively modulates the STED, but not the excitation beam (Wildanger et al. 2009).

Sample drift that occurs between image acquisition time points is a common problem when working with living tissue, but it can usually be corrected for by aligning the images afterward using standard image registry algorithms (e.g., in ImageJ), as long as the regions of interest are fully included in all of the image stacks.

DISCUSSION

Example of Application

Using the method described above, we recorded STED images with a resolution of ~70–80 nm. However, because of the refractive index mismatches between the oil objective, the coverslip, and the sample, the resolution was best close to the coverslip. Figure 2 shows a stretch of dendrite with spines, which was rendered from a STED image stack recorded with a pixel size of 29 × 29 × 500 nm. It illustrates well the potential of this technique to resolve subtle details in shape and size of living synaptic structures (I_{peak} = 400 MW/cm^2).

STED-Compatible Dyes and Laser Sources

A number of fluorophores have been shown to be suitable for STED microscopy; they include synthetic dyes and fluorescent proteins covering a wide spectral range (480–635 nm for excitation). A list of suitable fluorophores can be viewed at http://www.mpibpc.mpg.de/groups/hell/STED_Dyes.html.

In principle, laser sources other than the ones outlined above are also suitable for STED microscopy. Continuous-wave (CW) lasers can be used for STED microscopy of living YFP-labeled samples. They are easier to implement and maintain than a Ti:sapphire/OPO combination, and also considerably cheaper (Hein et al. 2008). However, CW STED on YFP samples requires about 10-fold more average STED power as compared with the 80-MHz pulsed system. Ongoing developments in pulsed supercontinuum lasers promise another interesting alternative, making it possible to use a single, spectrally flexible laser source for excitation and STED, significantly reducing cost and complexity of the setup (Wildanger et al. 2008). However, current systems operate at a repetition rate of only 1–2 MHz, which imposes relatively long recording times.

Future Perspective

The advent of nanoscopy is bound to break new ground in our understanding of cell-biological phenomena and their dynamics. STED microscopy has been shown to be suitable for live-cell imaging of molecular structures and cellular organelles at the nanoscale, using standard genetically encoded fluorescence markers such as YFP. Whereas the resolving capability of STED microscopy is well established by now, and this technique has come of age in many ways, future developments (especially with regard to fluorescent probes and laser sources) are bound to broaden its scope and versatility and to improve the resolution even further.

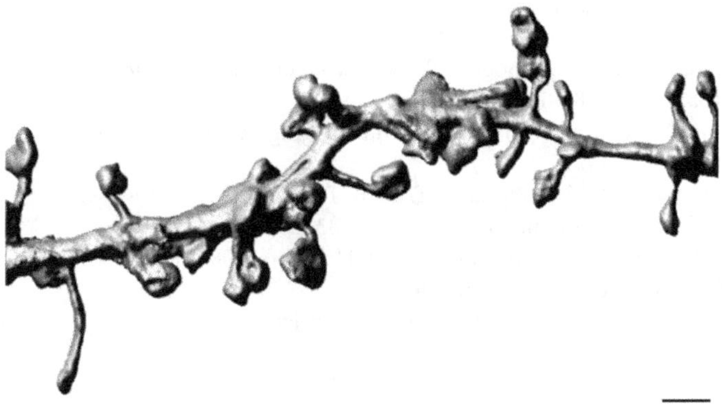

FIGURE 2. Bridging the gap between light and electron microscopy: 3D reconstruction of dendritic spines obtained by live-cell STED imaging in organotypic hippocampal slices from thy1-YFP transgenic animals. Scale bar, 1 µm, Δz = 0.5 µm. (Reprinted, with permission, from Nägerl et al. 2008.)

RECIPE

CAUTION: See Appendix 6 for proper handling of materials marked with <!>.
Recipes for reagents marked with <R> are included in this list.

Artificial Cerebrospinal Fluid (ACSF)

Reagent	Final concentration
NaCl	126 mM
KCl	2.5 mM
$CaCl_2$ <!>	2.5 mM
$MgCl_2$	1 mM
Glucose	10 mM
NaH_2PO_4	1.25 mM
$NaHCO_3$	26 mM

Adjust pH to 7.4.

REFERENCES

Betzig E, Patterson GH, Sougrat R, Lindwasser OW, Olenych S, Bonifacino JS, Davidson MW, Lippincott-Schwartz J, Hess HF. 2006. Imaging intracellular fluorescent proteins at nanometer resolution. *Science* **313:** 1642–1645.

Feng G, Mellor RH, Bernstein M, Keller-Peck C, Nguyen QT, Wallace M, Nerbonne JM, Lichtman JW, Sanes JR. 2000. Imaging neuronal subsets in transgenic mice expressing multiple spectral variants of GFP. *Neuron* **28:** 41–51.

Gähwiler BH. 1981. Organotypic monolayer cultures of nervous tissue. *J Neurosci Methods* **4:** 329–342.

Harke B, Keller J, Ullal CK, Westphal V, Schonle A, Hell SW. 2008. Resolution scaling in STED microscopy. *Opt Express* **16:** 4154–4162.

Harris KM, Kater SB. 1994. Dendritic spines: Cellular specializations imparting both stability and flexibility to synaptic function. *Annu Rev Neurosci* **17:** 341–371.

Hein B, Willig KI, Hell SW. 2008 Stimulated emission depletion (STED) nanoscopy of a fluorescent protein-labeled organelle inside a living cell. *Proc Natl Acad Sci* **105:** 14271–14276.

Hell SW. 2003. Toward fluorescence nanoscopy. *Nat Biotechnol* **21:** 1347–1355.

Hell SW. 2009. Microscopy and its focal switch. *Nat Methods* **6:** 24–32.

Hell SW, Wichmann J. 1994. Breaking the diffraction resolution limit by stimulated emission depletion fluorescence microscopy. *Opt Lett* **19:** 780–782.

Hess ST, Girirajan TP, Mason MD. 2006. Ultra-high resolution imaging by fluorescence photoactivation localization microscopy. *Biophys J* **91:** 4258–4272.

Kittel RJ, Wichmann C, Rasse TM, Fouquet W, Schmidt M, Schmid A, Wagh DA, Pawlu C, Kellner RR, Willig KI, et al. 2006. Bruchpilot promotes active zone assembly, Ca^{2+} channel clustering, and vesicle release. *Science* **312:** 1051–1054.

Nägerl UV, Willig KI, Hein B, Hell SW, Bonhoeffer T. 2008. Live-cell imaging of dendritic spines by STED microscopy. *Proc Natl Acad Sci* **105:** 18982–18987.

Rust MJ, Bates M, Zhuang X. 2006. Sub-diffraction-limit imaging by stochastic optical reconstruction microscopy (STORM). *Nat Methods* **3:** 793–795.

Sieber JJ, Willig KI, Heintzmann R, Hell SW, Lang T. 2006. The SNARE motif is essential for the formation of syntaxin clusters in the plasma membrane. *Biophys J* **90:** 2843–2851.

Westphal V, Rizzoli SO, Lauterbach MA, Kamin D, Jahn R, Hell SW. 2008. Video-rate far-field optical nanoscopy dissects synaptic vesicle movement. *Science* **320:** 246–249.

Wildanger D, Rittweger E, Kastrup L, Hell SW. 2008. STED microscopy with a supercontinuum laser source. *Opt Express* **16:** 9614–9621.

Wildanger D, Buckers J, Westphal V, Hell SW, Kastrup L. 2009. A STED microscope aligned by design. *Opt Express* **17:** 16100–16110.

Willig KI, Rizzoli SO, Westphal V, Jahn R, Hell SW. 2006. STED microscopy reveals that synaptotagmin remains clustered after synaptic vesicle exocytosis. *Nature* **440:** 935–939.

25 Two-Photon Uncaging Microscopy

Masanori Matsuzaki[1,2] and Haruo Kasai[1]

[1]*Laboratory of Structural Physiology, Center for Disease Biology and Integrative Medicine, Graduate School of Medicine, University of Tokyo, Tokyo, 113-0033, Japan;* [2]*Division of Brain Circuits, National Institute for Basic Biology, Okazaki, 444-8787, Japan*

ABSTRACT

Two-photon uncaging takes advantage of the inherent optical sectioning power of two-photon excitation to generate highly localized concentration increases of neurotransmitters such as glutamate. This can be used to activate isolated clusters of receptors and, thus, produce maps of receptor densities, or activate intracellular signal transduction under these receptors, in three dimensions and in complex structures such as hippocampal brain slices. Used in combination with two-photon imaging, two-photon uncaging provides a means to study the long-term structural and functional consequences of stimulation of structures such as dendritic spines. In this chapter, we describe the development of two-photon uncaging microscopy technology and provide an overview of its use in our laboratory.

INTRODUCTION

Background of the Development of Two-Photon Uncaging Microscopy

The photochemically initiated release of biochemically inert substrates (uncaging) is currently used by many laboratories to study a very diverse range of cell types (from plants, sea urchins, mammals, etc.). The history and general properties of caged compounds are described in more detail in Ellis-Davies (2011), Hess et al. (2011), and Zayat et al. (2011). The superb abilities of two-photon excitation for focal photochemistry were already realized in the original report describing two-photon microscopy (Denk et al, 1990). Shortly afterward, Denk showed two-photon uncaging of caged carbamolycholine, but found that the most widely used caging chromophore (*ortho*-nitrobenzyl) was fairly insensitive to two-photon excitation (Denk 1994). Denk had to use 640-nm light for 35 ms in each pixel, as the chromophore has a two-photon cross section of >0.0001 GM at 720 nm (1 GM = 10^{-50} cm^4 sec $photon^{-1}$) (Brown and Webb 1998). Lipp and Niggli (1998) used two-photon uncaging of calcium via irradiation of dimethoxy (DM)-nitrophen (Ellis-Davies and Kaplan 1988) at 720 nm to mimic calcium sparks, thus, initiating calcium waves in cardiac myocytes. Furuta et al. (1999) syn-

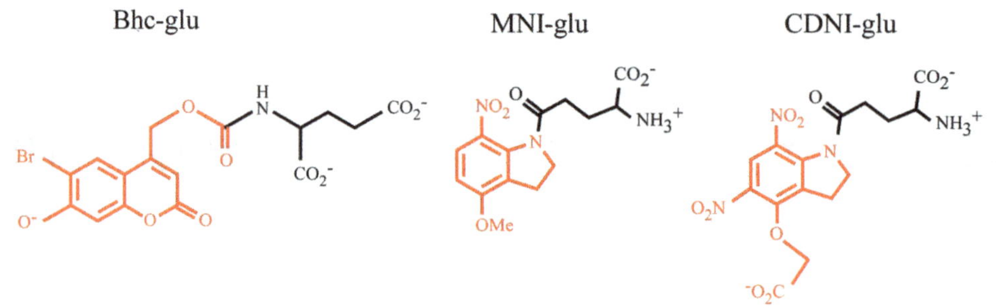

FIGURE 1. Structure of two-photon-sensitive caged glutamate molecules. Caging groups and glutamate are indicated in red and black, respectively.

thesized 6-bromo-7-hydroxycoumarin-4-ylmethoxycarbonyl-glutamate (Bhc-glu) as a caged neurotransmitter with a good two-photon cross section (~1 GM; Fig. 1).

Properties of Caged Glutamate for Two-Photon Photolysis

Caged neurotransmitters for two-photon uncaging must possess the following three chemical properties: (1) effective absorption of excitation light (i.e., good two-photon cross section); (2) high chemical resistance to aqueous hydrolysis at pH 7.4; and (3) rapid uncaging of substrate (ideally, the time constant should be in the microsecond range). As Bhc-glu possesses properties (1) and (2) but not property (3), it cannot mimic a single vesicle secretion in a single synapse. In 2001, methoxy-7-nitroindolinyl-glutamate (MNI-glu) was developed independently as a caged glutamate by two groups (Canepari et al. 2001; Matsuzaki et al. 2001; Papageorgiou and Corrie 2001;). Matsuzaki et al. (2001) were the first to show that MNI-glu satisfies all three design criteria, has a two-photon cross section of ~0.06 GM, and is effective for two-photon uncaging in physiological experiments. Since that time, many laboratories have performed two-photon uncaging of MNI-glu (Smith et al. 2003; Carter and Sabatini 2004; Araya et al. 2006; Beique et al. 2006; Harvey and Svoboda 2007) and also with a novel series of ruthenium-based caged compounds (Zayat et al. 2011). Recently, we developed 4-carboxymethoxy-5,7-dinitroindolinyl-glutamate (CDNI-glu) as a derivative of MNI-glu. Ellis-Davies et al. (2007) estimated that the two-photon cross section of CDNI-glu is about fourfold larger than that of MNI-glu (~0.24 GM; Fig. 1). However, as MNI-glu is now commercially available (whereas CDNI-glu is still not), in this chapter, we describe two-photon uncaging using MNI-glu.

Two-Photon Uncaging and Imaging

In general, higher energy is needed for the intramolecular cleavage of the caged compound between a caging group and a physiologically active molecule than for the excitation of fluorescent molecules from the ground state to the excited state. An excitation wavelength of 700–800 nm is used for two-photon uncaging, whereas an excitation wavelength of 800–1000 nm is used for fluorescence imaging. For example, MNI-glu is effectively uncaged at 720 nm, whereas commonly used fluorescent molecules, such as Alexa Fluor 488, 594, and the Oregon Green BAPTA (1,2-bis-(2-aminophenoxy)ethane-*N*,*N*,*N*′,*N*′-tetra-acetic acid) series, are excited at 800–870 nm much more efficiently than at 720 nm. Most fluorescent proteins, including green fluorescent protein (GFP), yellow fluorescent protein (YFP), and DsRed, are efficiently excited at wavelengths >900 nm. To examine how intracellular calcium, fluorescent small molecules and proteins, and fluorescent protein-fused functional proteins are affected dynamically by two-photon uncaging of MNI-glu, a second laser set at a different wavelength should be installed on the microscope. Two lasers can be introduced to single scan head, although imaging cannot be performed during uncaging (Carter and Sabatini 2004). In this chapter, we introduce a microscopy system comprising two scan heads, which is used to achieve two-photon uncaging and two-photon imaging simultaneously.

Protocol

Two-Photon Imaging and Uncaging of Dendritic Spines

This protocol gives an overview of the procedures used by the investigators for two-photon uncaging microscopy. It includes a detailed description of the development of a microscope that enables effective two-photon release of caged neurotransmitters. It also provides several examples of its use in cultured and acutely isolated hippocampal neurons.

IMAGING SETUP

Microscopy Setup

The two-photon uncaging microscope was constructed with minimal modifications to a commercial laser-scanning microscope (BX51WI and FluoView, Olympus, Tokyo, Japan). The two-photon uncaging microscope and two mode-locked Ti:sapphire lasers (Tsunami, pumped by 5-W Millenia, Spectra-Physics, Mountain View, CA) were fixed on an antivibration optical table (RS4000, New Port, San Jose, CA). The scan unit of the commercial laser-scanning microscope can be used for two-photon imaging, and practical guidance for the setup of microscopy for two-photon imaging is provided in Nikolenko and Yuste (2011), so, here, we focus on the setup for the two-photon uncaging microscope. For two-photon uncaging, a second scan unit is connected to a backport of the microscope (Fig. 2). The external photomultiplier tube (PMT; R3896, Hamamatsu, Japan) is also connected to the backport of the microscope. The optical pathway of the laser for uncaging is merged to that used for

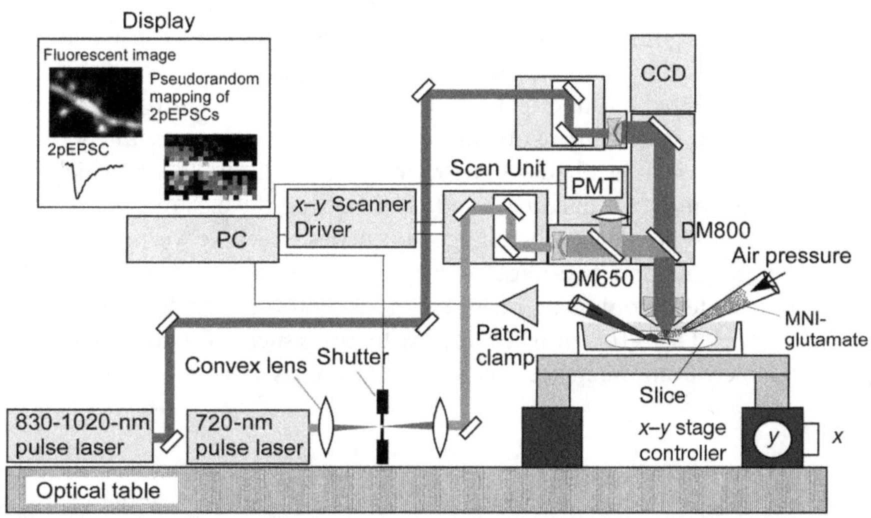

FIGURE 2. Schematic of a two-photon uncaging and imaging microscope. A second scan head (scan unit), which is used for uncaging, is added to a commercial laser-scanning microscope. Full control of the *x–y*-scanner mirrors is achieved via the use of an external scanner driver. A mode-locked laser beam is focused on a small aperture of a fast shutter using a pair of convex lenses, which are used for rapid gating. MNI-glu is superfused locally on a slice preparation on which whole-cell recording is performed. Automated mapping of glutamate currents is performed on a three-dimensional region of interest (ROI). CCD, charge-coupled device camera; DM, dichroic mirror; PC, personal computer; PMT, photomultiplier tube.

two-photon imaging via an 800-nm dichroic mirror. A 650-nm dichroic mirror separates the pathway of the emission fluorescence gathered into the PMT from the laser line used for uncaging. It is most important to achieve full control of the *x–y* scanner that determines the uncaging points; in this setup, external scanner drivers (CX6120, General Scanning, Inc., Watertown, MA) are connected to the driver ports of the scan unit (Fig. 2). The external scanner drivers are controlled by analog output of a digital-to-analog (DA) converter (NI 6030E and PCI-6713, National Instruments, Austin, TX). The output of the PMT is fed into an analog-to-digital (AD) converter (NI 6110E).

Two-photon uncaging and imaging microscopes with two scan heads can now be purchased from Olympus (FV-1000 MPE) and Prairie Technologies (Ultima; Middleton, WI).

Mode-Locked Laser Beam

A mode-locked Ti:sapphire laser is used at 720 nm for the uncaging of MNI-glu. This is the shortest wavelength at which consistent stable mode locking can be achieved. The use of shorter wavelengths can lead to increased photodamage (Denk, 1994). The peak wavelength and bandwidth of the laser are continuously monitored using a spectrometer (E201, IST-Rees, Horseheads, NY). A neutral-density filter wheel (VND-100, Sigma Koki, Tokyo, Japan) is placed at the input port of the microscope to attenuate the power at the specimen to a certain value (3–12 mW). The diameter of the beam must be reduced for rapid gating of the laser, as it is larger (3–5 mm) than the aperture of fast shutters (1.5 mm). A mechanical shutter (UHS1, Uniblitz, Rochester, NY) is placed at the focal point of two convex lenses (Fig. 2) to achieve a pulse duration of 0.6–5 msec. A Pockels cell (350-50, Conoptics, Danbury, CT) is placed between two beam expanders for shorter pulses (50 μsec).

The group velocity dispersion at the specimen is minimized by chirp compensation optics using two SF-10 prisms (New Focus) separated by a distance of ~1 m. Without chirp compensation, pulse duration at the specimen is expanded from 75 to 330 fsec, whereas compensation brings this value to only ~100 fsec. However, assuming that the majority of the photodamage is caused by two-photon absorption, chirp compensation is not necessary if there is enough laser power at the specimen.

The laser beam is aligned for optimal two-photon excitation at the specimen after a warm-up period of 1–2 h. The beam is aligned using two mirrors in front of two irises that are set for optimal laser alignment. First, a rough alignment is obtained using the two irises. Then, the alignment is improved by maximizing a reflection image of an *x–y* scanner mirror that appears in the absence of an infrared cut filter placed in front of the PMT. Third, two-photon fluorescence images of a fluorescent specimen are obtained, and the two mirrors are finely readjusted immediately before entering the scan head, to acquire homogeneous illumination (which is brightest at the center of image). The other two mirrors are adjusted so that the axis of the *x–z* image of a bead (diameter, 3 or 0.1 μm, Molecular Probes, Eugene, OR) is precisely vertical. For the performance of two-photon uncaging and two-photon imaging using two different lasers, the mirror adjustment should be repeated to obtain an image using the 720-nm laser that is identical to the image acquired using the laser at 800–1000 nm. If this adjustment is successful, the uncaging points or regions can be assigned solely from the image obtained using the laser at 800–1000 nm.

Software

We have written custom software to control the *x–y* scanner (DA channels 1 and 2), the shutter (DA channel 3), and the *z*-axis controller (DA channel 4), to determine the uncaging points. The software also acquires a current record from the patch-clamp amplifier (AD channel 1) and an output from the photomultiplier (AD channel 2). For the image acquisition mode, the *x* position is scanned using saw waves, and the *y* position is incremented at each cycle of saw waves. Because of a delay in the movement of the *x* scanner during image acquisition, a small offset of the *x* position occurs between the fluorescence image and the mapping experiment. This offset is constant and is readily corrected by precise positioning of the scanner during mapping to a corresponding position in the fluorescence image.

To eliminate the effects of desensitization of a receptor by previous uncaging at nearby pixels, it is essential to pseudorandomize the distribution of the positions of successive uncaging events during mapping experiments (Denk 1994). If one image is composed of $2Nx * 2Ny$ pixels, the position of the scanner $(X(i),Y(i))$ at the ith uncaging $(i = 0 ,..., 4 * Nx * Ny–1)$ is set,

$$X(i) = Nx * \mathrm{mod}[i,2] + Px(\mathrm{int}[\mathrm{mod}[i ,4 * Nx]/4]),$$

$$Y(i) = Ny * \mathrm{int}[\mathrm{mod}[i,4]/2] + \mathrm{int}[i/(4 * Nx)],$$

where Px is, for example, $Px(i) = (0, 4, 2, 6, 1, 5, 3, 7)$ when $Nx = 8$.

The x–y scanner is fixed to a certain pixel for 50–100 msec, during which time the sample is irradiated with the laser beam for a short period, and the whole-cell current and the PMT output at the time of irradiation are recorded. The laser is then repositioned to other pixels, and the process is repeated until all pixels are mapped. For example, it takes 26–52 sec to map an ROI consisting of 32 × 16 pixels. The lateral interval of a pixel is 0.33 µm. For three-dimensional mapping, x–y screens are performed sequentially for several z-axis positions at an interval of 0.8 µm. The error inherent to the sampling intervals is estimated as 0.1 (the coefficient of variation [CV]); the spatial resolution of glutamate mapping is 0.6 µm laterally and 1.4 µm axially, by means of recoding glutamate current responses at the soma of pyramidal cells. The fluorescence images obtained simultaneously are used to detect any structural alterations in the ROI during the experiments. For the fine analysis of structures, high-quality three-dimensional images with a pixel resolution of ~0.1 µm laterally and 0.2–0.5 µm axially are obtained before and after the mapping experiment.

Electrophysiology Setup

The position of the microscope stage (30 × 40 cm), can be finely adjusted using a mechanical x–y manipulator (Narishige Group, Tokyo, Japan) on the optical table (Fig. 2). Two mechanical manipulators on the stage control the headstage of the patch-clamp amplifier (Axon 200A, Axon Instruments, Inc., Foster City, CA) and a glass pipette for the application of caged compounds, respectively. The microscope system is shielded from electromagnetic waves by a Faraday cage covered with light-protective shielding cloth (Nihon Koden, Tokyo, Japan).

MATERIALS

Reagents

Alexa 594 (optional; see Step 5)

Brain slices

For experiments aimed at inducing structural plasticity of dendritic spines, we dissected hippocampal slices from P6 rats, incubated them at 35°C under 5% CO_2 for 5–6 d in vitro, and then transfected them with the use of a gene gun system (PDS-1000; Bio-Rad Laboratories, Hercules, CA), with a vector containing an enhanced green fluorescent protein (eGFP) cDNA under the control of the chicken β-actin gene promoter and the cytomegalovirus enhancer. Each slice was transferred to the recording chamber 2–5 d after transfection.

MNI-glutamate (MNI-glu)

MNI-glu is commercially available from Tocris Cookson, Ltd. (Bristol, UK). MNI-glu is dissolved to a final extracellular solution at a concentration of 5–12 mM on the experimental day. The preparing procedure should be performed under yellow-light illumination. To avoid the precipitation of Ca^{2+} phosphate in the solution during puff application, the solution should be bubbled with 5% CO_2 gas well before it is included in the puff pipette.

Equipment

Glass pipettes, 20-µm-diameter opening

Imaging setup, as described above

Pneumatic injector (for puff application)

EXPERIMENTAL METHOD

We have used this method to perform two-photon uncaging experiments with a single laser.

1. Place the hippocampal slice on the microscope stage.
2. Align the laser beam for two-photon uncaging of MNI-glu on the specimen.
3. Patch-clamp the neurons in a brain slice for whole-cell perfusion of fluorescent tracers and for the recording of glutamate-induced currents.
4. Acquire the fluorescence images of dendrites, and identify an appropriate ROI or a point.
5. Apply MNI-glu around the ROI or to the point via positive air pressure (0.1–0.2 psi), using a glass pipette placed just above the surface of the slice.

   The stability of the concentration of MNI-glu can be assessed by measuring the two-photon fluorescence of the polar tracer Alexa 594, which is also applied to a brain slice using a glass micropipette. In our experiments, the CV of these fluorescence values over time was <0.03.

6. Record both two-photon-induced excitatory postsynaptic-like currents (2pEPSCs) at the soma and fluorescence values at the time of the irradiation at each pixel within the ROI or at the point. For three-dimensional mapping, perform systematic alteration of the *z*-axis.

DISCUSSION

Examples of Applications

Imaging of AMPA Receptor Densities in Hippocampal Neurons

In this first example, we describe an application of two-photon uncaging microscopy without the imaging scan head. The efficacy of two-photon excitation of MNI-glu was shown in cultured hippocampal neurons (Fig. 3A–C). The caged compound did not activate any current at a concentration of 10 mM, nor did it affect the amplitude or the time course of miniature excitatory postsynaptic currents (mEPSCs), indi-

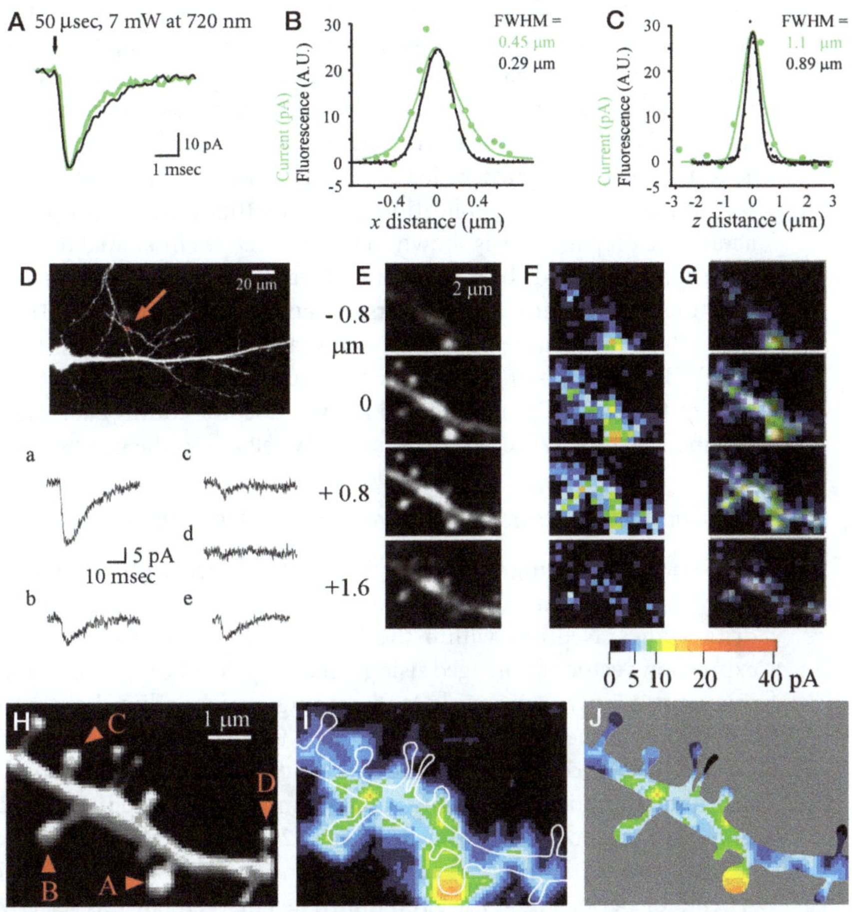

FIGURE 3. Functional mapping of glutamate sensitivity in hippocampal neurons. (*A*) Averaged current traces obtained from cultured hippocampal neurons in response to photolysis of MNI-glu for 50 µsec using a light source set at 7 mW and with a wavelength of 720 nm (green), and averaged mEPSCs (black) recorded from the same cell. Arrow indicates the start of light application. (*B* and *C*) Amplitude of the 2pEPSCs (green dots) along horizontal (*B*) or vertical (*C*) lines crossing the center of an AMPA-receptor cluster, and fluorescence profiles of a bead with a 0.1-µm diameter (black dots) along the horizontal (*B*) or vertical (*C*) axes. The smooth black lines (fluorescence) represent a Gaussian fitting of the data, and the smooth green lines (current) are theoretical predictions, as described (Matsuzaki et al. 2001). (*D–J*) Three-dimensional mapping of glutamate receptor densities in brain slices. (*D, top panel*) Fluorescence image of a CA1 pyramidal neuron within a hippocampal slice. The red arrow indicates the location of the ROI (*D, bottom panel, a–d*) 2pEPSCs induced at four different sites of the dendrite, which are indicated by red letters A–D in *H*. Traces were averages of data from four neighboring pixels. e shows an averaged trace of 10 mEPSCs recorded from the same neuron. (*E–G*) Fluorescence images at four z-axis sections, as indicated (*E*), two-dimensional pseudocolor map of 2pEPSC amplitude (*F*), and overlay images (*G*). (*H*) Three-dimensional reconstruction of the fluorescent images. (*I*) Maximal intensity plot of the glutamate sensitivity for the four sections shown in (*F*). The map was smoothed using linear interpolation. The white lines indicate the contours of the dendritic structures. (*J*) Glutamate-sensitivity map within the dendritic contours. (Reprinted, by permission of MacMillan Publishers Ltd., from Matsuzaki et al. 2004.)

cating that it does not exert any agonistic or antagonistic effect on AMPA (α-amino-3-hydroxy-5-methyl-4-isoxazolepropionic acid) receptors. It should be noted, however, that the high concentration (10 mM) of MNI-glu was antagonistic to GABA (γ-aminobutyric acid)-A receptor (Matsuzaki et al. 2010). Short-period (50-μsec) two-photon uncaging of MNI-glu in a region proximal to dendrites elicited a 2pEPSC (Fig. 3A, black), the time course of which was similar to those of spontaneous mEPSCs (Fig. 3A, green). As the MNI-glu uncaging period is probably in the submicrosecond range (Morrison et al. 2002), the extremely rapid onset (<100 μsec) of the two photon postsynaptic–like currents (PSC) suggests that the uncaged glutamate acts directly on glutamate receptors at the postsynaptic membrane. The current amplitude shows a supralinear (power exponent = 2.5 ± 0.6, mean ±S.D., $n = 7$) dependence on the radiation power (Matsuzaki et al. 2001). The time course of 2pEPSC can be predicted from the rapid photolysis (<50 μsec) of MNI-glu within a focal volume of two-photon excitation, measured using subresolution beads, with lateral and axial full width at half-maximal (FWHM) diameters of 0.29 μm and 0.89 μm, respectively (black lines in Fig. 3B,C). The FWHM diameters of glutamate receptor activation were 0.45 μm laterally and 1.1 μm axially (green lines in Fig. 3B,C), as assessed using an objective lens with a high numerical aperture (1.20) (UPlanApo60xW/IR, Olympus). A slight loss of resolution was caused by the diffusion of glutamate before its binding to the receptors. Nevertheless, two-photon excitation of MNI-glu permits the functional mapping of glutamate sensitivity with a spatiotemporal resolution similar to that achieved by activation of presynaptic terminals (~0.4 μm and ~0.1 msec) (Rusakov and Kullmann 1998).

An example of the application of our methodology to hippocampal CA1 pyramidal neurons in acute slice preparations is shown in Figure 3, D–J. The spatial resolution of AMPA-receptor mapping was slightly reduced (by 0.6 μm laterally and by 1.4 μm axially) as a result of the smaller numerical aperture (0.9) of the water-immersion lens (LUMPlanFI60xW/IR2, Olympus). Therefore, the study focused on dendrites with relatively sparse spines (Fig. 3E, ~1 spine/μm). The time course of 2pEPSC (Fig. 3D,a–d) was similar to that of spontaneous mEPSCs (Fig. 3D,e). Maps of the amplitude of the 2pEPSCs (represented by pseudocolor coding) are shown in Fig. 3F,G,I. The expression of functional AMPA receptors was closely related to the geometry of spines.

Induction of Structural Plasticity at Single Dendritic Spines

Two-photon uncaging microscopy, when combined with two-photon imaging, provides a means to study the long-term structural and functional consequences of glutamate stimulation at single dendritic spines. Neurons within the hippocampal slices were transfected biolistically with an eGFP expression vector and imaged using a second two-photon laser set at 910 nm. The power of the laser was set at ~5 mW on sample for uncaging, which induced currents through AMPA receptors with amplitudes in the range of those observed for mEPSCs. Repetitive stimulation of *N*-methyl-D-aspartate (NMDA) receptors at a single dendritic spine at a frequency of 1 Hz for 1 min was achieved using an external solution lacking Mg^{++}. This stimulation procedure induced a marked increase (>50%) in spine fluorescence (Fig. 4; Matsuzaki et al. 2004). This spine enlargement did not occur in the absence of MNI-glu. Furthermore, it was either transient, with spines recovering their original volume, or persistent, for periods of at least 100 min. The enlargement was selective toward stimulated spines (Fig. 4A,B); only 9% of neighboring spines (mean distance from stimulated spines, 1.2 μm) showed an enlargement >50%, and only 3% showed a long-lasting enlargement (Fig. 4C). Simultaneous two-photon imaging and two-photon uncaging revealed that spines were enlarged within 60 sec of stimulation and that the enlargement was first apparent within 10 sec of the initiation of the stimulus (Fig. 4D). This spine enlargement was triggered by a Ca^{2+} influx mediated by NMDA receptors. Together with the structural enlargement, AMPA receptor-mediated currents were also potentiated exclusively at stimulated spines (Matsuzaki et al. 2004; Harvey and Svoboda, 2007). Spine enlargement is now thought to be the structural basis of long-term potentiation that obeys the Hebbian rule.

Perspectives

The adjustment of the wavelength of the laser used for two-photon imaging allows the imaging of various synthetic small fluorescent molecules and fluorescent proteins. We and other groups have

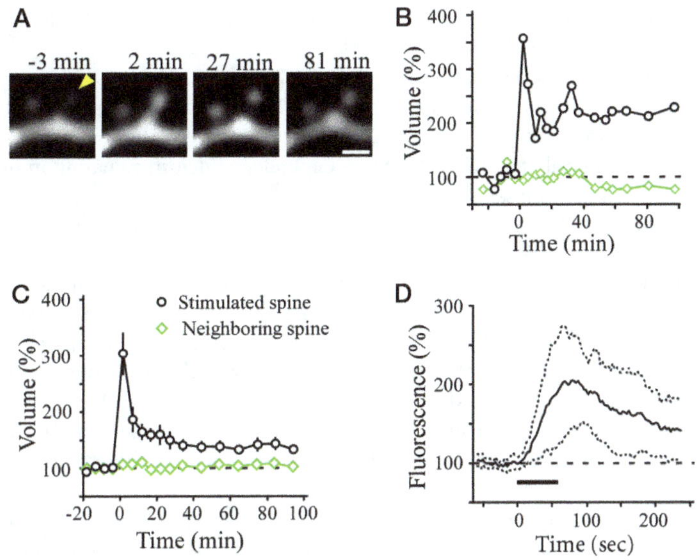

FIGURE 4. Spine-head enlargement induced by repetitive uncaging of MNI-glutamate. (*A*) Time-lapse images of the dendritic spines of a hippocampal CA1 pyramidal neuron expressing eGFP. The *arrowhead* indicates the spot of two-photon uncaging of MNI-glutamate, which was achieved via stimulation for 1 min at 1 Hz beginning at time 0 in a bathing solution lacking Mg^{++}. Images were stacked along the *z*-axis. Scale bar, 1 µm. (*B*) Time course of spine-head volume, estimated from stacked images, of the stimulated (black circles) and neighboring (green diamonds) spines shown in (*A*). (*C*) Averaged time courses of spine-head volume variation after two-photon uncaging of MNI-glu (circles, $n = 40$) and neighboring spines (green diamonds, $n = 34$). Error bars represent ± S.E.M. (*D*) Initial time course of spine fluorescence. The intensity in a single plane drawn through the center of the spine was sampled, and the average value of 11 spines is shown. Dotted lines show time courses corresponding to the mean ±S.D., and the horizontal bar indicates the period of repetitive uncaging. (Reprinted, by permission of MacMillan Publishers Ltd., from Matsuzaki et al. 2004.)

clarified the dynamics of Ca^{2+}, the elongation of actin filaments, and the spreading of active-form Ras and active-form calcium/calmodulin kinase II proteins (intracellular molecules that act downstream from glutamate receptors), all of which are activated by two-photon uncaging of MNI-glu (Carter and Sabatini 2004; Noguchi et al. 2005; Harvey et al. 2008; Honkura et al. 2008; Lee et al. 2009). Fast multipoint uncaging at dendritic spines induces supralinear depolarization and dendritic plasticity (Losonczy et al. 2008). Although the two-photon microscope was developed by Watt Webb and coworkers in 1990 (Denk et al. 1990), it is only now, with the development of new caged compounds and fluorescent molecules that can be excited by two-photon microscopy and the availability of lasers in the wavelength range of 700–1000 nm that the full potential of this invention aimed at the chemical probing of neuronal membranes can be fully exploited.

ACKNOWLEDGMENTS

This work was supported by Grants-in-Aid from the Japanese Ministry of Education, Culture, Sports, Science, and Technology and from the Japan Society for the Promotion of Science (to HK and MM), and by a research grant from the Human Frontier Science Program Organization (HK).

REFERENCES

Araya R, Eisenthal KB, Yuste R. 2006. Dendritic spines linearize the summation of excitatory potentials. *Proc Natl Acad Sci* **103:** 18799–18804.

Beique JC, Lin DT, Kang MG, Aizawa H, Takamiya K, Huganir RL. 2006. Synapse-specific regulation of AMPA receptor function by PSD-95. *Proc Natl Acad Sci* **103:** 19535–19540.

Brown EB, Webb WW. 1998. Two photon activation of caged calcium with submicron, submillisecond resolution. *Meth Enzymol* **201:** 356–380.

Canepari M, Nelson L, Papageorgiou G, Corrie JET, Ogden D. 2001. Photochemical and pharmacological evaluation of 7-nitroindolinyl-and 4-methoxy-7-nitroindolinyl-amino acids as novel, fast caged neurotransmitters. *J Neurosci Methods* **112:** 29–42.

Carter AG, Sabatini BL. 2004. State-dependent calcium signaling in dendritic spines of striatal medium spiny neurons. *Neuron* **44:** 483–493.

Denk W. 1994. Two-photon scanning photochemical microscopy: Mapping ligand-gated ion channel distributions. *Proc Natl Acad Sci* **91:** 6629–6633.

Denk W, Stricker JH, Webb WW. 1990. Two-photon laser scanning microscopy. *Science* **248:** 73–76.

Ellis-Davies GCR. 2011. Nitrobenzyl-based caged neurotransmitters. In *Imaging: A laboratory manual* (ed. Yuste R), pp. 889–896. Cold Spring Harbor Laboratory Press, Cold Spring Harbor, NY.

Ellis-Davies GCR, Kaplan JH. 1988. A new class of photolabile chelators for the rapid release of divalent cations: Generation of caged Ca and caged Mg. *J Org Chem* **53:** 1966–1969.

Ellis-Davies GCR, Matsuzaki M, Paukert M, Kasai H, Bergles DE. 2007. 4-Carboxymethoxy-5,7-dinitroindolinyl-Glu: An improved caged glutamate for expeditious ultraviolet and two-photon photolysis in brain slices. *J Neurosci* **27:** 6601–6604.

Furata T, Wang SSH, Dantzer JL, Dore TM, Bylee WJ, Callaway EM, Denk W, Tsien RY. 1999. Brominated 7-hydroxycoumarin-4-ylmethyls: Photolabile protecting groups with biologically useful cross-sections for two photon photolysis. *Proc Natl Acad Sci* **96:** 1193–1200.

Harvey CD, Svoboda K. 2007. Locally dynamic synaptic learning rules in pyramidal neuron dendrites. *Nature* **450:** 1195–1200.

Harvey CD, Yasuda R, Zhong H, Svoboda K. 2008. The spread of Ras activity triggered by activation of a single dendritic spine. *Science* **321:** 136–140.

Hess GP, Lewis RW, Chen Y. 2011. Caged neurotransmitters and other caged compounds: Design and application. In *Imaging: A laboratory manual* (ed. Yuste R), pp. 869–888. Cold Spring Harbor Laboratory Press, Cold Spring Harbor, NY.

Honkura N, Matsuzaki M, Noguchi J, Ellis-Davies GCR, Kasai H. 2008. The subspine organization of actin fibers regulates the structure and plasticity of dendritic spines. *Neuron* **57:** 719–729.

Lee SJ, Escobedo-Lozoya Y, Szatmari EM, Yasuda R. 2009. Activation of CaMKII in single dendritic spines during long-term potentiation. *Nature* **458:** 299–304.

Lipp P, Niggli E. 1998. Fundamental calcium release vents revealed by two-photon photolysis of caged calcium in guinea-pig cardiac myocytes. *J Physiol* **508:** 801–809.

Losonczy A, Makara JK, Magee JC. 2008. Compartmentalized dendritic plasticity and input feature storage in neurons. *Nature* **452:** 436–441.

Matsuzaki M, Ellis-Davies GCR, Nemoto T, Miyashita Y, Iino M, Kasai H. 2001. Dendritic spine geometry is critical for AMPA receptors expression in hippocampal CA1 pyramidal neurons. *Nature Neurosci* **4:** 1086–1092.

Matsuzaki M, Honkura N, Ellis-Davies GCR, Kasai H. 2004. Structural basis of long-term potentiation in single dendritic spines. *Nature* **429:** 761–766.

Matsuzaki M, Hayama T, Kasai H, Ellis-Davies GCR. 2010. Two-photon uncaging of γ-aminobutyric acid in intact brain tissue. *Nat Chem Biol* **6:** 255–257.

Morrison J, Wan P, Corrie JET, Papageorgiou G. 2002. Mechanisms of photorelease of carboxylic acids from 1-acyl-7-nitroindolines in solutions of varying water content. *Photochem Photobiol Sci* **1:** 960–969.

Nikolenko V, Yuste R. 2011. How to build a two-photon microscope with a confocal scan head. In *Imaging: A laboratory manual* (ed. Yuste R), pp. 111–116. Cold Spring Harbor Laboratory Press, Cold Spring Harbor, NY.

Noguchi J, Matsuzaki M, Ellis-Davies GCR, Kasai H. 2005. Spine-neck geometry determines NMDA receptor-dependent Ca^{2+} signaling in dendrites. *Neuron* **46:** 609–622.

Papageorgiou G, Corrie JET. 2001. Effects of aromatic substituents on the photocleavage of 1-acyl-7-nitroindolines. *Tetrahedron* **56:** 8197–8205.

Rusakov DA, Kullmann DM. 1998. Extrasynaptic glutamate diffusion in the hippocampus: Ultrastructural constraints, uptake, and receptor activation. *J Neurosci* **18:** 3158–3170.

Smith MA, Ellis-Davies GCR, Magee JC. 2003. Synaptic mechanisms of the distance-dependent scaling of Schaffer collateral synapses in CA1 pyramidal neurons. *J Physiol* **548:** 245–258.

Zayat L, Baraldo LM, Etchenique R. 2011. Uncaging with visible light: Inorganic caged neurotransmitters. In *Imaging: A laboratory manual* (ed. Yuste R), pp. 897–902. Cold Spring Harbor Laboratory Press, Cold Spring Harbor, NY.

26 Acousto-Optical Deflector–Based Patterned Ultraviolet Uncaging of Neurotransmitter for the Study of Neuronal Integration

Eugene F. Civillico, Shy Shoham, Daniel H. O'Connor, Dmitry V. Sarkisov, and Samuel S.-H. Wang

Department of Molecular Biology and Princeton Neuroscience Institute, Princeton University, Princeton, New Jersey 08544

ABSTRACT

This chapter describes an acousto-optical deflector (AOD)-based system for rapid ultraviolet (UV) photolysis in arbitrary spatial and temporal patterns (Shoham et al. 2005). Some basics on caged neurotransmitters and the theory of operation of AODs are covered, as are descriptions for implementing the system. Protocol A covers the construction of the UV beam path. Protocol B covers the alignment and calibration of patterned uncaging, including operation of the AODs. Protocol C describes how to prepare caged compounds and introduce them into experimental preparations.

INTRODUCTION

Because of the relative noninvasiveness of light and the ease with which it may be manipulated into spatiotemporally complex patterns, the method of patterned photoactivation is a natural fit for the study of dendritic integration. Photoactivatable molecules that influence a wide range of extracellular and intracellular neurophysiological functions are available. The choice of photosensitive molecules depends on the research question and will influence the design of the experimental apparatus (Civillico et al. 2010).

The first major choice is whether to use photolysis-activated "caged" diffusible molecules or one of the light-sensitive membrane proteins that have recently become available (Zhang et al. 2007; Sjulson and Miesenböck 2008). Caged compounds are preferred for the activation of native receptors with a time course matching endogenously occurring binding, unbinding, and biochemical kinetics. In general, newer optogenetic approaches seem attractive when "on–off" control of neuronal membrane potential or intracellular cascades is desired as a means of determining the downstream effects on other cells, or when it is desirable to manipulate function without an electrode (e.g., in multiple cells of a particular type).

Characteristics of Caged Compounds

A complete listing and discussion of available caged compounds is beyond the scope of this chapter, but reviews are available elsewhere (see Chapters 16 and 25; also Adams and Tsien 1993; Gurney 1994; Ellis-Davies 2007, 2011; Sarkisov and Wang 2007; Civillico et al. 2010). To be useful in biological experiments, caged compounds must meet a number of basic criteria: solubility, lack of biological activity (e.g., interference with receptors or toxic effects), and rapid release of ligand on illumination (Lester and Nerbonne 1982). In addition to these properties, one key parameter is the sensitivity to photoactivation. In this regard, a useful quantity is the uncaging index *U*, which is defined for conventional (one-photon) absorption as $U = \varepsilon\phi$, where ε is the extinction coefficient (typically in units of $\text{M}^{-1}\ \text{cm}^{-1}$) and ϕ is the quantum yield, or the probability that a group will be photolyzed after absorbing a photon. The relative merits of the different cage groups have been quantitatively well characterized for caged glutamates but less so for other compounds. *U* values for some commonly used compounds can be found in Adams and Tsien (1993), Gurney (1994), Ellis-Davies (2007), and Sarkisov et al. (2007).

One- and Two-Photon Uncaging

Caged compounds are typically excited by near-UV light (330–405 nm) using the "one-photon" approach described here. UV light is strongly scattered by brain tissue, leading to limited depth penetration (Shoham et al. 2005) and degradation of lateral resolution with increasing depth in the specimen. One-photon methods do not provide a way to restrict excitation to a single *z* focal plane, so molecules away from the plane of interest may also be excited.

Two-photon photolysis overcomes some of these disadvantages, resulting in decreased scattering of infrared light, a more localized excitation spot, and an intrinsic "optical section" around the plane of focus (Zipfel et al. 2003). Two-photon activation requires pulsed near-infrared (IR) laser light and is essential when single-spine resolution is desired (see Chapter 25). To obtain a diffraction-limited uncaging spot, it is necessary to use the full numerical aperture of the objective lens by completely filling the back aperture. The two-photon cross section of caged glutamate reached a usability threshold of 1 Göppert-Mayer unit with the synthesis of the Bhc (6-bromo-7-hydroxycoumarin-4-ylmethyl) (Furuta et al. 1999) and MNI (4-methoxy-7-nitroindole) (Matsuzaki et al. 2001) caging groups. The utility of light-gated proteins for multiphoton applications, as determined by chromophore properties and photocycle dynamics, is beginning to be explored (Rickgauer and Tank 2008).

Simultaneous two-photon calcium imaging and two-photon glutamate uncaging using two infrared lasers has recently been performed to study dendritic integration and excitability using one (Harvey and Svoboda 2007; Sobczyk and Svoboda 2007; Harvey et al. 2008) or two (Gasparini and Magee 2006; Gasparini et al. 2007) pairs of galvanometric scanning mirrors. It is also possible to perform simultaneous two-photon imaging and two-photon uncaging with a single scanned beam (Nikolenko et al. 2007).

The major disadvantage of pulsed infrared two-photon excitation relative to UV excitation is the cost of the light source. The need for two-photon excitation can sometimes be obviated by targeted application or expression of a photoactivable molecule. For one-photon uncaging of glutamate, we use a frequency-tripled *Q*-switched Nd:YVO_4 laser (355 nm, DPSS Series 3501). Considerably less expensive sources of UV and near-UV light include solid-state UV lasers (e.g., Oxxius violet), flash lamps (Ellis-Davies 2011; Kandler et al. 2000), and ultraviolet light-emitting diodes (LEDs) (available from Nichia Corporation; Bernardinelli et al. 2005; Venkataramani et al. 2007; Xu et al. 2008).

Acousto-Optic Deflectors

To explore the integrative properties of dendrites, it is desirable to have precise control over the timing and location of photoactivation in a preparation. Perhaps the most rapidly developing technology for beam-steering applications is the acousto-optic deflector (AOD), which allows inertia-free scanning (see also Chapter 53; also Duemani Reddy and Saggau 2011). For a comparison of AODs with other emerging technologies such as digital micromirror devices and spatial light modulators, see Civillico et al. (2010). The functionality of an AOD arises from a crystal that behaves as a tunable diffraction grating, whose grating constant can be adjusted by varying the frequency of sound propagating across the crystal; the sound is generated by a piezoelectric transducer interfaced with the crystal. AODs may be set up to direct most (>50%) of the input beam intensity into the first-order diffraction peak, generating a movable beam that can traverse a range of angles determined by the bandwidth of sound modulation. One AOD is required for each dimension of the coverage area.

The Spatial Resolution of an AOD-Based System

The number of resolvable points in the sample plane is determined by the number of resolvable deflection angles in the range of the scanning element. For a given range of angles $\Delta\theta$, a beam with an intrinsic divergence angle ϕ can occupy N resolvable angular locations; that is,

$$N = \frac{\Delta\theta}{\phi}. \tag{1}$$

For an AOD-based scanner, the range of deflection angles depends on the beam wavelength λ, the bandwidth of the driving acoustic frequency Δf, and the speed of sound in the AOD crystal medium V:

$$\Delta\theta = \frac{\lambda \Delta f}{V}. \tag{2}$$

The divergence of the beam, ϕ, is approximated (for a uniformly illuminated aperture) by dividing the beam wavelength by the beam diameter, which for AOD-based scanners is limited by the deflector aperture size D:

$$\phi = \frac{\lambda}{D}. \tag{3}$$

Substituting Equations 2 and 3 into Equation 1 and simplifying gives the number of resolvable spots, N, in terms of the acoustic propagation time across the crystal Δt:

$$N = \Delta f \cdot \Delta t. \tag{4}$$

This constraint is often referred to as a time-bandwidth product. For a Gaussian beam, which does not fill the aperture uniformly, N is an overestimate of the true performance. The most notable trade-offs are between spot size and scanning rate, and between bandwidth and number of resolvable spots (Bullen et al. 1997). Increasing D, for example, requires less beam expansion to maintain a full back aperture and therefore allows more resolvable points. A useful calculator for examining design trade-offs can be found at the website of MolTech GmbH (http://www.mt-berlin.com/frames_ao/acousto_frames.htm). The ongoing development of AODs with larger apertures (e.g., 13 mm; Kremer et al. 2008) will increase the field of view and number of resolvable points available with this scanning method.

Tellurium dioxide (TeO_2) crystals are particularly well suited for scanning applications with AODs. TeO_2 possesses a high acousto-optic figure of merit (Saleh and Teich 1991) with a slow acoustic velocity, conferring on it high transmission efficiency, high resolution, and a large scan angle for a relatively low radiofrequency (RF) bandwidth. TeO_2 transmits effectively over a wide spectral range including both infrared and near-UV. For UV applications, screening individual crystals for custom selection is advisable for achieving maximum transmission (Shoham et al. 2005). For pulsed IR applications the spectral bandwidth of optical pulses leads to pulse broadening by these birefringent crystals. Pulse dispersion compensation is often used to restore original optical pulse properties. AODs are available commercially from many sources (e.g., IsoMet, Brimrose Corporation, Noah Industries Inc., Crystal Technology Inc.).

IMPLEMENTATION OF AN AOD-BASED SYSTEM

In the protocols that follow, we describe the construction and operation of an AOD-based instrument for rapid UV photolysis of caged compounds in arbitrary spatiotemporal patterns. In Protocol A we describe the addition of an ultraviolet beam for uncaging to a home-built two-photon microscope. The fine alignment of the UV beam and the implementation of the AOD-based beam steering are described in Protocol B. Protocol C describes how to prepare caged compounds and introduce them into experimental preparations. Within the constraints of the components used, stimulation patterns are totally programmable. Editing them is as simple as writing an assignment statement in MATLAB, and can be made even simpler through the use of a basic custom graphical user interface (GUI). The design is ideal for incorporating into an existing two-photon microscope constructed around a vertical rail, and uses relatively inexpensive ultraviolet light steered by a pair of crosswise AODs. The system is particularly well suited for investigations of dendritic integration by neurons in brain slices.

Comparison with Two-Photon Uncaging Systems

Our system has a number of advantages relative to two-photon uncaging systems, including the low cost of ultraviolet light. An ultraviolet laser is approximately 1/20th the price of a femtosecond-pulse infrared laser suitable for two-photon photolysis. Recently developed ultraviolet LEDs are available for less than 1/100th the infrared cost. In addition, one-photon uncaging works with a wider toolbox of caged compounds and photolyzes a larger volume per site, allowing the stimulation of larger portions of cells or entire cells, without complex equipment specializations (Nikolenko et al. 2007). Finally, one-photon systems can make use of the advantages of AODs without requiring compensation for pulse broadening by the crystal.

One-photon advantages notwithstanding, the fact remains that ultraviolet light is strongly scattered by biological tissue. This degrades the axial resolution of one-photon uncaging. This disadvantage is reduced if the targeted elements are close to coplanar, as is the case in the dendrite of a Purkinje neuron.

Comparison with Other Beam-Steering Approaches

Other systems used for multispot patterned uncaging direct the beam with a pair of mirrors attached to galvanometric scanners. The choice between AOD-based and galvanometric scanning involves a tradeoff between addressing speed (where AODs are preferred) and range (where mirrors are preferred).

For AOD-mediated beam deflection, the travel time between two scanned points is determined by the propagation time of sound across the beam diameter at the crystal, usually tens of microseconds. This makes any two arbitrarily spaced points within the scan area accessible within one switching time. Compared with galvanometric mirrors, which typically span hundreds of mrads, AODs typically provide deflection at more than ~50 mrads. This disadvantage can be partially compensated by magnifying the beam between the scanner and the objective.

The multispot illumination provided by our system is able to stimulate 20 locations in 1 msec, less time than typical neuronal time constants. When activation of a large area of hundreds of μm^2 is desired, the number of points necessary will result in significant temporal spreading of the stimulus. For such an experimental design, a digital micromirror device (DMD) or spatial light modulator (SLM) may be the preferred beam-shaping technology. For more details on alternative choices, please see Civillico et al. (2010).

APPLICATION: USING PATTERNED UNCAGING OF GLUTAMATE TO SIMULATE NETWORK ACTIVITY IN VITRO

The spatial and temporal performance of this system allows the activation of multiple dendritic locations virtually simultaneously, both directly and monosynaptically via evoked action potentials in presynaptic neurons (Shoham et al. 2005). Figure 1 shows data from such an experiment. A cerebellar Purkinje neuron filled with dye and the presynaptic granule cell layer visible in negative stain are shown in Figure 1A. At each of 20 randomly selected locations (colored squares), uncaging was performed at a 3 × 3 array of points spaced 1.3–2.0 µm apart. At each responding location ($n = 18$ locations in four recordings; 718 responses total), averaged responses (Fig. 1B) had short onset delays from the first uncaging flash (12.6 ± 10.2 msec, mean ± S.D.) and peak amplitudes of 1–54 pA (mean, 19 pA; 25%–75% interquartile range, 7–28 pA). Responses were completely eliminated by addition of the sodium channel blocker lidocaine (Fig. 1B), which prevents propagation of action potentials. Lidocaine did not affect responses to uncaging pulses delivered directly to the Purkinje cell dendritic arbor (Fig. 1D). Thus, photolyzed glutamate triggered action potentials in granule cells and synaptic responses in Purkinje cells.

Amplitudes of excitatory postsynaptic currents (EPSCs) evoked with this system are comparable to measurements made using electrical stimulation of single granule cells (compare with 2–60 pA, mean 14 pA in, e.g., Barbour 1993). Multiple rising phases could often be seen in individual responses (Fig. 1B), indicating simultaneous activation of multiple granule cells; however, when the

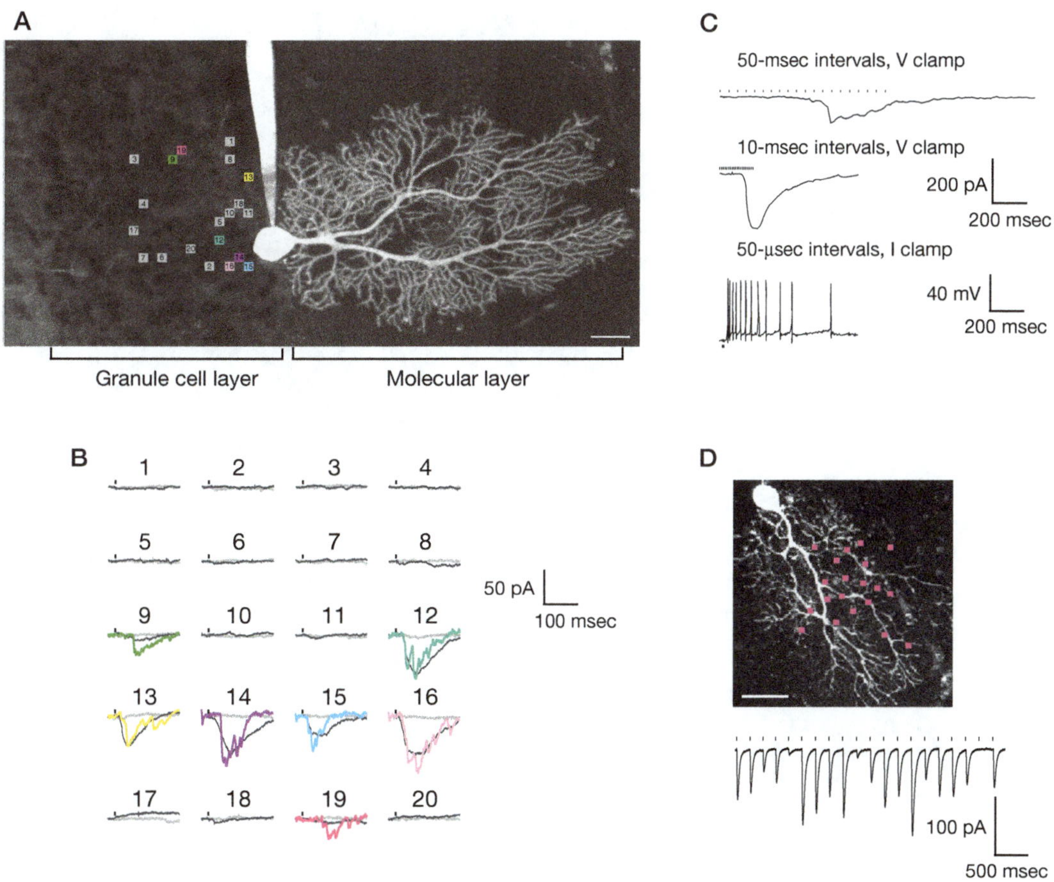

FIGURE 1. Patterned activation of the Purkinje cell dendritic arbor. *(A)* Maximum intensity projection of a cerebellar Purkinje neuron filled via a patch clamp recording electrode with fluorescent dye. Squares indicate the positions of 20 uncaging locations in the presynaptic granule cell layer. The locations were visited in the numbered order. Scale bar, 25 µm. *(B)* Postsynaptic voltage-clamp responses to uncaging at the locations shown in *A*, with 200-msec interlocation intervals. Vertical marks indicate times of uncaging. Shown in gray at each location are responses after addition of APV (400 µM) and lidocaine (500 µM). Single examples (colored traces) are also shown. *(C)* The same sequence of uncaging locations as in *B*, but with interlocation intervals of 50 msec (*top*), 10 msec (*middle*), or 50 µsec (*bottom* in current clamp). *(D)* Twenty uncaging locations over a Purkinje cell dendritic arbor (*top*; scale bar, 20 µm) were visited at 200-msec intergrid intervals in the presence of 500 µM lidocaine. Voltage clamp responses from the Purkinje cell are also shown (*bottom*). (Adapted, with permission, from Shoham et al. 2005.)

granule layer stimulus pattern was shifted 3 μm laterally, individual Purkinje cell responses were reduced by at least half in 72% of movements across eight responding locations, favoring the possibility that the multiple rising phases were caused by single granule cells firing bursts. When all 20 sites were activated within 9 msec, granule cell–Purkinje cell responses summated to produce a burst of spikes in the Purkinje cell (Fig. 1C; $n = 4$ cells). Experiments such as these are ongoing in our laboratory as we study synaptic integration, and particularly the importance of input synchrony, in the Purkinje cell dendritic arbor.

Protocol A

Integrating AOD-Based Patterned Uncaging with UV Light into a Two-Photon Microscope

Here we describe the addition of a UV beam for uncaging to a homebuilt two-photon microscope. The goal is to get ultraviolet light from the light source (laser) to the approximate center of the objective's back aperture, passing through a pair of perpendicularly oriented AODs along the way. The fine alignment of the UV beam and the implementation of AOD-based beam steering are described in Protocol B.

OVERVIEW OF THE UV BEAM PATH

Our system integrates a rapid-deflection ultraviolet (UV) laser-based patterned uncaging system into a two-photon laser-scanning microscope (Fig. 2; the numbered parts correspond to those described in the Imaging Setup section below). The UV light source is a frequency-tripled Nd:YVO_4 laser (1) that produces 50–60-nsec, λ = 355-nm pulses at a 100-kHz repetition rate. The laser beam is expanded using a variable-zoom UV beam expander (3), circularly polarized for improved transmission through the AODs using a $\lambda/4$ plate (4), and directed along the light path using UV mirrors (7) aligned to match the directions of movement of an *x–y* translational stage (XYR-8080, Danaher Motion) that positions the microscope (see Sarkisov et al. 2007).

Beam steering is performed using a two-dimensional TeO_2 AOD (6) attached to the microscope's vertical rail using a five-axis kinematic mount (7). Light energy reaching the specimen is reduced to the desired amount of energy with a neutral-density filter wheel (8) and by modulating the radiofre-

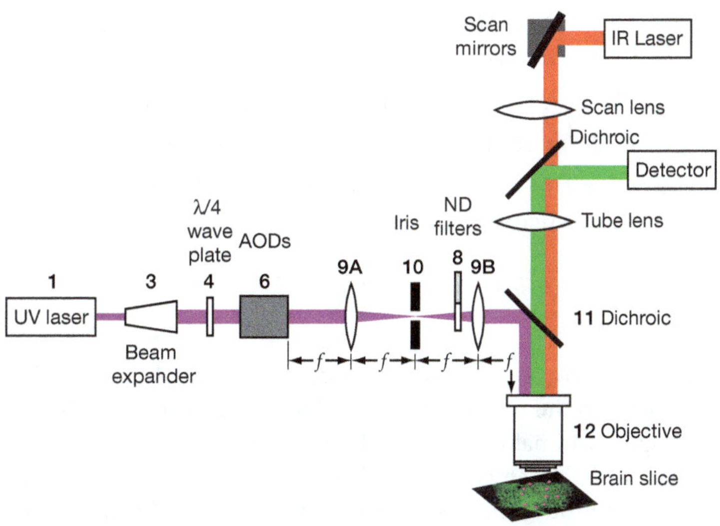

FIGURE 2. UV beam path for patterned uncaging integrated into a two-photon microscope. Part numbers in bold refer to the parts list. Parts 2, 5, and 7 omitted for clarity. Purple beam, UV light for photoactivation of MNI-glutamate. Red beam, scanning infrared beam for two-photon excitation of fluorescence. Green beam, emitted fluorescence, detected by a photomultiplier tube (Detector). (Modified, with permission, from Shoham et al. 2005.)

quency (RF) power. An analog output board is used to control the acousto-optical driver, as well as an electronic shutter (2) and a *Q*-switching input to the laser that provides microsecond level control of the laser's output. The AOD output is projected through two lenses (9A, 9B). An aperture (10) is placed in the image plane to block higher diffraction orders as well as the "parked" beam. A dichroic mirror (11) is used to merge the UV uncaging beam and the IR imaging beam, both of which are then directed through a 40x objective (12) into a single shared focal plane.

Imaging Setup

Components are listed in order along the beam path, starting with the UV laser and ending with the objective lens. We have tested and found both of the objective lenses listed (12) to be satisfactory. The UIS2 line of objectives from Olympus Corporation may also be suitable, although we have not tested them. Note that when used with short UV pulses, the effective neutral-density filter values (8) are typically ~80% of the specified values. Photographs of the components may be found in the Supplemental Methods to Shoham et al. (2005).

1. UV light source (frequency tripled Nd:YVO_4 laser; model 3504-150, DPSS Corp.)
2. Shutter (Uniblitz LS6ZM2, Vincent Associates; used with VMM-D1 driver)
3. Variable-zoom UV beam expander (model 4401.402.020, LINOS Photonics GmBH & Co.)
4. Quarter-wave plate (355-nm, mounted, zeroth-order; model WPQ05M-355, Thorlabs, Inc.)
5. Mirrors (UV MAXBRIte model 02 MLQ 011/007, Melles Griot)
6. Acousto-optic crystal (model 2DS-150-50-.364, Brimrose Corporation; used with acousto-optical driver model VFE-150-50-V-B1-F2-2Ch)
7. Kinematic mount (model 9082, New Focus)
8. Fused silica neutral-density filters: 03 FNQ 089, 03 FNQ 057, 03 FNQ 113, 03 FNQ 065 (Melles Griot; used with 12 Station Dual Filter Wheel FW2A, Thorlabs, Inc.)

9A. Plano-convex fused silica lens (150/25.4; LA4874, Thorlabs, Inc.) mounted in an *x–y* lens holder (111-0340, OptoSigma Corporation)

10. Lever-activated threaded iris (SM1D12, Thorlabs, Inc.) mounted on an adjustable-length lens tube (SM1V10, Thorlabs, Inc.), which is mounted in an *x–y* mount (111-0340, OptoSigma Corporation)

9B. Plano-convex fused silica lens (150/25.4; LA4874, Thorlabs, Inc.)

11. Dichroic mirror (DCLP 450, Chroma Technology)
12. Objective (HCX APO UVI 40x 0.8 numerical aperture [NA] or HCX APO UVI 20x 0.5 NA, Leica Microsystems)

Basic Control of the AODs

Basic control of the AODs is required for assembling the UV beam path. We recommend beginning to develop the routines to be used for AOD control in the complete system at this stage, using a PC, a digital-to-analog board, and instrument control software to generate the analog outputs. Our system is implemented in LabVIEW 5.1 and uses a National Instruments PCI-6713 board with eight analog outputs and two counter-timers. Six of the analog outputs are used to control the UV shutter, the laser *Q*-switching, and the two AODs (one frequency and one amplitude channel per AOD). More details are described in Protocol B.

ASSEMBLING THE UV BEAM PATH AND INCORPORATING THE AODS

The two-photon microscope is constructed around an X95 vertical rail (Newport Corporation) with the main optical axis oriented vertically from the scanning mirrors to the preparation. The ultraviolet beam is brought up from the optical table to a horizontal plane above the objective (this occurs between parts 4 and 6 of Fig. 2 and for clarity is not shown), and is introduced into the main vertical optical axis of the microscope by dichroic mirror (11). The following steps describe the basic alignment of the UV beam path.

1. Determine the center frequencies of the AOD crystals (i.e., the drive frequencies that produce a first-order output beam that is collinear with the input beam) by consulting the manufacturer or by empirical tests using a simple LabVIEW virtual instrument to drive them.

   The first AOD will produce at least three output beams owing to different diffraction orders, each of which will in turn result in at least three beams after the second AOD. All beams except the doubly deflected first-order beam will be excluded using an aperture in the functioning system.

2. Attach a horizontal rail to the vertical rail so that it is approximately vertically aligned with the desired location of the AOD output aperture. Bring the UV beam to the height of this rail by mounting mirrors as needed. Align the beam horizontally along the rail by attaching two post-mounted irises and closing them down so that the beam just passes through them.

   Note that ultraviolet and infrared light can cause retinal damage and blindness and are invisible to the naked eye. Wear protective eyewear and observe all laser safety practices.

3. Using laser viewing cards, adjust the last mirror to position the beam at the center of the objective back aperture, perpendicular to the vertical axis of the microscope.
4. Attach the AODs to the five-axis kinematic mount, and attach the mount to the vertical or horizontal rail. Drive the AODs with their center frequencies as determined in Step 1.
5. Adjust the mirrors before the AODs, and/or the AOD mount axes, so that the first-order doubly deflected beam passes through the pair of apertures as in Step 2. This will reestablish the position established in Step 3. If it does not, re-adjust the final mirror before the dichroic.
6. Introduce the two lenses (9A and B) which will relay the AOD pivot point to the back aperture of the objective. Introduce the lens closer to the objective first. Be sure that the position established in Step 4 stays constant. The completed system should be approximately a 4*f* system (Saleh and Teich 1991) as shown in Figure 2.
7. Introduce the iris between the lenses. Position it and adjust its diameter so as to exclude all beams except the doubly deflected first-order beam. Proceed to Protocol B for fine adjustment and alignment of uncaging.

Protocol B

Alignment and Implementation of Uncaging

Here we describe the optimization of the uncaging spot after passing through the AODs and the operation of the AODs themselves. We assume that the UV beam path has been constructed as in Protocol A (i.e., that it passes through two AODs and is directed to the back aperture of the objective). Performing the final alignment with the beam passing through the AODs will ensure that the system is optimized for the idiosyncrasies of the crystals.

MATERIALS

CAUTION: See Appendix 6 for proper handling of materials marked with <!>.

Reagents

Bovine serum albumin (BSA)
Caged fluorescein dextran
Cyanoacrylate glue (e.g., Krazy Glue) <!>
Fluorescein dye (F2456, Sigma-Aldrich)

Equipment

Coverslips
Glass slides
Parafilm
Two-photon laser-scanning microscope with UV laser <!> for uncaging, as described in Protocol A

EXPERIMENTAL METHOD

Preparation of the Caged Fluorescein Dried Sample

Steps 1–5 describe the preparation of a thin, immobile layer of caged dye on a glass slide. This slide may be examined at leisure after an uncaging flash or pattern as described in Step 9, and greatly simplifies alignment of an uncaging system.

1. Prepare a 10-mL solution of 1% BSA and 2 mg/mL caged fluorescein dextran in H_2O.
2. Evenly spread 5 μL of the solution on the coverslip. Let it dry on a horizontal surface for ~1 h, until the water evaporates.

   See *Troubleshooting.*
3. Firmly attach a layer of Parafilm to a glass slide.

   To allow better contact, the glass slide may be slightly heated.
4. Remove a patch of Parafilm from the center of the slide to make a hole. Position the coverslip on top of the hole with the caged dye layer facing down toward the slide.
5. To prevent water from leaking into the dried sample, the coverslip can be attached to the Parafilm with cyanoacrylate glue.

   The caged fluorescein slide will keep for several months at room temperature if stored in the dark.

Coarse Alignment

The alignment instructions that follow should be performed with the AODs on and the beam in the center position as described in Protocol A. Instructions for fine alignment suitable for higher-resolution systems may be found in Sarkisov and Wang (2006).

6. Pour a 10-mM solution of fluorescein (prepared in H_2O) into a chamber in front of the objective and illuminate it with UV light (<0.1 mW) or IR light. Avoid developing bubbles under the objective.

   Under these conditions, the cone of the excited fluorescein solution should be visible from the side (see Fig. 1 of Sarkisov and Wang 2006). If the observed cone is not axially symmetric with respect to the optical axis of the system, or does not coincide with the point at which the excitation light is focused, adjust the angle and position of the incoming UV light by adjusting the last mirrors in the beam path (after the AODs) until these conditions are met.

7. Observe fluorescein excitation through the eyepieces or camera. Make fine adjustments to the mirrors to move the fluorescent region to the middle of the field of view.

   Only low levels of UV light energy are needed for excitation of fluorescein. Do not increase UV light energy above 1 mW if fluorescence is not visible. The increased intensities, which are not visible, will damage your eyes or the detector. Instead, try making fresh fluorescein solution and darken the room.

8. Make slight changes to the beam expander setting to optimize the size and shape of the fluorescent spot, observed as in Step 7. If the position of the fluorescent spot moves laterally, readjust the beam as in Step 7. If the spot is distorted, reduce the beam expansion before the AODs.

9. Put the glass slide from Step 5 under the microscope and start imaging. Focus on the caged dye layer. Deliver an uncaging flash of 1–10 μJ energy. This may be performed with a simple program to control the shutter or *Q*-switching.

   A spot of photolyzed dye should become visible.

   See *Troubleshooting.*

10. Make adjustments to the UV beam expander to obtain the best shape and position of the uncaging spot, moving the slide to expose fresh regions of dye as necessary.

    From this point the implementation of patterned uncaging is a matter of delivering proper control signals to the AODs and optimizing the size and shape of uncaging spots across the uncaging zone. The dried sample of caged fluorescein will be useful in optimizing the beam size, AOD parameters, iris position, and telescope to obtain as large a field as possible of identical uncaging spots. This will be an iterative process of adjustment of the control signals and beam properties while monitoring the uncaging performance.

    See *Troubleshooting* and *Discussion.*

TROUBLESHOOTING

Problem (Step 2): It is impossible to spread the solution on the coverslip evenly.
Solution: The glass surface is hydrophobic. Increase the concentration of buffer protein (BSA), use coated coverslips, or treat the coverslip to make it hydrophilic.

Problem (Step 2): The layer of dried dye is not even; cracks are visible.
Solution: The concentration of carrier protein (BSA) may be too high. Decrease the concentration of BSA.

Problem (Step 9): The uncaging spot diffuses rapidly.
Solution: There may be a high concentration of water in the sample. Let the sample dry for a longer time period. Use a high-molecular-mass dextran conjugate. Perform the measurement in the linescan mode.

Problem (Step 9): The uncaging spot appears slowly.
Solution: There may be a low water concentration in the sample. Decrease the drying time.

Problem (Step 9): The uncaging spot is not bright.
Solution: There may be low uncaging power or low dye concentration. Increase ultraviolet-light power, or make a new sample with a higher dye concentration.

Problem (Step 9): There is high background fluorescence of the caged sample.
Solution: There may be a high concentration of uncaged dye. Prepare a fresh sample.

Problem (Step 9): It may be difficult to focus on the dried layer of caged dye, which usually has very low fluorescence.
Solution: Mark the coverslip and then focus on it using visible light. Alternatively, add a small concentration of fluorescein to the water between the objective and the coverslip. Focus above the coverslip and observe fluorescence. Move the objective down until the focus reaches the coverslip and the fluorescence disappears. Increase the IR laser power. Once you move down for another 0.17 mm (the thickness of a standard coverslip), the excitation focal point will reach the layer of caged dye on the other side of the coverslip, which can be detected as a slight increase of the fluorescence signal.

Problem (Step 10): The uncaging spots are asymmetric.
Solution: Reduce beam expansion before the AODs, or rotate the first lens after the AODs.

Problem (Step 10): The uncaging spots vary in intensity across locations.
Solution: Flatten the field by reducing the intensity control voltage in a location-dependent way for the brighter locations as described in Shoham et al. (2005).

Problem (Step 10): The uncaging performance is variable.
Solutions: Check that the UV laser power output is stable. Ensure the AODs are warmed up sufficiently. Calibrate daily.

DISCUSSION

Aligning Uncaging and Imaging Beams

Because objectives focus different wavelengths differently, collimated beams of uncaging (UV) and imaging (IR) light will focus to different planes if the objective is not properly corrected. For our AOD-based system, this problem is solved mainly by selecting objective lenses with wide-band correction for intrinsic variations in the index of refraction, and good transmittance at both UV and IR wavelengths (Leica UVI objectives; transmission at 355 nm is ~60% and at 830 nm is ~75%). In addition, the focal plane of uncaging can be shifted relative to the imaging plane by moving the first lens after the AODs along the optical axis. This adjustment may be used to bring the uncaging and imaging planes together or to position the uncaging plane in a different plane of focus.

For the AODs and objectives used in our system, fine alignment is less critical than for a single-point photolysis system in which the full numerical aperture of the objective is used (Sarkisov and Wang 2006, 2008). In practice, to prevent spatial distortion of the uncaging spot, the AODs require that we operate in a low-NA regime, using a beam width that underfills the objective back aperture. This degrades the axial resolution of the uncaging spot and ensures that it will overlap with the imaging plane. For this reason our system is optimal for two-dimensional (2D) or quasi-2D dendrites.

Implementing Beam Steering by Software Control of the AOD

Our AOD control is implemented using what is now referred to as the "Legacy" or "Traditional" Ni-DAQ drivers for control of National Instruments A/D conversion hardware. These drivers are no longer supported by National Instruments. Below we provide some design principles for low-level control, and our software is available on request; however, we recommend that investigators starting from scratch make use of the newer Ni-DAQmx drivers. Other software possibilities for interfacing with NI hardware include MATLAB (The Mathworks) and IGOR Pro with the Ni-DAQ Tools extension (WaveMetrics).

Analog Outputs

Each of the two AOD crystals is controlled by two voltages via the radiofrequency AOD with a nominal switching time of 50 µsec (channels: deflection in *x* and *y* and intensity in *x* and *y*). One voltage controls the wave frequency in a nonlinear manner over ~1–9 V applied control voltage; modulating this wave frequency between 125 and 175 MHz directs the first-order diffracted beam over a 1.7° range of angles. The second voltage is varied over 0–1 V to control the wave amplitude, which controls the intensity of this first-order beam.

In addition to the four signals that directly control the AODs, our software also controls the *Q*-switching of the laser and the shutter. This sets the timing of laser pulses so that they are only emitted after the AOD has settled to a new stable position. When *Q*-switching is first enabled, the first pulse in a series has higher energy than subsequent pulses, so it is not emitted by the laser. The software must account for the slight delay caused by this.

Parking the Beam

The iris is positioned and adjusted in such a way that the UV beam may be "parked" by directing it to a position at which it does not pass the iris. As an added mechanism for preventing unwanted uncaging, lasing is also stopped via the *Q*-switch control when the AODs are directed to this park point.

Stimulus Design

Stimuli are constructed with a custom MATLAB GUI that allows the specification of locations and timing for uncaging using a guide image. Delays between point stimuli are converted to "parked time." Stimulus files are saved out of MATLAB and read by the LabVIEW virtual instrument, which converts them to synchronized AOD, shutter, and *Q*-switch control waveforms for analog output. It is possible to specify in the software the number of UV pulses delivered to each uncaging "point," providing another level of control over uncaging energy.

Daily Recalibration

At the start of each experiment, we repeat Step 9 of the alignment procedure using a stimulus pattern consisting of a 10 x 10 grid tiled over the uncaging zone, with ~5 µm between points. Visualization of the uncaging zone in this way allows us to determine if it has drifted relative to the two-photon microscope's imaging field. If it has, we recenter it by adjusting the offsets to the scan waveforms in the two-photon imaging software. This drift is commonly <1 µm. In addition, our AOD control software allows for the possibility of adjusting the AOD drive signals each day using a calibration routine in which the actual uncaging spots produced by the 100-point grid are compared with the idealized grid and used to generate a correction file for that day. This procedure compensates for slight variations from day to day in the mapping of AOD control signals to deflection angles owing to, for example, variations in ambient temperature.

Protocol C

Preparing and Handling Caged Compound Solutions

Here we describe how to prepare caged compound solutions and discuss options for introducing caged compounds into an experimental preparation.

MATERIALS

Reagents

Artificial cerebrospinal fluid (ACSF)
MNI-glutamate (4-methoxy-7-nitroindolinyl-caged L-glutamate; MW = 323.3)

Equipment

Foil-covered box
Lyophilizer or SpeedVac (optional; see Step 2)
Microcentrifuge tubes

If using a reservoir such as the one depicted in Figure 3 (see Discussion), use microcentrifuge tubes that match the reservoir lid.

EXPERIMENTAL METHOD

Preparation of Caged Glutamate

Take care at all times to prevent the sample from being exposed to stray light, which could cause uncaging. Whether introducing caged compounds into an experimental preparation by bath perfusion (Step 4) or capillary application (Step 5), cover the reservoir during experiments. Although the exact guidelines vary depending on the stability of the compound in use, taking extra precautions will never hurt.

1. To a 10-mg vial of MNI-glutamate, add 1 mL of ACSF to make a 30 mM stock solution.
2. Aliquot the solution into 20 samples of 50 μL each.

   Aliquotting will minimize freezing and thawing cycles. If the compound will be stored for prolonged periods of time, aliquots can be dried in a lyophilizer or SpeedVac before freezing. In this case the solvent should be distilled water rather than ACSF, which would be concentrated by the drying process.
3. Store the aliquots in a foil-covered box at –20°C. They will be good for several months.

Bath Application

4. Use multiple aliquots depending on the bath volume, and dilute 12x to a final concentration of 2.5 mM in ACSF.

   See *Discussion.*

Capillary Application

5. Thaw one aliquot and dilute it 2x by adding 50 µL of oxygenated ACSF to obtain a 15 mM solution for local perfusion.

   See *Discussion*.

DISCUSSION

Application of Caged Compounds

For caged neurotransmitters and neuromodulators, the principal options for the introduction of caged compounds into an experimental preparation are bath perfusion and focal application through a glass capillary. Bath application has the advantage of relative simplicity and guarantees uniform concentration of the compound throughout the bath. Commonly, in vitro experiments using caged compounds are performed with caged compound solution pumped in a looped perfusion (Harvey and Svoboda 2007; Nikolenko et al. 2007). Caged intracellular ligands such as calcium and second messengers must be delivered by microinjection or via patch-clamp electrode, or else they must be made membrane-permeant by using AM esters to neutralize anionic moieties, an approach that is useful for delivering calcium chelators (Tsien 1989; Stosiek et al. 2003).

Focal Application through Capillary Tubing

Local pressure ejection of a caged compound solution (Furuta et al. 1999) has a number of advantages. First, the flow of the compound can be adjusted independently of the bath flow. This is particularly useful as a control to determine whether the caged compound itself affects biological function. Second, much smaller volumes are required, thus economizing on cost.

Suitable capillary glass tubing of various diameters can be purchased from Polymicro Technologies. Tubing must be carefully scored and broken to a useful length using a ceramic cleaving stone or capillary cutter. A clean flat break reduces turbulent flow. A schematic for a simple pressure ejector is shown in Figure 3.

If a droplet does not appear at the end of capillary tubing when gas pressure is applied, tighten the connections on reservoir-tubing adapter. Submerge the reservoir in a beaker of water to locate leaks. Apply more wax or adjust the wax position. If there are no leaks, then the capillary tube is clogged. In that case, cut a new capillary tube.

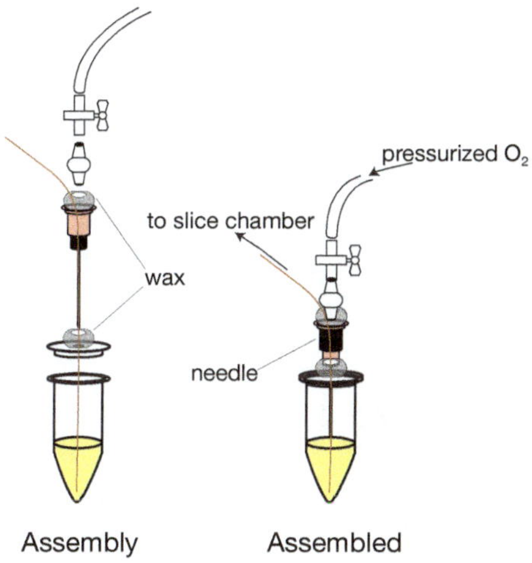

FIGURE 3. Assembly of a low-volume reservoir for focal elution of a caged compound solution into an experimental preparation. Bone wax is used to seal joints.

Capillary glass is autofluorescent at typical illumination wavelengths for calcium imaging experiments, so the application tube may be easily visualized in functional images. The flow of the applied solution may be visualized by replacing the caged compound solution with fluorescein.

At the conclusion of an experiment, run distilled water through the capillary for a few minutes and then run air through it until droplets no longer appear at the tip. This will ensure that it does not clog between experiments. If it does clog, use a capillary cutter to make a new end cut.

ACKNOWLEDGMENTS

EFC was supported by a Robert Leet and Clara Guthrie Patterson Postdoctoral Fellowship in Brain Circuitry. SSHW was a W.M. Keck Foundation Distinguished Young Investigator and was supported by National Institutes of Health grant NS045193 and a National Science Foundation CAREER award.

REFERENCES

Adams SR, Tsien RY. 1993. Controlling cell chemistry with caged compounds. *Annu Rev Physiol* **55:** 755–784.

Barbour B. 1993. Synaptic currents evoked in Purkinje cells by stimulating individual granule cells. *Neuron* **11:** 759–769.

Bernardinelli Y, Haeberli C, Chatton JY. 2005. Flash photolysis using a light emitting diode: An efficient, compact, and affordable solution. *Cell Calcium* **37:** 565–572.

Bullen A, Patel SS, Saggau P. 1997. High-speed, random-access fluorescence microscopy I. High-resolution optical recording with voltage-sensitive dyes and ion indicators. *Biophys J* **73:** 477–491.

Civillico EF, Rickgauer JP, Wang SS-H. 2010. Targeting and excitation of photoactivatable molecules: Design considerations for neurophysiology experiments. In *Photosensitive molecules for controlling biological function* (ed. Chambers J, Kramer R), pp. 5–28. Humana, New York.

Duemani Reddy G, Saggau P. 2011. High-speed two-photon imaging. In *Imaging: A laboratory manual* (ed. Yusti R), pp. 831–838. Cold Spring Harbor Laboratory Press, Cold Spring Harbor, NY.

Ellis-Davies GCR. 2007. Caged compounds: Photorelease technology for control of cellular chemistry and physiology. *Nat Methods* **4:** 619–628.

Ellis-Davies GCR. 2011. Nitrobenzyl-based caged neurotransmitters. In *Imaging: A laboratory manual* (ed. Yusti R), pp. 889–896. Cold Spring Harbor Laboratory Press, Cold Spring Harbor, NY.

Furuta T, Wang SS-H, Dantzker JL, Dore TM, Bybee WJ, Callaway EM, Denk W, Tsien RY. 1999. Brominated 7-hydroxycoumarin-4-ylmethyls: Photolabile protecting groups with biologically useful cross-sections for two photon photolysis. *Proc Natl Acad Sci* **96:** 1193–1200.

Gasparini S, Magee JC. 2006. State-dependent dendritic computation in hippocampal CA1 pyramidal neurons. *J Neurosci* **26:** 2088–2100.

Gasparini S, Losonczy A, Chen X, Johnston D, Magee JC. 2007. Associative pairing enhances action potential back-propagation in radial oblique branches of CA1 pyramidal neurons. *J Physiol* **580:** 787–800.

Gurney AM. 1994. Flash photolysis of caged compounds. *Microelectrode techniques. The Plymouth Workshop handbook*, 2nd ed. (ed. Ogden D), pp. 389–406. Company of Biologists, Cambridge.

Harvey CD, Svoboda K. 2007. Locally dynamic synaptic learning rules in pyramidal neuron dendrites. *Nature* **450:** 1195–1202.

Harvey CD, Yasuda R, Zhong H, Svoboda K. 2008. The spread of Ras activity triggered by activation of a single dendritic spine. *Science* **321:** 136–140.

Kandler K, Givens RS, Katz LC. 2000. Photostimulation with caged glutamate. In *Imaging neurons: A laboratory manual* (ed. Yusti R, Lanni F, Konnerth A), pp. 27.1–27.9. Cold Spring Harbor Laboratory Press, Cold Spring Harbor, NY.

Kremer Y, Léger JF, Lapole R, Honnorat N, Candela Y, Dieudonné S, Bourdieu L. 2008. A spatio-temporally compensated acousto-optic scanner for two-photon microscopy providing large field of view. *Opt Express* **16:** 10066–10076.

Lester HA, Nerbonne JM. 1982. Physiological and pharmacological manipulations with light flashes. *Annu Rev Biophys Bioeng* **11:** 151–175.

Matsuzaki M, Ellis-Davies GC, Nemoto T, Miyashita Y, Iino M, Kasai H. 2001. Dendritic spine geometry is critical for AMPA receptor expression in hippocampal CA1 pyramidal neurons. *Nat Neurosci* **4:** 1086–1092.

Nikolenko V, Poskanzer KE, Yuste R. 2007. Two-photon photostimulation and imaging of neural circuits. *Nat Methods* **4:** 943–950.

Rickgauer JP, Tank DW. 2009. Two-photon activation of channelrhodopsin-2 at saturation. *Proc Natl Acad Sci* **106:** 15025–15030.

Saleh BEA, Teich MC. 1991. *Fundamentals of photonics*. Wiley, New York.

Sarkisov DV, Wang SS-H. 2007. Combining uncaging techniques with patch-clamp recording and optical physiology. In *Patch-clamp analysis: Advanced techniques, Neuromethods,* 2nd ed. (ed. Walz W, et al.), Vol. 38, pp. 149–168. Humana Press, Totowa, NJ.

Sarkisov DV, Wang SS-H. 2006. Alignment and calibration of a focal neurotransmitter uncaging system. *Nat Protoc* **1:** 828–832.

Sarkisov DV, Wang SS-H. 2008. Order-dependent coincidence detection in cerebellar Purkinje neurons at the inositol trisphosphate receptor. *J Neurosci* **28:** 133–142.

Sarkisov DV, Gelber SE, Walker JW, Wang SS-H. 2007. Synapse specificity of calcium release probed by chemical two-photon uncaging of inositol 1,4,5-trisphosphate. *J Biol Chem* **282:** 25517–25526.

Shoham S, O'Connor DH, Sarkisov DV, Wang SS-H. 2005. Rapid neurotransmitter uncaging in spatially defined patterns. *Nat Methods* **2:** 837–843.

Sjulson L, Miesenböck G. 2008. Photocontrol of neural activity: Biophysical mechanisms and performance in vivo. *Chem Rev* **108:** 1588–1602.

Sobczyk A, Svoboda K. 2007. Activity-dependent plasticity of the NMDA-receptor fractional Ca^{2+} current. *Neuron* **53:** 17–24.

Stosiek C, Garaschuk O, Holthoff K, Konnerth A. 2003. In vivo two-photon calcium imaging of neuronal networks. *Proc Natl Acad Sci* **100:** 7319–7324.

Tsien RY. 1989. Fluorescent probes of cell signaling. *Annu Rev Neurosci* **12:** 227–253.

Venkataramani S, Davitt KM, Xu H, Zhang J, Song YK, Connors BW, Nurmikko AV. 2007. Semiconductor ultra-violet light-emitting diodes for flash photolysis. *J Neurosci Methods* **160:** 5–9.

Xu H, Zhang J, Davitt KM, Song Y, Nurmikko AV. 2008. Application of blue-green and ultraviolet micro-LEDs to biological imaging and detection. *J Phys London D Appl Phys* **41:** 94013.

Zhang F, Wang LP, Brauner M, Liewald JF, Kay K, Watzke N, Wood PG, Bamberg E, Nagel G, Gottschalk A, Deisseroth K. 2007. Multimodal fast optical interrogation of neural circuitry. *Nature* **446:** 633–639.

Zipfel WR, Williams RM, Webb WW. 2003. Nonlinear magic: Multiphoton microscopy in the biosciences. *Nat Biotechnol* **21:** 1369–1377.

27 Two-Photon Calcium Imaging of Dendritic Spines

Roberto Araya,[1] Jesse H. Goldberg,[2] and Rafael Yuste[1]

[1]*Howard Hughes Medical Institute, Department of Biological Sciences, Columbia University, New York, New York 10027;* [2]*Department of Brain and Cognitive Sciences, Massachusetts Institute of Technology, Cambridge, Massachusetts 02139*

ABSTRACT

Dendritic spines are key structural and functional components of the nervous system, yet are too small for direct electrophysiological probing. Two-photon microscopy has enabled, essentially for the first time, systematic imaging of spines from living, intact tissue. In this chapter, we describe the use of two-photon excitation to image calcium dynamics of spines and to quantify and analyze these measurements.

INTRODUCTION

In the mammalian neocortex, most excitatory synaptic input occurs in dendritic spines. This chapter describes two-photon calcium imaging in spines of living neurons that are physiologically and chemically activated. The technique can be applied to acute slices or organotypic neuronal cultures from the hippocampus, cerebellum, or neocortex (Yuste and Denk 1995; Holthoff et al. 2002; Sabatini et al. 2002; Araya et al. 2006). It uses one or two fluorescence detection channels to provide quantitative estimates of calcium concentration. A method for estimating calcium concentrations from fractional changes in fluorescent light intensity ($\Delta F/F$) is reviewed, along with two spine-stimulation methods for neurons loaded with calcium indicator. It should be emphasized that calcium indicators typically have significantly faster binding to and higher affinity for calcium than endogenous buffers. To minimize unavoidable distortions of natural calcium dynamics, we recommend that small (<50 μM) concentrations of indicator be used when possible.

Protocol

Two-Photon Calcium Imaging of Dendritic Spines

In this protocol, we describe the use of two-photon excitation for imaging calcium signals in dendritic spines, as well as methods to quantify and analyze these measurements.

MATERIALS

See the end of the chapter for recipes for reagents marked with <R>.

Reagents

Caged glutamate (e.g., MNI-glutamate, Tocris) (optional; see Step 5)

Calcium indicators (Invitrogen)

Use the potassium salt form of a calcium indicator in the green emission range (e.g., Fluo-4, Calcium Green-1 [CG-1], or Oregon Green BAPTA-1 [OGB-1]). If using only one (i.e., green emission) detection channel, Alexa Fluor 488 (Invitrogen) may be used in combination with one of these (see Step 1). For the red emission range, use Alexa Fluor 594.

Dextran, Alexa 594 dye (Invitrogen)

Internal recording solution <R>

Reagents for whole-cell patch-clamp recordings or intracellular bolus injection (see Steps 1–3 and Chapter 35)

Equipment

Equipment for whole-cell patch-clamp recordings or intracellular bolus injection (see Steps 1–3 and Chapter 35)

ISO-Flex stimulus isolator (A.M.P.I.) for extracellular stimulation of spines

Microforge (Narishige) (optional; see Step 4)

Microscope

Use a two-photon microscope (see Nikolenko and Yuste 2011) equipped with a recording chamber and, for two-channel calcium imaging, a dichroic mirror, which separates red and green light (e.g., 565 DCXR; Chroma Technology). At an angle of 45°, the dichroic mirror transmits ~80% of light between ~580 nm and 880 nm (IR from the laser to the sample and red Alexa 594 fluorescence back to the internal photomultiplier tube [PMT]; red channel) and reflects light below 530 nm (emission from calcium indicator; green channel) to the external PMT. We place an additional band-pass 510/40 filter (which transmits between 490 nm and 530 nm) in front of the external PMT to reduce scattered red light. Alternatively, a simplified, one-channel recording system, with one external PMT collecting green emission from both calcium indicator and Alexa dye, can be used.

Two-photon uncaging setup for single-spine uncaging of glutamate (optional; see Step 5 and Chapter 25)

EXPERIMENTAL METHOD

Loading Neurons with Calcium Indicators and Alexa Dyes

Individual cells can be loaded via patch pipette with calcium indicator and Alexa dyes in a recording chamber as described in Steps 1 and 2 (see also Chapter 35). Alternatively, if the dialysis that

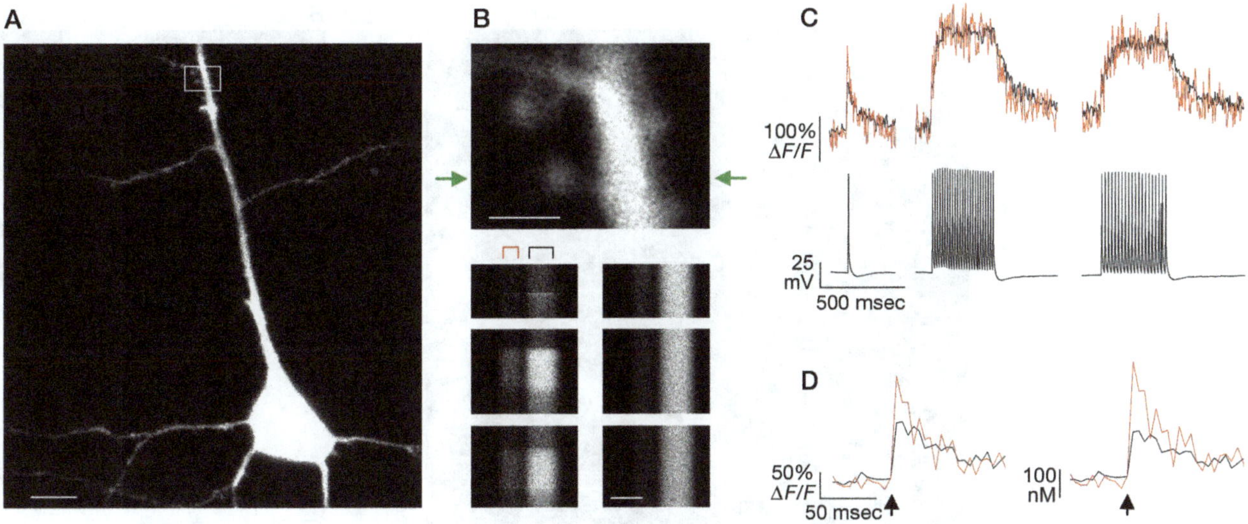

FIGURE 1. Analysis of two-channel calcium imaging of spines and dendrites of neurons dual-labeled with 25 µM Alexa Fluor 594 and 50 µM Fluo-4. (*A*) The morphology of an L2/3 pyramidal neuron from a p11 mouse visual cortex slice is visualized with the calcium-independent Alexa 594 fluorophore recorded on the red channel. (*B*) (*Top panel*) A higher magnification of the region of interest (boxed area in *A*). The green arrows indicate the spine and dendrite from which the readings were taken. (*Bottom, left panels*) Line scans (1 or 2 sec long) through the spine (red bracket) and dendrite (black bracket) performed using the calcium-dependent green channel during one AP (*top*), and during the 66-Hz (*middle*) and 50-Hz (*bottom*) trains. (*Bottom, right panels*) The Alexa Fluor 594 signal (red channel), which does not change in fluorescence, reflecting its insensitivity to calcium. (*C*) Calcium fluorescence signals acquired from the green channel over the spine (red lines) and dendrite (black lines) from *B* are plotted above time-locked APs. $\Delta F/F_{max}$ was estimated from the $\Delta F/F_{plateau}$ reached during the 66-Hz (v_2) (*middle*) and 50-Hz (v_1) (*right*) AP trains. (*D*) Calcium signal during a single AP is plotted as %ΔF/F (*left*) and as $\Delta[Ca^{2+}]$ (*right*). Arrowheads indicate timing of AP generation. Note how the spine $\Delta[Ca^{2+}]$ signal appears "stretched" compared to the %ΔF/F signal, reflecting how ΔF/F sublinearly reflects $\Delta[Ca^{2+}]$ as indicator saturation is approached. Scale bars, 10 µm (*A*); 1 µm (*B*).

occurs during sustained whole-cell recording is a concern (e.g., during spine motility experiments), calcium indicator can be loaded using intracellular bolus injection as described in Step 3 (Helmchen et al. 1996; Majewska et al. 2000).

1. Load the neurons using one of the following approaches.
   - When using two fluorescence detection channels (i.e., red and green emission channels) (Fig. 1), patch the neuron of choice with a whole-cell recording electrode filled with 20–100 µM green-emission calcium indicator and 20–100 µM Alexa Fluor 594 in standard internal recording solution.
   - If using only one (i.e., green emission) detection channel (Fig. 2A–C), patch the neuron with a pipette filled with >100 µM green emission calcium indicator (e.g., 200 µM Fluo-4) and 200 µM Alexa Fluor 488 in standard internal recording solution.

     This combination better resolves spines and dendrites. Alternatively, use only a calcium indicator, such as CG-1 (>100 µM), with strong fluorescence at resting calcium concentration; this is sufficient to resolve the structure of dendrites and spines.
2. After 20–30 min of recording, visualize the dendrites and spines (between 20 and 150 µm from soma) on the red channel or on the green channel. Use the green channel to image calcium changes in spines during back-propagation of action potentials after current injections in the soma, extracellular stimulation of spines, or two-photon uncaging of glutamate (see Steps 4 and 5).
3. For the intracellular bolus injection technique, prepare millimolar concentrations (e.g., 1–5 mM) of calcium indicator in internal recording solution, and hold the cell in whole-cell mode

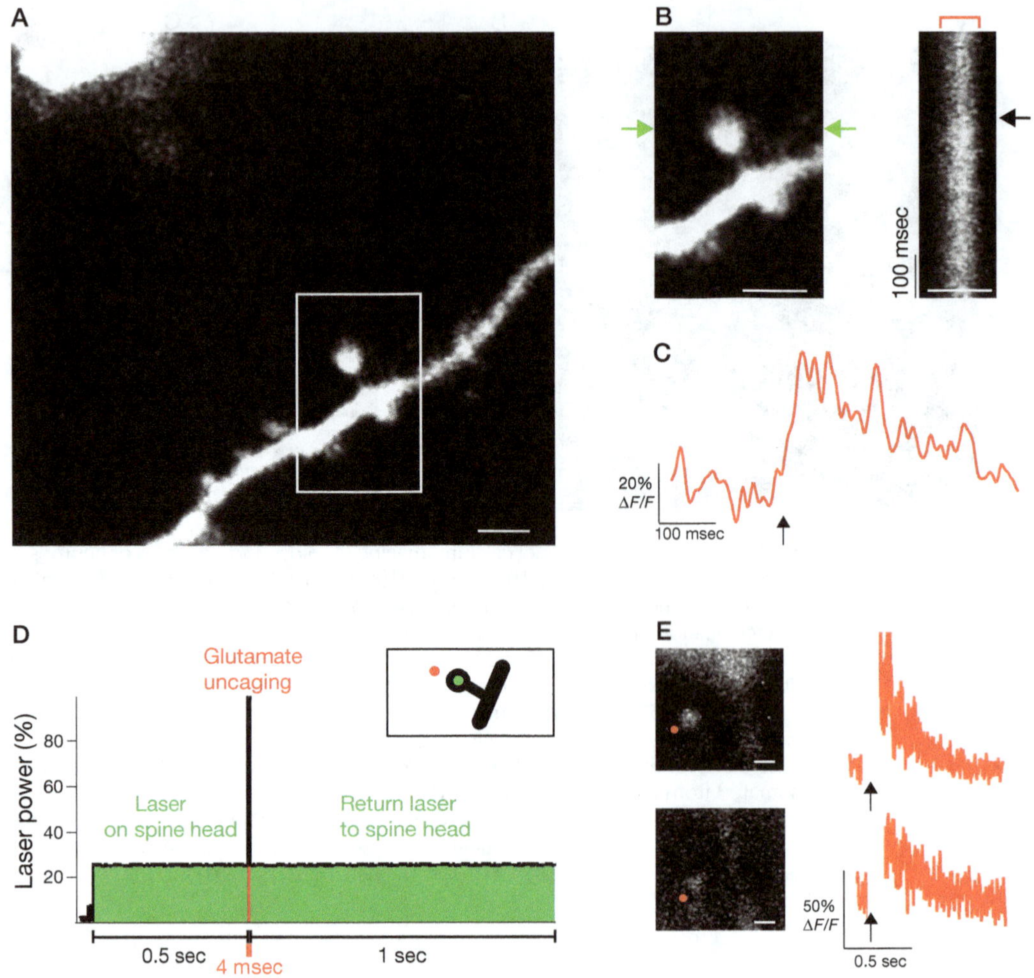

FIGURE 2. (*A–C*) Analysis of one-channel calcium imaging of spines and dendrites of neurons labeled with 200 µM Alexa Fluor 488 and 200 µM Fluo-4. (*A*) The morphology of a basal dendrite from a layer 5 pyramidal neuron from a p14 mouse visual cortex was imaged using the green emission detection channel. (*B*) (*Left panel*) A higher magnification of the region of interest (boxed area in *A*). Line scans through the spine head, at positions indicated by the green arrows, were recorded using the calcium-dependent green channel. The line scan in the *right panel* was generated before (above black arrow) and after (below black arrow) a 20-µsec extracellular stimulation delivered through an extracellular electrode (the electrode is shown in the *upper, left* corner of *A*). (*C*) A plot from the calcium fluorescence line scan image from *B*. The black arrow indicates the time of the extracellular stimulation of the spine. (*D*) The protocol for the measurement of intracellular calcium at the head of the spine before and after two-photon uncaging of glutamate. (*E*) Two examples of spines from basal dendrites of layer 5 pyramidal cell filled with 200 µM CG-1. The red traces corresponded to the average measurements of intracellular calcium in the head of the spine in response to two-photon uncaging of glutamate. Scale bars, 1 µm.

for 3 min. When patching the cell, target a region of the cell body that does not overlie the nucleus; this will ensure cell viability after electrode withdrawal.

Twenty minutes after withdrawal, the dendritic tree should be loaded satisfactorily and will be ready for calcium imaging.

Imaging Calcium Transients in Dendritic Spines

4. Electrically stimulate spines.

   i. Position a stimulating, high-resistance electrode (>20 mΩ) filled with dextran-Alexa 594 dye (10,000 MΩ, 50 µM) close (5–15 µm) to the spine head that you plan to record (Fig. 2A).

This will increase the probability of activating an axon contacting the selected spine. If desired, bend the tips of the stimulation pipettes by 50°–70° with a microforge to allow positioning of the stimulation pipette perpendicular to the slice surface; this will minimize tissue movement if simultaneous whole-cell recording is being performed.

ii. Deliver short stimulating pulses (100 µsec) to the pipette through an ISO-Flex stimulus isolator.

The calcium signals are recorded with either a high-sensitivity EM-CCD camera or by using a PMT to collect the fluorescence signal arising from a line scan intersecting the spine head (e.g., see the right panel of Fig. 2B).

5. Perform two-photon uncaging of glutamate (Fig. 2D,E).

i. Add caged glutamate to recirculating oxygenated medium at a final concentration of 2.5 mM for MNI-glutamate.

ii. Select the spine you want to image. Then, position the laser at the spine head for ~500 msec at low imaging power, and generate a 2–4-msec high-power uncaging laser pulse (25–30 mW on sample) at a position next to the spine head (see the red dots in Fig. 2D,E). Immediately reposition the laser beam for 1 sec at the same spine head location for calcium imaging again at low laser power.

The calcium fluorescence signals are collected with a PMT.

iii. Calculate the average calcium signals from spine heads before and after uncaging of glutamate, and use these values to determine the percentage change in basal fluorescence as $\Delta F/F$.

Estimating Changes in Calcium Concentration from $\Delta F/F$

Maravall et al. (2000) developed a method for estimating calcium concentrations that is ideal for use during two-photon excitation of calcium indicators with large dynamic ranges (such as Fluo-4 and OGB-1). Briefly, the approach takes advantage of the ability to estimate $\Delta F/F_{max}$ during experiments by achieving near-indicator saturation with strong stimulation, such as during high-frequency action potential (AP) trains.

By comparing the $\Delta F/F_{plateau}$ at two frequencies, $v_2 > v_1$, the percentage, x, of true indicator saturation ($\Delta F/F_{max}$) can be estimated using

$$x = 100 \cdot \frac{1 - Q_{v1/v2}}{1 - v_1/v_2},$$

where Q equals the $\Delta F/F$ plateau reached during v_2 divided by that reached during v_1 (see Fig. 1C). Using this equation it is possible to estimate the maximum fluorescence transient as

$$\Delta F_{max} = \Delta F_{plateau} \cdot 100/x.$$

Equipped with a derived value of $\Delta F/F_{max}$, solve for $\Delta[Ca^{2+}]$:

$$\Delta[\mathrm{Ca}] = K_d \left[\left((\Delta F/F)_{max} + 1 \right) \left(1 - R^{-1} \right) \left(\frac{\Delta F/F}{\left((\Delta F/F)_{max} - \Delta F/F \right) (\Delta F/F)_{max}} \right) \right]$$

where R is the dynamic range of the indicator and K is its dissociation constant (Maravall et al. 2000; Sabatini et al. 2002). An alternative method was pioneered by Feller and coworkers (Feller et al. 1996). For additional information on how to calibrate calcium indicators, see Helmchen et al. (1996) and Helmchen (2011).

DISCUSSION

Calcium imaging in small structures such as dendrites and spines is complicated by the fact that the calcium indicators, by binding calcium, effectively perturb calcium efflux and diffusion. To better approximate physiological conditions, exogenous buffer capacity must be reduced by using low-affinity indicators, such as Oregon Green BAPTA-6F, or, as described here, low indicator concentrations. One problem associated with low indicator concentrations is insufficient resting fluorescence and, hence, poor visualization of the dendritic tree. This problem can be solved by using a separate channel to visualize morphology with a calcium-insensitive fluorophore that has nonoverlapping emission.

An important consideration when working with low indicator concentrations, however, is indicator saturation. This can be advantageous when examining small calcium influxes, such as during single APs, because it allows estimation of $\Delta F/F_{max}$, and therefore $\Delta[Ca^{2+}]$. However, if large calcium signals, such as during *N*-methyl-D-aspartate (NMDA) receptor activation or release from intracellular stores, are of interest, we recommend the use of higher indicator concentrations, a simplified one-channel detection system, or low-affinity indicators (e.g., Oregon Green BAPTA-6F).

CG-1, OGB, and Fluo-4 have all been used successfully in single-channel imaging, and all three indicators are well excited by a mode-locked laser at 800 nm. However, each indicator is suited to different conditions. When imaging with one channel (using the indicator's resting fluorescence to visualize the dendritic tree), CG-1 and OGB (150–200 µM) are recommended for their strong fluorescence at resting calcium concentration. However, partly because of their strong fluorescence, their increase in intensity on binding calcium is compromised. These indicators are, therefore, unsuitable for detecting small signals. Fluo-4 has low fluorescence at resting calcium concentration. This means that higher concentrations (200–400 µM) are necessary for one-channel use, but Fluo-4 has a large dynamic range and is ideal for imaging small or heavily buffered signals, such as single APs in cortical interneurons (Goldberg et al. 2003). Fluo-4 is also a good indicator to use in tandem with Alexa 488 or 594 during one- or two-channel image acquisition.

RECIPE

Recipes for reagents marked with <R> are included in this list.

Internal Recording Solution

Reagent	Final Concentration
$KMeSO_4$	130 mM
KCl	5 mM
NaCl	5 mM
HEPES	10 mM
Mg-ATP	2.5 mM
GTP	0.3 mM

Titrate the solution to pH 7.3.

REFERENCES

Araya R, Jiang J, Eisenthal KB, Yuste R. 2006. The spine neck filters membrane potentials. *Proc Natl Acad Sci* **103:** 17961–17966.

Feller MB, Delaney KR, Tank DW. 1996. Presynaptic calcium dynamics at the frog retinotectal synapse. *J Neurophysiol* **76:** 381–400.

Goldberg JH, Tamas G, Yuste R. 2003. Ca^{2+} imaging of mouse neocortical interneurone dendrites: Ia-type K^+ channels control action potential backpropagation. *J Physiol* **551:** 49–65.

Helmchen F. 2011. Calibration of fluorescent calcium indicators. In *Imaging: A laboratory manual* (ed. Yuste R), pp. 403–416. Cold Spring Harbor Laboratory Press, Cold Spring Harbor, NY.

Helmchen F, Imoto K, Sakmann B. 1996. Ca^{2+} buffering and action potential–evoked Ca^{2+} signaling in dendrites of pyramidal neurons. *Biophys J* **70:** 1069–1081.

Holthoff K, Tsay D, Yuste R. 2002. Calcium dynamics of spines depend on their dendritic location. *Neuron* **33:** 425–437.

Majewska A, Tashiro A, Yuste R. 2000. Regulation of spine calcium compartmentalization by rapid spine motility. *J Neurosci* **20:** 8262–8268.

Maravall M, Mainen ZF, Sabatini BL, Svoboda K. 2000. Estimating intracellular calcium concentrations and buffering without wavelength ratioing. *Biophys J* **78:** 2655–2667.

Nikolenko V, Yuste R. 2011. How to build a two-photon microscope with a confocal scan head. In *Imaging: A laboratory manual* (ed. Yuste R), pp. 111–116. Cold Spring Harbor Laboratory Press, Cold Spring Harbor, NY.

Sabatini BL, Oertner TG, Svoboda K. 2002. The life cycle of Ca^{2+} ions in dendritic spines. *Neuron* **33:** 439–452.

Yuste R, Denk W. 1995. Dendritic spines as basic units of synaptic integration. *Nature* **375:** 682–684.

28 Imaging Calcium Waves and Sparks in Central Neurons

William N. Ross and Satoshi Manita[1]

Department of Physiology, New York Medical College, Valhalla, New York 10595

ABSTRACT

This chapter describes wide-field charge-coupled device (CCD) camera-based imaging methods that are used to detect the spatial and temporal aspects of calcium release from internal stores in dendrites of neurons in brain slice preparations. This approach is useful for revealing aspects of this signaling system, which is generally invisible to electrical recording. The changes in intracellular calcium ion concentrations, $[Ca^{2+}]_i$, sometimes occur as large-amplitude, propagating Ca^{2+} waves or as much smaller, localized events (sparks). Wide-field CCD camera systems have some advantages for detecting and analyzing these $[Ca^{2+}]_i$ changes because the waves are spatially extended and the sparks do not always occur at the same locations.

INTRODUCTION

Calcium waves and randomly generated Ca^{2+} release events (sparks [e.g., Cheng and Lederer 2008] or puffs [e.g., Sun et al. 1998]) are examples of signals that occur over spatially extended regions of cells including neurons. When and where they occur in the dendrites is not predictable. Single detectors (photodiodes or photomultipliers) have high temporal resolution but cannot follow the propagation of a wave or detect events at several locations. Confocal detectors, operated in the line-scan mode, also have high temporal resolution, but the "line" cannot easily track the complex geometry of the dendrites. Even video-rate confocals, which may have adequate temporal and spatial resolution for some of these measurements, are not ideal because the dendrites do not lie in the plane of a single *z* section. These instruments can only follow Ca^{2+} events in a restricted spatial region. In contrast, high-speed, cooled charge-coupled device (CCD) cameras, which can image spatially extended regions, are good instruments for examining these dendritic $[Ca^{2+}]_i$ changes. Here we describe the use of wide-field CCD camera–based imaging methods to detect the spatial and temporal aspects of calcium release from internal stores in dendrites of neurons in brain slice preparations.

[1]Current address: Brain Science Institute, RIKEN, Wako, Saitama, Japan.

Protocol

Loading Cells and Detecting $[Ca^{2+}]_i$ Changes

In this protocol, a cell is loaded with an indicator that responds to Ca^{2+}, waves or sparks are stimulated in the cell, and the spatial and temporal characteristics of calcium release from internal stores in the cell are detected using wide-field CCD camera–based imaging methods.

MATERIALS

CAUTION: See Appendix 6 for proper handling of materials marked with <!>.

Reagents

Ca^{2+} indicator, membrane-impermeant

Many different indicators can be used to stain the cytosol of the neurons of interest. There is usually a trade-off between an indicator with high sensitivity (high affinity [e.g., bis-Fura-2 or OGB-1]), which may buffer and distort transient changes in $[Ca^{2+}]_i$, and an indicator with low sensitivity (low affinity [e.g., Furaptra or OGB-5N]), which follows changes in $[Ca^{2+}]_i$ more accurately. Ca^{2+} waves are particularly sensitive to the buffering effects of indicators because they involve the regenerative release of Ca^{2+} from internal stores, so low-affinity indicators are preferred for detecting waves. On the other hand, high-affinity indicators are preferred for detecting small, localized events.

Intracellular solution

Typical intracellular (pipette) solution contains 145 mM potassium gluconate <!>, 4 mM NaCl, 4 mM Mg–ATP, 0.3 mM Na–GTP, 14 mM Na–phosphocreatine, and 10 mM HEPES, pH adjusted to 7.3 with KOH <!>. Final osmolarity typically is ~297 mOs. This solution is supplemented with Ca^{2+} indicators (see below). Note that there are no other added Ca^{2+} buffers.

Equipment

Camera and Software

A high-speed CCD camera and appropriate software are used to correlate electrophysiological and optical recordings. The Quantix 57 or Evolve 512 (Photometrics), both with frame-transfer CCD chips, and the NeuroCCD-SMQ (RedShirt Imaging) are recommended. These low-noise cameras have high quantum efficiency in the appropriate wavelength range and have sufficient spatial and temporal resolution for interesting measurements. Our laboratory custom-wrote the software for the Photometrics cameras (an early version is described in Lasser-Ross et al. 1991). For the NeuroCCD-SMQ, we use Neuroplex (RedShirt Imaging).

Electrical Recording Device

Most conventional bridge amplifiers (e.g., Axoclamp 900A, Multiclamp 700B, or equivalent) and analog-to-digital converters can be used to record the membrane current and voltage. It is important that the software be designed to synchronize the optical and electrical data streams.

Filter Sets

Standard Fura-2 or fluorescein filter sets can be used for the different indicators. Although there is some variation in the bandwidth peak transmission and sharpness of filters from different manufacturers, the choice of filter set is usually not a critical variable in obtaining good results.

Microscope

We use an Olympus BX50WI equipped with an epifluorescence port, computer-controlled shutter, 75-W xenon arc lamp, and 40× and/or 60× water-immersion lenses. Equivalent microscopes from other manufacturers could be used. The usual requirements for a slice patching rig are appropriate. The microscope should be stable, have good transmission characteristics, and have differential interference contrast (DIC) to improve cell visibility when patching.

Stimulating Electrode

A sharpened tungsten electrode (e.g., TM33B01KT, World Precision Instruments) or a blunt patch pipette filled with artificial cerebrospinal fluid (ACSF) works well for generating synaptic responses.

EXPERIMENTAL METHOD

1. Prepare a high-concentration stock solution of indicator in distilled H_2O and add an aliquot of the stock to the appropriate standard intracellular solution.

   Recommended final concentrations of some indicators are as follows: 100–250 μM bis-Fura-2, 25–100 μM OGB-1, 500 μM Furaptra, and 300 μM OGB-5N. Because the final concentration of the indicator is much lower than the concentrations of the other solution components, there is no need to correct for osmotic changes.

2. Use patch pipettes to load the neurons with the chosen Ca^{2+} indicator. Leave the loading electrode in place and use it to record the electrical response from the neuron.

   The use of patch pipettes for loading cells with Ca^{2+} indicators is now a standard technique (see Chapter 35). Patch pipettes are preferred to other loading methods because the concentration of indicator can be controlled. See Nakamura et al. (1999).

3. Place an extracellular stimulating electrode in the dendritic field near the cell and within 100 μm of the soma. Evoke Ca^{2+} waves by applying repetitive high-frequency stimulation, typically 100 Hz for 0.5 sec. Usually the current intensity can be kept below the threshold for evoking spikes in the patched neuron.

   Lower intensity and lower frequency stimulation will evoke isolated sparks. See Zhou and Ross (2002) for an analysis of effective stimulation parameters for waves. See Manita and Ross (2009) for effective parameters for sparks.

4. Detect a Ca^{2+} wave as a propagating change in fluorescence that corresponds to changes in $[Ca^{2+}]_i$. Apply standard corrections for indicator bleaching and background fluorescence.

   The magnitude of the $[Ca^{2+}]_i$ change can be estimated from the fractional change in fluorescence using standard formulas (see, e.g., Nakamura et al. 1999). Single-wavelength measurements are necessary if high time resolution and maximum signal-to-noise ratios (SNRs) are desired.

DISCUSSION

Camera Considerations

High speeds with CCD cameras are achieved with frame transfer chips and pixel binning or with the use of special lower-resolution chips. The camera shutter is never closed with this technique, maximizing the total amount of light detected and improving the SNR of the measurement. The low noise, linearity, and high digital accuracy that are characteristic of scientific-grade cameras are preserved. Developments in CCD camera technology since this technique was originally described (Lasser-Ross et al. 1991) have greatly improved the spatial and temporal resolution of the data.

CCD cameras can detect fluorescence changes over an extended area of the tissue slice with a greater depth of field than a laser scanning system. On the downside, this increased detection volume does lead to a loss of spatial resolution and, consequently, the possible misassignment of the signal to the wrong dendritic branch or the inability to localize a signal to a spine or a presynaptic terminal region.

Examples of Application

This approach has been used in our laboratory to detect calcium waves (Nakamura et al. 1999, 2002; Larkum et al. 2003) and sparks (Manita and Ross 2009) in pyramidal neurons from the rat hippocampus and somatosensory cortex. Similar approaches were used in other laboratories to detect waves in principal neurons from the amygdala (Power and Sah 2007), midbrain dopamine neurons (Morikawa et al. 2003), and prefrontal cortical neurons (Stutzmann et al. 2003; Hagenston et al. 2008).

Figure 1 shows a Ca^{2+} wave evoked in the dendrites of a neocortical pyramidal neuron by repetitive stimulation. The synaptic response generated a delayed, large-amplitude response that was not synchronous and which extended over a limited range of the dendritic field. The high temporal and spatial resolution of the measurements clearly show that the waves were initiated at several locations near branch points. The "line-scan" display is particularly useful for following wave propagation.

Figure 2A shows examples of localized Ca^{2+} release events in the dendrites of pyramidal neurons. The "line scan" through the apical dendrite shows a series of events at different locations and times.

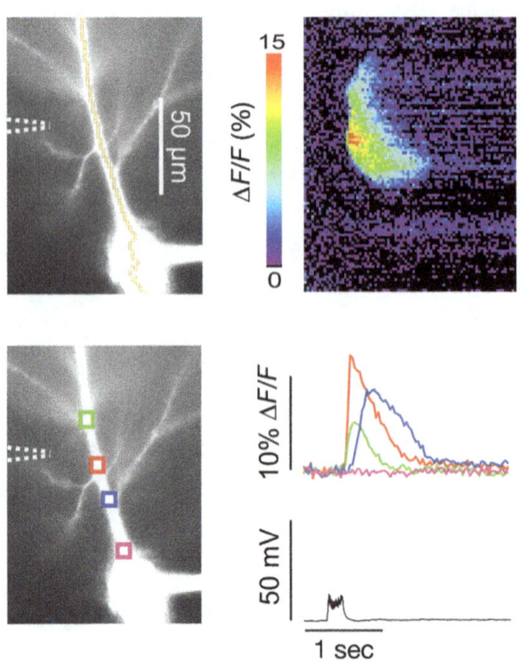

FIGURE 1. A synaptically activated Ca^{2+} wave in a neocortical pyramidal neuron. The cell, which was from a slice of the cortex from a 3-wk-old rat, was patched under DIC optics. It was loaded with the low-affinity indicator Furaptra (300 μM) via a patch electrode on the soma (visible to *right*), and stimulated via a tungsten electrode (dotted arrow) at 100 Hz for 0.25 sec. Two views of the resulting changes in $[Ca^{2+}]_i$ are shown. The *bottom panels* show the time course of the fluorescence changes at the regions of interest (ROIs) (indicated by colored rectangles). The time courses and amplitudes of the delayed responses differ at the four ROIs. The *top panels* show a "line scan" of the same data. A selected series of pixels along the somato-dendritic axis, as shown on the cell image, was chosen for analysis. The pseudocolor image shows the time course of the $[Ca^{2+}]_i$ changes at each of these locations. The timescale is the same as for the optical and electrical traces shown below. With this view, it is clear that two waves initiated at different locations in the dendrites (close to branch points) and propagated along the main dendrite in both directions. In this cell, the waves did not propagate into the soma.

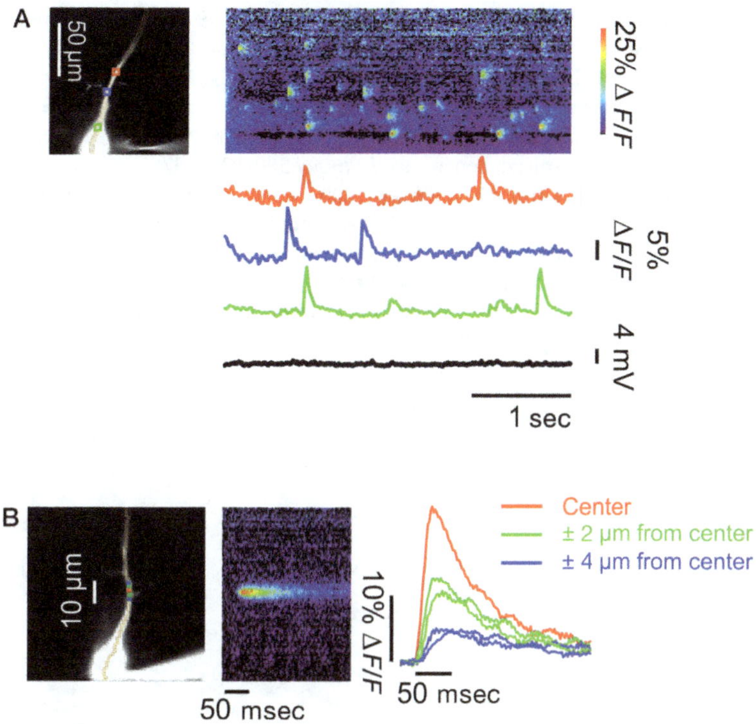

FIGURE 2. (*A*) Spontaneous Ca^{2+} release events occur in localized regions of the dendrites of hippocampal pyramidal neurons. The image shows a pyramidal neuron filled with 50 µM OGB-1. Three regions of interest (ROIs) are marked. The string of pixels indicates the locations of the line-scan images. In normal ACSF, spontaneous increases in $[Ca^{2+}]_i$ were detected asynchronously at the three locations. There was no corresponding change in membrane potential. The pseudocolor "line-scan" image shows the increases at all locations along the dendrite. (*B*) Averaged records of event signals from nearby locations on a pyramidal cell dendrite. The data were recorded at 500 Hz and spontaneous event signals from 11 events were aligned at the time of the start of the rising phase of the fluorescence transient. The resulting traces were filtered with a three-point interpolation algorithm. The pseudocolor image shows the "line scan" of the averaged signal. The traces show the signal at the center and neighboring (±2 µm, ±4 µm) locations. (Adapted, with permission, from Manita and Ross 2009.)

The signals at three selected regions of interest (ROIs) show the time course of some of these events. Figure 2B shows a spontaneous event at one location recorded at 500 Hz with the NeuroCCD-SMQ camera. The event initiated at one ~2 µm-long location and spread by diffusion over a larger region. The high speed and sensitivity of the camera was important for resolving the different time courses in the nearby pixels.

REFERENCES

Cheng H, Lederer WJ. 2008. Calcium sparks. *Physiol Rev.* **88:** 1491–1545.

Hagenston AM, Fitzpatrick JS, Yeckel MF. 2008. MGluR-mediated calcium waves that invade the soma regulate firing in layer V medial prefrontal cortical pyramidal neurons. *Cereb Cortex* **18:** 407–423.

Larkum ME, Watanabe S, Nakamura T, Lasser-Ross N, Ross WN. 2003. Synaptically activated Ca^{2+} waves in layer 2/3 and layer 5 rat neocortical pyramidal neurons. *J Physiol* **549:** 471–488.

Lasser-Ross N, Miyakawa H, Lev-Ram V, Young SR, Ross WN. 1991. High time resolution fluorescence imaging with a CCD camera. *J Neurosci Methods* **36:** 253–261.

Manita S, Ross WN. 2009. Synaptic activation and membrane potential changes modulate the frequency of spontaneous elementary Ca^{2+} release events in the dendrites of pyramidal neurons. *J Neurosci* **29:** 7833–7845.

Morikawa H, Khodakhah K, Williams JT. 2003. Two intracellular pathways mediate metabotropic glutamate receptor-induced Ca^{2+} mobilization in dopamine neurons. *J Neurosci* **23:** 149–157.

Nakamura T, Barbara J-G, Nakamura K, Ross WN. 1999. Synergistic release of Ca^{2+} from IP_3-sensitive stores evoked by synaptic activation of mGluRs paired with backpropagating action potentials. *Neuron* **24:** 727–737.

Nakamura T, Lasser-Ross N, Nakamura K, Ross WN. 2002. Spatial segregation and interaction of calcium signalling mechanisms in rat hippocampal CA1 pyramidal neurons. *J Physiol* **543:** 465–480.

Power JM, Sah P. 2007. Distribution of IP_3-mediated calcium responses and their role in nuclear signalling in rat basolateral amygdala neurons. *J Physiol* **580:** 835–857.

Stutzmann GE, LaFerla FM, Parker I. 2003. Ca^{2+} signaling in mouse cortical neurons studied by two-photon imaging and photoreleased inositol triphosphate. *J Neurosci* **23:** 758–765.

Sun XP, Callamaras N, Marchant JS, Parker I. 1998. A continuum of $InsP_3$-mediated elementary Ca^{2+} signalling events in *Xenopus* oocytes. *J Physiol* **509:** 67–80.

Zhou S, Ross WN. 2002. Threshold conditions for synaptically evoking Ca^{2+} waves in hippocampal pyramidal neurons. *J Neurophysiol* **87:** 1799–1804.

29 Dendritic Voltage Imaging

Knut Holthoff,[1] Marco Canepari,[2] Kaspar Vogt,[2] Arthur Konnerth,[3] and Dejan Zecevic[4]

[1]Hans-Berger-Klinik für Neurologie, Friedrich-Schiller-Universität Jena, Jena, Germany; [2]Division of Pharmacology and Neurobiology, Biozentrum–University of Basel, Basel, Switzerland; [3]Center for Integrated Protein Science and Institute of Neuroscience, Technical University Munich, Munich, Germany; [4]Department of Cellular and Molecular Physiology, Yale University School of Medicine, New Haven, Connecticut 06510

ABSTRACT

Understanding the biophysical properties and functional organization of single neurons and how they process information is fundamental to understanding how the brain works. Regional electrical properties of branching neuronal processes, which determine the input–output function of any neuron, are extraordinarily complex, dynamic, and, in the general case, impossible to predict in the absence of spatially well-resolved measurements. To obtain such a measurement, one would, ideally, like to be able to monitor, at multiple places, subthreshold events as they travel from the sites of origin on neuronal processes and dendritic spines and summate at particular locations to influence action potential initiation. It recently became possible to carry out this type of measurement from vertebrate neurons using high-resolution multisite optical recording of membrane potential changes with voltage-sensitive dyes. This chapter describes the method of voltage-sensitive dye recording from individual neurons, which allows one to monitor membrane potential transients from dendritic and axonal processes at single spine resolution.

INTRODUCTION

A central question in neuronal network analysis is how the interaction between individual neurons produces behavior and behavioral modifications. This task depends critically on exactly how signals are integrated by individual nerve cells functioning as complex operational units. Because the primary function of any nerve cell is to process electrical signals (i.e., membrane potential transients), usually from multiple sources, there is a need for detailed spatiotemporal analysis of electrical events in thin axonal and dendritic processes. The conventional patch-electrode measurements are limited in their capacity for assessing changing spatiotemporal patterns of signal initiation and propagation in complex dendritic and axonal processes. Moreover, many subcellular structures, including small-diameter terminal dendritic branches as well as dendritic spines and most axon terminals and axon collaterals, are inaccessible to electrodes. To overcome these limitations, it was highly desirable to complement the patch-electrode approach with technologies that permit extensive parallel recordings from all parts of a neuron with adequate spatial and temporal resolution. This can now be accomplished using optical recording of membrane potential changes with organic voltage-sensitive dyes (V_m imaging). The sensitivity of this measurement technique has recently reached a level that permits single-trial optical recordings of membrane potential transients from all parts of a neuron, including axon terminals and collaterals, terminal dendritic branches, and individual dendritic spines.

Areas of Application

Voltage imaging from individual nerve cells selectively stained by intracellular injection of fluorescent dyes has been used in the analysis of the electrical properties and the functional organization of identified neurons in invertebrate ganglia (Antic and Zecevic 1995; Zecevic 1996; Antic et al. 2000), as well as in experiments on several classes of vertebrate nerve cells, including neocortical pyramidal neurons (Antic et al. 1999; Antic 2003; Palmer and Stuart 2006, 2009; Zhou et al. 2008), mitral cells of the olfactory bulb (Djurisic et al. 2004, 2008), CA1 pyramidal neurons of the hippocampus (Canepari et al. 2007), and Purkinje cells in the cerebellum (Canepari and Vogt 2008).

IMAGING SETUP

Voltage imaging from individual vertebrate neurons in brain slices requires an optical monitoring apparatus that includes the equipment for patch-pipette recordings under infrared video microscopy. In our laboratory, the experimental setup is organized around a compound upright microscope (Olympus BX51WI) equipped with a moveable head; a stationary rigid stage; dark-field, bright-field, IR-DIC, and epifluorescence optics; three camera ports; and two micromanipulators mounted on the fixed stage.

One camera port has the conventional CCD camera (CCD-300; Dage-MTI) for visualizing neurons in slices under infrared-DIC illumination. The second camera port has a data acquisition CCD. We use the NeuroCCD-SM Imaging System (RedShirtImaging LLC) based on an 80 × 80-pixel, frame-shift, back-illuminated, cooled CCD characterized by 14-bit A-to-D resolution, well depth of 200,000 electrons/msec, frame shift time of 7 µsec, and read noise of 8 electrons rms at a 1-kHz frame rate. It is also possible to increase the frame rate from 2 kHz (full frame) to 3 kHz (2 × 2 binned), 5 kHz (partial readout; 80 × 28 pixels), and 10 kHz (partial readout; 80 × 12 pixels). The third camera port in our system has a CCD camera with high spatial resolution (1392 × 1024 pixels) mounted on a spinning-disk confocal scanner (Model CSU-10; Yokogawa) used to collect *z*-stacks of confocal images for detailed morphological reconstruction of the stained cell. The final refinement of the anatomy of dendrites and spines is obtained by sharpening the confocal image using deconvolution software (Huygens, Scientific Volume Imaging). The microscope has to be placed on the vibration isolation table to reduce vibration noise in optical recording. We use Model 350 BM-1 Nano-K vibration isolation systems (http://www.minusk.com). The analysis and display of data are made using the NeuroPlex program (RedShirtImaging, LLC) written in IDL (Interactive Data Language; ITT Visual Information Solutions) running on an IBM-PC computer. A small chamber with a brain slice is placed on the stage of the microscope, and the fluorescence image of the stained neuron is projected by a water-immersion objective onto the fast data acquisition CCD. Nikon 40×/0.9 NA or 60×/1.0 NA water-immersion objectives were used in different measurements. These objectives were selected as a compromise between imaging area, spatial resolution, and the signal-to-noise ratio (SNR). Initially, as in Figure 1, a 250-W xenon, short-gap arc lamp (Osram; XBO 250 W/CR ORF) powered by a low ripple power supply (Model 1700XT/A; Opti-Quip) was used, in combination with a 520/90 interference filter, as a source for excitation light. In recent experiments, as in Figure 2, the arc lamp has been replaced with a frequency-doubled 200-mW diode-pumped Nd:YVO_4 continuous wave (CW) laser emitting at 532 nm (Excelsior 532 single mode; Newport-Spectra-Physics). The laser beam is directed to a light guide coupled to the microscope via a single-port epifluorescence condenser (TILL Photonics GmbH) designed to overfill the back aperture of the objective. In this way, approximately uniform illumination of the object plane can be attained. High-intensity monochromatic excitation light from a laser at a wavelength close to the extreme red edge of the dye absorption spectrum increased the sensitivity of recording by a factor between 10 and 40. The best optical signals were obtained using a dichroic mirror with the central wavelength of 570 nm and a 610-nm barrier filter (Schott RG610).

Protocol

Staining Individual Vertebrate Neurons

To image dendritic events at single-spine resolution, neurons are infused with voltage-sensitive dyes that are subsequently viewed using infrared video microscopy coupled with patch-clamp recording. This protocol describes a method for staining neurons with a voltage-sensitive dye.

MATERIALS

CAUTION: See Appendix 6 for proper handling of materials marked with <!>.
See the end of the chapter for recipes for reagents marked with <R>.

Reagents

Halothane anesthesia <!>
Standard extracellular solution for vertebrate neurons <R>
Standard intracellular solution for vertebrate neurons <R>
Voltage-sensitive dye solution for vertebrate neurons

Dissolve 1 mg/mL voltage-sensitive dye JPW1114 (D6923; Invitrogen/Molecular Probes) in the standard intracellular solution. Before filling the electrodes, filter the dye solution to eliminate microscopic particles. Use Millex-GV4 filters with 0.22-μm pore size (Millipore).

Equipment

DIC video microscope (see Imaging Setup above for more details)
Incubator or heating apparatus set at 35°C
Patch pipettes, 5–7 MΩ
Vibratome <!> (e.g., Vibroslice VSLM1; WPI, Inc.)

EXPERIMENTAL METHOD

1. Use rat brain slices prepared as described in Stuart and Sakmann (1994). Briefly, decapitate rats following halothane anesthesia, and cut 300–400-μm-thick slices in ice-cold standard extracellular solution using a vibratome. Incubate slices for 30 min at 35°C, and thereafter maintain slices at room temperature (24°C–25°C).
2. For somatic whole-cell recordings, use 5–7-MΩ patch pipettes. Fill the patch electrodes from the tip with standard intracellular solution by applying negative pressure for ~20 sec. Backfill the electrodes with dye solution.
3. Patch individual neurons using DIC video microscopy. Allow 20–60 min of free diffusion of the dye from the pipette into the cell body to complete the intracellular staining.

   After the whole-cell configuration is formed, do not apply pressure to the patch electrode! Simple diffusion is sufficient to load the soma with an adequate amount of the dye.
4. Remove the patch electrode by forming an outside-out patch.
5. Incubate slices for an additional 2 h to allow the dye to spread from the soma into distal dendritic processes before making optical recordings.

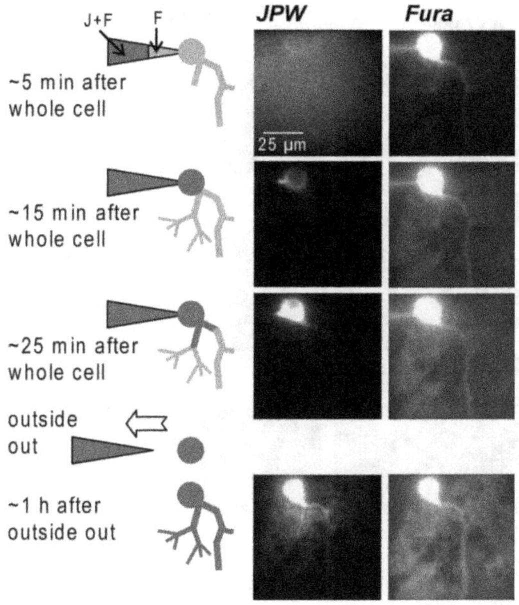

FIGURE 1. Intracellular application of a voltage-sensitive dye (JPW3028 [J], *left* column) and a Ca^{2+} indicator (bis-Fura-2 [F], *right* column) into a Purkinje neuron in the cerebellar slice. At the start, the internal solution in the tip of the patch electrode did not contain voltage-sensitive dye. Images were taken at 5, 15, and 25 min after whole-cell configuration and 1 h after the patch electrode was removed. (Reprinted, with permission of Wiley-Blackwell, from Canepari et al. 2008.)

The staining procedure as applied to a Purkinje neuron in a cerebellar slice is shown schematically in Figure 1. In this experiment the standard intracellular solution contained the calcium indicator bis-Fura-2. The electrode was backfilled with the voltage-sensitive dye solution. The leakage of bis-Fura-2 onto the slice from the tip of the electrode has no effect because this indicator is practically non-fluorescent at high calcium concentration in the extracellular medium.

See *Troubleshooting*.

TROUBLESHOOTING

Problem (Step 5): Background fluorescence is too high.

Solution: A major problem in staining vertebrate neurons from patch pipettes is dye leakage from the electrode. Patching technique requires positive pressure to be applied to the patch pipette during electrode manipulation through the tissue while approaching healthy neurons. This pressure ejects the solution from the electrode. To avoid extracellular deposition of the voltage-sensitive dye and the resulting large background fluorescence, the tip of the electrode must be filled with dye-free solution. It is possible, using this approach, to load neurons without any leakage of the dye to the surrounding tissue. An example of the selective staining of a neuron is shown in Figure 2A.

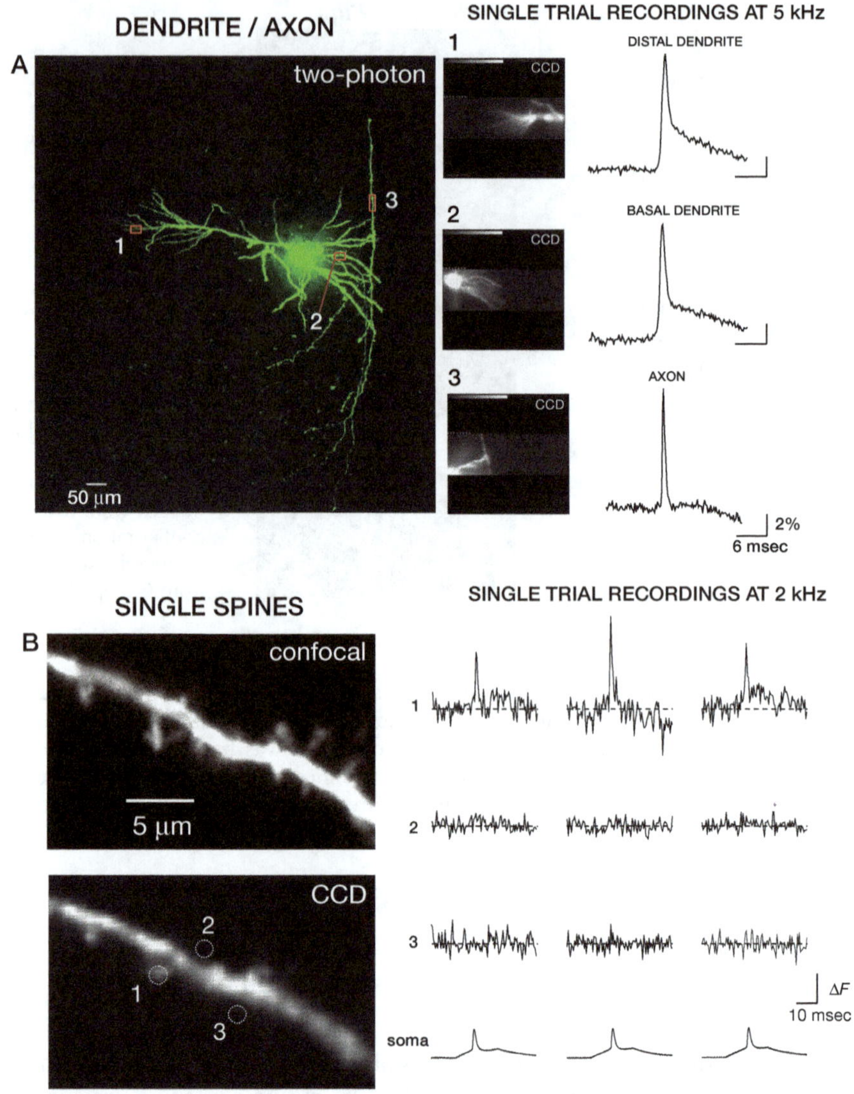

FIGURE 2. High-speed multisite recording of action potential signals from processes and dendritic spines of pyramidal neurons. (*A*) A voltage-sensitive dye-stained CA1 pyramidal neuron reconstructed from a *z*-stack of two-photon fluorescence images. Black and white images on the *right*: parts of the dendritic and axonal arbor projected to a CCD camera configured for a 5-kHz frame rate recording of fluorescence intensity changes from a subset of 28 x 80 pixels. Fluorescence intensity traces show the time course of the action potential (AP) signals from small regions on distal and basal dendrites (marked 1 and 2) and from one location on an axon collateral (location 3). Axonal AP is characterized by faster dynamics. An excellent signal-to-noise ratio is obtained in single-trial recordings. (*B*) Back-propagated AP (bAP) signals recorded optically from an individual dendritic spine of a mouse visual cortex layer 5 pyramidal neuron stained with the voltage-sensitive dye. (*Upper* micrograph) Anatomical reconstruction of a basal dendrite section in recording position obtained from a *z*-stack of spinning-disk confocal images. (*Lower* micrograph) Fluorescence images of the same dendritic spiny branch obtained with the CCD camera for voltage imaging. (*Right* panel) Fluorescence intensity traces from locations 1, 2, and 3 outlined on CCD image. Single-trial recordings are shown. (*Bottom* traces) Electrode recordings from the soma. The bAP signals, clearly recorded in spines, are absent from regions without spines (locations 2 and 3).

DISCUSSION

Example of Selectively Staining a Neuron with a Voltage-Sensitive Dye

A typical multisite recording of action potential signals from processes of mouse pyramidal neurons labeled with a voltage-sensitive dye is illustrated in Figure 2. In the experiments shown in Figure 2A, the extent of propagation and the time course of back-propagated action potential (bAP) signals were monitored at relatively low optical magnification from the axon as well as from basal and apical dendrites. In these measurements, the full CCD frame (80 x 80 pixels) corresponded to a 300 x 300-µm area in the object plane. At this magnification each pixel receives light from an ~4 x 4-µm area in the object plane, and the dendritic spines could not be resolved because their image was smaller than individual pixels. The dye-loaded pyramidal neuron was re-patched with a dye-free electrode, and spikes were elicited by depolarizing current pulses delivered from an electrode in the soma. The bAP signals were recorded optically at a frame rate of 5 kHz (partial readout of 80 x 28 pixels) from various locations by sequentially repositioning the field of view (Fig. 2A). As indicated in the figure, at all recording sites, bAPs could be monitored with excellent signal-to-noise ratio in single-trial measurements. The combined effect of an increase in light intensity and the use of near optimal excitation wavelength provided by a laser was a dramatic improvement in the sensitivity of voltage imaging by a factor between 10 and 40 compared to measurements using an arc lamp. The relatively high sensitivity of V_m imaging of bAP signals allowed direct comparison of spike waveforms from different regions on neuronal processes. Because the response time of the charge-shift voltage-sensitive dyes is much faster than the AP signal (in the microsecond range) (Rohr and Salzberg 1994) and because the relationship between fluorescence changes and membrane potential is strictly linear within a wide range (–100 mV to +100 mV) (Ross et al. 1977; Loew et al. 1992; Wu and Cohen 1993; Canepari et al. 2007), the dye signal tracks the membrane potential without distortion, and the time course of the signal is obtained directly from optical data. For example, there is a clear difference in waveform between axonal and dendritic action potentials shown in Figure 2A that can be quantified precisely. Because regional changes in the spike waveform are determined by the biophysical properties and distribution of voltage-gated channels, this type of information, when linked to a multicompartmental numerical simulation and to data on voltage-gated channel distribution from immunocytochemical analysis, will facilitate elucidating the functional significance of different signaling patterns in dendrites and axons of vertebrate neurons.

The sensitivity of the above measurements indicates that it should be possible to increase the optical magnification by a factor of 10 and monitor V_m transients from individual dendritic spines in wide-field epifluorescence microscopy mode. This approach is limited to neurons located near the surface of acute brain slices (<30 µm) because light scattering prevents resolution of individual spines in deeper layers. Figure 2B shows that the spine bAP signals can be clearly resolved in single-trial measurements. One concern with these measurements is the amount of scattered light from the parent dendrite that might contaminate the measurement from individual spines. To address this issue, dendritic signals from individual spines can be compared with recordings from analogous regions without spines. The results show that the signals from regions without spines were smaller than the noise in the measurement (Fig. 2B, locations 2 and 3). Thus, the interference from scattered light is insignificant in the superficial layers of the slice. Another concern in these measurements is the amount of photodynamic damage caused by the high-intensity excitation light. It was established, in a series of control experiments, that in the range tested (up to 50 recording trials) the first and the last bAP signals were identical, indicating that the photodynamic damage was not significant. The SNR in these measurements can be further improved by averaging a small number of trials (data not shown). The experimental analysis of electrical signaling in individual dendritic spines based on direct measurements is in its infancy, and numerous important questions remain open. The partial list of unresolved questions includes the extent of AP back-propagation as a function of spiking frequency under different conditions, the role of the spine neck in synaptic potential summation and plasticity, the composition and role of voltage-gated channels in spines, and the electrical structure of spines as a function of their dimensions and morphology. The V_m-imaging technique described above will facilitate these studies.

Advantages and Limitations

Voltage imaging from individual nerve cells has two fundamental advantages that make this approach unique. The first is the possibility of simultaneous measurements from many locations. The second is the ability to monitor membrane potential changes from neuronal processes that are not accessible to electrode recordings. Thin dendritic branches and individual dendritic spines that receive most of the synaptic inputs are likely to be a key compartment in which the input–output transform of an individual neuron is determined. These structures can now be probed for electrical signals using voltage imaging.

The major limitations of this technique are that the signal size is still relatively small at the spatial resolution of individual dendritic spines and that its application requires attention to many technical details. Nevertheless, it is plausible that the technology best suited for the investigation of signal processing at the level of axonal and dendritic processes of single neurons might ultimately be voltage imaging that allows extensive parallel recordings. Recent improvement in the sensitivity achieved by using laser excitation epifluorescence microscopy mode resulted in a method that is considerably more powerful. The pharmacological effects and photodynamic damage caused by the dye are possible additional limitations of this technique. The problem of damage is most effectively avoided by following strict and verified protocols regarding the amount of staining and the duration of intense illumination. Several previous studies have documented that the voltage-sensitive dyes used for intracellular labeling (JPW 1114 or its close analog JPW 3028) have little or no pharmacological effect and cause little or no photodynamic damage when applied at functional concentrations to both invertebrate (Antic and Zecevic 1995; Antic et al. 2000) and vertebrate (Antic et al. 1999; Antic 2003; Djurisic et al. 2004, 2008; Palmer and Stuart 2006; Canepari et al. 2007) neurons.

RECIPES

CAUTION: See Appendix 6 for proper handling of materials marked with <!>.
Recipes for reagents marked with <R> are included in this list.

Standard Extracellular Solution for Vertebrate Neurons

NaCl	125 mM
$NaHCO_3$	26 mM
Glucose	20 mM
KCl	2.3 mM
KH_2PO_4	1.26 mM
$CaCl_2$ <!>	2 mM
$MgSO_4$ <!>	1 mM

Adjust to pH 7.4 when bubbled with 95% O_2 <!> and 5% CO_2 <!>.

Standard Intracellular Solution for Vertebrate Neurons

Potassium gluconate <!>	115 mM
KCl	20 mM
Mg_2ATP	4 mM
Phosphocreatine	10 mM
Na-GTP	0.3 mM
HEPES	10 mM

Adjust to pH 7.3 with KOH <!>.

REFERENCES

Antic SD. 2003. Action potentials in basal and oblique dendrites of rat neocortical pyramidal neurons. *J Physiol* **550:** 35–50.

Antic SD, Zecevic D. 1995. Optical signals from neurons with internally applied voltage-sensitive dyes. *J Neurosci* **15:** 1392–1405.

Antic S, Major G, Zecevic D. 1999. Fast optical recordings of membrane potential changes from dendrites of pyramidal neurons. *J Neurophysiol* **82:** 1615–1621.

Antic SD, Wuskell JP, Loew L, Zecevic D. 2000. Functional profile of the giant metacerebral neuron of helix aspersa: Temporal and spatial dynamics of electrical activity in situ. *J Physiol (Lond)* **527:** 55–69.

Canepari M, Vogt KE. 2008. Dendritic spike saturation of endogenous calcium buffer and induction of postsynaptic cerebellar LTP. *PLoS ONE* **3:** e4011. doi: 10.1371/journal.pone.0004011.

Canepari M, Djurisic M, Zecevic D. 2007. Dendritic signals from rat hippocampal CA1 pyramidal neurons during coincident pre- and post-synaptic activity: A combined voltage- and calcium imaging study. *J Physiol* **580:** 463–484.

Canepari M, Vogt KE, Zecevic D. 2008. Combining voltage and calcium imaging from neuronal dendrites. *Cell Mol Neurobiol* **28:** 1079–1093.

Djurisic M, Antic S, Chen WR, Zecevic D. 2004. Voltage imaging from dendrites of mitral cells: EPSP attenuation and spike trigger zones. *J Neurosci* **24:** 6703–6714.

Djurisic M, Popovic M, Carnevale N, Zecevic D. 2008. Functional structure of the mitral cell dendritic tuft in the rat olfactory bulb. *J Neurosci* **28:** 4057–4068.

Loew LM, Cohen LB, Dix J, Fluhler EN, Montana V, Salama G, Wu J. 1992. A naphthyl analog of the aminostyryl pyridinium class of potentiometric membrane dyes shows consistent sensitivity in a variety of tissue, cell, and model membrane preparations. *J Membr Biol* **130:** 1–10.

Palmer LM, Stuart GJ. 2006. Site of action potential initiation in layer 5 pyramidal neurons. *J Neurosci* **26:** 1854–1863.

Palmer LM, Stuart GJ. 2009. Membrane potential changes in dendritic spines during action potentials and synaptic input. *J Neurosci* **29:** 6897–6903.

Rohr S, Salzberg BM. 1994. Multiple site optical recording of transmembrane voltage (MSORTV) in patterned growth heart cell cultures: Assessing electrical behavior, with microsecond resolution, on a cellular and subcellular scale. *Biophys J* **67:** 1301–1315.

Ross WN, Salzberg BM, Cohen LB, Grinvald A, Davila HV, Waggoner AS, Wang CH. 1977. Changes in absorption, fluorescence, dichroism, and birefrigence in stained giant axons: Optical measurements of membrane potential. *J Membr Biol* **33:** 141–183.

Stuart GJ, Sakmann B. 1994. Active propagation of somatic action potentials into neocortical pyramidal cell dendrites. *Nature* **367:** 69–72.

Wu JY, Cohen LB. 1993. Fast multisite optical measurement of membrane potential. In *Biological techniques: Fluorescent and luminescent probes for biological activity* (ed. Mason WT). Academic Press, New York.

Zecevic D. 1996. Multiple spike-initiation zones in single neurons revealed by voltage-sensitive dyes. *Nature* **381:** 322–325.

Zhou WL, Yan P, Wuskell JP, Loew LM, Antic SD. 2008. Dynamics of action potential backpropagation in basal dendrites of prefrontal cortical pyramidal neurons. *Eur J Neurosci* **27:** 923–936.

30 Two-Photon Sodium Imaging in Dendritic Spines

Christine R. Rose

Institute for Neurobiology, Heinrich-Heine University of Düsseldorf, 40225 Düsseldorf, Germany

ABSTRACT

Neuronal activity causes substantial postsynaptic Na^+ transients (Rose 2002). However, the physiological consequences of such Na^+ transients remain largely unknown. High-resolution Na^+ imaging is pivotal to study these questions, and, up to now, two-photon imaging with the fluorescent Na^+ indicator sodium-binding benzofuran isophthalate (SBFI) has been the method of choice. When introduced into individual neurons in acute tissue slices, SBFI enables measurements of $[Na^+]_i$ transients in dendrites and spines (Rose et al. 1999; Rose and Konnerth 2001). This technique is also suitable for the determination of $[Na^+]_i$ transients in other cell types, such as glial cells, with high spatial resolution, and is likely to be useful for measurement of Na^+ signals in vivo.

INTRODUCTION

In the early 1990s, two-photon imaging was introduced for the measurement of calcium transients in single spines (Denk et al. 1996). Several years later, the sodium-sensitive fluorescent dye SBFI was used with two-photon microscopy to image Na^+ within neurons demonstrating that $[Na^+]_i$ transients could be followed in living dendrites and spines within brain tissue for periods of several hours without any significant dye bleaching or photodamage (Rose et al. 1999; Rose and Konnerth 2001).

SBFI is a ratiometric UV-excited dye (similar to the well-known calcium-sensitive dye Fura-2) that has been used for conventional Na^+ imaging in many cell types (e.g., Donoso et al. 1992; Lasser-Ross and Ross 1992; Bennay et al. 2008; also see Chapter 21). With single-photon excitation, the optimal Na^+-sensitive excitation wavelength of SBFI is between 380 and 400 nm, and for two-photon absorption favorable excitation, wavelengths are 760–800 nm (Denk et al. 1995). The quantum efficiency of SBFI is about six times lower than for Fura-2, and its fluorescence change when binding Na^+ is less pronounced than that of Fura-2 binding Ca^{2+} (Minta and Tsien 1989). Thus, for measurements within fine cellular processes in tissue slices, a relatively high concentration of SBFI (i.e., 0.5–2 mM) must be used, and the membrane-impermeable form of SBFI must be loaded directly into cells using either a sharp microelectrode (Jaffe et al. 1992) or a patch pipette (Rose et al. 1999). Although early evidence indicated that SBFI does not discriminate sufficiently well between Na^+ and

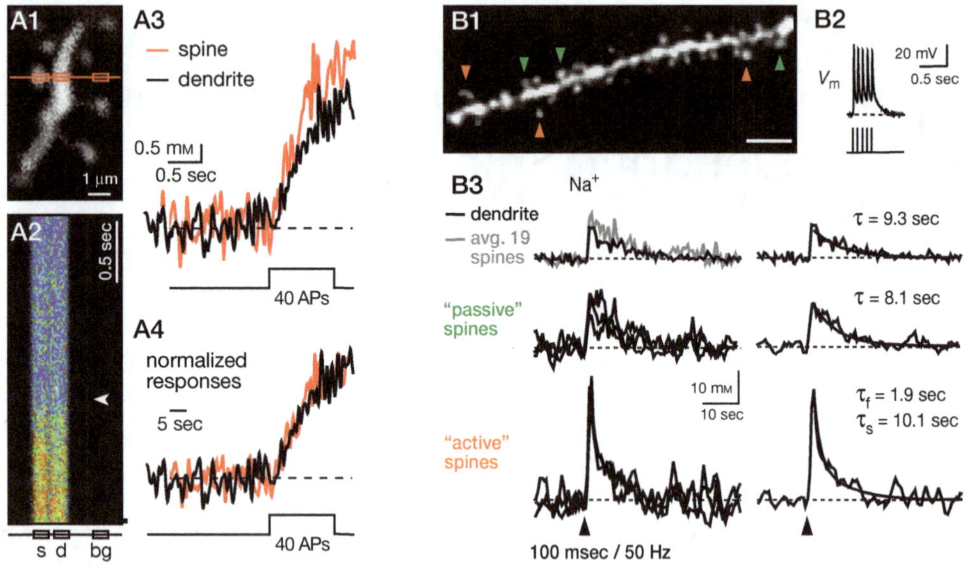

FIGURE 1. Activity-induced Na^+ transients in dendrites and spines. (*A*) Rising phase of action potential–induced $[Na^+]_i$ transients as recorded in the line scan mode (sampling frequency 160 Hz). (*A1*) Image of the spiny dendrite, and the line and regions of interest chosen for the experiment. (*A2*) Color-coded image of the action potential-induced changes in fluorescence in the different compartments; arrowhead indicates the start of the stimulation. (*A3*) $[Na^+]_i$ increase in the spine and the dendrite. (*A4*) Scaling the responses showed that the time to peak of $[Na^+]_i$ increases was similar in spines and dendrites. Data are the average of five consecutive stimulations, traces were low-pass filtered (80 Hz). (*B*) Synaptically induced sodium transients. (*B1*) Image of a spiny dendrite of a CA1 pyramidal cell. Arrowheads indicate spines from which measurements were taken. Scale bar, 5 μm. (*B2*) Electrical response of the neuron to a synaptic stimulation (five pulses, 50 Hz). (*B3*) *Left*: synaptically induced Na^+ transient in spines (average of 19 spines) and in the dendrite (*upper traces*). *Middle traces*: Na^+ transients in three single passive spines. *Lower traces*: Na^+ transients in three single active spines. *Right*: The recovery of dendritic Na^+ transients and passive spines could be fitted by a monoexponential decay with similar time constants (τ). Recovery in active spines followed a biexponential time course. Spine traces on the *right* represent averages of the single spine traces depicted on the *left*. (*A*, Reprinted, with permission of Springer Science+Business Media, from Rose et al. 1999; *B*, modified, with permission, from Rose and Konnerth 2001, ©Society for Neuroscience.)

K^+ (Minta and Tsien 1989), in practice, SBFI's sensitivity to changes in intracellular K^+ is negligible under most experimental conditions (Rose and Ransom 1996; Rose et al. 1999).

Two-photon Na^+ imaging and whole-cell patch-clamp recordings were used to measure activity-induced $[Na^+]_i$ transients in dendrites and spines of CA1 pyramidal neurons in rat hippocampal slices (Rose et al. 1999). A long-lasting burst of back-propagating action potentials, induced by brief depolarizing current pulses into the soma, evoked $[Na^+]_i$ transients in the low millimolar range throughout the proximal dendritic tree and in adjacent spines (Fig. 1A). The data suggested that accumulation of $[Na^+]_i$ in spines was caused by the opening of voltage-gated Na^+ channels on the spines themselves as previously demonstrated experimentally (Araya et al. 2007).

The same technique was used to determine synaptically induced Na^+ transients in CA1 pyramidal cells (Rose and Konnerth 2001). Although subthreshold postsynaptic depolarizations in response to short burst synaptic stimulation (five pulses at 50 Hz) did not elicit detectable $[Na^+]_i$ transients in dendrites and spines, suprathreshold synaptic activity, accompanied by three to five action potentials, caused local dendritic $[Na^+]_i$ transients of ~10 mM. Interestingly, two classes of spines could be distinguished based on the amplitude and time course of their $[Na^+]_i$ transients in response to suprathreshold stimulation. In the first class of spines (75% of all spines), the $[Na^+]_i$ level rose to values similar to those of adjacent dendrites. Moreover, the decay of the $[Na^+]_i$ transients in these spines and in dendritic regions was monoexponential with time constants of ~9 sec. In contrast, in the second class of spines (25%), activity-induced $[Na^+]_i$ transients were two to three times higher than those of the adjacent dendrite, reaching up to 35–40 mM. Recovery followed a biexponential decay with a fast time constant of <2 sec and a slow time constant that was very similar to the dendritic decay constant. It was concluded, therefore, that these spines were the sites of synaptic input and thus the sites of synaptic [Na^+] influx ("active" spines). In the other class of spines, in which the amplitude and the time course of $[Na^+]_i$ transients corresponded to that of the dendrites, the Na^+ signals were probably caused by diffusion of free and SBFI-bound [Na^+] ions from their site of entry ("passive" spines). In presumed active spines (those that most likely received synaptic input), activity-induced Na^+ transients reached up to 35–40 mM (Fig. 1B). It was found that the postsynaptic $[Na^+]_i$ transients were largely mediated by Na^+ entry through *N*-methyl-D-aspartate (NMDA) receptor channels. Indeed, blocking glutamate receptors of the NMDA type with APV strongly reduced the stimulus-induced $[Na^+]_i$ transient in the spines and the dendrite, while still eliciting a suprathreshold postsynaptic electrical response. The large amplitudes of the Na^+ transients and their location on dendritic spines suggested that this signal is an important determinant of electrical and biochemical spine characteristics.

Protocol

Calibration and Measurement of $[Na^+]_i$ within Neurons Using SBFI

This protocol provides guidelines and tips for measuring dynamic Na^+ concentrations within neurons, as well as an in situ calibration procedure. Because the basic properties of SBFI inside a cell differ strongly from the same properties in vitro, in situ calibrations are required to obtain meaningful experimental results.

IMAGING SETUP

For high-resolution imaging of sodium, we use a custom-built two-photon laser-scanning microscope based on an Olympus FV300 system, coupled to a Mai-Tai broadband laser (Newport Spectra-Physics). Laser intensity is controlled by a Pockels cell (350/80 LA Conoptics, coupled to a custom-built laser beam intensity controller). Objectives are 40x or 60x high-numerical-aperture, water-immersion lenses (e.g., Olympus LUMPlanFL/IR 40x/0.80 W or Nikon NIR APO 60x/1.0 W). Fluorescence emission of SBFI is collected by a custom-adapted channel photomultiplier (CPM 993, PerkinElmer) at wavelengths between 400 and 590 nm. Image analysis is performed using Igor Pro 6.0 (WaveMetrics) software.

MATERIALS

CAUTION: See Appendix 6 for proper handling of materials marked with <!>.
See the end of the chapter for recipes for reagents marked with <R>.

Reagents

Ethanol <!> (optional; see Troubleshooting)
Hippocampal brain slices
Intracellular saline containing SBFI <R>
Na^+ calibration solutions (see Step 3iii)
Na^+-free extracellular saline containing ionophores <R>
SBFI (Invitrogen)

Equipment

Equipment for whole-cell patch-clamp recording (see also Chapter 35)
Software for imaging analysis (see Imaging Setup)
Two-photon fluorescence microscope and related instruments (see Imaging Setup)

EXPERIMENTAL METHOD

1. Using a patch pipette, load single cells with 0.5–2 mM SBFI (the membrane-impermeable form) in intracellular saline. Alternatively, load the dye into single cells using electroporation. The objective is to obtain an acceptable signal-to-noise ratio.

   For CA1 pyramidal neurons in mouse hippocampal slices, when using a patch pipette, allow the dye to load for at least 45 min to ensure diffusion of the dye into distal dendrites. Relatively high

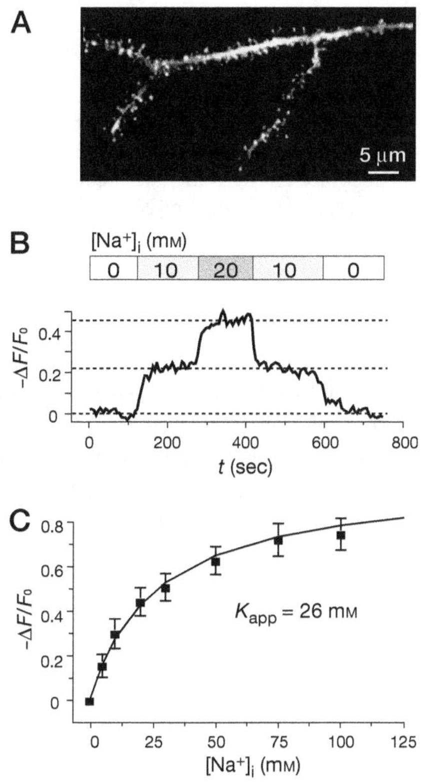

FIGURE 2. In situ calibration of SBFI fluorescence in a CA1 pyramidal neuron in an acute hippocampal tissue slice. (*A*) Image of the SBFI-filled spiny dendrite chosen for the experiment. (*B*) In saline containing ionophores (see text), changes in the extracellular $[Na^+]$ from 0 to 10 and to 20 mM evoked decreases in the SBFI fluorescence. Data are normalized to values at 0 mM $[Na^+]$ (F_0). (*C*) Relationship between SBFI fluorescence and $[Na^+]_i$, normalized to the fluorescence at 0 mM $[Na^+]_i$ (F_0). (Mean values of 6 measurements in 10 cells ± S.D.) Data points were fitted analogous to an equation following Michaelis–Menten kinetics. (Reprinted, with permission of Springer Science+Business Media, from Rose et al. 1999.)

dye concentrations are necessary to ensure sufficient fluorescence emission from small structures such as spines.

2. After dye loading, excite the cells with laser light between 760 and 810 nm wavelength.

   The cross section for two-photon imaging with SBFI has not been determined yet. Longer wavelengths might be suitable as well. Choose the lowest excitation intensity possible to prevent dye bleaching or phototoxicity. Increases in Na^+ are reflected by decreases in the fluorescence emission of SBFI; thus, do not set baseline fluorescence emission too low. Because the kinetics of Na^+ transients (decay times) are quite slow (when compared with Ca^{2+} transients), choose the lowest possible imaging frequency.

3. Calibrate the absolute $[Na^+]_i$ values in situ in CA1 pyramidal cells (Rose et al. 1999; see Fig. 2). This can be performed either after each physiological experiment or as a separate procedure.

   i. Fill the cells with SBFI using a patch electrode. Carefully withdraw the electrode to allow the cell membrane to reseal.

   ii. Perfuse the tissue slices with Na^+-free extracellular saline containing ionophores. This step induces equilibration of Na^+ across the plasma membrane.

   iii. Once a stable fluorescence emission is reached (which usually takes 20–30 min), perfuse the slices with calibration solutions of different Na^+ concentrations (e.g., 10, 20, 30, 50, 75, and 100 mM Na^+). Stepwise changes in Na^+ concentrations should result in stepwise changes in SBFI fluorescence (Fig. 2B).

   See *Troubleshooting*.

4. Plot normalized data from the calibration experiments.

   This generates a graph that follows Michaelis–Menten kinetics and should reveal an apparent K_d of 26 mM at room temperature (Fig. 2C). The calibrations also indicate that two photon–induced SBFI fluorescence reliably reports $[Na^+]_i$ changes between 0 and 75 mM $[Na^+]_i$ and is, therefore, well suited for the measurement of $[Na^+]_i$ alterations that occur during physiological conditions.

5. Use the in situ calibrations to express measured changes in SBFI fluorescence emission as $[Na^+]_i$ changes, assuming that the baseline $[Na^+]_i$ throughout the cell corresponds to the $[Na^+]$ of the pipette solution.

   This assumption probably does not hold true for cellular compartments far away from the soma. SBFI fluorescence, however, changes in a roughly linear fashion with changes in $[Na^+]_i$ between 10 and 30 mM. Thus, within this range, absolute changes in $[Na^+]_i$ can be determined with reasonable accuracy, even if baseline $[Na^+]_i$ levels differ from the $[Na^+]$ of the pipette.

TROUBLESHOOTING

Problem (Step 3): SBFI fluorescence from fine dendrites and spines is low, and high excitation intensities are required (that rapidly damage the cells) to obtain a sufficient signal-to-noise-ratio. Calibrations report significantly elevated $[Na^+]_i$ values in regions far away from the soma.

Solution: Ionophores, especially gramicidin, might have contaminated the perfusion system and tubing—thus, perturbing the integrity of the cellular membranes. Take great care to clean everything with 70% ethanol after working with these substances.

DISCUSSION

Generally, quantification of baseline $[Na^+]_i$ values and $[Na^+]_i$ transients is easy and straightforward. Still, it is important to bear in mind that reporters of ion concentrations that bind a particular ion will always act as a buffer and, therefore, will distort ion transients. Fortunately, when imaging Na^+ with SBFI, this problem is almost negligible because even dye concentrations of 2 mM are significantly below SBFI's K_d. The binding of Na^+ to SBFI, however, will alter Na^+ diffusion parameters. Whereas Na^+ moves with an apparent diffusion coefficient of ~13 x 10^{-6} $cm^2\,sec^{-1}$ inside cells (Pusch and Neher 1988), SBFI-bound Na^+ will diffuse much more slowly, at ~5 x 10^{-6} $cm^2\,sec^{-1}$ (assuming that the diffusion coefficient of SBFI is similar to that of Fura-2) (Pusch and Neher 1988).

To date, SBFI is the only Na^+-sensitive fluorescent dye routinely used for Na^+ imaging experiments (see also Bennay et al. 2008; Langer and Rose 2009). However, its low quantum efficiency and small fluorescence change when binding Na^+ result in poor signal-to-noise ratios. This makes the high-resolution measurement of Na^+ transients in fine processes a tedious task, and binning or averaging of several trials is often required to obtain satisfactory signals. However, even this will not permit the detection of Na^+ changes below the millimolar range. In addition, the kinetics of Na^+ transients are slow, making it necessary to record for long periods of time.

Several years ago, a new Na^+ indicator dye became available: CoroNa Green (CoroNa), with an absorbance maximum at 492 nm. CoroNa is suitable for the measurement of Na^+ in CA1 pyramidal cell dendrites (Meier et al. 2006) and rat optic nerve axons (Nikolaeva et al. 2005) using confocal systems. Moreover, CoroNa has recently been used for two-photon Na^+ imaging within axon initial segments (Bender and Trussell 2009). Although substantial transmembrane dye leakage and lower Na^+ sensitivity are clear disadvantages of CoroNa compared with SBFI (Meier et al. 2006), CoroNa may prove to be a valuable reagent for two-photon imaging of Na^+ in dendritic spines when large Na^+ elevations are expected that might exceed the dynamic range of SBFI.

Clearly, the development of more sensitive fluorescent indicators for Na^+ is required to enable the detection of small Na^+ transients, such as those that appear during subthreshold synaptic activation of AMPA receptors.

RECIPES

CAUTION: See Appendix 6 for proper handling of materials marked with <!>.
Recipes for reagents marked with <R> are included in this list.

Intracellular Saline Containing SBFI

1. Combine the following in 8 mL of ultrapure water:

170 mM	KCl <!>	126.7 mg
12.5 mM	HEPES buffer	29.8 mg
12.5 mM	NaCl	7.3 mg
5 mM	Mg-ATP	25.4 mg
0.5 mM	Na-GTP	2.6 mg

2. Adjust the pH to 7.3 with KOH <!>. The osmolarity should be 365 mOsmol/L. Adjust to a final volume of 10 mL using ultrapure water. Aliquot into 1-mL portions, and store at –20°C for up to 1 wk.

3. Prepare SBFI stock solution (10 mM) according to manufacturer's instructions, and store at –20°C.

4. Add SBFI stock solution to intracellular saline (1:4). Do not use the working solution for >1 d!

Na$^+$-Free Extracellular Saline Containing Ionophores

1. Combine the following in 8 mL of ultrapure water:

182.5 mM	KCl <!>	134.0 mg
12.5 mM	HEPES buffer	29.8 mg
5 mM	Mg-ATP	25.4 mg
0.5 mM	Na-GTP	2.6 mg

2. Adjust the pH to 7.3 with KOH <!>. The osmolarity should be 365 mOsmol/L. Adjust to a final volume of 10 mL using ultrapure water. Aliquot into 1-mL portions, and store at –20°C for up to 1 wk.

3. Just before use, add the following ionophores to the extracellular saline solution: 3 μM gramicidin, 10 μM monensin, and 100 μM ouabain.

REFERENCES

Araya R, Nikolenko V, Eisenthal KB, Yuste R. 2007. Sodium channels amplify spine potentials. *Proc Natl Acad Sci* **104:** 12347–12352.

Bender KJ, Trussell LO. 2009. Axon initial segment Ca^{2+} channels influence action potential generation and timing. *Neuron* **61:** 259–271.

Bennay M, Langer J, Meier SD, Kafitz KW, Rose CR. 2008. Sodium signals in cerebellar Purkinje neurons and Bergmann glial cells evoked by glutamatergic synaptic transmission. *Glia* **28:** 28.

Denk W, Piston D, Webb W. 1995. Two-photon molecular excitation in laser-scanning microscopy. in *Handbook of biological confocal microscopy* (ed. Pawley J), pp. 445–458. Plenum, New York.

Denk W, Yuste R, Svoboda K, Tank DW. 1996. Imaging calcium dynamics in dendritic spines. *Curr Opin Neurobiol* **6:** 372–378.

Donoso P, Mill JG, O'Neill SC, Eisner DA. 1992. Fluorescence measurements of cytoplasmic and mitochondrial sodium concentration in rat ventricular myocytes. *J Physiol* **448:** 493–509.

Jaffe DB, Johnston D, Lasser-Ross N, Lisman JE, Miyakawa H, Ross WN. 1992. The spread of Na^+ spikes determines the pattern of dendritic Ca^{2+} entry into hippocampal neurons. *Nature* **357:** 244–246.

Langer J, Rose CR. 2009. Synaptically induced sodium signals in hippocampal astrocytes in situ. *J Physiol* **587:** 5859–5877.

Lasser-Ross N, Ross WN. 1992. Imaging voltage and synaptically activated sodium transients in cerebellar Purkinje cells. *Proc Biol Sci* **247:** 35–39.

Meier SD, Kovalchuk Y, Rose CR. 2006. Properties of the new fluorescent Na^+ indicator CoroNa Green: Comparison with SBFI and confocal Na^+ imaging. *J Neurosci Methods* **155:** 251–259.

Minta A, Tsien RY. 1989. Fluorescent indicators for cytosolic sodium. *J Biol Chem* **264:** 19449–19457.

Nikolaeva MA, Mukherjee B, Stys PK. 2005. Na^+-dependent sources of intra-axonal Ca^{2+} release in rat optic nerve during in vitro chemical ischemia. *J Neurosci* **25:** 9960–9967.

Pusch M, Neher E. 1988. Rates of diffusional exchange between small cells and a measuring patch pipette. *Pflugers Arch* **411:** 204–211.

Rose CR. 2002. Na^+ signals at central synapses. *Neuroscientist* **8:** 532–539.

Rose CR, Konnerth A. 2001. NMDA receptor–mediated Na^+ signals in spines and dendrites. *J Neurosci* **21:** 4207–4214.

Rose CR, Ransom BR. 1996. Intracellular Na^+ homeostasis in cultured rat hippocampal astrocytes. *J Physiol* **491:** 291–305.

Rose CR, Kovalchuk Y, Eilers J, Konnerth A. 1999. Two-photon Na^+ imaging in spines and fine dendrites of central neurons. *Pflugers Arch* **439:** 201–207.

31 Transcranial Two-Photon Imaging of the Living Mouse Brain

Jaime Grutzendler,[1] Guang Yang,[2] Feng Pan,[2] Christopher N. Parkhurst,[2] and Wen-Biao Gan[2]

[1]Departments of Neurology and Physiology, Northwestern University Feinberg School of Medicine, Chicago, Illinois 60611; [2]Department of Physiology and Neuroscience, Molecular Neurobiology Program, Skirball Institute, New York University School of Medicine, New York, New York 10016

ABSTRACT

In this chapter, we describe a protocol for imaging the living mouse brain through a thinned skull using two-photon microscopy. This transcranial two-photon imaging method allows high-resolution imaging of fluorescently labeled neurons, microglia, astrocytes, and blood vessels, as well as subcellular structures such as dendritic spines and axonal varicosities. The surgical procedure for transcranial imaging can be performed in ~30 min and can be repeated multiple times, allowing brain cells and tissues to be studied in the same animals over short or long time intervals, depending on the design of the experiment. Two-photon imaging through a thinned and intact skull avoids side effects caused by skull removal and is a minimally invasive method for studying the living mouse brain under physiological and pathological conditions.

INTRODUCTION

In vivo imaging of the mouse brain using two-photon microscopy (TPM) (Denk et al. 1990) through a thinned skull is a minimally invasive method that allows for longitudinal studies of cortical structures at high optical resolution over intervals ranging from minutes to years (Grutzendler et al. 2002; Yoder and Kleinfeld 2002; Stosiek et al. 2003; Tsai et al. 2004; Davalos et al. 2005; Nimmerjahn et al. 2005; Zhang et al. 2005; Zuo et al. 2005a,b; Haynes et al. 2006; Majewska et al. 2006; Nishiyama et al. 2007; Xu et al. 2007; Kim et al. 2009; Wake et al. 2009; Wu et al. 2009; Yan et al. 2009; Yang et al. 2009). By creating a thinned skull (thickness ~20 µm) cranial window, it has been possible to image fluorescently labeled structures located up to ~400-µm deep within the cortex (Grutzendler et al. 2002; Tsai et al. 2004; Davalos et al. 2005; Zuo et al. 2005a; Yang et al. 2009). In recent years, the transcranial two-photon imaging approach has been used to study the development and plasticity of synaptic connections (Grutzendler et al. 2002; Tsai et al. 2004; Zuo et al. 2005a,b;

Majewska et al. 2006; Nishiyama et al. 2007; Xu et al. 2007; Wu et al. 2009; Yang et al. 2009), neuronal network activity (Stosiek et al. 2003), cerebral microvascular occlusion (Yoder and Kleinfeld 2002), amyloid plaque deposition (Christie et al. 2001; Tsai et al. 2004; Yan et al. 2009), and microglial dynamics and function (Davalos et al. 2005; Nimmerjahn et al. 2005; Haynes et al. 2006; Wake et al. 2009).

Protocol

Live Imaging of Mouse Brain Using Transcranial Two-Photon Microscopy

A method is provided that allows for long-term imaging of brain tissue using fluorescent reporters and two-photon microscopy. The surgical procedure that is required to allow imaging thins the cranium so that it becomes optically transparent. Once learned, the surgical procedure can be performed quickly, and imaging can follow immediately.

IMAGING SETUP

We use either a custom-built multiphoton microscope equipped with a mode-locked laser system or a Bio-Rad 2001 multiphoton microscope. For both systems, the laser system (Tsunami and Millenia Xs, Spectra-Physics) is tunable from a 690–1000-nm wavelength with an 80-MHz pulse repeat and a <100-fsec pulse width. The laser-scanning system is coupled to an upright fluorescence microscope. For the custom-built unit, an Olympus laser-scanning system is used, and detectors (photomultiplier tubes) are placed close to the objective to facilitate the signal detection. We generally use an Olympus 60x, numerical aperture (NA) 1.1 objective for adequate resolution of small structures, such as dendritic spines. A dissecting microscope coupled to a charge-coupled device (CCD) camera is used before TPM imaging for obtaining images of the superficial meningeal vasculature, which are then used as maps for future relocation of the cortical region of interest.

MATERIALS

CAUTION: See Appendix 6 for proper handling of materials marked with <!>.
See the end of the chapter for recipes for reagents marked with <R>.

Reagents

Alcohol preparation pads, sterile (e.g., Fisher 06-669-62)
Artificial cerebrospinal fluid (ACSF) <R>
Cyanoacrylate glue <!> (Fisher Scientific 11-999-24)
Dextran-rhodamine B MW 70,000, 10 mg/mL in phosphate-buffered saline (PBS; Molecular Probes D1841) (optional, see Step 13)
Ketamine/xylazine (KX) mixture <R>
Lubricating eye ointment, sterile (Del Pharmaceuticals, Inc.)
Transgenic mice expressing fluorescent proteins within a variety of cell types, such as layer 5 cortical neurons (e.g., thy1-YFP line) or microglia (e.g., CX3CR1-eGFP line)(Feng et al. 2000; Jung et al. 2000)

Equipment

Cotton pads
Custom-built plate

Attach two 18 × 18 × 18-mm aluminum blocks to a 14 x 10 x 0.1-cm aluminum plate. Place the blocks 3 cm from each other and ~2 cm from one of the short sides of the plate. Drill a hole with an internal thread on each block to accommodate a ¼-in screw (Fig. 1A).

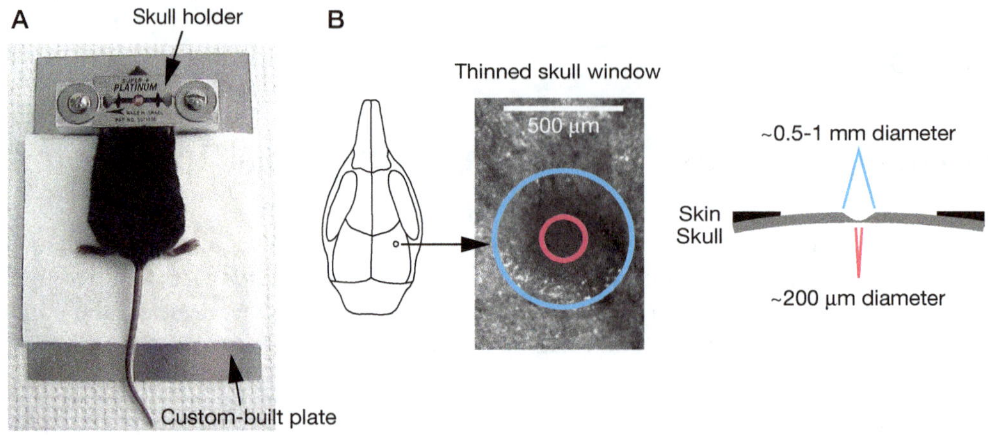

FIGURE 1. Schematic of thinned skull preparation. A head immobilization device including a custom-built plate and a skull holder is used for reducing movement artifacts during imaging (*A*). The skull holder/razor blades are glued onto the mouse skull such that the region of interest is exposed in the center of the circular hole of the razor blades. The skull holder/razor blades are then tightened on the aluminum blocks of the custom-built plate. A circular area of skull (typically ~0.5–1 mm in diameter, marked with a blue circle) over the region of interest is shaved (*B*). The thinnest region (marked with a pink circle) for two-photon laser-scanning microscope (TPLSM) imaging is ~20 µm in thickness and ~200 µm in diameter.

Dissecting instruments

Dissecting stereomicroscope (Olympus BX50WI)

Double-edge razor blades (CAMB Machine Knives International LLC CMK169S)

Fluorescence microscope

High-speed microdrill (Fine Science Tools 18000-17)

Make sure that the drill is adequately charged and in good working order because low drill speeds may result in nicking of the skull or uneven thinning.

Marker pen

Microdrill steel burrs (Fine Science Tools 19007-05 or 19007-07)

The drill burr should be replaced frequently to assure it performs properly.

Microsurgical blades (Surgistar, Inc. 6900)

The size, dimensions, and quality of the microsurgical blades are important for achieving proper thinning. For best results, use a new blade with each mouse.

Silk suture, 6-0 (LOOK M552830)

Skull holder

Cut the sharp edges of three double-sided razor blades, and then glue the blades to each other with cyanoacrylate to create a thick metal plate. The razor blades have circular holes in the middle for access to the skull. Use two spacers for stabilizing the skull holder (Fig. 1A).

Temperature regulator for mice (optional, see Step 1)

Two-photon microscope (TPM; see Imaging Setup)

EXPERIMENTAL METHOD

Thinned Skull Preparation

1. Anesthetize mice by intraperitoneal injection (5–6-µL/g body weight) of KX mixture. Allow sufficient time (5–10 min) for a surgical level of anesthesia to be achieved. Place the mouse on top of a cotton pad. For experiments involving long imaging periods, use a mouse temperature regulator.

   Experiments should comply with institutional and national animal care guidelines. Continuously monitor the depth of anesthesia by testing the animal's reflexes during the surgery (e.g., pinching the animal's foot with a blunt pair of forceps, and checking for the lack of withdrawal reflex). Inject more KX as necessary.

2. Lubricate both eyes with eye ointment.
3. Thoroughly shave the head over most of the scalp with a double-edge razor blade. Clean the scalp with a sterile alcohol preparation pad, and then make a midline scalp incision extending from the neck region between the ears to the frontal area between the eyes. Carefully disrupt the connective tissue located between the scalp and the underlying skull with a pair of spring scissors.
4. Localize the brain area to be imaged based on stereotactic coordinates, and label it with a marker pen.
5. Remove the connective tissue attached to the skull over the area to be imaged by gently scraping the skull with a blunt microsurgical blade.
6. Place a small amount of cyanoacrylate glue around the edges of the internal opening of the skull holder, and press the holder against the skull. Make sure that the area to be imaged is exposed in the center of the internal opening (Fig. 1A).

   Skull immobilization, which is, in part, determined by the quality of contact and adhesion between the skull holder and the skull, is critical to avoid movement artifacts during imaging.
7. Draw the loose skin gently up to the edges of the internal opening of the skull holder, and apply a small amount of glue to the perimeter of the skin to fix it to the skull.

   Gluing the skin around the internal opening of the skull holder and the skull aids in holding the artificial ACSF in place during imaging under a water-immersion objective.

   See *Troubleshooting.*
8. Wait several minutes until the skull holder is firmly glued to the skull. Attach the skull holder to the plate by inserting the lateral edges of the skull holder between the aluminum blocks and spacers/screws. Tighten the screws to completely immobilize the skull holder (Fig. 1A). Wash the opening of the skull holder several times with ACSF to remove remnants of nonpolymerized glue. This helps to prevent the glue from contaminating the microscope objectives during imaging.
9. Apply a few drops of ACSF to the skull surface during drilling to reduce overheating. Under a dissecting microscope, use a high-speed microdrill to thin a circular area (typically ~0.5–1 mm in diameter; Fig. 1B) over the skull region of interest. Perform drilling intermittently to avoid friction-induced overheating. Replace the ACSF periodically to wash away the bone debris. The mouse skull consists of two thin layers of compact bone (periosteum and endosteum) sandwiching a thick layer of spongy bone, which contains small cavities and canaliculi that carry blood vessels. Remove the external layer of compact bone and most of the spongy bone with the drill. Bleeding may occur during drilling of the spongy bone portion but usually stops spontaneously within a few minutes.

   Do not thin a large region (>1.5 mm) to a thin layer (<50 µm) as it is difficult to exercise fine control of the drilling process and may cause damage to the underlying cortex.
10. After removing the majority of the spongy bone, remaining concentric structures within the bone can usually be seen under the dissecting microscope, indicating that drilling is approaching the internal compact bone layer. At this stage, skull thickness is still >50 µm. Continue the skull-thinning procedure with a microsurgical blade to obtain a very thin (~20 µm) and smooth preparation (~200 µm in diameter) (Fig. 1B). During the thinning process, repeatedly examine the preparation with a conventional fluorescence microscope until fluorescently labeled structures, such as the dendrites and spines in the area of interest can clearly be visualized.

    Hold the microsurgical blade at an ~45° angle during thinning, and take great care not to push the skull downward against the brain surface or to break through the bone, as minor brain trauma or bleeding may induce inflammation and disruption of neuronal structures. In our experience, the best way to know when to stop thinning is by practice and by directly measuring the thickness of the skull from imaging skull autofluorescence with the TPM. For beginners, periodic measurement of skull thickness during thinning greatly helps in preventing skull overthinning.

It is critical not to thin a large surface area (>300 μm in diameter) to <15 μm in thickness as it may cause disruption of neuronal structures and activation of immune cells as evidenced by neurite blebbing, microglia process retraction, and thickening of epidural connective tissues. These events will reduce the quality of subsequent images and may confound experimental results.

Mapping the Imaging Area for Future Relocation

11. To identify the same imaged area at a later time point, take a high-quality picture of the brain vasculature with a CCD camera attached to a stereo dissecting microscope (Fig. 2A) or directly to the TPLSM setup.

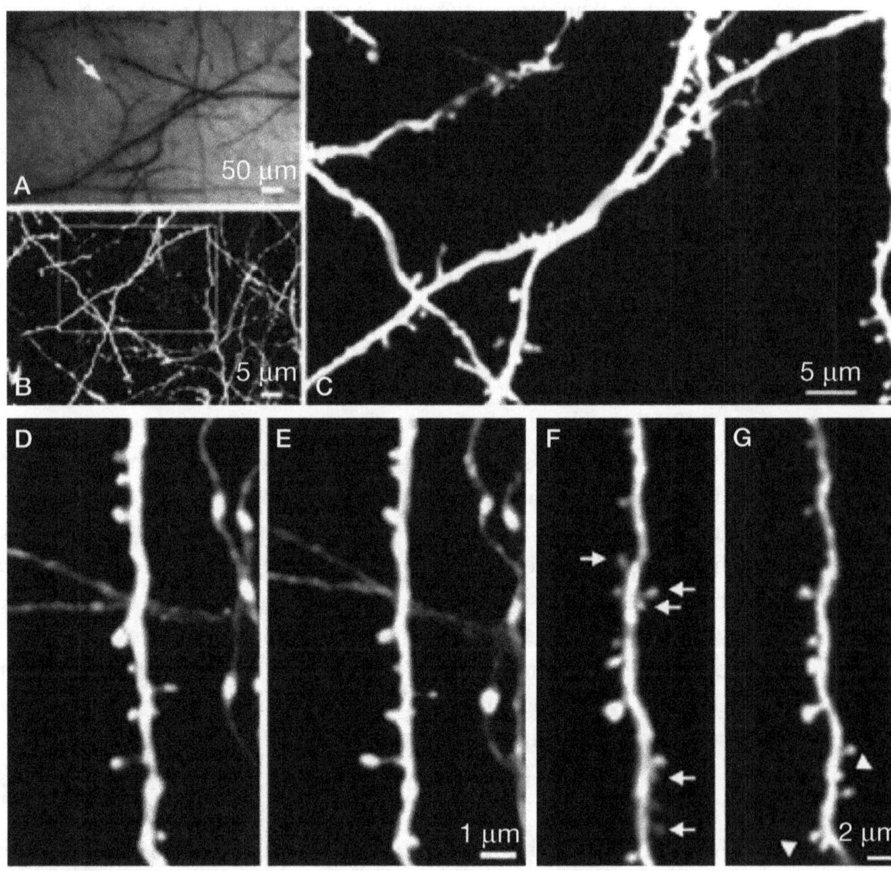

FIGURE 2. Long-term transcranial TPLSM imaging of fine neuronal structures. (*A*) CCD camera view of brain vasculature under the thinned skull. The cortical vasculature can clearly be seen through the thinned skull. The vasculature pattern remains stable over months to years and can be used as a landmark to relocate the imaged region at subsequent time points. Arrow indicates the region in which subsequent TPLSM images were obtained. (*B*) Two-dimensional (2D) projection of a three-dimensional (3D) stack of dendritic branches and axons in the primary visual cortex (60x, ~0.39 μm per pixel). The stack was 50-μm deep (2-μm step size). The boxed region was then imaged at higher magnification (*C*). High-power 2D projection of a 3D stack (60x, ~0.13 μm per pixel, 10-μm reconstructed, 0.70-μm step size) reveals clear neuronal structures including axonal varicosities, dendritic shafts, and dendritic protrusions. (*D,E*) Axonal and dendritic branches from two animals imaged 3 d apart show the same spines and boutons at the same locations. (Adapted, with permission, from Grutzendler et al. 2002.) (*F,G*) Dendritic branches imaged over 19 mo apart. The arrows indicate spines that are eliminated in the second view. The arrowheads indicate spines that are formed in the second view. Note that most spines in *F* persist in *G*. (Adapted, with permission, from Zuo et al. 2005a.) Two-dimensional projections of 3D image stacks containing dendritic segments of interest were used for *D–G*.

TPLSM Imaging of Neuronal or Glial Structures

12. Carefully move the mouse mounted on the custom-built plate to the TPM stage. Select an adequately thinned and uniform area for imaging under a fluorescence microscope, and carefully denote the selected area on the CCD brain vasculature map by observing the pattern of blood vessels adjacent to it (Fig. 2A).
13. If performing microglial imaging, inject 100 µL of rhodamine-dextran solution retro-orbitally for vessel labeling before application of eye ointment; alternatively, inject it through the tail vein.
14. Tune the TPM to the appropriate wavelength (e.g., 890 nm for green fluorescent protein [GFP] and 920 nm for YFP). Although we have used 40x (0.8 NA) and 60x (0.9 NA) in the past, we currently use higher-NA water-immersion objectives (e.g. 60x, 1.1 NA) for imaging small structures, such as dendritic spines.

    The objective lens should remain immersed in mouse ACSF at all times during imaging. If there is a sudden deterioration of imaging quality, check that the lens remains fully immersed in ACSF, and add more if necessary. Gradual leakage of ACSF is usually caused by inadequate gluing of the scalp around the skull holder openings (Step 7).

    For neuronal imaging, obtain a low-magnification stack of fluorescently labeled neuronal processes (e.g., 60x objective; 200 µm x 200 µm; 512 x 512 pixel; 2-µm step), which serves as a higher-resolution map for accurate relocation of the same region at later time points in conjunction with the CCD brain vasculature map (Fig. 2B). When performing microglial imaging, take a low-magnification (10x air objective) stack of the rhodamine dextran–labeled vasculature, and mark the selected area before switching to a 40x/60x objective, making sure that objectives (10x and 40x/60x) are cocentered.
15. Without changing the position of the stage, take high-magnification images (e.g., 66.7 x 66.7 µm; 512 x 512 pixel; 0.75-µm step: Fig. 2C) from the same area with TPM. The image stack is typically taken within ~100 µm below the pial surface for spine imaging and within ~200 µm for microglia imaging (Figs. 2 and 3).

    We typically use laser intensities in the range of 10 to 30 mW (measured at the sample) to minimize phototoxicity.

    See *Troubleshooting.*

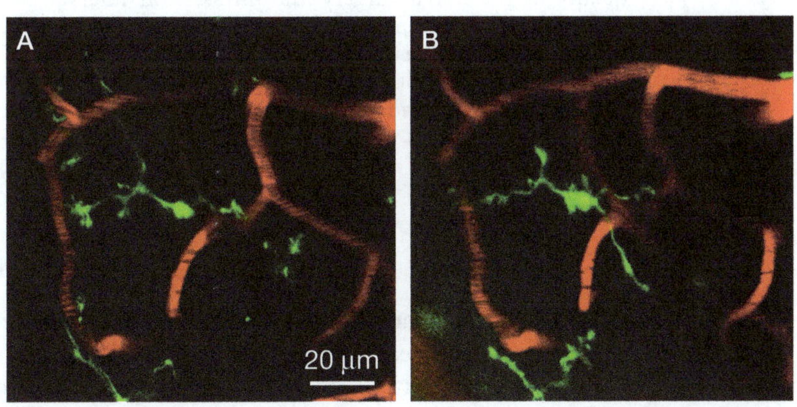

FIGURE 3. Transcranial TPLSM imaging of eGFP-labeled microglia vasculature within the mouse cerebellum. Two-dimensional projections of a 3D z-stack from the cerebellum (40x, digital zoom = 1) of a mouse harboring a single copy of the CX3CR1-eGFP allele driving eGFP expression in a subset of myeloid cells including central nervous system resident parenchymal microglia imaged 24 h apart (*A,B*). The cerebellar vasculature has been labeled in red by intravenous injection of a rhodamine-dextran conjugate solution. The stacks are 6 µm in depth (2-µm step size) and are representative of a section of cortex spanning 40–46 µm below the dural surface. eGFP-labeled microglia retain their characteristic highly branched morphology indicative of a resting state.

Recovery

16. When imaging is complete, detach the skull holder from the skull by gently pushing the skull away from the skull holder. When removed, the holder should take with it any remaining glue and be separated from the skin. If glue remains adherent to the skull or skin, gently remove any remaining dried glue with a pair of forceps. Suture the scalp with 6-0 silk, and leave the mouse in a separate cage until it is fully mobile. After recovery, return the mouse to its original housing cage until the next viewing.

Re-imaging

17. Begin by repeating Steps 1–8. Locate the previously thinned region based on the brain vasculature map (CCD for dendrite imaging, 10x rhodamine-dextran map for microglial imaging). If the second view is within 1 wk after the first view, carefully remove the connective tissue that has regrown on top of the thinned region using a microsurgical blade, and check the image quality with the TPM. If the image quality is poor (blurry features, high background fluorescence, or significantly reduced depth of penetration), use a microsurgical blade to carefully shave the skull (as in Step 10) until a clear image can be obtained. If the second view is >1 wk after the first view, repetition of Steps 9–10 will be required.

    The bone in the thinned area can regenerate substantially if the time interval between imaging sessions is >1 wk. The newly grown bone layer is optically less transparent for TPM imaging, so it is generally necessary to shave the bone so that it is slightly thinner than the previous imaging session.

    See *Troubleshooting.*

18. Roughly locate the previously imaged region under the fluorescence microscope. More precisely, align the region using TPM in accordance with the low magnification map, and then zoom in to high magnification to further align it. After the region is precisely aligned with the first view, take images as in Step 15.

    See *Troubleshooting.*

19. For multiple imaging sessions, repeat Steps 16–18.

    Although we have successfully imaged the same region with high optical clarity up to five times, the number and quality of imaging sessions is dependent on a variety of factors including the quality of the initial surgery, the intrasession interval, and the experience of the operator. Although extreme thinning of the skull (~15 μm) during the initial surgery may provide excellent clarity during the first imaging session, it will complicate future imaging sessions because of an increased propensity for compensatory bone regrowth. We recommend limiting the number of imaging sessions and skull-thinning procedures to preserve image quality and reduce the perturbation to the brain microenvironment.

TROUBLESHOOTING

Problem (Step 7): The skull holder does not hold ACSF during imaging.
Solution: It is important to seal the internal opening of the skull holder with the surrounding skin. First dry the skin by blotting off excess liquid, and then by gently drying with compressed air. Follow by pulling the skin around the skull holder and by fixing the skin in place with glue.

Problem (Step 15): There are excessive movements of fluorescently labeled structures during imaging.
Solution: Begin by checking the depth of anesthesia. If the animal begins to wake up, inject more KX, and wait for a few minutes for anesthesia to deepen. If the animal is under deep anesthesia,

check the attachment between the skull and the skull holder. The skull holder must be tightly glued to the skull. Detach the skull holder, and reglue it to the skull if the attachment is loose.

Problem (Steps 15 and 17): The fluorescent image is not clear under fluorescence microscope/TPM.
Solution: (1) Check the skull thickness with TPM. If the skull is too thick, carefully use either the drill or a microsurgical blade to thin the skull, and check again. (2) The skull is thin but uneven or marred by the drill bit or blunt microsurgical blades. An unevenly thinned or marred skull is a result of an improper thinning technique. It may be difficult to correct once part of the bone is already exceedingly thin. Use another animal for the experiment. (3) Red blood cells that have accumulated anywhere between the objective and the fluorescent structures can substantially degrade the image quality. If the bleeding is from the skull bone, wait until it stops spontaneously, and wash with several changes of ACSF to remove any coagulated blood. Bleeding underneath the bone indicates that the brain has been pushed or has been compromised during the thinning procedure. If this occurs, use another animal for the experiment.

Problem (Steps 15 and 17): Artifacts, such as neurite blebbing or microglial process retraction, appear.
Solution: Use another animal for the experiment. Neurite blebbing is an indicator that neuronal structures have been disrupted and is concurrent with high spine turnover. Microglial process retraction is indicative of an inflammatory response. One or both of these may appear if the skull is pushed downward against the brain surface during thinning, the skull has been overthinned, or the skull has been fractured.

Problem (Step 18): The previously imaged region cannot be located.
Solution: Check the CCD or 10x image of blood vasculature, and try to relocate to the previously imaged region with obvious patterns of blood vasculature and dendritic branches. Try to glue the skull holder such that the imaged region is in the center of the holder opening and parallel to the holder.

DISCUSSION

Studies Using Transcranial Two-Photon Imaging

The method of transcranial two-photon imaging has been used successfully to image individual dendritic spines and axonal varicosities in multiple cortical regions, over intervals of minutes up to 19 mo (Grutzendler et al. 2002; Zuo et al. 2005a). Figure 2, D and E, shows examples of images obtained over a 3-d interval in mice at 4 mo of age. Note the remarkable stability of the number and location of adult spines and axonal varicosities between the two views (Fig. 2D,E). Most adult spines persist even over a 19-mo interval (Fig. 2F,G). The noninvasive nature of the thinned skull technique also allows microglia to be imaged, which are highly sensitive to their environment and typically respond immediately after various perturbations (Davalos et al. 2005; Xu et al. 2007). Figure 3 shows images of GFP-expressing microglia cells in the cerebellum over 2 d with minimal changes of their ramified shape under a normal physiological state.

Comparison of Thinned Skull Imaging and Imaging through a Cranial Window

Two-photon imaging through a thinned and intact skull is a relatively straightforward and minimally invasive method for high-resolution study of structural and functional changes in the living cortex. In addition to thinned skull preparations, the cortex can also be visualized through open-skull windows after performing a craniotomy and replacing the skull with a cover glass and/or a layer of agarose on top of the dural layer (see Chap. 30; Svoboda et al. 1997; Trachtenberg et al. 2002; Holtmaat et al. 2005, 2006, 2009; De Paola et al. 2006; Lee et al. 2006; Keck et al. 2008; Hofer et al. 2009). It is important to point out that skull removal can induce a significant inflammatory reaction and could, thus, confound efforts to elucidate changes of neurons and glia in the intact brain (Xu et al. 2007; Holtmaat et al. 2009; Yan et al. 2009). The thinned skull window approach not only causes minimal perturbation, but also allows imaging of the cortex immediately after surgery, as opposed to the open-skull method in which optimal imaging quality is achieved many days after craniotomy (Grutzendler et al. 2002; Trachtenberg et al. 2002; Holtmaat et al. 2005; Zuo et al. 2005a,b; Yang et al. 2009). Therefore, we suggest that two-photon imaging through a thinned skull window, rather than through an open-skull window, should be the method of choice for studying structure and function of the living cortex. In experiments that cannot easily be performed without removing the skull, it is important to interpret the data carefully in the context of the confounding factors associated with open-skull preparations.

Challenges of Preparing an Animal for Transcranial TPM

The main limitations of the thinned skull technique are related to the fact that skull thickness is critical for image quality and that the optimal thickness exists within narrow limits. For example, a nonuniform skull thickness may cause significant spherical aberrations, resulting in decreased two-photon excitation and distortion of fluorescent structures located deep in the cortex (Helm et al. 2009). We have found that the skull thickness should ideally be ~20 µm to obtain high-resolution images of synapses (Grutzendler et al. 2002; Zuo et al. 2005a; Xu et al. 2007). Overthinning the skull to <15 µm carries the risk of mild cortical trauma resulting from pushing the skull downward with the drill bit or microsurgical blade. In such cases, the thinning process generally leads to activation of microglia as manifested by their amoeboid morphology and is often associated with axonal and dendritic blebbing. Injury of this type can be prevented by carefully using the microsurgical blade at an angle during the thinning procedure and by avoiding pushing the skull against the cortical surface. Because of this, achieving an optimal thickness in a consistent way requires significant surgical practice. Furthermore, although multiple imaging sessions within the first few days after the initial surgery can be easily performed and require only minimal removal of debris over the thinned cranium, repeated imaging over intervals >2–3 d requires skull rethinning, which may present a challenge for a beginner. Even with sufficient surgical skill, it is often difficult to perform more than five imaging sessions on the same animal, as the optical properties of the skull may gradually deteriorate with repeated thinning.

TPM imaging through a thinned skull window has emerged in recent years as an important tool for studying structural and functional changes in the living cortex in a minimally invasive manner (Grutzendler et al. 2002; Yoder and Kleinfeld 2002; Stosiek et al. 2003; Tsai et al. 2004; Davalos et al. 2005; Zhang et al. 2005; Zuo et al. 2005a,b; Haynes et al. 2006; Majewska et al. 2006; Nishiyama et al. 2007; Xu et al. 2007; Kim et al. 2009; Wake et al. 2009). With the growing availability of fluorescent reporters to probe the structure and function of the brain (Giepmans et al. 2006; Livet et al. 2007), we envision transcranial TPM imaging will greatly expand our current understanding of how neural circuits are assembled and are modified throughout life and how cells function in the living brain.

RECIPES

CAUTION: See Appendix 6 for proper handling of materials marked with <!>.
Recipes for reagents marked with <R> are included in this list.

ACSF

NaCl	119 mM
$NaHCO_3$	26.2 mM
KCl	2.5 mM
NaH_2PO_4	1 mM
$MgCl_2$	1.3 mM
Glucose	10 mM

Gas with 5% CO_2 <!>/95% O_2 for 10–15 min, then add 2.5-mM $CaCl_2$ <!>. Filter sterilize with a 0.22-µm filter apparatus, and store at 4°C. ACSF is stable for 3–4 wk. If overt contamination (solution becomes cloudy) or precipitation is apparent, discard, and make fresh ACSF.

KX Mixture

Mix 20 mg/mL ketamine <!> (Fort Dodge) and 3 mg/mL xylazine <!> (Shenandoah) in saline. Store at room temperature (20°C). Properly stored KX solution is stable for up to 1 yr.

ACKNOWLEDGMENTS

This work was supported by grants from the National Institutes of Health to WBG and JG. GY was supported by a postdoctoral fellowship from the NYS Spinal Cord Injury Research Program.

REFERENCES

Christie RH, Bacskai BJ, Zipfel WR, Williams RM, Kajdasz ST, Webb WW, Hyman BT. 2001. Growth arrest of individual senile plaques in a model of Alzheimer's disease observed by in vivo multiphoton microscopy. *J Neurosci* 21: 858–864.

Davalos D, Grutzendler J, Yang G, Kim JV, Zuo Y, Jung S, Littman DR, Dustin ML, Gan WB. 2005. ATP mediates rapid microglial response to local brain injury in vivo. *Nat Neurosci* **8:** 752–758.

Denk W, Strickler JH, Webb WW. 1990. Two-photon laser scanning fluorescence microscopy. *Science* **248:** 73–76.

De Paola V, Holtmaat A, Knott G, Song S, Wilbrecht L, Caroni P, Svoboda K. 2006. Cell type–specific structural plasticity of axonal branches and boutons in the adult neocortex. *Neuron* **49:** 861–875.

Feng G, Mellor RH, Bernstein M, Keller-Peck C, Nguyen QT, Wallace M, Nerbonne JM, Lichtman JW, Sanes JR. 2000. Imaging neuronal subsets in transgenic mice expressing multiple spectral variants of GFP. *Neuron* **28:** 41–51.

Giepmans BN, Adams SR, Ellisman MH, Tsien RY. 2006. The fluorescent toolbox for assessing protein location and function. *Science* **312:** 217–224.

Grutzendler J, Kasthuri N, Gan WB. 2002. Long-term dendritic spine stability in the adult cortex. *Nature* **420:** 812–816.

Haynes SE, Hollopeter G, Yang G, Kurpius D, Dailey ME, Gan WB, Julius D. 2006. The P2Y12 receptor regulates microglial activation by extracellular nucleotides. *Nat Neurosci* **9:** 1512–1519.

Helm PJ, Ottersen OP, Nase G. 2009. Analysis of optical properties of the mouse cranium—Implications for in vivo multi photon laser scanning microscopy. *J Neurosci Methods* **178:** 316–322.

Hofer SB, Mrsic-Flogel TD, Bonhoeffer T, Hubener M. 2009. Experience leaves a lasting structural trace in cortical circuits. *Nature* 457: 313–317.

Holtmaat AJ, Trachtenberg JT, Wilbrecht L, Shepherd GM, Zhang X, Knott GW, Svoboda K. 2005. Transient and persistent dendritic spines in the neocortex in vivo. *Neuron* **45:** 279–291.

Holtmaat A, Wilbrecht L, Knott GW, Welker E, Svoboda K. 2006. Experience-dependent and cell-type-specific spine growth in the neocortex. *Nature* **441:** 979–983.

Holtmaat A, Bonhoeffer T, Chow DK, Chuckowree J, De Paola V, Hofer SB, Hubener M, Keck T, Knott G, Lee WC, et al. 2009. Long-term, high-resolution imaging in the mouse neocortex through a chronic cranial window. *Nat Protoc* **4:** 1128–1144.

Jung S, Aliberti J, Graemmel P, Sunshine MJ, Kreutzberg GW, Sher A, Littman DR. 2000. Analysis of fractalkine receptor CX(3)CR1 function by targeted deletion and green fluorescent protein reporter gene insertion. *Mol Cell Biol* **20:** 4106–4114.

Keck T, Mrsic-Flogel TD, Vaz Afonso M, Eysel UT, Bonhoeffer T, Hubener M. 2008. Massive restructuring of neuronal circuits during functional reorganization of adult visual cortex. *Nat Neurosci* **11:** 1162–1167.

Kim JV, Kang SS, Dustin ML, McGavern DB. 2009. Myelomonocytic cell recruitment causes fatal CNS vascular injury during acute viral meningitis. *Nature* **457:** 191–195.

Lee WC, Huang H, Feng G, Sanes JR, Brown EN, So PT, Nedivi E. 2006. Dynamic remodeling of dendritic arbors in GABAergic interneurons of adult visual cortex. *PLoS Biol* **4:** e29.

Livet J, Weissman TA, Kang H, Draft RW, Lu J, Bennis RA, Sanes JR, Lichtman JW. 2007. Transgenic strategies for combinatorial expression of fluorescent proteins in the nervous system. *Nature* **450:** 56–62.

Majewska AK, Newton JR, Sur M. 2006. Remodeling of synaptic structure in sensory cortical areas in vivo. *J Neurosci* **26:** 3021–3029.

Nimmerjahn A, Kirchhoff F, Helmchen F. 2005. Resting microglial cells are highly dynamic surveillants of brain parenchyma in vivo. *Science* **308:** 1314–1318.

Nishiyama H, Fukaya M, Watanabe M, Linden DJ. 2007. Axonal motility and its modulation by activity are branch-type specific in the intact adult cerebellum. *Neuron* **56:** 472–487.

Stosiek C, Garaschuk O, Holthoff K, Konnerth A. 2003. In vivo two-photon calcium imaging of neuronal networks. *Proc Natl Acad Sci* **100:** 7319–7324.

Svoboda K, Denk W, Kleinfeld D, Tank DW. 1997. In vivo dendritic calcium dynamics in neocortical pyramidal neurons. *Nature* **385:** 161–165.

Trachtenberg JT, Chen BE, Knott GW, Feng G, Sanes JR, Welker E, Svoboda K. 2002. Long-term in vivo imaging of experience-dependent synaptic plasticity in adult cortex. *Nature* **420:** 788–794.

Tsai J, Grutzendler J, Duff K, Gan WB. 2004. Fibrillar amyloid deposition leads to local synaptic abnormalities and breakage of neuronal branches. *Nat Neurosci* **7:** 1181–1183.

Wake H, Moorhouse AJ, Jinno S, Kohsaka S, Nabekura J. 2009. Resting microglia directly monitor the functional state of synapses in vivo and determine the fate of ischemic terminals. *J Neurosci* **29:** 3974–3980.

Wu SH, Arevalo JC, Sarti F, Tessarollo L, Gan WB, Chao MV. 2009. Ankyrin repeat-rich membrane spanning/Kidins220 protein regulates dendritic branching and spine stability in vivo. *Dev Neurobiol* **69:** 547–557.

Xu HT, Pan F, Yang G, Gan WB. 2007. Choice of cranial window type for in vivo imaging affects dendritic spine turnover in the cortex. *Nat Neurosci* **10:** 549–551.

Yan P, Bero AW, Cirrito JR, Xiao Q, Hu X, Wang Y, Gonzales E, Holtzman DM, Lee JM. 2009. Characterizing the appearance and growth of amyloid plaques in APP/PS1 mice. *J Neurosci* **29:** 10706–10714.

Yang G, Pan F, Gan WB. 2009. Stably maintained dendritic spines are associated with lifelong memories. *Nature* **462:** 920–924.

Yoder EJ, Kleinfeld D. 2002. Cortical imaging through the intact mouse skull using two-photon excitation laser scanning microscopy. *Microsc Res Tech* **56:** 304–305.

Zhang ZG, Zhang L, Ding G, Jiang Q, Zhang RL, Zhang X, Gan WB, Chopp M. 2005. A model of mini-embolic stroke offers measurements of the neurovascular unit response in the living mouse. *Stroke* **36:** 2701–2704.

Zuo Y, Lin A, Chang P, Gan WB. 2005a. Development of long-term dendritic spine stability in diverse regions of cerebral cortex. *Neuron* **46:** 181–189.

Zuo Y, Yang G, Kwon E, Gan WB. 2005b. Long-term sensory deprivation prevents dendritic spine loss in primary somatosensory cortex. *Nature* **436:** 261–265.

32 Imaging Neocortical Neurons through a Chronic Cranial Window

Anthony Holtmaat,[1] Vincenzo de Paola,[2] Linda Wilbrecht,[3] Josh T. Trachtenberg,[4] Karel Svoboda,[5] and Carlos Portera-Cailliau[6]

[1]Department of Basic Neuroscience, University of Geneva, Geneva 1211, Switzerland; [2]MRC Clinical Sciences Centre, Imperial College London, London W12 0NN, United Kingdom; [3]Ernest Gallo Clinic and Research Center, University of California, San Francisco, Emeryville, California 94608; [4]Department of Neurobiology, University of California, Los Angeles, Los Angeles, California 90095; [5]Janelia Farm Research Campus, Ashburn, Virginia 20147; [6]Departments of Neurology and Neurobiology, David Geffen School of Medicine, University of California, Los Angeles, Los Angeles, California 90095

ABSTRACT

The rich structural dynamics of axonal arbors and neuronal circuitry can only be revealed through direct and repeated observations of the same neuron(s) over time, preferably in vivo. Here we describe a long-term, high-resolution method for imaging neocortical neurons in vivo, using a combination of two-photon laser scanning microscopy and a surgically implanted chronic cranial window. Using this approach, it is feasible to image the smallest neuronal structures in the superficial layers of the neocortex, such as dendritic spines and axonal boutons. Because the surface area of the craniotomy is relatively large, this technique is even suitable for use when labeled neurons are relatively uncommon. The surgery and imaging procedures are illustrated with examples from our studies of structural plasticity in the developing or adult mouse brain.

INTRODUCTION

Developmental and activity-dependent changes in the brain can take anywhere from milliseconds to weeks to occur. Historically, structural plasticity was inferred from comparisons of the average neuronal morphology in one group of fixed specimens to another. Not only was this imprecise and indirect, but this approach would also miss changes in steady state (e.g., when the gain and loss of dendritic spines is in balance). Therefore, to thoroughly understand adaptive changes in the brain, it is essential to monitor the structure and function of individual, identified neurons in the intact brain over long timescales.

Using two-photon excitation laser-scanning microscopy (2PLSM) of fluorescently labeled cortical neurons, synaptic structures can be monitored in vivo over periods of minutes or months (Denk et al. 1990; Svoboda and Yasuda 2006). Typically, cortical neurons can be visualized to a depth of ~500 µm, deep enough to include layers 1–4 of the mouse neocortex.

For imaging neurons in vivo with 2PLSM, the fluorophores of choice are the green fluorescent protein (GFP) and its derivatives (XFPs). The spectral characteristics of these proteins are comparable to some of the better synthetic fluorophores, they have large extinction ratios and quantum efficiencies over a broad spectral range (e.g., 830–1020 nm for EGFP), and they are quite resistant to photobleaching (Giepmans et al. 2006). One potential drawback is that high levels of fluorophore expression are necessary to visualize small cellular compartments, such as dendritic spines. It has been estimated that cytoplasmic GFP must be >1 µM to reliably detect dendritic spines (Niswender et al. 1995), which exceeds the concentration of the vast majority of endogenous proteins expressed by neurons.

Germline transgenesis has been the most common method for labeling neurons in vivo. Spatial and temporal control of fluorophore expression can be achieved using either a minimal promoter (Holtmaat et al. 1998) or engineered bacterial artificial chromosomes for transgenesis (Heintz 2004). The most commonly used transgenes for imaging neuronal structure in vivo use the *Thy1* promoter to drive high-level expression of the fluorophore (e.g., GFP-M, YFP-H, mGFP-L21) in subpopulations of neurons (Caroni 1997; Feng et al. 2000; De Paola et al. 2003). Electroporation techniques and viral vectors (e.g., Lentivirus and adeno-associated virus-based constructs) have also been increasingly useful to express foreign genes for long-term imaging of the adult brain (e.g., Gray et al. 2006; Stettler et al. 2006; Nishiyama et al. 2007).

The skull of most mammals is too opaque to allow high-resolution imaging of cortical neurons. Therefore, in all 2PLSM imaging studies in vivo, the overlying bone is partially or completely removed, but the dura mater is left intact. Here we describe the procedure to implant a chronic cranial window (Fig. 1) and its subsequent use for in vivo imaging and analysis (Fig. 2). See Mostany and Portera-Cailliau (2008) and Holtmaat et al. (2009) for detailed descriptions and discussions of this method.

Protocol

Imaging through a Chronic Cranial Window

This protocol is optimized for adult mice; we have used mice up to postnatal day 511 (P511). With minor modifications, it is possible to image neurons in rats and mice from P2 (Portera-Cailliau et al. 2005). Most of our studies have used the *Thy1* promoter to drive expression of fluorophores in subsets of cortical neurons (Feng et al. 2000; Trachtenberg et al. 2002; De Paola et al. 2003, 2006; Holtmaat et al. 2005, 2006; Portera-Cailliau et al. 2005).

IMAGING SETUP

In vivo images of fluorescently labeled neurons are best acquired using a custom-built 2PLSM (e.g., Lendvai et al. 2000; Helmchen and Denk 2005) with the animal positioned very close to the fluorescence detectors. We use a tunable Ti:sapphire laser (Tsunami; Spectra Physics), water-immersion objectives with a high NA (20x–60x magnification, 0.8–0.9 NA, optimized for IR light), and photomultiplier tubes with high quantum efficiencies and low dark currents (Hamamatsu R3896 or H7422P). ScanImage software (Pologruto et al. 2003), custom written in MATLAB (Mathworks), is used for image acquisition and processing. Off-the-shelf 2PLSM setups are also available from major microscope manufacturers.

MATERIALS

CAUTION: See Appendix 6 for proper handling of materials marked with <!>.
See the end of the chapter for recipes for reagents marked with <R>.

Reagents

All solutions should be sterile.

Agarose, 1.2% or 1.5% in cortex buffer (cooled to <37°C immediately before use)
Betadine (optional; see Step 4)
Buprenorphine, 0.3 mg/mL (optional; see Step 15)
Carprofen <!>, 50 mg/mL (optional; see Step 15)
Cortex buffer <R>
Cyanoacrylate (CA) glue <!> (e.g., Vetbond or 3M)
Dental acrylic or cement (e.g., Jet or Lang Dental)
Dexamethasone sodium phosphate <!>, 4 mg/mL (dexamethasone, final, 3 mg/mL)
Ethanol <!>, 70% (optional; see Step 4)
Eye ointment (sterile lubricant eye ointment [Stye, Del Pharmaceuticals] or optical gel [Nye Optical])
Isoflurane, <!> 2% (optional; anesthetic method)
Ketamine <!>, 13 mg/mL

Used in combination with xylazine for anesthesia.

Saline
Transgenic mice, expressing fluorophore in cells of interest

Xylazine <!>, 1 mg/mL

Used in combination with ketamine for anesthesia.

Xylocaine (lidocaine) <!>, 1%

Equipment

Confocal microscope, customized, and associated instruments (see Imaging Setup)
Cotton swabs
Dental drill, high-speed, pneumatic, 1/4-in bit (e.g., Dental Burs USA or Henry Schein)
Dissecting microscope
Gelfoam (e.g., Pharmacia, Upjohn)
Glass coverslips, # 1, 5-mm-diameter, or custom-made
Heating blanket
Razor

For shaving the scalp.

Stainless steel bars, 10 x 4 x 15-mm, custom, with screw holes
Stereotaxic frame (e.g., Stoelting)
Surgical instruments, including scalpel blade, dissection scissors, and fine forceps
Syringes and small-gauge injection needles

EXPERIMENTAL METHOD

Surgery

Humane treatment of animals must be observed at all times and should follow the local institutional guidelines for care and use of animals.

1. Anesthetize the mouse deeply with a mixture of ketamine and xylazine (intraperitoneal injection of 100 µg/g and 10 µg/g body weight, respectively) or with a continuous flow of gas anesthesia (e.g., 2% isoflurane).
2. Place the animal on a heating pad to maintain body temperature, and immobilize the head in a stereotaxic frame. Apply eye ointment to the eyes.

   It is important to protect the eyes from dehydration and irritation by dental acrylic.
3. Administer dexamethasone (~2 µg/g body weight) by intramuscular injection to the quadriceps muscle.

   This minimizes the cortical stress response during chronic experiments.
4. Shave the head using the razor, and wash the scalp with ethanol or Betadine. Using scissors, remove a 1-cm^2 flap of skin to reveal the skull overlying the barrel cortex of both hemispheres.

   All equipment and reagents should be sterile from this step in the procedure.
5. Apply 1% xylocaine to the skull and exposed muscles using cotton-swab applicators.
6. Using a scalpel blade, scrape away the fascia covering the skull. Separate the right lateral muscle (temporalis) from the bone and push it ventrally.

   It is important to avoid bleeding as much as possible, to prevent blood clots from obscuring the cranial window.
7. Mark an area of interest over the cortex. Surround this area with a thin layer of CA glue applied to the skull, muscle, and wound margins to stop bleeding and prevent seepage of serosanguineous fluid. After the glue has dried, apply a thin layer of dental acrylic on top of the glue.

   The layer of CA glue improves the adherence of the dental acrylic.
8. Use the dental drill to thin the circumference of an area of skull (2–3-mm-diameter) (Fig. 1B,

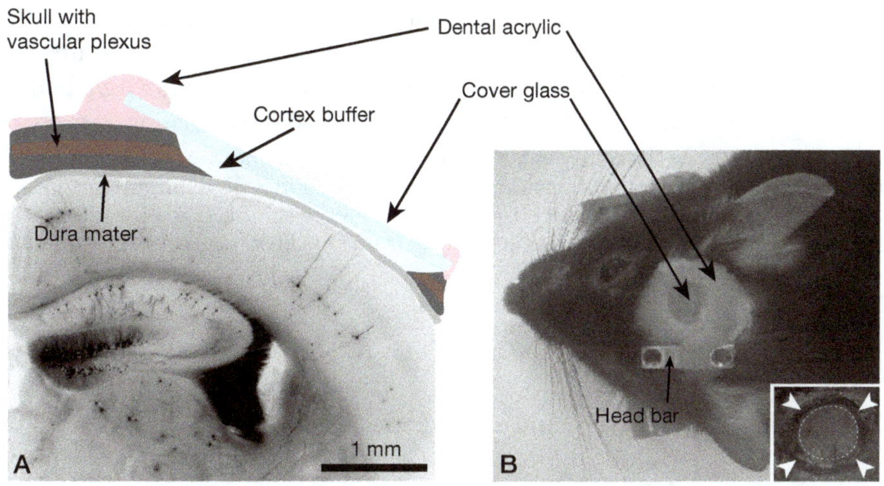

FIGURE 1. Experimental preparation for in vivo imaging. (*A*) Schematic view of the cranial window in cross section superposed on a histologic section of neocortex from a GFP-M transgenic mouse. (*B*) External view of a cranial window implant; (*inset*) the circle of thinned bone (arrowheads) around an island of skull (dotted circle) that will be removed during the craniotomy.

inset). Work slowly to avoid heating the bone and damaging the dura mater. Keep the skull moist with cortex buffer.

A circular shape is the simplest, with an island of skull left intact in the center. The cortical surface vasculature should become visible where the skull is thinned, and there should be minimal bleeding from the skull and none from the brain. Take care not to apply pressure to the skull and brain or to puncture the skull while drilling.

9. Gently perforate the thinned skull with the sharp point of a pair of forceps, and lift up the island of bone within the drilled circle to expose the dura.

   Do not exert any pressure on the brain. This removal is most successful if performed under a puddle of cortex buffer. Gelfoam is used to control any bleeding from the dura and skull and can be applied directly to the dura when wet. Prior to use, the Gelfoam should be cut into 7-mm^2 squares and soaked in cortex buffer.

10. Cover the skull opening with the cover glass, which should lie flush with the skull. To reduce movement during imaging, a thin layer of molten 1.2% agarose can be applied to the brain surface before implanting the cover glass.

    The coverslip should be ~100 µm from the brain surface and should be wider than the skull window. The agarose protects the brain from the air and the dental cement. The 1.2% agarose must be made fresh and sterile for each experiment and kept molten during the procedure. Cool the molten agarose to <37°C before applying it to the cortex.

11. If agarose was used, carefully remove any excess from the edges of the window once it is set, and dry the remaining liquid on the skull.

12. Seal the optical window to the skull with dental acrylic, covering all exposed areas of skull, wound margins, and edges of the cover glass (Fig. 1B). Do not allow any glue to contact the dura, or it will lose its transparency.

13. Embed the stainless steel bar into the dental acrylic over the intact hemisphere (Fig. 1B).

    The bar is used to stabilize the animal during imaging. Ensure that the screw threads are clear of acrylic, and allow the acrylic to set before manipulating the mouse or the bar (30 min is typically sufficient).

14. Imaging experiments may be performed for the duration of the anesthesia (typically up to 1 h after surgery using ketamine/xylazine anesthetic).

    See Image Acquisition, Step 17.

15. For chronic experiments, allow the mouse to recover on the heating pad before returning it to its cage. Allow 10–14 d for recovery before subsequent imaging sessions. Buprenorphine (0.1 µg/g body weight, subcutaneous) and Carprofen (5 µg/g body weight, intraperitoneal) may be used to reduce postoperative pain and inflammation, respectively.

    The clarity of the cranial window often improves during the recovery period and can remain clear for months without the use of antibiotics. To reduce stress for the mouse, minimize the size of the stainless steel bar so that the mouse can freely move in its cage. A large cage with a high ceiling can help prevent the mice from knocking the dental acrylic and losing a bar or the entire window and headcap.

Image Acquisition

16. Anesthetize animals with ketamine/xylazine (65 µg/g and 6.5 µg/g, respectively). Place the animals under the microscope on a heating pad, and stabilize the head using screws placed through the bars.
17. Trail 1.5% agarose around the edge of the imaging window, making a small enclosure to hold water for the objective during imaging.
18. In the first imaging session, acquire a bright-field image of the vasculature in the dura mater (Fig. 2A).

    This provides important anatomical landmarks that will be used to relocate previously imaged neurons in subsequent imaging sessions.

    See *Troubleshooting.*
19. Acquire images of apical dendritic tufts of fluorescently labeled pyramidal neurons using 2PLSM. To relocate previously imaged cells, use the unique vascular pattern near the cell to match fluorescent images to the bright-field image (Fig. 2A,B).

    For high-magnification spine imaging of second and higher-order branches, we typically select seven to 15 fields, each 50 x 50 µm, for each cell (Fig. 2C). Our image stacks typically consist of sections (512 x 512 pixels; ~0.09 µm/pixel, 2 msec/line) collected in 1-µm steps (three to five samples per resolution element).

    See *Troubleshooting.*
20. Collect low-magnification fluorescent images (512 x 512 pixels; ~0.3 µm/pixel; 3-µm steps) for overview (Fig. 2B). Take care to achieve close to identical fluorescence levels across imaged regions and imaging sessions by adjusting the excitation power.

Image Analysis

21. Use predetermined criteria to score the images, as follows.
    i. Align related images from the same time point using stable anatomical reference points, such as dendritic branch points.
    ii. Assign a single-blind code to the images to mask the experimental parameters (e.g., cortical region, developmental age, and experimental manipulation) before being scored by a naïve analyst.
    iii. Use length measurements to quantify the appearance and disappearance of dendritic spines and *terminaux* axonal boutons.
    iv. Use threshold intensity ratio measurements to determine changes in the volume of spines or *en passant* boutons that are continuous with the axonal shaft (Holtmaat et al. 2005; De Paola et al. 2006).

        Threshold intensity ratio measurements are relatively easy to acquire from 2PLSM images, because the point-spread function in the *z*-axis typically exceeds the size of small structures

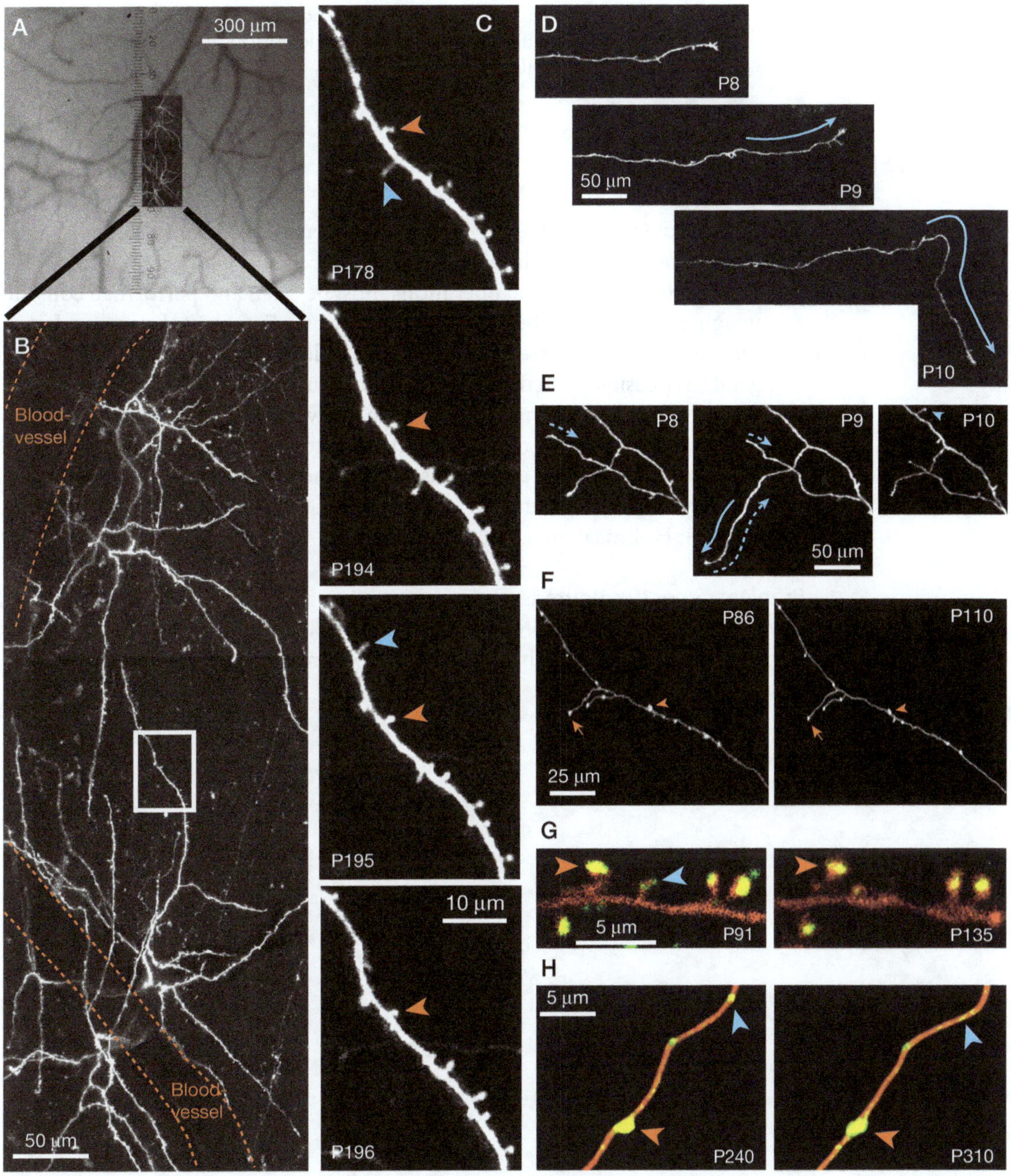

FIGURE 2. Long-term imaging of GFP-expressing pyramidal neurons in transgenic mice. (*A*) Composite bright-field image of the exposed dura mater and the high-resolution 2PLSM image in *B*, demonstrating how the vascular tree is used as a reference to locate specific neurons. (*B*) Top view of dendritic tufts on two pyramidal neurons in layer 5 (projection of a stack of optical sections). Blood vessels and a region of interest (white box) are indicated. (*C*) Long-term images revealing stable (orange) and transient (blue) dendritic spines from the region of interest (boxed in *A*). (*D,E*) Long-term images of postnatal thalamocortical neurons in the developing mouse neocortex showing axonal growth (solid arrows), retraction (dotted arrows), and a new *terminaux* bouton (blue arrowhead). (Modified, with permission, from Portera-Cailliau et al. 2005.) (*F*) Examples of stable branches (arrows) and boutons (arrowheads) on thalamocortical axons in the adult mouse neocortex. (Modified, with permission of Elsevier, from De Paola et al. 2006.) (*G*) Presumptive synaptic contacts indicated by the presence of PSD95-GFP (green) on DsRed-labeled dendritic spines (red); stable (orange arrowhead) and transient (blue arrowhead) synapses are indicated. (*H*) Synaptophysin-GFP (green) identifies synaptic contacts on axonal boutons expressing DsRed (red). Some synapses are stable (orange arrowhead), but other varicosities appear and disappear (blue arrowheads), possibly indicating synaptic changes. (*G* and *H*, Modified, by permission of Macmillan Publishers Ltd., from Holtmaat and Svoboda 2009.) In all images, (P) postnatal day. Panels *A–C* and *F* are GFP-M transgenics; *D* and *F* are mGFP-L21 transgenics; neurons in *F* and *G* were transfected in utero. All images are best projections, as described (Holtmaat et al. 2005; Portera-Cailliau et al. 2005; De Paola et al. 2006).

such as dendritic spines and axonal boutons. Therefore, the integrated fluorescence intensity of these microstructures in a single optical section is a good indicator of their volumes (Holtmaat et al. 2005).

TROUBLESHOOTING

Problem (Step 18): The imaging window is obscured or the image quality is poor.

Solution: Poor surgical technique is the most common source of problems with high-resolution chronic imaging. Surgical operations must be clean and performed consistently and quickly. The imaging window may become obscured by bleeding from the dura, edema of the neocortex, and infection. These adverse events can be avoided by keeping the operation sterile and by being as gentle as possible with dura and brain to minimize tissue damage and reaction. Regrowth of the skull is inevitable and may occur after a few weeks or months. Aberrant structural distortions and glial cell reactivity may also contribute to poor image quality and should be assessed at several time points after surgery using routine histological methods and immunostaining, and should be mild and transient (<2 wk for astrocytes, <1 wk for microglia) (Xu et al. 2007; Lee et al. 2008; Holtmaat et al. 2009).

Problem (Step 19): Boutons or dendritic spines are difficult to resolve.

Solution: Fluorophore expression may not be sufficiently high. Switch to a transgenic line with a stronger promoter to drive higher-level expression in the cell of interest.

DISCUSSION

High-resolution chronic imaging in vivo can be achieved using either a glass-covered cranial window (Levasseur et al. 1975; Brown et al. 2001; Trachtenberg et al. 2002) or a thin-skull preparation (Christie et al. 2001; Grutzendler et al. 2002). Both techniques yield comparable results (Majewska et al. 2006; Holtmaat et al. 2009; Livneh et al. 2009) but have their own advantages and disadvantages. The cranial window provides excellent optical access and exposes a relatively large area (~0.8–12 mm^2), and imaging can be repeated with unlimited frequency so long as the window does not become obscured, as discussed above. The cranial window is the method of choice for analysis of animals with a low proportion of fluorescently labeled cells (e.g., GFP-M; see above), cell reconstructions, and post hoc anatomy (Knott et al. 2006). In contrast, thinning of the skull (to ~20 µm) is usually performed over a very small area (~0.1–0.3 mm^2) to avoid contusion of the underlying cortex and is therefore more suitable for mice with a high proportion of labeled cells (e.g., YFP-H; see above). Although the thinning procedure must be repeated before every imaging session and can only be performed fewer than four times, it can be carried out at any time during the life of the animal.

Quantification of Morphological Changes

The cranial window provides excellent optical access to fluorescently labeled neurons. In control experiments, including direct comparisons with electron microscopy, we have confirmed that even the smallest dendritic spines can be reliably detected in vivo using this method, with certain limitations (Holtmaat et al. 2005). First, analysis should be limited to protrusions with fluorescence intensity that is greater than severalfold above background, based on the standard deviation of the background fluorescence close to the dendrite. Second, a major drawback of long-working-distance, water-immersion objectives is their poor resolving power in the (optical) *z*-axis, owing to their low NA. Therefore, analysis should be restricted to protrusions that emanate laterally from the dendritic shaft, irrespective of apparent shape, whereas those that project above or below the dendrite in the *z*-axis should not be scored. We use custom software to analyze dendritic spines and axonal boutons in three dimensions rather than in projections, discussed in detail in Holtmaat et al. (2005) and De Paola et al. (2006).

The reliability of measurements acquired in vivo should be assessed as follows: Estimates of spine and bouton densities acquired from in vivo images should be comparable with estimates from the same region in the fixed, dissected brain. Structural dynamics and spine or bouton densities should be stable over long time frames in naive adult mice (Trachtenberg et al. 2002; Holtmaat et al. 2005, 2006, 2009; De Paola et al. 2006; Lee et al. 2006, 2008; Keck et al. 2008; Hofer et al. 2009).

Chronic in vivo imaging of the neocortex has revealed how dendritic and axonal structures can be either remarkably dynamic or stable during development and experience-dependent learning in adults (Fig. 2C-F; Grutzendler et al. 2002; Trachtenberg et al. 2002; Mizrahi and Katz 2003; Holtmaat et al. 2005; Portera-Cailliau et al. 2005; Zuo et al. 2005a; De Paola et al. 2006; Lee et al. 2006; Majewska et al. 2006; Stettler et al. 2006; Nishiyama et al. 2007). In vivo 2PLSM can also be used to study the distributions of synaptic proteins (Fig. 2G,H; Gray et al. 2006). Structural dynamics in developing and young-adult tissue often result in net changes of neuronal complexity and synaptic density (Fig. 2D,E; Holtmaat et al. 2005; Portera-Cailliau et al. 2005; Zuo et al. 2005a), whereas in naive adults there is a dynamic balance between neurite growth and retraction, and synapse formation and loss (Fig. 2C,F; Grutzendler et al. 2002; Trachtenberg et al. 2002; Holtmaat et al. 2005; De Paola et al. 2006; Lee et al. 2006; Majewska et al. 2006). Because growth and retraction of spines and boutons are linked to synapse formation and elimination (De Paola et al. 2006; Knott et al. 2006), the structural dynamics associated with new experiences or changes in sensory input may alter connectivity within neuronal networks, at least temporarily (Zuo et al. 2005b; Holtmaat et al. 2006; Keck et al. 2008; Hofer et al. 2009).

RECIPE

CAUTION: See Appendix 6 for proper handling of materials marked with <!>.
Recipes for reagents marked with <R> are included in this list.

Cortex Buffer

NaCl	125 mM
KCl	5 mM
Glucose	10 mM
HEPES	10 mM
$CaCl_2$ <!>	2 mM
$MgSO_4$ <!>	2 mM

Adjust the pH to 7.4.

ACKNOWLEDGMENTS

This work was supported by HHMI, NIH, the Swiss National Science Foundation, the IRP Foundation, the Larry L. Hillblom Foundation, the March of Dimes Foundation, the FRAXA, and the Dana Foundation.

REFERENCES

Brown EB, Campbell RB, Tsuzuki Y, Xu L, Carmeliet P, Fukumura D, Jain RK. 2001. In vivo measurement of gene expression, angiogenesis and physiological function in tumors using multiphoton laser scanning microscopy. *Nat Med* **7:** 864–868.

Caroni P. 1997. Overexpression of growth-associated proteins in the neurons of adult transgenic mice. *J Neurosci Methods* **71:** 3–9.

Christie RH, Bacskai BJ, Zipfel WR, Williams RM, Kajdasz ST, Webb WW, Hyman BT. 2001. Growth arrest of individual senile plaques in a model of Alzheimer's disease observed by in vivo multiphoton microscopy. *J Neurosci* **21:** 858–864.

Denk W, Strickler JH, Webb WW. 1990. Two-photon laser scanning microscopy. *Science* **248:** 73–76.

De Paola V, Arber S, Caroni P. 2003. AMPA receptors regulate dynamic equilibrium of presynaptic terminals in mature hippocampal networks. *Nat Neurosci* **6:** 491–500.

De Paola V, Holtmaat A, Knott G, Song S, Wilbrecht L, Caroni P, Svoboda K. 2006. Cell type–specific structural plasticity of axonal branches and boutons in the adult neocortex. *Neuron* **49:** 861–875.

Feng G, Mellor RH, Bernstein M, Keller-Peck C, Nguyen QT, Wallace M, Nerbonne JM, Lichtman JW, Sanes JR. 2000. Imaging neuronal subsets in transgenic mice expressing multiple spectral variants of GFP. *Neuron* **28:** 41–51.

Giepmans BN, Adams SR, Ellisman MH, Tsien RY. 2006. The fluorescent toolbox for assessing protein location and function. *Science* **312:** 217–224.

Gray NW, Weimer RM, Bureau I, Svoboda K. 2006. Rapid redistribution of synaptic PSD-95 in the neocortex in vivo. *PLoS Biol* **4:** e370. doi: 10.1371/journal.pbio.0040370.

Grutzendler J, Kasthuri N, Gan WB. 2002. Long-term dendritic spine stability in the adult cortex. *Nature* **420:** 812–816.

Heintz N. 2004. Gene Expression Nervous System Atlas (GENSAT). *Nat Neurosci* **7:** 483. doi: 10.1038/nn0504-483.

Helmchen F, Denk W. 2005. Deep tissue two-photon microscopy. *Nat Methods* **2:** 932–940.

Hofer SB, Mrsic-Flogel TD, Bonhoeffer T, Hubener M. 2009. Experience leaves a lasting structural trace in cortical circuits. *Nature* **457:** 313–317.

Holtmaat A, Svoboda K. 2009. Experience-dependent structural synaptic plasticity in the mammalian brain. *Nat Rev Neurosci* **10:** 647–658 (Erratum **10:** 759).

Holtmaat AJ, Oestreicher AB, Gispen WH, Verhaagen J. 1998. Manipulation of gene expression in the mammalian nervous system: Application in the study of neurite outgrowth and neuroregeneration-related proteins. *Brain Res Brain Res Rev* **26:** 43–71.

Holtmaat AJ, Trachtenberg JT, Wilbrecht L, Shepherd GM, Zhang X, Knott GW, Svoboda K. 2005. Transient and persistent dendritic spines in the neocortex in vivo. *Neuron* **45:** 279–291.

Holtmaat A, Wilbrecht L, Knott GW, Welker E, Svoboda K. 2006. Experience-dependent and cell-type-specific spine growth in the neocortex. *Nature* **441:** 979–983.

Holtmaat A, Bonhoeffer T, Chow DK, Chuckowree J, De Paola V, Hofer SB, Hübener M, Keck T, Knott G, Lee W-CA. 2009. Long-term, high-resolution imaging in the mouse neocortex through a chronic cranial window. *Nat Protoc* **4:** 1128–1144.

Keck T, Mrsic-Flogel TD, Vaz Afonso M, Eysel UT, Bonhoeffer T, Hubener M. 2008. Massive restructuring of neuronal circuits during functional reorganization of adult visual cortex. *Nat Neurosci* **11:** 1162–1167.

Knott GW, Holtmaat A, Wilbrecht L, Welker E, Svoboda K. 2006. Spine growth precedes synapse formation in the adult neocortex in vivo. *Nat Neurosci* **9:** 1117–1124.

Lee WC, Huang H, Feng G, Sanes JR, Brown EN, So PT, Nedivi E. 2006. Dynamic remodeling of dendritic arbors in GABAergic interneurons of adult visual cortex. *PLoS Biol* **4:** e29. doi: 10.1371/journal.pbio.0040029.

Lee WC, Chen JL, Huang H, Leslie JH, Amitai Y, So PT, Nedivi E. 2008. A dynamic zone defines interneuron remodeling in the adult neocortex. *Proc Natl Acad Sci* **105:** 19968–19973.

Lendvai B, Stern E, Chen B, Svoboda K. 2000. Experience-dependent plasticity of dendritic spines in the developing rat barrel cortex in vivo. *Nature* **404:** 876–881.

Levasseur JE, Wei EP, Raper AJ, Kontos AA, Patterson JL. 1975. Detailed description of a cranial window technique for acute and chronic experiments. *Stroke* **6:** 308–317.

Livneh Y, Feinstein N, Klein M, Mizrahi A. 2009. Sensory input enhances synaptogenesis of adult-born neurons. *J Neurosci* **29:** 86–97.

Majewska AK, Newton JR, Sur M. 2006. Remodeling of synaptic structure in sensory cortical areas in vivo. *J Neurosci* **26:** 3021–3029.

Mizrahi A, Katz LC. 2003. Dendritic stability in the adult olfactory bulb. *Nat Neurosci* **6:** 1201–1207.

Mostany R, Portera-Cailliau C. 2008. A craniotomy surgery procedure for chronic brain imaging. *J Vis Exp* **12:** 680. doi: 10.3791/680.

Nishiyama H, Fukaya M, Watanabe M, Linden DJ. 2007. Axonal motility and its modulation by activity are branch-type specific in the intact adult cerebellum. *Neuron* **56:** 472–487.

Niswender KD, Blackman SM, Rohde L, Magnuson MA, Piston DW. 1995. Quantitative imaging of green fluorescent protein in cultured cells: Comparison of microscopic techniques, use in fusion proteins and detection limits. *J Microsc* **180:** 109–116.

Pologruto TA, Sabatini BL, Svoboda K. 2003. ScanImage: Flexible software for operating laser-scanning microscopes. *Biomed Eng Online* **2:** 13. doi: 10.1186/1475-925X-2-13.

Portera-Cailliau C, Weimer RM, Paola VD, Caroni P, Svoboda K. 2005. Diverse modes of axon elaboration in the developing neocortex. *PLoS Biol* **3:** e272. doi: 10.1371/journal.pbio.0030272.

Stettler DD, Yamahachi H, Li W, Denk W, Gilbert CD. 2006. Axons and synaptic boutons are highly dynamic in adult visual cortex. *Neuron* **49:** 877–887.

Svoboda K, Yasuda R. 2006. Principles of two-photon excitation microscopy and its applications to neuroscience. *Neuron* **50:** 823–839.

Trachtenberg JT, Chen BE, Knott GW, Feng G, Sanes JR, Welker E, Svoboda K. 2002. Long-term in vivo imaging of experience-dependent synaptic plasticity in adult cortex. *Nature* **420:** 788–794.

Xu HT, Pan F, Yang G, Gan WB. 2007. Choice of cranial window type for in vivo imaging affects dendritic spine turnover in the cortex. *Nat Neurosci* **10:** 549–551.

Zuo Y, Lin A, Chang P, Gan WB. 2005a. Development of long-term dendritic spine stability in diverse regions of cerebral cortex. *Neuron* **46:** 181–189.

Zuo Y, Yang G, Kwon E, Gan WB. 2005b. Long-term sensory deprivation prevents dendritic spine loss in primary somatosensory cortex. *Nature* **436:** 261–265.

33 Maintaining Live Cells and Tissue Slices in the Imaging Setup

Michael E. Dailey, Glen S. Marrs,[1] and Dana Kurpius

Department of Biology, University of Iowa, Iowa City, Iowa 52242

ABSTRACT

Live imaging provides exciting opportunities to study dynamic cellular events as they occur in real time. However, such experimental procedures present several challenges. This chapter discusses some of the major considerations relating to the maintenance of live biological samples during ex vivo imaging and presents some relatively simple low-cost solutions for keeping samples healthy.

INTRODUCTION

The development of new fluorescent probes and GFP (green fluorescent protein) reporter animal models, coupled with improved access to and convenience of imaging technologies, has generated a good deal of excitement in the field. One of the challenges that would-be users face is the application of these technologies to live specimens that must remain viable in the imaging setup. Although some preparation methods permit high-resolution imaging of live cells in vivo, in many cases, it is still necessary or desirable to isolate the cells and tissues so that they may be stained and, subsequently, observed over time. Such procedures can compromise the health of the cells and tissues. The length of time needed to make the required observations will clearly dictate the kinds of measures taken to maintain the specimens. Experimental observations made over the course of many hours require a different level of specimen maintenance than those made over only a few minutes. Under optimal conditions, it is possible to maintain viable tissues on the stage of a microscope for many hours—even days—while collecting valuable information on dynamic events in live cells and tissues.

How can biological samples be mounted and be maintained to last long enough to make meaningful observations? This chapter offers some practical advice based on the investigators' experiences imaging neurons (e.g., O'Rourke et al. 1992; Dailey and Smith 1994, 1996; Dailey et al. 1994; Marrs et al. 2001; Zha et al. 2005; Ahmed et al. 2006), astrocytes (Benediktsson et al. 2005), and microglia (Dailey and Waite 1999; Stence et al. 2001; Dailey 2002; Grossmann et al. 2002; Petersen and Dailey 2004; Kurpius et al. 2006, 2007) in rodent brain tissue slices, as well as neurons in excised chick neural retinal preparations (Marrs et al. 2006). Although the specific requirements of different samples

[1]Current address: Center for Neuroscience, West Virginia University, Morgantown, West Virginia 26506.

will vary, there are some general principles to consider when attempting to maintain viable specimens in the imaging setup (Dailey et al. 2006). The emphasis here is on simple low-cost approaches for maintaining live mammalian brain cells and tissue slices in imaging setups.

KEEPING THE SPECIMEN ALIVE IN THE IMAGING SETUP

There are several considerations for maintaining live cells and tissues in the imaging setup, including temperature, pH, gas exchange, and light exposure. The relative importance of each of these factors varies with the particular specimen, the spatial and temporal sampling requirements, and the expected length of observation.

Medium Considerations

The medium in which the specimen is mounted must support the cells and tissues for the duration of the observation. When working with cultured cells and tissues, it is necessary to consider the changes that may occur when the sample is taken from the environment of a culture incubator to ambient air. Most bicarbonate-based culture media are designed to buffer an air environment composed of ~5% CO_2, but in ambient air, these media rapidly (within ~5 min) become alkaline. Thus, by the time the sample is mounted, the gas content and the pH of the culture medium may have changed substantially. Such changes should be monitored. For determining pH changes, a pH indicator (phenol red) can be included in the chosen medium, or pH indicator strips (e.g., colorpHast; EMD Chemicals) may be used to test the medium at different stages in the mounting and imaging procedure. In the investigators' experience, a saline solution (e.g., phosphate-buffered saline) is sufficient to maintain neonatal mammalian brain tissue slices for a few hours. Longer observations usually require the more substantial nutritional support of culture media. In addition, long-term imaging of mammalian brain tissues derived from more mature animals (approximately postnatal day 7) may require continuous perfusion of oxygenated medium.

Mounting Live Specimens for Microscopic Observation

This section describes a simple mounting technique for rodent organotypic hippocampal slice cultures grown on filter culture membranes, a popular in vitro neural tissue slice preparation method (Stoppini et al. 1991; Gähwiler et al. 1997). These slice cultures are readily mounted and are imaged because the culture membrane provides a stable substratum that can be secured within an imaging chamber without applying pressure directly to the tissues. The investigators use translucent filter membrane inserts (Falcon 3090 or 3102) containing polyethylene terephthalate, track-etched porous membranes (1-µm pore size). Typically, two to five tissue slices are cultured on a single membrane insert. Although it is possible to image samples through a translucent filter membrane using low-magnification or long working–distance lenses, the highest resolution will be achieved by mounting samples in a way that allows direct imaging without an intervening filter membrane.

Technique for Chamber Construction and Sample Mounting

The mounting technique described here (Fig. 1A) is especially useful for imaging tissues through a coverslip on an inverted microscope system, but it could be adapted for an upright system in conjunction with saline or water-dipping lenses.

1. One or two drops of medium are placed on top of the membrane culture insert with attached organotypic tissue slices, and then the insert is removed from the culture well and is placed upside down on a flat surface. A piece of membrane with attached slices is excised from the culture insert by making a rectangular cut through the membrane and around the slices using a razor blade or a scalpel.

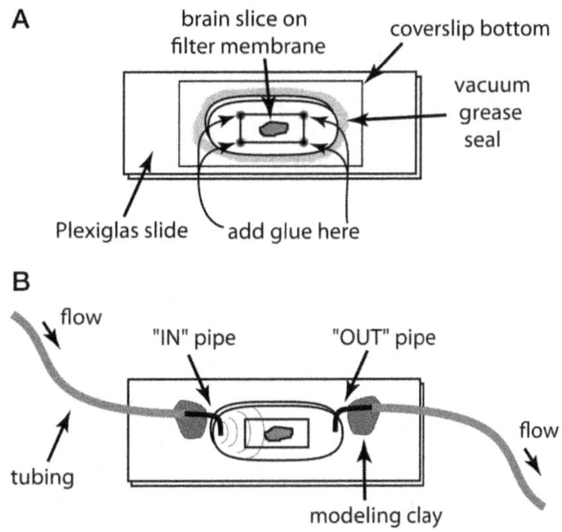

FIGURE 1. A simple specimen chamber for live cell or tissue imaging. (*A*) The specimen chamber is made of a rectangular coverslip (22 × 50-mm) and one or more 3-mm-thick Plexiglas slides with an elongated hole (1.5 × 3-cm) drilled in the center of each one (see text for details). The specimen is secured to the coverslip using a plasma/thrombin clot, netting, tape, glue, or a small weight, depending on the specimen. Here, a brain slice culture growing on a filter membrane is shown attached to a coverslip with cyanoacrylate glue. The chamber is partially filled with medium such that the bottom of the chamber is covered. The chamber opening may be covered with a second glass coverslip or may be kept uncovered for perfusion. (*B*) Perfusate is delivered by gravity feed to one side of the open chamber and withdrawn by vacuum suction from the other side. The tips of two syringe needles are cut off, bent, and inserted into the ends of the tubing to serve as small "in" and "out" pipes. The pipes, which stick down into the chamber well, are held in place with tape or modeling clay. The out pipe works best if the tip of the needle is beveled, or if a small filter-paper wick is inserted into the end of the needle. This allows continuous withdrawal of perfusate, rather than a pulsed removal that can induce severe focus jumps. The position of the out pipe must be adjusted (raised) so as not to suck the chamber dry.

2. The rectangular piece of membrane is then placed tissue-side down in the center of a dry 22 × 50-mm glass coverslip. To minimize air pockets and bubbles, the membrane is lowered slowly onto the coverslip starting at one edge. A very small drop of cyanoacrylate glue is then applied to each of the corners of the membrane. Care is taken to avoid placing glue too near the tissue slices or on top of them. The glue is allowed to dry for 10–20 sec.

3. A custom-made Plexiglas slide (~3-mm-thick) with an oblong hole (~1.5 × 3-cm) in the center is used to form a chamber around the tissue. First, a thin bead of silicone vacuum grease (e.g., Dow Corning) is laid down around the edge of the hole in the Plexiglas slide. (A 3-cc syringe, loaded from the back with vacuum grease, works well as a grease applicator. A 22-gauge tapered polyethylene tip [Small Parts, Inc., #FFT-22-05] works well to lay down a thin line of grease.) The greased Plexiglas slide is then inverted, is carefully positioned around the membrane rectangle, and is sealed to the coverslip by application of firm pressure. The vacuum grease forms a stable waterproof seal between the Plexiglas slide and the coverslip, thus, forming an open chamber containing the specimen. It is important that the piece of membrane and glue spots fit entirely inside the hole of the Plexiglas slide so that the grease ring is able to make a tight leak-resistant seal. A second custom-made Plexiglas slide may be similarly sealed on top of the first to create a chamber with greater volume.

4. HEPES-buffered culture medium (1–2-mL) is then added directly on top of the membrane to fill the chamber one-half to two-thirds full, with care taken to spread the medium across the entire floor of the well. Then the imaging chamber is transferred to the microscope stage.

5. If a perfusion setup is not used, a second coverslip (≥22 × 40-mm) is placed, without sealing, across the top of the well cavity to prevent evaporation of medium during imaging. Medium in the

chamber cavity should not directly contact the top coverslip as this can increase focal drift during time-lapse experiments. Also, because the chamber top is not sealed, it is possible to remove the top coverslip to add or replace the chamber medium during the course of the experiment. It is important to mount the samples swiftly to prevent evaporation and, thus, maintain proper salinity and pH of the medium.

Maintenance of Gas and pH Conditions after Mounting

For some short-term (<2-h) time-lapse experiments, it may be possible to maintain the specimen in a closed chamber configuration for the duration of the experiment. Larger volumes of chamber medium can support tissues for longer periods of time, although the pH of the chamber medium should be monitored. When large pieces (or several pieces) of tissue are confined to small volumes, the pH of the medium may significantly change (acidify) within a couple of hours as a consequence of tissue metabolism. This can cause rundown in cell behavior resulting, for example, in greatly reduced cell motility or migration. Mounting the specimen in a larger volume of medium may help, or periodic exchange of medium may adequately support the sample for many hours (O'Rourke et al. 1992).

An alternative solution for longer-term imaging experiments is to set up a continuous perfusion system. The rate of perfusion need not be very high: 1–2 mL of HEPES-buffered culture medium per hour can support developing rodent brain tissue slices and slice cultures on the stage of the microscope for many hours (Dailey and Smith 1996). The challenge is to set up a perfusion system that does not leak (thus, minimizing risk of damage to microscope components) and that does not induce focus jumps or drift. Dropwise delivery of perfusate to the specimen chamber, as well as periodic removal of medium from the chamber by suction, can cause pressure changes that translate into focus jumps. A rather simple and inexpensive gravity-fed perfusion setup can be constructed mostly from common laboratory supplies (Figs. 1B, 2). This can be integrated into a microscope setup with an external stage heater (Fig. 3). However, more elaborate chamber and perfusion systems can be constructed or can be purchased commercially (see Table 19.3 in Dailey et al. 2006). For a more thorough treatment of specimen chambers for live cell imaging, see Dailey et al. (2009).

For mature mammalian brain tissues, it is usually necessary to provide continuous perfusion of oxygenated medium. Figure 4 shows a simple perfusion arrangement that can be used to maintain mature brain tissue slices. Whatever the exact chamber configuration, for lengthy time-lapse experiments, it appears important that the imaging setup provide sufficient tissue oxygenation and removal of metabolic by-products such as CO_2.

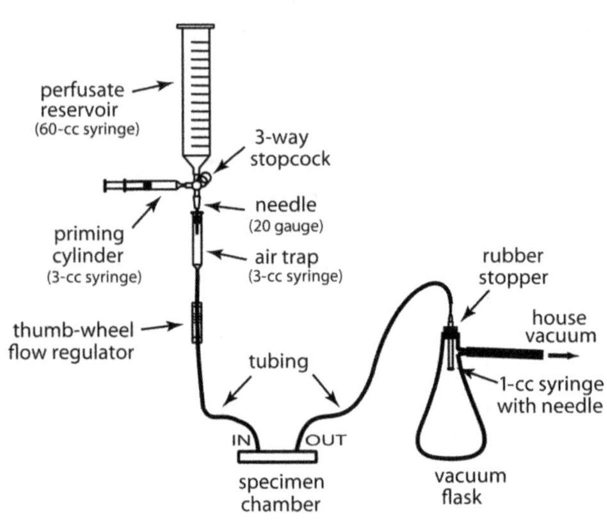

FIGURE 2. A simple gravity-feed perfusion system constructed from common laboratory supplies. Perfusate is added to the reservoir (e.g., a 60-cc syringe), and the system is primed with a 3-cc syringe to remove any air. The flow of perfusate is monitored by observing drip rates in the air trap (a second 3-cc syringe), and the rate is controlled by a thumb-wheel flow regulator taken from a standard medical intravenous set. A vacuum flask serves as the waste container.

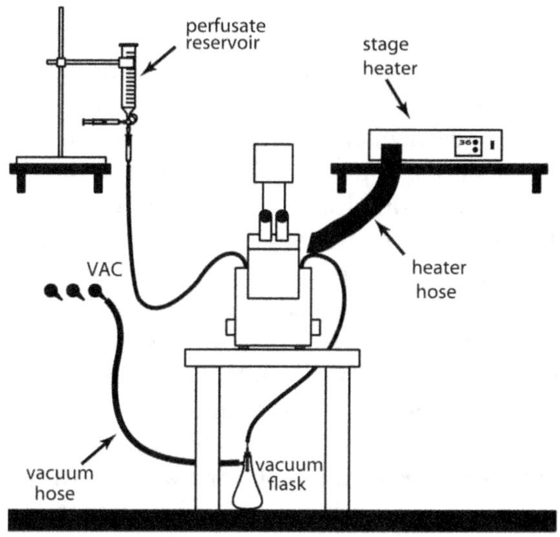

FIGURE 3. A simple integrated setup for live cell imaging. A gravity-feed perfusion system and forced air stage heater is shown in the context of an inverted microscope.

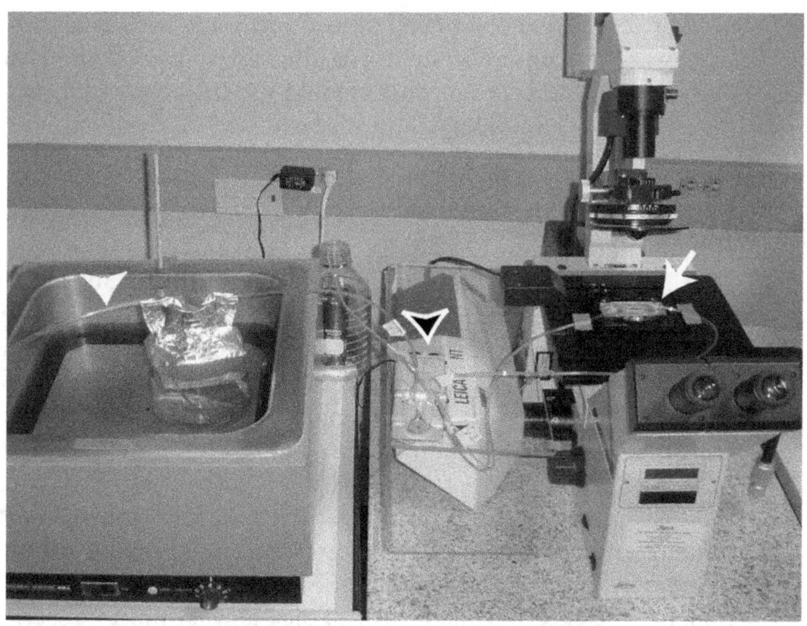

FIGURE 4. Perfusion setup for imaging adult rodent brain tissue slices on an inverted microscope platform. Warm perfusate is maintained in a covered reservoir (1-L glass bottle) sitting within a warm water bath. For perfusates requiring oxygenation, a tube (white arrowhead) is run from the oxygen tank (95% O_2, not shown) into the reservoir perfusate, and the perfusate is continuously aerated with the gas. Perfusate is delivered to and removed from the specimen chamber using a two-channel variable-speed pump (black arrowhead; e.g., Bioptechs) that maintains a constant rate of perfusate delivery and removal (~1 mL/min). The specimen chamber (white arrow) holds ~2 mL of perfusate. The tubes running to and from the specimen chamber are taped to the mobile portion of the microscope stage to minimize drag. Used perfusate is aspirated from the chamber and pumped into a waste container. Care must be taken to avoid dripping or spilling perfusate on any microscope or electronic components. Here, a custom-made Plexiglas shield (beneath the pump) was built to prevent perfusate from leaking onto electronic components in the confocal scan head.

Temperature Considerations

Maintaining the specimen at a physiological temperature during the imaging session increases experimenter confidence that the observed phenomena truly reflect native behaviors and not artifacts of the experimental conditions. Indeed, many dynamic cellular activities are highly sensitive to temperature. In the case of mammalian cells and tissues, it is often desirable to maintain a constant temperature of 34°C–36°C. This can be accomplished by perfusing warmed medium through the specimen chamber, by heating the stage, or by blowing warm air onto the preparation. There are very elaborate ways to accomplish this (see, e.g., Dailey et al. 2009), although it is also possible to put together a simple inexpensive arrangement that will adequately maintain constant temperature for many hours. For example, a standard inexpensive hair dryer can be used to blow warm air onto the microscope stage. The investigator must ensure, however, that the sample is protected from evaporation, which may occur when blowing warm nonhumidified air over the specimen chamber. Evaporation can be reduced or prevented by using a closed chamber configuration, or by continuous replacement (perfusion) of the chamber medium. Blowing warm air from the objective side of the chamber (on an inverted microscope) will help reduce airflow over the chamber and will also serve to more efficiently heat the objective, which may be a significant heat sink when using oil-coupled objectives. On inverted platforms, use of (nondipping) water- or saline-immersion lenses in combination with forced air stage heating can lead to rapid evaporation of the immersion medium and loss of the image, but this problem can be overcome by using a nondrying synthetic immersion medium that has an index of refraction (IR) matched to water (IR = 1.33; R.P. Cargille Laboratories, Cedar Grove, NJ).

Unfortunately, heating the specimen can introduce further complications such as increased rate of fluorescence photobleaching or focus drift due to thermal expansion across a chamber or a stage that is heated nonuniformly. Preheating the stage and the chamber can minimize temperature-induced focus drift. The medium should also be preheated to avoid the formation of bubbles within the chamber when cold medium is heated.

Collecting Images

How are images with sufficient spatial and temporal resolution collected without inducing changes in cell structure or physiology? Photodamage may occur when imaging live fluorescently labeled specimens. The cellular consequences of photodamage can appear within seconds, and the effects of a single photodamaging event can persist for many minutes or even hours. A compromise between the desired spatial and temporal resolution and the risk of photodamage to the sample must be reached. This is a particularly important issue when performing extended time-lapse imaging. Increasing the incident illumination (until fluorophore saturation) can improve the signal-to-noise ratio (SNR) of the image and can permit higher rates of image collection for a given SNR. However, most fluorescently stained samples cannot tolerate continuous light exposure for very long. The problem is particularly acute when attempting to sample a three-dimensional tissue volume over time (four-dimensional imaging). It is often necessary to lose some spatial and/or temporal resolution, but the experimental goals will dictate what must be sacrificed. For example, for fast physiological events (such as calcium spikes), it may be necessary to reduce exposure time and to sacrifice some spatial resolution to collect images at the fastest rate possible. On the other hand, for relatively slow events (such as long-distance migration of cells over periods of hours), the required long imaging sessions can be achieved only at the cost of reduced temporal or spatial resolution.

These considerations raise the question of whether it is better to keep the intensity of incident illumination low while increasing exposure time or to turn up the light and to reduce the exposure time. The incident illumination intensity will be a primary determinant of image quality and temporal sampling. In scanning imaging systems, image quality (SNR) can be improved by increasing pixel dwell time or by frame or line averaging. Unfortunately, there is no clear consensus as to whether it is better (for the specimen) to absorb less intense light over a longer period of time or to allow the sample sufficient time to recover between events of higher illumination intensity. In live specimens, movement of organelles and cells can smear the image when several frames are averaged over the period of a few seconds. This is less of an issue when imaging relatively slow events such as cell migration or axon and dendrite growth, but it is problematic when attempting to image fast physiological events such as calcium spikes, which tend to be averaged out. Most newer scanning imaging systems are capable of line averaging (resampling line by line) or slowing the scan rate (increasing pixel dwell time), which helps minimize the smearing of fast events while maintaining high SNR. Suffice it to say that the optimal imaging parameters vary with different specimens and fluorescent tags, and they need to be determined empirically for each biological and imaging system.

Photodynamic damage of the sample may be reduced somewhat by the inclusion of antioxidants such as ascorbate (vitamin C, 25–30-mM) in the chamber medium. However, the pH of the medium should be monitored because addition of significant amounts of ascorbate will acidify the medium.

KNOWING WHEN THE SPECIMEN IS UNHEALTHY

Deterioration of specimens in the imaging setup will, of course, limit the usefulness of observations. However, what criteria might be used to determine the health of the specimen? The primary consideration would seem to be whether the essential features of cell morphology, physiology, and dynamic behavior change under control conditions during the imaging session. If abnormal morphological or physiological features are apparent in the first images, then the culturing, staining, or mounting conditions must be altered. If abnormal features appear only after the imaging session has begun, one would suspect a problem with the imaging protocol or possibly the mounting procedure.

What are the indicators of ill health? Different cells vary in their responses to different conditions. Neurons are especially sensitive to changing conditions. When stained with fluorescent membrane dyes, nerve cell processes sometimes show abnormal features such as beading, blebbing, and triangulation of the branch points. The same cell types, in different samples, can show different patterns of beading and blebbing that may indicate different stress factors. Declining health can also affect the rates of cell growth and motility, usually resulting in a general decrease in activity. However, ill health is not always indicated by a decline in cellular activity. For example, compromised cells can show a marked increase in Brownian-like organelle motions. Likewise, increases in physiological activity (calcium spikes) are sometimes seen when imaging cells loaded with a fluorescent calcium indicator (e.g., Fluo-3).

Whatever morphological or physiological criteria are used, a critical question is whether the specimen looks and behaves in a similar way throughout the course of a control imaging session. For fluorescence imaging, it is important to remember that photodynamic damage can occur long before significant photobleaching is observed. To determine whether the light load is contributing to specimen deterioration, it can be helpful to compare the morphology or physiology of experimental specimens with control specimens that receive similar treatments (i.e., staining, mounting, drug application, etc.) but are imaged only at the end of a comparable experimental period. It can be especially instructive to sacrifice a few good-looking specimens to clearly define the sequence of changes that occur when the sample is subjected to phototoxic stress. Appropriate control experiments will help identify which cell behaviors are caused by the experimental treatment and which are consequences of suboptimal imaging conditions.

CONCLUSION

Even as imaging technologies improve, the problems encountered in live cell imaging can be daunting. Researchers are often limited by the photosensitivity of the biological sample, especially when fluorescent probes are used. Improvements in imaging system performance have allowed higher-quality images to be acquired with greater frequency, but the limits of the biological systems under study will always be tested. Increasingly, imaging involves the simultaneous use of two or three different fluorophores in individual cells. It is necessary, therefore, to become thoroughly familiar with the limitations of both the imaging and biological systems so that sound conclusions about the observed biological phenomena can be made.

ACKNOWLEDGMENTS

Development of the techniques described herein was facilitated by grants from the National Institutes of Health (NS37159, NS43468, and AA018823), the Whitehall Foundation (#S98-6), the Roy J. Carver Charitable Trust, the American Heart Association (0950160G), and a seed grant from the American Cancer Society (#IN-122R) administered through the University of Iowa Cancer Center.

REFERENCES

Ahmed R, Zha X-M, Green SH, Dailey ME. 2006. Synaptic activity and F-actin coordinately regulate CaMKIIα localization to dendritic postsynaptic sites in developing hippocampal slices. *Mol Cell Neurosci* **31:** 37–51.

Benediktsson AM, Schachtele SJ, Green SH, Dailey ME. 2005. Ballistic labeling and dynamic imaging of astrocytes in organotypic hippocampal slice cultures. *J Neurosci Methods* **141:** 41–53.

Dailey ME. 2002. Optical imaging of neural structure and physiology: Confocal fluorescence microscopy in live brain slices. In *Brain mapping: The methods*, 2nd ed. (ed. Toga A, Mazziotta J), pp. 49–76. Elsevier, San Diego.

Dailey ME, Smith SJ. 1994. Spontaneous Ca^{2+} transients in developing hippocampal pyramidal cells. *J Neurobiol* **25:** 243–251.

Dailey ME, Smith SJ. 1996. The dynamics of dendritic structure in developing hippocampal slices. *J Neurosci* **16:** 2983–2994.

Dailey ME, Waite M. 1999. Confocal imaging of microglial cell dynamics in hippocampal slice cultures. *Methods* **18:** 222–230.

Dailey ME, Buchanan J, Bergles D, Smith SJ. 1994. Mossy fiber growth and synaptogenesis in rat hippocampal slices in vitro. *J Neurosci* **14:** 1060–1078.

Dailey ME, Manders E, Soll D, Terasaki M. 2006. Confocal microscopy of live cells. In *Handbook of biological confocal microscopy*, 3rd ed. (ed. Pawley JB) Plenum, New York.

Dailey ME, Focht DC, Khodjakov A, Rieder CL, Spring KR, Claxton NS, Olenych SG, Griffin JD, Davidson MW. 2009. Culture chambers for live-cell imaging. http://microscopyu.com/articles/livecellimaging/culturechambers.html. Accessed 10/15/09.

Gähwiler BH, Capogna M, Debanne D, McKinney RA, Thompson SM. 1997. Organotypic slice cultures: A technique has come of age. *Trends Neurosci* **20:** 471–477.

Grossmann R, Stence N, Carr J, Fuller L, Waite M, Dailey ME. 2002. Juxtavascular microglia migrate along brain capillaries following activation during early postnatal development. *Glia* **37:** 229–240.

Kurpius D, Wilson N, Fuller L, Hoffman A, Dailey ME. 2006. Early activation, motility, and homing of neonatal microglia to injured neurons does not require protein synthesis. *Glia* **54:** 58–70.

Kurpius D, Nolley EP, Dailey ME. 2007. Purines induce directed migration and rapid homing of microglia to injured pyramidal neurons in developing hippocampus. *Glia* **55:** 873–884.

Marrs G, Green SH, Dailey ME. 2001. Rapid formation and remodeling of postsynaptic densities in developing dendrites. *Nat Neurosci* **4:** 1006–1013.

Marrs G, Honda T, Fuller L, Thangavel R, Balsamo J, Lilien J, Dailey ME, Arregui C. 2006. Dendritic arbors of developing retinal ganglion cells are stabilized by β1-integrins. *Molec Cell Neurosci* **32:** 230–241.

O'Rourke NA, Dailey ME, Smith SJ, McConnell SK. 1992. Diverse migratory pathways in the developing cerebral cortex. *Science* **258:** 299–302.

Petersen MA, Dailey ME. 2004. Diverse microglial motility behaviors during clearance of dead cells in hippocampal slices. *Glia* **46:** 195–206.

Stence N, Waite M, Dailey ME. 2001. Dynamics of microglial activation: A confocal time-lapse analysis in hippocampal slices. *Glia* **33:** 256–266.

Stoppini L, Buchs P-A, Muller D. 1991. A simple method for organotypic cultures of nervous tissue. *J Neurosci Methods* **37:** 173–182.

Zha X-M, Green SH, Dailey ME. 2005. Regulation of hippocampal synapse remodeling by epileptiform activity. *Mol Cell Neurosci* **29:** 494–506.

34 Calcium Imaging in Neuronal Endoplasmic Reticulum

Natasha Solovyova and Alexei Verkhratsky

School of Biological Sciences, The University of Manchester, Manchester M13 9PT, United Kingdom

ABSTRACT

This chapter describes a technique for the simultaneous monitoring of free calcium concentrations in the cytosol ($[Ca^{2+}]_i$) and within the lumen of the endoplasmic reticulum (ER) ($[Ca^{2+}]_L$) of cultured or freshly isolated dorsal root ganglia (DRG) neurons. The method uses two synthetic fluorescent Ca^{2+} probes (Fluo-3 and MagFura-2), in combination with fluorescence microscopy and a whole-cell patch-clamp technique.

INTRODUCTION

The approach described in this chapter (Fig. 1) has been used successfully in acutely isolated/cultured DRG neurons, in Purkinje neurons acutely isolated from cerebellar slices, and in cultured astrocytes. Similar techniques, using low-affinity Ca^{2+} indicators (MagFluo-4, MagFura-2, Fluo-3FF) have been used extensively for monitoring $[Ca^{2+}]_L$ dynamics in a variety of nonexcitable cells (e.g., in epithelial cells [Hofer and Machen 1993], pancreatic acinar [Mogami et al. 1998] and β-cells [Tengholm et al. 2001], endothelial cells [Dedkova and Blatter 2002], and in muscle cells [Shmigol et al. 2001]).

The two probes used in this method are a low-affinity Ca^{2+} indicator, MagFura-2 AM, and a high-affinity calcium indicator, Fluo-3 (in its pentapotassium salt form). MagFura-2 is a ratiometric probe (i.e., it shows spectral shift on Ca^{2+} binding) that was originally developed as an Mg^{++} indicator (Raju et al. 1989). Notwithstanding this history, MagFura-2 is ideally suited for $[Ca^{2+}]_L$ measurements because (1) it has a propensity to accumulate in subcellular compartments, particularly in the ER (see Thomas et al. 2000), and (2) its affinity for Ca^{2+} (K_d ~50 μM) is much higher than that for Mg^{++} (K_d ~ 2–5 mM) (see Hyrc et al. 2000) so that the probe can accurately report free Ca^{2+} values between tens and several hundreds of micromoles, within the characteristic range of $[Ca^{2+}]_L$. Fluo-3 has a K_d of ~325–390 nM, which allows it to measure cytosolic Ca^{2+} concentrations within the range of ~20 to ~2000 nM.

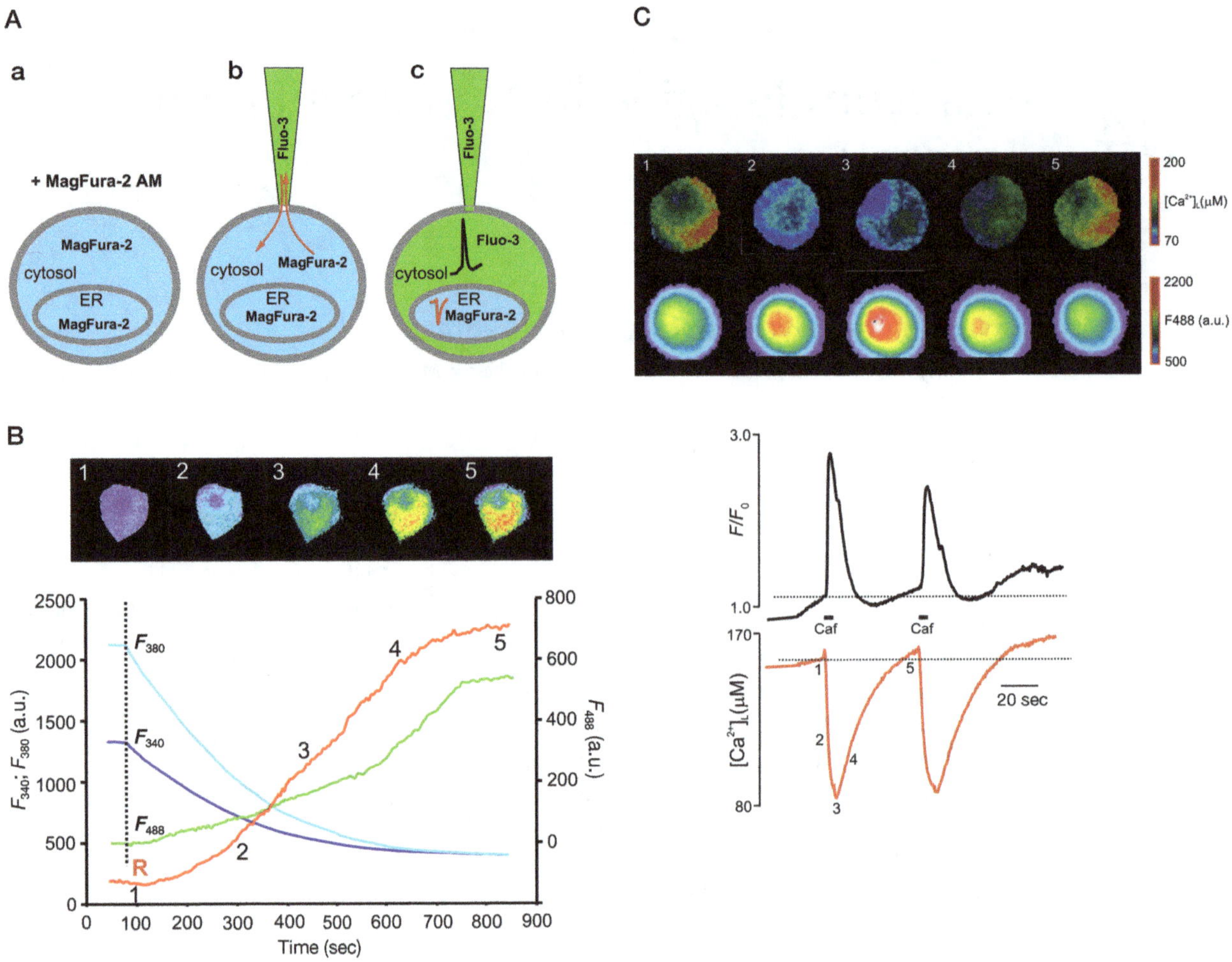

FIGURE 1. Monitoring of the intra-ER Ca^{2+} dynamics using MagFura-2 in combination with a whole-cell patch clamp. (*A*) Principle of the method. (*a*) Cells are stained by MagFura-2 AM in such a way that the probe is distributed both within the cytosol and within the ER lumen. (*b*) The cytoplasmic portion of the dye is then removed by intracellular dialysis via patch pipette. (*c*) The intrapipette solution is supplemented with the high-affinity Ca^{2+} indicator Fluo-3, which allows simultaneous, although independent, monitoring of Ca^{2+} changes in the cytosol and within the ER lumen. (*B*) Typical kinetics of washout of cytosolic MagFura-2 and wash-in of Fluo-3 into the cytosol. (*Bottom*) At the beginning of the experiment, the values of F_{340} (dark blue curve) and F_{380} (light blue curve) were high (fluorescent values were averaged over the whole cell), and *R* (red curve) was low, reflecting mostly cytosolic origin of the signal. After the breakthrough (indicated by dotted line) both F_{340} and F_{380} start to decrease and *R* increases, reflecting MagFura-2 washout from the cytosol. Simultaneously, an increase in F_{488} (green curve) is observed, which reflects dialysis of the cytosol with Fluo-3 (in its pentapotassium salt form). All three fluorescence signals stabilize after completion of dialysis, when the calculated $[Ca^{2+}]$ derived from MagFura-2 approaches 250 μM, reflecting a high free Ca^{2+} level in the cytosol. (*Top*) Selected ratio images corresponding to numbers on the graph in the bottom panel. (*C*) Example of simultaneous $[Ca^{2+}]_L$ and $[Ca^{2+}]_i$ recordings from a DRG neuron stained and patch-clamped as described above. (*Top*) Images of MagFura-2 ratios and Fluo-3 fluorescence; (*bottom*) corresponding $[Ca^{2+}]_L$ and $[Ca^{2+}]_i$ traces. The cell was challenged by 20-mM caffeine (as indicated by bars) which caused transient decreases in $[Ca^{2+}]_L$ and increases in $[Ca^{2+}]_i$. (Reprinted, with permission, from Solovyova and Verkhratsky 2002.)

Protocol

Use of Low- and High-Affinity Fluorescent Probes for Calcium Imaging in Neuronal ER

In this protocol, isolated neurons are first loaded with the membrane-permeant low-affinity Ca^{2+} indicator MagFura-2 AM, which preferentially, although not exclusively, accumulates in the ER. Cells are then loaded with the membrane-impermeant high-affinity calcium indicator Fluo-3 using a whole-cell patch-clamp configuration. This second loading removes the majority of cytosolic MagFura-2, replacing it with Fluo-3. MagFura-2 and Fluo-3 signals can be separated by virtue of their distinct excitation properties (340 and 380 nm for MagFura-2 and 488 nm for Fluo-3). $[Ca^{2+}]_L$ values are calculated using the 340/380-nm ratio.

IMAGING SETUP

Real-Time Video Imaging

This protocol was designed with the use of an Olympus IX70 inverted microscope (40x ultraviolet objective) equipped with a charge-coupled device (CCD)-cooled intensified camera (Pentamax Gene IV, Roper Scientific, UK). The illumination was provided by a monochromator (Polychrom IV, TILL Photonics, Germany). The equipment was linked to a Windows 98 workstation. Data were collected, stored, and analyzed using MetaFluor/MetaMorph software (Universal Imaging Corporation Ltd., USA).

Whole-Cell Voltage-Clamp Recordings

These recordings were made using an EPC-9 amplifier run by the PC-based PULS software (both from HEKA, Germany) (see Penner 1995).

MATERIALS

CAUTION: See Appendix 6 for proper handling of materials marked with <!>.
See the end of the chapter for recipes for reagents marked with <R>.

Reagents

Calibrating solutions <R>
DRG neurons, cultured or freshly isolated
Fluo-3 (in its pentapotassium salt form) (Molecular Probes) dissolved in the intrapipette solution
Intrapipette solution <R>
MagFura-2/AM, 1 mM in dimethylsulfoxide (DMSO) <!>
Standard extracellular saline <R>

Equipment

Incubator, preset to 37°C
Patch pipette

EXPERIMENTAL METHOD

Staining Neurons with MagFura-2 AM and Fluo-3

1. Dilute an aliquot of MagFura-2 AM stock solution (1 mM in DMSO) in standard extracellular saline to a final concentration of 5 µM.
2. Load cultured or freshly isolated cells with MagFura-2 by bathing them in the MagFura-2 AM extracellular saline solution and by incubating for 30 min at 37°C.
3. Wash the cells with prewarmed *dye-free* standard extracellular saline.
4. Incubate the cells in standard extracellular saline for 1 h at 37°C.

   This step ensures complete deesterification of the probe.
5. Establish whole-cell patch-clamp configuration on the cells (now loaded with MagFura-2) with a patch pipette, containing intrapipette solution supplemented with 50-µM Fluo-3.

   This step washes out the cytosolic portion of the MagFura-2 and introduces Fluo-3 in its place. For information on the kinetics of intracellular dialysis, see Solovyova and Verkhratsky (2002). Any remaining cytosolic fluorescence can be additionally quenched by adding Mn^{++} to the dialyzing solution (see Tse et al. 1994).

Real-Time Video Imaging

6. Illuminate the cells alternately at 340, 380, and 488 nm using a monochromator at a cycle frequency of 0.5–5 Hz.
7. Capture data at 510 ± 15 nm.

Intracellular Calibration of MagFura-2 Signals

8. Calculate $[Ca^{2+}]_L$ using the 340/380-nm ratio with the following equation:

$$[Ca^{2+}]_L = K^* (R - R_{min})/(R_{max} - R) \cdot R_{max},$$

   where R is the ratio obtained during measurement, R_{min} is the fluorescent ratio of Ca^{2+}-free dye, and R_{max} is the ratio for dye saturated by Ca^{2+}. The parameter K^* comprises the value of the affinity of dye to Ca^{2+} ions (K_d, the dissociation constant of the dye) and the constant β, which is determined by the optical properties of the experimental setup. The K_d, although known for each dye, may vary depending on the physical properties of the solution (e.g., viscosity, temperature, and pH). For practical reasons, K^* is usually determined empirically by measuring R for a given Ca^{2+} concentration and by using it in the formula with known values for R_{min} and R_{max}.
9. Determine the R_{min}, R_{max}, and K^* values by exposing the neurons to 10–20-µM ionomycin and three to four calibrating solutions with known Ca^{2+} concentrations (e.g., with $[Ca^{2+}]$ <10 nM, 100, 400 µM, and 10 mM).

DISCUSSION

Example of Application

Figure 2 shows an example of simultaneous $[Ca^{2+}]_L$ and $[Ca^{2+}]_i$ recordings from a single DRG neuron as it responds to a brief (5-sec) application of 20-mM caffeine or to the entry of Ca^{2+} into the cytoplasm via voltage-operated Ca^{2+} channels. The latter provides an example of genuine Ca^{2+}-induced Ca^{2+} release. Both treatments provoked the release of Ca^{2+} from intracellular stores, resulting in a transient increase of $[Ca^{2+}]_i$ and a transient decrease of $[Ca^{2+}]_L$.

Advantages and Limitations

Intraluminal free Ca^{2+} regulates numerous signaling cascades that converge in the ER (Berridge 2002; Bootman et al. 2002; Verkhratsky and Petersen 2002), giving rise to fast signaling events (e.g., Ca^{2+} release or Ca^{2+} uptake) and long-lasting adaptive responses (e.g., posttranslational protein folding controlled by Ca^{2+}-dependent chaperones). Therefore, the ability to make direct measurements of $[Ca^{2+}]_L$ is important for further understanding of the intimate mechanisms of ER-signaling processes.

The use of low-affinity Ca^{2+} indicators, such as MagFura-2, provides a relatively simple and reliable means of monitoring $[Ca^{2+}]_L$. Because MagFura-2 has a much higher affinity for Ca^{2+} than it does for Mg^{++}, the indicator should respond to changes in intracellular-free Ca^{2+} with little interference from changes in intracellular Mg^{++}. The combination of the MagFura-2/AM ester indicator compartmentalized within the ER lumen and Fluo-3 (loaded using a patch pipette), allows (1) simultaneous but independent monitoring of Ca^{2+} dynamics in the ER and in the cytosol (see Solovyova et al. 2002), and (2) the application of established electrophysiological protocols to the cell of interest. The use of a ratiometric probe for monitoring intra-ER Ca^{2+} permits accurate calibration and quantitative measurement of $[Ca^{2+}]_L$; it also reduces the effects of uneven dye loading, photobleaching, uneven cell thickness, and leakage of the dye.

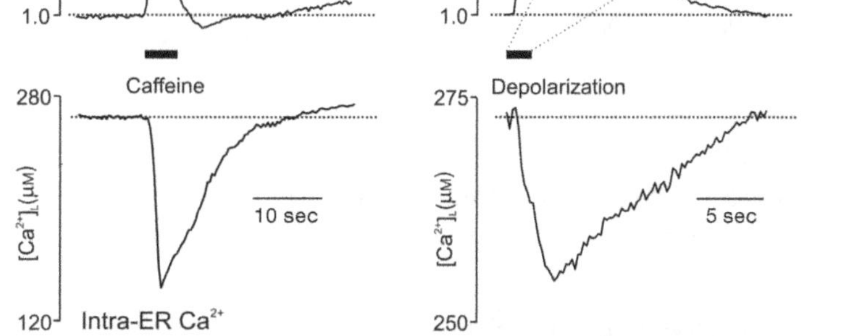

FIGURE 2. Simultaneous recordings of $[Ca^{2+}]_i$ (cytosolic Ca^{2+}) and $[Ca^{2+}]_L$ (intra-ER Ca^{2+}) in an isolated DRG neuron upon activation of Ca^{2+} release from the ER stores. Calcium release was activated by brief (5-sec) application of caffeine (*A*) or by Ca^{2+} entry through voltage-gated Ca^{2+} channels (*B*). In the latter case, the voltage-clamped neuron was depolarized from –70 to 0 mV for 2 sec. The *inset* shows the voltage protocol and the calcium current recorded in the presence of 1-μM tetrodotoxin. Both caffeine and Ca^{2+} entry resulted in transient decrease in $[Ca^{2+}]_L$, indicative of Ca^{2+} release from the ER.

The accumulation of the low-affinity Ca^{2+} probe in the ER lumen can be significantly improved by targeted expression of recombinant esterases into the ER lumen; this technique is described in detail in Rehberg et al. (2008). However, unlike ER-targeted cameleons (Yu and Hinkle 2000; Varadi and Rutter 2002) and aequorins (Montero et al. 1995), MagFura-2 cannot be targeted exclusively to the ER. Indeed, MagFura-2 is known to compartmentalize to a variety of organelles. The dye that accumulated in organelles other than the ER contributes to the overall MagFura-2 signal and can interfere with estimation of $[Ca^{2+}]_L$ levels. Yet, the dynamic range of synthetic fluorescent Ca^{2+} probes significantly exceeds that of cameleons; and, hitherto, low-affinity dyes remain the indicators of choice for measuring low-amplitude $[Ca^{2+}]_L$ signals in response to direct electrical stimulation of nerve cells under voltage/current clamp conditions.

ACKNOWLEDGMENTS

This research was supported by the Biotechnology and Biological Sciences Research Council and the Wellcome Trust.

RECIPES

CAUTION: See Appendix 6 for proper handling of materials marked with <!>.
Recipes for reagents marked with <R> are included in this list.

Calibrating Solutions

1. Prepare the following solution:
   NaCl 130 mM
   KCl 3 mM
   HEPES buffer/KOH <!> 10 mM
   Glucose 10 mM
   Ionomycin 0.01 mM
   pH to 7.3
2. Obtain the desired Ca^{2+} concentrations by adding Ca^{2+}/EGTA or Ca^{2+}/BAPTA <!> buffer to the above solution. Appropriate concentrations of Ca^{2+} and EGTA/BAPTA can be calculated using widely available calculators (e.g., http://www.stanford.edu/~cpatton/webmaxcE.htm).

Intrapipette Solution

CsCl <!> 122 mM
TEA-CL <!> 20 mM
Na_2ATP 3 mM
Fluo-3 (in its pentapotassium salt form) 0.05 mM
HEPES buffer/CsOH <!>, pH 7.3 10 mM

Standard Extracellular Saline

NaCl 135 mM
KCl 3 mM
$CaCl_2$ <!> 2 mM
Glucose 20 mM
HEPES buffer/NaOH <!>, pH 7.4 20 mM

REFERENCES

Berridge MJ. 2002. The endoplasmic reticulum: A multifunctional signalling organelle. *Cell Calcium* **32:** 235–249.

Bootman MD, Petersen OH, Verkhratsky A. 2002. The endoplasmic reticulum is a focal point for co-ordination of cellular activity. *Cell Calcium* **32:** 231–234.

Dedkova EN, Blatter LA. 2002. Nitric oxide inhibits capacitative Ca^{2+} entry and enhances endoplasmic reticulum Ca^{2+} uptake in bovine vascular endothelial cells. *J Physiol* **539:** 77–91.

Hofer AM, Machen TE. 1993. Technique for in situ measurement of calcium in intracellular inositol 1,4,5-trisphosphate-sensitive stores using the fluorescent indicator mag-fura-2. *Proc Natl Acad Sci* **90:** 2598–2602.

Hyrc KL, Bownik JM, Goldberg MP. 2000. Ionic selectivity of low-affinity ratiometric calcium indicators: Mag-Fura-2, Fura-2FF and BTC. *Cell Calcium* **27:** 75–86

Mogami H, Tepikin AV, Petersen OH. 1998. Termination of cytosolic Ca^{2+} signals: Ca^{2+} reuptake into intracellular stores is regulated by the free Ca^{2+} concentration in the store lumen. *EMBO J* **17:** 435–442.

Montero M, Brini M, Marsault R, Alvarez J, Sitia R, Pozzan T, Rizzuto R. 1995. Monitoring dynamic changes in free Ca^{2+} concentration in the endoplasmic reticulum of intact cells. *EMBO J* **14:** 5467–5475.

Penner R. 1995. A practical guide to patch-clamping. In *Single-channel recording* (ed. Shakmann B, Neher E), pp. 3–30. Plenum, New York.

Raju B, Murphy E, Levy LA, Hall RD, London RE. 1989. A fluorescent indicator for measuring cytosolic free magnesium. *Am J Physiol* **256:** 540–548.

Rehberg M, Lepier A, Solchenberger B, Osten P, Blum R. 2008. A new non-disruptive strategy to target calcium indicator dyes to the endoplasmic reticulum. *Cell Calcium* **44:** 386–399.

Shmigol AV, Eisner DA, Wray S. 2001. Simultaneous measurements of changes in sarcoplasmic reticulum and cytosolic $[Ca^{2+}]$ in rat uterine smooth muscle cells. *J Physiol* **531:** 707–713.

Solovyova N, Verkhratsky A. 2002. Monitoring of free calcium in the neuronal endoplasmic reticulum: An overview of modern approaches. *J Neurosci Methods* **122:** 1–12.

Solovyova N, Veselovsky N, Toescu EC, Verkhratsky A. 2002. Ca^{2+} dynamics in the lumen of the endoplasmic reticulum in sensory neurones: Direct visualisation of Ca^{2+}-induced Ca^{2+} release triggered by physiological Ca^{2+} entry. *EMBO J* **21:** 622–630.

Tengholm A, Hellman B, Gylfe E. 2001. The endoplasmic reticulum is a glucose-modulated high-affinity sink for Ca^{2+} in mouse pancreatic β-cells. *J Physiol* **530:** 533–540.

Thomas D, Tovey SC, Collins TJ, Bootman MD, Berridge MJ, Lipp P. 2000. A comparison of fluorescent Ca^{2+} indicator properties and their use in measuring elementary and global Ca^{2+} signals. *Cell Calcium* **28:** 213–223.

Tse FW, Tse A, Hille B. 1994. Cyclic Ca^{2+} changes in intracellular stores of gonadotropes during gonadotropin-releasing hormone-stimulated Ca^{2+} oscillations. *Proc Natl Acad Sci* **91:** 9750–9754.

Varadi A, Rutter GA. 2002. Dynamic imaging of endoplasmic reticulum Ca^{2+} concentration in insulin-secreting MIN6 cells using recombinant targeted cameleons: Roles of sarco(endo)plasmic reticulum Ca^{2+}-ATPase (SERCA)-2 and ryanodine receptors. *Diabetes* **51:** S190–S201.

Verkhratsky A, Petersen OH. 2002. The endoplasmic reticulum as an integrating signalling organelle: From neuronal signalling to neuronal death. *Eur J Pharmacol* **447:** 141–154.

Yu R, Hinkle PM. 2000. Rapid turnover of calcium in the endoplasmic reticulum during signaling. Studies with cameleon calcium indicators. *J Biol Chem* **275:** 23648–23653.

35 Dye Loading with Whole-Cell Recordings

Hartmut Schmidt and Jens Eilers

Carl-Ludwig-Institute of Physiology, University of Leipzig, 04103 Leipzig, Germany

ABSTRACT

This chapter describes the loading of individual cells with fluorescent probes via pipettes in the whole-cell patch-clamp configuration. The technique represents a versatile tool for combining electrophysiological and optical measurements at a quantitative level, and it can be used in almost any cell type in various preparations, including acutely dissociated cells, cultured cells, acute slices, or in vivo preparations.

INTRODUCTION

In comparison to other staining techniques, such as dye loading with membrane-permeable dyes (e.g., of the so-called AM [acetoxymethyl] type; Tsien 1981; see Chapters 13 and 49), dye injection via sharp electrodes (von Blankenfeld and Kettenmann 1992), or electroporation (see Chapter 50), the patch-clamp dye loading technique offers important advantages (see also Chapter 47 for a description of in vivo recording). First, all cell types that can be studied with the whole-cell patch-clamp technique are suitable for staining with patch-clamp dye loading. This is of crucial importance, especially when studying cells that are too small to tolerate impalement with a sharp electrode or neurons that are extensively covered by glia cells and, therefore, cannot be efficiently stained with AM dyes. Second, the intracellular dye concentration can be predicted rather precisely, either during the loading phase (Neher and Augustine 1992; Schneggenburger et al. 1993) or after an appropriate loading time, when the cytosolic dye concentration is close to that of the pipette solution (Rexhausen 1992). Thus, quantitative studies relying on accurate knowledge of cytosolic dye concentrations are possible (Neher 1995). Finally, loading of substances too large to pass through a sharp electrode (e.g., labeled proteins) is possible.

The patch-clamp methodology has been successfully used for single-cell dye labeling in cultured neurons, brain slices (Edwards et al. 1989), and in vivo preparations (Kitamura et al. 2008). A broad range of fluorescent probes can be used for this loading technique. Markers for morphological reconstruction (e.g., Lucifer yellow), functional indicator dyes for monitoring ion or second-messenger concentration changes (e.g., the calcium indicator Fura-2), and dye-labeled proteins for Förster resonance energy transfer [FRET], fluorescence correlation spectroscopy [FCS], and fluorescence recovery after photobleaching [FRAP] studies, respectively, are all suitable for patch-clamp loading. The most widespread application of this technique has been for Ca^{2+} imaging.

Protocol

Dye Loading with Patch Pipettes

This protocol describes a method for using the whole-cell patch-clamp recording configuration to load fluorescent dyes for imaging. Advice is provided on determining and obtaining the appropriate electrical resistance for the patch pipette, which is of particular importance for fast and efficient dye loading. The loading protocol presented here assumes familiarity with the standard procedures of whole-cell patch-clamp recordings.

IMAGING SETUP

In general, any imaging system can be used for combined fluorescence/patch-clamp recordings. Two aspects, however, require special consideration. First, for live cell recordings, the sensitivity of the system should be optimized. Phototoxic damage is a constant problem, which can only be reduced by limiting the intensity and duration of the illumination for fluorescence excitation and by increasing the detection sensitivity (see Yuste 2011, Section 1). Second, any parts of the imaging system that have to be moved or exchanged during the recordings (e.g., shutters, dichroic mirrors, barrier filters, magnifiers, objectives, *x–y* stages) must allow smooth vibration-free movements and stops. In particular, motorized systems can induce vibrations that easily couple to patch-clamp manipulators and patch electrodes and, thus, may rapidly deteriorate the whole-cell configuration.

MATERIALS

See the end of the chapter for recipes for reagents marked with <R>.

Reagents

Trace amounts of heavy metals or other contaminants either in the pipette or in the extracellular solution may severely affect the fluorescence properties of the fluorophores (Grynkiewicz et al. 1985). Therefore, it is highly recommended to use culture-grade water (e.g., W3500 from Sigma-Aldrich) for dissolving the dyes and for preparing the pipette solutions. For the extracellular solution, laboratory water processed by a (well-maintained) on-site purification system (e.g., from Millipore) will suffice. When in doubt, use medical-grade water from a commercial supplier (e.g., sterile water for irrigation, B. Braun Medical, Inc.).

Dye-containing pipette solution <R>
Extracellular solution <R>
Tissue source (e.g., cultured cells, acute slices)

Equipment

Imaging system (see above)
Microloaders (Eppendorf)

Alternatively, a loading device can be pulled from a 1-mL syringe. To do this, retract the plunger, carefully heat the middle of the barrel over a Bunsen burner until the plastic starts to melt, thin the barrel by pulling it apart, let the plastic cool down, reinsert the plunger, and cut the thinned barrel to a diameter that fits into the patch pipettes.

Pipettes

Use borosilicate pipettes (e.g., from Hilgenberg GmbH) rather than pipettes made out of soft soda glass. The latter release trace amounts of cations that may interfere with voltage-gated channels (Cota and Armstrong 1988) and, potentially, ion-sensitive dyes (Grynkiewicz et al. 1985).

Puller (e.g., PC-10 from Narishige Group or PIP 6 from HEKA)

Standard electrophysiology equipment for patch-clamp recordings

Amplifier, manipulator, pipette holder, recording chambers, oscilloscope, pressure meter, etc., are as described in detail, for example, in Penner (1995) or in The Axon Guide (http://www.moleculardevices.com/pdfs/Axon_Guide.pdf).

Syringe filter (e.g., 0.2-μm, Nalgene)

EXPERIMENTAL METHOD

1. Pull several patch pipettes of appropriate resistance.

   Depending on the experimental requirements and the cell type under study, the optimal pipette resistance may vary considerably.

   See *Troubleshooting*.

2. Prepare 1x dye-containing pipette solution from the stock solution (see Recipes), pass it through a syringe filter (e.g., 0.2 μm), and store on ice until use.

   Protect the solution from intense light.

3. Start up the imaging setup and the electrophysiological recording system. Place the preparation under the microscope and ensure an adequate supply of extracellular solution.

   Shutter the excitation light to prevent phototoxic damage throughout the following steps.

4. Select a cell of interest using bright-field illumination (optionally with contrast enhancement; see Dodt et al. 2011; Oshiro et al. 2011; and Ziv and Schiller 2011).

5. Use a microloader or a laboratory-made loading device to fill a patch pipette with 5–10 μL of the dye-containing solution.

   Do not use syringe needles for loading because these may release heavy metals.

6. Insert the pipette into the pipette holder and apply positive pressure to prevent dust floating on the surface from entering the pipette. Lower the pipette into the recording chamber containing the chosen cell. Once the pipette tip is in the bath, set the positive pressure to a minimal level that prevents the ejection of too much dye (see Fig. 1A). Move the pipette tip into the close vicinity (a few micrometers) of the chosen cell.

7. Move the pipette tip toward the cell, and apply a brief pulse of stronger pressure immediately before seal formation to clean the surface of the cell. Immediately afterward, apply gentle suction to establish the seal and the whole-cell configuration as per standard electrophysiological recording methods (Marty and Neher 1995). Release the suction immediately after the patch is ruptured.

   Depending on the strength and duration of the strong positive pressure pulse, the surrounding tissue will be stained to a variable extent. Therefore, limit the pulse to a minimum. Under certain conditions, debris on the cell surface may hinder seal formation.

   See *Troubleshooting*.

8. Allow the dye to diffuse to the cellular compartments of interest (e.g., distal dendrites and axons).

   Diffusion may take from 2 to 60 min. You may briefly check the state of loading of the cell by short-term inspection in the epifluorescence mode. Take every precaution to minimize the illumination period to prevent phototoxic damage. The time required to reach a certain cytosolic dye concentration will depend on the dye concentration in the pipette, the mobility of the dye molecule (Pusch and Neher 1988), the cell geometry, the distance between the pipette and the region of interest (Rexhausen 1992), and the series resistance (R_s) of the whole-cell configuration (Pusch

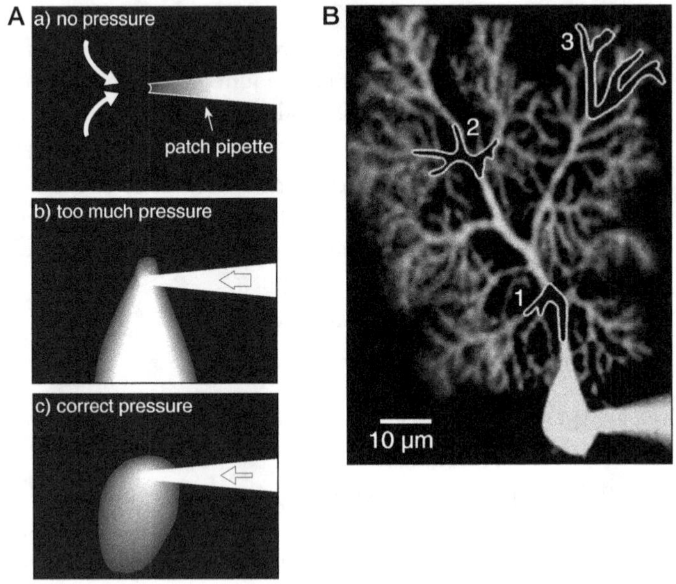

FIGURE 1. (*A*) A schematic showing the effect of different pressure settings, as viewed through fluorescence optics. Minimal pressure should be applied to the patch pipette while approaching the tissue. (*a*) If no pressure is applied, bath solution enters the pipette because of capillary forces. Thus, the composition of the pipette solution is modified, and successful seal formation may become difficult. (*b*) If the pressure applied to the pipette is too high, a strong jet of dye-containing solution is expelled from the pipette tip, and surrounding tissue may become heavily stained. (The bath solution flows from the top to the bottom.) (*c*) At the correct pressure setting, only a small amount of dye is ejected. (*B*) Dye loading of distal dendrites. A Purkinje neuron in a cerebellar slice preparation was loaded with the calcium indicator dye Fura-2 via a somatic patch pipette for 120 min (R_s = 5.5–6.7 MΩ). During the whole-cell recording, repeated measurements of the dendritic fluorescence at the calcium-insensitive wavelength were performed. From these data, the times of half-maximal loading were estimated to be ~2.5, ~16, and ~30 min for regions 1, 2, and 3, respectively. (Modified, with permission, from Rexhausen 1992.)

and Neher 1988; Rexhausen 1992).

During the loading period, R_s may gradually increase.

See *Troubleshooting*.

9. Start the imaging experiments when the regions of interest show a sufficient brightness in fluorescence. If knowledge about the approximate intracellular dye concentration is of importance, limit the recording time to an estimated or calculated (Pusch and Neher 1988; Rexhausen 1992) time window in which an appropriate dye level in the region of interest is reached.

   Note that molecules in the cytoplasm may be washed out on a timescale similar to the wash-in of the dye (see, e.g., Müller et al. 2005).

TROUBLESHOOTING

Problem (Step 1): The proper electrical resistance for the patch pipette must be determined and obtained.

Solution: The electrical resistance of the patch pipette R_P is of special importance for fast and efficient dye loading. R_P, which should not be confused with the series resistance or the access resistance (R_s), is determined by the conductivity of the pipette solution and the geometry of the pipette (i.e., by the length of the pipette taper and the width of the pipette opening). Depending on the experimental requirements and the cell type under study, the optimal R_P value may vary considerably. A low R_P permits a fast cell dialysis (Fig. 1B; see also Pusch and Neher 1988; Rexhausen 1992). However, it is more difficult to maintain long-lasting recordings with low-resistance pipettes. A higher R_P should be chosen if multiple measurements at a steadily increasing dye concentration are needed, as for measurements of the calcium-binding ratio with the calcium flux approach (Neher 1995). Furthermore, the cell size usually imposes limits on the diameter of the pipette tip and, consequently, on R_P. As a general rule, large cells tolerate larger tips than small cells or fine cellular processes (Sakmann and Stuart 1995).

Fine-tune R_P by changing the settings of the pipette puller (for details, see Penner 1995) until the resistance and the geometry of the pipette allows stable whole-cell recordings at a minimal access resistance (R_s, typically in the range of two to three times the R_P value). For instance, patch pipettes with resistances of 2.0–3.5 MΩ (measured with KCl-based pipette solution) are well suited for long-lasting recordings with R_s values of 5–10 MΩ from the large cerebellar Purkinje cell bodies (Fig. 1B), which have a diameter of 25–30 μm.

Problem (Step 7): In some preparations, cell debris, white matter, glia cells, and/or the perineuronal net hinder successful seal formation.

Solution: Cleaning of the cell membrane by means of a cleaning pipette, as described by Edwards et al. (1989), may solve this problem.

Problem (Step 8): A gradual increase in R_s is observed.

Solution: A difficulty frequently encountered, especially in brain slices (Edwards et al. 1989), is that R_s tends to increase gradually during whole-cell recordings to levels that seriously hinder dye loading and proper voltage control. Particularly for long-lasting experiments, the first minutes of the whole-cell recordings should be devoted entirely to the establishment of a low and stable series resistance by applying suction pulses to the pipette whenever R_s tends to increase. Although this approach may sometimes destroy the whole-cell configuration, it is preferable to lose a cell during the first minutes than to load it for a prolonged time with an unfavorable R_s and compromise on the degree of dye loading.

DISCUSSION

Once established in a laboratory, dye loading with patch pipettes represents an accurate, reliable, and straightforward technique that allows one to load a broad range of dyes and/or dye-labeled substances in almost any cell type. The key advantages are the well-defined intracellular dye concentration and the possibility to combine high-resolution electrophysiological and imaging experiments on single dye-loaded cells surrounded by unstained tissue. The drawbacks of the technique are associated with (1) the difficulties of the whole-cell patch-clamp technique, (2) the possible washout of cytosolic constituents during prolonged whole-cell recordings, and (3) the limitations in the size of dye-labeled substances that can be loaded via pipettes. Dye-labeled proteins have been successfully used up to a molecular weight of ~30 kDa (Schmidt et al. 2003; Müller et al. 2005) or even 150 kDa (Pusch and Neher 1988).

RECIPES

CAUTION: See Appendix 6 for proper handling of materials marked with <!>.
Recipes for reagents marked with <R> are included in this list.

Dye-Containing Pipette Solution

This recipe provides an example of a suitable solution.

Potassium gluconate <!>	140 mM
HEPES buffer	10 mM
NaCl	10 mM
$MgCl_2$	0.5 mM
Mg-ATP	4 mM
Na_3-GTP	0.4 mM
EGTA (*see note below*)	0.2 mM
Dye of choice	
pH to 7.3 (adjust with KOH <!>)	

Once prepared, dye-containing pipette solutions should be stored at –20°C until use. Fresh solutions should be prepared every 1–2 wk. For efficient use of expensive dyes with different pipette solutions, it is possible to prepare appropriately concentrated (e.g., 125%) dye-free pipette solutions and a separate stock solution of the dye that will yield the desired concentration after dilution. Mix (vortex) stock solutions just before use. Most dyes can be stored at –20°C for several months.

Important: When using Ca^{2+}-sensitive indicator dyes such as Fura-2, no additional substances that buffer Ca^{2+} [e.g., EGTA or BAPTA (1(,2-bis-(2-aminophenoxy)ethane-*N,N,N′,N′*-tetra-acetic acid) <!>] should be included in the pipette solution. The competition between the indicator dye and the additional buffer would seriously complicate the interpretation of the measurements; compare with Chapter 36.

External Solution

This recipe provides an example of a suitable solution.

NaCl	125 mM
KCl	2.5 mM
$CaCl_2$ <!>	2 mM

$MgCl_2$	1 mM
NaH_2PO_4	1.25 mM
$NaHCO_3$	26 mM
Glucose	20 mM

Aerate with 95% O_2 <!> and 5% CO_2 <!>. The resulting pH should be 7.3 at 21°C.

REFERENCES

Cota G, Armstrong CM. 1988. Potassium channel "inactivation" induced by soft-glass patch pipettes. *Biophys J* **53:** 107–109.

Dodt H-U, Becker K, Zieglgänsberger W. 2011. Infrared video microscopy, In *Imaging: A laboratory manual* (ed. Yuste R), pp. 79–88. Cold Spring Harbor Laboratory Press, Cold Spring Harbor, NY.

Edwards F, Konnerth A, Sakmann B, Takahashi T. 1989. A thin slice preparation for patch clamp recordings from neurones of the mammalian central nervous system. *Pflügers Arch* **414:** 600–612.

Grynkiewicz G, Poenie M, Tsien RY. 1985. A new generation of Ca^{2+} indicators with greatly improved fluorescence properties. *J Biol Chem* **260:** 3440–3450.

Kitamura K, Judkewitz B, Kano M, Denk W, Häusser M. 2008. Targeted patch-clamp recordings and single-cell electroporation of unlabeled neurons *in vivo*. *Nat Methods* **5:** 61–67.

Marty A, Neher E. 1995. Tight-seal whole-cell recordings. In *Single-channel recordings*, 2nd ed. (ed. Sakmann B, Neher E), pp. 31–52. Plenum, New York.

Müller A, Kukley M, Stausberg P, Beck H, Müller W, Dietrich D. 2005. Endogenous Ca^{2+} buffer concentration and Ca^{2+} microdomains in hippocampal neurons. *J Neurosci* **25:** 558–565.

Neher E. 1995. The use of fura-2 for estimating Ca buffers and Ca fluxes. *Neuropharmacology* **34:** 1423–1442.

Neher E, Augustine GJ. 1992. Calcium gradients and buffers in bovine chromaffin cells. *J Physiol* **450:** 273–301.

Oshiro M, Moomaw LA, Keller HE. 2011. Video microscopy, video cameras, and image enhancement. In *Imaging: A laboratory manual* (ed. Yuste R), pp. 57–70. Cold Spring Harbor Laboratory Press, Cold Spring Harbor, NY.

Penner R. 1995. A practical guide to patch-clamping. In *Single-channel recordings*, 2nd ed. (ed. Sakmann B, Neher E), pp. 3–30. Plenum, New York.

Pusch M, Neher E. 1988. Rates of diffusional exchange between small cells and a measuring patch pipette. *Pflügers Arch* **411:** 204–211.

Rexhausen U. 1992. "Bestimmung der Diffusionseigenschaften von Fluoreszenzfarbstoffen in verzweigten Nervenzellen unter Verwendung eines rechnergesteuerten Bildverarbeitungssystems." PhD thesis, University of Göttingen, Germany.

Sakmann B, Stuart GJ. 1995. Patch-pipette recordings from the soma, dendrites, and axon of neurons in brain slices. In *Single-channel recordings*, 2nd ed. (ed. Sakmann B, Neher E), pp. 199–211. Plenum, New York.

Schmidt H, Brown E, Schwaller B, Eilers J. 2003. Diffusional mobility of parvalbumin in spiny dendrites of cerebellar Purkinje neurons quantified by fluorescence recovery after photobleaching. *Biophys J* **84:** 2599–2608.

Schneggenburger R, Zhou Z, Konnerth A, Neher E. 1993. Fractional contribution of calcium to the cation current through glutamate receptor channels. *Neuron* **11:** 133–143.

Tsien RY. 1981. A non-disruptive technique for loading calcium buffers and indicators into cells. *Nature* **290:** 527–528.

von Blankenfeld G, Kettenmann H. 1992. Application of Lucifer yellow as intracellular marker in electrophysiology. In *Practical electrophysiological methods* (ed. Kettenmann H, Grantyn R), pp. 309–315. Wiley-Liss, New York.

Yuste R, ed. 2011. *Imaging: A laboratory manual.* Cold Spring Harbor Laboratory Press, Cold Spring Harbor, NY.

Ziv NE, Schiller J. 2011. Differential interference contrast imaging of living cells. In *Imaging: A laboratory manual* (ed. Yuste R), pp. 71–78. Cold Spring Harbor Laboratory Press, Cold Spring Harbor, NY.

36 A Single-Compartment Model of Calcium Dynamics in Nerve Terminals and Dendrites

Fritjof Helmchen[1] and David W. Tank[2]

[1]Department of Neurophysiology, Brain Research Institute, University of Zurich, 8057 Zurich, Switzerland;
[2]Department of Molecular Biology, Carl Ichan Laboratory, Princeton University, Princeton, New Jersey 08544

ABSTRACT

This chapter describes a single-compartment model of calcium dynamics that has been applied to fluorescence measurements of intracellular free calcium concentration ($[Ca^{2+}]_i$) changes in neurons. The model describes intracellular calcium handling under simplified conditions, for which analytical expressions for the amplitude and the time constants of $[Ca^{2+}]_i$ changes can be explicitly derived. In particular, it reveals the dependence of the measured $[Ca^{2+}]_i$ changes on the calcium indicator concentration. Applied to experimental data from small cells or subcellular compartments, the model equations have been extremely useful for obtaining quantitative information about essential parameters of Ca^{2+} influx, buffering, and clearance. The chapter also illustrates several changes that occur when the basic assumptions do not hold (e.g., when calcium diffusion, dye saturation, or kinetic effects become significant). Finally, we discuss how the changes in calcium dynamics, which are explained by the model, have been exploited for measuring properties of calcium-driven reactions, such as those regulating short-term synaptic enhancement, vesicle recycling, and adaptation. For a complementary treatment of calcium dynamics, see Neher (2011).

INTRODUCTION

Intracellular $[Ca^{2+}]_i$ changes are key to many fundamental cellular processes, such as muscle contraction, neurotransmitter release, or neuronal plasticity; and it is, therefore, desirable to quantitatively understand how exactly Ca^{2+} exerts these particular functions. The spatiotemporal dynamics of intracellular $[Ca^{2+}]_i$ changes is, however, a complicated function governed by many processes including Ca^{2+} influx, release, diffusion, binding to various proteins, and extrusion across the plasma membrane (Friel and Chiel 2008). In general, the corresponding differential equations cannot be solved analytically and, thus, need to be integrated numerically. Nonetheless, useful approximations can be made under certain conditions so that a simplified description of $[Ca^{2+}]_i$ dynamics can help researchers to understand key parameters of Ca^{2+} handling and guide their experimental measurements (Neher 1998). The single-compartment model described in this chapter is the simplest description of intracellular $[Ca^{2+}]_i$ under conditions relevant to the experimenter because buffering by intracellular Ca^{2+}-binding proteins as well as by the Ca^{2+} indicator introduced into the cells to achieve an optical $[Ca^{2+}]_i$ measurement is considered in the model.

The single-compartment model treats a cellular compartment as a homogenous well-mixed compartment, typically containing several pools of fast nonsaturating buffers (including the fluorescent indicator as one of them) as well as a simple linear extrusion mechanism. The model has been particularly useful in describing the buildup and decay of $[Ca^{2+}]_i$ produced by sodium action potentials in nerve terminals, dendrites, and dendritic spines (Regehr and Tank 1994; Sabatini et al. 2001). Rapid (<1 msec) and local (<0.1 µm) actions of Ca^{2+} near its entry site, such as the triggering of neurotransmitter release by calcium microdomains (Neher 1998; see Chapter 15), are not addressed by this approach. Rather, the model describes the dynamics of longer lasting (typically >10 msec) and spatially more homogenous $[Ca^{2+}]_i$ elevations—sometimes referred to as residual free calcium—that are commonly measured in dendrites and nerve terminals with high-affinity calcium indicators.

MATERIALS

Imaging Setup

Various types of microscope setups may be used for calcium imaging depending on the specific requirements. Charge-coupled device (CCD) camera–equipped fluorescence microscopes are useful for measurements from thick dendrites or large nerve terminals, whereas confocal microscopy or two-photon microscopy has been used to study thin dendrites, dendritic spines, and small axonal boutons. Most quantitative studies have been performed in acute brain slice preparations.

Calcium Indicator Dyes

The single-compartment model describes calcium dynamics in the presence of an exogenous calcium indicator dye. To date, most studies using the single-compartment model for quantitative analysis have been performed with synthetic calcium indicators. Commonly used high-affinity indicators (Ca^{2+}-binding dissociation constant $K_d < 1$ µM) include Fura-2, Calcium Green-1, Fluo-4, and Oregon Green BAPTA (1,2-bis-(2-aminophenoxy)ethane-*N,N,N′,N′*-tetra-acetic acid)-1 (Invitrogen). Commonly used low-affinity indicators ($K_d > 1$ µM) include MagFura-2, Calcium Green-5N, and Fluo-5F (Invitrogen).

For the theoretical background and experimental procedures for the calibration of these Ca^{2+} indicators in terms of absolute intracellular-free Ca^{2+} concentration, see Helmchen (2011). In addition to synthetic indicators, there now exist many variants of Ca^{2+}-sensitive fluorescent proteins (Mank and Griesbeck 2008; see Chapter 8). These genetically encoded Ca^{2+} sensors open new avenues for the study of intracellular $[Ca^{2+}]_i$ by enabling targeting to specific cell types or to subcellular organelles as well as long-term expression. A caveat for using the framework of the single-compartment model with genetically encoded calcium indicators is that these proteins often display cooperativity in Ca^{2+} binding, which makes the observed fluorescence changes more difficult to interpret (Palmer and Tsien 2006).

SINGLE-COMPARTMENT MODEL

Model Assumptions and Parameters

The single-compartment model is based on the assumption that Ca^{2+} gradients and diffusion can be neglected on the timescale of interest, making it possible to treat the cytosolic volume as a single

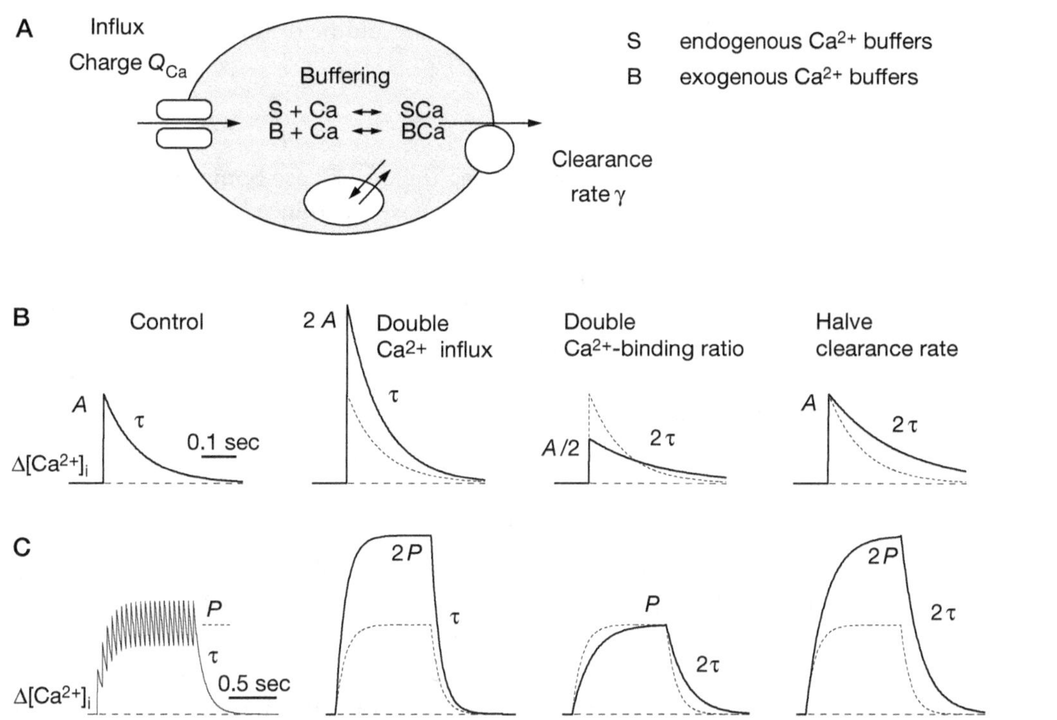

FIGURE 1. (*A*) Schematic of the single-compartment model. Ca^{2+} clearance from the cytosol includes extrusion via the plasma membrane, uptake into intracellular organelles, and binding to slow endogenous buffers. Two pools of endogenous rapid Ca^{2+} buffers (S) and exogenously introduced buffers (B) are considered. (*B*) A single, brief calcium influx leads to an exponential $[Ca^{2+}]_i$ transient (*left*). Changes in influx, buffering, or clearance parameters alter the shape of the transient (*middle* and *right*). For comparison, the control trace is drawn as dashed curves. (C) Summation of $[Ca^{2+}]_i$ transients during a train of stimuli for 1 sec at 20 Hz (*left*). The accumulation of the average $[Ca^{2+}]_i$ is described by an exponential increase to a steady-state plateau level *P* and an exponential decay. The dependence of the $[Ca^{2+}]_i$ accumulation on changes in influx, buffering, and clearance is shown as in *B*. Note that a change in the Ca^{2+}-binding ratio does not affect the plateau level.

homogenous compartment. However, when is this assumption of spatial homogeneity justified? For a compartment of radius r, a characteristic diffusion time is given by $t_d = r^2/(6D_{Ca})$, where D_{Ca} is an effective Ca^{2+} diffusion constant that takes the slowdown of Ca^{2+} diffusion caused by binding to cytosolic proteins into account. For $r = 1$ μm, the size of most mammalian nerve terminals and dendrites, and for $D_{Ca} \approx 10$–$20\ \mu m^2\ sec^{-1}$ (Gabso et al. 1997; Murthy et al. 2000), t_d is ~10 msec. Following the cessation of influx, Ca^{2+} gradients dissipate within this time period. $[Ca^{2+}]_i$ changes that occur on the timescale of 0.1 sec (widely observed in small neural processes) can be well described using a single-compartment model that considers Ca^{2+} influx, buffering, and clearance as the relevant biophysical processes (Fig. 1A). The approximations used to describe these processes are discussed below.

Influx

We limit our attention to pulses of calcium influx that are brief compared with the time required for calcium clearance. This condition is quite reasonable for short depolarizing voltage steps and, in particular, for the calcium influx through voltage-gated calcium channels during the repolarizing phase of an action potential, which lasts less than a millisecond (Borst and Helmchen 1998). Ca^{2+} clearance times in presynaptic terminals and dendrites are typically at least one order of magnitude longer. A brief Ca^{2+} pulse at time t_p can be described by a δ-function and the increase in total calcium concentration ($\Delta[Ca^{2+}]_T$):

$$j_{in} = \Delta[Ca^{2+}]_T\,\delta(t - t_p) = \frac{Q_{Ca}}{2FV}\,\delta(t - t_p), \tag{1}$$

where F is Faraday's constant, V is the volume of the cellular compartment, and Q_{Ca} is the calcium charge injected.

Buffering

Most of the calcium ions entering the cytosol are bound by endogenous calcium-binding proteins or exogenous calcium buffers that were introduced into the cell by the experimenter. The initial changes in free calcium concentration depend on the Ca^{2+}-buffering capacity of the rapid Ca^{2+} buffers. Usually, a pool S of rapid endogenous buffers and a pool B of exogenously introduced rapid buffers are considered (Fig. 1A). The Ca^{2+}-binding ratio of the exogenous buffer (κ_B), which usually is the fluorescent indicator itself, is defined as the ratio of the change in buffer-bound Ca^{2+} over the free Ca^{2+} change (Neher and Augustine 1992). κ_B depends on the total buffer concentration $[B]_T$, its dissociation constant K_d, and $[Ca^{2+}]_i$:

$$\kappa_B = \frac{\partial[BCa]}{\partial[Ca^{2+}]_i} = \frac{[B]_T K_d}{([Ca^{2+}]_i + K_d)^2} \approx \frac{[B]_T}{K_d} \quad ([Ca^{2+}]_i << K_d). \tag{2}$$

An analogous expression exists for the Ca^{2+}-binding ratio of the endogenous buffer (κ_S). The maximal Ca^{2+}-binding ratio ($[B]_T/K_d$) occurs at $[Ca^{2+}]_i$ levels that are well below the K_d. In its simplest version, the single-compartment model assumes that κ_S and κ_B are $[Ca^{2+}]_i$ independent, corresponding to the idea that there are only small $[Ca^{2+}]_i$ changes from a resting level during the experiment. Under these conditions, changes in free calcium concentration are proportional to changes in total calcium concentration (see Equation 6). A value of $\kappa_B = 100$ implies that, for each free calcium ion, there are 100 ions bound to the buffer B. Note that Ca^{2+}-binding ratios are dimensionless and that experimentally determined values for the endogenous buffers range from 10 to 2000 (Tables 1 and 2). Also note that Ca^{2+} indicators, which generally have fast kinetics, are potent exogenous Ca^{2+} buffers and, therefore, have to be included in pool B.

TABLE 1. Some representative experimental results for the basic parameters of calcium dynamics in dendrites and dendritic spines

Preparation	Amplitude A (nM)	Time constant τ (msec)	Ca^{2+}-binding ratio κ_S	Clearance rate γ (sec^{-1})	Reference(s)
Neocortex					
L5 pyramidal neurons	260	70	120	1700	Helmchen et al. 1996
Dendritic spines; L5 pyramidal neurons	1050	<50	20	~800	Cornelisse et al. 2007
L2/3 pyramidal neurons	190–240	60–90	110	1800	Koester and Sakmann 2000; Kaiser et al. 2001
Bitufted interneurons	140	200	300	1500	Kaiser et al. 2001
Hippocampus					
CA1 pyramidal neurons	150–240	90	60–180	700–2000	Helmchen et al. 1996; Maravall et al. 2000; Liao and Lien 2009
Cultured hippocampal excitatory neurons, room temperature	—	176	60	340	Lee et al. 2000a
Dendritic spines; CA1 pyramidal neurons, 34°C	1500–1700	12–15	20	1200–1600	Sabatini et al. 2002
Dendritic spines; CA1 pyramidal neurons, room temperature	1000–1100	35	11	350	Simons et al. 2009
Dendritic spines; CA2 pyramidal neurons, room temperature	230	34	41	1200	Simons et al. 2009
Dendritic spines; CA3 pyramidal neurons, room temperature	1300	24	9	420	Simons et al. 2009

(Continued)

TABLE 1. *(Continued)*

Preparation	Amplitude A (nM)	Time constant τ (msec)	Ca^{2+}-binding ratio κ_S	Clearance rate γ (sec^{-1})	Reference(s)
Dentate gyrus granule cells, room temperature	230 (P18–25) 190 (>P60)	1000 230	50–100 150–300	100 1000	Stocca et al. 2008
Cultured hippocampal inhibitory neurons, room temperature	—	429	130–150	305	Lee et al. 2000a
Fast-spiking basket cells	20–40	350–400 (room temperature) 200 (34°C) 500–600 (room temperature) 1100 (34°C)	200		Aponte et al. 2008
Dendrite-targeting interneurons	650	~300–500	20–30	~40–100	Liao and Lien 2009
Olfactory bulb					
Granule cells	250 (spines) 230 (dend.)	175	110–140 (spines) 80–90 (dend.)	870 (spines) 570 (dend.)	Egger and Stroh 2009
Lateral dendrites of mitral cells	130	190	360–400	3000	Egger and Stroh 2009
Cerebellum					
Cerebellar Purkinje cells	—	400	900–2000	—	Callewaert et al. 1996; Fierro and Llano 1996
SN					
Dopaminergic neurons in SN pars compacta	270	125	110 (P13–17) 180 (P25–32)	900 1400	Foehring et al. 2009

The reported amplitudes and time constants refer to single action potential–evoked calcium transients. Note that values depend on exogenous buffer conditions, animal age, and temperature (see original publications for specific conditions).

Dend, dendrites; SN, substantia nigra. (—) not determined.

TABLE 2. Some representative experimental results for the basic parameters of calcium dynamics in nerve terminals

Preparation	Amplitude A (nM)	Time constant τ (msec)	Ca^{2+}-binding ratio κ_S	Clearance rate γ (sec^{-1})	Reference(s)
Neocortex					
Neocortical pyramidal cell nerve terminals	500–1000	60	140	2600	Cox et al. 2000; Koester and Sakmann 2000
Hippocampus					
DG granule cell axonal boutons in the hilus, $\kappa_B = 0$	900	40	20	500	Jackson and Redman 2003
Hippocampal mossy fiber terminals, $\kappa_B > 0$	5–10	1000	—	—	Regehr et al. 1994
Hippocampal CA3–CA1 synapse	—	40	—	—	Sinha et al. 1997
Cerebellum					
Presynaptic boutons of cerebellar granule cells	390–860	24–54	34–86	700–3500	Brenowitz and Regehr 2007
Brain stem					
Calyx of Held nerve terminals in brain stem	400–500	20–90	40	400–900	Helmchen et al. 1997; Müller et al. 2007
Other					
Crayfish neuromuscular junction	5–10	4000	600	80–100	Tank et al. 1995
Frog retinotectal synapse	140	100	—	—	Feller et al. 1996
Synaptic terminals in *Drosophila* larvae	—	50–150	—	—	Lnenicka et al. 2006

The reported amplitudes and time constants refer to single action potential–evoked calcium transients. Note that values depend on exogenous buffer conditions, animal age, and temperature (see original publications for specific conditions).

(—) not determined. DG, dentate gyrus.

Clearance

Calcium ions are either sequestered into intracellular organelles or extruded via the plasma membrane until $[Ca^{2+}]_i$ decays back to a resting level of typically 30–100 nM (Helmchen et al. 1996; Maravall et al. 2000). Most single-compartment models used for the analysis of experimental imaging data have used a calcium clearance mechanism linearly dependent on the deviation of $[Ca^{2+}]_i$ from the resting calcium level with a clearance rate γ:

$$j_{out} = -\gamma\,\Delta[Ca^{2+}]_i = -\gamma\,([Ca^{2+}]_i - [Ca^{2+}]_{rest}). \tag{3}$$

This expression provides a good fit to experimental data in several systems (Neher and Augustine 1992; Tank et al. 1995; Helmchen et al. 1996, 1997). Nevertheless, Equation 3 is generally considered to be the low $[Ca^{2+}]_i$ limit of a saturable enzyme-driven extrusion mechanism (e.g., one following Michaelis–Menten kinetics). The good agreement with experimental data suggests that saturation effects might only be found under rather extreme stimulus conditions. In large presynaptic nerve terminals of the brain stem, Na^+–Ca^{2+} exchange is the predominant mechanism of calcium clearance at low $[Ca^{2+}]_i$ (with additional contribution by the plasma membrane Ca^{2+} ATPase), whereas mitochondrial uptake is recruited only at very high $[Ca^{2+}]_i$ levels when the Na^+–Ca^{2+} exchangers become saturated (Kim et al. 2005).

Dynamics of Single $[Ca^{2+}]_i$ Transients

Combining these kinetic descriptions for influx, buffering, and clearance, the differential equation for the temporal dynamics of $[Ca^{2+}]_i$ following a single brief calcium influx, is

$$\frac{d[Ca^{2+}]_T}{dt} = j_{in} + j_{out},$$
$$\frac{d[Ca^{2+}]_i}{dt}(1+\kappa_S+\kappa_B) = \Delta[Ca^{2+}]_T\,\delta(t-t_p) - \gamma\Delta[Ca^{2+}]_i. \tag{4}$$

This equation simply states that the change in total calcium equals the increase in calcium minus the clearance. Assuming constant Ca^{2+}-binding ratios, the analytical solution of Equation 4 is an exponential function with amplitude A and decay time constant τ (Fig. 1B). The relationships between A and τ and the model parameters are

$$A = \frac{Q_{Ca}/(2FV)}{(1+\kappa_S+\kappa_B)}, \tag{5}$$

$$\tau = \frac{(1+\kappa_S+\kappa_B)}{\gamma}. \tag{6}$$

Both A and τ depend on the Ca^{2+}-binding ratios. Note, however, that the product $A\tau$, which is the integral of the $[Ca^{2+}]_i$ transient (the area underneath the exponential curve), is independent of the Ca^{2+}-binding ratios. Figure 1B illustrates that changes in influx, total buffering capacity, and clearance rate lead to specific changes of the $[Ca^{2+}]_i$ transient shape (Sabatini and Regehr 1995). Although changes in influx or clearance only affect either A or τ, both magnitudes depend on the Ca^{2+}-buffering capacity. Because of this dependence, A and τ may be altered by introducing a calcium indicator dye. Note that these alterations are large (severalfold) for typical calcium indicator concentrations and, thus, are not negligible. A description of how exogenous calcium buffers can be exploited to obtain an estimate of the endogenous Ca^{2+}-binding ratio κ_S is given below.

Summation of $[Ca^{2+}]_i$ Transients during Repetitive Calcium Influx

If multiple calcium influx pulses occur on a timescale that is short compared with the $[Ca^{2+}]_i$-decay time, the amplitude of the $[Ca^{2+}]_i$ transient simply increases corresponding to the greater total calcium charge Q_{Ca}. For example, the amplitude of $[Ca^{2+}]_i$ transients has been shown to scale approximately linearly with the number of high-frequency action potentials in presynaptic terminals (Regehr et al. 1994), in Purkinje cell axons (Callewaert et al. 1996), in mammalian pyramidal cell dendrites in vivo (Svoboda et al. 1997), and in hippocampal inhibitory interneurons (Aponte et al. 2008; Liao and Lien 2009). Such a linear increase would be expected to break down for more stimuli when calcium influx per pulse changes or saturation of buffers and pumps occurs. A linear sum of individual transients can also approximate the buildup of $[Ca^{2+}]_i$ during repetitive calcium influx as it occurs during firing of a neuron at a lower constant frequency. We consider a train of stimuli starting at time $t = 0$ and with a time interval of Δt (frequency $f = 1/\Delta t$). The $[Ca^{2+}]_i$ level above resting level immediately before the $(n{+}1)$th stimulus is given by a geometric progression (Regehr et al. 1994):

$$\Delta[Ca^{2+}]_i(n\Delta t) = A\sum_{i=1}^{n} e^{-(i\Delta t)/\tau} = \frac{A}{(e^{\Delta t/\tau}-1)}(1-e^{-(n\Delta t)/\tau}), \tag{7}$$

where A is the amplitude and τ is the time constant of each individual transient. For stimulation frequencies $f < 1/(2\tau)$, there is little buildup, and individual spaced transients result; but for higher frequencies, the transients add up, and $[Ca^{2+}]_i$ exponentially reaches a steady state in which influx and clearance balance and $[Ca^{2+}]_i$ fluctuates around a plateau level (Fig. 1C). The time constant of the increase and the decay following the end of stimulation are both given by τ. The mean plateau level P reached at steady state is proportional to the frequency (Tank et al. 1995; Helmchen et al. 1996):

$$P = A\tau f = \frac{Q_{Ca}}{2FV\gamma} f, \tag{8}$$

with the integral of the single calcium transient ($A\tau$) as proportionality constant. P is independent of the Ca^{2+}-binding ratios, indicating that Ca^{2+} buffers can affect the transient dynamics of $[Ca^{2+}]_i$ but not its steady-state levels. Figure 1C summarizes the effects of changes in influx, buffering, and clearance on the $[Ca^{2+}]_i$ accumulation during repetitive stimulation. As in the case of the single $[Ca^{2+}]_i$ transient, each change in these biophysical parameters leads to a characteristic alteration of the overall shape of the $[Ca^{2+}]_i$ change (Tank et al. 1995).

APPLICATION OF THE SINGLE-COMPARTMENT MODEL

One of the main advantages of the single-compartment model is that it takes the effect of calcium indicator buffering into account. This provides a framework to estimate how severely the indicator perturbs calcium signaling and, in addition, it suggests systematic variation of the indicator concentration in the experiment as a tool to determine endogenous calcium handling parameters and to reveal unperturbed calcium dynamics.

Estimates of Endogenous Ca^{2+}-Binding Ratio and Clearance Rate

The endogenous Ca^{2+}-binding ratio κ_S and the clearance rate γ can be estimated by measuring changes in the $[Ca^{2+}]_i$ transient decay time produced by increasing concentrations of an exogenous Ca^{2+} buffer, such as BAPTA or Fura-2 (Fig. 2A). This follows directly from Equation 6, which can be rearranged to give the following:

$$\begin{aligned} \tau &= a_1\,\kappa_B + a_0, \\ a_1 &= 1/\gamma, \\ a_0 &= (1 + \kappa_S)/\gamma. \end{aligned} \tag{9}$$

The decay time constant τ should be linearly related to the exogenous buffer capacity κ_B (Tank et al. 1991). The inverse of the measured slope (a_1) provides the clearance rate, and the negative x-

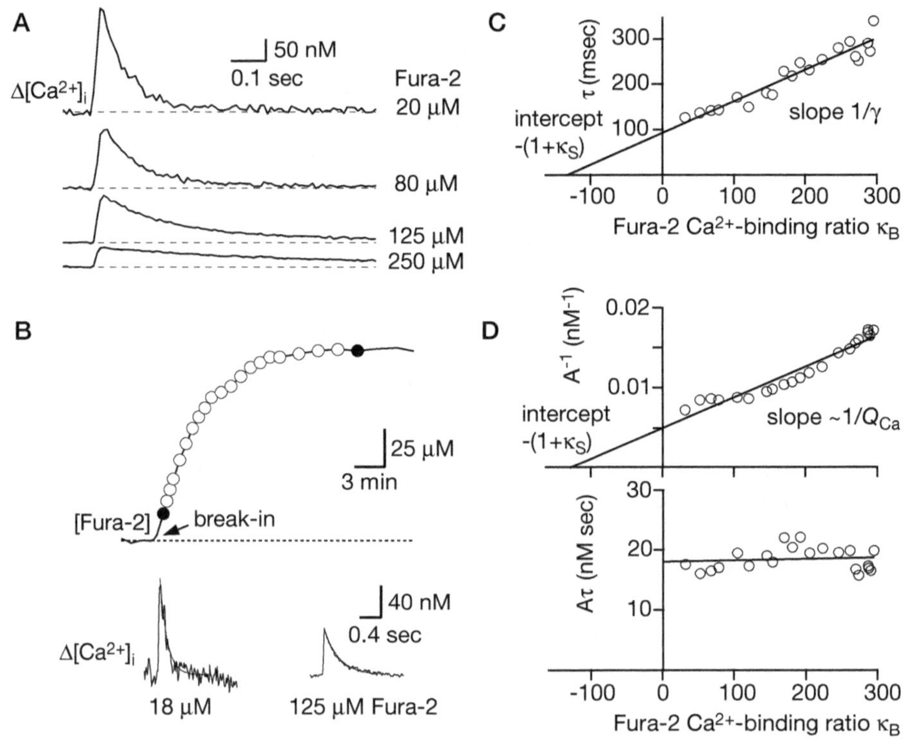

FIGURE 2. (*A*) Amplitude and decay time of calcium transients evoked by single action potentials in proximal apical dendrites of neocortical pyramidal neurons depend on the Fura-2 concentration (four different experiments). (*B*) This dependence can also be measured in single experiments by monitoring action potential–induced calcium transients during the loading of the indicator. Following break-in with a whole-cell patch pipette, the fluorescence exponentially reaches a steady-state plateau level, which represents equilibration with the patch pipette solution containing 125-µM Fura-2. Calcium transients were measured at 20–60-sec time intervals (circles). Two example calcium transients at low and high Fura-2 concentration, respectively (solid circles), are shown at the bottom. (*C*) Analysis of the time constant of calcium transients during Fura-2 loading. The endogenous Ca^{2+}-binding ratio κ_S and the clearance rate γ can be determined by plotting the decay time constant as a function of the exogenously introduced Ca^{2+}-buffering capacity. The extrapolation to zero Fura-2 Ca^{2+}-binding ratio yields an estimate of the time constant in the absence of an indicator. (*D*) Similarly, κ_S and the extrapolated calcium transient amplitude can be determined from a plot of the inverse of the measured amplitudes versus κ_B (*top*). As a control (*bottom*), the product of the amplitude and time constant is also plotted, which should not depend on κ_B. (Modified, with permission, from Helmchen et al. 1996.)

axis intercept ($-a_0/a_1$) provides an estimate of the endogenous buffer capacity κ_S (Neher and Augustine 1992). An example of this method, taken from a study of $[Ca^{2+}]_i$ transients in pyramidal cell dendrites, is shown in Figure 2C. Two methods have been used to provide known exogenous buffer concentrations: dialysis with patch pipettes (Neher and Augustine 1992; Helmchen et al. 1996; Maravall et al. 2000) and intracellular sharp microelectrode injection of a fluorescent buffer from which estimated concentrations can be calculated from the fluorescence intensity (Tank et al. 1995). Whenever possible, an elegant and effective method is to monitor the indicator buffering effect during the dye-loading process in individual experiments, avoiding cell-to-cell variability (Fig. 2B–D; Neher and Augustine 1992; Helmchen et al. 1996; Sabatini et al. 2002; Aponte et al. 2008; Foehring et al. 2009; Liao and Lien 2009). The actual indicator concentrations in this case are back-calculated from the estimated final concentration.

Estimates of Amplitude A and Total Calcium Charge Q_{Ca}

Compared with the unaltered $[Ca^{2+}]_i$ transients, the amplitude A that is obtained from a Ca^{2+} indicator measurement is reduced by a factor $(1 + \kappa_S)/(1 + \kappa_S + \kappa_B)$ (see Equation 5). Similar to the analysis of the decay time constant, the inverse of the amplitude (A^{-1}) is expected to depend linearly on the exogenous Ca^{2+}-buffer capacity κ_B (Equation 5), and this relationship can be used to obtain another estimate of κ_S (Fig. 2D). The absolute values of A can either be measured using ratiometric or nonratiometric calibration techniques (see Helmchen 2011) or be estimated from the degree of saturation of the fluorescent dye (see below). If the total buffering capacity of the cytoplasm is known, and with an estimate of the compartment volume from morphology, the total calcium charge Q_{Ca} can be calculated by rearranging Equation 5:

$$Q_{Ca} = 2FVA(1 + \kappa_S + \kappa_B). \tag{10}$$

Alternatively, Q_{Ca} can be measured directly using excessive dye loading of the compartment to obtain fluorescence changes proportional to Q_{Ca} (Neher 1995; Borst and Helmchen 1998). When individual calcium transients cannot be easily resolved, Q_{Ca} can be determined from the frequency dependence of the initial slope of the calcium buildup produced by a stimulus train (Tank et al. 1995).

Estimates of Unperturbed Calcium Dynamics

Because many cells have relatively low endogenous Ca^{2+}-buffer capacity (see below), even the lowest concentrations of indicator dyes that can be used for fluorescence measurements often can cause significant perturbation of calcium signals. Systematic variation of the exogenous Ca^{2+}-buffer capacity κ_B, in this case, also permits the estimation of the amplitude and time course of unperturbed calcium transients, yielding the best possible approximation of the intracellular $[Ca^{2+}]_i$ changes as they occur in cells under physiological conditions. The unperturbed amplitude and decay time constant are obtained by extrapolation of the regression lines from the plots of τ and A^{-1} against κ_B to the zero exogenous Ca^{2+}-binding ratio (Fig. 2C,D). Another method to approximate unperturbed dynamics is the use of low-affinity indicators, which introduce less exogenous Ca^{2+}-buffer capacity but usually require averaging of traces to resolve the small signals (Helmchen et al. 1997; Sabatini et al. 2002).

EXAMPLES OF APPLICATION

Changes in calcium dynamics similar to those illustrated (Fig. 1B,C) have been measured in small cell somata, presynaptic terminals, dendrites, dendritic spines, and axons. The results of these experiments provide evidence for the applicability of the single-compartment model as well as quantitative estimates for the model parameters. Tables 1 and 2 provide a comparison of observed parameters for dendritic compartments and nerve terminals, respectively, listing the corresponding references.

In general, relatively large and rapid calcium signals are associated with action potentials. In central nervous system (CNS) pyramidal cells, the physiological $[Ca^{2+}]_i$ changes in dendrites that are caused by single action potentials were estimated to have several hundred nanomolar amplitudes and to decay within 100 msec or less (Table 1). The reason for this fast signaling capacity is a relatively low endogenous Ca^{2+}-buffer capacity in the range of 50 to 180. For calcium handling in dendritic spines, the application of the single-compartment model revealed even lower κ_S values of ~10–30, which consequently result in fast spinous calcium transients with decay time constants of ~10–30 msec (Sabatini et al. 2002; Cornelisse et al. 2007; Simons et al. 2009). Interestingly, there appear to be regional differences (e.g., between different subfields in the hippocampus; Simons et al. 2009). Consistent with the large diversity of GABAergic interneurons, a wide range of κ_S values has been reported for interneuron subtypes ranging from relatively high Ca^{2+}-buffer capacities (Kaiser et al. 2001; Aponte et al. 2008) to surprisingly low values of κ_S in dendrite-targeting interneurons in the hippocampus (Liao and Lien 2009). The highest κ_S values so far have been reported for cerebellar Purkinje neurons (Callewaert et al. 1996; Fierro and Llano 1996). The presynaptic nerve terminals of the mammalian CNS that have been investigated so far also possess Ca^{2+}-binding ratios of ~20–140, and the decay time constant of single action potential–evoked calcium transients can be as fast as a few tens of milliseconds (Table 2).

EXTENSIONS TO THE SINGLE-COMPARTMENT MODEL

This section describes several extensions that can be added to the single-compartment model if certain assumptions do not hold. In addition, a description is given of how the model has helped to characterize calcium-dependent cellular processes.

Buffered Calcium Diffusion

In addition to buffering Ca^{2+}, fluorescent indicators may perturb intracellular calcium signals because of their mobility. This becomes relevant if Ca^{2+} influx occurs rather locally and subsequently spreads via diffusion (e.g., along a dendrite). The single-compartment model can be extended to such situations if the diffusional exchange occurs on a rather long spatial scale so that local chemical equilibrium can still be assumed. In this case, calcium diffusion is described by an apparent diffusion constant that depends on the diffusion constants D_m and the Ca^{2+}-buffer capacity κ_m of mobile buffers as well as the Ca^{2+}-buffer capacity of fixed buffers κ_f (Gabso et al. 1997). In general, immobile buffers tend to hold Ca^{2+} in place and thereby slow down diffusion, whereas highly mobile buffers such as Fura-2 or Calcium Green-1 facilitate the spread of Ca^{2+} (Neher 1998).

Deviations from Linear Behavior

When $[Ca^{2+}]_i$ changes are appreciable compared with the dissociation constant of the dominant intracellular buffers or when buffer kinetics are slow compared with the decay time constant, exponential decaying transients are not observed. Here, we describe expected deviations under these two circumstances.

Saturation of Buffers and Pumps

If buffer kinetics are still rapid but $[Ca^{2+}]_i$ changes become large compared with the K_d values of the intracellular buffers, then Equation 4 is still appropriate, but κ_B and κ_S are not constant, and the solution is not an exponential decay. Numerically calculated decay curves as well as measured transients show pronounced upward curvature (faster decays) at higher $[Ca^{2+}]_i$ levels on a semilogarithmic plot (Tank et al. 1995) because the effective buffering capacity is reduced at these levels caused by partial saturation of the buffers. An opposite effect would be expected from saturation of clearance mechanisms at very high $[Ca^{2+}]_i$ levels. Saturation in the clearance rate would be expected to also alter the linear frequency dependence of the plateau level during long stimulus trains (Equation 8). A related buffer saturation effect can be exploited to advantage to provide an estimate of the amplitude A of individual $[Ca^{2+}]_i$ transients when ratiometric measurements of absolute calcium concentration are not feasible (e.g., when fluorescence is measured from a large region of AM [acetoxymethyl-ester]-loaded brain slices or intact brains [Regehr and Atluri 1995; Feller et al. 1996]). Except when confocal or two-photon techniques for volume resolution are used, such measurements contain contaminating fluorescence from other structures. In these cases, A can be estimated from the degree of saturation of a high-affinity calcium indicator (Regehr and Atluri 1995; Sabatini and Regehr 1995; Feller et al. 1996). Figure 3A shows the fluorescence change of an indicator evoked by two subsequent brief calcium injections. The second injection evokes a smaller fluorescence change than the first one because of partial saturation of the indicator. Assuming that the change in calcium concentration is the same for the two stimuli, its magnitude can be estimated from the ratio α of the two fluorescence changes:

$$\Delta[\mathrm{Ca}^{2+}]_i = \frac{([\mathrm{Ca}^{2+}]_{rest} + K_d)(1-\alpha)}{2\alpha}, \tag{11}$$

where $\alpha = (F_3 - F_2)/(F_1 - F_0)$ and assuming $F_2 \approx F_1$ (Feller et al. 1996). This method, however, relies on reasonable estimates of the resting $[Ca^{2+}]_i$ level and the dissociation constant of the indicator.

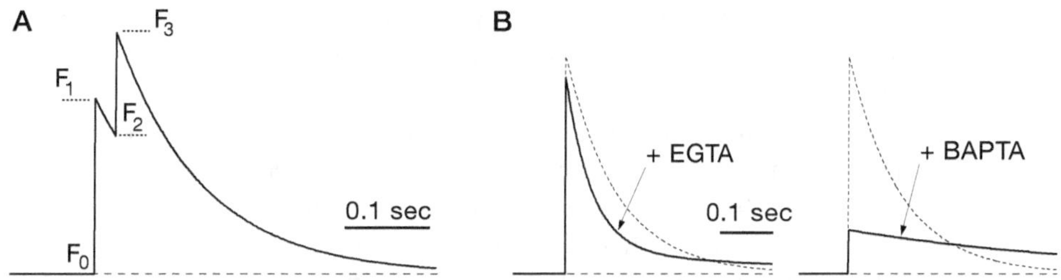

FIGURE 3. *(A)* The amplitude *A* of the $[Ca^{2+}]_i$ transient can be estimated from the degree of saturation of a fluorescent indicator with a known dissociation constant (see text). Schematic of the indicator fluorescence change on two brief calcium injections. The $[Ca^{2+}]_i$ change per stimulus is assumed to be equal, but the change in fluorescence *F* is smaller for the second stimulus. *(B)* Schematic of the effect of adding Ca^{2+} buffers with different kinetic properties on individual control calcium transients (dashed curves). The fast buffer, such as BAPTA, changes the $[Ca^{2+}]_i$ transient shape according to equilibrium considerations. A slow buffer, such as EGTA, in contrast, leads to a faster initial decay and a subsequent slow phase, which are explained by its slow association rate.

Slow Buffers

Radical departures in calcium dynamics from the expected results (Equations 4–6) have been observed when slow calcium buffers such as EDTA or EGTA were introduced as exogenous buffers into presynaptic nerve terminals (Atluri and Regehr 1996; Feller et al. 1996) and nerve cell dendrites (Markram et al. 1998). A comparison between the effects on a $[Ca^{2+}]_i$ transient of adding a fast buffer BAPTA versus a slow buffer EGTA is illustrated schematically in Figure 3B. Addition of BAPTA reduces the amplitude of the transient while prolonging its decay, consistent with what is expected from Equations 5 and 6. Addition of EGTA, however, has little effect on the amplitude *A* but changes the shape of the transient by providing a faster initial decay while also producing a slower subsequent phase. A similar effect was observed when chromaffin cells were loaded with parvalbumin, a calcium-binding protein with slow-binding kinetics (Lee et al. 2000b). As easily shown in numerical simulations, the early faster decay is produced by Ca^{2+} binding to EGTA (or parvalbumin) on a timescale longer than the response time of the Ca^{2+} indicator (Atluri and Regehr 1996; Feller et al. 1996; Markram et al. 1998; Lee et al. 2000b). The fast time constant can be approximated by adding the association rate of Ca^{2+} binding of EGTA to the clearance mechanisms (Atluri and Regehr 1996):

$$\tau_{\text{fast}} = \frac{(1+\kappa_S+\kappa_B)}{\gamma + k^{\text{on}}_{\text{EGTA}}[\text{EGTA}]}. \tag{12}$$

The longer time constant of the second decay phase is produced by the same mechanism as for BAPTA: More bound calcium must be removed per unit change of free calcium ion change. It is given by Equation 6 after addition of the Ca^{2+}-binding ratio κ_{EGTA}. Kinetic effects might be particularly relevant in small structures displaying rapid Ca^{2+} handling, such as dendritic spines because the sharing of Ca^{2+} between proteins will be a complicated function of their relative kinetics and the time course of calcium influx. Proteins with slow-binding kinetics could be bypassed following single action potentials and could be activated significantly only during prolonged or synaptic activation of Ca^{2+} influx (Sabatini et al. 2002). In large presynaptic terminals in the brain stem, a mobile calcium buffer with slow kinetics, presumably parvalbumin, was identified as an important intracellular factor that accelerates both spatially averaged calcium transients and paired pulse facilitation (Müller et al. 2007).

Single-Compartment Models and the Measurement of Calcium-Driven Reactions

The single-compartment models we have summarized here were developed to explain basic properties of calcium transients observed in imaging experiments on small structures such as presynaptic terminals and dendrites. The analysis also provides a set of tools to alter calcium dynamics, which has proven to be important in biophysical characterization of calcium-driven reactions. For example, the systematic prolongation of calcium transients with increasing exogenous buffer concentrations helped to show that the time constant of $[Ca^{2+}]_i$ decay was the rate-limiting step in the decay of calcium-driven short-term synaptic enhancement at an invertebrate synapse (Delaney and Tank 1994). In a mammalian synapse, exogenous buffer addition and calcium plateau manipulation by changes in stimulus frequency were used to alter calcium dynamics to determine kinetic rate constants of a calcium-driven reaction involved in short-term synaptic enhancement (Regehr et al. 1994). Similar approaches have been used in characterization of vesicle mobilization in chromaffin cells (Heinemann et al. 1993), of mitochondrial Ca^{2+} uptake and release in bullfrog sympathetic neurons (Friel and Tsien 1994), of adaptation of cellular excitability (Sobel and Tank 1994), and of synaptic facilitation in the cerebellum (Atluri and Regehr 1996).

To summarize, we have described the basic equations and parameters of a simple model of calcium dynamics that is applicable to action potential–evoked $[Ca^{2+}]_i$ transients in nerve terminals and dendrites. Methods have been introduced to quantify the main parameters of calcium influx, buffering, and clearance. We conclude that the single-compartment model is well suited as a starting point for a more detailed characterization of Ca^{2+} handling in small cellular compartments.

ACKNOWLEDGMENTS

We thank E. Neher for comments on this chapter.

REFERENCES

Aponte Y, Bischofberger J, Jonas P. 2008. Efficient Ca^{2+} buffering in fast-spiking basket cells of rat hippocampus. *J Physiol* **586:** 2061–2075.

Atluri PP, Regehr WG. 1996. Determinants of the time course of facilitation at the granule cell to Purkinje cell synapse. *J Neurosci* **16:** 5661–5671.

Borst JGG, Helmchen F. 1998. Calcium influx during an action potential. *Methods Enzymol* **293:** 352–371.

Brenowitz SD, Regehr WG. 2007. Reliability and heterogeneity of calcium signaling at single presynaptic boutons of cerebellar granule cells. *J Neurosci* **27:** 7888–7898.

Callewaert G, Eilers J, Konnerth A. 1996. Axonal calcium entry during fast 'sodium' action potentials in rat cerebellar Purkinje neurones. *J Physiol* **495:** 641–647.

Cornelisse LN, van Elburg RAJ, Meredith RM, Yuste R, Mansvelder HD. 2007. High speed two-photon imaging of calcium dynamics in dendritic spines: Consequences for spine calcium kinetics and buffer capacity. *PLoS ONE* **10:** e1073. doi: 10.1371/journal.pone.0001073.

Cox CL, Denk W, Tank DW, Svoboda K. 2000. Action potentials reliably invade axonal arbors of rat neocortical neurons. *Proc Natl Acad Sci* **97:** 9724–9728.

Delaney KR, Tank DW. 1994. A quantitative measurement of the dependence of short-term synaptic enhancement on presynaptic residual calcium. *J Neurosci* **14:** 5885–5902.

Egger V, Stroh O. 2009. Calcium buffering in rodent olfactory bulb granule cells and mitral cells. *J Physiol* **587:** 4467–4479.

Feller MB, Delaney KR, Tank DW. 1996. Presynaptic calcium dynamics at the frog retinotectal synapse. *J Neurophysiol* **76:** 381–400.

Fierro L, Llano I. 1996. High endogenous calcium buffering in Purkinje cells from rat cerebellar slices. *J Physiol* **496:** 617–625.

Foehring RC, Zhang XF, Lee JCF, Callaway JC. 2009. Endogenous calcium buffering capacity of substantia nigral dopamine neurons. *J Neurophysiol* **102:** 2326–2333.

Friel DD, Chiel HJ. 2008. Calcium dynamics: Analyzing the Ca^{2+} regulatory network in intact cells. *Trends Neurosci* **31:** 8–19.

Friel DD, Tsien RW. 1994. An FCCP-sensitive Ca^{2+} store in bullfrog sympathetic neurons and its participation in stimulus-evoked changes in $[Ca^{2+}]_i$. *J Neurosci* **14:** 4007–4024.

Gabso M, Neher E, Spira ME. 1997. Low mobility of the Ca^{2+} buffers in axons of cultured *Aplysia* neurons. *Neuron* **18:** 473–481.

Heinemann C, von Rüden L, Chow RH, Neher E. 1993. A two-step model of secretion control in neuroendocrine cells. *Pflügers Arch* **424:** 105–112.

Helmchen F. 2011. Calibration of fluorescent calcium indicators. In *Imaging: A laboratory manual* (ed. Yuste R), pp. 403–416. Cold Spring Harbor Laboratory Press, Cold Spring Harbor, NY.

Helmchen F, Imoto K, Sakmann B. 1996. Ca^{2+} buffering and action potential-evoked Ca^{2+} signaling in dendrites of pyramidal neurons. *Biophys J* **70:** 1069–1081.

Helmchen F, Borst JGG, Sakmann B. 1997. Calcium dynamics associated with a single action potential in a CNS presynaptic terminal. *Biophys J* **72:** 1458–1471.

Jackson MB, Redman SJ. 2003. Calcium dynamics, buffering, and buffer saturation in the boutons of dentate granule-cell axons in the hilus. *J Neurosci* **23:** 1612–1621.

Kaiser MMK, Zilberter Y, Sakmann B. 2001. Back-propagating action potentials mediate calcium signalling in dendrites of bitufted interneurons in layer 2/3 of rat somatosensory cortex. *J Physiol* **535:** 17–31.

Kim M-H, Korogod N, Schneggenburger R, Ho W-K, Lee S-H. 2005. Interplay between Na^+/Ca^{2+} exchangers and mitochondria in Ca^{2+} clearance at the calyx of Held. *J Neurosci* **25:** 6057–6065.

Koester HJ, Sakmann B. 2000. Calcium dynamics associated with action potentials in single nerve terminals of pyramidal cells in layer 2/3 of the young rat neocortex. *J Physiol* **529:** 625–646.

Lee S-H, Rosenmund C, Schwaller B, Neher E. 2000a. Differences in Ca^{2+} buffering properties between excitatory and inhibitory hippocampal neurons from the rat. *J Physiol* **525:** 405–418.

Lee S-H, Schwaller B, Neher E. 2000b. Kinetics of Ca^{2+} binding to parvalbumin in bovine chromaffin cells: Implications for $[Ca^{2+}]$ transients of neuronal dendrites. *J Physiol* **525:** 419–432.

Liao CW, Lien CC. 2009. Estimating intracellular Ca^{2+} concentrations and buffering in a dendritic inhibitory hippocampal interneuron. *Neuroscience* **164:** 1701–1711.

Lnenicka GA, Grizzaffi J, Lee B, Rumpal N. 2006. Ca^{2+} dynamics along identified synaptic terminals in *Drosophila* larvae. *J Neurosci* **26:** 12283–12293.

Mank M, Griesbeck O. 2008. Genetically encoded calcium indicators. *Chem Rev* **108:** 1550–1564.

Maravall M, Mainen ZF, Sabatini BL, Svoboda K. 2000. Estimating intracellular calcium concentrations and buffering without wavelength ratioing. *Biophys J* **78:** 2655–2667.

Markram H, Roth A, Helmchen F. 1998. Competitive calcium binding: Implications for dendritic calcium signaling. *J Comp Neurosci* **5:** 331–348.

Müller M, Felmy F, Schwaller B, Schneggenburger R. 2007. Parvalbumin is a mobile presynaptic Ca^{2+} buffer in the calyx of Held that accelerates the decay of Ca^{2+} and short-term facilitation. *J Neurosci* **27:** 2261–2271.

Murthy VN, Sejnowski TJ, Stevens CF. 2000. Dynamics of dendritic calcium transients evoked by quantal release at excitatory hippocampal synapses. *Proc Natl Acad Sci* **97:** 901–906.

Neher E. 1995. The use of fura-2 to estimate Ca buffers and Ca fluxes. *Neuropharmacology* **34:** 1423–1442.

Neher E. 1998. Usefulness and limitations of linear approximations to the understanding of Ca^{2+} signals. *Cell Calcium* **24:** 345–357.

Neher E. 2011. Quantitative aspects of calcium fluorimetry. In *Imaging: A laboratory manual* (ed. Yuste R), pp. 417–426. Cold Spring Harbor Laboratory Press, Cold Spring Harbor, NY.

Neher E, Augustine G. 1992. Calcium gradients and buffers in bovine chromaffin cells. *J Physiol* **450:** 273–301.

Palmer AE, Tsien RY. 2006. Measuring calcium signaling using genetically targetable fluorescent indicators. *Nat Protoc* **1:** 1057–1065.

Regehr WG, Atluri PP. 1995. Calcium transients in cerebellar granule cell presynaptic terminals. *Biophys J* **68:** 2156–2170.

Regehr WG, Tank DW. 1994. Dendritic calcium dynamics. *Curr Opin Neurobiol* **4:** 373–382.

Regehr WG, Delaney KR, Tank DW. 1994. The role of presynaptic calcium in short-term enhancement at the hippocampal mossy fiber synapse. *J Neurosci* **14:** 523–537.

Sabatini BL, Regehr WG. 1995. Detecting changes in calcium influx which contribute to synaptic modulation in mammalian brain slice. *Neuropharmacology* **34:** 1453–1467.

Sabatini BL, Maravall M, Svoboda K. 2001. Ca^{2+} signaling in dendritic spines. *Curr Opin Neurobiol* **11:** 349–356.

Sabatini BL, Oertner TG, Svoboda K. 2002. The life cycle of Ca^{2+} ions in dendritic spines. *Neuron* **33:** 439–452.

Simons SB, Escobedo Y, Yasuda R, Dudek SM. 2009. Regional differences in hippocampal calcium provide a cellular mechanism for limiting plasticity. *Proc Natl Acad Sci* **106:** 14080–14084.

Sinha SR, Wu L-G, Saggau P. 1997. Presynaptic calcium dynamics and transmitter release evoked by single action potentials at mammalian synapses. *Biophys J* **72:** 637–651.

Sobel EC, Tank DW. 1994. In vivo Ca^{2+} dynamics in a cricket auditory neuron: An example of chemical computation. *Science* **263:** 823–826.

Stocca G, Schmidt-Hieber C, Bischofberger J. 2008. Differential dendritic Ca^{2+} signalling in young and mature hippocampal granule cells. *J Physiol* **586:** 3795–3811.

Svoboda K, Denk W, Kleinfeld D, Tank DW. 1997. In vivo dendritic calcium dynamics in neocortical pyramidal neurons. *Nature* **385:** 161–165.

Tank DW, Regehr WG, Delaney KR. 1991. Modeling a synaptic chemical computation: The buildup and decay of presynaptic calcium. *Soc Neurosci Abstr* **17:** 578.

Tank DW, Regehr WG, Delaney KR. 1995. A quantitative analysis of presynaptic calcium dynamics that contribute to short-term enhancement. *J Neurosci* **15:** 7940–7952.

37 Imaging Action Potentials with Calcium Indicators

Rafael Yuste,[1] Jason MacLean,[2] Joshua Vogelstein,[3] and Liam Paninski[3]

[1]Department of Biological Sciences, Columbia University, New York, New York 10027; [2]Department of Neurobiology, The University of Chicago, Chicago, Illinois 60637; [3]Department of Statistics, Columbia University, New York, New York 10027

ABSTRACT

The understanding of neuronal circuits has been greatly advanced by the ability to simultaneously image action-potential generation within large populations of neurons. This chapter describes bulk loading of brain slices with acetoxymethyl (AM) ester calcium indicators to monitor action-potential activity in functional neuronal circuits. The imaging of calcium influx into neurons provides an indirect but accurate measure of action-potential generation within individual neurons. The key advantage of the technique is that it allows the researcher to densely sample the activity of a large population of neurons with single-cell resolution.

INTRODUCTION

Starting with Yuste and Katz (1991), this method of imaging action potentials has been successfully applied throughout the central nervous system including neocortex (e.g., Mao et al. 2001; Cossart et al. 2003; Stosiek et al. 2003), hippocampus (e.g., Tanaka et al. 2002), the cerebellum (e.g., Ghozland et al. 2002), striatum (e.g., Mao and Wang 2003), and spinal cord (e.g., Voitenko et al. 1999). Furthermore, this approach has been applied to cultured dissociated neurons, cultured slices, acute slices, and, most recently, to in vivo preparations. Although still fairly recent, calcium imaging of circuit activity has become very popular in many fields of neuroscience and has generated significant insights into the operation of the central nervous system.

Protocol

Imaging Action Potentials with Calcium Indicators

IMAGING SETUP

Both one- and two-photon excitation of calcium indicators can be used to monitor calcium changes in large populations of neurons. Two-photon imaging allows highly sensitive detection of changes in neuronal calcium concentration with relatively little bleaching and photodamage. Epifluorescence imaging of calcium indicators is sufficient to detect changes in neuronal calcium concentration and has the advantage that, using a fast camera, the detection of changes can be of higher temporal resolution than in conventional raster-scanning two-photon systems. However, wide-field fluorescent imaging of bulk-loaded slices is subject to rapid bleaching and is also prone to high background fluorescence. Finally, in our experience, spinning-disk confocals, together with fast cameras, can be used to image thousands of neurons simultaneously, without significant photobleaching and with good signal-to-noise ratios, over long periods of time.

MATERIALS

CAUTION: See Appendix 6 for proper handling of materials marked with <!>.

Reagents

Artificial cerebral spinal fluid (ACSF)
Cortical slices for imaging
Dimethylsulfoxide (DMSO) <!>
Fura-2 AM (Molecular Probes)

> Different AM ester calcium indicators under similar experimental conditions have different loading efficacies. Ranked from best loading to worst, in our experience: MagFura-2>Fura-2>Indo-1>Calcium Green-1>Fluo-4>Oregon-Green BAPTA-1>Calcium Orange>Calcium Crimson. Based on our experience, we have also ranked several of the indicators by their efficacy as detectors of action potentials: Fluo-4>Fura-2>MagFura-2. The majority of our experimental procedures use Fura-2 in brain slices. However, the methods described below can be used for any of the listed indicators.

Pluronic F-127, 20% (w/v) in DMSO (Molecular Probes)

Equipment

Fire-polished pipette, tip diameter ~30 μm (for Procedure 3)
Imaging setup (see options described above)
Loading chamber

EXPERIMENTAL METHOD

Choose Procedure 1, 2, or 3 for loading, depending on the characteristics of the slices to be loaded.

Procedure 1: Loading Embryonic and Neonatal Acute Cortical Slices

1. Dissolve 50 µg of Fura-2 AM (Molecular Probes) in 48 µL of DMSO and 2 µL of Pluronic F-127 (Molecular Probes) for a final concentration of 1 mM.

   Because the water solubility of calcium AM indicators is poor, vortex the solution for 10–15 min before use.

2. Add enough Fura-2 AM solution for a final concentration of 10 µM Fura-2 AM, and load the slices in oxygenated ACSF for 30–60 min.
3. Remove slices from the loading chamber, and place into an incubation chamber containing oxygenated ACSF. Imaging may begin after a 30-min recovery period.

Procedure 2: Loading Juvenile and Adult Acute Cortical Slices

1. Dissolve 50 µg of Fura-2 AM in 48 µL of DMSO and 2 µL of Pluronic F-127 for a final concentration of 1 mM.

   Because the water solubility of calcium AM indicators is poor, vortex the solution for 10–15 min before use.

2. Place slices in the loading chamber containing 2.5 mL of oxygenated ACSF.
3. Pipette 5–10 µL of Fura-2 AM solution on top of each slice.

   This results in a high initial concentration of Fura-2 AM. The concentration decreases as the Fura-2 AM diffuses away from the site of application, resulting in a final concentration in the entire chamber of ~10–20 µM.

4. Load slices in the dark for 20–30 min at 35°C–37°C with 95% O_2/5% CO_2 lightly ventilated into the chamber.

   As a rule of thumb, the loading time should be 10 min, plus as many minutes as the age of the animal in postnatal days.

5. Remove slices from the loading chamber, and place into an incubation chamber containing oxygenated ACSF for wash. The wash occurs through simple diffusion. Imaging may begin after a 30-min recovery period.

Procedure 3: Pressure Injection

Although local neural connections are maintained in the procedure described above (see Kozloski et al. 2001), long axonal projections, such as thalamocortical axons in thalamocortical slice preparations (Agmon and Connors 1991), may be compromised. A modification of Procedure 1 can be used to circumvent this problem (MacLean et al. 2005). The pressure injection of AM indicators is also used in Stosiek et al. (2003).

1. Dissolve 50 µg of Fura-2 AM in 13 µL of DMSO and 2 µL of Pluronic F-127, achieving a final concentration of 3.3 mM. Vortex the solution for 10–15 min.
2. Deposit the slices into the first loading chamber containing 2 mL of ACSF, ventilated with 95% O_2/5% CO_2, and place onto the microscope stage.
3. Fill a fire-polished pipette (tip diameter ~30 µm) with 7.5 µL of Fura-2 AM.
4. Insert the filled pipette into a standard patch-clamp electrode holder, with tubing attached, and, using a micromanipulator, place the pipette tip 100–200 µm above the surface of the slice. Apply 5–10 psi positive pressure to the pipette. Slowly (1–2 µL in 1 min) move the pipette across the surface of the slice using the manipulator, covering the area of interest with the dissolved Fura-2 AM.
5. Move the slices to the second loading chamber containing 2.5 mL of ACSF and the remaining 7.5 µL of Fura-2 AM.

6. Load the slices in the dark for 20–30 min (as a rule of thumb, 10 min plus the age of the animal in days) at 35°C–37°C with 95% O_2/5% CO_2 lightly ventilated into the chamber.
7. Remove the slices from the loading chamber, and transfer them to an incubation chamber containing oxygenated ACSF for washing. Imaging may begin after a 30-min recovery period.

DISCUSSION

Example of Application

Figure 1 illustrates two-photon imaging of Fura-2 as an indicator of neuronal spiking activity. Figure 1A illustrates several thousand loaded neurons in a neocortical slice imaged using a two-photon microscope. Figure 1B reveals the correlation between neuronal activity and a change in fluorescence as detected by epifluorescence. The fluorescence of Fura-2 decreases in the presence of calcium, thus, the negative fluorescent change in Figure 1Bii. As illustrated, even one-photon imaging of Fura-2 is sufficient to detect and to resolve action-potential generation in a single neuron within a field of loaded cells. Because of the rapid time course for the onset of its response, Fura-2 lends itself to the elucidation of network dynamics when simultaneously imaging large populations of neurons.

Inferring Spikes from Calcium Imaging

Given fluorescence traces from imaged neurons, such as those in Figure 1, we would like to be able to infer the underlying spike trains. This is a difficult statistical problem for a number of reasons, including the low temporal resolution of the imaging (often <30 Hz) and, in some cases, the limited signal-to-noise ratio of the fluorescence observations (e.g., caused by indicator saturation). A number of groups have devised algorithms for extracting spiking activity from fluorescence observations, including thresholding (Mao et al. 2001), template matching (Kerr et al. 2005), linear deconvolution (Yaksi and Friedrich 2006; Holekamp et al. 2008), support vector methods (Sasaki et al. 2008), nonlinear Bayesian deconvolution via sequential Monte Carlo (particle filtering) (Vogelstein et al. 2009), and fast nonnegatively constrained optimization methods (Vogelstein et al. 2010). Most of these approaches either explicitly or implicitly assume a model relating the spike trains to the fluorescence observations: The spike train is convolved with some linear filter (usually a simple exponential filter, although this may be generalized) to obtain the intracellular calcium concentration, and the fluorescence signal is essentially a noisy, possibly saturated version of the calcium. The parameters of this model may be estimated via standard statistical time-series methods if simultaneous observations of electrophysiological and imaging data are available or via the expectation-maximization algorithm if only imaging data are available (Vogelstein et al. 2009). These latter approaches have been shown to perform well in a variety of experimental preparations, including both in vitro and in vivo data sets (see Fig. 2 for an in vitro example).

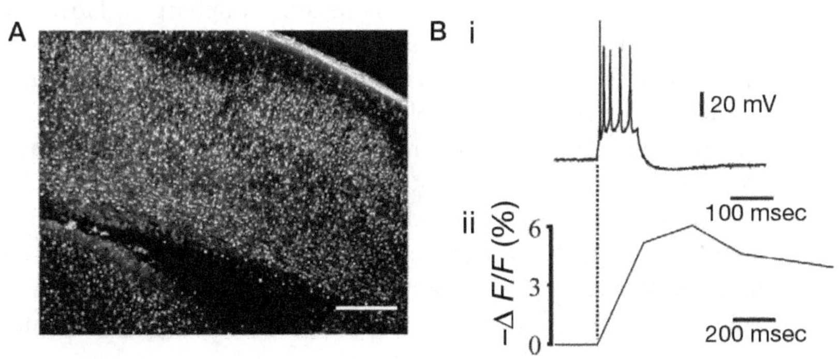

FIGURE 1. High-resolution calcium imaging of neuronal activity. *(A)* Two-photon image of transverse section of mouse neocortex loaded with Fura-2 AM. Scale bar, 250 µm. Note that thousands of neurons can be adequately visualized. *(B)* Correspondence between action potentials and somatic fluorescence change. *(i)* Whole-cell recording of a burst of action potentials in response to a single intracellular depolarizing current step (150 pA, 200 msec). *(ii)* Normalized fluorescence change for the recorded neuron imaged with a cooled charge-coupled device (CCD) camera (Micromax, Princeton Instruments). The recording pipette contained 50 µm of Fura pentapotassium salt (comparable to the intracellular concentration of Fura following bulk loading). Although individual action potentials cannot be resolved, the onset of neuronal activity is accurately imaged.

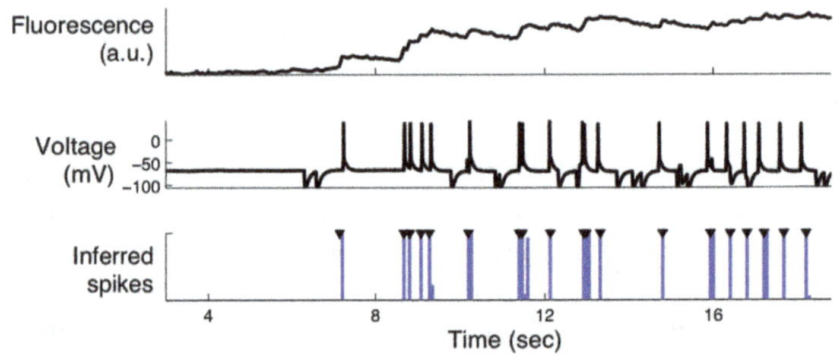

FIGURE 2. Automatic detection of action potentials. The fluorescence trace (*top panel*) was recorded from a pyramidal cell in the somatosensory cortex from a P14 mouse using Fura-2 (see Vogelstein et al. 2009 for more details). The true spike times were ascertained using simultaneous electrophysiological methods (*middle panel*). A nonlinear Bayesian deconvolution algorithm provides the probability of a spike occurring in each time bin (*bottom panel*), given the fluorescence observations. Although this fluorescence trace is highly saturated, this nonlinear approach performs well. Note that the parameters were estimated using only the observed fluorescence trace; that is, the electrophysiological data were not used to train the algorithm.

Advantages and Limitations

The use of calcium AM ester indicators provides a very robust method of imaging activity in large populations of neurons simultaneously. In our experience, bulk application of voltage-sensitive fluorescence or second-harmonic voltage-sensitive dyes does not enable the detection of activity in an individual cell within a population of neurons, such as a cortical slice. Their nonspecific staining pattern and poor signal strength currently make optical recording with single-cell resolution unfeasible in cortical slices (Yuste et al. 1997). On the other hand, calcium indicators (Tsien 1989) that can be bulk loaded into brain slices using their AM ester derivatives (Yuste and Katz 1991) act as very good, albeit indirect, measures of action-potential generation (Smetters et al. 1999). These still provide the best means of imaging activity in large populations of neurons when single-cell resolution is desirable.

The limitations of the technique are due to the properties of the dyes themselves. Calcium indicators, being charged molecules, do not easily cross the cell membrane and need, therefore, to be microinjected. To circumvent this problem, AM ester derivatives of the indicators were synthesized (Tsien 1981). The AM esters mask negative charges, making the indicator molecules more lipophilic and membrane permeant, thus allowing them to enter the cell. Once inside the cell, cytoplasmic esterases hydrolyze the acetyl ester linkage, releasing formaldehyde and free indicator, which then accumulates intracellularly as it is once again charged. However, the dependence on intracellular enzymatic cleavage makes this process cell dependent. This can result in differential loading efficiency in different neurons. In addition, the increased hydrophobicity of the AM ester derivatives of the indicators can cause problems in delivering sufficient amounts to their targets. This problem becomes significant in adult preparations or in vivo in which the pressure injection method appears to be the best loading strategy. Finally, because calcium dynamics are slower than action potentials and because the calcium signal can become saturated with increasing numbers of spikes, advanced quantitative approaches, as described above, are necessary to deconvolve the fluorescence change into the underlying spike train. Although significant advances have been made in this regard (see Vogelstein et al. 2009, 2010), continued improvement appears necessary.

REFERENCES

Agmon A, Connors BW. 1991. Thalamocortical responses of mouse somatosensory (barrel) cortex in vitro. *Neuroscience* **41:** 365–379.

Cossart R, Aronov D, Yuste R. 2003. Attractor dynamics of network UP states in the neocortex. *Nature* **423:** 283–288.

Ghozland S, Aguado F, Espinosa-Parrilla JF, Soriano E, Maldonado R. 2002. Spontaneous network activity of cerebellar granule neurons: Impairment by in vivo chronic cannabinoid administration. *Eur J Neurosci* **16:** 641–651.

Holekamp TF, Turaga D, Holy TE. 2008. Fast three-dimensional fluorescence imaging of activity in neural populations by objective-coupled planar illumination microscopy. *Neuron* **57:** 661–672.

Kerr JN, Greenberg D, Helmchen F. 2005. Imaging input and output of neocortical networks in vivo. *Proc Natl Acad Sci* **102:** 14063–14068.

Kozloski J, Hamzei-Sichani F, Yuste R. 2001. Stereotyped position of local synaptic targets in neocortex. *Science* **293:** 868–872.

MacLean, JN, Watson BO, Aaron, GB, Yuste R. 2005. Internal dynamics determine the cortical response to thalamic stimulation. *Neuron* **48:** 811–823.

Mao L, Wang JQ. 2003. Group I metabotropic glutamate receptor-mediated calcium signalling and immediate early gene expression in cultured rat striatal neurons. *Eur J Neurosci* **17:** 741–750.

Mao BQ, Hamzei-Sichani F, Aronov D, Froemke RC, Yuste R. 2001. Dynamics of spontaneous activity in neocortical slices. *Neuron* **32:** 883–898.

Sasaki T, Takahashi N, Matsuki N, Ikegaya Y. 2008. Fast and accurate detection of action potentials from somatic calcium fluctuations. *J Neurophysiol* **100:** 1668–1676.

Smetters DK, Majewska A, Yuste R. 1999. Detecting action potentials in neuronal populations with calcium imaging. *Methods* **18:** 215–221.

Stosiek C, Garaschuk O, Holthoff K, Konnerth A. 2003. In vivo two-photon calcium imaging of neuronal networks. *Proc Natl Acad Sci* **100:** 7319–7324.

Tanaka E, Uchikado H, Niiyama S, Uematsu K, Higashi H. 2002. Extrusion of intracellular calcium ion after in vitro ischemia in the rat hippocampal CA1 region. *J Neurophysiol* **88:** 879–887.

Tsien RY. 1981. A non-disruptive technique for loading calcium buffers and indicators into cells. *Nature* **290:** 527–528.

Tsien RY. 1989. Fluorescent probes of cell signaling. *Annu Rev Neurosci* **12:** 227–253.

Voitenko NV, Kostyuk EP, Kruglikov IA, Kostyuk PG. 1999. Changes in calcium signalling in dorsal horn neurons in rats with streptozotocin-induced diabetes. *Neuroscience* **94:** 887–890.

Vogelstein JT, Watson BO, Packer A, Yuste R, Jedynak B, Paninski L. 2009. Spike inference from calcium imaging using sequential Monte Carlo methods. *Biophys J* **97:** 637–656.

Vogelstein JT, Packer A, Machado TM, Sippy T, Babadi B, Yuste R, Paninski L. 2010. Fast non-negative deconvolution for spike train inference from population calcium imaging. *J Neurophysiol* **104:** 3691–3704.

Yaksi E, Friedrich RW. 2006. Reconstruction of firing rate changes across neuronal populations by temporally deconvolved Ca^{2+} imaging. *Nat Methods* **3:** 344–346.

Yaksi E, Judkewitz B, Friedrich RW. 2007. Topological reorganization of odor representations in the olfactory bulb. *PLoS Biol* **5:** e178. doi:10.1371/journal.pbio.0050178.

Yuste R, Katz LC. 1991. Control of postsynaptic Ca^{2+} influx in developing neocortex by excitatory and inhibitory neurotransmitters. *Neuron* **6:** 333–344.

Yuste R, Kleinfeld D, Tank DW. 1997. Functional characterization of the cortical microcircuit with voltage-sensitive dye imaging of neocortical slices. *Cereb Cortex* **7:** 546–558.

38 Structure–Function Analysis of Genetically Defined Neuronal Populations

Alexander Groh[1] and Patrik Krieger[2]

[1]*Institute for Neuroscience of the Technical University Munich, 80802 Munich, Germany;* [2]*Department of Neuroscience, Karolinska Institutet and Stockholm Brain Institute, SE-17177 Stockholm, Sweden*

ABSTRACT

Cell type–specific expression of green fluorescent protein can be used to visualize specific subsets of neurons in the brain or brain slice preparation. The advantages of genetically defined neurons are not only their homogeneity and suitability for systematic descriptions of networks but also their tremendous potential for the cell type–specific manipulation of neuronal networks in vivo. Here we describe some analytical tools and protocols for basic structure–function analysis, such as morphology and connectivity, of these genetically defined neurons.

INTRODUCTION

The morphological and functional classification of individual neurons is a crucial aspect of the characterization of neuronal networks. Systematic structural and functional analysis of individual neurons is now possible using transgenic mice with genetically defined neurons that can be visualized in vivo or in brain slice preparations. Genetically defined neurons are useful for studying a particular class of neurons, as well as more comprehensive studies of the neuronal content of a network. Furthermore, these tools allow the cell type–specific modulation of networks in vivo. The GENSAT Brain Atlas (www.gensat.org) and the Allen Brain Atlas (www.brain-map.org) are valuable resources that contain a growing number of transgenic lines with a different population of genetically tagged neurons (see Chapter 12). These neuronal populations can be studied using a multitude of techniques including imaging, electrophysiology, anatomy, gene profiling, and behavioral studies. This practical guide describes a selection of procedures to visualize and study cell type–specific properties of genetically defined neurons, such as electrical and morphological phenotype, and involvement in macrocircuits. Most of the methods described in this chapter have been used in a number of studies (Krieger et al. 2007; Groh et al. 2008, 2010) and are presented here as a practical guide.

Protocol

Single-Cell and Network Anatomy of Genetically Defined Neurons

Here we describe a protocol for a combined study of the anatomy and physiology of genetically defined neurons in transgenic mice. We provide information about basic equipment, reagents, procedures, and analysis approaches showing how to obtain 3D cell morphologies and how to determine the axonal input and output of genetically defined neurons. We exemplify with genetically labeled cortical neurons but the procedures apply to other brain regions with little or no alterations.

IMAGING SETUP

Fluorescently labeled cells in the living brain slice can be visualized using confocal laser scanning microscopy, two-photon laser scanning microscopy, or an epifluorescence microscope equipped with an infrared (IR) camera (from, e.g., Hamamatsu, PCO, Visitron). The IR camera is optimized for contrast imaging of the brain slice and is used to identify neurons and for visual guidance during whole-cell patch-clamp procedures. The IR camera can also be used to detect fluorescence. Images of large areas of brain or high resolution imaging of fluorescently labeled cells for morphological reconstructions is most efficiently acquired on a confocal microscope equipped with a three-axis, computer-controlled motorized table (e.g., Leica SP2 or SP5 with a motorized Märzhäuser stage).

The basic components required for bright-field three-dimensional (3D) reconstructions of neuronal morphology are (i) tracing software, (ii) a (color) digital camera, (iii) a three-axis, motorized stage, (iv) a microscope (e.g., Olympus BX-51), and (v) a computer and monitor. Complete packages, including hardware and software, are commercially available, such as Neurolucida (MicroBrightField) and FilamentTracer (Bitplane AG); however, software packages can also be obtained as freeware, such as Neuromantic (http://www.reading.ac.uk/neuromantic/), NeuronStudio (http://research.mssm.edu/cnic/tools-ns.html), NeuronIQ (http://www.cbi-tmhs.org/Neuroniq/), and ORION (http://cbl.uh.edu/ORION/) (not tested by the authors).

MATERIALS

CAUTION: See Appendix 6 for proper handling of materials marked with <!>.
See the end of the chapter for recipes for reagents marked with <R>.

Reagents

Biocytin (Sigma-Aldrich)

Diaminobenzidine (DAB) <!> stock solution (17.5 mg/mL in 0.1 M phosphate buffer [PB])

Wear gloves and prepare the solution in a fume hood. Store 1 mL aliquots at –20°C. Commercially prepared DAB solutions are also available.

Glycerol anhydrous (e.g., FLUKA 49767)

H_2O_2 (30%) <!>

Before use, make a 3% solution in 0.1 M PB and stir for 30 min.

ImmPACT DAB <!> (Vector Labs; optional)

Mowiol 4-88 (Kuraray) mounting medium <R>

Na_2HPO_4, 0.1 M

NaH_2PO_4, 0.1 M

Paraformaldehyde <!> (PFA; 4% in 0.1 M PB) <R>

4% PFA solution is also commercially available.

Phosphate buffer (PB; 0.1 M, pH 7.2)

For 1 L, mix 750 mL Na_2HPO_4 and 250 mL NaH_2PO_4.

Phosphate buffer with Triton X-100 (PBT)

10% triton X-100 is diluted with PB, as required (see below).

Streptavidin-conjugated ALEXA Fluor (Invitrogen, Molecular Probes)

Tris base <!>

Tris-HCl <!> (0.2 M; pH 8.5)

Use HCl to pH the solution.

Triton X-100 <!>, 10% in H_2O

VECTASTAIN Elite ABC Kit (Vector Labs)

Equipment

Beaker, glass, 1-L

Centrifuge and tubes, 50-mL

Flask, conical, 100-mL

Magnetic stirrer with hot plate

Stir bar

Tissue culture plates, 24-well plates (Gibco or Falcon)

Vials, glass or plastic, with lids, 1–2-mL

For storage of 1-mL aliquots of Mowiol at –20°C.

Water bath

Wire mesh, 0.4-mm (optional)

EXPERIMENTAL METHOD

All fixation and staining procedures are performed in individual wells of standard 24-well plates (e.g., Falcon, Gibco) and solutions are changed using 1-mL pipettes or pasteur pipettes with a narrow opening. High-throughput processing of many brain slices can be achieved using custom-built 24-well staining plates, which make solution changes less labor intensive. High-throughput plates are prepared by replacing the bottom of a standard 24-well plate with 0.4-mm wire mesh. During the procedure, the bottom of the plate is submerged in a custom-built trough (~1.4 cm height); 30 mL of solution (per step) is sufficient to cover the specimens.

Tissue Preparation, Staining, and Mounting

1. Prepare acute brain slices of ~300 µm thickness using standard methods (e.g., Edwards et al. 1989).
2. Following imaging and cell injections, brain slices are placed in a 24-well plate and fixed with PFA for 2 h at room temperature (RT), or up to 12 h at 4°C. Longer fixation is possible but may interfere with subsequent procedures, such as cytochrome *c* staining and immunostaining.
3. Wash 4 times for 10 min with 0.1 M PB, to remove PFA.
4. Visualize biocytin-filled neurons using one of the following methods.

   Using fluorophore-conjugated streptavidin:

   i. Permeabilize brain slice with 1% PBT for 1 h at RT.

   ii. Incubate with fluorophore-conjugated streptavidin (e.g., ALEXA 594) diluted to 1–10 µg/mL in 0.5% PBT for 2 h at RT.

iii. Wash 5 times for 10 min in PB.

iv. Mount slices in your preferred glycerol-based mounting medium for fluorescence samples.

Using Avidin-HRP:

Use 2 mL of solution per step per slice, unless indicated.

v. Incubate brain slice for 15–30 min in freshly prepared 3% H_2O_2 to quench endogenous peroxidases.

vi. Wash with PB at least 5 times for 10 min.

vii. Permeabilize tissue with 2% PBT for 1 h at RT.

viii. Incubate with VECTASTAIN ABC reagent at least 2 h at RT or overnight at 4°C.

ABC reagent is prepared as follows: Add 4 drops of component A to 20 mL 1% PBT, and then vortex. Add 4 drops of component B, and then mix well and let stand for 30 min. For economy, use 1 mL/well.

ix. Rinse 5 times for 10 min in PB.

HRP color development:

x. Prepare DAB development solution by adding 570 µL DAB stock solution and 6.7 µL 30% H_2O_2 to 20 mL PB.

Omit if using ImmPACT DAB.

xi. Add DAB development solution, or ImmPACT DAB, to brain slices and monitor color development under a dissection microscope.

Color development is determined empirically. For P20–P30 rats and mice, 3–5 min is usually sufficient.

xii. Stop the reaction by transferring the slice (e.g., using a paintbrush) into PB and rinse 5 times for 10 min in PB.

DAB is a suspected carcinogen. All DAB solutions and instruments that have come into contact with DAB must be appropriately decontaminated.

Mounting of Slices in Aqueous Medium

5. Transfer the brain slice onto a glass slide and gently remove excess buffer using a filter paper cut into triangular shape.
6. Add ~60 µL of Mowiol per 22 x 22-mm coverslip to the slice and coverslip the specimen, avoiding air bubbles. Allow Mowiol to harden for at least 1 h at RT.
7. View the samples or store at 4°C (Mowiol-embedded) or –20°C (glycerol-embedded).

   Specimens mounted in Mowiol can be kept for years at 4°C. For bright-field cell reconstructions a hardening mounting medium such as Mowiol is necessary to prevent tissue movement and ensure that the reconstructions are as comparable to the live cell size as possible.

EXAMPLES OF APPLICATIONS

Single-Cell Electrophysiology and Morphological Analysis of Genetically Defined Neurons

Identification of genetically defined neurons in vivo or in vitro is achieved by fluorescence imaging of enhanced green fluorescent protein (eGFP) or another fluorophore expressed under the control of the cell type–specific promoter (Groh et al. 2010; see also Chapter 12). Basic physiological analysis and dye-loading can be performed using a standard electrophysiology setup for acute brain slices (Penner 1995; also see Chapter 35). Use fluorescence imaging to identify somata of genetically labeled neurons in the region of interest. A common problem is identifying the fluorescent neuron after switching to gradient contrast optics for patch-clamping. There are several solutions to ensure that the correct (fluorescent) neuron is successfully patch-clamped. Ideally, the imaging setup (Fig. 1A) should allow the contrast image (Fig. 1B) and the fluorescence image (Fig. 1C) to be viewed simultaneously (Fig. 1D), allowing fluorescence-guided patch-clamping. One example of such a setup is the Leica SP2 system equipped with a two-photon laser and a Dodt tube for gradient contrast imaging (Wimmer et al. 2004). Alternatively, if using epifluorescence or confocal optics, identify a healthy, fluorescent neuron by toggling between contrast imaging (e.g., differential interference contrast [DIC] or Dodt optics) and fluorescence illumination, and directly mark the location of the soma on the TV monitor using a nonpermanent pen. If the images are digitized with a frame grabber, the fluorescence and contrast images can be overlaid using imaging software. If the fluorescence is too weak to be detected with an IR camera, fluorescence should be confirmed using confocal microscopy (one-photon or two-photon).

Once the genetically defined neuron of interest is identified, use standard techniques to obtain a whole cell recording of the identified neuron (Penner 1995; also see Chapter 35). Leakage of dye or biocytin from the pipette is a common cause of background staining, and is minimized by maintaining the pressure in the pipette tip to <30 mbar. In the whole-cell configuration, biocytin will passively diffuse into the cell without the need for current injection. Allow the dye to load the cell for ~30 min; during this time, electrophysiological recordings (e.g., firing patterns or ionic currents) can be made. The dye-loading time depends on the length of the neuronal processes and the access resistance of the recording. After dye loading, slowly retract the patch pipette under visual control. If more than one genetically labeled cell is to be patched, a fluorescent dye such as Alexa Fluor can be used to mark the position of the recorded cells during the experiment (Krieger et al. 2007). Furthermore, the fluorescent dye can also be used as an alternative to biocytin stain for 3D cell reconstructions.

Histological Reconstruction of Biocytin-Filled Cells

A major limitation in cell reconstructions is tissue shrinkage during fixation and histological staining procedures. One way to estimate shrinkage is to fill the live cell with a fluorescent dye (such as Alexa Fluor 594) and acquire confocal images of the cell to provide a scale comparison for the bright-field images of the cell after fixation and processing (Egger et al. 2008). Note that different histological procedures may cause different amounts of shrinkage; thus reconstructions from different histological methods might not be directly comparable.

Reconstruction of 3D Morphology

The 3D morphology can be reconstructed using commercially available (e.g., MBF Bioscience, Inc. and Bitplane scientific software) or free tracing software (we use Neurolucida [MBF Bioscience, Inc.]). Figure 2, A and B, shows 3D reconstructions of two types of genetically labeled layer 5 pyramidal neurons that were filled with biocytin (Groh et al. 2010). 3D morphology can be quantified using parameters such as apical and basal dendrite length (Fig. 2C). For bright-field reconstructions it is recommended to use a 100x oil objective (1.4 numerical aperture) so that fine processes can be resolved. The short working distance of high magnification objectives is a problem if the processes

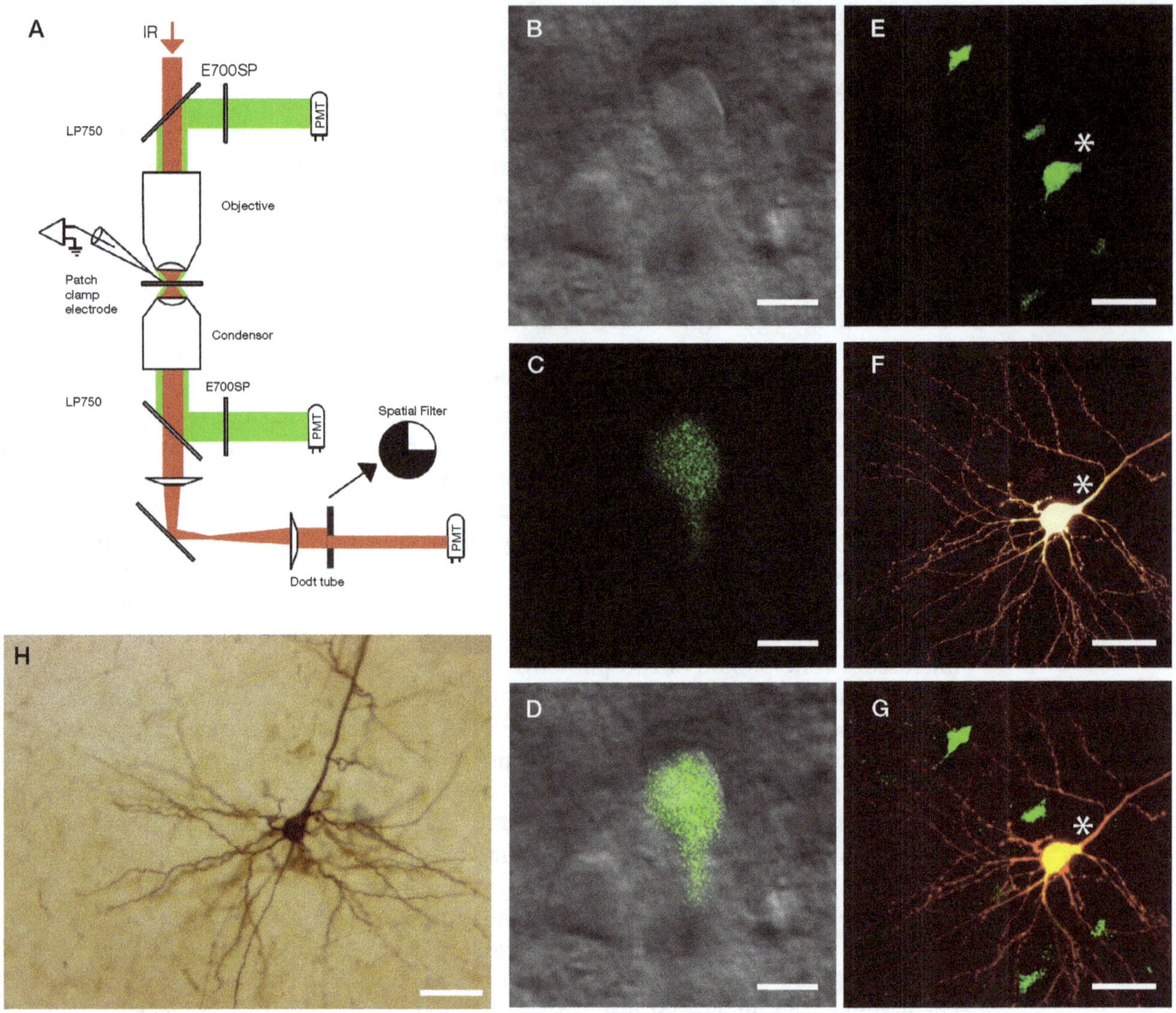

FIGURE 1. Identification of fluorescently labeled etv-expressing corticostriatal neurons in cortical layer 5A (Groh et al. 2010) using two-photon laser scanning and scanning-gradient contrast microscopy. (*A*) The microscope setup for simultaneous imaging of cell fluorescence and gradient contrast (for cell morphology) with two-photon microscopy to identify neurons expressing a cell type–specific genetic tag. Excitation light (IR laser illumination; red) passes through an LP750 excitation beam splitter onto the brain slice and via a Dodt gradient contrast tube onto photomultiplier tube (PMT) to generate a gradient contrast image (shown in *B*). Fluorescence emissions are reflected by a LP750 excitation beam splitter and pass through IR-block filters (E700SP), onto two external, non-descanned detectors behind the objective and condenser to generate the fluorescence image (*C*). (*B*) High-power, scanning-gradient contrast image of a brain slice preparation. (*C*) eGFP-fluorescence (green) identifies genetically labeled cell within the field shown in *B*. Laser illumination, 840 nm. (*D*) The cell body of the genetically defined cell is readily identified by overlaying the two images. The morphology of fluorescently labeled cells (green) can be reconstructed by filling them with biocytin and visualization using either streptavidin-conjugated Alexa Fluor (*E–G*) or avidin-HRP and a DAB color reaction (*H*). (*E*) Confocal image of eGFP-labeled genetically defined neurons (green). (*F*) One cell in *E* was filled with biocytin and visualized using streptavidin-conjugated Alexa Fluor 594 (red). (*G*) Overlay of the two images confirms the biocytin-injected cell is eGFP-labeled (yellow signal; asterisk marks the same position in *E–G*). (*H*) DAB staining of a biocytin filled neuron. Scale bars, 10 µm (*B–D*); 30 µm (*E–G*); 50 µm (*H*).

to be reconstructed are deep within the tissue. In this case, ultrathin cover glass (0.04–0.06 mm, Menzel #0) should be used. The most time-consuming cell structures to reconstruct are those that span several consecutive brain slices, such as long, fenestrated axons, especially for neurons filled in vivo.

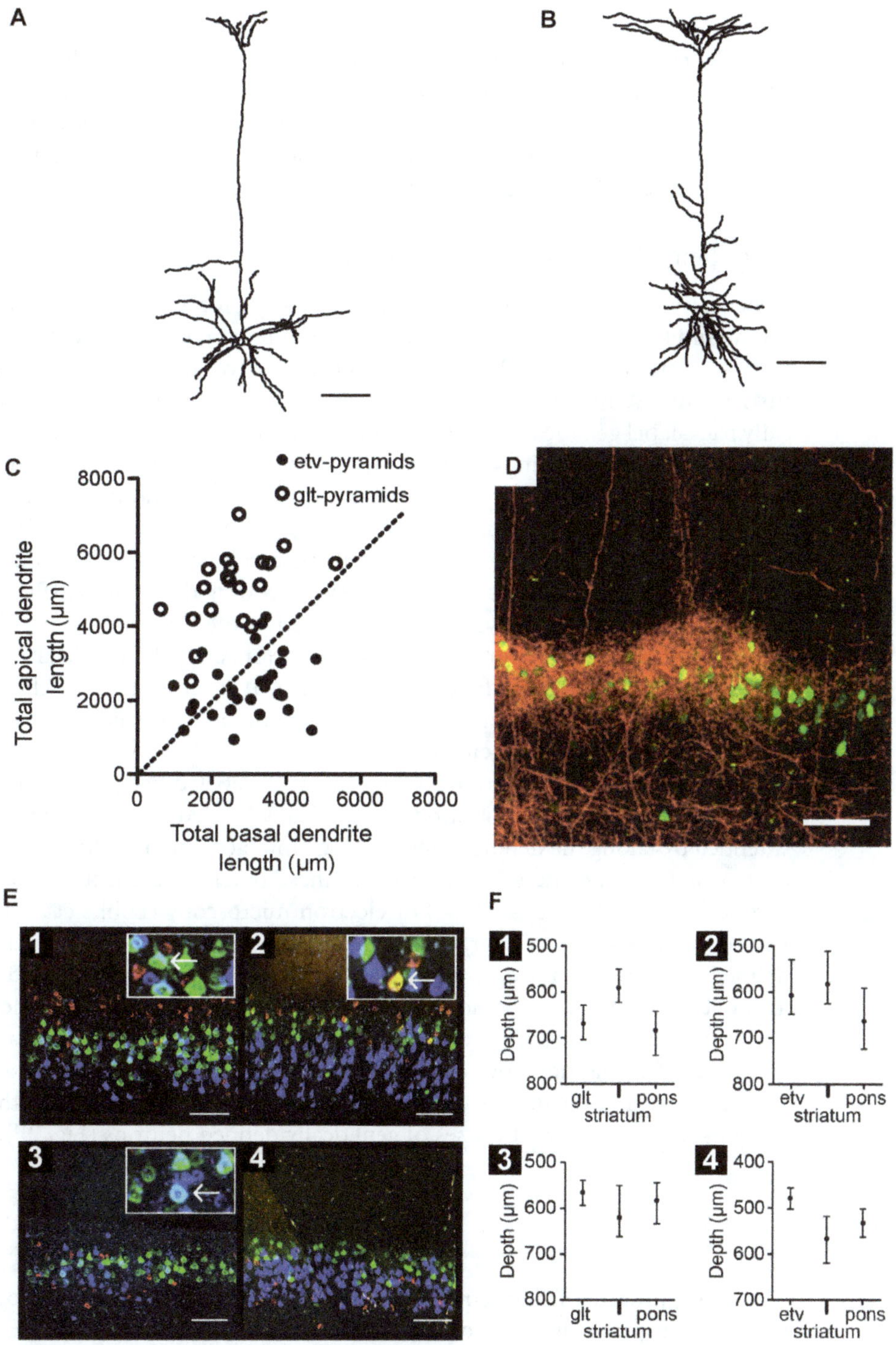

FIGURE 2. Characterization of the morphology, location, and connectivity of genetically labeled cortical pyramidal neurons in layers 5A (etv-pyramids) and 5B (glt-pyramids). etv and glt refer to cell type–specific markers; see Groh et al. (2010) for details. 3D morphological reconstruction of a typical layer 5A etv-pyramid (*A*) and a typical glt-pyramid (*B*) from the barrel field of the somatosensory cortex. (*C*) Quantification of neuronal morphology using apical and basal dendritic length. (*D*) Estimating inputs to basal dendrites of etv-pyramids. Thalamic axons were labeled by injection of an anterograde tracer (AAV-dTomato; red) into the POm thalamus. Labeled thalamic axons (red) overlap with the eGFP-labeled soma and basal dendrites of layer 5A etv-pyramids (green). (*E*) Use of retrograde labeling to study projection targets of genetically labeled neurons (green) in the somatosensory barrel cortex (*1,3*) and visual cortex (*2,4*). CTB-Alexa conjugate injected into the pons (blue) results in retrograde-labeling of glt-pyramidal neurons, whereas injection of CTB-Alexa conjugate into the striatum (red) results in retrograde-labeling of etv-pyramidal neurons. Insets show double-labeled neurons at higher magnification (arrow). (*F*) Distribution by depth measured from pia of double-labeled pyramidal neurons shown in *E*. Scale bars, 100 µm. (Reprinted, with permission, from Groh et al. 2010.)

If the equipment for bright-field reconstructions is not available, confocal image stacks may be used (Maravall et al. 2004; Losavio et al. 2008). However, this is not recommended for large, complex cells because the sample can become photobleached during the several hours it takes to acquire high resolution image stacks covering the whole neuron. Furthermore, unless scanning is performed on a motorized table, it is practically impossible to align the stacks correctly.

Circuit Analysis: Long-Range Projections and Input Mapping

Genetically defined neurons are especially useful for the characterization of cell type–specific inputs and long-range projections. Small amounts of retrograde (e.g., 100 nL of CTB 647, Invitrogen) or anterograde label can be injected in vivo using a stereotaxic apparatus (e.g., Kopf Instruments; Wimmer et al. 2004). A retrograde label in a putative projection target will specifically back-label genetically labeled neurons that form synapses in that target area. Choleratoxin B-Alexa (CTB-Alexa) conjugates have superior signal strength and specificity compared with other retrograde tracers (e.g., fluorescent beads or dextrans), although good results have been obtained with fluorescent beads (Morishima and Kawaguchi 2006; Hattox and Nelson 2007; Brown and Hestrin 2009). Anterograde labeling of axons can be achieved by viral expression of fluorescent proteins (e.g., adeno-associated virus [AAV] encoding dTomato; Groh et al. 2010), especially with synaptic proteins, such as synaptophysin, fused to the fluorophore (Groh et al. 2008). After sufficient survival time (>3 d for CTB-Alexa, >10 d for AAV) the whole animal is fixed by transcardial perfusion followed by postfixation for 1–12 h. It is important to choose fluorescent dyes with emissions that are well separated from the genetic tag (e.g., red or far red fluorescence in combination with eGFP-labeled neurons).

To evaluate the axonal input to the genetically labeled neurons overlay images of anterogradely labeled axons onto the distribution of genetically labeled neurons (Fig. 2D). This approach can be extended by taking the dendritic morphology into account and thereby determining spatial overlap of axonal and dendritic signals. To verify these potential contacts, high resolution imaging using confocal (Wouterlood et al. 2008) or electron microscopy (Lübke et al. 1996) must be performed. Figure 2E shows how retrograde labeling can be used to identify synaptic targets of genetically defined neurons. The retrograde tracer will back-label genetically defined neurons projecting to the area injected with the retrograde tracer. Thus, CTB-Alexa injected into the pons back-labels glt-pyramids in the barrel cortex but not etv-pyramids, which project to the striatum (Fig. 2E). These types of data can be quantified by counting the numbers of single-, double-, and triple-labeled neurons (Jafari-Mamaghani et al. 2010), and by comparing the average depths of the back-labeled projection neurons with the different classes of genetically defined neurons (Fig. 2F).

DISCUSSION

Classical approaches using retrograde tracers to tag and analyze groups of neurons are labor intensive and nonspecific in terms of the identities of the labeled cells (Morishima and Kawaguchi 2006; Hattox and Nelson 2007). Here we describe how eGFP expression under the control of cell type–specific promoters can be used to identify specific cell types within the brain so that they can be readily visualized using standard fluorescence optics. These genetically defined neurons can be used to characterize the structure and function of the neurons themselves and to systematically describe the contents, biophysical properties, and connectivity of neuronal networks. The ability to focus on homogeneous groups of neurons minimizes the variability in the data obtained when compared with analysis of random ("blind") samples of untagged neurons. A limitation of this approach at present is that all groups of neurons are not yet identifiable using specific transgenic tags. However, the number of transgenic lines with genetically defined neurons is steadily growing and transgenes that specifically label neurons in each cortical layer are available (O'Connor et al. 2009).

The methods described in this chapter are a starting point for the characterization of neuronal populations, and this characterization can be extended to functional studies of these neurons and their neuronal networks in vivo and in freely moving animals. Finally, Cre recombinase expression in genetically defined neurons (Gong et al. 2007) allows functional manipulations using chemical probes (such as calcium indicators) or light-sensitive ion channels (such as channelrhodopsin). The powerful combination of tools that are now becoming available opens the door to studies to define and manipulate specific groups of neurons and neuronal networks to determine their contribution to brain function and behavior.

RECIPES

CAUTION: See Appendix 6 for proper handling of materials marked with <!>.
Recipes for reagents marked with <R> are included in this list.

Mowiol 4-88 Mounting Medium

Makes 20 mL.

1. Weigh 6.0 g of glycerol into a 100 mL conical flask, add magnetic stirrer.
2. Add 2.4 g of Mowiol 4-88 and stir slowly for 1 h at RT.

Ensure that the Mowiol is always covered with the glycerol.

3. Add 6 mL distilled water and stir for 2 h at RT.
4. Add 12 mL 0.2 M Tris-HCl <!> pH 8.5 and heat to 50°C–60°C for 15 min using a water bath.

Make sure that also the Mowiol at the edges is dissolved.

5. Centrifuge at 5000*g* for 10 min at 10°C.
6. Prepare 1 mL aliquots and store at –20°C.

Paraformaldehyde (PFA), 4% in 0.1 M PB

PFA <!> is highly toxic and should be handled in a fume hood. Avoid breathing PFA dust and wear protective gloves.

For 1 L of 4% PFA, pH ~7.2:

1. Mix 600 mL PB, 20 drops 2 M NaOH <!>, and 40 g PFA in a 1-L glass beaker.
2. Heat to ~80°C on a hot plate while stirring until the solution is clear. Do not boil.
3. Transfer to an ice bath and allow to cool; fill to 1 L with 0.1 M PB.
4. Store at 4°C, or in aliquots at –20°C.

ACKNOWLEDGMENTS

We thank Marlies Kaiser who helped develop, teach, and refine many of the methods described here.

REFERENCES

Brown SP, Hestrin S. 2009. Intracortical circuits of pyramidal neurons reflect their long-range axonal targets. *Nature* **457:** 1133–1136.

Edwards FA, Konnerth A, Sakmann B, Takahashi T. 1989. A thin slice preparation for patch clamp recordings from neurones of the mammalian central nervous system. *Pflügers Arch* **414:** 600–612.

Egger V, Nevian T, Bruno RM. 2008. Subcolumnar dendritic and axonal organization of spiny stellate and star pyramid neurons within a barrel in rat somatosensory cortex. *Cereb Cortex* **18:** 876–889.

Gong S, Doughty M, Harbaugh CR, Cummins A, Hatten ME, Heintz N, Gerfen CR. 2007. Targeting Cre recombinase to specific neuron populations with bacterial artificial chromosome constructs. *J Neurosci* **27:** 9817–9823.

Groh A, de Kock CP, Wimmer VC, Sakmann B, Kuner T. 2008. Driver or coincidence detector: Modal switch of a corticothalamic giant synapse controlled by spontaneous activity and short-term depression. *J Neurosci* **28:** 9652–9663.

Groh A, Meyer HS, Schmidt EF, Heintz N, Sakmann B, Krieger P. 2010. Cell-type specific properties of pyramidal neurons in neocortex underlying a layout that is modifiable depending on the cortical area. *Cereb Cortex* **20:** 826–836.

Hattox AM, Nelson SB. 2007. Layer V neurons in mouse cortex projecting to different targets have distinct physiological properties. *J Neurophysiol* **98:** 3330–3340.

Jafari-Mamaghani M, Andersson M, Krieger P. 2010. Spatial point pattern analysis of neurons using Ripley's K-function in 3D. *Front Neuroinformatics* **4:** 9.

Krieger P, Kuner T, Sakmann B. 2007. Synaptic connections between layer 5B pyramidal neurons in mouse somatosensory cortex are independent of apical dendrite bundling. *J Neurosci* **27:** 11473–11482.

Losavio BE, Liang Y, Santamaría-Pang A, Kakadiaris IA, Colbert CM, Saggau P. 2008. Live neuron morphology automatically reconstructed from multiphoton and confocal imaging data. *J Neurophysiol* **100:** 2422–2429.

Lübke J, Markram H, Frotscher M, Sakmann B. 1996. Frequency and dendritic distribution of autapses established by layer 5 pyramidal neurons in the developing rat neocortex: Comparison with synaptic innervation of adjacent neurons of the same class. *J Neurosci* **16:** 3209–3218.

Maravall M, Koh IY, Lindquist WB, Svoboda K. 2004. Experience-dependent changes in basal dendritic branching of layer 2/3 pyramidal neurons during a critical period for developmental plasticity in rat barrel cortex. *Cereb Cortex* **14:** 655–664.

Morishima M, Kawaguchi Y. 2006. Recurrent connection patterns of corticostriatal pyramidal cells in frontal cortex. *J Neurosci* **26:** 4394–4405.

O'Connor DH, Huber D, Svoboda K. 2009. Reverse engineering the mouse brain. *Nature* **461:** 923–929.

Penner R. 1995. A practical guide to patch clamping. In *Single channel recording* (ed. Sakmann B, Neher E), pp. 3–30. Plenum, New York.

Wimmer VC, Nevian T, Kuner T. 2004. Targeted in vivo expression of proteins in the calyx of Held. *Pflügers Arch* **449:** 319–333.

Wouterlood FG, Boekel AJ, Kajiwara R, Belien JA. 2008. Counting contacts between neurons in 3D in confocal laser scanning images. *J Neurosci Methods* **171:** 296–308.

39 Infrared-Guided Neurotransmitter Uncaging on Dendrites

Hans-Ulrich Dodt,[1] Matthias Eder,[2] Anja Schierloh,[2] and Walter Zieglgänsberger[2]

[1]*Vienna University of Technology, 1040 Vienna, Austria;* [2]*Max-Planck-Institute of Psychiatry, 80804 Munich, Germany*

ABSTRACT

Photosensitive caged neurotransmitters have made it feasible to rapidly and precisely stimulate single neurons in brain-slice preparations (Callaway and Katz 1993; Katz and Dalva 1994; Wieboldt et al. 1994). Using a combination of patch-clamp recording and focused ultraviolet (UV) illumination to locally activate a neurotransmitter, it is relatively easy to determine, for example, the distribution of neurotransmitter receptors along a dendrite or to rapidly explore the connectivity of neuronal networks by photostimulating neurons in the vicinity of a patch-clamped neuron or even distant neurons that have a low probability of being connected.

INTRODUCTION

The characterization of neuronal networks would be greatly facilitated by a method that enabled large numbers of neurons to be surveyed. Although, in principle, this could be achieved by patch-clamp recording from multiple cells, such electrical stimulation requires great experimental skill and would necessarily be limited to only a few presynaptic inputs. Photoactivated caged neurotransmitters offer an elegant optical solution to this problem. Using the powerful combination of photolytic neurotransmitter activation and patch-clamp recording from a target neuron, we have gained new insights into neuronal function and connectivity through the ability to rapidly survey hundreds of presumptive presynaptic neurons for connectivity to the postsynaptic cell and by precisely mapping neurotransmitter sensitivity along a single dendrite.

Protocol

Infrared-Guided Laser Stimulation of Neurons in Brain Slices

Infrared-gradient contrast and a UV light spot is used to visualize and to photostimulate neurons in brain-slice preparations (Dodt and Zieglgänsberger 1994; Dodt et al. 1998).

IMAGING SETUP

Patch-clamp recording equipment is used in conjunction with an epifluorescence microscope equipped with a gradient contrast system, an infrared video camera, and a UV light source connected to the epifluorescence port via a quartz fiber optic. The UV light beam is reflected onto the tissue using a dichroic mirror and focused on a small spot in the focal plane of the objective. Fine control of the position of the quartz fiber is used to direct the light spot to the center of the image plane. A gradient contrast system is used to visualize neurons, and the neuron of interest is photolytically stimulated by illumination with the UV light spot. A high-pressure mercury burner is sufficient for light spots of ~10 μm in diameter; however, light spots of ~1 μm require a UV laser (e.g., Enterprise II, Coherent). Light flashes of a few milliseconds can be produced using an electronic shutter, and depolarizations are typically induced in the millivolt range (Dodt et al. 2002).

MATERIALS

CAUTION: See Appendix 6 for proper handling of materials marked with <!>.
See the end of the chapter for recipes for reagents marked with <R>.

Reagents

Brain slice
Caged neurotransmitter (Molecular Probes)

We have used γ-(α-carboxy-2-nitrobenzyl [CNB]-caged) L-glutamic acid (L-glutamic acid, γ-[CNB] ester, trifluoroacetic acid salt; Molecular Probes, Amsterdam) and *O*-(CNB-caged) GABA (γ-aminobutyric acid, CNB ester, trifluoroacetic acid salt; Molecular Probes).

Carbogen <!>, O_2, 95%; CO_2, 5%
Krebs-Ringer solution <R>
Patch-clamp pipette filling solution <R>

Equipment

Computer, equipped with data acquisition board (e.g., InstruTECH [HEKA]) and software to analyze pulse-clamp recordings (e.g., PULS [HEKA] Igor Pro [WaveMetrics] and Surfer [Golden Software])
Dodt gradient contrast system (Luigs & Neumann, Ratingen, Germany)
Electronic Shutter (e.g., UniBlitz)
Microscope, compound, epifluorescence (Zeiss Axioskop), equipped with a motorized three-dimensional stage (Luigs & Neumann) and a dichroic mirror
Microscope objective, 60x, 0.9-numerical-aperture (Olympus)

Patch-clamping equipment, including amplifier with appropriate bridge and capacitance compensation (e.g., SEC 1 l, NPI Electronic GmbH, Tamm, Germany)
Patch-clamp pipettes (direct current [DC] resistance 4–7 MΩ)
Peristaltic pump
Quartz optical fiber
Slice chamber
Teflon tubing, 0.3-mm
UV laser (e.g., Enterprise II, Coherent)
Vial, 5-mL or similar
Video camera, infrared, and monitor

EXPERIMENTAL METHOD

1. Prepare a concentrated solution (25–100 mM) of caged neurotransmitter in Krebs-Ringer solution, and store aliquots at –20°C in the dark.
2. Prepare brain slices, and maintain in a recirculating bath of Krebs-Ringer solution.
3. Place the slice chamber on the microscope stage, and visualize the location of the UV spot using a piece of white paper in the slice chamber. Mark the position of the UV spot on the screen of the video monitor.
4. Remove the paper, and replace it with a brain slice. Orient the slice such that the mark on the video monitor is positioned over the cell of interest.
5. Position the patch-clamp electrode on the target neuron.

   The UV illumination is controlled by a shutter. During opening the UV spot can become briefly visible because of autofluorescence of the tissue.
6. Add the caged neurotransmitter to the slice chamber medium.
   i. Using the Teflon tubing, connect the slice chamber to a vial containing 1.5 mL of Krebs-Ringer solution, continuously perfused with carbogen.
   ii. Circulate the medium in a closed loop system via a small peristaltic pump.
   iii. Add caged neurotransmitter to the vial to a final concentration of 0.25–1 mM.

      Allow 20 min for the caged neurotransmitter to permeate the brain slice.
7. Map localized neurotransmitter sensitivity by consecutively illuminating different segments along the dendrite of a patch-clamped pyramidal neuron by 3-msec UV flashes and recording the electrical response (Frick et al. 2001).

Applications of Targeted Stimulation

The above procedure revealed the differential distribution of $GABA_A$ and $GABA_B$ receptors on neocortical neurons (Eder et al. 2001). Photolytic stimulation and patch-clamp recording from pairs of neurons were used to identify neuronal networks in the barrel cortex (Dodt et al. 2003). The high spatial resolution of photolytic stimulation was also used to investigate the spatial extent of neurobiological phenomena involved in neuronal plasticity, such as long-term depression (LTD; Dodt et al. 1999; Eder et al. 2002). The highly localized nature of LTD was revealed by recording the response to low-frequency photostimulation glutamate activity (1-μM spot, 3-msec flash every 20 sec) before and after tetanic laser stimulation (5 Hz for 60 sec) at a single location. A LTD response was observed at the location of the tetanic stimulus but not 10 μM further away (Fig. 1B; 10 μM is the limit of the lateral spatial resolution of neurotransmitter photoactivation).

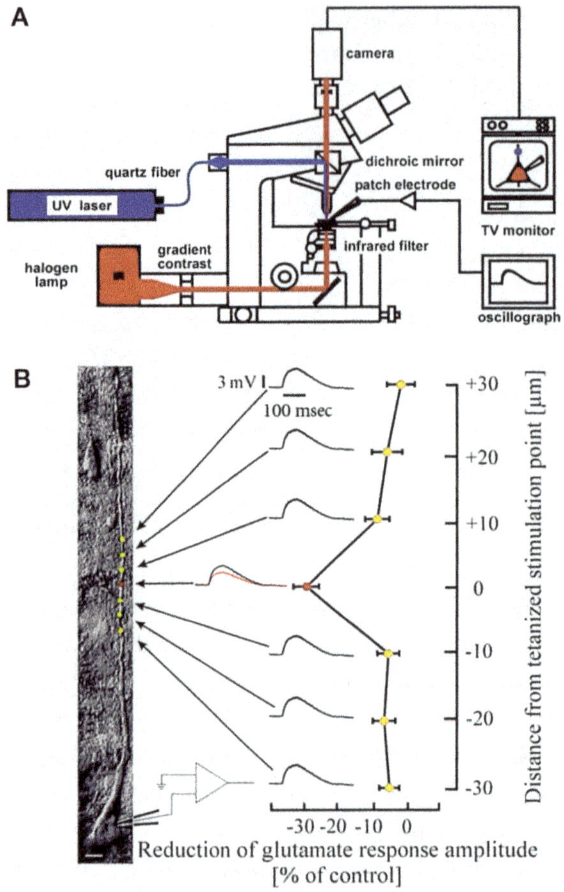

FIGURE 1. (*A*) Schematic of the setup for infrared video microscopy and laser stimulation. An epifluorescence compound microscope is augmented with a patch-clamp recording setup, an infrared video microscopy and a gradient contrast system, and a UV light source connected by a fiber optic through the epifluorescence port. (B) Use of photolytic conversion to show that LTD is spatially highly restricted. Example of a pyramidal neuron of neocortical layer 5 overlaid with seven UV-stimulation points on the dendrite (scale bar, 10 mm). Laser-induced LTD reduces the response to subsequent glutamate activation only in the immediate region of tetanic stimulation (5 Hz, indicated in red). At all other sites, no significant change in the response to glutamate uncaging was observed. Values were obtained by averaging five stimulations for each site on each neuron before and 10–20 min after the 5-Hz stimulation and by calculating the relative reduction.

Advantages and Limitations of Photolytic Neurotransmitter Activation

The experimental simplicity of laser stimulation now makes it feasible to map the neuropharmacology of subcellular compartments and the connectivity of neuronal networks. Mapping the neurotransmitter sensitivity of the neuronal membrane with targeted photostimulation can be performed with any neurotransmitter that is available in a photoactivated form, including glutamate, GABA, aspartate, dopamine, noradrenaline, and serotonin, among others. By combining photolytic neurotransmitter activation with a specific receptor blockade, the distribution of many subtypes of neurotransmitter receptors can be studied. This approach has significant logistical and technical advantages over microiontophoresis, which requires multiple micropipettes for patch-clamp recording and neurotransmitter delivery. In addition, because no active neurotransmitter is present before uncaging, the desensitization problems of iontophoresis are avoided.

One disadvantage of photostimulation, however, is its poor resolution in time. After uncaging, the active neurotransmitter diffuses over a relatively large area of neuronal membrane, compared with a neurotransmitter released from a single synaptic site. As a result, the time course of membrane depolarization induced by uncaging is about three times longer than that following synaptic stimulation. This makes spike timing on the millisecond timescale unfeasible. The low temporal resolution of photostimulation has been partially overcome by two-photon photostimulation (Niklenko et al. 2007). Using a femtosecond laser and a suitable caged glutamate (MNI [4-methoxy-7-nitroindolinyl]-caged L-glutamate), neurons can be driven to action-potential threshold within 5 msec of photolysis: The use of two-photon photostimulation also overcomes the lack of *z* resolution of one-photon photostimulation because of the inherent optical sectioning of two-photon excitation. To increase the effectiveness of the femtosecond laser, a diffractive optical element was used for beam multiplexing so that uncaging would occur over the whole area of the neuronal cell body. The temporal precision obtained using this method was 10-fold more precise than one-photon uncaging, although still slower than with paired patch-clamp recordings.

RECIPES

CAUTION: See Appendix 6 for proper handling of materials marked with <!>.
Recipes for reagents marked with <R> are included in this list.

Krebs-Ringer Solution

NaCl	125 mM
$NaHCO_3$	25 mM
Glucose	25 mM
KCl	2.5 mM
NaH_2PO_4	1.25 mM
$CaCl_2$ <!>	2 mM
$MgCl_2$	1 mM

Saturate with carbogen <!>. Adjust to pH 7.4.

Patch-Clamp Pipette Filling Solution

K-gluconate <!>	130 mM
KCl	5 mM
HEPES buffer	10 mM
EGTA	0.5 mM
Mg-ATP	2 mM
Glucose	5 mM

Adjust to pH 7.2.

REFERENCES

Callaway EM, Katz LC. 1993. Photostimulation using caged glutamate reveals functional circuitry in living brain slices. *Proc Natl Acad Sci* **90:** 7661–7665.

Dodt HU, Zieglgänsberger W. 1994. Infrared videomicroscopy: A new look at neuronal structure and function. *Trends Neurosci* **17:** 453–458.

Dodt HU, Frick A, Kampe K, Zieglgänsberger W. 1998. NMDA and AMPA receptors on neocortical neurons are differentially distributed. *Eur J Neurosci* **10:** 3351–3357.

Dodt HU, Eder M, Frick A, Zieglgänsberger W. 1999. Precisely localized LTD in the neocortex revealed by infrared-guided laser stimulation. *Science* **286:** 110–113.

Dodt HU, Eder M, Schierloh A, Zieglgänsberger W. 2002. Infrared-guided laser stimulation of neurons in brain slices. *Sci STKE* **120:** PL2.

Dodt HU, Schierloh A, Eder M, Zieglgänsberger W. 2003. Circuitry of rat barrel cortex investigated by infrared-guided laser stimulation. *Neuroreport* **14:** 623–627.

Eder M, Rammes G, Zieglgänsberger W, Dodt HU. 2001. $GABA_A$ and $GABA_B$ receptors on neocortical neurons are differentially distributed. *Eur J Neurosci* **13:** 1065–1069.

Eder M, Zieglgänsberger W, Dodt HU. 2002. Neocortical long-term potentiation and long-term depression: Site of expression investigated by infrared-guided laser stimulation. *J Neurosci* **22:** 7558–7568.

Frick A, Zieglgänsberger W, Dodt HU. 2001. Glutamate receptors form hot spots on apical dendrites of neocortical pyramidal neurons. *J Neurophysiol* **86:** 1412–1421.

Katz LC, Dalva MB. 1994. Scanning laser photostimulation: A new approach for analyzing brain circuits. *J Neurosci Methods* **54:** 205–218.

Nikolenko V, Poskanzer KE, Yuste R. 2007. Two-photon stimulation and imaging of neural circuits. *Nat Methods* **4:** 943–950.

Wieboldt R, Gee KR, Niu L, Ramesh D, Carpenter BK, Hess GP. 1994. Photolabile precursors of glutamate: Synthesis, photochemical properties, and activation of glutamate receptors on a microsecond time scale. *Proc Natl Acad Sci* **91:** 8752–8756.

40 Uncaging Calcium in Neurons

Kerry R. Delaney[1] and Vahid Shahrezaei[2]

[1]*Department of Biology, University of Victoria, Victoria V8W 3N5, Canada;* [2]*Department of Mathematics, Imperial College London, London SW7 2AZ, United Kingdom*

ABSTRACT

Changes in intracellular free calcium concentration ($\Delta[Ca^{2+}]_i$) driving physiological events such as neurotransmitter release or Ca^{2+}-dependent currents can be monitored using Ca^{2+}-sensitive fluorescent dyes. Although these dyes can correlate $\Delta[Ca^{2+}]_i$ with a physiological event, they cannot directly test for causality between changes in $[Ca^{2+}]_i$ and that event. Photolabile Ca^{2+} chelators are Ca^{2+}-binding molecules that can alter and, to a certain extent, control $[Ca^{2+}]_i$ in an inducible manner and with temporal and spatial resolution that surpasses microinjection or ionophore application. In this chapter, we discuss the properties of caged Ca^{2+} compounds as well as some practical considerations for their use in neuronal cells, where they have proven particularly effective.

INTRODUCTION

Unlike caged neurotransmitters, photolabile Ca^{2+} chelators (Ca^{2+} cages) are actually Ca^{2+} buffers whose affinity for Ca^{2+} is altered upon photolysis of a covalent bond somewhere within the compound. Several useful Ca^{2+} cages have been developed during the last 25 years, but only DM-nitrophen (DMNP), nitrophenyl-EGTA (NP-EGTA), and Diazo-2 are commercially available. Their properties, as well as other caged compounds that may be available from their inventors, are listed in Table 1.

CHEMICAL PROPERTIES OF CAGED CALCIUM CHELATORS

Photolysis of a covalent bond within the Ca^{2+} cage either increases (Diazo-2) (Adams et al. 1989) or decreases (DMNP and NP-EGTA) the Ca^{2+}-binding affinity of the chelator (Fig. 1). This change in affinity rapidly alters free $[Ca^{2+}]_i$ from close to resting level to >200 μM inside neurons (~0.1 msec with a pulsed laser; ~1 msec with an arc-based flash lamp; ~1 sec with an arc lamp). Because the photoproduct(s) have low affinity for Ca^{2+}, the free $[Ca^{2+}]$ after photolysis is determined by the rel-

TABLE 1. Properties of photolabile Ca^{2+} chelators

Cage	K_d Ca^{2+} intact (nM)	K_d Ca^{2+}, products (mM)	ΔK_d for Ca^{2+} (fold change)	K_d for Mg^{2+} (mM)	Quantum yield	Extinction coefficient (M^{-1} cm^{-1}) (two-photon cross sections)
Diazo-2[a]	2200	0.0066	0.033	~8		
DMNP	5	0.25[b]		0.0025	0.18	4300
		3[c]	50,000			(0.01–0.04)
			660,000			
NP-EGTA	80	1	12,500	9	0.20 +Ca	975
					0.23 –Ca	(0.001)
Azid-1	235	0.12	520	8	1	3.3×10^4
						(1)
DMNPE-4[d]	48	2	441,000	10	0.09	5140
						(0.01–0.04)
NDBF-EGTA[e]	100	2	20,000	15	0.7	18,400
						(0.6)

ΔK_d Ca^{2+} is expressed as fold-change and is derived from K_d Ca^{2+} products/K_d Ca^{2+} intact. ΔK_d >1.0 indicates decreased affinity for Ca^{2+}; ΔK_d <1.0 indicates increased affinity.

[a]Adams et al. 1989 at 0.15 M ionic strength.

[b]Ayer and Zucker 1999.

[c]Kaplan and Ellis-Davies 1988.

[d]Ellis-Davies 2008.

[e]Momotake et al. 2006.

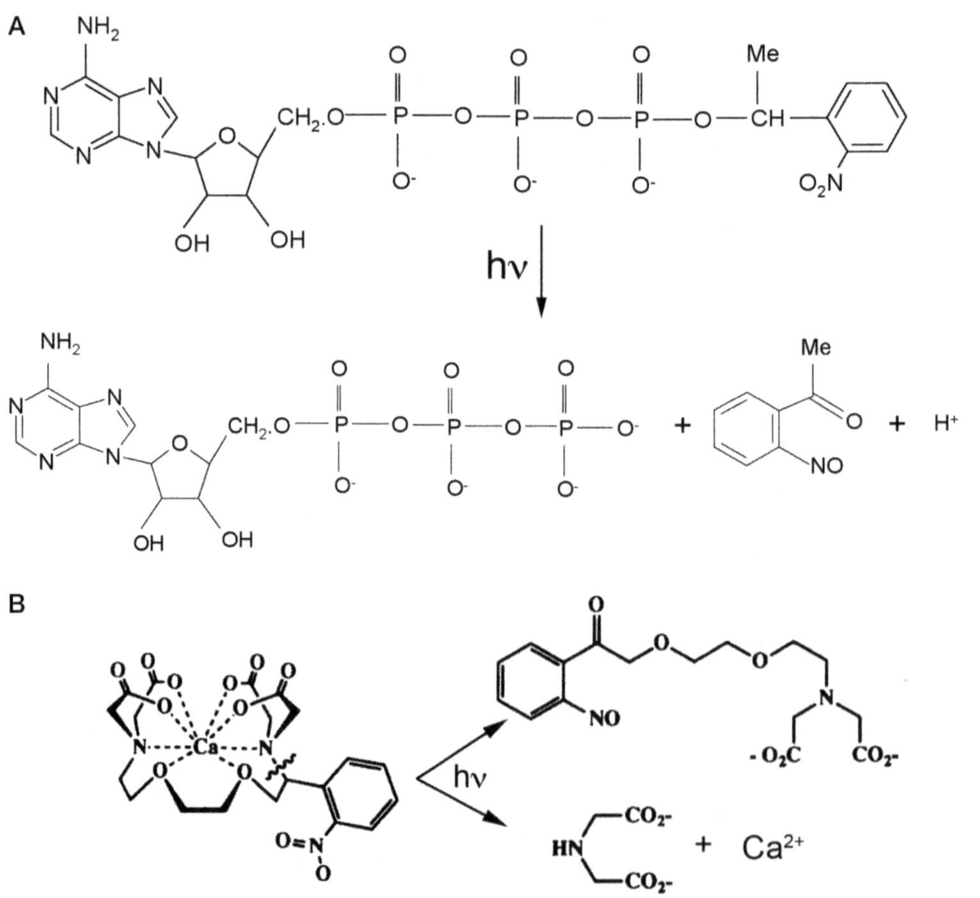

FIGURE 1. Structures and photoproducts of (*A*) DM-nitrophen and (*B*) NP-EGTA. Note that photolysis of DMNP will acidify unbuffered solutions, whereas photolysis of NP-EGTA has an alkalizing effect because of protonation of the amine nitrogens.

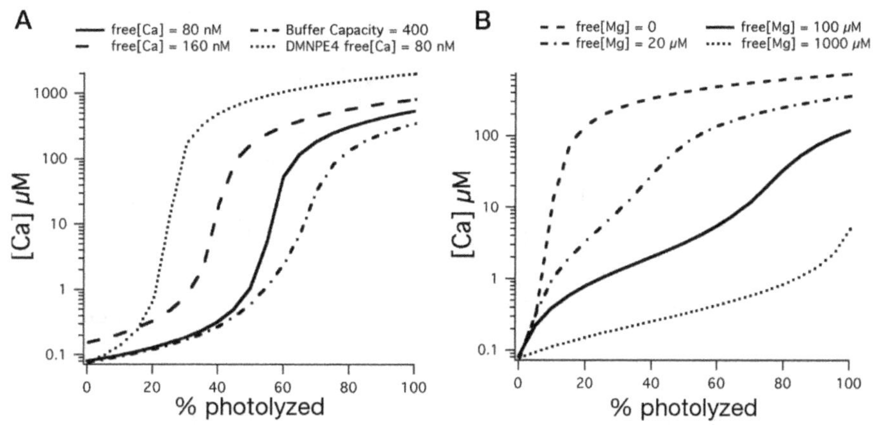

FIGURE 2. Equilibrium free $[Ca^{2+}]_i$ as a function of photolysis for NP-EGTA, DMNPE-4, and DM-Nitrophen (DMNP). (*A*) Photolysis is simulated under conditions in which 5 mM NP-EGTA or 5 mM DMNPE-4 (dotted line) are loaded into a neuron. For the dotted, dashed, and solid lines, NP-EGTA or DMNPE-4 is added to a neuron with an assumed endogenous buffer capacity of 100 (200 μM of a 2 μM K_d buffer). For the dot-dash line NP-EGTA is added to a neuron with an endogenous buffer capacity of 400. Simulations start with resting $[Ca^{2+}]_i$ of 80 nM (solid, dotted, and dot-dash lines) or 160 nM (dashed line) with resting $[Mg^{2+}]_i$ of 1 mM. (*B*) Free $[Ca^{2+}]_i$ as a function of DMNP photolysis for different concentrations of Mg^{2+}. For all simulations, 5 mM DMNP is added, free $[Ca^{2+}]$ before photolysis is 80 nM, and 200 μM of K_d 2 μM buffer is included. The free $[Mg^{2+}]_i$ is indicated next to each trace. Depending on the extent to which normal (i.e., ~1 mM) free Mg^{2+} concentration levels are achieved after addition of the cage, actual changes in $[Ca^{2+}]$ will lie between the 0 $[Mg^{2+}]$ (dashed line) and 1 mM $[Mg^{2+}]$ (dotted line) cases. For these simulations the K_d of photolyzed DMNP for Ca^{2+} was assumed to be 250 μM (Ayer and Zucker 1999) and properties of DMNPE-4 used for the simulation were those described by DelPrincipe et al. (1999).

ative proportions and Ca^{2+}-binding affinities of the photolyzed and unphotolyzed forms. The difference between the Ca^{2+}-binding affinity of the unphotolyzed and photolyzed forms of the chelator sets the potential range over which Ca^{2+} can be changed. In practice, the $\Delta[Ca^{2+}]_i$ that can be achieved with caged Ca^{2+} compounds is often determined by the affinity of the unphotolyzed cage relative to physiological "resting" $[Ca^{2+}]_i$, because this determines how much cage is bound (loaded) with Ca^{2+}, and thus how much Ca^{2+} is available for release. Figure 2 illustrates the change in free $[Ca^{2+}]$ produced as a function of percentage photolysis, assuming physiologically reasonable estimates of resting $[Ca^{2+}]_i$. Photolysis and release of Ca^{2+} occur on a microsecond-to-millisecond time scale for all the cages listed in Table 1 (McCray et al. 1992). Because a number of Ca^{2+} buffers are involved in setting the final equilibrium $[Ca^{2+}]_i$ after photolysis, each with its own forward binding rate, the time for $[Ca^{2+}]_i$ to reach the level predicted by the equilibrium solution for the relevant buffer equations can be quite complex (Fig. 3).

The Ca^{2+}-binding affinity of NP-EGTA, like its parent molecule EGTA, is pH-dependent and Mg^{2+}-independent. NP-EGTA can be used to raise $[Ca^{2+}]_i$ from resting level to several hundred micromolar (Fig. 2A). Starting from typical resting $[Ca^{2+}]_i$ and assuming moderate endogenous buffer capacity, ~40% photolysis will elevate steady-state $[Ca^{2+}]_i$ by several micromolar, and $\Delta[Ca^{2+}]_i$ accelerates rapidly above 50% photolysis (Fig. 2A). Based on the forward Ca^{2+}-binding rate, chemical equilibrium should be reached between unphotolyzed free NP-EGTA, its photoproducts, and Ca^{2+} within 10 msec of photolysis. If significant photolysis is achieved within 1 msec, a $[Ca^{2+}]$ transient will occur that is several times larger than the predicted equilibrium concentration and will recur until a significant portion of the available cage is photolyzed (Fig. 3). DMNPE-4 has similar properties to NP-EGTA but causes greater $\Delta[Ca^{2+}]_i$ per photolysis event because it becomes more heavily loaded with Ca^{2+} (owing to its significantly higher affinity for Ca^{2+}) and is more efficiently photolyzed (DelPrincipe et al. 1999). Unfortunately, DMNPE-4 is not commercially available at this time.

Based on its high affinity for Ca^{2+} (K_d = 5 nM) and large ΔK_d upon photolysis (Table 1), DMNP would appear to be an ideal caged compound. However, like its EDTA parent molecule, its significant affinity for Mg^{2+} complicates its use when Mg^{2+} is at physiological levels (see below, Estimating $\Delta[Ca^{2+}]_i$). If Mg^{2+} can be reduced or eliminated (e.g., by rapid exchange of low $[Mg^{2+}]$ solution near

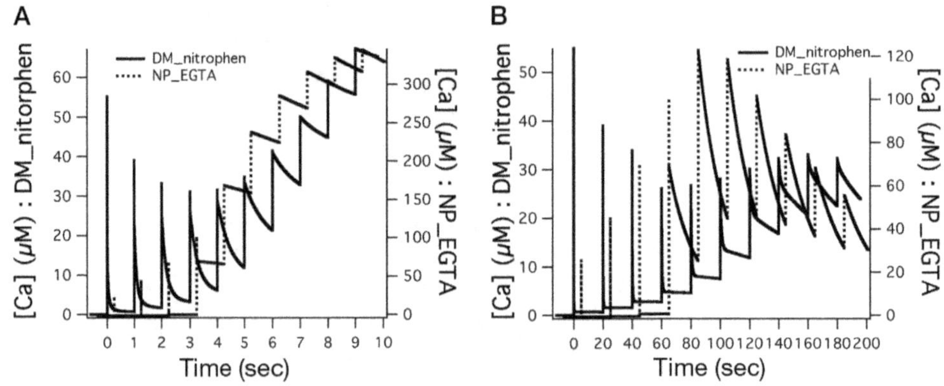

FIGURE 3. Simulation of [Ca^{2+}] dynamics after partial photolysis. These simulations assume 20% photolysis of the remaining unphotolyzed cage per flash with 5 mM cage loaded into a neuron with 80 nM initial free [Ca^{2+}], 100 μM initial free [Mg^{2+}], and an endogenous buffer capacity of 100 (i.e., 200 μM Ca^{2+} buffer, K_d = 2 μM). By the tenth flash, ~90% of the cage is photolyzed. We approximated the effect of Ca^{2+} surface pumps and sequestration by a simple exponential mechanism that returned free [Ca^{2+}] to resting level with a time constant of 5 sec in the absence of buffers and cage. (*A*) Photolysis at 1-sec intervals. For both cages, the earliest Ca^{2+} transient is due to rebinding of Ca^{2+} by endogenous buffers and intact cage. For DMNP, recovery of [Ca^{2+}] is initially dominated by the time required for equilibration between Ca^{2+}-bound and Mg^{2+}-bound DMNP. With NP-EGTA photolysis, the initial rebinding of free Ca^{2+} to intact cage and endogenous buffers occurs within a few milliseconds and is followed by "steps" of $\Delta[Ca^{2+}]_i$ with little recovery of [Ca^{2+}] between flashes. (*B*) Photolysis at 20-sec intervals. The full effects of Ca^{2+} buffering by endogenous mechanisms are seen more clearly. At 100 μM free [Mg^{2+}], very little NP-EGTA would be bound by Mg^{2+} because the affinity is so low (K_d = 9 mM); for simplicity, the binding affinity was assumed to be 0 μM.

the opening of a patch recording electrode), the full potential of DMNP to change $[Ca^{2+}]_i$ can be exploited (Fig. 2B; Bollmann et al. 2000). Because resting cytoplasmic free $[Mg^{2+}]_i$ is usually ~1 mM and the Mg^{2+} dissociation constant of DMNP is 5 μM, almost no DMNP will be bound to Ca^{2+} when the [DMNP] is less than free $[Mg^{2+}]_i$. Even if DMNP pre-loaded with Ca^{2+} is introduced into a neuron, cytoplasmic Mg^{2+} will displace the pre-bound Ca^{2+}.

However, if DMNP is introduced rapidly enough to overcome the cell's ability to restore the normal resting [Mg^{2+}], then a potentially large portion of the cage could be Ca^{2+}-bound owing to its higher affinity for Ca^{2+}. The proportion of DMNP that is bound by Mg^{2+} reduces its effectiveness at raising $[Ca^{2+}]_i$—because there is less "loaded" Ca^{2+} available for photolytic release, a higher degree of photolysis is required to achieve $\Delta[Ca^{2+}]$ in the micromolar range (Fig. 2B). Because Mg^{2+} is unlikely to equilibrate as rapidly as Ca^{2+} in neurons, this is an additional complication when estimating the degree of Ca^{2+} loading (and thus the Ca^{2+} that will be released by photolysis) when DMNP diffuses into distal dendritic compartments. Finally, because Mg^{2+} is also released upon photolysis of DMNP, the physiological effects of Mg^{2+}, as well as its potential to displace Ca^{2+} bound to intact cage, may have to be considered.

Rebinding of Ca^{2+} to unphotolyzed buffer occurs over a period of several milliseconds to >10 msec (Zucker 1993). Thus, partial photolysis of both DMNP and NP-EGTA can produce a large Ca^{2+} transient, between Ca^{2+} release and its chelation by unphotolyzed buffer. The slow off-rate of Mg^{2+} from DMNP delays the equilibration of Ca^{2+} with unphotolyzed DMNP (Fig. 3A). This delay can be a negative property of DMNP but can also be used creatively in some situations, because higher levels of Ca^{2+} are reached during continuous low-intensity photoconversion (Ayer and Zucker 1999). In physiological solutions, NP-EGTA has slower Ca^{2+} release kinetics upon photolysis (τ = 2.8–10 msec vs. 15 msec for DMNP) and faster rebinding of calcium to unphotolyzed buffer (Faas et al. 2005); therefore, under some conditions, the peak transient of [Ca^{2+}] after a brief flash to DMNP can be much higher compared to NP-EGTA (Fig. 3A), making DMNP particularly effective for driving Ca^{2+} release with a short light pulse. Because photoproducts of cages may change cytosolic pH and because Ca^{2+} binding by DMNP and NP-EGTA is pH-sensitive, it is advisable to increase endogenous pH buffering capacity when possible, for example, by introducing at least equal amounts of pH buffer along with the chelator via the injection or patch pipette.

After Ca^{2+} release by photolysis (or sequestration by Diazo-2), $[Ca^{2+}]_i$ is restored to normal resting levels by cellular processes such as surface pumping, leaks, and internal sequestration. When cage compounds are at millimolar concentrations, tens of seconds may be required to return $[Ca^{2+}]_i$ to resting levels. These temporal complexities of Ca^{2+} (and Mg^{2+}) binding by photolyzed and unphotolyzed cage, endogenous buffers, and cellular Ca^{2+} removal processes can sometimes be turned to experimental advantage, allowing $[Ca^{2+}]_i$ to be transiently elevated at intervals by repeated pulses of photolysis (Delaney and Zucker 1990; Zucker 1993). Examples of Ca^{2+} transients caused by repetitive, partial photolysis of DMNP or NP-EGTA are illustrated in Figure 3.

Photolabile Ca^{2+} chelators have been used to rapidly release or buffer Ca^{2+} to characterize the time course and voltage dependence of Ca^{2+}-activated K^+ currents and nonspecific cation currents (Gurney and Lester 1987; Lando and Zucker 1989; Hall and Delaney 2002), as well as inactivation of Ca^{2+} currents (Fryer and Zucker 1993; Johnson and Byerly 1993). Several studies have used Ca^{2+} cages to trigger neurotransmitter release to study autoreceptor activation, Ca^{2+} cooperativity, and Ca^{2+} sensitivity and voltage dependence of release (Delaney and Zucker 1990; Mulkey and Zucker 1991; Heidelberger et al. 1994; Lando and Zucker 1994; Bollmann et al. 2000; Schneggenburger and Neher 2000; Ma and Lowe 2007). Ca^{2+} cages have also been used to study Ca^{2+} buffer capacity (al-Baldawi and Abercrombie 1995), Ca^{2+} sequestration by mitochondria, and the effect of pH on Ca^{2+} extrusion rates (Sidky and Baimbridge 1997). Photolytic Ca^{2+} release (or scavenging with Diazo-2) has also been used to study the time course for induction or maintenance of various activity-dependent aspects of synaptic plasticity including facilitation, post-tetanic and long-term potentiation, and the localization of these processes to presynaptic or postsynaptic compartments (Malenka et al. 1992; Mulkey and Zucker 1993; Kamiya and Zucker 1994; Fischer et al. 1997).

Effective use of caged Ca^{2+} compounds requires a thorough understanding of the principles and processes of neuronal Ca^{2+} buffering and homeostasis (Sala and Hernandez-Cruz 1990; Neher and Augustine 1992; Roberts 1994; Neher 1995; Tank et al. 1995; see also Chapter 36). Adams and Tsien (1993), Zucker (1993), and Ellis-Davies (2008) are excellent reviews of the use of caged Ca^{2+} compounds.

PRACTICAL CHALLENGES TO THE USE OF CAGED CALCIUM COMPOUNDS IN NEURONS

Three major practical issues need to be addressed for the effective use of caged Ca^{2+} compounds: (i) loading the compound into the cell, (ii) photoconversion of the compound, and (iii) estimating the spatiotemporal aspects of Ca^{2+} changes resulting from photolysis.

Cell Loading

Caged Ca^{2+} chelators can be introduced into neurons by one of the following methods:

- iontophoresis through sharp electrodes
- pressure injection through beveled micropipettes (Lando and Zucker 1989; Delaney and Zucker 1990)
- bath application of acetoxymethyl (AM) ester derivatives
- diffusion from whole-cell recording pipettes (Neher and Augustine 1992; Schneggenburger and Neher 2000)

For photolysis to alter $[Ca^{2+}]_i$ sufficiently, cells must contain enough Ca^{2+}-bound cage so that the buffer capacity of the cage is severalfold greater than the capacity of endogenous Ca^{2+} buffers and removal of cytoplasmic Ca^{2+} by pumps. Because caged Ca^{2+} chelators are soluble to ~100 mM in physiological solutions and neurons can withstand injections of 10% of their volume, cage concentrations of 2–10 mM can be achieved when filling by pressure injection. However, because the cage is at high concentration in the injection solution, the sudden increase in Ca^{2+} buffering capacity may by itself interfere with the endogenous changes in $[Ca^{2+}]_i$ involved in, for example, neurotransmitter release and Ca^{2+}-dependent modulation of currents. Fortunately, these perturbations to the normally brief, endogenous Ca^{2+} transients that drive transmitter release are limited by the moderately slow forward binding rate of NP-EGTA and the particularly slow unbinding of Mg^{2+} from DMNP.

The preferred loading method is usually determined by the size of the cell and the method that will be used to monitor the consequences of photolysis. Resting $[Ca^{2+}]_i$ in neurons is maintained at ~100 nM or less by endogenous Ca^{2+} pumps and inward leaks. These processes also ensure that the change in free $[Ca^{2+}]_i$ resulting from addition of exogenous Ca^{2+} buffer (free or preloaded with Ca^{2+}) will be transient, and that the chelator will become Ca^{2+}-loaded in relation to the free $[Ca^{2+}]_i$ (and free $[Mg^{2+}]_i$, in the case of DMNP) and the dissociation constant (K_d) of the intact chelator. When using AM ester cage derivatives to load the cell, it is unlikely that influx into the cell and de-esterification will occur fast enough to perturb normal resting $[Ca^{2+}]_i$ or $[Mg^{2+}]_i$, even transiently. Consequently, DMNP in DMNP AM–loaded cells is almost entirely Mg^{2+}-bound at normal resting free $[Mg^{2+}]_i$ until [DMNP] exceeds several millimolar (Ellis-Davies 2006). When using whole-cell recording pipettes, the continuous exchange between cytoplasm and the pipette filling solution can be exploited to set $[Ca^{2+}]$ and/or $[Mg^{2+}]$ in the vicinity of the pipette tip before photolysis. Introducing mixtures of cage + $[Ca^{2+}]$ (DMNP and NP-EGTA) or pipette solutions lacking Mg^{2+} (Schneggenburger and Neher 2000) can increase the amount of Ca^{2+} released by a flash because the fractional loading of the cage is increased. However, although this approach can be effective near the pipette tip, the neuron's endogenous mechanisms of Ca^{2+} homeostasis will compete with diffusion of the Ca^{2+} bound cage to remove excess Ca^{2+} along the dendrite or axon so at the end of thin processes the fraction of cage bound to Ca^{2+} will be primarily determined by the equilibrium affinities of the cage for Ca^{2+} (and Mg^{2+} in the case of DMNP) and the natural resting $[Ca^{2+}]_i$ of the neuron.

The cytoplasmic concentration of chelator achieved by pressure injection can be estimated by co-injection of a long-wavelength fluorophore (e.g., Texas Red [10 µM]), mixed at a known ratio with the chelator. The fluorescence intensity of fluorophore/chelator mixtures can be calibrated using 20-µm path-length microcuvettes, or droplets under oil (see below). Determining the concentration of chelator after loading by AM ester or iontophoresis is more problematic and limits the use of these loading methods for quantitative studies.

Estimating the Efficiency of Photolysis

Because $\Delta[Ca^{2+}]$ is nonlinearly related to the proportion of cage that is photolyzed (Fig. 3), sufficient photolysis must be achieved to elevate $[Ca^{2+}]_i$ to an effective level before the cell buffers the released Ca^{2+}. A twofold increase in photolysis can be the difference between success and failure of an experiment using a Ca^{2+} cage. If more of the cage is loaded with Ca^{2+}, then less photolysis is required to achieve the same $\Delta[Ca^{2+}]$. Ca^{2+} loading can be increased by firing action potentials immediately prior to photolysis to generate a Ca^{2+} influx. Although this will cause a rise in free $[Ca^{2+}]_i$, the enhancement in $\Delta[Ca^{2+}]_i$ due to photolysis is more significant (Fig. 2A). For NP-EGTA (K_d = 80 nM), the difference in $\Delta[Ca^{2+}]$ achieved with 20%–40% photolysis is about threefold starting from a free $[Ca^{2+}]$ of 80 nM versus 160 nM (Fig. 1). Twice as much photolysis (50% vs. 25%) is required to elevate $[Ca^{2+}]$ to 1 µM starting from a free $[Ca^{2+}]$ of 80 nM versus 160 nM. Because the cytoplasmic cage concentration may be several millimolar, free $[Ca^{2+}]$ must be >100 nM to significantly load DMNP and NP-EGTA, and many action potentials or long voltage clamp steps may be required to cause a sufficient influx of Ca^{2+} (Tank et al. 1995).

Quantification of $\Delta[Ca^{2+}]_i$

Quantification of $\Delta[Ca^{2+}]$ is by far the most difficult issue to resolve and can be the main obstacle to the effective use of Ca^{2+} cages. It is difficult to predict the magnitude of $\Delta[Ca^{2+}]_i$ after photolysis

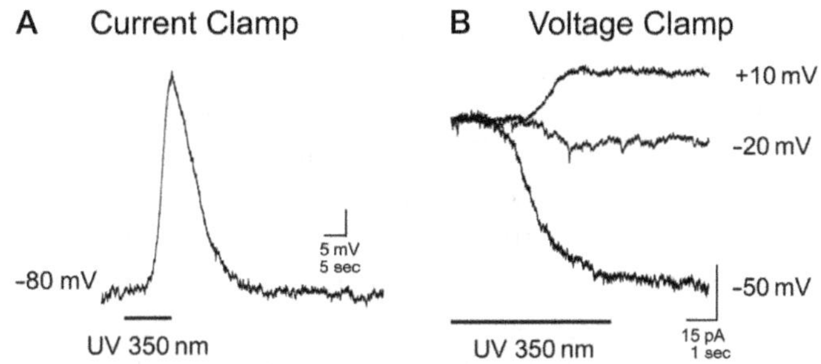

FIGURE 4. Whole-cell patch recording from an olfactory bulb granule cell in a frog. An intact olfactory bulb was maintained in artificial cerebrospinal fluid containing 20 μM CNQX, 100 μM APV, and 5 μM bicuculline methiodide to block major ionotropic receptor currents. Pipette solution contained cesium methane sulfonate, QX314, and TEA with 2 mM *o*-nitrophenol-EGTA (NP-EGTA; Invitrogen) added as the only Ca^{2+} buffer. Photolysis was achieved by exposure to ultraviolet light from a 75-W xenon arc lamp, 350 ± 6 nm as indicated, controlled by a switching monochrometer (Polychrome II, T.I.L.L. Photonics) through a 20x, 0.5 NA, quartz, water-immersion lens (Olympus). (*A*) Response to photolysis obtained ~10 min after obtaining whole-cell recording configuration. (*B*) In voltage clamp mode, repeated photolysis with the membrane held at the potentials indicated evokes a current that reverses near –10 mV (data corrected for a 10-mV junction potential), suggesting that a mixed-cation conductance is activated by released Ca^{2+}. To increase Ca^{2+} loading of the cage prior to photolysis, the neuron was depolarized to fire a train of approximately 10 action potentials (current clamp) or stepped to 0 mV for 1 sec (voltage clamp) ~2 sec prior to photolysis. Data provided by Dr. Benjamin Hall (see Hall and Delaney 2002).

because of the large number of factors involved, including the generation of one or more photoproducts with significant Ca^{2+} affinity; greater quantum yield for Ca^{2+}-bound versus free forms of the chelator; absorption of light by the cell; competition by endogenous buffers for released Ca^{2+}; and binding of protons and/or Mg^{2+} to the Ca^{2+} cage. Binding of photo-released Ca^{2+} to endogenous buffers or the intact cage occurs on a millisecond time scale, whereas removal of Ca^{2+} from the cytoplasm by pumps takes from 0.1 sec to several seconds.

In many situations, precise determination of $\Delta[Ca^{2+}]_i$ may not be required to differentiate between hypotheses (e.g., when Ca^{2+} is photolytically released in part of a neuron to test for localization of a Ca^{2+}-dependent process) or to dissociate $\Delta[Ca^{2+}]_i$ from transmembrane voltage (Fig. 4), or if temporal control of $[Ca^{2+}]_i$ is the main objective.

ILLUMINATION SYSTEMS

Light Sources

Any light source with significant energy in the near-UV range (~350 nm) can be used for photolysis, including conventional Xenon (Xe) and mercury (Hg) arc lamps, UV lasers, and frequency-doubled or -tripled long-wavelength lasers (Parker et al. 1997). Wavelengths up to 390 nm will photolyze most Ca^{2+} cages, but 320–355 nm is most effective; therefore, glass lenses and filters must be avoided along the illumination path. Two-photon excitation with long-wavelength light (Denk 1994) works very well for Azid-1 (Brown et al. 1999), NDPF-EGTA (Mototake et al. 2006), and reasonably well for DMNPE-4 (DelPrincipe et al. 1999), but it is difficult to achieve high levels of $[Ca^{2+}]$ using two-photon illumination without causing direct photodamage, because of the conversion efficiencies of commercially available cages (Lipp and Niggli 1998; DelPrincipe et al. 1999). Whichever UV light source is used, it is important to determine the level of illumination that by itself does not cause photodamage or induce physiological responses.

Xe or Hg arc lamps (75–150 W) operated in continuous mode can photolyze ~20% of the chelator in about a second, provided the arc is focused optimally. Brief (~1 msec), high-intensity discharges from Xe arc lamps can deliver the equivalent of ~1 sec of continuous illumination (up to 300 mJ in the 330–380-nm band) (Rapp and Güth 1988). Hg arc lamps can be briefly “pulsed” to deliver short UV light with intensities equivalent to Xe arcs (Denk 1997), but without the associated electromagnetic artifacts and acousto-mechanical vibrations that occur when using Xe flash lamps (Zucker 1994). Xe flash lamp systems are available from Rapp Optoelektronik, T.I.L.L. Photonics, and Cairns Inc; T.I.L.L. Photonics, Rapp Optoelektronik, and Cairns Inc. also sell complete systems. Rapp Optoelektronik offers customization of pulse width and high repetition rate upon request. T.I.L.L. Photonics, Rapp Optoelektronik, and Cairns Inc. offer fiber-optic condenser couplings to standard microscopes. The repetition rate of flash lamps is limited to <1 Hz at full power by the charging of the capacitors unless a second bank of capacitors is used. Lower-power, fast-repetition (e.g., 10–20 Hz) modes, essential for focusing and positioning the area of illumination, are provided in the commercial systems mentioned above.

Although more costly, lasers provide a convenient collimated beam that is easily focused to a small spot. Frequency-doubled ruby lasers and XeF excimer lasers with relatively long pulse durations (>10 nsec) can be pulsed at repetition rates >1 Hz and provide intensities comparable to flash lamps. Pulsed nitrogen lasers (Spectra Physics) are less expensive and operate at 337 nm but should be tested before purchase because their pulse durations (≤4 nsec) may be too short for efficient conversion of some compounds. Compact, quasi-continuous pulsed UV lasers are well suited for photolysis of Ca^{2+} cages (355 nm, 100 kHz repetition rates, 100 to 500 mW; DPSS Lasers Inc.) and can be gated with a temporal resolution of 10 μsec (Shepherd et al. 2003). Finally, modern high-power 365-nm LEDs have sufficient power when properly focused to uncage Ca^{2+} over small areas within <1 sec (Bernardinelli et al. 2005).

Setup and Accessories

For many purposes, a standard lamp housing attached to an epifluorescence microscope can be used, provided the illumination path is equipped with a shutter to control the duration of illumination. If visualization of fluorescence from Ca^{2+}-sensitive dyes is desired in combination with photolysis (Kao et al. 1989; Zucker 1993; Escobar et al. 1997), dichroic mirrors are required that have high reflectance in the UV and one or more regions of the visible spectrum (Chroma Corp. or Omega Optical). Caged FITC (Invitrogen) suspended in a thin layer of agarose is useful for initial alignment and to confirm the delivery of sufficient photolysis light prior to experiments.

If optical fibers are used to illuminate the specimen, these should be either quartz/fused silica or liquid light guides (e.g., Oriel Instruments). Liquid guides are less expensive and are available with a 3-mm diameter, which simplifies centering the arc on the input end of the fiber. Although light guides simplify the delivery of light to the specimen, the beam coupling must be carefully designed to avoid light loss. For laser coupling to microscopes, Oz Optics offers several affordable custom solutions. Custom-made tapered quartz fibers (Eberius and Schild 2001) can be positioned directly beside neurons to efficiently deliver photolysis light to small areas.

Provided there is sufficient space and pinpoint focusing is not required, the simplest way to deliver unfocused light for photolysis is direct illumination by a light source above the preparation. Elliptical reflectors simplify focusing the arc on the preparation because UV and visible wavelengths are focused to the same point, whereas singlet quartz lenses are not color-corrected and require some empirical compensation for depth. Elliptical reflectors that wrap around the lamp are generally more efficient for gathering light than lenses, especially considering the cost of large-diameter quartz lenses. High-quality, lens-based or elliptical reflector housings can be obtained from Rapp Optoelektronik or Photon Technology International. Large elliptical mirrors (100–200 mm focal length) suitable for building a lamp housing can be obtained from Oriel Instruments, Melles Griot, or other major optical supply companies for ~$500.

Most microscope illuminators use KG1 or KG2-type heat filters to block the IR emissions from arc lamps; however, these filters are unsuitable for UV photolysis as they transmit poorly below ~350 nm. UG11 filters are a good substitute because they transmit well between 300 and 370 nm and block effectively in the range that excites fluorescent indicators such as Fluo-4 or Ca-Green, which is useful for avoiding photobleaching when using an external flash lamp for photolysis.

Measuring the Effectiveness of Photolysis

The methods for calibrating photolysis of DMNP (Zucker 1993) are also suitable for other chelators. Either the change in the absorbance spectrum of the chelator following photolysis or the increase in free $[Ca^{2+}]_i$ resulting from photolysis is measured using a Ca^{2+} indicator dye and microcuvettes with path lengths similar to cells. Rectangular glass microcuvettes with 20–50-µm path lengths can be purchased from VitroCom Inc. Short lengths can be filled with chelator by capillary action, and submerged in light mineral oil or the ends sealed with grease, to prevent evaporation. This is suitable for calibrating arc lamps or weakly focused laser light, but high-NA lenses require shorter path lengths because of their limited focal depth. Two methods for creating shorter path lengths are to extrude small droplets of solution under mineral oil or place 5–10 µL of solution between two 25-mm square coverslips. Indicator dyes (e.g., Fluo-3) and chelators may interact (Zucker 1992); therefore, if quantitation is critical, indicators should be calibrated in the presence of the chelator. Many methods are available for calibration of photolysis in vitro using a variety of fluorescent Ca^{2+} indicators (Neher and Zucker 1993; Thomas et al. 1993; Xu et al. 1997; Kishimoto et al. 2001). For methods that estimate photolysis rate by the rise in free $[Ca^{2+}]$ after photolysis, absorption spectroscopy and reference to published values of extinction coefficients are used to accurately determine the concentration of chelator in the test solution; this must be performed before adding Ca^{2+} to the free $[Ca^{2+}]$ expected in neurons.

Other Considerations

If electrophysiological measurements are made along with photolysis, it is important to avoid illuminating AgCl reference wires, because this generates photochemical artifacts that may be misinterpreted as physiological responses to Ca^{2+} release.

Chelator concentration should be checked at the time of use. Commercially obtained compounds are typically ~90% pure, but batch-to-batch variation is common. Because chelators in solution can degrade over time and with freeze–thaw cycles, the chelator concentration must be checked prior to use. Purchase of 50-µg aliquots is recommended; however, larger quantities can be dissolved, aliquoted into single-use portions, and dried, preferably using a stream of dry N2. Solutions of chelator must be protected from light <400 nm at all times, including light from microscope illuminators using colored glass filters.

REFERENCES

Adams SR, Tsien RY. 1993. Controlling cell chemistry with caged compounds. *Annu Rev Physiol* **55:** 755–784.

Adams S, Kao J, Tsien R. 1989. Biologically useful chelators that take up Ca^{2+} upon illumination. *J Am Chem Soc* **111:** 7957–7968.

al-Baldawi NF, Abercrombie RF. 1995. Cytoplasmic calcium buffer capacity determined with Nitr-5 and DM-nitrophen. *Cell Calcium* **17:** 409–421.

Ayer RK Jr, Zucker RS. 1999. Magnesium binding to DM-nitrophen and its effect on the photorelease of calcium. *Biophys J* **77:** 3384–3393.

Bernardinelli YC, Haeberli C, Chatton JY. 2005. Photolysis using a light emitting diode: An efficient, compact and affordable solution. *Cell Calcium* **37:** 565–572.

Bollmann JH, Sakmann B, Borst JG. 2000. Calcium sensitivity of glutamate release in a calyx-type terminal. *Science* **289:** 953–957.

Brown EB, Shear JB, Adams SR, Tsien RY, Webb WW. 1999. Photolysis of caged calcium in femtoliter volumes using two-photon excitation. *Biophys J* **76:** 489–499.

Delaney KR, Zucker RS. 1990. Calcium released by photolysis of DM-nitrophen stimulates transmitter release at squid giant synapse. *J Physiol (Lond)* **426:** 473–498.

DelPrincipe F, Egger M, Ellis-Davies GC, Niggli E. 1999. Two-photon and UV-laser flash photolysis of the Ca^{2+} cage, dimethoxynitrophenyl-EGTA-4. *Cell Calcium* **25:** 85–91.

Denk W. 1994. Two-photon scanning photochemical microscopy: Mapping ligand-gated ion channel distributions. *Proc Natl Acad Sci* **91:** 6629–6633.

Denk W. 1997. Pulsing mercury arc lamps for uncaging and fast imaging. *J Neurosci Meth* **72:** 39–42.

Eberius C, Schild D. 2001. Local photolysis using tapered quartz fibres. *Pflugers Arch* **443:** 323–330.

Ellis-Davies GCR. 2006. DM-nitrophen AM is caged magnesium. *Cell Calcium* **39:** 471–473.

Ellis-Davies GCR. 2008. Neurobiology with caged calcium. *Chem Rev* **108:** 1603–1613.

Escobar AL, Velez P, Kim AM, Cifuentes F, Fill M, Vergara JL. 1997. Kinetic properties of DM-nitrophen and calcium indicators: Rapid transient response to flash photolysis. *Pflügers Arch* **434:** 615–631.

Faas G, Karacs K, Vergara JL, Mody I. 2005. Kinetic properties of DM-nitrophen binding to calcium and magnesium. *Biophys J* **88:** 4421–4433.

Fischer TM, Zucker RS, Carew TJ. 1997. Activity-dependent potentiation of synaptic transmission from L30 inhibitory interneurons of *Aplysia* depends on residual presynaptic Ca^{2+} but not on postsynaptic Ca^{2+}. *J Neurophysiol* **78:** 2061–2071.

Fryer MW, Zucker RS. 1993. Ca^{2+}-dependent inactivation of Ca^{2+} current in *Aplysia* neurons: Kinetic studies using photolabile Ca^{2+} chelators. *J Physiol* **464:** 501–528.

Gurney AM, Lester HA. 1987. Light-flash physiology with synthetic photosensitive compounds. *Physiol Rev* **67:** 583–617.

Hall BJ, Delaney KR. 2002. Contribution of a calcium-activated non-specific conductance to NMDA receptor-mediated synaptic potentials in granule cells of the frog olfactory bulb. *J Physiol* **543:** 819–834.

Heidelberger R, Heinemann C, Neher E, Matthews G. 1994. Calcium dependence of the rate of exocytosis in a synaptic terminal. *Nature* **371:** 513–515.

Johnson BD, Byerly L. 1993. Photo-released intracellular Ca^{2+} rapidly blocks Ba^{2+} current in *Lymnaea* neurons. *J Physiol* **462:** 321–347.

Kamiya H, Zucker RS. 1994. Residual Ca^{2+} and short-term synaptic plasticity. *Nature* **371:** 603–606.

Kao JPY, Harootunian H, Tsien RY. 1989. Photochemically generated cytosolic calcium pulses and their detection by Fluo-3. *J Biol Chem* **264:** 8179–8184.

Kaplan JH, Ellis-Davies GC. 1988. Photolabile chelators for the rapid photorelease of divalent cations. *Proc Natl Acad Sci* **85:** 6571–6575.

Kishimoto T, Liu TT, Ninomiya Y, Takagi H, Yoshioka T, Ellis-Davies GC, Miyashita Y, Kasai H. 2001. Ion selectivities of the Ca^{2+} sensors for exocytosis in rat phaeochromocytoma cells. *J Physiol* **533:** 627–637.

Lando L, Zucker RS. 1989. "Caged calcium" in *Aplysia* pacemaker neurons: Characterization of calcium-activated potassium and nonspecific cation currents. *J Gen Physiol* **93:** 1017–1060.

Lando L, Zucker RS. 1994. Ca^{2+} cooperativity in neurosecretion measured using photolabile Ca^{2+} chelators. *J Neurophysiol* **72:** 825–830.

Lipp P, Niggli E. 1998. Fundamental calcium release events revealed by two-photon excitation photolysis of caged calcium in guinea-pig cardiac myocytes. *J Physiol* **508:** 801–809.

Ma J, Lowe G. 2007. Calcium permeable AMPA receptors and autoreceptors in external tufted cells of rat olfactory bulb. *Neuroscience* **144:** 1094–1108.

Malenka RC, Lancaster B, Zucker RS. 1992. Temporal limits on the rise in postsynaptic calcium required for the induction of long-term potentiation. *Neuron* **9:** 121–128.

McCray JA, Fidler-Lim N, Ellis-Davies GC, Kaplan JH. 1992. Rate of release of Ca^{2+} following laser photolysis of the DM-nitrophen-Ca^{2+} complex. *Biochemistry* **31:** 8856–8861.

Momotake A, Lindegger N, Niggli E, Barsotti RJ, Ellis-Davies GC. 2006. The nitrodibenzofuran chromophore: A new caging group for ultra-efficient photolysis in living cells. *Nat Methods* **3:** 35–40.

Mulkey RM, Zucker RS. 1991. Action potentials must admit calcium to evoke transmitter release. *Nature* **350:** 153–155 (Erratum **351:** 419).

Mulkey RM, Zucker RS. 1993. Calcium released by photolysis of DM-nitrophen triggers transmitter release at the crayfish neuromuscular junction. *J Physiol (Lond)* **462:** 243–260.

Neher E. 1995. The use of fura-2 for estimating Ca buffers and Ca fluxes. *Neuropharmacology* **34:** 1423–1442.

Neher E, Augustine GJ. 1992. Calcium gradients and buffers in bovine chromaffin cells. *J Physiol* **450:** 273–301.

Neher E, Zucker RS. 1993. Multiple calcium-dependent processes related to secretion in bovine chromaffin cells. *Neuron* **10:** 21–30.

Parker I, Callamaras, N, Wier WG. 1997. A high-resolution confocal laser-scanning microscope and flash photolysis system for physiological studies. *Cell Calcium* **21:** 441–452.

Rapp G, Güth K. 1988. A low cost high intensity flash device for photolysis experiments. *Pflügers Arch* **411:** 200–203.

Roberts WM. 1994. Localization of calcium signals by a mobile calcium buffer in frog saccular hair cells. *J Neurosci* **14:** 3246–3262.

Sala F, Hernandez-Cruz A. 1990. Calcium diffusion modeling in a spherical neuron: Relevance of buffering properties. *Biophys J* **57:** 313–324.

Schneggenburger R, Neher E. 2000. Intracellular calcium dependence of transmitter release rates at a fast central synapse. *Nature* **406:** 889–893.

Shepherd GM, Pologruto, TA, Svoboda K. 2003. Circuit analysis of experience-dependent plasticity in the developing rat barrel cortex. *Neuron* **38:** 277–289.

Sidky AO, Baimbridge KG. 1997. Calcium homeostatic mechanisms operating in cultured postnatal rat hippocampal neurones following flash photolysis of nitrophenyl-EGTA. *J Physiol* **504:** 579–590.

Tank DW, Regehr WG, Delaney KR. 1995. A quantitative analysis of presynaptic calcium dynamics that contribute to short-term enhancement. *J Neurosci* **15:** 7940–7952.

Thomas P, Wong JG, Lee AK, Almers W. 1993. A low affinity Ca^{2+} receptor controls the final steps in peptide secretion from pituitary melanotrophs. *Neuron* **11:** 93–104.

Xu T, Naraghi M, Kang H, Neher E. 1997. Kinetic studies of Ca^{2+} binding and Ca^{2+} clearance in the cytosol of adrenal chromaffin cells. *Biophys J* **73:** 532–545.

Zucker RS. 1992. Effects of photolabile calcium chelators on fluorescent calcium indicators. *Cell Calcium* **13:** 29–40.

Zucker RS. 1993. The calcium concentration clamp: Spikes and reversible pulses using the photolabile chelator DM-nitrophen. *Cell Calcium* **14:** 87–100.

Zucker RS. 1994. Photorelease techniques for raising or lowering intracellular Ca^{2+}. *Methods Cell Biol* **40:** 31–63.

41 An Optical Fiber–Based Uncaging System

Karl Kandler,[1] Tuan Nguyen,[1] Jihyun Noh,[1] and Richard S. Givens[2]

[1]*Department of Otolaryngology, University of Pittsburgh, Pittsburgh, Pennsylvania 15213-2588;*
[2]*Department of Chemistry, The University of Kansas, Lawrence, Kansas 66045-7582*

ABSTRACT

This chapter describes the assembly and performance of a simple and inexpensive ultraviolet-flash system suitable for rapid focal photolysis of caged compounds in cultured neurons and brain slices. Advantages and limitations of this system are discussed. Examples are provided illustrating how this system can be used for stimulating neurons and mapping their functional inputs in brain slices.

INTRODUCTION

Use of laser-based scanning focal photolysis has become widespread for rapid and local stimulation of neurons (Callaway and Katz 1993; Pettit et al. 1997; Dantzker and Callaway 2000a; Nikolenko et al. 2007). However, for many applications, the high spatial resolution and scanning speed provided by a laser-based system are not required. For these cases, an inexpensive and easy to assemble and operate uncaging system is more desirable. Here, we describe a simple system composed of a mercury arc lamp for ultraviolet (UV) light generation and an optical fiber for light delivery, which can be implemented without modifications to the existing microscopy setup. This system has been used to stimulate small dendritic segments of neurons in the hippocampus (via caged glutamate) and in the auditory brain stem (via caged γ-aminobutyric acid [GABA]). In addition, it has been used to map functional connectivity patterns (Kandler et al. 1998; Kim and Kandler 2003). By selectively stimulating presynaptic inhibitory neurons, this approach has been combined with calcium imaging to measure postsynaptic responses in populations of neurons (Kullmann et al. 2002). The system is capable of rapidly photolyzing a variety of caged compounds and is applicable to a diversity of cell types and neuronal systems in slices as well as cultured neurons.

MATERIALS TO ASSEMBLE THE UNCAGING SYSTEM

The light source we use is a 100-W mercury arc lamp with a Series Q lamp housing and a UV grade fused silica condenser (Oriel, Newport Corporation). The power supply is from OPTI QUIP. The electronic shutter used to control light pulse width is a critical component. We use a Uniblitz Model LSG with $AlMgF_2$ coating (Vincent Associates), which can generate light pulses as brief as 5 msec. The shutter is triggered by a pulse generator (Master 8, A.M.P.I.). The uncaging system is mounted onto a breadboard. Required mounting components are a fiber positioner, fiber chucks, and mounting hardware (base plates, mounting rods, and screws) (Newport Corporation, New Focus, and Thorlabs, Inc.). The optical fibers are made from high-OH high UV–transmitting fused silica. We have used fibers with inner diameters ranging from 5 to 50 µm obtained from CeramOptec Industries, Inc. or Polymicro Technologies. Fibers are cut to the required length with a FITEL fiber cleaver from Fiber Instrument Sales, Inc.

The caged neurotransmitters are supplied by various vendors (Calbiochem, Dojindo Molecular Technologies, Inc., and Invitrogen) or can be synthesized in-house (e.g., Givens et al. 1997; Stensrud et al. 2009).

ASSEMBLY AND OPTIMIZATION

The uncaging system is illustrated in Figure 1A,B. For protection, the optical fiber is placed within small-diameter Tygon tubing. The light-emitting end of the fiber is attached to an aluminum rod that is mounted to a manual micromanipulator. For mapping experiments, we feed the fiber through a bent glass capillary tube (Fig. 1B inset) to ensure that the fiber is approximately perpendicular to the surface of the slice. The position of the optical fiber above the slice is monitored with the microscope. For off-line analysis, a microscope-mounted video camera and a video frame grabber are used to document the exact location of uncaging sites in the slice.

Assembly of the system requires no special knowledge and can be accomplished in 1–2 days. Attention should be paid to maximizing the amount of light launched into the fiber. To this end, the cut of the fiber has to be flat and smooth and should be checked under a stereomicroscope. A smooth cut is also necessary at the light-emitting end of the fiber. A quick way to check the quality of the cut at this end is to shine the light against a smooth surface: A circular light spot with well-defined edges indicates a good cut. Another crucial step is to correctly align the optical fiber: Both position (the end should be at the focal point of the condenser) and angular orientation are critical. Optimal alignment can be checked by measuring the relative light output power of the fiber with a UV-sensitive photodiode (Thorlabs, Inc.) and a UV-band-pass filter (Oriel, Newport Corporation). We also routinely test for optimal alignment while recording from a neuron using the amplitudes and rise times of membrane currents as sensitive indicators. If uncaging is unsuccessful, checking fiber alignment and the surfaces of the fiber cuts is the starting point for troubleshooting.

EXPERIMENTS UTILIZING THE OPTICAL FIBER UNCAGING SYSTEM

This system has been used to rapidly release glutamate or GABA from several different types of caged compounds (Givens et al. 1997; Kandler et al. 1998; Conrad et al. 2000; Kim and Kandler 2003; Stensrud et al. 2009). The two major neurobiological applications in which this system has been applied are focal dendritic stimulation with glutamate (Fig. 1C) or GABA (Kandler et al. 1998; Kim and Kandler 2003) and construction of functional connectivity maps in a GABA/glycinergic auditory brain-stem circuit (Kim and Kandler 2003). An example of an inhibitory input map for the MNTB to a single neuron in the LSO is shown in Figure 1D. In these experiments, we uncaged glu-

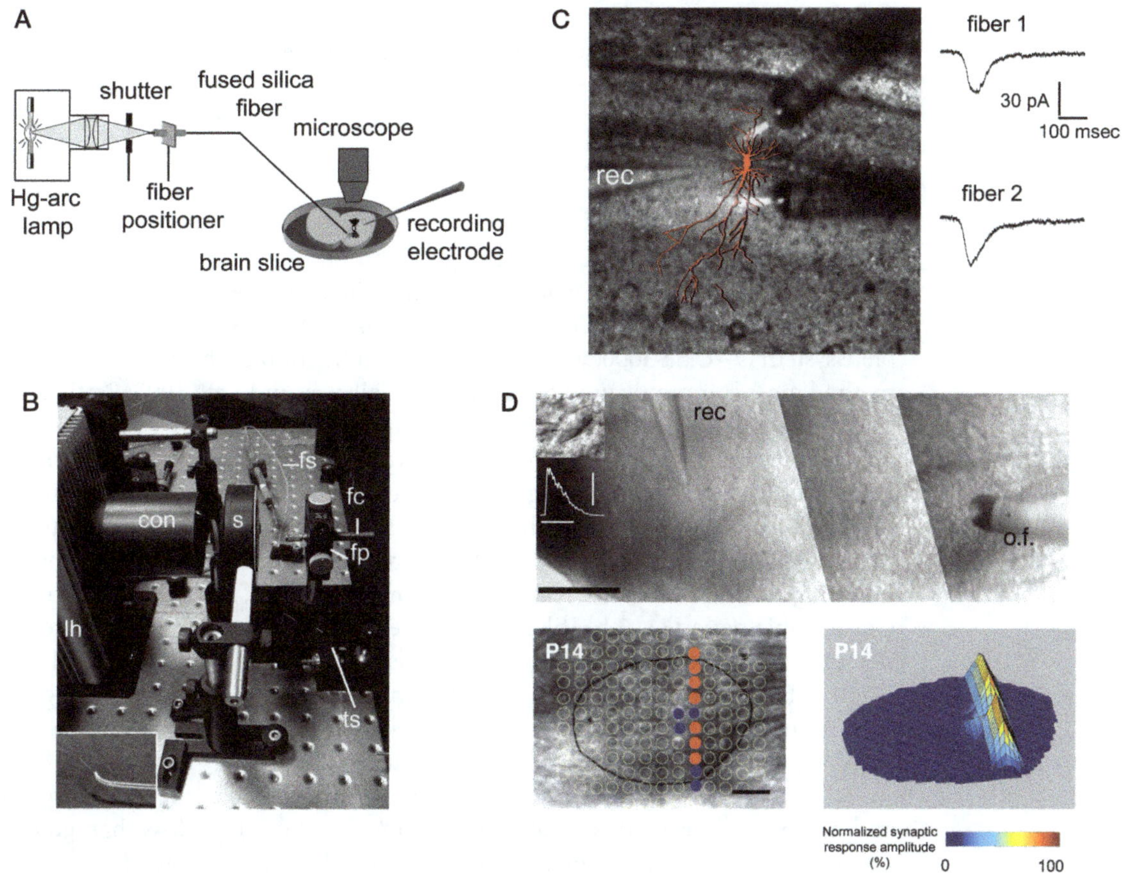

FIGURE 1. (*A*) Schematic of the uncaging system. (*B*) Photograph of lamp housing with optical-fiber mount. Inset (*lower left*) shows optical fiber protruding from curved glass capillary. con, condenser; fc, fiber chuck; fp, fiber positioner; fs, fiber sleeve; lh, lamp housing; s, shutter; ts, translational stage. (*C*) Stimulation of two dendritic areas with glutamate uncaging using two optical fibers. Post-hoc reconstruction of the recorded hippocampal CA3 pyramidal neuron (red) is overlaid on a photograph of the slice taken during stimulation. Current traces show responses elicited by applying UV flashes (20 msec) through either one of the fibers. (Reprinted, with permission, from Kandler et al. 1998.) (*D*) Mapping functional connections in an auditory brain-stem slice. The optical fiber (o.f.) is aimed at the medial nucleus of the trapezoid body (MNTB). A neuron in the lateral superior olive (LSO) is recorded in the whole-cell patch-clamp mode (rec). In the *lower panel*, each uncaging site in and around the MNTB (outlined by the black line) is indicated by a circle. Filled circles mark sites at which uncaging of glutamate elicited synaptic responses in the LSO neuron; open circles mark nonresponse sites. Scale bar, 100 µm. In the three-dimensional plot on the right, stimulation sites are encoded by the peak amplitude of corresponding synaptic responses. (Modified, with permission, from Kim and Kandler 2003.)

tamate from *p*-hydroxyphenacyl glutamate (150–200 µM) (Givens et al. 1997) in and around the MNTB at more than 100 discrete locations, separated by ~50 µm. Synaptic responses in the LSO neuron were elicited from only 12 uncaging locations (filled circles) that formed a sharp dorsoventrally oriented band mirroring the tonotopic organization of the MNTB–LSO pathway.

For correctly interpreting such mapping experiments, it is important to determine the spatial resolution of the uncaging system. This resolution is influenced by numerous factors, such as UV light power and flash duration, concentration of caged glutamate, UV transmittance of the brain area under investigation, and the geometry and physiological properties of the stimulated neurons. In our MNTB–LSO mapping experiments, we defined an effective, spike-eliciting resolution by recording from an MNTB neuron while uncaging glutamate in its vicinity. Under our conditions, action potentials were elicited only if the center of the uncaging spot was <25 µm from the cell body, corresponding to spatial resolution of 50 µm. Because light is diverging from the optical fiber, light can never be focused to a spot inside the slice and the intensity will decrease exponentially with distance from the fiber. Therefore, axial resolution of the system is expected to be significantly worse than the lateral resolution.

ADVANTAGES AND LIMITATIONS

The major advantages of this uncaging system are its low cost, its ease of use, and the minimal expertise in optics required. The system can be incorporated into any existing in vitro electrophysiology setup without the need for modifications to the existing microscope. The price of the complete system is ~U.S.$3000, which is only a small fraction of the cost of a UV laser–based system, the method most commonly used for uncaging (Callaway and Katz 1993; Pettit et al. 1997; Dantzker and Callaway 2000b). Additional optical fibers (up to four) can easily be added when using the Series Q arc lamp housing so that one setup can serve multiple electrophysiology rigs or allow for simultaneous multisite stimulation (Fig. 1C). The use of a continuous-wave light source permits a wide range of flash durations and high repetition rates, limited only by the shutter speed.

A limitation of the system described here is that the optical fiber is moved manually between uncaging positions, a rather time-consuming procedure, which requires stable recordings over a considerable period (1–3 h) if large input areas are to be mapped. If such recordings are difficult to obtain, use of a motorized micromanipulator under computer control should be considered. Another limitation arises from the fact that both the fiber and the recording electrode approach the slice from above. Consequently, the area beneath the recording electrode is obstructed, which, in some applications, can result in incomplete maps.

In principle, UV light from the arc lamp can be delivered to the microscope objective using lenses and mirrors. However, this would result in much greater light loss than guiding with small optical fibers, and, thus, would require longer flash durations, more powerful light sources, and/or higher concentrations of expensive caged compounds. With optical fibers, the uncaging spot size on the slice ultimately is restricted by the smallest diameter of the fiber core, which is 3–5 μm. Therefore, this system is less suitable for experiments that require extremely small uncaging areas as is necessary for mapping the subcellular distribution of receptors (Pettit et al. 1997; Frick et al. 2001) or for stimulating single spines (Matsuzaki et al. 2001; Smith et al. 2003).

ACKNOWLEDGMENTS

Work in our laboratory has been supported by the National Institute of Deafness and Other Communication Disorders (KK), the Center for Neural Basis of Cognition (TN, KK), and the National Institutes of Health grant GM72910 (RG).

REFERENCES

Callaway EM, Katz LC. 1993. Photostimulation using caged glutamate reveals functional circuitry in living brain slices. *Proc Natl Acad Sci* **90:** 7661–7665.

Conrad PG, Givens RS, Weber JF, Kandler K. 2000. New phototriggers: Extending the *p*-hydroxyphenacyl π–π* absorption range. *Org Lett* **2:** 1545–1547.

Dantzker JL, Callaway EM. 2000a. Laminar sources of synaptic input to cortical inhibitory interneurons and pyramidal neurons. *Nat Neurosci* **3:** 701–707.

Dantzker JL, Callaway EM. 2000b. Laminar sources of synaptic input to cortical inhibitory interneurons and pyramidal neurons. *Nat Neurosci* **3:** 701–707.

Frick A, Zieglgansberger W, Dodt HU. 2001. Glutamate receptors form hot spots on apical dendrites of neocortical pyramidal neurons. *J Neurophysiol* **86:** 1412–1421.

Givens RS, Jung A, Park CH, Weber J, Bartlett W. 1997. New photoactivated protecting groups. 7. *p*-hydroxyphenacyl: A phototrigger for excitatory amino acids and peptides. *J Am Chem Soc* **119:** 8369–8370.

Kandler K, Katz LC, Kauer JA. 1998. Focal photolysis of caged glutamate produces long-term depression of hippocampal glutamate receptors. *Nat Neurosci* **1:** 119–123.

Kim G, Kandler K. 2003. Elimination and strengthening of glycinergic/GABAergic connections during tonotopic map formation. *Nat Neurosci* **6:** 282–290.

Kullmann PH, Ene FA, Kandler K. 2002. Glycinergic and GABAergic calcium responses in the developing lateral superior olive. *Eur J Neurosci* **15:** 1093–1104.

Matsuzaki M, Ellis-Davies GC, Nemoto T, Miyashita Y, Iino M, Kasai H. 2001. Dendritic spine geometry is critical for AMPA receptor expression in hippocampal CA1 pyramidal neurons. *Nat Neurosci* **4:** 1086–1092.

Nikolenko V, Poskanzer KE, Yuste R. 2007. Two-photon photostimulation and imaging of neural circuits. *Nat Methods* **4:** 943–950.

Pettit DL, Wang SS, Gee KR, Augustine GJ. 1997a. Chemical two-photon uncaging: A novel approach to mapping glutamate receptors. *Neuron* **19:** 465–471.

Smith MA, Ellis-Davies GC, Magee JC. 2003. Mechanism of the distance-dependent scaling of Schaffer collateral synapses in rat CA1 pyramidal neurons. *J Physiol* **548:** 245–258.

Stensrud K, Noh J, Kandler K, Wirz J, Heger D, Givens RS. 2009. Competing pathways in the photo-Favorskii rearrangement and release of esters: Studies on fluorinated *p*-hydroxyphenacyl-caged GABA and glutamate phototriggers. *J Org Chem* **74:** 5219–5227.

42 Multiphoton Stimulation of Neurons and Spines

Hajime Hirase,[1,2] Volodymyr Nikolenko,[2] and Rafael Yuste [2]

[1]RIKEN Brain Science Institute, Wako, Saitama 351-0198, Japan; [2]Department of Biological Science, Columbia University, New York, New York 10027

ABSTRACT

This chapter describes an optical method to directly stimulate a neuron (i.e., without using any caged chemicals or genetic probes) using an infrared ultrafast mode-locked laser. This method can trigger action potentials in a targeted neuron when a laser beam is applied to the somatic membrane. Alternatively, it can mimic excitatory postsynaptic potentials (EPSPs) when applied to dendritic spines. The protocol has been applied successfully using juvenile (postnatal day 7–14) C57 mouse neocortical and hippocampal acute slices (~300-µm thickness). It can be used in conjunction with slices bulk loaded with calcium indicators, such as Fura-2 AM.

Protocol

Multiphoton Stimulation of Neurons and Spines

IMAGING SETUP

Two-Photon Microscope

This protocol was optimized using an ultrafast (repetition rate 76 MHz, pulse width ~120 fsec) laser beam, produced by a mode-locked Ti:sapphire laser (Mira 800, Coherent, Inc.), pumped by a solid-state Nd:YVO_4 laser (Verdi-5, Coherent, Inc.). Similar results were obtained using a Chameleon laser (Coherent, Inc.). The near-infrared laser beam was collimated and was modulated by a Pockels cell (350-50BK, Conoptics), driven by a high-voltage direct current power amplifier (Model 302, Conoptics). The output beam was then directed to a laser-scanning system (FluoView, Olympus) coupled to an upright microscope (BX50WI, Olympus). The Pockels cell was modulated by a signal generated using a square-wave signal generator (Master-8, A.M.P.I.). Transistor-to-transistor logic (TTL) signals marking the laser exposures (LINE ACTIVE and FRAME ACTIVE) were produced by the controller box (FV5-PSU) and were used to trigger the signal generator. The laser-scanning system and *z*-positioning of the microscope objective were controlled by dedicated software supplied by the vendor (FluoView software, Olympus). A more detailed description of the setup is available elsewhere (Majewska et al. 2000; Nikolenko et al. 2003).

MATERIALS

See the end of the chapter for recipes for reagents marked with <R>.

Reagents

Artificial cerebrospinal fluid <R>
Brain slices, acute
Fluorescent indicator (see Step 1)

Equipment

Two-photon microscope and associated instruments (see Imaging Setup)

EXPERIMENTAL METHOD

Direct Multiphoton Stimulation of Neurons and Spines

1. Prepare acute slices using a standard method. Bulk load the slice with a fluorescent indicator (e.g., with Fura-2 AM ester), as described in Chapter 37. Alternatively, label a neuron by intracellular recording (e.g., whole-cell patch-clamp method) with a fluorescent indicator (e.g., Alexa 488, Fura-2, or Calcium Green I, Molecular Probes) (Chapter 35).

2. Place the slice under the microscope objective (LUMPlanFL/IL 40x [numerical aperture (NA) = 0.8] or 60x [NA = 0.90], Olympus), and maintain in oxygenated artificial cerebrospinal fluid superfused through the recording chamber.
3. Scan the slice (x–y–t scan mode) with minimal laser intensity to visualize fluorescence-loaded cells. Choose a cell to be stimulated (target) from the visualized cell population.
4. Locate the soma of the target cell, and adjust the objective z-position such that the focal point is at the top surface of the somatic membrane, as observed by a diminished fluorescence signal.
5. Program the modulation signal to the Pockels cell in such a way that a high-intensity (>20 mW at the exit of the objective) excitation beam is applied only in the target cell area, keeping the rest of the scan with minimized laser intensity.
6. Repeated scanning (typically several times) of the membrane surface area of the target cell will lead to gradually increasing depolarizing responses of the target cell, eventually causing it to fire action potentials. Action potential generation can be monitored optically by calcium imaging of the soma, electrically via whole-cell intracellular recording, or by placing a glass pipette (~1 MΩ) close to the target cell's soma to record the extracellular spikes.

EXAMPLES OF APPLICATION TO SOMATA AND SPINES

To confirm the effectiveness of the stimulation protocol, the membrane potential of the target cell was recorded using the whole-cell patch-clamp method. Figure 1 shows an example of direct two-photon stimulation of a neuronal soma from a cerebral cortical slice. A neuron was loaded with a fluorescent dye (Alexa 488) through a patch pipette (Fig. 1A). When the laser beam was targeted at the intrasoma area, there was a very small change in the membrane potential (Fig. 1B; bottom trace). In contrast, the membrane potential depolarized in response to laser irradiation of the membrane and eventually showed bursts of action potentials (Fig. 1B; top trace).

The probability for successful optical stimulation depends heavily on the wavelength of excitation (Fig. 1C). Generally, a wavelength of ≤790 nm produces reliable optical stimulation. Wavelengths >800 nm did not produce reliable optical stimulation.

Figure 2 shows an example of a similar experiment in which the stimulated region is next to a dendritic spine (left panel, spot). Laser irradiation triggers a small depolarization (right panel), which resembles the amplitude and kinetics of an EPSP. This optically triggered depolarization can be repeatedly elicited without any significant changes in its amplitude and kinetics. It is possible that the depolarization is caused by the activation of the presynaptic terminal impinging on the spine. Regardless of its mechanisms, this method could be used to reliably stimulate any portion of the dendritic tree, without the use of caged compounds.

ADVANTAGES AND LIMITATIONS

One of the major limitations of the technique is the absence of the clear understanding of the biophysical mechanism of the direct optical stimulation. We found some evidence of an important role for free radical generation as well as possible transient breakage of the plasma membrane, perhaps caused by optical rectification (Hirase et al. 2002). Alternatively, temperature dependence of gating

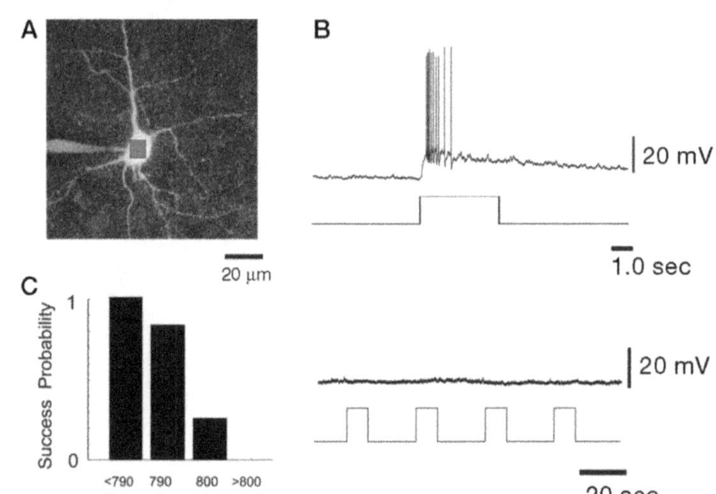

FIGURE 1. Optical stimulation of a cortical layer-5 pyramidal cell. (*A*) A layer-5 pyramidal cell was labeled with Alexa 488 through a whole-cell patch pipette. Laser irradiation was targeted at the dark square area. (*B*) (*Top*) When the membrane surface was targeted, the cell was depolarized in response to laser irradiation and produced a burst of action potentials. (*Bottom*) When the somatic interior was targeted with the laser, avoiding irradiation of the cell membrane, no change in membrane potential was recorded. (*C*) Bar chart indicating the relationship between probability of optical stimulation and excitation wavelength of the mode-locked laser beam. The probability of induction of optical stimulation is dependent on the excitation wavelength. (Modified, with permission, from Hirase et al. 2002.)

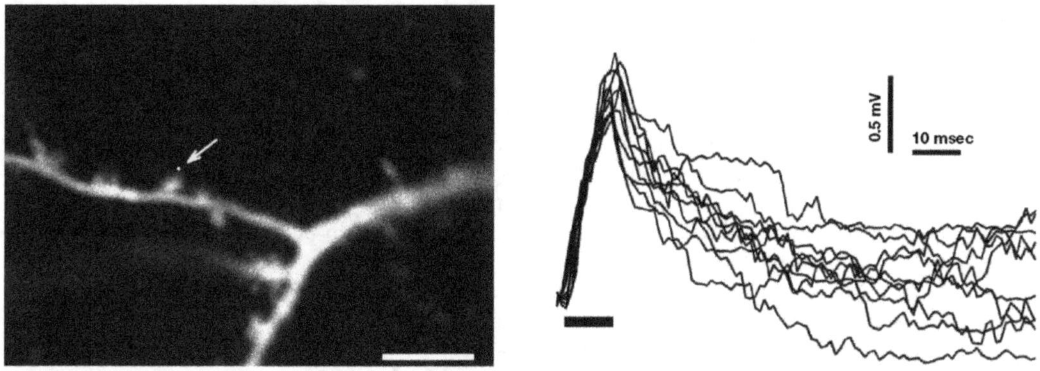

FIGURE 2. Optical stimulation of a spine. Experiment performed in a layer-5 large pyramidal neuron from a P13 mouse visual cortex. The neuron was labeled with Alexa 488 (250 µm) using a patch pipette, and the spot above the spine was irradiated for 10 msec with 80–90 mW of average power on the sample (*left* panel). Scale bar, 5 µm. (*Right* panel) Simultaneous electrophysiological recording from the soma during the experiment. (Black bar) Duration of the irradiation. The laser stimulation produces a depolarizing response that resembles an EPSP. This effect is produced repeatedly.

properties of intrinsic ion channels (Liang et al. 2009) can contribute to the phenomenon. See also more recent works (Liu et al. 2009; Zhao et al. 2009) for additional references and discussion.

Despite the limitations, the method of direct optical stimulation has some advantages. One is that it does not require the use of any extra chemicals (such as caged compounds) other than fluorescent markers used to label cell populations. This may be particularly important for in vivo experiments in which the application of exogenous chemicals is not straightforward. With recent advances in molecular biology, a wide variety of transgenic mice expressing fluorescent proteins in various cell types has become available. These allow a combination of optical stimulation, as described here, and optical probing (i.e., optically revealing postsynaptic targets) (Peterlin et al. 2000) to be used to map neuronal connectivity in various kinds of genetically labeled interneurons without electrophysiology or immunocytochemistry. In addition, optical stimulation of spines can be a useful tool for investigating dendritic biophysics. The stimulation regime presented in this chapter works more efficiently for wavelengths <800 nm. When designing these experiments, it is important that the indicator be effectively excited at the wavelength of the beam used for direct multiphoton excitation (see Yuste 2011).

RECIPE

CAUTION: See Appendix 6 for proper handling of materials marked with <!>.
Recipes for reagents marked with <R> are included in this list.

Artificial Cerebrospinal Fluid

NaCl	125 mM
KCl	3 mM
Glucose	10 mM
$NaHCO_3$	26 mM
NaH_2PO_4	1.1 mM
$CaCl_2$ <!>	2 mM
$MgSO_4$ <!>	1 mM

Adjust pH to 7.4, ~37°C.

REFERENCES

Hirase H, Nikolenko V, Goldberg JH, Yuste R. 2002. Multiphoton stimulation of neurons. *J Neurobiol* **51:** 237–247.

Liang S, Yang F, Zhou C, Wang Y, Li S, Sun CK, Puglisi JL, Bers D, Sun C, Zheng J. 2009. Temperature-dependent activation of neurons by continuous near-infrared laser. *Cell Biochem Biophys* **53:** 33–42.

Liu X, Lv X, Zeng S, Zhou W, Luo Q. 2009. Noncontact and nondestructive identification of neural circuits with a femtosecond laser. *Appl Phys Lett* **94:** 061113.

Majewska A, Yiu G, Yuste R. 2000. A custom-made two-photon microscope and deconvolution system. *Pflugers Arch* **441:** 398–408.

Nikolenko V, Nemet B, Yuste R. 2003. A two-photon and second-harmonic microscope. *Methods* **30:** 3–15.

Peterlin ZA, Kozloski J, Mao BQ, Tsiola A, Yuste R. 2000. Optical probing of neuronal circuits with calcium indicators. *Proc Natl Acad Sci* **97:** 3619–3624.

Yuste R, ed. 2011. *Imaging: A laboratory manual.* Cold Spring Harbor Laboratory Press, Cold Spring Harbor, NY.

Zhao Y, Zhang Y, Liu X, Lv X, Zhou W, Luo Q, Zeng S. 2009. Photostimulation of astrocytes with femtosecond laser pulses. *Opt Express* **17:** 1291–1298.

43 Circuit Mapping by Ultraviolet Uncaging of Glutamate

Gordon M.G. Shepherd

Department of Physiology, Feinberg School of Medicine, Northwestern University, Chicago, Illinois 60611

ABSTRACT

In laser photostimulation, small clusters of neurons in brain slices are induced to fire action potentials by focal glutamate uncaging, and synaptic connectivity between photoexcited presynaptic neurons and individual postsynaptic neurons is assessed by intracellular recording of synaptic events. With a scanner, this process can be repeated sequentially across a patterned array of stimulus locations, generating maps of neurons' local sources of synaptic inputs. Laser scanning photostimulation (LSPS) based on patterned glutamate uncaging offers an efficient, quantitative, optical-electrophysiological way to map synaptic circuits in brain slices.

INTRODUCTION

Less than 20 years ago, Callaway and Katz (1993) demonstrated that synaptic circuits in brain slices could be mapped by recording from a single postsynaptic neuron at one location and "photoexciting"—through UV photolysis of caged glutamate—a small number of presynaptic neurons at another. Since then, photostimulation by UV uncaging of glutamate has been used to map synaptic circuits across much of the central nervous system, including spinal cord (Kato et al. 2007), brain stem (Kim and Kandler 2003), superior colliculus (Pettit et al. 1999), thalamus (Lam and Sherman 2005; Deleuze and Huguenard 2006), hippocampus (Brivanlou et al. 2004), thalamocortical pathways (Bureau et al. 2006; Llano et al. 2009), and more. A number of neocortical areas have been examined with LSPS, particularly visual (Dantzker and Callaway 2000; Sawatari and Callaway 2000) and barrel cortex (Schubert et al. 2001; Shepherd et al. 2003; Bureau et al. 2006; Brill and Huguenard 2009; Xu and Callaway 2009) and recently also auditory (Barbour and Callaway 2008) and motor cortex (Weiler et al. 2008; Yu et al. 2008). LSPS has also proven useful for characterizing cortical "circuitopathies" in neurological disease models (Jin et al. 2006; Bureau et al. 2008; Wood et al. 2009). The efficacy of glutamate-based photostimulation for circuit mapping (in contrast to electrical stimulation) derives from the ability to stimulate neurons with high precision and speed, and without stimulating axons of passage. The interested reader is referred to the primary literature for further specific information about circuit mapping in different brain regions and species, and for detailed descriptions of recent technical innovations, including two-photon-based LSPS (Nikolenko et al. 2007; Matsuzaki et al. 2008), high-speed scanning (Boucsein et al. 2005; Shoham et al. 2005), and LSPS combined with imaging of activity-dependent intrinsic signals (Llano et al. 2009). This chapter presents a general approach and a set of guidelines for quantitative circuit mapping using "standard" LSPS methods based on single-photon glutamate uncaging using a UV laser, a pair of scanning mirror galvanometers, a patch-clamp setup, and open-source data acquisition software (Suter et al. 2010). For further details of the LSPS system described here, see Shepherd et al. (2003, 2005), Shepherd and Svoboda (2005), Shepherd (2006), and Weiler et al. (2008). For an early general review, see Katz and Dalva (1994).

Protocol

Circuit Mapping by UV Uncaging of Glutamate

This protocol describes the components, assembly, and operation of a laser scanning microscope for UV uncaging, along with experimental methods for mapping synaptic circuits in brain slices.

IMAGING SETUP

LSPS Hardware

Patterned glutamate uncaging requires that a focused beam of light be controlled in both space and time. A standard approach is to use voltage-controlled mirror galvanometers (Model 6210; Cambridge Technology, Inc.) to scan the beam of an ultraviolet (UV) laser (3500 series; DPSS Lasers, Inc.). To assemble the LSPS setup, place a laser and an upright microscope (BX series; Olympus) on a vibration isolation table. (*Caution:* Only qualified operators should use Class 4 UV lasers.) Place in the beam path (in order): Pockels cell (Conoptics), mechanical shutter (VMM-D1; Uniblitz, Vincent Associates), neutral density filter wheel (Edmund Optics), partial reflector (e.g., a glass slide) deflecting a small fraction of the beam to a UV-sensitive photodiode (Edmund Optics), scan mirrors, scan lens (focal length 100 mm, UV-grade, UV-coated, planoconvex lens; Edmund Optics), and dichroic mirror (380DRLP; Chroma Technology). The dichroic, mounted in a video adapter port (U-DPTS; Olympus) on top of the microscope, reflects the UV beam into the optical axis of the microscope and transmits visible light for capturing bright-field or epifluorescence images by a video camera mounted on the adapter. A camera with a large CCD array such as the Kodak 2020 chip (e.g., Retiga 2000 series; Q-Imaging) provides a suitable field of view for mapping circuits in cortical slices. Replace the microscope's tube lens with a UV-transmissive lens (focal length 180 mm, UV-grade, UV-coated, custom lens; CVI Melles Griot). Attach a low-magnification objective lens (4x, 0.16 NA, UPlanApo; Olympus). In this hardware configuration, the optics and microscope are fixed to the vibration isolation table, and the manipulators for the recording electrodes are attached to the stage, which is moved by an *x–y* stage translator (translator, Danaher; controller, Sutter). Alternatively, the microscope can be mounted on an *x–y* translator, with the beam coupled via an optical fiber; in this case, the laser and other optical components "upstream" to the scan mirrors can be moved off the vibration table. Install equipment for patch-clamp recording (amplifier, micromanipulators) and computer interface boards (National Instruments) for controlling devices and sampling signals. Build a small-volume recirculating perfusion system as described elsewhere (Shepherd and Svoboda 2005).

LSPS Software

Download and install Ephus, an open-source suite of software tools for laser-scanning and electrophysiology experiments. Ephus is freely available at http://openwiki.janelia.org, along with detailed information on software installation, calibration, and experimental procedures related to patterned photostimulation (see also Suter et al. 2010).

MATERIALS

CAUTION: See Appendix 6 for proper handling of materials marked with <!>.
See the end of the chapter for recipes for reagents marked with <R>.

Reagents

Artificial cerebrospinal solution (ACSF) <R>
Brain slices (see note to Step 3)
Caged glutamate stock solution <R>
Carbogen (95% O_2, 5% CO_2 mixture) <!>
Cutting solution <R>
High-divalent ACSF <R>
R-CPP (Tocris) or other NMDA antagonist

Equipment

Imaging instrumentation and software (see Imaging Setup)
Slice weights; short (2–4-mm) curved segments of gold wire (0.813-mm-diameter; Alfa Aesar) slightly flattened with a ball-peen hammer
Temperature-controlled recirculating water bath
Vibratome <!>

EXPERIMENTAL METHOD

Preparation of Brain Slices

1. Cut brain slices on a vibratome (0.3-mm-thick) in cutting solution.
2. Incubate the slices in prewarmed (35°C) oxygenated ACSF, for 15–30 min.
3. Store the slices at 22°C until used.

   Familiarity with brain slice procedures is assumed. Slice quality is important. The protocol described here is for mapping inputs to pyramidal neurons in mouse cortex, but it should be easy to adapt for other vertebrate and invertebrate preparations.

Mapping a Neuron's Synaptic Inputs

4. Add 7.5 mL of high-divalent ACSF to the recirculating bath system. Also add 0.01 mM CPP or other NMDA antagonist, and 0.2 mM MNI-glu. Set the water-bath temperature to room temperature (22°C).
5. Transfer a slice to the microscope's recording chamber, orient it approximately as desired for mapping, and anchor it with slice weights.
6. Using a high-power objective, select a neuron and establish a whole-cell recording. To record excitatory synaptic currents, put the amplifier in voltage clamp mode, and hold near the reversal potential for inhibitory conductances, typically near –70 mV.

   To record inhibitory inputs, hold at the reversal potential for glutamatergic conductances, typically ~0 mV. To record synaptic potentials, use current clamp mode.
7. Carefully switch to a low-power objective.

   It is important to keep the objectives clean, as salt accretions can block UV light.
8. Using the stage controller, carefully position the slice within the camera's field of view as desired for mapping (Fig. 1A).

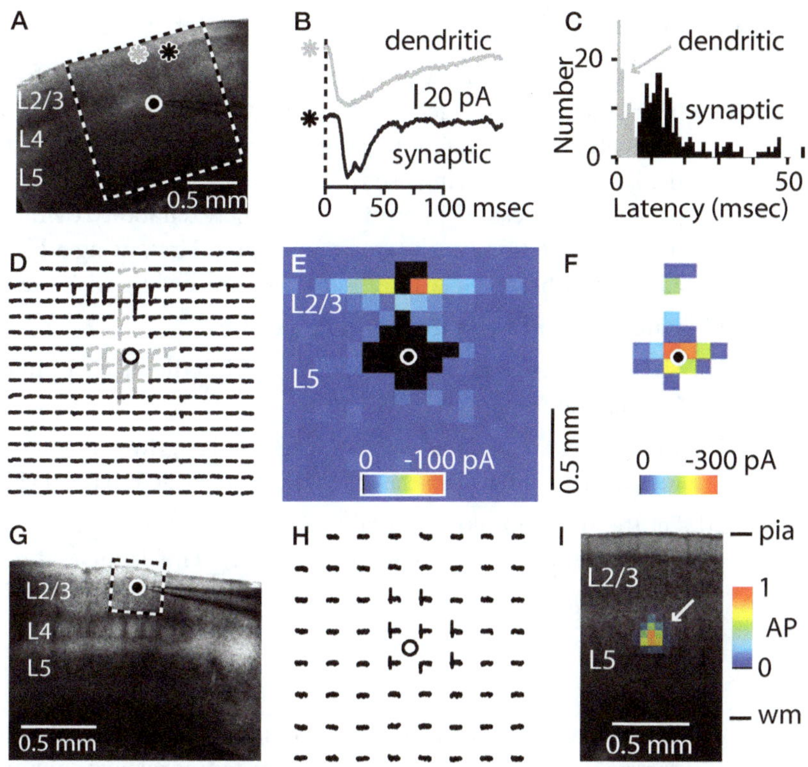

FIGURE 1. Synaptic input maps and excitation profiles. (*A*) Arrangement for recording a synaptic input map from an L5 neuron in a mouse motor cortex brain slice. (L) Cortical layers; (circle) recorded neuron; (asterisks) photostimulus locations; (dashed box) perimeter of 16 x 16 photostimulus grid, 100-µm spacing between sites. (*B*) Dendritic (*top*, gray) and synaptic (*bottom*, black) responses to photostimuli at the locations indicated by the asterisks in *A*. Note shorter onset latency of dendritic than synaptic response. (*C*) Histogram of event latencies. (Gray) Dendritic responses; (black) synaptic inputs. (*D*) Trace map of responses to patterned uncaging. (Gray) Dendritic responses; (black) synaptic or null responses. (*E*) Pixel map of synaptic input amplitude (black traces in *D*). (Black pixels) Dendritic response sites. (*F*) Pixel map of dendritic responses (gray traces in *D*). (*G*) Arrangement for recording an excitation profile from an L2/3 neuron in somatosensory barrel cortex; note L4 barrels. (Circle) Recorded neuron; (dashed box) perimeter of 8 x 8 stimulus grid, 50-µm spacing. (*H*) Excitation profile: trace map showing that action potentials were evoked by stimuli close to the soma (circle). (*I*) Average excitation profile for motor cortex pyramidal neurons (arrow), overlaid on slice image (to scale), illustrating spatial scale and amplitude of photoexcitation of neurons at a stimulus location.

9. Using software tools, select a 16 x 16 stimulus grid with 0.1-mm spacing between sites. Overlay it on the slice image, and align as desired with slice landmarks for mapping (Fig. 1A).
10. Deliver a 1.0-msec photostimulus to a site near the neuron's own apical dendrites. This should generate a rapid-onset, large-amplitude response (Fig. 1A–C).

    See *Troubleshooting*.
11. Deliver stimuli to other sites, away from the neuron's dendrites. At some locations this should evoke synaptic inputs, arriving with longer onset latencies than dendritic responses (Fig. 1A–C).

    See *Troubleshooting*.
12. Adjust the power to the desired mapping level (e.g., 20 mW in the specimen plane).
13. Set the interstimulus interval to 1.0 sec and acquire a map (Fig. 1D–F).
14. Change the map sequence, and acquire a second map.
15. Change the sequence again, and acquire a third map. This is the end of the synaptic input map acquisition.

Mapping a Neuron's Excitation Profile

16. The efficacy of photostimulation can vary greatly for different cell types, developmental ages, laser intensities, caged glutamate concentrations, and other experimental parameters. Therefore it is important to gauge the resolution and intensity of photostimulation for the particular conditions being used. This can be done using cell-attached recordings from neurons in presynaptic areas of interest to map the excitability of their somatodendritic membranes to direct stimulation by glutamate uncaging. First, using a high-power objective, select a neuron and establish a loose-seal (cell-attached) recording with the amplifier in voltage follower mode.
17. Carefully switch to a low-power objective.
18. Using the stage controller, carefully position the slice within the camera's field of view as desired for mapping (Fig. 1G).
19. Using software tools, select an 8 x 8 stimulus grid with 0.05-mm spacing. Overlay it on the slice image, and center it on the soma of the recorded neuron (Fig. 1G).
20. Set the interstimulus interval to 1.0 sec and acquire a map. This should show several perisomatic sites of photoexcitability (Fig. 1H). This is the end of the excitation profile acquisition.

    See *Troubleshooting*.

Analyzing Traces

21. To analyze synaptic input map traces, first exclude traces with dendritic responses. For mouse cortical circuits this can be done by temporal windowing, because dendritic and synaptic response latencies are bimodally distributed (Fig. 1C).
22. Calculate the mean poststimulus response over a short time interval (e.g., 50 msec). Pixels in synaptic input maps are color-coded based on a look-up table, or colormap, rendering dendritic responses as black pixels (Fig. 1E). Dendritic response maps (Fig. 1F) reflect the neuron's dendritic topography.

    Note that any parameter of interest can be mapped (e.g., peak, mean, latency, or number of events).
23. (Optional) Average the three maps to obtain a single map of the neuron's inputs. Averaging is primarily used as a noise-reduction image analysis technique.
24. For excitation profile traces, use spike-detection routines to identify the number and timing of action potentials in the excitation profile traces.
25. Analyze excitation profile data to estimate the intensity and resolution of the photostimulation (Shepherd et al. 2003; Shepherd and Svoboda 2005; Weiler et al. 2008).
26. (Optional) Visualize the intensity and resolution of photostimulation by averaging a group of excitation profiles for a given cell type (e.g., pyramidal neurons) and superimposing this on a slice image (Fig. 1I).

TROUBLESHOOTING

Problem (Steps 10 and 11): Responses are absent or weak.

Solution: Ensure that there is adequate laser intensity in the specimen plane. At least 20 mW should be available for mapping. If not, either the laser output or UV transmission of the optical system should be increased. Ensure that the slices are healthy.

Problem (Step 11): Synaptic responses seem too strong.

Solution: Ensure that synaptic driving is not causing reverberant excitatory activity in response to photostimulation, by performing excitation profiles and synaptic input maps for a range of stimulation intensities. The use of high divalent concentrations and NMDA blockade is usually sufficient to prevent overexcitation, but it may be necessary to work at low stimulus intensities if disynaptic events are evident.

Problem (Steps 10 and 11): Responses appear to be mismapped.
Solution: Shifting of the locations of dendritic responses relative to the location of the soma indicates that the imaging system needs recalibration. Use software tools to calibrate the correct voltage offsets and amplitudes for controlling the mirror angles.

Problem (Step 20): There are no action potentials in the excitation profiles.
Solution: Ensure that the lack of action potentials is not a false negative caused by a poor loose-seal recording. If no action potentials are evoked using the standard stimulus intensities, then increase the power and stimulate directly over the soma. If no action potentials are detected even at the highest powers, reject the neuron as a potentially false negative. If an action potential is recorded at higher but not standard powers, keep it as a true negative.

Problem (Step 20): Excitation profiles indicate different efficacies of photostimulation for different experimental conditions.
Solution: This is the reason for doing excitation profiles; if the efficacy of photostimulation differs for different conditions, such differences should be taken into consideration in the analysis. Otherwise, a change in apparent synaptic strength could be incorrectly ascribed to differences in synaptic connectivity, rather than to photoexcitability. Excitation profiles should be acquired for any new set of conditions.

DISCUSSION

The expected outcome of LSPS mapping of synaptic inputs is a rapidly acquired, highly reproducible set of "images" showing the local sources of synaptic input to individual neurons in brain slices. The expected outcome of excitation profile mapping is a data set that can be analyzed to obtain quantitative estimates for the resolution and intensity of photostimulation. LSPS microscopy based on patterned uncaging offers a unique tool for investigating the "electroanatomy" of synaptic circuits. The advantages of LSPS include (1) high spatial resolution for stimulating neurons near their cell bodies, because axons of passage are not excited; (2) rapid sampling across multiple sites in a stimulus pattern; and (3) no need for genetic manipulations. Its limitations include (1) lack of cell-type selectivity, in that any neurons at the stimulus location are potentially photostimulated; (2) selective stimulation of individually targeted presynaptic neurons is not routinely feasible; and (3) limited use for higher-frequency trains of stimuli at the same location, because of glutamate receptor desensitization and caged compound depletion. The general mapping approach described here can also be used for circuit mapping with two-photon glutamate uncaging (Nikolenko et al. 2007; Matsuzaki et al. 2008). Photostimulation based on channelrhodopsin-2 (ChR2) (Nagel et al. 2003) offers another way to selectively photostimulate (transfected) neurons (Petreanu et al. 2007; Wang et al. 2007). Channelrhodopsin-2-assisted circuit mapping (CRACM) (Petreanu et al. 2007) is closely related to the methods described here. A key difference is that CRACM by design excites ChR2-expressing axons, whereas LSPS based on patterned glutamate uncaging avoids axons of passage.

RECIPES

CAUTION: See Appendix 6 for proper handling of materials marked with <!>.
Recipes for reagents marked with <R> are included in this list.

ACSF (Artificial Cerebrospinal Fluid)

1. Prepare the stock solution as follows.

Reagent	Quantity (for 2 L)
NaCl	14.844 g
$NaHCO_3$	4.2 g
$NaH_2PO_4 \cdot H_2O$	0.345 g
KCl <!>	0.373 g
Glucose	9.01 g

Store at 4°C. Its shelf life is up to 1 wk.

2. Prepare the working solution, which contains standard divalent cation concentrations, 2 mM Ca^{2+} and 1 mM Mg^{2+}.

Reagent	Quantity (for 500 mL)
ACSF stock	500 mL
$CaCl_2$ <!>, 1 M	1.0 mL
$MgCl_2$, 1 M	0.5 mL

Store at 4°C. Its shelf life is up to 1 wk.

Caged Glutamate Stock Solution

Add H_2O to a vial of MNI-caged glutamate (Tocris) to a concentration of 10 mM.

Important: Prepare at room temperature, as the compound will precipitate if vials are placed on ice.

Important: The formula weight is batch specific. Aliquot into single-use aliquots.

Store at –20°C. Its shelf life is months to years.

Cutting Solution

1. Prepare the stock solution as follows.

Reagent	Quantity (for 2 L)
$NaHCO_3$	84.01 g
$NaH_2PO_4 \cdot H_2O$	137.99 g
KCl, 3 M	1.667 mL
$CaCl_2$ <!>, 1 M	1 mL
$MgCl_2 \cdot 6H_2O$, 1 M	14 mL
Glucose	9.01 g

Store at 4°C. Its shelf life is up to 1 wk.

2. Prepare the working solution.

Reagent	Quantity (for 500 mL)
Cutting solution stock	500 mL
Choline choride	7.68 g
Ascorbic acid	1.15 g
Pyruvic acid	0.17 g

Store at 4°C. Its shelf life is up to 1 wk.

High-Divalent ACSF

This contains high divalent cation concentrations (4 mM Ca^{2+} and 4 mM Mg^{2+}).

Reagent	Quantity (for 500 mL)
ACSF stock	500 mL
$CaCl_2$ <!>, 1 M	2 mL
$MgCl_2$, 1 M	2 mL

Store at 4°C. Its shelf life is up to 1 wk.

ACKNOWLEDGMENTS

This work was funded by the International Rett Syndrome Foundation, National Institutes of Health (grants NS061963 and EB001464), Simons Foundation, and Whitehall Foundation.

REFERENCES

Barbour DL, Callaway EM. 2008. Excitatory local connections of superficial neurons in rat auditory cortex. *J Neurosci* **28:** 11174–11185.

Boucsein C, Nawrot M, Rotter S, Aertsen A, Heck D. 2005. Controlling synaptic input patterns in vitro by dynamic photo stimulation. *J Neurophysiol* **94:** 2948–2958.

Brill J, Huguenard JR. 2009. Robust short-latency perisomatic inhibition onto neocortical pyramidal cells detected by laser-scanning photostimulation. *J Neurosci* **29:** 7413–7423.

Brivanlou IH, Dantzker JL, Stevens CF, Callaway EM. 2004. Topographic specificity of functional connections from hippocampal CA3 to CA1. *Proc Natl Acad Sci* **101:** 2560–2565.

Bureau I, von Saint Paul F, Svoboda K. 2006. Interdigitated paralemniscal and lemniscal pathways in the mouse barrel cortex. *PLoS Biol* **4:** e382. doi: 10.1371/journal.pbio.0040382.

Bureau I, Shepherd GM, Svoboda K. 2008. Circuit and plasticity defects in the developing somatosensory cortex of FMR1 knock-out mice. *J Neurosci* **28:** 5178–5188.

Callaway EM, Katz LC. 1993. Photostimulation using caged glutamate reveals functional circuitry in living brain slices. *Proc Natl Acad Sci* **90:** 7661–7665.

Dantzker JL, Callaway EM. 2000. Laminar sources of synaptic input to cortical inhibitory interneurons and pyramidal neurons. *Nat Neurosci* **3:** 701–707.

Deleuze C, Huguenard JR. 2006. Distinct electrical and chemical connectivity maps in the thalamic reticular nucleus: Potential roles in synchronization and sensation. *J Neurosci* **26:** 8633–8645.

Jin X, Prince DA, Huguenard JR. 2006. Enhanced excitatory synaptic connectivity in layer V pyramidal neurons of chronically injured epileptogenic neocortex in rats. *J Neurosci* **26:** 4891–4900.

Kato G, Kawasaki Y, Ji RR, Strassman AM. 2007. Differential wiring of local excitatory and inhibitory synaptic inputs to islet cells in rat spinal lamina II demonstrated by laser scanning photostimulation. *J Physiol* **580:** 815–833.

Katz LC, Dalva MB. 1994. Scanning laser photostimulation: A new approach for analyzing brain circuits. *J Neurosci Methods* **54:** 205–218.

Kim G, Kandler K. 2003. Elimination and strengthening of glycinergic/GABAergic connections during tonotopic map formation. *Nat Neurosci* **6:** 282–290.

Lam YW, Sherman SM. 2005. Mapping by laser photostimulation of connections between the thalamic reticular and ventral posterior lateral nuclei in the rat. *J Neurophysiol* **94:** 2472–2483.

Llano DA, Theyel BB, Mallik AK, Sherman SM, Issa NP. 2009. Rapid and sensitive mapping of long-range connections in vitro using flavoprotein autofluorescence imaging combined with laser photostimulation. *J Neurophysiol* **101:** 3325–3340.

Matsuzaki M, Ellis-Davies GC, Kasai H. 2008. Three-dimensional mapping of unitary synaptic connections by two-photon macro photolysis of caged glutamate. *J Neurophysiol* **99:** 1535–1544.

Nagel G, Szellas T, Huhn W, Kateriya S, Adeishvili N, Berthold P, Ollig D, Hegemann P, Bamberg E. 2003. Channelrhodopsin-2, a directly light-gated cation-selective membrane channel. *Proc Natl Acad Sci* **100:** 13940–13945.

Nikolenko V, Poskanzer KE, Yuste R. 2007. Two-photon photostimulation and imaging of neural circuits. *Nat Methods* **4:** 943–950.

Petreanu L, Huber D, Sobczyk A, Svoboda K. 2007. Channelrhodopsin-2-assisted circuit mapping of long-range callosal projections. *Nat Neurosci* **10:** 663–668.

Pettit DL, Helms MC, Lee P, Augustine GJ, Hall WC. 1999. Local excitatory circuits in the intermediate gray layer of the superior colliculus. *J Neurophysiol* **81:** 1424–1427.

Sawatari A, Callaway EM. 2000. Diversity and cell type specificity of local excitatory connections to neurons in layer 3B of monkey primary visual cortex. *Neuron* **25:** 459–471.

Schubert D, Staiger JF, Cho N, Kotter R, Zilles K, Luhmann HJ. 2001. Layer-specific intracolumnar and transcolumnar functional connectivity of layer V pyramidal cells in rat barrel cortex. *J Neurosci* **21:** 3580–3592.

Shepherd GMG. 2006. Imaging cortical circuits with laser scanning photostimulation. In *Visualizing large-scale patterns of activity in the brain: Optical and electrical signals* (ed. Buzsáki G), pp. 54–59. Society for Neuroscience, Washington, DC.

Shepherd GMG, Svoboda K. 2005. Laminar and columnar organization of ascending excitatory projections to layer 2/3 pyramidal neurons in rat barrel cortex. *J Neurosci* **25:** 5670–5679.

Shepherd GMG, Pologruto TA, Svoboda K. 2003. Circuit analysis of experience-dependent plasticity in the developing rat barrel cortex. *Neuron* **38:** 277–289.

Shepherd GMG, Stepanyants A, Bureau I, Chklovskii DB, Svoboda K. 2005. Geometric and functional organization of cortical circuits. *Nat Neurosci* **8:** 782–790.

Shoham S, O'Connor DH, Sarkisov DV, Wang SS. 2005. Rapid neurotransmitter uncaging in spatially defined patterns. *Nat Methods* **2:** 837–843.

Suter BA, O'Connor T, Iyer V, Petreanu LT, Hooks BM, Kiritani T, Svoboda K, Shepherd GMG. 2010. *Ephus:* Multipurpose data acquisition software for neuroscience experiments. *Front Neurosci Methods* **4:** 1–12.

Wang H, Peca J, Matsuzaki M, Matsuzaki K, Noguchi J, Qiu L, Wang D, Zhang F, Boyden E, Deisseroth K, et al. 2007. High-speed mapping of synaptic connectivity using photostimulation in Channelrhodopsin-2 transgenic mice. *Proc Natl Acad Sci* **104:** 8143–8148.

Weiler N, Wood L, Yu J, Solla SA, Shepherd GMG. 2008. Top-down laminar organization of the excitatory network in motor cortex. *Nat Neurosci* **11:** 360–366.

Wood L, Gray NW, Zhou Z, Greenberg ME, Shepherd GM. 2009. Synaptic circuit abnormalities of motor-frontal layer 2/3 pyramidal neurons in an RNA interference model of methyl-CpG-binding protein 2 deficiency. *J Neurosci* **29:** 12440–12448.

Xu X, Callaway EM. 2009. Laminar specificity of functional input to distinct types of inhibitory cortical neurons. *J Neurosci* **29:** 70–85.

Yu J, Anderson CT, Kiritani T, Sheets PL, Wokosin DL, Wood L, Shepherd GM. 2008. Local-circuit phenotypes of layer 5 neurons in motor-frontal cortex of YFP-H mice. *Front Neural Circuits* **2:** 1–8.

44 Two-Photon Mapping of Neural Circuits

Volodymyr Nikolenko, Elodie Fino, and Rafael Yuste
Howard Hughes Medical Institute, Department of Biological Sciences, Columbia University, New York, New York 10027

ABSTRACT

For the study of neural circuits, it would be ideal to be able to discern in a living preparation the interconnections among neurons. Working toward that aim, we have developed a method, using two-photon uncaging of glutamate in brain slices, by which we can detect neurons that are presynaptic to any particular cell. This method has single-cell resolution and can be applied systematically and quickly to test entire populations of neurons, making it possible to perform online functional analysis of connectivity.

INTRODUCTION

We have developed a method for the selective photostimulation of individual neurons that uses two-photon uncaging of chemically caged neurotransmitters. Molecules of caged (inactive) neurotransmitter are bath-applied to the sample, and two-photon absorption of near-infrared light triggers the release of biologically active transmitters. For example, spatially selective illumination of caged glutamate, an excitatory neurotransmitter, can locally release enough free glutamate to depolarize a specific neuron to its action potential firing threshold.

Conceptually similar experiments that also map neuronal circuitry have traditionally been performed using one-photon uncaging of caged glutamate irradiated with ultraviolet (UV) light (Callaway and Katz 1993; Dantzker and Callaway 2000; Schubert et al. 2001; Shepherd et al. 2003; and Chapter 43), as well as recently developed genetic photosensitive probes (see Chapter 76). Unfortunately, one-photon photostimulation suffers from limited spatial resolution of one-photon absorption, as well as extensive scattering of UV or visible light in thick biological tissue specimens. As a result, in practice, one-photon photostimulation techniques cannot provide single-cell specificity.

In contrast, two-photon photostimulation (specifically, two-photon uncaging of glutamate) provides outstanding spatial resolution and has been used to map α-amino-3-hydroxy-5-methyl-4-isoxazolepropionic acid (AMPA) receptor distribution (Matsuzaki et al. 2001, and Chapter 25) and for probing spine function (Araya et al. 2006; Matsuzaki et al. 2004). The method works well for

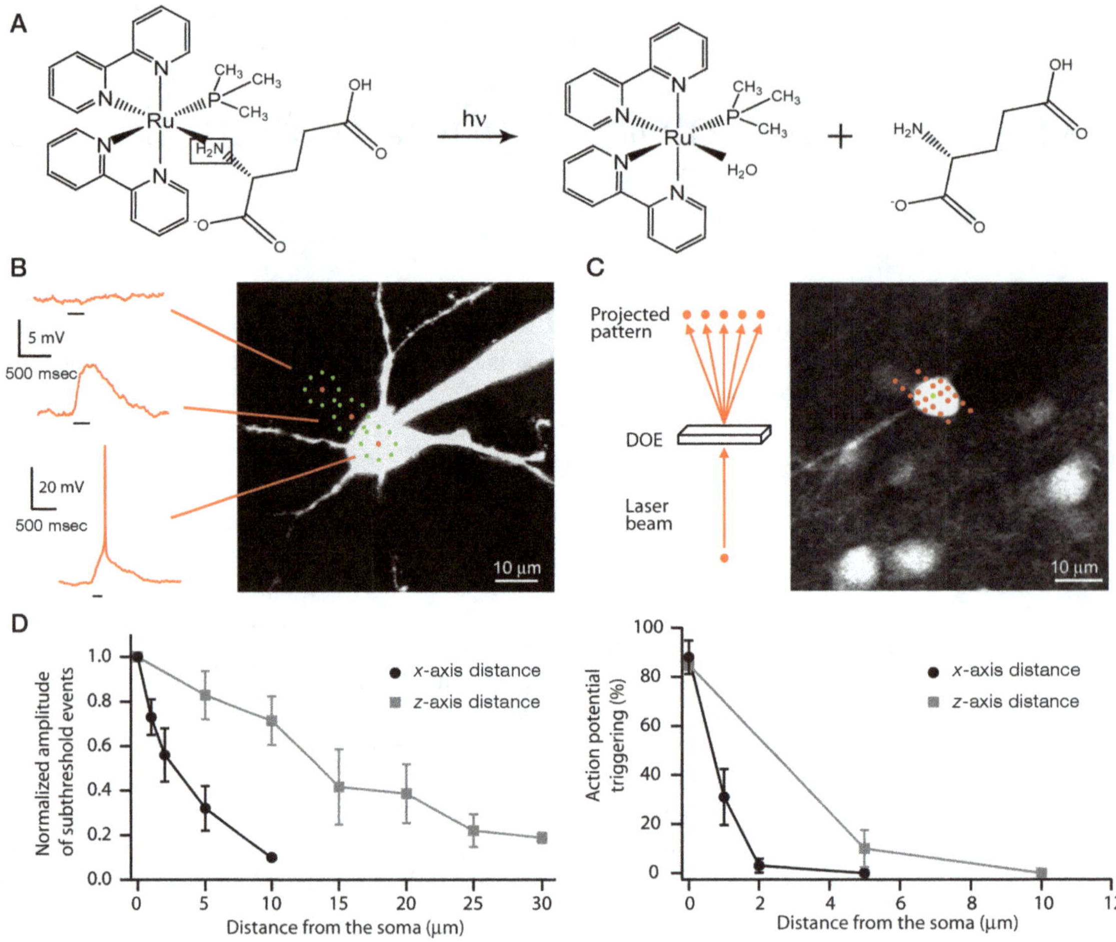

FIGURE 1. Optically activating neurons with two-photon uncaging of glutamate. (*A*) Structure of ruthenium-bipyridine-trimethylphosphine-glutamate (RuBi-glutamate) and glutamate photorelease. (*B*) Layer 2/3 pyramidal cells were filled with Alexa 594. Position of the uncaging targets (eight subtargets [color] encircling the central target [color, not stimulated]) to achieve temporal multiplexing. When placed on the somata, RuBi-glutamate uncaging elicited an action potential. When the targets were moved away from the soma, uncaging elicited only subthreshold depolarizations, and no response was evoked 10 μm away. (*C*) Neuron (loaded with AM-ester calcium indicator) stimulated with a linear diffractive optical element (DOE). The DOE splits the laser beam spatially, illuminating five spots simultaneously. In this example, the five beamlets are repositioned rapidly to four locations on the target cell soma. (*D*) Lateral (*x–y* plane) and axial (*z*-plane) resolution of RuBi-glutamate uncaging to elicit either supra- or subthreshold events. Note how subthreshold depolarizations were strongly reduced when targets were placed 10 μm away from the soma in the *x*-axis and 30 μm away in the *z*-axis (*left panel*). Uncaging did not elicit an action potential if targets were moved from the soma 2 μm away in the *x*-axis and 5 μm away in the *z*-axis (*right panel*).

stimulating a single dendritic compartment as a way to simulate excitatory postsynaptic potentials or currents (EPSPs or EPSCs). However, because of the small point spread function of two-photon absorption, it is difficult to photorelease enough glutamate to sufficiently depolarize the entire neuron so that it fires an action potential. As a practical solution to this problem, we developed a multiplexed uncaging technique (Nikolenko et al. 2007). In this technique the laser beam is split into separate closely spaced beamlets that create several uncaging centers spatially separated in the image plane near the neuron of interest (i.e., multiplex photostimulation in space). In addition, the laser beam is moved rapidly around the soma (i.e., multiplex photostimulation in time). As can be seen in Figure 1, this approach allows for the triggering of reliable action potentials with excellent axial and tangential spatial resolution of stimulation.

The laser beam is split by placing a diffractive optical element (DOE) into the plane optically conjugated to the back aperture of the microscope objective (and therefore the plane conjugated to the scanning mirrors). This is similar to the spatial light modulator (SLM) strategy described in Chapter 47, in which the full optical scheme is presented. In brief, the DOE basically serves as a simple single-element diffractive spatial light modulator with a fixed pattern. We used a DOE (model SLH-505Xa-(0.23)-(780) from StockerYale) that splits the laser into a linear pattern of five spots, although if a more powerful laser is used, the beam could be multiplexed into a larger number of beamlets. Multiplexing in time is a built-in feature of the custom software that is implemented by our group (a free version and a basic description are available at www.twophoton.com/software), and in the simplest case it constitutes "jumping" of the laser beam (or corresponding group of beamlets created by the DOE) around cell somata.

PRACTICAL EXAMPLES

Choice of Caged Glutamate

For this technique, several commercially available caged compounds were tested as well as compounds developed in collaboration with the Etchenique laboratory (Zayat et al. 2011). We compared the commercially available 4-methoxy-7-nitroindolinyl-caged-L-glutamate (MNI-caged glutamate) and our newly developed ruthenium-bipyridine-trimethylphosphine-glutamate (RuBi-glutamate).

MNI-glutamate, available commercially since 2003, is the form of caged glutamate used most often for two-photon uncaging. It can be used for one-photon stimulation in the UV range and two-photon stimulation at 725 nm. However, MNI-glutamate needs to be applied to the tissue at relatively high (i.e., millimolar) concentrations for effective two-photon uncaging and, unfortunately, it is a very effective antagonist of GABAergic transmission at these concentrations (Fino et al. 2009). As MNI-glutamate has been well characterized, we will focus on RuBi-glutamate and compare it to MNI-glutamate.

RuBi-glutamate is a new compound synthesized by the Etchenique laboratory (Fino et al. 2009; Salierno et al. 2010; Zayat et al. 2011). This caged glutamate is composed of a ruthenium center with six coordination positions, four of which are occupied by two bidentate bipyridines, the fifth by a trimethylphosphine, and the sixth by a molecule of L-glutamate, coordinated through its amino group (Fig. 1A). Following one- or two-photon excitation, RuBi-glutamate photocleaves within nanoseconds, releasing glutamate with a quantum yield of ~0.13 at pH 7 and an extinction coefficient of >4000 $\text{M}^{-1}\text{cm}^{-1}$ at 473 nm (Salierno et al. 2010). For two-photon uncaging, RuBi-glutamate needs to be illuminated at 800 nm and can be used at 300 µM, a much lower concentration than MNI-glutamate, but it requires more laser power than MNI-glutamate. At 300 µM, RuBi-glutamate still has a significant antagonist effect on endogenous GABAergic currents, but this effect (close to 50%) is much smaller compared with that of MNI-glutamate (~80% at 300 µM, ~100% at the effective concentration of 2.5 mM) (Fino et al. 2009).

Two-photon uncaging of RuBi-glutamate generated brief depolarizations, but with a single laser spot, the evoked depolarizations were too small to reliably cause the cell to fire action potentials. As mentioned above, the use of the temporal and spatial multiplexed uncaging technique permits reliable action potentials to be evoked in neurons (Fig. 1B). In this example, the beam of the laser is sequentially moved into eight subtargets. In addition, a DOE is used to spatially multiplex the laser beam into five closely spaced beamlets (Fig. 1C; Nikolenko et al. 2007). This beamlet array is then directed sequentially onto eight subtargets on top of the cell (Fig. 1B), releasing more glutamate and generating stronger cellular responses that lead to action potentials. Using this protocol, two-photon uncaging of RuBi-glutamate (on average, ~70 msec, 150–400 mW on the sample) was able to reliably generate action potentials (90.0 ± 2.6% successes, n = 11 cells; Fig. 1B,D) without causing damage to the neuron. In addition, the axial and tangential spatial resolution of stimulation was good because we observed a strong decrease in the occurrence of evoked action potentials as the uncaging targets were moved only 2 µm from the edge of the soma in the x–y plane (3.0 ± 2.8% of occurrence of an action potential, n = 7 neurons) and 5 µm above the soma in the z-plane (10.0 ± 7.6%, n = 4 neurons; Fig. 1D).

TABLE 1. Comparison of EPSPs evoked by Glu uncaging with spontaneous EPSPs

	Amplitude (mV)	Rate of rise (mV/msec)	Rise time, 10%–90% (msec)	Decay time (msec)
RuBi-glu (300 μM)	1.20 ± 0.09 (n = 65)	0.092 ± 0.01 (n = 65)	13.98 ± 1.13 (n = 65)	23.11 ± 1.79 (n = 59)
MNI-glu (2.5 mM)	0.92 ± 0.03 (n = 127)	0.04 ± 0.002 (n = 127)	19.08 ± 1.24 (n = 127)	53.43 ± 4.24 (n = 114)
Sr^{2+}-mEPSPs	0.55 ± 0.006 (n = 1317)	0.06 ± 0.001 (n = 1316)	8.20 ± 0.09 (n = 1316)	16.06 ± 0.21 (n = 1217)

Pairwise comparisons between MNI-glutamate (2.5 mM) and RuBi-glutamate, MNI-glutamate and mEPSP, and RuBi-glutamate and mEPSPs were statistically significant for all four variables with $p < 0.001$. Values expressed as mean ± S.E.M.

MNI-glutamate has also been used to activate single spines (Araya et al. 2006). RuBi-glutamate allows two-photon stimulation of single spines as well, with very good spatial resolution. The EPSPs evoked with RuBi-glutamate uncaging are similar in amplitude to events recorded after two-photon uncaging of MNI-glutamate. Nevertheless, the kinetics are faster and more closely resemble the kinetics of spontaneous mEPSPs (Table 1).

Mapping Connections in Brain Slices

We have used this photostimulation technique to map the synaptic inputs to different neurons within a brain slice by performing current-clamp recordings on selected neurons while sequentially photostimulating individual neurons in the surrounding area of a brain slice (Fig. 2). We usually used neocortical slices bulk-loaded with one of the AM ester calcium and magnesium indicators to visualize cell somata and target photostimulation following automatic recognition of cell contours. Typically, we load with MagIndo-1 AM, because it produces very robust staining and the dye can be excited for imaging by the same 725-nm beam that is used for MNI-glutamate uncaging. MagFura-2 produces similarly strong staining and is effectively excited by 800-nm light that is optimal for uncaging of RuBi-glutamate. A second approach is to recognize cell contours using transgenic animals, in which certain (or all) neurons are labeled by one of the fluorescent proteins. A third strategy avoids fluorescent imaging altogether and relies instead on cell soma recognition in differential interference contrast (DIC) images of the brain slice.

In a typical experiment with a 10x objective, automatic cell recognition software detects approximately 700–5000 neurons in a single focal plane of a P14 mouse brain slice, encompassing all six cortical layers. One could easily stimulate every single detected cell, as we do in some experiments. However, in a sequential stimulation protocol, it is often desirable to use sufficiently long waiting periods between stimulations of different cells (such as several seconds to minimize effects of plasticity). Thus, to complete the protocol in a reasonable time it makes sense to either restrict the number of stimulated neurons in a pseudorandom fashion, stimulate only a preselected subgroup of neurons (for example labeled by a genetic marker), or use a protocol with more beamlets.

In typical mapping experiments when the conditions are close to physiological, two distinct classes of postsynaptic responses are found to time-lock to the uncaging pulses (Fig. 2A,B, and Table 1). Note that this analysis excludes hyperpolarizing/inhibitory events that can be observed when RuBi-glutamate is used, but that cannot be observed when concentrations of MNI-caged glutamate are used that are sufficient to block GABA transmission as described above. These events are similar to the two types of events described during one-photon glutamate uncaging:

- compound monosynaptic excitatory postsynaptic potentials (EPSPs) induced by action potentials in presynaptic photostimulated cells (i.e., true positives; Fig. 2A)
- larger, slower depolarizations with shorter latencies, caused by direct uncaging onto distal den-

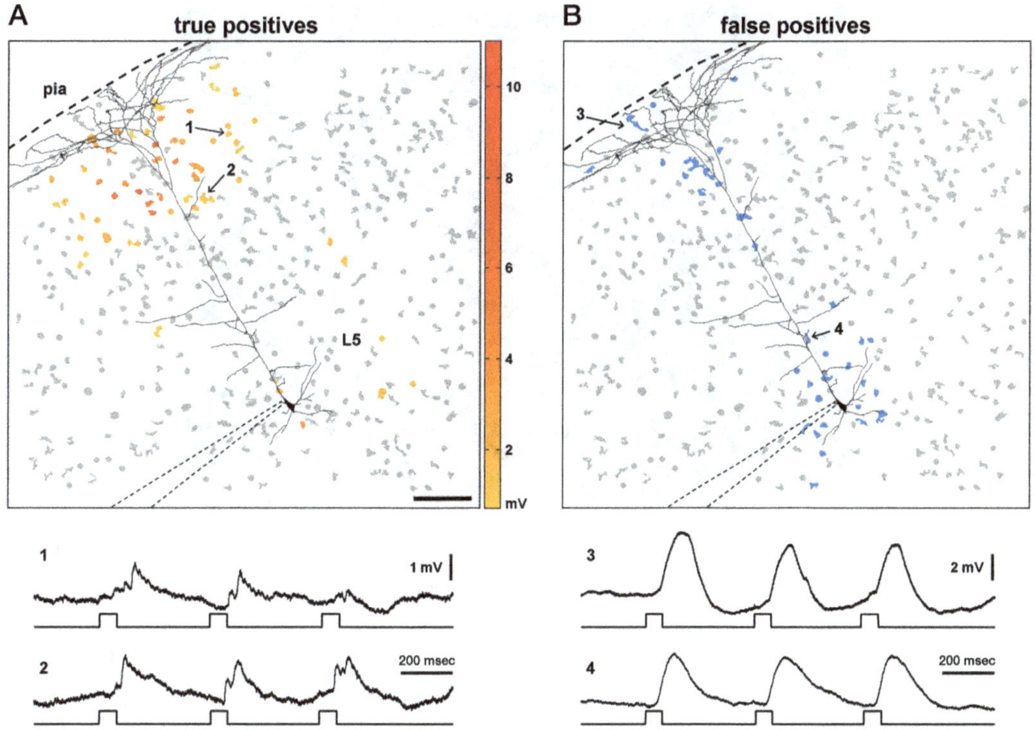

FIGURE 2. Mapping connections in brain slices with single-cell resolution. (*A*) Input map for a layer-5 pyramidal neuron and superimposed morphological reconstruction of its dendritic tree. Gray areas outline stimulated neurons. Colored areas are outlines of all "true positive" cells, color-coded according to peak EPSP amplitude. Scale bar, 100 µm. Dotted outline is the patch pipette. (*Lower panel*) Voltage recordings (*upper traces*) from the patched cell during uncaging (laser pulses in *lower traces*) in locations labeled by arrows in input map. Holding potential –65 mV. (*B*) Map of false positive signals for the same neuron (blue areas). Note that they track the morphology of the dendritic tree. (*Lower panel*) Recordings of false positive signals during uncaging in corresponding locations labeled by arrows. Note the difference in kinetics and amplitude between true and false positive signals.

drites of the patch-clamped cells (i.e., false positives; Fig. 2B), in cases when those dendrites cross near a stimulated area and thus are stimulated directly

It is easy to distinguish true positives from false ones based on their statistically different onset kinetics and latencies to the uncaging pulse (Table 2). Moreover, it is possible to definitively confirm the manual or semiautomatic sorting of true and false positive responses by applying the sodium channel blocker tetrodotoxin (TTX, 1 mM). TTX blocks action potential generation in the brain slice and prevents all evoked synaptic release, revealing solely false positive responses. Alternatively, one

TABLE 2. Analysis of true and false positive responses (manual sorting)

	True positives				False positives			
	Amplitude first EPSP (mV)	Peak amplitude (mV)	Latency after laser pulse (msec)	Rise speed (mV/msec)	Amplitude first EPSP (mV)	Peak amplitude (mV)	Latency after laser pulse (msec)	Rise speed (mV/msec)
Mean	1.04	3.21	57.85	0.066	5.69	5.76	30.00	0.056
Standard deviation	0.26	1.16	6.46	0.021	2.00	1.83	9.35	0.016
Standard error	0.11	0.52	2.89	0.009	0.89	0.82	4.18	0.007
Two-tail *p*-value	1.77×10^{-13}	2.84×10^{-5}	9.89×10^{-20}	0.0365				

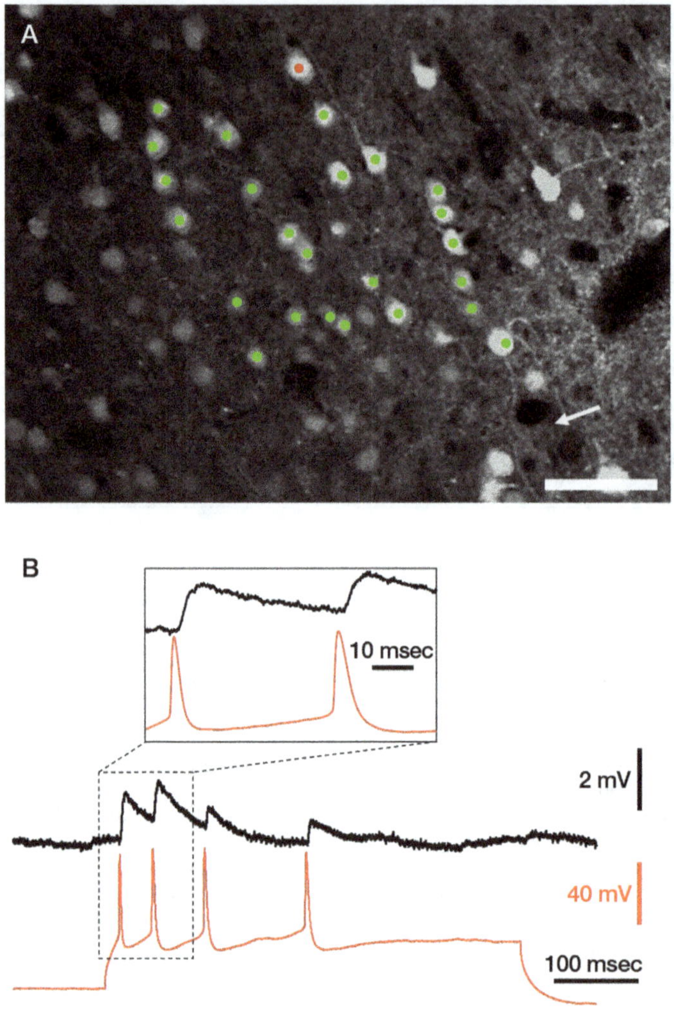

FIGURE 3. Confirmation of a connected pair of neurons. (*A*) True positive signals are monosynaptic connections. A postsynaptic neuron was patched (not in focus; arrow) and all neurons labeled by green dots were sequentially stimulated. The neuron labeled by a red dot gave a true positive signal and was subsequently patch-clamped. Note the significant distance between both neurons. Scale bar, 50 µm. (*B*) Dual recording from these two neurons reveals monosynaptic connection. Action potentials induced by current injection in the putative presynaptic neuron (red trace); neuron labeled by the red dot in *A* causes time-locked EPSPs in the postsynaptic cell (black trace). *Inset*: note correspondence between onset of action potentials and EPSPs. Both neurons were held at –63 mV. In this series of experiments, the average delay was 1.83 ± 0.50 msec, ranging from 0.79 to 3.49 msec for the connected pairs (*n* = 5).

can repeat the mapping protocol in low Ca^{2+} conditions (described in Kotter et al. 2005) to eliminate synaptic transmission by inhibiting presynaptic release, and also to definitively separate true positive from false positive signals.

Finally, one can use dual whole-cell recordings to confirm that detected true positive signals are indeed generated by photostimulated distal neurons monosynaptically connected to the patch-clamped postsynaptic neuron. Figure 3 shows an example of such a confirmation experiment: action potentials elicited in putative presynaptic neurons reliably triggered excitatory postsynaptic responses in the postsynaptic cell with short delays corresponding to monosynaptic connections.

SUMMARY

The described method of two-photon circuit mapping, using a space–time multiplexed photostimulation beam, is a highly specific tool for revealing neuronal connectivity and possibly for the development of novel therapeutic approaches based on highly selective optical stimulation of the individual neurons in the living brain.

ACKNOWLEDGMENTS

We thank Kira E. Poskanzer and Adam Packer for additional control experiments and further development of the protocol.

REFERENCES

Araya R, Jiang J, Eisenthal KB, Yuste R. 2006. The spine neck filters membrane potentials. *Proc Natl Acad Sci* **103:** 17961–17966.

Callaway EM, Katz LC. 1993. Photostimulation using caged glutamate reveals functional circuitry in living brain slices. *Proc Natl Acad Sci* **90:** 7661–7665.

Dantzker JL, Callaway EM. 2000. Laminar sources of synaptic input to cortical inhibitory interneurons and pyramidal neurons. *Nat Neurosci* **3:** 701–707.

Fino E, Araya R, Peterka DS, Salierno M, Etchenique R, Yuste R. 2009. RuBi-Glutamate: Two-photon and visible-light photoactivation of neurons and dendritic spines. *Front Neural Circuits* **3:** 2.

Kotter R, Schubert D, Dyhrfjeld-Johnsen J, Luhmann HJ, Staiger JF. 2005. Optical release of caged glutamate for stimulation of neurons in the in vitro slice preparation. *J Biomed Opt* **10:** 11003.

Matsuzaki M, Ellis-Davies GC, Nemoto T, Miyashita Y, Iino M, Kasai H. 2001. Dendritic spine geometry is critical for AMPA receptor expression in hippocampal CA1 pyramidal neurons. *Nat Neurosci* **4:** 1086–1092.

Matsuzaki M, Honkura N, Ellis-Davies GC, Kasai H. 2004. Structural basis of long-term potentiation in single dendritic spines. *Nature* **429:** 761–766.

Nikolenko V, Poskanzer KE, Yuste R. 2007. Two-photon photostimulation and imaging of neural circuits. *Nat Methods* **4:** 943–950.

Salierno M, Marceca E, Peterka DS, Yuste R, Etchenique R. 2010. A fast Ruthenium polypyridine cage complex photoreleases glutamate with visible or IR light in one and two photon regimes. *J Inorg Biochem* **104:** 418–422.

Schubert D, Staiger JF, Cho N, Kotter R, Zilles K, Luhmann HJ. 2001. Layer-specific intracolumnar and transcolumnar functional connectivity of layer V pyramidal cells in rat barrel cortex. *J Neurosci* **21:** 3580–3592.

Shepherd GM, Pologruto TA, Svoboda K. 2003. Circuit analysis of experience-dependent plasticity in the developing rat barrel cortex. *Neuron* **38:** 277–289.

Zayat L, Baraldo LM, Etchenique R. 2011. Uncaging with visible light: Inorganic caged compounds. In *Imaging: A laboratory manual* (ed. Yuste R), pp. 897–902. Cold Spring Harbor Laboratory Press, Cold Spring Harbor, NY.

45 All-Optical In Situ Histology of Brain Tissue with Femtosecond Laser Pulses

Philbert S. Tsai,[1] Pablo Blinder,[1] Jeffrey A. Squier,[2] and David Kleinfeld[1,3]

[1]*Department of Physics, University of California at San Diego, La Jolla, California 92093;* [2]*Department of Physics, Colorado School of Mines, Golden, Colorado 80401;* [3]*Graduate Program in Neurosciences, University of California at San Diego, La Jolla, California 92093*

ABSTRACT

This chapter describes the application of laser pulses to image and ablate neuronal tissue for the purpose of automated histology. The histology is accomplished in situ using serial two-photon imaging of labeled tissue and removal of the imaged tissue with amplified, femtosecond pulses. Together with the use of endogenous fluorescent indicators and/or deep penetration of antibody labels and organic dyes, this method may be used to automatically image, reconstruct, and vectorize structures of interest across millimeter to centimeter regions of brain with micrometer resolution.

INTRODUCTION

We describe the coupling of plasma-mediated ablation, driven by laser pulses of ~100 fsec, with two-photon microscopy for the purpose of performing serial histology of neuronal tissue (Tsai et al. 2003). This combined procedure, together with related labeling and analysis techniques, extends the histological use of two-photon microscopy to image tissue throughout arbitrarily thick samples.

The ablation of tissue with femtosecond infrared laser pulses requires large optical fluences, in excess of 1 J/cm^2 (Tsai et al. 2009a). Such high fluences can be achieved with commercially available ~100-MHz laser oscillators with ~100-fsec pulse duration and a pulse energy of ~10 nJ when the beam is focused with high numerical aperture objectives to a 1-µm diameter spot. Such beams have been used for fine-scale ablation of subcellular structures (Tirlapur and Konig 2002; Yanik et al. 2004). For large-scale ablation, practical issues suggest the use of microjoule-scale pulse energies. These are obtained by amplifying individual nanojoule seed pulses with an optical amplifier to form pulses with femtosecond duration, but with energies of 1–100 µJ, albeit at repetition rates of only 1–100 kHz. Amplified pulses have been used to ablate a wide variety of tissues, including cornea (Loesel et al. 1996; Oraevsky et al. 1996; Juhasz et al. 1999; Lubatschowski et al. 2000; Maatz et al. 2000), dental tissue (Loesel et al. 1996; Neev et al. 1996), skin (Frederickson et al. 1993), and brain

(Loesel et al. 1996, 1998; Suhm et al. 1996; Goetz et al. 1999; Tsai et al. 2003). The focus in our laboratory is on data acquisition from brain tissue that has been structurally labeled. Examples in this chapter are taken from results in our laboratory.

Protocol

Preparation of Mouse Brain Tissue and Imaging Using All-Optical In Situ Histology with Femtosecond Laser Pulses

IMAGING SETUP

Integration of the imaging and ablation optomechanical setup is shown schematically in Figure 1A. The individual components are as follows.

Imaging laser source: One of two commercially available Ti:sapphire laser oscillators are used. The Mira-F900, pumped by a Verdi V-10 (Coherent Inc.) delivers 10-nJ 100-fsec pulses at a repetition rate of 78 MHz. Alternatively, the Tsunami HP, pumped by a 15-W Millennia (Spectra-Physics) delivers 25-nJ 100-fsec pulses at a repetition rate of 80 MHz. Each system has a separate pump and oscillators, instead of all-in-one systems, so that the pump beam focus in the Ti:sapphire rod may be realigned to compensate for potential deterioration in laser power over many months of operation.

Ablation laser source: A commercially available Ti:sapphire amplified laser system is used. The Libra is an all-in-one system that includes a Vitesse seed oscillator and a regenerative Ti:sapphire amplifier pumped with an Evolution Q-switched Nd:YLF laser (Coherent Inc.). This system delivers 200-µJ 100-fsec pulses at a repetition rate of 5 kHz. A potential alternative ablation source is the RegA 9000 amplifier system (Coherent Inc.). This system provides a 200-kHz train of pulses with a few microjoules of energy, which is sufficient for plasma-mediated ablation, although it is slightly less energy than we use for our current experimental configurations.

Two-photon laser scanning microscope: A custom-built microscope is used based on a design derived from the original in vivo system, which was constructed by Denk (Svoboda et al. 1997). The custom design (Tsai and Kleinfeld 2009) allows integration of the imaging and ablation beams through a polarizing beam splitter (Fig. 1B). The microscope is operated by a custom-designed control software suite, MPScope (Nguyen et al. 2006, 2009), that incorporates a scripting language for automation of the multiple steps in the all-optical histology sequence (Fig. 1C). Imaging is performed using a dipping objective with a high numerical aperture (NA; 0.5–1.0).

Fast programmable translation stages: The tissue is mounted on a set of fast translation stages, capable of speeds up to 8 mm/sec with micrometer accuracy (XYR-4040, Danaher Motion). The stages are interfaced to the computer system through a set of Galil DMC-4040 controllers (Galil Motion Control Inc.) that raster the tissue across the focus of the objective lens. Centimeter-sized tissue volumes may be ablated, and multiple fields of view may be imaged using two-photon laser-scanning microscopy (TPLSM) (Fig. 1A).

MATERIALS

CAUTION: See Appendix 6 for proper handling of materials marked with <!>.
See the end of the chapter for recipes for reagents marked with <R>.

Reagents

Blocking solution <R>
Cyanoacrylate cement <!> (Superbonder 49550; Loctite)

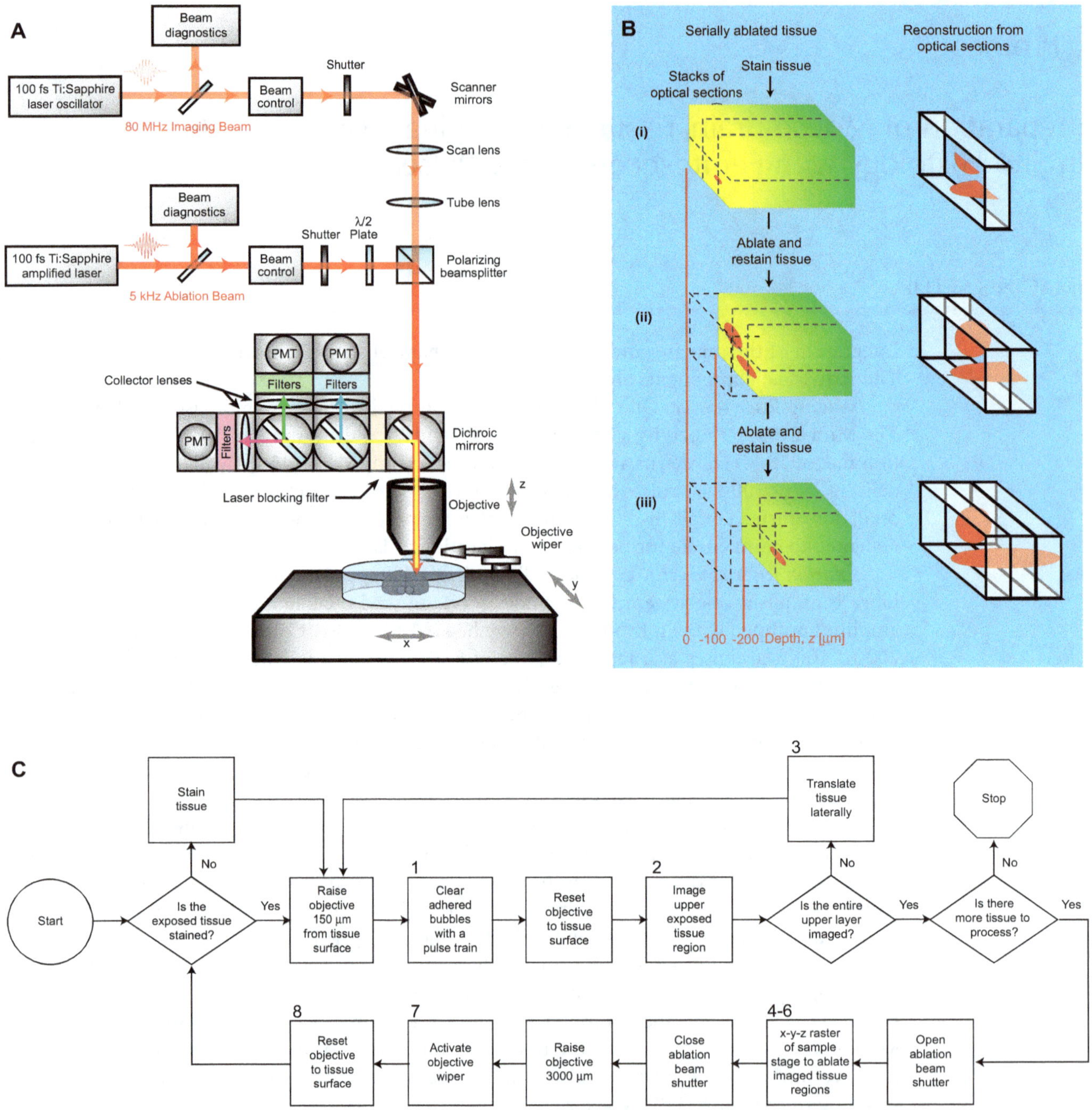

FIGURE 1. Overview of all-optical histology. (*A*) Schematic of laser and imaging systems. An 80-MHz train of 100-fsec pulses from an optically pumped Ti:sapphire oscillator are directed with scanning mirrors to image a preparation placed on a programmable translation stage. A 5-kHz train of amplified 100-fsec pulses is coupled into the microscope with a polarizing beam splitter and directed to the sample for ablation. Beam control of both beams includes intensity and beam profile modifications. The tissue is moved laterally by the programmable translation stages to allow ablation over large areas as well as imaging over multiple fields of view. (*B*) Schematic illustration of the iterative process of all-optical histology. (*i*) The tissue sample (*left column*) containing two fluorescently labeled structures is imaged by conventional two-photon laser scanning microscopy to collect optical sections. Sections are collected until scattering of the incident light reduces the signal-to-noise ratio below a useful value; typically this occurs at ~150 μm in fixed tissue. Labeled features in the resulting stack of optical sections are digitally reconstructed (*right column*). (*ii*) The top of the now-imaged region of tissue is completely ablated with amplified femtosecond laser pulses to expose a new surface for imaging. The sample is again imaged down to a maximal depth, and the new optical sections are added to the previously stored stack. (*iii*) The process of ablation and imaging is again repeated so that the structures of interest can be fully sectioned and reconstructed. (*C*) Flowchart of the all-optical histology process.

4′,6-diamidino-2-phenylindole (DAPI) dilactate <!> (D9564, Sigma-Aldrich)
Fluorescent antibodies
Fluorescent gel solution <R>

Warm the solution to 40°C before the perfusion.

Mice (see Discussion for examples of strains that have been used)
Paraformaldehyde <!>, 4% (w/v), in PBS
Phosphate buffered saline (PBS; Sigma-Aldrich)

Equipment

All-optical histology rig (see Imaging Setup)
Charge-coupled device (CCD) camera (Pantera 1M60, Dalsa Corp.)
Coverslips
Diode array capable of emitting 630-nm light
Perfusion equipment (see Step 4)
Piezoelectric device
Razor blade mounted to a precision translation stage

EXPERIMENTAL METHOD

We focus on the region of cortex that codes the sensory input from the rodent vibrissae, but the procedures are applicable for all brain areas.

Mapping the Region of Interest

Localize and identify barrels in the mouse somatosensory cortex with maps of the intrinsic optical signal in response to single-vibrissa stimulation (Frostig et al. 1993).

1. Stimulate a single vibrissa with a piezoelectric device while the skull is illuminated with 630-nm light from a diode array (Drew and Feldman 2009).
2. Capture the backscattered light on a CCD camera.

   The response is defined as the optical signal averaged over the 0.5–1.5-sec epoch after stimulation, whereas baseline is defined as the signal averaged over the 1.0-sec epoch that starts 1.0 sec before the onset of stimulation.

FIGURE 1. *(Continued)*
The following notes apply: (1) Hydrolysis bubbles formed by the ablation process can adhere to the tissue surface, and diminish subsequent imaging quality. The application of a single burst of 500 amplified laser pulses targeted to the solution above the tissue surface acts to clear the bubbles from the tissue surface. (2) Laser power should be increased approximately exponentially with depth to compensate for scattering and absorption losses. (3) The translation stages are programmed to move to an adjoining lateral field of view for imaging. It is useful to allow roughly 10%–20% spatial overlap with previously imaged fields of view to verify proper registration. (4) The translation stages are programmed to perform a raster scan across the entire lateral extent of the tissue. Parameters for ablation can be varied to trade off between speed and precision. Higher pulse energies can allow faster translation speed and larger step sizes, but may increase the surface roughness of the tissue, thereby potentially reducing the quality of subsequent images. (5) To maximize spatial overlap, so as to minimize surface roughness, the lines of ablation in successive planes can be interlaced by offsetting the start position in each plane by one-half of the lateral step size between lines. (6) The depth of tissue to be removed per ablation round will depend on the tissue scattering properties and the strength of the fluorescent signal. It is important to ablate less depth than was imaged in the previous imaging round, as a means to verify registration. (7) Hydrolysis bubbles resulting from the ablation process can drift upward and adhere to the front surface of the objective. These bubbles are removed by raising the objective and sweeping a lens-tissue wiper across the objective with an automated servo motor. (8) On occasion, the dipping of a dry objective into immersion solution can result in the trapping of a large air bubble under the objective surface. This can be ameliorated by dipping the objective twice without a wipe step between dips.

3. Divide the response signal by the baseline signal at each pixel to generate a map of the functional representation of the stimulated whisker. The response is registered relative to the surface vasculature.

Perfusing with Fluorescent Gel

4. Perfuse rodent transcardially at 0.3 mL/sec with 50 mL of saline (0.9% NaCl) at 37°C.
5. Switch perfusion to 50 mL of 4% (w/v) paraformaldehyde (PFA) in PBS, followed again by 50 mL of saline.
6. Perfuse the animal with a fluorescent gel solution.
7. Extract the brain and postfix it for 20 h in 4% (w/v) PFA in PBS at 4°C.

   Fixation of the fluorescently labeled tissue keeps the label from leaking after slabs of tissue are prepared (in Step 8).

Blocking and Staining

8. Guided by the results from the intrinsic optical signal to define the region of interest, cut an ~1.5-mm-thick coronal slab of tissue from the extracted brain using a razor blade mounted to a precision translation stage.
9. Incubate the slab of tissue for 10–12 h in blocking solution.

   This improves the permeability of the tissue.
10. Incubate in 1 µg/mL of DAPI dilactate and the specific antibodies of choice that are directly conjugated to a fluorophore (Tsai et al. 2009b).

    For example, we have stained with anti-NeuN conjugated to Alexa-594 (custom order, Chemicon); the incubation lasted for 4 d at room temperature with the goal of complete staining throughout the thick slabs. Finally slabs were washed in PBS.

Subblocking and Mounting

11. Subblock the thick coronal slab over the region of interest.

    In our work this is centered on the vibrissa barrel field.
12. Mount the block to a coverslip using a thin layer of cyanoacrylate cement.
13. Once the cement has cured, immerse the entire sample in PBS and transfer it to the all-optical histology rig.
14. Image the sample using a dipping objective having a high numerical aperture (i.e., 0.5–1.0) (Fig. 1A).

Iterative Imaging and Ablation

The iterative process of imaging and ablation are laid out as a flowchart in Figure 1C, and technical details pertaining to particular steps are provided in the legend. Examples of cutting parameters that have been used for sample ablation with 1 and 5 kHz trains of 100-fsec pulses are provided in Table 1. Note that these parameters represent cutting speeds that are overly conservative by a factor ranging from 2 to 10.

DISCUSSION

Laser Alignment

One technical issue concerns the centering and alignment of the imaging and ablation beams. The collimation and alignment of the ablation beam is adjusted by telescopes and routing mirrors located

TABLE 1. Ablation parameters that have been used with 100-fsec pulses

Objective NA	Pulse repetition rate (kHz)	Energy (μJ)	Translation speed (μm/sec)	Lateral step size (μm)	Axial step size (μm)	Volume rate (mm^3/sec)
0.8 Dipping	5	4	4.0	10	10	4×10^{-4}
0.5 Dipping	1	2	4.0	2.5	10	1×10^{-4}
0.5 Dipping	1	8	4.0	10	10	4×10^{-4}
0.3 Dipping	1	22	4.0	10	50	20×10^{-4}
0.2 Air	1	33	1.5	10	20	3×10^{-4}

within the amplified beam path. The microscope objective is temporarily replaced by a 45° mirror that deflects both imaging and ablation beams to a target at least 30 cm away. We align the ablation beam to the imaging beam at both the distant target and the back focal plane of the objective. This ensures colocalization of the foci of the two beams at the center of the imaging field. Alignment can be verified by ablating and imaging the top of a glass slide immersed in a fluorescent dye.

Example Applications

We have used all-optical histology to visualize the three-dimensional structure of the cortical vasculature in mice and the location and identity of surrounding cells. Figure 2A shows the raw overlay of four iterations of imaging and ablation on cortical tissue from a transgenic mouse expressing cyan fluorescent protein in endothelial cells. The technique has been extended to tissue volumes in excess of a cubic millimeter, as shown in Figure 2B for a C57Bl6 mouse with gel-perfused vasculature. Here, each individual imaging field was 256 x 256 μm wide, extended 120 μm into the tissue, and overlapped with neighboring fields by 56 μm laterally and 80 μm axially. A 5 x 5 x 25 array of imaging fields was merged to reproduce the full tissue volume. Additionally, we simultaneously localized all of the cell nuclei using DAPI and identified neurons and non-neuronal cells using an Alexa-conjugated antibody to the neuronal specific antigen NeuN (Fig. 2C).

To show the ability to visualize fine neuronal processes, we have applied all-optical histology to layer 5 pyramidal neurons from mouse neocortex obtained from the transgenic mouse strain B6.Cg-TgN(thy1-YFPH)2Jrs (Feng et al. 2000), available from The Jackson Laboratory. The result of 24 iterations of imaging and ablation along the radial axis of neocortex is shown in Figure 2D. The imaging in each iteration progressed in the dorsal to ventral direction (left to right in the figure), over a single 200 x 200 μm field of view through at least 100 μm of tissue depth. The ablation also progressed from the dorsal to ventral direction, removing an average of 60 μm per iteration. Each image stack was then projected along the coronal direction and overlaid to produce the composite image.

Advantages and Limitations

The primary advantage of the all-optical histological technique is the ability to image large volumes of unfrozen and unembedded soft tissue in an automated, unsupervised fashion. The potential for tissue distortion and misalignment owing to the generation of thin physical sections is eliminated, allowing for straightforward image registration and generation of three-dimensional maps and reconstructions.

The technique is limited to fluorescently labeled tissues whose optical properties are amenable to two-photon microscopy; that is, the scattering depth must be large compared with the axial extent of the focal volume. The speed of TPLSM imaging is currently the time-limiting aspect of this technique (Tsai et al. 2003). At the imaging speeds used in our experiments, acquisition of 1 mm^3 of image data takes 30 min to 5 h, depending on the degree of imaging overlap desired for registration. By comparison, at the ablation speeds used in our experiments, removal of 1 mm^3 of tissue requires 40 min.

Both imaging and ablation times may be shortened in the future. To calculate the theoretical limit for the speed of imaging, we consider the imaging of cellular nuclei labeled with DAPI. At a 1:5

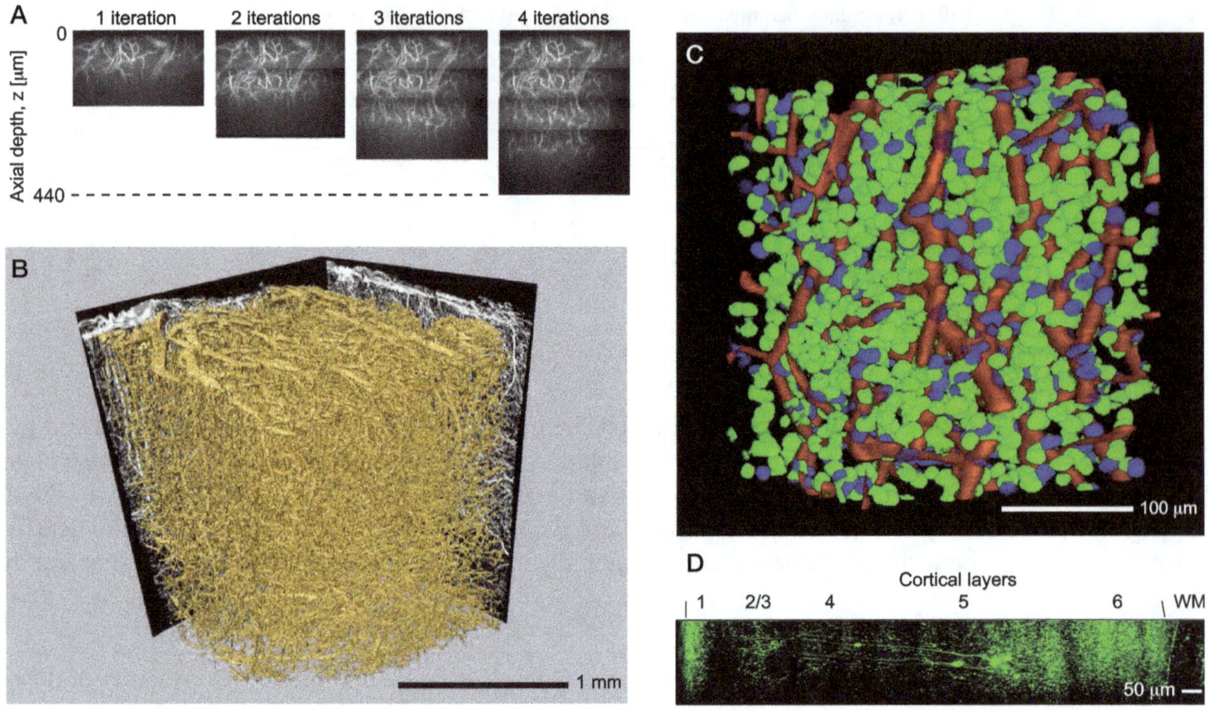

FIGURE 2. Examples of the application of all-optical histology. (*A*) Serial reconstruction of vasculature in a block of neocortex of a transgenic mouse in which actin filaments in the vascular wall are endogenously labeled with cyan fluorescent protein. Four successive cutting and imaging cycles are shown. The laser was focused onto the cut face with an intermediate numerical aperture (0.5 NA) objective and a raster pattern was used to optically ablate successive planes at a depth of 10 µm each with total thicknesses of 70 µm per cut. The energy per pulse varied from 0.4 to 1.7 µJ. Each stack of images represents a maximal side projection of all accumulated optical sections. The sharp breaks in the images shown in successive panels demarcate the cut boundaries. (Adapted from Tsai et al. 2003). (*B*) Large volume (1.5 mm^3) reconstruction of block-face data obtained from mouse neocortex. The vasculature was filled with fluorescein-labeled gel. Ablation was performed with pulse energies of 5 µJ and rastering of the translation stage, as in *A*. The image data consisted of overlapping stacks, 256 x 256 x 120 µm, that were registered into a single volume by means of cross-correlation over nearest neighbor features in the overlap region. (*C*) A 200 x 200 x 150-µm volume extracted from a larger reconstruction of mouse cortical vasculature and surrounding cell nuclei. Volume-accurate vascular reconstructions are shown in red, neuronal nuclei are shown in green, and non-neuronal nuclei are shown in blue. (*D*) Iterative processing of a block of neocortex of a transgenic mouse with neurons labeled by yellow fluorescent protein. Twenty-four successive cutting and imaging cycles are shown. The laser was focused onto the cut face with a 20x magnification 0.5 NA dipping objective, and a raster pattern was performed at 4 mm/sec to optically ablate successive planes at a depth of 10 µm each with total thicknesses between 40 and 70 µm per cut. The energy per pulse was maintained at 8 µJ. Each stack of images represents a maximal side projection of all accumulated optical sections obtained using TPLSM at λ = 920 nm. Anatomical layers are labeled above the figure, including the cortical layers (1 to 6) and the white matter (WM).

binding ratio of DAPI to DNA base pairs, and assuming that the 3 million base pairs in the mouse genome are uniformly distributed within a 100 µm^3 nuclear volume, each 1 µm^3 voxel would contain 6000 fluorophores. Operating at 20% of the dye saturation, and assuming a 10% collection efficiency for a 0.8 NA objective, and 1% total transmission and photoconversion efficiency for the microscope (Tsai and Kleinfeld 2009), each laser pulse would generate ~10 photoelectrons, and thus a single-voxel signal-to-noise ratio of ~3. It is therefore reasonable to consider the imaging limit of one laser pulse per voxel. For an 80-MHz laser system, this results in a imaging time of 3 h for a 0.5 cm^3 whole mouse brain at 1 µm^3 sampling. With regard to ablation, the current time could be substantially reduced by the use of structured illumination (Oron et al. 2005). Thus, if one considers only strong fluorophores so that ablation is the rate limiting step, the rate of all-optical histology comes to ~20 h per 0.5 cm^3 mouse brain.

RECIPES

CAUTION: See Appendix 6 for proper handling of materials marked with <!>.
Recipes for reagents marked with <R> are included in this list.

Blocking Solution

Normal goat serum, 10% (v/v)
Triton X-100 <!>, 2% (v/v)
Sodium azide <!>, 0.1% (w/v)

Mix contents in PBS.

Fluorescent Gel Solution

Fluorescein labeled albumin (A9771; Sigma-Aldrich)
PBS
Porcine skin gelatin type A (G1890; Sigma-Aldrich)

1. Prepare a 2% (w/v) solution of gelatin by boiling in PBS followed by cooling to <50°C.
2. Combine with 1% (w/v) fluorescein-albumin, maintained with stirring at 40°C.
3. Filter through filter paper (no. 1 Whatman) that has been pre-moistened with PBS.
4. Warm the filtered solution to 40°C before the perfusion.

ACKNOWLEDGMENTS

We thank Beth Friedman for help with the deep antibody staining and John P. Kaufhold for his work on associated analysis of the histology data. This work was funded by the National Institutes of Health (grants EB003832, MH72570, and MH085499 to DK) and a ISF-Bikura postdoctoral fellowship to PB.

REFERENCES

Drew PJ, Feldman DE. 2009. Intrinsic signal imaging of deprivation-induced contraction of whisker representations in rat somatosensory cortex. *Cereb Cortex* **19:** 331–348.

Feng G, Mellor RH, Bernstein M, Keller-Peck C, Nguyen QT, Wallace M, Nerbonne JM, Lichtman JW, Sanes JR. 2000. Imaging neuronal subsets in transgenic mice expressing multiple spectral variants of GFP. *Neuron* **28:** 41–51.

Frederickson KS, White WE, Wheeland RG, Slaughter DR. 1993. Precise ablation of skin with reduced collateral damage using the femtosecond-pulsed, terawatt titanium-sapphire laser. *Arch Dermatol* **129:** 989–993.

Frostig RD, Dory Y, Kwon MC, Masino SA. 1993. Characterization of functional organization within rat barrel cortex using intrinsic signal optical imaging through a thinned skull. *Proc Natl Acad Sci* **90:** 9998–10002.

Goetz MH, Fischer SK, Velten A, Bille A, Strum V. 1999. Computer-guided laser probe for ablation of brain tumours with ultrashort laser pulses. *Phys Med Biol* **44:** N119–N127.

Juhasz T, Loesel HL, Kurtz RM, Horvath C, Bille JF, Mourou G. 1999. Corneal refractive surgery with femtosecond lasers. *IEEE J Sel Top Quant Electron* **5:** 902–910.

Loesel FH, Niemez MH, Bille JF, Juhasz T. 1996. Laser-induced optical breakdown on hard and soft tissues and its dependence on the pulse duration: Experiment and model. *IEEE J Quant Electron* **32:** 1717–1722.

Loesel FH, Fischer JP, Gotz MH, Horvath C, Juhasz T, Noack F, Suhm N, Bille JF. 1998. Non-thermal ablation of neural tissue with femtosecond laser pulses. *Appl Phys B* **66:** 121–128.

Lubatschowski H, Maatz G, Heisterkamp A, Hetzel U, Drommer W, Welling H, Ertmer W. 2000. Application of ultrashort laser pulses for intrastromal refractive surgery. *Graefe's Arch Clin Exp Opthamol* **238:** 33–39.

Maatz G, Heisterkamp A, Lubatschowski H, Barcikowski S, Fallnich C, Welling H, Ertmer W. 2000. Chemical and physical side effects at application of ultrashort laser pulses for intrastromal refractive surgery. *J Opt A* **2:** 59–64.

Neev J, Carrasco WA, Armstrong WB, Da Silva LB, Feit MD, Matthews DL, Perry MD, Rubenchik AM, Stuart BC. 1996. Applications of ultrashort pulse lasers for hard tissue surgery. *IEEE J Sel Top Quant Electron* **2:** 790–800.

Nguyen Q-T, Tsai PS, Kleinfeld D. 2006. MPScope: A versatile software suite for multiphoton microscopy. *J Neurosci Methods* **156:** 351–359.

Nguyen Q-T, Dolnick EM, Driscoll J, Kleinfeld D. 2009. MPScope 2.0: A computer system for two-photon laser scanning microscopy with concurrent plasma-mediated ablation and electrophysiology. In *Methods for in vivo optical imaging*, 2nd ed. (ed. Frostig RD), pp. 117–142. CRC Press, Boca Raton, FL.

Oraevsky A, Da Silva L, Rubenchik A, Feit M, Glinsky M, Perry M, Mammini B, Small W, Stuart B. 1996. Plasma mediated ablation of biological tissues with nanosecond-to-femtosecond laser pulses: Relative role of linear and nonlinear absorption. *IEEE J Sel Top Quant Electron* **2:** 801–809.

Oron D, Tal E, Silberberg Y. 2005. Scanningless depth-resolved microscopy. *Opt Express* **13:** 1468–1476.

Suhm N, Gotz MH, Fischer JP, Loesel F, Schlegel W, Sturm V, Bille JF, Schroder R. 1996. Ablation of neural tissue by short-pulsed lasers—A technical report. *Acta Neurochir* **138:** 346–349.

Svoboda K, Denk W, Kleinfeld D, and Tank DW. 1997. In vivo dendritic calcium dynamics in neocortical pyramidal neurons. *Nature* **385:** 161–165.

Tirlapur UK, Konig K. 2002. Targeted transfection by femtosecond laser light. *Nature* **418:** 290–291.

Tsai PS, Kleinfeld D. 2009. In vivo two-photon laser scanning microscopy with concurrent plasma-mediated ablation: Principles and hardware realization. In *Methods for in vivo optical imaging*, 2nd ed. (ed. Frostig RD), pp. 59–115. CRC Press, Boca Raton, FL.

Tsai PS, Friedman B, Ifarraguerri AI, Thompson BD, Lev-Ram V, Schaffer CB, Xiong Q, Tsien RY, Squier JA, Kleinfeld D. 2003. All-optical histology using ultrashort laser pulses. *Neuron* **39:** 27–41.

Tsai PS, Blinder P, Migliori BJ, Neev J, Jin Y, Squier JA, Kleinfeld D. 2009a. Plasma-mediated ablation: An optical tool for submicrometer surgery on neuronal and vascular systems. *Curr Opin Biotechnol* **20:** 90–99.

Tsai PS, Kaufhold J, Blinder P, Friedman B, Drew P, Karten HJ, Lyden PD, Kleinfeld D. 2009b. Correlations of neuronal and microvascular densities in murine cortex revealed by direct counting and colocalization of nuclei and vessels. *J Neurosci* **29:** 14553–14570.

Yanik MF, Cinar H, Cinar HN, Chisholm AD, Jin Y, Ben-Yakar A. 2004. Functional regeneration after laser axotomy. *Nature* **432:** 822.

46 Ballistic Delivery of Dyes for Structural and Functional Studies of the Nervous System

Wen-Biao Gan,[1] Jaime Grutzendler,[2] Rachel O. Wong,[3] and Jeff W. Lichtman[4]

[1]Department of Physiology and Neuroscience, Skirball Institute of Biomolecular Medicine, New York University School of Medicine, New York, New York 10016; [2]Departments of Neurology and Physiology, Northwestern University Feinberg School of Medicine, Chicago, Illinois 60611; [3]Department of Biological Structure, University of Washington, Seattle, Washington 98195; [4]Department of Molecular and Cellular Biology, Harvard University, Cambridge, Massachusetts 02138

ABSTRACT

This chapter describes a detailed protocol for rapid labeling of cells in a variety of preparations by means of particle-mediated ballistic (gene gun) delivery of fluorescent dyes. This method has been used for rapid labeling of cells with either lipid- or water-soluble dyes in a variety of preparations. In particular, carbocyanine lipophilic dyes, such as DiI, have been used to obtain Golgi-like labeling of neurons and glia in fixed and live cell cultures and brain slices, as well as fixed postmortem human brain. Water-soluble calcium indicators, such as Calcium Green-1 dextran have been used to image calcium dynamics in living brain slices and retinal explants. Thus, this ballistic labeling technique is useful for studying the structure and function of neurons and glia in both living and fixed specimens.

INTRODUCTION

Fluorescent indicators are essential tools for studying the structure and function of the nervous system (Feng et al. 2000; Gan et al. 2000; Giepmans et al. 2006). For example, carbocyanine dyes, because of their high extinction coefficients and lipophilic properties, have been very useful for labeling neurons in their entirety and for tracing long-range neuronal connections. These dyes are maintained within the cell membrane and can be imaged for extended periods of time without obvious cytotoxicity (Honig and Hume 1986; Liu and Westerfield 1990). In addition, fluorescent calcium indicators have been critical for monitoring calcium dynamics in neuronal processes and studying the role of calcium signaling in neuronal plasticity (Giepmans et al. 2006; Garaschuk et al. 2007).

Loading fluorescent indicators into individual cells is often accomplished with sharp microelectrodes or patch pipettes (Neher and Almers 1986; Liu and Westerfield 1990; Gan and Macagno 1995a,b; Gan et al. 1999). Such dye-loading approaches are technically demanding and time consuming, and only a small number of cells can be labeled at any one time. Here, we present a detailed protocol for particle-mediated ballistic delivery of fluorescent dyes into individual cells. This technique allows rapid labeling of entire neurons or glia in a Golgi-like manner after membranes of individual cells are contacted by particles coated with lipophilic dyes (Gan et al. 2000). Many cells can be labeled by dyes of different colors at controlled densities to facilitate the study of structural interactions between cells (Gan et al. 2000; Rockhill et al. 2002; Strettoi et al. 2002; Sun et al. 2002; Kong et al. 2005; Coombs et al. 2007; Li et al. 2010; Yu et al. 2010). Furthermore, in conjunction with other histochemical labeling methods, the technique can be used to study neuronal/glial structures associated with pathologic processes in animal models or the postmortem human brain (Alifragis et al. 2002; Grutzendler et al. 2003; Moolman et al. 2004; Oberheim et al. 2008; Neely et al. 2009). In addition to lipophilic dyes, water-soluble molecules, such as calcium indicators, can also be delivered efficiently with this technique (Kettunen et al. 2002; Lohmann et al. 2002; Lohmann and Bonhoeffer 2008). Thus, the ballistic delivery of indicators into individual cells provides a useful approach to study the structure and function of the nervous system.

Protocol

Ballistic Delivery of Dyes into Neurons and Glial Cells

A method is provided that allows for rapid labeling of neurons and glia in a variety of preparations with either lipid- or water-soluble fluorescent dyes.

IMAGING SETUP

A confocal microscope is recommended for imaging fluorescently labeled cells. We frequently use an oil-immersion objective (60x 1.4 numerical aperture) for high-resolution imaging of neuronal structures, such as dendritic spines.

MATERIALS

CAUTION: See Appendix 6 for proper handling of materials marked with <!>.

Reagents

Glycerol (80% in 0.2-M phosphate-buffered saline [PBS] or 100%; see Step 17)
Lipophilic dyes: DiI, DiO, or DiD (D-282, D-275, D-307, Molecular Probes)
Methylene chloride <!>
Nervous system tissue

Fixing the tissue may be desirable or necessary to obtain good quality labeling of neuronal structures. For mouse brain slices, perfuse the animal with 10 mL of isosmotic PBS followed by rapid perfusion with 40 mL of 4% paraformaldehyde <!> in PBS (20 mL/min). Remove the brain quickly from the skull, and post fix it in 4% paraformaldehyde for 10 min. Using a vibratome, cut the brain into slices 200–300-μm thick.

PBS, 0.2 M (containing monobasic 0.2 M NaH_2PO_4-H_2O; dibasic 0.2 M Na_2HPO_4 (anhydrous) and 0.9% NaCl [all from Sigma-Aldrich]) (see Step 16)
Polyvinylpyrrolidone ([PVP], 10 mg/mL in distilled water, from Sigma-Aldrich or Bio-Rad)
Water-soluble dye, such as calcium indicators (e.g., Calcium Green-1 dextran, Molecular Probes)

Equipment

Biolistic device: Helios Gene Gun System (Bio-Rad) or a custom-made ballistic device (see http://thalamus.wustl.edu/nonetlab/ResourcesF/genegun/Genegun.htm)
Compressed air
Glass slides
Membrane filter holder

Construct a filter holder by cutting a central hole (~20-mm diameter) in the bottom of a plastic Petri dish, and cover it with a metal mesh (~1-mm pore). Glue the mesh to the Petri dish. Additionally, cut a 20-mm diameter hole in the cover of the Petri dish.

Membrane filters, Isopore polycarbonate with 3-μm pore size and 8.0 x 10^5 pores/cm^2 (TSTP04700, Millipore)
Microfuge tubes
Microscope (see Imaging Setup)
Razor blades

Sonicator
Syringe for injecting and withdrawing solutions through the Tefzel tubing
Tefzel tubing (165-2441, Bio-Rad)
Test tubes, 10 mL
Tubing preparation station (165-2418, Bio-Rad)
Tungsten particles (1.0–1.7-µm-diameter, Bio-Rad)
Weighing paper

EXPERIMENTAL METHOD

Procedure 1: Preparation and Delivery of Lipophilic Dyes

Preparation of Lipophilic Dyes

Perform either Step 1 or Steps 2–4.

1. *For single color labeling*, dissolve 3 mg of lipophilic dye such as DiI, DiO, or DiD in 100 µL of methylene chloride.
2. *For multicolor labeling*, dissolve 7 mg of DiI, DiO, and DiD in 70 µL of methylene chloride in three separate microfuge tubes.
3. Mix various proportions of the different dyes to obtain, for example, seven different dye combinations in seven tubes as follows.

Tube #	DiI (µL)	DiO (µL)	DiD (µL)
1	15	15	—
2	15	—	15
3	—	15	15
4	10	10	10
5	30	—	—
6	—	30	—
7	—	—	30

4. Add 70 µL of methylene chloride to each tube. The final concentration of each dye combination is 3 mg dye/100 µL methylene chloride.

Coating Tungsten Beads

5. Spread 50–100 mg of tungsten particles evenly on a glass slide by mixing them with a few drops of methylene chloride.
6. Add 100 µL of dye solution to the particles on the glass slides. The methylene chloride evaporates within 1 min, causing the lipophilic dye to precipitate onto the tungsten particles.
7. Scrape the dye-coated particles off of the glass slide using a razor blade, and transfer them to a 10-mL test tube.

   For multicolor labeling, coat particles separately with different combinations of dyes, and then combine them in a 10-mL test tube.
8. Resuspend the particles in 3–10 mL of distilled H_2O, and sonicate for 5–10 min to prevent formation of large clusters.

Preparation of Bullets

9. Inject 10-mg/mL PVP in distilled H_2O into a 762-mm long piece of Tefzel tubing. Rapidly drain to precoat the tubing before injecting the dye solutions.

   This step improves attachment of the beads to the tube walls.

10. Inject sonicated solutions of particles of single color or mixtures of colors into the precoated Tefzel tube.
11. Insert the tube into a tubing preparation station, and allow the dye-coated particles to precipitate and to settle onto the tube wall for ~3 min (depending on the desired labeling density) before slowly withdrawing the remaining liquid. To spread the beads evenly, rotate the tube on the preparation station for ~5 min while drying with a constant flow of air.
12. Cut the air-dried tube into 13-mm pieces (the bullets), and store at room temperature, protected from light, for future use. The storage time could be up to several months.

Particle Delivery

13. Insert a membrane filter into the filter holder located between the gene gun and the preparation. Specifically, place a filter on top of the metal mesh, and hold it in place with a Petri dish cover that also has a central opening of ~20 mm in diameter (Fig. 1).

    The filter prevents large clusters of particles from landing on the tissue, causing nonuniform labeling and interfering with single-cell resolution (Fig. 1C). The use of a membrane filter also protects the tissue from the air shock wave generated by the gun.

14. Place the filter holder on top of the target tissue, and align it with the gene gun barrel before shooting (Fig. 1).
15. Deliver dye-coated particles to the preparation by firing the bullets from the gene gun.

Dye Diffusion, Tissue Mounting, and Imaging

16. After delivering the dye-coated particles, keep tissues, such as brain slices, in solution (e.g., 0.2 M PBS) for several hours to allow the dye to diffuse along neuronal processes. One option is to postfix the tissue for 2–12 h at room temperature, which further preserves structures while also allowing for dye diffusion.

    Labeling takes longer in fixed tissue than in living slices, presumably because dye diffusion is slower along the fixed plasma membrane.

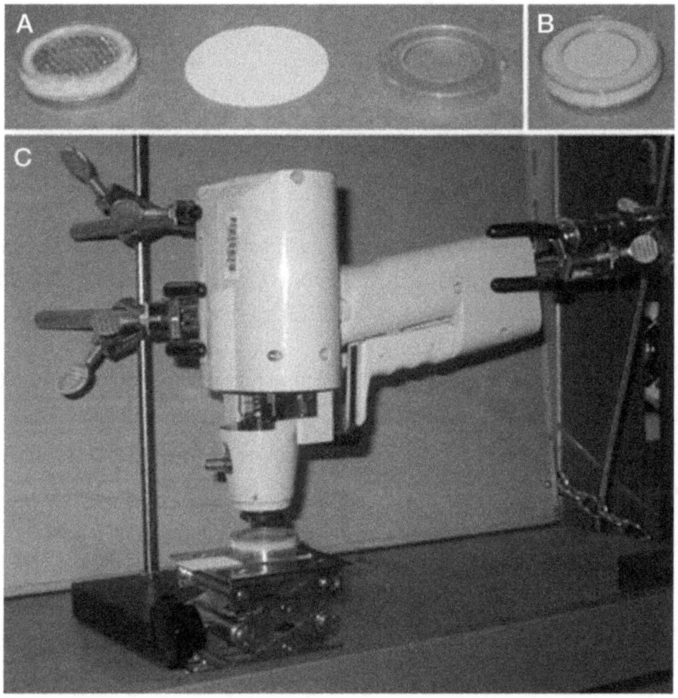

FIGURE 1. Setup for ballistic dye delivery. (*A*) Membrane filter holder. From *left to right*: metal mesh glued to a Petri dish with 20-mm-diameter opening, membrane filter, a Petri dish cover with ~20-mm-diameter opening. (*B*) Membrane filter is placed on top of metal mesh and is held in place by the Petri dish cover. (*C*) Membrane filter holder is placed on top of tissue and is aligned with the gene gun barrel.

17. Following dye diffusion, mount tissues on glass slides using 80% glycerol in 0.2 M PBS or 100% glycerol.
18. Image the fluorescently labeled cells.

    See *Troubleshooting*.

Procedure 2: Labeling Cells with Water-Soluble Dyes

When using water-soluble dyes, labeling density needs to be approximately fivefold higher than with lipophilic dyes to increase the probability of successful labeling because dyes need to enter the cell soma and diffuse into the cytoplasm rather than diffuse along the cell membrane as with lipophilic dyes. Because of this, labeling with water-soluble dyes works best in preparations containing a high density of cell somata, such as the ganglion cell layer of the retina in which 15% of cells can be routinely labeled.

1. Dissolve 1–2 mg of an H_2O-soluble dye, such as Calcium Green-1 dextran in 12–25 µL of distilled H_2O.
2. Spread 25–50 mg of tungsten particles (1.3- or 1.7-µm diameter) evenly onto a clean microscope glass slide. Use a razor blade to dissociate large clusters of particles until a fine powder results.
3. Pipette the dissolved dye solution uniformly onto the tungsten particles, and air dry.
4. Scrape the dye-coated particles onto a piece of weighing paper with the razor blade.
5. Precoat a 15-cm-long piece of Tefzel tubing with 10-mg/mL PVP in distilled water.
6. Pour the powder of dye-coated particles into the Tefzel tubing.

    Note that, unlike with the preparation of bullets containing lipophilic dyes, the particles coated with water-soluble dye are added to the tubing in a dry form.
7. Gently rotate and shake the tube until the particles spontaneously adhere to the inner wall of the tubing.
8. Cut the particle-coated tube into 13-mm pieces, and store them in a dark dry container at 4°C until use. The storage time could be up to several months.
9. For particle delivery into tissues and subsequent imaging, proceed with Steps 13–18 above.

TROUBLESHOOTING

Problem (Step 18): No fluorescently labeled cells were found.
Solutions:

1. Check whether dye-coated tungsten particles were delivered to the preparation. If not, shoot the tissue again varying the conditions based on the following. The density and depth of labeling can be controlled by varying the gas pressure (60–200 psi of helium gas), by changing the distance between the gun and the preparation (10–25 mm), by altering the density of particles per bullet, by using a membrane filter having a different pore density, or by shooting the tissue multiple times. A lower gas pressure and a larger distance between the gun and the tissue lead to lower labeling densities. Higher gas pressures may tear the membrane filter, resulting in tissue injury and nonuniform labeling.
2. If tungsten particles are present in the tissue, check that the tungsten particles were coated with enough fluorescent dye. If necessary, prepare new bullets having more dye per particle.

Problem (Step 18): Neuronal processes are not clearly labeled.
Solution: Evaluate the fixation conditions. The duration of fixation before ballistic particle delivery is important for obtaining good quality labeling. Overfixation before shooting (>1 h) often causes the dye to spread between cells or nearby processes, thus, reducing the number of cells

that can be individually resolved. However, after ballistic labeling, the duration of postfixation does not appear to cause dye spread between nearby cells, suggesting that dye spread and decreased labeling is probably caused by failure of initial dye transfer from the particle to surrounding individual cells. The problem of dye spread between cells caused by overfixation is less severe in young animals, presumably because of greater extracellular space.

Fixation conditions are also critical when labeling the postmortem human brain. It appears that labeling is best when small pieces of brain are fixed with 4% paraformaldehyde within 24 h postmortem. Because lipophilic dye labeling requires the presence of an intact and continuous cell membrane, it is important to avoid using fixatives that can dissolve membrane lipids, such as methanol, which is contained in standard formalin. In addition, frozen specimens cannot be labeled with lipophilic dyes because of the membrane disruption that occurs during the freezing and thawing process.

DISCUSSION

Examples of Application

The ballistic technique has been used to label neurons in both live and fixed tissues, ranging from isolated neurons in culture to whole brain and brain slices (Fig. 2). When labeling at low densities, we typically observed only one dye-coated particle contacting either the soma or a branch of the dendritic or axonal arbor of a labeled cell, indicating that a single particle contains enough dye to label one cell through retrograde and anterograde diffusion (Fig. 2A,B). Figure 2C shows multiple pyramidal neurons labeled in a fixed mouse cortical brain slice by delivering a mixture of particles coated with different combinations of dyes. Many neurons and their processes, including dendritic spines, appeared completely labeled and can be distinguished from each other because of the multicolor labeling. Labeled dendritic arbors and axons can often be followed for hundreds of micrometers. In living tissues, labeled dendritic processes appeared rapidly (<5 min) after particles contacted the tissue (Gan et al. 2000; Kettunen et al. 2002; Grutzendler et al. 2003). This technique is especially useful for morphological analysis of neurons in retinal explants (Gan et al. 2000; Rockhill et al. 2002; Strettoi et al. 2002; Sun et al. 2002; Kong et al. 2005; Coombs et al. 2007) and has also been used for analysis of dendritic spines in different regions of the brain (Grutzendler et al. 2003; Tsai et al. 2004; Wu et al. 2004a,b; O'Brien and Unwin 2006). Furthermore, it has been successfully used for morphological analysis of astrocytic domains in both mouse and postmortem human brain (Oberheim et al. 2008, 2009).

In addition to lipophilic dye labeling, the ballistic technique can be used to label cells with water-soluble compounds, such as calcium indicators (Kettunen et al. 2002; Lohmann et al. 2002; Lohmann and Bonhoeffer 2008). Figure 2D shows a cortical pyramidal cell from an acute brain slice labeled with the calcium indicator, Calcium Green-1 delivered on tungsten particles. The time required for complete filling of dendritic arbors occurred within 1 min. The intensity of labeling varied from cell to cell, most likely because particles carried differing amounts of indicator. In addition, these labeled cells showed an increase in intracellular calcium levels on stimulation with KCl, indicating that they remained viable after labeling (Kettunen et al. 2002; Lohmann et al. 2002).

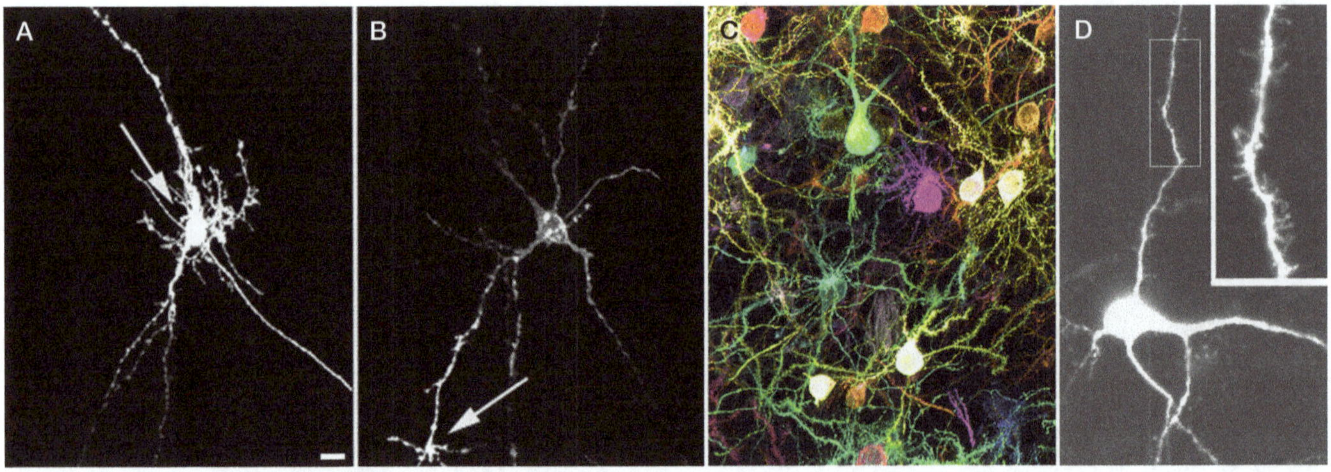

FIGURE 2. Ballistic labeling of cells in fixed and living tissues. An individual cell in a P10 mouse brain slice is labeled with lipophilic dye with a single dye-coated particle (arrow) landed on either the cell soma (*A*) or a dendritic arbor (*B*). (*C*) Labeling of many pyramidal neurons from a fixed brain slice from a P20 mouse that was shot with a combination of seven different lipophilic dyes. The image represents a collapsed view of 50 confocal planes covering ~25 µm of depth. Many of the cells had processes that extended beyond the upper or lower surface of the slice and, hence, have dendritic or axonal branches that end abruptly. (*D*) Labeling of a pyramidal neuron and spines (*inset*) with Calcium Green-1 dextran from an acute slice preparation of a P12 rat. Scale bar (in *A*), 10 µm for *A–D*.

Advantages of the Ballistic Delivery Method

The ballistic method allows rapid delivery of fluorescent indicators into multiple cell types in both fixed and living preparations, ranging from embryonic tissue to adult postmortem human brain. This labeling technique provides several advantages over existing labeling methods.

Efficient labeling of multiple cells: In comparison to intracellular injection of indicators using micropipettes, this approach allows many neurons to be labeled within minutes rather than targeting cells individually. Dendrites in many cells are often labeled from a single indicator-coated particle, and the labeling process appears not to cause a significant perturbation of cellular structure or function (Gan et al. 2000; Kettunen et al. 2002; Lohmann et al. 2002). In addition, ballistic delivery can be very useful for labeling immature cells, which are often too fragile to survive labeling by dye injection through micropipettes. Finally, using various combinations of carbocyanine dyes, adjacent neurons and glia can be labeled in different colors (Gan et al. 2000; Oberheim et al. 2008). Combined with high-density labeling, this multicolor feature is potentially useful for studying complicated neuronal networks.

Rapid labeling: The ballistic technique permits immediate imaging of neuronal tissues as soon as the preparation is made, unlike green fluorescent protein transfection techniques that require many hours for gene expression to occur (Lo et al. 1994). This is particularly important for brain slice experiments in which the health of the preparation can be maintained for only a limited period of time. In addition, the technique allows extensive dendritic arbors to be labeled rapidly without dialyzing cellular contents as seen in patch-pipette loading (Majewska et al. 2000).

Labeling a variety of cell types in live and fixed tissues: The labeling technique involves passive dye transfer and diffusion that is independent of gene transcription and protein synthesis. Thus, any type of cell in fixed or living tissue can be labeled nonselectively. Furthermore, more complete labeling of fine neuronal structures, such as dendritic spines is achieved with lipophilic dyes compared with immunolabeling.

Labeling with various indicators and molecules: In addition to delivering lipophilic dyes and calcium indicators as described here, the ballistic technique is potentially useful for delivering other indicators and molecules, such as voltage-sensitive dyes, dextran-conjugated pH indicators, ion indicators, and pharmacological reagents. This potential has been exploited in recent studies using a rhodamine analog to investigate cell migration in embryonic brains (Alifragis et al. 2002). It is also possible to simultaneously deliver more than one indicator by co-coating the particles with multiple substances (O'Brien and Lummis 2004).

Combining the ballistic labeling technique with other labeling approaches: The ballistic labeling technique in combination with other fluorescent dyes such as Thioflavin-S has been used to study dendritic abnormalities near amyloid plaques in a mouse model of Alzheimer's disease and in the human brain (Grutzendler et al. 2003). In addition, if immunostaining does not require extensive tissue permeabilization and lipid extraction, antibody labeling can also be used in conjunction with lipophilic dye labeling (Alifragis et al. 2002; Neely et al. 2009).

Disadvantages of the Ballistic Delivery Method

Despite many advantages, ballistic delivery of indicators also has some drawbacks. First, this approach is highly variable because dye crystal size, density, and penetration are difficult to control. Using the commercially available Bio-Rad gene gun, the depth of penetration of dye-coated particles is generally ~20–30 µm and up to 60 µm. Therefore, most of the labeled cells are usually located near the surface. However, because of retrograde transfer of dyes, cells located deeper can also be labeled, particularly with lipophilic dyes. A potentially useful modification to the gene gun that can increase the depth of penetration substantially was reported (O'Brien et al. 2001). Such a modification may allow penetration of particles up to ~300 µm in brain slices. Second, because water-soluble dye-coated particles have to be in or near the cell somata for labeling to occur, it is difficult to get high-density labeling without showering the tissue with many particles and potentially leading to some tissue damage. Third, whereas labeling of dendritic structures is very good, especially with lipophilic dyes, labeling of axonal arbors over extensive distances has not been successful. This is likely because each particle carries a limited amount of dye on its surface. In the future, it may be possible to improve labeling by designing special beads that can carry a larger amount of dye. A fourth limitation of the technique is the difficulty in targeting a particular cell or region. However, using a mask to cover parts of the preparation during shooting can prevent labeling of undesired areas and can target a smaller region of interest.

Summary

To summarize, we describe, in detail, a ballistic technique that delivers fluorescent dyes and indicators into multiple cell types, providing rapid labeling of neurons, glia, and their processes both in vivo and in vitro. This approach should be useful for studying neuronal connectivity, function, and pathology in the nervous system of experimental animals as well as in postmortem human tissue specimens.

REFERENCES

Alifragis P, Parnavelas JG, Nadarajah B. 2002. A novel method of labeling and characterizing migrating neurons in the developing central nervous system. *Exp Neurol* **174:** 259–265.

Coombs JL, Van Der List D, Chalupa LM. 2007. Morphological properties of mouse retinal ganglion cells during postnatal development. *J Comp Neurol* **503:** 803–814.

Feng G, Mellor RH, Bernstein M, Keller-Peck C, Nguyen QT, Wallace M, Nerbonne JM, Lichtman JW, Sanes JR. 2000. Imaging neuronal subsets in transgenic mice expressing multiple spectral variants of GFP. *Neuron* **28:** 41–51.

Gan WB, Macagno ER. 1995a. Developing neurons use a putative pioneer's peripheral arbor to establish their terminal fields. *J Neurosci* **15:** 3254–3262.

Gan WB, Macagno ER. 1995b. Interactions between segmental homologs and between isoneuronal branches guide the formation of sensory terminal fields. *J Neurosci* **15:** 3243–3253.

Gan WB, Bishop DL, Turney SG, Lichtman JW. 1999. Vital imaging and ultrastructural analysis of individual axon terminals labeled by iontophoretic application of lipophilic dye. *J Neurosci Methods* **93:** 13–20.

Gan WB, Grutzendler J, Wong WT, Wong RO, Lichtman JW. 2000. Multicolor "DiOlistic" labeling of the nervous system using lipophilic dye combinations. *Neuron* **27:** 219–225.

Garaschuk O, Griesbeck O, Konnerth A. 2007. Troponin C-based biosensors: A new family of genetically encoded indicators for in vivo calcium imaging in the nervous system. *Cell Calcium* **42:** 351–361.

Giepmans BN, Adams SR, Ellisman MH, Tsien RY. 2006. The fluorescent toolbox for assessing protein location and function. *Science* **312:** 217–224.

Grutzendler J, Tsai J, Gan WB. 2003. Rapid labeling of neuronal populations by ballistic delivery of fluorescent dyes. *Methods* **30:** 79–85.

Honig MG, Hume RI. 1986. Fluorescent carbocyanine dyes allow living neurons of identified origin to be studied in long-term cultures. *J Cell Biol* **103:** 171–187.

Kettunen P, Demas J, Lohmann C, Kasthuri N, Gong Y, Wong RO, Gan WB. 2002. Imaging calcium dynamics in the nervous system by means of ballistic delivery of indicators. *J Neurosci Methods* **119:** 37–43.

Kong JH, Fish DR, Rockhill RL, Masland RH. 2005. Diversity of ganglion cells in the mouse retina: Unsupervised morphological classification and its limits. *J Comp Neurol* **489:** 293–310.

Li M, Cui Z, Niu Y, Liu B, Fan W, Yu D, Deng J. 2010. Synaptogenesis in the developing mouse visual cortex. *Brain Res Bull.* **81:** 107–113.

Liu DW, Westerfield M. 1990. The formation of terminal fields in the absence of competitive interactions among primary motoneurons in the zebrafish. *J Neurosci* **10:** 3947–3959.

Lo DC, McAllister AK, Katz LC. 1994. Neuronal transfection in brain slices using particle-mediated gene transfer. *Neuron* **13:** 1263–1268.

Lohmann C, Bonhoeffer T. 2008. A role for local calcium signaling in rapid synaptic partner selection by dendritic filopodia. *Neuron* **59:** 253–260.

Lohmann C, Myhr KL, Wong RO. 2002. Transmitter-evoked local calcium release stabilizes developing dendrites. *Nature* **418:** 177–181.

Majewska A, Tashiro A, Yuste R. 2000. Regulation of spine calcium dynamics by rapid spine motility. *J Neurosci* **20:** 8262–8268.

Moolman DL, Vitolo OV, Vonsattel JP, Shelanski ML. 2004. Dendrite and dendritic spine alterations in Alzheimer models. *J Neurocytol* **33:** 377–387.

Neely MD, Stanwood GD, Deutch AY. 2009. Combination of diOlistic labeling with retrograde tract tracing and immunohistochemistry. *J Neurosci Methods* **184:** 332–336.

Neher E, Almers W. 1986. Patch pipettes used for loading small cells with fluorescent indicator dyes. *Adv Exp Med Biol* **211:** 1–5.

O'Brien J, Lummis SC. 2004. Biolistic and diolistic transfection: Using the gene gun to deliver DNA and lipophilic dyes into mammalian cells. *Methods* **33:** 121–125.

O'Brien J, Unwin N. 2006. Organization of spines on the dendrites of Purkinje cells. *Proc Natl Acad Sci* **103:** 1575–1580.

O'Brien JA, Holt M, Whiteside G, Lummis SC, Hastings MH. 2001. Modifications to the hand-held gene gun: Improvements for in vitro biolistic transfection of organotypic neuronal tissue. *J Neurosci Methods* **112:** 57–64.

Oberheim NA, Tian GF, Han X, Peng W, Takano T, Ransom B, Nedergaard M. 2008. Loss of astrocytic domain organization in the epileptic brain. *J Neurosci* **28:** 3264–3276.

Oberheim NA, Takano T, Han X, He W, Lin JH, Wang F, Xu Q, Wyatt JD, Pilcher W, Ojemann JG, et al. 2009. Uniquely hominid features of adult human astrocytes. *J Neurosci* **29:** 3276–3287.

Rockhill RL, Daly FJ, MacNeil MA, Brown SP, Masland RH. 2002. The diversity of ganglion cells in a mammalian retina. *J Neurosci* **22:** 3831–3843.

Strettoi E, Porciatti V, Falsini B, Pignatelli V, Rossi C. 2002. Morphological and functional abnormalities in the inner retina of the rd/rd mouse. *J Neurosci* **22:** 5492–5504.

Sun W, Li N, He S. 2002. Large-scale morphological survey of mouse retinal ganglion cells. *J Comp Neurol* **451:** 115–126.

Tsai J, Grutzendler J, Duff K, Gan WB. 2004. Fibrillar amyloid deposition leads to local synaptic abnormalities and breakage of neuronal branches. *Nat Neurosci* **7:** 1181–1183.

Wu CC, Chawla F, Games D, Rydel RE, Freedman S, Schenk D, Young WG, Morrison JH, Bloom FE. 2004a. Selective vulnerability of dentate granule cells prior to amyloid deposition in PDAPP mice: Digital morphometric analyses. *Proc Natl Acad Sci* **101:** 7141–7146.

Wu CC, Reilly JF, Young WG, Morrison JH, Bloom FE. 2004b. High-throughput morphometric analysis of individual neurons. *Cereb Cortex* **14:** 543–554.

Yu DM, Tang WC, Wu P, Deng TX, Liu B, Li MS, Deng JB. 2010. The synaptic remodeling between regenerated perforant pathway and granule cells in slice culture. *Cell Mol Neurobiol* **30:** 309–316.

47 Two-Photon Targeted Patching and Electroporation In Vivo

Michael Häusser[1] and Troy W. Margrie[2]

[1]*Wolfson Institute for Biomedical Research and Department of Neuroscience, Physiology and Pharmacology, University College London, London WC1E 6BT, United Kingdom;* [2]*Division of Neurophysiology, The National Institute for Medical Research, London NW7 1AA, United Kingdom*

ABSTRACT

By combining patch-clamp methods with two-photon microscopy, it is possible to target recordings to specific classes of neurons in vivo. Here we describe methods for imaging and recording from the soma and dendrites of neurons identified using genetically encoded probes such as green fluorescent protein (GFP) or functional indicators such as Oregon Green BAPTA-1. Extracellular perfusion of membrane-impermeable dyes, resulting in a neuron-negative staining pattern, can also be used to discern cell types based on their morphology and target them for electrical recordings ("shadowpatching"). We describe how these approaches can be adapted for single-cell electroporation to manipulate specific cells genetically. These approaches thus permit the recording and manipulation of rare genetically, morphologically, and functionally distinct subsets of neurons in the intact nervous system.

INTRODUCTION

Whole-cell recordings are useful for recording synaptic and spiking activity of neurons in vivo because the formation of a high-resistance seal between pipette and membrane ensures stable recordings for prolonged periods and is an approach that can even be used in awake, head-fixed, or behaving animals (Margrie et al. 2002; Lee et al. 2006). However, "blind" (i.e., nontargeted) recordings suffer from the problem that cell sampling is generally based on anatomical predominance, with the largest and/or the most numerous cell types being more readily sampled than the smallest or least abundant cells. This is, of course, dependent on the overall architecture of the brain area of interest, but in many regions that contain anatomically, genetically, and functionally diverse populations of neurons, such as the neocortex, reliable sampling has not been possible for many subtypes of cells.

Transgenic mouse lines paved the way for targeted recordings from specific cells expressing a fluorescent marker (Fig. 1). This has been made possible by combining in vivo two-photon microscopy

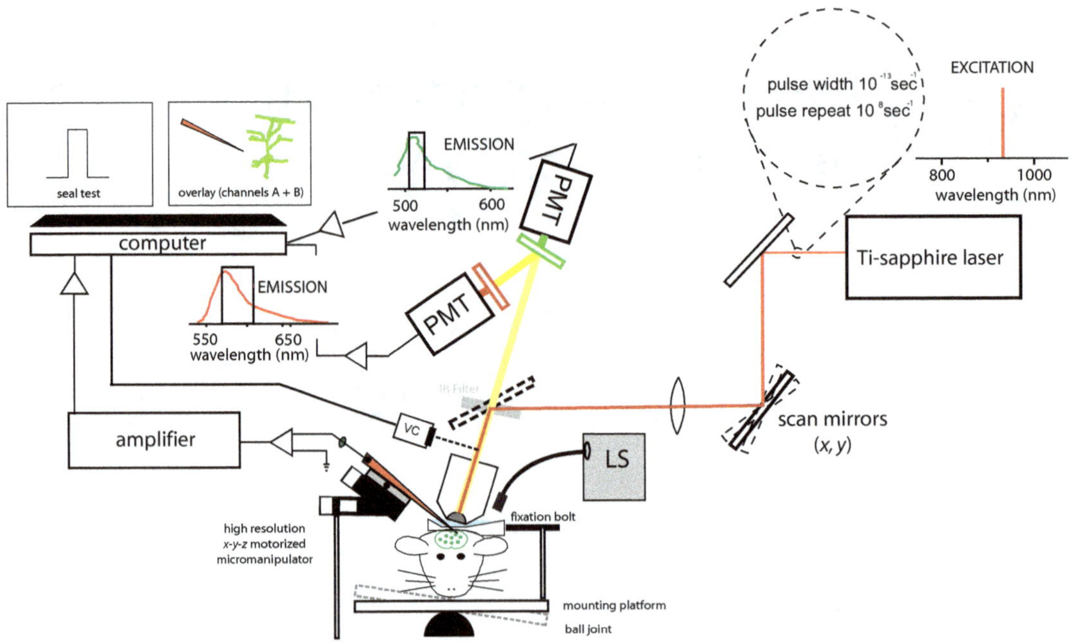

FIGURE 1. Schematic of the imaging setup for the targeted recording approach. Neurons are visualized using a two-photon microscope operated in conjunction with a mode-locked Ti:sapphire laser. Dual-channel detection is used to collect emitted light selected by an interference filter for monitoring the signals from Alexa 594 (the pipette; 610/BP40) and enhanced green fluorescent protein (eGFP) (expressed by the targeted cell type; 520/BP30). Under high magnification, water-immersion conditions can be achieved by loading the imaging window of the head-plate chamber (which is fixed to the cranium) with Ringer's solution. For added ease of use, the stereotaxic platform may be fixed to an *x–y* table or a ball-and-socket mounting plate.

(Denk et al. 1990; Svoboda et al. 1997) with the use of transgenic mice expressing GFP (Margrie et al. 2003), and more recently using viral vectors designed to drive expression via cell-specific promoters (Dittgen et al. 2004). Two-photon targeted patching can also be adapted for use with wild-type brains by filling the extracellular space with a nonpermeant dye to visualize the cells by their negative image, a technique termed "shadowpatching" (Kitamura et al. 2008). Furthermore, by prelabeling populations of neurons with a fluorescent Ca^{2+} indicator for action potential imaging (Stosiek et al. 2003; Garaschuk et al. 2006) or by expressing genetically encoded Ca^{2+} sensors, functionally defined neurons can be targeted for recording (Kerr et al. 2005; Greenberg et al. 2008; Wallace et al. 2008; Rochefort et al. 2009; Schultz et al. 2009; Tian et al. 2009; Lütcke et al. 2010). Finally, extracellular stimulation via patch pipettes targeted to individual cells has also made in vivo targeted electroporation of dyes (Nevian and Helmchen 2007) and DNA (Kitamura et al. 2008) possible. This chapter describes the steps involved in this multi-applicable technique that permits targeted in vivo recordings from the soma and dendrites of specific cells, as well as genetic manipulation of specific cells in vivo.

Protocol

Targeting Single Cells with Patch Pipettes In Vivo

This protocol describes the use of patch-clamp methods together with two-photon microscopy to image and record from the soma and dendrites of specific cells in vivo. We present detailed instructions for electrophysiological manipulations, shadowpatching, and electroporation to deliver dye or DNA into selected neurons.

MATERIALS

CAUTION: See Appendix 6 for proper handling of materials marked with <!>.
See the end of the chapter for recipes for reagents marked with <R>.

Reagents

Biological sample

The animal (typically a rat or transgenic mouse) must be anesthetized and maintained according to the appropriate local and national ethical guidelines. The animal's head should be secured using a stereotaxic frame or a custom head mount (e.g., Judkewitz et al. 2009).

Cyanoacrylate glue (or dental acrylic) <!>
Fluorescent dye (choose according to study of interest; see Step 2)
Internal recording solution <R>
Ketamine/xylazine mixture <!> (used as anesthetic; Sigma-Aldrich)

Equipment

Analog-to-digital (A-D) converter (Instrutech)
Electroporation device (e.g., Axoporator 800A; Molecular Devices)
Motorized *x–y–z* micromanipulator (e.g., Luigs & Neumann)
Pipette puller (e.g., from Narishige or Sutter)
Stereotaxic frame
Temperature controller (FHC)
Two-photon microscope for in vivo imaging

This can be a commercial instrument (e.g., from Sutter Instruments or Prairie Technologies) or a custom-made design (e.g., see Tsai et al. 2002), and it requires a mode-locked Ti:sapphire laser (e.g., Coherent or Spectra-Physics).

Voltage/current amplifier (e.g., MultiClamp 700B; Molecular Devices)
Water-immersion high-NA objective (e.g., 40x, 0.8-NA Nikon)

EXPERIMENTAL METHOD

1. Place the anesthetized animal in a stereotaxic frame, and perform the craniotomy according to standard surgical procedures used for in vivo patch-clamp preparations. Because optical clarity is essential, it is imperative that little or no bleeding occurs in the region where imaging and targeted patching will be performed. Before attaching the head plate used for stabilization, allow the skull to dehydrate to ensure robust fixation of the head plate (Fig. 1). Apply either cyanoacrylate glue or dental acrylic to the skull and gently fix the head plate. This step may also be carried out in reverse; that is, the craniotomy can be created through the center hole of the head plate once it is fixed to the cranium. This has the advantages that it negates the need for the stereotaxic frame, fixative cannot leak into the craniotomy, and it reduces the overall preparation time.

   For many preparations, removal of the dura is essential in order to prevent the pipette from clogging as it enters the brain. However, in mice it is not always necessary.

2. To visualize electrodes, fill patch pipettes with internal recording solution containing a fluorescent dye (e.g., Alexa 594) or other indicator that can be clearly distinguished from the reporter probe (e.g., GFP) based on their different emission spectra (Fig. 1).

3. While illuminating the cortical surface with an external light source, coarsely position the pipette over the region of interest using a low-power objective (4X–20X) under either video guidance or visualization via the binocular tube (Fig. 1). The patch pipette should be guided to the center of the field of view, near the cortical region of interest, such that the forward diagonal motion (see Step 5) will place the pipette in the center of the region of interest.

4. Replace the low-power objective with the water-immersion, high-NA objective that is needed for targeted recordings (Fig. 1). Now, using two-photon excitation, the dye-filled patch pipette can be visualized. While the pipette is in the bath above the preparation, low positive pressure (20–40 mbar) should result in a bright plume of fluorescent dye ejected from the pipette. Next, position the pipette just over the surface of the brain. If the dura is intact, its bright autofluorescence can be used to easily locate the correct depth. Apply a higher positive pressure (200–400 mbar) and enter the brain by moving the pipette in a diagonal motion. Once the pipette has cleanly entered the brain, with or without the dura, dye should be flowing freely out of the tip. Quickly reduce the positive pressure back to 20–40 mbar to prevent increasing the background fluorescence level in the neuropil.

   Note that a single low-power, high-NA objective can be used for both Steps 3 and 4.

5. Find a cell that is suitable for targeting. This can be done by visualizing GFP-positive neurons in transgenic animals (or by virally transfecting neurons to express GFP) (Dittgen et al. 2004). Alternatively, ejection of Alexa dye from the patch pipette into the extracellular space in wild-type brains can be used to visualize neurons via the "negative image" of the unfilled cells (Fig. 2; Kitamura et al. 2008). Once the cell has been selected, position it in the center of the field of view. Determine the depth of the cell from the surface of the brain, then retract the pipette from the craniotomy by a known distance and lower the manipulator in the *z*-axis according to the distance of the cell from the brain surface. For successful navigation of the patch pipette toward GFP-labeled cells, it is helpful to overlay the two emission channels (Fig. 1).

6. In voltage-clamp mode, move the pipette rapidly down toward the cell while monitoring the tip resistance using a test pulse. Moving the pipette back along the diagonal axis, position it inside the brain to ~50–100 μm from the cell of interest. A plume of fluorescent dye around the pipette tip should be observed—this is a good sign, as it indicates that the electrode tip has not been clogged or obstructed.

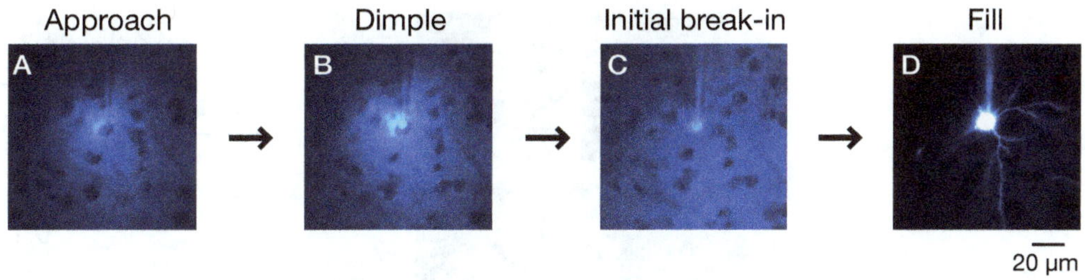

FIGURE 2. Shadowpatching of unlabeled cells in vivo using two-photon microscopy. (*A*) A water-soluble, low-molecular-weight dye is included in the pipette solution and fills the neuropil without entering cell bodies, which consequently appear as dark shadows. This way, morphological cues can be used to target recordings and the pipette can be optimally positioned for gigaseal formation. (*B*) Once a dimple forms in the surface of the cell (a layer 2/3 pyramidal neuron 150 μm deep in mouse visual cortex), it is immediately clear in the image, and standard procedures are used to establish the whole-cell patch-clamp configuration. After releasing positive pressure, the dye in the neuropil quickly diffuses away. (C) Following break-in, the cell begins to fill with dye and the neuropil continues to fade. (*D*) Low series resistance whole-cell recordings permit rapid dialysis with pipette solution, demonstrated here by a bright fill. (Unpublished data from Spencer Smith.)

7. When approaching the cell, reduce the positive pressure to ~20–40 mbar to reduce the background fluorescence, ensuring that the pipette resistance remains constant. This should be verified using both optical and electrical readouts (see Fig. 1). If >20–40 mbar of pressure is required to observe a plume out of the pipette tip, then the tip may be clogged, even if little to no increase in tip impedance is observed.
8. Navigate toward the target cell while monitoring the electrode resistance continuously. Lateral excursions of the pipette are feasible but should be limited to <50 μm. Position the electrode tip in a focal plane ~5–10 μm directly away from the cell. While visually tracking the tip and monitoring the electrode resistance simultaneously, move toward the center of the cell body. This may be done by either moving along the diagonal axis or by positioning the tip directly above the center of the cell and moving vertically. Observe that once the tip of the electrode comes into contact with the cell membrane, the resistance of the pipette will increase and the seal test amplitude will fluctuate at heartbeat frequencies (Margrie et al. 2002). In some instances, contact with the cell can been recognized via a dimple in the cell membrane (Fig. 2).
9. Remove the positive pressure and take the command potential to –70 mV. Some slight negative pressure may be required to achieve gigaohm seal formation. This can take from 5–10 sec up to 1 min. Noninvasive recordings of activity by recording cell-attached spikes can be carried out at this point. If seal formation fails, the pipette should be changed; the same cell may be attempted more than once. In cell-attached mode, increase the test pulse amplitude and apply a slow ramp of negative pressure to obtain the whole-cell configuration.
10. Verify that targeting is successful using two different approaches.
    i. Electrophysiologically: Verify resting membrane potential, assess access resistance, and determine the firing properties.
    ii. Optically: Observe filling of the target cell with dye (Figs. 2 and 3), and when targeting GFP-positive neurons, observe initial mixing of the GFP and dye signals, followed by a gradual washout of GFP signal and replacement with the internal dye.
11. This approach can be adapted for electroporation of dye or DNA into selected neurons, which should be included in the pipette solution. At Step 9, release the positive pressure and apply a train of voltage pulses (e.g., 50 square –12-V pulses of 0.5 msec duration at 50 Hz works well

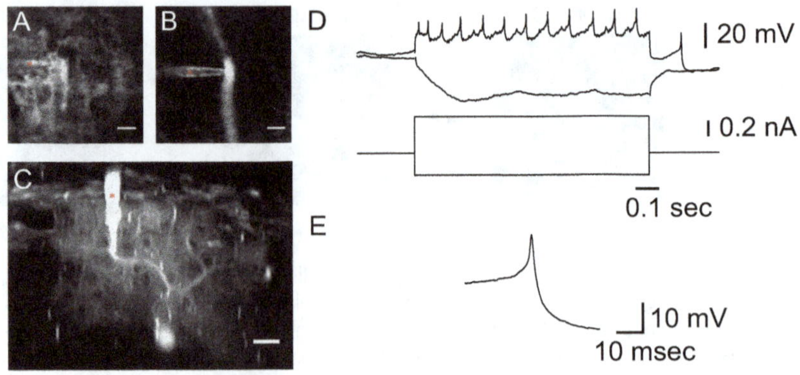

FIGURE 3. Targeted dendritic recording from cerebellar Purkinje cell in vivo. (*A*) Individual dendrites can be visualized using the shadow imaging technique. Red dot indicates the patch pipette. Scale bar, 5 µm. (*B*) Following break-in, the targeted dendrite is filled with dye. (*C*) *x*–*z* projection image of the entire cell after filling. Scale bar, 20 µm. (*D*) Voltage responses to current pulses (–0.3 and 0.9 nA). (*E*) Enlarged waveform of a simple spike at the dendritic recording site. (Unpublished results by Kazuo Kitamura.)

for layer 2/3 cortical neurons) via the electroporation device. Juxtacellular (or loose cell-attached) recordings of spontaneous or stimulus-evoked spikes can be made before the electroporation to characterize the targeted cell functionally. Successful electroporation can be visualized directly via the Alexa 594 included in the pipette solution by monitoring its injection into the cell, and subsequently by imaging expression of GFP (if plasmid DNA for GFP is also included in the pipette) (Fig. 4).

PRACTICAL CONSIDERATIONS

- An outstanding preparation is crucial for successful targeted recordings. Any preparation that exhibits visible blood, edema, and/or pulsation can prevent adequate visualization of the cell, seal formation, and/or maintenance of the recording. Great care should be taken when performing craniotomies, removing dura, and mounting and securing recording chambers to minimize damage (for detailed procedures and troubleshooting tips, see Komai et al. 2006; Judkewitz et al. 2009).
- Brain movement must be rigorously minimized to maximize the stability of the preparation and duration of recordings. This can be achieved by covering the brain with 200 µL of 1.5% (w/v)

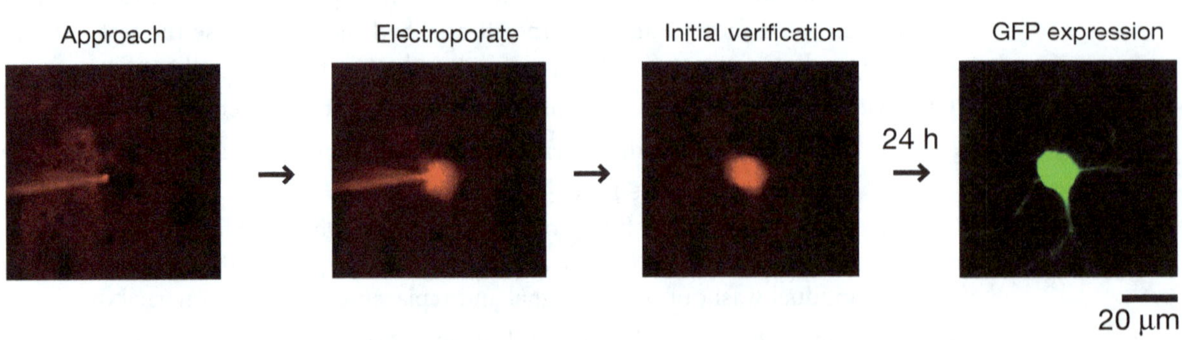

FIGURE 4. Targeted electroporation in vivo. Negative contrast image of a mouse neocortical layer 2/3 pyramidal neuron before electroporation, during electroporation, and a few minutes after electroporation (filled with Alexa 594). On the *far right* is an image of the same neuron taken 24 h later, showing bright eGFP fluorescence throughout the dendritic tree. Scale bar, 20 µm. (Unpublished data from Benjamin Judkewitz.)

agarose and a glass coverslip (optional) in conjunction with an appropriate chamber (e.g., see Judkewitz et al. 2009).

- Clogged or contaminated electrodes can prevent gigaohm seal formation. This should be avoided by rigorously filtering pipette solutions (both for storage and on day of use), ensuring that the pipette enters the brain under high pressure, and verifying that a plume of solution is exiting the mouth of the pipette at all times. Clogged pipettes can sometimes be unclogged using a brief pulse of high pressure (>200 mbar), although this may dramatically raise background fluorescence in the tissue, limiting visibility.
- As in two-photon microscopy in general, laser illumination should be minimized, particularly during seal formation, which can be compromised by high laser powers (Komai et al. 2006). The laser power used should be the minimum consistent with adequate resolution of the pipette and targeted cell.

EXAMPLES OF APPLICATIONS

Targeted patch-clamp recordings have been used to make recordings from specific cell types in intact mammalian networks. For example, recordings have been targeted to genetically identified interneurons (Margrie et al. 2003; Liu et al. 2009) and genetically modified neurons labeled with GFP (Dittgen et al. 2004; Komai et al. 2006). Furthermore, this approach can be used to record from neurons that have been characterized functionally using synthetic (Greenberg et al. 2008; Rochefort et al. 2009; Schultz et al. 2009) or genetically encoded (Wallace et al. 2008) Ca^{2+} sensors. Such an approach is particularly valuable for calibration of Ca^{2+} signals in terms of the underlying temporal spiking patterns, which are notoriously difficult to extract quantitatively from Ca^{2+} signals alone and require sophisticated algorithms that must be optimized based on combined imaging and electrophysiological data (Yaksi and Friedrich 2006; Greenberg et al. 2008; Vogelstein et al. 2009).

One application that may prove to be particularly useful for the targeted approach is in vivo dendritic recording. There are only a few examples in the literature of "blind" recordings from dendrites in vivo (e.g., Llinas and Nicholson 1971; Helmchen et al. 1999; Larkum and Zhu 2002; Waters et al. 2003) owing in part to the small caliber of dendrites. As shown in Figure 3, it is possible using the targeted approach to make recordings directly from dendrites of visually identified neurons at defined locations in the dendritic tree (Kitamura et al. 2008). This approach may allow a more systematic investigation of the properties of dendrites in vivo.

Finally, using essentially the same approach, it is possible to electroporate visually and/or genetically identified neurons. This can be useful for injecting Ca^{2+} dyes into defined groups of neurons (Nevian and Helmchen 2007), providing more control over cell type and number than the use of genetically encoded or bulk-loaded indicators. Moreover, this approach can be used to genetically alter identified neurons via electroporation of plasmid DNA (Fig. 4; Kitamura et al. 2008; Judkewitz et al. 2009). This should allow the function of specific genes to be identified on the single-cell level in the absence of the more widespread deleterious effects normally associated with the use of transgenic mice or viral approaches. One of the great advantages of the single-cell electroporation approach is that it can be targeted to neurons identified by their functional profile by using the same electrode to make electrophysiological recordings from neurons before electroporation.

ADVANTAGES AND LIMITATIONS

Targeted recording allows the experimenter to select the cell type to explore. Thus, with the use of GFP and the appropriate promoters, genetically defined cell types can be targeted for electrophysiological recordings. If neuronal types have a distinctive morphological signature or location (e.g., in specific cell layers), shadowpatching can also be used to discriminate different cell types. Furthermore, the introduction or expression of activity-dependent probes (such as Ca^{2+} dyes) allows the functional signature of individual cells to be used as a criterion for targeting. Together, these approaches allow recordings to be targeted to rare or distinctive cell types based on their genetic, morphological, or functional signature, which would be extremely difficult or impossible using blind recording techniques.

The ability to visualize the cell before and during recording provides several additional advantages. First, cell properties can be assessed morphologically or functionally (using activity-dependent probes) before recording from the cell of interest. Second, the process of seal formation can be visualized directly: Not only can the position of the electrode on the cell be optimized (e.g., to allow dendritic recordings or target the pipette to the center of the soma), but dimple formation, often a valuable "signature" preceding an effective gigaohm seal, can be monitored and terminated at precisely the right moment before applying suction to form a seal. This can dramatically improve the success rates and quality of patch-clamp recordings. Third, the position of the electrode can be monitored during the recording and corrected, for example, for drift or movement. Finally, introduction of morphological markers or activity-dependent dyes via the patch pipette can allow morphology to be assessed in vivo during the experiment and also permit simultaneous measurement of electrical and chemical signaling in the same cell. Furthermore, it allows the experimenter to return to the same cell at later time points to assess changes in function following manipulations, either extrinsic or following genetic manipulation via electroporation (Kitamura et al. 2008).

The main limitation of the targeted approach, aside from requiring an expensive two-photon microscopy setup, is that it is limited to cells that can be visualized using two-photon microscopy. Currently this is possible down to a depth of up to ~1000 µm (Theer et al. 2003; Helmchen and Denk 2005; Theer and Denk 2006), which in principle allows targeted recording from most cortical layers in rodents but precludes targeting neurons in deeper brain structures. In the future, it may be possible to combine visualization of deep neurons, such as those in hippocampus, using microendoscopy (Jung et al. 2004) combined with patch-clamp recording, although this will require a strategy for positioning the patch pipette within the field of view of the microendoscope. Targeted patch-clamp recordings have also thus far been restricted to anesthetized, head-fixed animals. However, given recent progress in two-photon imaging from awake, head-fixed (Dombeck et al. 2007; Greenberg et al. 2008), and freely moving animals (Helmchen et al. 2001; Flusberg et al. 2005; Engelbrecht et al. 2008), as well as in patch-clamp recording from freely moving animals (Lee et al. 2006, 2009), one can envisage the possibility of targeted recordings from awake, head-fixed, and possibly even freely moving animals in the near future.

RECIPE

CAUTION: See Appendix 6 for proper handling of materials marked with <!>.
Recipes for reagents marked with <R> are included in this list.

Internal Recording Solution

Reagent	Final concentration
Potassium gluconate <!>	130 mM
Sodium gluconate	10 mM
HEPES	10 mM
Phosphocreatine	10 mM
MgATP	4 mM
GTP	0.3 mM
NaCl	4 mM
Biocytin (Sigma-Aldrich)	0.3%–1%
Alexa 594 (Molecular Probes)	0.06 mM

Adjust the pH to 7.2. Filter through a 0.45-µm syringe filter (Millipore) and store at –20°C. The same pipette solution can be used for electroporation. Alternatively, HEPES-buffered artificial cerebrospinal fluid or phosphate-buffered saline containing 50 µM Alexa 594 can also be used.

ACKNOWLEDGMENTS

We thank our colleagues in our respective laboratories for providing figures and Benjamin Judkewitz, Spencer Smith, and Christian Wilms for their comments on the manuscript. Work in our laboratories is funded by the Wellcome Trust and Gatsby Charitable Foundation (MH) and the Medical Research Council (TWM).

REFERENCES

Denk W, Strickler JH, Webb WW. 1990. Two-photon laser scanning fluorescence microscopy. *Science* **248:** 73–76.

Dittgen T, Nimmerjahn A, Komai S, Licznerski P, Waters J, Margrie TW, Helmchen F, Denk W, Brecht M, Osten P. 2004. Lentivirus-based genetic manipulations of cortical neurons and their optical and electrophysiological monitoring in vivo. *Proc Natl Acad Sci* **101:** 18206–18211.

Dombeck DA, Khabbaz AN, Collman F, Adelman TL, Tank DW. 2007. Imaging large-scale neural activity with cellular resolution in awake, mobile mice. *Neuron* **56:** 43–57.

Engelbrecht CJ, Johnston RS, Seibel EJ, Helmchen F. 2008. Ultra-compact fiber-optic two-photon microscope for functional fluorescence imaging in vivo. *Opt Express* **16:** 5556–5564.

Flusberg BA, Jung JC, Cocker ED, Anderson EP, Schnitzer MJ. 2005. In vivo brain imaging using a portable 3.9 gram two-photon fluorescence microendoscope. *Opt Lett* **30:** 2272–2274.

Garaschuk O, Milos RI, Konnerth A. 2006. Targeted bulk-loading of fluorescent indicators for two-photon brain imaging in vivo. *Nat Protoc* **1:** 380–386.

Greenberg DS, Houweling AR, Kerr JN. 2008. Population imaging of ongoing neuronal activity in the visual cortex of awake rats. *Nat Neurosci* **11:** 749–751.

Helmchen F, Denk W. 2005. Deep tissue two-photon microscopy. *Nat Methods* **2:** 932–940.

Helmchen F, Svoboda K, Denk W, Tank DW. 1999. In vivo dendritic calcium dynamics in deep-layer cortical pyramidal neurons. *Nat Neurosci* **2:** 989–996.

Helmchen F, Fee MS, Tank DW, Denk W. 2001. A miniature head-mounted two-photon microscope. High-resolution brain imaging in freely moving animals. *Neuron* **31:** 903–912.

Judkewitz B, Rizzi M, Kitamura K, Häusser M. 2009. Targeted single-cell electroporation of mammalian neurons in vivo. *Nat Protoc* **4:** 862–869.

Jung JC, Mehta AD, Aksay E, Stepnoski R, Schnitzer MJ. 2004. In vivo mammalian brain imaging using one- and two-photon fluorescence microendoscopy. *J Neurophysiol* **92:** 3121–3133.

Kerr JN, Greenberg D, Helmchen F. 2005. Imaging input and output of neocortical networks in vivo. *Proc Natl Acad Sci* **102:** 14063–14068.

Kitamura K, Judkewitz B, Kano M, Denk W, Häusser M. 2008. Targeted patch-clamp recordings and single-cell electroporation of unlabeled neurons in vivo. *Nat Methods* **5:** 61–67.

Komai S, Denk W, Osten P, Brecht M, Margrie TW. 2006. Two-photon targeted patching (TPTP) in vivo. *Nat Protoc* **1:** 647–652.

Larkum ME, Zhu JJ. 2002. Signaling of layer 1 and whisker-evoked Ca^{2+} and Na^{+} action potentials in distal and terminal dendrites of rat neocortical pyramidal neurons in vitro and in vivo. *J Neurosci* **22:** 6991–7005.

Lee AK, Manns ID, Sakmann B, Brecht M. 2006. Whole-cell recordings in freely moving rats. *Neuron* **51:** 399–407.

Lee AK, Epsztein J, Brecht M. 2009. Head-anchored whole-cell recordings in freely moving rats. *Nat Protoc* **4:** 385–392.

Liu BH, Li P, Li YT, Sun YJ, Yanagawa Y, Obata K, Zhang LI, Tao HW. 2009. Visual receptive field structure of cortical inhibitory neurons revealed by two-photon imaging guided recording. *J Neurosci* **29:** 10520–10532.

Llinas R, Nicholson C. 1971. Electrophysiological properties of dendrites and somata in alligator Purkinje cells. *J Neurophysiol* **34:** 532–551.

Lütcke H, Murayama M, Hahn T, Margolis DJ, Astori S, Zum Alten Borgloh SM, Göbel W, Yang Y, Tang W, Kügler S, et al. 2010. Optical recording of neuronal activity with a genetically-encoded calcium indicator in anesthetized and freely moving mice. *Front Neural Circuits* **4:** 9.

Margrie TW, Brecht M, Sakmann B. 2002. In vivo, low-resistance, whole-cell recordings from neurons in the anaesthetized and awake mammalian brain. *Pflugers Arch* **444:** 491–498.

Margrie TW, Meyer AH, Caputi A, Monyer H, Hasan MT, Schaefer AT, Denk W, Brecht M. 2003. Targeted whole-cell recordings in the mammalian brain in vivo. *Neuron* **39:** 911–918.

Nevian T, Helmchen F. 2007. Calcium indicator loading of neurons using single-cell electroporation. *Pflugers Arch* **454:** 675–688.

Rochefort NL, Garaschuk O, Milos RI, Narushima M, Marandi N, Pichler B, Kovalchuk Y, Konnerth A. 2009. Sparsification of neuronal activity in the visual cortex at eye-opening. *Proc Natl Acad Sci* **106:** 15049–15054.

Schultz SR, Kitamura K, Post-Uiterweer A, Krupic J, Häusser M. 2009. Spatial pattern coding of sensory information by climbing fiber-evoked calcium signals in networks of neighboring cerebellar Purkinje cells. *J Neurosci* **29:** 8005–8015.

Stosiek C, Garaschuk O, Holthoff K, Konnerth A. 2003. In vivo two-photon calcium imaging of neuronal networks. *Proc Natl Acad Sci* **100:** 7319–7324.

Svoboda K, Denk W, Kleinfeld D, Tank DW. 1997. In vivo dendritic calcium dynamics in neocortical pyramidal neurons. *Nature* **385:** 161–165.

Theer P, Denk W. 2006. On the fundamental imaging-depth limit in two-photon microscopy. *J Opt Soc Am A Opt Image Sci Vis* **23:** 3139–3149.

Theer P, Hasan MT, Denk W. 2003. Two-photon imaging to a depth of 1000 µm in living brains by use of a Ti:Al_2O_3 regenerative amplifier. *Opt Lett* **28:** 1022–1024.

Tian L, Hires SA, Mao T, Huber D, Chiappe ME, Chalasani SH, Petreanu L, Akerboom J, McKinney SA, Schreiter ER, et al. 2009. Imaging neural activity in worms, flies and mice with improved GCaMP calcium indicators. *Nat Methods* **6:** 875–881.

Tsai PS, Nishimura N, Yoder EJ, Dolnick EM, White GA, Kleinfeld D. 2002. Principles, design and construction of a two-photon laser-scanning microscope for in vitro and in vivo brain imaging. In *In vivo optical imaging of brain function* (ed. Frostig RD), pp. 113–171. CRC Press, Boca Raton, FL.

Vogelstein JT, Watson BO, Packer AM, Yuste R, Jedynak B, Paninski L. 2009. Spike inference from calcium imaging using sequential Monte Carlo methods. *Biophys J* **97:** 636–655.

Wallace DJ, Meyer zum Alten Borgloh S, Astori S, Yang Y, Bausen M, Kügler S, Palmer AE, Tsien RY, Sprengel R, Kerr JN, et al. 2008. Single-spike detection in vitro and in vivo with a genetic Ca^{2+} sensor. *Nat Methods* **5:** 797–804.

Waters J, Larkum M, Sakmann B, Helmchen F. 2003. Supralinear Ca^{2+} influx into dendritic tufts of layer 2/3 neocortical pyramidal neurons in vitro and in vivo. *J Neurosci* **23:** 8558–8567.

Yaksi E, Friedrich RW. 2006. Reconstruction of firing rate changes across neuronal populations by temporally deconvolved Ca^{2+} imaging. *Nat Methods* **3:** 377–383.

48 Imaging with Voltage-Sensitive Dyes
Spike Signals, Population Signals, and Retrograde Transport

Bradley Baker,[1] Xin Gao,[2] Brian S. Wolff,[2] Lei Jin,[1] Lawrence B. Cohen,[1,3] Chun X. Bleau,[1,3] and J.-y. Wu[2,4]

[1]Department of Cellular and Molecular Physiology, Yale University School of Medicine, New Haven, Connecticut 06520; [2]Department of Physiology and Biophysics, Georgetown University, Washington, D.C. 20007; [3]RedShirtImaging LLC, Decatur, Georgia 30030; [4]WuTech, Instruments, Inc., Gaithersburg, Maryland 20877

ABSTRACT

Optical recording of membrane potential allows simultaneous measurements to be taken from many different locations in the nervous system. This is important in the studies of the nervous system in which simultaneous activity can occur at the regional, cellular, and subcellular levels. This chapter discusses the use of fluorescence and absorption probes of membrane potential. (For a discussion of second-harmonic generation probes, see Wilson et al. 2011, and for a discussion of intrinsic signals, see Chapters 60 and 83.) We describe techniques for optimizing the signal-to-noise ratio in these measurements and then present protocols addressing (1) single-neuron recordings from an invertebrate preparation, (2) population signals from mammalian preparations, and (3) retrograde staining. The chapter ends with a discussion of future directions.

INTRODUCTION

Optical recording of membrane potential allows simultaneous measurements to be taken from many different locations in the nervous system, an important issue in nervous system studies in which simultaneous activity can occur at the regional, cellular, and subcellular levels. A voltage-sensitive dye signal reflects the changes in transmembrane potential while ignoring the currents in the tissue caused by volume conductance, thus allowing a more accurate localization of the source of the activity in comparison to local field potential measurements (Zochowski et al. 2000; Huang et al. 2004).

Our discussion is limited to "fast" signals (Cohen and Salzberg 1978). These signals occur only at the wavelengths related to the dye chromophore and are presumed to arise from the membrane-bound dye. They follow changes in membrane potential with time courses that are rapid compared with the rise time of an action potential. Protocols addressing three types of experiments will be presented: (1) action potential signals from individual neurons, (2) population signals in which light from thousands of neurons falls on each pixel, and (3) signals from hydrophobic dyes that have been retrogradely transported from axon to the cell body. A fourth application of voltage-sensitive dye imaging, intracellularly injected dye, is described in Chapter 29.

All of the results presented here are from the use of "organic" voltage-sensitive dyes. The development of genetically encoded voltage sensors (fluorescent protein [FP] voltage sensors) is discussed briefly at the end of this chapter and in Chapter 7.

Measuring Small Signals

Voltage-sensitive dye signals from bath-applied dye have been used in many different preparations, including individual neurons in tissue culture (Parsons et al. 1991), neurons in ganglia (London et al. 1987; Nakashima et al. 1992; Obaid et al. 1999), population signals from brain slices (e.g., Grinvald et al. 1982; Albowitz and Kuhnt 1993; Komuro et al. 1993; Sato et al. 1998; Wu et al. 2001; Bao and Wu 2003; Huang et al. 2004; Hirata and Sawaguchi 2008), and in vivo preparations (Orbach et al. 1985; Prechtl et al. 1997; Lam et al. 2003; Lippert et al. 2007; Xu et al. 2007). In addition, signals have been recorded from retrogradely and anterogradely transported dyes in embryonic chick spinal cord (Tsau et al. 1996), lamprey cord (Hickie et al. 1996), and zebrafish olfactory neurons (Friedrich and Korsching 1998).

As a result of these experiments, it is possible to anticipate the fractional intensity changes that might be obtained from available dyes in a new experiment. Quite often these fractional changes are small; the $\Delta I/I$, or fractional change, of the signal can be $\leq 10^{-3}$. To record such signals, the noise must be 10 times smaller. Two examples of such signals are shown in Figure 1. Figure 1A shows the outputs of four photodiodes in a measurement of absorption signals from individual neurons of an *Aplysia* abdominal ganglion. Figure 1B shows population signals from a cortical slice illustrating spontaneous oscillations. In these examples, the fractional changes in the signal intensity are in the range 10^{-4}–10^{-3}, and the noise is $<10^{-4}$. Because these signals are not tightly time-locked to a stimulus, multiple trial signal averaging could not be used to improve the signal-to-noise ratio. Obviously, making a measurement in which the noise is $<10^{-4}$ of the total light requires that all sources of noise must be $<10^{-4}$. Five noise sources are briefly discussed in the following sections.

Noise in the incident light: The light emission from a tungsten filament lamp is easily stabilized to one part in 10^{-5}. Light-emitting diode sources (e.g., OptoLED, Cairn Research Ltd., Faversham, UK) are also stable and they avoid the vibrations associated with opening of a mechanical shutter. If a higher-intensity source is needed, a 75- or 150-W xenon arc lamp and power supply (Cairn Research Ltd.) have intensity fluctuations that are not too much larger than 10^{-4} of the incident light. For even higher intensities, a laser source can be used (Chapter 29). Higher intensities are usually associated with shorter recording times because of the bleaching and phototoxic effects of the dyes.

Vibrational noise: Reducing the vibrational noise to $<10^{-4}$ requires some effort. The power spectrum of the signals can help to determine the source of the noise; for example, 20–50-Hz vibrations are usually caused by ventilation fan motors attached to the building structure; vibrations below 5 Hz may come from movement of the building in the wind. A number of precautions for reducing vibrational noise can be taken (Salzberg et al. 1977; London et al. 1987). A coverslip or water-immersion lens will reduce vibration noise from the air–water interface. As a starting point for vibration isolation, try air-filled, soft rubber tubes (Newport Corporation). For more severe vibration problems, a BM-1 bench top isolation table sold by Minus K Technology (Inglewood, CA; www.minusk.com) will help. With their very low resonant frequency (~0.5 Hz) in both vertical and horizontal directions, a large reduction of vibration noise in the frequency range of biological signals can be obtained.

Dark noise: Noise in the apparatus that occurs in the absence of illumination can result from electron generation in the photodetector in the dark or from amplifier nosie. This dark noise will be

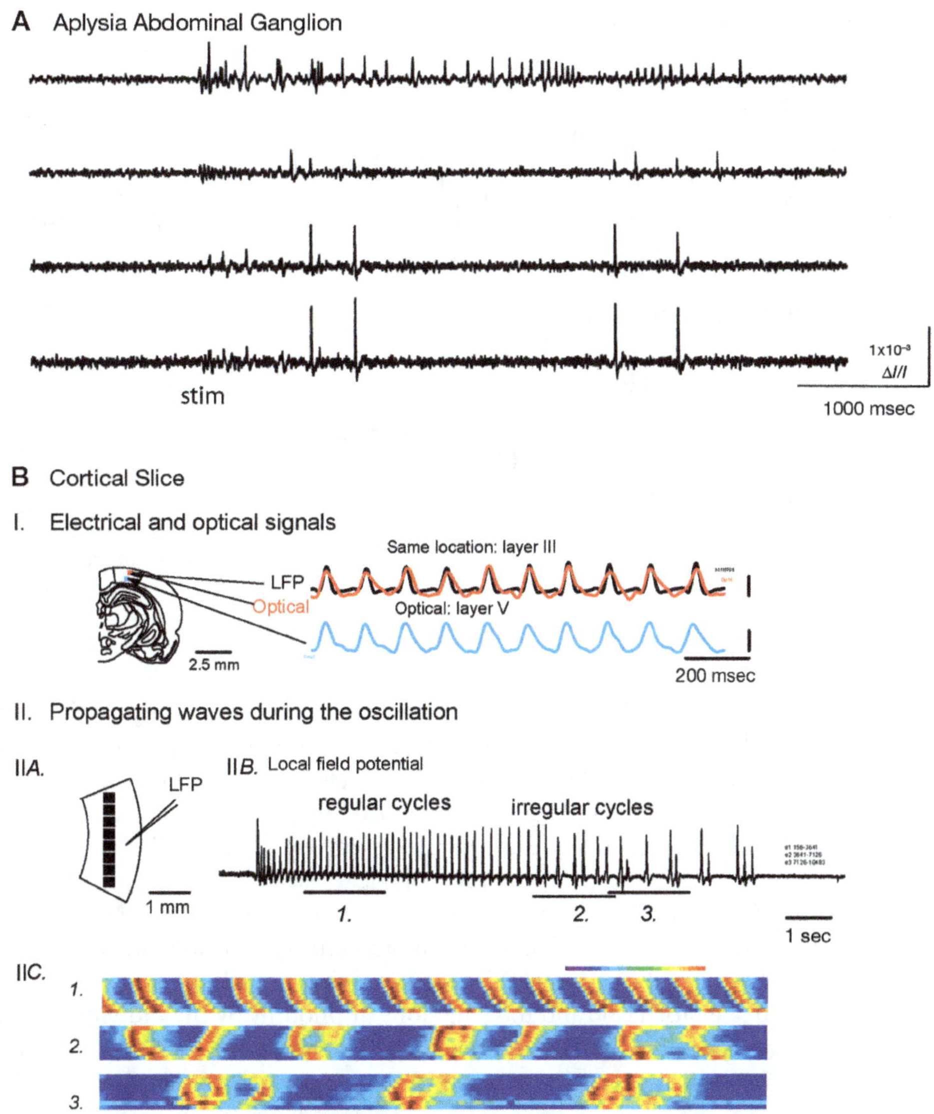

FIGURE 1. Voltage-sensitive dye signals from two different preparations. In both, the signal represents a fractional intensity change of $\leq 10^{-3}$ and the noise in the measurement is $\leq 10^{-4}$. The staining must be sufficiently intense, the optics sufficiently efficient, and the array and amplifiers good enough to provide this kind of noise performance. (*A*) Action potential signals from four photodiodes receiving light from four areas of an *Aplysia* abdominal ganglion. An absorption dye (an oxonol, RH155) and transillumination were used. A light touch to the siphon skin was applied at the time labeled "stim"; vibrational noise from the stimulus can be seen in the second trace. Spike signals are seen on all four detectors (from data collected for the experiments in Falk et al. 1993). (*B*) Population oscillations in a rat cortical slice. Absorption dye (an oxonol, RH482) and transillumination were used. (*I, left*) The recording was performed in a slice containing occipital cortex. A microelectrode recorded the local field potential (LFP) from cortical layers 2/3. Optical signals from the tissue surrounding the electrode (red square) and from layer 5 of the same column (blue square) were simultaneously recorded with the LFP. (*I, right*) In layers 2/3, simultaneous electrical (black) and optical (red) recordings had a similar waveform during the oscillations. In layer 5, the optical signal (blue) has the same polarity. Calibration bar, –50 μV for LFP and 10^{-4} for optical recordings. (Reprinted, with permission, from Bao and Wu 2003.) (*IIA*) Eight optical detectors (black squares) were horizontally distributed in the deep cortical layers. Their signals were used in the images in *IIC*. A local field potential microelectrode was placed in layers 2/3. (Modified, with permission, from Bao and Wu 2003.) (*IIB*) An oscillation epoch contains regular and irregular cycles. Optical signals during the periods 1, 2, and 3 (marked under the trace) are shown in *IIC*. (*IIC*) Pseudocolor images from optical signals recorded by the eight detectors. The optical signal from each detector was normalized to the maximum on that particular detector during that particular period and normalized values were assigned to colors according to a color scale (*top right*). The *x* direction of the images is time (total time ~12 sec) and the *y* direction is space in the horizontal direction in deep cortical layers, ~2.6 mm. *IIC1* are propagating waves and *IIC2* and *IIC3* are irregular patterns. (Modified, with permission, from Bao and Wu 2003.)

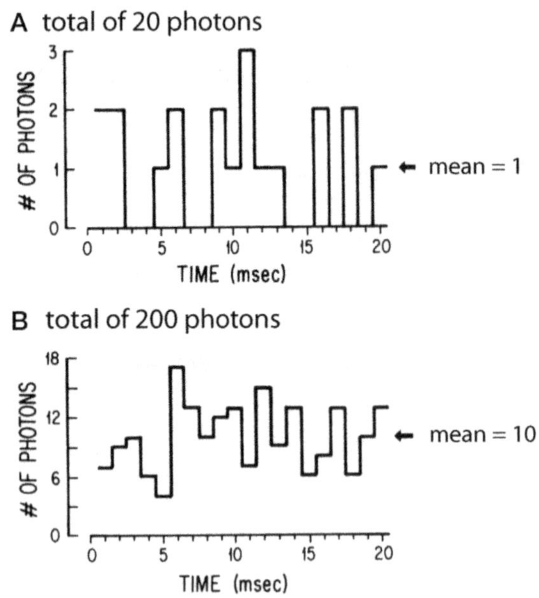

FIGURE 2. Plots of the results of using a table of random numbers to distribute 20 photons (*A*) or 200 photons (*B*) into 20 time bins. The result illustrates the fact that when more photons are measured the noise relative to the average intensity (signal-to-noise ratio) is improved. (Modified from Cohen and Lesher 2003.)

of concern in situations in which the light level is low (e.g., a single neuron is stained and only a small fraction of that neuron is imaged onto one pixel [Chapter 29]). In other circumstances (e.g., the experiments in Fig. 1), the shot noise (see below) is much larger than the dark noise. If dark noise is a major contributor, it can be reduced by using a cooled charge-coupled device (CCD) camera. Very dim images that require an image intensifier or a CCD chip with a gain stage (e.g., e2v CCD60) are rarely encountered in neurobiological research.

Shot noise: The limit of accuracy with which light can be measured is set by the shot noise arising from the statistical nature of photon emission from light sources and fluorophores. (In filament and arc lamps the interval between emission of photons from an individual element of the light source is stochastic and the emissions of photons by the individual elements are independent.) In a shot noise–limited measurement, the signal-to-noise ratio is proportional to the square root of the number of measured photons and the bandwidth of the photodetection system (Malmstadt et al. 1974). The basis for the square root dependence on intensity is illustrated in Figure 2. Figure 2A shows the result of using a random number table to distribute 20 photons into 20 time windows and Figure 2B shows the same procedure being used to distribute 200 photons into the same 20 bins. Clearly, relative to the average light level, there is more noise in the top trace (20 photons) than in the bottom trace (200 photons). In Figure 2, a signal-to-noise ratio of ~3 was obtained with a photon count of 10 photons/bin. Following the square root relationship, a signal-to-noise ratio of 10^{-4} will require a photon count of 10^8 photons/bin.

This result places two constraints on the apparatus. First, the optics must be efficient and the fluorescent dyes bright enough to generate 10^8 photons/time bin per region of interest at the imaging device. In epi-illumination measurements, both the excitation light and the emitted light pass through the objective; the intensity reaching the photodetector is proportional to the fourth power of numerical aperture (NA). Unfortunately, conventional microscope optics have low NAs at low magnifications. Salama (1988), Ratzlaff and Grinvald (1991), and Kleinfeld and Delaney (1996) took advantage of the high NA that can be achieved by using a camera or video lens in place of a microscope objective. At a magnification of 4x, a Macroscope (RedShirtImaging) based on a 25-mm focal length, *f*/0.95 camera lens had an image intensity 100 times greater than a conventional 4x, 0.16-NA microscope lens. Second, the imaging apparatus must have a dynamic range (>80 db or >13 bits) large enough to measure a signal of 10^4 photons in a background of 10^8 photons.

Choice of cameras: Several factors affect the determination of the optimal camera. These include dynamic range, spatial resolution, temporal resolution, and dark (read) noise. The photodiode array used in the protocols described here has a very large dynamic range but also a relatively large dark noise and a limited spatial resolution. In Protocol D efforts were made to increase the measured intensity so that the noise was not dominated by the dark noise of the photodiode array. Although the spatial resolution of the photodiode array is limited, in in vitro (Protocol B) and in vivo (Protocol C) preparations the actual spatial resolution is limited by the light scattering of the preparation rather than the photodiode array camera. If additional spatial resolution is needed along with a large dynamic range, fast CMOS (complementary metal oxide semiconductor) cameras are available (e.g., RedShirtImaging or SciMedia USA Ltd.)

In situations where the signals are larger ($\geq 10^{-3}$) but the light intensity is smaller (measurements from individual dendrites, spines, or axons), a camera with lower dark noise and increased spatial resolution is optimal. Some CCD cameras meet this requirement.

Protocol A

Recording Action Potentials from Individual Neurons in Invertebrate Ganglia

The original motivation for developing optical methods was to record all of the action potential activity of all of the neurons in simple invertebrate ganglia during behaviors. In the first attempt to use voltage-sensitive dyes in ganglia (Salzberg et al. 1973), we were fortunate to be able to monitor activity in a single neuron because of severe photodynamic damage and a low signal-to-noise ratio. Now, however, with better dyes and methods, the action potential activity of hundreds of individual neurons can be recorded simultaneously (Zecevic et al. 1989; Nakashima et al. 1992; Obaid et al. 1999; Frost et al. 2010). In the measurements described below, the action potential activity of about half of the 1000 cells in the *Aplysia* abdominal ganglion is measured.

IMAGING SETUP

Place a photodiode array (NeuroPDA, RedShirtImaging) in the image plane formed by a 25x 0.4-NA microscope objective. Because the very irregular image of a ganglion is formed on a rectilinear diode array, there is no simple correspondence between images of cells and individual pixels. The light from a larger cell body will fall on several pixels and its activity will be recorded as simultaneous events on neighboring pixels (e.g., the signals on the bottom two traces in Fig.1A). In addition, because these preparations are multilayered, most detectors will receive light from several cells (e.g., the two different signals on the second trace in Fig. 1A). As a result, a sorting step is required to determine the activity in neurons from the spike signals on individual pixels. This sorting can be performed using a computer program assisted by the human pattern recognition system or by using independent component analysis (ICA; Brown et al. 2001). Additional suggestions for improving the signal-to-noise ratio of recordings from invertebrate ganglia can be found in Frost et al. (2010).

MATERIALS

Reagents

Aplysia intact ganglia (see Nakashima et al. 1992)
Bathing solution (artificial seawater; Zecevic et al. 1989.)
Oxonol dye RH155 (or its diethyl analog)

EXPERIMENTAL METHOD

Dye Staining

Considerable effort has been made to optimize dye staining to increase the signal while causing minimal pharmacological effects on the gill-withdrawal behavior.

1. Incubate intact ganglia in a 0.15 mg/mL solution of the oxonol dye RH155 (or its diethyl analog) using a protocol developed by Nakashima et al. (1992). Note that oxonol dyes are moder-

ately hydrophobic. More hydrophobic dyes do not penetrate through the connective tissue sheath, whereas a less hydrophobic dye destains too rapidly.

2. Once the ganglia are stained, wash the dye out of the bathing solution.
3. Record signals for several hours using the imaging setup described above.

DISCUSSION

The result of a complete analysis of one data set showed that 135 *Aplysia* neurons were activated by a light touch to the siphon (Wu et al. 1994). Similar large numbers of activated neurons were obtained by Nakashima et al. (1992) and Brown et al. (2001) in *Tritonia*. The *Aplysia* recording was estimated to be only ~50% complete. Thus, the actual number of activated neurons in the abdominal ganglion during the gill-withdrawal reflex was ~300. It is as if the *Aplysia* nervous system is designed such that every cell in the abdominal ganglion "cares" about this (and perhaps every modest to strong) sensory stimulus. In addition, more than 1000 neurons in other ganglia are activated by this touch (Tsau et al. 1994). Clearly, information about this very mild and localized stimulus is propagated widely in the *Aplysia* nervous system.

Protocol B

Voltage-Sensitive Dye Imaging of Population Signals in Brain Slices

In a bright-field measurement from a vertebrate brain stained by superfusing a solution of the dye over the surface, each pixel in a camera receives light from a substantial number (thousands) of neurons and neuronal processes. Because of scattering and out-of-focus light, this will be true even if the pixel size corresponds to a small area of the brain. In this situation, the voltage-sensitive dye signal will be a population average of the change in membrane potential of all of these neurons and processes.

During the last 30 years, many investigators have published voltage-sensitive dye imaging studies of population activities in brain slices. Their methods, including choice of dyes, illumination intensity, and imaging device, vary across a large spectrum. We present two protocols for visualizing spatiotemporal patterns in rodent neocortex. Detecting these patterns requires high-sensitivity imaging in single trials, because averaging will obscure the complex dynamics of the spatiotemporal patterns.

IMAGING SETUP

Imaging Chamber

During imaging, the brain slice is mounted in a perfusion chamber on a microscope stage. The slice is set on a mesh in the center of the chamber such that both sides of the slice are well perfused. The holding solution (artificial cerebrospinal fluid [ACSF], see below) is warmed before perfusion with fine Tygon tubing wound on a temperature-controlled heating block. The solution is heated slowly from room temperature to 28°C–32°C while it flows through the 100-cm-long tubing. Perfusing with high flow rate (3–6 mL/min) and on both sides of the slice is important for maintaining network activities in submerged chambers (Wu et al. 2005; Hajos and Mody 2009).

Imaging Apparatus

The optical signal from absorption dyes in brain slices ranges from 10^{-5} to 10^{-3}. We used a photodiode array (WuTech Instruments, Gaithersburg, MD) for imaging. The array consists of 464 individual photodiodes, each glued to an end of an optical fiber (750 µm in diameter). The optical fibers are bundled coherently into a 19-mm hexagonal aperture. Photocurrent from each diode is fed into a two-stage amplifier circuit. The output of the first-stage amplifier has a very small signal of 0.05–5 mV riding on top of a large 5-V resting light signal. The resting voltage is removed by a hardware DC (resting voltage) subtraction circuit between the first- and second-stage amplifiers. The second stage has a 100x–1000x voltage gain. As a result, the two-stage amplifier construction can extend the dynamic range to ~19 bits (Cohen and Lesher 1986; Wu and Cohen 1993). We use a 14-bit data acquisition board (Microstar Laboratories, Bellevue, WA) installed in a desktop personal computer. A sampling speed of 1000–2000 frames/sec is adequate for imaging oscillations in brain slices.

Microscope, Light Source, and Optical Filters

For imaging with absorption dyes, inexpensive low-NA objectives (e.g., 5x 0.12-NA, Zeiss) and a conventional 100-W halogen–tungsten filament lamp can provide excellent results. The stained slice has a peak of absorption of ~670 nm, which stays the same after a long wash with dye-free ACSF. The light transmission at the absorption peak (670 nm) is reduced to 1/10–1/50 of that of the unstained slice.

The dye signal ($\Delta I/I$) and signal-to-noise ratio are maximal at 705 nm. The signal decreases below 690 nm and reaches a 0 at ~675 nm. The signal polarity reverses at wavelengths shorter than 670 nm (Jin et al. 2002). The signal reaches a second maximum at 660 nm. For nonratiometric measurement, band-pass filtering around 705 nm should yield the largest signal; an alternating illumination of 700 and 660 nm may be used for ratiometric measurements. Illumination at 675 nm results in a minimum voltage-sensitive dye signal that can be used for distinguishing dye signals from broadband intrinsic signals. The bandwidth of the filter depends on other factors: When the signal is small and shot noise is a limiting factor, a wider band pass (e.g., 715 ± 30 nm) allows more light and thus increases the signal-to-noise ratio. When measuring large signals and when long total recording time is needed, a narrower band filter of 705 ± 15 can be used for reducing dye bleaching. The heat filter (infrared filter) in the incident light path should be removed.

Data Acquisition and Analysis

NeuroPlex, developed by Avrum Cohen and Chun Bleau of RedShirtImaging, is used for acquiring and analyzing imaging data. The program can display the data in the form of traces for numerical analysis and pseudocolor images for studying the spatiotemporal patterns. Figure 1B shows examples of data display in which signal from the local field potential (LFP) electrode and that of an optical detector viewing the same location are plotted together. The spatiotemporal patterns of the activity are often presented with pseudocolor maps, where the signals from each individual detector are normalized to their own maximum amplitude.

MATERIALS

See the end of the chapter for recipes for reagents marked with <R>.

Reagents

Brain, rodent
Holding artificial cerebrospinal fluid <R>
Modified artificial cerebrospinal fluid <R>, used at 4°C
Oxonol dye NK3630 (available from Nippon Kankoh-Shikiso Kenkyusho, Japan or Charkit Chemical Corporation, Darien, CT)

Equipment

Holding chamber
Magnetic stir bar
Staining chamber
Vibratome <!> stage (e.g., 752M Vibroslice, Campden Instruments Ltd., Sarasota, FL)

EXPERIMENTAL METHOD

Slice Preparation

For imaging population events, local circuits need to be well preserved with a large portion of viable neurons and functional synapses. Thin slices usually have a poor signal-to-noise ratio, owing to the smaller number of cells and to larger relative cell damage. We prefer slices with a thickness of 400–600 µm. Brain slices can be cut in several directions to conserve different aspects of cortical circuitry. Oblique directions can preserve the thalamocortical connections (Agmon and Connors 1992; Metherate and Cruikshank 1999; MacLean et al. 2006). Slices cut tangentially can best preserve the horizontal connections in layers 2/3 to observe two-dimensional waves such as spirals (Huang et al. 2004).

1. Harvest the whole brain carefully and chill for 90 sec in cold (0°C–4°C), modified ACSF.
2. Block the brain and slice it on a vibratome stage.
3. Place the brain slices into a holding chamber that contains holding ACSF. Holding chambers with a large volume (500 mL) of ACSF can better preserve local circuits.
4. Bubble the holding solution with a mixture of 95% O_2 and 5% CO_2, and slowly circulate it with a magnetic stir bar. Slices can be kept viable in this setting for 8–12 h.

   Stirring provides a slow but steady convection around the slice for delivering oxygen and washing out noxious molecules produced by the cold shock and the trauma of slicing.

Dye Staining

Absorption signals are larger than fluorescent signals in brain slices, as predicted and measured (Cohen and Lesher 1986; Jin et al. 2002).

5. Prepare the staining chamber containing ~50 mL of holding ACSF, bubbled slowly with a mixture of 95% O_2 and 5% CO_2 (about 1 bubble/sec) and circulated by a magnetic stir bar.

   Dyes have a tendency to be concentrated on the air–water interface, and slower bubbling with large bubbles can reduce the depletion of the dye from the solution.

6. In the example described in this section (see Fig. 1), the slices were stained with 5–20 µg/mL of an oxonol dye, NK3630 (first synthesized by R. Hildesheim and A. Grinvald as RH482; see Momose-Sato et al. 1995 for molecular structure). We prefer a lower dye concentration and longer staining time (1–2 h) because it results in more even staining through the thickness of the tissue.

Sensitivity, Phototoxicity, Bleaching, and Total Recording Time

The sensitivity of absorption measurements in a slice can be similar to that of local field potential measurements from the same preparation (Fig. 1B) (Jin et al. 2002). Imaging with high sensitivity is important not only for detecting small population activity signals but also for detecting spatiotemporal patterns without averaging across trials. Phototoxicity or photodynamic damage (Cohen and Salzberg 1978) in cortical slices appears to be irreversible even after a long dark recovery period. Absorption dyes have much less phototoxicity than the fluorescent dye RH795 in brain slice experiments (Jin et al. 2002). Intermittent light exposure also significantly reduces phototoxicity. With the same intensity (10^{12} photons/msec per mm^2), the phototoxicity using NK3630 became less significant when the exposure was broken into 2-min sessions with 2-min dark intervals; the LFP signals decreased <30% after 60 min of total exposure time. Optical signals, however, decreased ~90%. We refer to the reduction in optical signal while LFP signals remain unchanged as "bleaching." After a long light exposure, bleaching is visible even by eye as loss of color in the exposed area. Optical signals cannot be recovered in the bleached area even if the preparation is restained.

In the example shown in Figure 1B, oscillations induced by carbachol were manifested as propagating waves. Each oscillation cycle was associated with one propagating wave (Fig. 1B, IIC, image 1). This one-cycle one-wave pattern was seen during all oscillation cycles. Some cycles are "irregular," with the initiation site of the wave and the propagating velocity varying from cycle to cycle (Fig. 1B, IIC, images 2 and 3). Complex patterns, such as collision and reflection of the waves, were also observed. In one of the cycles, the wave was apparently reflected at the edge of the tissue and the reflected waves propagated back and collided in the center, forming an "O" pattern (Fig. 1B, IIC, image 3).

Protocol C

In Vivo Voltage-Sensitive Dye Imaging of Mammalian Cortex Using "Blue" Dyes

New "blue" dyes, developed by Amiram Grinvald's group (Shoham et al. 1999; Chapter 74), are a great advance for in vivo voltage-sensitive dye imaging of mammalian cortex. The blue dyes are excited by red light (630 nm) that does not overlap with light absorption of hemoglobin (510–590 nm). This virtually eliminates the heart pulsation artifact.

IMAGING SETUP

The cortex is imaged by a 5× macroscope with a field of view of ~4 mm in diameter. A halogen tungsten filament lamp (12 V, 100 W, Zeiss) is used for illumination. Light is filtered with a 630 ± 15-nm interference filter (Chroma Technology) and then reflected down onto the cortex via a 655-nm dichroic mirror (Chroma Technology). Köhler illumination is achieved through the macroscope. The fluorescence of the dye (~700 nm, Optical Imaging) from the stained cortex is collected via the macroscope, filtered through a 695-nm long-pass filter (RG-695, Edmund Scientific), and projected onto the fiber optic aperture of the photodiode array. Each detector of the array receives light from a cortical area of ~160 μm in diameter.

MATERIALS

CAUTION: See Appendix 6 for proper handling of materials marked with <!>.

Reagents

Atropine sulfate <!> (40 μg/kg)
Dexamethasone sulfate <!> (1 mg/kg)
Dye RH 1691 (Optical Imaging)
Holding artificial cerebrospinal fluid
Isoflurane <!>, 2%–3%, in air
Sprague Dawley and Long Evans rats, 250–400-g

Equipment

Calibrating multigas monitor (BCI 9100)
Catheter, 16-gauge, over-the-needle
Perfusion pump, custom-made (see London et al. 1989)
Silicone valve grease
Silicon oil, high-viscosity, 60,000 centistokes (Sigma-Aldrich DMPS-60M)
Small animal respirator (Harvard Apparatus)
Stereotaxic frame with a regulated heating pad
Surgical adhesive (Vetbond, 3M)
Surgical drill

EXPERIMENTAL METHOD

Surgery

1. Pretreat Sprague Dawley and Long Evans rats (250–400 g) with atropine sulfate (40 µg/kg intraperitoneally [i.p.]) ~30 min before anesthetic induction to reduce mucus secretion.
2. Anesthetize the animals with 2%–3% isoflurane in air.
3. Perform a tracheostomy by inserting a 16-gauge over-the-needle catheter into the trachea.
4. Ventilate the animal using a small animal respirator (Harvard Apparatus) with isoflurane in room air. Adjust the respiration rate (60–100 /min) and volume (2–4 mL) so that inspiratory pressure is ~5 cm H_2O and the end-tidal carbon dioxide is maintained at ~26–28 mm Hg with a calibrating multigas monitor.
5. Place the animal in a stereotaxic frame with a regulated heating pad. As soon as the animal is secure, reduce the isoflurane to 1.5%–2% for craniotomy.
6. Clean the cranial surface thoroughly of soft tissue, and then apply a thin layer of surgical adhesive to prevent dye leakage during staining.
7. Drill a cranial window (~5 mm in diameter). Immerse the drill tip in ACSF to diffuse heat during drilling. Great care must be taken to avoid irritation of the cortex by heat or pressure during this procedure.
8. Carefully separate the bone from the dura. Leave the dura intact to reduce the heart beat movement artifact during optical recording (London et al. 1989; Lippert et al. 2007).

   Irritated brain can appear reddish owing to increased blood flow, which leads to poor staining. Dexamethasone sulfate (1 mg/kg i.p.) may be given 6–24 h before to the experiment to reduce the inflammatory response of the dura.

Staining

9. Dissolve the voltage-sensitive dye RH 1691 at 2 mg/mL in ACSF solution.
10. Wash the surface of the dura with ACSF, and then remove the fluid thoroughly by suction. Allow the dura to dry completely for 3–4 min, with gentle air flow, until the tissue becomes very transparent and "glassy." This "glassy" appearance is essential for good staining. Drying the dura before staining increases its permeability to the dye.
11. Staining is performed through the intact dura. Construct a temporary staining chamber on the surrounding bone using silicone valve grease.
12. Use ~200 µL of dye solution (~2 mg/mL of RH 1691) to stain an area of the dura 5 mm in diameter. During staining, circulate the dye solution continuously using a custom-made perfusion pump (London et al. 1989).

    The pump has a battery-powered gear motor that gently presses the rubber nipple of a pasteur pipette once every few seconds. The tip of this pipette is placed in the staining solution and performs a gentle, back-and-forth circulation of a small amount of dye (~100 µL). Such circulation is necessary because CSF slowly exudes from the dura and dilutes the dye concentration locally at the dye/dura interface.

13. Stain for 90 min, and then wash the dura with dye-free ACSF for >15 min.

    Physical stabilization of the cortex can be further improved by application of high-viscosity silicone oil (60,000 centistokes; Sigma-Aldrich DMPS-60M) on top of the dura after the ACSF wash. This silicon oil is replaced every 30 min to 1 h, because exuded CSF can accumulate underneath and hinder the dampening effect.

Sensitivity and Total Recording Time

14. Image as described in imaging setup, above.

DISCUSSION

When the cortex is properly stained, the sensitivity of the voltage-sensitive dye recording can be comparable to that of LFP recordings (Lippert et al. 2007). When the LFP is simultaneously recorded with dye imaging, by comparing these two signals we found most of the peaks in the LFP showed a corresponding event in the imaging signal. There is a high correlation between optical traces that is not caused by light scattering or optical blurring because there is a small timing difference between locations caused by waves propagating in spatiotemporal patterns (Slovin et al. 2002; Petersen et al. 2003; Lippert et al. 2007; Xu et al. 2007). At the light intensity for a good signal-to-noise ratio for sleeplike waves, the total recording time can be up to 1000 sec, broken into 80–100 trials of 10 sec, which is sufficient for many types of experiment.

Protocol D

Retrograde Staining with Voltage-Sensitive Dyes

One limitation of the population signals described above is that all cell types in the region are stained and it is difficult to determine which cell type contributes what to the overall signal. One method for obtaining cell type–specific staining has been explored in embryonic chick and in lamprey spinal cords (Wenner et al. 1996). An identified cell class (motor neurons) was stained selectively by injecting a slurry of a very concentrated solution-cum-suspension of a hydrophobic styryl dye (dioctyl or didodecyl dye) into the ventral root. After waiting for 4–12 h some of this dye traveled to the motor neuron cell bodies in the chick spinal cord. This dye was reasonably efficient in signaling changes in membrane potential. Later, in lamprey experiments, spike signals from individual neurons were measured, but not reliably.

IMAGING SETUP

Same as Protocol A.

MATERIALS

In vitro chick embryo spinal cord preparation (see Landmesser and O'Donovan 1984)
Styryl dye, Di-8-ANEPEQ (see Tsau et al. 1996)
Tyrodes bathing solution (see Wenner et al. 1996)

EXPERIMENTAL METHOD

Imaging

The photodiode array has a relatively large dark noise that can be the dominant noise source in a fluorescence measurement where the number of dye molecules is relatively small. The following four modifications were made to increase the measured light intensity in order to reduce the relative contribution of dark noise.

1. The light source was a 150-W xenon short-arc lamp (Osram, XBO 150 WrCR ORF). The light intensity from this arc lamp was about three times brighter than that of a tungsten-halogen source. Noise from arc wander was not detected.
2. A wide bandwidth incident light filter (520 ± 45 nm) was used that passes wavelengths that include most of the long-wavelength excitation band of the dye and thus was optimal for a signal that results from a shift in the peak wavelength of the absorption (Loew et al. 1985). Smaller signals were obtained with narrower bandpass filters.
3. We attempted to use lenses with a large numerical aperature (NA) because the light intensity in an epifluorescence measurement is proportional to the fourth power of the NA.
4. The wavelengths passed by the RG-160 secondary filter include all of the long-wavelength portion of the emission spectrum.

Staining

Recordings were obtained using the in vitro chick embryo spinal cord preparation (Landmesser and O'Donovan 1984). Voltage-sensitive dyes were introduced into the ventral roots by pressure injection of a concentrated ethanol, chloroform, or dimethylsulfoxide suspension of the dye from a broken-tipped microelectrode. Staining with crystals of the dye was less effective. The dye concentrations were in the range of 10–20 mg/mL. The optical measurements were made after allowing a period of up to 24 h (usually overnight) for the dye to reach the motor neuron cell bodies. We presume that the concentrated solvent will damage the neuronal membranes with which it comes in contact. Controls for pharmacological effects of the dye injection suggested that this procedure does not damage the motor neuron cell bodies. Experiments were performed at room temperature because higher temperatures reduce the viability of the isolated cord when it is maintained for prolonged periods.

Analysis

The analysis consisted of measurements of the signal-to-noise ratio in the optical recordings. Both spatial and temporal smoothing were used.

FUTURE DIRECTIONS

Because the voltage-sensitive dye measuring apparatus is already close to optimal, any improvement in sensitivity will need to come from the development of better dyes. The vast majority of those previously synthesized organic voltage-sensitive dyes are of the general class named polyenes (Hamer 1964), a group that is used to extend the wavelength response of photographic film. The best of these styryl and oxonol dyes have fluorescence changes of 10%–20%/100 mV in a bilayer. Gonzalez and Tsien (1995) introduced a new scheme for generating voltage-sensitive signals using two chromophores and energy transfer. Although these fractional changes were also initially in the range of 10%/100 mV, more recent results are nearer 30% (T. Gonzalez and R.Y. Tsien, pers. comm.). However, to obtain fast responses, one of the chromophores must be hydrophobic, and molecules of such hydrophobicity will not penetrate mammalian brain tissue. Thus, it has not been possible to measure signals in intact mammalian tissues using this combination of dyes (T. Gonzalez and R.Y. Tsien, pers. comm.; A. Obaid and B.M. Salzberg, pers. comm.).

Second-harmonic generation provides a novel alternative to fluorescence-based voltage imaging (Wilson et al. 2011).

Neuron Type–Specific Staining

There are new developments of methods for neuron type–specific staining. The original approach was the use of anterograde/retrograde staining procedures (see above). The development of a more general approach was pioneered by Siegel and Isacoff (1997), who constructed a genetically encoded chimera of a potassium channel and green fluorescent protein (FlaSh, an FP voltage sensor; Chapter 7). In frog oocytes, FlaSh had a relatively slow voltage-dependent signal with a fractional fluorescence change of 1%–5%. Unfortunately, in mammalian cells, FlaSh and two other first-generation FP-voltage sensors are only minimally expressed in the extracellular membrane and no signals were detected (Baker et al. 2007). Thomas Knöpfel and his collaborators (Dimitrov et al. 2007) found that using the *Ciona* voltage-sensitive phosphatase (Murata et al. 2005) as the host protein led to a construct that was expressed well in mammalian cells and had large ($\Delta F/F \sim 10\%$) but slow responses to changes in membrane potential. It is hoped that improvements in response speed will lead to useful FP voltage sensors that provide cell type specificity for monitoring membrane potential in mammalian systems.

Chanda et al. (2005) (see also Bradley et al. 2009) have developed a hybrid approach that combines a genetically encoded fluorescent protein (membrane-anchored green fluorescent protein) together with dipicrylamine, a charged hydrophobic molecule that partitions into the membrane and moves when membrane potential changes. It is not clear if this method will be generally useful because it is not known how well dipicrylamine penetrates into mammalian brain tissue.

Optical recording with voltage-sensitive dyes already provides unique insights into brain activity and organization. Improvements in sensitivity or selectivity would make these methods even more powerful.

RECIPES

CAUTION: See Appendix 6 for proper handling of materials marked with <!>.
Recipes for reagents marked with <R> are included in this list.

Holding Artificial Cerebrospinal Fluid (ACSF)

NaCl	126 mM
KCl	2.5 mM
$CaCl_2$ <!>	2 mM

$MgSO_4$ <!>	2 mM
NaH_2PO_4	1.25 mM
$NaHCO_3$	26 mM
Dextrose	10 mM

Modified Artificial Cerebrospinal Fluid (ACSF)

Sucrose	250 mM
KCl	3 mM
$CaCl_2$ <!>	2 mM
$MgSO_4$ <!>	2 mM
NaH_2PO_4	1.25 mM
$NaHCO_3$	27 mM
Dextrose	10 mM

Saturated with 95% O_2 <!> and 5% CO_2<!>.

ACKNOWLEDGMENTS

The authors are indebted to their collaborators Vicencio Davila, Amiram Grinvald, Kohtaro Kamino, David Kleinfeld, Ying-wan Lam, Les Loew, Bill Ross, Brian Salzberg, Matt Wachowiak, Alan Waggoner, Dejan Zecevic, and Michal Zochowski for numerous discussions about optical methods. The experiments performed in our laboratories were supported by National Institutes of Health grants DC05259, NS057631, NS036447, and NS059034.

REFERENCES

Agmon A, Connors BW. 1992. Correlation between intrinsic firing patterns and thalamocortical synaptic responses of neurons in mouse barrel cortex. *J Neurosci* **12:** 319–329.

Albowitz B, Kuhnt U. 1993. Spread of epileptiform potentials in the neocortical slice: Recordings with voltage-sensitive dyes. *Brain Research* **631:** 329–333.

Baker BJ, Lee H, Pieribone VA, Cohen LB, Isacoff EY, Knopfel T, Kosmidis, E.K. 2007. Three fluorescent protein voltage sensors exhibit low plasma membrane expression in mammalian cells. *J Neurosci Methods* **161:** 32–38.

Bao W, Wu J-Y. 2003. Propagating wave and irregular dynamics: Spatiotemporal patterns of cholinergic theta oscillations in neocortex in vitro. *J Neurophysiol* **90:** 333–341.

Bradley J, Luo R, Otis TS, DiGregorio DA. 2009. Submillisecond optial reporting of membrane potential in situ using a neuronal tracer dye. *J Neurosci* **29:** 9197–9209.

Brown GD, Yamada S, Sejnowski TJ. 2001. Independent components analysis (ICA) at the neural cocktail party. *Trends Neurosci* **24:** 54–63.

Chanda B, Blunck R, Faria LC, Schweizer FE, Mody I, Bezanilla F. 2005. A hybrid approach to measuring electrical activity in genetically specified neurons. *Nat Neurosci* **8:** 1619–1626.

Cohen LB, Lesher S. 1986. Optical monitoring of membrane potential: Methods of multisite optical measurement. *Soc Gen Physiol Ser* **40:** 71–99.

Cohen LB, Salzberg BM. 1978. Optical measurement of membrane potential. *Rev Physiol Biochem Pharmacol* **83:** 35–88.

Dimitrov D, He Y, Mutoh H, Baker BJ, Cohen L, Akemann W, Knöpfel T. 2007. Engineering and characterization of an enhanced fluorescent protein voltage sensor. *PLoS One* **2:** e440. doi: 10.1371/journal.pone.0000440.

Falk CX, Wu J-Y, Cohen LB, Tang AC. 1993. Nonuniform expression of habituation in the activity of distinct classes of neurons in the *Aplysia* abdominal ganglion. *J Neurosci* **13:** 4072–4081.

Friedrich RW, Korsching SI. 1998. Chemotopic, combinatorial, and noncombinatorial odorant representations in the olfactory bulb revealed using a voltage-sensitive axon tracer. *J Neurosci* **18:** 9977–9988.

Frost WN, Wang J, Brandon CJ, Moore-Kochlacs C, Sejnowski TJ, Hill ES. 2010. Use of fast-responding voltage-sensitive dyes for large-scale recording of neuronal spiking activity with single-cell resolution. In *Membrane poten-*

tial imaging in the nervous system: Methods and applications (ed. Canepari M, Zecevic D), pp. 53–60. Springer-Verlag, Berlin.

Gonzalez JE, Tsien RY. 1995. Voltage sensing by fluorescence energy transfer in single cells. *Biophysical J* **69:** 1272–1280.

Grinvald A, Manker A, Segal M. 1982. Visualization of the spread of electrical activity in rat hippocampal slices by voltage-sensitive optical probes. *J Physiol (Lond)* **333:** 269–291.

Hajos N, Mody I. 2009. Establishing a physiological environment for visualized in vitro brain slice recordings by increasing oxygen supply and modifying aCSF content. *J Neurosci Methods* **183:** 107–113.

Hamer FM. 1964. *The cyanine dyes and related compounds.* Wiley, New York.

Hickie C, Wenner P, O'Donovan M, Tsau Y, Fang J, Cohen LB. 1996. Optical monitoring of activity from individual and identified populations of neurons retrogradely labeled with voltage-sensitive dyes. *Abs Soc Neurosci* **22:** 321.

Hirata Y, Sawaguchi T. 2008. Functional columns in the primate prefrontal cortex revealed by optical imaging in vitro. *Neurosci Res* **61:** 1–10.

Huang X, Troy WC, Yang Q, Ma H, Laing CR, Schiff SJ, Wu J-Y. 2004. Spiral waves in disinhibited mammalian neocortex. *J Neurosci* **24:** 9897–9902.

Jin WJ, Zhang RJ, Wu JY. 2002. Voltage-sensitive dye imaging of population neuronal activity in cortical tissue. *J Neurosci Methods* **115:** 13–27.

Kleinfeld D, Delaney K. 1996. Distributed representation of vibrissa movement in the upper layers of somatosensory cortex revealed with voltage-sensitive dyes. *J Comp Neurol* **375:** 89–108.

Komuro H, Momose-Sato Y, Sakai T, Hirota A, Kamino K. 1993. Optical monitoring of early appearance of spontaneous membrane potential changes in the embryonic chick medulla oblongata using a voltage-sensitive dye. *Neuroscience* **52:** 55–62

Lam Y-W, Cohen LB, Michal R, Zochowski MR. 2003, Effect of odorant quality on the three oscillations and the DC signal in the turtle olfactory bulb. *Eur J Neurosci* **17:** 436–446.

Landmesser LT, O'Donovan MJ. 1984. The activation patterns of embryonic chick motoneurons projecting to inappropriate muscles. *J Physiol* **347:** 205–224.

Lippert MT, Takagaki K, Xu W, Huang X, Wu J-Y. 2007. Methods for voltage-sensitive dye imaging of rat cortical activity with high signal-to-noise ratio. *J Neurophysiol* **98:** 502–512.

Loew LM, Cohen LB, Salzberg BM, Obaid AL, Bezanilla F. 1985. Charge-shift probes of membrane potential. Characterization of aminostyrylpyridinium dyes on the squid giant axon. *Biophys J* **47:** 71–77.

London JA, Zečević D, Cohen LB. 1987. Simultaneous optical recording of activity from many neurons during feeding in *Navanax*. *J Neurosci* **7:** 649–661.

London JA, Cohen LB, Wu J-Y. 1989. Optical recordings of the cortical response to whisker stimulation before and after the addition of an epileptogenic agent. *J Neurosci* **9:** 2182–2190.

MacLean JN, Fenstermaker V, Watson BO, Yuste R. 2006. A visual thalamocortical slice. *Nat Methods* **3:** 129–134.

Malmstadt HV, Enke CG, Crouch SR, Harlick G. 1974. Electronic measurements for scientists. Benjamin, Menlo Park, CA.

Metherate R, Cruikshank SJ. 1999. Thalamocortical inputs trigger a propagating envelope of γ-band activity in auditory cortex in vitro. *Exp Brain Res Experimentelle Hirnforschung* **126:** 160–174.

Momose-Sato Y, Sato K, Sakai T, Hirota A, Matsutani K, Kamino K. 1995. Evaluation of optimal voltage-sensitive dyes for optical monitoring of embryonic neural activity. *J Membr Biol* **144:** 167–176.

Murata Y, Iwasaki H, Sasaki M, Inaba K, Okamura Y. 2005. Phosphoinositide phosphatase activity coupled to an intrinsic voltage sensor. *Nature* **435:** 1239–1243.

Nakashima M, Yamada S, Shiono S, Maeda M, Sato F. 1992. 448-detector optical recording system: Development and application to *Aplysia* gill-withdrawal reflex. *IEEE Trans Biomed Eng* **39:** 26–36.

Obaid AL, Koyano T, Lindstrom J, Sakai T, Salzberg BM. 1999. Spatiotemporal patterns of activity in an intact mammalian network with single-cell resolution: Optical studies of nicotinic activity in an enteric plexus. *J Neurosci* **19:** 3073–3093.

Orbach HS, Cohen LB, Grinvald A. 1985. Optical mapping of electrical activity in rat somatosensory and visual cortex. *J Neurosci* **5:** 1886–1895.

Parsons TD, Salzberg BM, Obaid AL, Raccuia-Behling F, Kleinfeld D. 1991. Long-term optical recording of patterns of electrical activity in ensembles of cultured *Aplysia* neurons. *J Neurophysiol* **66:** 316–333.

Petersen CC, Grinvald A, Sakmann B. 2003. Spatiotemporal dynamics of sensory responses in layer 2/3 of rat barrel cortex measured in vivo by voltage-sensitive dye imaging combined with whole-cell voltage recordings and neuron reconstructions. *J Neurosci* **23:** 1298–1309.

Prechtl JC, Cohen LB, Peseran B, Mitra PP, Kleinfeld D. 1997. Visual stimuli induce waves of electrical activity in turtle cortex. *Proc Nat Acad Sci* **94:** 7621–7626.

Ratzlaff EH, Grinvald A. 1991. A tandem-lens epifluorescence microscope: Hundred-fold brightness advantage for wide-field imaging. *J Neurosci Methods* **36:** 127–137.

Salama G. 1988. Voltage-sensitive dyes and imaging techniques reveal new patterns of electrical activity in heart and cortex. *SPIE Proc* **94:** 75–86.

Salzberg BM, Davila HV, Cohen LB. 1973. Optical recording of impulses in individual neurones of an invertebrate central nervous system. *Nature* **246:** 508–509.

Salzberg BM, Grinvald A, Cohen LB, Davila HV, Ross WN. 1977. Optical recording of neuronal activity in an invertebrate central nervous system: Simultaneous monitoring of several neurons. *J Neurophysiol* **40:** 1281–1291.

Sato K, Momose-Sato Y, Hirota A, Sakai T, Kamino K. 1998. Optical mapping of neural responses in the embryonic rat brainstem with reference to the early functional organization of vagal nuclei. *J Neurosci* **18:** 1345–1362.

Shoham D, Glaser DE, Arieli A, Kenet T, Wijnbergen C, Toledo Y, Hildesheim R, Grinvald A. 1999. Imaging cortical dynamics at high spatial and temporal resolution with novel blue voltage-sensitive dyes. *Neuron* **24:** 791–802.

Siegel MS, Isacoff EY. 1997. A genetically encoded optical probe of membrane voltage. *Neuron* **19:** 735–741.

Slovin H, Arieli A, Hildesheim R, Grinvald A. 2002. Long-term voltage-sensitive dye imaging reveals cortical dynamics in behaving monkeys. *J Neurophysiol* **88:** 3421–3438.

Tsau Y, Wu JY, Hopp HP, Cohen LB, Schiminovich D, Falk CX. 1994. Distributed aspects of the response to siphon touch in *Aplysia*: Spread of stimulus information and cross-correlation analysis. *J Neurosci* **14:** 4167–4184.

Tsau Y, Wenner P, O'Donovan MJ, Cohen LB, Loew LM, Wuskell JP. 1996. Dye screening and signal-to-noise ratio for retrogradely transported voltage-sensitive dyes. *J Neurosci Meth* **70:** 121–129.

Wenner P, Tsau Y, Cohen LB, O'Donovan MJ, Dan Y. 1996. Voltage sensitive dye recording using retrogradely transported dye in the chicken spinal cord: Staining and signal characteristics. *J Neurosci Methods* **70:** 111–120.

Wilson SA, Millard A, Lewis, Loew LM. 2011. Monitoring membrane potential with second-harmonic generation. In *Imaging: A laboratory manual* (ed. Yuste R), pp. 721–734. Cold Spring Harbor Press, Cold Spring Harbor, NY.

Wu J.-Y, Cohen LB. 1993. Fast multisite optical measurement of membrane potential. In *Fluorescent probes for biological activity* (ed. Mason WT), pp. 389–404. Academic, New York.

Wu J-Y, Falk CX, Cohen LB. 1994. Neuronal activity during different behaviors in *Aplysia*: A distributed organization? *Science* **263:** 820–823.

Wu J-Y, Guan L, Bai L, Yang Q. 2001. An evoked all-or-none population activity in rat sensory cortical slices. *J Neurophysiol* **86:** 2461–2474

Wu C, Luk WP, Gillis J, Skinner F, Zhang L. 2005. Size does matter: Generation of intrinsic network rhythms in thick mouse hippocampal slices. *J Neurophysiol* **93:** 2302–2317.

Xu W, Huang X, Takagaki K, Wu J-Y. 2007. Compression and reflection of visually evoked cortical waves. *Neuron* **55:** 119–129.

Zecevic D, Wu JY, Cohen LB, London JA, Hopp HP, Falk CX. 1989. Hundreds of neurons in the *Aplysia* abdominal ganglion are active during the gill-withdrawal reflex. *J Neurosci* **9:** 3681–3689.

Zochowski M, Wachowiak DM, Falk CX, Cohen LB, Lam Y-W, Antic S, Zecevic D. 2000. Imaging membrane potential with voltage-sensitive dyes. *Biol Bull* **198:** 1–21.

WWW RESOURCE

http://www.minusk.com Minus K Technology, Inglewood, CA.

49 In Vivo Two-Photon Calcium Imaging Using Multicell Bolus Loading of Fluorescent Indicators

Nathalie L. Rochefort, Christine M. Grienberger, and Arthur Konnerth

Institute of Neuroscience, Technical University Munich, 80802 Munich, Germany

ABSTRACT

Imaging the activity of neuronal populations allows the simultaneous mapping of spontaneous and evoked responses of numerous neurons and is critical for determining the function of neuronal circuits. This chapter describes a simple and effective method called "multicell bolus loading" that has been developed to deliver a calcium indicator in vivo to a large population of cortical neurons. Cells are stained by a brief injection of a membrane-permeant Ca^{2+}-indicator dye into the extracellular space. The injected dye diffuses into the cells of interest, where it is hydrolyzed by intracellular esterases. The activity-dependent calcium signals from multiple cells can then be imaged with two-photon microscopy at single-cell resolution. Calcium imaging of neuronal circuits was first developed in slices, in which it was shown that calcium transients correspond to action potentials. This method has been successfully applied to in vivo imaging of calcium signals during spontaneous or evoked activity in various brain regions, including the somatosensory cortex, the visual cortex, the olfactory bulb, the spinal cord, the cerebellum, and the hippocampus. The same protocol can be used in different species and at different developmental ages, for example, in zebrafish larvae, mice, rats, cats, ferrets, and monkeys.

Protocol

Multicell Bolus Loading of Fluorescent Indicators for In Vivo Two-Photon Ca^{2+} Imaging

The following protocol describes an approach for the staining and subsequent functional imaging of a large population of cortical neurons in vivo by using two-photon microscopy. This approach can be adapted for various indicator dyes and various target brain regions as well as for different species of choice.

IMAGING SETUP

In vivo Ca^{2+} imaging is performed by using a custom-built two-photon microscope based on a Ti:sapphire pulsing laser (model, Chameleon; repetition rate, 80 MHz; pulse width, 140 fsec; Coherent) and resonant galvo-mirror (8 kHz; GSI) system. The scanner is mounted on an upright microscope (BX51WI, Olympus, Tokyo, Japan) equipped with a water-immersion objective (40x/0.8-numerical-aperture (NA) Nikon, Japan or 60x/1.0-NA Nikon). Emitted photons are detected by two detection channels equipped with photomultiplier tubes (H7422-40; Hamamatsu), a "green" channel for OGB-1-dependent calcium recordings (480–560 nm) and a "red" channel for the Alexa-594-generated fluorescence (580–680 nm). Full-frame images at 480x400 pixel resolution are acquired at 30 Hz by custom-programmed software based on LabVIEW (version 8.2; National Instruments).

MATERIALS

CAUTION: See Appendix 6 for proper handling of materials marked with <!>.
See the end of the chapter for recipes for reagents marked with <R>.

Reagents

Agarose, low-melting-point (Invitrogen GmbH)
Anesthetic agent (e.g., isoflurane <!>, ketamine/xylazine <!>, or urethane <!>)
Calcium-indicator dye, membrane-permeable (e.g., Oregon Green <!> 488 BAPTA-1 <!> acetoxymethyl (AM) ester (OGB-1 AM), Calcium Green-1 AM, Fura-2 AM, Fura-PE3 AM, Fluo-4 AM, or Indo-1 AM from Molecular Probes or TefLabs)
Cyanoacrylic glue <!>
Experimental animal
Local anesthetic agent (e.g., lidocaine <!>)
20% Pluronic F-127 in dimethylsulfoxide (DMSO) <!> (e.g., 2 g Pluronic F-127 in 10 mL of DMSO; Sigma-Aldrich)
Standard external saline <R>
Standard pipette solution <R>
Sulforhodamine-101 (SR101; Sigma-Aldrich)

Equipment

Anesthesia-monitoring system to measure the blood pressure, arterial oxygen concentration, body temperature, and respiratory and heart rates of the animal (e.g., ADI Instruments)

Anesthesia unit, including a chamber for pre-anesthetic medication, a flow meter, and a vaporizer (the latter items are for volatile anesthetic agents only)

Consult the literature (e.g., Flecknell 2000) for the best choice of anesthesia for each species.

Custom-made recording chamber from a standard tissue-culture dish (see Fig. 1A, below).

Glass capillaries (e.g., from Hilgenberg GmbH)

Manipulator and pressure-application device

The LN-Mini manipulator (Luigs & Neumann GmbH) and either Picospritzer II (General Valve) or the pneumatic drug-ejection system (NPI) are used to inject the staining solution into the brain in this protocol.

Patch-clamp amplifier (e.g., from HEKA) to monitor the resistance of the pipette during dye injection

Monitoring of the pipette resistance is not mandatory, but it allows recognizing pipette clogging and estimating when the pipette touches the cortical surface.

Pipette puller (e.g., PP830 from Narishige)

Surgical equipment, including a stereotaxic instrument, drill, and warming blanket to ensure a constant body temperature (e.g., TSE Systems)

EXPERIMENTAL METHOD

Staining Neurons with a Calcium Indicator Dye

1. Anesthetize the animal (e.g., isoflurane inhalation anesthesia 1%–1.5% [vol/vol] in pure O_2) and position it on a heating plate (38°C). Monitor the level of anesthesia regularly (e.g., by testing the pinch withdrawal and the eyelid reflex).
2. Inject ~50 µL of a local anesthetic agent (e.g., 2% lidocaine) subcutaneously at the location where the skin is to be removed.
3. Remove the skin above the desired brain area. Gently clean and dry the surface of the skull.
4. Use cyanoacrylic glue to adhere the custom-made recording chamber (see Fig. 1A, below) to the skull so that the opening of the chamber is located above the desired brain region.
5. Thin the bone carefully above the brain area of interest using dental drills. Thinning can be stopped as soon as the pial vasculature can be visualized clearly and the bone displays flexibility when touched lightly with a forceps.
6. Transfer the animal into the recording setup and place it onto a warming plate (38°C). Start the anesthesia monitoring system (including respiratory and heart rates).
7. Perfuse the recording chamber with a warm (37°C) standard external saline.
8. Perform a small (0.5–1 mm) craniotomy above an area devoid of large blood vessels.
9. Dissolve the AM ester of the chosen calcium indicator dye in 20% Pluronic F-127 in DMSO (e.g., 2 g of Pluronic in 10 mL of DMSO) to yield a dye concentration of 10 mM. Dilute this solution 1/10 to 1/100 with standard pipette solution to obtain the final staining solution.
10. Fill a micropipette (pipette resistance 3–3.5 MΩ) with the staining solution. Insert the pipette into the cortex and advance it along its axis until it reaches the desired depth (Fig. 1B) while monitoring the pipette resistance. Apply a pressure pulse (1 min, 0.7 bar) to eject the staining solution near the cells of interest. Remove the pipette.
11. Wait for an hour to achieve a stable maximal fluorescence level in the stained cells (Stosiek et al. 2003). This protocol yields a stained area with a diameter of 200–400 µm.

Two-Photon Imaging of Stained Cells

12. The wavelength of the excitation light depends on the indicator dye used (see two-photon excitation spectra of common Ca^{2+}-sensitive dyes). Excitation light of 800 nm can be used for the initial visualization of neurons stained with the following indicator dyes: Fura-PE3 AM, Fura Red AM, Indo-1 AM, Calcium Green-1 AM, OGB-1 AM, Fluo-4 AM, and Magnesium Green AM. The average power of the excitation light has to be measured under the objective. For imaging, the excitation power should be raised carefully to avoid dye bleaching. With the dye application pipette located 150–200 µm below the cortical surface, all cortical cells between the surface and 400-µm depth are stained (Stosiek et al. 2003). When imaging through the thinned skull, individual cells can be well resolved up to 250 µm below the cortical surface. Removing the skull above the imaged field further improves depth resolution. The depth limit for single-cell imaging in the cortex is ~500 µm (see Theer et al. 2003; Helmchen and Denk 2005). Although most imaging results have been obtained so far with scanners that are based on conventional galvo-scanners (e.g., Stosiek et al. 2003; Ohki et al. 2005), improved recordings were obtained recently by using high-speed scanning involving a resonant galvo-mirror (Rochefort et al. 2009).

    See *Troubleshooting.*

TROUBLESHOOTING

Problem (Step 12): Cells are not stained or are stained poorly.
Solution: This can be caused by a failure in dye injection. It is useful to monitor the pipette resistance during dye injection (a value >5 MΩ indicates that the pipette is clogged). In addition, Ca^{2+}-indicator dyes should be dissolved immediately before use and filtered (e.g., with a Millipore filter; pore diameter, 0.45 µm). Another possible source of poor staining is damaged brain tissue. The skull opening should be performed carefully, avoiding damage to blood vessels, controlling the temperature of bath solution, checking the anesthesia monitoring parameters, and increasing the speed of dissection and staining (<2 h).

Problem (Step 12): Movement artifacts occur in the imaged area.
Solution: It should be stressed that the stability of recordings depends critically on the diameter of the craniotomy. Thus, openings >1 mm in diameter are often accompanied by movement artifacts occurring at heartbeat frequency. In addition, movement artifacts are more profound in regions with a high density of blood vessels. To reduce these artifacts, the skull opening can be made, if possible, above a region without big blood vessels or a low density of blood vessels. Then, the opened region can be covered with 2% low-melting-point agarose (e.g., 20 mg in 1 mL of standard external saline).

DISCUSSION

Application Examples

Figures 1 and 2 show examples of in vivo two-photon calcium imaging experiments in mouse cortical areas (somatosensory and visual cortices). The calcium dye (OGB-1 AM) was injected in cortical layer 2/3 (Fig. 1B) of the sensory area of interest. In all experiments, the correct location of the imaged neurons can be confirmed in vivo by their sensory-evoked responses and/or post hoc by imaging the stained area in cortical slices (e.g., Fig. 1C). After in vivo injection of the dye, a stable

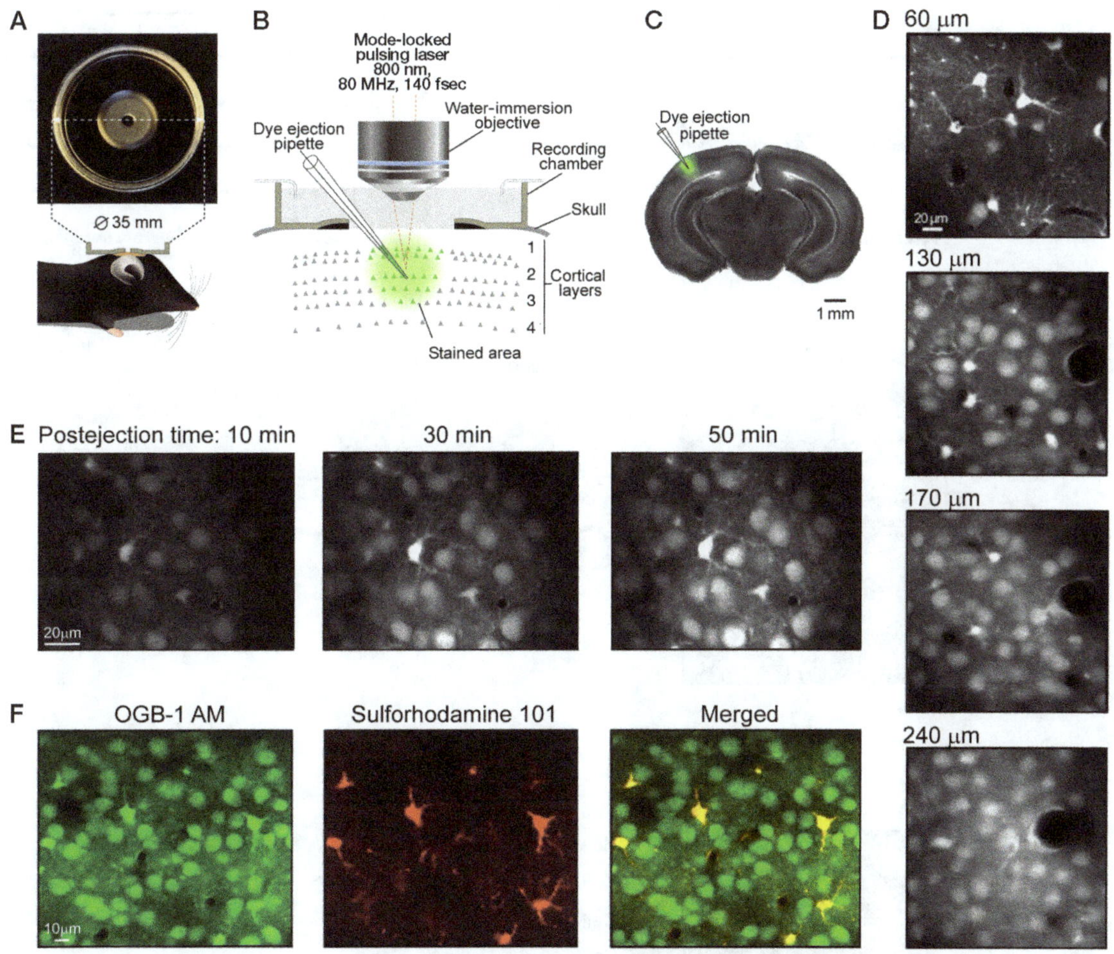

FIGURE 1. In vivo staining of neuronal populations with a Ca^{2+}-indicator dye. (*A,B*) Experimental setup for in vivo staining with Ca^{2+}-indicator dyes. (*A*) Photograph of the recording chamber (*upper panel*) and schematic drawing of its cross section (*lower panel*) on the mouse skull. (*B*) The custom-made recording chamber is glued to the skull to allow head fixation and the use of a water-immersion objective. The chamber is perfused with warm (37°C) standard external saline. The stained area (represented in green) has a diameter of 200–400 µm. (*C*) Combined fluorescence/transmitted light micrograph of a coronal brain slice from an 11-d-old (P11) C57BL/6 mouse taken after an in vivo experiment. The area stained in vivo with OGB-1 is shown in green. (*D*) High-magnification images of the barrel cortex of a P13 mouse taken at different depths. (*E*) Consecutive images taken in vivo at 90-µm depth 10, 30, and 50 min after injection (1 min) of 1 mM OGB-1 AM into the cortex of a P14 mouse. (*F*) Microphotographs of layer 2/3 cells in the mouse visual cortex (P30) stained simultaneously using a dye mixture containing OGB-1 AM and the glial marker sulforhodamine 101 (SR101). The fluorescence of OGB-1 AM was directed to the green channel (*left*) and the fluorescence of SR101 was directed to the red channel (*middle*). The merged image on the *right* shows neurons in green and glial cells in yellow. (*A,F,* Reprinted, with permission, from Garaschuk et al. 2006; *B,D,E,* reprinted, with permission, from Stosiek et al. 2003; *C,* reprinted, with permission, from Rochefort et al. 2009.)

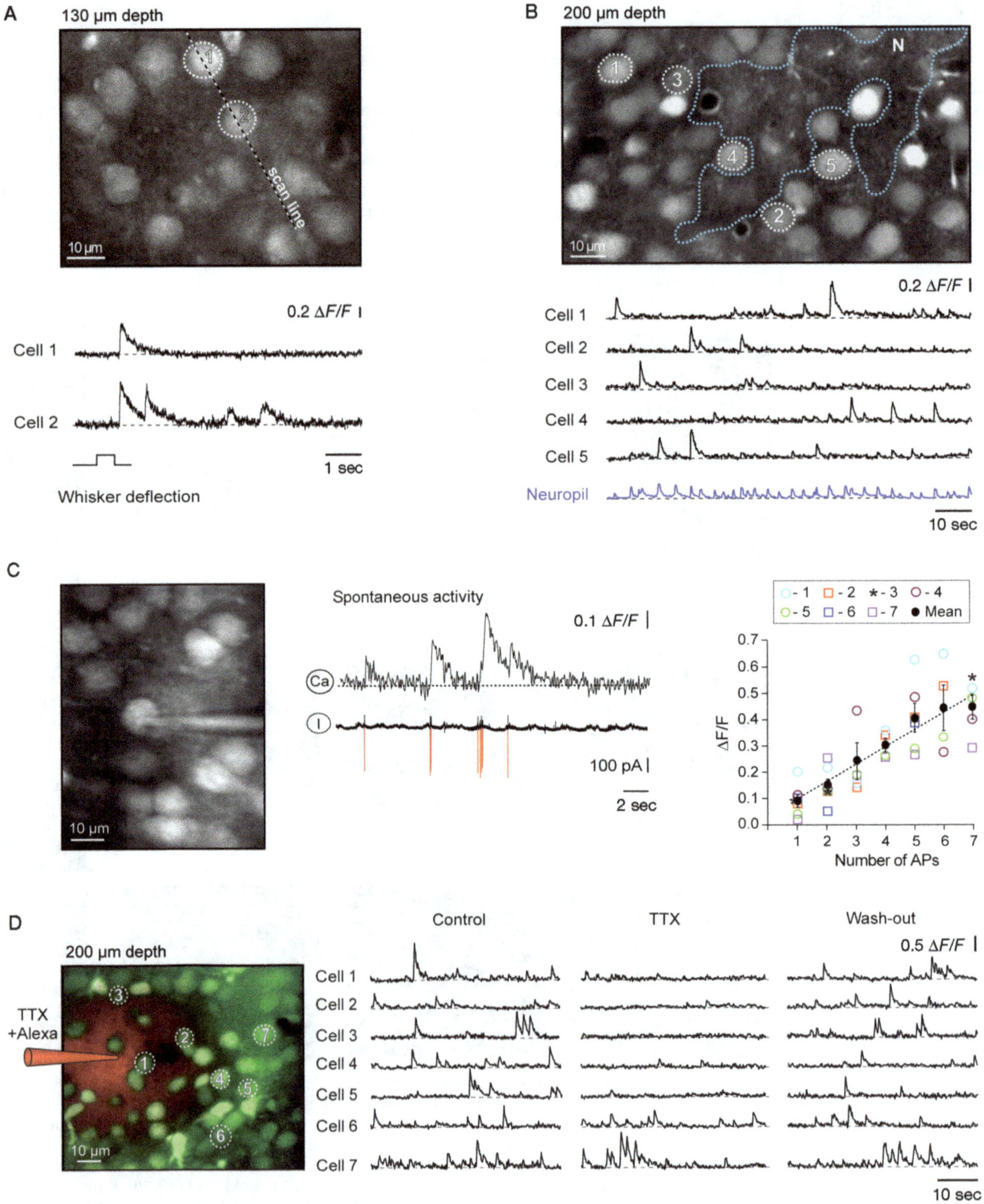

FIGURE 2. In vivo Ca^{2+} imaging of neuronal populations in the mouse sensory areas (somatosensory and visual cortex). (*A*) In vivo recordings of calcium transients evoked by whisker deflection, in the mouse somatosensory cortex. Two-photon line-scan recordings of Ca^{2+} transients (*lower panel*) evoked in two-layer 2/3 neurons by whisker deflection on the contralateral side of the mouse's snout (P13 mouse). The position of the scanned line and the cells analyzed are indicated in the upper panel. (*B*) In vivo recordings of spontaneous calcium transients in the visual cortex of a P26 C57BL/6 mouse. Spontaneously occurring Ca^{2+} transients (*lower panel*) in five individual layer 2/3 neurons and in surrounding neuropil, marked in the *upper* panel with corresponding numbers and the dashed blue line, respectively. Images were acquired at 30 Hz.

level of fluorescence within stained cells was reached in ~1 h (Fig. 1E). There was no major difference in the time course of dye loading in vivo and in vitro (Stosiek et al. 2003). It is possible to image in vivo calcium transients for 4–6 h without any significant dye leakage and/or bleaching. At the beginning of the imaging session, we recommend that the objective be focused on the brain surface. The following structures appear when focusing deeper in the cortex (Fig. 1D): large epithelial cells on the top of the cortex; layer 1 cells up to a depth of 100–200 μm, depending on the age of the animal; and layer 2/3 cells. Removing the skull above the imaging field improves depth resolution, allowing the detection of individual cells up to 300–400 μm below the cortical surface. The multicell bolus loading method can be combined with other indicators to distinguish specific cell types. For example, it is possible to combine OGB-1 AM with sulforhodamine 101 staining, which specifically labels astroglia (Nimmerjahn et al. 2004; see Chapter 63). From these experiments, it appears that just OGB-1 AM–labeled astrocytes can be identified unambiguously based on their OGB-1 AM labeling alone, because of their brighter staining and their specific morphology with thick processes (see, e.g., the bright cells in Fig. 1D,E).

The advantage of the method described in this chapter is that it allows the simultaneous recording of the activity of multiple neurons in vivo. Figure 2 shows examples of calcium transients recorded in layer 2/3 neurons of the barrel (Fig. 2A) and visual (Fig. 2B) cortices of the mouse in vivo. Both spontaneous (Fig. 2B) and evoked (Fig. 2A) activities can be imaged for several hours with a resolution of individual somata. By combining electrical recordings with Ca^{2+} imaging, it was shown that Ca^{2+} transients recorded in the somata of cortical neurons result exclusively from neuronal firing (Smetters et al. 1999; Stosiek et al. 2003; Kerr et al. 2005; Greenberg et al. 2008; Rochefort et al. 2009). Using high-speed two-photon imaging and cell-attached recordings, single action potential–evoked Ca^{2+} transients can be detected reliably and the amplitude of Ca^{2+} transients correlates with the number of action potentials (Fig. 2C). Another way to show that spontaneous Ca^{2+} transients in vivo result from action potential firing is to apply the voltage-gated Na^{+} channels blocker TTX (Fig. 2D). TTX produces a specific blockade of the Na^{+} channels required for action potential generation. To test the effect of a drug on neuronal activity in vivo, it is possible to apply the drug in the bath solution in the recording chamber (see Fig. 1B). However, under the condition of bath application, the primary site of drug action remains unclear. Therefore, it is often useful to apply the drug focally near the cells of interest. The reliability and accuracy of this approach is greatly improved when the drug is coapplied with an inert fluorescent dye. In the experiment illustrated in Fig. 2D we used Alexa Fluor 594 and show, as an example, that local TTX application effectively and reversibly blocked the Ca^{2+} transients only in the region where the Alexa 594 was also detected. At remote sites at which the Alexa 594 signal was absent, the TTX effect was not observed (cells 6 and 7 in Fig. 2D).

Advantages and Limitations

The approach described here, known as multicell bolus loading (Stosiek et al. 2003), allows the simultaneous monitoring of Ca^{2+} levels in many individual neurons. The major difference between multicell bolus loading and other staining methods using AM indicator dyes is that the indicators

FIGURE 2. (*Continued*) (*C*) Cellular properties of neuronal Ca^{2+} signals. Simultaneous recordings of spontaneous Ca^{2+} transients and underlying action potential firing in loose-seal cell-attached configuration (*middle panel*) in a layer 2/3 neuron (*left panel*) in the primary visual cortex of a C57BL/6 15-d-old mouse. Downward deflections (in red) are spontaneously occurring action potentials (APs). (*Right panel*) Relation between the number of APs and the amplitude ($\Delta F/F$) of the corresponding Ca^{2+} transient (P14-P16 C57BL/6 mice, seven cells as indicated). Dotted line indicates the linear least-squares fit to the data. (*D*) In vivo recording of spontaneous Ca^{2+} transient of seven layer 2/3 neurons before, during, and after local drug application (tetrodotoxin [TTX]). Note the local blockade of Ca^{2+} transients at the site of local TTX (2.5 μM in the pipette) application (P17 mouse). For the identification of the region of TTX action, Alexa Fluor 594 (500 μM, red channel) was added to the pipette solution. The fluorescence of OGB-1 was detected in the green channel. (*A*, Reprinted, with permission, from Stosiek et al. 2003; *B*, modified, with permission, from Rochefort et al. 2009; *C*, reprinted, with permission, from Rochefort et al. 2009.)

are delivered directly to the target cells for a short period. In particular, this approach improves the staining of neurons in the adult brain, which are, in general, not stained by AM indicator dyes bath-applied to brain slices. It is important to remember that good-quality in vitro brain slices are mostly obtained at relatively early developmental stages. In contrast, in vivo two-photon calcium imaging experiments are feasible and produce good results in both young and aged mice (e.g., see Eichhoff et al. 2008).

The bulk-loading approach of calcium indicators is well suited for the analysis of somatic activity in neural circuits, but this method has limitations when it comes to the imaging of subcellular compartments, like dendrites, because the image contrast is reduced because of the staining of many fine processes in the surrounding neuropil. Furthermore, the dye concentration in multicell bolus loaded cells is lower than in cells stained using a microelectrode (Svoboda et al. 1997; Stosiek et al. 2003). These limitations also reduce the depth resolution of the recordings (200–300 µm under the cortical surface as compared with >500 µm when imaging cell dendrites of microelectrode-loaded cells) (Svoboda et al. 1997). Another limitation of the multicell bolus loading method is that it results in a nonspecific labeling of all cells, including all neuronal as well as glial cells in the stained region. This limitation can be partially overcome when using additional specific indicators in parallel. For example, calcium signaling has been studied in cortical astrocytes labeled by the fluorescent dye sulforhodamine 101 (Winship et al. 2007) and in γ-aminobutyric acid-producing (GABAergic) interneurons expressing enhanced green fluorescent protein (eGFP) (Sohya et al. 2007). Another method of labeling a specific neuronal population is to inject a dextran-conjugated form of dye into axon bundles that is then transported anterogradely and retrogradely over hours or days (Hirase et al. 2004). This method has been used fruitfully in the olfactory bulb (Oka et al. 2006) and in the olivary-cerebellar system (Kreitzer et al. 2000).

Until recently, all in vivo imaging studies were performed in head-fixed and anesthetized animals. But lately, the multicell bolus loading technique has been combined with a new technique for two-photon fluorescence imaging with cellular resolution in awake, behaving head-fixed mice (Dombeck et al. 2007) (see Chapter 76), allowing the study of brain functions in unanesthetized animals. In addition, a miniaturized head-mounted multiphoton microscope has been used for recording of Ca^{2+} transients from neuronal somata in the visual cortex of awake, freely moving rats (Sawinski et al. 2009; see Chapter 78). Another approach for monitoring brain functions in awake animals involves the use of fiber optics-based microendoscopes (Adelsberger et al. 2005) (see Chapter 81). Contrary to head-fixed setups, this latter approach permits imaging in freely moving animals. Such a fiber can be used for signal detection with a high temporal resolution in deep brain structures, but has a restricted field of view and low spatial resolution. To achieve imaging with cellular resolution through such a microendoscope, one approach is based on the use of microelectromechanical scanners coupled with gradient-index lenses coupled to the optical fibers (Mehta et al. 2004; Flusberg et al. 2005). This approach, combined with the multicell bolus loading technique described here, will allow the recording of population activities in deep brain regions, such as the hippocampus or the thalamus, in the near future.

In conclusion, the approach described here allows in vivo two-photon Ca^{2+} imaging of large neuronal circuits at a single-cell level of resolution. It has a wide range of applications from in vitro recordings to in vivo imaging of neuronal population activities. Combined with two-photon microscopy and/or optical fibers, it makes possible the study of spontaneous and evoked activities in any brain region in anesthetized as well as in awake and behaving animals. Furthermore, in combination with the focal application of pharmacological drugs, multicell bolus loading is a powerful approach for the exploration of mechanisms underlying network function in vivo on the level of individual neurons. The recent generation of a wide range of transgenic mice, including neurological disease models and/or labeling of specific cell types, further increases the application fields of this technique.

RECIPES

CAUTION: See Appendix 6 for proper handling of materials marked with <!>.
Recipes for reagents marked with <R> are included in this list.

Standard External Saline

NaCl	125 mM
KCl	4.5 mM
$NaHCO_3$	26 mM
NaH_2PO_4	1.25 mM
$CaCl_2$ <!>	2 mM
$MgCl_2$	1 mM
Glucose	20 mM

pH 7.4, when bubbled with 95% O_2 <!> and 5% CO_2 <!>.

Standard Pipette Solution

NaCl	150 mM
KCl	2.5 mM
HEPES	10 mM

ACKNOWLEDGMENTS

This research was supported by grants of the Deutsche Forschungsgemeinschaft (SFB 596, GA 654/1-1, HO 2156/2-1) and the Bundesministerium für Bildung und Forschung (NGFN-2). AK is a Carl von Linde Senior Fellow of the Institute for Advanced Study of the Technische Universität München.

REFERENCES

Adelsberger H, Garaschuk O, Konnerth A. 2005. Cortical calcium waves in resting newborn mice. *Nat Neurosci* **8:** 988–990.

Dombeck DA, Khabbaz AN, Collman F, Adelman TL, Tank DW. 2007. Imaging large-scale neural activity with cellular resolution in awake, mobile mice. *Neuron* **56:** 43–57.

Eichhoff G, Busche MA, Garaschuk O. 2008. In vivo calcium imaging of the aging and diseased brain. *Eur J Nucl Med Mol Imaging* (suppl. 1) **35:** S99–S106.

Flecknell P. 2000. *Laboratory animal anaesthesia*. Academic, San Diego.

Flusberg BA, Jung JC, Cocker ED, Anderson EP, Schnitzer MJ. 2005. In vivo brain imaging using a portable 3.9 gram two-photon fluorescence microendoscope. *Opt Lett* **30:** 2272–2274.

Garaschuk O, Milos RI, Konnerth A. 2006. Targeted bulk-loading of fluorescent indicators for two-photon brain imaging in vivo. *Nat Protoc* **1:** 380–386.

Greenberg DS, Houweling AR, Kerr JN. 2008. Population imaging of ongoing neuronal activity in the visual cortex of awake rats. *Nat Neurosci* **11:** 749–751.

Helmchen F, Denk W. 2005. Deep tissue two-photon microscopy. *Nat Methods* **2:** 932–940.

Hirase H, Creso J, Singleton M, Bartho P, Buzsaki G. 2004. Two-photon imaging of brain pericytes in vivo using dextran-conjugated dyes. *Glia* **46:** 95–100.

Kerr JN, Greenberg D, Helmchen F. 2005. Imaging input and output of neocortical networks in vivo. *Proc Natl Acad Sci* **102:** 14063–14068.

Kerr JN, de Kock CP, Greenberg DS, Bruno RM, Sakmann B, Helmchen F. 2007. Spatial organization of neuronal population responses in layer 2/3 of rat barrel cortex. *J Neurosci* **27:** 13316–13328.

Kreitzer AC, Gee KR, Archer EA, Regehr WG. 2000. Monitoring presynaptic calcium dynamics in projection fibers by in vivo loading of a novel calcium indicator. *Neuron* **27:** 25–32.

Mehta AD, Jung JC, Flusberg BA, Schnitzer MJ. 2004. Fiber optic in vivo imaging in the mammalian nervous system. *Curr Opin Neurobiol* **14:** 617–628.

Nimmerjahn A, Kirchhoff F, Kerr JN, Helmchen F. 2004. Sulforhodamine 101 as a specific marker of astroglia in the neocortex in vivo. *Nat Methods* **1:** 31–37.

Ohki K, Chung S, Ch'ng YH, Kara P, Reid RC. 2005. Functional imaging with cellular resolution reveals precise micro-architecture in visual cortex. *Nature* **433:** 597–603.

Oka Y, Katada S, Omura M, Suwa M, Yoshihara Y, Touhara K. 2006. Odorant receptor map in the mouse olfactory bulb: In vivo sensitivity and specificity of receptor-defined glomeruli. *Neuron* **52:** 857–869.

Rochefort NL, Garaschuk O, Milos RI, Narushima M, Marandi N, Pichler B, Kovalchuk Y, Konnerth A. 2009. Sparsification of neuronal activity in the visual cortex at eye-opening. *Proc Natl Acad Sci* **106:** 15049–15054.

Sawinski J, Wallace DJ, Greenberg DS, Grossmann S, Denk W, Kerr JN. 2009. Visually evoked activity in cortical cells imaged in freely moving animals. *Proc Natl Acad Sci* **106:** 19557–19562.

Smetters D, Majewska A, Yuste R. 1999. Detecting action potentials in neuronal populations with calcium imaging. *Methods* **18:** 215–221.

Sohya K, Kameyama K, Yanagawa Y, Obata K, Tsumoto T. 2007. GABAergic neurons are less selective to stimulus orientation than excitatory neurons in layer II/III of visual cortex, as revealed by in vivo functional Ca^{2+} imaging in transgenic mice. *J Neurosci* **27:** 2145–2149.

Stosiek C, Garaschuk O, Holthoff K, Konnerth A. 2003. In vivo two-photon calcium imaging of neuronal networks. *Proc Natl Acad Sci* **100:** 7319–7324.

Svoboda K, Denk W, Kleinfeld D, Tank DW. 1997. In vivo dendritic calcium dynamics in neocortical pyramidal neurons. *Nature* **385:** 161–165.

Theer P, Hasan MT, Denk W. 2003. Two-photon imaging to a depth of 1000 μm in living brains by use of a $Ti{:}Al_2O_3$ regenerative amplifier. *Opt Lett* **28:** 1022–1024.

Winship IR, Plaa N, Murphy TH. 2007. Rapid astrocyte calcium signals correlate with neuronal activity and onset of the hemodynamic response in vivo. *J Neurosci* **27:** 6268–6272.

50 In Vivo Local Dye Electroporation for Ca^{2+} Imaging and Neuronal-Circuit Tracing

Shin Nagayama,[1] Max L. Fletcher,[1] Wenhui Xiong,[1] Xiaohua Lu,[1] Shaoqun Zeng,[2] and Wei R. Chen[1]

[1]Department of Neurobiology and Anatomy, University of Texas Health Science Center at Houston, Houston, Texas 77030; [2]The Key Laboratory of Biomedical Photonics of the Ministry of Education, Huazhong University of Science and Technology, Wuhan 430074, China

ABSTRACT

A major technical challenge for using optical imaging to analyze neuronal circuit functions is how to effectively load synthetic Ca^{2+} dyes or neural tracers into the brain. This chapter introduces a simple but versatile approach to label many neurons and clearly visualize their axonal and dendritic morphology. The method uses a large-tip patch pipette filled with dextran-conjugated Ca^{2+} dyes or fluorescent tracers. By inserting the pipette into a targeted brain area and passing microampere current pulses, dyes or tracers are electroporated into dendrites and axons near the pipette tip. The dyes are then transported to reveal the entire cell morphology, suitable for both functional Ca^{2+} imaging and neuronal circuit tracing.

INTRODUCTION

Two major challenges persist in our long-lasting endeavor of understanding how the brain functions. One is to reveal the entire brain-wiring diagram completely, and the other is to understand how information is encoded and processed by populations of neurons embedded in intricate networks. To address the first issue of hardware circuitry, it is essential to label individual neuron morphology in its entirety and in a sparse manner. The traditional Golgi silver staining method does this, but it is performed on fixed dead tissues and is thus not suited for carrying out long-range circuit tracking analysis. With respect to the second challenge, neural coding and processing usually take the form of complex spatiotemporal activity patterns in large neuron populations that are sophisticatedly interconnected. Understanding this process requires not only monitoring the functional dynamics of many neurons simultaneously but also interpreting the measured activity patterns in the context of neuronal circuitry. One good example of this approach is the analysis of odor coding-pattern transformation within the olfactory bulb. Odors are initially represented as spatial patterns of activated olfactory glomeruli (Rubin and Katz 1999; Fletcher et al. 2009), which then break down into scattered mitral cell ensemble codes (Nagayama et al. 2007). To understand such an exquisite coding-pattern transformation, the mitral cell activity pattern should be analyzed with reference to its specific connectivity with individual olfactory glomeruli.

This chapter describes a simple but versatile dye-loading method (local electroporation of dextran-conjugated dyes) that can be applied to address both challenges effectively. Electroporation of the cell membrane for loading exogenous substances such as DNA plasmids or fluorescent dyes into a cell has been practiced for many years. Initially, a large electric field set up by a pair of planar electrodes was used for loading many cells simultaneously. Later, a focal electric field was applied to achieve single-neuron labeling (Pinault 1996; Haas et al. 2001; Nevian and Helmchen 2007). By controlling the tip size of a glass pipette, as well as the amplitude of electroporation current, we reported recently that an intermediate number of neurons could be labeled in vivo with a pattern and resolution similar to those of Golgi silver staining (Nagayama et al. 2007).

Protocol

Local Electroporation of Dextran-Conjugated Dyes

The idea of this electroporation labeling method is quite simple. (1) Short electrical pulses are applied to the brain sample using a glass pipette, which probably induces small apertures in the neural membrane. (2) The dextran-conjugated dyes are also pushed into the neuronal cytosol by the electrical pulses. (3) The dyes would then be transferred throughout the cytosol of the neurons, allowing clear visualization of their structure. This process requires basic physiological equipment normally contained within a physiological laboratory.

IMAGING SETUP

In vivo two-photon imaging can be performed either with a custom-built system or with a commercially available system. We use a mode-locked laser operated at a 100-fsec pulse width, 80-MHz pulse-repeating frequency, and 810–830-nm wavelength (Tsunami and Millennia Xs, Spectra Physics). To reduce tissue photodamage, the laser power delivered into the brain should be <120 mW, which is usually enough to achieve an imaging depth of ~500 µm. Calcium imaging is carried out either in line-scan mode with a rate up to 2 msec/line or in frame-scan mode at 2–4 frames/sec. Image analysis is carried out by deriving $\Delta F/F_0 = (F - F_0)/F_0$, where ΔF is the change in fluorescence signal F and F_0 is the baseline fluorescence level.

MATERIALS

CAUTION: See Appendix 6 for proper handling of materials marked with <!>.
See the end of the chapter for recipes for reagents marked with <R>.

Reagents

Agarose (1.5%) dissolved in Ringer solution
Dextran-conjugated Alexa Fluor 488 or 594 dyes (10,000 MW) for long-range circuit tracing
Dextran-conjugated calcium indicators (dextran-conjugated Oregon Green 488 BAPTA-1 <!> [10,000 MW] and dextran-conjugated Calcium Green-1 [3,000 or 10,000 MW])
Mouse
Ringer solution, Ca^{2+}-free <R>
Urethane <!> (1.2 g/kg body weight) or Nembutal <!> (50 mg/kg body weight)

Equipment

Current delivery circuit consisting of an electric stimulator connected to an isolator set at current-output mode (Master-8 and Iso-Flex, A.M.P.I.)
Dental drill
Electrode puller (Sutter P-97)
Glass pipette
Heating pad
Metal clip to ground mouse
Micromanipulator
Oscilloscope

EXPERIMENTAL METHOD

1. Anesthetize the mouse to be used for in vivo dye electroporation using intraperitoneal (i.p.) injection of either urethane or Nembutal. The animal is mounted in a stereotaxic frame and kept on a heating pad to maintain body temperature throughout the entire procedure.
2. Thin the skull with a dental drill and then open a small hole of ~100 μm diameter (without removing the dura) for inserting a dye-loading glass pipette.

   A small hole is critical for reducing brain pulsation, which affects the image quality significantly. In some experiments involving a large craniotomy, 1.5% agarose dissolved in Ringer solution can be applied to reduce brain pulsation.
3. Fabricate a glass pipette using a Sutter P-97 electrode puller to make a tip size of 2.5–4 μm inner diameter.
4. Dissolve 10% (mass/volume) of dextran-conjugated Ca^{2+} indicators or fluorescent tracers in Ca^{2+}-free Ringer solution. Backfill the dye-loading pipette with the dye solution.
5. Insert the pipette into a targeted brain region using a micromanipulator.
6. Carry out local electroporation by applying positive current pulses to the dye-filled glass pipette. See Figure 1A for the current delivery circuit setup; the current delivery circuit consists of an electric stimulator connected to an isolator set at current-output mode (Master-8 and Iso-Flex, A.M.P.I.). Set the current-pulse amplitude at 3–5 μA (Nagayama et al. 2007). The frequency and total number of electroporation pulses are usually 2 Hz and 1200 pulses, respectively.

   A pulse width of 25 msec is found to be the minimal pulse duration that can maintain loading efficiency comparable to that of longer pulses, although it can reduce background fluorescence at the electroporation site.
7. Monitor the actual waveform and amplitude of the injected current pulses on an oscilloscope by measuring the voltage drop across a 100-kΩ series resistor in the circuit.
8. The brain, if alive and healthy, can remove excess dyes very efficiently. Usually, at ~2 h after the termination of dye electroporation, background fluorescence is already cleared away, with local neuronal circuits well labeled for Ca^{2+} imaging. However, if there are concerns about temporary tissue damage caused by electroporation, a longer waiting time is recommended.

   One useful variant of the method is to use local electroporation of dextran-conjugated fluorescent tracers to analyze the fine organization of long-range projection circuits. In such a case, a much longer waiting period is required. We usually wait for 3 d for dextran dyes to be transported from the olfactory bulb glomerulus all the way to the piriform cortex, which covers a distance of ~10 mm.

APPLICATION EXAMPLES

Figures 2 and 3 demonstrate the effectiveness of in vivo dye electroporation for functional analysis of local neuronal circuits. In Figure 2, many neurons in layers 2/3 of the mouse barrel cortex were labeled for Ca^{2+} imaging. Figure 3 shows Ca^{2+} responses of 10 labeled mitral cells to a series of odor stimuli. Apparently, dye electroporation can achieve the labeling of local circuits with a pattern and resolution similar to those of the classical Golgi silver staining but in living brain tissues. This method should expedite in vivo functional analysis of subcellular structures such as dendritic spines (as shown in Fig. 2B) and axonal terminals. It also makes possible the study of neuronal population codes in the context of local circuitry (as demonstrated in Fig. 3 by tracing mitral cell glomerular projections).

Because of the small and restricted dye-loading area in in vivo local electroporation (as illustrated in Fig. 1C), targeted labeling of neural elements can be carried out with unprecedented precision, making it possible to trace some exquisitely organized long-range projection pathways. Figure 4 illustrates examples of this application by analyzing the projection pathway of a single olfactory glomerulus.

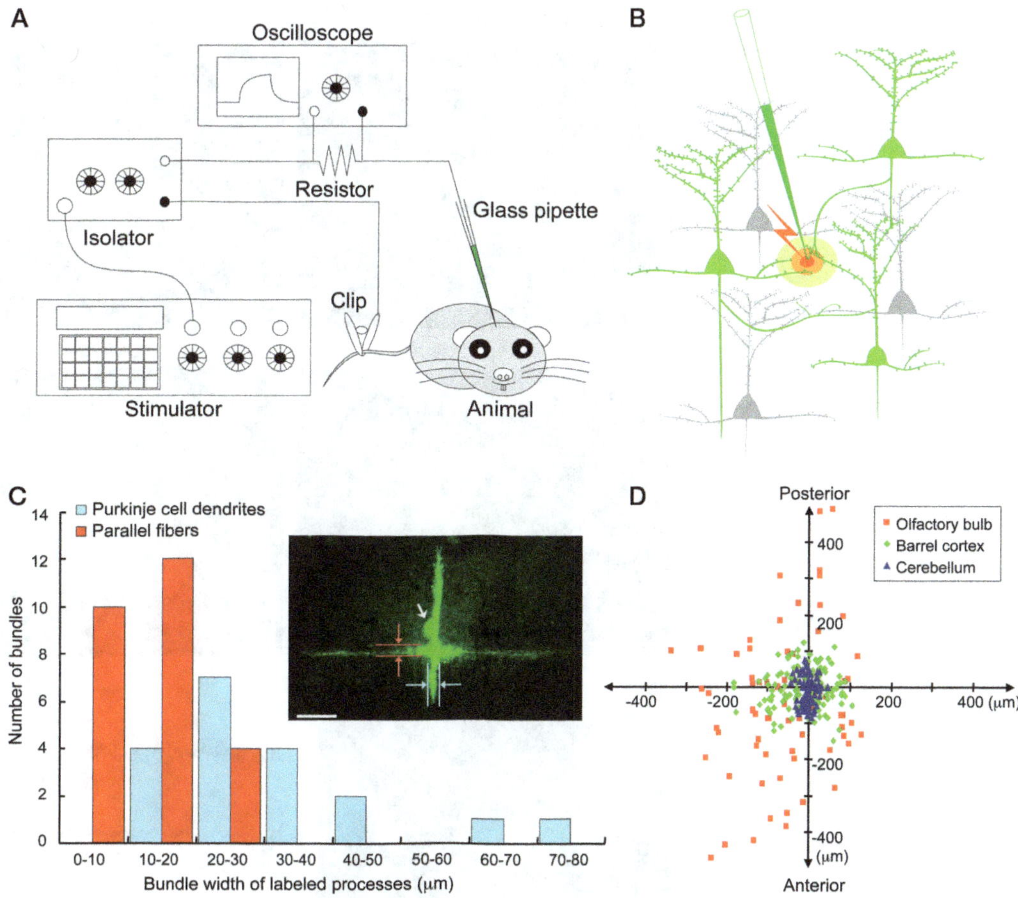

FIGURE 1. Experimental setup, dye-loading mechanism, size of effective dye-loading area, and dependence of labeled-neuron distribution on local circuit characteristics of each brain region. (*A*) Schematic diagram showing the experimental setup for in vivo local electroporation dye loading. An electric circuit, consisting of a pulse generator, a current-output isolator, a series resistor, a dye-filled glass pipette, an anesthetized mouse, and a metal clip attached to the animal tail is used. The waveform and amplitude of current pulses are monitored on an oscilloscope by measuring the voltage drop across a 100-kΩ resistor. (*B*) Schematic illustration showing how dextran-conjugated fluorescent dyes are loaded to visualize local neuronal circuits. A cell membrane in close vicinity of the pipette tip is ruptured transiently by electric current pulses and dyes are electrophoresed into the axons and dendrites that pass through a small effective dye-loading area. The loaded dyes are then transported or simply diffuse along axons and dendrites to enable visualization of the entire neuron morphology. (*C*) Estimation of the size of effective dye-loading area by taking advantage of the unique organization of cerebellar neuronal circuits. (Red) Width distribution of labeled parallel fibers; (blue) distribution of labeled Purkinje cell dendrites. (*Inset*) Example of labeling. The vertical fluorescent bundle represents Purkinje cell dendrites. In this specific case, only one Purkinje cell was labeled; its soma is indicated by a white arrow. The horizontal bundle represents labeled parallel fibers. Also indicated is how the widths of labeling were measured. Scale bar, 50 μm. (*D*) Summary of the distribution patterns of labeled neuronal somata in three different brain structures. The dye-injection sites are aligned at the center of the diagram. (Blue triangles) Distribution of labeled cells in the cerebellum; (green diamonds) labeling in the barrel cortex; (red squares) labeling in the olfactory bulb. The vertical and horizontal axes correspond, respectively, to the rostrocaudal and mediolateral directions. (Modified, with permission, from Nagayama et al. 2007, ©Elsevier.)

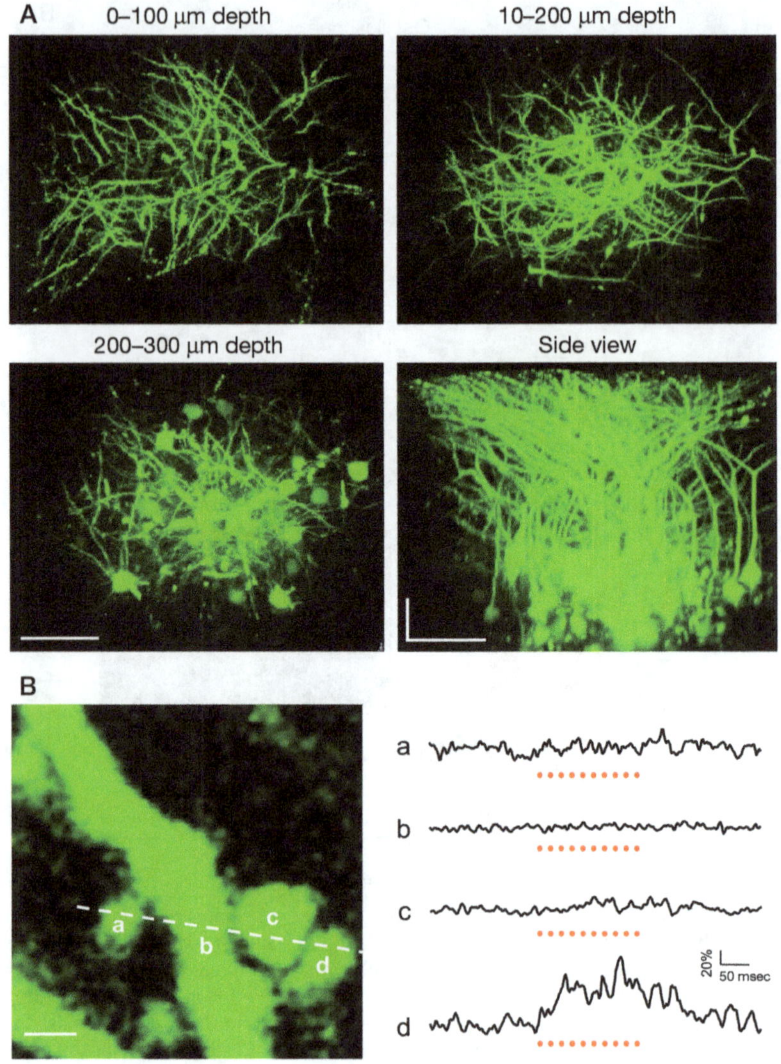

FIGURE 2. In vivo two-photon imaging of neuronal labeling in the neocortex, as well as Ca^{2+} responses in the dendritic shaft and spines of a labeled pyramidal cell. (*A*) Top and side views of labeled neurons in the mouse barrel cortex. The three top-view panels were obtained by projecting along the *z*-axis of the image stacks captured consecutively from three different ranges of depth. Scale bars, 50 µm. The side view is horizontally stretched and thus carries two different scale bars. (*B*) In vivo Ca^{2+} imaging of pyramidal cell dendritic spines. A line-scan recording was made along a dashed line that covered three neighboring spines (a, c, d) and their parental dendrite (b). Only one out of three spines showed a Ca^{2+} response to 10 electric stimuli shown as red dots. No Ca^{2+} increase was observed in the parental dendritic shaft, thus providing in vivo evidence that spines are independent chemical microcompartments. Red dots below the traces indicate individual electric stimuli. Scale bar, 1 µm. (Modified, with permission, from Nagayama et al. 2007, ©Elsevier.)

TROUBLESHOOTING

Problem: There is no labeling at all.

Solution: The correct amount of electric current flowing through the tip of the loading pipette is critical for the effectiveness of electroporation dye loading. If there is no labeling at all, it is important to check if there is current leakage in the dye-loading electric circuit or whether there is a short circuit. In our experience, caution should be taken to completely isolate the animal and stereotaxic instrument electrically from the general equipment ground.

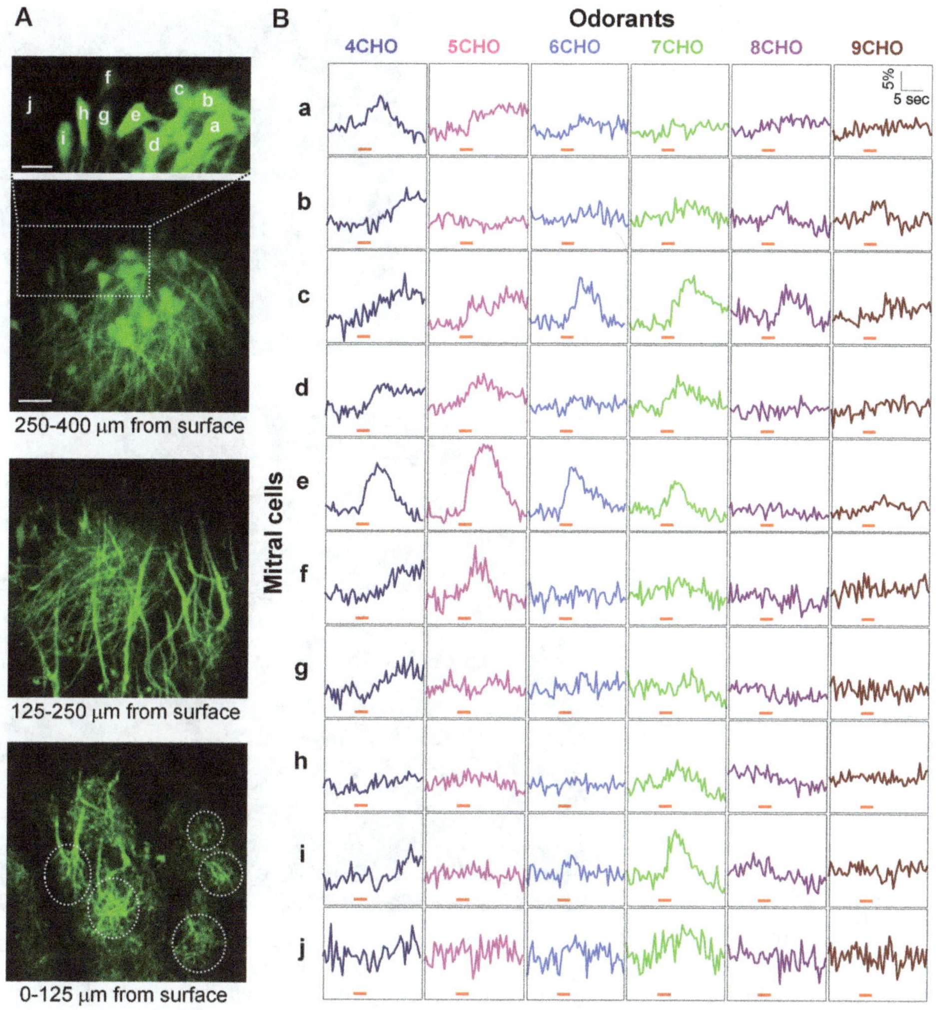

FIGURE 3. In vivo two-photon imaging of olfactory ensemble codes from 10 labeled mitral cells. (*A*) The overall labeling of mitral cells in the mouse olfactory bulb. (*Top two panels*) Ten different mitral cells with their somata labeled by local electroporation of dextran-conjugated Oregon Green 488 BAPTA-1 Ca^{2+} indicator. (*Upper panel*) Magnified view of the area outlined by a white dashed box in the second panel. (*Lower two panels*) Images obtained at two more superficial imaging depths, illustrating the feasibility of tracing the mitral cell apical dendrites from their somata all the way into individual olfactory glomeruli (outlined by white dashed circles). Scale bar, 25 μm (for the *top panel*) and 50 μm (for the *second panel*). (*B*) The response spectra of the 10 mitral cells to a series of aliphatic aldehydes varying systematically in their backbone carbon-chain length. Each row shows the responses of a certain mitral cell to six different aldehydes, and each column represents the responses of the 10 mitral cells to a given aldehyde. (Red bars) Timing of odor delivery. (Modified, with permission, from Nagayama et al. 2007, ©Elsevier.)

Problem: Dye washout after electroporation is poor, leading to strong fluorescence background.

Solution: This usually occurs when brain tissues are not healthy. Using a microscope, check whether regional blood circulation is still normal, slows down, or even stops. In the case of normal blood circulation, this problem may be due to having applied too much dye. Consider reducing the total number of electroporation pulses or decreasing the inner diameter of the pipette tip.

Problem: Neural fluorescence labeling is weak.

Solution: The glass pipette may be clogged; check by directly observing the pipette tip in Ringer solution while delivering the current pulses. If it is not clogged, increase the current amplitude of electroporation. Sometimes, moving the dye-filled glass pipette through Ringer solution to

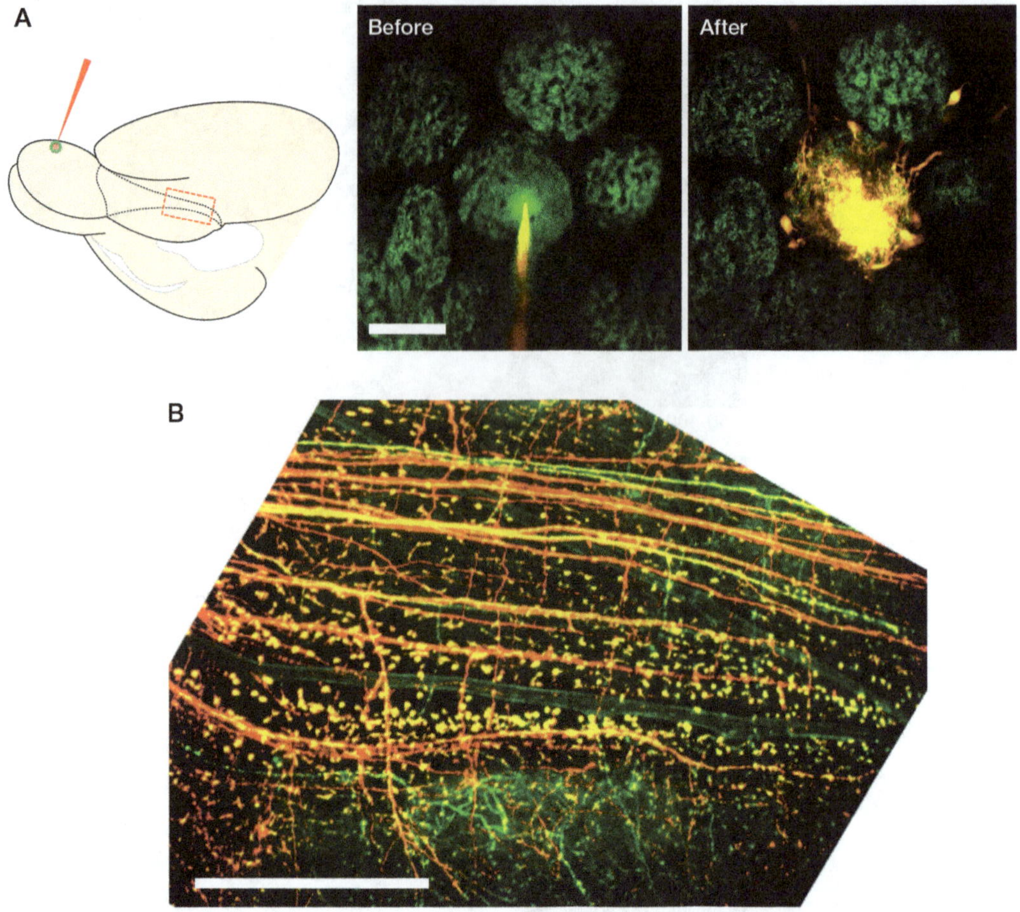

FIGURE 4. Precisely targeted dye loading for tracing exquisitely organized long-range projection pathways. (*A*) Selective labeling of single olfactory glomerulus by local dye electroporation in an OMP (olfactory marker protein)-synapto-pHluorin transgenic mouse. The diagram on the *left* illustrates loading dextran-conjugated fluorescent tracers into an individually targeted glomerulus and then waiting for 5 days for the tracers to be transported over a long distance to visualize the single-glomerulus projection pattern in the olfactory cortex. The two image panels on the *right* demonstrate the precision and confinement of successful dye loading targeted to a single glomerulus. (*B*) Labeling of mitral/tufted cell axonal projections in the lateral olfactory tract in fixed whole brain sample and their sending off collateral branches perpendicularly toward the piriform cortex (upward) and olfactory tubercle (downward). The location of the imaged area is roughly indicated by a red dashed rectangle box in *A*, but the data shown in *B* were from a different animal experiment. Scale bars, 50 µm (*A*); 500 µm (*B*).

reach brain tissue can cause significant dye dilution within the pipette. This can happen especially when using a large pipette tip. To correct this, first drain the Ringer solution that covers the brain, position the dye-filled pipette, and insert it into brain tissue. Then refill the Ringer solution over the brain surface.

Problem: Neuronal Ca^{2+} response is poor.

Solution: This could also be due to poor health of the brain tissue. First, check the animal's condition and local blood circulation. Slowing down or stopping the local blood flow can dramatically affect imaging quality and depth as well as neuronal responses. Also, check the morphology of stained neurons. Obvious signs of cell damage include blebbing of processes or a bright fluorescence level due to Ca^{2+} overload.

DISCUSSION

An alternative approach to carrying out in vivo Ca^{2+} imaging of neuron populations is bulk loading with the acetoxymethyl-ester form of Ca^{2+} indicator dyes (see Chapter 49). Unlike the local electroporation method, AM-dye bulk loading can essentially label all neurons within 200–400 µm of the pipette, leading to a much larger number of neurons being labeled (Stosiek et al. 2003). This can be both an advantage and a disadvantage for this method. The advantage is that up to thousands of neurons (Ohki et al. 2005, 2006) can be imaged simultaneously for functional analysis of the entire local neuron population. The disadvantage is that labeling everything (including glial cells) can obscure signal origins and thus prevent effective signal separation and proper data interpretation. Such universal staining of axons and dendrites makes tracing local neuronal circuits essentially impossible. The electroporation loading method, on the other hand, labels neurons in a sparse manner, allowing individual axonal and dendritic processes to be imaged clearly (Göbel and Helmchen 2007; Fujiwara et al. 2009). In this regard, electroporation has a labeling quality as good as traditional intracellular or whole-cell recording, and, at the same time, it avoids the tediousness commonly encountered in the latter two techniques.

Targeted single-cell electroporation has recently been reported both in vitro (Nevian and Helmchen 2007) and in vivo (Judkewitz et al. 2009). With visual guidance, a dye-loading pipette was positioned at a neuronal soma, and current pulses of microampere level were delivered. Although the same principle of electroporation was applied, the location for dye delivery may be different between our approach and these single-cell electroporation methods. In our method, membrane electroporation and dye delivery are more likely to occur in dendrites and axons rather than at the soma. This might cause much less cell damage, as the major metabolic apparatus is located in the cell body. With the electroporation current amplitudes comparable to those used in our experiments, cell viability after single-cell electroporation is reported to be ~85% (Nevian and Helmchen 2007). Such a high viability provides a very positive perspective to the application of our approach.

RECIPE

CAUTION: See Appendix 6 for proper handling of materials marked with <!>.
Recipes for reagents marked with <R> are included in this list.

Ca^{2+}-Free Ringer Solution

Reagent	Quantity
NaCl	140 mM
KCl	5 mM
EDTA <!>	0.01 mM
HEPES	10 mM
Glucose (pH 7.5)	10 mM

ACKNOWLEDGMENTS

This publication was supported by National Institutes of Health Grant Nos. DC003918 and DC009666 to Dr. Wei Chen, Grant No. DC010057 to Dr. Shin Nagayama, and Grant No. DC009853 to Dr. Max Fletcher.

REFERENCES

Fletcher ML, Masurkar AV, Xing J, Imamura F, Xiong W, Nagayama S, Mutoh H, Greer CA, Knopfel T, Chen WR. 2009. Optical imaging of postsynaptic odor representation in the glomerular layer of the mouse olfactory bulb. *J Neurophysiol* **102:** 817–830.

Fujiwara T, Kazawa T, Haupt SS, Kanzaki R. 2009. Ca^{2+} imaging of identifiable neurons labeled by electroporation in insect brains. *Neuroreport* **20:** 1061–1065.

Göbel W, Helmchen F. 2007. New angles on neuronal dendrites in vivo. *J Neurophysiol* **98:** 3770–3779.

Haas K, Sin WC, Javaherian A, Li Z, Cline HT. 2001. Single-cell electroporation for gene transfer in vivo. *Neuron* **29:** 583–591.

Judkewitz B, Rizzi M, Kitamura K, Häusser M. 2009. Targeted single-cell electroporation of mammalian neurons in vivo. *Nat Protoc* **4:** 862–869.

Nagayama S, Zeng S, Xiong W, Fletcher ML, Masurkar AV, Davis DJ, Pieribone VA, Chen WR. 2007. In vivo simultaneous tracing and Ca^{2+} imaging of local neuronal circuits. *Neuron* **53:** 789–803.

Nevian T, Helmchen F. 2007. Calcium indicator loading of neurons using single-cell electroporation. *Pflugers Arch* **454:** 675–688.

Ohki K, Chung S, Ch'ng YH, Kara P, Reid RC. 2005. Functional imaging with cellular resolution reveals precise microarchitecture in visual cortex. *Nature* **433:** 597–603.

Ohki K, Chung S, Kara P, Hubener M, Bonhoeffer T, Reid RC. 2006. Highly ordered arrangement of single neurons in orientation pinwheels. *Nature* **442:** 925–928.

Pinault D. 1996. A novel single-cell staining procedure performed *in vivo* under electrophysiological control: Morpho-functional features of juxtacellularly labeled thalamic cells and other central neurons with biocytin or neurobiotin. *J Neurosci Methods* **65:** 113–136.

Rubin BD, Katz LC. 1999. Optical imaging of odorant representations in the mammalian olfactory bulb. *Neuron* **23:** 499–511.

Stosiek C, Garaschuk O, Holthoff K, Konnerth A. 2003. In vivo two-photon calcium imaging of neuronal networks. *Proc Natl Acad Sci* **100:** 7319–7324.

51 In Vivo Two-Photon Calcium Imaging in the Visual System

Kenichi Ohki[1,2] and R. Clay Reid[1]

[1]*Department of Neurobiology, Harvard Medical School, Boston, Massachusetts 02115;*
[2]*Department of Molecular Physiology, Kyushu University, Fukuoka 812-8582, Japan*

ABSTRACT

Two-photon imaging of calcium-sensitive dyes in vivo has become a common tool used by neuroscientists, largely because of the development of bolus loading techniques, which can label every neuron in a local circuit with calcium-sensitive dye. Like multielectrode recordings, two-photon imaging paired with bolus loading provides a method for monitoring many neurons at once, but, in addition, it provides a means for determining the precise location of every neuron. Thus, it is an ideal method for studying the fine-scale functional architecture of the cortex and guiding the experimenter to individual neurons that can be targeted for further anatomical study. Two-photon calcium imaging enables study of the fine structure of functional maps in the visual cortex in cats and rodents. In mice, it can allow the characterization of specific cell types when paired with transgenic or retrograde labeling.

INTRODUCTION

Multicell bolus loading labels thousands of cortical neurons in a sphere of ~300-µm diameter (Stosiek et al. 2003; see Chapter 49). These neurons can then be imaged using two-photon excitation to monitor the activity of the population at single-cell resolution, up to a depth of ~350 µm in cats, ~400 µm in rats, and 550 or 400 µm in mice with or without a durotomy, respectively. In our laboratory, we have used the technique to study the fine-scale spatial organization of orientation and direction maps in area 18 of the cat visual cortex (Ohki et al. 2005, 2006) and the lack of such maps in the rat visual cortex (Ohki et al. 2005). In other laboratories, this method has been used to study the plasticity of ocular dominance in mice (Mrsic-Flögel et al. 2007), the lack of orientation tuning in inhibitory interneurons in the mouse (Sohya et al. 2007), the plasticity of direction selectivity in ferrets (Li et al. 2008), and the structure of binocular disparity maps in cats (Kara and Boyd 2009).

Two-photon calcium imaging fills a niche that permits experiments that would otherwise be impossible. It has a much higher spatial resolution and signal-to-noise ratio than intrinsic optical imaging. Thus, multicell bolus loading may be more akin to electrophysiology than optical imaging because the calcium signal more directly reflects spiking activity (Smetters et al. 1999; Stosiek et al. 2003; Kerr et al. 2005; Sato et al. 2007).

However, there are several spatial and temporal limitations to calcium imaging in neural tissue labeled via multicell bolus loading. First, the area that can be imaged is limited by the extent of the dye spread (sphere of ~300 μm^2). By making multiple partially overlapping multicell bolus loading injections and using new low-magnification, high-numerical-aperture (NA) objectives, we can image areas up to almost 1 mm^2. However, because most functional maps in the visual cortex of primates or carnivores have a periodicity of ~1 mm, it can be useful to combine this method with conventional optical imaging to see the overall functional architecture. Thus, to study sparse features of a map, such as orientation pinwheels, we have performed a two-stage experiment that comprises intrinsic-signal imaging followed by multicell bolus loading at targeted regions in the intrinsic-signal map (Ohki et al. 2006).

Second, achieving high temporal resolution of many cells over a wide field of view is a challenge. The dominant source of noise is photon shot noise; thus, at higher scan rates, there is a trade-off between high signal-to-noise ratio and tissue damage due to high laser power. Moreover, owing to the kinetics of the calcium indicator dye, the calcium signal is a low-pass filtered version of the spike train. Although some timing information is present in the rapid (~50 msec) rising phase of the signal with Oregon Green BAPTA-1 (OGB-1) AM dye, the decay of the fluorescence transient is much slower. Thus, methods such as white-noise mapping are quite difficult, and the best results are obtained by using stimuli that yield robust responses and produce sustained spiking over a period of several seconds on a quiescent background. We have had the best success with very simple stimuli, such as oriented gratings that are presented for several seconds, interspersed with unstimulated periods of roughly the same length (Ohki et al. 2005, 2006).

Protocol

In Vivo Two-Photon Calcium Imaging in the Visual System

IMAGING SETUP

We use water-immersion objectives (16x, 20x, 25x, 40x with an NA of 0.8–1.05). The imaging laser is a mode-locked laser, such as the Chameleon Ultra (Coherent) or MaiTai Deep See (Spectra Physics, with prechirper), with a wavelength or 800 nm or 920 nm for orange-green-blueOGB-1 imaging. The filter sets are green filter (517–560 or 500–550 nm) for OGB-1; red filter (603–655 or 570–650 nm) for sulforhodamine 101 (SR101) or Alexa 598; and blue filter (432–482 or 460–500 nm) for enhanced green fluorescent protein (eGFP).

MATERIALS

CAUTION: See Appendix 6 for proper handling of materials marked with <!>.
See the end of the chapter for recipes for reagents marked with <R>.

Reagents

Agarose
Antibiotic cream
Artificial cerebrospinal fluid (ACSF) <R>
Atropine <!> (for cats, monkeys)
Dexamethasone <!>
Dimethylsulfoxide (DMSO) <!>
Isoflurane <!>
Ketamine <!>/xylazine <!>/acepromazine <R>
Lactated Ringer's + 5% dextrose
Lidocaine <!>
Nitrous oxide <!>
Oregon Green BAPTA-1 AM <!> (or other calcium indicator, such as Fluo-4, Fura-2) <R>
Oxygen <!>
Phenylephrine <!> (for cats)
Pluronic
Silicon oil (for rodents)
Vecuronium <!> (0.1 mg/mL) in 5% dextrose

Equipment

Anesthesia control and monitor
Catheter (cat)
CO_2 monitor (cat)
Contact lens
Dissecting microscope

Electrocardiograph (ECG) monitor
Electroencephalograph (EEG) monitor
Gas mask (cat)
Glass pipette
Head plate
Heat blanket
Isoflurane vaporizer
Kimwipes
Lens
Light source (cat)
Microloader (Eppendorf 5242 956.003)
Micromanipulator (Sutter MP-285)
Millipore Ultrafree-MC HV Centrifugal filter (UFC30HV00)
Needle (30-gauge) or tungsten electrode
Nose cone (rodent)
Ophthalmoscope (conventional, slit)
Picospritzer
Pipette puller
Spongel
Stereotaxic frame
Surgical microscope
Surgical tools, including clipper, dental drill, dental cement, forceps, scalpels, scissors, sutures
Trachea tube (cat)
Ventilator (cat)

EXPERIMENTAL METHOD

Induction of Anesthesia (Rodent)

See also Troubleshooting point 10.

1. Weigh the animal to determine anesthesia dose.
2. Inject a ketamine/xylazine (ketamine:xylazine:saline 1:0.05:8.95 in volume) cocktail (dose, 0.05 cc [cocktail]/10 g of body weight) intraperitoneally. This is the initial anesthesia.
3. Inject atropine solution (atropine:saline 1:9 in volume; dose, 0.02 cc/10 g of body weight) intraperitoneally.

   After administering atropine at the start of the procedure (1/10 dilution), also administer it if the animal's heart rate becomes too slow.
4. Start 2% isoflurane with oxygen from a nose cone.

   The recommended doses of isoflurane are as follows: If you induce with ketamine/xylazine, use 1.5%–2% isoflurane during surgery and 1% during imaging; if you induce with isoflurane alone, use 2%–2.5% isoflurane during surgery and 1%–1.5% during imaging.
5. Place the animal on a heating pad.
6. Apply eye cream (or suture eyes to make sure that eyes are not dried).

Induction of Anesthesia (Cat)

See also Troubleshooting point 10.

7. Prepare the mask, drugs (lidocaine, atropine, dexamethasone, ketamine), trachea tube and silk sutures, intravenous catheter, and saline.
8. Weigh the animal to determine anesthesia dose.

9. Inject a 1 mL/kg of body weight ketamine/acepromazine cocktail (20 mg/mL ketamine, 0.2 mg/mL acepromazine) intramuscularly.
10. Inject 0.5 mL/kg of body weight atropine (0.54 mg/mL) intramuscularly.
11. Inject 0.5 mL/kg of body weight dexamethasone (4 mg/mL) intramuscularly.
12. Shave the hair of the head.
13. Administer isoflurane (3%) from a mask.
14. Place the animal on a heating pad.
15. Perform a tracheotomy and suture the trachea tube in place.
16. Connect the isoflurane circuit to the trachea tube.
17. Reduce the isoflurane to 1.5%.
18. Insert an intravenous catheter to the femoral vein to administer the lactated Ringer and later paralytics.
19. Connect the electrocardiogram leads.
20. Administer the eye drops (phenylephrine, atropine), and then insert contact lenses into the eyes.
21. Start intravenous lactated Ringer's + 5% dextrose.

Initial Surgery (Rodent)

22. Cut the whiskers and shave the skull hair.
23. Position the mouse in the earbars of the stereotaxic frame.
24. Open the scalp and remove the temporal and posterior muscles. Scrape the surface of the skull to remove membranes and tissue, and let it dry.
25. Draw the position of the craniotomy onto the skull surface (see Fig. 1A).
26. Affix the head plate with dental cement by filling the space between the skull bone and head plate with cement. Let dry.

    Place the head plate at an angle so as to improve clearance for the contralateral eye.
27. Tighten the head plate onto the pillars of the stereotaxic frame.
28. Make a large well around the craniotomy area with glue or dental cement that can be filled with water for the water–immersion objective.

    Make sure the shape of the well will fit your objective and injection pipette.

Initial Surgery (Cat)

29. Shave the skull hair.
30. Open the scalp and remove the temporal muscles. Let the skull dry.
31. Draw the position of the craniotomy onto the skull surface.
32. Position the head plate in place with dental cement. Fill the space between the bone and head plate with cement, and let it dry.
33. Place the head plate onto the pillars of the sterotaxic frame.
34. Implant the electroencephalograph screws. Then lift up the eyelids by sutures to keep eyes widely open. Insert contact lenses (see Step 38).
35. Connect the electroencephalograph leads and check the electroencephalograph.
36. Make a large well around the craniotomy area with glue or dental cement to fill water for the water-immersion objective.

Eyes

See also Troubleshooting point 11.

37. For rodents, insert contact lenses or silicon oil. There is no correction.
38. For cats, insert contact lenses with correction (see also the Visual Stimulation section for the correction method).

Craniotomy (Rodent)

39. Perform the craniotomy: 3 x 3 mm around region V1. (See Fig. 1 for mice.) Open the skull 1.5–4.5 mm laterally from the midline, along the posterior sinus, which is often 0.5–1 mm behind the suture and 3 mm anterior from the sinus (see also Chapter 60).
40. Stop drilling when the bone starts to crack. Make sure to have cracks all along the craniotomy.
41. Tap the bone with forceps to test whether it has detached. Pour artificial cerebrospinal fluid into the craniotomy, and wait for a few minutes. Then pull the bone very gently. Do not exert strong force, and do not tilt the bone flap.
42. Wash the dura surface with artificial cerebrospinal fluid until the bleeding, if there is any, stops completely. If the bleeding attaches to the dura, use Spongel to wipe out the blood. Wet the Spongel with artificial cerebrospinal fluid, and gently wipe the dura.

Craniotomy (Cat)

43. Perform the craniotomy around A2-6 mm, 1–3.5 mm ML (Horsley–Clarke coordinate), for area 18. Make sure not to make the craniotomy too medially and avoid the sinus. (See also Troubleshooting point 13.)
44. Pour artificial cerebrospinal fluid into the craniotomy well several times during the procedure to dissipate the heat created by drilling.
45. Wash the dura with artificial cerebrospinal fluid.
46. Start the paralytics (vecuronium 0.1 mg/mL in 5% dextrose) at 2.8 cc/kg of body weight to start and 2.8 cc/h per kilogram of body weight for maintenance.

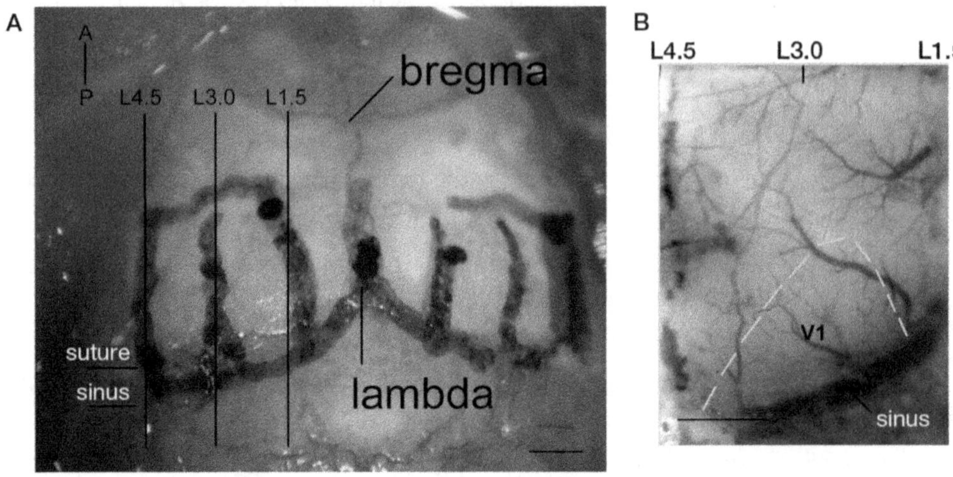

FIGURE 1. Location of craniotomy and primary visual cortex in mouse. (*A*) Location of craniotomies over mouse V1 on both hemispheres. (*B*) Location of primary visual cortex on left hemisphere (see Wang and Burkhalter 2007). Inject around L3.0 mm and as posteriorly as possible. Scale bars, 1 mm.

Durotomy (Rat, Cat)

47. Bend a 30-gauge needle or tungsten electrode.
48. Hook up dura with the bent needle and tear the dura with fine forceps, or cut the dura with fine scissors.

    It is better to keep the durotomy small to minimize brain motion (see Troubleshooting point 12).

    A durotomy is not necessary for mice because the murine dura is transparent enough to see up to 400 µm in depth with two-photon microscopy, and glass pipettes can penetrate through the dura. However, when you need more depth penetration (up to 500 µm) or clearer images, durotomy is helpful even in mice.
49. Fill the craniotomy well with artificial cerebrospinal fluid.

    Applying dexamethasone (10X diluted) onto the brain may help to keep the brain clear, but dexamethasone makes it harder to stop bleeding.

Dye Loading (All Species)

Preparation for Injection

50. Use glass capillary of 1.5 mm outside diameter (OD), 0.85 mm inside diameter (ID), and pull pipettes with filament. The pipette should have a long taper (1:5–10 tapering at the tip).
51. Using a Kimwipe, break the tip of the pipette to 3–4 µm. Inspect it under a conventional light microscope; it should have a sharp, beveled break.
52. Make the fill solution with calcium indicator dye (such as Oregon Green BAPTA-1 AM) and Alexa 594/sulforhodamine 101 (see Chapter 49).
53. Load the pipettes with a microloader and store them vertically.

    Use clean glass pipettes and microloaders (see Troubleshooting point 2).

    Make sure that there are no air bubbles in the pipette.

    Double-check for debris that may occlude pipettes under a conventional microscope (see Troubleshooting point 2).
54. Mix agarose (type III-A; 2%–3%, ~20 mL) with artificial cerebrospinal fluid, and then boil the mixture until it is transparent.
55. Set the micromanipulator angle at 30°.

    It is advisable to use a motorized manipulator for injection to avoid vibration (see Troubleshooting point 5). A four-axes manipulator (*x*–*y*–*z* + diagonal axis; e.g., Sutter MP-285) is very helpful for targeted injection (see below).

Target Injection Site

56. Find the imaging and injection site under the microscope. From here on, do not move the *x*–*y* stage on which the animal and manipulator are mounted; otherwise, you may miss the target site.

    If you do not have any specific target, avoid large vessels. Those vessels will cast a shadow during imaging.
57. Move the objective up to create clearance for the pipette.
58. Attach the dye-loaded pipette to the micromanipulator.
59. Find the tip of the pipette with fluorescence by moving the pipette in any direction and the objective in the *z* direction.
60. Make sure that the pipette is not clogged while the pipette is still outside of the cortex by sending a short pulse (0.1 sec) of low pressure (5 psi) (see Troubleshooting point 2).

61. Lower the pipette tip to the cortical surface at the site you want to penetrate. White light illumination is better for seeing both the pipette and the cortex. Again, do not move the *x–y* stage.
62. A small amount of agarose on the surface of the cortex will suppress the pulsation of the cortex and will also seal the pipette. This will help to prevent backflow of injected dye along the pipette. For rats, cats, and monkeys, it is advisable to add agarose here. For mice, it depends on the size of pulsation of the cortex. Make sure that the pipette is not clogged by the agarose (see Troubleshooting point 5). If you want to add agarose now, do the following.
    i. Move the objective up.
    ii. Suction off the artificial cerebrospinal fluid and make sure that the cortex is dry.
    iii. Start positive pressure.
    iv. Apply agarose at 44°C and wait for 3 min.
    v. Fill the craniotomy well with artificial cerebrospinal fluid.
    vi. Move the objective down.

More Precise Targeting (see Fig. 2)

63. Find the imaging and injection sites under microscope.
64. Focus at the injection site. If you want to inject 200 µm deep from the surface, focus there.
65. Focus up 2 mm.
66. Find the tip of the pipette with fluorescence, without moving the focus or the animal. Move only the pipette.
67. Withdraw the pipette 2 mm, diagonally (30° from the horizontal plane). This will move the pipette 1 mm in *z*, 1.7 mm in *x*.
68. Move the pipette down 2 mm in *z*.
69. Focus slightly above the surface of the target site.
70. Advance the pipette 2 mm diagonally (30°). When you move the pipette ~1.5 mm, you should start to see the pipette.
71. Make sure that the pipette does not damage surface blood vessels. Otherwise, follow the same procedure as above (e.g., positive pressure).

    Untargeted injections are much easier because you can carefully avoid surface vasculature. It is advisable to achieve good success rates in making untargeted injections before attempting an experiment that requires targeted injections.

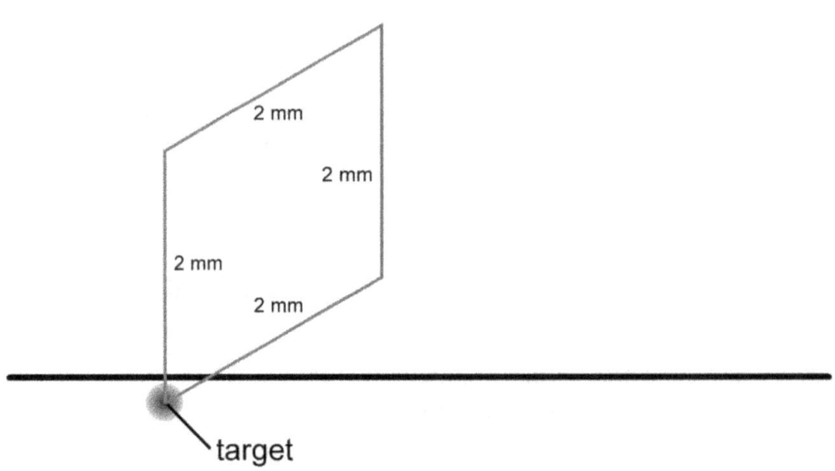

FIGURE 2. Method of precise targeting of injection. See Steps 63–71.

Calcium Indicator Injection under Two-Photon Imaging

72. Start scanning using two-photon excitation. Set the laser power to the optimal power recommended below (see Optimal Laser Power).
73. Under two-photon imaging with a low-magnification objective (16–20X), penetrate the pipette diagonally (~30°) into the target site.
74. Advance the pipette 200–250 μm in the *z* direction and 400–500 μm in the *x* direction. To avoid clogging of the pipette, apply short pulses (0.1 sec) of positive pressure (5–10 psi) every 5–10 sec or apply continuous very weak pressure (<5 psi).
75. At every advance of the pipette (25–50 μm for each advance), make sure that the dye is coming out from the pipette.

    If the pipette is clogged during penetration, see Troubleshooting point 2.
76. Check that the pipette tip is really around at the 200–250 μm depth. Move up the focus, find the surface, and recalibrate the depth.
77. If the depth is correct, inject the dye at, for example, 10 psi, 1 sec, every 10 sec, 20–50 pulses, or at 5–10 psi continuous pressure for several minutes. (Find your favorite parameters.)

    See Troubleshooting point 3 for how to estimate the amount of injected dye.

    See Troubleshooting points 4 and 5 for good injection and how to prevent backflow of the injection.
78. Wait 30 min. If the loading is correct, withdraw the pipette. If not, repeat the injection.

    If cells are not loaded by the calcium indicator, see Troubleshooting point 2.

    If loading looks hazy, see Troubleshooting point 6.

    Dye loading is easier in younger animals, but it may be harder to get stable anesthesia and visual responses. In older animals, dye loading can take more time. Wait for another 1–3 h if necessary.

Before Imaging

79. Put 2% agarose and a coverslip over the craniotomy area to minimize brain motion. A thicker 3% agarose layer without a coverslip also works for mice.

    If there is excessive brain motion, see Troubleshooting point 12.

    If the brain starts to sink below the skull, see Troubleshooting point 13.
80. Test the light shield objective by presenting visual stimuli and record with photomultiplier tubes.

Optimal Laser Power

See also Troubleshooting point 7.

81. Check parameters to avoid photodamage from high laser power. Photodamage depends on many factors, including imaging depth, transparency of the cortex (age, species), imaging size (field of view [FOV]), whether the laser pulses are prechirped, and the objective (NA, aberration, etc.). At typical parameters, the FOV is 300 μm and the objective's NA is 0.8–1.05. If the laser is prechirped (<100 fsec), the parameters are 10 mW at 130–180 μm and 15 mW at 180–230 μm. If the laser is not prechirped, then the parameters are 30 mW at 130–180 μm and 45 mW at 180–230 μm (at the objective tip).

    See Troubleshooting point 7 for how to estimate the optimal laser power.

Visual Stimulation

82. To obtain a visual response, many parameters of visual stimuli must be optimized for the imaged neurons. These are retinotopy, orientation, and spatial frequency. To find the optimal parameters, the easiest way is to use a time course of the entire field of view, including all the

cells and neuropil. This time course will tell you the average properties of imaged neurons, with good signal-to-noise ratio. You usually do not have to repeat runs.

The best visual stimuli are those that produce sustained spiking over a period of several seconds. We have had the best success with very simple stimuli, such as oriented gratings that are presented for 4–8 sec, interspersed with unstimulated periods of 4–8 sec. Drifting gratings produce sustained spiking activity.

83. For mouse and rat:
    i. Examine retinotopy: six positions, cover 0°–90° in azimuth, ±45° in elevation; 40°/cycle, 1 Hz, any orientation; center receptive field.
    ii. Examine orientation tuning: four or six orientations (eight or 12 directions), 40°/cycle, 1 Hz.
    iii. Examine spatial frequency tuning: 7°, 10°, 14°, 20°, 28°, 40°, 56°, 80°/cycle, square wave, 1–1.5 Hz, with the best orientation obtained above.

        Sampling could be sparser; it usually peaks around 20°–40°/cycle; 30°/cycle would be the best choice if you want to use only one.
    iv. After optimizing retinotopy, orientation, and spatial frequency, go to real scan. Usually repeat 10 times.

        Be careful that eyes could move, which would move receptive fields.
    v. If orientation selective cells are sparse, check retinotopy again.
84. For cats:
    i. Find the optic disc and area centralis by reflecting the retina.
    ii. Put contact lenses on the eyes to focus the retina image on the screen.
    iii. Place the monitor centering on the area centralis.
    iv. Examine orientation tuning: four or six orientations (eight or 12 directions), 10°/cycle, 2 Hz for area 18; 2°/cycle, 2 Hz for area 17.
    v. Examine spatial frequency tuning with the best orientation obtained above: 2.5°, 3.5°, 5°, 7°, 10°, 14°, 20°, 28°/cycle for area 18; 0.5°, 0.7°, 1°, 1.4°, 2°, 2.8°, 4°, 5.6°/cycle for area 17.
    vi. After optimizing orientation and spatial frequency, go to real scan. Usually repeat 10 times.
85. Any species:
    i. Keep collecting data until the calcium indicator is extruded from the cells, from several hours to >24 h (see Troubleshooting point 8).

        If there is no or only sparse visual response, see Troubleshooting points 9 and 10.

Somatic Signal from Neurons

86. In the steady state, the mean calcium concentration at soma is proportional to the instantaneous firing rate up to tens of hertz (Helmchen et al. 1996). However, high-affinity calcium-sensitive dyes (such as OGB-1) saturate at a high calcium concentration that is triggered by a high firing rate of neurons. Endogenous or exogenous (calcium indicator itself) calcium buffer can alter the peak amplitude and decay time constant of the calcium signal (Helmchen et al. 1996; see Chapter 36). However, with sustained activation, integration of calcium signal does not depend on the calcium buffering (Helmchen et al. 1996). Deconvolution of the calcium signal is possible (Yaksi and Friedrich 2006). Without knowing the size of unitary events (single action potential–evoked calcium transients), only the relative firing rate can be obtained, not the absolute firing rate. At a low firing rate and a high signal-to-noise ratio, single action potentials can be detected (Kerr et al. 2005; Sato et al. 2007).

Signal from Astrocytes

87. With a large volume of OGB-1 injection, astrocytes are also loaded by OGB-1 (Nimmerjahn et al. 2004). Sulforhodamine 101 (S101) is an astrocyte-specific marker; thus, neurons and astrocytes are easily distinguished by co-injecting OGB-1 and S101 (Nimmerjahn et al. 2004; see Chapter 63). Astrocytes show a slower time course to sensory stimulation than neurons (Wang et al. 2006; Schummers et al. 2008; but see Winship et al. 2007). Astrocytes show sharp orientation tuning in cats (see Fig. 3B; Kelly and Van Essen 1974; K. Ohki and R.C. Reid, unpubl.) and ferrets (Schummers et al. 2008). In the rodent barrel cortex, they show a large calcium increase in response to whisker stimulation (Wang et al. 2006). In the rodent visual cortex, they do not show much response to visual stimulation (see Fig. 3A; K. Ohki and R.C. Reid, unpubl.).

Neuropil Signal

88. The neuropil signal shows no orientation tuning in rodents (see Fig. 3A; Ohki et al. 2005). It shows orientation tuning in cats, but the tuning width is broader than neurons (see Fig. 3B;

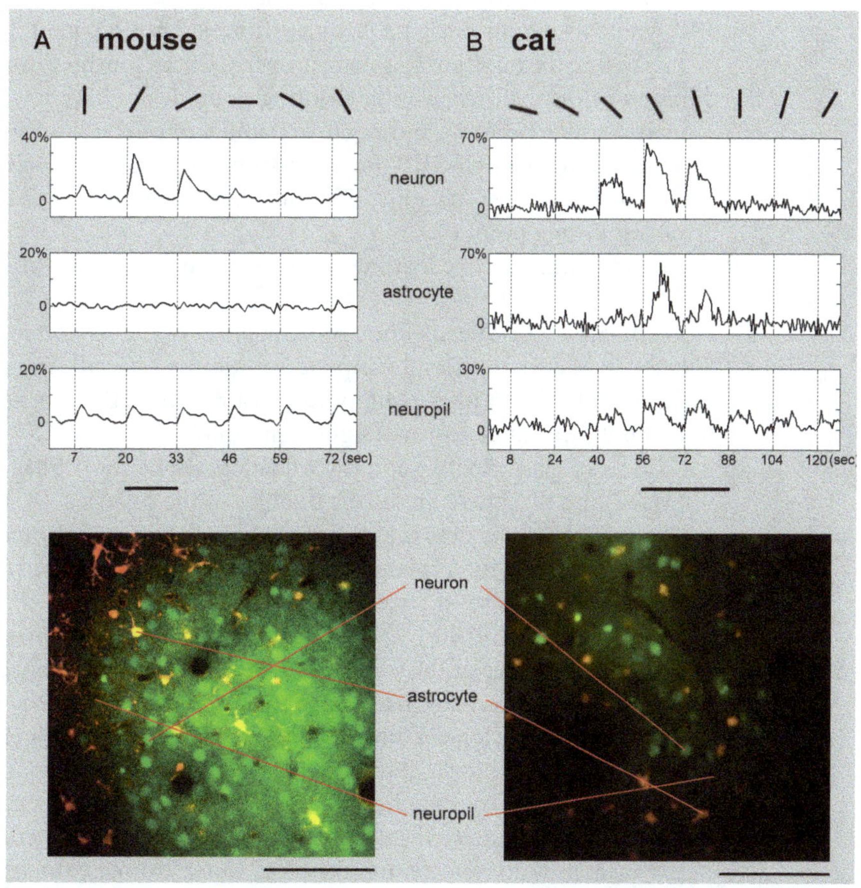

FIGURE 3. Time courses of neurons, astrocytes, and neuropil in the rodent and cat visual cortex. (*A*) Time courses of a neuron, an astrocyte, and neuropil in mouse visual cortex (averaged across 10 trials). Six different orientations separated by 30° were presented for 6 sec. (*Bottom* figure) Regions of interest for the neuron, astrocyte, and neuropil displayed *above*. (Green) Oregon Green BAPTA-1 488; (red) sulforhodamine 101. (*B*) Time courses of a neuron, an astrocyte, and neuropil in the cat visual cortex (averaged across 10 trials). Eight different orientations separated by 15° were presented for 8 sec. Note that the full range of orientations was not presented: for example, horizontal orientation was not presented because neither neurons, astrocytes, nor neuropil responded to those orientations (data not shown). Even neuropil was tuned to an orientation range of <90°. Scale bars (at the *bottoms* of time courses), 30°.

Ohki et al. 2005). The neuropil signal can contaminate the soma signal of neurons owing to the point-spread function of two-photon excitation and out-of-focus fluorescence (for discussion, see Helmchen and Denk 2005; Göbel and Helmchen 2007). The out-of-focus fluorescence from the cortical surface can be reduced by a targeted small injection around the recording site, and the point-spread function can be made smaller by correcting spherical aberration with a recently developed objective (Olympus XLPLN25XWMP).

TROUBLESHOOTING

1. *Problem (Step 78):* Cells are not loaded by the calcium indicator.
   Solutions: Make sure that the calcium indicator has not degraded. After dissolving with DMSO, the OGB-1 mixture should look slightly yellowish. It also helps to sonicate the solution again just before the injection to make the DMSO droplets smaller.

   Make sure enough calcium indicator was injected. Often, pipettes are clogged during penetration and injection (see Troubleshooting point 2), and enough calcium indicator is not injected. The amount of injected dye can be assessed by observing the injection sphere of SR101 (or Alexa 594) with two-photon imaging (see also Troubleshooting point 3).

   Make sure the injection was done properly. Check for three things during the injection: isotropic diffusion of the dye (see Troubleshooting point 4), no backflow of the dye along the pipette (see Troubleshooting point 5), and negative staining of cell bodies (see Troubleshooting point 6).

   Check if you can see S101-loaded astrocytes. If so, only calcium indicator was not loaded for some of the above reasons. If not, the cortex may not be transparent enough (see Troubleshooting point 6).

   In older animals, dye loading can take more time. Wait for another 1–3 h.

2. *Problem (Step 75):* The injection pipettes are clogged very often during penetration or injection.
   Solutions: Make sure that the pipette solution is filtered. Any debris in the solution will cause clogging. A 0.45-µm filter (Millipore Ultrafree-MC HV Centrifugal filter; UFC30HV00) works best. A 0.23-µm filter removes significant amounts of dye.

   Use clean glass pipettes and microloaders (Eppendorf 5242 956.003).

   Make sure that there are no air bubbles in the pipette.

   Double-check for debris in the pipette under a conventional light microscope.

   Make sure that the pipette is not clogged when the pipette is still outside of the cortex by applying a short pulse (0.1 sec) of low pressure (5 psi).

   Apply short pulses (0.1 sec) of positive pressure (5–10 psi) during penetration, every 5–10 sec. At every advance of the pipette (25–50 µm for each advance), make sure that the dye comes out from the pipette. Alternatively, you can use continuous low pressure (<5 psi).

   If the pipette is clogged during injection, withdraw the pipette to the surface, and apply high pressure (20–40 psi) to clear up the pipette. If this does not work, replace the pipette with a new one. (Do not apply too much pressure inside the cortex.)

   Avoid blood vessels. Vessel walls are sticky, and the pipette can become clogged easily. In larger animals (cats, ferrets, monkeys) or older animals, the tissue tends to stick to the pipette more. Consider using a larger tip size (~5 µm).

3. *Problem (Step 77):* How do you estimate the amount of dye injection?
   Solution: Use the same concentration of SR101 (or Alexa 594), laser power, objective, and photomultiplier tube (PMT) gain for every injection. Inject around the same depth (e.g., 200–250 mm). Then, the intensity of SR101 (or Alexa 594) will be proportional to the amount of injected dye.

   The injection sphere will be typically 300 µm in diameter, and you should see a very bright

SR101 (or Alexa 594) signal after each pulse. Importantly, cells should show up as negative staining by SR101.

After injection, check the OGB-1 signal by epifluorescence (excitation, blue; emission, green). The OGB-1 signal from the entire volume (mostly in neuropil) will develop relatively quickly (15–30 min). You can tell the approximate amount of total injection even before OGB-1 signal in the soma is fully developed, and it can be used to calibrate the injection. If the injection was done properly (see also Troubleshooting point 1), cells will be eventually loaded by OGB-1. If not, go to Troubleshooting point 1. Even when enough dye was injected, cells will not be loaded if there was backflow or leakage to the surface (see Troubleshooting point 5).

If too much dye was injected, you will see a darkening of the OGB-1 signal around the injection tip, which probably shows damaged (or dead) cells.

The amount of injected dye cannot be determined solely by tip size, pressure, and duration of injection. It largely depends on the condition of the pipette tip (whether it is clear or partially blocked) and whether the tip is in a soft or a tough tissue. Thus, it is important to always check injection under the two-photon microscope.

4. *Problem (Step 77):* How do you know a good injection was performed?
   Solution: The dye should diffuse to the extracellular space slowly and almost by passive diffusion. The injection flow should be isotropic and with low pressure, which can be checked by observing the injection under two-photon imaging.

   If too much dye is injected at once, or if the pressure is too high, it will backflow along the pipette (see Troubleshooting point 5), or it may leak to the surface through vessel walls or breaking the tissue. If this occurs, the loading is not good.

   Avoid injecting around blood vessels. The injected dye can escape to the surface along the vessel walls.

5. *Problem (Step 77):* Injected calcium indicator backflows along the pipette.
   Solution: The flow of the injection should be slow (or the amount of each pulse of injection should be small). If it is too fast, it may backflow along the pipette, or it may leak to the surface through vessel walls or breaking the tissue (see also Troubleshooting point 4).

   The pipette should be tightly sealed to the cortical tissue. Vibration of the pipette or pulsation of the cortex creates a space between the pipette and the tissue. Motorized manipulators help decrease vibration during the injection.

   Put a small amount of agarose on the surface of the cortex. It will suppress the pulsation of the cortex and will also help to seal the pipette. (But make sure that the pipette is not clogged by the agarose.)

6. *Problem (Step 78):* The loading looks hazy.
   Solution: The cortex may have become opaque. This can be checked in two ways. (1) During injection, SR101 (or Alexa 594) fills up the extracellular space, and cells should look dark on a bright background (i.e., negatively stained). If no cells are visible at this stage, the cortex is not very transparent. (2) After loading, even if the cells are not loaded by OGB-1, astrocytes should be loaded by SR101 (astrocytes take up SR101 more readily than neurons take up OGB-1). If SR101 staining of astrocytes looks clear, the cortex is transparent, and potential problems are likely due to OGB-1 loading. If astrocytes also look hazy, the cortex is not transparent.

   If the cortex does not look transparent, check the cortical surface by epifluorescence or white light.

   If the cortex is transparent, then there was likely a problem with the loading (go to Troubleshooting point 1).

7. *Problem (Step 81):* What is the optimal laser power to use?
Solution: If you see a steady decrease or increase of baseline signal of calcium indicator, the laser power may be too high. Note that the decrease could also happen from leakage of water.
Even when the calcium baseline is stable, high laser power could suppress the response of neurons. If you see that responses of neurons disappear after ~1 min of scanning, consider reducing laser power. However, note that attenuation of responses could also occur as a result of adaptation.
Examples of optimal laser power are listed in the section Optimal Laser Power. However, note that optimal laser power depends on many factors, including (1) imaging depth, (2) transparency of cortex (age, species), (3) imaging size (field of view), (4) whether laser pulses are prechirped, and (5) objective (NA, aberration, etc.).
If the laser power is optimal, then you can keep scanning continuously until the dye disappears (typically from several hours to >24 h; see Troubleshooting point 8).

8. *Problem (Step 85):* How long can we image?
Solution: As mentioned in Troubleshooting point 7, if the laser power is optimal, the calcium indicator does not bleach very much, even with continuous scanning. However, the calcium indicator will be extruded from cells after a while, ranging from several hours to >24 h. The extrusion time seems to depend on the animal's age. Extrusion seems to be faster in younger animals.
The cortex can also become opaque over time (12–24 h). This possibility of opacity will limit the imaging time. Opacity can be suppressed by intravenous injection of dexamethasone (~0.1 mg/h per kilogram of body weight).

9. *Problem (Step 85):* There is no visual response in calcium signal.
Solution: There are many causes for no response, including (1) poor loading of calcium indicator (see Troubleshooting point 1); (2) deep anesthesia (see Troubleshooting point 10); (3) low body temperature (see Troubleshooting point 10); (4) unhealthy cortex; (5) cataract of eyes (see Troubleshooting point 11); (6) unfocused eyes (see Troubleshooting point 11); (7) the stimulus is not optimized for neurons' receptive fields (see Visual Stimulation); (8) the imaging site is outside the target area (such as primary visual cortex); and (9) the laser power is too high (Troubleshooting point 7).

10. *Problem (Step 85):* How can anesthesia be controlled so that a good visual response results? (Also see Troubleshooting point 9.)
Solution: Sensory responses crucially depend on the type and dose of anesthesia. We recommend using isoflurane for mice, rats, and cats.
A higher dose of anesthesia is needed during surgery to make sure that animals are deeply anesthetized. During imaging, you may need to reduce the dose to obtain strong responses. The depth of anesthesia should be monitored by toe pinch and electroencephalography. Our recommended dose is described in the Anesthesia section.
Responsiveness correlates with core body temperature. Because anesthesia disrupts temperature regulation, it is important to maintain body temperature at 37°C–38°C with a heat blanket. Insert the temperature probe into the anus and measure the core temperature.
Responsiveness of neurons also correlates with heart rate. Make sure that the heart rate is high enough: 400–600 beats per minute (bpm) for mice, 300–400 bpm for rats, and 150–300 bpm for cats. If the heart rate is too low, decrease the dose or consider switching to a different anesthetic.
If the heart rate starts to be arrhythmic or bradycardic, inject atropine intramuscularly.
In rodents, xylazine helps to avoid brain swelling during craniotomy. However, it has a long half-life and suppresses heart rate and neural response. Using isoflurane alone is better for

getting neural response, but the craniotomy will be more difficult because of high blood pressure, bleeding, and brain swelling. Increase the isoflurane during the craniotomy to help reduce swelling and bleeding.

11. *Problem (Steps 37 and 38):* How do you take care of and focus the eyes?
Solution: For vision experiments, it is very important to keep eyes healthy. If they dry out, the cornea will develop a cataract and visual responses will be much poorer.
For rodents, apply eye drops to moisturize the eyes at the very beginning of the procedure. Place antibiotic cream on the eyes during surgery, or, alternatively, suture the eyes closed during the surgery to protect the eyes from light and to keep them from drying out.
Once you finish the surgery, put contact lenses on both eyes. For mice, you may need to cut the eyelids to put on contact lenses.
If you do not have contact lenses, you can apply silicon oil to the eyes. The oil may dry out in a few hours. Check eyes every hour, and add oil if it dries out. Because silicone oil requires a good deal of maintenance, contact lenses are better for long experiments.
In rodents, it is not easy to focus eyes accurately. It is better not to dilate pupils by applying atropine eye drops.
For cats and other large animals (ferrets, monkeys), apply eye drops to moisturize the eyes at the beginning of the procedure. Once you finish the induction of anesthesia, place contact lenses in both eyes.
Apply atropine eye drops to dilate the pupils. Then focus the eyes on the screen by using appropriate contact lenses or ophthalmic lenses. For cats, you can find the best lenses by reflecting the retina on the screen.

12. *Problem (Step 79):* The brain is pulsating more than several micrometers.
Solution: The craniotomy should not be too big (at maximum, 3 × 3 mm). The durotomy should also be minimal (1 × 2 mm).
Apply 2%–3% agarose on the cortex with a cover glass. Before applying the agarose, the cortex (or dura if it exists) should be dry so that the agarose will attach to the cortex tightly. Use a relatively large size of cover glass (e.g., 5–8 mm diameter). After pouring the agarose, put on the cover glass immediately. Let it dry for 3–5 min.
Horizontal (x–y) motions can be corrected by registration across frames. z motions cannot be corrected.

13. *Problem (Step 79):* Brain starts to sink below the skull.
Solution: In cats, particularly in young cats, the brain sometimes sinks far below the skull (a few millimeters), disrupting imaging of the cortex with the two-photon microscope.
Keep the general condition of the animals very healthy from the beginning. If the blood pressure drops, there is a higher risk of the brain sinking.
Do not position the head too high above the rest of the body. It will decrease the blood pressure in the brain.
Minimize the durotomy. The dura is often loosely attached to the skull and the cortex, and it lifts up the cortex. If a large area of the dura is removed, the cortex loses this support and sinks. Particularly, avoid opening the very medial side of the dura.

14. *Problem:* What other calcium indicators are recommended?
Solution: In addition to Oregon Green BAPTA-1, Fluo-4 and Fura-2 are useful for specific purposes. Fluo-4 has a larger percent change of signals and is helpful to detect single action potentials (Sato et al. 2007). Fura-2 has a shorter wavelength for two-photon excitation than OGB-1, which is helpful for separating its signal from eGFP (Sohya et al. 2007).
However, loading of these dyes is more difficult than OGB-1, particularly in adult animals, and loading is often more sparse.

DISCUSSION

The ability to examine the receptive-field properties of almost every neuron in a local volume opens up a new realm of experimental investigations of the cerebral cortex. From its inception, the idea of functional architecture has been considered both at a macroscopic scale, hundreds of micrometers to several millimeters, and at a microscopic scale, minicolumns that are 50 μm wide. Until recently, the microscopic scale could only be studied with microelectrodes, which typically sample neurons sparsely and without precise localization. Single-cell calcium imaging in vivo promises to elucidate the relationship between anatomy and physiology in neuronal ensembles at an unprecedented level of completeness. Combined with a characterization of the sensory or motor function of single neurons, this imaging method could permit the study of microfunctional architecture throughout the cerebral cortex and reveal other examples of maps that have single-cell precision.

Stronger correlations between anatomy and physiology could be achieved by labeling cells to reveal their transmitters or other molecular markers, their projection patterns, or their detailed dendritic and axonal morphology. The evolution of genetic tools in mice has fostered an unprecedented opportunity to study neurons in living animals. The ability to stably label specific subsets of neurons with genetically encoded reporters allows the dissection of cell type–specific function in the dauntingly complex cortical circuit. Several groups have generated mouse lines with subsets of neurons labeled with fluorescent proteins using various promoters. By combining the power of two-photon calcium imaging with the cell type–specific labeling afforded by mouse genetics, the complexity of the cortical network can be studied by examining cell type–specific function in intact cortical circuits (Sohya et al. 2007). This approach can lead to a more complete understanding of sensory physiology of interneurons and their role in cortical information processing.

RECIPES

CAUTION: See Appendix 6 for proper handling of materials marked with <!>.
Recipes for reagents marked with <R> are included in this list.

Anesthetic for Cats

1. Prepare ketamine/acepromazine cocktail (10× diluted):

Ketamine (20 mg/mL) <!>	1 cc
Acepromazine (0.2 mg/mL)	0.1 cc
Saline	8.9 cc

2. Add vecuronium <!> (0.1 mg/mL) in 5% dextrose.
3. Add dexamethasone <!> for cortex (0.4 mg/mL; diluted by ACSF).

Anesthetic for Rodents

1. Ketamine/xylazine (1:0.05) cocktail (10× diluted):

Ketamine <!> (20 mg/mL)	1 cc
Xylazine <!> (2 mg/mL)	0.05 cc
Saline	8.95 cc

2. Atropine (10× diluted):

Atropine <!> (0.54 mg/mL)	1 cc
Saline	9 cc

Artificial Cerebrospinal Fluid (ACSF)

NaCl	124 mM
KCl	2.5 mM
$MgSO_4$ <!>	2.0 mM
KH_2PO_4	1.25 mM
$NaHCO_3$	26 mM
Glucose	10 mM
Sucrose	4 mM
$CaCl_2$ <!>	2.5 mM

The ACSF solution may be slightly milky when the components are first mixed, but it should clear when aerated with 95% O_2 <!> and 5% CO_2 <!>, after which the pH of the medium should be 7.4.

Dye-Making Recipe

OGB-1 <!>	
Alexa 594 <!>	2 mM
DMSO <!> + Pluronic	20% by weight; keep at 60°C for 30 min
ACSF	
R101 <!>	2 mM

Millipore Ultrafree-MC HV Centrifugal filter (UFC30HV00)

1. Keep the OGB-1 dry.
2. Thaw the OGB-1 with your gloved hand.
3. Dissolve the OGB-1 with 4 mL of DMSO (containing 20% of Pluronic). After dissolving with DMSO, the OGB-1 should look slightly yellowish.
4. Vortex the mixture for 1 min.
5. Add 35 µL of ACSF.
6. Add 1 µL of 2 mM Alexa 594.
7. Vortex the mixture for 1 min.
8. Seal the top of the tube with Parafilm.
9. Sonicate the mixture for 5 min.
10. Filter the dye mixture through a 0.45-µL mesh using a centrifuge.

ACKNOWLEDGMENTS

We especially thank Lindsey Glickfeld for her valuable comments on the manuscript; Tim Cherry for reading the manuscript; Aaron Kerlin, Mark Histed, and Vincent Bonin for discussions; and other members of the Reid laboratory.

REFERENCES

Göbel W, Helmchen F. 2007. In vivo calcium imaging of neural network function. *Physiology* **22:** 358–365.

Helmchen F, Denk W. 2005. Deep tissue two-photon microscopy. *Nat Methods* **2:** 932–940.

Helmchen F, Imoto K, Sakmann B. 1996. Ca^{2+} buffering and action potential-evoked Ca^{2+} signaling in dendrites of pyramidal neurons. *Biophys J* **70:** 1069–1081.

Kara P, Boyd JD. 2009. A micro-architecture for binocular disparity and ocular dominance in visual cortex. *Nature* **458:** 627–631.

Kelly JP, Van Essen DC. 1974. Cell structure and function in the visual cortex of the cat. *J Physiol* **238:** 515–547.

Kerr JN, Greenberg D, Helmchen F. 2005. Imaging input and output of neocortical networks in vivo. *Proc Natl Acad Sci* **102:** 14063–14068.

Li Y, Van Hooser SD, Mazurek M, White LE, Fitzpatrick D. 2008. Experience with moving visual stimuli drives the early development of cortical direction selectivity. *Nature* **456:** 952–956.

Mrsic-Flögel TD, Hofer SB, Ohki K, Reid RC, Bonhoeffer T, Hubener M. 2007. Homeostatic regulation of eye-specific responses in visual cortex during ocular dominance plasticity. *Neuron* **54:** 961–972.

Nimmerjahn A, Kirchhoff F, Kerr JN, Helmchen F. 2004. Sulforhodamine 101 as a specific marker of astroglia in the neocortex in vivo. *Nat Methods* **1:** 31–37.

Ohki K, Chung S, Ch'ng YH, Kara P, Reid RC. 2005. Functional imaging with cellular resolution reveals precise micro-architecture in visual cortex. *Nature* **433:** 597–603.

Ohki K, Chung S, Kara P, Hubener M, Bonhoeffer T, Reid RC. 2006. Highly ordered arrangement of single neurons in orientation pinwheels. *Nature* **442:** 925–928.

Sato TR, Gray NW, Mainen ZF, Svoboda K. 2007. The functional microarchitecture of the mouse barrel cortex. *PLoS Biol* **5:** e189. doi: 10.1371/journal.pbio.0050189.

Schummers J, Yu H, Sur M. 2008. Tuned responses of astrocytes and their influence on hemodynamic signals in the visual cortex. *Science* **320:** 1638–1643.

Smetters D, Majewska A, Yuste R. 1999. Detecting action potentials in neuronal populations with calcium imaging. *Methods* **18:** 215–221.

Sohya K, Kameyama K, Yanagawa Y, Obata K, Tsumoto T. 2007. GABAergic neurons are less selective to stimulus orientation than excitatory neurons in layer II/III of visual cortex, as revealed by in vivo functional Ca^{2+} imaging in transgenic mice. *J Neurosci* **27:** 2145–2149.

Stosiek C, Garaschuk O, Holthoff K, Konnerth A. 2003. In vivo two-photon calcium imaging of neuronal networks. *Proc Natl Acad Sci* **100:** 7319–7324.

Wang Q, Burkhalter A. 2007. Area map of mouse visual cortex. *J Comp Neurol* **502:** 339–357.

Wang X, Lou N, Xu Q, Tian GF, Peng WG, Han X, Kang J, Takano T, Nedergaard M. 2006. Astrocytic Ca^{2+} signaling evoked by sensory stimulation in vivo. *Nat Neurosci* **9:** 816–823.

Winship IR, Plaa N, Murphy TH. 2007. Rapid astrocyte calcium signals correlate with neuronal activity and onset of the hemodynamic response in vivo. *J Neurosci* **27:** 6268–6272.

Yaksi E, Friedrich RW. 2006. Reconstruction of firing rate changes across neuronal populations by temporally deconvolved Ca^{2+} imaging. *Nat Methods* **3:** 377–383.

52 Three-Dimensional Imaging of Neuronal Network Activity

Björn M. Kampa, Werner Göbel, and Fritjof Helmchen
Department of Neurophysiology, Brain Research Institute, CH-8057 Zurich, Switzerland

ABSTRACT

Neural tissue is organized in three-dimensional (3D) networks of neuronal and glial cell populations. To understand the functional organization of these networks, it is desirable to achieve 3D activity measurements from large cell populations in intact tissue and with high temporal resolution. Repeated acquisition of image stacks with standard laser-scanning microscopes is too slow for this purpose. This chapter presents a protocol for fast 3D calcium imaging in the living brain using mechanical laser scanning with standard galvanometric mirrors and a piezoelectric focusing element. With a sinusoidal swinging microscope objective as a constraint, 3D scan trajectories are generated that sample fluorescence signals from the majority of cells within a cuboidal volume. Using this technique, measurements with 10-Hz temporal resolution can be achieved for population calcium signals from several hundreds of identified neurons and glial cells within cuboids with side lengths of ~250 µm. As an example, we present a cellular 3D orientation map of the mouse visual cortex. This 3D laser-scanning technique enables direct observation of in vivo neural network dynamics in cell populations of substantial size.

INTRODUCTION

Most physiological processes take place in three-dimensionally organized dynamic networks of cells. In particular, 3D spatiotemporal activation patterns in excitable neuronal networks are fundamental to brain function. Therefore, a major goal in neuroscience is to read out neuronal population activity in the intact brain. Electrophysiological approaches using multiple electrodes partially achieve this goal by providing simultaneous measurements of tens to hundreds of neurons (Buzsáki 2004). However, the subset of sampled cells in these recordings is necessarily widely dispersed because of the size and spacing of electrodes.

Fluorescence imaging using functional indicators is an alternative approach for large-scale population readout. Specifically, deep tissue imaging with two-photon laser-scanning microscopy (Denk et al. 1990; Helmchen and Denk 2005) enables in vivo calcium measurements to probe electrical excitation of neuronal networks indirectly. Standard laser-scanning approaches are limited, however, in their ability to collect 3D data rapidly. Typically, to obtain 3D information, stacks of images are acquired by raster-like, line-by-line two-dimensional (2D) scanning and stepwise focus changes. Repeated collection of image stacks—so-called four-dimensional (4D) imaging—typically requires stack-to-stack intervals of at least a few seconds and, thus, permits measurements of only slow 3D dynamics (e.g., morphological changes). A higher temporal resolution is required for analyzing neural excitation flow using calcium imaging.

Calcium transients evoked by action potentials typically have durations of a few hundred milliseconds and, thus, are likely to be missed even at 1-Hz sampling rates. Improved acquisition speeds can be achieved by deviating from raster-like scanning and optimizing scan trajectories for data collection from the structures of interest (Göbel et al. 2007; Göbel and Helmchen 2007a; Lillis et al. 2008). This chapter presents a simple 3D scanning scheme that uses mechanical scanners and a free line-scanning approach (Göbel et al. 2007). It describes the experimental setup, discusses the relevant technical aspects, and shows how 3D imaging is applied in combination with population staining techniques for in vivo calcium imaging from neuronal networks (see Chapter 49; for reviews, see Garaschuk et al. 2006; Göbel and Helmchen 2007b; Grewe and Helmchen 2009). 3D network imaging in the mouse visual cortex is presented as an example. Finally, we briefly compare this approach to other 3D imaging techniques and provide a future outlook.

Protocol

Measuring Neuronal Population Activity Using 3D Laser Scanning

The purpose of 3D laser scanning is to create a 3D line scan that samples relatively homogenously from a particular observation volume. Clearly, the spatial resolution of this approach is low, except along the line. However, the main goal is not to resolve subcellular structures, but rather to hit as many cell bodies as possible within the volume. In this manner, local network activity can be inferred from the somatic calcium signals of a significant fraction of the cell population. This protocol describes key technical aspects of implementing fast 3D laser scanning for measuring in vivo neural network dynamics as described in the original paper (Göbel et al. 2007).

IMAGING SETUP

Any mirror-based two-photon laser-scanning microscope can, in principle, be adapted for 3D mechanical scanning. However, because custom control software is required, it probably is easier to adapt a custom-built, rather than a commercial microscope for 3D scanning. We use a custom-built two-photon laser-scanning microscope with two standard galvanometric scan mirrors for *x–y* scanning. A Ti:sapphire laser system (e.g., Tsunami, Newport Spectra-Physics) provides ultrashort laser pulses. For fluorescence excitation of the calcium indicator Oregon Green BAPTA-1 (OGB-1) a wavelength of 870 nm is used, and laser intensity is adjusted using a Pockels cell. The only special device required is a piezoelectric focusing element for rapid movement of the objective in the *z* direction (i.e., along the optical axis; Fig. 1A). Almost any standard microscope objective can be mounted to a piezoelectric focusing element that allows travel ranges of 400 µm or more. We mainly use 40× (e.g., 40× LUMPlanFl/IR, 0.8 numerical aperture [NA], mass 75 g; Olympus) or 20× (NA 0.95, mass 235 g; Olympus) water-immersion objectives. Mirror scanning and movement of the piezofocusing element are controlled by custom software written in LabVIEW (National Instruments); the core routines can be downloaded at http://www.nature.com/nmeth/journal/v4/n1/extref/nmeth989-S9.zip.

MATERIALS

CAUTION: See Appendix 6 for proper handling of materials marked with <!>.
See the end of the chapter for recipes for reagents marked with <R>.

Reagents

Agarose, type III-A (1%–1.5%, prepared in normal rat Ringer solution (NRR); Sigma-Aldrich)
Calcium indicator dye for population staining (e.g., acetoxymethyl [AM] ester form of Oregon Green BAPTA-1 <!> [OGB-1 AM; Invitrogen])
NRR <R>
Rats, 3–4-wk-old
Sulforhodamine 101 (SR101; e.g., from Sigma-Aldrich)
  Sulforhodamine is used for specific counterstaining of cortical astrocytes (Nimmerjahn et al. 2004; see also Chapter 63).
Urethane <!>

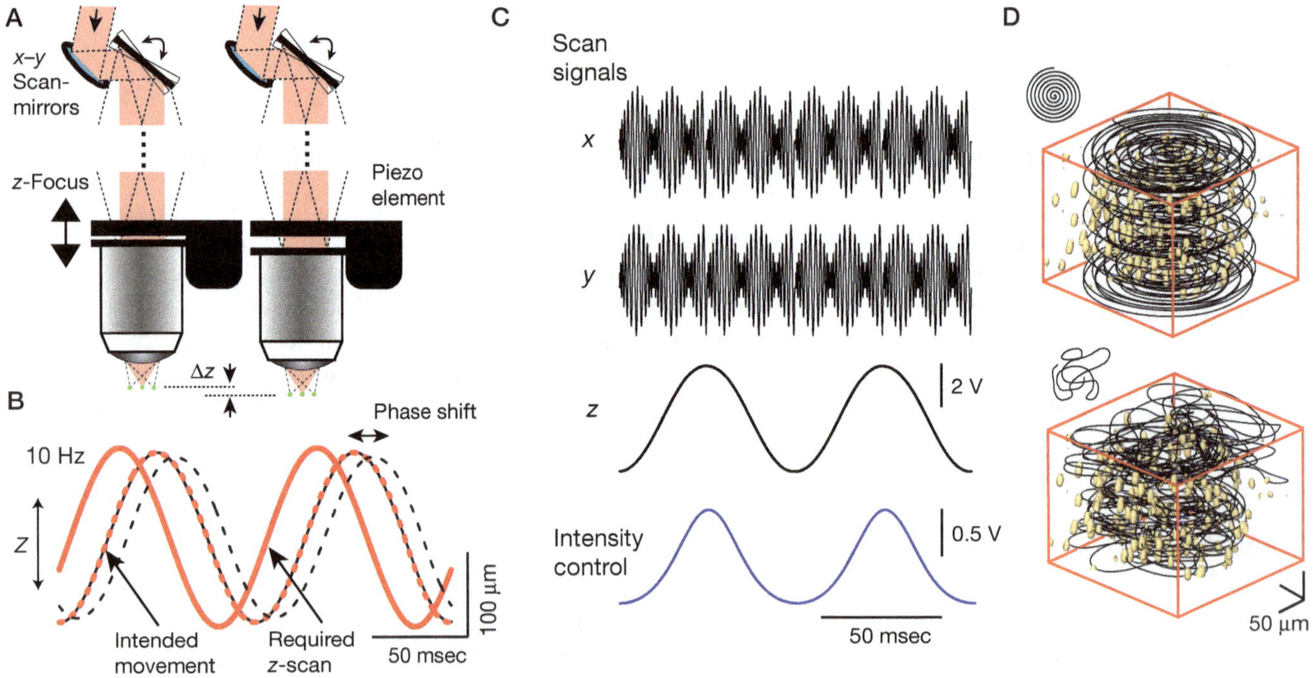

FIGURE 1. Principles of fast 3D laser scanning. (*A*) Mechanical 3D laser scanning: A pair of standard, galvanometric scan mirrors is used for *x–y* deflection of the laser beam. Fast *z* scanning is performed with a piezoelectric focusing element that continuously displaces the microscope objective along the optical axis, shifting the focal plane by the same amount. (*B*) Generation of the intended objective's *z* movement by appropriate adjustment of the drive signal. Providing the intended drive signal (solid black curve) to the piezoelement results in *z* motion with phase shift and amplitude reduction (dashed black curve). A drive signal with an opposing phase shift and enhanced amplitude (solid red curve) produces the intended movement of the piezoelement (dashed red curve). (*C*) Exemplary *x–y–z* scan signals for 3D spiral scanning (black). Depending on scan depth, the laser intensity is modulated with an appropriate signal provided to the Pockels cell (blue) to ensure even illumination with depth. (*D*) 3D line scan trajectories (black) in spiral (*top*) and user-defined modes (*bottom*) overlaying the 3D-rendered view of 10-µm beads from the reference stack (beige). 3D scan paths were reconstructed based on position feedback signals. Note the rather homogenous scan density for the spiral mode. (*D*, Adapted, with permission, from Göbel et al. 2007.)

Equipment

Coverslips, glass

Optical *x–y* scanner, galvanometer based (e.g., model 6210 or 6215, Cambridge Technology)

We use mirrors for a 3-mm beam diameter. Note that larger mirrors have larger inertias and, thus, are more limited in speed.

Piezoelectric focusing element for *z* scanning (e.g., model P-725.4CD PIFOC, equipped with E-665 control electronics, Physik Instrumente; alternatively MIPOS 500, Piezosystem Jena, Inc.).

Pockels cell (e.g., Model 302; Conoptics)

The Pockels cell is used for rapid laser intensity modulation.

Surgical instruments

Three-dimensional visualization software (e.g., Amira, Visage Imaging, Inc.)

EXPERIMENTAL METHOD

Creating 3D Scan Trajectories

1. Drive the piezoelement with a sinusoidal voltage such that a mechanical vibration of the microscope objective is induced along the optical axis (Fig. 1A,B).

   This moves the focal plane up and down continually with a certain amplitude.

2. A critical issue in establishing trajectories is the inertia of the objective, which causes a frequency-dependent phase shift and amplitude reduction (Fig. 1B). With a standard 40× objective and using 10-Hz motion, the deviation of the real *z* movement from the intended motion is small but nonetheless needs to be taken into account.

   i. Determine the phase shift and amplitude reduction for a given objective and frequency.

   ii. Calculate backward to determine the drive signal required to obtain the intended motion (Fig. 1B).

      Practically, the inertial properties have to be measured by comparing the actual position of the objective—which requires that the piezoelement provide a position-output signal—and the position-drive signal. Note that such a calibration is necessary for each particular objective.

3. With the swinging motion of the objective as a constraint, various 3D trajectories can be realized by driving the *x–y* scan mirrors appropriately.

   Sharp trajectory edges should be avoided because they can exceed the frequency range of the scan mirrors and, thus, would be mechanically smoothed. To correctly assign the fluorescence signals to the identified cells (see Steps 21 and 22), it is essential to always know the exact position of the laser focus along the trajectory. This information is obtained by always measuring the feedback position signals from the scan mirrors and the piezoelectric focusing element together with the fluorescence signal.

   i. A spiral pattern in the *x–y* plane is very useful to create a 3D set of opening and closing spirals, which can cover a large portion of the sampled volume (Fig. 1C,D).

      Although a predefined spiral trajectory does not require preselection of cells, it can nevertheless achieve a hit rate of up to 90% of the cells in the imaged volume (Göbel et al. 2007). The use of predefined scan patterns can, thus, eliminate the time-consuming selection step during an experiment. Nevertheless, a reference stack is still required for analysis because, independently of the detailed scan mode, the fluorescence signals of the pixels along the scan line need to be correctly assigned to individual cells (Steps 21 and 22).

   ii. Different regular scan patterns and even user-defined 3D trajectories are possible as well (Fig. 1D; Göbel et al. 2007).

4. Fine-tune the disribution of scan points along the 3D scan trajectory. In the original implementation, the generation of scan points and the acquisition of fluorescence signals were performed in parallel using a fixed predefined pixel dwell time, Δt (typically, 10 µsec; Göbel et al. 2007). To relate fluorescence intensity values from different pixels to each other, adjacent scan points were spaced equidistantly in 3D, which means that the scan trajectory was run at constant speed.

      Because a continuous sinusoidal waveform is used for the objective's motion in the *z* direction, the speed of *z* movement changes during the vibration cycle, requiring a correction of pixel distances along the scan line. This correction is achieved by first independently defining lateral pixel position and *z* motion, respectively; then the actual 3D distances are calculated according to the position of the objective along the *z*-axis and the amount by which lateral pixel coordinates are shifted to equalize their distances in 3D space (Göbel et al. 2007).

5. For user-defined line scans:

   i. Divide the line scan into several segments (depending on the distribution of cells).

   ii. Slow down or accelerate the laser focus between segments, as needed. On each segment run the trajectory at constant speed as described in Step 4. A detailed algorithm for generation of user-defined trajectories is provided in Göbel et al. (2007).

      In general, modulation of the scan speed along the line can prove useful for optimizing dwell time on the cell bodies (Lillis et al. 2008).

Laser Intensity Adjustment

The laser intensity for fluorescence excitation decreases rapidly with depth in brain tissue as a result of scattering. For neocortical tissue, an exponential decrease with a scattering length λ_S of 200–250 µm can be assumed (Kleinfeld et al. 1998; Oheim et al. 2001).

6. To compensate for the loss of excitation, the laser power needs to be adjusted continually according to the focal depth (Fig. 1C). For automatic adjustment, a fast laser modulator, such as a Pockels cell, is required. Modulate the signal applied to the Pockels cell according to

$$I(t) = I_0 \cdot e^{O_z(t)/\lambda_S}, \tag{1}$$

where I_0 denotes the laser intensity at the z position $O_z(0)$ of the objective at the start of the 3D scan. The parameter λ_S can be empirically adjusted with the goal to achieve similar mean-cellular fluorescence intensities near the top and bottom of the imaged volume.

Animal Preparation and 3D Imaging Procedure

Experimental procedures for animal preparation and dye labeling of cell populations are similar to procedures described in detail in Chapters 49 and 51. Perform all procedures in accordance with the guidelines and with the approval of your local animal facility.

7. Anesthetize the rat or mouse with urethane (1–1.5-g/kg body weight).
8. Perform a craniotomy above the cortical area under study (e.g., the visual cortex).
9. Carefully remove the dura.
10. Superfuse the exposed cortex with NRR.
11. Bolus load neurons in layer 2/3 of the rat visual cortex with a calcium-sensitive fluorescent dye (e.g., OGB-1 AM) (Stosiek et al. 2003; Göbel et al. 2007; see also Chapter 49).
12. To distinguish neurons from astrocytes, counterstain the latter by application of SR101 to the exposed brain surface for several minutes (Nimmerjahn et al. 2004; see also Chapter 63).
13. To dampen heartbeat- and breathing-induced motion, fill the cranial window with 1%–1.5% agarose. Cover with a glass coverslip.
14. Place the animal under the custom-built two-photon microscope adapted for 3D laser scanning.
15. Search for well-stained neurons in your area of interest. Choose an appropriate zoom factor.

    Stained neurons and astrocytes are visualized using the green and red detection channels, respectively.

16. Before 3D scanning, acquire a reference stack at high spatial resolution using a standard 2D frame scan approach at increasing depths (Fig. 2A).
17. Define a 3D scan trajectory as described in Steps 3–5.

    For predefined 3D scan pattern this can be done immediately. For user-defined trajectories, target cells have to be selected manually from the reference stack and their 3D coordinates stored as a list of coordinates. From these manually selected cell positions a user-defined 3D scan trajectory is created (see Step 5).

18. Adjust the laser intensity at the initial imaging plane and turn on the automatic laser intensity adjustment. Set λ_S to a reasonable value (e.g., 200 μm).
19. Run the 3D laser scanning to acquire spontaneous or stimulus-evoked calcium signals and store data to disk.
20. Fluorescence data can be prescreened by displaying the temporal evolution of fluorescence along the entire trajectory as line-scan image (space-time plot; Fig. 2B).

    See *Troubleshooting.*

Signal Assignment and 3D Visualization of Network Dynamics

21. From the reference stack (Step 16), define volumes of interest (e.g., as ellipsoids surrounding cell bodies).

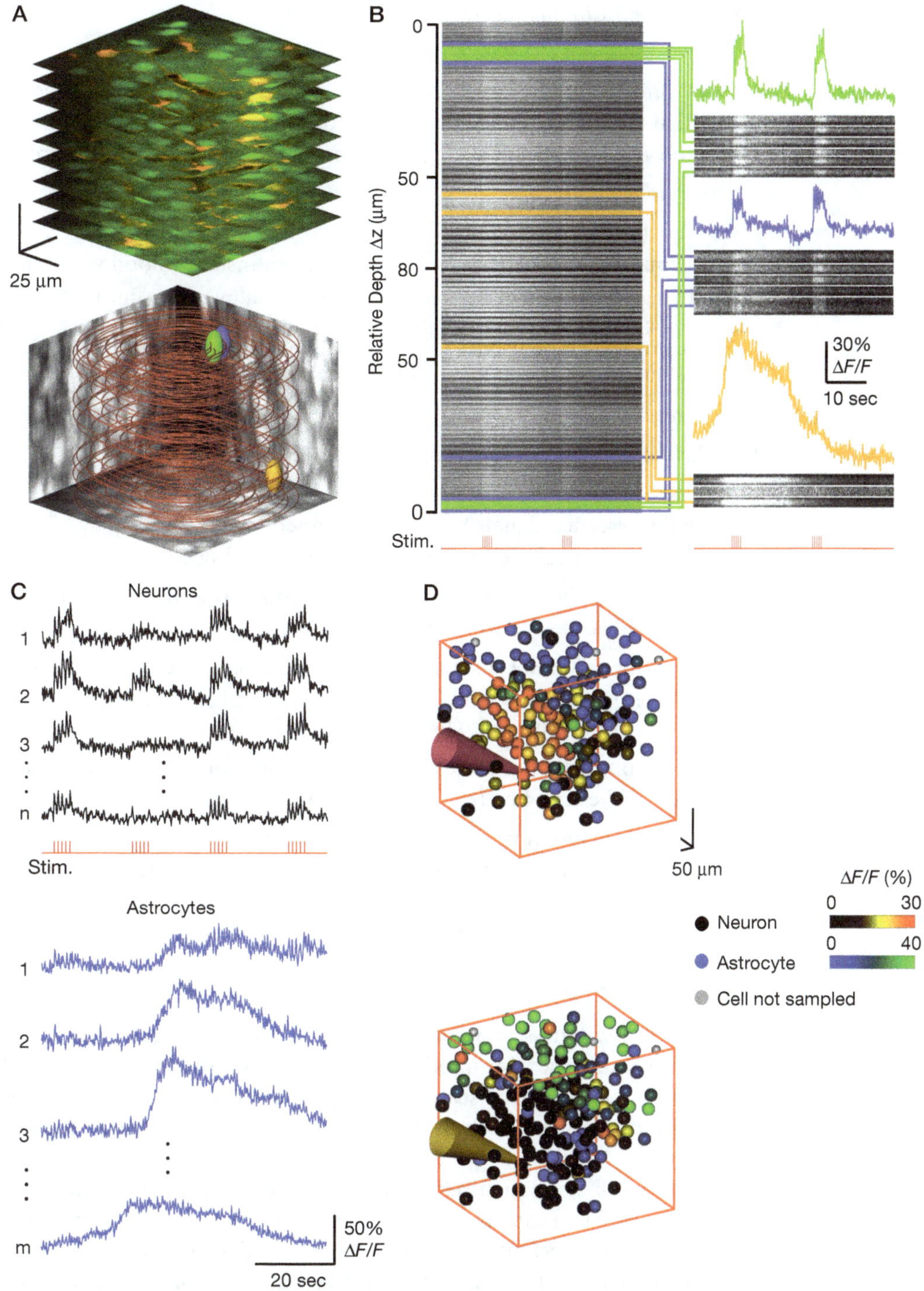

FIGURE 2. Assigning fluorescence signals to the cells within the scan volume. (*A*) (*Top*) Reference image stack of cells in layer 2 of rat neocortex stained with OGB-1 (green) and SR101 (red). Dual-stained astrocytes appear yellow. (*Bottom*) Spiral 3D scan trajectory. Three ellipsoids (green, blue, and yellow) exemplify positions and volumes of interest for three cell soma that were hand selected from the reference stack. (*B*) (*Left*) Entire set of line-scan raw data. The unfolded 3D line is presented as a vertical line with the temporal dimension to the *right*. The three selected cells were hit multiple times during one 3D scan period. (*Right*) Extracted line-scan segments belonging to the three cells selected in (*A*). For each cell, the fluorescence signals of all segments were averaged to obtain the time course of the calcium signal on local electrical stimulation through an extracellular micropipette (five bursts at 1 Hz of 10 pulses at 100 Hz every 20 sec). (*C*) Another example of simultaneously measured fluorescence traces for a small subset of neurons (*top*, 4 of 116) and of identified astrocytes (*bottom*, 4 of 89) from a different experiment using the 3D spiral scanning and simulation protocols. Note the distinct characteristics of neuronal and glial calcium dynamics. (*D*) Reconstructed 3D calcium dynamics for the entire cell population for the experimental recording shown in C. Different lookup tables were used for neurons and astrocytes. Two snapshots of the activity state are shown, one during a stimulus-evoked response of neuronal subsets (*top*) and one during a slow calcium wave in layer 1 (*bottom*). (Modified, with permission, from Göbel et al. 2007.)

22. Based on the simultaneously recorded 3D position signals, find all points of the 3D scan trajectory that are inside a particular volume of interest (typically a cell but it could also be a segment of neuropil). Assign the average fluorescence signal of all these points to the respective object.

    See *Troubleshooting.*

23. Convert fluorescence signals into relative fluorescence changes ($\Delta F/F$) using an estimate of the background fluorescence (obtained, e.g., from inside a blood vessel lumen).

    Using 3D spiral scanning, cells often are hit multiple times during one scan cycle, which permits all segments of the scan line belonging to a particular cell to be averaged, thus, improving the signal-to-noise ratio (Göbel et al. 2007). Ultimately, one ends up with a data set consisting of a list of fluorescence traces for all of the sampled cells.

    See *Troubleshooting.*

24. If particular cell types are counterstained (e.g., astrocytes with SR101), sort cells into respective subtypes (Fig. 2C).

25. Once fluorescence signals have been assigned to all cells, display 3D dynamics using an appropriate software package (e.g., Amira, Visage Imaging).

    i. Most simply, display population dynamics as a 3D plot of spheres that are pseudocolor labeled according to their activity state.

    ii. For movies, indicate the temporal evolution of activity states by the changing of colors.

        Figure 2D uses two different color lookup tables for neurons and astrocytes, allowing their dynamic changes to be visualized simultaneously.

    iii. For completeness, indicate cells that are identified in the reference stack but obviously not hit by the scan line (e.g., as small gray spheres; Fig. 2D).

    iv. Another option is to color code spheres in a still image according to a certain feature that has been extracted in the analysis (e.g., orientation tuning; see below, Fig. 3).

TROUBLESHOOTING

Problem (Step 20): Line-scan images (Fig. 2B) show apparent uneven illumination with depth.
Solution: The automatic laser intensity adjustment is not set appropriately. If cells at greater depths are too dim, decrease the scattering length parameter, which should result in enhanced excitation with depth. Conversely, if deep cells are too bright, increase the scattering length.

Problem (Step 22): The assignment of signals to cells appears to be incorrect.
Solution: This problem can have several causes. First check whether the scan position signals are correctly measured. Second, note that the inertia-induced delay of the actual scan position from the intended movement also needs to be taken into account for acquisition of the reference stack. Stack acquisition typically relies on sawtooth-like scan signals, for which actual scan positions deviate with a phase delay and slight amplitude reduction. Without compensation, pixel coordinates from the reference stack cannot be related to 3D line-scan segments because these coordinates would not reflect true scan position. Consequently, cell positions identified from the reference stack might not exactly match with respective 3D scan line segments. Ideally, position signals from the *x–y* scanners are recorded during acquisition of the reference stack so that coordinate matching can be based solely on true scan positions. Use 5–10-µm fluorescent beads embedded in agarose at a density comparable to the cell density as a test to verify correct signal assignment.

Problem (Step 23): A significant percentage of the cells in the volume is missed by the 3D spiral scan line.

Solution: The homogeneity of the 3D spiral scan trajectory might not be set optimally. Check the parameters of the spiral scan (turns per cycle, number of spirals, etc.). Alternatively, the volume could be so large that the density of the line scan is not sufficiently high to ensure sampling of all or most cells. Note that for a given sampling density, 3D scanning volume trades off against scan line length and, thus, against total 3D sampling rate. Reduce either the volume size or the volume sampling rate.

EXAMPLE OF APPLICATION: ORIENTATION PREFERENCE IN 3D NEURAL NETWORKS OF VISUAL CORTEX

In the visual cortex, neurons respond best to moving grating stimuli presented to the animal's eyes. The neuronal responses are sustained throughout the duration of the stimulus and can reach high firing rates. Moving grating therefore results in large calcium signals that can be detected with in vivo two-photon imaging. Whereas neurons in the cat visual cortex have been shown to be highly organized into cortical columns (Ohki et al. 2006), the orientation selectivity of neurons in rat visual cortex is more randomly dispersed (Ohki et al. 2005). To assess the three-dimensional organization of visual responses we used the fast 3D scanning technique to identify a complete set of neurons in a supragranular cube of 200-µm side length. Neurons in layer 2/3 of the rat visual cortex were bolus loaded with OGB-1 (Stosiek et al. 2003; Göbel et al. 2007) and astrocytes were counterstained with SR101 (Nimmerjahn et al. 2004). Animals were placed under a custom-built two-photon microscope and neuronal calcium transients were measured using 3D spiral scanning in response to visual stimulation with moving gratings of different orientations (Fig. 3A). The location of the neurons in the imaged volume was manually determined from a reference stack and small volumes of interest were placed around the neurons for assigning signals to individual neurons. Neurons with a clear orientation preference could be recognized by a peak response during the preferred stimulus (Fig. 3B). Neuropil signals that were analyzed from segments of the 3D scan trajectory in between cells at various depths showed little orientation preference (Fig. 3B). For visualization of a 3D map of orientation tuning, the imaged neuronal population was reconstructed according to their spatial position and neurons were schematically color-coded in green, blue, red, or yellow according to their orientation preference (Fig. 3C). Color saturation in addition indicated the orientation selectivity index (OSI), which was calculated as OSI = $(A_{max} - A_{max,90°})/(A_{max} + A_{max,90°})$, where A_{max} refers to the maximal $\Delta F/F$ response amplitude and $A_{max,90°}$ to the $\Delta F/F$ amplitude for the perpendicular orientation. These results show a noncolumnar organization of orientation tuned neurons in rat visual cortex and demonstrate the fidelity of the fast 3D scanning technique to detect neuronal tuning properties in populations of several hundred neurons simultaneously.

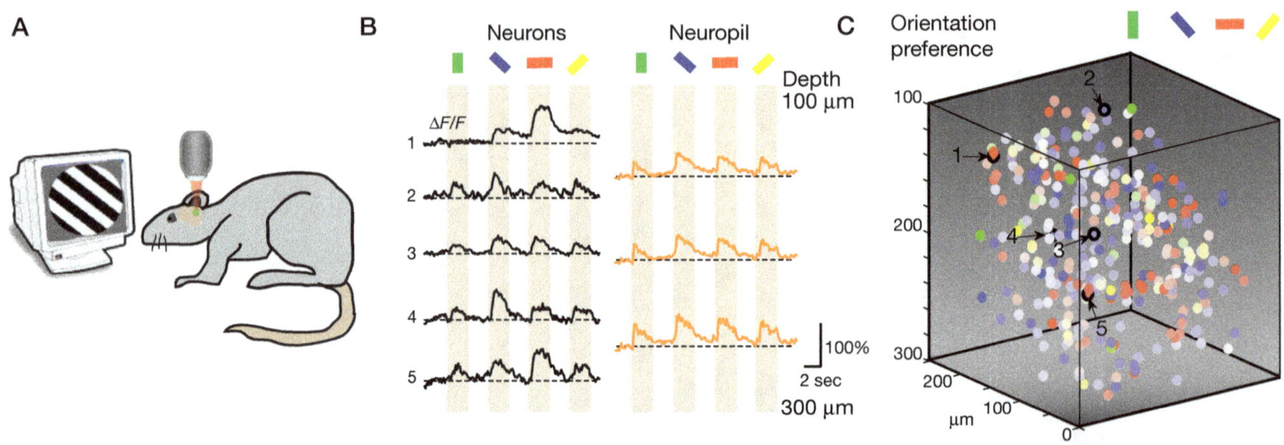

FIGURE 3. 3D imaging of orientation tuning in rat visual cortex. (*A*) Calcium transients in OGB-1-labeled neurons were measured in response to visual stimuli presented on a monitor in front of the contralateral eye. High-contrast gratings at four different orientations were used. (*B*) Examples of traces from five neurons (*left*) and neuropil activity (*right*) obtained at different depths (325 neurons in total). Colored vertical bars indicate stimulation periods; the orientation of the moving grating is shown at the top. (*C*) 3D visualization of orientation preference. Neurons are shown at their respective positions within the volume with their orientation-tuning color coded as indicated at the top. Increasing color saturation indicates stronger orientation selectivity. The five sample cells from *B* are marked by arrows. White cells possess no orientation preference.

DISCUSSION

For dynamic measurements on the subsecond timescale, standard laser-scanning microscopes are mostly limited to planar line scans or 2D frame scanning. Therefore, the number of cells that can be imaged is usually restricted by the lateral field of view. By enabling volume sampling, the fast 3D scanning approach described here increases the number of simultaneously imaged cells to a few hundred, still offering a sufficient temporal resolution to resolve action potential–driven calcium transients. This approach allows the rapid recording of the orientation tuning of several hundred neurons, which might have taken about 10 times longer using a sequential 2D scanning approach.

The main advantages of the mechanical scanning approach are (1) the simplicity of implementation, which is based on a single additional mechanical component to the microscope; (2) the lack of additional optical components that could lead to transmission losses, restrictions of spectral windows, or optical aberrations; and (3) the flexibility of creating scan lines specifically tuned to one's needs. A current limitation is that the achievable temporal resolution is constrained by the speed of the piezoelectric element. In our experiments, large *z* movements of up to 400 µm were possible at 10 Hz, but higher oscillation frequencies resulted in reduced amplitude of the microscope movement. Although we have thus far applied the piezoelement in the closed-loop mode, it might be conceivable to use the open-loop mode instead, in which case the electronic circuits would not be frequency limiting (although the stability of the trajectory would need to be verified). Together with more powerful piezoelectric devices, this could allow 3D scanning at video rate or even faster.

Inertia-induced limitations scale with the mass of the objective mounted to the piezoelement. Hence, reducing the mass of the object moved would also enable a significant increase in scanning frequency. This improvement is, in principle, possible by shifting *z* scanning to a plane optically conjugate to the focal plane of the microscope objective (Botcherby et al. 2008). In this scheme, *z* scanning would be performed with small lightweight mirrors and—as an additional advantage—the microscope objective would remain stationary during *z* scanning, avoiding mechanical interference with recording pipettes under the objective.

Acousto-optic deflectors (AODs) are an alternative to mechanical scanners. AOD scanning enables random access scanning, which means that the laser focus can jump from cell to cell without wasting time on the path in between, efficiently increasing recording speed without compromising the signal-to-noise ratio (see Grewe et al. 2010; Duemani Reddy and Saggau 2011; Chapter 53). In principle, this approach is extendable to volumes, targeting points of interest in 3D space at kilohertz rates (Duemani Reddy et al. 2008; Kirkby et al. 2010). However, the volume that can be scanned is relatively small and may be limited to an octahedral shape. Because several AODs in series are required, the optical pathway is more complicated. So far, 3D–AOD scanning has been used to record multisite dendritic calcium transients, but its use in vivo for measuring neuronal network activity remains a challenge. As a final alternative, scanless techniques are being developed in which the combination of spatial multiplexing of the laser with fast cameras can enable the simultaneous acquisition of fluorescence signals from populations of neurons (Bansal and Saggau 2011; Nikolenko et al. 2011; Oron and Silberberg 2011).

The 3D scanning technique for population imaging described here will benefit from other advances in the field. Limitations in imaging depth could be partially overcome with more powerful lasers and by replacing the blood with a nonabsorbing solution (Haiss et al. 2009). Moreover, the use of genetically encoded calcium indicators (see Chapters 8 and 57) instead of synthetic dyes is highly promising, enabling cell type–specific targeting as well as repeated measurements of the same cells over long time spans (Mank et al. 2008; Wallace et al. 2008; Tian et al. 2009). Overall, 3D imaging of neural network activity is likely to become an important tool for investigating the principles of local neural circuit dynamics in the living brain.

RECIPE

CAUTION: See Appendix 6 for proper handling of materials marked with <!>.
Recipes for reagents marked with <R> are included in this list.

NRR

NaCl	135 mM
KCl	5.4 mM
HEPES buffer	5 mM
$CaCl_2$ <!>	1.8 mM

Adjust pH to 7.2 with NaOH <!>.

ACKNOWLEDGMENTS

This work was supported by the Center for Neuroscience Zurich, the International Human Frontier Science Program Organization (HFSP), and the Swiss National Science Foundation (Grants 3100A0-114624 [FH] and 31-120480 [BMK]).

REFERENCES

Bansal V, Saggau P. 2011. Digital micromirror devices: Principles and applications in imaging. In *Imaging: A laboratory manual* (ed. Yuste R), pp. 839–848. Cold Spring Harbor Laboratory Press, Cold Spring Harbor, NY.

Botcherby EJ, Juskaitis R, Booth MJ, Wilson T. 2008. An optical technique for remote focusing in microscopy. *Opt Commun* **281:** 880–887.

Buzsaki G. 2004. Large-scale recording of neuronal ensembles. *Nat Neurosci* **7:** 446–451.

Denk W, Strickler JH, Webb WW. 1990. Two-photon laser scanning fluorescence microscopy. *Science* **248:** 73–76.

Duemani Reddy G, Saggau P. 2011. High-speed two-photon imaging. In *Imaging: A laboratory manual* (ed. Yuste R), pp. 831–838. Cold Spring Harbor Laboratory Press, Cold Spring Harbor, NY.

Duemani Reddy G, Kelleher K, Fink R, Saggau P. 2008. Three-dimensional random access multiphoton microscopy for functional imaging of neuronal activity. *Nat Neurosci* **11:** 713–720.

Garaschuk O, Milos RI, Grienberger C, Marandi N, Adelsberger H, Konnerth A. 2006. Optical monitoring of brain function in vivo: From neurons to networks. *Pfluegers Arch* **453:** 385–396.

Göbel W, Helmchen F. 2007a. In vivo calcium imaging of neural network function. *Physiology* **22:** 358–365.

Göbel W, Helmchen F. 2007b. New angles on neuronal dendrites in vivo. *J Neurophysiol* **98:** 3770–3779.

Göbel W, Kampa BM, Helmchen F. 2007. Imaging cellular network dynamics in three dimensions using fast 3D laser scanning. *Nat Methods* **4:** 73–79.

Grewe BF, Helmchen F. 2009. Optical probing of neuronal ensemble activity. *Curr Opin Neurobiol* **19:** 520–529.

Grewe BF, Langer D, Kasper H, Kampa BM, Helmchen F. 2010. High-speed in vivo calcium imaging reveals spike trains in neuronal networks with near-millisecond precision. *Nat Methods* **7:** 399–405.

Haiss F, Jolivet R, Wyss MT, Reichold J, Braham NB, Scheffold F, Krafft MP, Weber B. 2009. Improved in vivo two-photon imaging after blood replacement by perfluorocarbon. *J Physiol* **587:** 3153–3158.

Helmchen F, Denk W. 2005. Deep tissue two-photon microscopy. *Nat Methods* **2:** 932–940.

Kirkby PA, Srinivas Nadella KM, Silver RA. 2010. A compact acousto-optic lens for 2D and 3D femtosecond based 2-photon microscopy. *Opt Express* **18:** 13721–13745.

Kleinfeld D, Mitra PP, Helmchen F, Denk W. 1998. Fluctuations and stimulus-induced changes in blood flow observed in individual capillaries in layers 2 through 4 of rat neocortex. *Proc Natl Acad Sci* **95:** 15741–15746.

Lillis KP, Eng A, White JA, Mertz J. 2008. Two-photon imaging of spatially extended neuronal network dynamics with high temporal resolution. *J Neurosci Methods* **172:** 178–184.

Mank M, Santos AF, Direnberger S, Mrsic-Flogel TD, Hofer SB, Stein V, Hendel T, Reiff DF, Levelt C, Borst A, et al. 2008. A genetically encoded calcium indicator for chronic in vivo two-photon imaging. *Nat Methods* **5:** 805–811.

Nikolenko V, Peterka DS, Araya R, Woodruff A, Yuste R. 2011. Spatial light modulator microscopy. In *Imaging: A laboratory manual* (ed. Yuste R), pp. 849–860. Cold Spring Harbor Laboratory Press, Cold Spring Harbor, NY.

Nimmerjahn A, Kirchhoff F, Kerr JN, Helmchen F. 2004. Sulforhodamine 101 as a specific marker of astroglia in the neocortex in vivo. *Nat Methods* **1:** 31–37.

Oheim M, Beaurepaire E, Chaigneau E, Mertz J, Charpak S. 2001. Two-photon microscopy in brain tissue: Parameters influencing the imaging depth. *J Neurosci Methods* **111:** 29–37.

Ohki K, Chung S, Ch'ng Y, Kara P, Reid R. 2005. Functional imaging with cellular resolution reveals precise micro-architecture in visual cortex. *Nature* **433:** 597–603.

Ohki K, Chung S, Kara P, Hübener M, Bonhoeffer T, Reid RC. 2006. Highly ordered arrangement of single neurons in orientation pinwheels. *Nature* **442:** 925–928.

Oron D, Silberberg Y. 2011. Temporal focusing microscopy. In *Imaging: A laboratory manual.* (ed. Yuste R), pp. 861–868. Cold Spring Harbor Laboratory Press, Cold Spring Harbor, NY.

Stosiek C, Garaschuk O, Holthoff K, Konnerth A. 2003. In vivo two-photon calcium imaging of neuronal networks. *Proc Natl Acad Sci* **100:** 7319–7324.

Tian L, Hires SA, Mao T, Huber D, Chiappe ME, Chalasani SH, Petreanu L, Akerboom J, McKinney SA, Schreiter ER, et al. 2009. Imaging neural activity in worms, flies and mice with improved GCaMP calcium indicators. *Nat Methods* **6:** 875–881.

Wallace DJ, Borgloh SM, Astori S, Yang Y, Bausen M, Kugler S, Palmer AE, Tsien RY, Sprengel R, Kerr JN, et al. 2008. Single-spike detection in vitro and in vivo with a genetic Ca^{2+} sensor. *Nat Methods* **5:** 797–804.

53 High-Speed Two-Photon Calcium Imaging of Neuronal Population Activity Using Acousto-Optic Deflectors

Benjamin F. Grewe and Fritjof Helmchen

Department of Neurophysiology, Brain Research Institute, University of Zurich, CH-8057 Zurich, Switzerland

ABSTRACT

Two-photon calcium imaging of neuronal populations allows optical measurements of spiking activity in living animals. However, laser-scanning microscopes with galvanometric scan mirrors are too slow to capture population activity on a millisecond timescale. This chapter describes a two-photon microscope that is based on two-dimensional laser scanning with acousto-optic deflectors (AODs), enabling in vivo recording of neuronal population activity at temporal resolutions of several hundred hertz. We provide a detailed construction plan of the AOD-based microscope and introduce a novel random-access pattern scanning (RAPS) technique for fast population measurements. AOD-based RAPS can measure calcium transients in neocortical neuronal populations, revealing spike trains with near-millisecond precision. The current limitations of the AOD-based microscope are discussed, and we provide an outlook of its future applications.

INTRODUCTION

The characterization of neuronal network dynamics is essential to understanding brain function. Neuronal networks process incoming information as coordinated spike sequences on a millisecond-to-second timescale. Investigation of these activity patterns requires fast readout methods, which in the future will need to be combined with unambiguous identification of neuronal cell types, cell positions, and the functional connectivity within the network organization. Although electrical recordings provide excellent temporal resolution and can reveal spike trains in widely dispersed neuronal populations (Buzsáki 2004), they cannot comprehensively resolve activity patterns in the local microcircuits. The alternative is high-resolution optical imaging, which permits population measurements from many well-identified neurons (for review, see Grewe and Helmchen 2009). In particular, two-photon laser scanning microscopy enables imaging to a depth of several hundred micrometers in optically thick, light-scattering tissue in vivo (for review, see Helmchen and Denk 2005; Kerr and Denk 2008). However, using frame scanning with mechanical scan mirrors in a standard two-photon microscope, the temporal resolution of population measurements has been limited maximally to 15–30 Hz (Kerr et al. 2005; Rochefort et al. 2009). Hence, in vivo optical recordings of network dynamics on the millisecond timescale have not been possible so far.

One alternative to mechanical laser scanning is scanning with acousto-optical deflectors (AODs) (Reddy and Saggau 2011). AOD scanners enable transitions of the laser focus within several microseconds between any two positions in the scanning area. Such fast transitions avoid the waste of scanning time on background regions and enable random-access scanning from preselected points of interest only (e.g., neurons). AODs are, however, limited by their low laser diffraction efficiency (~55% transmission for an *x–y* pair of AODs) and the large optical dispersion of the acousto-optic crystal, which causes laser pulse broadening and spatial deformation of the laser focus. Using relatively long laser pulses (0.5–1.8 psec) can minimize dispersive effects (Iyer et al. 2006; Otsu et al. 2008; Reddy et al. 2008), although this reduces two-photon fluorescence excitation. Alternatively, adding diffractive elements can compensate for the dispersion (Zeng et al. 2006, 2007; Kremer et al. 2008; Kirkby et al. 2010). For a broader discussion of fast-scanning and scanless laser microscopes, see Bansal and Saggau (2011), Liebling (2011), Nikolenko et al. (2011), Oron and Silberberg (2011), and Reddy and Saggau (2011).

Several AOD-based two-photon microscopes have been applied to calcium imaging from neuronal dendrites in brain slices (Iyer et al. 2006; Otsu et al. 2008; Reddy et al. 2008), but their application to measure network dynamics in the intact brain has been pending. This chapter provides detailed instructions and recommendations for the assembly of a recently introduced high-speed AOD-based two-photon microscope optimized for in vivo imaging applications (Grewe et al. 2010).

Protocol

Random-Access Pattern Scanning of Neuronal Population Activity Using AOD-Based Two-Photon Microscopy

This protocol describes how to assemble an acousto-optic two-photon microscope and details all optimization procedures for rendering the microscope suitable for high-speed in vivo measurements of neuronal population activity.

IMAGING SETUP

We first provide a list of the hardware components that are necessary to build the microscope and then discuss several design considerations.

Laser and Accessories

Beam expander, variable (S6ASS2075, Sill Optics)
Infrared (IR) viewer, miniature, 350–1700 nm (IRV1-1700, Newport Corporation)
Laser intensity modulator (i.e., Pockels cell; #M350-80, Conoptics)
Laser system, Ti:sapphire, capable of producing an average power >2.8 W at 850 nm (e.g., Chameleon Ultra II, Coherent, Inc.)
Mirror mounts, kinematic, metric (KMS/M, Thorlabs, Inc.)
Mirrors, broadband, dielectric, 750–1100-nm, 1-in and 2-in-diameter (BB1-E03 and BB2-E03, respectively, Thorlabs, Inc.)
Power and energy meter console (PM100D, Thorlabs, Inc.)
Thermal power sensor, C-Series (S350C, Thorlabs, Inc.)

Microscope Stand

Optical table, equipped with four stabilizers (RS2000 series and S-2000, respectively, Newport Corporation)
Precision linear stage (i.e., *z* stage; model PMT 160-DC, Feinmess Dresden GmbH)
Rail system (e.g., Profile X95-640 column, baseplate, and six to eight clamps [LINOS Photonics])
x–y stage, compact, motorized (KT310, Feinmess Dresden GmbH)
x–y–z stage controller (Galil)

Scanning Apparatus and Dispersion Compensation Unit

AOD-alignment platform (e.g., pitch and yaw platform w/micrometers; PY003, Thorlabs, Inc.)
AODs, high-resolution, large-scan-angle (DTSXY-A12-850, AA Opto-Electronic)
Compensation prism, N-SF11, uncoated, apex angle 60° (e.g., equilateral dispersive prism; Thorlabs, Inc.)

Alternatively, use an uncoated custom prism (N-SF66, apex angle 60°).

Computer, equipped with quad-core processor (e.g., HP Compaq dc7900, Hewlett Packard)
Frequency generators (two) (e.g., CMOS 300 MSPS Complete DDS digital synthesizer; AD9852, Analog Devices)
Half-wave ($\lambda/2$) plate modules, zeroth-order (two) (RZBH-850, Thorlabs, Inc.)
Lenses, cylindrical telescope, achromatic, N-BK7, focal lengths f_1 = 100 mm and f_2 = 50 mm (Thorlabs, Inc.) (for use with N-SF11 prism [Thorlabs, Inc.] only)

Power amplifiers (two) (AMPA-B-36, AA Opto-Electronic)
Rotation mount, prism, high-precision (PR01/M, Thorlabs, Inc.)
Rotation mounts, λ/2 plate (two) (e.g., rotation mount for 25.4-mm-diameter optics; RSP1/M, Thorlabs, Inc.)
Scan control board, field-programmable gate array (FGPA) (PXI-7813R, National Instruments)
Scan lens, achromatic, near-infrared (NIR) coating, f_{SL} = 300 mm (Thorlabs, Inc.)
Scan software, custom, written in LabVIEW 2009 (National Instruments)
Tube lens, achromatic, f_{TL} = 180 mm (Olympus)

Detection Pathway

Analog/digital (A/D) converter, 40-MHz (custom design, CSEM)
Beam splitter, two-photon (cold light mirror KS 93/45°, LINOS Photonics GmbH)
Data acquisition board, FPGA (NI PXI-7813R, National Instruments)
Dichroic mirror, DC green (LINOS Photonics GmbH)
Emission filter, green (HQ 535/50, Chroma Technology)
Emission filter, red (720SP, AHF Tübingen)
Microscope objectives (40× LUMPlanFl/IR, numerical aperture (NA) 0.8; 10× UPLFLN 10×2, NA 0.3 [Olympus])
The 10× objective is used for large-field overviews of preparations.
Photomultiplier tube (PMT) preamplifier (XPG-ADC-PREAMP, Sigmann-Elektronik)
PMTs (e.g., R6357, Hamamatsu)

Design Considerations

The AOD microscope described here was designed as a "fixed-stage" microscope that can be moved over the sample. This design provides a large space below the objective lens and supports simultaneous stable electrophysiological recordings during in vivo imaging experiments using a table-mounted manipulator. The complete microscope is built around an X95 aluminum column mounted on a motorized *x–y* stage (Fig. 1A). Several important issues must be taken into account. The AOD specifications, together with the choice of objective and lenses, determine the image resolution and the field-of-view (FOV) size. Because the maximal AOD scan angle (Θ) is limited (typically 47 mrad), the number of resolvable spots (N) is limited too, depending on the laser beam size at the AOD aperture (Fig. 1B). For a Gaussian laser beam of full $1/e^2$ width D and wavelength λ, $N = \pi/4 \times \Theta \times D/\lambda$ (Kremer et al. 2008). If one is interested in the best possible optical resolution, the laser beam should be expanded to fill the AOD aperture (to maximize N) and the FOV size should be maximally $S \times N$, where S is the diffraction-limited spatial resolution as determined by the effective NA of the microscope objective. If one is rather interested in a large FOV (e.g., to functionally image neuronal populations), one has to accept that the image resolution may be limited by the AOD pointing precision (to ~1 µm in the FOV center in our setup; Grewe et al. 2010). To obtain a certain FOV size, choose a combination of scan and tube lens for a particular microscope objective (i.e., first choose the desired FOV size, then the objective, and finally the scan lens). The tube lens should be selected so that objective magnification is maintained (e.g., f_{TL} = 180 mm for Olympus or Nikon objectives). Because the combination of the scan and tube lenses acts as a telescope—thus changing the laser beam diameter—the corresponding demagnification (or magnification) should match the factor required to image the laser beam at the AOD aperture into the objective's back aperture. The system described here produces a maximum FOV of 305 µm × 305 µm with a 40× objective (scan lens–tube lens telescope demagnification of 0.6).

The second key optical component is the dispersion compensation unit (DCU) (Fig. 1C). A single-prism compensation approach (Zeng et al. 2006) was modified and optimized for laser beam transmission and scan resolution over a large FOV. Two λ/2-wave plates before the prism and the AODs, respectively, turn laser beam polarization along the pathway and thereby maximize transmission through the prism (85%) and the AOD pair (55%). Overall, 45% of the incoming light is

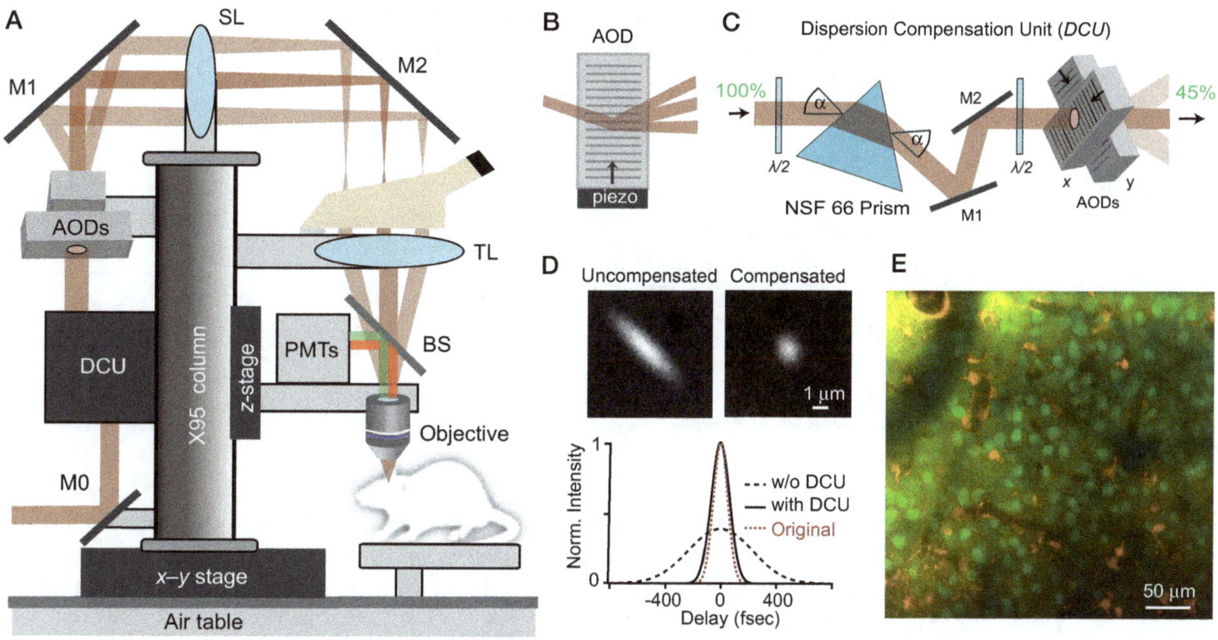

FIGURE 1. Microscope setup for acousto-optic deflector (AOD)-based high-speed calcium imaging. (*A*) Schematic of the two-photon microscope setup mounted on an air table. A motorized *x–y* stage supports movements of the entire microscope over the sample. After the dispersion compensation unit (DCU), two orthogonal AODs are used for *x–y* scanning. Scan (SL) and tube (TL) lenses provide a large field of view (FOV) (~300 µm) with a 40× objective. BS, beam splitter; M0, M1, M2, mirrors. (*B*) Schematic of laser beam scanning with an AOD crystal. A standing sound wave is produced by a piezoelectric transducer and acts as a diffraction grating. The angle of laser beam deflection is proportional to the piezo-oscillation frequency. (*C*) Schematic of the optical path through the single-prism DCU consisting of two λ/2 plates and a custom prism oriented at 45° to the AOD axes. If an SF11 prism is used, an additional cylindrical telescope (not shown) is required to compensate for beam ellipticity caused by mismatch of the input and output angle at the prism. (*D*) Compensation of spatial and temporal dispersion. (*Top*) Two-photon images of 500-nm fluorescent beads without (*left*) and with (*right*) the DCU. Note that compensation abolishes ellipticity. (*Bottom*) Laser pulse width with and without the DCU. The single-prism DCU nearly restores the original pulse width under the objective. (*E*) Two-photon fluorescence image of neocortical L2/3 cell populations in intact mouse neocortex. Cells were bulk labeled with the calcium indicator OGB-1; astrocytes were counterstained with SR101.

transmitted by the DCU and AOD scanners. With a standard SF11 prism the beam path is not symmetrical through the prism, which makes the beam elliptical after the prism. This effect can be corrected, however, by using a cylindrical telescope in front of the prism, restoring a nearly round laser beam profile at the AOD aperture (Grewe et al. 2010). Alternatively, the SF11 prism can be replaced with a custom N-SF66 prism for which the beam path is symmetric, eliminating the need for the cylindrical telescope.

MATERIALS

CAUTION: See Appendix 6 for proper handling of materials marked with <!>.
See the end of the chapter for recipes for reagents marked with <R>.

Reagents

Agarose, type III-A (1%, prepared in normal rat Ringer [NRR] solution; Sigma-Aldrich)

Calcium indicator dye, membrane-permeable (e.g., acetoxymethyl [AM] form of Oregon Green 1,2-bis-(2-aminophenoxy)ethane-*N*,*N*,*N′*,*N′*-tetra-acetic acid (BAPTA)-1 <!> [OGB-1; Invitrogen])

Mice, wild-type

Normal rat Ringer (NRR) solution <R>

Sulforhodamine 101 (SR101; e.g., from Sigma-Aldrich)

Sulforhodamine is used for specific counterstaining of cortical astrocytes (Nimmerjahn et al. 2004; see also Chapter 63).

Urethane <!>

Equipment

Beads, fluorescent (≤500 nm)

Microscope, two-photon, AOD-based

See Imaging Setup for a detailed list of required parts.

EXPERIMENTAL METHOD

Microscope Setup

To set up the optical light path of the microscope, use a "backward" approach: (1) Mount the microscope objective, (2) install the tube and scan lenses, (3) mount the AOD scanners and the DCU, and (4) proceed with the beam alignment. For direct viewing of the samples, mount a binocular in the beam path (Fig. 1A).

1. Set up the laser, the shutter, the Pockels cell, and the variable beam expander.
2. Fix the motorized *x–y* stage on the optical table.
3. Mount the X95 column with two base plates attached on the *x–y* stage.
4. Using a clamping carrier, mount the linear *z* stage onto the x95 column.
5. Construct a custom objective holder. Attach it to the *z* stage.

   We use a distance of ~20 cm of the optical axis from the column to facilitate the alignment procedures and to provide plenty of space beneath the objective (e.g., for a living animal).
6. Design and construct a fluorescence detection pathway that includes a sliding two-photon beam splitter, the dichroic mirrors, fluorescence filters, and PMTs.

   This system uses two channels but, in principle, more channels can be added.
7. Attach the fluorescence detection unit (from Step 6) to the linear *z* stage directly above the objective.
8. Install the tube lens at the appropriate distance (i.e., f_{TL}) above the objective.
9. Mount the scan lens at the correct distance from the tube lens ($f_{TL} + f_{SL}$) on top of the X95 column.
10. Use a 45° mirror (BB2-E03) to redirect the laser beam (M2 in Fig. 1A).
11. Mount the AODs at the correct distance from the scan lens (f_{SL}) on the opposite side of the column.
12. Use a 45° mirror (BB2-E03) to redirect the laser beam to the AOD scanners (M1 in Fig. 1A).
13. Insert a λ/2-wave plate in a rotation mount. Use a clamping carrier to mount this assembly before the AODs.
14. Mount the prism rotation mount (with the compensation prism) on a clamping carrier.

    Use a custom-built holder to install the compensation prism at an angle of 45° with respect to the AOD axes (Fig. 1C).
15. Attach mirror mounts for two mirrors after the prism that redirect the laser beam to the AODs (M1 and M2 in Fig. 1C).
16. Install a λ/2-wave plate (in a rotation mount) on a clamping carrier in front of the compensation prism.
17. Mount a 45° mirror at the X95 column base to reflect the laser beam into the microscope (M0 in Fig. 1A).

18. Mount two mirrors oriented 90° to each other onto the two linear moving parts of the x–y motorized stage (i.e., in the x- and y-directions, respectively; mirrors not shown in Fig. 1A).

    These mirrors are used to steer the fixed input laser beam to the microscope while avoiding laser beam runoff during stage movements.

Alignment and Optimization Procedures

The microscope design is kept relatively simple, with only two key components that have to be properly aligned (the AOD scanners and the DCU prism). Given ~45% transmission through these two components, the total transmission efficiency of the entire microscope should not go below ~25% (assuming an 850-nm wavelength and the use of a 40× objective). Because the DCU prism and the AODs form a pair of dispersive elements that provide negative group velocity dispersion, they can be used to compensate for temporal pulse broadening in the light path. Optimal compensation is achieved by adjusting the distance between the DCU prism and the AODs, so that minimal pulse lengths are achieved behind the objective (Fig. 1D). Pulse width can be measured directly with an autocorrelator; alternatively, the obtained fluorescence signal can be maximized (see below). To correct for spatial dispersion, the aim is to obtain a uniformly round focus shape in the FOV center with a full-width at half-maximum of ~1 µm. Performance is best evaluated by imaging small (100–500 nm) fluorescent beads (Fig. 1D). Note that optimal compensation of spatial dispersion is only possible in the FOV center and that lateral resolution and beam ellipticity worsen toward the edges and corners of the FOV.

19. Remove the AODs, λ/2-wave plates, and the DCU by removing the clamps from the X95 column.
20. With the help of an iris, adjust the two in-coupling mirrors (from Step 18) that steer the laser beam into the microscope (via M0 in Fig. 1A) to ensure a stable laser beam alignment during stage movements.
21. Use two or more irises to center the beam on all lenses and the objective. Also, center the laser beam on the mirrors (M1 and M2 in Fig. 1A).
22. Remount the AOD scanners.
23. Drive the AODs with the center frequency. Center the laser beam by shifting the AODs laterally with respect to the laser beam.
24. Adjust the pitch/yaw platform below the AODs to optimize the incident angle and ensure maximum laser beam diffraction efficiency, as measured using the power meter.
25. Reinstall the DCU. Adjust the incident angle of the laser beam to ~53° for the N-SF11 prism (or ~69° for the N-SF66 prism).
26. Realign the laser beam after the DCU with respect to the AODs.

    A distance of ~35 cm between the AODs and the DCU prism is a good starting point for nearly restoring the original laser pulse width (Fig. 1D). The optimal distance depends on the specific optical components used in the beam path and can be found empirically (see Step 30).
27. Reinstall and adjust the first λ/2-wave plate before the prism to ensure maximum laser beam transmission through the prism, as measured using the power meter.
28. Reinstall the second λ/2-wave plate before the AODs. Turn the polarization to maximize AOD diffraction efficiency, as measured using the power meter.
29. Position small (≤500 nm) round fluorescent beads in the FOV. Adjust the incident angle of the laser beam at the prism to achieve a homogeneous bead resolution (Fig. 1D).

    The laser beam needs to be realigned with respect to the AODs by using the two parallel mirrors of the DCU each time the prism angle is changed.

    See *Troubleshooting*.
30. To optimize compensation of the temporal dispersion (i.e., to achieve the shortest pulses below the objective), shift the complete DCU parallel to the X95 column until the fluorescence signal is maximal.

Perform this optimization during AOD scanning. Alternatively, measure the laser pulse width below the objective with an autocorrelator. Realign the laser beam for maximal signal each time after shifting the DCU.

31. During regular maintenance, be sure to check the alignment of the two key components (i.e., the AODs and the DCU prism), which can be adjusted slightly without any major laser beam runoff.

Animal Preparation

Experimental procedures for animal preparation and dye labeling of cell populations are similar to procedures described in detail in Chapters 49 and 51. Perform all procedures in accordance with the guidelines, and with the approval, of your local animal facility.

32. Anesthetize mice, for example, by intraperitoneal administration of urethane (1.2–1.5 g/kg body weight).
33. Perform a craniotomy above the brain region of interest (e.g., the barrel cortex) and carefully remove the dura if needed.
34. Superfuse the exposed cortex with NRR solution.
35. Bolus load neurons in the superficial layers of the cortex in the area under study with a calcium-sensitive fluorescent dye (e.g., OGB-1) (Stosiek et al. 2003; see also Chapter 49).
36. To distinguish neurons from astrocytes, counterstain the latter by application of SR101 to the exposed brain surface for several minutes (Nimmerjahn et al. 2004; see also Chapter 63).
37. To dampen heartbeat- and breathing-induced motion, fill the cranial window with 1% agarose. Cover with a glass coverslip.
38. Place animals under the custom-built AOD two-photon microscope and start two-photon imaging in the frame-scan mode.

Random Access Pattern Scanning of Neuronal Populations

The main advantage of using AOD laser scanners is the possibility to rapidly shift the focal excitation spot between any two positions within the scanning field within only a few microseconds (10 µsec in our setup). Therefore, stained cell populations can be scanned in a "random-access" manner avoiding waste of time on background regions. To optimize dwell times on neuronal somata and to minimize effects of photobleaching and phototoxicity, we devised a random-access pattern scanning (RAPS) modus, in which a small stereotype point pattern is sequentially run on each neuron in the population (Fig. 2A).

39. Search for an area of interest that contains well-stained neurons. Choose an appropriate zoom factor to clearly identify neuronal somata.

    Stained neurons and astrocytes are visualized in the green and red detection channels, respectively.

    See *Troubleshooting.*

40. Before random-access scanning, acquire a reference image at high spatial resolution using the standard 2D frame scanning mode with the AODs (Fig. 1E).
41. Choose your stereotype scan pattern, defined by the number of points ($n = 1$–9; we mostly used five points) and the geometrical arrangement (e.g., as a cross or a circle).
42. Define the random-access pattern scanning pathway by manually selecting neuronal somata from the previously acquired reference image. From the positions of the selected cells (and the selected pattern), a user-defined list of all scan positions is created and sent to the AOD driving electronics.

    Choose at least one background region (e.g., the lumen of an unstained blood vessel) that provides an estimate for background fluorescence that should be subtracted. Note that the number of cells selected trades off against effective cell sampling rate. For example, with our settings and using 5-point RAPS, 100 cells are sampled at 167 Hz, 50 cells at 334 Hz, or 30 cells at 556 Hz.

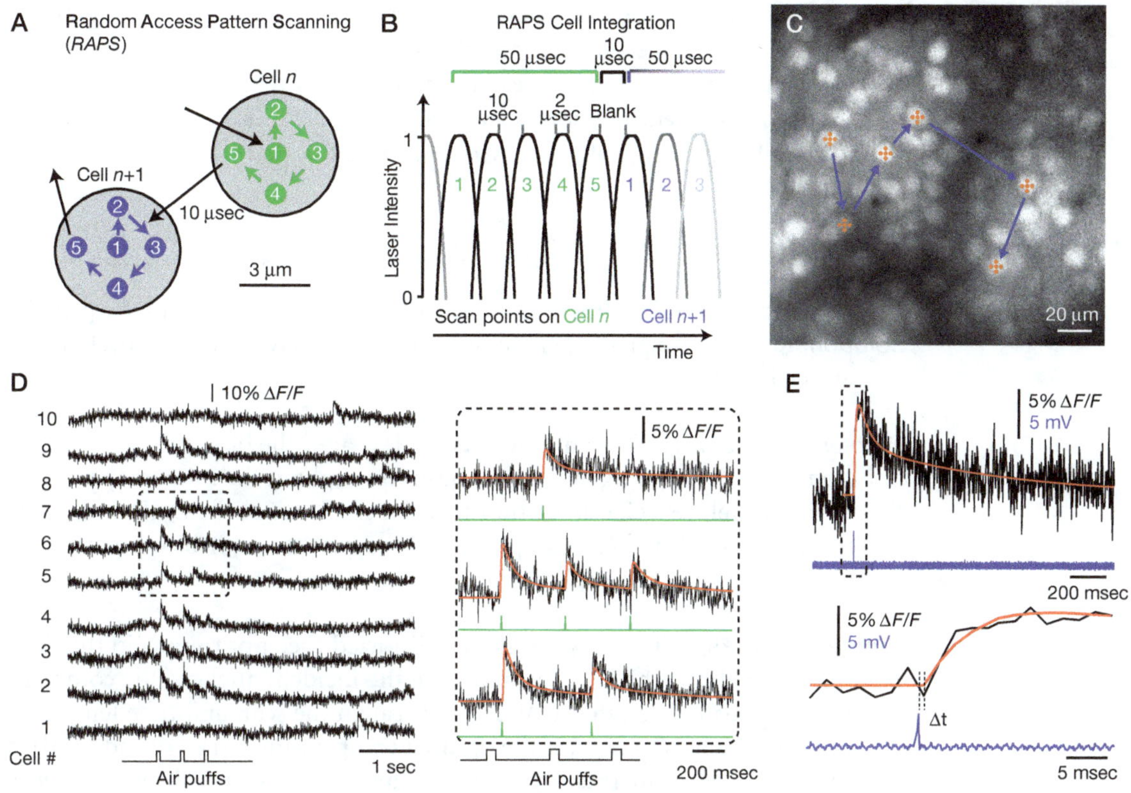

FIGURE 2. High-speed random-access recording of calcium signals in neocortical networks. (*A*) The principle of random-access pattern scanning (RAPS) is exemplified by a five-point pattern that is sequentially scanned on two cell somata. (*B*) Signal integration protocol for five-point RAPS. Cellular fluorescence signals are integrated over the entire five-point scan periods except for the 10-µsec transition between cells. (*C*) L2/3 neuronal population in mouse barrel cortex labeled with OGB-1 (depth = 210 µm). High-speed imaging using a 314-Hz sampling rate per cell was performed using five-point RAPS on 53 cells manually preselected from a reference image. Five of the RAPS transitions are shown in the image. (*D*) Calcium transients in 10 of the 53 selected L2/3 neurons in response to whisker stimulation by three air puffs at 2.5 Hz (*left*). Close-up of the responses of three cells (as indicated in the box at *left*) to air-puff stimulation (*right*). Model traces (red) are reconstructed by an automated spike detection algorithm. (*E*) Example of a single action-potential-evoked calcium transient (measured at 490 Hz) fitted with a model trace (red) (*top*). A simultaneous in vivo juxtacellular recording reveals the spike eliciting the calcium transient. Expanded plot of the onset of the single action-potential-evoked calcium transient and its temporal relationship to the simultaneously recorded spike (*bottom*). In this case, the difference (Δt) between actual spike time and the spike time estimate obtained from the onset fit (red) was 0.7 msec.

43. During the pattern scan on each cell continuously integrate the fluorescence signal, interrupting only during the brief transition periods when the laser focus is transiting to the next neuron.

    In our setup we used a five-point pattern and a total signal integration time of 50 µsec/cell (Fig. 2B). Although the 10-mm aperture size of our AODs theoretically supports a point-to-point transition time of 15.4 µsec (sound velocity 650 m/sec), this can be shortened without any measurable effects to 10 µsec (Otsu et al. 2008), in particular when the AOD aperture is underfilled. For frame scanning, an additional pixel dwell time of 2–3 µsec results in an acquisition time of ~800 msec for a 256 × 256-pixel frame.

44. For optimizing the signal-to-noise ratio (SNR) of the fluorescence recordings from neuronal somata, you can either increase the signal integration time per neuron or slightly increase the laser intensity.

45. Start random-access pattern scanning of the neuronal cell population and acquire spontaneous or stimulus-evoked fluorescence signals.

    See *Troubleshooting*.

46. For each neuron, subtract background fluorescence and convert the raw fluorescence signal to relative percentage fluorescence changes ($\Delta F/F$). From the onsets of the calcium transients, action potential timing can be determined with near-millisecond precision (Grewe et al. 2010). For converting entire fluorescence traces to spike trains, use the "peeling" algorithm described in detail in the original paper (Grewe et al. 2010).

TROUBLESHOOTING

Problem (Step 23): AOD drivers do not work properly (e.g., driving frequencies are not generated).
Solution: Reset the frequency generators of the AOD driver circuit. Check the frequency output with an oscilloscope.

Problem (Step 29): Laser beam power is too low after the objective.
Solution: Look for clipping of the laser beam throughout the microscope. Check proper functioning of the Pockels cell. Check the transmission efficiency of the DCU and the AODs. If transmission is <40%, check the λ/2-wave plates and readjust the laser beam incident angle of the AODs by adjusting the pitch/yaw platform.

Problem (Step 29): Fluorescent beads are distorted (i.e., not round) in the FOV center.
Solution: Recalibrate the DCU by adjusting the incident angle of the compensation prism. Realign the laser beam through the AODs and realign the scan and tube lenses. If installed, realign the cylindrical lens telescope. If the optimal compensation (i.e., round bead) is offset to the side of the FOV's center, increase or reduce the amount of spatial dispersion by changing the incident angle between the laser beam and the compensation prism to ensure optimal compensation in the FOV center.

Problem (Step 29): Astigmatic effects are apparent during axial movement of the focal plane.
Solution: Verify whether the laser beam is centered on all optical components. If a cylindrical telescope is used, check that the lens distance is the exact sum of the focal lengths of the cylindrical lenses. The focusing planes of both lenses must fit properly.

Problem (Step 39): There is no measurable fluorescence signal in any channel.
Solution: Use an IR viewer to see if the laser illuminates the sample. If not, check if and where the alignment of the excitation beam might be off. Make sure all the electronics (AOD drivers and amplifiers, shutters, Pockels cell) are turned on. Check the laser beam alignment. If the laser does illuminate the sample, check all the components of the detection pathway (PMTs, preamplifiers, A/D converters, dichroic mirrors). Try to locate the area of dye injection within the craniotomy. Check the microscope system using fluorescent test samples (e.g., fluorescent beads or pollen grains).

Problem (Step 39): The staining of the neurons is weak and insufficient to measure neuronal activity.
Solution: If dye loading is not optimal, repeat the labeling procedure. Multicell bolus loading critically depends on the quality of the preparation (Garaschuk et al. 2006). If the labeling is sufficient but other factors are limiting (e.g., obstruction of illumination by blood vessels), reduce the scanning speed and increase the cell integration time by raising the number of points scanned on each cell; this will help to maintain a good SNR.

Problem (Step 39): The brightness of the fluorescence signal is inhomogeneous and falls off toward the FOV edges.

Solution: This effect occurs because the AOD deflection efficiency is maximal for the AOD center frequency but changes as a function of the AOD frequency and thus of the scan angle. As a result, the laser illumination is inhomogeneous across the FOV. Because the fast FPGA board controls both AOD driving frequency and the sound wave amplitude (which directly modulates diffraction efficiency), it is possible to compensate for this effect by slightly increasing the sound wave amplitude toward the FOV edges. Try adjusting the AOD sound wave amplitude for each deflection angle until a given transmission efficiency is reached. Perform this calibration for each AOD (x and y) separately.

Problem (Step 45): Motion-induced artifacts during RAPS experiments make recordings difficult.

Solution: During imaging experiments motion artifacts can be caused by tissue pulsation resulting from animal breathing and/or heartbeats. Damp brain tissue pulsation as much as possible by adjusting the animal's posture and by renewing the agarose and coverslip on top of the craniotomy.

EXAMPLE OF APPLICATION

To directly demonstrate RAPS recording we measured fluorescence traces from a neuronal population of 53 cells in mouse barrel cortex at a 314-Hz cell sampling rate (Fig. 2C–E). Repetitive stimulation of the contralateral whiskers evoked calcium transients with rapid onsets and $\Delta F/F$ amplitudes in the range of 5%–20% (Fig. 2D). With a peeling algorithm for the automated spike reconstruction algorithm (Grewe et al. 2010), we were able to extract a model of action potential–evoked calcium transients that closely followed the experimental data and thereby provide an estimate of the underlying spike train. The accuracy of spike time reconstruction was verified using simultaneous juxtacellular recordings from single neurons together with the population RAPS measurement and analyzing the time difference Δt between the spike times predicted from the calcium transient onsets and the true spike times as revealed by juxtacellular recording (Fig. 2E). This analysis showed that the high cell sampling rates allowed a reliable and precise action potential reconstruction from the optical recordings with near-millisecond precision (Grewe et al. 2010).

DISCUSSION

By scanning small point patterns on each individual neuron the RAPS mode enables measurements with high SNR and enables a post hoc reconstruction of spike times with near-millisecond accuracy (Grewe et al. 2010). The imaging depth of the AOD microscope is sufficient to visualize neuronal activity of L2/3 neurons in mouse neocortex down to 300 µm below the pia. At present it will be difficult to make recordings at much greater depths because of the slightly reduced spatial resolution of the AOD system and laser power limitations. Compared to in vivo imaging studies using standard galvanometric scan mirrors (1–30-Hz frame rate; Kerr et al. 2005; Ohki et al. 2005; Sato et al. 2007; Rochefort et al. 2009), the AOD-based imaging approach provides up to a 500-Hz cell sampling rate on populations of comparable size, hence gaining a factor of 16–500 in temporal resolution. Although user-defined line-scanning approaches can also be used to increase sampling speed (Göbel et al. 2007; Lillis 2008; Rothschild et al. 2010), it is unclear whether they can reach a comparable SNR as the AOD-based RAPS measurements.

All fluorescence traces shown in Figure 2 are raw traces without any filtering. The excellent SNR at the high sampling rates is explained by the rather long dwell times per cell and the efficient fluorescence collection. AOD-based scanning thus opens new ways to measure spatiotemporal spiking dynamics in the living brain across neuronal populations of 10 to 100 cells with a few milliseconds resolution. We expect this method to facilitate in vivo imaging studies, for which a high temporal resolution is essential. This in particular concerns questions related to spike timing within neuronal populations (e.g., phase relationships of spiking activity relative to global oscillations) or to circuit plasticity, where relative spike timing is crucial for modulation of synaptic strength.

There are several promising future directions for AOD-based microscopy. First, although our in vivo measurements so far were restricted to a single focal plane, they should be extendable to 3D by using either mechanical *z* scanning in combination with AOD-based lateral scanning (Göbel et al. 2007) or chirped acoustic waves in a series of AODs (Vucinic and Sejnowski 2007; Reddy et al. 2008; Kirkby et al. 2010). Extending the method volumetrically would allow the measurement of neural dynamics in local 3D networks. Second, the combination of high-speed AOD-based imaging with genetically encoded calcium indicators (Mank et al. 2008; Wallace et al. 2008; Tian et al. 2009; Lütcke et al. 2010) should allow for repeated long-term functional interrogation of neuronal networks. A third goal is to perform high-speed AOD-based calcium imaging in awake animals (Dombeck et al. 2007; Greenberg and Kerr 2009). This faces particular challenges because the AOD-RAPS mode is especially sensitive to motion artifacts, which are expected under awake conditions. Finally, improvements in algorithms for inferring neuronal spiking from noisy fluorescence traces (Kerr et al. 2007; Vogelstein et al. 2009; Grewe et al. 2010) are possible and could increase the reliability and precision of spike train reconstruction. In summary, the combination of fast optical recording with AODs, a high SNR, and a reliable automated spike reconstruction will facilitate the spatiotemporal analysis of neuronal population activity on the level of local microcircuits and thus help to reveal fundamental aspects of neural coding.

RECIPE

CAUTION: See Appendix 6 for proper handling of materials marked with <!>.
Recipes for reagents marked with <R> are included in this list.

Normal Rat Ringer (NRR) Solution

NaCl	135 mM
KCl	5.4 mM
HEPES	5 mM
$CaCl_2$ <!>	1.8 mM
$MgCl_2$	1 mM

Adjust pH to 7.2 with NaOH <!>.

ACKNOWLEDGMENTS

This work was supported by a Forschungskredit of the University of Zurich (BG), and by grants to FH from the Swiss National Science Foundation (Grant 3100A0-114624), the EU-FP7 program (SPACE-BRAIN project 200873), and the Swiss SystemsX.ch initiative (project 2008/2011-Neurochoice).

REFERENCES

Bansal V, Saggau P. 2011. Digital micromirror devices: Principles and applications in imaging. In *Imaging: A laboratory manual* (ed. Yuste R), pp. 839–848. Cold Spring Harbor Laboratory Press, Cold Spring Harbor, NY.

Buzsáki G. 2004. Large-scale recording of neuronal ensembles. *Nat Neurosci* **7:** 446–451.

Dombeck DA, Khabbaz AN, Collman F, Adelman TL, Tank DW. 2007. Imaging large-scale neural activity with cellular resolution in awake, mobile mice. *Neuron* **56:** 43–57.

Garaschuk O, Milos R-I, Konnerth A. 2006. Targeted bulk-loading of fluorescent indicators for two-photon brain imaging in vivo. *Nat Protoc* **1:** 380–386.

Göbel W, Kampa BM, Helmchen F. 2007. Imaging cellular network dynamics in three dimensions using fast 3D laser scanning. *Nat Methods* **4:** 73–79.

Greenberg DS, Kerr JN. 2009. Automated correction of fast motion artifacts for two-photon imaging of awake animals. *J Neurosci Methods* **176:** 1–15.

Grewe BF, Helmchen F. 2009. Optical probing of neuronal ensemble activity. *Curr Opin Neurobiol* **19:** 520–529.

Grewe BF, Langer D, Kasper H, Kampa BM, Helmchen F. 2010. High-speed in vivo calcium imaging reveals neuronal network activity with near-millisecond precision. *Nat Methods* **7:** 399–405.

Helmchen F, Denk W. 2005. Deep tissue two-photon microscopy. *Nat Methods* **2:** 932–940.

Iyer V, Hoogland TM, Saggau P. 2006. Fast functional imaging of single neurons using random-access multiphoton (RAMP) microscopy. *J Neurophysiol* **95:** 535–545.

Kerr JND, Denk W. 2008. Imaging in vivo: Watching the brain in action. *Nat Rev Neurosci* **9:** 195–205.

Kerr JND, Greenberg D, Helmchen F. 2005. Imaging input and output of neocortical networks in vivo. *Proc Natl Acad Sci* **102:** 14063–14068.

Kerr JND, de Kock CPJ, Greenberg DS, Bruno RM, Sakmann B, Helmchen F. 2007. Spatial organization of neuronal population responses in layer 2/3 of rat barrel cortex. *J Neurosci* **27:** 13316–13328.

Kirkby PA, Srinivas Nadella KM, Silver RA. 2010. A compact acousto-optic lens for 2D and 3D femtosecond based 2-photon microscopy. *Opt Express* **18:** 13721–13745.

Kremer Y, Léger J-F, Lapole R, Honnorat N, Candela Y, Dieudonné S, Bourdieu L. 2008. A spatio-temporally compensated acousto-optic scanner for two-photon microscopy providing large field of view. *Opt Express* **16:** 10066–10076.

Liebling M. 2011. Imaging the dynamics of biological processes via fast confocal microscopy and image processing. In *Imaging: A laboratory manual* (ed. Yuste R), pp. 823–830. Cold Spring Harbor Laboratory Press, Cold Spring Harbor, NY.

Lillis KP, Eng A, White JA, Mertz J. 2008. Two-photon imaging of spatially extended neuronal network dynamics with high temporal resolution. *J Neurosci Methods* **172:** 178–184.

Lütcke H, Murayama M, Hahn T, Margolis DJ, Astori S, zum Alten Borgloh SM, Göbel W, Yang Y, Tang W, Kügler S, et al. 2010. Optical recording of neuronal activity with a genetically-encoded calcium indicator in anesthetized and freely moving mice. *Front Neural Circuits* **4:** 9.

Mank M, Santos AF, Direnberger S, Mrsic-Flogel TD, Hofer SB, Stein V, Hendel T, Reiff DF, Levelt C, Borst A, et al. 2008. A genetically encoded calcium indicator for chronic in vivo two-photon imaging. *Nat Methods* **5:** 805–811.

Nikolenko V, Peterka DS, Araya R, Woodruff A, Yuste R. 2011. Spatial light modulator microscopy. In *Imaging: A laboratory manual* (ed. Yuste R), pp. 849–860. Cold Spring Harbor Laboratory Press, Cold Spring Harbor, NY.

Nimmerjahn A, Kirchhoff F, Kerr JND, Helmchen F. 2004. Sulforhodamine 101 as a specific marker of astroglia in the neocortex in vivo. *Nat Methods* **1:** 31–37.

Ohki K, Chung S, Ch'ng YH, Kara P, Reid RC. 2005. Functional imaging with cellular resolution reveals precise microarchitecture in visual cortex. *Nature* **433:** 597–603.

Oron D, Silberberg Y. 2011. Temporal focusing microscopy. In *Imaging: A laboratory manual* (ed. Yuste R), pp. 861–868. Cold Spring Harbor Laboratory Press, Cold Spring Harbor, NY.

Otsu Y, Bormuth V, Wong J, Mathieu B, Dugué GP, Feltz A, Dieudonné S. 2008. Optical monitoring of neuronal activity at high frame rate with a digital random-access multiphoton (RAMP) microscope. *J Neurosci Methods* **173:** 259–270.

Reddy GD, Saggau P. 2011. High-speed two-photon imaging. In *Imaging: A laboratory manual* (ed. Yuste R), pp. 831–839. Cold Spring Harbor Laboratory Press, Cold Spring Harbor, NY.

Reddy GD, Kelleher K, Fink R, Saggau P. 2008. Three-dimensional random access multiphoton microscopy for functional imaging of neuronal activity. *Nat Neurosci* **11:** 713–720.

Rochefort NL, Garaschuk O, Milos R-I, Narushima M, Marandi N, Pichler B, Kovalchuk Y, Konnerth A. 2009. Sparsification of neuronal activity in the visual cortex at eye-opening. *Proc Natl Acad Sci* **106:** 15049–15054.

Rothschild G, Nelken I, Mizrahi A. 2010. Functional organization and population dynamics in the mouse primary auditory cortex. *Nat Neurosci* **13:** 353–360.

Sato TR, Gray NW, Mainen ZF, Svoboda K. 2007. The functional microarchitecture of the mouse barrel cortex. *PLoS Biol* **5:** e189.

Stosiek C, Garaschuk O, Holthoff K, Konnerth A. 2003. In vivo two-photon calcium imaging of neuronal networks. *Proc Natl Acad Sci* **100:** 7319–7324.

Tian L, Hires SA, Mao T, Huber D, Chiappe ME, Chalasani SH, Petreanu L, Akerboom J, McKinney SA, Schreiter ER, et al. 2009. Imaging neural activity in worms, flies and mice with improved GCaMP calcium indicators. *Nat Methods* **6:** 875–881.

Vogelstein JT, Watson BO, Packer AM, Yuste R, Jedynak B, Paninski L. 2009. Spike inference from calcium imaging using sequential Monte Carlo methods. *Biophys J* **97:** 636–655.

Vučinić D, Sejnowski TJ. 2007. A compact multiphoton 3D imaging system for recording fast neuronal activity. *PLoS ONE* **2:** e699.

Wallace DJ, zum Alten Borgloh SM, Astori S, Yang Y, Bausen M, Kügler S, Palmer AE, Tsien RY, Sprengel R, Kerr JND, et al. 2008. Single-spike detection in vitro and in vivo with a genetic Ca^{2+} sensor. *Nat Methods* **5:** 797–804.

Zeng S, Lv X, Zhan C, Chen WR, Xiong W, Jacques SL, Luo Q. 2006. Simultaneous compensation for spatial and temporal dispersion of acousto-optical deflectors for two-dimensional scanning with a single prism. *Opt Lett* **31:** 1091–1093.

Zeng S, Lv X, Bi K, Zhan C, Li D, Chen WR, Xiong W, Jacques SL, Luo Q. 2007. Analysis of the dispersion compensation of acousto-optic deflectors used for multiphoton imaging. *J Biomed Opt* **12:** 024015.

54 Calcium Imaging in the *Drosophila* Olfactory System with a Genetic Indicator

Cory M. Root, Allan M. Wong, Jorge Flores, and Jing W. Wang
Division of Biological Sciences, University of California San Diego, La Jolla, California 92093

ABSTRACT

Insects show sophisticated odor-mediated behaviors controlled by an olfactory system that is genetically and anatomically simpler than that of vertebrates, providing an attractive system to investigate the mechanistic link between behavior and odor perception. Advances in neuroscience have been facilitated by modern optical imaging technologies—both in instrumentation and in probe design—that permit the visualization of functional neural circuits. This chapter describes an imaging system to monitor neural activity in the *Drosophila* antennal lobe that couples two-photon microscopy with the specific expression of the calcium-sensitive green fluorescent protein G-CaMP and discusses the practical considerations of this imaging system.

INTRODUCTION

The yeast Gal4 transcriptional activator and the upstream activating sequence (UAS) have been exploited to create a binary gene expression system in *Drosophila* to express transgenes in a spatially restricted manner (Brand and Perrimon 1993). Many Gal4 lines exist that label defined subpopulations of the fly central nervous system, providing a means to genetically dissect the function of neural circuits. To study the olfactory response in the antennal lobe, we used GH146–Gal4, which expresses Gal4 in approximately 90 of the 200 projection neurons that connect the antennal lobe to the mushroom body and the protocerebrum (Stocker et al. 1997). The G-CaMP-coding sequence was fused to the UAS promoter to generate UAS–G-CaMP transgenic flies (Wang et al. 2003). Multiple insertions of the G-CaMP transgene and simultaneous translation of multiple transgenes by the internal ribosome entry site (IRES) (Jang et al. 1988; Pelletier and Sonenberg 1988) were used to increase G-CaMP expression levels. Flies harboring the GH146–Gal4 and UAS–G-CaMP transgenes express G-CaMP in projection neurons (Fig. 1B), such that changes in calcium concentration—an indicator of neural activity—can be monitored (Yuste and Katz 1991; see also Chapter 37).

The buffering capacity of G-CaMP is an important consideration in the imaging system. The fluorescence intensity of labeled neurons is proportional to the concentration of the fluorescent probe. Low [G-CaMP] results in a poor signal-to-noise ratio (SNR), but high concentrations can alter the

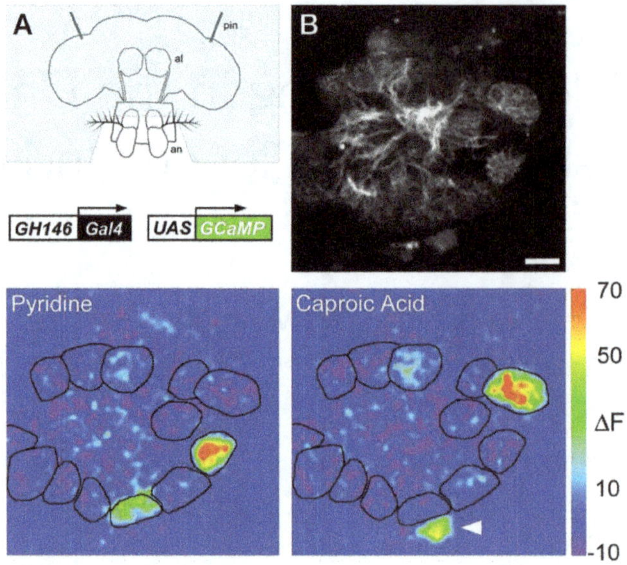

FIGURE 1. (*A*) Schematic of the antennae–brain preparation. This preparation allows airborne odorants to reach the antennae (an) and imaging of the antennal lobe (al). The gray area represents the part of the preparation covered with agarose. (*B*) Flies harboring GH146–Gal4 and UAS–G-CaMP transgenes express G-CaMP in projection neurons and were analyzed by two-photon imaging in the antennal lobe. The grayscale prestimulation image (*top*) reveals glomerular structure (scale bar, 10 μm). (*Bottom*) Peak glomerular response to pyridine and caproic acid. Odorants (16% saturated vapor) were delivered for 2 sec and false color images (128 x 128 pixels) were collected at 4 frames/sec. In this plane, pyridine activates the VM2 and VA3 glomeruli and caproic acid activates the DM3 glomerulus and a cell body (arrowhead).

temporal dynamics of the calcium influx (see Chapter 36). If the buffering capacity of G-CaMP approaches that of the endogenous buffers, increasing [G-CaMP] should alter the amplitude and decay time constant of the fractional change in fluorescence intensity ($\Delta F/F$). Genetic manipulation can be used to alter [G-CaMP] by varying the transgene copy number. A comparison of the peak $\Delta F/F$ response of projection neurons to a range of stimulus intensities showed no significant differences in the peak or the decay time constant of $\Delta F/F$ (Fig. 2A,B), suggesting that the buffering capacity of G-CaMP is negligible when compared with that of the endogenous buffers. Similar G-CaMP expression levels have negligible effects on odor-evoked action potential firing in projection neurons (Jayaraman and Laurent 2007), supporting the notion that the genetic expression of G-CaMP does not alter the buffering capacity of the projection neurons.

SNR is another important factor that affects the sensitivity of the imaging system. The SNR increases proportionally to $N^{1/2}$, in which N is the number of detected photons (Yasuda et al. 2004). There are two different ways to improve the SNR. First, increasing laser power generates more photons, but it can also cause excessive photobleaching: In two-photon imaging, the rate of photobleaching as a function of excitation power increases faster than that of photon emission (Patterson and Piston 2000; Chen et al. 2002). Increased laser power also is associated with increased photodamage (Koester et al. 1999; Hopt and Neher 2001). Alternatively, one can increase the [G-CaMP], which should improve the SNR proportional to $[\text{G-CaMP}]^{1/2}$ (Yasuda et al. 2004; Hires et al. 2008). An examination of photodamage and photobleaching using different G-CaMP concentrations shows significantly more photobleaching using a low [G-CaMP] when the SNR is kept constant (Fig. 2C,D). Similarly, samples with low [G-CaMP] show severe photodamage with a significant reduction in the peak $\Delta F/F$ after prolonged laser illumination (Fig. 2E). Therefore, high [G-CaMP] (e.g., four to six copies in projection neurons) is necessary to provide an adequate SNR without the side effects of photobleaching and photodamage that is associated with high laser power.

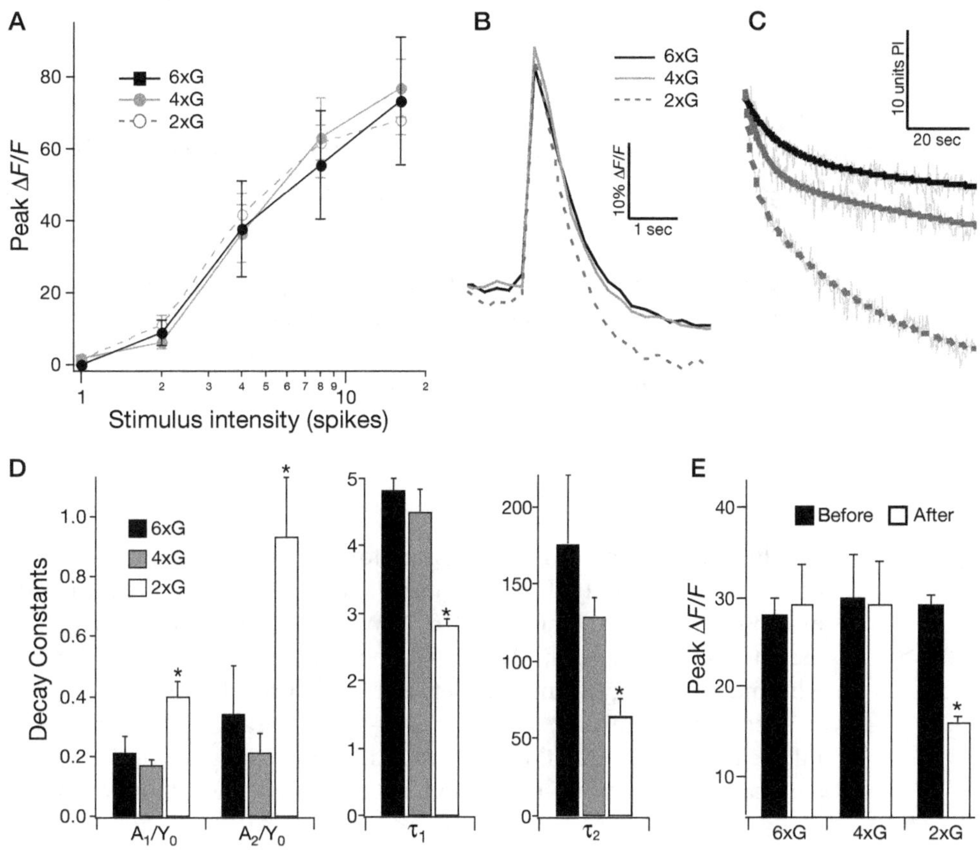

FIGURE 2. (*A*) Two-photon imaging of dendritic calcium activity of a DM2 projection neuron in response to electrical stimulation of the olfactory nerve across a range of stimulus intensities. The level of G-CaMP expression was altered by varying the transgene copy number. 2xG: two copies of UAS–G-CaMP; 4xG: four copies of UAS–G-CaMP; 6xG: two copies of UAS–G-CaMP–IRES–G-CaMP–IRES–G-CaMP). (*B*) Change in fluorescence over time in response to stimulation of the olfactory nerve. Traces show the average of three trials from different preparations. (*C*) Photobleaching was assayed by measuring G-CaMP fluorescence over 75 sec of high-intensity illumination (1.4x that used in *B*). Units of pixel intensity (PI) are plotted as a function of time. Representative traces are overlaid with the fitting from a double exponential equation: $Y = Y_0 + A_1e^{-t/\tau_1} + A_2e^{-t/\tau_2}$. (*D*) Average values for the curve-fitting constants for the photobleaching experiments in *C*. (*E*) Photodamage was assayed by comparing projection neuron response to electrical stimulation of the olfactory nerve before and after a 75-sec period of high-intensity illumination. $n = 3–5$. Error bars show S.E.M. $p \leq 0.05$ by *t*-test.

Protocol

Calcium Imaging in the *Drosophila* Olfactory System with a Genetic Indicator

Imaging calcium activity in genetically defined populations of neurons provides an important tool for investigating the function of neural circuits. The following protocol describes a two-photon imaging setup and antennae–brain preparation that can be used to measure odor-evoked calcium activity in the *Drosophila* antennal lobe.

IMAGING SETUP

For this protocol, calcium imaging is performed using a custom-made two-photon microscope (Denk et al. 1990; Denk 2011). The system uses a Mira 900 Ti:sapphire laser (Coherent, Inc.) at a wavelength of ~930-nm, pumped by a Verdi V10 10-W solid-state laser (Coherent, Inc.) as its light source. A blocking filter (E700SP-special, Chroma Technology) in conjunction with a band-pass filter (HQ525/50m, Chroma Technology) is inserted between the focusing lens and the photomultiplier tube light detector to select for the emission spectrum of G-CaMP (peak at 509 nm; Nakai et al. 2001) and to block the infrared laser light and other background fluorescent light (for details, see Wang et al. 2003). Samples are imaged using a water-immersion objective (Zeiss, IR-Achroplan, 40×, NA 0.80).

MATERIALS

CAUTION: See Appendix 6 for proper handling of materials marked with <!>.
See the end of the chapter for recipes for reagents marked with <R>.

Reagents

Agarose (2%, prepared in adult hemolymph-like saline [AHLS]; Sigma-Aldrich)
CO_2 <!>
Drosophila, adult, transgenic

Use adult flies (within 48 h of eclosion) carrying four copies of the UAS–G-CaMP transgene for this procedure. Older flies should be avoided because of their high background fluorescence.

Drosophila AHLS <R>

Prepare separate solutions with and without calcium.

Mineral oil (optional; see Step 10)
Odorants
Sylgard

Equipment

Bottles, 100-mL
Capillary tubes, glass
Coverslips, glass
Insect pins, 0.1-mm diameter
Filter paper
Forceps, #55
Microscope slides

Olfactometer

This technique uses a simple custom-made odor delivery system that permits quantitative control of odor stimulation by diluting saturated vapor up to 1000-fold and delivering a precise amount of odorant to the brain preparation at a specific time during imaging. For details, see Wang et al. (2003) and Root et al. (2008).

Petri dish, Sylgard-coated

Stimulator (Grass Technologies)

Wires, tungsten, 20-μm-diameter (California Fine Wire Company)

EXPERIMENTAL METHOD

Separate procedures are presented for imaging upon olfactory stimulation (Steps 1–10) and direct electrical stimulation of the olfactory nerve (Steps 11–17).

Antennae–Brain Preparation

This procedure provides a *Drosophila* antennae–brain preparation that accommodates a water-immersion objective lens for imaging and airborne odor stimulation (Fig. 1A). Imaging should be performed within 4 h of dissection.

1. Briefly anesthetize flies with CO_2. Decapitate them.
2. Use an insect pin to affix one of the heads through the proboscis onto a Sylgard-coated plate. Flood the plate with cold calcium-free AHLS.

   Cold calcium-free AHLS suppresses synaptic transmission during dissection, thereby maintaining the integrity of the preparation. Preparations dissected in normal AHLS (i.e., containing 2 mM $CaCl_2$) are much less responsive to odor stimulation.
3. Using forceps, expose the brain by removing the head capsule and air sacs.
4. Remove the maxillary palps.

   This leaves the third antennal segment as the sole sensory organ that provides input to the antennal lobes.
5. Transfer the preparation onto a thin Sylgard slab on a microscope slide. Pin the antennae and brain in position with tungsten wires.
6. Replace the calcium-free AHLS with normal AHLS.
7. To minimize movement, replace the AHLS with 2% agarose maintained at 38°C.
8. Before imaging, carefully remove the agarose covering the antennae to allow odor stimulation. Place a glass coverslip on top of the brain region.

Olfactory Stimulation and Imaging

9. Place the preparation under the two-photon microscope.
10. Deliver constant carrier airflow (1 L/min) of the odorant of choice to the antennae via a pipe (12 mm in diameter).

    Deliver odors by mixing a defined percentage of the carrier air with air redirected through 100-mL bottles containing 20 μL of odorant on a piece of filter paper. In some cases, it might be necessary to first dilute the odorants in mineral oil. For more information on olfactory stimulation and imaging, see Wang et al. (2003) and Root et al. (2008).

Nerve Stimulation Preparation

This procedure allows electrical stimulation of the olfactory nerve while accommodating a water-immersion objective lens for imaging.

11. Dissect the brain as described in Steps 1–3.
12. Pin the antennae and brain in position with fine tungsten wires (Fig. 1A).
13. Use forceps to cut the olfactory nerves near the base of the antennae.
14. Replace the calcium-free AHLS with normal AHLS.
15. Prepare a glass suction electrode for stimulation:
    i. Pull glass capillary tubing to a fine tip and break the tip using forceps.
    ii. Fire-polish the tip to achieve a diameter that is ~1.5x the diameter of the olfactory nerve (8–10 μm).
16. Position the preparation and electrode under the two-photon microscope. Gently aspirate the nerve as a loop into the electrode.
17. Use a Grass stimulator to deliver pulses at 100 Hz, 1 msec in duration and 10 V in amplitude.

    Varying the stimulus duration from 10–160 msec allows for the delivery of 1–16 spikes.

DISCUSSION

At 2 and 4 h after dissection, the odor responsivity is the same as that at 1 h after dissection: In response to isoamyl acetate (8% saturated vapor), the $\Delta F/F$ in the VM3 glomerulus were 107 ± 10, 102 ± 14, and 101 ± 17 at 1, 2, and 4 h after dissection, respectively (mean ± S.D., $n = 4$). Images in response to olfactory stimulation with pyridine and caproic acid are shown in Figure 1B.

Advantages and Limitations

The optical imaging system described here is capable of simultaneously monitoring neural activity in a genetically defined population of neurons. It uses two-photon microscopy with the specific expression of the genetic calcium-sensitive probe, G-CaMP. The Gal4/UAS gene expression system targets G-CaMP expression in a population of sensory or projection neurons, and the power of two-photon microscopy to perform optical sections permits neural activity from individual glomeruli to be monitored in the three-dimensional antennal lobe. At present, this imaging system is limited by the availability of Gal4 lines that label neurons of interest.

The dissociation constant of G-CaMP is slightly greater than the resting calcium concentration, making it possible to detect small calcium influxes evoked by odors at concentrations likely to be encountered by insects in their natural habitats. Although calcium influx is correlated with the number of action potentials, the measurement of calcium influx is not an adequate reflection of fast neural dynamics in millisecond resolution, an intrinsic problem of all calcium probes that can be mitigated by concurrent electrical recordings (Wang et al. 2003; Jayaraman and Laurent 2007; Root et al. 2007; see also Chapter 37). Nevertheless, this calcium imaging system should be useful to study higher-order neurons in the olfactory system to decipher the olfactory code, and it might also permit the study of other sensory modalities in the fly.

RECIPE

CAUTION: See Appendix 6 for proper handling of materials marked with <!>.
Recipes for reagents marked with <R> are included in this list.

Drosophila Adult Hemolymph-like Saline (AHLS)

$CaCl_2$ <!>	2 mM
KCl	5 mM
HEPES	5 mM
$MgCl_2$	8.2 mM
NaCl	108 mM
$NaHCO_3$	4 mM
NaH_2PO_4	1 mM
Sucrose	10 mM
Trehalose	5 mM

Adjust to pH 7.5. Filter. Store at 4°C.

Measure this solution using a freezing-point osmometer. The osmolarity should be 265 mOsm, very close to the value of adult *Drosophila* hemolymph (Singleton and Woodruff 1994). Use $CaCl_2$ and $MgCl_2$ solutions instead of powders to avoid precipitation in the saline.

REFERENCES

Brand AH, Perrimon N. 1993. Targeted gene expression as a means of altering cell fates and generating dominant phenotypes. *Development* **118:** 401–415.

Chen T-S, Zeng S-Q, Luo Q-M, Zhang Z-H, Zhou W. 2002. High-order photobleaching of green fluorescent protein inside live cells in two-photon excitation microscopy. *Biochem Biophys Res Commun* **291:** 1272–1275.

Denk W. 2011. Introduction to mulitphoton-excitation fluorescence microscopy. In *Imaging: A laboratory manual* (ed. Yuste R), pp. 105–110. Cold Spring Harbor Laboratory Press, Cold Spring Harbor, NY.

Denk W, Strickler JH, Webb WW. 1990. Two-photon laser scanning fluorescence microscopy. *Science* **248:** 73–76.

Hires SA, Tian L, Looger LL. 2008. Reporting neural activity with genetically encoded calcium indicators. *Brain Cell Biol* **36:** 69–86.

Hopt A, Neher E. 2001. Highly nonlinear photodamage in two-photon fluorescence microscopy. *Biophys J* **80:** 2029–2036.

Jang SK, Kräusslich H-G, Nicklin MJH, Duke GM, Palmenberg AC, Wimmer E. 1988. A segment of the 5′ nontranslated region of encephalomyocarditis virus RNA directs internal entry of ribosomes during in vitro translation. *J Virol* **62:** 2636–2643.

Jayaraman V, Laurent G. 2007. Evaluating a genetically encoded optical sensor of neural activity using electrophysiology in intact adult fruit flies. *Front Neural Circuits* **1:** 3.

Koester HJ, Baur D, Uhl R, Hell SW. 1999. Ca^{2+} fluorescence imaging with pico- and femtosecond two-photon excitation: Signal and photodamage. *Biophys J* **77:** 2226–2236.

Nakai J, Ohkura M, Imoto K. 2001. A high signal-to-noise Ca^{2+} probe composed of a single green fluorescent protein. *Nat Biotechnol* **19:** 137–141.

Patterson GH, Piston DW. 2000. Photobleaching in two-photon excitation microscopy. *Biophys J* **78:** 2159–2162.

Pelletier J, Sonenberg N. 1988. Internal initiation of translation of eukaryotic mRNA directed by a sequence derived from poliovirus RNA. *Nature* **334:** 320–325.

Root CM, Semmelhack JL, Wong AM, Flores J, Wang JW. 2007. Propagation of olfactory information in *Drosophila*. *Proc Natl Acad Sci* **104:** 11826–11831.

Root CM, Masuyama K, Green DS, Enell LE, Nässel DR, Lee C-H, Wang JW. 2008. A presynaptic gain control mechanism fine-tunes olfactory behavior. *Neuron* **59:** 311–321.

Singleton K, Woodruff RI. 1994. The osmolarity of adult *Drosophila* hemolymph and its effect on oocyte–nurse cell electrical polarity. *Dev Biol* **161:** 154–167.

Stocker RF, Heimbeck G, Gendre N, de Belle JS. 1997. Neuroblast ablation in *Drosophila* P[GAL4] lines reveals origins of olfactory interneurons. *J Neurobiol* **32:** 443–456.

Wang JW, Wong AM, Flores J, Vosshall LB, Axel R. 2003. Two-photon calcium imaging reveals an odor-evoked map of activity in the fly brain. *Cell* **112:** 271–282.

Yasuda R, Nimchinsky EA, Scheuss V, Pologruto TA, Oertner TG, Sabatini BL, Svoboda K. 2004. Imaging calcium concentration dynamics in small neuronal compartments. *Sci STKE* **2004:** l5.

Yuste R, Katz LC. 1991. Control of postsynaptic Ca^{2+} influx in developing neocortex by excitatory and inhibitory neurotransmitters. *Neuron* **6:** 333–344.

55 Calcium Imaging in the Intact Olfactory System of Zebrafish and Mouse

Rainer W. Friedrich

Friedrich Miescher Institute for Biomedical Research, CH-4058 Basel, Switzerland

ABSTRACT

Calcium imaging methods are important tools used to analyze the encoding and processing of odor information by distributed and dynamic activity patterns in the olfactory bulb and higher brain areas. Odor-evoked patterns of sensory inputs across the array of glomeruli can be measured with high spatial and temporal resolutions by selective loading of Ca^{2+} indicators into sensory afferents. Spatiotemporal activity patterns across neurons within the olfactory bulb have also been measured by two-photon Ca^{2+} imaging after bolus loading of acetoxymethyl (AM)-ester dyes.

INTRODUCTION

Odors are first detected by olfactory sensory neurons (OSNs) and evoke stimulus-specific patterns of activation across the input channels of the olfactory bulb (OB), the glomeruli. The output of the OB consists of spatiotemporal activity patterns across mitral/tufted cells that are conveyed to multiple pallial and subpallial target areas. In the main olfactory system of vertebrates, as well as in the olfactory system of insects, odor information is encoded by distributed patterns of activity across a large number of glomeruli or neurons. Ca^{2+} imaging has therefore become an important approach used to analyze the encoding and processing of olfactory information by populations of glomeruli or neurons. Experiments in the intact olfactory system are important to maintain the integrity of the system, to analyze activity patterns evoked by natural odors, and to examine the influence of active sampling strategies, such as sniffing in mammals.

Odor-evoked activity patterns across glomeruli have been visualized by a wide range of methods, including Ca^{2+} imaging (Friedrich and Korsching 1997; Wachowiak and Cohen 1999, 2001), voltage-sensitive dyes (Cinelli et al. 1995; Friedrich and Korsching 1998; Lam et al. 2000; Spors and Grinvald 2002), intrinsic optical signals (Rubin and Katz 1999), functional magnetic resonance imaging (Yang et al. 1998), synapto-pHluorin (Bozza et al. 2004), and others. Similar approaches have also been applied in the insect analog of the OB, the antennal lobe (Joerges et al. 1997; Ng et al. 2002; Wang et al. 2003). This chapter focuses primarily on imaging of glomerular Ca^{2+} signals after loading of an

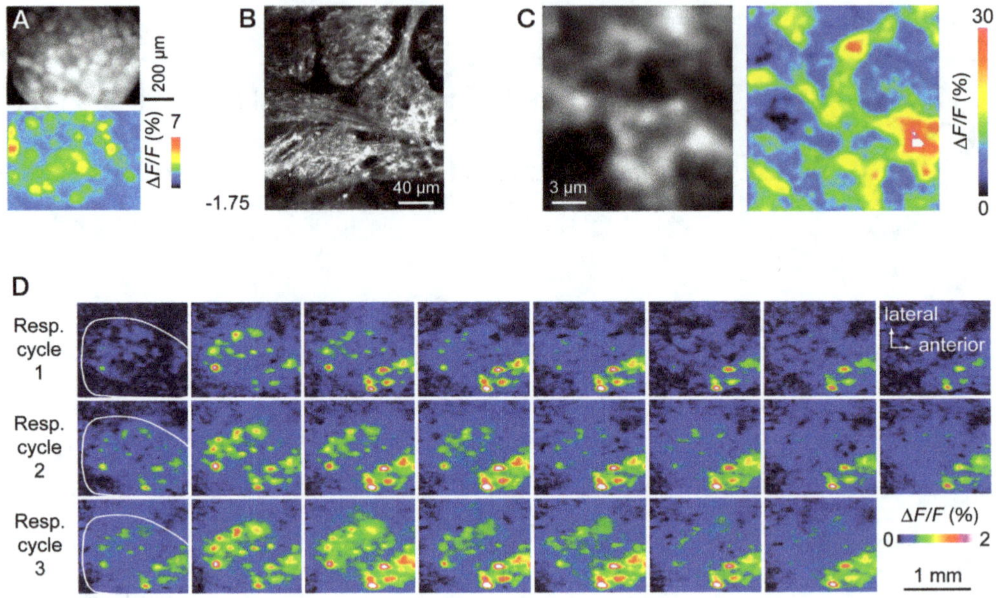

FIGURE 1. Imaging of glomerular responses performed through the thinned skull of a mouse after loading of olfactory sensory neurons with Calcium Green dextran (10 kDa). (*A*) Wide-field fluorescence image of glomeruli (*top*) and response pattern to odor stimulation with 2% 2-hexanone (*bottom*). (*B*) Multiphoton image of glomeruli. (*C*) Multiphoton imaging of fluorescence within a glomerulus (*left*) and map of odorant-evoked Ca^{2+} signals within the same region (*right*). (*D*) Wide-field imaging of response dynamics across respiration cycles (77 msec/frame). Note the phase differences of responses between glomeruli. (*A–C*, Modified, with permission, from Wachowiak et al. 2004, ©National Academy of Sciences; *D*, modified, with permission, from Spors et al. 2006, ©Society for Neuroscience.)

indicator into OSNs (Friedrich and Korsching 1997; Wachowiak and Cohen 2001) (Fig. 1). The advantages of this approach are (1) the selectivity of imaging of sensory input to glomeruli, (2) sufficient temporal resolution to reveal dynamics of activity patterns (e.g., within sniff cycles [Spors et al. 2006]), (3) the potential for high-resolution mapping of Ca^{2+} signals within individual glomeruli by multiphoton microscopy (Wachowiak et al. 2004), and (4) the potential for applications in awake animals (Verhagen et al. 2007; Wesson et al. 2008). Downstream from glomeruli, odor-evoked activity has been measured by multiple imaging approaches, including bolus loading of AM-ester dyes (Yaksi and Friedrich 2006; Yaksi et al. 2007; Petzold et al. 2009) and genetically encoded activity probes (Fiala et al. 2002; Ng et al. 2002; Wang et al. 2003; Hasan et al. 2004; Yu et al. 2004; Shang et al. 2007).

LOADING OF CALCIUM INDICATORS

Selective Loading of Olfactory Sensory Neurons Projecting to Glomeruli

The loading of dextran-coupled Ca^{2+} indicators into vertebrate OSNs is based on the reversible ablation of cilia and microvilli by a nasal lavage with a low concentration of Triton X-100 (Cancalon 1983; Adamek et al. 1984), which allows a dextran-coupled Ca^{2+} indicator to enter OSNs. Subsequently, cilia and microvilli regenerate and the dye spreads throughout the axon; the dextran prevents extrusion (Friedrich and Korsching 1997; Wachowiak and Cohen 2001). This approach has been used successfully in a variety of vertebrate species, including zebrafish (Friedrich and Korsching 1997), turtle (Wachowiak et al. 2002), mouse (Wachowiak and Cohen 2001; Wachowiak et al. 2004; Spors et al. 2006), and rat (Verhagen et al. 2007; Wesson et al. 2008). Odor-evoked Ca^{2+} signals within glomeruli were observed between 2 and 8 d after dye loading and most likely reflect Ca^{2+} influx near synaptic release sites (Wachowiak et al. 2004). The highest signal-to-noise ratios were obtained with Oregon Green BAPTA-1 and Calcium Green-1 coupled to a 10-kDa dextran.

Bolus Loading of Neurons Downstream from Glomeruli

In vertebrates, synthetic Ca^{2+} indicators can be efficiently loaded into populations of neurons by AM-ester bolus loading (Brustein et al. 2003; Stosiek et al. 2003). Careful optimization of the loading procedure can be beneficial, albeit time-consuming. Good results are usually achieved by determining the bolus injection when a predetermined target fluorescence intensity is reached. The target intensity is determined separately for different brain areas by finding dye and light levels that do not cause detectable toxicity. Two simple methods to assess toxicity are described here; further tests for toxicity include electrophysiology. First, a fluorescent marker for necrotic cells (e.g., SYTOX Green or sulforhodamine 101) can be co-injected with the Ca^{2+} indicator to detect cell death inflicted mechanically or by the imaging light (Yaksi and Friedrich 2006 [see Supplementary Methods]). Second, under multiphoton optics, the laser power can be adjusted to keep fluorescence intensity in the absence of stimulation constant over extended periods of time (~20 sec). If intensity slowly increases or decreases, the excitation light can exert direct effects on neurons and should be reduced.

Genetically Encoded Calcium Indicators and Other Probes

In insects, bolus loading by injection of AM-ester dyes has not yet been described. Instead, Ca^{2+} indicators were introduced by injections into axon tracts (Sachse and Galizia 2002) or by expression of genetically encoded probes (Fiala et al. 2002; Wang et al. 2003; see Chapter 54). Genetically encoded Ca^{2+} indicators have also been used to measure odor-evoked activity in vertebrates (Hasan et al. 2004; Li et al. 2005).

IMAGING METHODS

Wide-Field Fluorescence Imaging

In most species, conventional epifluorescence microscopy using low-noise charge-coupled device (CCD) cameras is sufficient to achieve single-glomerulus resolution. Because the field of view is usually large, the total light intensity is often high. Because many CCD cameras use chips with low well depths that are optimized for low-light applications, it may be desired to select cameras with fast frame rates and/or large well depths to collect all available photons.

Multiphoton Imaging

High-resolution optical methods, such as multiphoton microscopy, are used to resolve subglomerular structures and to measure Ca^{2+} signals from individual neurons. In rodents, mitral cells are located relatively deep and imaging is often performed conveniently through the thinned skull. For such applications the recommendation is to use objectives with high numerical apertures (NAs) and large fields of view (e.g., 20×, NA 0.95; Olympus; 20×, NA 1.0, Zeiss; 16×, NA 0.8, Nikon), and to optimize detection optics for the collection of scattered photons (Oheim et al. 2001).

Accessory Methods

Efficient animal preparation and odor delivery methods are key to the success of Ca^{2+} imaging experiments in the olfactory system. Procedures have been described for zebrafish (Friedrich and Korsching 1997; Yaksi and Friedrich 2006), anesthetized rodents (Rubin and Katz 1999; Wachowiak and Cohen 2001), awake rats with immobilized heads (Verhagen et al. 2007), turtles (Wachowiak and Cohen 1999; Lam et al. 2000), *Drosophila* (Fiala et al. 2002; Yu et al. 2004; Asahina et al. 2009), and other insects (Joerges et al. 1997; Sachse and Galizia 2002).

Protocol

Loading of Dextran-Coupled Calcium Indicators into Olfactory Sensory Neurons

Separate procedures are described for zebrafish (Friedrich and Korsching 1997) and mice (Wachowiak and Cohen 2001). Expect some variability between animals. Differences between batches of dye can be another source of variability.

IMAGING SETUP

The zebrafish OB can be imaged in an explant preparation of the entire brain with nose attached (Friedrich et al. 2006) using an inverted or upright microscope. The entire OB can be imaged using a 10x objective; in many cases, subregions are imaged using 20x or 40x objectives. Whole-field imaging is usually done with sensitive CCD cameras (e.g., CoolSnap HQ, Photometrics). Light intensities are in the low range and do not usually saturate most CCD cameras. Typical pixel resolutions used after on-chip binning are 128 x 128 or 170 x 170 pixels. Typical frame rates used for acquisition are 2–20 Hz.

Wide-field imaging of the dorsal OB of the mouse through a craniotomy is usually done using an upright microscope with low-magnification lenses (e.g., 2.5x–10x) or a tandem photo lens system ("macroscope," e.g., from RedShirtImaging). It is advantageous to use a camera with high frame rate and/or well depth (e.g., NeuroCCD models, RedShirtImaging) because light levels are substantially higher than in zebrafish and can saturate cameras optimized for low-light applications. Typical pixel resolutions used are between 80 x 80 and 256 x 256 pixels. Typical frame rates are between 2 and 200 Hz. High frame rates are sometimes used to avoid camera saturation and reduced before storage to disk to reduce file size.

Light sources for wide-field imaging should be stablized arc lamps (e.g., 150 W Xenon with stabilized power supply, OptiQuip) or halogen lamps (e.g., 100 W with stabilized power supply, Kepco). To avoid bleaching and phototoxicity, it may be necessary to reduce light levels by neutral density filters to 1.5%–50% of full intensity.

Multiphoton imaging in zebrafish and mice is performed using mode-locked Ti:Sapphire lasers (e.g., Cameleon, Coherent; MaiTai, Spectra Physics) and long-distance water-immersion lenses (e.g., 16x, NA 0.8, Nikon; 20x, NA 0.95, Olympus; 20x, NA 1.0, Zeiss) in an upright microscope. It is recommended to use external detection optics with high collection efficiency for scattered photons, particularly for deep imaging in the mouse. Appropriate microscope designs may be offered by commercial suppliers (e.g., Sutter Instruments, ThorLabs, Prairie Technologies) or custom-built. Custom solutions can be based on components of commercially available microscopes (e.g., the body of BX-51, Olympus) and have the advantage that they can be optimized for specific needs. A variety of free software packages are available for the control of custom microscopes (e.g., Scanimage, http://research.janelia.org/wiki/display/ephus/ScanImage%20 [Pologruto et al. 2003]; MPScope, http://physics.ucsd.edu/neurophysics/links.html [Nguyen et al. 2006]). Because odor-evoked glomerular activation patterns can be sparse, it is often very useful to first localize activated glomeruli in wide-field mode and then zoom into the regions of interest in multiphoton mode. A microscope that can be switched quickly between wide-field and multiphoton modes is therefore highly recommended. Typical pixel resolutions and frame rates are between 64 x 64 and 256 x 256 pixels and between 2 and 12 Hz, respectively. Frame rates can be further increased by reducing the field of view (number of pixels) along one dimension (e.g., 8 x 64 pixels at 62.5 Hz).

MATERIALS

CAUTION: See Appendix 6 for proper handling of materials marked with <!>.
See the end of the chapter for recipes for reagents marked with <R>.

Reagents

Experimental animals (e.g., mice, zebrafish)
Ketamine <!> (for mouse studies only)
OSN dye solution <R>
   Prepare for mice or zebrafish, as appropriate.
Petroleum jelly (e.g., Vaseline) (for mouse studies only)
Tricaine methanesulfonate <!> (MS-222; 0.01%) (for zebrafish studies only)
   Adjust the pH as necessary.
Xylazine <!> (for mouse studies only)

Equipment

Heating pad or lamp (for mouse studies only)
Microscope, dissection
Paper towels (for zebrafish studies only)
Pipette tips, 200-μL (for zebrafish studies only)
Pipette tips, GELoader (Eppendorf)
Syringe, Hamilton, 10-μL

EXPERIMENTAL METHOD

Loading of Dextran-Coupled Calcium Indicators into Mouse OSNs

1. Anesthetize a mouse with ketamine (90 mg/kg) and xylazine (10 mg/kg).
2. Place dorsal-side down under a dissection microscope. Keep the animal warm using a heating pad or lamp.
3. Attach a GELoader pipette tip (Eppendorf) coated with a thin layer of petroleum jelly to a Hamilton syringe (10 μL). Make a mark about 10 mm from the tip. Load 10 μL of dye solution.
4. Insert the pipette tip into one nostril as far as possible, then retract by ~1 mm.
   The mark should be close to the nostril.
5. At 1-min intervals, inject 2 μL of dye solution (five injections total).
6. Position the head such that the snout is roughly vertical to ensure spreading of the solution within the nasal cavity.
7. Return the mouse to its cage when it starts to recover from anesthesia (~15 min).
8. Allow at least 4 d for regeneration of cilia and microvilli.
9. Perform odor stimulation and Ca^{2+} imaging, following appropriate procedures (see, e.g., Wachowiak and Cohen 2001; Wachowiak et al. 2004; Spors et al. 2006).

Loading of Dextran-Coupled Calcium Indicators into Zebrafish OSNs

10. Anesthetize a zebrafish with 0.01% MS-222.
11. Place under dissection microscope on a stage inclined by ~20°.

12. Perfuse the gills with fish water containing 0.01% MS-222 via a cannula (e.g., a 200-µL pipette tip) inserted in the mouth.
13. Keep fish moist in a wet paper towel and stabilize laterally. Avoid water flow over the nostrils.
14. Cut a GELoader pipette tip at an angle of ~45° to ~1.5 cm in length. Load 1 µL of dye solution.
15. Insert the pipette tip under the skin over the nasal epithelium from the posterior through the outflow opening. Inject the dye.

    Verify visually that the dye is in the nasal epithelium.
16. Repeat Step 15 on the contralateral side if desired.
17. Return the fish to its tank after 5 min.
18. Allow at least 3 d for regeneration of cilia and microvilli.
19. Perform odor stimulation and Ca^{2+} imaging following appropriate procedures (see, e.g., Friedrich and Korsching 1997).

DISCUSSION

Alternative approaches for the visualization of glomerular activity patterns include imaging of intrinsic signals and the genetically encoded reporter of transmitter release, synapto-pHluorin. Both these methods have lower temporal resolution than Ca^{2+} imaging but do not require dye delivery. Synapto-pHluorin has also been used to measure synaptic activity downstream from glomeruli in the olfactory system of insects (Ng et al. 2002; Yu et al. 2004; Shang et al. 2007). Voltage-sensitive dyes applied to the OB surface allow for high temporal resolution, but the signal might not be specific for glomeruli (Cinelli et al. 1995; Lam et al. 2000; Spors and Grinvald 2002). This can be overcome by tracing OSN axons with a voltage-sensitive dye from the nose using dye application procedures similar to those described above but without detergent (Friedrich and Korsching 1998).

Bolus loading of AM-ester dyes and multiphoton imaging (Yaski et al. 2007) allows the topology and dynamics of odor-evoked activity patterns across thousands of individual neurons to be analyzed. Using temporal deconvolution, the measured Ca^{2+} signals can be converted into firing rate changes, which increases the effective temporal resolution and allows for a more direct assessment of neuronal activity (Yaksi and Friedrich 2006). It should be noted that this procedure requires a very high signal-to-noise ratio in the optical data. Cell type identification can be achieved by combining the red fluorescent Ca^{2+} indicator Rhod-2 AM with green or yellow fluorescing marker proteins (Yaksi and Friedrich 2006). The use of Rhod-2 requires careful optimization of loading protocols but provides an excellent signal-to-noise ratio and dynamic range.

Genetically encoded Ca^{2+} indicators have been used extensively in invertebrates and are expected to become increasingly more important for studies in vertebrates. Obvious advantages of genetically encoded probes are the ability to target selected neurons, the potentially noninvasive loading procedure, and the potential for long-term studies.

RECIPE

CAUTION: See Appendix 6 for proper handling of materials marked with <!>.
Recipes for reagents marked with <R> are included in this list.

OSN Dye Solution

Reagent	Mouse	Zebrafish
NaCl	1–10 mM	1–10 mM
Oregon Green 488 BAPTA-1 dextran (10 kDa)	2%–6% (w/v)	2%–6% (w/v)
Triton X-100 <!>	0.25%	0.1%

This solution can be kept frozen for at least 1 wk.

REFERENCES

Adamek GD, Gesteland RC, Mair RG, Oakley B. 1984. Transduction physiology of olfactory receptor cilia. *Brain Res* **310:** 87–97.

Asahina K, Louis M, Piccinotti S, Vosshall LB. 2009. A circuit supporting concentration-invariant odor perception in *Drosophila*. *J Biol* **8:** 9.

Bozza T, McGann JP, Mombaerts P, Wachowiak M. 2004. In vivo imaging of neuronal activity by targeted expression of a genetically encoded probe in the mouse. *Neuron* **42:** 9–21.

Brustein E, Marandi N, Kovalchuk Y, Drapeau P, Konnerth A. 2003. "In vivo" monitoring of neuronal network activity in zebrafish by two-photon Ca^{2+} imaging. *Pflügers Arch* **446:** 766–773.

Cancalon P. 1983. Influence of a detergent on the catfish olfactory mucosa. *Tissue Cell* **15:** 245–258.

Cinelli AR, Hamilton KA, Kauer JS. 1995. Salamander olfactory bulb neuronal activity observed by video rate, voltage-sensitive dye imaging. III. Spatial and temporal properties of responses evoked by odorant stimulation. *J*

Neurophysiol **73:** 2053–2071.

Fiala A, Spall T, Diegelmann S, Eisermann B, Sachse S, Devaud J-M, Buchner E, Galizia CG. 2002. Genetically expressed cameleon in *Drosophila melanogaster* is used to visualize olfactory information in projection neurons. *Curr Biol* **12:** 1877–1884.

Friedrich RW, Korsching SI. 1997. Combinatorial and chemotopic odorant coding in the zebrafish olfactory bulb visualized by optical imaging. *Neuron* **18:** 737–752.

Friedrich RW, Korsching SI. 1998. Chemotopic, combinatorial and noncombinatorial odorant representations in the olfactory bulb revealed using a voltage-sensitive axon tracer. *J Neurosci* **18:** 9977–9988.

Hasan MT, Friedrich RW, Euler T, Larkum ME, Giese G, Both M, Duebel J, Waters J, Bujard H, Griesbeck O, et al. 2004. Functional fluorescent Ca^{2+} indicator proteins in transgenic mice under TET control. *PLoS Biol* **2:** e163.

Joerges J, Küttner A, Galizia CG, Menzel R. 1997. Representations of odours and odour mixtures visualized in the honeybee brain. *Nature* **387:** 285–288.

Lam Y-W, Cohen LB, Wachowiak M, Zochowski MR. 2000. Odors elicit three different oscillations in the turtle olfactory bulb. *J Neurosci* **20:** 749–762.

Li J, Mack JA, Souren M, Yaksi E, Higashijima S, Mione M, Fetcho JR, Friedrich RW. 2005. Early development of functional spatial maps in the zebrafish olfactory bulb. *J Neurosci* **25:** 5784–5795.

Ng M, Roorda RD, Lima SQ, Zemelman BV, Morcillo P, Miesenböck G. 2002. Transmission of olfactory information between three populations of neurons in the antennal lobe of the fly. *Neuron* **36:** 463–474.

Nguyen QT, Tsai PS, Kleinfeld D. 2006. MPScope: A versatile software suite for multiphoton microscopy. *J Neurosci Methods* **156:** 351–359.

Oheim M, Beaurepaire E, Chaigneau E, Mertz J, Charpak S. 2001. Two-photon microscopy in brain tissue: Parameters influencing the imaging depth. *J Neurosci Methods* **111:** 29–37.

Petzold GC, Hagiwara A, Murthy VN. 2009. Serotonergic modulation of odor input to the mammalian olfactory bulb. *Nat Neurosci* **12:** 784–791.

Pologruto TA, Sabatini BL, Svoboda K. 2003. ScanImage: Flexible software for operating laser scanning microscopes. *BioMed Eng OnLine* **2:** 13.

Rubin BD, Katz LC. 1999. Optical imaging of odorant representations in the mammalian olfactory bulb. *Neuron* **23:** 499–511.

Sachse S, Galizia CG. 2002. Role of inhibition for temporal and spatial odor representation in olfactory output neurons: A calcium imaging study. *J Neurophysiol* **87:** 1106–1117.

Shang Y, Claridge-Chang A, Sjulson L, Pypaert M, Miesenböck G. 2007. Excitatory local circuits and their implications for olfactory processing in the fly antennal lobe. *Cell* **128:** 601–612.

Spors H, Grinvald A. 2002. Spatio-temporal dynamics of odor representations in the mammalian olfactory bulb. *Neuron* **34:** 301–315.

Spors H, Wachowiak M, Cohen LB, Friedrich RW. 2006. Temporal dynamics and latency patterns of receptor neuron input to the olfactory bulb. *J Neurosci* **26:** 1247–1259.

Stosiek C, Garaschuk O, Holthoff K, Konnerth A. 2003. In vivo two-photon calcium imaging of neuronal networks. *Proc Natl Acad Sci* **100:** 7319–7324.

Verhagen JV, Wesson DW, Netoff TI, White JA, Wachowiak M. 2007. Sniffing controls an adaptive filter of sensory input to the olfactory bulb. *Nat Neurosci* **10:** 631–639.

Wachowiak M, Cohen LB. 1999. Presynaptic inhibition of primary olfactory afferents mediated by different mechanisms in lobster and turtle. *J Neurosci* **19:** 8808–8817.

Wachowiak M, Cohen LB. 2001. Representation of odorants by receptor neuron input to the mouse olfactory bulb. *Neuron* **32:** 723–735.

Wachowiak M, Cohen LB, Zochowski MR. 2002. Distributed and concentration-invariant spatial representations of odorants by receptor neuron input to the turtle olfactory bulb. *J Neurophysiol* **87:** 1035–1045.

Wachowiak M, Denk W, Friedrich RW. 2004. Functional organization of sensory input to the olfactory bulb glomerulus analyzed by two-photon calcium imaging. *Proc Natl Acad Sci* **101:** 9097–9102.

Wang JW, Wong AM, Flores J, Vosshall LB, Axel R. 2003. Two-photon calcium imaging reveals an odor-evoked map of activity in the fly brain. *Cell* **112:** 271–282.

Wesson DW, Carey RM, Verhagen JV, Wachowiak M. 2008. Rapid encoding and perception of novel odors in the rat. *PLoS Biol* **6:** e82.

Yaksi E, Friedrich RW. 2006. Reconstruction of firing rate changes across neuronal populations by temporally deconvolved Ca^{2+} imaging. *Nat Methods* **3:** 377–383.

Yaksi E, Judkewitz B, Friedrich RW. 2007. Topological reorganization of odor representations in the olfactory bulb. *PLoS Biol* **5:** e178.

Yang X, Renken R, Hyder F, Siddeek M, Greer CA, Shepherd GM, Shulman RG. 1998. Dynamic mapping at the laminar level of odor-elicited responses in rat olfactory bulb by functional MRI. *Proc Natl Acad Sci* **95:** 7715–7720.

Yu D, Ponomarev A, Davis RL. 2004. Altered representation of the spatial code for odors after olfactory classical conditioning: Memory trace formation by synaptic recruitment. *Neuron* **42:** 437–449.

56 Calcium Imaging in Populations of Olfactory Neurons by Planar Illumination Microscopy

Timothy E. Holy

Department of Anatomy and Neurobiology, Washington University in St. Louis School of Medicine, St. Louis, Missouri 63110

ABSTRACT

Neurons in the olfactory system display extraordinary functional diversity, which at the level of sensory neurons arises from the expression of one out of several hundred distinct receptor types. To cope with this diversity, one approach is to use techniques that can record sensory responses from many neurons simultaneously. We have developed a form of light-sheet microscopy, called objective-coupled planar illumination (OCPI) microscopy, that is well suited to recording at high signal-to-noise ratios from large neuronal populations. This chapter describes the theory, advantages, and practical implementation of planar illumination and briefly discusses its application to neuronal recording in the mouse vomeronasal organ.

INTRODUCTION

Theory

Imaging of neuronal activity is predominantly based on fluorescence microscopy. The two most commonly used types of instruments for imaging are wide-field (Fig. 1A) and laser-scanning (Fig. 1B) microscopes. Both of these techniques have been enormously productive, yet each has its own drawbacks. For three-dimensional samples, the main weakness of wide-field microscopy is that out-of-focus portions of the sample also emit light, reducing the ability to resolve the in-focus signal. This motivates laser-scanning microscopy, typified by confocal or two-photon microscopy, in which the illumination is focused to a point and an image is assembled sequentially by scanning this point over the sample (point scanning). However, point scanning is relatively slow over large fields of view: Whereas there have been many strategies for "fast" point scanning, optophysiological studies have been remarkably uniform in devoting ~5–10 µsec to each voxel (three-dimensional pixel) (Nguyen et al. 2001; Lechleiter et al. 2002; Salome et al. 2006; Yaksi and Friedrich 2006; Göbel et al. 2007; Vucinic and Sejnowski 2007; Duemani Reddy et al. 2008). This allows for fast imaging over small numbers of pixels, but a 1-megapixel image would require 5–10 sec, a period that is long on the timescale of neuronal activity.

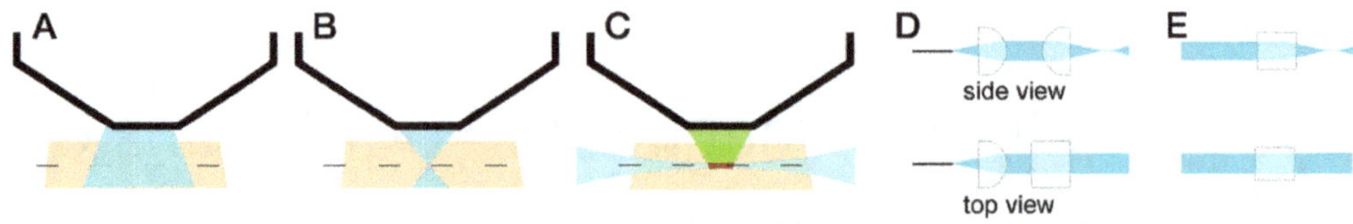

FIGURE 1. Illumination methods in fluorescence microscopy. (*A*) Epifluorescence. (*B*) Laser-scanning microscopy. (*C*) Planar illumination microscopy. In all images, the dashed gray line indicates the focal plane of the objective. (*D*) A design for a fiber-coupled illumination system. A plano-convex lens and cylindrical lens (approximately matched in terms of radius of curvature, to minimize aberrations) act to collimate and focus the light sheet, respectively. (*E*) A design for a system using a collimated beam input. A collimated beam impinges upon a uniaxial gradient index (GRIN) lens.

Why does fluorescence, at least for the purpose of measuring neuronal activity, seem to be limited to per-voxel integration times in the microsecond range? In fact, point-scanning microscopes have been built that can acquire frames at video rates (Nguyen et al. 2001). However, the resulting images are rather noisy, making it difficult, for example, to quantify fluorescence intensity with enough fidelity to observe physiologically relevant changes by calcium imaging. By integrating over a longer time for each voxel, one reduces the noise, eventually to a level that produces useful images.

The main source of these noisy images is the randomness of photon emission, which can be accurately described as a Poisson process. The expected number of photons, N, will be observed with a variance equal to the mean, or a standard deviation of $\sqrt{N}$. Consequently, to reliably measure a 10% change in fluorescence, one needs a minimum of 100 photons; to detect a 1% change in fluorescence, one needs at least 10,000 photons.

The limit on speed arises because it takes time to collect these photons. Inside a single voxel, the fluorophores undergo many excitation-emission cycles. The rate of excitation is a function of the light intensity, so under some circumstances one can increase the brightness of the image by increasing the laser intensity. However, in the focal spot of a point-scanning laser, it is common to saturate the fluorophores with light, so further increases in laser intensity do not make the image brighter. Under these conditions, emission is the rate-limiting step. A theoretical analysis (Holekamp et al. 2008) indicates that for optophysiological imaging at the diffraction limit, minimum dwell times should be ~1 µsec. This suggests that current point-scanning fluorescence microscopes may be within an order of magnitude of their maximum performance.

Imaging Multiple Voxels Simultaneously: From Epifluorescence to Planar Illumination

To circumvent this physical limit on imaging speed, the most common solution is to illuminate multiple voxels at once. Epifluorescence acquires the entire image in parallel. If one requires optical sectioning, there are a growing number of approaches, such as spinning disk and slit-scanning confocal microscopy (Pétran et al. 1968; Lichtman et al. 1989; Masters and Thaer 1994), deconvolution (Conchello and Lichtman 2005), and structured illumination (Neil et al. 1997; Kner et al. 2009). These are quite powerful, but all share an important limitation: Light bathes the entire preparation, rather than just the in-focus plane. Consequently, photobleaching and phototoxicity will occur at rates faster than one would experience if illumination could be restricted to the in-focus plane. This problem is resolved by using multiphoton effects while illuminating multiple points in parallel (Bewersdorf et al. 1998; Nikolenko et al. 2008). This idea has many advantages and will likely play a growing role in applications. However, the illumination power required for multiphoton excitation is nontrivial, especially when multiplied by many focal spots, and it seems likely that this could serve as a constraint.

To overcome the individual disadvantages of epifluorescence and confocal microscopy, several groups (Seidentopf and Zsigmondy 1903; Voie et al. 1993; Fuchs et al. 2002; Huisken et al. 2004; Dodt et al. 2007; Holekamp et al. 2008) have independently conceived a sort of hybrid approach: to illuminate the entire field of view using a thin sheet of light projected along the focal plane of the objective (Fig. 1C). This approach, alternately called light-sheet microscopy or planar illumination microscopy, provides optical sectioning without the speed trade-offs of confocal microscopy. In the nervous system, planar illumination has been used both to image structure (Dodt et al. 2007) and to record activity via fluorescent calcium indicators (Holekamp et al. 2008).

The light sheet may be formed by passing a collimated beam through a lens that focuses along only one axis, such as a cylindrical lens (Huisken et al. 2004; Dodt et al. 2007; Holekamp et al. 2008) or a uniaxial gradient-index (GRIN) lens (Turaga and Holy 2008). This light sheet should be positioned in the focal plane of the objective. In our implementations (Holekamp et al. 2008; Turaga and Holy 2008), the optics that form the light sheet are coupled to the objective, thereby allowing the light sheet to remain in the focal plane as the objective is moved up and down. To image large samples, the microscope is tilted relative to the sample so that both excitation and emission occur at an ~45° angle to the tissue surface (Holekamp et al. 2008). The 45° tilt is used to minimize the distance that the illumination has to travel through tissue.

Examples Demonstrating the Properties of OCPI Microscopy

The main advantage of OCPI microscopy is its speed. At the theoretical limit, one still requires a minimum integration time of 5–10 µsec/frame, but this implies a maximum of 100,000 frames/sec. Crucially, the speed of OCPI microscopy comes without cost to signal-to-noise ratio: We have detected fluorescence changes ($\Delta F/F$) near 0.1% at hundreds of (subregion) frames per second. This has allowed us to simultaneously record putative action potentials, following each event in a 20-Hz burst, over a field of approximately 100 neurons (Fig. 2A).

By coupling the illumination to the objective, it becomes possible to quickly image a three-dimensional volume by moving both together with a piezoelectric objective positioner. We have used OCPI to image volumes hundreds of micrometers on a side every few seconds, collecting the activity from ~700 neurons simultaneously (Fig. 2B,C).

Aside from speed, the other major advantage of planar illumination is that, as in two-photon microscopy, little photobleaching or phototoxicity occurs in out-of-focus regions of the specimen (Fig. 2D). Given that phototoxicity constrains the total number of frames that can be acquired from the preparation, this represents a significant advantage.

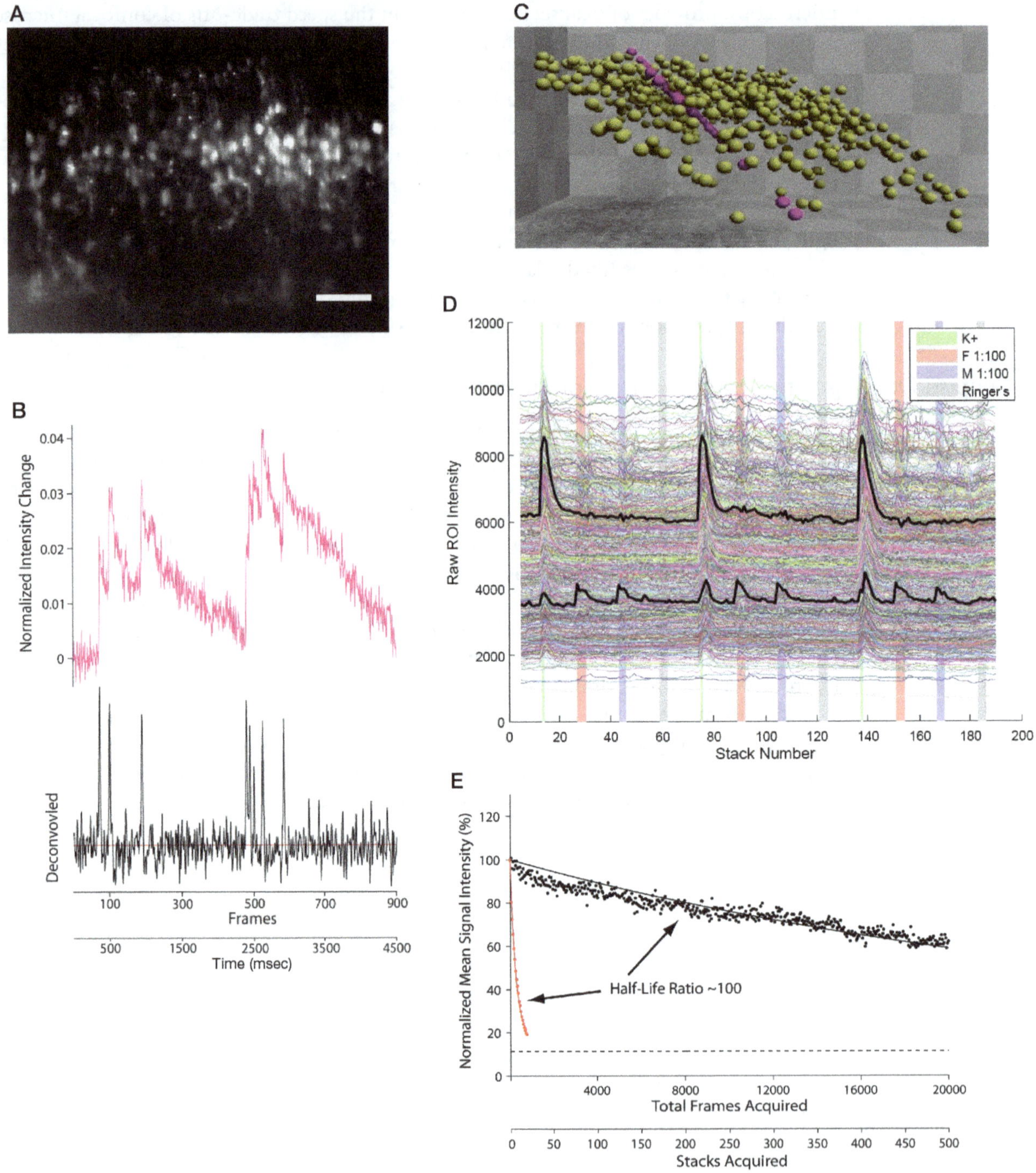

FIGURE 2. Fast, low-phototoxicity calcium imaging in vomeronasal neurons by OCPI microscopy. (*A*) An image of the vomeronasal organ labeled with Oregon Green BAPTA-1. (*B*) Fluorescence in a single neuron recorded at 200 Hz; 87 other neurons were simultaneously recorded. Temporal deconvolution reveals putative action potentials. Note that the noise level is ~0.1%. (*C*) A three-dimensional region containing nearly 700 neurons, the soma of each indicated by a sphere; the purple spheres lie in a single focal plane. (*D*) Fluorescence intensity in these neurons over time, during presentation of different olfactory stimuli. K+ is high-potassium (50 mM, equimolar substitution for sodium) Ringer's solution. F 1:100 and M 1:100 are mouse urine diluted 1:100 in Ringer's solution from females and males, respectively. (*E*) Fluorescein bead intensity as a function of stack number for both confocal microscopy (red dots) and OCPI microscopy (black dots). Photobleaching occurs much faster with confocal microscopy.

Protocol

Calcium Imaging in Populations of Olfactory Neurons by Planar Illumination Microscopy

This protocol describes a method for imaging olfactory neurons by OCPI microscopy. Because OCPI microscopy illuminates the entire field simultaneously, it allows fast imaging without compromising field of view. At current camera speeds, pixels can be acquired more than 100-fold faster than by point-scanning fluorescence microscopy. In this protocol, the animal is pretreated with dye for 1 wk to allow labeling of the sensory neurons prior to stimulation and imaging.

IMAGING SETUP

There are many possible variations on the principle of OCPI microscopy. The most basic implementation is relatively inexpensive and requires only a modest number of custom parts. The necessary components are listed in Table 1.

The microscope proper consists of two major components: the objective and the tube lens. These can be mounted on a stereomicroscope boom stand and (if tilting is desired) may require simple custom-built mounting hardware. We connect the objective to a piezoelectric positioner, which mounts to a custom block that can accept the barrier filter, which then connects to the tube lens, which is screwed into the C-mount of the camera.

Most of the custom work is devoted to the illumination. The greatest flexibility is obtained with a fiber-coupled system. Some care needs to be devoted to the optics forming the light sheet, both to control the geometry of the sheet and to minimize aberrations that would otherwise widen the sheet. The minimum thickness of the sheet is determined by the numerical aperture (NA) of the sheet-forming lens (for details, see Holekamp et al. 2008). For a field a few hundred micrometers in size, an illumination NA near 0.05 is appropriate. To limit aberrations, the collimating lens should be matched to the "cylindrical" lens. For example, if one plans to use an ordinary glass cylindrical lens to form the sheet, one should choose a simple plano-convex lens to collimate the output of the fiber (Holekamp et al. 2008). The two lenses should be made of the same material, should have similar

TABLE 1. Components needed for OCPI microscopy

Commercially available components	Custom components
Vibration-isolation table	Light-sheet-forming optics
Stereomicroscope boom stand	Objective coupler
Microscope objective (water-immersion, long-working-distance)	Mounting hardware: piezo to tube lens (perhaps via emission filter mounting block)
Tube lens (e.g., Infinity, Edmund Optics, Thorlabs, Inc.)	Mounting hardware: system to boom stand
Camera (CCD or CMOS) (good low-cost choice: an uncooled camera based on Sony's ICX285 sensor)	
Laser coupled to a single-mode fiber (with shutter, and control over delivered power over a range from ~10 µW to ~1 mW)	
Piezoelectric objective positioner (e.g., Physik Instrumente, Piezosystem Jena)	
Emission filter	

OCPI, objective-coupled planar illumination; CCD, charge-coupled device; CMOS, complementary metal oxide semiconductor.

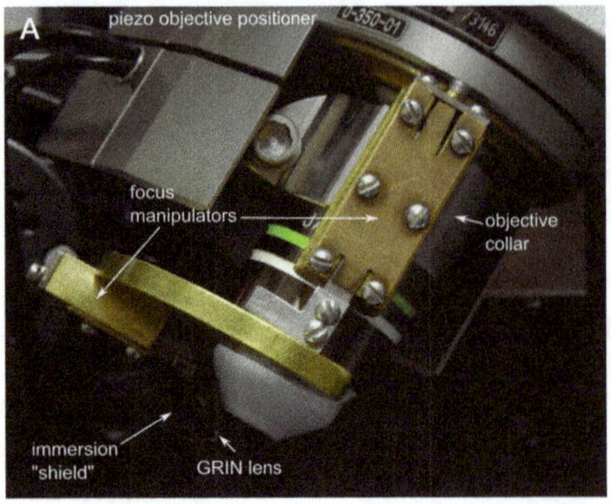

FIGURE 3. (*A*) Photograph of a coupler design for OCPI microscopy. This example takes as input the collimated beam from a laser. Two manipulators, controlling the position relative to the central focal spot of the objective, are visible (brass). The GRIN lens that forms the illumination sheet is barely visible projecting beyond the black shield, whose purpose is to prevent the immersion liquid from flowing to cover the back side of the GRIN lens. The piezoelectric manipulator allows scanning along the z-axis. (*B*) Schematic of the GRIN lens and immersion shield attached to the mount, which in turn is movable by the focus manipulators.

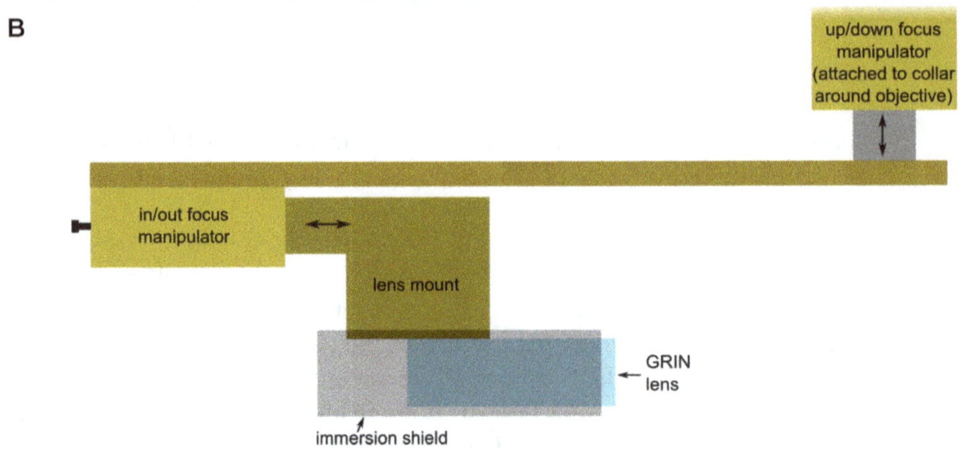

radii of curvature, and should be placed close together; in this way, the aberrations of these two unsophisticated lenses will largely cancel. Alternatively, if the optical fiber is terminated with a GRIN-lens collimator, one should use a higher-quality lens to form the sheet; a uniaxial GRIN lens (cylindrical type) is a good choice here (Turaga and Holy 2008). A further advantage of the GRIN lens is its small size; however, one must be alert for reflections off the internal faces of the lens.

Ideally, the custom coupler should possess manipulators to position the light sheet (Fig. 3). Tuning these manipulators is best done while examining a sample of fluorescent 200-nm beads in agarose. One first positions the light sheet in the focal plane simply by adjusting it to produce the best-looking image with the tightest focus. The position of the waist along the axis of propagation is tuned by minimizing the thickness of the light sheet; to determine the thickness, move the objective with the piezoelectric objective positioner and measure the full width at half-maximum of a single bead's intensity. Once set, these manipulators do not need to be adjusted again in normal operation.

We have used both electron-multiplying charge-coupled device (EMCCD) cameras (Andor Technology) and non-EMCCD cameras (Point Grey) for imaging. At signal-to-noise ratios sufficient to detect few-percent changes in $\Delta F/F$, it seems that electron multiplying is not a major advantage, because the number of photons that must be collected makes a readout noise of less than 10 electrons largely irrelevant. At high speeds, cooling is likewise a nonissue. Consequently, even relatively inexpensive (but fast) cameras are now suitable for OCPI microscopy. We typically store the acquired data to drives in a hot-swappable serial ATA (SATA) drive tower, which have the capacity for the hundreds of gigabytes that can be generated in a single experiment.

MATERIALS

CAUTION: See Appendix 6 for proper handling of materials marked with <!>.

Reagents

Dextran-conjugated Oregon Green BAPTA-1 <!>
Mice for imaging
Olfactory stimuli to be tested
Triton X-100 <!>

Equipment

Flow chamber (see Step 2)
Imaging setup (see description above)

EXPERIMENTAL METHOD

1. Prelabel the sensory neurons in vivo by injecting a mixture of 6% dextran-conjugated Oregon Green BAPTA-1 and 0.25% Triton X-100 into the lumen of the vomeronasal organ.
2. After a 1-wk recovery period, kill the animal, excise the tissue from the nose, and place in a custom flow chamber with walls high enough to allow immersion of both the objective and illumination lenses.

   The entire microscope is tilted at an angle (Holekamp et al. 2008) to allow horizontal superfusion of the tissue.

3. Apply olfactory stimuli in the superfusion and record activity in neuronal populations.

   We typically apply stimuli in full three-dimensional volumes (Fig. 2) with stacks of 40 images collected every few seconds.

DISCUSSION

OCPI microscopy's main advantages are (1) it circumvents the trade-off between speed and signal-to-noise ratio, allowing very high signal-to-noise imaging at high speeds, and (2) it is relatively efficient with photons, so there is little in the way of unnecessary photobleaching or phototoxicity. These two advantages are substantial, particularly for neuronal imaging. Although the technique is still relatively new, it seems likely that development and applications of OCPI microscopy will continue to grow.

OCPI microscopy is not without its disadvantages. The major ones are (1) the need for access from two perpendicular faces is restrictive for some potential applications, although we have found (D. Turaga and H.E. Holy, unpubl. results) that this does not prevent the technique from being applied in vivo (see also Dunsby 2008); (2) optical aberrations are induced by imaging through tilted surfaces (but see Turaga and Holy 2008; D. Turaga and H.E. Holy, unpubl. results); and (3) because OCPI microscopy is a single-photon technique, scattering is likely to prevent it from imaging as deeply as two-photon microscopy.

ACKNOWLEDGMENTS

Our imaging research has been supported by the Mallinckrodt Foundation, a McKnight Technological Innovation in Neuroscience award, the Leila Y. and Harold G. Mathers Foundation, and National Institutes of Health (NIH)/NINDS/NIAAA (National Institute of Neurological Disorders and Stroke/National Institute on Alcohol Abuse and Alcoholism).

REFERENCES

Bewersdorf J, Pick R, Hell SW. 1998. Multifocal multiphoton microscopy. *Opt Lett* **23:** 655–657.

Conchello JA, Lichtman JW. 2005. Optical sectioning microscopy. *Nat Methods* **2:** 920–931.

Dodt H-U, Leischner U, Schierloh A, Jahrling N, Mauch CP, Deininger K, Deussing JM, Eder M, Zieglgansberger W, Becker K. 2007. Ultramicroscopy: Three-dimensional visualization of neuronal networks in the whole mouse brain. *Nat Methods* **4:** 331–336.

Duemani Reddy G, Kelleher K, Fink R, Saggau P. 2008. Three-dimensional random access multiphoton microscopy for functional imaging of neuronal activity. *Nat Neurosci* **11:** 713–720.

Dunsby C. 2008. Optically sectioned imaging by oblique plane microscopy. *Opt Express* **16:** 20306–20316.

Fuchs E, Jaffe JS, Long RA, Azam F. 2002. Thin laser light sheet microscope for microbial oceanography. *Opt Express* **10:** 145–154.

Göbel W, Kampa BM, Helmchen F. 2007. Imaging cellular network dynamics in three dimensions using fast 3D laser scanning. *Nat Methods* **4:** 73–79.

Holekamp TF, Turaga D, Holy TE. 2008. Fast three-dimensional fluorescence imaging of activity in neural populations by objective-coupled planar illumination microscopy. *Neuron* **57:** 661–672.

Huisken J, Swoger J, Del BF, Wittbrodt J, Stelzer EH. 2004. Optical sectioning deep inside live embryos by selective plane illumination microscopy. *Science* **305:** 1007–1009.

Kner P, Chhun BB, Griffis ER, Winoto L, Gustafsson MG. 2009. Super-resolution video microscopy of live cells by structured illumination. *Nat Methods* **6:** 339–342.

Lechleiter JD, Lin DT, Sieneart I. 2002. Multi-photon laser scanning microscopy using an acoustic optical deflector. *Biophys J* **83:** 2292–2299.

Lichtman JW, Sunderland WJ, Wilkinson RS. 1989. High-resolution imaging of synaptic structure with a simple confocal microscope. *New Biol* **1:** 75–82.

Masters BR, Thaer AA. 1994. In vivo human corneal confocal microscopy of identical fields of subepithelial nerve plexus, basal epithelial, and wing cells at different times. *Microsc Res Tech* **29:** 350–356.

Neil MAA, Juskaitis R, Wilson T. 1997. Method of obtaining optical sectioning by using structured light in a conventional microscope. *Opt Lett* **22:** 1905–1907.

Nguyen QT, Callamaras N, Hsieh C, Parker I. 2001. Construction of a two-photon microscope for video-rate Ca^{2+} imaging. *Cell Calcium* **30:** 383–393.

Nikolenko V, Watson BO, Araya R, Woodruff A, Peterka DS, Yuste R. 2008. SLM microscopy: Scanless two-photon imaging and photostimulation with spatial light modulators. *Front Neural Circuits* **2:** 5.

Pétran M, Hadravsky M, Eggar MD, Galambos R. 1968. Tandem-scanning reflected-light microscopy. *J Opt Soc Am* **58:** 661–664.

Salomé R, Kremer Y, Dieudonné S, Léger JF, Krichevsky O, Wyart C, Chatenay D, Bourdieu L. 2006. Ultrafast random-access scanning in two-photon microscopy using acousto-optic deflectors. *J Neurosci Methods* **154:** 161–174.

Siedentopf H, Zsigmondy R. 1903. Uber Sichtbarmachung und Grossenbestimmung ultramikroskopischer Teilchen, mit besonderer Anwendung auf Goldrubinglaser. *Ann Phys* **10:** 1.

Turaga D, Holy TE. 2008. Miniaturization and defocus correction for objective-coupled planar illumination microscopy. *Opt Lett* **33:** 2302–2304.

Voie AH, Burns DH, Spelman FA. 1993. Orthogonal-plane fluorescence optical sectioning: Three-dimensional imaging of macroscopic biological specimens. *J Microsc* **170:** 229–236.

Vucinic D, Sejnowski TJ. 2007. A compact multiphoton 3D imaging system for recording fast neuronal activity. *PLoS One* **2:** e699.

Yaksi E, Friedrich RW. 2006. Reconstruction of firing rate changes across neuronal populations by temporally deconvolved Ca^{2+} imaging. *Nat Methods* **3:** 377–383.

57

Functional Neuron-Specific Expression of Genetically Encoded Fluorescent Calcium Indicator Proteins in Living Mice

Matthias Heindorf and Mazahir T. Hasan

Department of Molecular Neurobiology, Max Planck Institute for Medical Research, 69120 Heidelberg, Germany

ABSTRACT

Genetically encoded fluorescent calcium (Ca^{2+}) indicator proteins (FCIPs) are promising tools for long-term population recordings of neuronal activity in living systems. With FCIPs, it has now become possible to detect neuronal activity with single-cell, single-action-potential resolution both ex vivo and in living mice. Because the signal-to-noise ratio crucially depends on the protein concentration of the Ca^{2+} sensors within a cell, we have found that with the current FCIPs, it is necessary to reach high expression levels in neurons to reliably detect functional responses in vitro and in vivo. This chapter provides relevant information and protocols for generating transgenic mice with tetracycline (Tet)–controlled genetic switches and recombinant adeno-associated viruses (rAAVs) to achieve high FCIP expression in neurons. We also provide examples of Ca^{2+} activity responses with FCIPs both in vitro and in vivo.

INTRODUCTION

Development of new imaging technologies has been heralded since the green fluorescent protein (GFP) was first sequenced (Prasher et al. 1992; Prasher 1995) and functionally expressed in living cells (Chalfie et al. 1994). The first genetically encoded fluorescent Ca^{2+} indicator protein (FCIP; Miyawaki et al. 1997), also called a genetically encoded Ca^{2+} indicator (GECI), was engineered by linking yellow fluorescent protein (YFP) and the cyan fluorescent protein (CFP) via a Ca^{2+}-sensing domain, calmodulin (CaM), and its interacting partner, M13, derived from the myosin light chain kinase (MLCKp). These ratiometric FCIPs, called Yellow Cameleons (YCs), report Ca^{2+} transients by harnessing a phenomenon called Förster resonance energy transfer (FRET). During FRET, an electronically excited donor chromophore—in this case, CFP—transfers energy nonradiatively to an acceptor chromophore—here, YFP. This results in a decrease in fluorescence of the donor and an increase in fluorescence of the acceptor. As FRET strongly depends on the distance between the donor and the acceptor (sixth-power dependency), the efficiency of FRET dramatically increases when a Ca^{2+}-induced conformational change in the Ca^{2+}-sensing domain brings the two GFP variants closer together.

Previous YCs could successfully report Ca^{2+} transients in worms (Kerr et al. 2000; Suzuki et al. 2003), fruit fly (Fiala et al. 2002; Reiff et al. 2002; Liu et al. 2003; Wang et al. 2003; Yu et al. 2003), and zebrafish (Higashijima et al. 2003). However, results were disappointing in mammalian neurons and remained unreported. Therefore, large efforts were made to optimize FCIP performance and to improve different features of GFP and the Ca^{2+} sensor domain. Both by random and rational designs, GFP variants were generated with increased brightness, photostability, reduced sensitivity to H^+ and Cl^- ions, and rapid chromophore maturation at physiological temperatures (Hires et al. 2008). In addition, re-engineering of a Ca^{2+}-sensing domain employing split CaM and a different CaM-interacting partner, CCLp, improved the dynamic range somewhat (Truong et al. 2001). In addition, efforts were undertaken to (1) improve FPs to increase FRET efficiency in responses to Ca^{2+} and (2) reduce interactions between CaM and endogenous cellular interacting factors.

A major breakthrough was the surprising finding that a more sensitive FCIP can be engineered by inserting short DNA sequences at specific positions (typically at position 144) in FPs without compromising the fluorescence signal (Baird et al. 1999). More important, breaking the FPs at these specific positions and rejoining the C and N termini of the GFP variants generated circularly permuted FPs (cpFPs) with bright fluorescence (Baird et al. 1999; Griesbeck et al. 2001). These cpFPs were also more sensitive to fluctuation in pH and temperature and thus served as better substrates to develop more sensitive FCIPs. For YCs, a YFP variant, Venus, was circularly permuted to generate a new variant, cpVenus (Nagai et al. 2004). Pairing of cpVenus with ECFP in a FRET-based FCIP (YC3.60) increased the Ca^{2+}-induced dynamic range of FRET change by up to 560% (Nagai et al. 2004).

To reduce interaction of CaM with endogenous CaM-interacting factors, two different approaches were undertaken. In the first case, CaM/M13 binding interfaces were mutated to strongly reduce their interaction with cellular targets while preserving intramolecular interaction between them (Palmer et al. 2006). These YCs of the "Design" series include D3cpv (Palmer et al. 2006). In the second case, at about the same time, a Ca^{2+}-sensing domain from the skeletal muscle protein troponin (TN) (Heim and Griesbeck 2004) was used to generate novel FCIPs that include TN-XL (Mank et al. 2006) and TN-XXL (Mank et al. 2008). All these FCIPs (YC3.60, D3cpv, and TN-XXL) can record sensory-evoked neuronal responses (Mank et al. 2008; Wallace et al. 2008; Lütcke et al. 2010). Interestingly, the performance of YC3.60 (Lütcke et al. 2010), a predecessor of D3cpv, is quite comparable to that of D3cpv (Wallace et al. 2008) and TN-XXL (Mank et al. 2008) both in vitro and in vivo. In addition, YC3.60, like D3cpv, can reliably detect Ca^{2+} transients in response to sensory stimulation in vivo with single-cell, single-action-potential (AP) resolution (Lütcke et al. 2010). Moreover, unlike D3cpv (Wallace et al. 2008), YC3.60 has a faster decay kinetics, and it can report up to at least 10 APs without saturation of signal amplitude (Lütcke et al. 2010). It is plausible that the Ca^{2+}-dependent large dynamic range of D3cpv, TN-XXL, and YC3.60 is largely due to the use of circular permuted YFP variants (cp173Venus in YC3.60 and D3cpv and Citrizine-cp174 in TN-XL and TN-XXL) as FRET partners with CFP (Nagai et al. 2004; Mank et al. 2008; Wallace et al. 2008). However, with FRET-based indicators, one potential concern is that ATP might be acting as a FRET quencher (Willemse et al. 2007). It is conceivable that variable cellular ATP levels might compromise the FRET dynamic range in individual cells. This would likely also compromise the reliability of single-AP detection in labeled neurons under different metabolic states.

For single GFP-based FCIPs, insertion of CaM at specific positions in the sequence of GFP variants and the use of CaM/M13 in cpFPs resulted in sensor designs of the camgaroo-1/camgaroo-2 (Baird et al. 1999; Griesbeck et al. 2001) and pericams/GCaMPs (Nakai et al. 2001; Nagai et al. 2004) types, respectively, which can detect functional Ca^{2+} signals. The advantages of GCaMPs are high signal-to-noise ratios (Nakai et al. 2001; Tallini et al. 2006). Rational mutagenesis has further improved the GCaMP series with a highly promising FCIP, GCaMP3, as a very sensitive reporter of neuronal activity (Tian et al. 2009). However, one major benefit of ratiometric FCIPs over single-GFP-based FCIPs is their reduced sensitivity to brain pulsation–related motion artifacts. Thus they are likely to be more suitable for imaging neuronal activity in freely moving animals.

STRATEGIES FOR FCIP EXPRESSION

To achieve robust long-term FCIP expression in mammalian neurons in vivo, two different expression strategies can be deployed: (1) classical mouse transgenesis by pronuclei DNA injection using Tet-controlled genetic switches and (2) viral delivery using recombinant adeno-associated viruses (rAAVs). In the following sections, we describe in detail these two expression strategies and provide protocols for their use. We conclude this chapter by discussing potential future developments.

FCIP Expression Using Mouse Transgenesis

The tetracycline (Tet) transactivator (tTA) and reverse tTA (rtTA) systems allow for Dox-controlled, regulated gene expression in the brain of living mice (Hasan et al. 2001; Sprengel and Hasan 2007). In addition, the Tet systems also allow for expression amplification in neurons (Hasan et al. 2004; Sprengel and Hasan 2007) and are therefore highly recommended for generating transgenic mice. The Tet-inducible system requires three components for inducible and reversible control of gene expression: (1) a potent transcriptional activator protein, either tTA or the rtTA; (2) a minimal Tet-promoter (P_{tet}) or a bidirectional Tet-promoter (P_{tet}bi) to express one or more responder genes; and (3) tetracycline or one of its derivatives such as doxycycline (Dox) as an inducer. Both tTA and rtTA can be expressed under control of either a ubiquitous or a cell type–specific promoter that, upon binding to P_{tet}/P_{tet}bi, robustly activates the promoter (Sprengel and Hasan 2007). As a standard procedure, it is advisable to generate activator (tTA) and Tet-responder transgenic mice (P_{tet}/P_{tet}bi) separately. Double transgenic animals (tTA and P_{tet}/P_{tet}bi) can then be obtained by mating. One potential concern is that whereas low tTA/rtTA levels in neurons may be insufficient to activate responder genes (via either P_{tet} or P_{tet}bi), very high levels might be toxic (Barton et al. 2002). Therefore, several independent tTA/rtTA transgenic mouse lines should be generated to identify a line that has optimal tTA/rtTA levels in neurons for functional expression of responder genes without toxicity. However, we have not observed any obvious abnormalities in the brain histologically or in animal behavior when three different transgenic FCIPs (inverse pericam, camgaroo-2, and YC3.12) were mated with transgenic mice that express tTA specifically in excitatory neurons of the forebrain (Hasan et al. 2004). Additionally, long-term high FCIP expression in certain cell types can become a problem (Tallini et al. 2006). In such a case, when using the tTA system, gene expression can be switched off by doxycycline (Dox) during the nonexperimental periods and switched on again by removing Dox during the recording period (Fig. 1A). To switch off gene expression in the brain, mice can be treated with Dox in the drinking water (2 mg/mL plus 5% sucrose) (Hasan et al. 2001). Noninvasive in vivo bioluminescence imaging of gene expression can be performed with the firefly luciferase gene (Fig. 1B). With the tTA system, it takes 5 d of Dox treatment to strongly reduce gene expression and 10 d of Dox withdrawal to fully restore gene expression (Fig. 1C; Hasan et al. 2001). Because the firefly luciferase activity can be recorded noninvasively through an intact skull in living mice (Hasan et al. 2001), we recommend using bioluminescence imaging for long-term monitoring of Dox-controlled gene expression. In vivo bioluminescence imaging of the firefly luciferase activity can be a useful strategy to indirectly monitor the expression status of the second gene of interest (e.g., a gene that could perturb neuronal signaling). This strategy thus should allow, in the same animal over time, reliable analysis of gene activity on cell physiology (by FCIPs) and animal behavior.

Previously, we established transgenic mouse lines that express camgaroo-2 and inverse pericam selectively in the forebrain under control of the tTA expression system (Fig. 2A; Hasan et al. 2004). Cell type–specific tTA expression under control of the αCaMKII promoter ensured that FCIP expression amplification is restricted to excitatory neurons (Fig. 2A). The bidirectional Tet promoter (P_{tet}bi) can serve to drive two different genes, such as an FCIP and a Cre recombinase. Expression of the Cre gene, for example, provides the possibility to inducibly delete a "floxed" gene (a gene flanked by *lox* sites), if desired, while FCIP expression can be used as a sensitive reporter of neuronal activity. With

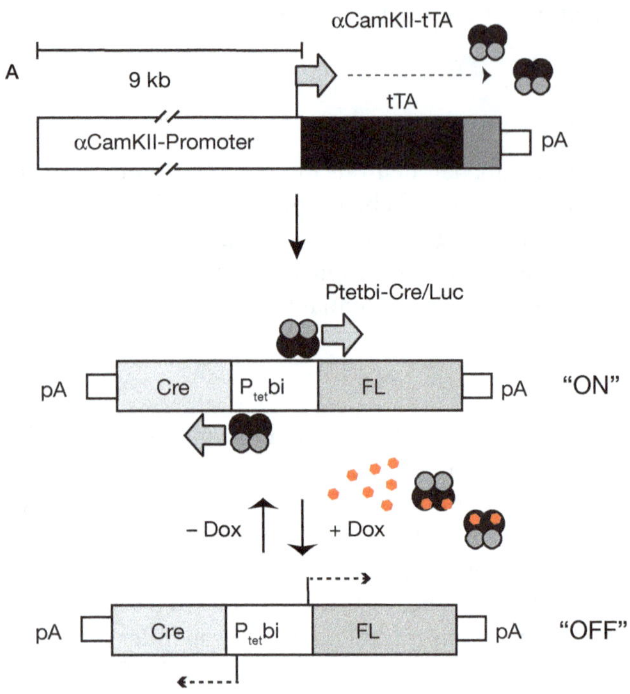

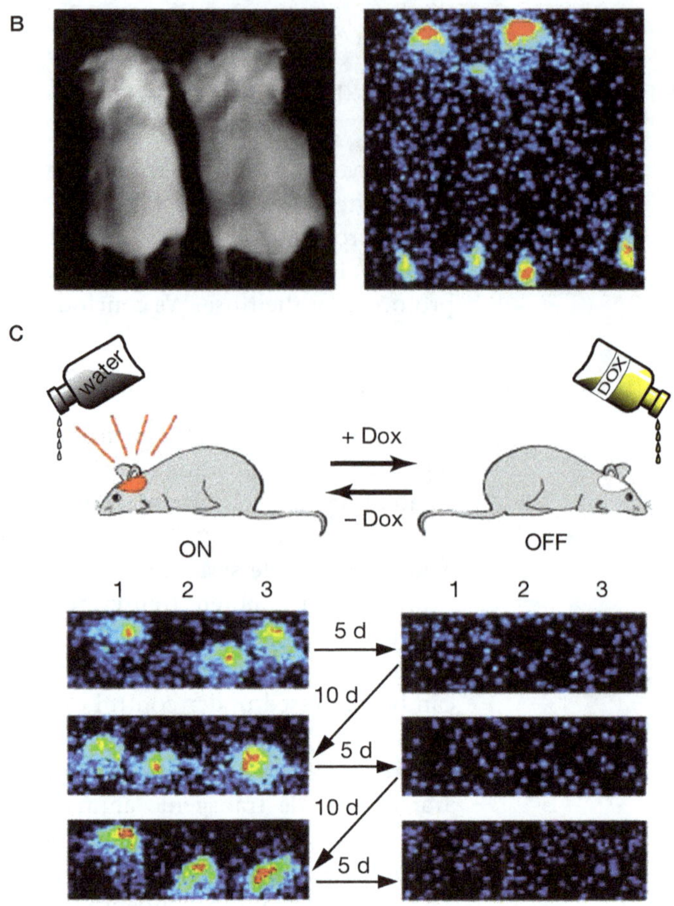

FIGURE 1. Long-term, in vivo bioluminescence imaging. (*A*) Tetracycline-controlled transactivators (tTA) bind to a bidirectional Tet promoter (P_{tet}bi) in the absence of doxycycline (Dox) and activate the expression of the firefly luciferase gene (FL) as a reporter and Cre as a second gene of interest (Cre/Luc). Dox-controlled, neuron-specific regulation of Cre/Luc is achieved by driving the expression of the tTA gene with the αCamKII promoter ($P_{\alpha CaMKII}$). (*B*) In vivo brain imaging of firefly luciferase gene activity in double-transgenic mice (αCamKII-tTA/P_{tet}bi-Cre/Luc) allows coexpression of luciferase and Cre genes in a tTA-dependent manner in the absence of Dox. Luciferase expression is restricted to the brain and foot pads (exposure time 5 min). (*C*) In vivo cycles (in days) of luciferase gene activation and deactivation via Dox (2 mg/mL in drinking water) were monitored in three animals.

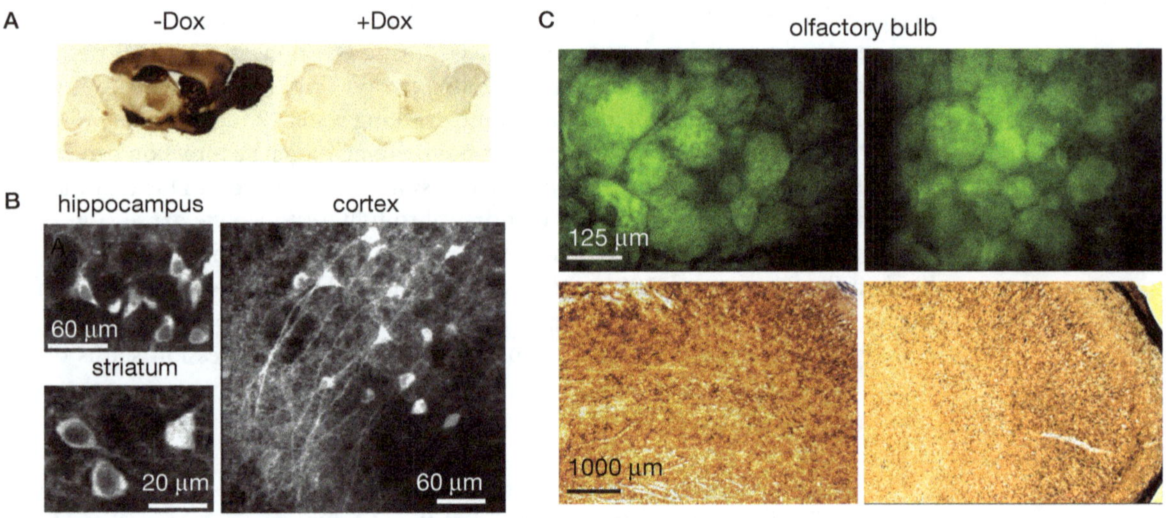

FIGURE 2. Fluorescent Ca^{2+} indicator protein (FCIP) in transgenic mice. (*A*) Dox-controlled regulated FCIP expression visualized in brain slices by immunohistology with the rabbit polyclonal anti-GFP antibodies/peroxidase-DAB system. In a double-positive mouse (αCamKII-tTA/P_{tet}bi-FCIP), forebrain excitatory neuron–specific tTA expression drives P_{tet}bi-mediated FCIP expression (*left* panel). Gene expression can be switched off by Dox (*right* panel). (*B*) Two-photon images of acute brain slices with neurons in both CA1 (*left-upper* panel) and striatum (*left-lower* panel). Fluorescence is usually excluded from the nucleus. (*Right* panel) Maximum-intensity projection of two-photon 3D-stacks taken from a cortical brain slice. (*C*) Camgaroo-2 (*left-top* and *left-bottom* panels) and inverse pericam (*right-top* and *right-bottom* panels) expression in olfactory epithelial neurons (glomeruli) and the main olfactory bulb. (*Continued*.)

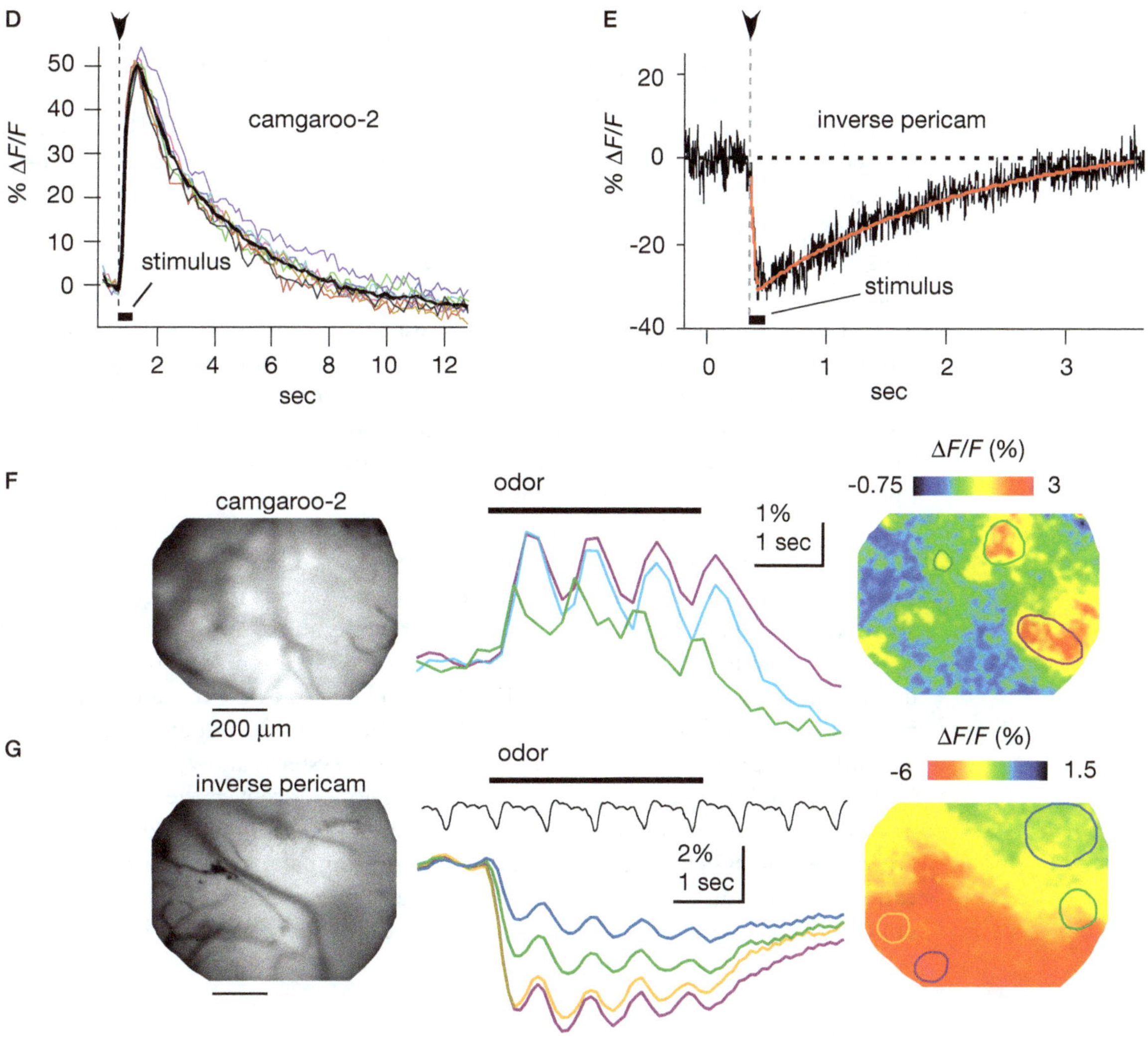

FIGURE 2. (*Continued.*) (*D,E*) Two-photon imaging of Ca^{2+} responses evoked by synaptic stimulation in the Schaffer collateral pathway of the hippocampus. Camgaroo-2 (*D*) and inverse pericam (*E*) were stimulated with 100 pulses and 20 pulses, respectively, both at 200 Hz. Single-trial responses and an average response are shown for both FCIPs (average traces; [black] camgaroo-2; [red] inverse pericam). (*F,G*) Odor-evoked activity maps in transgenic mice with camgaroo-2 (*F*) and inverse pericam (*G*). A raw fluorescence image (*left* panels) and the time course of fluorescence signal (*center* panels) in different regions of interest (*right* panels). Color-coded maps show the relative change in fluorescence evoked by different odors in each pixel during the first second of the odor response (*right* panels). The black trace in *G* shows respiratory activity. As expected, camgaroo-2 and inverse pericam show positive- and negative-going activity signals, respectively.

the Tet-inducible system, we achieved robust, Dox-controlled FCIP expression in forebrain excitatory neurons (Fig. 2A,B). Two-photon imaging of acute brain slices showed FCIP-labeled individual neurons in the hippocampus, the striatum, and the cortex (Fig. 2B). Strong expression was also readily visible in the olfactory bulb (Fig. 2C), being largely restricted to olfactory receptor neurons and granule cells. Acute brain slices derived from camgaroo-2 and inverse pericam transgenic mice showed large fluorescence changes in response to strong synaptic stimulation of the Schaffer collateral pathway (Fig. 2D,E). Moreover, wide-field imaging of the olfactory bulb revealed odor-specific activity maps (Fig. 2F,G), which were a sum of activity responses from both glomeruli and granule cells. In vivo two-photon imaging of layer 2/3 cortical neurons and the olfactory bulb, however, failed to reveal activity signals with good signal-to-noise ratios, likely because of low FCIP levels in neurons.

Viral Delivery Using rAAV

An alternative method for gene delivery and expression is to introduce FCIPs into neurons via rAAV vectors using constitutive and cell type–specific promoters (Lu 2004; Shevtsova et al. 2005). The AAV genome is ~4.8 kb long and is composed of single-stranded DNA (ssDNA), either positive or negative sense. The genome comprises inverted terminal repeat (ITR) sequences of 145 bases each at both ends of the DNA strand. ITRs are required in *cis* to the gene of interest, which replaces structural (*cap*) and packaging (*rep*) genes. The Rep proteins are required for the AAV life cycle, and the capsid proteins (VP1, VP2, and VP3) self-organize to form a capsid of an icosahedral symmetry. After infection of host cells, AAVs form episomal concatamers in the nucleus. In nondividing cells, these concatamers remain intact throughout the life of the cell. Random integration of AAV DNA into the host genome is quite rare. This is advantageous because it reduces the chances of mutagenesis due to random DNA insertion. In dividing cells, AAV DNA is diluted every cell division because the episomal DNA is unable to replicate in host cells. The other major advantage of the rAAVs is that they can be targeted to selected cell types at any time in the animal's life, thereby allowing detailed investigation of gene function without interference during development as is the case with traditional transgenic mice derived by pronuclei DNA microinjections. This not only obviates the use of conditional transgenics, but also easily expands the use of transgenics to species other than rodents and allows the use of multiple viruses to express multiple genes simultaneously, even in different locations. Maximum transgene expression can be observed in a matter of weeks and typically persists throughout life.

We have found that the rAAV expression systems are highly suitable for imaging neuronal activity both in vitro and in vivo. Purified rAAVs can be used to express an FCIP in dissociated neurons, in cultured brain slices, and in vivo. FCIPs can be expressed under control of either constitutive or Tet-inducible promoters (Fig. 3A). Both expression systems allow for fast and efficient gene delivery in different cell types and long-term gene expression. Moreover, they can also be applied to different animal species, from rats (Xu et al. 2001) to monkeys (Stettler et al. 2006). This makes rAAV vectors and expression systems valuable tools for creating animal models of diseases, for gene therapy applications, and for functional genomics. To prepare high-quality rAAVs, we describe below two different purification methods, using (1) an iodixanol gradient and (2) a heparin column (Wallace et al.

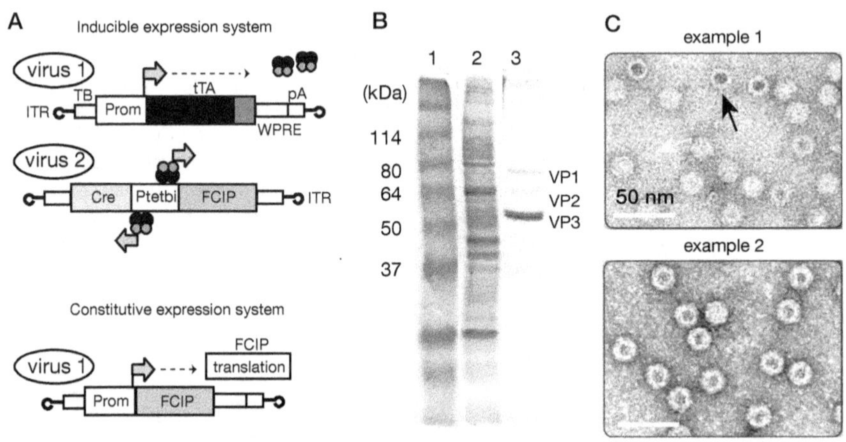

FIGURE 3. Recombinant adeno-associated virus (rAAV)–mediated FCIP expression in vitro and in vivo. (*A*) Inducible and constitutive expression systems. (*Top* panel) In the inducible expression system, two rAAVs are required: one to express tTA under control of the human synapsin promoter (virus 1) and the other to inducibly express FCIP and Cre genes under control of a P_{tet}bi (virus 2) in a tTA-dependent manner. (*Bottom* panel) In the constitutive expression system, the human synapsin promoter drives constitutive expression of FCIP. (TB) Synthetic transcription blocker; (WPRE) a woodchuck hepatitis virus post-transcriptional control element; (pA) polyadenylation signal; (ITR) inverted terminal repeats of AAV2. (*B*) Coomassie-stained gel (SDS-PAGE) with capsid proteins (VP1, VP2, VP3) from a purified virus (lane *3*), unpurified virus in cell lysate (lane *2*), and protein molecular weight markers (lane *1*). (*C*) Electron microscopic images of virus particles visualized by negative staining with uranyl acetate. Empty virus particle (black arrow) and DNA-filled virus particle (white arrow) are indicated in example 1. Bars, 50 nm. (*Continued*.)

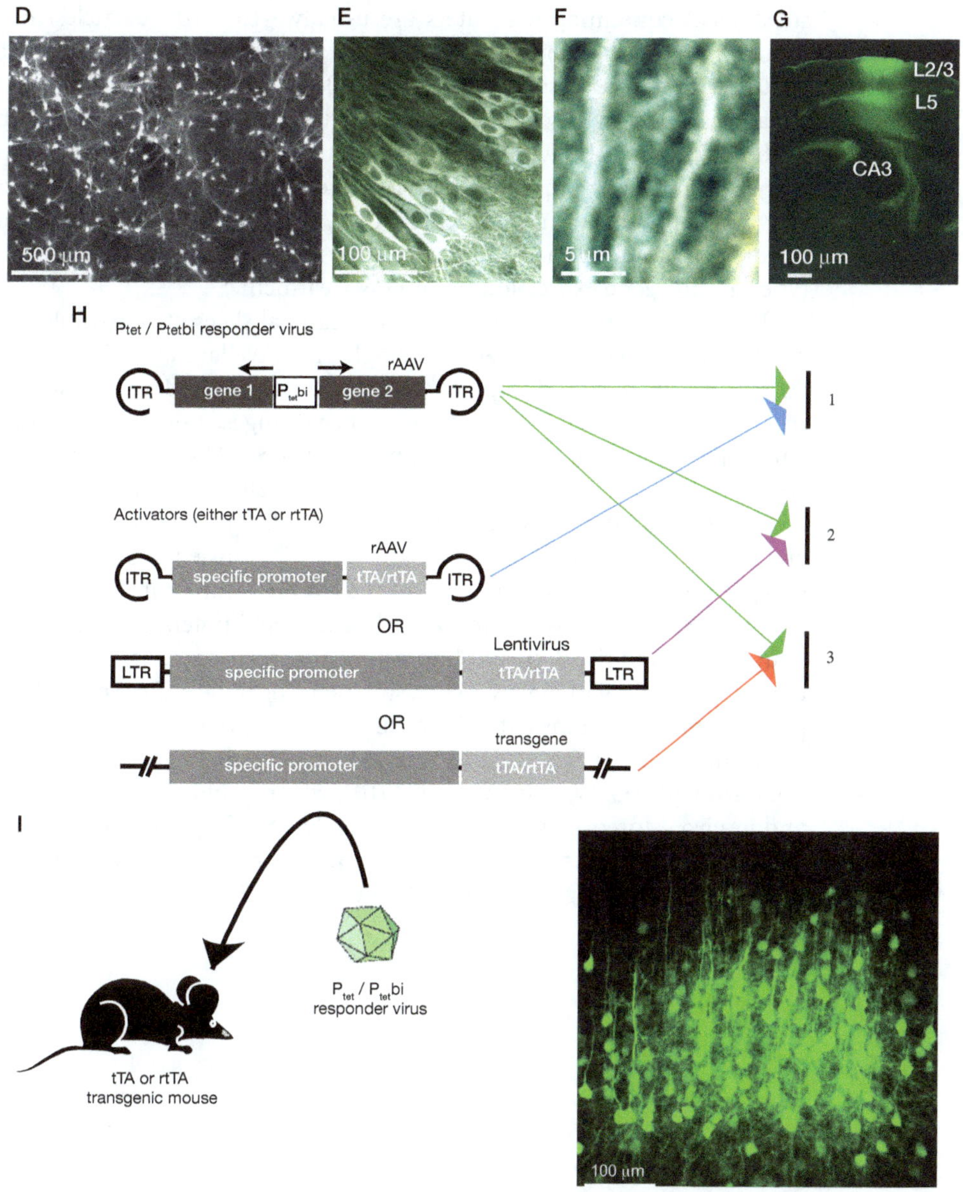

FIGURE 3. (*Continued*.) (*D*) FCIP expression in dissociated cultured neurons. (*E,F*) A tetracycline-inducible promoter expresses an FCIP (D3cpv) in hippocampal organotypic slices acquired at low and high zoom (both images are collapsed two-photon image stacks). (*G*) Human synapsin promoter (P_{hSYN})-driven FCIP (D3cpv) expression in vitro. Fluorescence in a fixed brain slice was detected by wide-field imaging with AAV injections at depths of 250 µm and 1.5 mm. (CA3) Hippocampal layer CA3; (L2/3) layer 2/3; (L5) layer 5. (*H*) A versatile combinatorial genetic approach for rAAV-mediated P_{tet}bi responder gene expression in tTA/rtTA-expressing cells delivered via three different approaches: (1) rAAV, (2) lentivirus, and (3) transgenic mice. (*I*) Example of an rAAV equipped with a P_{tet}bi responder. When injected into a tTA or rtTA transgenic mouse (*left* panel), this allows for forebrain-specific, neuron-specific gene expression in the presence of doxycycline (*right* panel).

2008). After the virus is purified, it is analyzed for purity (Fig. 3B,C) and infectivity both in vitro and in vivo (Fig. 3D–G). It is important to determine that the purified rAAVs show distinct capsid proteins on a gel (Fig. 3B, lane 3). Electron microscopic images of different virus preparations help to determine the fraction of empty and DNA-filled virus particles (Fig. 3C). It is highly recommended that a cloned DNA fragment between the two ITRs not be >4.6 kb in length. With larger insert sizes, the packaging efficiency drops sharply; thus empty virus particles dominate the fraction (Fig. 3C). Depending on the quality of the DNA transfection, packaging efficiency, and virus purification steps, DNA-containing AAV titers are usually ~10^{12}–10^{13} particles/mL with particle-to-infectivity ratios of <10, although with much effort a near-perfect infectivity is also possible (Zeltner et al. 2010). With both constitutive and inducible systems, we have observed robust FCIP expression ~2 wk after applying viruses to dissociated cultured neurons, rat hippocampal organotypic slices, and in the living animal (Fig. 3D–G; Wallace et al. 2008). To achieve cell type–specific FCIP expression amplification, rAAVs equipped with a Tet-responsive promoter (P_{tet}bi) can be combined with transactivators provided via rAAVs, lentiviruses, and transgenic mice (Fig. 3H; Zhu et al. 2007). (See Fig. 3I for an example of a combination of rAAV-P_{tet}bi and rtTA-expressing transgenic mice for cell type–specific gene expression upon Dox treatment.)

In CA3 neurons of cultured rat hippocampal slices, it is possible to detect fluorescence changes reliably with D3cpv in response to single APs (Wallace et al. 2008). During whole-cell recordings, current was injected in D3cpv-expressing neurons (Fig. 4A), and simultaneous optical two-photon recording allowed detection of fluorescence changes, a decrease in CFP and an increase in YFP signal, respectively (Fig. 4B). For one, two, and three APs, the average peak amplitudes of Ca^{2+} transients (%$\Delta R/R$) were 8.3 ± 0.9, 15.0 ± 2.4, and 20.0 ± 4.3, respectively (Fig. 4C). In the superficial layer of the somatosensory cortex (Fig. 4D,E), D3cpv detected spontaneous (%$\Delta R/R$ = 3.6 ± 0.16) and whisker-evoked (%$\Delta R/R$ = 3.2 ± 0.16) Ca^{2+} transients (Fig. 4F; Wallace et al. 2008). In layer 2/3 cortical neurons, whisker stimulation evoked Ca^{2+} transients both in cell somata and in the adjacent neuropil. Simultaneous recording of the local field potential (LFP) and fluorescence changes in the neuropil showed that LFP and fluorescence changes correlated well. Moreover, in vivo targeted cell-attached recording of D3cpv-expressing neurons in layer 2/3 showed good correlation between single AP and Ca^{2+} transients. Recordings from seven cells (four mice) showed an average Ca^{2+} transient %$\Delta R/R$ = 3.5 ± 0.2 in response to a single AP (Wallace et al. 2008). Similar results were obtained with YC3.60 (Lütcke et al. 2010), which could also detect single AP responses both in vitro and in vivo. Moreover, with a fiber-optic telescope implanted on top of a thinned skull (Fig. 5A), it was possible to detect spontaneous and behaviorally related changes in YFP fluorescence (YC3.60) even in freely moving mice (Fig. 5B; Lütcke et al. 2010).

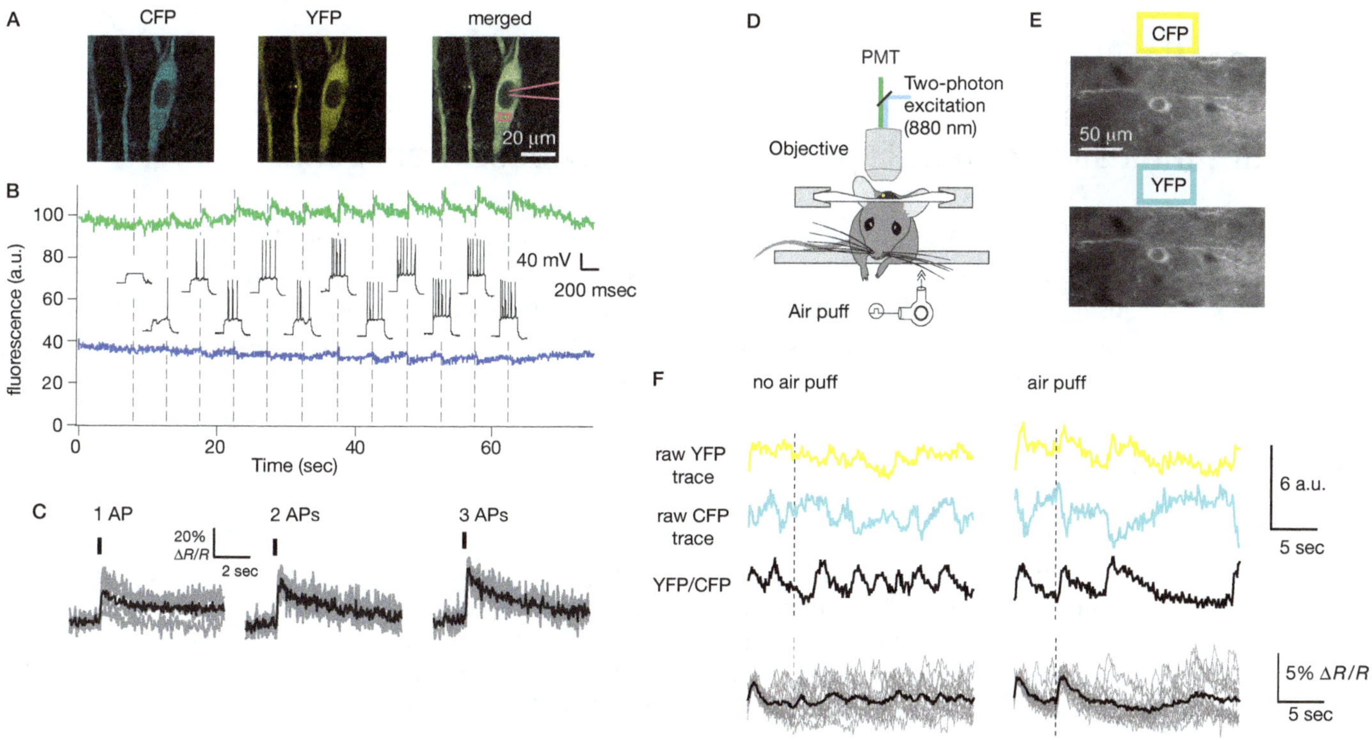

FIGURE 4. In vitro and in vivo functional responses with D3cpv. (*A*) Whole-cell recordings in organotypic slices of CA3 pyramidal cells shown as YFP, CFP, and merged images. The recording pipette is indicated in the merged image. (*B*) Time courses for YFP (green trace) and CFP (blue trace) channels show that there are increases in fluorescence changes (both in CFP and YFP raw traces) with increasing number of action potentials (APs); (a.u.) arbitrary units. (*C*) Single-trial and averaged ratios of CFP and YFP fluorescence in response to one, two, and three APs. (*D*) In vivo two-photon imaging experimental setup for recording neuronal activity in response to whisker stimulation by air puffs. (*E*) Mouse somatosensory cortex layer 1 neuropil (CFP and YFP channels). (*F*) Spontaneous and whisker-stimulation-induced fluorescence transients (examples of single-trial raw traces for CFP [cyan], YFP [yellow], and YFP/CFP [black]). An average of 15 trials across the whole frame without and in response to air-puff whisker stimulation; (black) averages over trials; (gray) single trials. (Adapted from Wallace et al. 2008.)

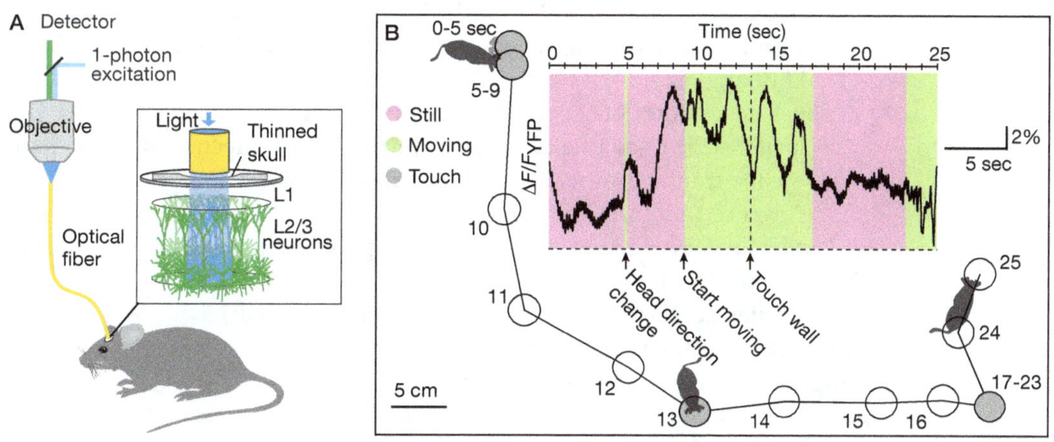

FIGURE 5. Fiber-optic recording of neuronal activity in freely moving mice. (*A*) The tip of an optical fiber is placed on the cortical surface with fluorescence excitation and emission collected from the same fiber. (*B*) Bulk Ca^{2+} signals were recorded in somatosensory cortex through a single-core optical fiber with fluorescence changes in the YFP-channel over a period of 25 sec. The fluorescence changes are shown here together with the position of the mouse in an open field box. Animal behavior (sitting still, moving, touches, or having contact with the wall) is indicated for each time period. (Adapted from Lütcke et al. 2010.)

Protocol A

Calcium Indicator Protein Expression Using Mouse Transgenesis

This protocol describes methods for regulated expression of FCIP using Tet-responsive transgenic mice. To ensure a high level of FCIP expression in neurons, transgenic founder mice are screened using an ear fibroblast culture method to identify those that are responsive to Dox treatment prior to use in experiments. The protocol describes use of Dox to regulate gene expression and provides a short description of in vivo recording of luciferase activity.

IMAGING SETUP

For in vivo bioluminescence imaging, images are digitized with the Argus 20 image processor (Hamamatsu). The intensity of acquired images is calculated as relative luminescent units per minute (rlu/min). The emitted light is acquired by a photon-counting camera (two-stage ICCD C2400-47) fitted with a Nikon lens (35 mm/f1.2) and a computer with image analysis capabilities. Body images are recorded as reference in white light. An in vivo luciferase imaging system can be purchased from either Hamamatsu or Caliper IVIS Imaging System.

MATERIALS

CAUTION: See Appendix 6 for proper handling of materials marked with <!>.
See the end of the chapter for recipes for reagents marked with <R>.

Reagents

Collagenase type IA (Sigma-Aldrich, C-2674)
D-luciferin
DNA constructs
- Promoter for expression of tTA: Examples include the alpha-Ca^{2+}/Calmodulin Kinase II (αCamKII) promoter (Mark Mayford, Scripps Research Institute, La Jolla, CA) and the Thy-1 promoter (Guoping Feng, Department of Neurobiology, Duke University Medical Center, Durham, NC).
- Constructs of the Tet expression system: A bidirectional Tet promoter (Ptetbi) with the firefly luciferase (FL) as a reporter gene (pBI-L Tet; Clontech 631005) as well as an optimized tetracycline transactivator (Tet-OFF Advanced Vector; Clontech 631070) and reverse tTA (Tet-ON Advanced Vector; Clontech 631069).
- *Renilla* luciferase mammalian expression plasmid (pRL-CMV vector; Promega)

Doxycycline (Dox) <!>
Dual-Luciferase Reporter Assay System (Promega, E1910)
FCS, certified tetracycline-free (Gibco)
Fungizone or nystatin (Sigma-Aldrich)
Ketamine <!>
Lipofectamine 2000 (Invitrogen)
Mice, transgenic

To generate transgenic mice by pronuclei injection, we recommend using the sucrose gradient procedure for purifying a transgene (Mann and McMahon 1993; Hasan et al. 2004). We inject transgenic DNA constructs at a concentration of 10 ng/mL. Our efficiency of transgenesis is usually >25% with the transgene inserted into the mouse genome at a low copy number. This method for generating transgenic mice helps to recapitulate faithful promoter activity patterns and reduces the risk associated with gene silencing induced by high copy number.

A list of transgenic mouse strains available from Jackson Labs (http://www.jax.org) and other sources is available in Table 1.

NaCl
Penicillin/streptomycin <!>/glutamine (100x, cat. no. 15970-063; Invitrogen Life Technologies-Gibco)
Phosphate-buffered saline (PBS), 1x <R>
RPMI-1640 (cat. no. 21870-092; Invitrogen Life Technologies-Gibco)
Sucrose
Xylazine <!>

Equipment

Culture dishes, 3.5-cm-diameter
Glass slide, sterile
Imaging setup, as described above
Light-proof chamber for mice
Luminometer (e.g., BioFix Lumi-10, Machery-Nagel)
Pipettes, 25-mL, 10-mL, 5-mL, and 1-mL, sterile
Plates, six-well
Razor blade
Syringes, for intraperitoneal injections

EXPERIMENTAL METHOD

Screening for Transgenic Mouse Founders with High FCIP Expression

To establish highly efficient transgene-expressing P_{tet}/P_{tet}bi mouse lines, it is necessary to screen a large number of transgenic Tet responders using the ear fibroblast method described below (Hasan et al. 2004). We recommend using a bidirectional Tet promoter (P_{tet}bi) with the firefly luciferase (FL; pBI-L Tet) as a reporter gene for quantitative measurement of gene activity. This early screen should help to save both time and resources.

Preparation of Mouse Ear Fibroblasts and Quantitative Gene Expression Assay

1. Take a biopsy of an ear (4 x 4-mm) and place it into a sterile tube with 1 mL of RPMI-1640 medium supplemented with 30% FCS and 300 U/mL penicillin/streptomycin/glutamine (P/S/G). To avoid fungal contamination, also add Fungizone or nystatin to a final concentration of 2.5 µg/mL or 250 U/mL, respectively. For best results, proceed immediately; however, one may keep the material at room temperature for several hours.
2. Cut the ear tip into small pieces on a sterile glass slide with a razor blade, and transfer the material into a 3.5-cm dish containing 2 mL of RPMI-1640 supplemented with 30% FCS, P/S/G, and 1 mg/mL collagenase type IA. Digest overnight for 12–18 h at 37°C.
3. Disaggregate the pieces by careful pipetting, resuspend in 10 mL of medium, centrifuge at 2000*g* for 5 min, resuspend pellet in 3 mL of fresh medium, transfer to a 3.5-cm culture dish, and allow the cells to adhere and grow for 24–36 h.
4. Wash adherent cells carefully with 1x PBS to remove debris, and add fresh medium.
5. Grow cells to 80% confluency (optimally this will take 3–5 d).

TABLE 1. Transgenic mice

Mouse line	Promoter	Transgene and expression pattern	Source	Reference
Tet-activator (tTA) mice				
Tg(Camk2a-tTA)1Mmay/J	αCaMKII	tTA in forebrain	Jackson Labs Stock 003010	Mayford et al. 1996
Tg(mod-Camk2a-tTA)	Modified αCaMKII	tTA mainly in dentate gyrus, CA1, pyriform cortex	Upon request from Rolf Sprengel[a] (line CN12)	Shimshek et al. 2005
Tg(mod-Camk2a-tTA)	Modified αCaMKII	tTA mainly in dentate gyrus, sparsely in CA1	Upon request from Rolf Sprengel[a] (line CN10)	Niewoehner et al. 2007
Tg(Eno2-tTA)5030Nes/J	enolase (Eno2)	tTA in rat neurons in striatum, lesser in cerebral cortex and hippocampus	Jackson Labs Stock 003763	Chen et al. 1998 (line A)
Tg(Eno2-tTA)5021Nes/J	Eno2	tTA in rat neurons in striatum, cerebellum, CA1, deep neo-cortical layers	Jackson Labs Stock 003767	Chen et al. 1998 (line B)
Tg(Pcp2-tTA)3Horr/J	Regulatory region of murine Pcp2	tTA in Purkinje cells	Jackson Labs Stock 005625	Zu et al. 2004
Tg(Prnp-tTA/F959)	Prion promoter	tTA in cortex and hippocampus, lesser in thalamus, hypothala-mus, and brainstem	Upon request from Stanley B. Prusiner[b]	Tremblay et al. 1998
Tg(P_{nestin}-tTA)	Nestin	tTA in a subset of proliferating SVZ-cells, rostral migratory stream; periglomerular (PG) neurons positive for calretinin, calbindin, GAD67, and tyrosine hydroxylase, but not granule cells	Upon request from Robert D. Beech[c]	Beech et al. 2004
Tg(hGFAP-tTA)110Pop/J	Human glial fibrillary lacidic protein (hGFAP)	tTA in astrocytes	Jackson Labs Stock 005964	Hitoshi et al. 2008
Tg(hGFAP-tTA)	hGFAP	tTA in astrocytes	Upon request from Philip J. Haydon[d]	Pascual et al. 2005
Tet-responder (P_{tet}/P_{tet}bi) mice				
Tg(MTH-Cg2-19)	P_{tet}bi	Camgaroo-2 and firefly luciferase activation in tTA-expressing cells	Upon request from Mazahir T. Hasan[e]	Hasan et al. 2004
Tg(MTH-IP-12)	P_{tet}bi	Inverse pericam and firefly luciferase activation in tTA-expressing cells	Upon request from Mazahir T. Hasan[e]	Hasan et al. 2004
Tg(MTH-YC3.12-1)	P_{tet}bi	YC3.12 and firefly luciferase activation in tTA-expressing cells	Upon request from Mazahir T. Hasan[e]	Hasan et al. 2004
Tg($P_{m\text{-}aMHC(tetO)7}$-GCaMP2)	Weakened α-myosin heavy chain promoter (ΔαMHC) + 7 tet-operator sequences (a P_{tet} variant)	GCaMP2 activation in tTA-expressing cells	Upon request from Michael I. Kotlikoff[f]	Tallini et al. 2006
Transgenic mice with constitutive promoters driving FCIP expression				
Tg(P_{CAGGS}-YC3.0)	β-actin + cytomegalo-virus (CMV) enhancer	Mosaic expression of YC3.0 observed in neocortex and cerebellar Purkinje cells	Upon request from Roger Y. Tsien[g]	Tsai et al. 2003
Tg(P_{CAGGS}-YC3.60pm)	β-actin + CMV enhancer	Membrane targeted YC3.60 for ubiquitous expression	Upon request from Atsushi Miyawaki[h]	Nagai et al. 2004
Tg(P_{Thy1}-CerTN-L15)	Thyl promoter fragment	Expression of CerTN-L15, a calcium sensor, in hippocampal and cortical neurons	Upon request from Oliver Griesbeck[i]	Heim et al. 2007

Contact information: [a]Rolf.Sprengel@mpimf-heidelberg.mpg.de; [b]stanley@ind.ucsf.edu; [c]robert.beech@yale.edu; [d]philip.haydon@tufts.edu; [e]Mazahir.Hasan@mpimf-heidelberg.mpg.de; [f]mik7@cornell.edu; [g]rtsien@ucsd.edu; [h]matsushi@brain.riken.go.jp; [i]griesbeck@neuro.mpg.de.

6. Split the culture into the six wells of a six-well plate. Keep two wells as a stock to expand the cultures for repeating the experiment. The other four wells can be used for plasmid transfection using Lipofectamine 2000, following the transfection protocol recommended by the vendor. For transfection per well, use the following mammalian expression plasmids: 0.5 µg of P_{CMV}-rtTA (Tet-ON) and 0.5 µg of P_{CMV}-*Renilla* luciferase (pRL-CMV).
7. With the rtTA system, treat matching fibroblast cell cultures without and with 1 µg/mL doxycycline (Dox). With the tTA system, use 10 ng/mL Dox in the culture medium.
8. After 48 h, wash cells once with PBS and incubate in 0.5 mL of lysis buffer (from the Promega Dual-Luciferase Reporter Assay System kit) on ice. Aliquots from each lysate can be tested for FL activity and RL activity in a luminometer using the Promega Dual-Luciferase Reporter Assay System. The ratio of FL to RL activity helps to correct for DNA transfection efficiency.

Dox Administration

9. To define the start point of Dox treatment, inject Dox by a single intraperitoneal (i.p.) injection (10 mg/mL in 0.9% NaCl; 100 µg Dox/10 µL per gram of mouse body weight).
10. Continue Dox treatment in drinking water (2 mg/mL Dox supplemented with 5% sucrose) for the remaining time. Change Dox water every 3 d.

    Note that Dox is light-sensitive. Therefore, it is important to provide animals with Dox water in brown bottles.

In Vivo Bioluminescence Imaging

The firefly luciferase is an ATP-dependent enzyme that catalyzes a light-emitting reaction in the presence of ATP and oxygen by converting luciferin into oxyluciferin. Luciferin is a low-molecular-weight organic compound composed of a benzothiazole moiety attached to a thiazole carboxylic acid moiety.

11. Anesthetize mice expressing the firefly luciferase gene with ketamine/xylazine (ketamine, 100 mg/kg body weight; and xylazine, 10 mg/kg body weight) by i.p. injections.
12. Once animals are anesthetized, administer D-luciferin (300 µg/g mouse body weight) by i.p. injection, and place animals into a light-proof chamber.
13. About 10 min after luciferin injection, close the chamber and start the photon counting program for 1–5 min. Collected photons over time generate images that later can be converted to pseudocolor, digitized, and stored in a computer. (See Imaging Setup for equipment and analysis details.)

DISCUSSION

Identifying tTA/rtTA-responsive transgenic founder mice from their ear-fibroblast cell cultures eliminates the tedious procedure of mating them with tTA-expressing mice, genotyping, Dox treatment, brain slicing, and preparation of tissue lysates for quantitative gene expression analysis. With ear-fibroblast cells derived from transgenic founder mice, we can identify high-expressing $P_{tet}/P_{tet}bi$ founders within a week (Hasan et al. 2004). Therefore, this method helps to save both time and resources. We have estimated that ~10% of the founders show Dox-controlled gene expression (Hasan et al. 2004). One potential matter of concern is that different genetic backgrounds can affect the responsiveness of $P_{tet}/P_{tet}bi$ transgenic mice (Robertson et al. 2002). In addition, the $P_{tet}/P_{tet}bi$ transgenes are quite susceptible to epigenetic silencing if they are inactive during development (Zhu et al. 2007). To prevent $P_{tet}/P_{tet}bi$ transgenes from becoming silenced, one possibility is to express tTA at high levels in neurons (Zhu et al. 2007).

Because the P_{tet}bi construct allows for simultaneous and nearly equal levels of dual gene expression, this system is also highly suitable to indirectly report the expression status of a second gene of interest (Hasan et al. 2001). To reversibly manipulate gene activity in selective cell types, the Dox-induced, tTA-dependent system can be used for either gene knockdown (by siRNA) or genetic interference by overexpression of wild-type, dominant-active, or dominant-negative mutants. Simultaneous imaging of neuronal activity with an FCIP can still be performed under control of a strong, constitutive neuron-specific promoter. Such a combinatorial genetic approach should help investigate gene function(s) in neuronal activity dynamics, experience-dependent plasticity, and even behavior.

Protocol B

Calcium Indicator Protein Expression Using rAAVs

This protocol describes the use of rAAVs to express FCIPs in the brain for imaging. HEK 293 cells are first transfected using calcium phosphate. rAAV is then prepared using either an iodixanol gradient or a heparin column. After the virus is purified, its quality is assessed by sodium dodecyl sulfate polyacrylamide gel electrophoresis (SDS-PAGE), estimation of genomic and functional virus titers by quantitative PCR (qPCR), and expression in dissociated neurons. Mice are injected with rAAV using a stereotactic instrument and can be imaged ~3 wk later.

IMAGING SETUP

In this protocol, two different types of imaging experiments are described. The imaging setup for each one is detailed in this section.

In Vivo Imaging in the Olfactory Bulb

Images are collected at rates of 5–15 Hz with a cooled charge-coupled device (CCD) camera (CoolSnapHQ, Photometrics) mounted on a custom-built upright fluorescence microscope with a 20x, 0.95-NA water-immersion objective (Olympus).

In Vivo Two-Photon Imaging

Two-photon excitation is achieved with a mode-locked Ti:sapphire laser (Coherent Mira F900; 930 nm [or 880 nm for CFP excitation], 100 fsec, 78 MHz) coupled into a custom-built imaging system. Images are acquired with a 4x, 0.8-NA (Nikon) or a 20x, 0.95-NA, 0.95 water-immersion lens (Olympus). A photomultiplier-based whole-field detector detects emitted light that is optimized for either YFP alone or ratiometric imaging using two fluorophores (YFP and CFP). For functional imaging, we normally acquire images at a frame scan rate of 15.6 or 8.3 Hz (128 x 64 pixels).

MATERIALS

CAUTION: See Appendix 6 for proper handling of materials marked with <!>.
See the end of the chapter for recipes for reagents marked with <R>.

Reagents

Reagents specific to particular options in the protocol are indicated by the corresponding step numbers.

Benzonase (Sigma-Aldrich; 25,000 U, cat. no. E1014), 300 U/μL stock solution
BES-buffered saline (BBS), 2x <R>
$CaCl_2$ <!>, 2 M stock solution, filter-sterilized
Coomassie Brilliant Blue solution <R>
Destaining solution <R>
Dissociated cultured neurons, for estimation of functional titer (Step 31)
Fetal calf serum (FCS), heat-inactivated (Life Technologies-Gibco)
Human embryonic kidney 293 cells (HEK 293 cells; American Tissue Culture, ATCC no. CRL-1573)
For transfection and virus production, HEK 293 cells of less than 30 passages are recommended.

Ketamine <!>
Minimal Essential Medium (MEM; Invitrogen Life Technologies-Gibco, cat. no. 111090-099)
NaCl
PBS, 1x, filter-sterilized <R>
PBS-MK, 1x <R>
Penicillin/streptomycin <!>/glutamine, 100x (Invitrogen Life Technologies-Gibco, cat. no. 15970-063)
Pentobarbital <!> (for imaging of olfactory bulb; Steps 45–47)
Plasmids <R>

Choose one of the two sets of packaging plasmids for use in the experiment. See Recipes for AAV plasmid preparation method.

AAV packaging plasmids:

Set 1—pDP1, pDP2 (Jürgen A. Kleinschmidt, German Cancer Research Center, Heidelberg, Germany)
Set 2—pFdelta6, pNLrep, and pH21 (Matthias J. During, Auckland, New Zealand)

AAV plasmids:

pAAV-hSYN-eGFP (pAAV-6P-SEWB)—A plasmid equipped with a human synapsin promoter to drive gene expression (Sebastian Kügler and Mazahir T. Hasan)
pAM-CAG-pL-WPRE-βGH-pA—A plasmid equipped with a CMV enhancer and a chicken β-actin promoter to drive gene expression (Matthias J. During)
pAAV-P_{tet}bi—A plasmid for simultaneous expression of two different genes (Mazahir T. Hasan and Rolf Sprengel)
pAAV-hSYN-tTA and pAAV-hSYN-rtTA—Two plasmids for expressing tTA and rtTA, respectively, under control of a human synapsin promoter (Mazahir T. Hasan and Rolf Sprengel)

Reagents for in vivo two-photon imaging (Steps 48–52)
Ringer buffer
Urethane <!> (for anesthesia)
Reagents for rAAV purification with a heparin column (Steps 32–41)
Cell lysis buffer 2 (150 mM NaCl, 20 mM Tris•HCl at pH 8)
Solution A (100 mM NaCl, 20 mM Tris•HCl at pH 8)
Solution B (200 mM NaCl, 20 mM Tris•HCl at pH 8)
Solution C (300 mM NaCl, 20 mM Tris•HCl at pH 8)
Solution D (400 mM NaCl, 20 mM Tris•HCl at pH 8)
Solution E (450 mM NaCl, 20 mM Tris•HCl at pH 8)
Solution F (500 mM NaCl, 20 mM Tris•HCl at pH 8)
Reagents for rAAV purification with an iodixanol gradient (Steps 17–31)
Cell lysis buffer 1 (150 mM NaCl, 50 mM Tris•HCl at pH 8)
Iodixanol (Progen Biotechnik, OptiPrep, cat no. 1114542)
Real-time PCR system (e.g., Applied Biosystems 7500 Fast Real-Time PCR System)

This system provides a comprehensive protocol, reagents, and equipment for quantitative analysis of DNA concentration in a sample. Analysis of DNA-containing viruses can be performed as described previously (Rohr et al. 2005).

SDS-PAGE gel <R>
SDS-PAGE 10x SDS running buffer <R>
SDS-PAGE Laemmli buffer (5x) <R>
Size markers for DNA (Biolabs)
Size markers for protein (Bio-Rad)
Sodium deoxycholate (NaDOC) <!> (Sigma-Aldrich, 5 g, D5670-5G), 10% stock
Xylazine <!>

Equipment

Equipment specific to particular options in the protocol is indicated by the corresponding step numbers.

Acrodisc syringe filters (for filtering small volumes), 0.22 μm (Millipore; Milliex-GV, cat. no. SLGV0113SL)
Cell scraper, disposable (Greiner Bio-One GmbH, 541070)
Centrifuge tubes, PC-26.3, polycarbonate bottle assembly (six per pack; Beckman cat. no. 355618)
Equipment for in vivo two-photon imaging (Steps 48–52)
  Cyanoacrylate glue (UHU) <!>
  Heating blanket (FHC)
  Rectal temperature probe
  Removable head plate with a cranial window of 3 mm in diameter (W. Denk and M.T. Hasan, unpubl.)
Equipment for rAAV purification by heparin column (Steps 32–41)
  HiTrap heparin columns (Amersham)
  Syringe pump (World Precision Instruments, SP200IZ)
Equipment for rAAV purification by iodixanol column (Steps 17–31)
  Liquid nitrogen <!>
  Needle, 20-G x 2¾ inches, bent at a 90° angle
Equipment for SDS-PAGE (e.g., gel casting plates, electrophoresis unit, power supply)
Flow dilution olfactometer (for imaging of olfactory bulb; Steps 45–47)
  We use a custom-built unit (see Spors et al. 2006).
Glass capillary (tip size 8–12 μm)
Imaging setup (as described above)
Incubators, 37°C, one with 3% CO_2 and one with 5% CO_2
Luminescent image analyzer (Fuji LAS-3000) for quantifying SDS-PAGE gel bands
Pipettes, 25-mL, 10-mL, 5-mL, and 1-mL, sterile
Stereotactic instrument (963 Ultra Precise Small Animal; David Kopf Instruments)
Test tubes, 50-mL conical (Falcon)
Ultracentrifuge and rotor (Beckman 70-TI/70.1-TI)
Virus concentrator (Amicon-Ultra centrifugal filters; Millipore, UFC910024)

EXPERIMENTAL METHOD

Calcium Phosphate–Mediated Gene Transfection of HEK 293 Cells for Producing rAAVs

1. Plate cells on 15-cm plates with 25 mL of culture medium (MEM supplemented with 1x penicillin/streptomycin/glutamine and 10% FCS). On the day of transfection, the cells should be 40% confluent.

2. Prepare a master mix of the plasmids for transfection.

   Packaging plasmids: Use either set 1 or set 2.

   Set 1—12.5 μg of pDP1, 12.5 μg of pDP2, and 25 μg of pAAV (i.e., one of the plasmids listed under "AAV Plasmids" in the Materials list)

   Set 2—25 μg of pFdelta6, 6 μg of pNLrep, 6.25 μg of pH21, and 12.5 μg of pAAV

   Amounts given are per 15-cm plate; scale up the master mix depending on the number of plates.

   For virus preparation, we recommend 20 to 40 plates per construct. Because quality of plasmid DNA seriously affects transfection efficiency, we recommend using high-quality plasmid DNA.

3. Thaw transfection reagents at least 2 h before transfection. To ensure that the concentration of each solution is homogeneous, mix well.

4. At 2–3 h before transfection, remove growth medium from each plate and replace with 25 mL of pre-warmed fresh medium.
5. Pre-warm H_2O, 2 M $CaCl_2$, and 2x BBS to 37°C.
6. Add pre-warmed water to the plasmid mixture (prepared in Step 2) to a volume of 437.5 µL.
7. Add 62.5 µL of 2 M $CaCl_2$, vortex it, and add 500 µL of 2x BBS dropwise while gently mixing; incubate for 10 min at room temperature.
8. Transfect 15-cm plates with appropriate constructs (50 µg total DNA/plate).
9. Add dropwise (~100 µL per drop) transfection solution on the plate without medium change; swirl the plate so that the DNA-containing transfection solution is mixed well in the medium.
10. Place cells in 3% CO_2 incubator for 24 h.
11. Change medium and incubate for another 24 h in 5% CO_2.
12. At 48–60 h post-transfection, check for fraction of transfected cells (red fluorescence). It should be >80%. If this is the case, proceed to the next step.
13. Scrape transfected cells from the plate without removing the medium. This is important because transfected cells tend to detach and start floating in the medium. Transfer the cell suspension to 50-mL Falcon tubes. For 10 x 15–cm plates, use 4 x 50–mL Falcon tubes.
14. Pellet cells at 1000*g* for 5 min at 4°C.
15. Discard supernatant and resuspend cell pellets in 10 mL of PBS (filter-sterilized) and transfer into a single 50-mL Falcon tube. Spin again at 2000*g* for 10 min to pellet.
16. Proceed to virus purification using either an iodixanol gradient (Step 17) or a heparin column (Step 32) as described below.

Purification of rAAV

Purify the rAAV using either an iodixanol gradient (Steps 17–31) or a heparin column (Steps 32–41). After purification, continue with Step 42. Commercial services for rAAV purification are also available (http://www.vectorbiolabs.com/vbs/index.html and http://www.PlasmidFactory.com).

Purification of rAAV by an Iodixanol Gradient

17. Resuspend the cell pellet (0.5 mL) from Step 15 above in 9 mL of lysis buffer 1, add 500 µL of NaDOC (10%) to yield a final concentration of 0.5%, and mix well by pipetting up and down several times.
18. Add Benzonase to a final concentration of 50 U/mL, and mix well by pipetting up and down several times to ensure that Benzonase has thoroughly mixed with the lysate. Incubate the lysate for 1 h at 37°C.

    Optional: The crude lysate can be frozen at –70°C at this step.
19. To thaw, incubate the lysate for 20 min at 37°C. Remove cell debris by centrifugation at 4000*g* for 15 min at 4°C.

    Optional: The crude lysate can be frozen at –70°C at this step.
20. To thaw, incubate the lysate for 20 min at 37°C. Add 584 mg of NaCl to the lysate and incubate for 10 min at room temperature. Do three freeze–thaw cycles in liquid nitrogen and 37°C, respectively. It takes ~15 min for each freeze–thaw cycle. Spin at 3000*g* for 15 min, and transfer the clear crude lysate into a fresh 50-mL Falcon tube. If the lysate is still not clear, follow the re-centrifugation step once or twice more.
21. Prepare 15%, 25%, 40%, and 54% iodixanol solutions with 1x PBS-MK (iodixanol comes as a 60% solution).

22. Load 6 mL of 15% iodixanol in Beckman centrifuge tubes. Underlay the 15% iodixanol with 4 mL of 25% iodixanol, followed by 3 mL of 40% iodixanol, and finally 3 mL of 54% iodixanol. Four discernible phases should be visible. With a permanent pen marker, label the boundary between the phases of 40% and 54% iodixanol.
23. Gently load the cell lysate (~9 mL) on top of the gradient. Set aside ~100 µL of the unpurified lysate for analysis on SDS-PAGE.
24. Spin in an ultracentrifuge at 60,000 rpm for 1.5 h at 18°C.
25. The virus should be extracted from the 40%–54% iodixanol interphase as well as the majority of the 40% iodixanol phase using a 5-mL syringe attached to a 20-G x 2¾-inch needle bent at a 90° angle. Make sure that the beveled side of the needle is facing downward. Avoid removing the "white film" above and below the 40% iodixanol layer.
26. Put the collected fraction in an Amicon concentrator (100 kDa), and fill the remaining volume with 1x PBS-MK, mix well, and centrifuge at 3000*g* for 5 min. Discard the flow-through, refill the reservoir with 1x PBS-MK again, and repeat the whole procedure twice more.
27. After the last spin, transfer the dialyzed and concentrated virus suspension (250–300 µL) into a microcentrifuge tube, and filter-sterilize it using an Acrodisc filter (0.22 µm) that is 13 mm in diameter. Do not use larger filter devices!
28. Store viruses at –70°C as 10-µL aliquots in small microcentrifuge tubes.
29. Analyze unpurified and purified viruses on SDS-PAGE.
    i. Prepare an SDS-PAGE gel (see Recipes).
    ii. Take 10 µL of virus mix and add 2.5 µL of loading buffer (5x Laemmli buffer).
    iii. Heat samples for 5 min at 99°C, then centrifuge briefly at 1000*g*. Load virus samples and molecular weight markers in gel slots. Run samples in the stacking gel at 80 V and in the separation gel at 150 V.
    iv. Disassemble the gel, stain in Coomassie Brilliant Blue solution for 1 h, and destain the gel with the destaining solution once for 10 min and one more time for 45 min. If gel has destained well, wash twice in water for 10 min. To further destain the background, leave the gel in water overnight at room temperature on a tray set on a shaker. Gel bands can be quantified with a luminescent image analyzer.

        The purified virus should show three predominant bands that correspond to the capsid proteins (VP1, VP2, and VP3) (Fig. 1C).
30. Perform a quantitative polymerase chain reaction (qPCR) for estimating viral genome particles per milliliter.

    We use the Applied Biosystems 7500 Fast Real-Time PCR System, following the manufacturer's instructions.
31. To estimate functional titer, infect dissociated cultured neurons with serially diluted purified rAAVs. Two weeks after infection, calculate the fraction of GFP-positive neurons.

Purification of rAAV by a Heparin Column

The heparin column purification method is suitable for serotype-2-containing viruses. Serotype-2 viruses use the heparan sulfate proteoglycan as its primary cell surface attachment receptor for entering cells.

32. Resuspend the cell pellet from Step 15 above in 50 mL of lysis buffer 2, split into two 50-mL Falcon tubes to help handle the volume adequately. Into each tube, add 1.25 mL of NaDOC (10%) to give a final concentration of 0.5% NaDOC and mix well by pipetting up and down several times.

33. Add Benzonase to a final concentration of 50 U/mL, and mix well by pipetting up and down several times to ensure that Benzonase has thoroughly mixed with the lysate. Incubate the lysate for 1 h at 37°C.

    Optional: The crude lysate can be frozen at –70°C at this step.

34. To thaw, incubate lysate for 20 min at 37°C. Remove cell debris by centrifugation at 4000*g* for 15 min at 4°C.

    Optional: The crude lysate can be frozen at –70°C at this step.

35. To thaw, incubate lysate for 20 min at 37°C. Do three freeze–thaw cycles in liquid nitrogen and 37°C, respectively. It takes ~15 min for each freeze–thaw cycle. Spin at 3000*g* for 15 min, and transfer the clear crude lysate into a fresh 50-mL Falcon tube. If the lysate is still not clear, follow the recentrifugation step once or twice more. Freeze the crude lysate at –70°C and prepare for heparin column purification steps.

36. Equilibrate a heparin column with 10 mL of cell lysis buffer 2 (150 mM NaCl, 20 mM Tris•HCl at pH 8). Use a 60-mL syringe pump with the attachment provided and pump solution at a flow rate of 1 mL/min.

37. Disconnect the heparin column, and fill the 60-mL syringe with 50 mL of crude clear lysate. Reconnect the heparin column to the syringe, and pump lysate through it at a flow rate of 1 mL/min. Set aside ~100 µL of unpurified lysate for analysis on SDS-PAGE.

38. Again disconnect the heparin column, and fill the 60-mL syringe with 20 mL of Solution A for a first wash. Reconnect the heparin column to the syringe, and pump Solution A through it at a flow rate of 1 mL/min.

39. The next steps can be done manually (without using the pump). Disassemble the heparin column, connect it to a 1-mL syringe, and gently and slowly wash the heparin column first with 1 mL of Solution B and finally with 1 mL of Solution C.

40. For the elution steps, use a 3-mL syringe. Elute successively with 1.5 mL of Solution D, 3 mL of Solution E, and 1.5 mL of Solution F.

41. Dialyze and concentrate the 6 mL of eluted volume, and analyze as described above for the iodixanol gradient purification method (Steps 26–31).

Stereotactic Injection of rAAV

For FCIP expression in vivo, we have successfully used 50–250 nL of high-titer rAAV to label neurons in selective brain regions. Virus suspension with 10% D-mannitol solution for in vivo injection (Mastakov et al. 2001) increased the infection diameter and even expression levels. Systematic control of virus concentration, amounts, and the use of D-mannitol should allow one to label neurons both sparsely and densely and with small and large expression diameters in the brain. We have been able to label neurons in vivo with expression diameters of 0.3–2 mm.

42. Anesthetize a mouse with ketamine plus xylazine by i.p. injection (ketamine, 100 mg/kg body weight; xylazine, 10 mg/kg body weight), and fix it on a stereotactic instrument.

43. Load 150–300 nL of purified virus suspension into a glass capillary (tip size 8–12 µm), and gently lower the glass pipette into the brain to a depth of ~250 µm, over which a small craniotomy has been performed, 1.5 mm posterior to the bregma and 3 mm lateral to the midline for the whisker-related somatosensory cortex.

44. Follow the stereotactic procedure for virus injection as described previously (Cetin et al. 2006).

    The mice can be used for functional analyses ~3 wk after the rAAV injection.

In Vivo Imaging

The methods below provide examples of two different imaging experiments.

In Vivo Imaging in the Olfactory Bulb

45. Anesthetize mice with pentobarbital, and perform surgical procedures as described previously (Wachowiak and Cohen 2001).
46. Deliver odors through a flow dilution olfactometer.
47. Collect images at rates of 5–15 Hz (see Imaging Setup for equipment details).

In Vivo Two-Photon Imaging

48. Anesthetize mice with urethane (1–2 g/kg), and monitor anesthesia by pinching one of the paws or the tail. Monitor core body temperature by a rectal probe and maintain at 37°C by a heating blanket (FHC). Apply local anesthesia (lidocaine, as recommended by the vendor) underneath the head skin before beginning surgical procedures (Cetin et al. 2006).
49. Glue a removable head plate with a cranial window of 3 mm in diameter to the top of the skull using cyano-acrylate, and attach it to a fixed bar before thinning the skull.

    The combination of a rigid head plate and a thinned skull reduces respiration motion as well as cardiac-pulsation-induced brain motion.
50. Thin the exposed skull through the cranial window (50 µm in thickness) using a dental drill.

    After the head-plate chamber is filled with Ringer's buffer, blood vessels become clearly visible under the binocular.
51. Position the two-photon microscope objective so that the optical axis is perpendicular to the surface of the cortex.
52. Perform two-photon imaging (see Imaging Setup for equipment details).

    Although we image mice for hours, each field of view is usually imaged in short time blocks. In whisker-stimulation experiments, we imaged mice through the thinned skull at cortical depths of 50 µm and 250 µm with laser powers of ~50 mW and ~150 mW, respectively, after the objective.

DISCUSSION

A variety of DNA expression vectors are available to generate recombinant adeno-associated viruses for cell type–specific, constitutive (Xu et al. 2001; Shevtsova et al. 2005; Xu et al. 2005) and inducible (Zhu et al. 2007) gene expression both in vitro and in vivo. These viruses are powerful tools for investigating the role of a particular gene product in cellular signaling and animal behavior. For functional brain imaging, high-titer, high-quality viruses are required for recording functional Ca^{2+} signals without tissue damage. The first critical step to produce high-quality viruses is to ensure that the AAV-plasmid transfection efficiency in HEK 293 cells is >80%. Once this is accomplished, a suitable virus purification procedure should be selected. Although there are different methods for virus purification (Hermens et al. 1999; Auricchio et al. 2001; Zolotukhin et al. 2002; Hauck et al. 2003; Zeltner et al. 2010), here we recommend virus purification by either iodixanol gradient or heparin column.

We recommend that virus titers be determined in dissociated neurons and later functionally tested for Ca^{2+} transients in cultured hippocampal organotypic slices. This saves both time and resources. We usually infect hippocampal organotypic slices with viruses by applying them in the medium (2.5–5 µL/mL of medium) and can achieve robust FCIP expression without toxicity. Alternatively, viruses can be applied directly on top of hippocampal organotypic slices. This also allows one to test for toxicity. If no toxic effects are detected in cultured brain slices, we proceed with virus injection into the cortex for in vivo imaging. With a functional virus titer of $>2–5 \times 10^8$ transducing units per microliter, injection of 50–200 nL of virus in 10% D-mannitol can provide high expression levels in neurons, which is suitable for long-term functional in vivo two-photon imaging. In our experience, 10–60 µM fluorescein equivalent units (FEUs) of FCIP in neurons should be sufficient to detect single action potentials in vivo (Wallace et al. 2008).

FUTURE PROSPECTS

With the recently improved FCIPs and expression strategies, it is now possible to record sensory-evoked neuronal activity from a large number of cells in vivo, both acutely (Hasan et al. 2004; Mank et al. 2008; Wallace et al. 2008; Tian et al. 2009; Lütcke et al. 2010) and chronically (Mank et al. 2008; Tian et al. 2009), and in freely moving mice (Lütcke et al. 2010). Both spontaneous and sensory-evoked Ca^{2+} transients can be detected in neurons in vivo with single-cell, single-AP resolution. However, new challenges lie ahead including further optimization of FCIP design and their targeting strategies. To improve signal-to-noise ratios for detecting single-AP responses both in vitro and in vivo, it is necessary to express FCIPs at high levels in neurons. However, high FCIP levels in neurons can be problematic: (1) binding of intracellular free Ca^{2+} to a large number of FCIP molecules could lower the peak amplitude of the stimulus-induced Ca^{2+} transient; and (2) high FCIP levels might contribute to Ca^{2+} buffering and affect cell physiology. The other issue relates to Ca^{2+}-binding affinity and high cooperativity, both of which contribute to rapid sensor saturation to increasing numbers of APs. Although high cooperativity with GCaMPs may be suitable for detection of single APs with high signal-to-noise ratios, FCIPs that respond more linearly to increasing numbers of APs would be preferred to detect variable spike rates.

To overcome these shortcomings, efforts should be focused to generate an "ideal" FCIP that should be bright, photostable, and also of low affinity and which, when expressed in neurons even at low levels, should still be able to detect single APs with high SNRs both in vitro and in vivo. The ideal FCIP should furthermore linearly increase in response amplitude with increasing numbers of APs and also with faster rise and decay kinetics to detect high-frequency spike rates. Such FCIPs would also tolerate longer imaging time, for example, by in vivo two-photon imaging, with little photobleaching.

RECIPES

CAUTION: See Appendix 6 for proper handling of materials marked with <!>.
Recipes for reagents marked with <R> are included in this list.

AAV Plasmid Preparation

Reagents

SURE 2 super-competent cells (Stratagene, cat. no. 200152) transformed with AAV plasmid
HiSpeed Plasmid Maxi Kit (QIAGEN, cat. no. 12662)
Ampicillin <!> (Sigma-Aldrich, cat. no. A0166)
Terrific Broth (TB):

TB-A:

Yeast extract	24 g
Tryptone	12 g
Glycerol	4 g

Bring final volume to 900 mL with MilliQ water; autoclave.

TB-B:

KH_2PO_4	2.3 g (0.17 M)
K_2HPO_4	16.4 g (0.72 M)

Bring final volume to 100 mL with MilliQ water; autoclave.

Method

To grow bacteria:

1. Mix 900 mL of TB-A with 100 mL of TB-B.
2. Add ampicillin (final concentration 100 µg/mL).
3. Inoculate with 1–3 mL of bacterial culture (AAV plasmid-transformed SURE 2 cells).
4. Grow 1 L of bacterial culture for 24 h in a 4-L autoclaved flask with constant shaking and aeration at 37°C.
5. Isolate plasmids using the HiSpeed Plasmid Maxi Kit according to the manufacturer's instructions.

BES-Buffered Saline (BBS), 2x

N,N-bis(2-hydroxyethyl)-2-amino-ethanesulfonic acid (BES), pH 6.95	50 mM
NaCl	280 mM
Na_2HPO_4	1.5 mM

With 2 M NaOH <!> and 2 M HCl <!>, adjust the pH to 6.95. Filter-sterilize by passing through 0.2-µm nitrocellulose filters (Nalge), and store aliquots at –20°C.

Coomassie Brilliant Blue Staining Solution

1. Prepare 1 L of staining solution:

   1.06% Coomassie Brilliant Blue <!>
   50% methanol <!>
   10% acetic acid <!> in H_2O

2. Mix well.
3. Filter through Whatman #1 paper.

Destaining Solution

Prepare 50% methanol <!>, 7.5% acetic acid <!> in H_2O. Mix well.

PBS, 10x and 1x

10.9 g of anhydrous Na_2HPO_4, 3.2 g of anhydrous NaH_2PO_4, and 9 g of NaCl, raised to a final volume of 1000 mL with distilled water (pH 7.2). To make 1 L of 1x PBS, thoroughly mix 100 mL of 10x PBS with 900 mL of distilled water.

PBS-MK, 1x

Prepare filter-sterilized 1x PBS supplemented with 10 mM $MgCl_2$ and 25 mM KCl.

SDS-PAGE Gel

1. Prepare the separation gel (10%). Mix in the following order:

H_2O	4.1 mL
Acrylamide/bis <!> (30% 37.5:1; Bio-Rad)	3.3 mL
Tris·HCl <!> (1.5 M, pH 8.8)	2.5 mL
SDS <!>, 10%	100 µL
N,N,N′,N′-tetramethylethylene-diamine (TEMED) <!> (Bio-Rad)	10 µL
Ammonium persulfate (APS) <!>, 10%	32 µL

After adding TEMED and APS to the SDS-PAGE separation gel solution, the gel will polymerize quickly, so add these two reagents when ready to pour.

2. Pour gel, leaving ~2 cm below the bottom of the comb for the stacking gel. Make sure to remove bubbles.
3. Layer the top of the gel with isopropanol <!>. This will help to remove bubbles at the top of the gel and will also keep the polymerized gel from drying out. In ~30 min, the gel should be completely polymerized.
4. Remove the isopropanol and wash out the remaining traces of isopropanol with distilled water.
5. Prepare the stacking gel (4%). Mix in the following order:

H_2O	6.1 mL
Acrylamide/bis (30%, 37.5:1) <!>	1.3 mL
Tris·HCl (0.5 M, pH 6.8)	2.5 mL
SDS, 10%	100 µL
TEMED <!>	10 µL
Ammonium persulfate (APS) <!>, 10%	100 µL

6. Pour stacking gel on top of the separation gel.
7. Add combs to make wells. In ~30 min, the stacking gel should become completely polymerized.
8. Clamp gel into apparatus, and fill both buffer chambers with gel running buffer according to the instructions for the specific apparatus.
9. Load samples and molecular weight protein markers into wells for separation by electrophoresis.

SDS-PAGE 10x SDS Running Buffer

Tris base	30.3 g
Glycine	144.4 g
SDS <!>	10 g

Dissolve in 1 L of MilliQ-filtered H_2O.

SDS-PAGE Laemmli Buffer (5x)

Tris·HCl (pH 6.8)	312.5 mM
SDS <!>	10%
Glycerol	25%
Bromophenol blue <!>	0.5%

Add 10% β-mercaptoethanol <!> before use.

ACKNOWLEDGMENTS

This work was supported by the Max Planck Society, the Schloessmann Foundation, Collaborative Research Center (SFB 488), Fritz Thyssen Stiftung, and Deutsche Forschungsgemeinschaft.

REFERENCES

Auricchio A, Hildinger M, O'Connor E, Gao GP, Wilson JM. 2001. Isolation of highly infectious and pure adeno-associated virus type 2 vectors with a single-step gravity-flow column. *Hum Gene Ther* **12:** 71–76.

Baird GS, Zacharias DA, Tsien RY. 1999. Circular permutation and receptor insertion within green fluorescent proteins. *Proc Natl Acad Sci* **96:** 11241–11246.

Barton MD, Dunlop JW, Psaltis G, Kulik J, DeGennaro L, Kwak SP. 2002. Modified GFAP promoter auto-regulates tet-activator expression for increased transactivation and reduced tTA-associated toxicity. *Brain Res Mol Brain Res* **101:** 71–81.

Beech RD, Cleary MA, Treloar HB, Eisch AJ, Harrist AV, Zhong W, Greer CA, Duman RS, Picciotto MR. 2004. Nestin promoter/enhancer directs transgene expression to precursors of adult generated periglomerular neurons. *J Comp Neurol* **475:** 128–141.

Cetin A, Komai S, Eliava M, Seeburg PH, Osten P. 2006. Stereotaxic gene delivery in the rodent brain. *Nat Protoc* **1:** 3166–3173.

Chalfie M, Tu Y, Euskirchen G, Ward WW, Prasher DC. 1994. Green fluorescent protein as a marker for gene expression. *Science* **263:** 802–805.

Chen J, Kelz MB, Zeng G, Sakai N, Steffen C, Shockett PE, Picciotto MR, Duman RS, Nestler EJ. 1998. Transgenic animals with inducible, targeted gene expression in brain. *Mol Pharmacol* **54:** 495–503.

Fiala A, Spall T, Diegelmann S, Eisermann B, Sachse S, Devaud JM, Buchner E, Galizia CG. 2002. Genetically expressed cameleon in *Drosophila melanogaster* is used to visualize olfactory information in projection neurons. *Curr Biol* **12:** 1877–1884.

Griesbeck O, Baird GS, Campbell RE, Zacharias DA, Tsien RY. 2001. Reducing the environmental sensitivity of yellow fluorescent protein. Mechanism and applications. *J Biol Chem* **276:** 29188–29194.

Hasan MT, Schonig K, Berger S, Graewe W, Bujard H. 2001. Long-term, noninvasive imaging of regulated gene expression in living mice. *Genesis* **29:** 116–122.

Hasan MT, Friedrich RW, Euler T, Larkum ME, Giese G, Both M, Duebel J, Waters J, Bujard H, Griesbeck O, et al. 2004. Functional fluorescent Ca^{2+} indicator proteins in transgenic mice under TET control. *PLoS Biol* **2:** e163. doi: 10.1371/journal.pbio.0020163.

Hauck B, Chen L, Xiao W. 2003. Generation and characterization of chimeric recombinant AAV vectors. *Mol Ther* **7:** 419–425.

Heim N, Griesbeck O. 2004. Genetically encoded indicators of cellular calcium dynamics based on troponin C and green fluorescent protein. *J Biol Chem* **279:** 14280–14286.

Heim N, Garaschuk O, Friedrich MW, Mank M, Milos RI, Kovalchuk Y, Konnerth A, Griesbeck O. 2007. Improved calcium imaging in transgenic mice expressing a troponin C-based biosensor. *Nat Methods* **4:** 127–129.

Hermens WT, ter Brake O, Dijkhuizen PA, Sonnemans MA, Grimm D, Kleinschmidt JA, Verhaagen J. 1999. Purification of recombinant adeno-associated virus by iodixanol gradient ultracentrifugation allows rapid and reproducible preparation of vector stocks for gene transfer in the nervous system. *Hum Gene Ther* **10:** 1885–1891.

Higashijima S, Masino MA, Mandel G, Fetcho JR. 2003. Imaging neuronal activity during zebrafish behavior with a genetically encoded calcium indicator. *J Neurophysiol* **90:** 3986–3997.

Hires SA, Tian L, Looger LL. 2008. Reporting neural activity with genetically encoded calcium indicators. *Brain Cell Biol* **36:** 69–86.

Hitoshi Y, Harris BT, Liu H, Popko B, Israel MA. 2008. Spinal glioma: Platelet-derived growth factor B–mediated oncogenesis in the spinal cord. *Cancer Res* **68:** 8507–8515.

Kerr R, Lev-Ram V, Baird G, Vincent P, Tsien RY, Schafer WR. 2000. Optical imaging of calcium transients in neurons and pharyngeal muscle of *C. elegans*. *Neuron* **26:** 583–594.

Liu L, Yermolaieva O, Johnson WA, Abboud FM, Welsh MJ. 2003. Identification and function of thermosensory neurons in *Drosophila* larvae. *Nat Neurosci* **6:** 267–273.

Lu Y. 2004. Recombinant adeno-associated virus as delivery vector for gene therapy—A review. *Stem Cells Dev* **13:** 133–145.

Lütcke H, Murayama M, Hahn T, Margolis DJ, Astori S, Zum Alten Borgloh SM, Gobel W, Yang Y, Tang W, Kugler S, et al. 2010. Optical recording of neuronal activity with a genetically-encoded calcium indicator in anesthetized and freely moving mice. *Front Neural Circuits* **4:** 9. doi: 10.3389/fncir.2010.00009.

Mank M, Reiff DF, Heim N, Friedrich MW, Borst A, Griesbeck O. 2006. A FRET-based calcium biosensor with fast signal kinetics and high fluorescence change. *Biophys J* **90:** 1790–1796.

Mank M, Santos AF, Direnberger S, Mrsic-Flogel TD, Hofer SB, Stein V, Hendel T, Reiff DF, Levelt C, Borst A, et al. 2008. A genetically encoded calcium indicator for chronic in vivo two-photon imaging. *Nat Methods* **5:** 805–811.

Mann JR, McMahon AP. 1993. Factors influencing frequency production of transgenic mice. *Methods Enzymol* **225:** 771–781.

Mastakov MY, Baer K, Xu R, Fitzsimons H, During MJ. 2001. Combined injection of rAAV with mannitol enhances gene expression in the rat brain. *Mol Ther* **3:** 225–232.

Mayford M, Bach ME, Huang YY, Wang L, Hawkins RD, Kandel ER. 1996. Control of memory formation through regulated expression of a CaMKII transgene. *Science* **274:** 1678–1683.

Miyawaki A, Llopis J, Heim R, McCaffery JM, Adams JA, Ikura M, Tsien RY. 1997. Fluorescent indicators for Ca^{2+} based on green fluorescent proteins and calmodulin. *Nature* **388:** 882–887.

Nagai T, Yamada S, Tominaga T, Ichikawa M, Miyawaki A. 2004. Expanded dynamic range of fluorescent indicators for Ca^{2+} by circularly permuted yellow fluorescent proteins. *Proc Natl Acad Sci* **101:** 10554–10559.

Nakai J, Ohkura M, Imoto K. 2001. A high signal-to-noise Ca^{2+} probe composed of a single green fluorescent protein. *Nat Biotechnol* **19:** 137–141.

Niewoehner B, Single FN, Hvalby O, Jensen V, Meyer zum Alten Borgloh S, Seeburg PH, Rawlins JN, Sprengel R, Bannerman DM. 2007. Impaired spatial working memory but spared spatial reference memory following functional loss of NMDA receptors in the dentate gyrus. *Eur J Neurosci* **25:** 837–846.

Palmer AE, Giacomello M, Kortemme T, Hires SA, Lev-Ram V, Baker D, Tsien RY. 2006. Ca^{2+} indicators based on computationally redesigned calmodulin-peptide pairs. *Chem Biol* **13:** 521–530.

Pascual O, Casper KB, Kubera C, Zhang J, Revilla-Sanchez R, Sul JY, Takano H, Moss SJ, McCarthy K, Haydon PG. 2005. Astrocytic purinergic signaling coordinates synaptic networks. *Science* **310:** 113–116.

Prasher DC. 1995. Using GFP to see the light. *Trends Genet* **11:** 320–323.

Prasher DC, Eckenrode VK, Ward WW, Prendergast FG, Cormier MJ. 1992. Primary structure of the *Aequorea victoria* green-fluorescent protein. *Gene* **111:** 229–233.

Reiff DF, Thiel PR, Schuster CM. 2002. Differential regulation of active zone density during long-term strengthening of *Drosophila* neuromuscular junctions. *J Neurosci* **22:** 9399–9409.

Robertson A, Perea J, Tolmachova T, Thomas PK, Huxley C. 2002. Effects of mouse strain, position of integration and tetracycline analogue on the tetracycline conditional system in transgenic mice. *Gene* **282:** 65–74.

Rohr UP, Heyd F, Neukirchen J, Wulf MA, Queitsch I, Kroener-Lux G, Steidl U, Fenk R, Haas R, Kronenwett R. 2005. Quantitative real-time PCR for titration of infectious recombinant AAV-2 particles. *J Virol Methods* **127:** 40–45.

Shevtsova Z, Malik JM, Michel U, Bahr M, Kugler S. 2005. Promoters and serotypes: Targeting of adeno-associated virus vectors for gene transfer in the rat central nervous system in vitro and in vivo. *Exp Physiol* **90:** 53–59.

Shimshek DR, Bus T, Kim J, Mihaljevic A, Mack V, Seeburg PH, Sprengel R, Schaefer AT. 2005. Enhanced odor discrimination and impaired olfactory memory by spatially controlled switch of AMPA receptors. *PLoS Biol* **3:** e354. doi: 10.1371/journal.pbio.0030354.

Spors H, Wachowiak M, Cohen LB, Friedrich RW. 2006. Temporal dynamics and latency patterns of receptor neuron input to the olfactory bulb. *J Neurosci* **26:** 1247–1259.

Sprengel R, Hasan MT. 2007. Tetracycline-controlled genetic switches. *Handb Exp Pharmacol* **2007:** 49–72.

Stettler DD, Yamahachi H, Li W, Denk W, Gilbert CD. 2006. Axons and synaptic boutons are highly dynamic in adult visual cortex. *Neuron* **49:** 877–887.

Suzuki H, Kerr R, Bianchi L, Frokjaer-Jensen C, Slone D, Xue J, Gerstbrein B, Driscoll M, Schafer WR. 2003. In vivo imaging of *C. elegans* mechanosensory neurons demonstrates a specific role for the MEC-4 channel in the process of gentle touch sensation. *Neuron* **39:** 1005–1017.

Tallini YN, Ohkura M, Choi BR, Ji G, Imoto K, Doran R, Lee J, Plan P, Wilson J, Xin HB, et al. 2006. Imaging cellular signals in the heart in vivo: Cardiac expression of the high-signal Ca^{2+} indicator GCaMP2. *Proc Natl Acad Sci* **103:** 4753–4758.

Tian L, Hires SA, Mao T, Huber D, Chiappe ME, Chalasani SH, Petreanu L, Akerboom J, McKinney SA, Schreiter ER, et al. 2009. Imaging neural activity in worms, flies and mice with improved GCaMP calcium indicators. *Nat Methods* **6:** 875–881.

Tremblay P, Meiner Z, Galou M, Heinrich C, Petromilli C, Lisse T, Cayetano J, Torchia M, Mobley W, Bujard H, et al. 1998. Doxycycline control of prion protein transgene expression modulates prion disease in mice. *Proc Natl Acad Sci* **95:** 12580–12585.

Truong K, Sawano A, Mizuno H, Hama H, Tong KI, Mal TK, Miyawaki A, Ikura M. 2001. FRET-based in vivo Ca^{2+} imaging by a new calmodulin-GFP fusion molecule. *Nat Struct Biol* **8:** 1069–1073.

Tsai PS, Friedman B, Ifarraguerri AI, Thompson BD, Lev-Ram V, Schaffer CB, Xiong Q, Tsien RY, Squier JA, Kleinfeld D. 2003. All-optical histology using ultrashort laser pulses. *Neuron* **39:** 27–41.

Wachowiak M, Cohen LB. 2001. Representation of odorants by receptor neuron input to the mouse olfactory bulb. *Neuron* **32:** 723–735.

Wallace DJ, Zum Alten Borgloh SM, Astori S, Yang Y, Bausen M, Kugler S, Palmer AE, Tsien RY, Sprengel R, Kerr JN, et al. 2008. Single-spike detection in vitro and in vivo with a genetic Ca^{2+} sensor. *Nat Methods* **5:** 797–804.

Wang JW, Wong AM, Flores J, Vosshall LB, Axel R. 2003. Two-photon calcium imaging reveals an odor-evoked map of activity in the fly brain. *Cell* **112:** 271–282.

Willemse M, Janssen E, de Lange F, Wieringa B, Fransen J. 2007. ATP and FRET—A cautionary note. *Nat Biotechnol* **25:** 170–172.

Xu R, Janson CG, Mastakov M, Lawlor P, Young D, Mouravlev A, Fitzsimons H, Choi KL, Ma H, Dragunow M, et al. 2001. Quantitative comparison of expression with adeno-associated virus (AAV-2) brain-specific gene cassettes. *Gene Ther* **8:** 1323–1332.

Xu R, Rahimi M, Ma H, Fung P, Chang C, Xu S, During M. 2005. Stability of infectious recombinant adeno-associated viral vector in gene delivery. *Med Sci Monit* **11:** BR305–BR308.

Yu D, Baird GS, Tsien RY, Davis RL. 2003. Detection of calcium transients in *Drosophila* mushroom body neurons with camgaroo reporters. *J Neurosci* **23:** 64–72.

Zeltner N, Kohlbrenner E, Clement N, Weber T, Linden RM. 2010. Near-perfect infectivity of wild-type AAV as benchmark for infectivity of recombinant AAV vectors. *Gene Ther* **17:** 872–879.

Zhu P, Aller MI, Baron U, Cambridge S, Bausen M, Herb J, Sawinski J, Cetin A, Osten P, Nelson ML, et al. 2007. Silencing and un-silencing of tetracycline-controlled genes in neurons. *PLoS ONE* **2:** e533. doi: 10.1371/journal.pone.0000533.

Zolotukhin S, Potter M, Zolotukhin I, Sakai Y, Loiler S, Fraites TJ Jr, Chiodo VA, Phillipsberg T, Muzyczka N, Hauswirth WW, et al. 2002. Production and purification of serotype 1, 2, and 5 recombinant adeno-associated viral vectors. *Methods* **28:** 158–167.

Zu T, Duvick LA, Kaytor MD, Berlinger MS, Zoghbi HY, Clark HB, Orr HT. 2004. Recovery from polyglutamine-induced neurodegeneration in conditional SCA1 transgenic mice. *J Neurosci* **24:** 8853–8861.

58 Chronic Calcium Imaging of Neurons in the Visual Cortex Using a Troponin C–Based Indicator

Alexandre Ferrão Santos and Mark Hübener
Max Planck Institute of Neurobiology, D-82152 Martinsried, Germany

ABSTRACT

This chapter describes the use of the genetically encoded troponin C–based calcium indicator TN-XXL to chronically monitor the functional properties of single neocortical neurons in vivo, using two-photon microscopy. We demonstrate the application of this technique for long-term in vivo imaging of stimulus response properties in the mouse visual cortex. Beyond providing functional information, long-term imaging of TN-XXL–labeled neurons also enables the simultaneous monitoring of structural properties down to the level of single dendritic spines.

INTRODUCTION

Since their introduction a decade ago (Miyawaki et al. 1997), genetically encoded calcium indicators have held great promise as convenient tools for imaging neuronal activity. In fact, they have proven to be very useful in several parts of the nervous system, in a variety of different species (Higashijima et al. 2003; Wang et al. 2003; Hasan et al. 2004; Reiff et al. 2005). Only recently, however, novel indicators have made it possible to reliably record sensory evoked signals from single neurons in the mammalian neocortex (Mank et al. 2008; Wallace et al. 2008). Here we describe the use of one such indicator, TN-XXL (Mank et al. 2008), for long-term in vivo imaging of neuronal response properties in the mouse visual cortex.

TN-XXL belongs to a class of Förster resonance energy transfer (FRET)–based calcium indicators, which consist of two chromophores—cyan fluorescent protein (CFP) and a circularly permutated citrine—sandwiched between a modified calcium-binding moiety derived from chicken skeletal muscle troponin C (Mank and Griesbeck 2008). In the absence of calcium binding, the calcium-sensing domain adopts an elongated conformation that largely prevents nonradiative energy transfer from CFP to citrine. Upon calcium binding, the calcium-sensing moiety adopts a new conformation that brings CFP and citrine into close proximity, thereby increasing FRET efficiency. As a result, upon excitation with the appropriate wavelength, CFP fluorescence emission decreases whereas that of citrine increases (Fig. 1A). The ratio of citrine to CFP emission thus provides a relative measure of the variations in calcium concentration within neurons expressing TN-XXL.

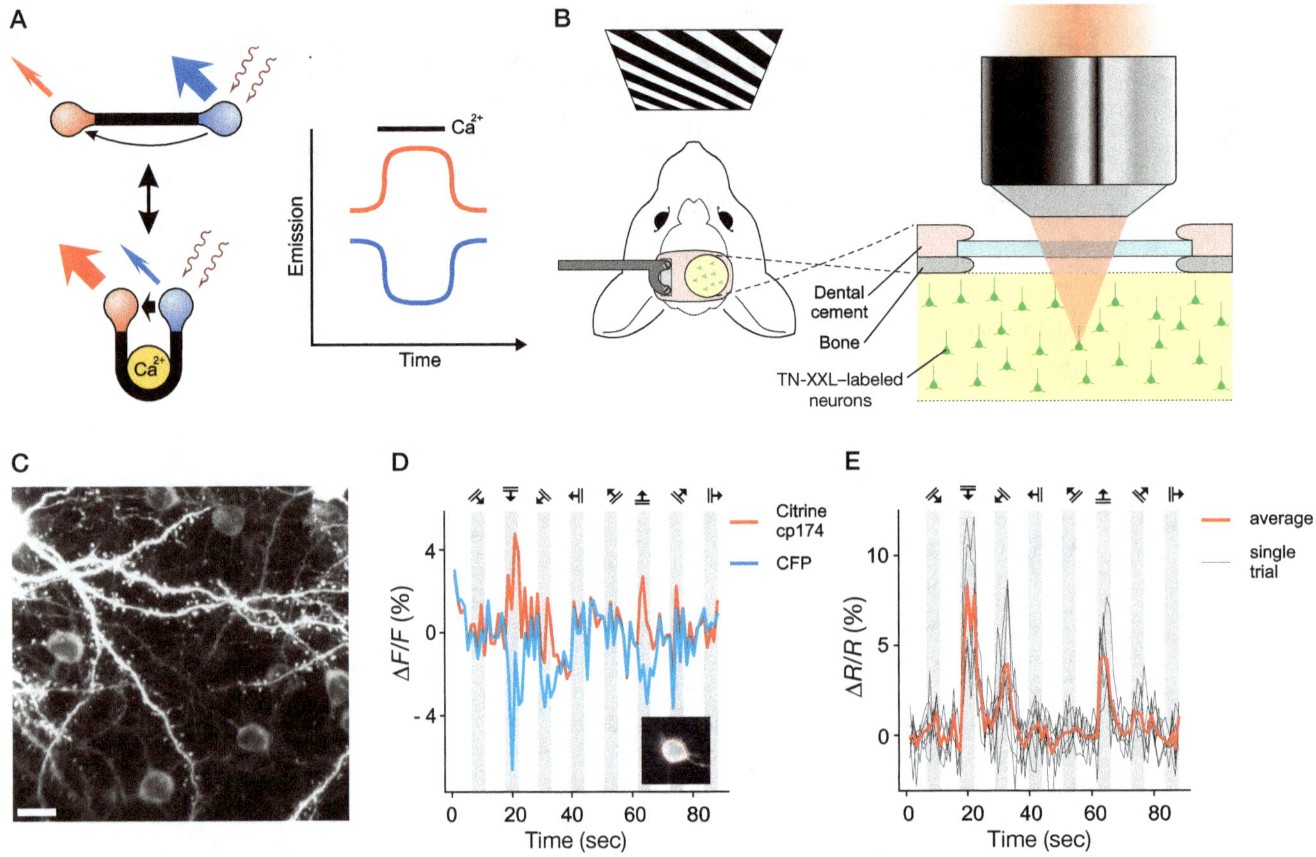

FIGURE 1. Imaging of calcium signals using TN-XXL. (*A*) Schematic depicting the principle of FRET-based calcium indicators. (*Left*) In the absence of bound calcium, TN-XXL adopts an elongated conformation that minimizes nonradiative energy transfer between CFP (blue) and citrine (red) following two-photon excitation of CFP. Upon calcium binding, TN-XXL changes its conformation, thus bringing both chromophores closer together, which increases FRET efficiency. (*Right*) Changes in emission of both chromophores are plotted over time during application and removal of calcium. (*B*) Schematic of the experimental setup. (*Left*) The mouse is facing the stimulus display with its head fixed by a head bar. TN-XXL–expressing neurons are imaged by two-photon microscopy through a cranial window implanted over the visual cortex. (*C*) In vivo two-photon imaging of TN-XXL–expressing cortical neurons labeled by in utero electroporation. Maximum-intensity projection of a 65-µm *z*-stack. Bar, 10 µm. (*D*) Average (six trials), normalized change in fluorescence (Δ*F*/*F*) of citrine and CFP during visual stimulation (gray bars). Stimulus orientation and direction are indicated by the symbols at the top. Note opposite sign of signal change in both channels during stimulation with horizontal gratings. The region of interest measured is indicated with a red dashed line in the inset. (*E*) Single-trial (black) and average (red) time courses of the ratiometric signal (Δ*R*/*R*) during visual stimulation. The apparent signal increase briefly before stimulus onset in some instances is due to jitter caused by the slow frame rate (1 Hz).

TN-XXL has a K_d value of 800 nM, comparable to the previously described troponin C–based indicator TN-L15 (710 nM; Heim et al. 2007). TN-XXL differs, however, from TN-L15 by its higher sensitivity between 100 and 500 nM free calcium (Mank et al. 2008), thus making it well suited for detecting changes in the range of physiologically relevant calcium concentrations. In cultured hippocampal slices, TN-XXL reliably reports single action potentials, although sensitivity in vivo might be lower.

IN VIVO EXPRESSION OF TN-XXL

Long-term expression of TN-XXL in pyramidal neurons of the cerebral cortex can be achieved by viral infection, by in utero electroporation (IUE) (Saito and Nakatsuji 2001), or by using transgenic mouse lines expressing the indicator under the control of neuron-specific promoters (e.g., Thy-1.2

[Feng et al. 2000]). For initial in vivo as well as in vitro testing of TN-XXL, we used infection with Semliki Forest virus (Lundstrom and Ehrengruber 2003), resulting in rapid (12 h), high-level expression of the indicator in cortical neurons. Infected neurons showed clear FRET signal changes, in response to either action potentials elicited with a patch pipette in slice cultures or stimulation of sensory inputs in vivo (Mank et al. 2008). However, long-term imaging is not possible using this virus, as neurons cease responding 4–5 d after infection and die soon thereafter. Therefore, for chronic imaging over weeks, infection with less cytotoxic viruses such as adeno-associated virus or lentiviruses is recommended.

Although stable transgenic expression of TN-XXL under control of the Thy-1.2 promoter seems ideally suited for long-term imaging, this approach is often hampered by the rather capricious expression patterns seen with this promoter. Consequently, many mouse lines have to be generated and screened to identify those with an appropriate distribution and density of labeled neurons. In the future, the availability of specific Cre driver mouse lines (Gong et al. 2007) in combination with "floxed" indicator mouse lines will hopefully make the transgenic approach much more straightforward.

We found IUE ideally suited to achieve high level expression of TN-XXL in sparse populations of neurons in the visual cortex (Fig. 1C; a detailed description of the procedure is described in the Supplementary Methods of Mank et al. 2008). When IUE was carried out at embryonic day 16, labeled cells were mostly confined to layers 2/3. Typically, dozens of neurons in the visual cortex expressed the sensor, thus allowing imaging from several neurons in one optical section. At the same time, the density of labeled neurons was sufficiently low to allow the isolation of signals from individual cell bodies without contamination from the surrounding neuropil, a potential problem with bolus loading of synthetic calcium indicator dyes (Göbel and Helmchen 2007). Levels of TN-XXL expression were very heterogeneous among labeled neurons, allowing resolution of fine anatomical details like dendritic spines in the brightest cells (Fig. 1C). There was no relationship between labeling intensity and frequency of recorded responses, suggesting that neurons with higher levels of TN-XXL expression do not have altered response profiles.

Protocol

Repeated Two-Photon Imaging of TN-XXL–Labeled Neurons in the Visual Cortex

This protocol describes a method for repeated two-photon imaging of neuronal activity with the calcium indicator TN-XXL. A cranial window is implanted over the brain; then, several days later, the visual cortex is mapped and photographed to facilitate repeated imaging of the same region. The first imaging may be done ~2 wk after the window is implanted.

IMAGING SETUP

Two-Photon Imaging

Imaging of neuronal calcium signals was performed using a custom-built two-photon microscope (Denk et al. 1990) equipped with a 40× long working distance, water-immersion objective (numerical aperture [NA] = 0.8, Olympus). A mode-locked Ti:sapphire laser (Mai Tai, Spectra-Physics) delivered laser pulses at 860 nm, with an average power <50 mW after the objective. A long-pass dichroic mirror (505 nm) was used to split emitted CFP and citrine fluorescence. Additional bandpass filters (AHF, Tübingen, extended transmission multiphoton) were used for the CFP (peak 480 nm, bandwidth 40 nm) and citrine channels (peak 535 nm, bandwidth 30 nm), which also contained an infrared (IR) blocking filter. The microscope was equipped with a movable *x–y* stage to which a mouse head holder was attached.

MATERIALS

CAUTION: See Appendix 6 for proper handling of materials marked with <!>.
See the end of the chapter for recipes for reagents marked with <R>.

Reagents

Anesthetics <!>
- Fentanyl (Hexal)
- Medetomidine <!> (Pfizer)
- Midazolam (Roche)

Anti-inflammatory agents and analgesics
- Carprofen <!> (Pfizer)
- Dexamethasone <!> (Sigma-Aldrich)

Bone wax dyed with black ink powder (Ivory Black, Kremer Pigments, New York)

Cortex buffer <R>

Mice for imaging

Long-term expression of TN-XXL in pyramidal neurons of the cerebral cortex can be achieved by viral infection, by in utero electroporation, or by using transgenic mouse lines expressing the indicator under the control of neuron-specific promoters. See the earlier section In Vivo Expression of TN-XXL, for details.

Ophthalmic cream (Isoptomax, Alcon)

Virus solution (if inducing expression by viral infection; see Step 4)

For initial in vivo as well as in vitro testing of TN-XXL, we used infection with Semliki Forest virus. For chronic imaging, we recommend infection with less cytotoxic viruses such as adeno-associated virus or lentiviruses. See the earlier section In Vivo Expression of TN-XXL for details.

Equipment

Coverslip, 5-mm-diameter
Dental acrylic cement (Palladur, Heraeus Kulzer)
Heating blanket
Imaging setup for two-photon imaging, as described above
LCD (liquid crystal display) monitor (e.g., 19-in) for presenting visual stimuli
Metal head bar
Micropipette, borosilicate glass (Harvard Apparatus, GC150F-10) (optional; see Step 5)
Opaque cone to fit cranial window
Stereotaxic apparatus

EXPERIMENTAL METHOD

Cranial Window Implantation

For repeated two-photon imaging, we implant a cranial glass window over one cortical hemisphere of juvenile or adult C57BL/6 mice (Trachtenberg et al. 2002; Holtmaat et al. 2009; see Chapter 32 for a detailed description of the procedure).

1. Induce anesthesia with a mixture of fentanyl (0.05 mg/kg body weight), midazolam (5.0 mg/kg), and medetomidine (0.5 mg/kg) injected i.p. Inject dexamethasone s.c. (25.0 mg/kg) to prevent cerebral edema and carprofen s.c. (5 mg/kg) to reduce pain and inflammation.
2. Affix the mouse to the stereotaxic apparatus, then expose the posterior skull by removing an 8 × 8-mm skin flap.
3. Perform a 4–5 mm diameter trepanation above the visual cortex, centered on a point 3 mm lateral from the midline and 1.5 mm anterior to the lambdoid suture, such that the craniotomy extends approximately two-thirds anterior and one-third posterior of the lambdoid suture (Fig. 1B). Keep the exposed cortical surface moist with cortex buffer.
4. (Optional) To induce TN-XXL expression by viral infection, gradually (3 min) pressure-inject (0.8 psi) 0.5 µL virus solution into the visual cortex at a depth of 50–300 µm with a borosilicate glass micropipette.
5. Insert a glass coverslip into the craniotomy such that it sits directly on the moist surface of the cortex, avoiding any air bubbles under the coverslip. Use dental acrylic cement to seal the coverslip and mount a metal head bar onto the skull to immobilize the mouse during imaging.

Mapping of the Visual Cortex

Several days after implantation of the cranial window, the mouse may be reanesthetized to map the location and extent of the monocular and binocular regions of the primary visual cortex using optical imaging of intrinsic signals (see Chapters 60 and 83).

6. While the cortex is illuminated with red light of 707 nm, acquire images from a plane 200–300 µm below the cortical surface using a charge-coupled device (CCD) camera.

   In our experiments, visual stimuli were small patches of gratings, presented at 24 positions in the visual field in random order. Use computer-controlled shutters to obtain independent stimulation of either eye.

7. Compute overall retinotopic maps for the contra- and ipsilateral eye from single-position maps based on 12 stimulus repetitions.
8. Photograph the superficial blood vessels of the region of interest under green illumination to provide topographical references during subsequent two-photon imaging sessions.

Two-Photon Imaging

See Imaging Setup, above, for equipment details.

To allow clearing of the cranial window and to minimize potential adverse effects of the implantation procedure on neuronal responses, it is advisable to commence two-photon calcium imaging no earlier than 2 wk after implant surgery.

9. Anesthetize the mouse using the fentanyl/midazolam/medetomidine mixture at the dosages given in Step 1. If necessary, prolong anesthesia by reinjecting one-third of the initial dose every 90 min.
10. Place the mouse on a heating blanket and affix the mouse under the microscope by screwing the head bar to a metal holder connected to a movable *x–y* stage under the microscope (Fig. 1B). Rotate the head such that the cranial window is oriented perpendicular to the objective.
11. Mount an opaque cone on top of the cranial window to minimize light from the visual stimulus screen entering the objective and to provide a fluid reservoir for the water-dipping lens.

    Bone wax dyed with black ink powder provides a temporary light- and water-tight seal that can be easily removed after each imaging session without damaging the implant.

12. Throughout all procedures, protect the eyes of the mouse with ophthalmic cream to avoid irritation of the cornea by dehydration or mechanical stress. Remove the cream (except for a thin film) prior to imaging.

Imaging Visually Driven Calcium Signals

Because only a fraction of TN-XXL–expressing pyramidal neurons in layer 2/3 responds reliably to visual stimulation (~22%; Mank et al. 2008), we often use the first imaging session to screen the visual cortex for responsive neurons using a limited set of visual stimuli, in particular in mice with sparse expression. Because we found a neuron's responsiveness to be typically stable over weeks, in subsequent imaging sessions these responsive neurons could then be scrutinized with a more complete set of visual stimuli. In most experiments we used frame rates of 1 or 2 Hz at 512 × 512 pixel per frame.

13. For visual stimulation, place a LCD monitor at a 15-cm distance from the mouse eye such that the central portion of the visual field is covered (Fig 1B).

    Use eye shutters to obtain independent stimulation of either eye; adjust carefully to avoid mechanical interference.

14. Present visual stimuli.

    In our experiments, visual stimuli were moving, whole-field, high-contrast, square-wave gratings, with a spatial frequency of ~0.03 cycles/degree, and a temporal frequency of 2 Hz.

    i. To measure a neuron's orientation tuning curve, present gratings at four or eight different orientations and moving in both directions.

       In our experiments, each individual stimulus cycle consisted of 6.6 sec of stationary grating, followed by 4.4 sec of drift. This standing-moving sequence was chosen to minimize stimulus-on responses, which might obscure orientation tuning.

    ii. To determine ocular dominance (OD), present the grating stimuli at the neuron's preferred orientation and use the shutters to record responses evoked through either eye.

       In our experiments, the stimulus cycle consisted of 16.6 sec of an average-luminance gray stimulus, followed by 4.4 sec of drifting grating. In particular for measuring OD, we found the extended interval between stimulus presentations important, because it substantially reduced response adaptation.

Repeated Imaging

The superficial cortical blood vessels stay largely stable even over extended periods of time. Distinctive features of their pattern can be used as convenient landmarks to relocalize imaged neurons in subsequent sessions. This approach usually works if labeling is sparse, but can fail if expression is dense. In such cases, acquire a *z*-stack of the region of interest and use the local blood vessel architecture to reference precisely the position of the imaged neuron. In general, the distinct somadendritic architecture of labeled neurons allows for the unequivocal identification of imaged cells over several imaging sessions.

DATA ANALYSIS

To measure visually evoked changes in calcium concentration, the fluorescence intensities of both chromophores of the TN-XXL FRET pair have to be determined and compared by deriving their ratio: $R = F(\text{citrine})/F(\text{CFP})$.

Individual neuronal cell bodies were defined as regions of interest (ROI), either by hand or by using a set of morphological filters (Ohki et al. 2005). For each frame, and both channels, all pixel values within the ROI were averaged. In cases of sparse labeling, the ROIs do not need to precisely conform to the cell body outline, and small motion artifacts are irrelevant because their effect is essentially removed by simultaneous imaging of both channels and the subsequent formation of the ratio. However, if labeling is dense, it might be necessary to align the frames of a time series to avoid signal contamination from processes of neighboring labeled cells caused by lateral movements during data acquisition. Alignment can be done either by using rigid x–y shifts (e.g., with the "RegisterROI" plug-in of ImageJ software [Abramoff et al. 2004]) or by using more complex algorithms that allow correction for in-frame motion (Dombeck et al. 2007; Greenberg and Kerr 2009). Background subtraction can be carried out, if appropriate and needed. To this end, ROIs devoid of any labeled structures are chosen, and their averaged fluorescence intensities are subtracted, for both channels, from those of labeled neurons. Because we were primarily interested in comparing the relative response amplitudes of a neuron to different visual stimuli, background subtraction was not critical.

We expressed response strength as the relative change of the citrine/CFP ratio, $\Delta R/R$. The averaged ratio values from the three frames preceding stimulus onset of each cycle were used as baseline in order to generate a $\Delta R/R$ time course. Response amplitude was calculated from the averages of at least five stimulus repetitions as the average $\Delta R/R$ value of the three (1-Hz frame rate) or six (2-Hz frame rate) frames immediately after stimulus onset, centered around the stimulus-evoked response maximum.

The $\Delta R/R$ values can be directly used to generate the orientation tuning curve of imaged neurons. To determine the preferred orientation, we used vector averaging. For further quantitative analysis, and to obtain a measure for tuning width, the tuning curves were fitted with the sum of two circular Gaussian functions (von Mises distribution: $A \exp\{-k[1 - \cos(j - j_0)]\}$, where A is the height of the peak, k is inversely related to its width, and j_0 is the peak's center direction; Swindale et al. 2003).

As a measure for OD, we computed an OD score, defined as the ratio of the ipsilateral eye response to the summed response of both eyes (Mrsic-Flogel et al. 2007). An OD score of 0 or 1 indicates exclusive dominance by the contra- or ipsilateral eye, respectively.

EXAMPLES OF APPLICATION

Figure 1 shows imaging of sensory-evoked calcium signals using TN-XXL. Neurons were labeled by IUE (Fig. 1C) and the responses to drifting gratings were measured (Fig. 1D,E).

Figure 2 shows repeated monitoring of neuron's sensory-evoked response properties. Orientation tuning curves from the same neuron were obtained at successive time points over the course of several days (Fig. 2A,B). Note that orientation tuning is largely stable over this period. Another example demonstrates repeated two-photon calcium imaging of ocular dominance (Fig 2C,D). Neurons were imaged three times over a 2-wk period. In both cells, the OD score remained relatively stable.

FUTURE PERSPECTIVES

Two-photon calcium imaging using TN-XXL allows for the repeated assessment of stimulus-driven response properties in the mammalian visual cortex. Although the signal amplitudes of this and

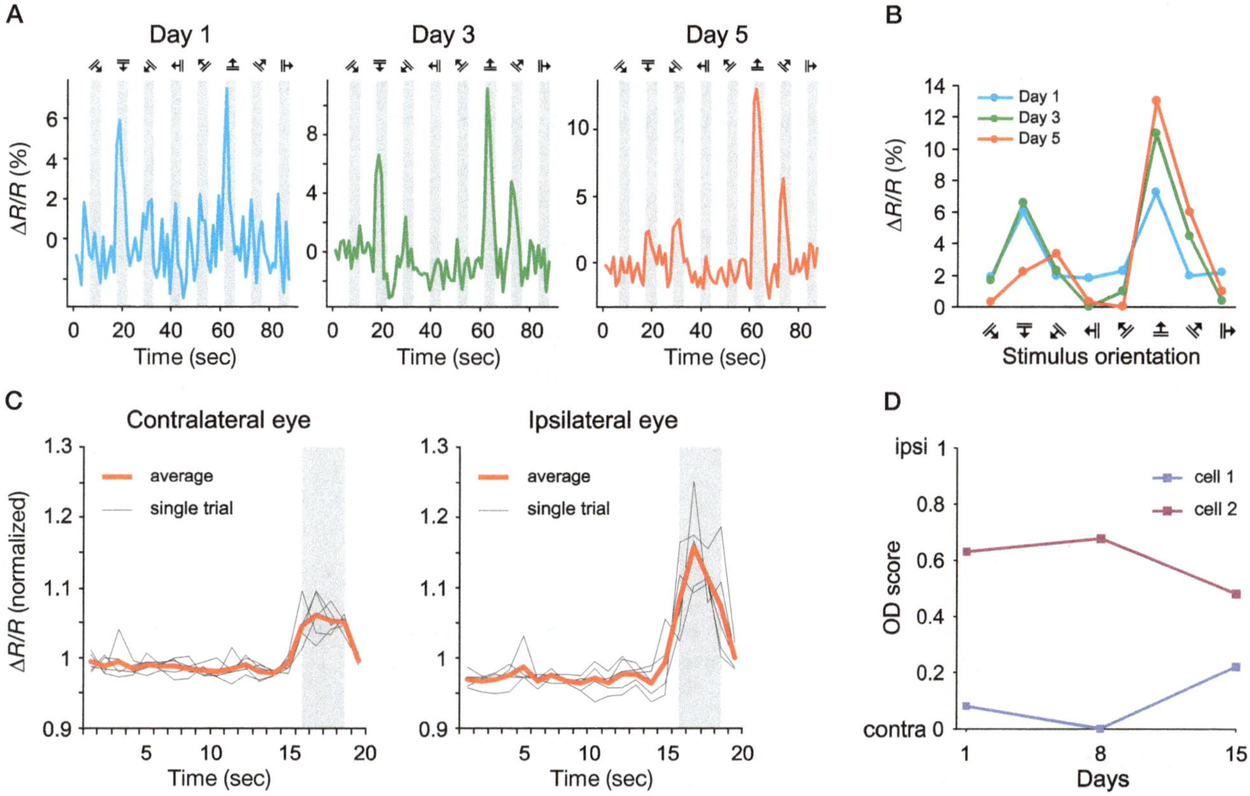

FIGURE 2. Repeated imaging of calcium signals using TN-XXL. (*A*) Repeated imaging of orientation-selective responses in a single neuron in the visual cortex. (*B*) Orientation tuning curves from the neuron in *A* measured on different days after start of the experiment. (*C*) Repeated imaging of ocular dominance (OD) using TN-XXL. Average (red) and single-trial (black) responses of the normalized ratiometric signal. The neuron was stimulated (gray regions) with moving gratings of the preferred orientation, presented independently to the contra- and ipsilateral eye. (*D*) Repeated imaging of a binocular (cell 2) and a contralaterally dominated (cell 1) neuron. In both cells, the OD score remained largely constant over a 2-wk period.

other genetically encoded indicators are currently still smaller than those of synthetic calcium dyes like OGB-1 AM (Oregon Green 488 BAPTA-1 acetoxymethyl; Stosiek et al. 2003), new improved protein based sensors will almost certainly alleviate this shortcoming (Tian et al. 2009; Lütcke et al. 2010). Already it is possible to follow the functional properties of single cortical neurons over time, thus putting us in a position to monitor changes in stimulus-specific responses of individual neurons caused by development, learning, or pathological processes. Targeted viral infection, in combination with appropriate Cre driver lines, will allow for region-specific and even cell type–specific expression of TN-XXL.

The development of techniques for long-term two-photon imaging of the fine structure of cortical neurons (Grutzendler et al. 2002; Trachtenberg et al. 2002) has provided us with rich insights into the dynamic nature of cortical circuitry. Genetically encoded calcium indicators like TN-XXL now make it possible to complement these structural studies with long-term imaging of neuronal function. Ultimately, what we would like to achieve is monitoring both structure and function of the same cortical neurons over time, thus allowing us to correlate functional and structural changes following modifications of sensory input or learning at the level of individual neurons. Because labeling neurons with TN-XXL yields not only functional data, but also exquisite anatomical detail (Fig. 1C), we are already close to achieving this goal.

RECIPE

CAUTION: See Appendix 6 for proper handling of materials marked with <!>.
Recipes for reagents marked with <R> are included in this list.

Cortex Buffer

NaCl	125 mM
KCl	5 mM
Glucose	10 mM
HEPES buffer	10 mM
$MgSO_4$	2 mM
$CaCl_2$ <!>	2 mM

Adjust pH to 7.4.

ACKNOWLEDGMENTS

We thank Oliver Griesbeck and Volker Scheuss for comments on the manuscript. Work in the authors' laboratory is supported by the Max Planck Society.

REFERENCES

Abramoff MD, Magelhaes PJ, Ram SJ. 2004. Image processing with ImageJ. *Biophoton Int* **11:** 36–42.

Denk W, Strickler JH, Webb WW. 1990. Two-photon laser scanning fluorescence microscopy. *Science* **248:** 73–76.

Dombeck DA, Khabbaz AN, Collman F, Adelman TL, Tank DW. 2007. Imaging large-scale neural activity with cellular resolution in awake, mobile mice. *Neuron* **56:** 43–57.

Feng GP, Mellor RH, Bernstein M, Keller-Peck C, Nguyen QT, Wallace M, Nerbonne JM, Lichtman JW, Sanes JR. 2000. Imaging neuronal subsets in transgenic mice expressing multiple spectral variants of GFP. *Neuron* **28:** 41–51.

Göbel W, Helmchen F. 2007. In vivo calcium imaging of neural network function. *Physiology* **22:** 358–365.

Gong S, Doughty M, Harbaugh CR, Cummins A, Hatten ME, Heintz N, Gerfen CR. 2007. Targeting Cre recombinase to specific neuron populations with bacterial artificial chromosome constructs. *J Neurosci* **27:** 9817–9823.

Greenberg DS, Kerr JN. 2009. Automated correction of fast motion artifacts for two-photon imaging of awake animals. *J Neurosci Methods* **176:** 1–15.

Grutzendler J, Kasthuri N, Gan WB. 2002. Long-term dendritic spine stability in the adult cortex. *Nature* **420:** 812–816.

Hasan MT, Friedrich RW, Euler T, Larkum ME, Giese G, Both M, Duebel J, Waters J, Bujard H, Griesbeck O, et al. 2004. Functional fluorescent Ca^{2+} indicator proteins in transgenic mice under TET control. *PLoS Biol* **2:** 763–775.

Heim N, Garaschuk O, Friedrich MW, Mank M, Milos RI, Kovalchuk Y, Konnerth A, Griesbeck O. 2007. Improved calcium imaging in transgenic mice expressing a troponin C–based biosensor. *Nat Methods* **4:** 127–129.

Higashijima S, Masino MA, Mandel G, Fetcho JR. 2003. Imaging neuronal activity during zebrafish behavior with a genetically encoded calcium indicator. *J Neurophysiol* **90:** 3986–3997.

Holtmaat A, Bonhoeffer T, Chow DK, Chuckowree J, De Paola V, Hofer SB, Hübener M, Keck T, Knott G, Lee WC, et al. 2009. Long-term, high-resolution imaging in the mouse neocortex through a chronic cranial window. *Nat Protoc* **4:** 1128–1144.

Lundstrom K, Ehrengruber MU. 2003. Semliki Forest virus (SFV) vectors in neurobiology and gene therapy. *Methods Mol Med* **76:** 503–523.

Lütcke H, Murayama M, Hahn T, Margolis DJ, Astori S, Meyer zum Alten Borgloh S, Göbel W, Yang Y, Tang W, Kügler S, et al. 2010. Optical recording of neuronal activity with a genetically-encoded calcium indicator in anesthetized and freely moving mice. *Front Neural Circuits* **4:** 1–12.

Mank M, Griesbeck O. 2008. Genetically encoded calcium indicators. *Chem Rev* **108:** 1550–1564.

Mank M, Santos AF, Direnberger S, Mrsic-Flogel TD, Hofer SB, Stein V, Hendel T, Reiff DF, Levelt C, Borst A, et al. 2008. A genetically encoded calcium indicator for chronic in vivo two-photon imaging. *Nat Methods* **5:** 805–811.

Miyawaki A, Llopis J, Heim R, McCaffery JM, Adams JA, Ikura M, Tsien RY. 1997. Fluorescent indicators for Ca^{2+} based on green fluorescent proteins and calmodulin. *Nature* **388:** 882–887.

Mrsic-Flogel TD, Hofer SB, Ohki K, Reid RC, Bonhoeffer T, Hübener M. 2007. Homeostatic regulation of eye-specific responses in visual cortex during ocular dominance plasticity. *Neuron* **54:** 961–972.

Ohki K, Chung S, Ch'ng YH, Kara P, Reid RC. 2005. Functional imaging with cellular resolution reveals precise micro-architecture in visual cortex. *Nature* **433:** 597–603.

Reiff DF, Ihring A, Guerrero G, Isacoff EY, Joesch M, Nakai J, Borst A. 2005. In vivo performance of genetically encoded indicators of neural activity in flies. *J Neurosci* **25:** 4766–4778.

Saito T, Nakatsuji N. 2001. Efficient gene transfer into the embryonic mouse brain using in vivo electroporation. *Dev Biol* **240:** 237–246.

Stosiek C, Garaschuk O, Holthoff K, Konnerth A. 2003. In vivo two-photon calcium imaging of neuronal networks. *Proc Natl Acad Sci* **100:** 7319–7324.

Swindale NV, Grinvald A, Shmuel A. 2003. The spatial pattern of response magnitude and selectivity for orientation and direction in cat visual cortex. *Cereb Cortex* **13:** 225–238.

Tian L, Hires SA, Tianyi M, Huber D, Chiappe ME, Chalasani SH, Petreanu L, Akerboom J, McKinney SA, Schreiter ER, et al. 2009. Imaging neural activity in worms, flies and mice with improved GCaMP calcium indicators. *Nat Methods* **6:** 875–881.

Trachtenberg JT, Chen BE, Knott GW, Feng GP, Sanes JR, Welker E, Svoboda K. 2002. Long-term in vivo imaging of experience-dependent synaptic plasticity in adult cortex. *Nature* **420:** 788–794.

Wallace DJ, Meyer zum Alten BS, Astori S, Yang Y, Bausen M, Kügler S, Palmer AE, Tsien RY, Sprengel R, Kerr JN, et al. 2008. Single-spike detection in vitro and in vivo with a genetic Ca^{2+}sensor. *Nat Methods* **5:** 797–804.

Wang JW, Wong AM, Flores J, Vosshall LB, Axel R. 2003. Two-photon calcium imaging reveals an odor-evoked map of activity in the fly brain. *Cell* **112:** 271–282.

59 Two-Photon Chloride Imaging Using MQAE In Vitro and In Vivo

Yury Kovalchuk and Olga Garaschuk

Institute of Physiology II, University of Tübingen, 72074 Tübingen, Germany

ABSTRACT

This chapter describes a technique for high-resolution chloride imaging of living cells using a quinoline-based Cl^- indicator dye, MQAE [*N*-(6-methoxyquinolyl) acetoethyl ester]. Bath-applied to acute brain slices, MQAE provides high-quality labeling of neuronal cells and their processes. In living anesthetized mice, cortical cells are labeled using the multicell bolus loading procedure. In combination with two-photon microscopy, this procedure enables in vivo visualization of cell bodies of neurons and astrocytes as well as some astrocytic processes and allows monitoring changes in the intracellular chloride concentration in dozens of individual cells.

INTRODUCTION

Almost every aspect of brain function involves multiple excitatory and inhibitory interactions between participating cells. Whereas excitatory interactions can be successfully monitored optically either directly (using voltage-sensitive dyes, see Chapters 29 and 48) or indirectly (using indicators for Na^+, Ca^{2+}, and second messengers; see Chapters 27, 30, and 49–58), monitoring inhibitory interactions is still rather difficult. It is largely accepted that the best way to visualize inhibitory synaptic signals is to measure changes in the intracellular Cl^- concentration ($[Cl^-]_i$). Because the major inhibitory neurotransmitters GABA (γ-aminobutyric acid) and glycine open ion channels highly permeable to chloride, intracellular Cl^- levels are expected to change even in the case of so-called shunting inhibition (Andersen et al. 1980), causing very little change in membrane potential. On the other hand, because, in many cells, the Cl^- equilibrium potential (E_{Cl}) is very close to the resting membrane potential, synaptically driven transmembrane Cl^- fluxes generate rather small changes in $[Cl^-]_i$. This imposes stringent requirements on the sensitivity of Cl^- sensors.

Two classes of fluorescent molecules are currently used to measure intracellular Cl^- levels in living cells: (i) small-molecule quinoline-based Cl^- indicator dyes and (ii) genetically encoded sensor proteins. Genetically encoded Cl^- sensors exploit the ability of yellow fluorescent protein (YFP) to be quenched by halides (Wachter and Remington 1999). They include Clomeleon (Kuner and Augustine

2000; see Chapter 9) and Cl sensor (Markova et al. 2008). The advantages of genetically encoded indicators include (i) their ratiometric nature (both indicators are based on Förster resonance energy transfer [FRET] between YFP and Cl^--insensitive cyan fluorescent protein [CFP]) and (ii) the ability to be targeted to different cell types/cellular subcompartments of interest. However, both proteins are pH sensitive (Kuner and Augustine 2000; Markova et al. 2008) and have slow binding kinetics (e.g., Clomeleon binds Cl^- over a time course of 100 msec [Jayaraman et al. 2000]). In addition, Clomeleon is much less Cl^- sensitive than the small-molecule quinoline-based Cl^- indicator dyes MQAE or MEQ (6-methoxy-*N*-ethylquinolinium chloride) (Verkman et al. 1989; Biwersi and Verkman 1991; Schwartz and Yu 1995).

The family of quinoline-based Cl^- indicator dyes includes three members: 6-methoxy-*N*-(3-sulphopropyl)quinolinium (SPQ), MQAE, and MEQ. Whereas MQAE and MEQ are membrane permeable, SPQ is usually loaded into cells either via a patch pipette or by transient hypotonic shock permeabilization. The advantages of quinoline-based Cl^- indicator dyes (especially MQAE) include their insensitivity to changes in bicarbonate concentration and pH as well as their very fast kinetics. The disadvantage of quinoline-based indicator dyes lies in the fact that they are prone to unusually strong bleaching (Inglefield and Schwartz-Bloom 1997; Nakamura et al. 1997). In the past, this significantly restricted the duration of the measurements and permitted only low data-acquisition rates (0.2–2 frames/min) (Inglefield and Schwartz-Bloom 1997; Fukuda et al. 1998). Fortunately, the bleaching is largely reduced when using two-photon microscopy (Marandi et al. 2002). Furthermore, it turned out that MQAE is also suitable for two-photon fluorescence lifetime imaging (Kaneko et al. 2004; Funk et al. 2008; Hille et al. 2009). This recording mode analyzes the lifetime of the MQAE fluorescence, rather than its intensity. Because chloride ions decrease the decay time of the emitted light due to the dynamic fluorescence quenching, this approach allows ratiometric imaging and, thus, the determination of the intracellular chloride concentration.

In this chapter, we will describe an approach for two-photon Cl^- imaging of living cells using MQAE. We refer the reader to other recent articles (Berglund et al. 2008; Markova et al. 2008; Mukhtarov et al. 2008) for descriptions of the technique utilizing genetically encoded Cl^- indicators.

Protocol

Two-Photon Chloride Imaging Using MQAE In Vitro and In Vivo

IMAGING SETUP

Any commercially available two-photon imaging system can be used. Such systems are available from several providers (e.g., Carl Zeiss AG, Jena, Germany; Olympus, Tokyo, Japan; Leica Microsystems, Wetzlar, Germany; Prairie Technologies, Middleton, WI). We currently use a custom-built setup based on a mode-locked Ti:sapphire laser with automated dispersion compensation (Mai Tai HP DeepSee, Spectra-Physics, Mountain View, CA) and a laser-scanning system (Olympus Fluoview, Olympus, Tokyo, Japan) coupled to an upright microscope (BX51WI, Olympus) and equipped with a 60x, 1.0-numerical-aperture (NA) water-immersion objective (Fluor 60x, Nikon, Tokyo, Japan). Such a custom-built system can be assembled following the instructions described in Majewska et al. (2000) and Nikolenko and Yuste (2005). We excite MQAE at 750–770 nm and collect the fluorescence between 400 and 720 nm. The acquired images are then background corrected and analyzed offline with the ImageJ program (http://rsb.info.nih.gov/ij/) and a LabVIEW-based software package (National Instruments, Austin, TX).

MATERIALS

CAUTION: See Appendix 6 for proper handling of materials marked with <!>.
See the end of the chapter for recipes for reagents marked with <R>.

Reagents

For in vitro staining via bath application:
- Cultured mouse cells or brain slices
- MQAE <!> (Invitrogen, Karlsruhe, Germany)
- Standard external saline <R>

For in vivo staining:
- Mice of desired strain
- Pipette staining solution, freshly prepared <R>
- Surgical and anesthesia reagents as described in Chapter 49

In vitro calibration solutions <R>
MQAE cuvette calibration solutions <R>

Equipment

For in vitro staining via bath application:
- Incubator, preset to 37°

For in vivo staining:
- Glass capillaries (e.g., from Hilgenberg GmbH, Malsfeld, Germany)
- LN-Mini manipulator (Luigs & Neumann GmbH, Ratingen, Germany)

Pipette puller (e.g., DMZ-Universal Puller, Zeitz-Instrumente GmbH, Martinsried, Germany)
Pressure application system (Picospritzer II, General Valve Corp., Fairfield, NJ)
Surgical and anesthesia equipment as described in Chapter 49
Ultrafree-MC centrifugal filter, pore diameter 0.45 μm (Millipore)

Imaging setup (see above)

Recording chamber with central access opening: custom made from a standard tissue-culture dish (diameter 35 mm; Garaschuk et al. 2006).

EXPERIMENTAL METHOD

Perform MQAE staining either in vitro (Steps 1–2) or in vivo (Steps 3–7), then proceed to imaging (Step 8).

Staining Cells in Culture or Tissue Slices via Bath Application of MQAE

This protocol allows high-quality staining of the upper 70–120 μm of a slice so that different types of neurons can be identified based on their morphology (Marandi et al. 2002; see also Fig. 1).

1. Dissolve MQAE in standard external saline to a final concentration of 6 mM.
2. Incubate cultured cells or brain slices with this solution for 10 min at 37°C, and then rinse them with dye-free saline for 10–15 min.

   See *Troubleshooting.*

In Vivo Staining of Neurons and Glia Using Multicell Bolus Loading

This protocol allows staining of both neurons and glia cells (see Fig. 2) within a spherical volume with a diameter of ~200 μm.

3. Conduct surgical procedure as described in Chapter 49.
4. Prepare fresh pipette staining solution, and filter it using an Ultrafree-MC centrifugal filter.
5. Fill a pipette similar to a patch pipette with the staining solution (the resistance of the filled pipette 3–6 MΩ), and position it under optical control at the desired depth within the brain tissue using an LN-Mini manipulator. Use the imaging system to continuously monitor the position of the pipette.
6. Pressure-eject MQAE into the brain using a brief (only 500 msec) ejection pulse (ejection pressure 34.5 kPa). Repeat the ejection two to four times at an interpulse interval of 1–3 min.
7. Inspect the quality of the obtained staining ~10 min after the last ejection pulse.

   This short waiting time is necessary to allow wash-out of the MQAE from the extracellular space (most probably due to microcirculation). In contrast to membrane–permeant calcium indicator dyes, MQAE is not undergoing deesterification inside the cells; and, therefore, no additional waiting time is required.

   See *Troubleshooting.*

Two-Photon Imaging of Cells Stained with MQAE

8. Perform two-photon imaging (see Imaging Setup, above).

   With one-photon excitation, MQAE is excited at wavelengths of 320–400 nm and has an emission maximum at 460 nm (Verkman et al. 1989). With two-photon imaging, MQAE is excited efficiently at ~740–770 nm. It is also possible to excite MQAE at longer wavelengths (up to 800 nm), but the intensity of the emitted light is lower (Marandi et al. 2002). Using our imaging system, it was not possible to excite MQAE at excitation wavelengths of 960–990 nm.

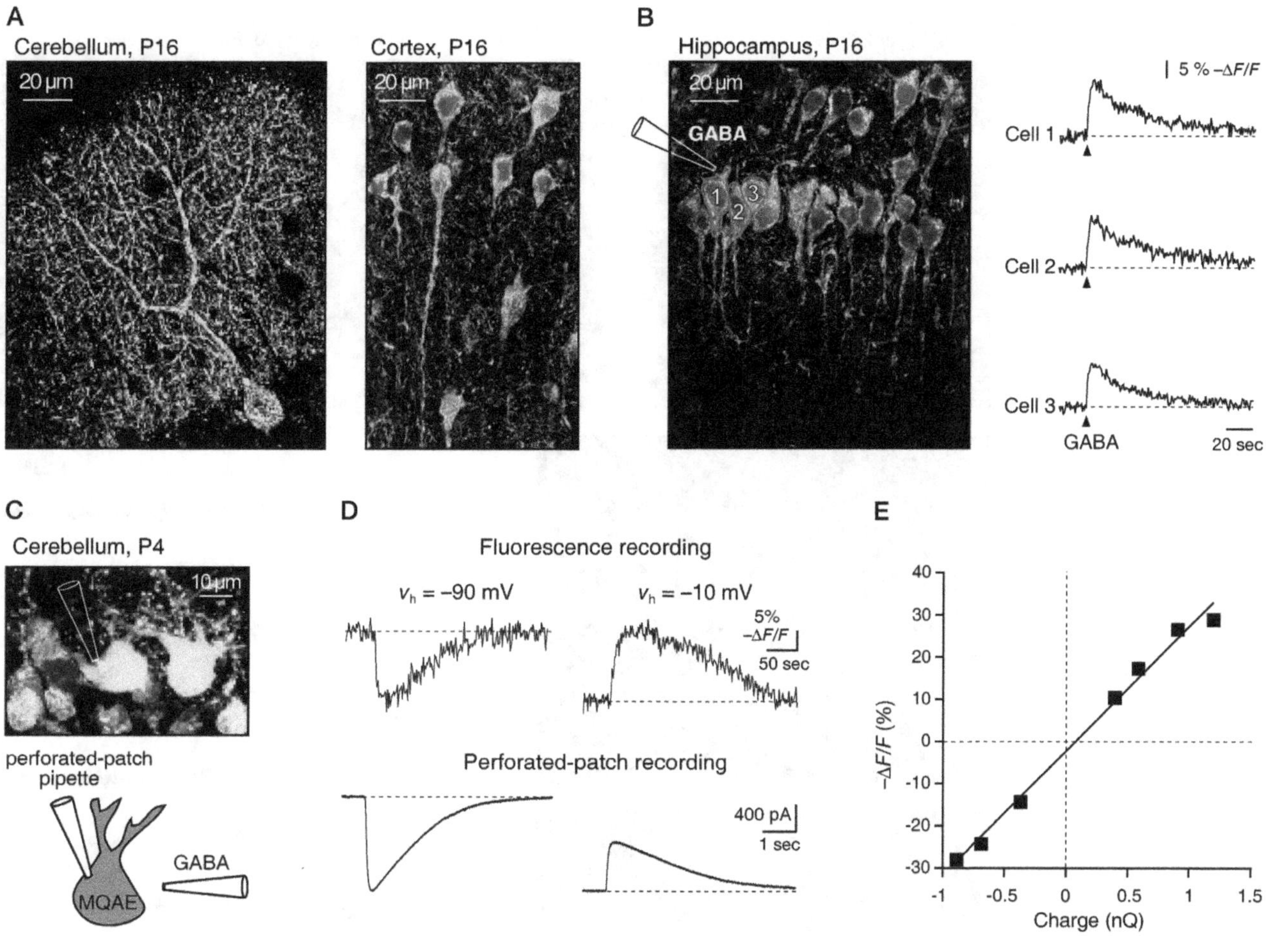

FIGURE 1. High-resolution MQAE-based Cl^- imaging in brain slices. (*A*) An image of a cerebellar Purkinje cell (*left*) and of cortical neurons (*right*) in slices obtained from 16-d-old rats. Each image represents a maximal projection of seven to nine images taken with 1-µm steps. (*B*) GABA-mediated Cl^- transients in somata of CA1 pyramidal neurons (*right*) labeled with corresponding numbers on the image (*left*). (*C*) A high-resolution image of cerebellar Purkinje cells in a slice from a 4-d-old rat (*top*) and a schematic of the experimental arrangement for the gramicidin-based perforated patch recordings (*bottom*). (*D*) Opposite directions of GABA-evoked Cl^- fluxes at holding potentials (V_h) of −90 mV (*left*) and −10 mV (*right*). Changes in fluorescence (*top*) and corresponding transmembrane currents (*bottom*) are caused by a 150-msec pressure application of GABA. (*E*) Relationship between the amplitude of MQAE-mediated Cl^- transients and corresponding transmembrane charges in another experiment using the gramicidin-based perforated patch technique. (Modified, with permission, from Marandi et al. 2002, ©Springer Science+Business Media.)

Intracellular Calibration of MQAE

The efficiency of quenching of quinolinium-based Cl^- indicators by Cl^- depends on the viscosity and/or the polarity of the solvent (Jayaraman and Verkman 2000) and may, therefore, be different inside cells compared with in the cuvette tests. The calibration protocol introduced by Krapf et al. may be used for calibration of Cl^- levels in neurons in slices (Krapf et al. 1988; Marandi et al. 2002).

9. Prepare in vitro calibration solutions (as described in Reagents) containing different amounts of Cl^- (e.g., 0, 10, 20, 30, 40 mM).
10. Add the Cl^-/OH^- antiporter tributyltin chloride (10 µM) and the K^+/H^+ ionophore nigericin (10 µM) to each of these solutions. This treatment will break down the Cl^- gradient across the cell membrane and will ascertain that the cytosolic Cl^- concentration ($[Cl^-]_i$) is equal to that of the corresponding calibration solution.

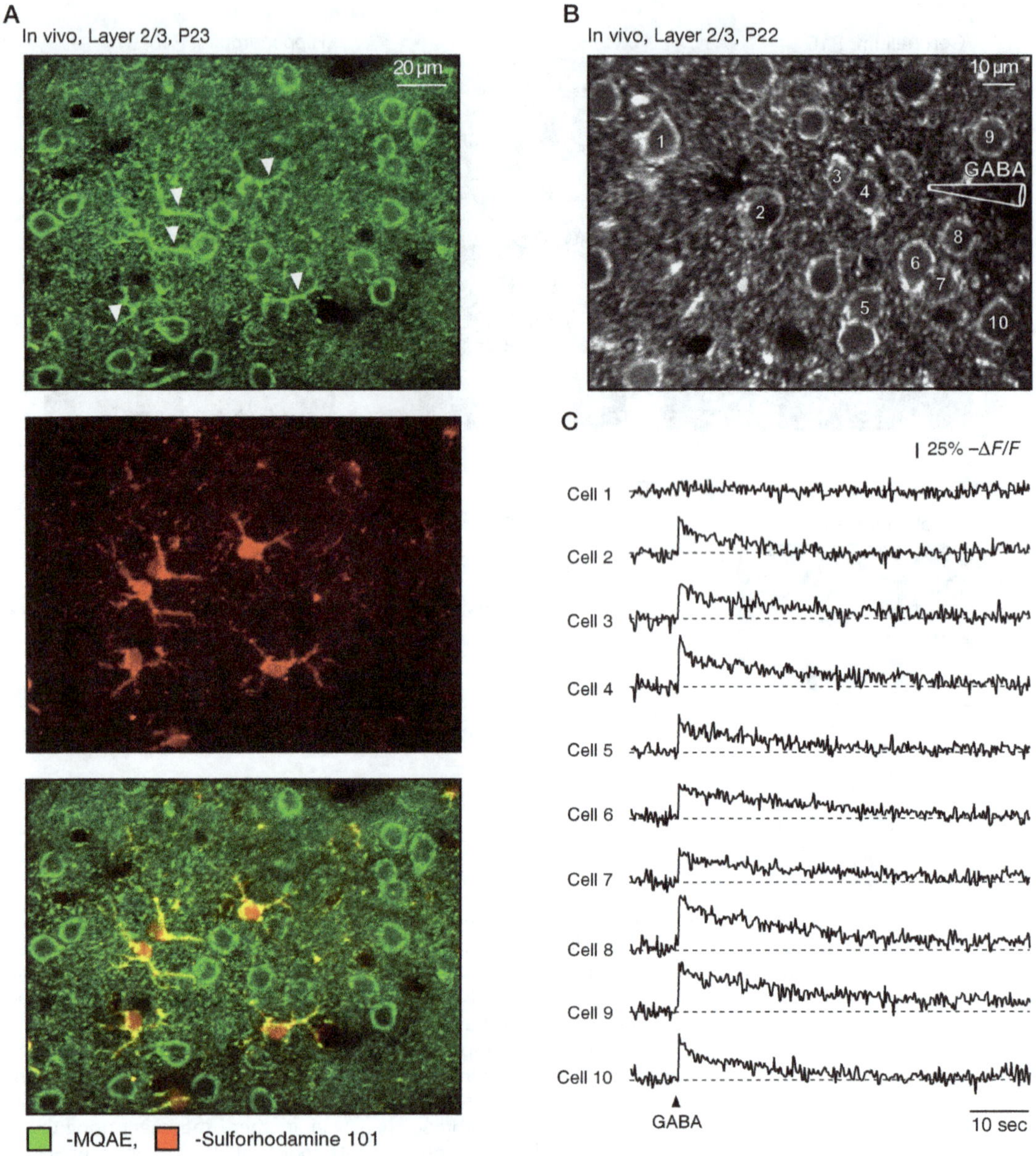

FIGURE 2. In vivo Cl^- imaging using MQAE. (*A*) Microphotographs of the cortical layer 2/3 cells stained both with MQAE (*top*, green) and sulforhodamine 101 (*middle*, red) using multicell bolus loading (MCBL). The merged image (*bottom*) shows that MQAE labels both neurons and astrocytes. Arrowheads point to MQAE-labeled glial processes. Each image represents a maximal projection of seven images taken with 1-µm steps. (*B*,*C*) Cl^- transients (*C*) caused by local pressure application of GABA (0.5 mM in the application pipette, 100 msec, 55 kPa), recorded from cells marked with the corresponding numbers in *B*.

11. Apply the in vitro calibration solutions sequentially, and measure the intracellular steady-state fluorescence levels. The mean-fluorescence level in the Cl^--free solution is defined as F_0. The F values for each calibration solution are plotted as F_0/F versus the corresponding $[Cl^-]_i$ (a so-called Stern–Volmer plot). The slope of the regression line (Stern–Volmer constant K_{SV}) is the reciprocal of an apparent dissociation constant (K_d).

    In our calibration experiments, the K_d of MQAE was 13 mM in the cuvette and 40 mM ($K_{SV} = 24.7$ M^{-1}) in vitro, inside neurons in brain slices (Marandi et al. 2002). In other tissues, K_{SV} values varied between 3 and 26 M^{-1} (Lau et al. 1994; Bevensee et al. 1997; Maglova et al. 1998; Eberhardson et al. 2000; Gilbert et al. 2007; Hille et al. 2009).

TROUBLESHOOTING

Problem (Steps 2 and 7): Poor staining is observed.

Solution: The MQAE staining itself is relatively easy and reliable. It requires, however, a high-quality slice/in vivo preparation. To obtain good-quality in vivo preparations, please follow the suggestions described in Chapter 49. In addition, we do not recommend delivering large amounts of MQAE at once. It looks like MQAE is washed out from the extracellular space less effectively than membrane-permeant calcium indicator dyes, and accumulation of a large amount of the dye leads to a bleary (low-contrast–high-brightness) staining.

DISCUSSION

Example of Application

Figure 1 shows examples of the two-photon Cl^- imaging experiments in vitro, in different types of brain slices (Marandi et al. 2002). Note that in vitro the low fluorescence of the extracellular space allows resolution of the secondary and tertiary dendritic branches of MQAE-stained neurons (Fig. 1A,B). The neurons reliably responded to brief (150–500 msec) GABA applications with Cl^- transients. The combination of two-photon Cl^- imaging with gramicidin-based, perforated patch-clamp recordings (Fig. 1C,D) reveals a clear correlation between the amplitudes of GABA-mediated transmembrane Cl^- fluxes and the amplitudes of MQAE-mediated fluorescence changes. The relation is linear and has the same slope for either direction of charge movement, indicating that MQAE is an accurate reporter of the transmembrane Cl^- charge transfer.

Figure 2 shows an in vivo image of layer 2/3 cortical neurons stained with MQAE using the MCBL technique (Stosiek et al. 2003; see Chapter 49). Similar to in vitro (Marandi et al. 2002), the indicator dye in vivo brightly stains the cell cytosol but is largely excluded from the nucleus. Both neuronal and glial cells are labeled, as confirmed by a costaining with the astrocyte-specific marker sulforhodamine 101 (Nimmerjahn et al. 2004). The quality of images obtained does not provide contrast sufficient for imaging individual neuronal dendrites (not shown). However, somewhat thicker processes of glial cells can be clearly distinguished (arrowheads in Fig. 2A). A brief (100-msec) application of GABA to the cells of interest causes reversible quenching of MQAE fluorescence (e.g., Cl^- transients) in the cells situated in the vicinity of the application pipette (Fig. 2B,C). Thus, MQAE enables measurements of GABA-mediated Cl^- transients at single-cell resolution in vitro and in vivo.

Advantages and Limitations

MQAE provides easy and fast staining of neurons in vitro and in vivo with satisfactory fluorescence levels in cell bodies. In brain slices, it also allows one to image neuronal dendrites, whereas in vivo, only glial processes can clearly be discerned. This discrepancy is most probably due to slow/incomplete wash-out of the dye from the extracellular space under the in vivo conditions.

Compared with other Cl^- indicators (see the Introduction), the advantages of MQAE include relatively high sensitivity and selectivity for Cl^-, insensitivity to changes in bicarbonate concentration and pH, and the possibility of prolonged continuous measurements when using two-photon excitation. It is also important to mention that MQAE is quenched rapidly by Cl^- (<1 msec; Verkman et al. 1989) and is thus well suited for monitoring physiological changes in $[Cl^-]_i$, often occurring in the millisecond-to-second range. Furthermore, MQAE is quenched by a collisional quenching mechanism, which does not involve binding of Cl^- to the indicator dye molecule (Verkman 1990). MQAE, therefore, does not buffer Cl^-, and an increase in the intracellular dye concentration improves the signal-to-noise ratio without disturbing the time course of Cl^- transients. Furthermore, when using fluorescence lifetime imaging (see the Introduction), MQAE becomes a ratiometric dye, allowing quantitative Cl^- measurements.

The major limitation of MQAE is the loss of the intracellular dye through leakage. The leakage rate seems to be preparation specific, ranging from 3%/h in liposomes (Verkman et al. 1989) to 30%/h in brain slices (Marandi et al. 2002). As can be expected for a lipophilic compound, its leakage rate is temperature dependent. Therefore, the leakage of the dye is very prominent in vivo. This restricts the duration of in vivo Cl^- measurements to ~2 h after the staining procedure.

RECIPES

CAUTION: See Appendix 6 for proper handling of materials marked with <!>.
Recipes for reagents marked with <R> are included in this list.

In Vitro Calibration Solutions

KCl	0–40 mM
HEPES buffer	10 mM
Na^+-gluconate	10 mM

Adjust osmolarity to 310 mOsmol/L with K^+-gluconate <!>; adjust pH to 7.4 with KOH <!>.

Add 10 μM tributyltin chloride <!> and 10 μM nigericin <!> as described in the protocol.

MQAE Cuvette Calibration Solutions

KCl	0–160 mM
HEPES buffer	30 mM
MQAE <!>	5 mM

Adjust osmolarity to 310 mOsmol/L with sucrose; adjust pH to 7.4 with KOH <!>.

Pipette Staining Solution

MQAE <!>	10–50 mM
KCl	2.5 mM
HEPES buffer	10 mM
pH 7.4	

Adjust osmolarity to 310 mOsmol/L with NaCl.

Standard External Saline

NaCl	125 mM
KCl	4.5 mM
$NaHCO_3$	26 mM
NaH_2PO_4	1.25 mM
$CaCl_2$ <!>	2 mM
$MgCl_2$	1 mM
Glucose	20 mM

pH 7.4 when bubbled with 95% O_2 <!> and 5% CO_2 <!>.

ACKNOWLEDGMENTS

This work was supported by grants from the Deutsche Forschungsgemeinschaft (SFB 596, GA 654/1-1) to OG.

REFERENCES

Andersen P, Dingledine R, Gjerstad L, Langmoen IA, Laursen AM. 1980. Two different responses of hippocampal pyramidal cells to application of γ-amino butyric acid. *J Physiol* **305:** 279–296.

Berglund K, Schleich W, Wang H, Feng G, Hall WC, Kuner T, Augustine GJ. 2008. Imaging synaptic inhibition throughout the brain via genetically targeted Clomeleon. *Brain Cell Biol* **36:** 101–118.

Bevensee MO, Apkon M, Boron WF. 1997. Intracellular pH regulation in cultured astrocytes from rat hippocampus. II. Electrogenic Na/HCO3 cotransport. *J Gen Physiol* **110:** 467–483.

Biwersi J, Verkman AS. 1991. Cell-permeable fluorescent indicator for cytosolic chloride. *Biochemistry* **30:** 7879–7883.

Eberhardson M, Patterson S, Grapengiesser E. 2000. Microfluorometric analysis of Cl^- permeability and its relation to oscillatory Ca^{2+} signalling in glucose-stimulated pancreatic β-cells. *Cell Signal* **12:** 781–786.

Fukuda A, Tanaka M, Yamada Y, Muramatsu K, Shimano Y, Nishino H. 1998. Simultaneous optical imaging of intracellular Cl^- in neurons in different layers of rat neocortical slices: Advantages and limitations. *Neurosci Res* **32:** 363–371.

Funk K, Woitecki A, Franjic-Wurtz C, Gensch T, Mohrlen F, Frings S. 2008. Modulation of chloride homeostasis by inflammatory mediators in dorsal root ganglion neurons. *Mol Pain* **4:** 32.

Garaschuk O, Milos RI, Konnerth A. 2006. Targeted bulk-loading of fluorescent indicators for two-photon brain imaging in vivo. *Nat Protoc* **1:** 380–386.

Gilbert D, Franjic-Wurtz C, Funk K, Gensch T, Frings S, Mohrlen F. 2007. Differential maturation of chloride homeostasis in primary afferent neurons of the somatosensory system. *Int J Dev Neurosci* **25:** 479–489.

Hille C, Lahn M, Lohmannsroben HG, Dosche C. 2009. Two-photon fluorescence lifetime imaging of intracellular chloride in cockroach salivary glands. *Photochem Photobiol Sci* **8:** 319–327.

Inglefield JR, Schwartz-Bloom RD. 1997. Confocal imaging of intracellular chloride in living brain slices: Measurement of $GABA_A$ receptor activity. *J Neurosci Methods* **75:** 127–135.

Jayaraman S, Verkman AS. 2000. Quenching mechanism of quinolinium-type chloride-sensitive fluorescent indicators. *Biophys Chem* **85:** 49–57.

Jayaraman S, Haggie P, Wachter RM, Remington SJ, Verkman AS. 2000. Mechanism and cellular applications of a green fluorescent protein–based halide sensor. *J Biol Chem* **275:** 6047–6050.

Kaneko H, Putzier I, Frings S, Kaupp UB, Gensch T. 2004. Chloride accumulation in mammalian olfactory sensory neurons. *J Neurosci* **24:** 7931–7938.

Krapf R, Berry CA, Verkman AS. 1988. Estimation of intracellular chloride activity in isolated perfused rabbit proximal convoluted tubules using a fluorescent indicator. *Biophys J* **53:** 955–962.

Kuner T, Augustine GJ. 2000. A genetically encoded ratiometric indicator for chloride: Capturing chloride transients in cultured hippocampal neurons. *Neuron* **27:** 447–459.

Lau KR, Evans RL, Case RM. 1994. Intracellular Cl^- concentration in striated intralobular ducts from rabbit mandibular salivary glands. *Pflugers Arch* **427:** 24–32.

Maglova LM, Crowe WE, Smith PR, Altamirano AA, Russell JM. 1998. Na^+-K^+-Cl^- cotransport in human fibroblasts is inhibited by cytomegalovirus infection. *Am J Physiol* **275:** C1330–C1341.

Majewska A, Yiu G, Yuste R. 2000. A custom-made two-photon microscope and deconvolution system. *Pflugers Arch* **441:** 398–408.

Marandi N, Konnerth A, Garaschuk O. 2002. Two-photon chloride imaging in neurons of brain slices. *Pflugers Arch* **445:** 357–365.

Markova O, Mukhtarov M, Real E, Jacob Y, Bregestovski P. 2008. Genetically encoded chloride indicator with improved sensitivity. *J Neurosci Methods* **170:** 67–76.

Mukhtarov M, Markova O, Real E, Jacob Y, Buldakova S, Bregestovski P. 2008. Monitoring of chloride and activity of glycine receptor channels using genetically encoded fluorescent sensors. *Philos Transact A Math Phys Eng Sci* **366:** 3445–3462.

Nakamura T, Kaneko H, Nishida N. 1997. Direct measurement of the chloride concentration in newt olfactory receptors with the fluorescent probe. *Neurosci Lett* **237:** 5–8.

Nikolenko V, Yuste R. 2005. How to build a two-photon microscope using a confocal scan head. In *Imaging in neuroscience and development: A laboratory manual* (ed. Yuste R, Konnerth A), pp. 75–78. Cold Spring Harbor Laboratory Press, Cold Spring Harbor, NY.

Nimmerjahn A, Kirchhoff F, Kerr JND, Helmchen F. 2004. Sulforhodamine 101 as a specific marker of astroglia in the neocortex in vivo. *Nat Methods* **1:** 31–37.

Schwartz RD, Yu X. 1995. Optical imaging of intracellular chloride in living brain slices. *J Neurosci Methods* **62:** 185–192.

Stosiek C, Garaschuk O, Holthoff K, Konnerth A. 2003. In vivo two-photon calcium imaging of neuronal networks. *Proc Natl Acad Sci* **100:** 7319–7324.

Verkman AS. 1990. Development and biological applications of chloride-sensitive fluorescent indicators. *Am J Physiol* **259:** C375–C388.

Verkman AS, Sellers MC, Chao AC, Leung T, Ketcham R. 1989. Synthesis and characterization of improved chloride-sensitive fluorescent indicators for biological applications. *Anal Biochem* **178:** 355–361.

Wachter RM, Remington SJ. 1999. Sensitivity of the yellow variant of green fluorescent protein to halides and nitrate. *Curr Biol* **9:** R628–R629.

60 Intrinsic Optical Imaging of Functional Map Development in Mammalian Visual Cortex

Tobias Bonhoeffer and Mark Hübener
Max-Planck Institute of Neurobiology, 82152 München-Martinsried, Germany

ABSTRACT

In this chapter, we describe how intrinsic imaging of the visual cortex can be used as a chronic technique. This method allows the time course of the development of cortical maps to be studied in a single animal during a period of up to one year.

INTRODUCTION

Optical imaging based on intrinsic signals takes advantage of the fact that neuronal activity is reflected in the intrinsic optical properties of brain tissue (see Chapter 83). It can be used to measure brain activity over large areas, with high spatial resolution (100 μm), in a relatively noninvasive manner. These properties have made intrinsic imaging ideal for experiments in which different cortical maps are obtained simultaneously from the same cortical area or studies in which map development is followed over several weeks in the same animal. This chapter concentrates on the latter kind of experiments and shows how recent progress has facilitated chronic imaging not only in cats and ferrets, but also in the genetically amenable mouse.

The general principle of the technique is not new. It has been known for decades that brain activity can be observed directly, without the use of extrinsic probes, by taking advantage of optical signals that are intrinsic to the brain and that reflect neuronal activity (Hill and Keynes 1949; Chance et al. 1962; Cohen 1973; Jobsis 1977). It was only much later, however, that hemoglobin-based intrinsic signals were used to monitor cortical activity in the living brain (Grinvald et al. 1986).

An advantage of intrinsic optical imaging is that, after an initial chamber implantation (see below), it is completely noninvasive. This is particularly helpful in developmental studies following cortical maps over time in individual animals. Earlier techniques allowed only interanimal comparisons of functional maps at a given age (e.g., see Singer et al. 1981). These studies could not address how cortical maps develop and change over time in a single animal. Consequently, it was not possible to determine, for example, whether orientation or ocular dominance maps change during early development or whether they remain stable once established. For these and similar questions, the development of intrinsic imaging presents a major advance. Another advantage of the technique is that it allows for a quick assessment of overall neuronal activity, such that changes following sensory deprivation can be determined in a simple manner. Finally, it should be pointed out that these intrinsic signals are mechanistically related to the hemodynamic changes that form the bases of functional MRI (see Chapter 83).

BASICS OF INTRINSIC IMAGING

Many publications deal extensively with the various sources of the different components of the intrinsic optical signal (for reviews on this topic, see Chapter 83 and Bonhoeffer and Grinvald 1996). Understanding the origin of the different components of the signal is essential for optimizing the protocol for different applications. The basic procedures used in intrinsic imaging experiments are also described in detail elsewhere (see Chapter 83) (Bonhoeffer and Grinvald 1996; Bonhoeffer 2000); this chapter concentrates exclusively on issues of chronic imaging.

CHRONIC INTRINSIC IMAGING

Chronic intrinsic imaging benefits greatly from the noninvasiveness of the method and allows observation of the development of functional maps in single animals. The first chronic imaging experiments were performed in cats (Kim and Bonhoeffer 1994; Gödecke and Bonhoeffer 1996) and ferrets (Chapman et al. 1996; Chapman and Bonhoeffer 1998), but in the meantime the technique has also been successfully applied to monkeys (Shtoyerman et al. 2000; Chen et al. 2002), rats (Polley et al. 1999), and mice (Hofer et al. 2006; Kaneko et al. 2008).

Chronic recordings necessitate a number of changes to the procedures used for acute experiments. First, in young animals, the bone of the skull may be too soft for the chamber to be firmly attached. In these cases, agar can be used to stabilize the brain (Chapman et al. 1996) or a cranial glass window can be attached to the skull (Keck et al. 2008; Hofer et al. 2009; see Chapter 32). Alternatively, recordings can be made through the thinned (Frostig et al. 1990; Bosking et al. 1997) or intact skull (Schuett et al. 2002). In any case, for good spatial resolution, the brain must be sufficiently stable to provide a good optical interface.

Second—perhaps not unexpectedly for an experiment that relies strongly on the oxy/deoxy transition of hemoglobin—proper respiration of the animal is essential for successful imaging. Although it has been shown that good-quality intrinsic signal imaging maps can be obtained in a spontaneously breathing animal (Rubin and Katz 1999; Meister and Bonhoeffer 2001), artificial respiration has proven considerably more reliable. Despite the technical challenges, intubation is highly recommended for intrinsic signal imaging experiments, even for animals weighing as little as 20 g.

Another concern is the anesthesia regime during chronic imaging experiments, particularly in those involving very young animals. A wide range of anesthetics, including halothane, isoflurane, ketamine, fentanyl, and urethane, are available. In our hands, gas anesthesia using isoflurane or halothane is most successful. It provides a high level of control over the animal's anesthetic state while not interfering noticeably with the source of the intrinsic signal (i.e., the coupling of neural activity to blood flow and oxygenation).

In chronic imaging experiments using very young animals, the physiological state of the animal significantly affects the duration and success of the experiment. Sterility of the surgical environment is paramount, and routine implementation of additional (nutritional) measures is highly recommended. Offering the animals additional food (milk) and limiting the number of siblings being fed by the same mother have proven to be extremely beneficial to the animals' health and strength in mice, rats, ferrets, and cats.

Exemplary data from a chronic imaging experiment on the development of orientation domains in ferret visual cortex are shown in Figure 1. These data illustrate the entire time course of the development of orientation domains in ferret visual cortex, from the time that the domains are first observed (postnatal day [PND] 32) to the time at which they are fully developed (PND 43). In these

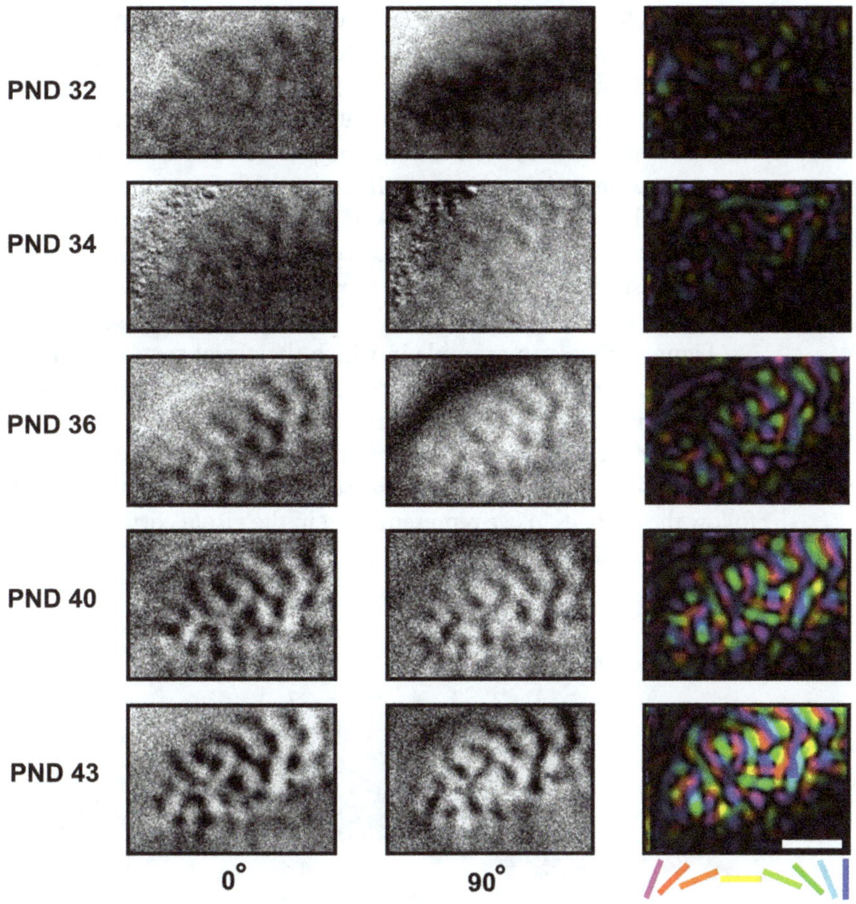

FIGURE 1. Chronic recording of developing orientation maps. Orientation maps obtained from ferret visual cortex (area 17) over a period of 11 d. Postnatal age of the ferret is shown at the left of each row. The two left-hand columns show single-condition maps in response to different orientations (0° and 90°). The right-hand column shows a "polar map" of orientation preference. In this map, color indicates the preferred angle for every cortical location. In addition, the brightness of the color indicates the strength of orientation preference. Scale bar, 2 mm.

experiments, four different orientation maps were acquired for each age studied (only two maps are shown for each age). To better display the overall layout of the iso-orientation domains and their development, "polar maps" are shown in the column on the right. In these maps, different colors indicate different orientations, and the brightness of the color indicates the degree of orientation tuning (for more details on the calculation of such maps, see Bonhoeffer and Grinvald 1993, 1996). Both the single-orientation maps and the polar maps in Figure 1 show that, between PND 32 and PND 43, the overall structure of the orientation domains changes very little. Patch-by-patch comparison of the maps reveals that, when the first hint of an orientation map is detected at PND 32, the structure of that map, although immature, is already remarkably similar to that seen in the mature map (PND 43). The polar maps shown in Figure 1 illustrate particularly well how, over the period of 11 d, a functional map, with only very weak orientation preferences (dark colors), develops into a crisp, clear orientation preference map while maintaining its overall structure.

By now intrinsic imaging is also a standard tool in rodents such as rats (e.g., see Masino et al. 1993; Dowling et al. 1996; Rubin and Katz 1999; Uchida et al. 2000; Meister and Bonhoeffer 2001) and mice (Schuett et al. 2002; Kalatsky and Stryker 2003; Hofer et al. 2006; Kaneko et al. 2008; Keck et al 2008). Intrinsic imaging in mice promises to be very useful. In that respect, imaging retinotopic maps in mouse visual cortex (Schuett et al. 2002) and superior colliculus (Mrsic-Flogel et al. 2005) holds particular promise. It allows for the use of genetics to scrutinize molecules thought to be important for pathfinding in the visual system with respect to their influence on the formation of functional maps (Cang et al. 2005, 2008). Intrinsic imaging studies in the olfactory system of mice and rats (Rubin and Katz 1999, 2001; Uchida et al. 2000; Meister and Bonhoeffer 2001; Soucy et al. 2009) open another interesting avenue: The olfactory system is particularly suitable for studying the intricate interplay between function and development in the rodent brain.

Another technical advance, which directly affects the ability to do chronic recordings, has been introduced by Kalatsky and Stryker (2003). These investigators showed that a Fourier-based data acquisition and analysis method can speed up the process of data acquisition considerably. Because data acquisition time is often the limiting factor when working on young and therefore fragile animals, this method can greatly improve the quality and usefulness of the data.

Figure 2 shows an example in which the visual cortex of a mouse was mapped in the traditional way, using 21 different stimulus locations in visual space. The figure shows the meticulous and precise mapping of the visual field onto the visual cortex.

Another area in which chronic optical imaging in mice has proven to be useful is the adaptation of the cerebral cortex to changes in the sensory environment. Using this approach, it became possible to follow the changes in eye representation in the visual cortex after monocular deprivation and recovery in individual animals (Hofer et al. 2006; Kaneko et al. 2008). Similarly, chronic optical imaging has

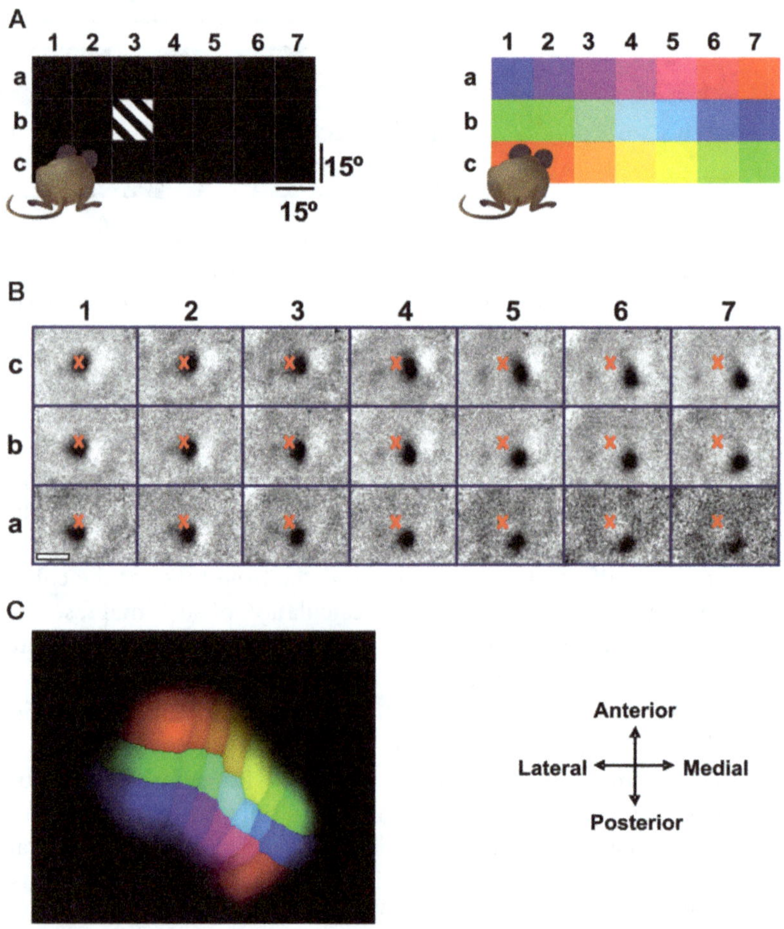

FIGURE 2. Mapping retinotopy in mouse visual cortex. (*A*) Experimental setup: Small visual stimuli are presented in random order at different locations in the visual field. (*B*) Activity maps are derived for each stimulus location. Note the systematic shift of the activated region relative to the red cross, which was placed at the same position in all maps. Scale bar, 2 mm. (*C*) A complete map of retinotopy is generated by assigning a color to each pixel, based on the stimulus location to which it best responded (see *A*). The response strength is encoded by pixel intensity, such that nonresponsive regions of the cortex appear dark.

been used to monitor at high spatial resolution the functional reorganization in the visual cortex following the induction of small retinal lesions (Keck et al. 2008). Such experiments are particularly useful in elucidating the role of molecular factors involved in cortical plasticity and reorganization (Kaneko et al. 2008).

CONCLUSIONS

Intrinsic optical imaging allows chronic experiments in which changes in cortical maps can be observed over time in single animals. These experiments have been extended to animals amenable to genetic manipulation. Chronic experiments like these are crucial for the understanding of cortical map development and reorganization. The fact that such experiments can now be performed in the cerebral cortex of the mouse has the important implication that the influence of single genes on the formation (or reorganization) of maps can be studied in great detail. Together with new data acquisition tools, this has further enhanced the power of chronic intrinsic imaging. Once at the forefront of imaging technologies in neuroscience, optical imaging of intrinsic signals has now become a standard tool in many laboratories. It is powerful, yet straightforward and provides in an easy way a comprehensive overview of developmental events and functional properties of the brain areas under investigation.

REFERENCES

Bonhoeffer T. 2000. Intrinsic signal optical imaging as a tool to visualize the development of functional maps in the mammalian visual cortex. In *Imaging neurons: A laboratory manual* (ed. Yuste R, et al.), pp. 46.1–46.10. Cold Spring Harbor Laboratory Press, Cold Spring Harbor, NY.

Bonhoeffer T, Grinvald A. 1993. The layout of iso-orientation domains in area 18 of cat visual cortex. Optical imaging reveals a pinwheel-like organization. *J Neurosci* **13:** 4157–4180.

Bonhoeffer T, Grinvald A. 1996. Optical imaging based on intrinsic signals: The methodology. In *Brain mapping: The methods* (ed. Toga AW, Mazziotta JC), pp. 55–97. Academic, San Diego, CA.

Bosking WH, Zhang Y, Schofield B, Fitzpatrick D. 1997. Orientation selectivity and the arrangement of horizontal connections in tree shrew striate cortex. *J Neurosci* **17:** 2112–2127.

Cang J, Kaneko M, Yamada J, Woods G, Stryker MP, Feldheim DA. 2005. Ephrin-As guide the formation of functional maps in the visual cortex. *Neuron* **48:** 577–589.

Cang J, Wang L, Stryker MP, Feldheim DA. 2008. Roles of ephrin-As and structured activity in the development of functional maps in the superior colliculus. *J Neurosci* **28:** 11015–11023.

Chance B, Cohen P, Jobsis F, Schoener B. 1962. Intracellular oxidation–reduction states in vivo. *Science* **137:** 499–508.

Chapman B, Bonhoeffer T. 1998. Overrepresentation of horizontal and vertical orientation preferences in developing ferret area 17. *Proc Natl Acad Sci* **95:** 2609–2614.

Chapman B, Stryker MP, Bonhoeffer T. 1996. Development of orientation preference maps in ferret primary visual cortex. *J Neurosci* **16:** 6443–6453.

Chen LM, Heider B, Williams GV, Healy FL, Ramsden BM, Roe AW. 2002. A chamber and artificial dura method for long-term optical imaging in the monkey. *J Neurosci Methods* **113:** 41–49.

Cohen LB. 1973. Changes in neuron structure during action potential propagation and synaptic transmission. *Physiol Rev* **53:** 373–418.

Dowling JL, Henegar MM, Liu D, Rovainen CM, Woolsey TA. 1996. Rapid optical imaging of whisker responses in the rat barrel cortex. *J Neurosci Methods* **66:** 113–122.

Frostig RD, Lieke EE, Ts'o DY, Grinvald A. 1990. Cortical functional architecture and local coupling between neuronal activity and the microcirculation revealed by in vivo high-resolution optical imaging of intrinsic signals. *Proc Natl Acad Sci* **87:** 6082–6086.

Gödecke I, Bonhoeffer T. 1996. Development of identical orientation maps for two eyes without common visual experience. *Nature* **379:** 251–254.

Grinvald A, Lieke EE, Frostig RD, Gilbert CD, Wiesel TN. 1986. Functional architecture of cortex revealed by optical imaging of intrinsic signals. *Nature* **324:** 361–364.

Hill DK, Keynes RD. 1949. Opacity changes in stimulated nerve. *J Physiol* **108:** 278–281.

Hofer SB, Mrsic-Flogel TD, Bonhoeffer T, Hübener M. 2006. Prior experience enhances plasticity in adult visual cortex. *Nat Neurosci* **9:** 127–132.

Hofer SB, Mrsic-Flogel TD, Bonhoeffer T, Hübner M. 2009. Experience leaves a lasting structural trace in cortical circuits. *Nature* **457:** 313–317.

Jobsis FF. 1977. Noninvasive, infrared monitoring of cerebral and myocardial oxygen sufficiency and circulatory parameters. *Science* **198:** 1264–1266.

Kalatsky VA, Stryker MP. 2003. New paradigm for optical imaging: Temporally encoded maps of intrinsic signal. *Neuron* **38:** 529–545.

Kaneko M, Stellwagen D, Malenka RC, Stryker MP. 2008. Tumor necrosis factor-α mediates one component of competitive, experience-dependent plasticity in developing visual cortex. *Neuron* **58:** 673–680.

Keck T, Mrsic-Flogel TD, Vaz Afonso M, Eysel UT, Bonhoeffer T, Hübener M. 2008. Massive restructuring of neuronal circuits during functional reorganization of adult visual cortex. *Nat Neurosci* **11:** 1162–1167.

Kim DS, Bonhoeffer T. 1994. Reverse occlusion leads to a precise restoration of orientation preference maps in visual cortex. *Nature* **370:** 370–372.

Masino SA, Kwon MC, Dory Y, Frostig RD. 1993. Characterization of functional organization within rat barrel cortex using intrinsic signal optical imaging through a thinned skull. *Proc Natl Acad Sci* **90:** 9998–10002.

Meister M, Bonhoeffer T. 2001. Tuning and topography in an odor map on the rat olfactory bulb. *J Neurosci* **21:** 1351–1360.

Mrsic-Flogel TD, Hofer SB, Creutzfeldt C, Cloëz-Tayarani I, Changeux JP, Bonhoeffer T, Hübener M. 2005. Altered map of visual space in the superior colliculus of mice lacking early retinal waves. *J Neurosci* **25:** 6921–6928.

Polley DB, Chen-Bee CH, Frostig RD. 1999. Two directions of plasticity in the sensorydeprived adult cortex. *Neuron* **24:** 623–637.

Rubin BD, Katz LC. 1999. Optical imaging of odorant representations in the mammalian olfactory bulb. *Neuron* **23:** 499–511.

Rubin BD, Katz LC. 2001. Spatial coding of enantiomers in the rat olfactory bulb. *Nat Neurosci* **4:** 355–356.

Schuett S, Bonhoeffer T, Hübener M. 2002. Mapping retinotopic structure in mouse visual cortex with optical imaging. *J Neurosci* **22:** 6549–6559.

Shtoyerman E, Arieli A, Slovin H, Vanzetta I, Grinvald A. 2000. Long-term optical imaging and spectroscopy reveal mechanisms underlying the intrinsic signal and stability of cortical maps in V1 of behaving monkeys. *J Neurosci* **20:** 8111–8121.

Singer W, Freeman B, Rauschecker JP. 1981. Restriction of visual experience to a single orientation affects the organization of orientation columns in cat visual cortex. A study with deoxyglucose. *Exp Brain Res* **41:** 199–215.

Soucy ER, Albeanu DF, Fantana AL, Murthy VN, Meister M. 2009. Precision and diversity in an odor map on the olfactory bulb. *Nat Neurosci* **12:** 210–220.

Uchida N, Takahashi YK, Tanifuji M, Mori K. 2000. Odor maps in the mammalian olfactory bulb: Domain organization and odorant structural features. *Nat Neurosci* **3:** 1035–1043.

61 Imaging Astrocytic Calcium and Vacuole-Like Vesicles

Jian Kang

Department of Cell Biology and Anatomy, New York Medical College, Valhalla, New York 10595

ABSTRACT

The study of astrocytic function is critically dependent on the imaging of fluorescent markers of activity because astrocytes are mostly electrically silent. In addition to their traditional supporting and nutritive functions in local neuronal circuits, astrocytes also actively participate in modulation of neuronal activity by releasing gliotransmitters, a process that occurs in response to astrocytic Ca^{2+} signals or as a result of vesicular fusion. Imaging of astrocytic Ca^{2+} and vesicles in acute (live) brain slices by two-photon microscopy combined with patch-clamp techniques is one of the best approaches to invetigate gliotransmission. Over the past few years, our laboratory has used a combination of patch-clamp and fluorescence imaging techniques to detect astrocytic Ca^{2+}, vacuole-like vesicles, and glutamate release. In this chapter, we discuss our experience studying astrocytes in hippocampal slices.

INTRODUCTION

Astrocytes account for one-third of the mass of the nervous system and outnumber neurons 10-fold (Privat et al. 1995). They form an extensive network that is highly coupled through gap junctions and intimately associated with neuronal and vascular elements (Ransom 1995). Traditionally, astrocytes have been considered to be supporting and nutritive cells in neuronal circuits. However, studies in the past decade have suggested that astrocytes can release gliotransmitters, such as glutamate, adenosine triphosphate (ATP), and D-serine, to modulate synaptic transmission (Parpura et al. 1994; Kang et al. 1998; Araque et al. 2001; Nedergaard et al. 2002; Liu et al. 2004; Fiacco et al. 2007).

Activity-dependent Ca^{2+} oscillations in individual astrocytes and Ca^{2+} waves between astrocytes are thought to be factors important for astrocytic release of gliotransmitters. Our laboratory has found that fusion of a vacuole-like vesicle is the mechanism underlying astrocytic glutamate release (Kang et al. 2005; Xu et al. 2007). The use of two-photon microscopy combined with patch-clamp techniques allows us to detect astrocytic Ca^{2+} signals/vesicular fusion and modulation of neuronal activity simultaneously.

IMAGING SETUP

Our imaging setup is a customized two-photon Olympus BX61WI microscope with a 40x or 60x/0.90 numerical aperture (NA) water-immersion lens and infrared differential inference contrast (IR-DIC). A Mai-Tai laser (Solid-State Laser Co., Mountain View, CA) is used for excitation (830 nm for Fluo-4 or 890 nm for FM1-43). Image acquisition is controlled by Olympus FluoView FV300 software (Olympus, Melville, NY). In the transfluorescence pathway, a 565-nm dichroic mirror is used to separate green and red fluorescence, and HQ525/50 and HQ605/50 high-pass filters are placed in the "green" and "red" pathways, respectively, to eliminate transmitted or reflected excitation light (Chroma Technology, Rockingham, VT). Alexa Fluor 594 or FM1-43 fluorescence is detected via the red pathway with the HQ605/50 filter. Fluo-4 fluorescence is detected using the green pathway. Baseline fluorescence (F_0) is the average of four images acquired during prestimulus baseline imaging, and $\Delta F/F$ is calculated as $(\Delta F/F)(t) = [F(t) - F_0]/F_0$. The position shift in the x–y section during 5 min scanning is 0.5 ± 0.3 µm (mean ± standard deviation [SD]; $n = 30$ cells).

IDENTIFICATION OF ASTROCYTES AND NEURONS IN HIPPOCAMPAL SLICES

Using DIC Microscopy

During electrophysiological recordings, the standard approach is to identify astrocytes as cells that have highly negative resting membrane potentials and do not show action potentials on depolarization (Walz and MacVicar 1988). However, in calcium imaging studies, many cells are visualized simultaneously and it is not feasible to obtain recordings from them all. The use of more traditional approaches for identifying astrocytes in fixed tissues, including immunoreactivity against glial fibrillary acidic protein (GFAP) (Dani et al. 1992) or biocytin staining (Robinson et al. 1993), is not an option in live tissues. Our goal was, therefore, to establish whether astrocytic morphology under IR-DIC microscopy allows the investigator to identify astrocytes and thus discriminate between astrocytes and neurons during imaging experiments. We have primarily studied astrocytes in hippocampal slices, but similar observations have been made in cortical slices. For information on DIC, see Chapter 1. In our studies we used an Olympus BX51WI upright microscope equipped with IR-DIC optics. Slices in a thin 1.5-mL chamber mounted on the microscope stage were perfused with a slice solution. Cells were visualized with a 40x or 60x water-immersion lens. Figure 1A illustrates a characteristic field of the stratum radiatum containing both astrocytes and neurons. Interneurons are characterized by large and bright somata (diameter 16.1 ± 2.7 µm; mean ± SD) with large dendrites, whereas astrocytic somata are small and round (diameter 8.1 ± 1.7 µm; mean ± SD) without large processes.

Confirming the Morphology Revealed by DIC

The identification of astrocytes revealed by DIC microscopy can be confirmed by electrophysiological recordings (Fig. 1B), biocytin staining (Fig. 1C), and GFAP staining (Fig. 1D). The use of slices prepared from transgenic mice expressing green fluorescence protein (GFP) under the control of the GFAP promoter is an excellent approach for identifying astrocytes in live tissue (Fig. 1E) (Zhuo et al. 1997; Nolte et al. 2001). Recently, sulforhodamine 101 (SR101) has been used as a specific marker

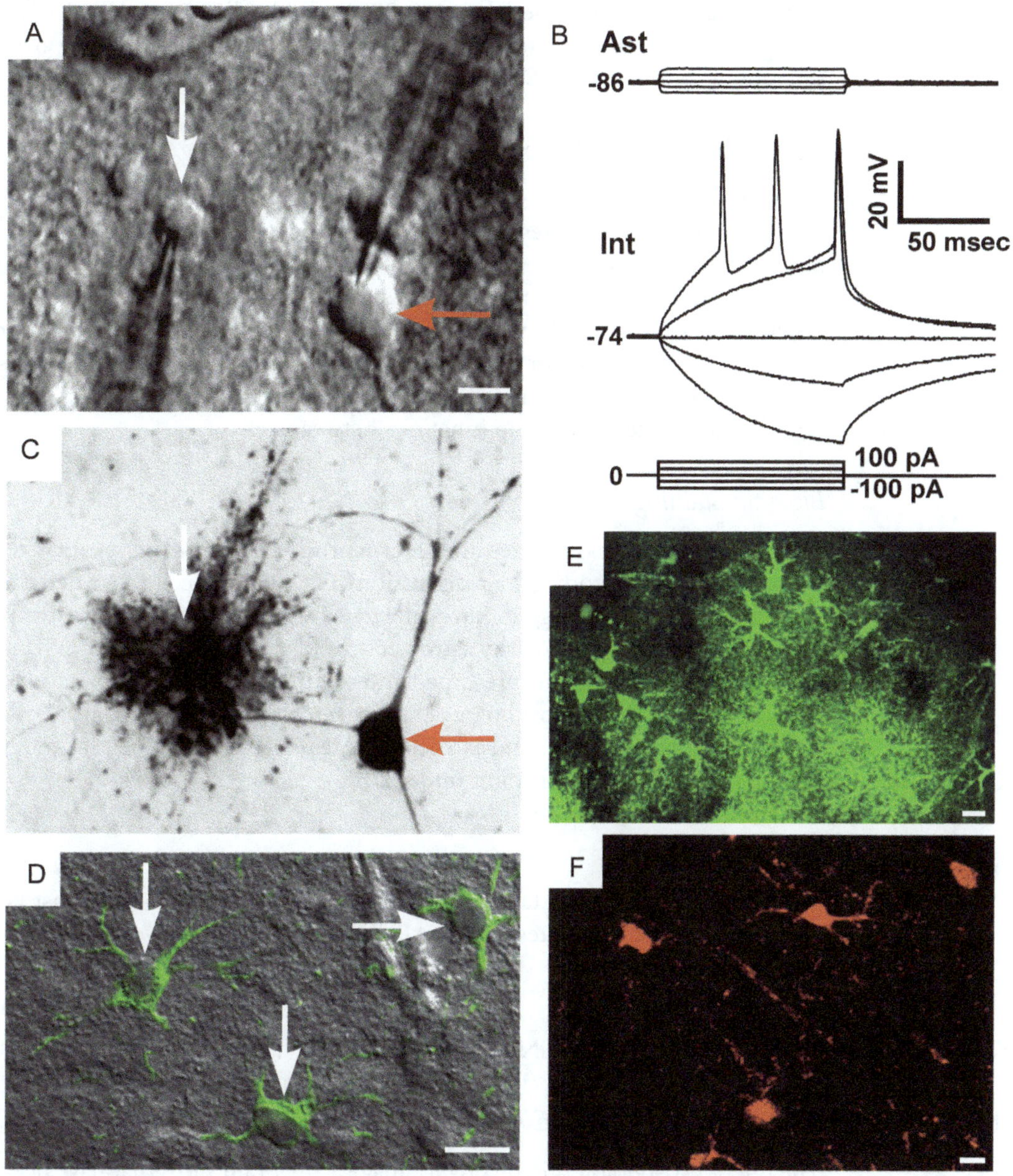

FIGURE 1. Identification of interneurons and astrocytes in hippocampal slices under differential interference contrast (DIC) optics. (*A*) An astrocyte (white arrow; characterized by a small, rounded cell body and poorly defined, irregular processes) and an interneuron (red arrow; characterized by a large soma with large dendrites that project in all directions) located in the stratum radiatum were patched by two pipettes. (*B*) Electrophysiological properties of interneurons (Int) and astrocytes (Ast). The *bottom panel* indicates depolarization currents delivered to patch pipettes. Astrocytes do not fire upon depolarization and have a low input resistance. (*C*) Biocytin staining of the astrocyte and interneuron shown in *A*. Biocytin (0.5%) was injected into both of these cells to compare DIC with cellular morphology. The cell identified by DIC as an astrocyte (white arrow) was characterized by an extensive array of branched short processes, which are phenotypical for astrocytes. In contrast, the presumed interneuron (red arrow) had few, but evenly sized, less branched processes, a morphology that is stereotypical for neurons. The size of the biocytin-stained astrocyte seems larger than that of the same cell viewed under DIC optics. This might result from the mixture of short processes with the cell body. (*D*) Overlapping of DIC optics (background) and GFAP staining (green) in the same field confirms that three cells (white arrows) that were identified as astrocytes under DIC optics are GFAP-positive. (*E*) GFP-labeled astrocytes in a slice prepared from a GFAP-eGFP mouse. (F) Sulforhodamine 101 (SR101) staining of astrocytes in a slice. SR101 (1 μM) was added to the superfusate. Scale bars, 10 μm.

of astroglia in vivo (Nimmerjahn et al. 2004) and in slices (Fig. 1F) (Kafitz et al. 2008). When using SR101 in experiments that test neuronal functions, a caution should be given so that SR101 enhances synaptic activation of NMDA receptors and neuronal excitability (Kang et al. 2010).

Electrophysiological Properties

Electrophysiological properties can confirm the DIC morphological criteria for identifying interneurons and astrocytes. In the study shown in Figure 1, recordings were taken from a total of 205 cells with DIC morphology typical for astrocytes, and none fired action potentials on depolarization (Fig. 1B, Ast). Recordings were also taken from 107 interneurons, all of which fired action potentials (Fig. 1B, Int). Astrocytes also have a more negative resting membrane potential (-86 ± 0.5 mV; mean $\pm$ standard error of the mean [S.E.M.]; $n = 87$) than interneurons (-74 ± 0.6 mV; mean $\pm$ S.E.M.; $n = 90$) and the input resistance of astrocytes is low (15.0 ± 1.1 MΩ; $n = 10$), reflecting both high ion conductance and gap junctions between cells.

Biocytin Staining

DIC morphology of astrocytes and interneurons can be further evaluated by biocytin staining (see Protocol A). In the study under discussion (Fig. 1C), the cell identified by DIC as an astrocyte was shown by biocytin staining to have an extensive array of branched short processes, which are phenotypical for astrocytes. Nearby astrocytes could also be stained through gap junctions (Robinson et al. 1993). In contrast, the presumed interneuron was shown to have few evenly sized and less branched processes, a morphology that is stereotypical for neurons. In this experiment, a total of eight pairs of interneurons and astrocytes were injected with biocytin, and in all cases, biocytin staining confirmed cell identification under DIC.

GFAP Staining

In a further experiment, staining of astrocytes with an antibody against GFAP showed overlapping of DIC-identified astrocytes and GFAP-staining (Fig. 1D).

DETECTION OF ASTROCYTIC Ca^{2+} SIGNALS IN SLICES BY TWO-PHOTON MICROSCOPY

Detection of Ca^{2+} Signals from Multiple Astrocytes in Slices

The use of two-photon microscopy to image acute brain slices that have been preincubated with Fluo-4 acetoxymethyl ester (AM) (see Protocol B) has special advantages for detecting astrocytic Ca^{2+} signals.

- The excellent penetration of long-wavelength excitation with a two-photon laser allows us to detect Ca^{2+} signals in deep-layer cells (50–150 µm).
- The low level of phototoxic damage associated with infrared excitation allows us to monitor Ca^{2+} signals in healthy astrocytes deep within the slice over a prolonged period of time.
- In brain slices prepared from rats older than PND (postnatal day) 10, astrocytes preferentially load with Fluo-4 AM (Fig. 2A) (Liu et al. 2004). The mechanism underlying the preferential loading of astrocytes is unknown, but it is specific to intact brain tissue because neurons, endothelial cells, and pericytes load well with AM dyes in cultures.
- Astrocytes are not patched and the intracellular content is intact.

It is important to note, however, that the time resolution in these experiments is limited because of the large scanning area. Also, photobleaching can occur during two-photon microscopy imaging, but to a lesser extent than after one-photon excitation. Very long exposure times can damage astrocytes, and this is indicated by a continuing elevation of Ca^{2+}.

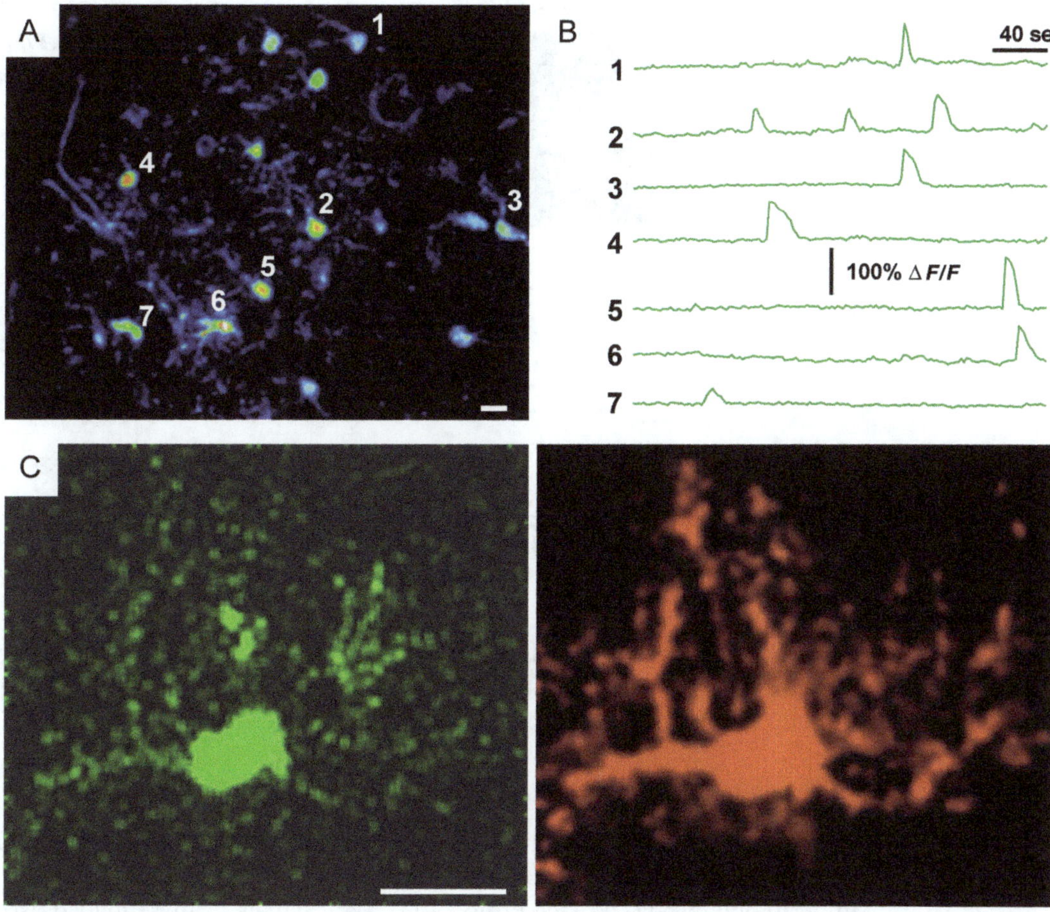

FIGURE 2. Two-photon imaging of astrocytic Ca^{2+}. *(A)* A two-photon microscopy image of astrocytes (indicated by numbers 1–7) loaded with the Ca^{2+} indicator Fluo-4 AM. *(B)* Time course of Ca^{2+} signals in the soma of astrocytes indicated in *A*. Spontaneous Ca^{2+} oscillations occurred in the astrocytes. (C) An astrocyte patched by a pipette containing 50 μM Fluo-4-potassium (green). *(D)* The same astrocyte patched by a pipette containing 100 μM Alexa Fluor 594 (red; *right*). Scale bar, 10 μm.

Detection of Ca^{2+} Signals from Astrocytes Infused with Membrane-Impermeable Ca^{2+} Indicators

Ca^{2+} signals in a single astrocyte and its processes can be detected by two-photon imaging of astrocytes filled with membrane-impermeable fluorescence Ca^{2+} indicators, like Fluo-4, which are infused into astrocytes through patch pipettes (Fig. 2C) (see Protocol C). The morphology of patched astrocytes can be monitored by infusing the fluorescent marker Alexa Fluor 594 (Fig. 2D). This approach has several advantages.

- Fine astrocytic processes are clearly outlined by Alexa Fluor 594 and therefore subcellular changes in Ca^{2+} can be visualized.
- The patch-clamp recording also provides information regarding cell membrane potential and ionic currents.
- By changing the contents of the pipette solution, it is possible to introduce other reagents, such as the Ca^{2+} chelator BAPTA (1,2-bis-(aminophenoxy)ethane-*N*,*N*,*N*′,*N*′-tetra-acetic acid), intracellularly.
- Gap-junction coupling can be detected by the diffusion of fluorescent dyes into neighboring cells.

There are some disadvantages to this approach. Patching stimulates astrocytes and often causes prolonged increases in cytosolic Ca^{2+}. These cells are also highly sensitive to mechanical stimulation,

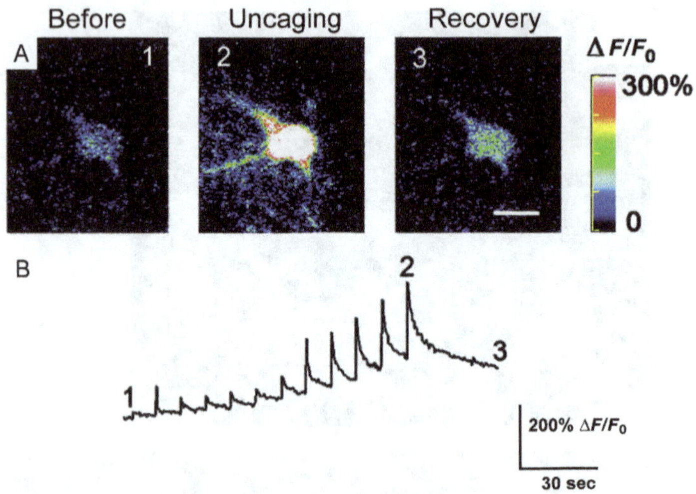

FIGURE 3. Ca^{2+} uncaging in astrocytes. (*A*) An astrocyte was stimulated with a train of 12 UV laser pulses (0.1 Hz) to uncage NP-EGTA. Images were taken before (1), during (2), and after (3) the UV pulses. Scale bar, 10 µm. (*B*) Time course of changes in calcium fluorescence ($\Delta F/F_0$) during the experiment shown in *A*.

and many groups have noted that astrocytes fail to respond to external stimulation after patching. Moreover, astrocytes are relatively small cells and their cytosolic content is quickly dialyzed (within minutes of breaking the membrane).

Detection of Ca^{2+} Signals from Photolysis of Caged Compound in a Single Astrocyte

Astrocytes can be loaded with caged Ca^{2+} either by incubating brain slices with membrane-permeable NP-EGTA-AM (o-nitrophenyl-EGTA-AM) or through a patch pipette filled with membrane-impermeable NP-EGTA (see Protocol D). Ca^{2+} signals in these loaded astrocytes are triggered by stimulation with ultraviolet (UV) light, which uncages Ca^{2+} from NP-EGTA (Fig. 3A,B). The advantage of this technique is that Ca^{2+} levels in a single astrocyte can be elevated by application of a laser beam and Ca^{2+}-driven release of transmitters can be tested (Lui et al. 2004). Ca^{2+} signals can also be triggered by uncaging IP_3, which evokes Ca^{2+} release from intracellular stores. Uncaging IP_3 triggers the release of Ca^{2+} from specific intracellular stores, whereas uncaging Ca^{2+} may reflect Ca^{2+} signals from several different sources. When focusing on the functions of Ca^{2+} release from intracellular stores, uncaging IP_3 is the preferred triggering method. When studying the overall effects of intracellular Ca^{2+} increase, uncaging NP-EGTA is the better choice.

DETECTION OF ASTROCYTIC VACUOLE-LIKE VESICLES BY TWO-PHOTON MICROSCOPY

A transient astrocytic glutamate release that activates a slowly decaying transient inward current (SIC) has been observed in hippocampal and thalamic neurons (Parri and Crunelli 2001; Angulo et al. 2004; Fellin et al. 2004; Kang et al. 2005; Xu et al. 2007; Navarrete and Araque 2008; Shigetomi et al. 2008). An astrocytic transient current (aTC) is associated with SICs in hippocampal neurons (Kang et al. 2005; Xu et al. 2007). We have shown that SICs are the result of fusion of a very large vesicle to the astrocytic membrane, which results in glutamate release. The size (2–6 µm) and electron microscopy ultrastructure of these large vesicles (Xu et al. 2007) suggest that they are vacuole-like vesicles.

Using Fluo-4 Potassium and Alexa Fluor 594 to Detect Vacuole-Like Vesicles

Vacuole-like vesicles can be loaded with Fluo-4 and Alexa Fluor 594 when these compounds are infused into astrocytes through patch pipettes. To elicit SICs or aTCs in astrocytes, high intracellu-

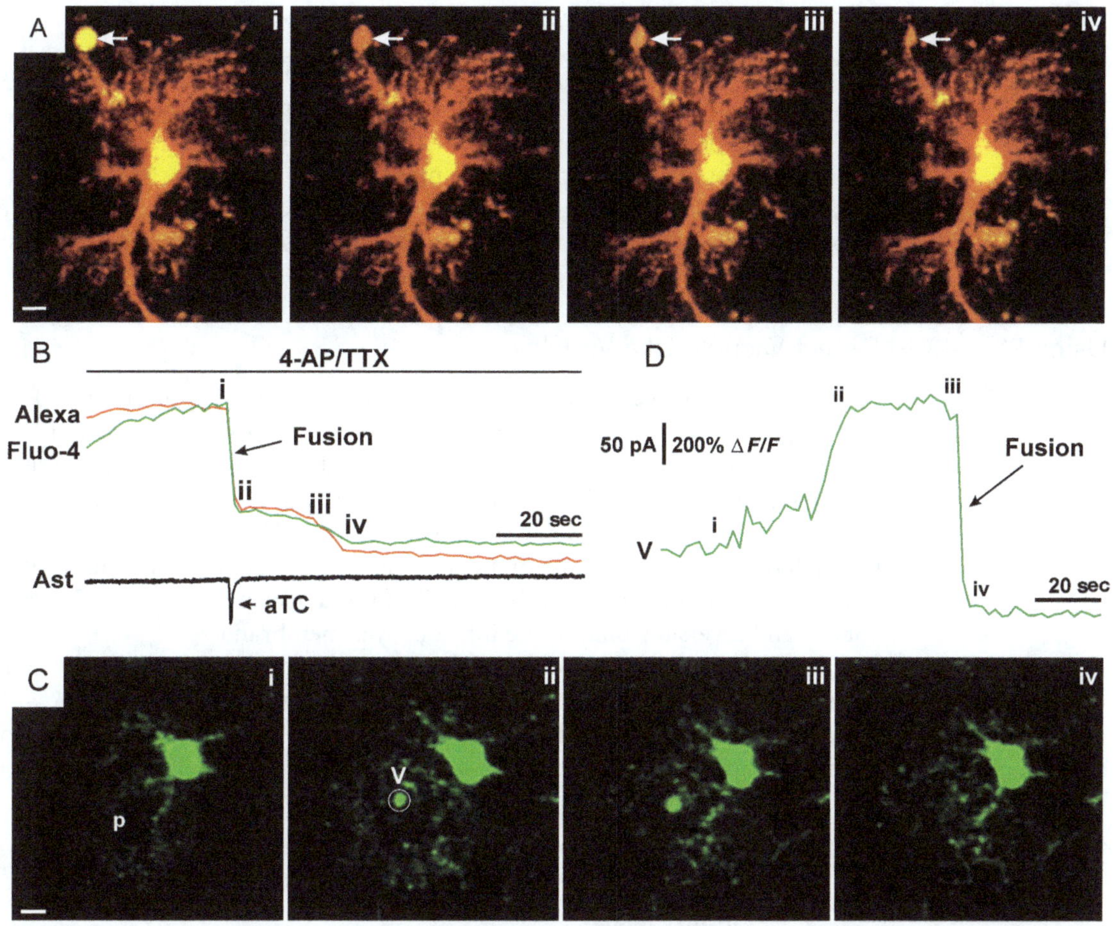

FIGURE 4. Astrocytic vacuole-like vesicles loaded with Fluo-4 and Alexa Fluor 594. (*A*) Two-photon microscopy images of overlapped Fluo-4 and Alexa Fluor 594 fluorescence from an astrocyte patched with a pipette solution containing 50 mM glutamate, 100 μM Alexa Fluor 594, and 50 μM Fluo-4 potassium at the time indicated in *B*. The intensity of Fluo-4 and Alexa Fluor 594 fluorescence in a vacuole-like vesicle (arrowhead) suddenly decreased, indicating fusion. (*B*) Alexa Fluor 594 (red) and Fluo-4 (green) traces in the vesicle indicated in *A* and whole-cell recording in the astrocyte (Ast trace) indicate that fusion of the vacuole-like vesicle coincided with an astrocyte transient current (aTC). (*C*) Two-photon microscopy images of an astrocyte that was preloaded with Fluo-4 AM at the time indicated in *B*. (*i*) An image after the first puff showing the position of the puffer pipette (p). A Fluo-4-loaded vacuole-like vesicle (V) appeared after the second puff with artificial cerebrospinal fluid (*ii*) and then suddenly disappeared (*iii–iv*). (*D*) Fluo-4 fluorescence in the circled area (V) indicated in C. Arrow indicates fusion. Scale bars, 5 μm. 4-AP, 4-aminopyridine; TTX, tetrodotoxin.

lar glutamate concentrations (>10 mM) are required. In an experiment illustrated in Figure 4, an astrocyte was patched with a pipette containing Fluo-4 potassium (50 μM), Alexa Fluor 594 (100 μM), and glutamate (50 mM), and vacuole-like vesicles could be observed during the perfusion of slices with 4-aminopyridine and tetrodotoxin (Fig. 4A). Vacuole-like vesicles have highly intense Fluo-4 and Alexa Fluor 594 fluorescence, indicating that Fluo-4, Ca^{2+}, and Alexa Fluor 594 are actively transported into these vesicles. Fusion of vacuole-like vesicles coincided with aTCs (Fig. 4B, aTC) that are mixtures of glutamate transport currents and ionotropic glutamate receptor currents, suggesting glutamate release from vacuole-like vesicles. It is important to note that vacuole-like vesicles can be induced by variety of stimulations, such as the K^+ channel blocker 4-aminopyridine, mechanical stimulation, and hypotonic stimulation. Also, when intracellular glutamate concentration in astrocytes is <5 mM, such as puffing ACSF into the extracellular space (see Fig. 5C), fusion of vacuole-like vesicles can still be induced, but their fusion does not elicit SICs or aTCs.

Using Fluo-4 AM to Detect Vacuole-Like Vesicles

Because vacuole-like vesicles can load with intracellularly applied Fluo-4 and Ca^{2+}, they are also stained with Fluo-4 and Ca^{2+} in Fluo-4 AM-pretreated astrocytes. If slices are preincubated with Fluo-4 AM (5 μM) for 40 min similar to the experiment described in Fig. 2A, Fluo-4-positive vesicles can be observed following weak mechanical stimulation (e.g., puffing with artificial cerebrospinal fluid) (Fig. 4C,D). The advantages of this approach are that the intracellular content of astrocytes is not changed, Ca^{2+} levels in astrocytes are normal and can be monitored, and the astrocytic location of vacuole-like vesicles can be shown.

Using FM1-43 to Detect Vacuole-Like Vesicles

Vacuole-like vesicles can also be detected by the extracellular application of FM1-43. FM1-43, which is an amphiphilic styryl dye developed by Betz and co-workers (Betz et al. 1992; see Chapter 18), has a hydrophilic head with two positive charges, which limits its translocation through the lipid bilayer, and a four-carbon hydrophobic tail, which has a high affinity for hydrophobic lipids (see Fig. 5A) (Cochilla et al. 1999; Cousin and Robinson 1999). FM1-43 fluorescence is low in aqueous solution and high when intercalated within membranes. These characteristics allow FM1-43 to bind and stay on the exposed nonpolar region (lipid tail) of fused bilayer membranes (Fig. 5A, red), and only occasionally and reversibly bind to the intact bilayer membrane.

FM1-43 is extensively used in exocytosis of synaptic vesicles. Generally, synaptic vesicles are first loaded with FM1-43 by endocytosis following depolarization stimulation with high KCl, and then stimulation-induced reduction of loaded vesicles (FM1-43 fluorescence) is monitored by two-photon microscopy. Because of the large size of astrocytic vacuole-like vesicles, here we can monitor the

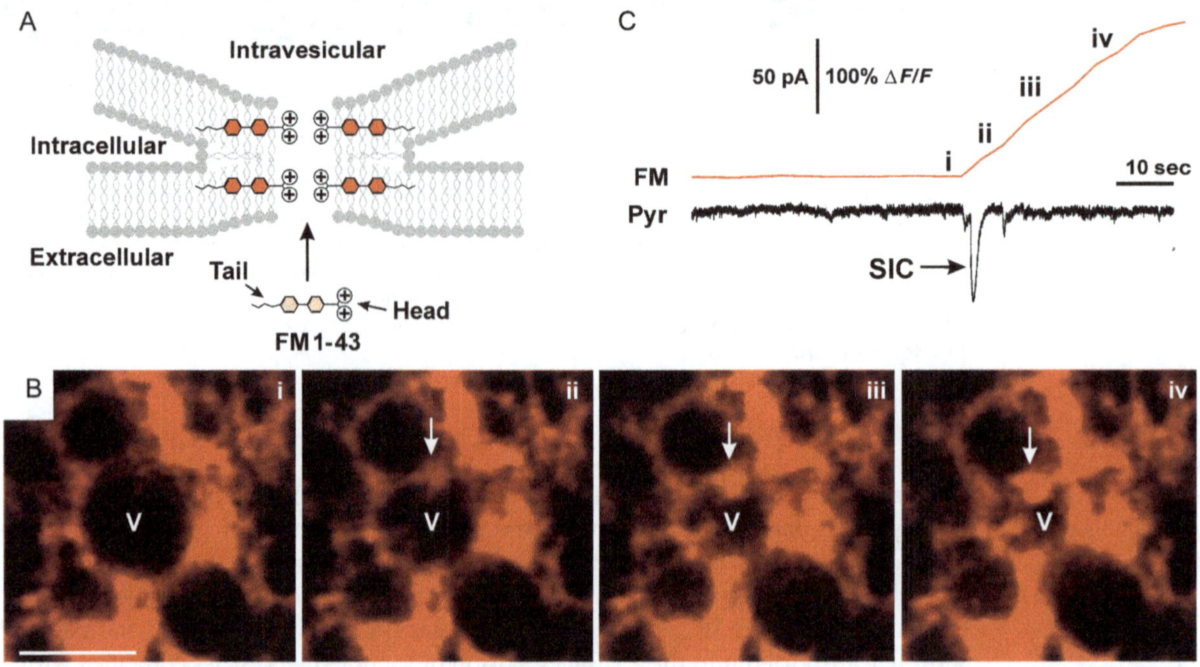

FIGURE 5. Fusion of vacuole-like vesicles coincides with SICs. (*A*) Schematic diagram indicating the detection of vacuole-like vesicles by the extracellular application of FM1-43. (*B*) A vacuole-like vesicle that is FM1-43-negative (V) is shown as it collapses (panels *i–iv*). A high FM1-43 fluorescence punctum (arrow) forms as the vesicle collapses. A mixture of FM1-43 (20 μM) and glutamate (50 mM) was puffed into the stratum radiatum using a puffer pipette. Scale bar, 5 μm. (*C*) Accumulation of FM1-43 in the vesicle shown in *B* (red line) and simultaneous whole-cell recording from a pyramidal neuron with apical dendrites passing through the imaging area. The start of vesicular collapse coincides with an SIC.

FM1-43 influx (loading process) into the vacuole-like vesicle that is collapsing. FM1-43 (20 μM) was added to the puffer pipette solution and puffed into the extracellular space. FM1-43 fluorescence quickly declined after the first puff, but higher fluorescence background was observed after repetitive puffs. Repetitive puffs induced multiple FM1-43-negative large vesicles (V in Fig. 5B). An FM1-43-negative large vesicle was identified as a newly appearing round structure with low fluorescence surrounded by the high FM1-43 fluorescence. As glutamate (20 or 50 mM) is included in the puffer pipette solution, fusion (collapse) of FM1-43-negative vacuole-like vesicles (Fig. 5B) is associated with SICs in whole-cell recording from pyramidal neurons (Fig. 5C). When a vacuole-like vesicle starts to collapse, a small FM1-43 punctum with high fluorescence intensity appears in the vesicular wall (Fig. 5B). This FM1-43 bright punctum indicates the position of the fusion pore because the membrane around the fusion pore is fused and binds to FM1-43 with high affinity (Fig. 5A). The vesicular lumen remains at low fluorescence intensity as vesicular collapse (V in Fig. 5B), suggesting that FM1-43 endocytosed in the vacuole-like vesicle binds to the intact internal vesicular membrane still with low affinity.

FM1-43 imaging has several advantages. The appearance of a high fluorescence punctum indicates FM1-43 influx into a vacuole-like vesicle through the fusion pore, and exocytotic fusion can be shown. Also, the morphological collapse of vesicles gives more information than does dye diffusion (Fig. 4). The disadvantages are that the astrocytic location of vacuole-like vesicles is not indicated by FM1-43 imaging and image screening of vacuole-like vesicles limits the time resolution for vesicular collapse.

It is important to note that FM1-43 fluorescence in the extracellular space quickly declines after puffing and the fluorescence intensity of two-photon microscopy imaging decreases with time. When analyzing two-photon microscopy imaging, offline compensation of time-dependent loss of the fluorescence intensity by the ImageJ software is needed. Morphological collapse of vesicles is slower than diffusion of the vesicular content, and the time course of vesicular collapse does not represent the gliotransmitter release time.

Ca^{2+}-dependent fusion of small microvesicles similar to synaptic vesicles has been suggested to be the mechanism underlying astrocytic glutamate release (Montana et al. 2004; Zhang et al. 2004; Jourdain et al. 2007; Marchaland et al. 2008). Recent studies also suggest that lysosomes are the major vesicular compartment undergoing Ca^{2+}-regulated exocytosis from astrocytes (Zhang et al. 2007; Li et al. 2008). Although our data cannot exclude the mechanism of microvesicle or lysosome exocytosis, SICs are unlikely to come from these small compartments because of the large quantal size of SICs and dispersed distribution of these small compartments.

Protocol A

Biocytin Staining Procedure

This technique is used to postexperimentally label morphology of glia and neurons through the patch pipette.

MATERIALS

CAUTION: See Appendix 6 for proper handling of materials marked with <!>.
See the end of the chapter for recipes for reagents marked with <R>.

Reagents

Albumin
Biocytin hydrochloride (Sigma-Aldrich) (0.5% in pipette solution)
    Prepare immediately before use.
Brain slices from Sprague-Dawley rats (PND 8–20)
Fixative <R>
Iodine <!> (0.3% in 0.1 M PBS)
Permount <!> (Fisher Scientific)
Phosphate-buffered saline (PBS), 0.1 M (Mediatech, Inc.)
Pipette solution for biocytin staining <R>
Slice solution <R>
Triton X-100 <!>
Vectastain Elite ABC kit (Vector Labs)

Equipment

DIC microscope
Patch pipettes and other electrophysiology equipment
Recording chamber, custom-made

EXPERIMENTAL METHOD

1. Prepare brain slices from 8–20-day-old (PND 8–20) Sprague-Dawley rats of either sex.
2. Place the slices in a recording chamber (1.5 mL) and perfuse with slice solution.
3. Using two patch pipettes, load biocytin (0.5% in pipette solution) into each cell type under study. First, patch an interneuron with a biocytin-containing pipette and then patch an astrocyte with another biocytin-containing pipette.
4. Without moving the patch pipettes, perfuse the slices with fixative for 10 min and then carefully withdraw the pipettes.
5. Postfix the slices for 10–12 h. Wash four times in 0.1 M PBS.
6. Incubate in 0.3% iodine (in 0.1 M PBS) at room temperature for 15 min. Wash four times in PBS.
7. Incubate in 0.2% Triton X-100/0.2% albumin in 0.1 M PBS for 45 min at room temperature. Wash four times in PBS.
8. Stain the fixed slices with the Vectastain Elite ABC kit. Mount with Permount.

Protocol B

Two-Photon Imaging of Acute Brain Slices Preincubated with Fluo-4 AM

This technique is used to detect Ca^{2+} signals from multiple astrocytes in hippocampal slices. Fluo-4 AM is a membrane-permeable Ca^{2+} indicator dye.

MATERIALS

CAUTION: See Appendix 6 for proper handling of materials marked with <!>.
See the end of the chapter for recipes for reagents marked with <R>.

Reagents

Artificial cerebrospinal fluid (ACSF) <R>
Fluo-4 AM (Invitrogen-Molecular Probes)
Gas mixture of 95% O_2 <!> and 5% CO_2 <!>
Hippocampal slices from Sprague-Dawley rats (PND 13–20)
Slice solution <R>

Equipment

Incubation tube, 5-ml
Recording chamber
See *Imaging Setup.*

EXPERIMENTAL METHOD

1. Prepare acute hippocampal slices from the brains of PND 13–20 rats and allow the slices to recover for 30 min at room temperature.
2. Incubate the slices in slice solution containing 5 µM Fluo-4 AM for 40 min at 30°C–32°C. Use a 5-mL incubation tube and gas with 95% O_2 and 5% CO_2.
3. Wash the slices with ACSF for 20 min and then transfer them to a recording chamber for microscopy.
4. Perform two-photon microscopy imaging with a 40x or 60x lens through the green pathway (525 nm).

   The excitation typically used in two-photon imaging of Fluo-4 AM is 830 nm. Ca^{2+} oscillations in the somata and processes, depending on magnification, of individual astrocytes can be detected (Fig. 2A). The time course of Ca^{2+} signals in individual astrocytes (Fig. 2B) can be obtained offline using the FluoView FV300 software or the ImageJ program.

Protocol C

Two-Photon Imaging of Astrocytes Filled with Fluo-4 Introduced by Patch Pipettes

This technique is used to detect Ca^{2+} signals from a single astrocyte that has been filled with a membrane-impermeable Ca^{2+} indicator such as Fluo-4.

MATERIALS

See the end of the chapter for recipes for reagents marked with <R>.

Reagents

Alexa Fluor 594 (Invitrogen-Molecular Probes)
Brain slices
Fluo-4 potassium (Invitrogen-Molecular Probes)
Pipette solution for Fluo-4 staining <R>

Equipment

DIC microscope
Patch pipettes and other electrophysiology equipment
See *Imaging Setup*.

EXPERIMENTAL METHOD

1. Astrocytes are identified under DIC optics as described in Figure 1.
2. Patch individual astrocytes with patch pipettes filled with pipette solution containing 100 μM Alexa Fluor 594 and 50 μM Fluo-4 potassium. Form the whole-cell recording configuration and proceed with two-photon microscopy imaging of patched astrocytes.

   The positive pressure applied to the pipettes before patching the astrocytes should be smaller than that applied when patching neurons because astrocyte membranes are softer than neuron membranes.
3. Pyramidal neurons are identified under DIC optics as described in Figure 1. Pyramidal neurons with apical dendrites that pass though the imaging area are selected and patched in the whole-cell voltage-clamp configuration.

Protocol D

Two-Photon Imaging of Astrocytes Filled with a Photolabile Caged Ca^{2+} Compound

This technique is used to detect Ca^{2+} signals from a single astrocyte that has been filled with a compound such as NP-EGTA, which releases Ca^{2+} after stimulation with UV light. Astrocytes can be loaded with such caged Ca^{2+} either by incubating brain slices with membrane-permeable NP-EGTA-AM or by introduction through a patch pipette filled with membrane-impermeable NP-EGTA.

MATERIALS

Reagents

Brain slices
NP-EGTA or NP-EGTA-AM (Invitrogen-Molecular Probes)

Equipment

Nitrogen-pulsed laser (VSL-337ND; Laser Science)
Photolysis system (Prairie Technologies)
See also *Imaging Setup*.

EXPERIMENTAL METHOD

1. Use a photolysis system to uncage NP-EGTA and release calcium. Focus a UV laser beam (337 nm) from a nitrogen-pulsed laser through a DIC 40x water-immersion objective to give an optical spot 10 µm in diameter. Couple a red spot from a weak 635-nm diode laser with the UV beam to guide the position of uncaging.
2. Perform fluorescence imaging using two-photon microscopy. Use the 488-nm line of the argon laser for excitation and a 510-nm emission filter. Analyze fluorescence intensities by using ImageJ or other analysis software.

   Changes in $[Ca^{2+}]_i$ are given as $\Delta F/F$, where F is the baseline fluorescence level and ΔF is determined by the relative change from the base level.

CONCLUSION

Using DIC microscopy, astrocytes in hippocampal slices are identified by their small, round somata, which lack large processes. Electrophysiological properties, biocytin staining, and GFAP staining can be used to confirm the DIC morphological criteria for identifying astrocytes. Astrocytes in slices prepared from PND 13–20 rats load well with Fluo-4 AM, and Ca^{2+} signals in astrocytes can be detected by two-photon microscopy. Fusion of vacuole-like vesicles in astrocytes can be monitored by two-photon microscopy imaging of astrocytes infused with Fluo-4 and Alexa Fluor 594 or subjected to puffs of FM1-43. A combination of two-photon microscopy imaging and patch-clamp techniques is an excellent approach to test gliotransmitter release by fusion of vacuole-like vesicles.

RECIPES

CAUTION: See Appendix 6 for proper handling of materials marked with <!>.
Recipes for reagents marked with <R> are included in this list.

Artificial Cerebrospinal Fluid (ACSF)

Reagent	Final concentration
NaCl	125 mM
$NaHCO_3$	26 mM
NaH_2PO_4	1.25 mM
KCl	2.5 mM
$MgCl_2$	1.0 mM
$CaCl_2$ <!>	2.0 mM
Glucose	25 mM

Bubble 95% O_2 <!> and 5% CO_2 <!> through the solution.

Fixative

Add 40 mL of saturated picric acid <!> (Sigma-Aldrich) and 80 mL of 25% glutaraldehyde <!> (Fisher Scientific) to 9.88 mL of 0.2 M phosphate-buffered saline (Sigma-Aldrich). Dilute this solution 1:1 with 8% paraformaldehyde solution <R>. Adjust the pH to 7.4 with NaOH <!>. The final concentrations are 4% paraformaldehyde, 0.2% picric acid, and 0.1% glutaraldehyde.

Paraformaldehyde Solution (8%)

Add 40 g of paraformaldehyde <!> (Sigma-Aldrich) to 400 mL of distilled water. Heat to 60°C and add 2–3 drops of 1 N NaOH <!>. Adjust the final volume to 500 mL with distilled water. Store this stock solution at 4°C.

Pipette Solution for Biocytin Staining

Reagent	Final concentration
Potassium gluconate <!>	120 mM
KCl	10 mM
$MgCl_2$	1 mM
HEPES	10 mM
EGTA	0.1 mM
$CaCl_2$ <!>	0.01 mM
ATP	1 mM
Guanosine triphosphate (GTP)	0.2 mM
Glucose	4 mM

Adjust pH to 7.2 with KOH <!>.

Pipette Solution for Fluo-4 Staining

Reagent	Final concentration
Potassium gluconate <!>	123 mM
KCl	10 mM
$MgCl_2$	1 mM
HEPES	10 mM
EGTA	5 mM
$CaCl_2$ <!>	0.5 mM
ATP	1 mM
GTP	0.2 mM
Glucose	4 mM

Adjust pH to 7.2 with KOH <!>.

Slice Solution

Reagent	Final concentration
NaCl	126 mM
KCl	2.5 mM
NaH_2PO_4	1.25 mM
$MgCl_2$	2 mM
$CaCl_2$ <!>	2 mM
Glucose	10 mM
NaH_2CO_3	26 mM

The pH should be 7.4 when the solution is gassed with 5% CO_2 <!> and 95% O_2 <!> at room temperature (23°C–24°C).

REFERENCES

Angulo MC, Kozlov AS, Charpak S, Audinat E. 2004. Glutamate released from glial cells synchronizes neuronal activity in the hippocampus. *J Neurosci* **24:** 6920–6927.

Araque A, Carmignoto G, Haydon PG. 2001. Dynamic signaling between astrocytes and neurons. *Annu Rev Physiol* **63:** 795–813.

Betz WJ, Mao F, Bewick GS. 1992. Activity-dependent fluorescent staining and destaining of living vertebrate motor nerve terminals. *J Neurosci* **12:** 363–375.

Cochilla AJ, Angleson JK, Betz WJ. 1999. Monitoring secretory membrane with FM1-43 fluorescence. *Annu Rev Neurosci* **22:** 1–10.

Cousin MA, Robinson PJ. 1999. Mechanisms of synaptic vesicle recycling illuminated by fluorescent dyes. *J Neurochem* **73:** 2227–2239.

Dani JW, Chernjavsky A, Smith SJ. 1992. Neuronal activity triggers calcium waves in hippocampal astrocyte networks. *Neuron* **8:** 429–440.

Fellin T, Pascual O, Gobbo S, Pozzan T, Haydon PG, Carmignoto G. 2004. Neuronal synchrony mediated by astrocytic glutamate through activation of extrasynaptic NMDA receptors. *Neuron* **43:** 729–743.

Fiacco TA, Agulhon C, Taves SR, Petravicz J, Casper KB, Dong X, Chen J, McCarthy KD. 2007. Selective stimulation of astrocyte calcium in situ does not affect neuronal excitatory synaptic activity. *Neuron* **54:** 611–626.

Jourdain P, Bergersen LH, Bhaukaurally K, Bezzi P, Santello M, Domercq M, Matute C, Tonello F, Gundersen V, Volterra A. 2007. Glutamate exocytosis from astrocytes controls synaptic strength. *Nat Neurosci* **10:** 331–339.

Kafitz KW, Meier SD, Stephan J, Rose CR. 2008. Developmental profile and properties of sulforhodamine 101–labeled glial cells in acute brain slices of rat hippocampus. *J Neurosci Methods* **169:** 84–92.

Kang J, Jiang L, Goldman SA, Nedergaard M. 1998. Astrocyte-mediated potentiation of inhibitory synaptic transmission. *Nat Neurosci* **1:** 683–692.

Kang N, Xu J, Xu Q, Nedergaard M, Kang J. 2005. Astrocytic glutamate release-induced transient depolarization and epileptiform discharges in hippocampal CA1 pyramidal neurons. *J Neurophysiol* **94:** 4121–4130.

Kang J, Kang N, Yu Y, Zhang J, Petersen N, Tian GF, Nedergaard M. 2010. Sulforhodamine 101 induces long-term potentiation of intrinsic excitability and synaptic efficacy in hippocampal CA1 pyramidal neurons. *Neuroscience* **169:** 1601–1609.

Li D, Ropert N, Koulakoff A, Giaume C, Oheim M. 2008. Lysosomes are the major vesicular compartment undergoing Ca^{2+}-regulated exocytosis from cortical astrocytes. *J Neurosci* **28:** 7648–7658.

Liu QS, Xu Q, Arcuino G, Kang J, Nedergaard M. 2004. Astrocyte-mediated activation of neuronal kainate receptors. *Proc Natl Acad Sci* **101:** 3172–3177.

Marchaland J, Calì C, Voglmaier SM, Li H, Regazzi R, Edwards RH, Bezzi P. 2008. Fast subplasma membrane Ca^{2+} transients control exo–endocytosis of synaptic-like microvesicles in astrocytes. *J Neurosci* **28:** 9122–9132.

Montana V, Ni Y, Sunjara V, Hua X, Parpura V. 2004. Vesicular glutamate transporter-dependent glutamate release from astrocytes. *J Neurosci* **24:** 2633–2642.

Navarrete M, Araque A. 2008. Endocannabinoids mediate neuron-astrocyte communication. *Neuron* **57:** 883–893.

Nedergaard M, Takano T, Hansen AJ. 2002. Beyond the role of glutamate as a neurotransmitter. *Nat Rev Neurosci* **3:** 748–755.

Nimmerjahn A, Kirchhoff F, Kerr JN, Helmchen F. 2004. Sulforhodamine 101 as a specific marker of astroglia in the neocortex in vivo. *Nat Methods* **1:** 31–37.

Nolte C, Matyash M, Pivneva T, Schipke CG, Ohlemeyer C, Hanisch U-K, Kirchhoff F, Kettenmann H. 2001. GFAP-promoter-controlled EGFP-expressing transgenic mice: A tool to visualize astrocytes and astrogliosis in living brain tissue. *Glia* **33:** 72–86.

Parpura V, Basarsky TA, Liu F, Jeftinija K, Jeftinija S, Haydon PG. 1994. Glutamate-mediated astrocyte-neuron signalling. *Nature* **369:** 744–747.

Parri HR, Crunelli V. 2001. Pacemaker calcium oscillations in thalamic astrocytes in situ. *NeuroReport* **12:** 3897–3900.

Privat A, Gimenez-Ribotta M, Ridet J. 1995. Morphology of astrocytes. In *Neuroglia* (ed. Kettenmann H, Ransom BR), pp. 3–22. Oxford University Press, New York.

Ransom B.R. 1995. Gap junctions. In *Neuroglia* (ed. Kettenmann H, Ransom BR), pp. 299–318. Oxford University Press, New York.

Robinson SR, Hampson EC, Munro MN, Vaney DI. 1993. Unidirectional coupling of gap junctions between neuroglia. *Science* **262:** 1072–1074.

Shigetomi E, Bowser DN, Sofroniew MV, Khakh BS. 2008. Two forms of astrocyte calcium excitability have distinct effects on NMDA receptor–mediated slow inward currents in pyramidal neurons. *J Neurosci* **28:** 6659–6663.

Walz W, MacVicar B. 1988. Electrophysiological properties of glial cells: Comparison of brain slices with primary cultures. *Brain Res* **443:** 321–324.

Xu J, Peng H, Kang N, Zhao Z, Lin JH, Stanton PK, Kang J. 2007. Glutamate-induced exocytosis of glutamate from astrocytes. *J Biol Chem* **282:** 24185–24197.

Zhang Q, Pangrsic T, Kreft M, Krzan M, Li N, Sul JY, Halassa M, Van Bockstaele E, Zorec R, Haydon PG. 2004. Fusion-related release of glutamate from astrocytes. *J Biol Chem* **279:** 12724–12733.

Zhang Z, Chen G, Zhou W, Song A, Xu T, Luo Q, Wang W, Gu XS, Duan S. 2007. Regulated ATP release from astrocytes through lysosome exocytosis. *Nat Cell Biol* **9:** 945–953.

Zhuo L. Sun B, Zhang CL, Fine A, Chiu SY, Messing A. 1997. Live astrocytes visualized by green fluorescent protein in transgenic mice. *Dev Biol* **187:** 36–42.

62 Monitoring Exocytosis in Astrocytes with Total Internal Reflection Fluorescence Microscopy

Paola Bezzi and Andrea Volterra

Department of Cell Biology and Morphology (DBCM), University of Lausanne, 1005 Lausanne, Switzerland

ABSTRACT

Increasing evidence indicates that astrocytes, the most abundant glial cell type in the brain, are competent for releasing chemical transmitters (also referred to as gliotransmitters) via regulated exocytosis of heterogeneous classes of organelles. By this process, astrocytes exert modulatory influences on neighboring cells and are thought to participate in the control of synaptic circuits and cerebral blood flow. Optical methods for monitoring vesicle exocytosis and recycling by use of specific fluorescent reporters and different types of illumination (such as epifluorescence and evanescent wave illumination) have recently been developed. Studying the properties of exocytosis in astrocytes constitutes a challenge because the cell biological basis of this process is very incompletely defined in these cells. For example, researchers have only recently started to understand that astrocytic exocytosis relies on multiple populations of secretory vesicles, which calls for the definition of adequate criteria to recognize the distinct populations and study them in isolation. In this chapter, we discuss emerging experimental strategies to recognize and follow regulated exocytosis and recycling processes of individual populations of secretory organelles in cultured astrocytes.

INTRODUCTION

Regulated exocytosis is a process by which a specialized secretory cell releases materials to the outside in response to a specific stimulus, such as elevation of the cytoplasmic calcium concentration ($[Ca^{2+}]_i$). Materials, including transmitters, are discharged via the fusion of secretory vesicles to the cell membrane. In the brain, neurons undergo regulated exocytosis when they release neurotransmitters. Astrocytes, in contrast, have long been considered to be unspecialized and incompetent for regulated exocytosis. In the past few years, however, optical imaging experiments in cultured astrocytes have revealed that these cells respond promptly (on a millisecond timescale) to $[Ca^{2+}]_i$ elevation by releasing secretory vesicles (Volterra and Meldolesi 2005). Various types of organelles have been proposed to underlie regulated exocytosis of transmitters in astrocytes, including synaptic-like microvesicles (SLMVs; Bezzi et al. 2004; Crippa et al. 2006; Jourdain et al. 2007), dense-core granules (DCGs; Coco et al. 2003; Striendinger et al. 2007), and lysosomes (Zhang et al. 2007; Li et al. 2008). Most of the early studies of astrocyte exocytosis could not distinguish such heterogeneity as they used generic exocytosis reporters (such as synthetic fluorescent dyes). Thus, these studies did not allow an accurate description of the dynamics and properties of astrocytic secretion because they mixed contributions by more than one exocytic organelle population.

IDENTIFICATION AND STAINING OF DISTINCT POPULATIONS OF SECRETORY ORGANELLES IN ASTROCYTES

In a typical secretory cell, the task of identifying a specific population of secretory organelles is already difficult; in the astrocyte, in which information on most of the basic properties of the secretory vesicles is missing, the challenge is even greater. A good experimental strategy involves trying to specifically live-stain a selected population (e.g., by transfecting the cells with a population-specific fluorescent reporter). This approach must, however, be validated by subsequent colocalization analysis with markers of the intracellular organelles, which will provide additional useful information on the nature of the stained organelles.

In Protocols A–C we summarize the available information for identifying individual populations of secretory organelles in astrocytes and present experimental procedures for staining such populations specifically, or at least to discriminate a population of interest from the other populations.

Protocol A

Identification and Staining of Synaptic-Like Microvesicles (SLMVs)

IDENTIFICATION CRITERIA

Confocal and electron microscopy evidence indicates that the gliotransmitters glutamate and D-serine accumulate in small and clear SLMVs, which are present in cultured astrocytes and in the perisynaptic processes of hippocampal astrocytes in situ (Bezzi et al. 2004; Mothet et al. 2005; Panatier et al. 2006; Jourdain et al. 2007; Martineau et al. 2008; Bergersen and Gundersen 2009). The identification of SLMVs can presently be made based on the following criteria.

- Ultrastructural characteristics: SLMVs have a clear and round appearance (they are not electron-dense), they have a small diameter (30–80 nm), they contain low-molecular-weight chemical transmitters (glutamate and/or D-serine), and they are distributed in proximity to or in direct contact with the plasma membrane.
- Expression of (1) proteins responsible for the vesicular uptake of glutamate (vesicular glutamate transporters type 1 or 2 [VGLUT1 or VGLUT2]) or D-serine (still to be identified); and (2) v-*N*-ethylmaleimide sensitive factor attachment receptor (SNARE) proteins involved in vesicular fusion, in particular members of the vesicle-associated membrane protein (VAMP) family that are substrates for the clostridial toxins, tetanus neurotoxin (TeNT) and botulinum toxin B (BoNTB), notably VAMP2/synaptobrevin or VAMP3/cellubrevin.
- Absence of expression of (1) dense-core granules (chromogranins, secretogranins, phogrin); (2) *trans*-Golgi network (tgn38, GM130); (3) early and recycling endosomes (EAA1, transferrin receptor); (4) late endosomes, multivesicular bodies, and lysosomes (lamp1, CD63); or (5) VAMP7 and VAMP8 isoforms.

Two methods for specifically staining populations of SLMVs are presented below.

MATERIALS

CAUTION: See Appendix 6 for proper handling of materials marked with <!>.
See the end of the chapter for recipes for reagents marked with <R>.

Reagents

ADVASEP-7 (Sigma-Aldrich)
α,β-methyleneadenosine 5′-triphosphate (2MeSADP) (Sigma-Aldrich)
Astrocytes, cortical or hippocampal, in culture (see Step 1)
Constructs: VGLUT1/2-eGFP (from Andrea Volterra, UNIL, Switzerland), VGLUT1/2-mCherry and VGLUT1-pHluorin (from Robert Edwards, UCSF)
FM1-43 (Molecular Probes)
FM4-64 (Molecular Probes)
FuGENE 6 transfection reagent <!> (Roche Applied Science)
HEPES-KRH buffer <R>
(*RS*)-3,5-dihydrophenylglycine (DHPG) (Tocris)

EXPERIMENTAL METHODS FOR STAINING SLMVs

Method 1: Transfection of Astrocytes with Constructs of VGLUT1 or VGLUT2 Tagged with a Green (eGFP) or Red (mCherry) Fluorescent Protein and Labeling with FM Dyes

1. Transfect cortical or hippocampal astrocytes in culture (0.1–0.2 x 10^4 cells/35-mm glass coverslip, plated 5 d before transfection) with VGLUT1/2-eGFP or VGLUT1/2-mCherry (0.5 µg/µL) (Bezzi et al. 2004; Marchaland et al. 2008) using FuGENE 6 (3 µL).
2. Wait 2–5 d before labeling the transfected cells with an FM dye. Use FM4-64 for VGLUT1/2-eGFP-expressing cells and FM1-43 for VGLUT1/2-mCherry-expressing cells. Superfuse the coverslips for 30 sec with HEPES-KRH buffer containing 15 µM of the FM dye and an agonist of an astrocyte G protein–coupled receptor (GPCR) coupled to glutamate exocytosis (e.g., DHPG [100 µM] or 2MeSADP [10 µM]) (Bezzi et al. 2004; Domercq et al. 2006).

   The latter agent is used to trigger exocytosis and favor rapid uptake of the FM dye in the fusing and recycling vesicles.
3. Wash with HEPES-KRH buffer for 1 min. Leave the FM dye in the extracellular solution for a minimum of 5 min and a maximum of 30 min to allow full staining of individual vesicles.
4. At the end of the incubation period, remove the FM dye present in the extracellular medium by superfusing for 5–20 min in HEPES-KRH buffer containing 0.1–0.2 mM of ADVASEP-7, an agent that reduces the background staining of plasma membranes with FM dyes.

Method 2: Transfection of Astrocytes with Constructs of VGLUT Tagged with Superecliptic pHluorin

pHluorin is a green fluorescent protein (GFP) with modified pH sensitivity that is particularly useful for studying the dynamics of exo/endocytosis processes of acidic organelles (see Chapter 3). When pHluorin is attached to the luminal domain of a protein (e.g., VGLUT) resident in an acidic organelle (pH 5.6), its fluorescence signal is completely quenched. During fusion of the organelle, pHluorin is externalized to pH 7.4, the quenching is relieved, and fluorescence rapidly increases, marking the fusion event. Fluorescence becomes quenched again when, upon endocytosis, the organelle undergoes re-acidification.

5. Transfect cortical or hippocampal astrocytes in culture (0.1–0.2 x 10^4 cells/35-mm glass coverslip, plated 3–5 d before transfection) with VGLUT1-pHluorin (0.5 µg/µL) (Voglmaier et al. 2006) or double transfect with VGLUT1-pHluorin and VGLUT1/2-mCherry (0.25 µg/µL for each) (Marchaland et al. 2008) using FuGENE 6 (3 µL).
6. Wait 2–5 d before using the transfected cells for imaging experiments.

Protocol B

Identification and Staining of Dense-Core Granules (DCGs)

IDENTIFICATION CRITERIA

Little information on the characteristics of DCGs in astrocytes is presently available and that is limited to cultured cells in which accumulation of secretogranin II (SgII) and neuropeptide Y (NPY) in large DCGs has been shown at the electron microscopic level (Calegari et al. 1999; Ramamoorthy and Whim 2008). Therefore, there are few established criteria for identifying DCGs in astrocytes and these are outlined below.

- Ultrastructural characteristics: DCGs have an electron-dense, often round core, contain peptides like NPY, and have a large and variable diameter (100–500 nm).
- Expression of (1) proteins of the granin family, notably secretogranins; and (2) proteins involved in vesicular fusion, notably the v-SNARE protein, VAMP2/synaptobrevin.
- Absence of expression markers for (1) SLMVs (e.g., VGLUT1 or VGLUT 2); (2) *trans*-Golgi network (tgn38, GM130); (3) early and recycling endosomes (EAA1 and transferrin receptor); (4) late endosomes, multivesicular bodies, and lysosomes (lamp1, CD63); or (5) VAMP7 and VAMP8 isoforms.
- Based on analogy with DCGs in neuroendocrine cells, additional markers could be phogrin, which is an integral glycoprotein of the DCG vesicular membrane (Taupenot et al. 2002), and chromogranins (Cg), which are also members of the granin family (Malosio et al. 2004; Meldolesi et al. 2004).

Two methods for specifically staining populations of DCGs are presented below.

MATERIALS

CAUTION: See Appendix 6 for proper handling of materials marked with <!>.

Reagents

Astrocytes, cortical or hippocampal, in culture (see Step 1)

Constructs: NPY-mRFP and CgA-eGFP (Romano Regazzi, UNIL, Switzerland), SgII-eGFP (Maité Corel, UNIRouen, France), phogrin pHluorin (Romano Regazzi, UNIL, Switzerland)

FuGENE 6 transfection reagent <!> (Roche Applied Science)

EXPERIMENTAL METHODS FOR STAINING DCG

Method 1: Transfection of Astrocytes with Constructs of DCG Cargoes Tagged with a Green or Red Fluorescent Protein

1. Transfect cortical or hippocampal astrocytes in culture ($0.1–0.2 \times 10^4$ cells/35-mm glass coverslip, plated 3–5 d before transfection) with constructs such as NPY-mRFP, CgA-eGFP, or SgII-eGFP (0.25 µg/µL) (Ramomoorthy and Whim 2008; Ravier et al. 2008) using FuGENE 6 (3 µL).

2. Wait 2–5 d before using the transfected cells for imaging experiments.

Method 2: Transfection of Astrocytes with a Construct of Phogrin Tagged with Superecliptic pHluorin

3. Transfect cortical or hippocampal astrocytes in culture (0.1–0.2 x 10^4 cells/35-mm glass coverslip, plated 3–5 d before transfection) with a construct formed by phogrin and superecliptic pHluorin (0.25 µg/µL) (Ravier et al. 2008) using FuGene 6 (3 µL).
4. Wait 2–5 d before using the transfected cells for imaging experiments.

Protocol C

Identification and Staining of Lysosomes

IDENTIFICATION CRITERIA

Very little information is available on the characteristics of lysosomes in astrocytes. Recent confocal studies in cultured cells have reported an accumulation of adenosine triphosphate (ATP) in large cisternae carrying specific markers for lysosomes (lamp1, CD63) (Zhang et al. 2007; Li et al. 2008). Previously, researchers had found this nucleotide in DCGs (Calegari et al. 1999). Without established criteria for identifying astrocytic lysosomes, one needs to rely on the criteria used in other cell types.

- Ultrastructural characteristics: Lysosomes have an electron-dense, often round core, and they have a very large and variable diameter (>500 nm).
- Expression of (1) lysosome marker proteins such as lamp1 and CD63; and (2) proteins involved in vesicular fusion, notably the v-SNARE protein VAMP7, which is an isoform that is typically insensitive to clostridial toxins such as TeNT and BoNTB.
- Absence of expression of markers for (1) SLMVs (e.g., VGLUT1 or VGLUT2); (2) *trans*-Golgi network (tgn38, GM130); or (3) early and recycling endosomes (EAA1, transferrin receptor).

A method for specifically staining populations of lysosomes is presented below.

MATERIALS

See the end of the chapter for recipes for reagents marked with <R>.

Reagents

ADVASEP-7 (Sigma-Aldrich)
Astrocytes, cortical or hippocampal, in culture (see Step 1)
Fetal bovine serum (FBS) (PAA)
FM1-43 (Molecular Probes)
FM4-64 (Molecular Probes)
HEPES-KRH buffer <R>
Minimum essential medium (MEM) <R>

EXPERIMENTAL METHOD FOR STAINING LYSOSOMES

Loading of Astrocytes with FM Dyes

1. Stain cortical or hippocampal astrocytes in culture (0.1–0.2 × 10^4 cells/35-mm glass coverslip, plated 5 d before transfection) with 4–15 μM of an FM dye (FM4-64 or FM1-43) in MEM containing 10% FBS for >1 h at 37°C.
2. At the end of the incubation period, remove the FM dye present in the extracellular medium by superfusing for 5–20 min with HEPES-KRH buffer containing 0.1–0.2 mM of ADVASEP-7, an agent that reduces the background staining of plasma membranes with FM dyes.

IMAGING EXOCYTOSIS AND RECYCLING OF SLMVs IN ASTROCYTES

Optical imaging techniques are well suited for following the dynamics of physiological processes in living cells. Among these techniques, total internal reflection fluorescence (TIRF) microscopy based on evanescent wave illumination (EWi) allows spectacular, real-time visualization of individual vesicle movements, fusions, and retrievals occurring during exo/endocytosis processes at the cell surface (i.e., within 100 nm of the plasma membrane; for details, see Chapter 19). TIRF microscopy is an ideal approach for studying the properties of exocytosis and recycling in cultured astrocytes, particularly because these cells have a rather flat surface and contain secretory vesicles with sparse distribution. The kinetics of exo/endocytosis can be studied by transfecting astrocytes with a fluorescent marker of the SLMV membrane together with a fluorescent reporter of the fusion event. Use of dual-wavelength TIRF then permits one to follow individual doubly fluorescent SLMVs undergoing exo/endocytosis and to distinguish between these processes occurring via the "kiss and run" or the "total fusion" mode (Marchaland et al. 2008) (Protocol D). Applying another approach, which uses SLMV-tagged pHluorin as reporter and a combination of EWi and epifluorescence illumination (EPIi), one can comparatively analyze fusion events taking place in the submembrane compartment and at the whole-cell level (Protocol E). In addition, by normalizing the pHluorin fluorescence curves obtained under EWi and EPIi, one can define the specific kinetics of exocytosis, endocytosis, and re-acidification of SLMVs (Marchaland et al. 2008).

Protocol D

Imaging Exocytosis at the Single Vesicle Level with EWi

To monitor exocytosis of individual SLMVs in astrocytes, two different strategies can be used: (1) transfection of astrocytes with the VGLUT1/2-eGFP (or VGLUT1/2-mCherry) construct and labeling with a FM dye; or (2) double transfection with VGLUT1-pHluorin and VGLUT1/2-mCherry constructs (Calì et al. 2008; Marchaland et al. 2008). In transfected cells the nature of the stained vesicles needs to be verified by colocalization analysis with markers of the different classes of intracellular organelles. Finally, imaging of single vesicle exocytosis under EWi is performed using the method described below.

IMAGING SETUP

Optical Design for Dual-Shutter Prismless TIRF Microscopy

The setup consists of a commercial inverted microscope (e.g., Axiovert 200B Zeiss) modified for EPIi and EWi (Bezzi et al. 2004; Calì et al. 2008; Marchaland et al. 2008). For TIRF experiments, the expanded beam of a 488/568-nm argon/krypton multiline laser (20 mW) passes through an acoustic-optic tunable filter (AOTF) wavelength selector synchronized with a SNAP-HQ charge-coupled device camera under MetaFluor software control and is introduced to the cell culture coverslip from the high-numerical-aperture objective lens (α-plan FLUAR 100x, 1.45 NA; Zeiss). Light enters the coverslip and undergoes total internal reflection at the glass–cell interface. Penetration depth of EWi, calculated according to Mattheyses and Axelrod (2006) with some modification (Marchaland et al. 2008), is 92.12 nm. In single-wavelength experiments (488 nm), the laser is filtered via a Zeiss filter set 10 (Ex BP 450–490, Em BP 515–565); in dual-wavelength EWi (488/568 nm), laser beams are combined by a Zeiss filter set 24 (Ex DBP 485/20 + 578/14, Em BP 515–540 + LP 610). Images are acquired at 20–40 Hz by MetaFluor software at binning 2. (The pixel size was 126 nm.)

MATERIALS

See the end of the chapter for recipes for reagents marked with <R>.

Reagents

α,β-methyleneadenosine 5′-triphosphate (2MeSADP) (Sigma-Aldrich)
HEPES-KRH buffer <R> (Sigma-Aldrich)
(*RS*)-3,5-dihydrophenylglycine (DHPG) (Tocris)
SDF-1α (R & D Systems Europe Ltd.)
Transfected astrocyte cultures (see Protocol A, B, or C)

Equipment

Microperfusion fast-step instrument (SF-77B perfusion fast step, Warner Instrument Corporation)
Open laminar flow perfusion indicator set at 37°C
See *Imaging Setup*.

EXPERIMENTAL METHOD

1. Before starting, verify by looking at the sample with EPI fluorescence illumination that transfection and/or FM dye labeling of the cells was successful.
2. Mount the cell culture coverslip in an open laminar flow perfusion incubator set at 37°C on the microscope stage.
3. Perfuse the preparation with HEPES-KRH buffer for 5 min at a rate of 1–1.5 mL/min.
4. Using EPIi, select a cell showing good transfection and/or FM labeling. Exclude any cell whose size exceeds the field of view.
5. Move to TIRF laser illumination and perform EWi. Select a cell with clear dark "footprint" (see Chapter 19). Focus the 100x objective on the footprint.
6. Make sure that SLMVs near the cell surface are in focus and start the recording (frame rate: 20–40 Hz).
7. Stimulate exocytosis via short (2-sec) local perfusion of an astrocyte GPCR agonist (e.g., DHPG [100 μM], 2MeSADP [10 μM], or the chemokine stromal-cell-derived factor SDF-1α [3 nM]) via a software-controlled microperfusion fast-step device (100 μL/min).

IMAGE ANALYSIS

Recognizing the Fluorescent Signal of SLMVs from Other Fluorescent Spots

To select and study fluorescent spots corresponding to a homogeneous vesicle population (i.e., SLMVs), one needs to estimate the size of the individual spots appearing in the EWi field during the imaging period. This is done by calculating the "radial sweep," which measures, for any given spot, the decay in fluorescence intensity with the distance from the center of the spot (Zenisek et al. 2002).

Recognizing a Fusion Event from the Movement of a Fluorescent Vesicle In and Out of the EWi Field

Once spots are identified that correspond to individual SLMVs, one can study the kinetics of their individual fusion events by positioning a circle on top of the fluorescent spot and a concentric annulus around the circle (Zenisek et al. 2000; Bezzi et al. 2004; Calì et al. 2008); thereby one can follow and quantify over time the diffusion into the plasma membrane of the fluorescent marker for fusion events during the fusion process (for the FM dye signal, refer to Calì et al. 2008; for the pHluorin signal, refer to Marchaland et al. 2008). In our experiments, we consider that a fluorescence change corresponds to a "fusion event" (and not to a movement) when the pHluorin (or FM dye) fluorescence signal of a single SLMV increases over basal by ≥4-fold. Fluorescence changes need to be expressed as background-subtracted fluorescence $\Delta F/F_0$, where F_0 represents the fluorescence level before stimulation and ΔF represents the change in fluorescence during the fusion process.

Calculating Temporal Distribution of Fusion Events

The entire profile of an astrocyte appearing in the field of view is subdivided in squared, equal-sized regions of interest (ROIs; diameter 5 pixels with the binning 2). The number of fusion events in each ROI should be counted and plotted against time.

Monitoring the Mode of Exo/Endocytosis

To obtain this type of information, astrocytes need to be doubly transfected and individual SLMVs will express both VGLUT1-pHluorin and VGLUT1/2-mCherry fluorescent proteins. The dynamics of the two signals from an SLMV are monitored in a circle positioned on top of the spot and in a concentric annulus, as above. Figure 1A shows the occurrence of a fusion event in the "kiss and run" mode, characterized by the abrupt appearance in the EWi field of the pHluorin fluorescence signal, which remains constant during the fusion event. VGLUT1-mCherry fluorescence, already present in the EWi field at the appearance of the pHluorin signal, also remains constant during the fusion event and then disappears

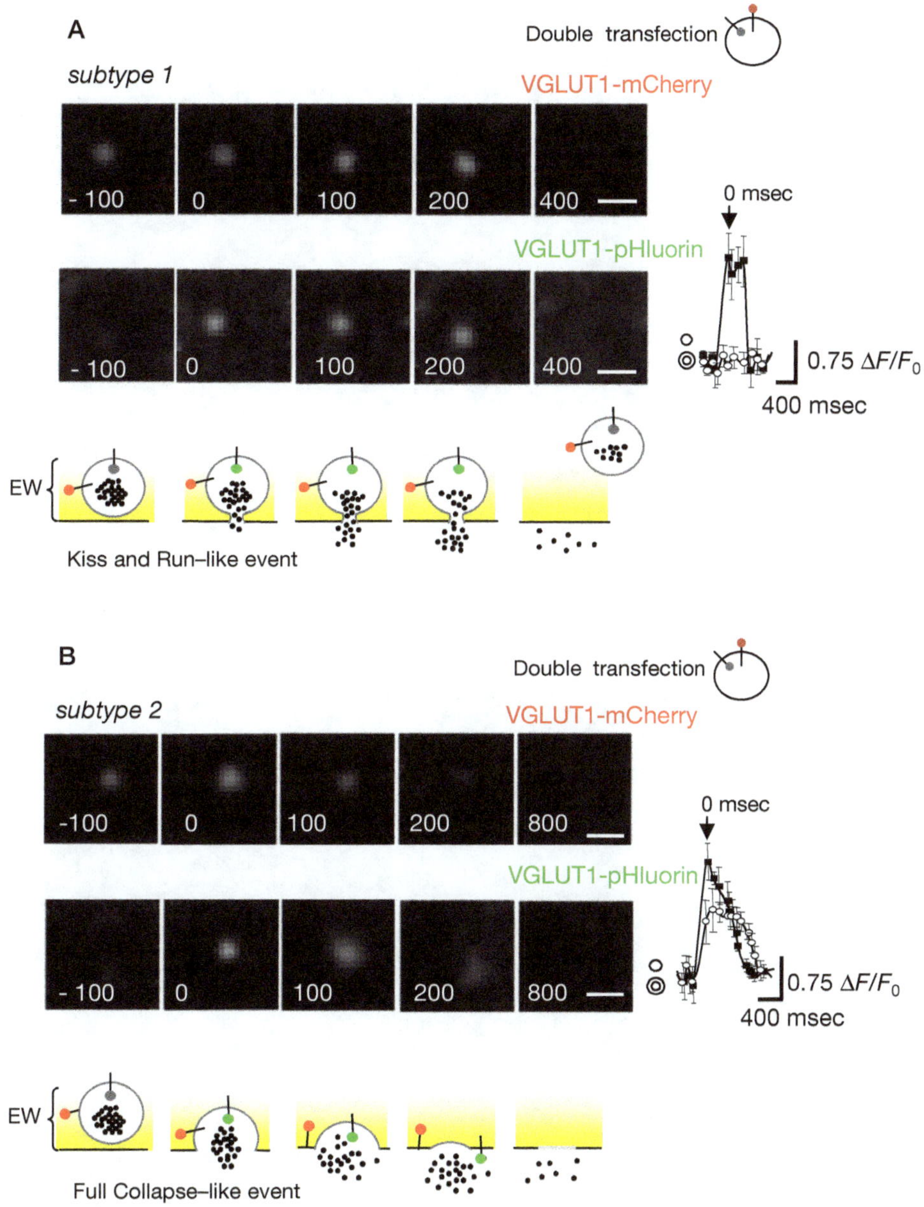

FIGURE 1. Two modes of exo/endocytosis of SLMVs in astrocytes. Experiments were performed under evanescent wave illumination (EWi). (*A*) Images and schematic diagram illustrate a fusion event occurring in the "kiss and run" mode. The first two sequences of images from *top* to *bottom* represent the dynamics of VGLUT1-mCherry and VGLUT1-pHluorin fluorescence signals taken in parallel from the same SLMV undergoing a fusion event. (*Right*) A time course of pHluorin signal changes (average of 112 different SLMVs; $n = 26$ cells) in a small circle enclosing the fluorescent spot (filled circles) and in a concentric annulus around the circle (open circles). The scheme below summarizes the behavior of the two fluorescent markers during "kiss and run" fusion. The color code for the pHluorin signal is gray when the signal is off and green when it is on. Scale bars: 0.5 µm. (*B*) Images and schematic diagram illustrate a fusion event occurring in the "full collapse" mode. The figure is organized as in *A*, but the timing of the images, the time courses of the pHluorin signal, and the behavior of the two fluorescent markers in the scheme are different. (Reprinted, with permission, from Marchaland et al. 2008.)

together with the pHluorin signal. The simultaneous disappearance of the two fluorescent markers corresponds to the movement of the SLMV out of the EWi field. Figure 1B shows instead an example of a fusion event occurring in the "full collapse" mode. The event is characterized by the diffusion of the pHluorin signal in the external annulus, followed by its dilution until complete disappearance in the plasma membrane; in parallel, the mCherry signal progressively decreases in intensity. These two combined events suggest that the SLMV loses its identity by mixing with the plasma membrane.

Protocol E

Imaging Exocytosis and Recycling at the Whole-Cell Level

The population kinetics of SLMV exocytosis, endocytosis, and re-acidification in response to astrocyte GPCR stimulation can be studied by taking advantage of a strategy recently developed for studying the dynamics of glutamatergic vesicles at synapses (Atluri and Ryan 2006; Voglmaier et al. 2006). This consists of monitoring fusion events under both EPIi and EWi using VGLUT1-pHluorin as reporter (Marchaland et al. 2008).

MATERIALS

See the end of the chapter for recipes for reagents marked with <R>.

Reagents

α,β-methyleneadenosine 5′-triphosphate (2MeSADP) (Sigma-Aldrich)
Astrocyte cultures transfected with VGLUT-1-pHluorin (see Protocol A)
HEPES-KRH buffer <R> (Sigma-Aldrich)
(*RS*)-3,5-dihydrophenylglycine (DHPG) (Tocris)
SDF-1α (R & D Systems Europe Ltd.)
Transfected astrocyte cultures (see Protocol A, B, or C)

Equipment

Microperfusion fast-step instrument (SF-77B perfusion fast step, Warner Instrument Corporation)
Open laminar flow perfusion indicator set at 37°C
See *Imaging Setup*.

IMAGING SETUP

See Protocol D.

EXPERIMENTAL METHOD

1. Before starting the imaging experiment, verify that transfection of astrocyte cultures with VGLUT1-pHluorin (see Protocol A) was successful. Take into account that, under basal conditions, the pHluorin signal is low both under EWi and EPIi, and that just a small number of fluorescent spots is visible, representing SLMVs undergoing spontaneous discharge.
2. Perform Steps 2–7 as in Protocol D.

IMAGE ANALYSIS

The pHluorin Signal at the Whole-Cell Level Under EPIi and EWi

The kinetics of the pHluorin fluorescence signal evoked by application of a GPCR agonist can be measured at the whole-cell level under both EPIi and EWi by placing an ROI on the cell perimeter (i.e., an ROI that comprises the whole astrocyte). The fluorescence curve that is obtained needs to be interpreted by taking into account that the pHluorin signal always represents the net balance of miniature fluorescence changes caused by hundreds or thousands of SLMVs undergoing exocytosis followed by endocytosis and re-acidification (Sankaranarayanan and Ryan 2001; Atluri and Ryan 2006). In addition, the components that generate the signal differ depending on the field of view. Under EPIi, the pHluorin signal ($F_{\mathrm{EPI}}[t]$) is given by the difference between two competing processes, exocytosis and endocytosis ($F_{\mathrm{EPI}}[t] = \mathrm{Exo}[t] - \mathrm{Endo}'[t]$). Under EWi, the signal ($F_{\mathrm{EW}}[t]$) is instead given by the difference between exocytosis ($\mathrm{Exo}[t]$) and two different processes that characterize the endocytic pathway: (1) fluorescence decay after fusion pore closure and re-acidification and (2) fluorescence decay due to movement of the vesicles out of the EWi field ($F_{\mathrm{EW}}[t] = \mathrm{Exo}[t] - \mathrm{Endo}'[t] - \mathrm{Mov}[t]$).

"Alkaline Trapping" Experiments Block Re-Acidification of Recycling Vesicles and Allow Calculation of the Exocytosis Kinetics

The experimental strategy of "alkaline trapping" (Fig. 2A) is used to block re-acidification of recycling vesicles on exocytosis (Sankaranarayanan and Ryan 2001). Astrocytes are incubated with the V-type ATPase inhibitor, bafilomycin A1 (Baf A1; 5 μM; for 1–2 min before GPCR agonist stimulation), which blocks the proton pump and traps the lumen of recycling vesicles at alkaline pH. This eliminates all changes of the pHluorin signal due to the re-acidification component and reveals pure exocytosis. Thus, in the presence of Baf A1: (1) under EPIi, the pHluorin signal represents only exocytosis ($F_{\mathrm{EPIBaf}}[t] = \mathrm{Exo}[t]$) (Fig. 2B2); and (2) under EWi, it represents the difference between exocytosis and movement out of the EWi field ($F_{\mathrm{EWBaf}}[t] = \mathrm{Exo}[t] - \mathrm{Mov}[t]$).

Normalization of the pHluorin Signal Curves

For further calculations, all the pHluorin fluorescence curves obtained in a given cell (under EPIi and EWi, in the presence and absence of Baf A1) need to be normalized to maximum fluorescence for that cell. In the case of EPIi, maximum fluorescence corresponds to the peak of the EPIi trace evoked by the stimulus in the presence of Baf A1; in the case of EWi, as a fraction of vesicles can still move out from the EWi field, images obtained after the stimulus need to be superposed with Photoshop by using the "exclusion" method (Marchaland et al. 2008).

Calculation of the Re-Acidification Kinetics

Subtraction of the normalized curve obtained with EPIi in the absence of Baf A1 from the curve obtained in the presence of the drug gives the "difference curve" representing the kinetics of the re-acidification processes evoked by the stimulus (Fig. 2B4).

Calculation of the Endocytosis Kinetics

The difference between the normalized curves obtained with EPIi and EWi in the presence of Baf A1 (Fig. 2B5) corresponds to the kinetics of the movement of SLMVs out of the EWi (Fig. 2B6), representing the first step of endocytosis.

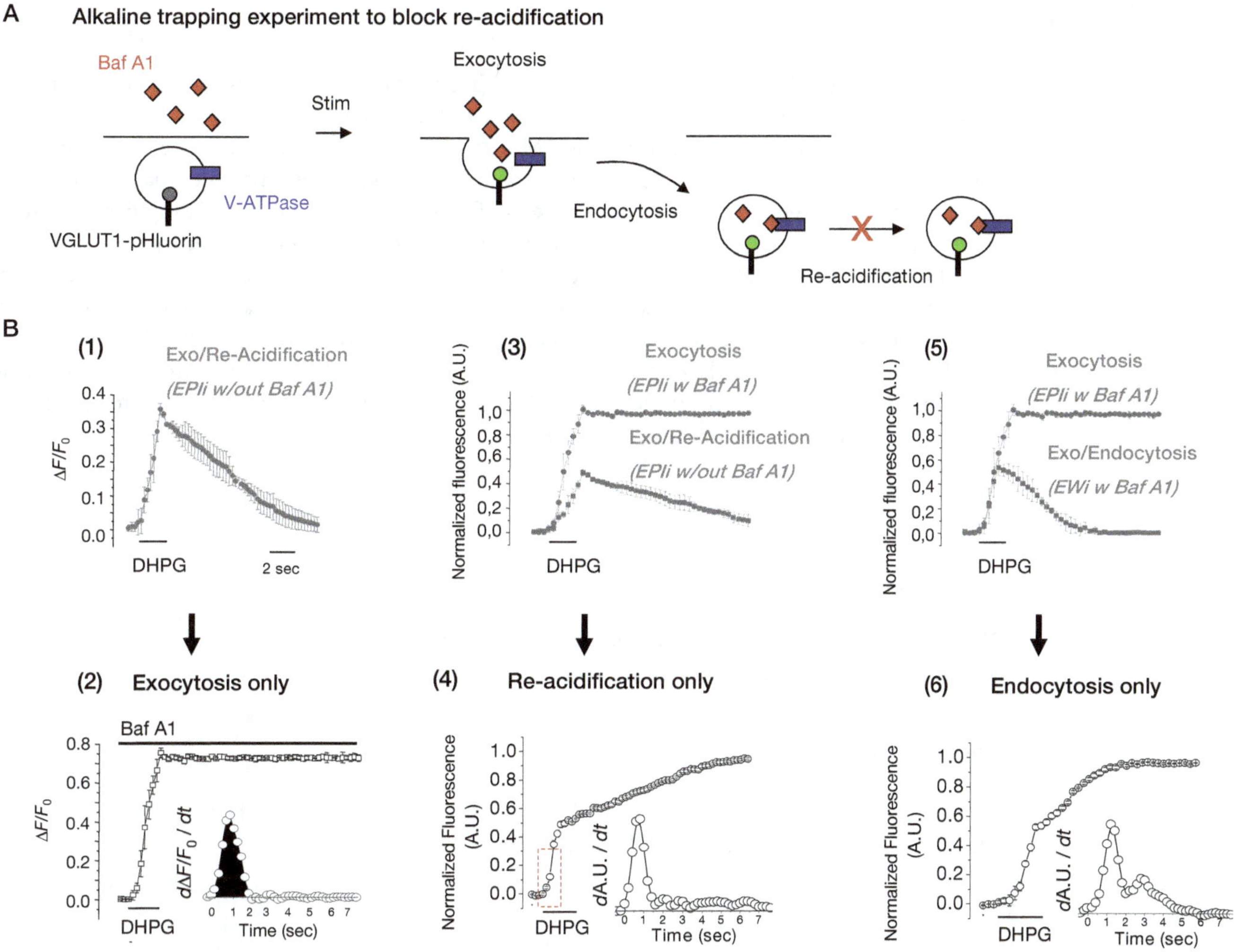

FIGURE 2. Kinetics of SLMV exocytosis, endocytosis, and re-acidification in astrocytes. (*A*) Bafilomycin A1 (Baf A1), a blocker of the V-type ATPase proton pump, prevents SLMV re-acidification upon exocytosis and traps VGLUT1-pHluorin and the vesicle lumen in the alkaline state. (*B*) Imaging experiment in which astrocytes are stimulated with the GPCR agonist DHPG (100 μM) and imaged under both EPIi and EWi. (*1*) Curve of the DHPG-evoked pHluorin fluorescence signal under EPIi, representing the net balance of exocytosis (Exo) and re-acidification. (2) Same curve as in *1*, but in the presence of Baf A1 (5 μM). The curve represents pure exocytosis. The *inset* shows the first derivative of the curve indicating the temporal distribution of the exocytic events. (*3–6*) Analysis using pHluorin fluorescence curves obtained under EPIi and EWi and normalized for maximum fluorescence (see text). (*3*) Superposition and (*4*) subtraction of curves obtained under EPIi in the presence (EPIi w Baf A1) and in the absence (EPIi w/out Baf A1) of Baf A1. The subtraction curve represents the kinetics of re-acidification. The *inset* shows the first derivative of the curve indicating the temporal distribution of the re-acidification processes. (*5*) Superposition and (*6*) subtraction of curves obtained under EPIi and EWi in the presence of Baf A1 (EPIi w Baf A1 and EWi w Baf A1, respectively). The subtraction curve represents movement of SLMVs out of the EWi field—that is, the first step of endocytosis. The *inset* shows the first derivative of the curve indicating the temporal distribution of the endocytic events. (Adapted, with permission, from Marchaland et al. 2008.)

Temporal Distribution of Exocytosis, Endocytosis, and Re-Acidification

By calculating the first derivative of each curve, one can appreciate the speed at which the three processes take place over time (Fig. 2B2,B4,B6).

CONCLUSIONS

Astrocytes exchange signaling molecules with their environment. Studies in cell culture highlight their capacity to release chemical transmitters such as glutamate, D-serine, and ATP via exocytosis of multiple populations of secretory organelles. More work is needed to characterize the molecular, cell biological, and functional properties of such organelles. Most of the dynamic imaging studies performed to date used generic fluorescent markers of exocytosis that did not distinguish among populations. This may have led to inaccurate information that needs verification by new studies with population-specific markers. We have presented here a number of emerging criteria and experimental approaches to recognize, label, and dynamically study individual populations of secretory organelles in cultured astrocytes. Among all the populations, we focused on SLMVs and illustrated how TIRF microscopy using EWi is useful to study exocytosis and recycling at the single vesicle level and, combined with EPIi, can provide detailed information on the kinetics of exocytosis, endocytosis, and re-acidification at the whole-cell level.

RECIPES

CAUTION: See Appendix 6 for proper handling of materials marked with <!>.
Recipes for reagents marked with <R> are included in this list.

HEPES-KRH Buffer

Reagent	Final concentration
NaCl	116 mM
KCl	4 mM
$MgCl_2$	1 mM
$CaCl_2$ <!>	1.8 mM
Glucose	25 mM
HEPES acid	10 mM

Adjust pH to 7.4.

MEM

MEM (Sigma-Aldrich)	500 mL
Glutamine (Sigma-Aldrich)	2 mM
Penicillin and streptomycin <!> (PAA)	100 U/mL

ACKNOWLEDGMENTS

Research in the AV laboratory is supported by the Swiss National Foundation (grant 3100A0-120398) and that in the PB laboratory by the University of Lausanne (grant FBM 2006) and the Foundation Novartis No. 26077772.

REFERENCES

Atluri PP, Ryan TA. 2006. The kinetics of synaptic vesicle reacidification at hippocampal nerve terminals. *J Neurosci* **26:** 2313–2320.

Bergersen LH, Gundersen V. 2009. Morphological evidence for vesicular glutamate release from astrocytes. *Neuroscience* **158:** 260–265.

Bezzi P, Gundersen V, Galbete JL, Seifert G, Steinhäuser C, Pilati E, Volterra A. 2004. Astrocytes contain a vesicular compartment that is competent for regulated exocytosis of glutamate. *Nat Neurosci* **7:** 613–620.

Calegari F, Coco S, Taverna E, Bassetti M, Verderio C, Corradi N, Matteoli M, Rosa P. 1999. A regulated secretory pathway in cultured hippocampal astrocytes. *J Biol Chem* **274:** 22539–22547.

Calì C, Marchaland J, Regazzi R, Bezzi P. 2008. SDF 1-α (CXCL12) triggers glutamate exocytosis from astrocytes on a millisecond time scale: Imaging analysis at the single-vesicle level with TIRF microscopy. *J Neuroimmunol* **198:** 82–91.

Coco S, Calegari F, Pravettoni E, Pozzi D, Taverna E, Rosa P, Matteoli M, Verderio C. 2003. Storage and release of ATP from astrocytes in culture. *J Biol Chem* **278:** 1354–1362.

Crippa D, Schenk U, Francolini M, Rosa P, Verderio C, Zonta M, Pozzan T, Matteoli M, Carmignoto G. 2006. Synaptobrevin2-expressing vesicles in rat astrocytes: Insights into molecular characterization, dynamics and exocytosis. *J Physiol* **570:** 567–582.

Domercq M, Brambilla L, Pilati E, Marchaland J, Volterra A, Bezzi P. 2006. P2Y1 receptor-evoked glutamate exocytosis from astrocytes: Control by tumor necrosis factor-α and prostaglandins. *J Biol Chem* **41:** 30684–30696.

Jourdain P, Bergersen LH, Bhaukaurally K, Bezzi P, Santello M, Domercq M, Matute C, Tonello F, Gundersen V, Volterra A. 2007. Glutamate exocytosis from astrocytes controls synaptic strength. *Nat Neurosci* **10:** 331–339.

Li D, Ropert N, Koulakoff A, Giaume C, Oheim M. 2008. Lysosomes are the major vesicular compartment undergoing Ca^{2+}-regulated exocytosis from cortical astrocytes. *J Neurosci* **28:** 7648–7658.

Malosio ML, Giordano T, Laslop A, Meldolesi J. 2004. Dense-core granules: A specific hallmark of the neuronal/neurosecretory cell phenotype. *J Cell Sci* **117:** 743–749.

Marchaland J, Calì C, Voglmaier SM, Li H, Regazzi R, Edwards RH, Bezzi P. 2008. Fast subplasma membrane Ca^{2+} transients control exo-endocytosis of synaptic-like microvesicles in astrocytes. *J Neurosci* **28:** 9122–9132.

Martineau M, Galli T, Baux G, Mothet JP. 2008. Confocal imaging and tracking of the exocytotic routes for D-serine-mediated gliotransmission. *Glia* **56:** 1271–1284.

Mattheyses AL, Axelrod D. 2006. Direct measurement of the evanescent field profile produced by objective-based total internal reflection fluorescence. *J Biomed Opt* **11:** 014006.

Meldolesi J, Chieregatti E, Luisa Malosio M. 2004. Requirements for the identification of dense-core granules. *Trends Cell Biol* **14:** 13–19.

Mothet JP, Pollegioni L, Ouanounou G, Martineau M, Fossier P, Baux G. 2005. Glutamate receptor activation triggers a calcium-dependent and SNARE protein-dependent release of the gliotransmitter D-serine. *Proc Natl Acad Sci* **102:** 5606–5611.

Panatier A, Theodosis DT, Mothet JP, Touquet B, Pollegioni L, Poulain DA, Oliet SH. 2006. Glia-derived D-serine controls NMDA receptor activity and synaptic memory. *Cell* **125:** 775–784.

Ramamoorthy P, Whim MD. 2008. Trafficking and fusion of neuropeptide Y-containing dense-core granules in astrocytes. *J Neurosci* **28:** 13815–13827.

Ravier MA, Tsuboi T, Rutter GA. 2008. Imaging a target of Ca^{2+} signalling: Dense core granule exocytosis viewed by total internal reflection fluorescence microscopy. *Methods* **46:** 233–238.

Sankaranarayanan S, Ryan TA. 2001. Calcium accelerates endocytosis of vSNAREs at hippocampal synapses. *Nat Neurosci* **2:** 129–136.

Sankaranarayanan S, De Angelis D, Rothman JE, Ryan TA. 2000. The use of pHluorins for optical measurements of presynaptic activity. *Biophys J* **4:** 2199–2208.

Striedinger K, Meda P, Scemes E. 2007. Exocytosis of ATP from astrocyte progenitors modulates spontaneous Ca^{2+} oscillations and cell migration. *Glia* **55:** 652–662.

Taupenot L, Harper KL, Mahapatra NR, Parmer RJ, Mahata SK, O'Connor DT. 2002. Identification of a novel sorting determinant for the regulated pathway in the secretory protein chromogranin A. *J Cell Sci* **115:** 4827–4841.

Voglmaier SM, Kam K, Yang H, Fortin DL, Hua Z, Nicoll RA, Edwards RH. 2006. Distinct endocytic pathways control the rate and extent of synaptic vesicle protein recycling. *Neuron* **1:** 71–84.

Volterra A, Meldolesi J. 2005. Astrocytes, from brain glue to communication elements: the revolution continues. *Nat Rev Neurosci* **6:** 626–640.

Zenisek D, Steyer JA, Almers W. 2000. Transport, capture and exocytosis of single synaptic vesicles at active zones. *Nature* **6798:** 849–854.

Zenisek D, Steyer JA, Feldman ME, Almers W. 2002. A membrane marker leaves synaptic vesicles in milliseconds after exocytosis in retinal bipolar cells. *Neuron* **35:** 1085–1097.

Zhang Z, Chen G, Zhou W, Song A, Xu T, Luo Q, Wang W, Gu XS, Duan S. 2007. Regulated ATP release from astrocytes through lysosome exocytosis. *Nat Cell Biol* **9:** 945–953.

63 In Vivo Labeling of Cortical Astrocytes with Sulforhodamine 101

Axel Nimmerjahn[1,3] and Fritjof Helmchen[2]

[1]Departments of Biology and Applied Physics, James H. Clark Center for Biomedical Engineering and Sciences, Stanford University, Stanford, California 94305; [2]Department of Neurophysiology, Brain Research Institute, University of Zurich, 8057 Zurich, Switzerland

ABSTRACT

Fluorescent markers that stain particular cell types in the intact brain are essential tools for fluorescence microscopy because they enable studies of structure and function of thus identified cells. Although cell type–specific fluorescence staining can be achieved through promoter-driven expression of fluorescent proteins, this genetic approach is generally labor- and cost-intensive. Alternative viral approaches for targeted fluorophore expression are relatively invasive. For astrocytes, there is a simple alternative. This chapter describes an easy and robust method for rapid and high-contrast staining of astrocytes in defined regions of the intact rodent cortex using the synthetic, red fluorescent dye sulforhodamine 101 (SR101). Selective staining is achieved through local uptake and gap junction–mediated spread of SR101 following its topical application or injection into tissue. Given its simplicity and reliability, SR101 staining is a valuable tool for the study of astrocyte function in the intact brain and for in vivo fluorescence microscopy at large.

INTRODUCTION

Astrocytes are one of the most numerous cell types in the brain. For a long time, these cells were thought to be merely passive elements performing mainly supportive functions. It is now becoming increasingly clear that astrocytes make crucial contributions to normal brain physiology. However, most studies on astrocytes and their interactions with other cells in the brain have been conducted in vitro. Although highly informative, studies in cell culture or brain slices have several limitations; these systems lack vascular dynamics, long-range modulatory input, normal synaptic activity, and metabolic demand. Investigating astrocyte physiology in live animals is therefore essential, but it requires selective markers that complement existing imaging techniques.

[3]Present address: Salk Institute for Biological Studies, La Jolla, California 92037.

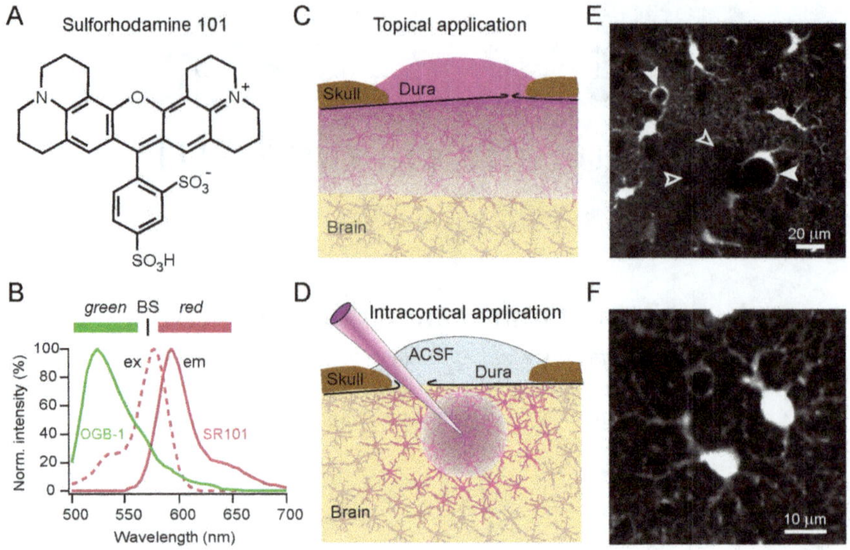

FIGURE 1. In vivo staining of cortical astrocytes with sulforhodamine 101 (SR101). (*A*) Chemical structure of SR101. In the fixable SR101 analog, Texas Red hydrazide (TRH), the SO_3H group is replaced by SO_2NHNH_2. (*B*) Single-photon excitation (ex; dashed purple line) and emission spectrum (em; solid purple line) of SR101. For comparison, the emission spectrum of Oregon Green 488 BAPTA-1 (OGB-1), which is similar to that of Calcium Green-1 and yellow fluorescent protein (YFP), is shown (solid green line). A beam splitter (BS) at ~570 nm in combination with appropriate band-pass filters can be used to separate SR101 and OGB-1 fluorescence. (*C,D*) Schematic diagrams showing two different routes of SR101 application for staining cortical tissue in vivo. (*C*) Topical route of SR101 application. (*D*) Intracortical route of application using a micropipette. Artificial cerebrospinal fluid (ACSF) serves to prevent dehydration. (*E*) An example of a two-photon image showing SR101-stained astrocytes in neocortical layer 2 of a live mouse. Astrocytic end feet surrounding blood vessels (closed arrow heads) and unstained neuronal cell bodies (open arrow heads) are indicated. (*F*) Higher magnification two-photon image of SR101-stained cortical astrocytes.

Specific staining of astrocytes or subsets thereof is possible through expression of fluorescent proteins using transgenic (Nolte et al. 2001; Zuo et al. 2004; Hirrlinger et al. 2005; Cahoy et al. 2008; Atkin et al. 2009) or viral (Colin et al. 2009; Hoogland et al. 2009) approaches. However, one drawback of these approaches is that in rodents they are relatively laborious and cost intensive. In contrast, nongenetic staining methods such as multicell bolus loading (Garaschuk et al. 2006; see Chapter 49) or in vivo electroporation (Nagayama et al. 2007) are comparatively straightforward and inexpensive but generally lack specificity.

A few years ago, we discovered that sulforhodamine 101 (SR101) stains astrocytes in the intact rodent cortex (Nimmerjahn et al. 2004) (Fig. 1). Using counterimmunostains in fixed tissue and taking advantage of transgenic mice expressing fluorescent proteins in different cell types, we verified that SR101 stains cortical astrocytes in a highly specific manner. Although the precise mechanism of SR101 uptake is still unknown, we showed that staining of astrocytes is facilitated by rapid diffusional spread of the dye through gap junctions (Nimmerjahn et al. 2004). Since then, SR101 staining has proven highly useful in a variety of in vivo imaging applications, particularly those that have used cell-unspecific green fluorescent indicators. This chapter describes two procedures that result in robust, rapid (within minutes), and selective staining of astrocytes by SR101 in the intact rodent cortex. Applications, technical pitfalls, and limitations of the SR101-staining technique are discussed.

Protocol

Staining of Astrocytes in the Intact Rodent Cortex with SR101

The procedures described here were developed for two-photon imaging in the cortex of live mice and rats (Nimmerjahn et al. 2004, 2009), but the methods can be readily adapted to other tissue preparations, brain areas, and animal species (Kafitz et al. 2008; Schummers et al. 2008). Details regarding surgical preparation and equipment can be found elsewhere (e.g., Chapters 69 and 87). All procedures were approved by the Cantonal Veterinary Office, Zurich and the Institutional Animal Care and Use Committee of the Max Planck Society and Stanford University.

IMAGING SETUP

Imaging Equipment

For in vivo imaging, any two-photon microscope may be used. We use a custom-built two-photon microscope equipped with an ultrafast Ti:sapphire laser, a 10-W all-solid-state pump laser (Spectra-Physics), a motorized micromanipulator-controlled (Sutter Instruments) *x–y* table (New England Affiliated Technologies), a motorized objective-focusing nosepiece (e.g., using a Vexta stepping motor), and one or more fluorescence detection channels (filter sets depend on the application). Other equipment required for imaging includes the following:

- beam shutter (e.g., item #SH05 from ThorLabs, Inc.)
- cold light source (e.g., KL 1500, Schott)
- data acquisition hardware (e.g., items NI PCI-6110, NI BNC-2110, and NI SHC68-68-EP from National Instruments)
- data acquisition software (e.g., MPScope; see Nguyen et al. 2006)
- electro-optic laser power controller (e.g., model 302RM and 350-80-LA from Conoptics)
- galvanometer-based optical scanner (e.g., item #6215H from Cambridge Technology, Inc.)
- high-numerical-aperture (NA) water-immersion objective (e.g., Zeiss Achroplan 40x/0.80-NA or Olympus XLUMPlanFl 20x/0.95-NA)
- high-speed digital signal integrator (custom-made as described in Nguyen et al. 2006)
- infrared viewer (e.g., FIND-R-SCOPE from FJW Optical Systems, Inc.) or viewing card (ThorLabs, Inc.)
- optical power meter (e.g., item #FM203 from ThorLabs, Inc.)
- variable-gain, low-noise current amplifiers (e.g., DLPCA-200 from FEMTO)

Imaging Parameters

SR101 is a red fluorescent dye with single-photon excitation and emission spectra similar to those of X-Rhod-1 and Alexa Fluor 568 (i.e., $\lambda_{max, ex} \sim 580$ nm and $\lambda_{max, em} \sim 595$ nm; Fig. 1B). Its two-photon excitation spectrum is relatively broad with SR101 being reasonably well excited at wavelengths between 800 and 900 nm. This is useful in experiments that require additional fluorophores to be excited simultaneously, such as the green fluorescent Ca^{2+} indicator Oregon Green 488 BAPTA-1 acetoxymethyl (OGB-1 AM) (see Fig. 3). Recordings can be made up to several hours after a single

SR101 application. However, long-term experiments (days to weeks) require tissue restaining. The number of images, frame rate, averaging, and sampling interval are determined according to the imaging application, biological process investigated, or volume sampled.

Image Processing

Image-processing software can be custom-written (e.g., using IGOR Pro from WaveMetrics) and is also available in the public domain (e.g., ImageJ) or as a commercial package (e.g., Amira or Imaris from Visage Imaging and Bitplane, respectively).

MATERIALS

See the end of the chapter for recipes for reagents marked with <R>.

Animal Models

Any wild-type or transgenic animal

Successful staining of cortical astrocytes by SR101 has been confirmed in cortical regions of adult mice, rats, and ferrets.

Reagents

Artificial cerebrospinal fluid <R> (ACSF; custom-made or from Harvard Apparatus)

SR101 (e.g., from Invitrogen or Sigma-Aldrich)

SR101 is a water-soluble but non-fixable red fluorescent dye.

Texas Red hydrazide (TRH; Invitrogen)

TRH is a fixable analog of SR101 useful for staining astrocytes when immunohistochemical analysis is required (see Figs. 1 and 2).

Equipment

Capillary glass (e.g., item #596000 from A-M Systems)
Centrifugal filter devices (e.g., item #UFC30GV25 Ultrafree-MC; Millipore)
Electrode/pipette holder (e.g., item #64-1232 from Warner Instruments)
Hypodermic needle, 27- or 30-gauge
Microloader pipettes (e.g., item #5242956.003 from Eppendorf) (for backfilling pipettes)
Micropipette puller (e.g., P-97 from Sutter Instruments)
Motorized micromanipulator (e.g., MP-285 from Sutter Instruments)
Pressure application system (e.g., Picospritzer III from Parker)
Stereomicroscope
Surgical eye spears (e.g., item #1556455 from Henry Schein Medical)

EXPERIMENTAL METHOD

Astrocytes can be stained with SR101 using either of two alternative procedures: topical application (Steps 1–5) or injection into tissue (Steps 6–9).

Option 1: Topical Application

This option provides efficient staining of cortical astrocytes while minimizing potential preparation-induced disturbances of tissue physiology or integrity (Fig. 1C). This procedure takes ~10 min.

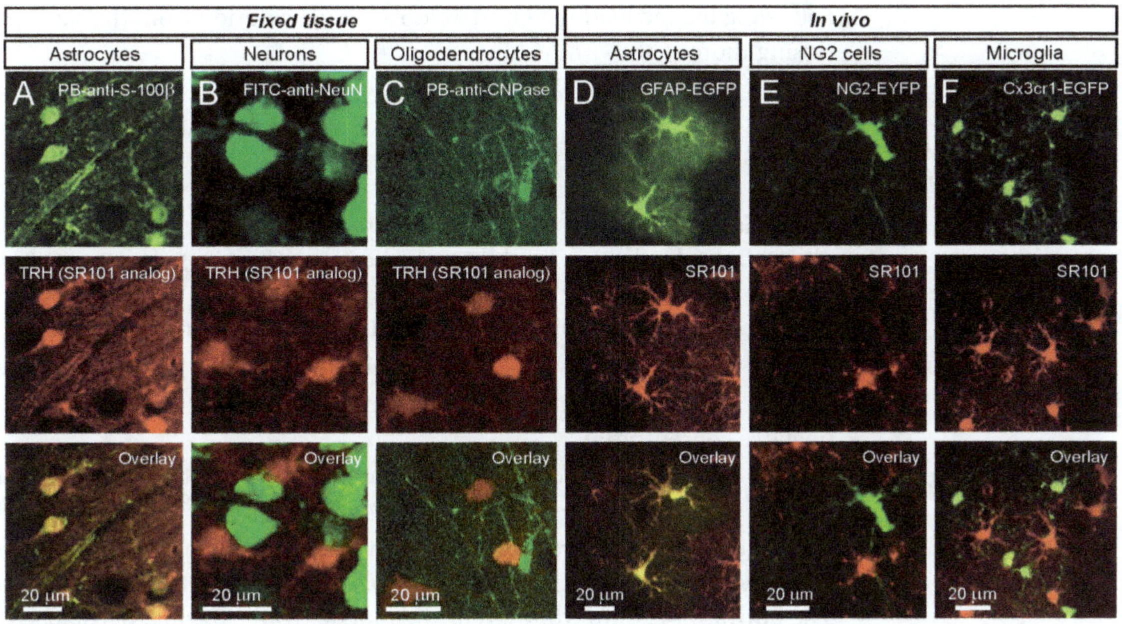

FIGURE 2. Specificity of SR101 staining. *(A–F) Top* row: Two-photon images of cortical tissue stained with cell type–specific markers for astrocytes, neurons, oligodendrocytes, NG2 cells, and microglia in vitro and in vivo. *Center* row: The same tissue regions costained with either SR101 or its fixable analog Texas Red hydrazide (TRH). *Bottom* row: Overlay of *top* and *center* row fluorescence images. Images in *A–C* were obtained in para-formaldehyde-fixed tissue, with astrocytes labeled by topical TRH application in vivo before transcardial perfusion. Images in *D–F* were obtained in vivo using transgenic mice and topical application of SR101. *(A)* Pacific Blue (PB) fluorescence of anti-S100β-labeled astrocytes and TRH fluorescence from the same tissue region. *(B)* Fluorescein isothiocyanate (FITC) fluorescence of neurons labeled with anti-NeuN and TRH fluorescence from the same region. *(C)* PB fluorescence of oligodendrocytes labeled with anti-CNPase. TRH staining as in *A* and *B*. *(D)* In vivo two-photon images of astrocytes expressing enhanced green fluorescent protein (eGFP) under control of the human GFAP promoter and cells labeled by SR101 in the same tissue region. Note that the GFAP transgene is active in only a subset of cortical astrocytes and that SR101 stains additional cells with similar morphology. *(E)* In vivo two-photon images of enhanced yellow fluorescent protein (eYFP)-expressing NG2 cells in mutant mice. No overlap with SR101-labeled cells is observed. *(F)* eGFP-expressing microglial cells in heterozygous Cx3cr1-eGFP mice. SR101- and eGFP-labeled cells in the same tissue region are distinct. *(A–D,F,* Modified, with permission, from Nimmerjahn et al. 2004.)

1. Prepare SR101 solution (typically ~200–300 μM in ACSF) either fresh or from a stock solution.

   Be careful not to use unnecessarily high dye concentrations or volumes as they may result in adverse side effects (see Discussion).

2. Perform a craniotomy above the brain region of interest (see Chapters 49, 51, and 69). Carefully perforate the dura locally to allow dye penetration into cortical tissue (Fig. 1C). Do not remove the dura unless absolutely necessary (see Discussion). If dura removal is required, perform SR101 staining before dura removal. To perforate the dura, use a stereomicroscope at high magnification and ample ACSF immersion. Use the tip of a new hypodermic needle (e.g., 27- or 30-gauge) to make a small (~0.5-mm diameter) incision away from surface blood vessels near one edge of the craniotomy. To minimize damage to the underlying tissue, hold the needle at a shallow angle and use upward-directed lateral movement to make the incision. Immediately resolve any bleeding.

3. Remove excess ACSF from the surface of the skull (e.g., using surgical eye spears), while keeping a thin layer of the liquid on top of the brain. Apply SR101 solution topically and incubate for 3–5 min.

   SR101 uptake occurs locally at sites where the dura has been perforated, but the dye will also reach astrocytes in distant (millimeters away) regions of the craniotomy via astrocytic gap junctions (see Discussion) (Nimmerjahn et al. 2004).

4. Following the incubation period, remove SR101 solution from the skull and cortical surface by rinsing thoroughly with ACSF.
5. Proceed with surgical preparation as necessary.

Option 2: Tissue Injection

This option is recommended whenever the experiment requires tissue injection(s) (e.g., in the case of multicell bolus loading or local drug application) (Fig. 1D). This procedure takes <10 min per injection.

6. Prepare SR101 stock solution (e.g., 1 mM in ACSF) and injection pipettes (e.g., patch-pipette-like glass electrodes with a tip diameter of 2–3 µm).
7. Add appropriate amount of SR101 stock solution to the pipette solution to yield the desired final dye concentration.

   The smaller the injected volume, the higher the final concentration of SR101 should be. For multicell bolus loading of a Ca^{2+} indicator, we typically use a final SR101 concentration of ~300–500 µM.

8. With the help of a micromanipulator and applying slight positive pressure (100–300 mbar), carefully insert the dye-filled pipette into a region of the brain that is devoid of blood vessels. The outward pressure will prevent the pipette from clogging, particularly when it is advanced through a layer of agar, which is often used to cover the brain surface for dampening of tissue pulsations. Advance the pipette slowly until its tip has reached the desired location within the brain. Increase pipette pressure as needed: We recommend using ≤300 mbar to avoid pressure-induced tissue damage. Verify dye ejection online using a 10x or 40x objective and two-photon microscopy.

   Note that the required injection duration depends on the applied pressure, the pipette tip diameter, and the concentration of SR101 in the pipette.

9. Proceed with tissue preparation as necessary.

Two-Photon Imaging

SR101 uptake occurs rapidly (within minutes) (Nimmerjahn et al. 2004). High-contrast and widespread staining of astrocytes is achieved within ~1–1.5 h after dye application (Fig. 1E,F). Two-photon imaging can be performed for up to several hours after a single SR101 application.

10. If the experiment requires simultaneous excitation of additional fluorophores, tune the laser to the appropriate excitation wavelength (e.g., ~950 nm for simultaneous excitation of eGFP and SR101; ~800 nm for simultaneous excitation of OGB-1 AM and SR101).
11. Make sure appropriate filter sets are installed (minimizing fluorescence bleed-through; Fig. 1B).
12. Perform imaging as necessary. If possible, restrict imaging to areas distant to the dye injection/tissue entry site because damaged or injured cells near that region may also take up SR101 (Thompson et al. 2006, 2008).

    See *Troubleshooting*.

TROUBLESHOOTING

Problem (Step 12): SR101 tissue staining is faint or absent.
Solution: The speed of SR101 uptake and the extent of gap junction–mediated spread may differ between brain regions and astrocytic subtypes (e.g., between radial astrocytes in folded regions of the mouse cerebellum and star-shaped astrocytes in the unfolded regions of the mouse neocortex) (Matyash and Kettenmann 2009). Allow ~1–1.5 h for dye uptake and spread before

imaging. If tissue staining remains locally restricted, perform a larger or an additional dura incision in future experiments. If tissue staining remains faint, insufficient amounts of the dye may have penetrated into the tissue. If possible reapply/reinject SR101 solution. Tissue staining may also appear faint owing to blood at the tissue surface. Hemoglobin has a large two-photon absorption cross section and may interfere with fluorescence imaging (Clay et al. 2007).

Problem (Step 12): In addition to astrocytes, other cells appear labeled by SR101.

Solution: Near the site of application, SR101 is taken up primarily by cells positive for the established astrocyte markers glial fibrillary acidic protein (GFAP; ~97%) and S100β (~95%) (Nimmerjahn et al. 2004) (Fig. 2). However, extracellular SR101 can also be taken up by damaged or injured cells (Thompson et al. 2006, 2008), a property that can be leveraged to assess tissue health. To minimize staining of non-astrocytic cells, keep dura incisions small, and pipette pressure and speed of pipette advancement low. Perform imaging in regions distant to the application/injection site. In these regions, SR101 staining is primarily owing to dye spread through astrocytic gap junctions.

Problem (Step 12): Over the course of several hours, SR101 fluorescence levels gradually decline and staining becomes grainy.

Solution: If possible, reapply/reinject SR101 solution. In future preparations, consider using a higher initial SR101 concentration or a longer incubation/injection duration. This may extend the time available for high-contrast imaging (see Discussion).

EXAMPLE OF APPLICATION

Figure 3 illustrates several applications of SR101 staining. SR101 staining, for example, allows the study of astrocytic structural changes as well as of spontaneous and evoked blood vessel diameter changes in vivo. For the latter purpose, the combination with a blood plasma stain further helps to reveal vessel outlines (Fig. 3A). In addition, this simple dual-labeling approach is highly useful for testing the in vivo performance of a newly set up two-photon microscope (e.g., in terms of achievable depth penetration).

In combination with multicell bolus loading of cell-permeant Ca^{2+} indicators (Garaschuk et al. 2006; see Chapter 49), SR101 labeling can be used to obtain functional information about astrocytes (Fig. 3B,C). This approach has helped reveal, for example, differences in Ca^{2+} dynamics between neocortical neurons and SR101-labeled astrocytes (Fig. 3C) as well as novel forms of astrocytic Ca^{2+} dynamics in various cortical areas and species (Nimmerjahn et al. 2004, 2009; Wang et al. 2006; Göbel et al. 2007; Schummers et al. 2008). Note that SR101 colabeling is particularly useful when astrocytic processes need to be identified unambiguously in regions containing a dense meshwork of cellular processes such as the molecular layer in cerebellar cortex (Fig. 3D; see Chapter 69).

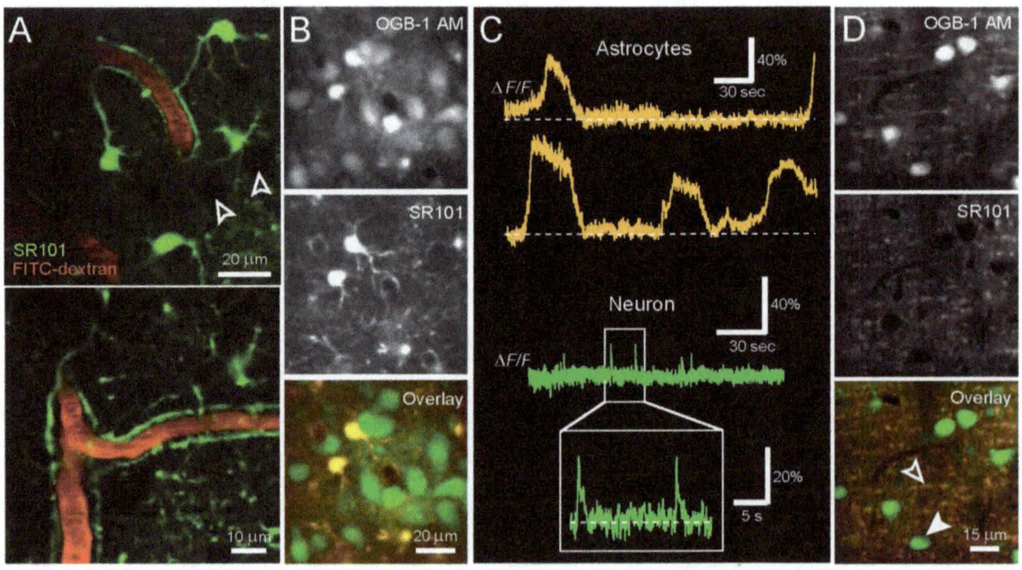

FIGURE 3. Examples showing in vivo applications of SR101 staining. (*A*) Two-photon images of SR101-labeled astrocytes (green) and blood vessels (red) in rat neocortex. Blood plasma was stained by tail-vein injection of FITC-labeled dextran. The unstained gap between astrocytic end feet and vessel lumen corresponds to endothelial cells and vessel-wall-associated pericytes. Unstained parenchymal entities represent neuronal cell bodies (open arrowheads). This negative stain of neurons facilitates, for example, targeted neuronal patch-clamp recordings. The extensive stain of astrocytic endfeet enables quantification of blood vessel diameter changes and potentially disease-related structural alterations. Note that for illustration purposes, blood plasma and astrocytes are shown in red and green pseudocolor, respectively. (*B*) Two-photon images showing cells in layer 2 of rat neocortex labeled by bolus injection of the cell-permeant, acetoxymethyl-ester form of the green fluorescent Ca^{2+} indicator Oregon Green 488 BAPTA-1 (OGB-1 AM) (*top*) and the red fluorescent dye SR101 (*center*). As astrocytes are stained with both dyes they appear orange in the overlay (*bottom*). (C) Spontaneous Ca^{2+} dynamics in two astrocytes (orange traces) and one neuron (green trace) in neocortical layer 2 of an anesthetized rat. Neuronal Ca^{2+} transients, presumably reflecting action potential–induced Ca^{2+} influx, are shown on an expanded timescale (*bottom*). Traces show relative fluorescence change, $\Delta F(t)/F$. Note the difference in amplitude and time course of astrocytic and neuronal transients. (*D*) Two-photon images showing fluorescence labeling in the molecular layer of mouse cerebellar cortex following multicell bolus loading of OGB-1 AM (*top*) and topical application of SR101 (*center*). The fine processes of Bergmann glia, astrocytes of the cerebellar cortex, are stained with both dyes (open arrowhead) and thus appear orange in the overlay (*bottom*). Interneuons (green; open arrowhead) are stained by OGB-1 AM but not SR101. (*A–C*, Modified, with permission, from Nimmerjahn 2009; *D*, reprinted, with permission, from Nimmerjahn et al. 2009.)

DISCUSSION

The synthetic, red fluorescent dye SR101 enables fast and reliable staining of astrocytes in defined cortical regions of live rodents and potentially other species. Following a single, brief, and minimally invasive application, high-contrast fluorescence images of astrocytes and their processes can be obtained for durations up to several hours. Long-term investigations (days to weeks) are possible by means of repeated SR101 application, although typically at the cost of potentially harmful repeated tissue access (Nimmerjahn et al. 2004). As only a few microliters of SR101 are required per application, the presented staining technique is remarkably low cost.

In the intact rodent brain, SR101 serves primarily as a marker of astrocyte morphology. Additionally, the extensive staining of astrocytic somata and processes by SR101 facilitates targeted patch-clamp recordings from defined cell types in the intact brain. It can also be used for motion artifact correction in functional recordings from anesthetized or awake behaving animals (Winship and Murphy 2008; Nimmerjahn et al. 2009). In combination with ion indicators, functional information about astrocyte dynamics can be obtained (Fig. 3).

Astrocytes in acute brain slices can also be stained by SR101 (Kafitz et al. 2008; Meier et al. 2008). However, successful in vitro staining appears to depend on various parameters. For example, although room temperature incubation of acute rat hippocampal slices may result in unspecific surface labeling, highly selective uptake of SR101 by astrocytes was observed for more physiological temperatures (≥34°C). Additionally, SR101 staining appears to depend on animal age: Although the vast majority of astrocytes in slices prepared from animals 15 d or older are SR101 positive, only a fraction of cells with glial morphology are stained in younger animals (Kafitz et al. 2008). This developmental dependence of SR101 staining appears plausible given the pronounced structural and functional changes astrocytes undergo during postnatal development (Bushong et al. 2004; Xu et al. 2009). Further, electrophysiological and optical studies in acute brain slices indicate that SR101 staining (1 µM; 20 min incubation) does not interfere with astrocyte physiology (Kafitz et al. 2008) and that astrocytes stained by SR101 in juvenile and adult animals represent classical astrocytes (i.e., cells with passive membrane properties, glutamate transporter activity, and a high rate of gap junction coupling) (Meier et al. 2008). However, a potential transient effect of SR101 on neuronal excitability during the period of dye uptake by astrocytes from the extracellular space cannot be excluded (Kang et al. 2010).

Some questions regarding SR101 uptake and staining still remain. What is the precise mechanism responsible for SR101 uptake? Are reactive astrocytes that typically show reduced gap junction coupling (Xu et al. 2009) also labeled by SR101? How can the differences in SR101 staining between rodent cortex and other central nervous system regions or species be reconciled (Keifer et al. 1992; Ehinger et al. 1994; Lavallee and Pflieger 2009)? Independent of the answers to these questions, SR101 has already proven and will continue to be a valuable tool in the study of astrocyte structure and function in the intact rodent brain and for a number of in vivo applications.

RECIPE

CAUTION: See Appendix 6 for proper handling of materials marked with <!>.
Recipes for reagents marked with <R> are included in this list.

Artificial Cerebrospinal Fluid (ACSF)

Reagent	Final concentration
NaCl	125 mM
KCl	5 mM
D-glucose	10 mM
HEPES	10 nM

$MgSO_4$ <!>	2 mM
$CaCl_2$ <!>	2.0 mM

Adjust pH to ~7.3 with NaOH <!>. Osmolarity should be ~300.

ACKNOWLEDGMENTS

The authors apologize to all colleagues whose important work was not directly cited owing to space limitations. This work was supported by the Max Planck Society, pre- and postdoctoral fellowships to AN from the Boehringer Ingelheim Fonds, the Alexander von Humboldt-Foundation, and the International Human Frontier Science Program Organization (HFSP), and by research grants to FH from HFSP and the Swiss National Science Foundation.

REFERENCES

Atkin SD, Patel S, Kocharyan A, Holtzclaw LA, Weerth SH, Schram V, Pickel J, Russell JT. 2009. Transgenic mice expressing a cameleon fluorescent Ca^{2+} indicator in astrocytes and Schwann cells allow study of glial cell Ca^{2+} signals in situ and in vivo. *J Neurosci Methods* **181:** 212–226.

Bushong EA, Martone ME, Ellisman MH. 2004. Maturation of astrocyte morphology and the establishment of astrocyte domains during postnatal hippocampal development. *Int J Dev Neurosci* **22:** 73–86.

Cahoy JD, Emery B, Kaushal A, Foo LC, Zamanian JL, Christopherson KS, Xing Y, Lubischer JL, Krieg PA, Krupenko SA, et al. 2008. A transcriptome database for astrocytes, neurons, and oligodendrocytes: A new resource for understanding brain development and function. *J Neurosci* **28:** 264–278.

Clay GO, Schaffer CB, Kleinfeld D. 2007. Large two-photon absorptivity of hemoglobin in the infrared range of 780–880 nm. *J Chem Phys* **126:** 025102.

Colin A, Faideau M, Dufour N, Auregan G, Hassig R, Andrieu T, Brouillet E, Hantraye P, Bonvento G, Deglon N. 2009. Engineered lentiviral vector targeting astrocytes in vivo. *Glia* **57:** 667–679.

Ehinger B, Zucker CL, Bruun A, Adolph A. 1994. In vivo staining of oligodendroglia in the rabbit retina. *Glia* **10:** 40–48.

Garaschuk O, Milos R-I, Konnerth, A. 2006. Targeted bulk-loading of fluorescent indicators for two-photon brain imaging in vivo. *Nat Protoc* **1:** 380–386.

Göbel W, Kampa BM, Helmchen F. 2007. Imaging cellular network dynamics in three dimensions using fast 3D laser scanning. *Nat Methods* **4:** 73–79.

Hirrlinger PG, Scheller A, Braun C, Quintela-Schneider M, Fuss B, Hirrlinger J, Kirchhoff F. 2005. Expression of reef coral fluorescent proteins in the central nervous system of transgenic mice. *Mol Cell Neurosci* **30:** 291–303.

Hoogland TM, Kuhn B, Göbel W, Huang W, Nakai J, Helmchen F, Flint J, Wang SS. 2009. Radially expanding transglial calcium waves in the intact cerebellum. *Proc Natl Acad Sci* **106:** 3496–3501.

Kafitz KW, Meier SD, Stephan J, Rose CR. 2008. Developmental profile and properties of sulforhodamine 101–labeled glial cells in acute brain slices of rat hippocampus. *J Neurosci Methods* **169:** 84–92.

Kang J, Kang N, Yu Y, Zhang J, Petersen N, Tian GF, Nedergaard M. 2010. Sulforhodamine 101 induces long-term potentiation of intrinsic excitability and synaptic efficacy in hippocampal CA1 pyramidal neurons. *Neuroscience* **169:** 1601–1609.

Keifer J, Vyas D, Houk JC. 1992. Sulforhodamine labeling of neural circuits engaged in motor pattern generation in the in vitro turtle brain-stem cerebellum. *J Neurosci* **12:** 3187–3199.

Lavallee A, Pflieger JF. 2009. Developmental expression of spontaneous activity in the spinal cord of postnatal opossums, *Monodelphis domestica*: An anatomical study. *Brain Res* **1282:** 1–9.

Matyash V, Kettenmann H. 2009. Heterogeneity in astrocyte morphology and physiology. *Brain Res Rev* **63:** 2–10.

Meier SD, Kafitz KW, Rose CR. 2008. Developmental profile and mechanisms of GABA-induced calcium signaling in hippocampal astrocytes. *Glia* **56:** 1127–1137.

Nagayama S, Zeng SQ, Xiong WH, Fletcher ML, Masurkar AV, Davis DJ, Pieribone VA, Chen WR. 2007. In vivo simultaneous tracing and Ca^{2+} imaging of local neuronal circuits. *Neuron* **53:** 789–803.

Nguyen QT, Tsai PS, Kleinfeld D. 2006. MPScope: A versatile software suite for multiphoton microscopy. *J Neurosci Methods* **156:** 351–359.

Nimmerjahn A. 2009. Astrocytes going live: Advances and challenges. *J Physiol* **587:** 1639–1647.

Nimmerjahn A, Kirchhoff F, Kerr JND, Helmchen F. 2004. Sulforhodamine 101 as a specific marker of astroglia in the neocortex in vivo. *Nat Methods* **1:** 31–37.

Nimmerjahn A, Mukamel EA, Schnitzer MJ. 2009. Motor behavior activates Bergmann glial networks. *Neuron* **62:** 400–412.

Nolte C, Matyash M, Pivneva T, Schipke CG, Ohlemeyer C, Hanisch UK, Kirchhoff F, Kettenmann H. 2001. GFAP promoter-controlled EGFP-expressing transgenic mice: A tool to visualize astrocytes and astrogliosis in living brain tissue. *Glia* **33:** 72–86.

Schummers J, Yu HB, Sur M. 2008. Tuned responses of astrocytes and their influence on hemodynamic signals in the visual cortex. *Science* **320:** 1638–1643.

Thompson RJ, Zhou N, MacVicar BA. 2006. Ischemia opens neuronal gap junction hemichannels. *Science* **312:** 924–927.

Thompson RJ, Jackson MF, Olah ME, Rungta RL, Hines DJ, Beazely MA, MacDonald JF, MacVicar BA. 2008. Activation of pannexin-1 hemichannels augments aberrant bursting in the hippocampus. *Science* **322:** 1555–1559.

Wang XH, Lou NH, Xu QW, Tian GF, Peng WG, Han XN, Kang J, Takano T, Nedergaard M. 2006. Astrocytic Ca^{2+} signaling evoked by sensory stimulation in vivo. *Nat Neurosci* **9:** 816–823.

Winship IR, Murphy TH. 2008. In vivo calcium imaging reveals functional rewiring of single somatosensory neurons after stroke. *J Neurosci* **28:** 6592–6606.

Xu G, Wang W, Kimelberg HK, Zhou M. 2009. Electrical coupling of astrocytes in rat hippocampal slices under physiological and simulated ischemic conditions. *Glia* **58:** 481–493.

Zuo Y, Lubischer JL, Kang H, Tian L, Mikesh M, Marks A, Scofield VL, Maika S, Newman C, Krieg P, Thompson WJ. 2004. Fluorescent proteins expressed in mouse transgenic lines mark subsets of glia, neurons, macrophages, and dendritic cells for vital examination. *J Neurosci* **24:** 10999–11009.

64 In Vivo Imaging of Structural and Functional Properties of Astrocytes

Takahiro Takano, Daniel Christensen, and Maiken Nedergaard

Center for Translational Neuromedicine, Department of Neurosurgery, University of Rochester Medical Center, Rochester, New York 14642

ABSTRACT

Recent advances in two-photon microscopy have enabled direct observation of astrocytes in vivo. These imaging studies have revolutionized our view on astrocytes by demonstrating that astrocytes are actively sensing neural activity and display increases in cytosolic calcium in response to both sensory input and motor behavior. In turn, astrocytes are thought to modulate neural network activity. Here, we discuss novel approaches used in the imaging of astrocytic structure, organization, metabolic state, and Ca^{2+} signaling in vivo.

INTRODUCTION

Experimental neuroscience has historically relied on electrical recordings of brain activity when studying live animals. As a consequence, our understanding of the function of largely electrically nonexcitable cell types lags behind that of neurons. A renewed interest in the functional roles of nonneuronal cells in the central nervous system has sparked an interest in adapting imaging approaches to live animals. More recent technological advances in functional imaging, such as positron emission tomography (PET) and functional magnetic resonance imaging (fMRI), have provided a wealth of insight on brain activity in real time. However, spatial resolution is limited and cellular resolution has yet to be achieved. Two-photon microscopy is currently the tool of choice for in vivo astrocyte imaging (Fig. 1A) (Tian et al. 2006; Wang et al. 2009). Cellular and even subcellular levels of spatial resolution at a depth of 1–300 µm can be visualized by optical imaging through a cranial window using two-photon microscopy. Two-photon microscopy eliminates the critical single-photon problem of photodamage while maintaining an image resolution within an order of traditional confocal microscopy. Additionally, both the signal-to-noise ratio and depth penetration of two-photon microscopy are considerably better than confocal microscopy. Briefly, two-pho-

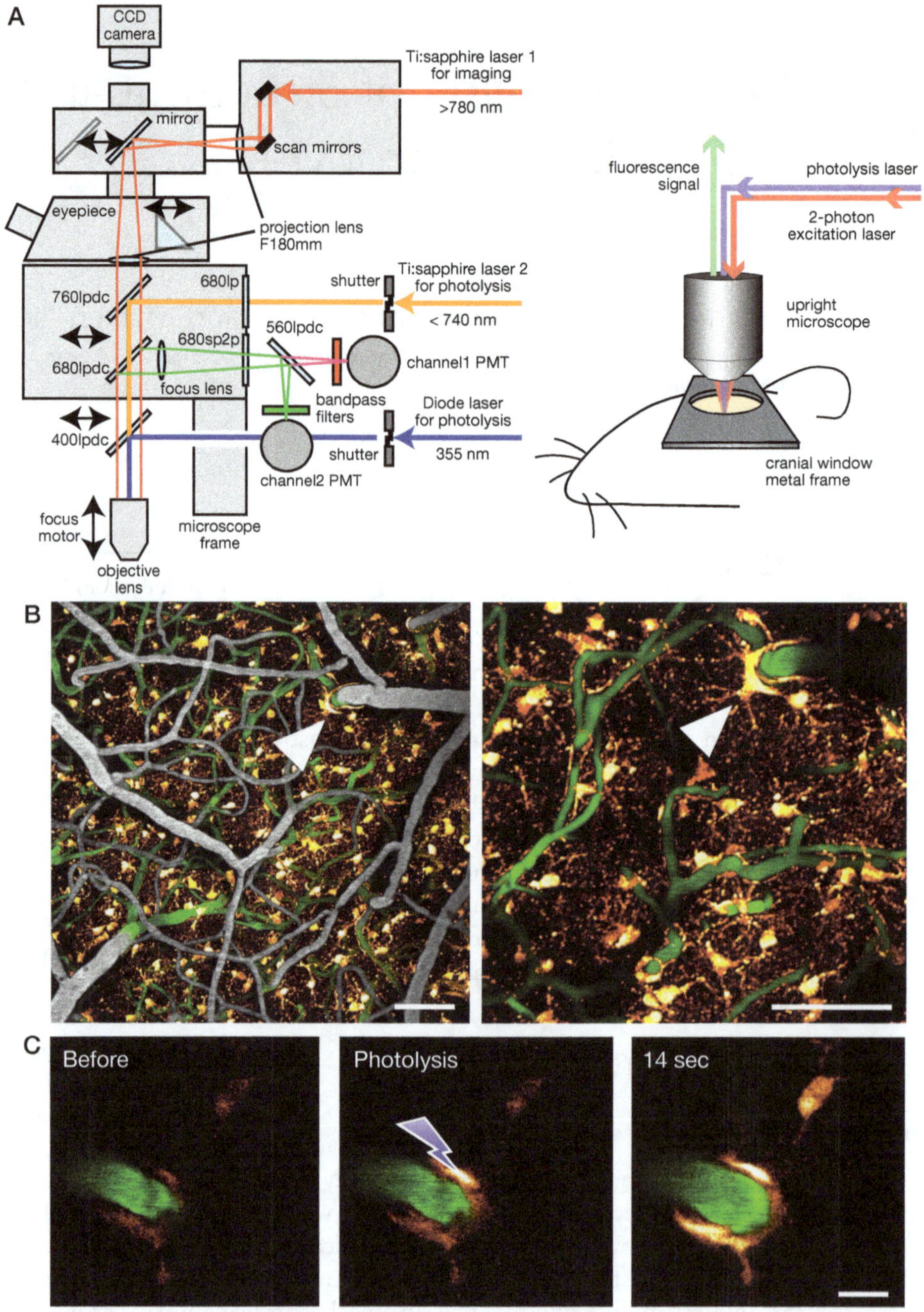

FIGURE 1. In vivo two-photon laser-scanning microscopy. (*A*) (*Left*) A schematic of the microscope design. For fluorescence excitation, a femtosecond pulsed laser is guided by a pair of scanning mirrors. The emission signal through the objective lens passes through emission bandpass filters and is detected by photomultiplier tubes (PMTs). Additionally, photolysis of caged compounds can be achieved by another laser, a Ti:sapphire laser for two-photon photolysis or an ultraviolet diode laser for single-photon photolysis, guided to the excitation laser path. A photolysis flash of 10^{-5} sec duration can be controlled by a mechanical shutter. (*Right*) Preparation of rodent cranial window. A stainless steel frame with a 5-mm hole in the middle is glued to the exposed skull using dental cement. The animal is secured to the microscope stage by screwing the metal frame to the stage at both edges. The cranial window may be made by just thinning the skull, or by removing the skull and covering the excision with agarose and a cover glass. (*B*) An *x–y–z*-projection image (at two magnifications) of a mouse cortex through the cranial window. The vasculature is stained by systemically administered fluorescein isothiocyanate (FITC)-dextran. Gray indicates the vessels on the surface of the cortex, whereas green indicates the vessels inside the cortex. Cortical astrocytes are loaded with a Ca^{2+} indicator dye, Rhod-2 AM (red). The white arrowhead points to a penetrating artery, which is surrounded by thick astrocytic endfeet. Scale bars, 50 µm. (*C*) Arterial vasodilation induced by an induction of Ca^{2+} increase after photolysis of caged Ca^{2+} compound. Photolysis laser is targeted to an endfoot of an astrocyte surrounding the artery. The photolysis causes Ca^{2+} increase in the endfoot and arterial vasodilation. Scale bar, 10 µm.

ton microscopy employs multiple, essentially simultaneously arriving photons of low energy to accomplish the molecular excitation of a single, high-energy photon (Yasuda 2011). This allows a very high excitation rate to be achieved at the focus of the objective while minimizing the fluorescence excitation in surrounding areas. Furthermore, because the photons have a lower energy, their mean free path through the sample is increased, leading to greater depth penetration with less scattering.

IN VIVO IMAGING OF FINE STRUCTURAL DETAILS OF ASTROCYTES IN LIVE ANIMALS

The best approach to study structural and organizational changes in astrocytes is to use transgenic mice expressing enhanced green fluorescent protein (eGFP), or its spectral variants, under astrocyte-specific promoters. However, although glial fibrillary acidic filament (GFAP) is a classic astrocyte marker protein in histological sections, GFAP is not an ideal promoter for visualizing mouse cortical astrocytes. In mice, <10% of protoplasmic astrocytes express GFAP, and eGFP labeling is often weak in the few astrocytes that express GFAP (Oberheim et al. 2008). In other regions, such as the hippocampus and brain stem, astrocytes express higher levels of GFAP; consequently the eGFP labeling is both more intense and more inclusive. Another example is the cerebellum, in which Bergmann glial cells express high levels of GFAP and are intensively labeled with eGFP driven by the GFAP promoter. Other very useful lines include transgenic mice expressing eGFP under the glutamate transporter 1 (GLT1) promoter (Regan et al. 2007). These mice exhibit specific and robust labels of almost all astrocytes in cortex and deeper regions. The promoter for aldehyde dehydrogenase, Aldh1L1, recently identified as a specific marker for astrocytes, also displays intense eGFP fluorescence and is commercially available from the Gene Expression Nervous System Atlas (GENSAT) (Cahoy et al. 2008). Mice expressing membrane-targeted eGFP under astrocyte-specific promoters are not yet available. Although astrocytes are readily labeled by adenoviral gene transfer and membrane-targeted eGFP is technically feasible, to our knowledge no studies have yet visualized astrocytes and their processes by membrane-targeted eGFP.

This chapter describes the application of in vivo imaging methods using dyes and intrinsic fluorescence to reveal morphology, Ca^{2+} activity, and redox state in mouse cortical astrocytes.

Protocol A

In Vivo Imaging of Astrocytes Stained with SR101

Sulforhodamine 101 (SR101) is a water-soluble amphoteric rhodamine that is taken up selectively by astrocytes when applied at the surface of the exposed cortical surface (Nimmerjahn et al. 2004; see also Chapter 63). SR101 reveals the detailed morphology of astrocytes in intact live brain 2–3 h after surface loading and concentration within the cytosol. The cell body and vascular endfeet of cortical astrocytes are clearly outlined when imaging as deep as 250 μm in adult mice. However, the spatial resolution of two-photon microscopy does not allow visualization of the fine filopodic processes of astrocytes, which appear as an even background staining in astrocytes tagged with either eGFP or SR101. Morphological changes in astrocytes, the most important of which is swelling, can be quantified in serial optical sections (*x–y–z* scans). SR101 leaks out of astrocytes within minutes after energy failure, rendering it difficult to use SR101 in studies of astrocytes in the setting of ischemia or hypoxia (M. Nedergaard, unpubl.).

MATERIALS

CAUTION: See Appendix 6 for proper handling of materials marked with <!>.

Reagents

Agar, 1.5%
Dental cement (or cyanoacrylate glue) <!>
Ketamine <!>
Mouse, adult
Physiological saline (0.9% NaCl)
Sulforhodamine 101 (SR101) (Invitrogen)
Xylazine <!>

Equipment

Cover glass (#1)
Dental drill (used for craniotomy)
Metal frame (thin, with 5-mm hole in the middle; used for craniotomy)
Surgical instruments
Ti:sapphire laser (e.g., Spectra-Physics MaiTai)
Upright microscope, preferably designed for electrophysiology (e.g., Olympus BX61WI)
Water-immersion objective lens (e.g., Olympus LUMPlanF1/IR 40x/0.80)

EXPERIMENTAL METHOD

1. Anesthetize an adult mouse with ketamine (60 mg/kg animal; intraperitoneal [IP] injection) and xylazine (10 mg/kg; IP).
2. Expose the skull and attach a thin metal frame with 5-mm hole in the middle to the skull using dental cement (Fig. 1A).

3. Secure the animal using the attached frame. Use a drill to remove the skull exposed in the middle hole of the frame.

   The dura mater may be left intact.

4. Dissolve SR101 in saline to give a final concentration of 1–100 μM. Apply the solution to the exposed cortical surface for 5–15 min.
5. Rinse the excess solution for 15 min with running prewarmed saline.
6. Fill the craniotomy with 1.5% agar and cover with a #1 cover glass.
7. Secure the animal to the microscope stage by screwing the metal frame to the stage at both edges. Use a water-immersion objective lens and fill the gap between the lens and cranial window with saline. Image with excitation at 740 or 860 nm. Collect emission at 620 nm.

Protocol B

Imaging of Ca^{2+} Dynamics in Cortical Astrocytes: Loading with Ca^{2+} Indicator Dyes

Ca^{2+} indicator dyes that are routinely used in cell cultures, such as Fluo-4 and Rhod-2 acetoxymethyl (AM) esters, can be used in vivo to visualize astrocytic Ca^{2+} activity (Takano et al. 2006; Wang et al. 2009). When introduced topically, these AM dyes are preferentially taken up by astrocytes (Fig. 1B). Esterases hydrolyze these compounds leaving the Ca^{2+}-sensitive indicator within the cytosol of astrocytes while neurons stay unloaded. Ca^{2+} waves can be observed in stimulated cortical astrocytes coupled to regional neuronal activity (Wang et al. 2006), as well as in astrocytes within the olfactory bulb (Chaigneau et al. 2007) and in Bergmann glia in the cerebellum (Hoogland et al. 2009; Nimmerjahn et al. 2009). Curiously, when these dyes are bulk-loaded with microinjection, all cells in the area appear to take up the dye, although astrocytes in most areas take up the dye better than other cell types (Stosiek et al. 2003).

MATERIALS

CAUTION: See Appendix 6 for proper handling of materials marked with <!>.
See the end of the chapter for recipes for reagents marked with <R>.

Reagents

Artificial cerebrospinal fluid (ACSF) <R>
Ca^{2+} indicator dye in AM ester form
Craniotomy reagents (see Protocol A)
Dimethylsulfoxide (DMSO) <!>
Microinjection solution <R>
Mouse, adult
Pluronic F-127

Equipment

Craniotomy materials (see Protocol A)
Micropipette pressure injection system (Picospritzer III, Parker Hannifin Corp.)
Upright microscope, preferably designed for electrophysiology (e.g., Olympus BX61WI)

EXPERIMENTAL METHOD

1. Perform craniotomy on mouse as described in Protocol A.
2. Combine 5 µL of DMSO, 2 µL of pluronic F-127, and 50 µg of Ca^{2+} indicator dye in AM ester form in a tube. Vortex for 15 min. Add 43 µL of ACSF. Vortex for 15 min.
3. Apply 10 µL of the solution to the cranial window and incubate for 40–50 min before removing the excess.
4. To administer the indicator using a micropipette pressure injection system, dissolve the dye in microinjection solution at a concentration in the range of 25–100 µM. Apply using a micropipette under 5–10 psi pressure.
5. For Fluo-4 AM, image with excitation at 800–820 nm and emission at 515 nm. For Rhod-2 AM, image with excitation at 840–860 nm and emission at 620 nm.

Protocol C

Imaging of Ca^{2+} Dynamics in Cortical Astrocytes: Loading with Photolabile Caged Ca^{2+} Compounds

The nondisruptive strategy for loading Ca^{2+} indicator dyes described in Protocol B can be adapted to load photolabile caged Ca^{2+} probes, including NP-EGTA and DMNPE-4 (Ellis-Davies and Kaplan 1994; Ellis-Davies and Barsotti 2006). As with the Ca^{2+} indicator dyes, surface loading results in preferential uptake of the esterified, cell-permeant form of caged Ca^{2+} by astrocytes, which can then be targeted by either ultraviolet or two-photon photolysis (Brown et al. 1999). Photolysis of loaded caged Ca^{2+} in vivo induces an increase of Ca^{2+} levels at the endfeet of astrocytes surrounding an artery and results in vessel dilation (Fig. 1C) (Takano et al. 2006).

MATERIALS

CAUTION: See Appendix 6 for proper handling of materials marked with <!>.
See the end of the chapter for recipes for reagents marked with <R>.

Reagents

Artificial cerebrospinal fluid (ACSF) <R>
Ca^{2+} indicator dye in AM ester form
Caged Ca^{2+} compound (NP-EGTA or DMNPE-4)
Craniotomy reagents (see Protocol A)
Dimethylsulfoxide (DMSO) <!>
Pluronic F-127

Equipment

Craniotomy materials (see Protocol A)
Diode laser, 355-nm
Master-8 pulse generator (A.M.P.I.)
Mechanical shutter (Uniblitz)
Upright microscope, preferably designed for electrophysiology (e.g., Olympus BX61WI)

EXPERIMENTAL METHOD

1. Perform craniotomy on mouse as described in Protocol A.
2. Combine 5 µL of DMSO, 2 µL of pluronic F-127, and 50 µg of Ca^{2+} indicator dye in tube 1. Simultaneously, combine 5 µL of DMSO, 2 µL of pluronic F-127, and 50 µg of caged Ca^{2+} compound in tube 2. Vortex each tube for 15 min.
3. Add 38 µL of ACSF to tube 1. Add 24.6 µL of ACSF to tube 2. Vortex each tube for 15 min.
4. Add 5 µL of the solution in tube 2 to tube 1. Vortex for 15 min.
5. Apply 10 µL of the solution in tube 1 to the cranial window. Wait 40–50 min before microscopy.
6. For photolysis of NP-EGTA AM, use a 355-nm diode laser. Generate a train of short pulses of the laser using a Uniblitz mechanical shutter controlled by a Master-8 pulse generator. Three 10-msec pulses at 10-msec intervals are delivered to the target position to induce photolysis.

Protocol D

In Vivo Imaging of Redox State Using NADH Intrinsic Fluorescence with Cellular Resolution

Astrocytes reside at the interface between the vasculature and metabolically active synapses. Many lines of work have suggested that astrocytes play a key role in maintaining the energy supply to neurons. The ability of astrocytes to deliver lactate by increasing the rate of glycogenolysis and glycolysis in response to increased energy demand from surrounding neurons has been intensively debated (Chih and Roberts 2003; Voutsinos-Porche et al. 2003; Dienel and Cruz 2008). Given the complete coverage of cerebral blood vessels by astrocytic processes, it has also been speculated that astrocytes mediate local blood flow increases in response to neuronal activity. NADH and its oxidized form NAD^+ are involved in redox reactions in metabolism. In particular, NADH is converted to NAD^+ in oxidative respiration in mitochondria, and conversely NADH is produced during glycolysis. Besides the function as a redox coenzyme, NADH possesses a unique property in that NADH is naturally fluorescent whereas NAD^+ is not. Therefore, a change in NADH fluorescence reflects a shift of NAD^+/NADH ratio, which in turn reflects a shift in metabolism inside the cells. Thus, NADH intrinsic signaling has been utilized in functional imaging for brain tissue metabolic activity. NADH can be excited at ~700–740 nm in two-photon microscopy (Kasischke et al. 2004), enabling visual observation of metabolic activity in cellular or subcellular resolution with minimal invasion (i.e., no dye loading is necessary).

In acutely isolated slices, NADH intrinsic signaling depicts a high basal level of NADH signal in the soma of astrocytes compared to surrounding cells (Kasischke et al. 2004). Curiously, the higher intensity of NADH labeling of astrocytes is not replicated when the same tissue is imaged in vivo (Fig. 2A). The NADH signal is rather uniform in cortical tissue of live mice, with somewhat higher levels in selected compartments, such as some astrocytic endfeet and pyramidal neuron soma (Fig. 2A,B). Cell identification can be achieved by using transgenic mice expressing eGFP under cell-specific promoters. Tissue redox state can be manipulated by altering CO_2 or O_2 concentration in inhalation gas. Imaging of NADH during waves of cortical spreading depression revealed that tissue O_2 can decrease to hypoxic levels in response to increases in energy demand, despite a lack of decreases in cerebral blood flow. NADH imaging documented the existence of microwatershed areas with increases in NADH during the peak of cortical spreading depression (Fig. 2C) (Takano et al. 2007). A possible explanation is that O_2 demand exceeds O_2 delivery in situations where energy consumption is increased in response to tissue depolarization.

MATERIALS

Reagents

NADH
See Protocol B

Equipment

Upright microscope, preferably designed for electrophysiology (e.g., Olympus BX61WI)

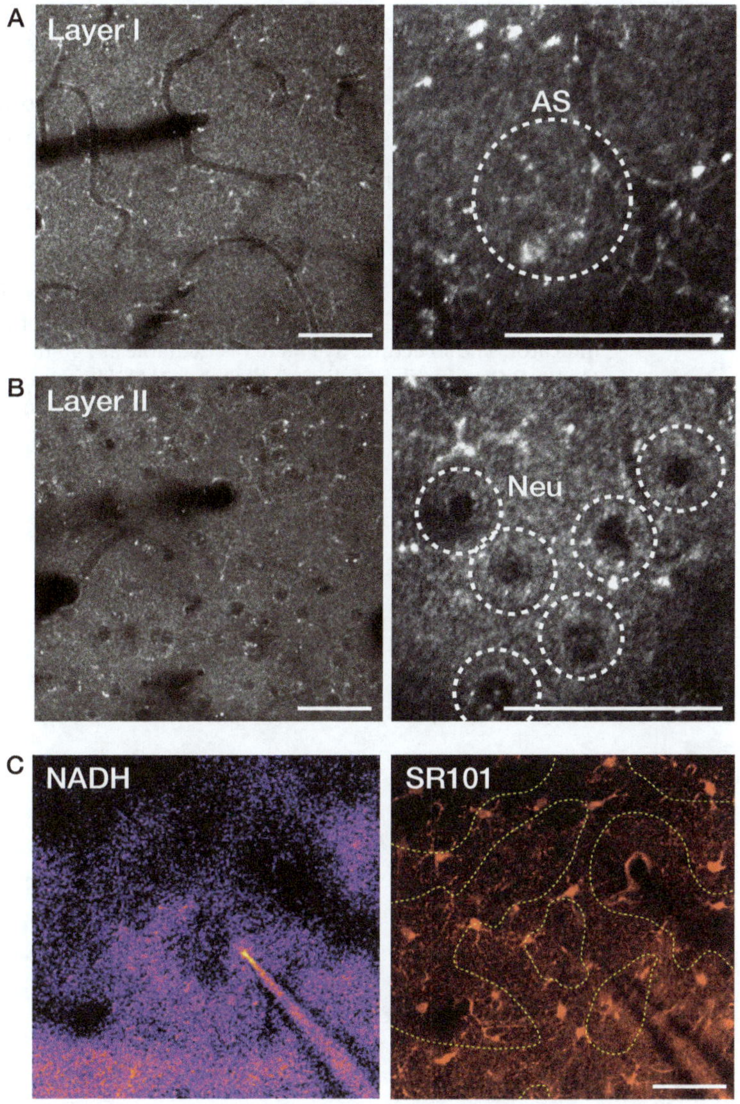

FIGURE 2. NADH intrinsic fluorescence imaging. (*A*) NADH imaging by two-photon laser scanning microscopy at mouse cortical layer 1. Dark areas with elongated shapes are blood vessels. The NADH signal is generally higher along the vessels, presumably astrocytic endfeet. Otherwise NADH does not show much of a structure. (*Right*) Higher magnification view. White circle indicates an astrocyte (AS). (*B*) NADH imaging at cortical layer 2. Dark holes are nuclei of pyramidal neurons. (*Right*) Higher magnification view. White circles indicate pyramidal neuron (Neu) somas. (*C*) NADH imaging (*left*, pseudocolor) with astrocytic staining by SR101 (*right*, red) during cortical spreading depression. NADH can be simultaneously imaged with other dyes to identify the cellular identity of the NADH signaling. During a cortical spreading depression wave, NADH shows a unique pattern of increase and decrease that is independent of cellular distribution. Yellow dotted lines indicate the area where NADH decreased. Scale bars, 50 µm.

EXPERIMENTAL METHOD

NADH Imaging

1. Perform craniotomy on mouse as described in Protocol A. Omit the dye-loading procedure.
2. Image NADH with excitation at 700–740 nm and emission at 460 nm.

   The intrinsic fluorescence from NADH is typically weaker than the fluorescence from most dyes. Use high-quality objective lenses and beware of excitation laser power. Laser power that is too high causes photodamage to the tissue (typically >50 mW, but depending on the depth, radiation duration, frequency, and loaded fluorophores).

NADH Imaging in Combination with Ca^{2+} Imaging

NADH imaging can also be combined with intracellular Ca^{2+} imaging using Ca^{2+} indicator dyes. However, great care must be taken to ensure that emission from the Ca^{2+} indicator does not leak into the detection of the NADH signals. NADH signal is typically weak compared to the loaded dyes.

3. Perform craniotomy on mouse as described in Protocol A. Load with Ca^{2+} indicator dye as described in Protocol B.
4. Image NADH and Ca^{2+} indicator with excitation at 740 nm.

   At this wavelength, Ca^{2+} indicator excitation is not optimal but is still visible. An emission filter of 460/40 nm is used to select for NADH, and a filter of longer wavelength is used for the Ca^{2+} indicator. The structural dye SR101 is excitable at 740 nm, and you must lower the concentration (1–10 µM) so it will not disturb the weak NADH signal. For GFP, simultaneous imaging cannot be performed unless GFP expression level is highly intense.

CURRENT STATUS OF IN VIVO IMAGING OF ASTROCYTES

Several practical issues inherently limit the data quality of in vivo imaging of astrocytes. In addition to the common problems associated with live animal imaging (light scattering, movement artifact, and poor or unspecific labeling), the complexity of neural and glial networks leads to unique difficulties in studies of astrocytes. For example, it is not possible to stimulate astrocytes selectively by local application of a receptor agonist. No known receptor agonists selectively activate astrocytes owing to the overlap in receptor expression between astrocytes and neurons. Photolysis of caged compound remains the most attractive approach for selectively targeting astrocytes because the time and localization of the stimulation are chosen by the investigator.

As with most in vivo studies, movements and vibrations on both the macro- (animal head and body) and micro- (cells and vessels) scales necessitate both mechanical (framing) and pharmacological (anesthetic) stabilization. This is significant because it is well known that both sleep and anesthetics have numerous effects on normal, active brain function. Recent progress has been made in designing compact, head-mounted, two-photon microscopes, which may enable imaging of astrocytes while the animal is awake, active, and not anesthetized (Engelbrecht et al. 2008).

It is currently difficult to monitor the communications of glia–glia, glia–neuron, glial–vascular, and neural networks. Most common commercially available techniques for functional imaging require that a relatively large (hundreds of micrometers) field of view be raster scanned over multiple depths to create a three-dimensional image of the network of interest. Typically this restricts the ability to collect line scans in one dimension. This is particularly troublesome because the vascular and neural networks are orientated vertically in the cortex. Collection of vertical line scans is too slow to acquire meaningful biological information.

FUTURE DEVELOPMENTS

There are a few emerging optical techniques that hold great promise for improving the temporal and spatial resolution of astrocyte imaging. The recent adaptation of acoustic-optical deflectors (AODs) to microscope scanning systems in random access multiphoton microscopy (RAMP) is a new approach for overcoming the obstacle of imaging large fields (Iyer et al. 2006; Duemani Reddy et al. 2008). In RAMP, the excitation laser beam is moved by an acoustically modulated crystal, which is not subject to the same inertial restrictions posed on a typical mirror galvanometer. Because the scan position of the beam is proportional to the acoustic modulation of the crystal, the beam can be placed at any position on the sample without raster scanning the entire field (see Chapter 53). Furthermore, the properties of acoustic crystal modulation allow the AOD to scan not only laterally but also axially, and thereby limit sampling to a specific region of interest within a three-dimensional space. Because virtually no time is spent imaging irrelevant areas, kilohertz functional imaging rates of complex communication networks are possible. In addition, spatial light modulators (SLMs) can also be used to greatly increase imaging speed without actually scanning the sample (Nikolenko et al. 2011). Finally, several nanoscopic, wide-field imaging techniques (photoactivated localization microscopy [PALM], stochastic optical reconstruction microscopy [STORM], and stimulated emission depletion [STED]) are currently under development (Bates et al. 2011; Zhong 2011). In particular, STED microscopy has recently been adapted for application to living cells (Kittel et al. 2006; Hein et al. 2008; Nagerl et al. 2008), and it is expected that this technique at some point will enable visualization of astrocytic processes in live mice. By using an optical trick to circumvent Abbe's diffraction limit, fluorescence from a sample can be localized to a region much smaller than that of conventional confocal microscopy. Because of the high powers currently associated with STED imaging, live cell neuronal imaging has been restricted to brain slices.

RECIPES

CAUTION: See Appendix 6 for proper handling of materials marked with <!>.
Recipes for reagents marked with <R> are included in this list.

ACSF

Reagent	Final concentration
NaCl	126 mM
KCl	2.5 mM
NaH_2PO_4	1.25 mM
$MgCl_2$	2 mM
$CaCl_2$ <!>	2 mM
Glucose	10 mM
$NaHCO_3$	26 mM

pH 7.4; gassed with 95% O_2 <!> and 5% CO_2 <!>.

Microinjection Solution

0.9% NaCl, pH 7.2.

REFERENCES

Bates M, Jones SA, Zhuang X. 2011. Stochastic optical reconstruction microscopy (STORM): A method for super-resolution fluorescence imaging. In *Imaging: A laboratory manual* (ed. Yuste R), pp. 547–576. Cold Spring Harbor Laboratory Press, Cold Spring Harbor, NY.

Brown EB, Shear JB, Adams SR, Tsien RY, Webb WW. 1999. Photolysis of caged calcium in femtoliter volumes using two-photon excitation. *Biophys J* **76:** 489–499.

Cahoy JD, Emery B, Kaushal A, Foo LC, Zamanian JL, Christopherson KS, Xing Y, Lubischer JL, Krieg PA, Krupenko SA, et al. 2008. A transcriptome database for astrocytes, neurons, and oligodendrocytes: A new resource for understanding brain development and function. *J Neurosci* **28:** 264–278.

Chaigneau E, Tiret P, Lecoq J, Ducros M, Knopfel T, Charpak S. 2007. The relationship between blood flow and neuronal activity in the rodent olfactory bulb. *J Neurosci* **27:** 6452–6460.

Chih CP, Roberts EL Jr 2003. Energy substrates for neurons during neural activity: A critical review of the astrocyte-neuron lactate shuttle hypothesis. *J Cereb Blood Flow Metab* **23:** 1263–1281.

Dienel GA, Cruz NF. 2008. Imaging brain activation: Simple pictures of complex biology. *Ann NY Acad Sci* **1147:** 139–170.

Duemani Reddy G, Kelleher K, Fink R, Saggau P. 2008. Three-dimensional random access multiphoton microscopy for functional imaging of neuronal activity. *Nat Neurosci* **11:** 713–720.

Ellis-Davies GC, Barsotti RJ. 2006. Tuning caged calcium: Photolabile analogues of EGTA with improved optical and chelation properties. *Cell Calcium* **39:** 75–83.

Ellis-Davies GC, Kaplan JH. 1994. Nitrophenyl-EGTA, a photolabile chelator that selectively binds Ca^{2+} with high affinity and releases it rapidly upon photolysis. *Proc Natl Acad Sci* **91:** 187–191.

Engelbrecht CJ, Johnston RS, Seibel EJ, Helmchen F. 2008. Ultra-compact fiber-optic two-photon microscope for functional fluorescence imaging in vivo. *Opt Express* **16:** 5556–5564.

Hein B, Willig KI, Hell SW. 2008. Stimulated emission depletion (STED) nanoscopy of a fluorescent protein-labeled organelle inside a living cell. *Proc Natl Acad Sci* **105:** 14271–14276.

Hoogland TM, Kuhn B, Gobel W, Huang W, Nakai J, Helmchen F, Flint J, Wang SS. 2009. Radially expanding transglial calcium waves in the intact cerebellum. *Proc Natl Acad Sci* **106:** 3496–3501.

Iyer V, Hoogland TM, Saggau P. 2006. Fast functional imaging of single neurons using random-access multiphoton (RAMP) microscopy. *J Neurophysiol* **95:** 535–545.

Kasischke KA, Vishwasrao HD, Fisher PJ, Zipfel WR, Webb WW. 2004. Neural activity triggers neuronal oxidative metabolism followed by astrocytic glycolysis. *Science* **305:** 99–103.

Kittel RJ, Wichmann C, Rasse TM, Fouquet W, Schmidt M, Schmid A, Wagh DA, Pawlu C, Kellner RR, Willig KI, et al. 2006. Bruchpilot promotes active zone assembly, Ca^{2+} channel clustering, and vesicle release. *Science* **312:** 1051–1054.

Nagerl UV, Willig KI, Hein B, Hell SW, Bonhoeffer T. 2008. Live-cell imaging of dendritic spines by STED microscopy. *Proc Natl Acad Sci* **105:** 18982–18987.

Nikolenko V, Peterka DS, Araya R, Woodruff, Yuste R. 2011. Spatial light modulator microscopy. In *Imaging: A laboratory manual* (ed. Yuste R), pp. 849–860. Cold Spring Harbor Laboratory Press, Cold Spring Harbor, NY.

Nimmerjahn A, Kirchhoff F, Kerr JN, Helmchen F. 2004. Sulforhodamine 101 as a specific marker of astroglia in the neocortex in vivo. *Nat Methods* **1:** 31–37.

Nimmerjahn A, Mukamel EA, Schnitzer MJ. 2009. Motor behavior activates Bergmann glial networks. *Neuron* **62:** 400–412.

Oberheim NA, Tian GF, Han X, Peng W, Takano T, Ransom B, Nedergaard M. 2008. Loss of astrocytic domain organization in the epileptic brain. *J Neurosci* **28:** 3264–3276.

Regan MR, Huang YH, Kim YS, Dykes-Hoberg MI, Jin L, Watkins AM, Bergles DE, Rothstein JD. 2007. Variations in promoter activity reveal a differential expression and physiology of glutamate transporters by glia in the developing and mature CNS. *J Neurosci* **27:** 6607–6619.

Stosiek C, Garaschuk O, Holthoff K, Konnerth A. 2003. In vivo two-photon calcium imaging of neuronal networks. *Proc Natl Acad Sci* **100:** 7319–7324.

Takano T, Tian GF, Peng W, Lou N, Libionka W, Han X, Nedergaard M. 2006. Astrocyte-mediated control of cerebral blood flow. *Nat Neurosci* **9:** 260–267.

Takano T, Tian GF, Peng W, Lou N, Lovatt D, Hansen AJ, Kasischke KA, Nedergaard M. 2007. Cortical spreading depression causes and coincides with tissue hypoxia. *Nat Neurosci* **10:** 754–762.

Tian GF, Takano T, Lin JH, Wang X, Bekar L, Nedergaard M. 2006. Imaging of cortical astrocytes using 2-photon laser scanning microscopy in the intact mouse brain. *Adv Drug Deliv Rev* **58:** 773–787.

Voutsinos-Porche B, Bonvento G, Tanaka K, Steiner P, Welker E, Chatton JY, Magistretti PJ, Pellerin L. 2003. Glial glutamate transporters mediate a functional metabolic crosstalk between neurons and astrocytes in the mouse developing cortex. *Neuron* **37:** 275–286.

Wang X, Lou N, Xu Q, Tian GF, Peng WG, Han X, Kang J, Takano T, Nedergaard M. 2006. Astrocytic Ca^{2+} signaling evoked by sensory stimulation in vivo. *Nat Neurosci* **9:** 816–823.

Wang X, Takano T, Nedergaard M. 2009. Astrocytic calcium signaling: Mechanism and implications for functional brain imaging. *Methods Mol Biol* **489:** 93–109.

Yasuda R. 2011. Imaging intracellular signaling using two-photon fluorescent lifetime imaging. In *Imaging: A laboratory manual* (ed. Yuste R), pp. 443–450. Cold Spring Harbor Laboratory Press, Cold Spring Harbor, NY.

Zhong H. 2011. Photoactivated localization microscopy (PALM): An optical technique for achieving ~10-nm resolution. In *Imaging: A laboratory manual* (ed. Yuste R), pp. 533–546. Cold Spring Harbor Laboratory Press, Cold Spring Harbor, NY.

65 Imaging Calcium Waves in Cerebellar Bergmann Glia

Michael Beierlein

Department of Neurobiology and Anatomy, University of Texas Health Science Center, Houston, Texas 77030

ABSTRACT

This chapter describes methods for recording synaptically evoked Ca^{2+} waves from individual Bergmann glia (BG) in slices of cerebellar cortex. Unlike protoplasmic, star-shaped astrocytes, whose thin processes pose a serious challenge to stable Ca^{2+} measurements, BG are large radial cells, with several main processes that run over distances of several hundred micrometers toward the pia. Stimulation of parallel fiber synapses with brief bursts can trigger long-lasting Ca^{2+} responses in BG processes, which can be reliably recorded using a cooled charge-coupled device (CCD) camera.

INTRODUCTION

Astrocytes in a number of regions of the central nervous system show complex patterns of Ca^{2+} increases in response to local synaptic activity. The most widely described mechanism for such increases involves release of Ca^{2+} from the endoplasmic reticulum. The basic steps that lead to intracellular Ca^{2+} signals typically involve the activation of glial G-protein-coupled receptors by neurotransmitters and phospholipase C (PLC), and the production of inositol trisphosphate (IP_3), which then triggers Ca^{2+} release from IP_3-sensitive internal stores (Fiacco and McCarthy 2006). The resulting increases in Ca^{2+} can be highly localized and transient. However, they can also propagate as waves some distance from their point of origin. Ca^{2+} waves can remain confined to individual processes or cells, or they can propagate intercellularly through tens to hundreds of neighboring glia.

Bergmann glia (BG) cells are radial glia whose elaborate processes in the molecular layer ensheathe thousands of parallel fiber (PF) synapses, which are formed between granule cells and Purkinje cells. The close proximity of PF synapses and glial processes, together with the uniquely simple cellular architecture of the cerebellar cortex, make BG an attractive model system to examine neuron–glia interactions. BG express both ionotropic and metabotropic receptors as well as glutamate transporters. Thus, transmitter release at activated PF synapses or exogenous agonist application can lead to an increase in glial Ca^{2+} concentration for several seconds, an event that is triggered by the activation of metabotropic glutamate and purinergic receptors and PLC, and that is mediated by Ca^{2+} release from IP_3-sensitive internal stores (Beierlein and Regehr 2006; Piet and Jahr 2007).

Ca^{2+} increases initiate near activated synapses before propagating as Ca^{2+} waves along cellular processes at speeds of ~50 μm/sec. Ca^{2+} waves in individual BG can be readily characterized by loading the cells with a high-affinity Ca^{2+} indicator such as Fura-2 and by imaging the stimulus-evoked Ca^{2+} response using a cooled charge-coupled device (CCD) camera.

Protocol

Imaging Stimulus-Evoked Ca^{2+} Responses in BG

The protocol described here was developed to enable measurements of Ca^{2+} waves in individual BG for up to 2 h. Because BG recorded in slices rarely display spontaneous (i.e., tetrodotoxin [TTX]-sensitive) or intrinsic Ca^{2+} transients, Ca^{2+} waves can be evoked repeatedly and reliably, which permits quantitative studies using pharmacological tools. Fluorescence measurements obtained using CCD technology offer a straightforward means of characterizing the mechanisms and potential functional consequences of widespread and long-lasting, store-mediated Ca^{2+} increases in astrocytes.

IMAGING SETUP

We use a standard upright, fixed-stage microscope (Olympus BX51WI) equipped with a 60x/0.9-NA water-immersion objective (Olympus LUMPlanFl). For fluorescence excitation of Fura-2 at 360 and 380 nm, we use a monochromator (Polychrome IV from TILL Photonics). Excitation light is fed via a quartz light guide into an epifluorescence condenser (TILL Photonics). To measure fluorescence changes, we use a high-resolution cooled CCD camera (Imago QE; 1376 x 1040 pixels; TILL Photonics). The speed and sensitivity of most measurements are increased by on-chip binning of 4 x 4 or 8 x 8 pixels.

MATERIALS

See the end of the chapter for recipes for reagents marked with <R>.

Reagents

Artificial cerebrospinal fluid (ACSF) <R>

Fura-2 (potassium salt) (Invitrogen)

> Ca^{2+} increases mediated by release from internal stores are long-lasting and typically remain below 1 μM, which justifies the use of a high-affinity indicator such as Fura-2.

Equipment

Polylysine-coated circular coverslips

> Circular coverslips (Fisher) should be cleaned in ethanol, coated with 50–100 μL poly-L-lysine solution (Sigma-Aldrich), and then allowed to dry overnight. We find that slice fixation using coverslips, rather than hold-down devices such as harps, permits extremely stable long-term measurements even during high rates of perfusion (4–5 mL/min).

Recording chamber (Warner Instruments or custom design)

EXPERIMENTAL METHOD

1. Prepare parasagittal or transverse cerebellar slices (300 μm) from rat brain using standard techniques (Atluri and Regehr 1996; Beierlein and Regehr 2006).

   The choice of slice orientation should be determined by the particular aims of the experiment. BG processes tend to follow the dendrites of Purkinje cells, which are oriented in a parasagittal plane. Therefore, slices cut in this manner will allow for easier isolation of individual BG processes, with limited signal contribution from out-of-focus processes. However, the reliable activation of PF inputs can be challenging in this orientation and placing the stimulus electrode near the glial process can lead to artifacts caused by direct stimulation of the process (see *Troubleshooting*). Transverse slices permit the reliable stimulation of PF inputs at distances of 50–100 µm. In this case, the isolation of individual branches, particular along the *z*-axis, can be more difficult, potentially leading to misassignment of the measured fluorescent signal to the wrong glial process.

2. Secure the slices on polylysine-coated circular coverslips and place them in a recording chamber. To obtain stable recordings at near-physiological temperatures (33°C–35°C), superfuse the slices with ACSF at a flow rate of at least 4 mL/min.

3. Load patch pipettes (2.5–3.0 MΩ resistance) with a potassium gluconate–based internal solution supplemented with the Ca^{2+}-indicator Fura-2 (50–100 µM).

   The red Ca^{2+}-insensitive dye Alexa 594 can also be included (100 µM) to improve the visualization of cell morphology. It is a good idea to tip-fill pipettes with internal solution free of indicator and dye to limit dye spill near the recorded cell.

4. Before patching, visualize BG using either standard bright-field or differential interference contrast (DIC) optics. BG cell bodies appear in the Purkinje cell layer and are distinctly larger than those of granule cells located in the adjacent granule cell layer. In whole-cell mode, BG resting membrane potential lies between –80 and –85 mV. When filled with Alexa 594, the cells should display two or more elaborate processes that reach through the molecular layer toward the pia where they form endfeet. Depending on the access resistance of the recording electrode, as well as the distance of the recorded process from the cell body, Ca^{2+}-indicator loading will take ~10–15 min.

   BG are strongly coupled by gap junctions. However, in our experience, there is only modest dye coupling; that is, little dye spreads from the recorded cell to its neighbors over the course of an experiment (<1 h). Occasionally, the somata of adjacent BG show increases in dye concentration over a time period of ≥60 min, but most processes remain essentially free of indicator.

5. To stimulate PF synapses, place a patch pipette (2–3 MΩ) filled with ACSF in the outer half of the molecular layer, near a well-isolated process (sagittal slice) or at a lateral distance of 50–100 µm at the same depth in the slice as the process (transverse slice).

6. At near physiological temperatures, Ca^{2+} responses elicited by brief stimulus trains (three to 10 stimuli; 20–50 Hz) will have a latency of several hundred milliseconds and last for tens of seconds. Therefore, individual fluorescence exposures during a trial should be brief (25–50 msec at frequencies of 5–10 Hz) and fluorescence intensity should be kept at levels that avoid significant photobleaching during trials. Photobleaching, as estimated by comparing images collected at 359 nm (isosbestic wavelength for Fura-2) before and after acquisition, should be <1%.

   Stimulation of PF at high enough intensities will evoke complex postsynaptic currents in BG, mediated in part by the activation of AMPA receptors and electrogenic glutamate transporters. However, we find that for stimulation sites in the outer half of the molecular layer, a short stimulus train at a given intensity can readily trigger Ca^{2+} increases without evoking a measurable postsynaptic current.

   See *Troubleshooting*.

TROUBLESHOOTING

Problem (Steps 1 and 6): Stimulus-evoked Ca^{2+} signals are not sensitive to TTX.
Solution: Stimulus electrodes need to be positioned at sufficient distance from the recorded neuron to avoid direct activation of glial processes.

Problem (Step 6): There is no clear stimulus-evoked Ca^{2+} transient.

Solution: We find that under our experimental conditions, every cell will show a Ca^{2+} increase in response to a PF burst. Recorded cells should have stable resting membrane potentials (–80 to –85 mV; measured with a potassium-gluconate-based internal solution) and low resting Ca^{2+} levels. Confirm that cells do not generate spontaneous transients, which would interfere with stimulus-evoked responses. In our experience, spontaneous Ca^{2+} transients only appear in unhealthy slices or in cells that experience light-induced photodamage.

Problem (Step 6): Ca^{2+} waves are unstable or rapidly fatigue over several trials.

Solution: Ca^{2+} waves are mediated by Ca^{2+} release from internal stores, and complete refilling of the stores can take several minutes. Thus, to obtain stable responses, intertrial intervals should be kept at 5–7 min. Note that the time necessary for store refilling is strongly temperature dependent. Therefore, experiments conducted at room temperature will require longer intertrial intervals.

EXAMPLE OF APPLICATION

Figure 1 shows data from an experiment in which synaptically evoked changes in glial Ca^{2+} are monitored using a cooled CCD camera. PF synapses associated with a distal glial process are activated by a nearby glass electrode (Fig. 1A), leading to long-lasting Ca^{2+} increases (Fig. 1B). The initial Ca^{2+} transient is limited to a dendritic region near the activated synapses (Δt = 1 sec) before spreading through the entire glial process at speeds of 50 µm/sec (Fig. 1C). Ca^{2+} increases are then uniform throughout the glial process and gradually return to resting levels. Thus, synaptically evoked Ca^{2+} signals in BG propagate slowly from a local initiation site.

DISCUSSION

BG of the cerebellum are an attractive model for studying synaptically evoked Ca^{2+} waves (Beierlein and Regehr 2006). They extend radial processes in near-planar fashion over several hundred micrometers, from the Purkinje cell layer to the pial surface, where they terminate as characteristic endfeet. These main processes can be readily loaded with dye-filled patch pipettes and Ca^{2+} measurements can be performed using standard CCD technology. Under our experimental conditions, we find that even after multiple trials, the likelihood of spontaneous Ca^{2+} increases remains very low, suggesting that even long-lasting fluorescence exposures lead to little photodamage. Furthermore, Ca^{2+} waves can be evoked repeatedly and reliably, suggesting that even long-lasting recordings do not lead to washout of the cytosolic components necessary for the generation of Ca^{2+} waves.

Although CCD cameras are ideal detectors for slow, spatially extended Ca^{2+} waves, they do not allow for the characterization of Ca^{2+} signals in small appendages termed microdomains, which extend from the main radial fibers (Grosche et al. 1999). Furthermore, it is possible that fast, local Ca^{2+} increases mediated by Ca^{2+} influx through Ca^{2+}-permeable AMPA receptors remain undetected (Piet and Jahr 2007). Thus, for measurements of local Ca^{2+} increases, two-photon microscopy is more suitable.

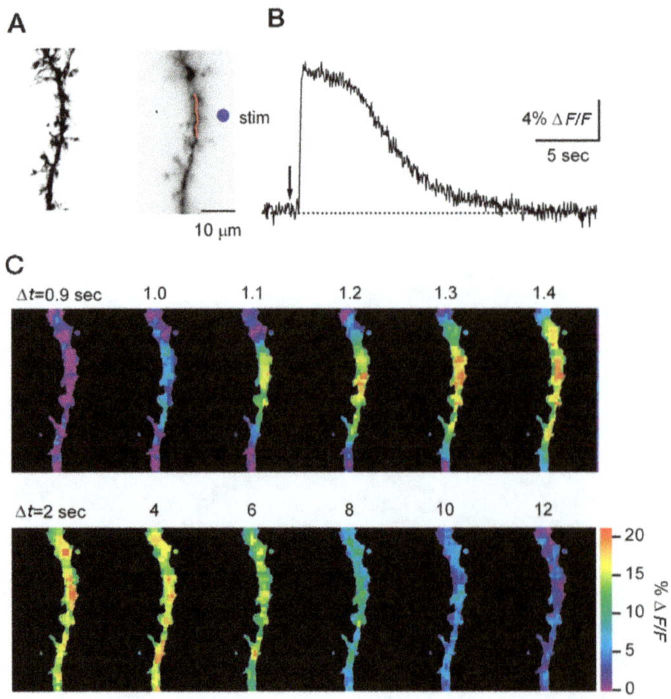

FIGURE 1. Spatiotemporal characteristics of Bergmann glia (BG) Ca^{2+} signals. (*A*) Images of BG processes obtained with a two-photon laser scanning microscope (*left*) and a charge-coupled device (CCD) camera (*right*). The blue dot indicates the location of the stimulus electrode. (*B*) Ca^{2+} response evoked by parallel fiber stimulus burst (10 pulses; 50 Hz) in the region indicated by the red line in *A*. The arrow indicates onset of stimulus train. (*C*) Spatial spread of Ca^{2+} wave. The sequence of fluorescence CCD images (50-msec frame duration) shows the initiation and spread of the Ca^{2+} response (*top row*), followed by a more uniform decay (*bottom row*). Times indicated are relative to the onset of the stimulus train. The color bar indicates fluorescence change (%$\Delta F/F$). (Reprinted, with permission, from Beierlein and Regehr 2006.)

RECIPE

CAUTION: See Appendix 6 for proper handling of materials marked with <!>.
Recipes for reagents marked with <R> are included in this list.

Artificial Cerebrospinal Fluid (ACSF)

Reagent	Final concentration
NaCl	125 mM
$NaHCO_3$	26 mM
NaH_2PO_4	1.25 mM
KCl	2.5 mM
$MgCl_2$	1.0 mM
$CaCl_2$ <!>	2.0 mM
Glucose	25 mM

Bubble 95% O_2 <!> and 5% CO_2 <!> through the solution.

ACKNOWLEDGMENTS

I would like to thank Wade Regehr, in whose laboratory these methods were developed.

REFERENCES

Atluri PP, Regehr WG. 1996. Determinants of the time course of facilitation at the granule cell to Purkinje cell synapse. *J Neurosci* **16:** 5661–5671.

Beierlein M, Regehr WG. 2006. Brief bursts of parallel fiber activity trigger calcium signals in Bergmann glia. *J Neurosci* **26:** 6958–6967.

Fiacco TA, McCarthy KD. 2006. Astrocyte calcium elevations: Properties, propagation, and effects on brain signaling. *Glia* **54:** 676–690.

Grosche J, Matyash V, Moller T, Verkhratsky A, Reichenbach A, Kettenmann H. 1999. Microdomains for neuron–glia interaction: Parallel fiber signaling to Bergmann glial cells. *Nat Neurosci* **2:** 139–143.

Piet R, Jahr CE. 2007. Glutamatergic and purinergic receptor-mediated calcium transients in Bergmann glial cells. *J Neurosci* **27:** 4027–4035.

66 In Vivo Calcium Imaging of Cerebellar Astrocytes with Synthetic and Genetic Indicators

Bernd Kuhn, Tycho M. Hoogland, and Samuel S.-H. Wang

Department of Molecular Biology and Princeton Neuroscience Institute, Princeton University, Princeton, New Jersey 08544

ABSTRACT

The cerebellar cortex contains two astrocyte types: the Bergmann glia of the molecular layer and the velate protoplasmic astrocytes of the granule cell layer. In vivo, these cell types generate both subcellular calcium transients and transglial calcium waves. In the molecular layer, transglial waves have unusual radial symmetry, encompassing many astrocyte processes in a near-ellipsoidal volume without filling any one cell in full. Here, we describe two protocols to study glial signals in vivo using synthetic and genetically encodable calcium indicators. Multicell bolus loading (MCBL) in the molecular layer of the synthetic calcium indicators Fluo-5F AM and Fluo-4 AM preferentially labels Bergmann glia. Because bolus loading tends to additionally label neurons, we have also used the fluorescent calcium indicator protein G-CaMP2. The use of a replication-incompetent recombinant adenovirus containing a cytomegalovirus (CMV) immediate-early (IE) promoter confines expression of G-CaMP2 to astrocytes. Expression is sufficiently high to allow calcium signals to be recorded in Bergmann glial processes as well as the processes and somata of velate protoplasmic astrocytes. To obtain structural information, G-CaMP2 fused with the brighter chromophore DsRed allows three-dimensional reconstruction of cells. G-CaMP2 expression lasts for at least 3 wk, enabling long-term functional imaging in both anesthetized and awake animals.

INTRODUCTION

Recent work adds to a growing body of evidence that astrocytes are active players in brain function (Allen and Barres 2009). Astrocytes regulate blood flow (Iadecola and Nedergaard 2007), reflect neuronal activity after whisker stimulation in barrel cortex (Wang et al. 2006), show orientation selectivity in visual cortex (Schummers et al. 2008), regulate sleep homeostasis (Halassa et al. 2009), participate in circadian rhythmicity in *Drosophila* (Suh and Jackson 2007) and perhaps mammals

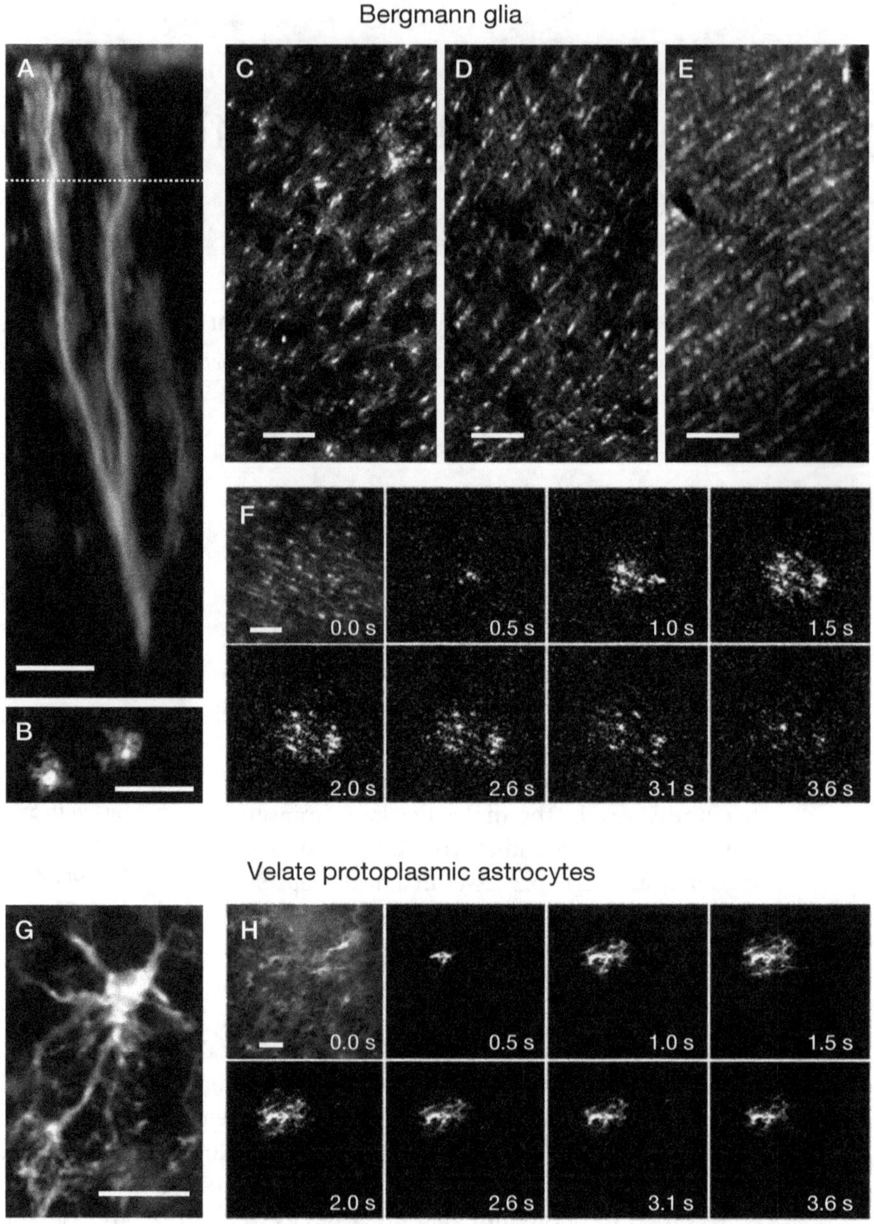

FIGURE 1. Astrocytes of the cerebellar cortex generate transglial calcium waves in vivo. (*A*) Bergmann glia cell expressing a fusion protein of the fluorescent calcium indicator protein (FCIP) G-CaMP2 and DsRed under the CMV IE promoter after adenoviral gene transfer. (*B*) Image showing the *x–y* plane indicated by the dotted line in *A*. The use of FCIP also allows imaging of glial microdomains when cells are sparsely infected or (*C*) cell populations at higher infection rates, which are almost comparable to (*D*) transgenic mice expressing green fluorescent protein (GFP) under the Bergmann glia-specific glial fibrillary acidic protein (GFAP) promoter. (*E*) MCBL with the synthetic calcium dye Fluo-5F AM shows preferential Bergmann glia labeling if injected superficially into the molecular layer of the cerebellar cortex. (*F*) In vivo Bergmann glia show radially expanding transglial calcium waves measured with G-CaMP2. The first image shows resting fluorescence; the following images show fluorescence changes relative to the first image. (*G*) Velate protoplasmic astrocyte of the granule cell layer expressing G-CaMP2 after infection by recombinant adenovirus. (*H*) Velate astrocytes also generate transglial calcium waves in vivo. The first image shows resting fluorescence; in subsequent images, fluorescence changes relative to the first image. Scale bars, 20 µm.

(Prolo et al. 2005), and are necessary for sensory organ function in *Caenorhabditis elegans* (Bacaj et al. 2008).

In the cerebellar cortex, Bergmann glia (Fig. 1A–E) of the molecular layer and velate protoplasmic astrocytes (Fig. 1G) of the granule cell layer may play active signaling roles. Using in vivo calcium imaging with two-photon microscopy, we have found (Hoogland et al. 2009) that both types of astrocytes show spontaneous compartmental calcium signals that fill small parts of individual cells, as they do in brain slices (Grosche et al. 1999), as well as transcellular calcium waves (Fig. 1F,H). These signals have been confirmed in awake animals (Nimmerjahn et al. 2009). Waves are particularly novel because they span many astrocyte processes in approximately ellipsoidal domains without fully encompassing any one cell. These waves spread at rates consistent with diffusion and depend on the P2 purine receptor, whose native agonists include adenosine triphosphate (ATP), adenosine diphosphate (ADP), and adenosine. The role of these waves in cerebellar function may include neutral protection or regional regulation of excitability or blood flow.

In this chapter, three protocols are provided for in vivo calcium imaging in cerebellar astrocytes. Protocol A describes the injection of a replication-incompetent recombinant adenovirus for gene transfer of the FCIP G-CaMP2. Protocol B presents a cerebellar craniotomy procedure, which is used days after virus injection to image astrocyte activity. The same procedure is used in Protocol C to prepare an animal for preferential MCBL (Stosiek et al. 2003) of Bergmann glia with Fluo-5F or Fluo-4 (Fig. 1E).

Protocol A

Injection of Recombinant Adenovirus for Gene Transfer of FCIP G-CaMP2 into Astrocytes of the Cerebellar Cortex

MATERIALS

CAUTION: See Appendix 6 for proper handling of materials marked with <!>.

Reagents

Buprenorphine (e.g., 0.3 mg/mL Buprenex; for mice, dilute 20-fold with saline to a final concentration of 15 μg/mL)
Hair removal cream or a hair trimmer for small animals
Ketamine <!>/xylazine <!> hydrochloride solution (K113, Sigma-Aldrich)

For mice, dilute 10-fold with saline to a final concentration of 8 mg/mL ketamine and 1.2 mg/mL xylazine.

Mannitol, 25% in H_2O (optional, see Step 5; heat to dissolve)
Mice or other experimental animals

All procedures must be approved by the local institutional animal care and use committee.

Recombinant adenovirus containing the FCIP sequence (in this protocol, it is G-CaMP2 fused to DsRed)
Saline (0.9% NaCl in H_2O)

Equipment

Absorption triangles (Sugi Saugtupfer, Kettenbach GmbH)
Box for storing quartz pipettes
Cotton-tipped applicators
Cyanoacrylate glue (e.g., Super Glue) <!>
Dissection microscope (e.g., Leica MZ16)
Electrode beveler (Sutter Instruments or World Precision Instruments [WPI])
Electrode puller (Sutter Instruments or WPI) suitable for preparing sharp pipettes, optimally from quartz tubing (Sutter Instruments)
Heating blanket
Low vibration dental drill (High-speed microdrill, Fine Science Tools [FST]; or EXL-M40, Osada, Inc.) with fine (0.5-mm) carbide drill burs
Marker, permanent fine tip
Micromanipulator with electrode holder, tubing, syringe for pressure control, and pressure gauge mounted in surgery area
Needle, 30-gauge
Parafilm
Pressurized air
Quartz tubing, outer diameter 1.0 mm, inner diameter 0.3 mm, no capillary

Alternatively, use thick-walled borosilicate glass without capillary. If volumes >200 nL are injected, a wider inner diameter should be used. The smaller surface-to-volume ratio reduces surface absorption of virus.

Stereotaxic apparatus
Surgery area, optimally on a vibration-isolated table, biosafety level 2 (BSL-2) approved
Surgical tools including scissors, microcurette, scalpel, and forceps #5 and #55 (FST)
Temperature controller
Temperature probe, rectal

EXPERIMENTAL METHOD

1. Pull sharp (>10 MΩ) electrodes from quartz or glass tubing, and bevel at an angle of ~20° to an inner tip diameter of 5–10 µm. Calibrate, and control the diameter of the tip during beveling by applying pressure in the range of 5–10 psi (0.35–0.7 bar) to the pipette.

   Quartz tubing is preferred rather than glass tubing because the resulting electrode shanks are stiffer.

2. Using a permanent marker, draw tick marks at 1-mm intervals on the beveled pipettes.

   These marks allow control over the injected volume. For electrodes with an inner diameter of 0.3 mm, the marks correspond to ~70-nL increments.

3. Keep pipettes in a clean closed box to prevent contamination with dust.
4. Anesthetize an animal with an intraperitoneal injection of a ketamine/xylazine solution (80 mg/mL and 12 mg/mL, respectively). Inject 1 mL/kg body weight. For mice, dilute the anesthetic 10-fold in saline to achieve a more precise dosage, and inject 16 mL/kg body weight.
5. (Optional) To infect a wider area of cells, inject mannitol intraperitoneally, 30 mL of 25% mannitol/kg body weight, and wait 15 min before starting surgery.

   This increases the extracellular space and promotes better virus diffusion (Burger et al. 2005).

6. Remove hair over the craniotomy site using either hair removal cream or a hair trimmer.
7. Immobilize the head of the animal in a stereotaxic apparatus such that the intended injection site is accessible. Insert a rectal temperature probe, and turn on a heating blanket to keep the animal's temperature at 37°C. Work on a vibration-isolated surgery table with a dissection microscope.
8. Using scissors, open the skin over the brain area of interest to a length of ~1 cm, and pull the skin to the side using #5 forceps.
9. If muscles cover the skull over the brain area of interest, carefully loosen tendons from the underlying bone with a microcurette or scalpel.

   Avoid damaging the muscles.

10. Dry the skull with cotton-tipped applicators or absorption triangles and a gentle application of pressurized air.
11. Drill a 0.75-mm circular region in the skull, leaving a thin layer of bone intact. Remove the last layer of bone by cutting a slit with the sharp edge of a 30-gauge needle tip, and then remove the disk of thinned bone with #55 forceps. Keep the dura dry.

    Do not push the bone into the underlying brain tissue. Make sure that the dura mater stays intact and no cerebrospinal fluid leaks out.

12. Mount a beveled pipette on a pipette holder. Attach tubing with a pressure meter and syringe to the back of the pipette.
13. Under visual control through the dissection microscope, tip load the beveled pipette with virus stock by putting a 2–5-µL droplet of virus stock on Parafilm, immersing the tip of the pipette, and applying suction until the pipette has taken up the virus stock. Follow progress by looking at the tick marks.

    Make sure that the electrode does not draw in any air.

14. Under visual control through the dissection microscope, approach the craniotomy with the pipette tip, applying 0.3 psi (0.02 bar) of positive pressure before touching. When the pipette tip reaches the dura, zero the micromanipulator position indicator.

    A dry dura will minimize the amount of spilled virus.

15. Enter the brain at an angle of ~45° to the skull surface. Slowly (10 µm/sec) penetrate in the direction of the pipette by 240 µm, which corresponds to a depth of ~170 µm below the dura mater.
16. Increase the pressure to ~0.5 psi (0.035 bar). Using the tick marks drawn on the pipette for guidance, inject 50–400 nL of virus stock over a period of ~15 min. Check the progress frequently, and adjust the pressure in a range of 0.1–1 psi (0.007–0.07 bar).

    Low pressure and low volume injection will give a more local infection; conversely, high pressure and large volume injection will give more widespread infection.

    See *Troubleshooting.*

17. Reduce the pressure to 0.1 psi (0.007 bar), and retract the pipette slowly. Keep the pipette mounted as it can be used for additional injections.

    We usually inject six to eight animals in a session.

18. Remove the mouse from the stereotaxic apparatus, wet the skull with saline, and close the skin by sealing it with Super Glue.
19. For postoperative analgesia, give buprenorphine at 20–50 µg/kg body weight. Keep the mouse on a heating blanket at 37°C until it recovers. Check it regularly afterward, and administer more buprenorphine if necessary.
20. Maintain the mouse for 1–25 d after virus injection to allow for protein expression.
21. To prepare the animal for in vivo imaging, use Protocol B.
22. For two-photon calcium imaging of G-CaMP2, use an excitation wavelength of 920 nm with an emission filter of 500–550 nm. For cell reconstruction with DsRed, use an excitation wavelength of 990–1050 nm and an emission filter of 550–650 nm. Objectives of high numerical aperture should be used.

    See *Troubleshooting.*

TROUBLESHOOTING

Problem (Step 16): The pipette with the virus solution is clogged.

Solution: This is the most critical problem as the virus stock is, in many cases, a precious substance and is difficult to recover from a pipette. While the tip is exposed to air, apply high pressure of up to 20 psi (1.4 bar) until a droplet comes out of the tip. If this fails, tissue or blood cells may have clogged the tip. In this case, immerse the tip in saline for a few seconds at low pressure, retract the tip, and again increase pressure until a droplet is expelled. Remove the droplet with absorbent triangles to prevent spillage of virus on the dura.

Problem (Step 22): No infected cells are found during in vivo imaging.

Solution: Make sure that the craniotomy for imaging was performed at the correct coordinates. The skull should recover almost completely after 10 d, especially when virus has been injected into young adult mice (16–25 d). Take good notes of the injection site, including drawing (or photographing) dura blood vessels after wetting the skull with saline. Remember that the angle of injection was 45° causing the center of infection to be away from the center of the original drilling. If all injected animals fail to show infected cells, check the potency of the adenoviral stock by in vitro infection of HEK293 cells and titration of the virus by agar plug assay (Luo et al. 2007).

Protocol B

Cerebellar Craniotomy for In Vivo Imaging

This protocol is used to prepare a virus-injected animal (Protocol A) for in vivo imaging. It is also used to prepare an animal for MCBL (Protocol C).

MATERIALS

CAUTION: See Appendix 6 for proper handling of materials marked with <!>.
See the end of the chapter for recipes for reagents marked with <R>.

Reagents

Artificial cerebrospinal fluid, HEPES-buffered <R>
Cyanoacrylate glue (e.g., Super Glue) <!>
Dental acrylic (e.g., MetaBond, Parker Inc.)
Hair removal cream or a hair trimmer for small animals
Isoflurane <!> (optional, see Step 1)
Ketamine <!>/xylazine <!> hydrochloride solution (K113, Sigma-Aldrich)

For mice, dilute 10-fold with saline to a final concentration of 8 mg/mL ketamine and 1.2 mg/mL xylazine.

Lidocaine hydrochloride topical solution (4%) <!>
Mice or other experimental animals

All procedures must be approved by the local institutional animal care and use committee.

Equipment

Absorption triangles (Sugi Saugtupfer, Kettenbach GmbH & Co. KG)
Cotton-tipped applicators
Dissection microscope (e.g., Leica MZ16)
Heating blanket
Isoflurane vaporizer (optional, see Step 1) with O_2 tank, plastic box for initial anesthesia, animal mask, and filter
Low vibration dental drill (High-speed microdrill, FST, or EXL-M40, Osada, Inc. dental drill) with fine (0.5-mm) carbide drill burs
Mounting frame with custom head plate
Needle, 30-gauge
Pressurized air
Stereotaxic apparatus
Surgery area, optimally on a vibration-isolated table
Surgical tools, including scissors, microcurette, scalpel, and forceps #5 and #55 (FST)
Temperature controller
Temperature probe, rectal
Two-photon microscope

EXPERIMENTAL METHOD

1. Anesthetize animal with an intraperitoneal injection of a ketamine/xylazine solution (80 mg/mL and 12 mg/mL, respectively). Inject 1 mL/kg body weight. For mice, dilute the anesthetic 10-fold in saline to achieve a more precise dosage, and inject 16 mL/kg body weight. If isoflurane is used, place the animal in a sealed plastic box, and connect to 2% isoflurane in O_2.
2. Remove hair over the craniotomy site using either hair removal cream or a hair trimmer.
3. Mount the animal on a stereotaxic apparatus. Insert a rectal temperature probe, and use a heating blanket to maintain an internal body temperature of 37°C. Work on a vibration-isolated surgery table with dissection microscope. If isoflurane is used, mount an animal mask, and reduce the isoflurane to 1.0–1.5% in O_2.
4. Make an incision with scissors over the brain region of interest. Pull the skin to the side with #5 forceps, and, with cotton-tipped applicators, apply 4% lidocaine solution topically onto the incision. If necessary, scrape muscle tendons and connective tissue from the skull with a microcurette or scalpel. Clean the skull with saline and cotton-tipped applicators. Gently dry the skull with pressurized air or absorbent triangles.
5. Mount a metal head plate to the skull with a thin dried layer of Super Glue followed by dental acrylic.
6. Transfer the animal from the stereotaxic apparatus to a frame that holds the head plate in a stable position.
7. Drill a 1–2-mm wide circular region defining the outline of the craniotomy. When the thinned bone is soft and transparent, stop drilling. Use pressurized air to blow away bone dust.

   Take care not to put excessive pressure on the skull. Constantly move the drill to avoid drilling-induced local heating of the brain.

   See *Troubleshooting.*
8. Puncture the thinned bone along the edge of the circle with the tip of a 30-gauge needle or #55 forceps.
9. Put a drop of saline on the area of the drilled region, and lift the bone patch with forceps.

   Avoid putting pressure on the underlying brain area. If slight dura bleeding occurs, rinse immediately with saline until bleeding stops. If the saline is cloudy, remove it with absorption triangles, and replace with fresh saline. The dura mater should still be intact with no brain swelling. Keep saline on the exposed craniotomy.
10. For experiments in which drugs are topically applied to the brain surface, the dura must be removed. Make a small slit in the dura leaving the arachnoid mater and pia mater intact. With one pair of forceps on each side of the slit, lift the dura simultaneously from both sides, and gently tear apart toward the edge of the craniotomy, taking care to exert force on dura against dura, not the brain.

    This step requires practice. If the brain swells after dura removal, not only the dura but also parts of the arachnoid or pia mater were probably removed. In this case, the experiment should be terminated.
11. Move the animal from the surgery area to the two-photon microscope setup. If more anesthesia is needed, inject additional ketamine/xylazine (50% of initial dose). When using isoflurane anesthesia, move quickly from the surgery area to the imaging setup.

TROUBLESHOOTING

Problem (Step 7): You break through the bone or there is bleeding while drilling the craniotomy.
Solution: The skull over the cerebellum varies in thickness, and thicker parts have blood-filled

sinuses. For this reason, perform a test surgery to become familiar with the bone thickness over the brain area of interest. Drill carefully and slowly when drilling in thin skull areas. If you drill in a thick part of the skull, slow drilling of the bone will stop sinus bleeding because the resulting bone–blood mixture coagulates quickly. Be careful with the use of cauterizers because these can easily cause damage to underlying brain structures, especially in mice.

Protocol C

Preferential Loading of Bergmann Glia with Synthetic Acetoxymethyl Calcium Dyes

MCBL has been described in previous chapters (see Chapters 49 and 59). In the cerebellum, a degree of cell-type specificity can be achieved by varying the depth of injection. Here, we give a short description of our particular loading procedure after a craniotomy has been performed as described in Protocol B.

MATERIALS

CAUTION: See Appendix 6 for proper handling of materials marked with <!>.
See the end of the chapter for recipes for reagents marked with <R>.

Reagents

Agarose, 1.5% in saline (Type III-A, Sigma-Aldrich)

Heat the agarose until it is completely dissolved, and keep in a 45°C water bath.

Calcium indicators: Fluo-5F AM (50 µg, Invitrogen) or Fluo-4 AM (50 µg, Invitrogen)
Dimethylsulfoxide (DMSO) <!>, desiccated by molecular sieves (3-Å pore size)
Mice or other experimental animal (prepared surgically as per Protocol B)

All procedures must be approved by the local institutional animal care and use committee.

Pluronic F-127 (Invitrogen or BASF)
Saline (0.9% NaCl in H_2O) or artificial cerebrospinal fluid, HEPES-buffered <R> without $CaCl_2$

Equipment

Borosilicate glass tubing for pipettes
Coverslips, glass #1 (~5 x 5-mm)
Electrode beveler (Sutter Instruments or WPI)
Electrode puller suitable for preparing patch pipettes (Sutter instruments or WPI)
Microloader (Eppendorf)
Micromanipulator with a position counter, an electrode holder with tubing, syringe, and pressure gauge on the two-photon microscope system
Ultrafiltration device, 0.22-µm pore size (e.g., Ultrafree-MC, Millipore)

EXPERIMENTAL METHOD

1. For the preparation of the dye stock solution, add 5 µL of 20% Pluronic F-127 freshly dissolved in H_2O-free DMSO to 50 µg of Fluo-5F AM or Fluo-4 AM. Vortex, and then add 80 µL of saline. Vortex, and filter the dye stock solution through a low-volume 0.22-µm filter.

   We find that Fluo-5F AM as well as Fluo-4 AM generally allow for measurements with better signal to noise in astrocytes than does OGB-1 AM. Fluo-5F (K_D ~ 1 µM) (Sarkisov and Wang 2008) is especially well suited for measuring calcium signals of the magnitude produced in transglial waves.

2. Pull a patch pipette, and backfill it with the dye solution using a microloader. Bevel the pipette at an angle of 20° to a resistance of 3–4 MΩ.
3. Mount the beveled pipette in the pipette holder on the micromanipulator, and apply 0.5 psi (0.035 bar) positive pressure to the pipette before entering the saline over the craniotomy.
4. Approach the dura under visual control by entering the saline over the craniotomy, and zero the manipulator when reaching the brain surface. Check under fluorescence viewing that the dye is coming out of the pipette.
5. Advance the pipette into the surface of the brain. After passing the dura, reduce the pressure to 0.3 psi (0.02 bar), and advance to a depth of ~50 µm below the dura. Verify that dye is ejected from the pipette by visualizing its fluorescence. To label Bergmann glia, eject the dye solution at ~2 psi (0.14 bar). A similar injection into the granule cell layer (in mice >140 µm below pia mater) will result in relatively increased labeling of Purkinje cells. Apply pressure for 5 min.
6. Reduce the pressure to 0.1 psi (0.007 bar), and slowly retract the electrode.
7. Add 1.5% agarose at 45°C over the craniotomy, and seal it with a coverslip while the agarose is still liquid.
8. Wait for ~30 min for dye uptake.
9. Image using an excitation wavelength of 840 or 920 nm and a 500–550-nm band-pass filter for detection.

DISCUSSION

We have described two protocols to monitor astrocytic calcium signals in the intact cerebellum. MCBL, which is reliable and quick, preferentially labels Bergmann glia if the dye is injected into the superficial molecular layer (Fig. 1E). When dye is injected into the granule cell layer, we have been able to simultaneously observe activity in Bergmann glia, interneurons, and Purkinje cells (Sullivan et al. 2005). The disadvantage of bolus loading is low contrast so that fine processes in the neuropil cannot be resolved; three-dimensional (3D) reconstructions of Bergmann glia are blurry, and signals from small structures cannot be assigned to particular cell types.

To overcome the problem of nonspecific labeling, we use attenuated nonreplicating adenovirus to express the FCIP G-CaMP2 in Bergmann glia and velate protoplasmic astrocytes. Compared with synthetic dyes such as Oregon Green BAPTA-1, the current generation of FCIPs has slower on and off calcium binding rates and smaller relative fluorescence changes (Hendel et al. 2008). However, neither constraint appears to be a limiting factor in the study of astrocytic calcium signals, which last for seconds and reach micromolar levels. Fluorescence changes in FCIPs are highly nonlinear with respect to calcium concentration, with Hill coefficients >1 (Hendel et al. 2008) so that measurements are, to a degree, a thresholded function of the actual change in free calcium. Signals can be recorded in vivo from even the finest structures, such as Bergmann glia microdomains (Grosche et al. 1999) or velate processes of protoplasmic astrocytes. FCIPs are also well suited for long-term imaging (days or longer), whereas dye-loaded cells lose their labeling after a few hours. Additionally, functionally imaged cells can be reconstructed in 3D in vivo with the help of two-photon microscopy (Fig. 1A).

The virus-based method presented in this chapter is one of several methods currently available for expressing FCIPs in neurons and astrocytes in situ. Other methods include transgenic animal generation (Hasan et al. 2004; Heim et al. 2007), in utero electroporation (Mank et al. 2008; Mao et al. 2008), single-cell electroporation (Judkewitz et al. 2009), and the use of other recombinant viruses (Mank et al. 2008; Wallace et al. 2008). As adenoviral vectors have long been known to be useful for gene transfer to astrocytes (Le Gal La Salle et al. 1993), in our protocol we used the replication-incompetent recombinant adenovirus AdEasy-1, which has low cytotoxic side effects (He et al. 1998). However, it should be mentioned that other viruses or virus serotypes also infect glia with high efficiency. Examples are baculovirus (Wang and Wang 2006), single- and double-stranded adeno-associated viruses (AAVs) of specific serotypes (Shevtsova et al. 2005; Lowery et al. 2009), lentivirus (Croci et al. 2006; Colin et al. 2009), and Semliki Forest virus (Mank et al. 2008).

In vitro, adenoviruses have a much higher probability of infecting glia than neurons (Sato et al. 2004). In addition, the CMV IE promoter results in significantly higher expression levels in astrocytes than in neurons (Boulos et al. 2006). Consistent with these two observations, the approach outlined here leads to expression of G-CaMP2 in Bergmann glia and velate protoplasmic astrocytes. Adenoviruses do not integrate into the genome, and calcium imaging is limited to ~1–25 d after virus injection. Sparse expression allows single cells and their processes to be studied (Fig. 1A,B), whereas high-density labeling across regions of several hundred micrometers can label all astrocytes (Fig. 1C), comparable to a transgenic approach (Fig. 1D).

In vivo, G-CaMP2 has low resting fluorescence comparable to wild-type GFP (Tallini et al. 2006; Mao et al. 2008). To improve detection of expression, we used a fusion protein of G-CaMP2 with DsRed, facilitating detection and 3D reconstruction using an excitation wavelength optimal for DsRed (990 nm). Two-channel detection also allows for ratiometric measurements and, thus, compensation of motion artifacts, an important consideration when recording from awake behaving animals.

RECIPES

CAUTION: See Appendix 6 for proper handling of materials marked with <!>.
Recipes for reagents marked with <R> are included in this list.

Artificial Cerebrospinal Fluid, HEPES-Buffered

NaCl	135 mM
KCl	5.4 mM
Na-HEPES buffer	5 mM
$CaCl_2$ <!>	1.8 mM
$MgCl_2$	1 mM

Adjust to pH 7.3 with HCl <!>.

Recombinant Adenovirus Containing the FCIP Sequence

The recombinant replication-incompetent adenovirus AdEasy-1 is commercially available and supplied as a viral backbone and shuttle vector for gene insertion (JHU-23, ATCC). Recombinant viruses can be produced with basic knowledge of molecular cloning by following the protocol in Luo et al. (2007). For virus amplification, a BSL-2 cell culture hood and incubator are required. The viral vectors can be harvested from HEK293 cells after lysis by freeze/thawing and centrifugation. Lysate can be used directly from the supernatant after 0.5-µm filtration or further purification and concentration using a $CsCl_2$ gradient or a commercially available kit (Vivapure AdenoPACK, Sartorius Stedim Biotech S.A.). The titer of infectious particles should be in the range of 10^8–10^9/mL. FCIP constructs are typically generously supplied by their developers, including the laboratories of R.Y. Tsien, J. Nakai, O. Griesbeck, and others. The construct we used, a fusion protein of G-CaMP2 and DsRed, was designed and made by J. Nakai (Saitama University, Japan).

ACKNOWLEDGMENTS

The work described in this chapter was supported by grants to SW from the National Institutes of Health (R01 NS045193), the National Science Foundation, the W. M. Keck Foundation, the Human Frontier Science Project, and the New Jersey Governor's Council for Medical Research and Treatment of Autism.

REFERENCES

Allen NJ, Barres BA. 2009. Neuroscience: Glia—More than just brain glue. *Nature* **457:** 675–677.

Bacaj T, Tevlin M, Lu Y, Shaham S. 2008. Glia are essential for sensory organ function in *C. elegans*. *Science* **322:** 744–747.

Boulos S, Meloni BP, Arthur PG, Bojarski C, Knuckey NW. 2006. Assessment of CMV, RSV and SYN1 promoters and the woodchuck post-transcriptional regulatory element in adenovirus vectors for transgene expression in cortical neuronal cultures. *Brain Res* **1102:** 27–38.

Burger C, Nguyen FN, Deng J, Mandel RJ. 2005. Systemic mannitol-induced hyperosmolality amplifies rAAV2-mediated striatal transduction to a greater extent than local co-infusion. *Mol Ther* **11:** 327–331.

Colin A, Faideau M, Dufour N, Auregan G, Hassig R, Andrieu T, Brouillet E, Hantraye P, Bonvento G, Deglon N. 2009. Engineered lentiviral vector targeting astrocytes in vivo. *Glia* **57:** 667–679.

Croci C, Fasano S, Superchi D, Perani L, Martellosio A, Brambilla R, Consalez G, Bongarzone ER. 2006. Cerebellar neurons and glial cells are transducible by lentiviral vectors without decrease of cerebellar functions. *Dev Neurosci* **28:** 216–221.

Grosche J, Matyash V, Moller T, Verkhratsky A, Reichenbach A, Kettenmann H. 1999. Microdomains for neuron–glia interaction: Parallel fiber signaling to Bergmann glial cells. *Nat Neurosci* **2:** 139–143.

Halassa MM, Florian C, Fellin T, Munoz JR, Lee SY, Abel T, Haydon PG, Frank MG. 2009. Astrocytic modulation of sleep homeostasis and cognitive consequences of sleep loss. *Neuron* **61:** 213–219.

Hasan MT, Friedrich RW, Euler T, Larkum ME, Giese G, Both M, Duebel J, Waters J, Bujard H, Griesbeck O, et al.

2004. Functional fluorescent Ca^{2+} indicator proteins in transgenic mice under TET control. *PLoS Biol* **2:** e163. doi: 10.1371/journal.pbio.0020163.

He TC, Zhou S, da Costa LT, Yu J, Kinzler KW, Vogelstein B. 1998. A simplified system for generating recombinant adenoviruses. *Proc Natl Acad Sci* **95:** 2509–2514.

Heim N, Garaschuk O, Friedrich MW, Mank M, Milos RI, Kovalchuk Y, Konnerth A, Griesbeck O. 2007. Improved calcium imaging in transgenic mice expressing a troponin C–based biosensor. *Nat Methods* **4:** 127–129.

Hendel T, Mank M, Schnell B, Griesbeck O, Borst A, Reiff DF. 2008. Fluorescence changes of genetic calcium indicators and OGB-1 correlated with neural activity and calcium in vivo and in vitro. *J Neurosci* **28:** 7399–7411.

Hoogland TM, Kuhn B, Gobel W, Huang W, Nakai J, Helmchen F, Flint J, Wang SS-H. 2009. Radially expanding transglial calcium waves in the intact cerebellum. *Proc Natl Acad Sci* **106:** 3496–3501.

Iadecola C, Nedergaard M. 2007. Glial regulation of the cerebral microvasculature. *Nat Neurosci* **10:** 1369–1376.

Judkewitz B, Rizzi M, Kitamura K, Hausser M. 2009. Targeted single-cell electroporation of mammalian neurons in vivo. *Nat Protoc* **4:** 862–869.

Le Gal La Salle G, Robert JJ, Berrard S, Ridoux V, Stratford-Perricaudet LD, Perricaudet M, Mallet J. 1993. An adenovirus vector for gene transfer into neurons and glia in the brain. *Science* **259:** 988–990.

Lowery RL, Zhang Y, Kelly EA, Lamantia CE, Harvey BK, Majewska AK. 2009. Rapid, long-term labeling of cells in the developing and adult rodent visual cortex using double-stranded adeno-associated viral vectors. *Dev Neurobiol* **69:** 674–688.

Luo J, Deng ZL, Luo X, Tang N, Song WX, Chen J, Sharff KA, Luu HH, Haydon RC, Kinzler KW, et al. 2007. A protocol for rapid generation of recombinant adenoviruses using the AdEasy system. *Nat Protoc* **2:** 1236–1247.

Mank M, Santos AF, Direnberger S, Mrsic-Flogel TD, Hofer SB, Stein V, Hendel T, Reiff DF, Levelt C, Borst A, et al. 2008. A genetically encoded calcium indicator for chronic in vivo two-photon imaging. *Nat Methods* **5:** 805–811.

Mao T, O'Connor DH, Scheuss V, Nakai J, Svoboda K. 2008. Characterization and subcellular targeting of GCaMP-type genetically-encoded calcium indicators. *PLoS One* **3:** e1796. doi: 10.1371/journal.pone.0001796.

Nimmerjahn A, Mukamel EA, Schnitzer MJ. 2009. Motor behavior activates Bergmann glia networks. *Neuron* **62:** 400–412.

Prolo LM, Takahashi JS, Herzog ED. 2005. Circadian rhythm generation and entrainment in astrocytes. *J Neurosci* **25:** 404–408.

Sarkisov DV, Wang SS-H. 2008. Order-dependent coincidence detection in cerebellar Purkinje neurons at the inositol trisphosphate receptor. *J Neurosci* **28:** 133–142.

Sato Y, Shiraishi Y, Furuichi T. 2004. Cell specificity and efficiency of the Semliki forest virus vector- and adenovirus vector-mediated gene expression in mouse cerebellum. *J Neurosci Methods* **137:** 111–121.

Schummers J, Yu H, Sur M. 2008. Tuned responses of astrocytes and their influence on hemodynamic signals in the visual cortex. *Science* **320:** 1638–1643.

Shevtsova Z, Malik JM, Michel U, Bahr M, Kugler S. 2005. Promoters and serotypes: Targeting of adeno-associated virus vectors for gene transfer in the rat central nervous system in vitro and in vivo. *Exp Physiol* **90:** 53–59.

Stosiek C, Garaschuk O, Holthoff K, Konnerth A. 2003. In vivo two-photon calcium imaging of neuronal networks. *Proc Natl Acad Sci* **100:** 7319–7324.

Suh J, Jackson FR. 2007. *Drosophila* ebony activity is required in glia for the circadian regulation of locomotor activity. *Neuron* **55:** 435–447.

Sullivan MR, Nimmerjahn A, Sarkisov DV, Helmchen F, Wang SS-H. 2005. In vivo calcium imaging of circuit activity in cerebellar cortex. *J Neurophysiol* **94:** 1636–1644.

Tallini YN, Ohkura M, Choi BR, Ji G, Imoto K, Doran R, Lee J, Plan P, Wilson J, Xin HB, et al. 2006. Imaging cellular signals in the heart in vivo: Cardiac expression of the high-signal Ca^{2+} indicator GCaMP2. *Proc Natl Acad Sci* **103:** 4753–4758.

Wallace DJ, Meyer zum Alten Borgloh S, Astori S, Yang Y, Bausen M, Kugler S, Palmer AE, Tsien RY, Sprengel R, Kerr JN, et al. 2008. Single-spike detection in vitro and in vivo with a genetic Ca^{2+} sensor. *Nat Methods* **5:** 797–804.

Wang CY, Wang S. 2006. Astrocytic expression of transgene in the rat brain mediated by baculovirus vectors containing an astrocyte-specific promoter. *Gene Ther* **13:** 1447–1456.

Wang X, Lou N, Xu Q, Tian GF, Peng WG, Han X, Kang J, Takano T, Nedergaard M. 2006. Astrocytic Ca^{2+} signaling evoked by sensory stimulation in vivo. *Nat Neurosci* **9:** 816–823.

67 Two-Photon Imaging of Neurons and Glia in the Spinal Cord In Vivo

Heinz Steffens,[1,2] Fabien Nadrigny,[1,3] and Frank Kirchhoff[1,4,5]

[1]Department of Neurogenetics, Max-Planck-Institute of Experimental Medicine, 37075 Göttingen, Germany; [2]Institute of Physiology, Georg August University, 37073 Göttingen, Germany; [3]University of Bordeaux, Neurocentre Magendie, 33077 Bordeaux, France; [4]DFG Research Centre Molecular Physiology of the Brain, 37073 Göttingen, Germany; [5]Department of Molecular Physiology, Institute of Physiology, University of Saarland, 66421 Homburg, Germany

ABSTRACT

Two-photon imaging of the nervous system is now used extensively for visualizing brain dynamics and signal activities. To date, scientists have focused on the analysis either of gray matter forebrain structures, such as the cortex and cerebellum, or they have investigated muscle innervation of peripheral nerves. The spinal cord represents an ideal structure to use for imaging central nervous system white matter. The dorsal columns formed by myelinated sensory axons are located directly at the surface of the spinal cord underneath the pia mater. This chapter presents three experimental protocols for imaging neuronal fibers and neighboring glial cells in transgenic mice using cell type–specific fluorescent protein expression.

INTRODUCTION

Two-photon laser-scanning microscopy (2pLSM) is routinely used to image brain structures to a depth of more than several hundred micrometers. Because it is possible to record neuronal and astroglial network activities in a cortical three-dimensional space of >250 µm side length, 2pLSM is the method of choice for the analysis of signal processing in the central nervous system (CNS).

In contrast to gray matter, white matter tracts are thought to show fewer structural changes. There are, however, several brain diseases in which the white matter is known to be affected. This has been well documented for multiple sclerosis, and schizophrenic patients show pathological alterations of hemisphere-connecting fiber tracts. Accidental dissection of spinal fiber tracts (such as can be caused by motor bike accidents) almost always results in paralysis of the lower extremities. Because effective therapies are lacking for white matter diseases and injuries, basic research on the molecular and cellular interactions and dynamics is essential for successful development of such therapies.

The dorsal columns of the spinal cord are the most suitable parts of the CNS for 2pLSM imaging of white matter because the myelinated axons of sensory ganglion neurons run directly underneath the bones of the vertebral column (see Fig. 1). In our experimental setup, the spines of the vertebrae can be held securely in place without injuring the nervous tissue. Cells of transgenic mice are labeled using cell type–specific expression of various fluorescent proteins (see Fig. 2).

The following protocols provide a detailed description of how to perform 2pLSM imaging in the spinal cord of living mice. The first two protocols describe the anesthesia and surgical procedures necessary to perform single and repetitive imaging, respectively. The third protocol explains how to mount the spinal column for imaging and the imaging procedure itself.

Protocol A

Preparation of the Mouse Spinal Column for Single Imaging Using 2pLSM

The procedures in this protocol are ideal for single-imaging experiments in which the spinal cord needs to be imaged at optimal spatial resolution with minimal motion artifacts. Before injecting the tissue for imaging, a tracheotomy is performed to allow for intubation and artificial ventilation. Mice are sacrificed after the imaging session. In all of our experiments the mouse is placed with the head to the right. We refer to the half of the body that is next to the person performing the experiment as ipsilateral. Determine the body weight of the mouse first. Barbiturates and morphines require precise dosage to avoid respiratory depression.

MATERIALS

CAUTION: See Appendix 6 for proper handling of materials marked with <!>.
See the end of the chapter for recipes for reagents marked with <R>.

Reagents

Artificial cerebrospinal fluid (ACSF) <R>
Bepanthen
Methohexital (i.e., Brevital sodium), 6.67 mg/mL in 0.9% NaCl containing 0.7 mg/mL glucose
Mice, genetically modified
- TgN(Thy1-eYFP)TYFF-Fki, yellow fluorescent axons (Hirrlinger et al. 2005)
- TgN(GFAP-eCFP)GCFD-Fki, cyan fluorescent astrocytes (Hirrlinger et al. 2005)
- TgH(CX3CR1-eGFP), green fluorescent microglia (Jung et al. 2000)

All experiments require approval by the local regulatory body responsible for animal experimentation.

Norepinephrine <!> to strengthen heart function
Pancuronium bromide <!>

Dilute 1 volume of the muscle relaxant in 9 volumes of 0.9% NaCl.

Pentobarbital <!>, sodium salt (5 mg/mL in 0.9% [w/v] NaCl)
Respiration gas mixture: 50% N_2, 47.5% O_2 <!>, 2.5% CO_2 <!>
Vaseline

Equipment

Anesthesia equipment: injection pump for application of small volumes (e.g., Programmable Syringe Pump AL-1000, World Precision Instruments), 5-mL syringe with Luer lock connection fitting and rubber piston, intravenous cannula (polyethylene, 0.28-mm-ID, 0.61-mm-OD) for cannulation of the jugular vein, hypodermic needle (0.3-mm-OD) to fit the intravenous cannula
Binocular dissection microscope with adjustable magnification and sufficient working distance
Differential preamplifier (for ECG recording)
Filter tips for syringes to sterilize injected fluids and ACSF
Hypodermic needles: 0.3 mm (OD) to fit the polyethylene microtube; 0.4 mm (OD) for intraperitoneal (i.p.) or subcutaneous (s.c.) injections
Oscilloscope
Pasteur pipettes to apply ACSF to the operation field
Platinum wire (0.3 mm in diameter) as ECG electrodes

Power supply for the heat pad (12-V, 1-A)
Rectal temperature sensor and display
Respiration equipment: high pressure gas bottle with reduction valve for respiration of gas mixture (see above), control logic for solenoid valve, fittings, flow meter to measure and adjust respiration gas flow (Platon GTF 1 BHD, for air, 20–250-mL/min), solenoid valve (ASCO Scientific 4224) for artificial respiration, tracheal tube (T-shaped, custom-made), pressure tubing to connect reduction valve and flow meter, silicon tubing (1-mm-ID, 3-mm-OD) to connect flow meter with tracheal tube, silicon tubing (2-mm-ID, 4-mm-OD) to connect tracheal tube with solenoid valve
Shaver, electric
Spongostan sponges to stop bleedings
Support for the mouse, heated by a flat heat pad with rail bars to fix two vertebral clamps for the backbone and the volatile anesthesia equipment
Surgical instruments: Curved forceps (e.g., FST 11051-10), dissector scissors (e.g., FST 14081-09), Dumont forceps (#5 Biology, #3c, and #7), Michel suture clips (FST 12040-01) and forceps (FST 12018-12), microcurette (e.g., FST 10081-10), microscissors, needle holder (e.g., FST 12002-12) and needles (FST 12050-03), scalpel handle (e.g., FST 10007-12) and blades (10010-00 or 10011-00), scissors type Wecker (FST 15010-11), silk suture (FST 18020-30), straight scissors (e.g., FST 14088-10)
Syringes: 1-mL, graduation 100/10 μL for i.p. injection of chemicals; 5-mL Luer lock with rubber piston, graduation 1/0.2 mL for continuous infusion of anesthetics; 0.5-mL with needle (0.3-mm-diameter) for intravenous (i.v.) injections into tail vein
Tape (Leukosilk, Beiersdorf)
Thin cotton sewing thread
Vacuum cleaner

EXPERIMENTAL METHOD

Anesthetizing and Intubating the Mouse

1. To initially anesthetize the mouse, inject i.p. 77 mg of sodium pentobarbital per kilogram of body weight.

   This should be enough anesthetic to permit placement of the central vein catheter in the jugularis vein for i.v. delivery of anesthetic. If necessary, the dose of pentobarbital may be increased by 10%. Indications that the mouse is sufficiently anesthetized are: respiration is slow and calm (2 breaths/sec), eyelid closure reflex is absent, pinna reflex is absent, and body temperature is falling.

2. Use a rectal probe (coat first with vaseline) to measure body temperature. Adjust body temperature to 36°C–37°C with the heating pad on which the mouse is lying.
3. Fix the mouse in a supine position to the heated support using tape at the limbs and tail.
4. Stretch the mouse neck with a loop of thread placed around the upper front teeth.
5. Cover the eyes with Bepanthen to protect them from drying out.
6. Shave the ventral skin of the neck. Carefully remove all of the shaved hair with a vacuum cleaner.

   Conscientious preparation at this step greatly reduces the formation of allergies against mice.

7. Make a longitudinal incision in the ventral skin along the midline of the neck. This provides access to the ipsilateral jugularis vein and to the trachea.
8. Fill a PE microtube catheter with sodium methohexital. Using a sharp scalpel cut the end of the catheter.
9. Insert the catheter into the jugular vein and fix it in place with two threads. Also suture the catheter at the skin of the neck.

10. Begin the flow of anesthetic, being careful to avoid injection of air bubbles. Adjust the flow rate of sodium methohexital to ~55 mg/kg/h.

    When the effect of the sodium pentobarbital fades, it might be necessary to double the dose of methohexital, but only for brief periods of time (because there is the danger of persistent bradycardia). To monitor the state of the heart, set up an electrocardiogram (ECG) by inserting platinum wires as electrodes into both forelimbs and connecting them via a differential amplifier to an oscilloscope.

    See *Troubleshooting.*

11. To access the trachea, separate the salivary glands in the middle and the sternohyoideus muscle at the midline. Because one of the main veins had been occluded, suture the ipsilateral carotid artery before inserting the tracheal tube, thereby avoiding accumulation of blood in the head. Take care not to include the sympathetic nerve in the carotid occlusion.
12. For intubation, place two loops of thread around the trachea, perform a transversal incisure between the two loops, cut the caudal cartilage ring, insert the tracheal tube and secure it with the two threads. Use the most caudal accessible part of the trachea; otherwise there is the danger of injuring the well perfused thyreoidea.
13. Close the wound with suture clips.

Preparing the Spinal Cord for Imaging

14. Fix the mouse in a prone position; shave the back and clean it with a vacuum cleaner.
15. Make an incision in the skin along the rostrocaudal axis (Fig. 1D).

    The dorsal process of the sixth lumbar spine can be recognized by the insertion of the last pair of tendons of the back muscles. In the lumbar spinal cord the main region of interest may be the lumbar intumescence (i.e., the region where the motor neurons for the legs are situated, and where the roots of the efferent and afferent innervation for the legs originate or enter, respectively). This region is related to the spinal cord segments L3 to L5, mainly L4. Because of long spinal roots, you will find spinal cord segment L4 under the spinal arch L1, which can be identified by counting the dorsal spinal processes from L6.

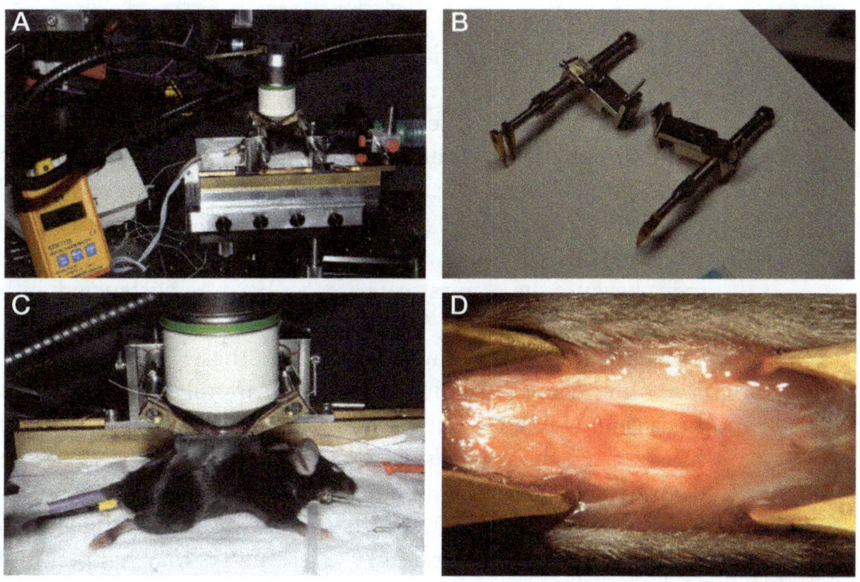

FIGURE 1. Preparation of mice for spinal cord imaging. (*A*) Mouse receives a volatile anesthetic, a rectal probe (see also C) monitors temperature, and imaging is performed with a high numerical aperture, 20x objective. (*B*) Clamps to fix the vertebral spines. (*C*) Mouse is artificially ventilated through a tracheal tube. Clamps fix the vertebral column. Skin flaps form a reservoir for artificial cerebrospinal fluid (ACSF) in which the objective is immersed. (*D*) Exposed spinal cord after a laminectomy.

16. To prepare the field of view, use microscissors or a microcurette to remove tendons and muscle tissue from the dorsal spinal processes and arches. Keep the preparation field moist with ACSF.

    In this and subsequent steps, stop occasional bleeding with Spongostan.

17. Use Dumont forceps #7 to dissect the spinal arches (i.e., a laminectomy) while holding the spinal column with a curved anatomical forceps. Although it is difficult and will take practice, take care to leave the dura mater intact. Imaging can be performed through the intact dura mater. In addition, lasers can be used to induce lesions through the intact dura mater.

    You may remove material from spinal arch L4 up to L1 depending on your aims and needs. When imaging the lateral spinal cord, the dorsal root ganglia or the ventral roots are required, thus remove spinal arches down to the vertebral bodies ipsilaterally.

18. Prepare the mouse for mounting the spinal column.

    i. Inject 100 μL of pancuronium:NaCl i.p. (repeat this every 60–90 min throughout the experiment).

    ii. Begin artificial ventilation.

        Feed the respiration gas mixture (pressure set to 2.5 bar) through the flow meter at a rate of 70 mL/min. The flow meter is connected to the short arm of the T-shaped tracheal tube. The large lateral arm is connected to a solenoid valve that closes for 100 msec per breath, thereby providing ~120 μL of inspiration volume. After closing for 100 msec, the valve opens for 400 msec, while the mouse is expiring passively. The resulting respiration rate is therefore 2/sec. The high fraction of O_2 allows for this rather flat respiration. As a result, you will be able to record images with almost no motion artifacts.

    iii. Monitor the heartbeat (should be 7–8/sec) and any changes in body temperature as indicators of the level of anesthesia.

    iv. Optionally remove the dura mater if necessary (e.g., for superfusion with drugs or for dye injections with micropipettes) with Dumont forceps #5 biology and microscissors.

19. Proceed with Protocol C to image spinal column neuronal tissue.

TROUBLESHOOTING

Problem (Step 10): Mice are insufficiently anesthetized with pentobarbital sodium.
Solution: Although the dosage of methohexital provided is precise with respect to body weight, mice, as with humans, are differentially susceptible to barbiturates. Thus, the dosage given here is a minimum and will guarantee no anesthesia related deaths. Some mice, however, will probably need higher dosages. The absence of central reflexes (corneal reflex, pinna reflex) is a sign of sufficient anesthesia. But be aware that flexor reflexes are mediated via the spinal cord and are still intact, albeit with lower amplitude. On the other hand, an overdose of methohexital leads to apnea because the respiratory passages become obstructed with mucus.

Problem (Step 10): Mice die while under anesthesia.
Solution: Monitoring the body temperature carefully can prevent unwanted animal loss due to anesthesia (Fig. 1A,C). Do not use automatic feedback control of the heating pad; adjust it manually instead. Once the animal's body temperature reaches a steady state while it is on the heating pad, a subsequent drop in body temperature indicates deeper anesthesia, whereas a rising temperature indicate insufficient anesthesia. However, although methohexital is not very critical in dosage, spontaneous respiratory arrest always may happen under the action of barbiturates. It is therefore always a good idea to connect the respiration gas mixture from the flow meter with the tracheal tube as soon as the mouse is in prone position and to switch the control of the solenoid valve on without connecting it to the tracheal tube. If a spontaneous respiratory arrest happens, tubing from tracheal tube to the solenoid valve can be connected immediately, and the animal will be stabilized.

Protocol B

Preparation of the Mouse Spinal Column for Repetitive Imaging Using 2pLSM

This protocol (referred to as a repetitive-imaging experiment) is used when the spinal cord is to be imaged repetitively over multiple days, with time for the mouse to recover between imaging sessions. In this case, only a volatile anesthesia is administered, without a tracheotomy and jugular vein catheter (unlike in Protocol A).

MATERIALS

CAUTION: See Appendix 6 for proper handling of materials marked with <!>.
See the end of the chapter for recipes for reagents marked with <R>.

Reagents

ACSF <R>
Bepanthen
Buprenorphine
Ethanol <!>, 70%
Isoflurane <!>
Mice, genetically modified
- TgN(Thy1-eYFP)TYFF-Fki, yellow fluorescent axons (Hirrlinger et al. 2005)
- TgN(GFAP-eCFP)GCFD-Fki, cyan fluorescent astrocytes (Hirrlinger et al. 2005)
- TgH(CX3CR1-eGFP), green fluorescent microglia (Jung et al. 2000)

All experiments require approval by the local regulatory body responsible for animal experimentation.

Nitrous oxide (N_2O) <!>
Octenisept
Oxygen (O_2) <!>
Vaseline

Equipment

Anesthesia equipment: Adjustable flow meters for N_2O and O_2 (0–1 L/min), anesthesia box, anesthesia mask, fittings, scavenge pump for waste anesthesia gas, tubes with three-way selector valve to switch from anesthesia box to anesthesia mask, vaporizer (Surgivet) for isoflurane
Binocular dissection microscope with adjustable magnification and sufficient working distance
Filter tips for syringes to sterilize injected fluids and ACSF
Glass beaker for 70% ethanol to keep instruments sterile
Hypodermic needles, 0.4-mm (OD) for i.p. or s.c. injections
Power supply for the heat pad
Rectal temperature sensor and display
Shaver, electric
Spongostan sponges
Support for the mouse, heated by a flat heat pad (12-V, 1-A) with rail bars to fix two vertebral clamps for the backbone and the volatile anesthesia equipment

Surgical instruments: Curved forceps (e.g., FST 11051-10), dissector scissors (e.g., FST 14081-09), Dumont forceps (#5 Biology, #3c, and #7), Michel suture clips (FST 12040-01) and forceps (FST 12018-12), microcurette (e.g., FST 10081-10), microscissors, needle holder (e.g., FST 12002-12) and needles (FST 12050-03), scalpel handle (e.g., FST 10007-12) and blades (10010-00 or 10011-00), scissors type Wecker (FST 15010-11), silk suture (FST 18020-30), straight scissors (e.g., FST 14088-10)

Syringes for filter tips to apply sterile ACSF to the operation field

Tape (Leukosilk, Beiersdorf)

Thin cotton sewing thread

Vacuum cleaner

EXPERIMENTAL METHOD

Anesthesia

1. Place the mouse into the anesthesia box that is connected to the vaporizer, and rapidly fill the box with 1:1 O_2:N_2O loaded with 5% isoflurane at 1000 mL/min. Within a few minutes the mouse will become atonic with a reduced respiration rate.
2. Remove the mouse from the anesthesia box and place it in front of the anesthesia mask that is fixed to the heated mouse support. Switch the gas flow from the anesthesia box to the mask using the three-way valve.
3. Reduce the gas flow to 150 mL/min N_2O and 200 mL/min O_2.
4. Reduce the percentage of isoflurane to 2%.
5. Set the respiration rate during surgery to ~1.5 breaths/sec by slightly modifying the isoflurane percentage. There should be no withdrawal reflexes when the mouse is pinched.
6. Control the body temperature and keep it at ~36.5°C by adjusting the temperature of the heated support.

Surgery

Surgery for repetitive imaging is similar to that for single-imaging experiments. Muscle and tendon tissue is removed only minimally. All procedures should be performed aseptically.

7. Shave the back and sterilize the preparation field with Octenisept. All surgical instruments should be sterilized with 70% ethanol.
8. Remove muscle and tendon tissue only between two neighboring spinal arcs. It might be necessary to bend the back of the mouse using the clamps, which fix the spinal column (see below).
9. Proceed with Protocol C and image through the prepared cleft between the vertebrae (e.g., L1 and L2). Small laser-induced lesions can be applied through the intact dura within the spinal tissue. The field of view is 500 x 500 μm.

   This procedure does not lead to any disability of the mice after recovery.

Protocol C

In Vivo Two-Photon Imaging of Mouse Spinal Cord

IMAGING SETUP

Any commercial or custom-made two-photon laser-scanning microscope can be used. It should be capable of low-magnification epifluorescence and bright field imaging in a separate channel that is used for overviews (e.g., Zeiss HBO 103 and a 5×, 0.15-NA objective). A CCD camera is required for recording these images.

For 2pLSM, a long-distance 20× objective with high NA is essential (e.g., Zeiss W Plan-Apochromat 20×, 1.0 NA DIC). For excitation, use a femtosecond-pulsed titanium-sapphire laser (e.g., Chameleon Ultra II, Coherent). To optimize the spatial resolution and the detection efficiency, fill 80% of the 17-mm-wide back aperture of the 20× objective. In addition, use two lenses to focus the fluorescence signal into the photomultiplier tubes. For deep tissue penetration, glass dispersion compensation is recommended (either as a separate module or integrated into the system (e.g., Coherent Vision II). Control the laser intensity with a Pockels cell (Conoptics). Scanimage version 3.5 is the image control software that we use (Pologruto et al. 2003).

MATERIALS

See the end of the chapter for recipes for reagents marked with <R>.

Reagents

ACSF <R>
Agar, 2% in ACSF
Mice, prepared surgically as in either Protocol A or B
Texas Red-dextran (70-kDa, 1.25% [w/v], Invitrogen)

Equipment

Anesthesia equipment (use appropriate equipment based on either Protocol A or B)
Cotton swabs
Power supply for the heat pad (12-V, 1-A)
Rectal temperature sensor and display
Scalpel
Scissors
Support for the mouse, heated by a flat heat pad with rail bars to fix two vertebral clamps for the backbone and the volatile anesthesia equipment
Thread or suture material
Two-photon laser scanning microscope (see Imaging Setup)
Vertebral clamp holders (two) for the backbone (Narishige or custom-made)

EXPERIMENTAL METHOD

Mounting of the Vertebral Column

For nearly motionless imaging, the vertebral column has to be mounted to two vertebral clamps (Fig. 1A,B).

1. Fix the lateral joint processes of the vertebrae next to the prepared region rostrally and caudally.
2. Cut holes into the tendon layers above the back muscles to gain access to the lateral joint processes.
3. Fix the clamps at an angle that leaves enough space for the imaging objective (Fig. 1C,D).

   For repetitive imaging it is best to mount the vertebral column before preparing the field of view.

   Imaging lateral aspects of the spinal cord (e.g., the ventral roots or the dorsal root ganglia) requires clamps that can be tilted. After tilting the mouse by ~80°, the lateral spinal cord and parts of the ventral roots can be exposed by moving the dorsal roots dorsally with a fine glass hook or a 0.3 mm metal hook. Care should be taken not to impact the spinal cord surface, the roots or the blood vessels. Dorsal root ganglia are accessible without shifting the dorsal roots.

Preparation for Imaging

4. Create a reservoir to hold the ACSF, to be able to properly use the water-immersion objective.

   For single-imaging experiments:

   i. Form walls for an ACSF reservoir by lifting the skin flaps with threads (Fig. 1A,C).

      If you image more laterally with a tilted mouse, form reservoir walls with skin flaps and small pieces of paper.

   ii. Fill this pool with warm 2% agar in ACSF. Keep the region of interest free of agar by covering it with a piece of cotton soaked in ACSF.

   iii. After the agar has hardened, cut a cone-shaped hole into the agar, making it large enough to remove the cotton and to insert the objective lens.

      The surrounding agar will help to reduce undesirable movement at the region of interest.

   For repetitive-imaging experiments:

   i. Much less muscle and tendon are removed (Protocol B vs. Protocol A), so the remaining tissue will suffice as walls for the ACSF reservoir.

   ii. Only ACSF sterlized with filter tips should be used for repeated-imaging experiments.

5. Fill the reservoir with ACSF.

Two-Photon Microscopy and Image Acquisition

6. Take a bright-field image from the surface of the spinal cord (Fig. 2A).

   Use the pattern of blood vessels to image from compatible regions in different animals or to find the same location during repetitive imaging.

7. Set the laser to 920 nm, a wavelength in which most fluorescent proteins (e.g., eCFP, eGFP, and eYFP) and sulforhodamine 101 (TexasRed) can be excited (Fig. 2B–D).
8. Separate eGFP and eYFP by using a 520-nm dichroic mirror (Semrock) and 500 ± 12-nm and a 542 ± 27-nm emission filters (Semrock) (Fig. 2C).
9. Separate the fluorescence of Texas Red from an eGFP signal with a 560-nm long-pass filter (Carl Zeiss) and an HQ575 emission filter (AHF Analysentechnik) (Fig. 2B).

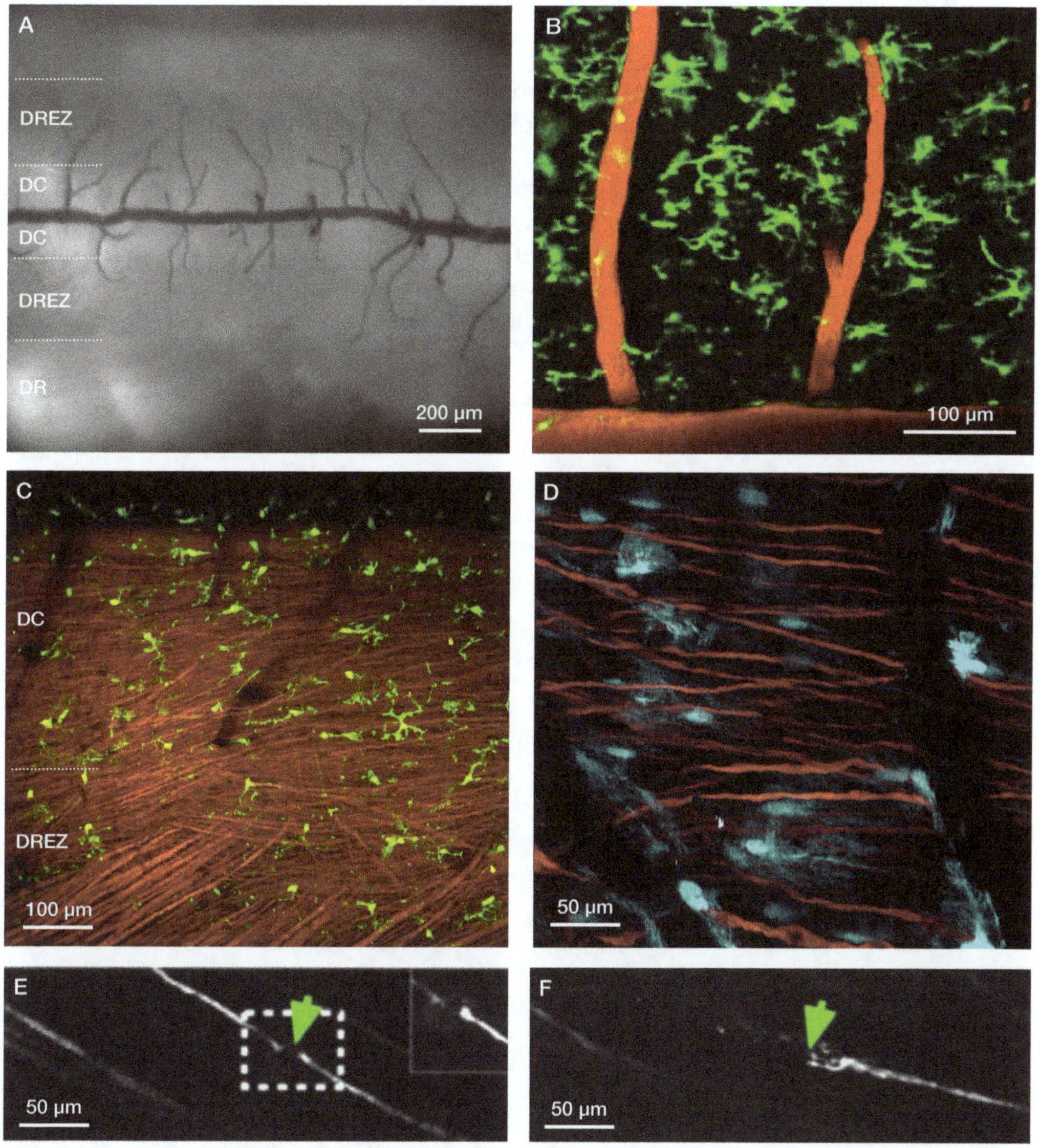

FIGURE 2. Images of the mouse spinal cord. (*A*) Bright-field overview of the spinal cord. Note the pattern of dark blood vessels. (*B*) In vivo 2pLSM of green fluorescent microglia. The spinal capillaries are labeled with Texas Red-dextran by injection into the tail vein. (*C*) Recording from a double-transgenic mouse (TgH(CX3CR1-eGFP)) x (TgN(Thy1-eYFP)) with green fluorescent microglia and yellow fluorescent axons (shown in red). Note the oblique axons from the roots in contrast to the spinal axons running in parallel. (*D*) Recording from a double-transgenic (TgN(GFAP-eCFP)) x (TgN(Thy1-eYFP)) mouse. The thin lamellipodia of the cyan-fluorescent astrocytes are faint but clearly present. (*E,F*) Repetitive imaging of single spinal axons in TgN(Thy1-eGFP) mice. (*E*) A single axon was dissected by a high-power laser pulse. Laser lesion was carried out on a square field of 7 x 7 μm, laser power being set to 50% (1 W), for one image (i.e., ~0.5 sec). (*F*) Three days later, the dissected axon shows some sprouting.

10. Use a 460/80BP emission filter (Semrock) for eCFP signals (Fig. 2D).
11. Record regions of the dorsal columns as uniformly spaced planes (1–2 μm) of fields of view from 170 x 170 to 600 x 600 μm as *z*-stacks of images (256 x 256 to 1024 x 1024 pixels in size). Voxel sizes should range from 0.24 x 0.24 x 1.5 to 0.72 x 0.72 x 2.0 μm for the *x*-, *y*-, and *z*-axis, respectively. The total acquisition time for a stack of 16 to 24 images will range from 1 to 2 min.

    See *Troubleshooting*.

TROUBLESHOOTING

Problem (Step 11): Image resolution is not optimal.

Solution: A very stable field of view is required to obtain high-resolution images. Thus, the holder should be placed as close as possible to the structure of interest (see the position of the clamps in Fig. 1). In most cases, this arrangement is sufficient for high-resolution imaging. It is also critical that the water-immersion objective pool built from the skin flaps of the wound is leakproof. If necessary, add a layer of 1.5% agarose to seal the pool against leaks. The agarose will also stabilize the spinal cord, which could improve image quality. Too much motion could also be caused by a lack of muscle relaxant (Pancuronium). Therefore re-inject Pancuronium in time.

DISCUSSION

Using the above protocols, we have obtained stable recordings continuously over several hours during a single imaging session. We have also been able to follow the progression of white matter scar formation by imaging repeatedly over a period of >6 mo. In addition, cell type–specific expression of eCFP, eGFP, and eYFP could be seen by simultaneously recording with four emission channels and subsequently performing spectral unmixing (Nadrigny et al. 2006; Neher et al. 2009).

A location that has optimal imaging conditions of white matter is found in the dorsal columns (DC) close to the dorsal central vein of the spinal cord. Here, we focused primarily on recordings of the lumbar segments receiving peripheral inputs from lower limbs, whereas others investigated spinal cord injuries at thoracic levels (Davalos et al. 2008; Dray et al. 2009). In our preparation, access to the DC can be achieved without a more invasive laminectomy. After careful removal of the dorsal parts of intervertebral discs and slight bending of the back, axons and glial cells of the white matter can be imaged repeatedly (Fig. 2E,F; Ylera et al. 2009).

More laterally, within the dorsal root entry zone (DREZ) region of the dorsal roots, ascending fibers and lamina I of the dorsal horn can be readily accessed (Fig. 2A,C). Johannssen and Helmchen (2010) achieved recording of stimulation-evoked Ca^{2+} transients from superficial neurons of this region. The pattern of the blood vessels and the consistency of the axons allow imaging of single cells and axons for several days and even months (Fig. 2).

RECIPE

CAUTION: See Appendix 6 for proper handling of materials marked with <!>.
Recipes for reagents marked with <R> are included in this list.

ACSF

NaCl	125 mM
$NaHCO_3$	25 mM
KCl	2.5 mM
KH_2PO_4	1.25 mM
$MgCl_2$	1 mM
$CaCl_2$ <!>	2 mM
Glucose	10 mM

Bubble with carbogen (5 vol% CO_2 <!>/95% O_2 <!>) to generate and stabilize a pH of 7.3.

REFERENCES

Davalos D, Lee JK, Smith WB, Brinkman B, Ellisman MH, Zheng B, Akassoglou K. 2008. Stable in vivo imaging of densely populated glia, axons and blood vessels in the mouse spinal cord using two-photon microscopy. *J Neurosci Meth* **169:** 1–7.

Dray C, Rougon G, Debarbieux F. 2009. Quantitative analysis by in vivo imaging of the dynamics of vascular and axonal networks in injured mouse spinal cord. *Proc Natl Acad Sci* **106:** 9459–9464.

Hirrlinger PG, Scheller A, Braun C, Quintela-Schneider M, Fuss B, Hirrlinger J, Kirchhoff F. 2005. Expression of reef coral fluorescent proteins in the central nervous system of transgenic mice. *Mol Cell Neurosci* **30:** 291–303.

Johannssen HC, Helmchen F. 2010. In vivo Ca^{2+} imaging of dorsal horn neuronal populations in mouse spinal cord. *J Physiol* **588:** 3397–3402.

Jung S, Aliberti J, Graemmel P, Sunshine MJ, Kreutzberg GW, Sher A, Littman DR. 2000. Analysis of fractalkine receptor CX_3CR1 function by targeted deletion and green fluorescent protein reporter gene insertion. *Mol Cell Biol* **20:** 4106–4114.

Nadrigny F, Rivals I, Hirrlinger PG, Koulakoff A, Personnaz L, Vernet M, Allioux M, Chaumeil M, Ropert N, Giaume C, Kirchhoff F, Oheim M. 2006. Detecting fluorescent protein expression and co-localisation on single secretory vesicles with linear spectral unmixing. *Eur Biophys J* **35:** 533–547.

Neher RA, Mitkovski M, Kirchhoff F, Neher E, Theis FJ, Zeug A. 2009. Blind source separation techniques for the decomposition of multiply labeled fluorescence images. *Biophys J* **96:** 3791–3800.

Pologruto TA, Sabatini BL, Svoboda K. 2003. ScanImage: Flexible software for operating laser scanning microscopes. *Biomed Eng Online* **2:** 13.

Ylera B, Erturk A, Hellal F, Nadrigny F, Hurtado A, Tahirovic S, Oudega M, Kirchhoff F, Bradke F. 2009. Chronically CNS-injured adult sensory neurons gain regenerative competence upon a lesion of their peripheral axon. *Curr Biol* **19:** 930–936.

68 Imaging Microglia in Brain Slices and Slice Cultures

Michael E. Dailey, Ukpong Eyo, Leah Fuller, John Hass, and Dana Kurpius
Department of Biology, University of Iowa, Iowa City, Iowa 52242-1324

ABSTRACT

This chapter describes methods for imaging fluorescently labeled parenchymal microglia (MG) in excised neonatal or adult rodent brain tissue slices. Using multichannel confocal or two-photon time-lapse imaging, the approach affords real-time analyses of MG behaviors, including motility, migration, chemotaxis, proliferation, and phagocytosis in live brain tissues. The methods are applicable to acutely prepared tissue slices from developing and adult rodents and to slice cultures derived from neonatal rodents, including transgenic and green fluorescent protein reporter mice. A variety of fluorescent tags can be used to study the structure and physiology of MG in these preparations. Moreover, bath application of reagents (such as ATP [adenosine triphosphate]) can establish spatial and temporal gradients that induce chemokinesis- and chemotaxis-like MG migration in tissue slices. Thus, the approach is useful for dissecting the molecular basis of MG behaviors and testing whether candidate reagents alter MG behavior and function in semi-intact central nervous system tissue preparations.

IMAGING SETUP

For imaging microglia (MG) in thick tissue slices and slice cultures (50–500 µm thick), optical sectioning with a confocal or two-photon imaging system is recommended. We use commercially available laser-scanning confocal or two-photon microscopes equipped with multiple lasers, computer-driven focus, and automated image capture (e.g., Leica SP2 AOBS confocal imaging system on an inverted microscope platform, or a Leica SP5 two-photon system on an upright platform). Stage heating is accomplished with a simple, custom-made heater modified from a commercially available hair dryer. Manufacturer's software is used to control the microscope and for image capture. To view dynamic activities of MG in live tissues, images are captured at intervals of a few seconds to minutes, over total periods of tens of minutes to several hours. A 20×, 0.7-numerical-aperture (NA) dry objective lens is sufficient to observe many dynamic activities of MG for several hours at a time, at submicron pixel resolution, with an incident illumination power of <20 µW at the specimen. To monitor cell movements in three dimensions using a 20× objective lens, optical sections are typically collected at 2–3 µm *z*-step intervals over 30–40 µm of tissue depth at time intervals of 3–4 min. Images can be captured more frequently for higher time resolution. For playback and analysis, image stacks are compiled and combined into extended focus time-series images using the microscope manufacturer's software or ImageJ. Because some lateral *x–y* drift of the field of view is common, it is very useful to perform postprocessing registration of images in a time series using ImageJ plug-ins (e.g., Stackreg). Time stamps and scale bars are also easily added to time series using ImageJ plug-ins. Time series are played back in ImageJ using the stack tools, or are exported from ImageJ for playback in QuickTime.

Protocol

Imaging Microglia in Brain Slices and Slice Cultures

MATERIALS

CAUTION: See Appendix 6 for proper handling of materials marked with <!>.
See the end of the chapter for recipes for reagents marked with <R>.

Reagents

Dissection medium (DM) <R>
Filter culture medium (FCM) <R>
Green fluorescent protein (GFP) and yellow fluorescent protein (YFP) reporter mouse lines
HEPES-buffered culture medium (HCM) <R>
Isolectin B4 (IB4 or GS-IB4) derived from *Griffonia simplicifolia* seeds <R>
Live- or dead-cell fluorescent markers (e.g., Sytox Orange, ToPro3, YoPro1)
Modified balanced salt solution (MBSS) <R>
pHrodo-labeled *Escherichia coli* bioparticles
Vacuum grease

Equipment

Confocal or two-photon microscope
Coverslips
Imaging chamber, Plexiglas (or similar)
Microscope stage heater
Nylon mesh (110-µm window, 47% open area; Small Parts, Inc., CMN-0112-C)
Syringe, 1-mL
Tissue culture dishes, six-well
Tissue culture incubator (5% CO_2 <!>, 36°C, 100% humidity)
Tissue slicer, manual (Stoelting Co. 51425)

EXPERIMENTAL METHOD

Tissue Slice Preparation

This should take 5–15 min.

1. Prepare tissue slices from developing or adult animals for immediate imaging (i), or prepare slices from neonatal animals for culturing and later imaging (ii).

   A more detailed protocol for preparation of rodent hippocampal slices and slice cultures is available elsewhere (Fuller and Dailey 2007).

   i. Prepare acutely isolated hippocampal tissue slices from neonatal (PND5–7) or adult (6–8-wk-old) Sprague Dawley rats or C57BL/6 mice (Harlan) as described in Dailey and Waite (1999). Briefly, rats or mice are decapitated, and brains are removed and placed in cold dissection medium (DM). Hippocampi are excised and sectioned transversely (300–400-µm-thick slices) using a manual tissue slicer.

ii. For organotypic slice cultures, prepare slices from neonatal (PND5–7) animals following the static filter culture method originally described by Stoppini et al. (1991). Slices are placed on porous membrane inserts in six-well plates and carried in FCM in a humidified tissue culture incubator (36°C, 5% CO_2).

Labeling Microglia in Tissue Slices with Fluorescent Lectin

This should take 45–60 min.

2. Label MG in neonatal tissue slices and slice cultures by incubating for 45–60 min at 36°C in HCM (ambient air) or FCM (5% CO_2 environment) containing 5 µg/mL of fluorescently conjugated IB4 lectin. For hippocampal slices from mature rodents, label MG by incubation in MBSS containing 5 µg/mL IB4. Incubate free-floating slices in a 35-mm dish or multiwell plate and then rinse with dye-free medium for 5–10 min. To label organotypic slice cultures, cut out membranes with attached slices using a scalpel and place membrane-attached tissue slices into a dish with IB_4-containing HCM. Make sure the slices are submerged and do not dry out.

   As an alternative to lectin labeling, live slices or slice cultures may be prepared from heterozygous mice expressing GFP in MG (see above). GFP expression levels vary somewhat with MG activation state. However, GFP expression (in MG from Cx3cr1 mice) is strong enough for good visualization of both activated and quiescent MG in living or formaldehyde-fixed tissue slices.

   Lectin labeling (or GFP expression) can be combined with other fluorescent labeling techniques to evaluate phagocytosis of dead or dying cells (nuclei of compromised cells labeled with Sytox Orange or ToPro3) or bacterially derived bioparticles (pHrodo-labeled *E. coli* bioparticles). Alternatively, reporter lines expressing GFP in MG can be crossed to available lines expressing YFP in neurons (e.g., Thy1-YFP; Jackson Labs; see Fig. 2).

Mounting Tissue Slices for Imaging

This should take 5–10 min. Methods of mounting slices for imaging vary with the tissue preparation. It is important to secure the tissue slices to prevent movement during imaging while retaining sufficient gas exchange and access to media. Procedures for mounting acutely prepared slices are given in Steps 3–7. Details for mounting organotypic slice cultures affixed to membranes are given elsewhere (Chapter 33).

3. To mount acutely prepared (free-floating) tissue slices from neonatal rodents, construct a simple chamber from a custom-made Plexiglas slide (75 x 25 x 3-mm) with oblong hole (35 x 13-mm) cut out in the center to form the chamber (see Chapter 33). Briefly, steps for constructing the chamber are as follows.

   i. Secure a 50-mm glass coverslip with vacuum grease to the underside of a Plexiglas slide that contains a hole in the center. The coverslip forms the chamber floor.

   ii. Apply two thin lines of vacuum grease ~5 mm long onto the coverslip floor, ~5 mm apart. Apply the vacuum grease using a custom-made "grease gun," a 1-mL syringe with flame-tapered tip.

4. Transfer one to three tissue slices with medium to the chamber and position the slices carefully between the lines of vacuum grease.

5. Lay a piece of nylon mesh over the slices and the grease lines, then gently press it into the grease with forceps to secure the slices in place under the mesh. Take care not to press too firmly as this will compress the tissue and inhibit MG cell movements.

6. (Optional) Layer a second Plexiglas slide, attached by vacuum grease, to the top of the first slide to increase the chamber volume.

7. Add medium to cover the tissue slices, and then lay a second coverslip over the top of the chamber to prevent evaporation.

Focal drift can be reduced if the chamber is not completely filled with medium (i.e., leave an air pocket in the chamber so that the top coverslip is not tightly sealed).

Confocal and Two-Photon Imaging

This should take 30 min–24 h.

8. To capture a large field of view (~500 x 500-µm) containing several (typically 10 to 30) MG cells, collect images using a 20x, 0.7-NA dry Plan-Apochromat objective lens.
9. Capture images using a confocal or two-photon microscope (see Imaging Setup).

   See Chapter 33 for a discussion on maintaining live tissue slices on the microscope stage.
10. Process and analyze images using Photoshop or ImageJ.

    See *Troubleshooting*.

TROUBLESHOOTING

Problem (Step 10): IB4 labeling of MG appears weak relative to labeled blood vessels.

Solution: The IB4 lectin labels endothelial cells of blood vessels in addition to MG, and lectin staining of MG varies somewhat with species and developmental age. IB4 lectin staining is stronger in tissues from rat than from mouse. In mouse tissues, lectin staining is stronger in tissues from developing mice than from adult mice. However, in fixed tissues, IB4 lectin staining can be significantly improved in mouse tissues if they are fixed with methanol as opposed to formaldehyde. Finally, IB4 lectin becomes internalized in active MG, so visibility of the full MG cell structure is somewhat reduced after 10 h. Mouse lines in which GFP is expressed in the microglia (Jung et al. 2000; Hirasawa et al. 2005) provide superior visibility, but they are less readily available.

Problem (Step 10): MG movements appear sluggish in time-lapse movies.

Solution: MG motility and migration in live tissue slices from developing and mature rodents are sensitive to temperature. Stage heating (preferably above 32°C) is necessary for maximal movement. Although MG in tissues from neonatal rodents are relatively resistant to hypoxia, MG in tissue slices from mature rodents are sensitive to oxygen levels. Adult tissues require oxygenated medium. Finally, MG movements appear to be hindered if tissue slices are mechanically compressed during mounting.

DISCUSSION

The present methods should yield highly visible MG that show spontaneous movements and that rapidly respond to exogenously applied reagents. Cells and tissues should be viable for hours without evidence of photobleaching or rundown induced by poor chamber conditions or phototoxicity.

Using this approach, we have shown that MG undergo rapid changes in cell morphology and motility following activation in live rodent brain tissues (Stence et al. 2001). Within minutes, MG in acutely isolated hippocampal tissue slices begin to activate and subsequently accumulate near injured pyramidal neurons, a process we refer to as "homing" (Kurpius et al. 2006). MG show a chemotaxis-like response to exogenously applied purines, including ATP and ADP (see Fig. 1 and

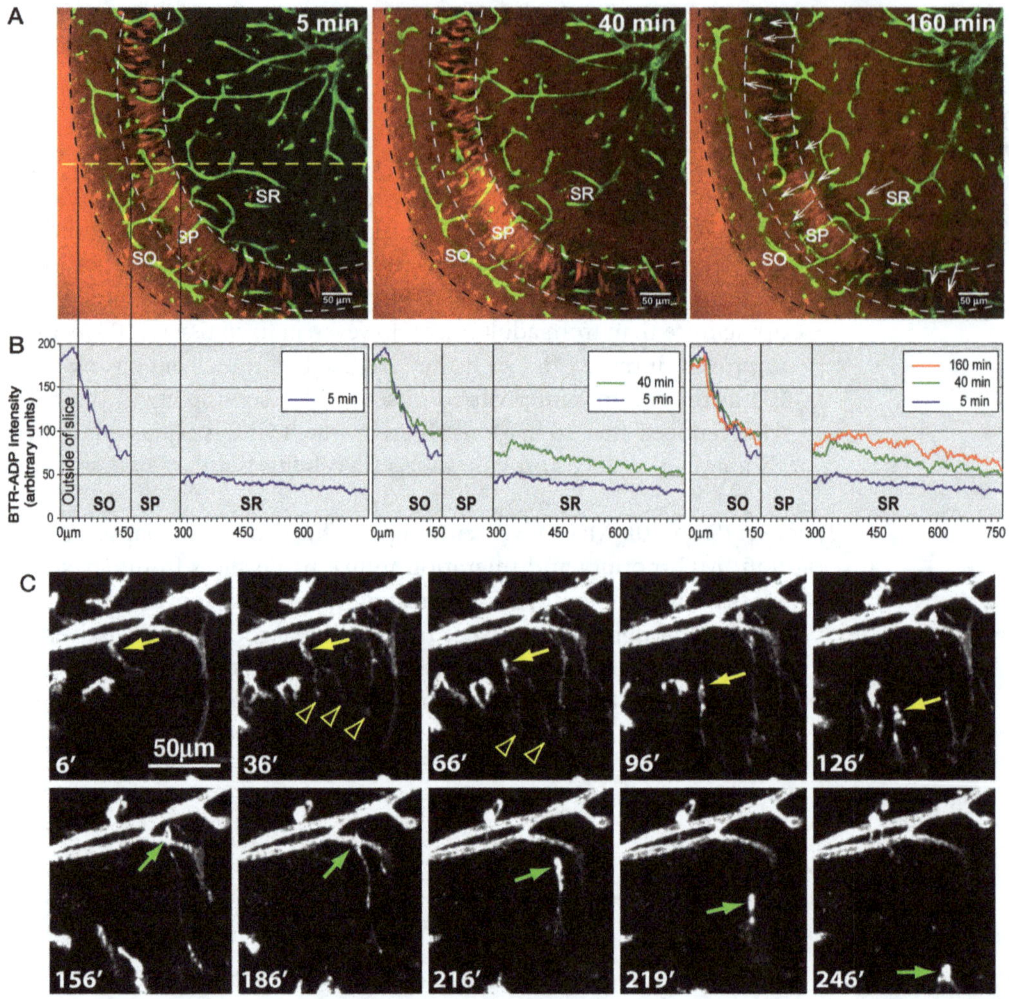

FIGURE 1. Migration of microglia (MG) in rat and mouse tissue slices following application of exogenous ADP (adenosine diphosphate). (*A,B*) Rat hippocampal slice was stained with Alexa Fluor 488-IB4 (green) and a mixture of ADP (1 mM) and fluorescently tagged ADP (BodipyTR-ADP, 10 μM; red) was added to the bathing medium. Selected images from a time-lapse confocal imaging session show diffusion of BodipyTR-ADP into the slice. Scale bars, 50 μm. (*B*) Measurement of BodipyTR-ADP fluorescence across the dashed yellow line for each of the three time points shown in *A*. Note the steep concentration gradient across the *stratum oriens* (SO). Levels of BodipyTR-ADP continue to rise in the interior of the slice (*stratum radiatum,* SR) over the 3-h time period imaged. During this time, many MG cells migrated up the gradient toward the edge of the slice. Arrows indicate the net movement of individual MG over the 3-h period. See Movie 68.1. (*C*) Time-lapse sequence in a different slice showing MG branch extension followed by cell migration in response to bath application of exogenous ADP (1 mM). ADP was applied to the slice immediately after excision, and imaging commenced 1 h later. When ADP is first applied at early stages of activation (as here), MG (arrows) first extend long branches (open arrowheads) directed toward the source of ADP, followed by MG cell migration. Two migrating MG are highlighted (yellow and green arrows, respectively) in this 4-h time sequence. See Movie 68.2.

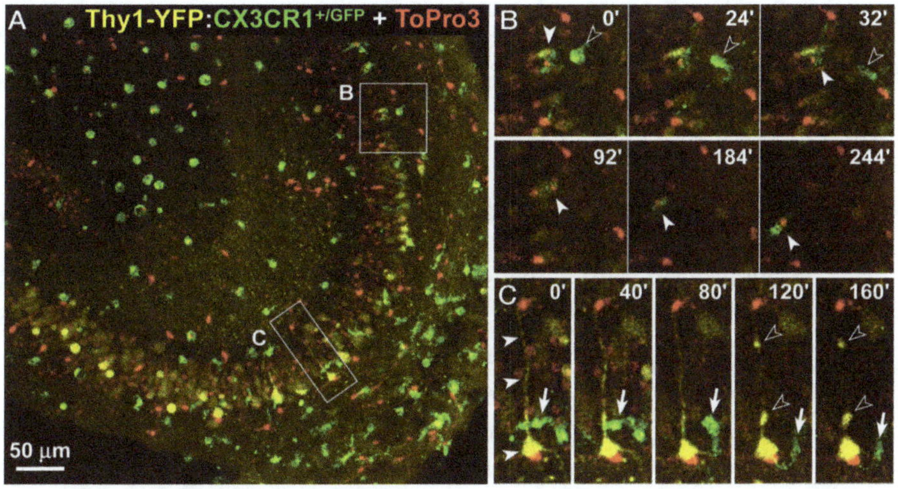

FIGURE 2. Three-channel, time-lapse confocal imaging in a live tissue slice from a double-reporter mouse expressing green fluorescent protein (GFP) in MG (green) and yellow fluorescent protein (YFP) in pyramidal neurons (yellow). The slice was counterstained with ToPro3 to label nuclei of dead cells (red). (*A*) Low-magnification view of hippocampal area CA3. (*B*) Selected higher magnification images from a time-lapse series shows movement of two MG cells (arrowheads), one (solid arrowhead) containing engulfed ToPro3-labeled dead cell nuclei. (*C*) A GFP-expressing MG cell (arrow) passes by a YFP-expressing pyramidal neuron with labeled cell body and apical dendrite (solid arrowheads). Several hours after injury, the neuron shows signs of degeneration, evident as dendrite blebs (open arrowheads at 120′ and 160′). The present methods enable dynamic analysis of cell interactions between MG and other cells in central nervous system tissues.

Kurpius et al. 2007), and work in tissues from knockout mice have shown that this is dependent on P2Y12-type purinoceptors (Haynes et al. 2006). Moreover, use of cultured tissue slices has shown that MG remain active for several days after tissue injury and play important roles in phagocytic uptake and clearance of dead cell debris (Petersen and Dailey 2004). Thus, the combination of acutely excised rodent tissue slices and cultured tissue slices has provided excellent ex vivo models for studying MG activation and motility in a central nervous system (CNS) tissue environment.

When GFP-expressing tissues are unavailable or undesirable, fluorescent conjugates of IB4 lectin provide a simple, one-step method for labeling MG. IB4 penetrates tissues efficiently and selectively stains MG and endothelial cells lining blood vessels by binding to terminal α-D-galactosyl residues of glycoproteins and glycolipids on these cells (Streit and Kreutzberg 1987). This selective surface staining provides good structural definition of MG. The variety of fluorescent IB4 conjugates allows flexibility for double or triple labeling with other fluorescent markers, including markers of live or dead cell nuclei (see Fig. 2 and Dailey and Waite 1999).

Because IB4 is nontoxic to live cells, it is amenable to time-lapse imaging in living tissues. However, some limitations must be considered. First, it is essential to reduce illumination intensity to minimize phototoxic damage or bleaching. Fortunately, IB4-labeled MG are relatively hardy and can be subjected to confocal or two-photon imaging conditions over many hours without ill effects. Second, membrane turnover in activated MG reduces surface labeling over the course of a lengthy time-lapse imaging session (>10 h). This can result in a reduced fluorescent signal and apparent “fading” of the MG over the course of time. Finally, imaging MG in adult tissues is more challenging because, unlike neonatal tissues, the adult tissues require continuous perfusion with oxygenated media, and they are prone to showing signs of reduced movement under hypoxic conditions. Moreover, IB_4 lectin staining of MG is weaker in adult mouse tissues in comparison to younger mouse or rat tissues. Some of these limitations can be overcome by using GFP reporter mouse lines (see Fig. 2). In summary, the present methods afford opportunities to study dynamic MG cell events as they occur in a brain tissue environment on a timescale of minutes, hours, and even days.

RECIPES

CAUTION: See Appendix 6 for proper handling of materials marked with <!>.
Recipes for reagents marked with <R> are included in this list.

Dissection Medium

Supplement Hanks' balanced salt solution (HBSS) with 6 mg/mL dextrose.

Filter Culture Medium (FCM)

50% modified Eagle's medium (MEM), 25% HBSS, 25% horse serum, 2 mM glutamine, 0.044% $NaHCO_3$, and 10 units/mL penicillin-streptomycin <!>.

HEPES-Buffered Culture Medium (HCM)

50% modified Eagle's medium (MEM), 25% HBSS, 25% horse serum, 6 mg/mL D-glucose, and 25 mM HEPES.

Isolectin B4 (IB4) Derived from Griffonia simplicifolia *Seeds*

Several fluorescent IB4 conjugates are commercially available (FITC-conjugated IB4 from Sigma-Aldrich; and Alexa Fluor 488-, 568-, 594-, or 647-conjugated GS-IB4 from Invitrogen). They all yield strong microglial staining and serve as excellent markers of resting or activated parenchymal MG in rodent brain tissues (Streit and Kreutzberg 1987). The Alexa Fluor 647 fluorophore is especially suitable for work in thick tissue slices because longer wavelength excitation and emission light show less tissue scatter. IB4 labeling can be performed by diluting the lectin to 5 µg/mL in any of a variety of solutions, including FCM, HCM, MBSS, or phosphate-buffered saline (PBS).

Modified Balanced Salt Solution (MBSS)

NaCl	125 mM
KCl	2.5 mM
$CaCl_2$<!>	2 mM
$MgSO_4$ <!>	1.3 mM
NaH_2PO_4	1.25 mM
$NaHCO_3$	25 mM
Glucose	10 mM

Oxygenate MBSS by bubbling it with 95% O_2 <!>.

ACKNOWLEDGMENTS

This work was supported by grants to MED from the National Institutes of Health (NS-43468, AA018823), the Roy J. Carver Charitable Trust, and the American Heart Association (0950160G).

REFERENCES

Dailey ME, Waite M. 1999. Confocal imaging of microglial cell dynamics in hippocampal slice cultures. *Methods* **18:** 222–230.

Fuller L, Dailey ME. 2007. Preparation of rodent hippocampal slice cultures. *Cold Spring Harb Protoc* doi: 10.1101/pdb.prot4848.

Haynes SE, Hollopeter G, Yang G, Kurpius D, Dailey ME, Gan W-B, Julius D. 2006. The P2Y12 receptor regulates microglial activation by extracellular nucleotides. *Nat Neurosci* **9:** 1512–1519.

Hirasawa T, Ohsawa K, Imai Y, Ondo Y, Akazawa C, Uchino S, Kohsaka S. 2005. Visualization of microglia in living tissues using Iba1-EGFP transgenic mice. *J Neurosci Res* **81:** 357–362.

Jung S, Aliberti J, Graemmel P, Sunshine MJ, Kreutzberg GW, Sher A, Littman DR. 2000. Analysis of fractalkine receptor CX3CR1 function by targeted deletion and green fluorescent protein reporter gene insertion. *Mol Cell Biol* **20:** 4106–4114.

Kurpius D, Wilson N, Fuller L, Hoffman A, Dailey ME. 2006. Early activation, motility, and homing of neonatal microglia to injured neurons does not require protein synthesis. *Glia* **54:** 58–70.

Kurpius D, Nolley EP, Dailey ME. 2007. Purines induce directed migration and rapid homing of microglia to injured pyramidal neurons in developing hippocampus. *Glia* **55:** 873–884.

Petersen M, Dailey ME. 2004. Diverse microglial motility behaviors during clearance of dead cells in hippocampal slices. *Glia* **46:** 195–206.

Stence N, Waite M, Dailey ME. 2001. Dynamics of microglial activation: A confocal time-lapse analysis in hippocampal slices. *Glia* **33:** 256–266.

Stoppini L, Buchs P-A, Muller D. 1991. A simple method for organotypic cultures of nervous tissue. *J Neurosci Methods* **37:** 173–182.

Streit WJ, Kreutzberg GW. 1987. Lectin binding by resting and reactive microglia. *J Neurocytol* **16:** 249–260.

MOVIE LEGENDS

Movies are freely available online at www.cshprotocols.org/imaging.

MOVIE 68.1. Migration of microglia (MG) cells in a rat hippocampal tissue slice following bath application of exogenous adenosine diphosphate (ADP) (1 mM). The red fluorescence is BodipyTR-ADP.

MOVIE 68.2. Branch extension and subsequent migration of IB4 lectin-labeled microglia in rat hippocampal tissue slices following application of exogenous adenosine diphosphate (ADP) (1 mM).

69 Two-Photon Imaging of Astrocytic and Neuronal Excitation in Cerebellar Cortex of Awake Mobile Mice

Axel Nimmerjahn[1,3] and Mark J. Schnitzer[1,2]

[1]*James H. Clark Center for Biomedical Engineering & Sciences, Stanford University, Stanford, California 94305;*
[2]*Howard Hughes Medical Institute, Stanford University, Stanford, California 94305*

ABSTRACT

Fluorescence imaging has enabled minimally invasive studies of Ca^{2+} dynamics within large networks of astrocytes and neurons in the intact brain. However, most recordings have been made in anesthetized animals. Because anesthesia alters normal brain function and precludes animal behavior, relationships between cellular Ca^{2+} excitation and animal behavior have remained elusive. This protocol describes procedures and techniques for monitoring astrocytic and neuronal Ca^{2+} excitation in the cortex of awake head-restrained mice allowed to behave on an exercise ball. The procedure is performed in several stages, over multiple days, and involves head-plate implantation, animal habituation, optical window preparation, tissue staining, and fluorescence imaging. Head-plate implantation can be completed in ~1.5 h, animal habituation in 3–5 d, and optical window preparation, tissue staining, and fluorescence imaging in ~4–5 h. In the cerebellar cortex, this protocol has helped reveal novel forms of astrocytic Ca^{2+} dynamics and spatiotemporal features of Ca^{2+} spiking within and across microzones of cerebellar Purkinje cells. The general methodology is widely applicable to the study of various cell types, brain regions, and genetically modified mouse lines, including mouse models of disease. We therefore expect new findings to be obtained using this protocol regarding the interrelationships between animal behavior, network excitation, and cellular dynamics in the normal and diseased mammalian brain.

[3]Present address: Salk Institute for Biological Studies, La Jolla, California 92037.

INTRODUCTION

Some of the central issues in neuroscience concern how animal behavior both arises from and influences cellular-level dynamics in the brain. To address such issues, it is crucial to be able to record cellular excitation patterns with high spatiotemporal resolution in the intact brain. One approach that can provide sufficient resolution is intravital fluorescence microscopy (Wilt et al. 2009), which has allowed minimally invasive studies of cells and their dynamics in the intact cortex of multiple species. However, most intravital microscopy studies have been performed in anesthetized animals, which have distinct brain dynamics from awake subjects.

Faced with these limitations, we developed fluorescence microscopy techniques that permit stable optical recordings of astrocytic and neuronal Ca^{2+} excitation in the brains of awake behaving mice (Flusberg et al. 2008; Mukamel et al. 2009; Nimmerjahn et al. 2009). Ca^{2+} excitation is thought to be a primary component of astrocyte signaling dynamics (Nimmerjahn 2009) and a useful signature of neuronal electrical activity (Stosiek et al. 2003). This set of four protocols describes methods for two-photon Ca^{2+} imaging in the cortex of awake, head-restrained mice that are allowed to move on an exercise ball. Using this approach, we have studied Ca^{2+} dynamics in Bergmann glia and Purkinje cells of the cerebellar cortex and examined how these dynamics depend on wakefulness and motor behavior. We have also adapted the methodology to other accessible brain areas and combined it with established electrophysiological, pharmacological, and laser Doppler flowmetry techniques, with the goal of obtaining new information regarding cellular and molecular mechanisms related to locomotion (Mukamel et al. 2009; Nimmerjahn et al. 2009). Given its broad applicability, we expect our approach to provide insights into the roles of various cell types in brain physiology and mammalian behavior.

Protocol A

Surgical Implantation of a Head Plate in Mice in Preparation for Two-Photon Imaging in the Cerebellar Cortex

MATERIALS

CAUTION: See Appendix 6 for proper handling of materials marked with <!>.

Reagents

Adult male mice (note that different mouse strains and sexes show different autonomic stress responses; Jones et al. 1998; Van Bogaert et al. 2006)
Analgesic (e.g., buprenorphine)
Dental acrylic (e.g., Coltène Whaledent, item H00335)
Eye ointment (e.g., Puralube Vet Ointment, PharmaDerm)
Hair removal cream (e.g., Nair, Church & Dwight Co.)
Local anesthetic (1% lidocaine <!>)
Physiologic saline/lactated Ringer's solution (Henry Schein Medical)
Skin disinfectant (e.g., Techni-Care, Care-Tech Laboratories)
Surface disinfectant (e.g., Virkon-S, DuPont)
Tissue adhesive (e.g., Vetbond, 3M)

Equipment

Aseptic instruments/surgical tools (Fine Science Tools)
Aseptic working field
Balance (e.g., Mettler Toledo PG503-S), for weighing animals (particularly important if injectable instead of gaseous anesthetics are used)
Flexible-arm magnetic-base holder (e.g., McMaster-Carr, item 20755A23)
Glass bead sterilizer (e.g., model BS-500, Dent-Eq)
Head holder, custom-made, 6.4 x 76 x 6.4-mm (W x L x H), stainless steel
Head-plate assembly, custom-made; including an anodized 9 x 15 x 1-mm (W x L x H) aluminum head plate and a C-shaped 7.5 mm diameter phosphor bronze metal ring (Fig. 1B)
Laboratory animal anesthesia system (e.g., VetEquip Inc., item 901806), anesthetic gas (e.g., isoflurane <!>, Southmedic, Inc.), carrier gas tank (e.g., medigrade oxygen from Praxair); recommended: waste anesthetic gas system (e.g., VetEquip Inc., item 933101)
Pressurized air container for removal of skull/hair shavings (e.g., Ultrajet Air Duster System, Chemtronics)
Stereotaxic apparatus/dissecting setup (custom-made or from David Kopf Instruments)
Surgical eye spears (e.g., Henry Schein Medical, item 1556455)
Temperature regulation system (e.g., FHC Inc., items 40-90-8, 40-90-5, 40-90-2-07)
Waste liquid suction line, custom-made

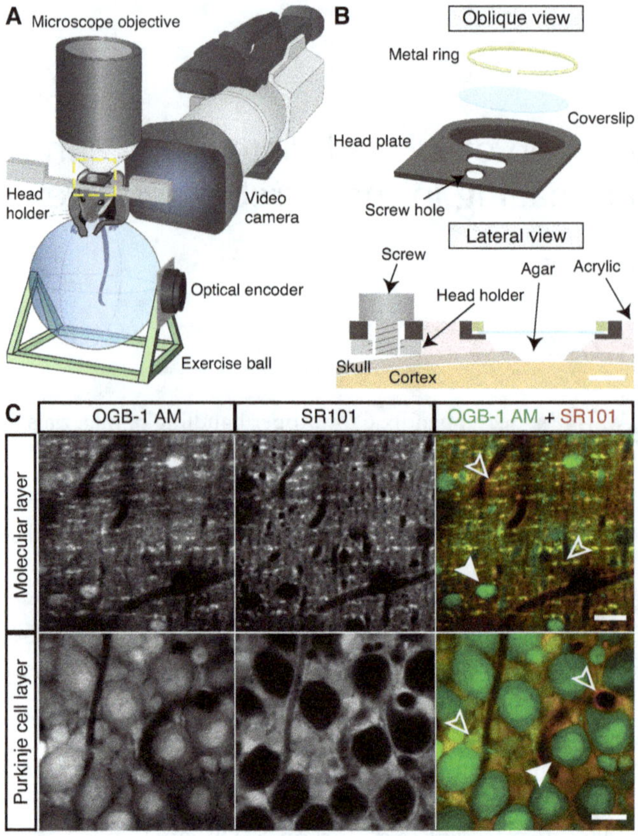

FIGURE 1. Experimental approach for two-photon fluorescence imaging in the cortex of awake behaving mice. (*A*) Schematic view of the experimental setup. A mouse is head restrained atop an exercise ball. Locomotion is tracked by an optical encoder monitoring the rotation of the ball. In addition, mouse behavior is recorded on video. The yellow-dashed box encloses the region shown in *B*. (*B*) Chronic implantation of a lightweight head plate allows repeatable head fixation for habituation and imaging. (*Top*) Schematic view of head-plate assembly. (*Bottom*) Lateral view showing how the skull, head plate, and head holder interface with each other. Scale bar, 2 mm. (*C*) Sulforhodamine-101 (SR101) co-staining allows online discrimination of astrocytes (yellow) and neurons in tissue bulk loaded with Ca^{2+} indicator Oregon Green-488 BAPTA-1 acetoxymethyl (OGB-1 AM). (*Left*) OGB-1 AM staining pattern in the cerebellar molecular layer (*top*) and Purkinje cell layer (*bottom*). (*Center*) SR101 staining in the same mouse. (*Right*) Overlay images. Bergmann glia fibers, endfoot processes, and somata are indicated (open arrow heads). Molecular layer interneurons and Purkinje cells (closed arrow heads in *top* and *bottom* rows, respectively) are stained by OGB-1 AM, but not SR101. Scale bars, 15 µm. (*A,C*, Reprinted, with permission, from Nimmerjahn et al. 2009.)

EXPERIMENTAL METHOD

All procedures described in this and the following protocols were approved by Stanford's Institutional Animal Care and Use Committee (IACUC). Some steps involve the handling of animals and the use of drugs. Exposure to laboratory animal allergens can cause allergies (Figler 2004). Improper use of drugs may constitute a health hazard. Wearing appropriate personal protective equipment (PPE) and carefully reviewing material safety data sheets (MSDS) are therefore recommended. As tissue health is of utmost importance during in vivo imaging experiments, surgery should be performed under conditions that are as sterile as possible. Use proper disinfectants to thoroughly clean surgical tools and equipment. This surgical procedure takes ~1.5 h.

1. Weigh the animal. Fluid and drug amounts to be administered during the experiment are based on body mass. For example, to prevent dehydration of the animal, inject ~0.25–0.5 mL of warm fluids (physiologic saline or lactated Ringer's solution) per 15–25 g body mass per hour.

2. Transfer the animal to an anesthesia induction chamber. Anesthetize the animal with gaseous isoflurane (3% in 1 L/min O_2).
3. Place the animal in a stereotaxic apparatus equipped with a nose cone for maintaining isoflurane anesthesia (typically 1%–2% in 1 L/min O_2 during surgery). Maintain the animal's body temperature (37°C) using a rectal probe and heating pad. Fix the animal's head with ear bars.
4. Protect the animal's eyes with ophthalmic ointment.
5. Remove the fur from the top of the head (from the back of the neck up to the eyes) using hair removal cream.
6. Disinfect the shaved scalp (using, e.g., Techni-Care surgical scrub).
7. Inject a local anesthetic (e.g., 1% lidocaine) subcutaneously into the scalp.
8. Begin surgery only after the animal is nonresponsive, as assessed by the absence of paw withdrawal to a toe pinch. Monitor breathing rate, mucous membrane color, and withdrawal reflex of the animal periodically and adjust the anesthetic level as necessary.
9. Using aseptic instruments, make an incision along the midline of the scalp from the level of the eyes to the ears. Remove the skin on both sides of the incision down to the ridge of the temporal bone where the jaw and neck muscles attach. Clean the surface of the skull (by clearing all periosteum and resolving any bleeding at the skull surface).
10. Carefully retract cervical muscle, as necessary, for imaging or head-plate attachment.
11. Mark the position for the center of the craniotomy (6.5 mm posterior and 0.5 mm lateral of bregma for lobules 4 and 5 of the cerebellar vermis in an adult C57BL/6J mouse; see Paxinos and Franklin 2003).
12. Clean scalp edges and attach to the skull using tissue adhesive.
13. Thoroughly dry the skull surface (using surgical eye spears). This is critical for firm head-plate attachment.
14. Connect the head plate to the head holder with an appropriate screw. Using a flexible-arm magnetic-base holder to hold the head holder by one of its distal ends, center the head plate over the marked position. Lower the head plate until its lower surface is level with the skull surface. Tilt the head plate until it is parallel to the underlying brain surface at the marked position.

    Note that the skull over the cerebellum has a variable thickness, so the brain and the overlying skull surface may not be parallel.
15. Attach the head plate to the skull using dental acrylic while keeping the area over which the craniotomy will be performed uncovered. Dental acrylic should cover all wound margins but should not extend beyond the upper head-plate surface to facilitate head alignment and avoid interference with short working-distance objectives.
16. Before awakening the animal, inject warm fluids (physiologic saline or lactated Ringer's solution; ~0.25–0.5 mL per 15–25 g body mass) and analgesic (e.g., buprenorphine at 0.05–0.1 mg/kg) subcutaneously.
17. Remove ear bars and terminate anesthesia.
18. Transfer the animal to a clean cage without bedding. Provide an external heat source (such as a heating pad) to prevent loss of body heat. Monitor the animal until it has recovered satisfactorily from anesthesia (i.e., normal respiration, able to stand and walk). Assess animal condition (activity level, coat, posture, etc.) periodically until the day of imaging. Monitor tissues adjacent to the surgical incisions for any signs of infection or lesion. Treat as necessary. For more details on recommended postoperative care, see Hayward et al. (2006).
19. Thoroughly disinfect all areas and instruments (using a glass bead sterilizer to autoclave tips of surgical instruments).

Protocol B

Habituation of Mice to Head Restraint, Exercise Ball, and Two-Photon Microscope Setup

MATERIALS

CAUTION: See Appendix 6 for proper handling of materials marked with <!>.

Reagents

Adult male mice (head plate implanted; see Protocol A)

Equipment

Head holder, custom-made, 6.4 x 76 x 6.4-mm (W x L x H), stainless steel
Lab jack (e.g., Newport 281; for height/legroom adjustment)
Laboratory animal anesthesia system (e.g., VetEquip Inc., item 901806), anesthetic gas (e.g., isoflurane <!>, Southmedic, Inc.), carrier gas tank (medigrade oxygen from Praxair); recommended: waste anesthetic gas system (e.g., VetEquip Inc., item 933101)
Spherical treadmill (e.g., mini exercise ball with stand, Habitrail) and a standard copper mesh to encase the exercise ball and help avoid foot slips

EXPERIMENTAL METHOD

Animal habituation is performed in several sessions of increasing duration and one to two sessions per day. Depending on the individual, the mouse should be ready for imaging while awake after 3–5 days of habituation.

1. Prepare setup (typically the microscope setup) for habituation. Mount an exercise ball on top of a lab jack of adjustable height. Attach a head holder to fixed posts on both sides of the exercise ball.
2. Briefly anesthetize the animal (see Protocol A, Step 2).
3. Transfer the animal onto an exercise ball inside the habituation setup. Fix the head plate to the head holder (Fig. 1). Adjust the head and ball position such that the animal's posture during walking/running is as natural as possible under the conditions required for imaging.

   See *Troubleshooting.*
4. Habituate the animal to head-restrained behavior and provide conditions mimicking those during imaging (shutter/scanner noise, etc.). Habituate in several sessions (one to two sessions per day over 3–5 d) of increasing duration (first session ~15 min) until the animal walks/runs effortlessly on the ball and is used to being head fixed for the durations required for imaging.
5. Return the animal to its home cage at the end of each session.
6. Thoroughly disinfect all areas and instruments.

TROUBLESHOOTING

Problem (Step 3): Animal behaves clumsily on the exercise ball and sits still for extended periods of time.

Solution: This is normal during the first habituation session when the animal learns to ambulate on the exercise ball during head fixation. Walking/running behavior will become more effortless and forward directed in subsequent sessions. Locomotor activity, however, varies from mouse to mouse and between mouse strains. Some mice run frequently and for extended periods of time; others run only sporadically. Persistent inept locomotor behavior and lack of grooming may be a sign of improper habituation conditions. For example, ensure that the animal has sufficient legroom and that its head is held in as natural a position as possible.

Problem (Step 3): Head plate detaches during habituation.

Solution: This is an infrequent scenario that normally occurs in <2% of implanted animals and may be caused by improper head-plate attachment due to insufficient drying of the skull surface in Step 13 of Protocol A or animal handling. Careful handling is particularly important during the first habituation session when the animal is more likely to react to unfamiliar stimuli with sudden movements. If head-plate detachment occurs, immediately anesthetize the animal and treat according to local facility guidelines for care and use of animals.

Protocol C

Optical Window Preparation and Tissue Staining in Mice in Preparation for Two-Photon Imaging

MATERIALS

CAUTION: See Appendix 6 for proper handling of materials marked with <!>.
See the end of the chapter for recipes for reagents marked with <R>.

Reagents

Adult male mice (head plate implanted and habituated; see Protocols A and B)
Agarose, Type III-A (Sigma-Aldrich)
Artificial cerebral spinal fluid (ACSF; custom-made or, e.g., from Harvard Apparatus) <R>
Dental acrylic (e.g., Coltène/Whaledent, Inc., item H00335)
Dimethylsulfoxide (DMSO) <!>, anhydrous, 99.9% (Sigma-Aldrich; for preparing Ca^{2+} indicator solution)
Eye ointment (e.g., Puralube Vet Ointment, PharmaDerm)
Oregon Green-488 BAPTA-1 acetoxymethyl (OGB-1 AM; Invitrogen)
Physiologic saline/lactated Ringer's solution (Henry Schein Medical)
Pluronic F-127 (Invitrogen; for preparing Ca^{2+} indicator solution)
Skin disinfectant (e.g., Techni-Care; Care-Tech Laboratories, Inc.)
Sulforhodamine-101 (SR101; Invitrogen)
Surface disinfectant (e.g., Virkon-S; DuPont)

Equipment

Aseptic instruments/surgical tools (Fine Science Tools)
Aseptic working field
Balance (e.g., Mettler Toledo PG503-S; for weighing animals; particularly important if injectable instead of gaseous anesthetics are used)
Capillary glass (e.g., A-M Systems, item 596000)
Centrifugal filter devices (e.g., Millipore Ultrafree-MC, item UFC30GV25; for preparing Ca^{2+} indicator solution)
Coverslips, #1 thickness, 7-mm-diameter, round (from Thermo Fisher Scientific)
Electrode/pipette holder (e.g., Warner Instruments, item 64-1232)
Gelfoam (Henry Schein Medical)
Glass bead sterilizer (e.g., Dent-Eq, model BS-500)
Head holder, custom-made, 6.4 x 76 x 6.4-mm (W x L x H), stainless steel
Heat block or water bath (from VWR)
Laboratory animal anesthesia system (e.g., VetEquip Inc., item 901806), anesthetic gas (e.g., isoflurane, Southmedic, Inc.), carrier gas tank (e.g., medigrade oxygen from Praxair); recommended: waste anesthetic gas system (e.g., VetEquip Inc., item 933101)
Microdrill (e.g., Osada EXL-M40)
Microloader pipettes (e.g., Eppendorf Microloader, item 5242956.003; for backfilling pipettes)

Micropipette puller (e.g., Sutter Instrument, P-97)
Microwave (for agarose gel preparation)
Motorized micromanipulator (e.g., Sutter Instrument, MP-285)
Pressure application system (e.g., Parker, Picospritzer III)
Pressurized air container (e.g., Ultrajet Air Duster System, Chemtronics; for removal of skull shavings)
Stereomicroscope (e.g., Leica MZ12.5)
Stereotaxic apparatus/dissecting setup (custom-made or David Kopf Instruments)
Surgical eye spears (Henry Schein Medical, item 1556455)
Temperature regulation system (e.g., FHC Inc., items 40-90-8, 40-90-5, 40-90-2-07)
Waste liquid suction line, custom-made

EXPERIMENTAL METHOD

Cranial window preparation can be performed in ~30 min. Tissue staining takes ~1.5 h.

1. Prepare for multicell bolus loading (Garaschuk et al. 2006; see Chapter 49).
   i. Mix the synthetic Ca^{2+} indicator of your choice. For the green fluorescent OGB-1 AM, combine dye and a solution of 20% Pluronic F-127 in DMSO to final dye and DMSO concentrations of ~500 µM and 2%–5%, respectively; filter final solution using centrifugal filter devices. Keep solution on ice while not in use.
   ii. Prepare SR101, the red fluorescent astrocyte marker (typically a final concentration of <300 µM in ACSF is sufficient for in vivo imaging). Keep solution on ice while not in use.
      SR101 can either be co-injected (see Step 11) or applied topically if the dura mater will be removed (incubation duration, <5 min; see Step 13; see Chapter 63).
   iii. Prepare patch pipette–like glass electrodes for intracortical dye ejections (tip diameter, 2–3 µm; pressure, 3–4 psi/200–300 mbar; duration, 1–2 min per ejection at a given depth; injection spacing, 100–150 µm. Use a new pipette for each injection).
2. Prepare 1.5%–2% agar in ACSF for optical window creation. Keep the agar on a heat block or in a water bath set to >37°C while not in use.
3. Transfer the animal to an induction chamber. Anesthetize it with gaseous isoflurane (3% in 1 L/min O_2 for induction).
4. Place the animal in a stereotaxic apparatus equipped with a nose cone for maintaining isoflurane anesthesia (typically 1%–2% in 1 L/min O_2 during surgery). Maintain the animal's body temperature (37°C) using a rectal probe and heating pad. Fix the animal's head by attaching the head plate to a head holder via the screw hole (Fig. 1B).
5. Protect the eyes with ophthalmic ointment.
6. Clean and disinfect the head-plate surface and exposed skull areas.
7. Begin surgery only after the animal is nonresponsive, as assessed by the absence of paw withdrawal to a toe pinch. Monitor the breathing rate, mucous membrane color, and withdrawal reflex of the animal periodically and adjust the anesthetic level as necessary.
8. Under a stereomicroscope and using a microdrill (20,000–40,000 rpm) equipped with a new drill bit (0.5-mm-diameter), begin thinning the bone in a ~1.5-mm-diameter region surrounding the marked target point. Drill under ACSF immersion to dissipate heat. Exchange the ACSF frequently to remove bone shavings. Immediately resolve any bleeding from bone vessels (e.g., using wet Gelfoam or drilling toward the bleeding site to remove the bleeding vessel). Continue thinning gently until large blood vessels at the brain surface can be seen through the thinned skull under ACSF immersion. Some brain regions may not require skull thinning to see surface blood vessels.

9. Then, drill carefully around the edges of the desired target area. Ideally, this area should be devoid of large blood vessels and bone sutures. Drilling across these structures can cause substantial bleeding. Continue drilling until fine blood vessels at the brain surface can be seen clearly through the thinned skull. Avoid exerting pressure onto the skull.
10. Using aseptic forceps, lift away the center bone in one piece. Alternatively, use a 30-gauge needle to perforate the edges (the needle should penetrate the thinned skull easily without much force). To minimize damage to the underlying cortex, hold the needle at a shallow angle and point its tip away from the area to be imaged. Immediately resolve any bleeding. Bleeding must be stopped before tissue labeling.
11. Perform multicell bolus loading as described in Garaschuk et al. (2006) and in Chapter 49.
12. (Optional) Unless otherwise required, leave the dura mater intact. Dura opening or removal may be required in experiments in which simultaneous electrophysiological measurements or pharmacological interventions are made (Komai et al. 2006; Nimmerjahn et al. 2009). To remove the dura, follow this procedure.
    i. While viewing through a stereomicroscope at high magnification and with ample ACSF immersion, use the tip of a new hypodermic needle (27- or 30-gauge) to make a small incision in the dura away from surface blood vessels near one edge of the craniotomy. To minimize damage to the underlying tissue, hold the needle at a shallow angle and use upward-directed lateral movement to make the incision.
    ii. Extend the incision along one edge of the craniotomy.
    iii. Using aseptic forceps with very fine tips (Dumont No. 5), grab the dura at the incision and gently pull it away. Because of the increased risk of causing cortical damage, avoid removing the dura near regions where the cerebellar folium bends. Proceed with the next step as quickly as possible to minimize edema-related tissue damage.
13. Create an optical window for imaging. To do so, first resolve any remaining surface bleeding. Hemoglobin has a large two-photon absorption cross section and may interfere with imaging (Clay et al. 2007). Remove excess ACSF from the surface of the skull using surgical eye spears while keeping a thin layer of the liquid on top of the brain. Do not touch the cortex. Apply agar while holding a coverslip of desired size in place. To avoid heat-related damage to the brain, let the agar cool down until it has nearly solidified (close to physiological temperature) before applying. Let the agar harden (~1 min). Remove excess gel around and on top of the coverslip. Fix the coverslip using a small metal ring that fits snugly within the head plate (Fig. 1B). Clean the exposed coverslip surface.
14. (Optional) Seal coverslip edges with dental acrylic.

    Sealing creates an airtight window and shields the underlying tissue but also prevents access to the brain for delivery of micropipettes or drugs. A sealed optical window preparation has been used, for example, for high-speed one-photon imaging in freely moving mice (Flusberg et al. 2008).
15. Inject the animal with warm fluids (physiological saline or lactated Ringer's solution; ~0.25–0.5 mL per 15–25 g body mass) before awakening.
16. Thoroughly disinfect all areas and instruments using a glass bead sterilizer to autoclave tips of surgical instruments.

Protocol D

Two-Photon Imaging of Astrocytic and Neuronal Excitation in Cerebellar Cortex of Awake Mobile Mice

IMAGING SETUP

Equipment

- Two-photon microscope (custom-made) equipped with an ultrafast Ti:sapphire laser, a 10-W all-solid state pump laser (Spectra-Physics), a motorized micromanipulator-controlled (Sutter Instrument Company) *x–y* table (New England Affiliated Technologies, Inc.), a motorized objective focusing nosepiece (e.g., using a Vexta stepping motor) and two fluorescence detection channels (e.g., Chroma; emission filters: HQ525/50M, HQ645/75M; beam splitter: 565DCXR)
- Low-magnification, high-NA water-immersion objective (Olympus XLUMPlanFl 20X, 0.95-NA)
- Electro-optic laser power controller (e.g., Conoptics, Inc., model 302RM and 350-80-LA)
- Galvanometer-based optical scanner (e.g., Cambridge Technology, Inc., item 6215H)
- Beam shutter (e.g., Thorlabs, item SH05)
- Nanofocusing system (e.g., Physik Instrumente, item PIFOC400D; for 3D imaging)
- Variable-gain low-noise current amplifiers (e.g., Femto, DLPCA-200)
- Data acquisition software (e.g., MPScope; see Nguyen et al. 2006)
- Data acquisition hardware (e.g., National Instruments, items NI PCI-6110, NI PCI-6711, NI BNC-2110, NI SHC68-68-EP)
- High-speed digital signal integrator (custom-made as per Nguyen et al. 2006)
- Cold light source (e.g., Schott, KL 1500)
- Optical power meter (e.g., Thorlabs, item FM203)

Imaging Parameters

- Excitation wavelength: ~800 nm (this wavelength is optimal for OGB-1AM [Kuba and Nakayama 1998] and sufficient for SR101 excitation)
- Excitation intensity at sample surface: typically <25 mW (depending on details of the optical setup and surgical preparation and on imaging depth)
- Images/frame rate: 256 x 256-pixel (10.17 Hz), 128 x 128-pixel (20.4 Hz)
- Section/stack rate (3D imaging): five to 20 sections, 5–10-µm axial spacing, 1–4-Hz stack acquisition rate
- Recording duration: typically 5–7 min for a given field of view

Image Processing

- ImageJ plug-ins for lateral image alignment, temporal downsampling, $\Delta F(t)/F$ calculation
- Matlab (custom-written software, e.g., for bleed-through correction, basic image analysis, and automated extraction of cellular Ca^{2+} signals based on an independent components analysis [Mukamel et al. 2009; Nimmerjahn et al. 2009])

MATERIALS

See the end of the chapter for recipes for reagents marked with <R>.

Reagents

ACSF <R>
Adult male mice (head plate implanted, habituated, prepared with optical window, and stained with fluorescent indicator(s); see Protocols A–C)

Equipment

Black cloth
Brightness adjustable LED flashlight (from Orion)
Cold light source (e.g., Cuda Products Corp., model I-150)
Head holder, custom-made, 6.4 x 76 x 6.4-mm (W x L x H), stainless steel
Lab jack (e.g., Newport 281; for height/legroom adjustment)
Light-emitting diode (LED) (Future Electronics; triggered by beam shutter to allow synchronization of video records and imaging data)
Optical encoder kit (e.g., US Digital, item E7PD-720-118)
Optical power meter (e.g., Thorlabs, item FM203)
Spherical treadmill (e.g., mini exercise ball with stand, Habitrail) and a standard copper mesh to encase the exercise ball and help avoid foot slips
Standard digital camcorder (e.g., Sony DCR-VX2000)
Two-photon microscope (see Imaging Setup)

EXPERIMENTAL METHOD

Imaging in awake mice typically takes ~2–3 h, but will take longer when combined with other procedures, such as electrophysiology or pharmacological intervention.

1. Prepare imaging setup for head-restraint recordings. Fix the exercise ball, with the optical encoder mounted for readouts of angular displacement, on top of the lab jack underneath the microscope's objective. Attach the head holder to fixed posts on both sides of the exercise ball (Fig. 1A). Use optically opaque material, such as a black cloth, to create an area around the exercise ball shielding the microscope's detectors from ambient light. Allow access for the objective from the top and a video camera from the front.
2. Using an optical power meter measure the average laser power exiting the microscope objective for different beam deflector settings.

   Excessive illumination of tissue can cause pathological Ca^{2+} dynamics (Wang et al. 2006). Photodamage threshold depends on various parameters, including imaging depth, pulse duration, and exposure time (Koester et al. 1999; Hopt and Neher 2001) and therefore needs to be determined experimentally, such as by exposing the stained tissue to various illumination conditions and then quantifying any changes in the frequency of Ca^{2+} transients over the duration of light exposure. Typically, this is performed in a separate set of experiments. Tissue well-stained with OGB-1 AM can be imaged with <25 mW average laser power at the sample surface at 800 nm excitation wavelength.
3. Carefully transfer the anesthetized animal onto the exercise ball. Fix the animal's head via a head plate and head holder, avoiding any unnecessary mechanical forces on the animal's head or neck region.
4. Adjust the tilt angle of the head holder as necessary. Adjust the distance between the apex of the ball and the head holder as the animal recovers from anesthesia and requires more legroom for walking and running.

5. Using the microscope's ocular (or camera) and cold light illumination, locate the brain surface by translating the microscope stage and/or objective. Switch to fluorescence mode.
6. To record mouse behavior on the ball, position the digital camcorder and beam shutter-triggered LED (for synchronization of video and imaging data). Use a brightness adjustable LED flashlight to illuminate the animal's torso and exercise ball but not the microscope's objective or its water immersion. Alternatively, an infrared (IR) camera and lights can be used to record animal behavior.
7. Let the animal recover from anesthesia (typically ~1 h). Adequate recovery time depends on depth and duration of anesthesia before awakening.
8. Apply ACSF/water immersion and start imaging using minimal excitation intensity.

   See *Troubleshooting.*
9. While staying below the photodamage threshold, acquire imaging and analog input data such as encoder signals as necessary (Figs. 1C and 2). For each measurement, record microscope settings as well as other parameters of interest.
10. At the end of the experiment detach the head plate from the head holder and return the animal to its cage. Proceed with further steps, such as transcranial perfusion, as necessary.
11. Clean the objective and record average laser power exiting the microscope objective for all beam deflector settings used during the experiment.
12. Thoroughly disinfect all areas and instruments. Remove/store blackout material, ball, head holder, LED, and camera.

TROUBLESHOOTING

Problem (Step 8): Optical recordings are unusually noisy, showing little or no Ca^{2+} transients.

Solution: Make sure that the mouse's head and neck region are not illuminated by the LED flashlight (Step 6). Check whether appropriate emission filters are being used. The signal-to-noise ratio of Ca^{2+} transients may be improved by increasing excitation intensity. Be careful, however, not to exceed the photodamage threshold (see Discussion). Lack of Ca^{2+} transients may be due to tissue damage. Tissue health can be assessed via the fluorescence stain. Processes of damaged/dead cells show blebbing and fluorescence intensities larger than surrounding healthy tissue as damaged/dead neurons tend to accumulate dye present in the extracellular space. If fluorescent dye preparation and tissue staining is suspected to be a problem, see Garaschuk et al. (2006) for troubleshooting (e.g., use and regularly exchange anhydrous DMSO for dye preparation).

Problem (Step 8): Large motion artifacts occur during imaging, particularly during animal locomotion.

Solution: Motion artifacts during imaging can be caused by movements of the skull relative to the microscope objective or the brain relative to the skull. The former can be reduced by anchoring the head plate to posts on both sides of the animal's head via a rigid head holder; the latter can be controlled to some extent by adjustments to the surgical preparation with a small craniotomy, an intact dura mater, and more viscous agarose (~2%), all of which help reduce motion artifacts. Lateral motion artifacts remaining in recorded images can be largely corrected offline by image registration algorithms; correction of axial motion artifacts remains challenging. See Dombeck et al. (2007), Greenberg and Kerr (2009), and Nimmerjahn et al. (2009) for example algorithms.

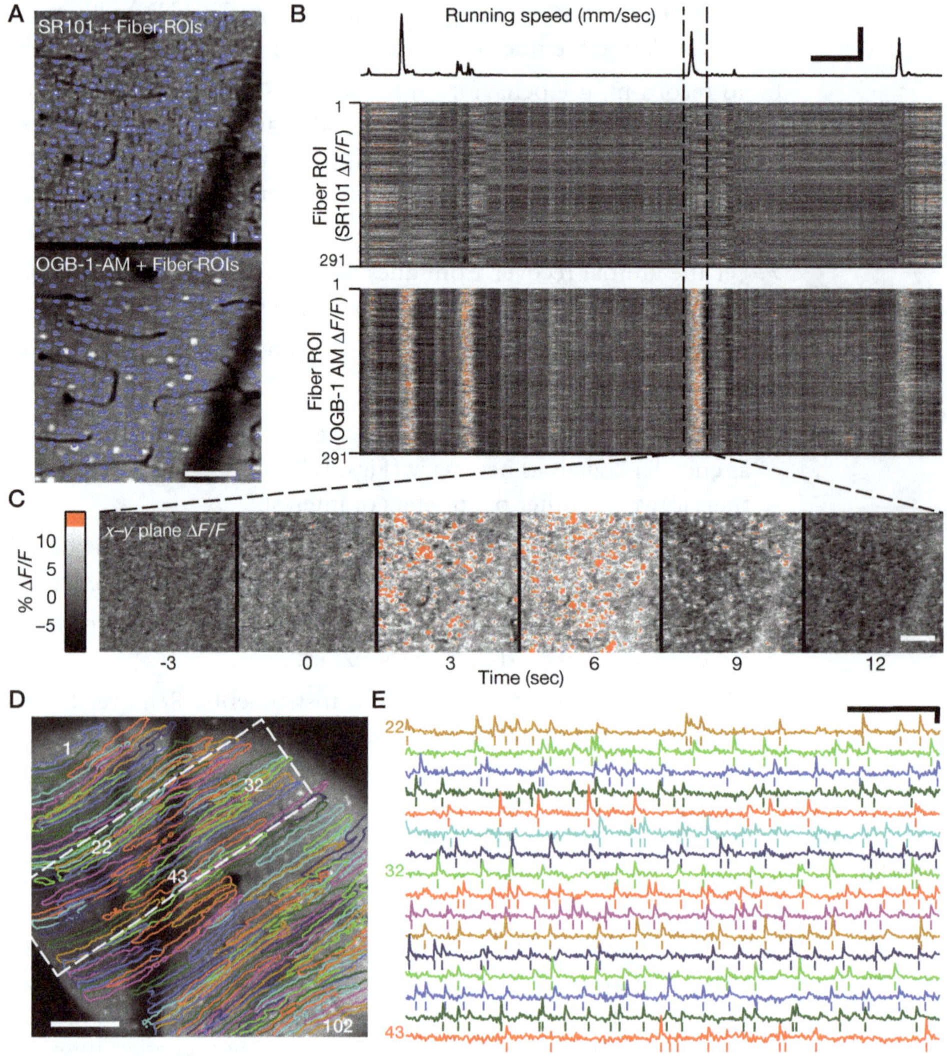

FIGURE 2. Example data showing Ca^{2+} excitation in Bergmann glia and Purkinje cell networks of alert mice. (*A–C*) Motor behavior–evoked Ca^{2+} excitation across Bergmann glial networks. (*D,E*) Ca^{2+} spiking activity in Purkinje cell dendrites of an alert, restrained mouse. (*A*) (*Top*) Sulforhodamine-101 (SR101) image showing cellular structures in a superficial region of the cerebellar molecular layer. Regions of interest (ROIs) outlining Bergmann glial palisade fibers are indicated (blue). Bottom, the same palisade ROIs are used to analyze glial Ca^{2+} dynamics in Oregon Green-488 BAPTA-1 acetoxymethyl (OGB-1 AM) images. Scale bar, 50 µm. (*B*) (*Top*) Encoder trace of the exercise ball's rotational speed, showing multiple running bouts. Scale bars, 50 sec (horizontal) and 20 mm/sec (vertical). (*Center*) SR101 $\Delta F(t)/F$ traces from 291 palisade fiber ROIs during the same time interval. (*Bottom*) The same ROIs in OGB-1 AM images show concerted $\Delta F(t)/F$ increases during running. Signals of similar magnitude and time course are absent in $\Delta F(t)/F$ traces of the nonfunctional indicator SR101. $\Delta F(t)/F$ scale bar in *C* also applies to *B*. ROIs were sorted by *x* coordinate. (*C*) Image frames of OGB-1 AM $\Delta F(t)/F$ increases relative to movement onset during the time interval indicated in *B* (dashed lines) showing the spatial extent of behaviorally evoked Ca^{2+} excitation. Scale bar, 50 µm. (*D*) OGB-1 AM image showing an optical cross section through cellular structures in the cerebellar molecular layer. Spatial contours of 102 active Purkinje cell dendrites identified by independent component analysis are indicated. Scale bar, 100 µm. (*E*) Example time courses of 16 of the independent component signals indicated in *D*. Tick marks indicate the times of estimated Ca^{2+} spikes. Scale bars, 10 sec (horizontal); 5 S.D. (vertical). (Reprinted and modified slightly, with permission, from Nimmerjahn et al. 2009 and Mukamel et al. 2009.)

DISCUSSION

Using the procedures detailed in these four protocols, stable optical recordings of neuronal and astrocytic Ca^{2+} excitation can be obtained in the cortex of awake behaving mice. The first requirement for obtaining such recordings is the ability to make clean cranial window preparations, with minimal bleeding and/or surface damage. This requires considerable practice. The second requirement is successful tissue staining. Once Ca^{2+} transients are reliably obtained in vivo, users should perform extensive control experiments, most of which can be performed in anesthetized animals. For example, to ensure that Ca^{2+} transients recorded by two-photon imaging represent physiological signals and not artifacts evoked by laser illumination (Wang et al. 2006), the photodamage threshold should be assessed. This threshold can be estimated in anesthetized animals as the point at which the intensity or illumination duration of the laser beam causes alterations in the amplitude or frequency of Ca^{2+} signals. Combined Ca^{2+} imaging and electrophysiology is recommended to determine which forms of neuronal excitation can be detected by optical imaging. For example, cortical cerebellar tissue stained with OGB-1 AM allows reliable detection of climbing fiber but not parallel fiber-evoked activity in Purkinje cell dendrites, at least without applied electrical stimulation (Sullivan et al. 2005; Mukamel et al. 2009). Finally, the third requirement is mechanical stability of the preparation. Reducing motion artifacts by, for example, firmly fixing the implanted head plate to posts on both sides of the animal's head is crucial in obtaining recordings of high quality. Control experiments probing whether mechanical forces acting on the animal's head during locomotion influence normal cortical dynamics are recommended (Nimmerjahn et al. 2009).

One limitation of the methodology is that it requires restraining the animal's head in a position suitable for imaging under water immersion. Restraining the mouse's head in a position uncomfortable to the animal will likely affect normal behavior. Given this limitation, brain regions close to the skull's midline and unfettered by overlying skeletal muscle (such as the primary visual, somatosensory, and motor cortices) appear better suited for imaging in awake head-restrained mice than other areas. Another shortcoming of the approach is its limitation to short-term experiments. However, this is a constraint arising from the use of the multicell bolus loading technique, in which dye extrusion from cells occurs over the course of several hours. Viral or genetic expression of genetically encoded Ca^{2+} indicators should help improve the suitability of this protocol to long-term studies involving repeated imaging sessions (Tian et al. 2009). Established surgical procedures for long-term imaging studies already exist. Despite these limitations our approach is applicable to many cortical areas and cell types therein (Mukamel et al. 2009; Nimmerjahn et al. 2009). Deeper lying regions, such as the hippocampus, may also be studied but require different surgical and optical procedures. For example, study of cells in hippocampal regions may require the use of microendoscopes (Jung et al. 2004; Levene et al. 2004) and might be enhanced by use of a virtual reality environment (Wilt et al. 2009). Use of an air-supported free-floating ball may permit bi-directional movement of the animal (Dombeck et al. 2007). However, the pressurized air to support the ball and animal may create a background noise that is also subtly modulated in response to the animal's own movements. A rodent exercise ball with stand and optical encoder avoids any such potential effects and is inexpensive and straightforward to implement and disinfect.

Alternative methods for the study of cellular excitation patterns in behaving mice include the use of miniature one- and two-photon head-mounted fluorescence microscopes (reviewed in Wilt et al. 2009). These miniature microscopes provide broader behavioral freedom to the animal, and some of these microscopes permit fast frame rates, up to ~100 Hz (Flusberg et al. 2008). In comparison to conventional microscopy, miniature microscopes are presently inferior in spatial resolution, depth penetration, volumetric imaging capability, and imaging duration, but over time these limitations are likely to relax. Miniature microscope approaches have already enabled the study of neuronal excitation and blood flow changes in freely moving rodents (Helmchen et al. 2001; Flusberg et al. 2008). For the study of some behaviors, such as basic forms of motor conditioning, the behavioral constraints provided by the head-fixed preparation may be preferred over having the rodent behave freely.

No matter which imaging approach is used to record network excitation in behaving animals, analysis of large-scale imaging data requires the use of efficient algorithms for automated extraction of cell locations and cell activity traces. Similarly, algorithms for automated analysis of behavioral records are needed to correlate animal behavior to cellular excitation. Semiautomated and automated techniques for extracting Purkinje cell dynamics have been documented (Ozden et al. 2008; Mukamel et al. 2009). Detailed information about computational analysis of Bergmann glial Ca^{2+} excitation and encoder signals has also been described (Nimmerjahn et al. 2009).

In summary, two-photon Ca^{2+} imaging in the brain of awake head-restrained mice moving on an exercise ball is a potent tool in the study of relationships between animal behavior, the dynamical excitation of networks of neurons and astrocytes, and the cellular and molecular mechanisms underlying these phenomena.

RECIPE

CAUTION: See Appendix 6 for proper handling of materials marked with <!>.
Recipes for reagents marked with <R> are included in this list.

Artificial Cerebral Spinal Fluid (ACSF)

NaCl	125 mM
KCl	5 mM
D-glucose	10 mM
HEPES	10 mM
$MgSO_4$ <!>	2 mM
$CaCl_2$ <!>	2 mM

Adjust pH to ~7.3 with NaOH <!>. Osmolarity should be ~300.

ACKNOWLEDGMENTS

This work was supported by postdoctoral fellowships to AN from the Alexander von Humboldt-Foundation and the International Human Frontier Science Program Organization, and grants to MJS from the National Institutes of Health's Nanomedicine Development Center (NDC) for Optical Control of Biological Function, National Institute on Deafness and Other Communication Disorders (NIDCD), the Stanford Bio-X program, and the Packard Foundation. We thank our collaborator Eran Mukamel for data analysis that contributed to Figure 2 (Mukamel et al. 2009; Nimmerjahn et al. 2009).

REFERENCES

Clay GO, Schaffer CB, Kleinfeld D. 2007. Large two-photon absorptivity of hemoglobin in the infrared range of 780–880 nm. *J Chem Phys* **126:** 025102.

Dombeck DA, Khabbaz AN, Collman F, Adelman TL, Tank DW. 2007. Imaging large-scale neural activity with cellular resolution in awake, mobile mice. *Neuron* **56:** 43–57.

Figler N. 2004. Laboratory animal allergies: Overview of causation and prevention. *Lab Anim* **33:** 25–27.

Flusberg BA, Nimmerjahn A, Cocker ED, Mukamel EA, Barretto RPJ, Ko TH, Burns LD, Jung JC, Schnitzer MJ. 2008. High-speed, miniaturized fluorescence microscopy in freely moving mice. *Nat Methods* **5:** 935–938.

Garaschuk O, Milos R-I, Konnerth A. 2006. Targeted bulk-loading of fluorescent indicators for two-photon brain imaging in vivo. *Nat Protoc* **1:** 380–386.

Greenberg DS, Kerr JN. 2009. Automated correction of fast motion artifacts for two-photon imaging of awake animals. *J Neurosci Methods* **176:** 1–15.

Hayward AM, Lemke LB, Bridgeford EC, Theve EJ, Jackson CN, Cunliffe-Beamer TL, Marini RP. 2006. Biomethodology and surgical techniques. In *The mouse in biomedical research* (ed. Fox JG, et al.), pp. 437–488. Academic, New York.

Helmchen F, Fee MS, Tank DW, Denk W. 2001. A miniature head-mounted two-photon microscope: High-resolution brain imaging in freely moving animals. *Neuron* **31:** 903–912.

Hopt A, Neher E. 2001. Highly nonlinear photodamage in two-photon fluorescence microscopy. *Biophys J* **80:** 2029–2036.

Jones BC, Sarrieau A, Reed CL, Azar MR, Mormede P. 1998. Contribution of sex and genetics to neuroendocrine adaptation to stress in mice. *Psychoneuroendocrinology* **23:** 505–517.

Jung JC, Mehta AD, Aksay E, Stepnoski R, Schnitzer MJ. 2004. In vivo mammalian brain Imaging using one- and two-photon fluorescence microendoscopy. *J Neurophysiol* **92:** 3121–3133.

Koester HJ, Baur D, Uhl R, Hell SW. 1999. Ca^{2+} fluorescence imaging with pico- and femtosecond two-photon excitation: Signal and photodamage. *Biophys J* **77:** 2226–2236.

Komai S, Denk W, Oste, P, Brecht M, Margrie TW. 2006. Two-photon targeted patching (TPTP) in vivo. *Nat Protoc* **1:** 648–653.

Kuba K, Nakayama S. 1998. Two-photon laser-scanning microscopy: Tests of objective lenses and Ca^{2+} probes. *Neurosci Res* **32:** 281–294.

Levene MJ, Dombeck DA, Kasischke KA, Molloy RP, Webb WW. 2004. In vivo multiphoton microscopy of deep brain tissue. *J Neurophysiol* **91:** 1908–1912.

Mukamel EA, Nimmerjahn A, Schnitzer MJ. 2009. Automated analysis of cellular signals from large-scale calcium imaging data. *Neuron* **63:** 747–760.

Nguyen QT, Tsai PS, Kleinfeld D. 2006. MPScope: A versatile software suite for multiphoton microscopy. *J Neurosci Methods* **156:** 351–359.

Nimmerjahn A. 2009. Astrocytes going live: Advances and challenges. *J Physiol* **587:** 1639–1647.

Nimmerjahn A, Mukamel EA, Schnitzer MJ. 2009. Motor behavior activates Bergmann glial networks. *Neuron* **62:** 400–412.

Ozden I, Lee HM, Sullivan MR, Wang SS. 2008. Identification and clustering of event patterns from in vivo multiphoton optical recordings of neuronal ensembles. *J Neurophysiol* **100:** 495–503.

Paxinos G, Franklin KBJ. 2003. *The mouse brain in stereotaxic coordinates*. Academic, New York.

Stosiek C, Garaschuk O, Holthoff K, Konnerth A. 2003. In vivo two-photon calcium imaging of neuronal networks. *Proc Natl Acad Sci* **100:** 7319–7324.

Sullivan MR, Nimmerjahn A, Sarkisov DV, Helmchen F, Wang SS. 2005. In vivo calcium imaging of circuit activity in cerebellar cortex. *J Neurophysiol* **94:** 1636–1644.

Tian L, Hires SA, Mao T, Huber D, Chiappe ME, Chalasani SH, Petreanu L, Akerboom J, McKinney SA, Schreiter ER, et al. 2009. Imaging neural activity in worms, flies and mice with improved GCaMP calcium indicators. *Nat Methods* **6:** 875–881.

Van Bogaert MJV, Groenink L, Oosting RS, Westphal KGC, van der Gugten J, Olivier B. 2006. Mouse strain differences in autonomic responses to stress. *Genes Brain Behav* **5:** 139–149.

Wang XH, Lou NH, Xu QW, Tian GF, Peng WG, Han XN, Kang J, Takano T, Nedergaard M. 2006. Astrocytic Ca^{2+} signaling evoked by sensory stimulation in vivo. *Nat Neurosci* **9:** 816–823.

Wilt BA, Burns LD, Ho ETW, Ghosh KK, Mukamel EA, Schnitzer MJ. 2009. Advances in light microscopy for neuroscience. *Annu Rev Neurosci* **32:** 435–506.

70 Automated Imaging and Analysis of Behavior in *Caenorhabditis elegans*

Eviatar Yemini,[1] Rex A. Kerr,[2] and William R. Schafer[1]

[1]*Medical Research Council Laboratory of Molecular Biology, Cambridge CB2 0QH, United Kingdom;*
[2]*Howard Hughes Medical Institute, Janelia Farm Research Campus, Ashburn, Virginia 20147*

ABSTRACT

Neurobiological research in genetically tractable organisms relies heavily on robust assays for behavioral phenotypes. The simple body plan of the nematode *Caenorhabditis elegans* makes it particularly amenable to the use of automated microscopy and image analysis to describe behavioral patterns quantitatively. This chapter outlines general features and worm-handling protocols for both single-worm and multiworm tracking systems. In addition, we present protocols for building and using a multiworm tracker for high-throughput phenotypic analysis.

INTRODUCTION

The analysis of behavior is an important aspect of neuroscience, in particular for simple genetically tractable organisms such as *Caenorhabditis elegans*. Behavioral assays are the principal tool for identifying and characterizing the effects of individual genes and neurons on nervous system function. In cases in which the relevant circuitry is well defined, behavior can often serve as an effective surrogate for the activity of particular neurons or for the output of specific molecular pathways.

Although most early work on *C. elegans* behavior was based on real-time observation, in recent years there has been increasing use of automated microscopy and video imaging analysis systems (known as "worm trackers") for behavioral assays. These approaches offer a number of advantages over nonautomated methods. For example, many behavioral phentoypes have initially been described only in qualitative, imprecise terms (e.g., "weak kinker," "loopy," "coiler"). More quantitative assays lend precision to these descriptions and make it feasible to break behaviors down into components that can be related to particular neural functions. Another advantage is increased sensitivity, especially with regard to behaviors that occur over long timescales. A further advantage is the potential for standardization between laboratories; if the same algorithms are used to score a behavior, data collected at different locations can in principle be compared as long as recording conditions are reproducibly controlled. Finally, an automated system (in particular a multiworm tracking microscope) offers the possibility of conducting high-throughput screens for mutants with complex behavioral phenotypes.

Worm trackers fall into two general types. Single-worm trackers follow individual animals at relatively high magnification, keeping them in the field of view by moving the camera or a motorized stage. Such systems have been used to study aspects of locomotion such as turning and speed (Pierce-Shimomura et al. 1999) as well as body posture and morphology (Karbowski et al. 2008). Multiworm trackers track multiple individuals in populations, by necessity at lower magnification, and usually focus on locomotion features. Generally speaking, single-worm trackers allow one to measure a greater number of parameters, whereas multiworm trackers offer much higher efficiency and throughput.

Principles of Single-Worm Tracking

The benefit of single-worm over multiworm tracking is that it delivers videos with much higher spatial resolution—resolutions that enable neuronal tracking as well. Moreover, single-worm tracking opens up the door for stimulus targeting to specific locations of a worm upon specific behavioral conditions (although these types of results have yet to be published). As such, image processing and computer vision to discover the worm must run as fast as possible. Among the simplest methods for worm localization, and the one seen most frequently in tracking software, is the following.

1. Incoming video is reduced to grayscale as worms present no additional information within color channels.
2. Video frames are dealt with individually and, using either a fixed or adaptive pixel value threshold, they are converted to a binary image of foreground and background pixels.
3. A connected-components algorithm (also known as particle or blob analysis) is run to label each group of connected foreground pixels with a separate numerical descriptor. The term "connected" usually refers to either 4-connected pixels that touch only adjacently or 8-connected pixels that are permitted to touch diagonally as well.
4. Once the connected foreground pixels have been labeled, the "worm" is chosen to be either the largest group of connected pixels or a group that has an appropriate size and/or shape.
5. This "worm" is then tracked by moving the stage to keep it in view.

In general, refinements on the discovery algorithm described in point 3 include subtracting frames to determine motion and moving the stage to keep this motion in view. Motion tracking is necessary when illumination or other occlusions prevent accurate worm discovery using the more traditional approach of connected-components analysis. Additional refinements involve image processing to determine the worm's contour, skeleton, or other descriptors that allow one to determine worm body part locations and/or behavior. These refinements are currently ad hoc as they can quickly approach the limit of what modern computers can perform in real time.

Concurrent behavioral and neural imaging has been achieved and this field is seeing rapid development (Chronis et al. 2007; Clark et al. 2007; Zhang et al. 2008; Ben Arous et al. 2010). *C. elegans*' hard cuticle and internal hydrostatic pressure have precluded the use of nongenetic neuronal-activity reporters. Presently, two types of genetically encoded calcium reporters form the de facto standard for neuronal imaging in *C. elegans*. Cameleon-based reporters use a Förster resonance energy transfer (FRET) between two nearby fluorescent proteins to indicate activity. Upon binding calcium, cameleon reporters change the ratio between the excitation intensities of their two fluorescent proteins, thereby providing a ratiometric measure of calcium presence and, as a result, of neuronal activity. GCaMP-based reporters, in contrast, are composed of a single fluorescing protein and simply change the brightness of their excitation upon binding calcium. GCaMP reporters are not ratiometric and, therefore, can be confounded by movements out of the focal plane that reduce excitation intensity out of tune with the actual calcium changes. In practice, this is a minor problem as there are many ways to determine and correct for out-of-focus effects on excitation intensity. Presently, GCaMP calcium sensors have an advantage over cameleon reporters in brightness and dynamic range from baseline to their calcium bound state (Tian et al. 2009; see Chapter 8).

EXPERIMENTAL METHODS

The first two protocols describe methods for preparing substrates for growing and maintaining worms during video analysis and an approach for obtaining uniform illumination during tracking. Although these protocols were developed for use in a single-worm tracker, they address factors important for the generation of reproducible, standardized images in all systems. The final protocol describes the use of the software package Multi-worm Tracker, and discusses possible applications for genetic analyses of behavior; the hardware setup, including stage, camera, and light source, is shown in Figure 1. Neuronal tracking uses precisely the same algorithm to search for labeled neurons. The few minor changes this protocol may require involve using higher magnification to view tracked neurons and inverting the binary thresholding to discover bright neurons on a dark background (worms are usually imaged in bright-field, where they are darker than their background).

Examples of single-worm trackers can be found in several published studies (e.g., Waggoner et al. 1998; Pierce-Shimomura et al. 1999; Hardaker et al. 2001; Feng et al. 2004; Geng et al. 2004; Cronin et al. 2005; Stephens et al. 2008). Another single-worm system can be found on our website, http://www.mrc-lmb.cam.ac.uk/wormtracker/. In addition to the system described below, multi-worm trackers also have been described (Ramot et al. 2008; Tsechpenakis et al. 2008).

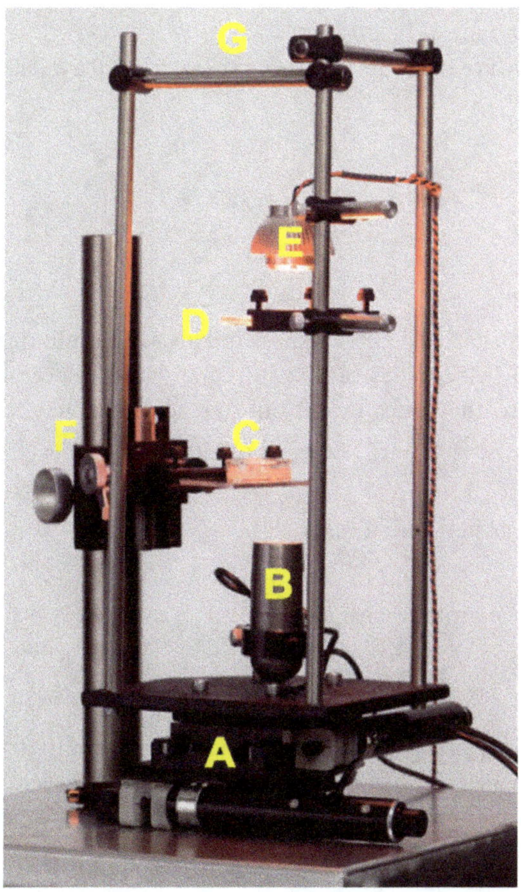

FIGURE 1. Hardware: (A) The stage, composed of *x*- and *y*-axis actuators and linear translation stages; (B) the camera; (C) the sample, a worm on an agar Petri dish; (D) a diffuser above a collimating Fresnel lens, to provide uniform illumination; (E) a high-intensity red light-emitting diode (LED) for illumination; (F) course and fine focusing knobs to focus the sample in view of the camera; (G) a rigid cage to center the illumination above the camera and eliminate vibrations caused by moving the stage.

Protocol A

Preparation of Samples for Worm Tracking

This protocol describes the preparation and use of media for growing and maintaining worms for tracking. The second part of the protocol describes how to prepare a single young adult worm for recording during video analysis.

MATERIALS

See the end of the chapter for recipes for reagents marked with <R>.

Reagents

Nematode adult worms
Nematode Growth Medium (NGM) agar medium <R>
NGM agar, low-peptone <R>
OP50 *Escherichia coli* bacteria

Equipment and Imaging Setup

Petri dishes, ≥5-cm
Stage with platform to support the sample
Worm tracking hardware and software package (e.g., Worm Tracker 2.0 available at http:www.mrc-lmb.cam.ac.uk/wormtracker/)

EXPERIMENTAL METHOD

Maintaining Strains

1. Prepare and pour fresh NGM agar into Petri dishes, and allow them to dry for 2 d with the lid on. After drying, these NGM Petri dishes can be stored at 4°C for several days to retain their moisture; although, before use, they must be left out to reach room temperature.
2. Seed each dish with three drops (~150 µL) of fresh OP50 bacteria. Leave them to dry, for 2 d, with the lid on.
3. Select six adult worms and transfer onto a dish.
4. Incubate the plate at 22°C.
5. Repeat Steps 1–4 every 3 d to maintain your strains.

Tracking

6. Prepare and pour fresh NGM agar into Petri dishes, and allow them to dry for 2 d with the lid on. After drying, these NGM Petri dishes can be stored at 4°C for several days to retain their moisture; although, before use, they must be left out to reach room temperature.

7. Seed each dish with three drops (~150 µL) of fresh OP50 bacteria. Leave them to dry, for 2 d, at room temperature with the lid on.
8. Approximately 16–18 h before tracking, select L4 worms from plates prepared in Steps 1–5 and transfer them to an NGM agar dish.
9. Store the plates at 22°C.
10. At 2 d before the experiment, prepare freshly poured low-peptone NGM dishes (these produce thin food lawns that produce less distortion in worm images). Leave them, for 2 d, to dry with the lid on.
11. At 30 min before each tracking experiment, seed each low-peptone dish with 20 µL of fresh OP50 bacteria. Leave them to dry with the lid on (30 min).
12. Select and transfer one worm to a plate (prepared in Step 11). Move the plate under the tracker. Wait 30 min for the worm to habituate, and record for 15 min at room temperature.
13. Record at least 20 worms (5 h of data). When performing multiple experiments on multiple trackers, they should be randomized in time and in choice of tracker. This helps avoid inconsistent spatial and temporal cues influencing behavior by driving their statistical effects into noise rather than amplifying them as signal. For similar reasons, it is best to space experiments out over several days using different experimenters.

Protocol B

Illumination for Worm Tracking and Behavioral Imaging

The following approach should be used to obtain uniform illumination during tracking. Good lighting can be more of an art than a science. Once the system is set up, it will be necessary to play with it, testing your results after each tweak to ensure that the analysis software is able to clearly identify the worm and its boundaries.

EXPERIMENTAL METHOD

1. Choose the appropriate wavelength(s) (e.g., use the red spectrum to avoid stimulating lite-1 receptor, blue for the converse; cameleon, GCaMP, and various neuronal reporters have published constraints for their excitation and emission spectra; and channelrhodopsins and halorhodopsins have published constraints for their excitation spectra).
2. Obtain your light source. Bright, single light-emitting diodes (LEDs) and arrays of LEDs are available at a variety of wavelengths, and are inexpensive. Alternatively, a selection of bulbs (e.g., halogen, mercury, xenon) exist that have high-intensity emission much of the way from ultraviolet, across visible light, and into the infrared spectrum. If you need an intense, tight spectrum unavailable through LEDs or filtered bulbs, lasers provide a great, albeit expensive, option.
3. Your lighting must remain uniform despite camera movement. If your camera and light source move independently from each other, several companies make light boxes that achieve uniform light over large areas. Unfortunately, inexpensive light boxes can be weak; therefore, one needs to check that their illumination is sufficient to provide the desired video frame rates.
4. Purchase appropriate filter sets to achieve tight spectral bands when necessary. Multiband excitation and emission will likely require one or more dichroic mirrors, in addition to your filter set, to differentiate between emission bands.
5. Your light source intensity significantly affects the camera exposure time. Test the two in conjunction to achieve fast image captures free from motion blur.
6. Good contrast means that the worm edges are significantly distinguished (much darker or much lighter) from the background. This includes the head and tail, which are often difficult to delineate from their background, as well as worm tracks, which can often be as dark as the worm itself, leading to a confusion between body and background. Therefore, the following factors must be considered to achieve good contrast.
   i. The distance of the light source from the worm: Greater distances yield more directional lighting, improving edge contrast at the expense of intensity.
   ii. Using a tight filter grating (e.g., a laptop privacy filter) to restrict angled light: Similar to distance, the grating leads to more directional light and, therefore, better contrast; but, once again, this is at the expense of intensity.
   iii. Using a diffusing filter (e.g., ground or colored glass) to achieve uniform lighting within your camera's field of view: Vignetting, due to nonuniform lighting, can lead to image

edges that are as dark as or darker than worm edges and, therefore, to confusion between the worm and its background.

iv. Using a collimating lens to achieve uniform, directional light: As the previous factors show, this greatly increases worm edge contrast.

v. Maintaining an unobstructed view of the worm: Among other suggestions, this can be achieved by resting the worm plate on a transparent platform and flipping it such that the worm faces the video camera directly as opposed to being filmed through the agar; please note that worms on agar appear behaviorally unaffected when crawling upside down.

Protocol C

Tracking Movement Behavior of Multiple Worms on Food

Forward genetic screens and screens of drug libraries require high-throughput phenotyping, a task traditionally incompatible with manual scoring of quantitatively varying behaviors. High-throughput automated analysis of *C. elegans* movement behavior, however, is now possible with several different tracking software packages. The Multi-worm Tracker (MWT) described here is designed especially for high-throughput analysis: it can record dozens of worms simultaneously at 30 frames per second for hours or days at a time. This is accomplished by performing all image analysis in real time, saving only the worm centroid, bearing, and outline data to the disk. To simplify the required image processing, the system focuses only on worms that have moved, and detects and discards worms that are touching rather than trying to isolate them computationally. Because the software is entirely automated, protocols can run unattended once the worms have been placed and the software has been started. The MWT does not save images for later analysis but behavior can be validated manually with a companion analysis tool that replays recorded body postures. This protocol describes a basic basal movement assay on food using the MWT; similar protocols apply to related assays and to similar multiple-animal trackers. This basic protocol can be extended to a variety of assays ranging from tap response to chemotaxis.

MATERIALS

See the end of the chapter for recipes for reagents marked with <R>.

Reagents

C. elegans, Bristol N2, your adult hermaphrodites
NGM agar medium <R>
NGM media <R>
OP50 *E. coli*

Equipment and Imaging Setup

Computer with at least 2GB RAM and 1680 x 1050 or 1600 x 1200 monitor
Dalsa Falcon 4M30 camera (2352 x 1728 @ 31 Hz) with M42 to F-Mount lens adapter
Multi-worm Tracker software (http://sourceforge.net/projects/mwt)
National Instruments NI-1427 PCI image capture card
Petri dishes, ≥5-cm
Plate holder, ~15 cm above light, ~30 cm below camera
Rodenstock 60 mm, f/4.0 Rodagon lens with MFB F-Mount adapter, modular focus block (25 mm), M39 x 0.75 lens adaptor
Schott ACE light source with fiber-optic backlight (4 x 4.88-in)

EXPERIMENTAL METHOD

1. Set up the recording system, the Multi-worm Tracker, as described in its User Manual, and use the Measurement and Automation Explorer tool to set the camera to 8 bit resolution.

2. Measure the size of a pixel in millimeters: View a ruler with the camera and compute, for example, [number of pixels in one inch]/25.4.
3. Prepare worms for tracking.
   i. Seed one plate of NGM with 50 μL of NGM media, spread to cover center of plate.
   ii. Leave plate overnight.
   iii. Select five young adult hermaphrodites and transfer them onto the NGM plate. Incubate them for 3 h to lay eggs, and then remove the adult hermaphrodites.
   iv. Allow the eggs to grow to young adults (72–96 h).
4. Place the plate in position for imaging with the lid on, making sure that the worms are in focus with the camera, that the brightness across the field of view varies by no more than ~40%, and that the brightest pixels are below saturation (a good target is 200–220 out of 255).

   See *Troubleshooting.*
5. Run the tracker.
   i. In the MWT software, select a target directory, an experiment title, and a duration to record.
   ii. Verify in the image window that the MWT is detecting the worms.

      See *Troubleshooting.*
   iii. Press "Go," and wait for capture to complete.
6. Analyze the resulting data with the companion command-line Choreography program as described in the MWT User Manual.
   i. Supply the pixel size measured in Step 2 with the -p option (e.g., -p 0.024 for a setup with 24 μm/pixel, which is approximately what it should be with the imaging setup described).
   ii. Request the output parameters desired (e.g., -o Ns will produce a table containing the number of worms and the mean speed of those worms at each time point).

      See the Users Guide for more detail.

TROUBLESHOOTING

Problem (Steps 4 and 5ii): The worms are obscured by condensation on the lid of the plate.
Solution: Check the temperature in the room; if the lid of the plate is colder than the base, condensation will appear. The light source box may need to be moved farther away or downwind instead of upwind; the light plate itself should heat the base enough to cause fogging.
Solution: Remove the lid from the Petri dish. Be aware, however, that worms are highly sensitive to drying, and long assays (>1 h) will be measuring predominantly a drying-related response unless the humidity is very high.

Problem (Step 5ii): The MWT fails to show any image of worms.
Solution: Check that the Measurement and Automation Explorer can find the camera and view images from it; then close the Explorer and try the MWT again.
Solution: The National Instrument Vision Runtime License has not been activated and the trial period has expired. Contact National Instruments to purchase a runtime license.

Problem (Step 5ii): The tracker cannot find the worms.
Solution: Make sure there is good contrast between the worm and the background. With the optical setup as described, contrast should be ~50%.

Solution: Make sure the lighting is bright but does not saturate the camera.
Solution: Alter the tracking parameters such as intensity threshold within the MWT.

Problem (Step 5ii): The tracker finds many tiny worms as well as the adult worms.
Solution: Increase the minimum object size in the MWT.
Solution: Raise the worms at a colder temperature or run the experiment before any eggs have hatched.
Solution: Select and transfer adult worms to the plate instead of waiting for them to grow from eggs. Allow at least 2 h to elapse between transfer and recording, if possible.

DISCUSSION

A variety of methods for recording many worms simultaneously have been recently developed. Image analysis methods used on single worms can be applied to high-resolution images of whole plates of worms to give similar data: some individual measurements are noisier or impossible, but collecting large sample sizes is much faster. Ramot et al. (2008) have created the Parallel Worm Tracker, a suite of three related open-source programs that acquire image data, perform basic image processing to identify worms and produce tracks of the worms's centroid positions, and analyze those tracks to compute linear and angular velocity. Similarly, Roussel et al. (2007) and Tsechpenakis et al. (2008) have created software that analyzes stored image data with more sophisticated algorithms that account for noise and adjacent worms and can provide basic postural data in addition to centroid position. Other efforts at simultaneous worm tracking have focused on worms in 96-well plates with no (Simonetta and Golombek 2007) or very minimal (Buckingham and Sattelle 2009) image processing, or on robustly detecting pairs of worms in contact for detailed assays. The Multiworm Tracker discussed here is conceptually very similar to the trackers described by Ramot et al. (2008) and Tsechpenakis et al. (2008), save for its focus on real-time image processing.

Multiworm tracking is possible with a wide variety of cameras and imaging speeds. For the MWT, the primary two requirements are that worms must overlap over at least 50% of their bodies from one frame to the next, and must be large enough to cause a clear change in brightness when moving into a new area of the plate. Thus, frame rates slower than ~0.5 Hz for crawling or ~10 Hz for swimming (because the worm swings side to side) will tend to fail because of excessive movements; magnifications that result in pixels that span more than ~60 µm also tend to give substandard results.

Converting a single-worm manual assay to an automated assay can be a considerable amount of work. The MWT does not automatically detect qualitative phenotypes often scored manually (e.g., "kinker"), so a quantitative proxy must be discovered. For example, for tap habituation assays, reversal distance is a typical metric (Broster and Rankin 1994), but is relatively difficult to extract computationally. However, for longer interstimulus tap intervals, increase of speed of animals on food is a good alternative metric for response to tap (Fig. 2A; the transient increase in speed due to manipulation of the plate before the beginning of the experiment is typical). When beginning a new assay, it is valuable to compare manually scored behaviors with automated results. To assist in this, the Choreography program contains a --map option that brings up a map browser interface, which allows one to replay the movements of the tracked worms (Fig. 2B). The analysis parameters and exclusion criteria used in the results shown in Figure 2 are listed in Table 1. Imaging output from Worm Tracker 2.0.3.1 is shown in Figure 3.

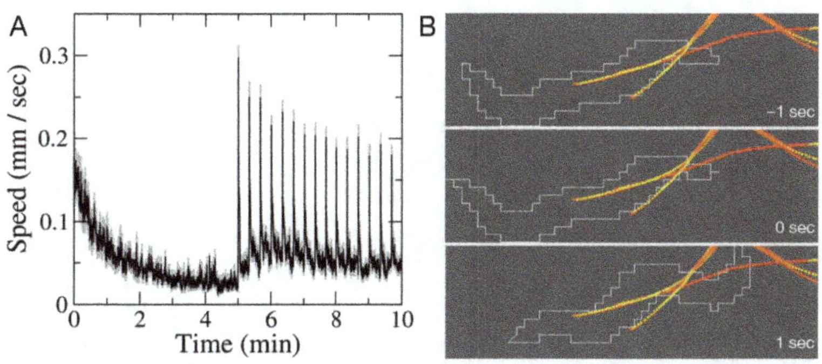

FIGURE 2. A tap habituation assay performed on the Multi-worm Tracker. (*A*) Average speed of ~*N* worms over 10 min with plate tap delivered at 5 min and 3x/min thereafter (15 taps total). Black line is mean; standard error in gray. *N* = 45–50 animals per time point (one plate). (*B*) Manual validation of tap response with worm position 1.0 sec before (*top*), at (*center*), and 1.0 sec after (*bottom*) a tap. Colored dots are centroid positions colored by speed, where red is slower and yellow faster; one dot is drawn every 33 msec.

TABLE 1. Choreography analysis parameters used in the example results shown in Figure 2

Pixel size	Speed averaging window	Object min duration	Object min movement	Output quantities
0.0243 mm/sec	0.2 sec	20 sec	2 body lengths	ss*s# (speed: mean, standard deviation, number of samples)

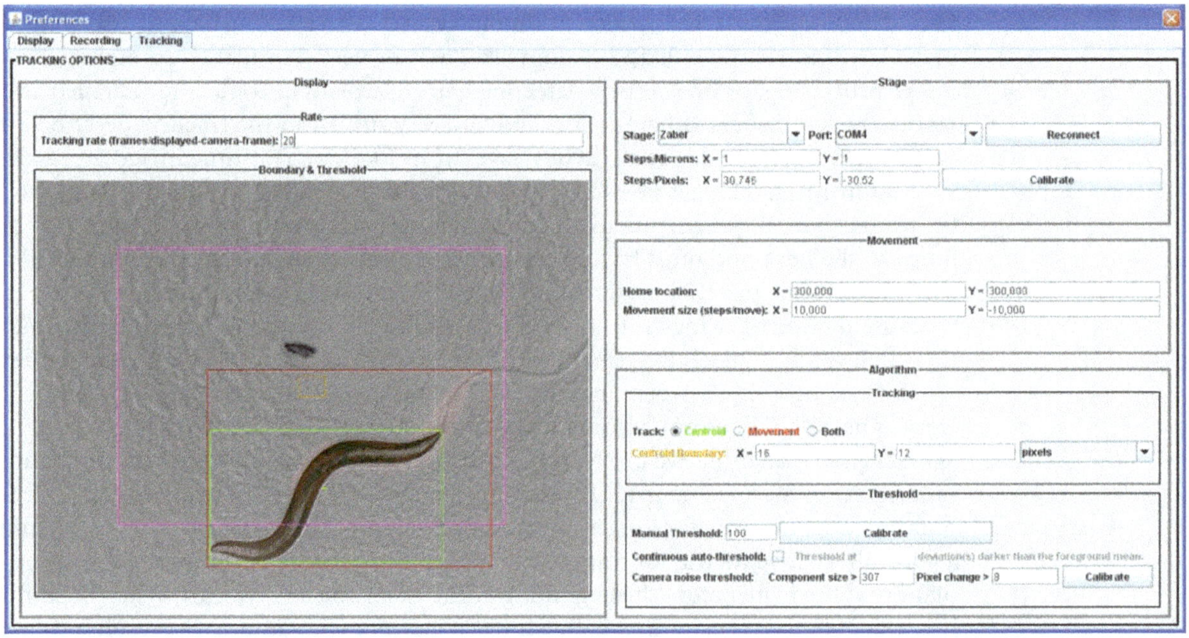

FIGURE 3. Single-worm tracking. This picture was taken from Worm Tracker 2.0.3.1. The largest connected component, the worm, is labeled in green. Motion is labeled in red. The tracker is attempting to keep the worm's centroid within the orange boundary and the worm's motion within the magenta boundary

RECIPES

CAUTION: See Appendix 6 for proper handling of materials marked with <!>.
Recipes for reagents marked with <R> are included in this list.

NGM Agar Medium

Reagent	Amount
NaCl	3 g
Agar	17 g
Peptone	2.5 g
Cholesterol (5 mg/mL in ethanol <!>)	1 mL
$CaCl_2$ <!>, 1 M	1 mL
$MgSO_4$ <!>, 1 M	1 mL
K_2HPO_4, 1 M (pH 6)	25 mL

1. Combine the first four reagents in 975 mL of water, mix well, and autoclave at 120 for 20 min. Wait for the solution to cool to 50°C before adding the remaining reagents.
2. Maintaining sterile technique, add the remaining three reagents, mixing after each addition.
3. Allow the solution to cool slightly, then pour into Petri dishes (≥5 cm).

Note that for NGM liquid media, you should omit the agar from the recipe.

NGM Agar Medium, Low-Peptone

Reagent	Amount
NaCl	3 g
Agar	20 g
Peptone	0.13 g
Cholesterol (5 mg/mL in ethanol <!>)	1 mL
$CaCl_2$ <!>, 1 M	1 mL
$MgSO_4$ <!>, 1 M	1 mL
K_2HPO_4, 1 M (pH 6)	25 mL

1. Combine the first four reagents in 975 mL of water, mix well, and autoclave at 120 for 20 min. Wait for the solution to cool to 50°C before adding the remaining reagents.
2. Maintaining sterile technique, add the remaining three reagents, mixing after each addition.
3. Allow the solution to cool slightly, then pour into Petri dishes (≥5 cm).

REFERENCES

Ben Arous J, Tanizawa Y, Rabinowitch I, Chatenay D, Schafer WR. 2010. Automated imaging of neuronal activity in freely behaving *Caenorhabditis elegans*. *J Neurosci Methods* **187:** 229–234.

Broster BS, Rankin CH. 1994. Effects of changing interstimulus interval during habituation in *Caenorhabditis elegans*. *Behav Neurosci* **108:** 1019–1029.

Buckingham SD, Sattelle DB. 2009. Fast, automated measurement of nematode swimming (thrashing) without morphometry. *BMC Neurosci* **10:** 84.

Chronis N, Zimmer M, Bargmann CI. 2007. Microfluidics for in vivo imaging of neuronal and behavioral activity in *Caenorhabditis elegans*. *Nat Methods* **4:** 727–731.

Clark DA, Gabel CV, Gabel H, Samuel AD. 2007. Temporal activity patterns in thermosensory neurons of freely moving *Caenorhabditis elegans* encode spatial thermal gradients. *J Neurosci* **27:** 6083–6090.

Cronin CJ, Mendel JE, Mukhtar S, Kim YM, Stirbl RC, Bruck J, Sternberg PW. 2005. An automated system for measuring parameters of nematode sinusoidal movement. *BMC Genet* **6:** 5.

Feng Z, Cronin CJ, Wittig JH Jr, Sternberg PW, Schafer WR. 2004. An imaging system for standardized quantitative analysis of *C. elegans* behavior. *BMC Bioinformatics* **5:** 115.

Geng W, Cosman P, Berry CC, Feng Z, Schafer WR. 2004. Automatic tracking, feature extraction and classification of *C. elegans* phenotypes. *IEEE Trans Biomed Eng* **51:** 1811–1820.

Hardaker LA, Singer E, Kerr R, Zhou G, Schafer WR. 2001. Serotonin modulates locomotory behavior and coordinates egg-laying and movement in *Caenorhabditis elegans*. *J Neurobiol* **49:** 303–313.

Karbowski J, Schindelman G, Cronin CJ, Seah A, Sternberg PW. 2008. Systems level circuit model of *C. elegans* undulatory locomotion: Mathematical modeling and molecular genetics. *J Comput Neurosci* **24:** 253–276.

Pierce-Shimomura JT, Morse TM, Lockery SR. 1999. The fundamental role of pirouettes in *Caenorhabditis elegans* chemotaxis. *J. Neurosci.* **19:** 9557–9569.

Ramot D, Johnson BE, Berry TL Jr, Carnell L, Goodman MB. 2008. The Parallel Worm Tracker: A platform for measuring average speed and drug-induced paralysis in nematodes. *PLoS One* **3:** e2208.

Roussel N, Morton CA, Finger FP, Roysam B. 2007. A computational model for *C. elegans* locomotory behavior: Application to multiworm tracking. *IEEE Trans Biomed Eng* **54:** 1786–1797.

Simonetta SH, Golombek DA. 2007. An automated tracking system for *Caenorhabditis elegans* locomotor behavior and circadian studies application. *J Neurosci Methods* **161:** 273–280.

Stephens GJ, Johnson-Kerner B, Bialek W, Ryu WS. 2008. Dimensionality and dynamics in the behavior of *C. elegans*. *PLoS Comput Biol* **4:** e1000028.

Tian L, Hires SA, Mao T, Huber D, Chiappe ME, Chalasani SH, Petreanu L, Akerboom J, McKinney SA, Schreiter ER, et al. 2009. Imaging neural activity in worms, flies and mice with improved GCaMP calcium indicators. *Nat Methods* **6:** 875–881.

Tsechpenakis G, Bianchi L, Metaxas D, Driscoll M. 2008. A novel computational approach for simultaneous tracking and feature extraction of *C. elegans* populations in fluid environments. *IEEE Trans Biomed Eng* **55:** 1539–1549.

Waggoner LE, Zhou GT, Schafer RW, Schafer WR. 1998. Control of alternative behavioral states by serotonin in *Caenorhabditis elegans*. Neuron **21:** 203–214.

Zhang M, Chung SH, Fang-Yen C, Craig C, Kerr RA, Suzuki H, Samuel AD, Mazur E, Schafer WR. 2008. A self-regulating feed-forward circuit controlling *C. elegans* egg-laying behavior. *Curr Biol* **18:** 1445–1455.

71 In Vivo Dendritic Calcium Imaging in the Fly Visual System

Alexander Borst,[1] Winfried Denk,[2] and Jürgen Haag

[1]Max-Planck-Institute for Neurobiology, D-82152 Martinsried, Germany; [2]Max-Planck-Institute for Medical Research, D-69120 Heidelberg, Germany

ABSTRACT

This chapter discusses the essential features of the blowfly lobula-plate tangential cell experimental system and presents practical guidelines, as well as recent technical improvements, for studying visually induced calcium dynamics in fine dendritic branches. These signals can be interpreted with respect to the presynaptic mechanism of local motion detection. The results exactly fit the predictions of the correlation-detector model of Reichardt (1961).

INTRODUCTION

The large motion-sensitive interneurons (lobula-plate tangential cells) of the fly represent a unique system for the in vivo study of dendritic integration and calcium signaling in individual neurons. In the blowfly, *Calliphora vicina*, there exist approximately 60 of these neurons per lobula plate. Each can be identified by its visual response characteristics, such as receptive field location and preferred direction of motion, as well as by its typical anatomy (Hausen 1984). The large dendrites of these neurons receive signals from thousands of retinotopically organized elements that are sensitive to the local direction of visual motion. Tangential neurons have been studied in detail with respect to the question of how their active and passive membrane properties (Borst and Haag 1996; Haag et al. 1997) give rise to their visual response properties (Haag and Borst 1996; Single et al. 1997; Haag et al. 1999). Lobula-plate tangential cells are particularly well suited for optical recording in vivo (Borst and Egelhaaf 1992). These cells are located directly beneath the rear surface of the brain and are directly accessible upon opening the head capsule. They are visible in whole mount after fluorescent labeling, even in tissue that has not been cleared.

EXPERIMENTAL APPROACHES AND GUIDELINES

Whole-Mount Preparation

In our system, female blowflies were briefly anesthetized with CO_2 and mounted ventral side up with wax on a small preparation platform. The head capsule was opened from behind; the trachea and air sacs, which normally cover the lobula plate, were removed. To eliminate movements of the brain caused by peristaltic contractions of the esophagus, the proboscis of the animal was cut away and the gut was pulled out. This preparation allowed stable intracellular recordings and reduced the occurrence of movement artifacts during imaging.

Cell Labeling

For electrophysiological recordings and intracellular injection of the calcium indicator Calcium Green (Ca-Green I hexapotassium salt; Molecular Probes), electrodes were pulled on a Brown-Flaming micropipette puller (P-97; Sutter Instruments) using thin-wall glass capillaries with an outer diameter of 1 mm (GC100TF-10; Clark). When filled with 2 M KAc/0.5 M KCl/8.8 mM Ca-Green, they had resistances of ~30–40 MΩ. Cells were labeled by applying a hyperpolarizing current of –3 nA for ~5 min. An SEC-10L amplifier (npi-electronics) was used throughout the experiments, which was operated in the Bridge or DCC mode.

Imaging Setup

We used a custom-built two-photon microscope and a novel stimulation device (Fig. 1). Briefly, the microscope consists of the following components: a 5W-pumped Ti:sapphire laser (MaiTai; Spectra Physics), a Pockels cell (Conoptics), scan mirrors including drivers (Cambridge Technology), a scan lens (4401-302; Rodenstock), a tube lens (MXA 22018; Nikon), a dichroic mirror (DCSPR 25.5x36; AHF Tuebingen), and a 40× water-immersion lens (Zeiss). The lens can move along all three axes by a stepper motor–driven micromanipulator (MP285-3Z; Sutter Instruments). Emitted light is filtered

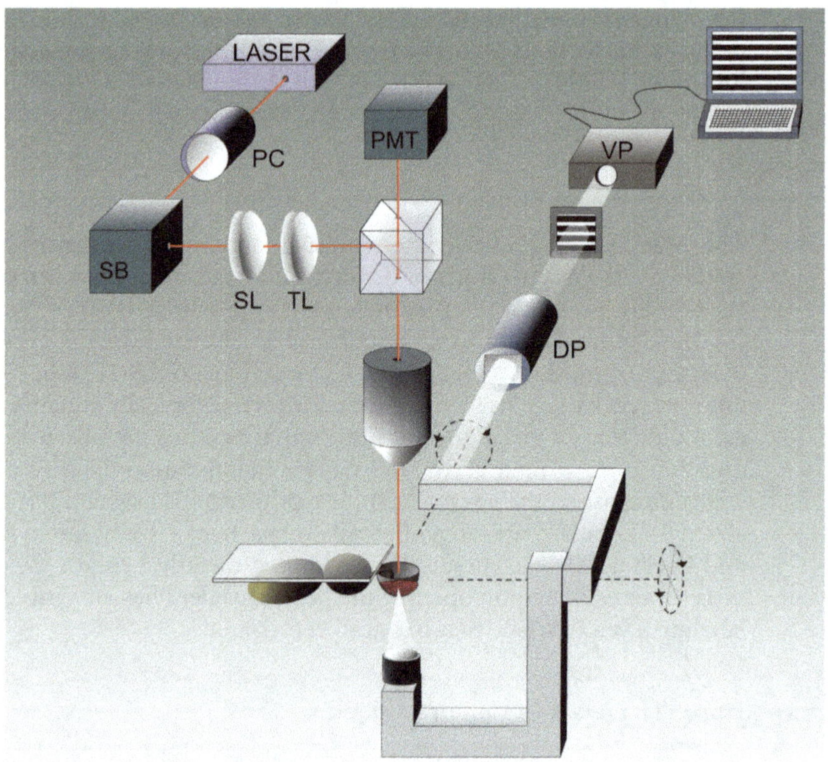

FIGURE 1. Schematic drawing of the setup used for two-photon calcium imaging and visual stimulation in vivo. The fly is shown under an objective lens, looking down onto a moving grating. From the top, a femtosecond laser pulse excites a fluorophore in the visual interneuron with the emitted photons being collected by a photomultiplier. On the *right* side, the stimulus light path is shown: A moving grating is generated by a computer and fed through a dove prism into the optics leading to the eyepiece. Abbreviations: PC, Pockels cell; SB, scan box; SL, scan lens; TL, tube lens; PMT, photomultiplier tube; DP, dove prism; VP, video projector.

by a band pass (HQ 535/50; Chroma) and collected by a multialkali photomultiplier (R6357; Hamamatsu). Visual stimuli are provided by a computer via a video projector (3M), filtered by a short pass (10SWF-450), and imaged by a total of three lenses in the path so that the pattern is focused into the reticule plane of the eyepiece. An adjustable aperture in the beam path was focused onto the center of the eye-surface curvature. The image can be rotated in two axes around the fly by an articulated arm-like arrangement, which is driven by two stepper motors (OWIS). The whole system is controlled by custom-written software.

Imaging

Despite the advantages that fly visual interneurons offer for optical recording in general, specific problems arise when one is trying to record visually evoked responses. This difficulty is because some of the intense light needed for fluorescence excitation almost inevitably reaches the photoreceptor and thus interferes with visual stimulation. This is true even if care is taken to illuminate only a small field of view since the neuropile of the lobula plate strongly scatters light. With careful experimental design (Borst and Single 2000; Elyada et al. 2009), visually evoked responses can be recorded, whereas rather bright visual stimulation must be used to compete successfully with photoreceptor activation by fluorescence excitation light. This problem can, however, be virtually eliminated by the use of two-photon microscopy (Denk et al. 1990). There pulsed infrared light is used to cause two-photon excitation, which, owing to the quadratic intensity dependency, is limited to the focal region and thereby provides strong optical sectioning. This approach has been successfully employed for calcium recordings in the mammalian retina (Denk and Detwiler 1999; Euler et al. 2002) and is currently being applied to the visual system of *Drosophila* using genetically encoded calcium indicators (Guerrero et al. 2005; Hendel et al. 2008; Mank et al. 2008). The lack of photoreceptor activation even by intense infrared light shone into the back of the fly's head was confirmed using electrophysiological recordings, where no membrane voltage change is seen in response to opening the shutter of the two-photon excitation laser.

To avoid generating a false signal in the fluorescence channel, one needs to prevent the stimulus light from reaching the photomultiplier tube. This can be achieved by using appropriate nonoverlapping wavelength filters for fluorescence and stimulation light. In our case, we used short-pass filters limiting the spectral range of the stimulus light to <450 nm. The remaining short-wavelength light stimulates the photoreceptors strongly enough because of the ultraviolet sensitivity of the fly photo pigments (Hardie 1979). In order not to tax the finite blocking ability of the filters, it is sometimes necessary to concentrate the stimulus light as much as possible. This can be achieved, for example, as follows (Fig. 1): Instead of using a screen onto which the visual pattern is projected, the fly is looking at the pattern through an eyepiece with the eye point (the illumination aperture) at the center of its eye-surface curvature. This minimizes the number of scattered photons and ensures that the light of the stimulus path is entering the fly visual system very efficiently.

[Ca^{2+}] MEASUREMENTS

Figure 2 gives an example of an in vivo experiment on a neuron that is sensitive to vertical image motion, a so-called VS2 cell. In Figure 2C, an image stack (50 images, 256 x 256 pixels, 1-μm steps, 200-μm image width) is combined into a single image ("extended focus"). The axon can be seen running from the center of the image to the left. The main dendrite extends from the upper-right to the lower-left corner of the image. In Figure 2A, the relative change in fluorescence, a measure of the cytosolic calcium concentration ([Ca^{2+}]), is shown during visual motion (bar below response traces) along the preferred direction of the cell. The optical signal is derived from a small dendritic branchlet of the cell. The [Ca^{2+}] rises steeply at the onset of motion, is modulated at the stimulus frequency in addition to the offset increase, and decays slowly after motion has stopped. Previous analyses of the mechanisms underlying such changes in free cytosolic calcium in lobula-plate tangential cells have revealed that Ca^{2+} enters the cytosol from outside of the cell, rather than being released from intracellular stores (Oertner et al. 2001). This influx is mainly voltage-dependent (Haag and Borst

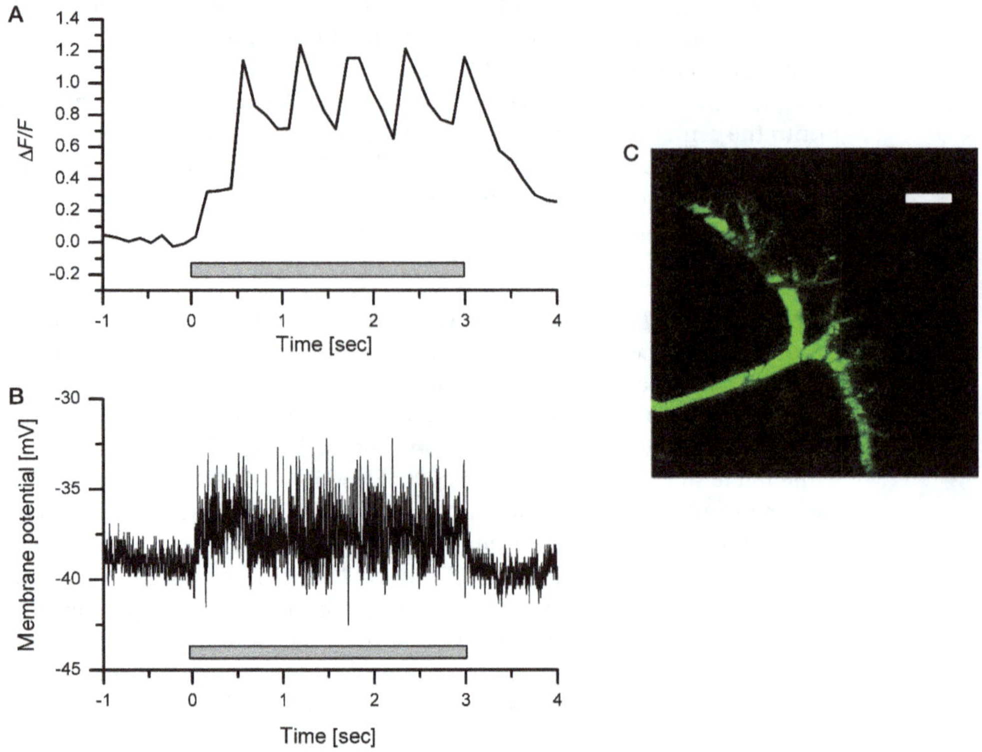

FIGURE 2. Example recording from a motion-sensitive neuron in the lobula plate (VS2 cell). (*A*) Calcium signals during motion along the preferred direction of the cell. Fluorescence was stimulated at 850 nm. (*B*) Axonal membrane potential recorded simultaneously. (*C*) Projected two-photon image stack ("extended focus"). Scale bar, 20 μm. Stimulus parameters: mean luminance, 500 cd/m^2; contrast, 98%; spatial pattern wavelength, 8.3°; stimulus velocity, 10.8°/sec; and temporal frequency, 1.3 Hz.

2000) but has a minor component that enters through nicotinic acetylcholine receptors (Single and Borst 2002). Acetylcholine is the main excitatory neurotransmitter in the insect central nervous system, as has also been demonstrated for the tangential cells (Brotz and Borst 1996). In Figure 2B, the axonal membrane potential is displayed, together with the simultaneously recorded optical signal. As can be seen, the neuron fires a barrage of spikes in addition to being slightly depolarized (Haag and Borst 1996). Unlike the optical signal, the electrical signal is not modulated with the temporal stimulus frequency (i.e., the grating velocity divided by the spatial wavelength). This is exactly as predicted by the correlation type of motion detector model (Reichardt 1961; Borst and Egelhaaf 1989): According to this model, local signals reflect both the direction of visual motion as well as local pattern properties. Only after spatial integration do modulations cancel owing to the phase shifts across the receptive field of the neuron, leaving a purely directionally selective signal in the axon of the cell (Egelhaaf et al. 1989; Single and Borst 1998). The great advantage of our current instrumentation over previous systems (Borst and Single 2000) lies in the fact that stimulus parameters can be chosen at will. In the experiment shown in Figure 2, the mean luminance level was ~500 cd/m^2. Furthermore, in contrast to one-photon excitation, no reduction of stimulus contrast due to stray excitation light occurs.

REFERENCES

Borst A, Egelhaaf M. 1989. Principles of visual motion detection. *Trends Neurosci* **12:** 297–306.

Borst A, Egelhaaf M. 1992. In vivo imaging of calcium accumulation in fly interneurons as elicited by visual motion stimulation. *Proc Natl Acad Sci* **89:** 4139–4143.

Borst A, Haag J. 1996. The intrinsic electrophysiological characteristics of fly lobula plate tangential cells. I. Passive membrane properties. *J Comput Neurosci* **3:** 313–336.

Borst A, Single S. 2000. In vivo calcium imaging in blowfly CNS. In *Imaging neurons: A laboratory manual* (ed. Yuste R, et al.), pp. 44.1–44.6. Cold Spring Harbor Laboratory Press, Cold Spring Harbor, NY.

Brotz T, Borst A. 1996. Cholinergic and GABAergic receptors on fly tangential cells and their role in visual motion detection. *J Neurophysiol* **76:** 1786–1799.

Denk W, Detwiler PB. 1999. Optical recording of light-evoked calcium signals in the functionally intact retina. *Proc Natl Acad Sci* **96:** 7035–7040.

Denk W, Strickler JH, Webb WW. 1990. Two-photon laser scanning fluorescence microscopy. *Science* **248:** 73–76.

Egelhaaf M, Borst A, Reichardt W. 1989. Computational structure of a biological motion detection system as revealed by local detector analysis in the fly's nervous system. *J Opt Soc Am A* **6:** 1070–1087.

Elyada YM, Haag J, Borst A. 2009. Different receptive fields in axons and dendrites underlie robust coding in motion-sensitive neurons. *Nat Neurosci* **12:** 327–332.

Euler T, Detwiler PB, Denk W. 2002. Directionally selective calcium signals in dendrites of starburst amacrine cells. *Nature* **418:** 845–852.

Guerrero G, Reiff DF, Agarwal G, Ball RW, Borst A, Goodman CS, Isacoff EY. 2005. Heterogeneity in synaptic transmission along a *Drosophila* larval motor axon. *Nat Neurosci* **8:** 1188–1196.

Haag J, Borst A. 1996. Amplification of high-frequency synaptic inputs by active dendritic membrane processes. *Nature* **379:** 639–641.

Haag J, Borst A. 2000. Spatial distribution and characteristics of voltage-gated calcium currents within visual interneurons. *J Neurophysiol* **83:** 1039–1051.

Haag J, Theunissen F, Borst A. 1997. The intrinsic electrophysiological characteristics of fly lobula plate tangential cells. II. Active membrane properties. *J Comput Neurosci* **4:** 349–368.

Haag J, Vermeulen A, Borst A. 1999. The intrinsic electrophysiological characteristics of fly lobula plate tangential cells. III. Visual response properties. *J Comput Neurosci* **7:** 213–234.

Hardie RC. 1979. Electrophysiological analysis of fly retina. I: Comparative properties of R1-6 and R7 and 8. *J Comp Physiol* **129:** 19–33.

Hausen K. 1984. The lobula-complex of the fly: Structure, function and significance in visual behaviour. In *Photoreception and vision in invertebrates* (ed. Ali MA), pp. 523–559. Plenum Press, New York.

Hendel T, Mank M, Schnell B, Griesbeck O, Borst A, Reiff DF. 2008. Fluorescence changes of genetic calcium indicators and OGB-1 correlated with neural activity and calcium in vivo and in vitro. *J Neurosci* **28:** 7399–7411.

Mank M, Ferrão Santos A, Direnberger S, Mrsic-Flogel TD, Hofer SB, Stein V, Hendel T, Reiff DF, Levelt C, Borst A, et al. 2008. A genetically encoded calcium indicator for chronic in vivo two photon imaging. *Nat Methods* **5:** 805–811.

Oertner TG, Brotz T, Borst A. 2001. Mechanisms of dendritic calcium signaling in fly neurons. *J Neurophysiol* **85:** 439–447.

Reichardt W. 1961. Autocorrelation, a principle for the evaluation of sensory information by the central nervous system. In *Sensory communication* (ed. Rosenblith WA), pp. 303–317. M.I.T. Press, Cambridge, MA.

Single S, Borst A. 1998. Dendritic integration and its role in computing image velocity. *Science* **281:** 1848–1850.

Single S, Borst A. 2002. Different mechanisms of calcium entry within different dendritic compartments. *J Neurophysiol* **87:** 1616–1624.

Single S, Haag J, Borst A. 1997. Dendritic computation of direction selectivity and gain control in visual interneurons. *J Neurosci* **17:** 6023–6030.

72 Imaging Neuronal Activity and Motor Behavior in Zebrafish

Germán Sumbre[1] and Mu-Ming Poo[2]

[1]*Section de Neurosciences, Institute de Biologie de l'Ecole Normale Supérieure, Paris 75005, France;*
[2]*Division of Neurobiology, Department of Molecular and Cell Biology, Helen Wills Neuroscience Institute, University of California, Berkeley, California 94720*

ABSTRACT

To understand how motor behaviors are controlled by the nervous system it is necessary to monitor large portions of the brain, still with single-cell resolution, in relatively simple behaving animal models. Here we describe a method that enables monitoring large neuronal circuits in an intact behaving vertebrate. For this purpose we chose the zebrafish larva, a small lower vertebrate with transparent skin, as the experimental model. To monitor the activity of large neuronal networks, the optic tectum of the larva was bolus injected with an acetoxymethyl (AM) ester-type fluorescent calcium indicator dye. Following a period of incubation, the larva was restrained under a microscope's objective by embedding only its head in a drop of low-melting agarose, leaving its tail free to move. The labeled tectal neurons were imaged using a two-photon fluorescence scanning microscope. This technique enabled us to simultaneously monitor visually induced or spontaneous neuronal activities and motor behaviors. Using this method, we could show how intervals, on the order of seconds, between rhythmic visual stimuli are represented, memorized, and predicted in the nervous system of a lower vertebrate.

INTRODUCTION

One of the long-standing goals in neuroscience is to characterize the activation patterns of neural networks underlying neuronal computations and motor behaviors. For this purpose it is necessary to simultaneously monitor the activity of large populations of neurons with single-cell resolution over a large area of the brain, in a behaving organism.

The zebrafish larva offers these possibilities. During the early larval stage, zebrafish breathe mainly through the skin and can be restrained under the microscope using a drop of low-melting agarose, without the use of anaesthetics or paralyzing drugs. Moreover, the transparency of the skin, the small diameter of the neurons (4–5 μm), and the high-neuronal density enable the use of in vivo noninvasive imaging techniques to monitor neuronal activities of up to ~500 cells within the central nervous system, still with single-cell resolution, in an intact behaving vertebrate (Brustein et al. 2003; Niell and Smith 2005; Ramdya and Engert 2008; Sumbre et al. 2008). Finally, the zebrafish model is suitable for genetic tools to monitor and manipulate the activities of specific cell types or large neuronal networks (Burrone et al. 2002; Hua et al. 2005; Szbobota et al. 2007; Wang et al. 2007; Zhang et al. 2007; Dougglas et al. 2008; Arrenberg et al. 2009; Wyart et al. 2009).

This chapter describes a method for simultaneously monitoring spontaneous and visually evoked activities of large populations of neurons in the optic tectum of the zebrafish larva, using a synthetic calcium dye (Oregon Green BAPTA-1 AM) and a conventional confocal or two-photon scanning fluorescence microscope, together with a method for measuring the tail motor behavior of the head-immobilized zebrafish larva.

RELATED INFORMATION

This technique has been used to determine how short-term perceptual memories of time intervals on the order of seconds are represented in the zebrafish optic tectum, the highest and largest brain region in teleost fish (Sumbre et al. 2008).

IMAGING SETUP

Calcium Imaging of Tectal Neuronal Population Activities

To image the Ca^{2+} dynamics of the labeled cells, we used a custom-built two-photon microscope system, with a 40x water-immersion objective (numerical aperture 0.8) and a Ti:sapphire laser tuned to 790 nm (Mai Tai, Newport-Spectra Physics). Using this method it is possible, in nacre zebrafish larvae (see Protocol), to monitor the activities of up to ~500 cells simultaneously with clear single-cell resolution at 1–2.7 Hz (Fig. 1A,D). Owing to pigmentation, wild-type larvae were imaged with a conventional Zeiss confocal system at 488 nm, under low laser power (<2 mW after the objective). If wild-type larvae have to be used, the fish should be treated with PTU (see Protocol), to avoid the black melanophores absorbing the IR laser heat and burning the skin of the larva. Otherwise, wild-type larvae can be imaged using a conventional confocal microscope and a 488-nm laser line (use no more than 2 mW after the objective). Because 488-nm light excites the zebrafish retina, small Ca^{2+} transients in tectal cells could be evoked by the onset of the 488-nm laser light; therefore it is necessary to discard data collected during the first 2–4 sec period after the onset of the laser, and scan without modulating the laser intensity during the galvanometers fast-backward nonacquiring periods. The small diameter of the periventricular-layer tectal cells (diameter ~4–5 µm) and their high density with very narrow extracellular space (Fig. 1A) ensures that virtually all pixels scanned carry relevant information about neuronal activities.

Visual Stimulation

Whole-field visual stimulation: Whole-field light flashes (200 msec) were delivered via the opening of a fast shutter (Uniblitz), controlled by a computer. The light source was a 150-W lamp connected to the shutter via an optic fiber. To avoid interference of the visual stimulation with the Ca^{2+} indicator emission, green light from the wide-spectrum light source was filtered with a Kodak 32 filter.

Patterned visual stimuli: Moving bars, looming circles, and slow-moving shapes were generated using custom-made software written in MATLAB and Psychtoolbox (Brainard 1997; Pelli 1997). The stimuli were presented to the larvae via a miniature 14 x 9-mm LCD screen (PLM-A35, Sony). The LCD screen was filtered by a Kodak 32 filter. The miniature LCD monitor was positioned against a coverslip that served as the window of the recording chamber, 7 mm away from the eye contralateral to the Ca^{2+} dye-injected tectum, covering a visual field of 90° x 65° (Fig. 1B). At this distance the lens

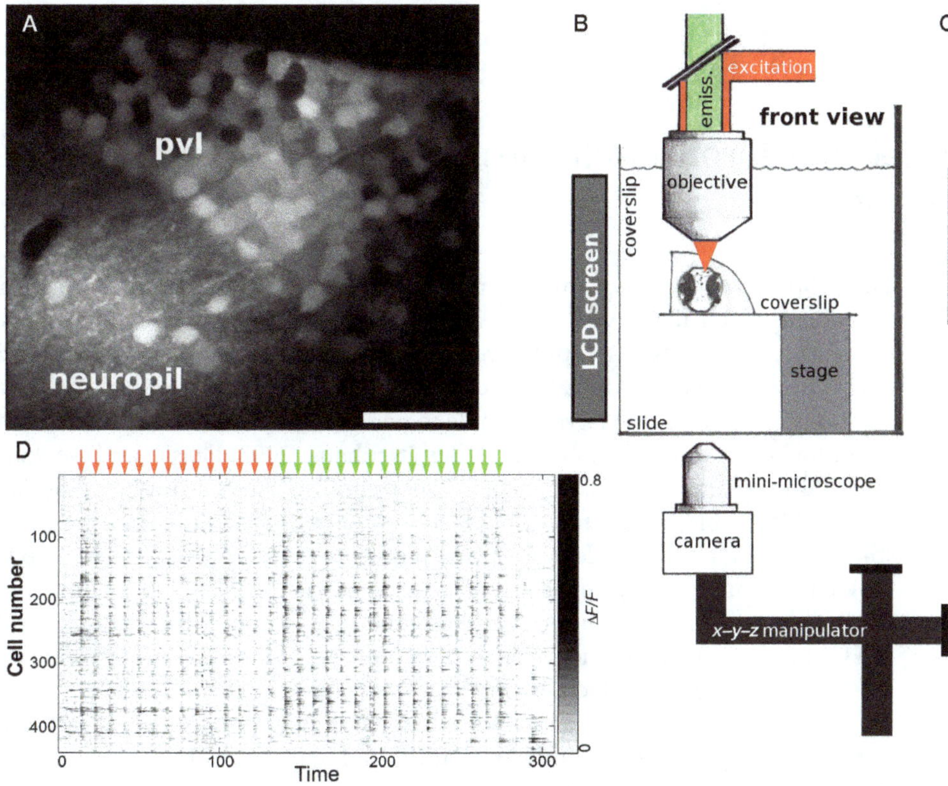

FIGURE 1. Two-photon imaging of the Ca^{2+} dynamics of labeled cells. (*A*) An optical section of zebrafish larva optic tectum (8 d postfertilization), labeled with Ca^{2+} indicator OGB-1 AM and imaged with a two-photon scanning fluorescence microscope. The tectal neuropil and periventricular layer (pvl) can be clearly distinguished. Scale bar, 20 µm. (*B*) The recording chamber showing the two-photon microscope objective (*top*), the mini custom-made microscope to monitor tail motor behavior (*bottom*), and the liquid crystal display (LCD) screen for visual stimulation (gray bar positioned next to the chamber). (*C*) A top view of the positioned larva in the recording chamber. The agarose (gray) is removed from the distal third of the tail. A sharp cut orthogonal to the eye contralateral to the labeled tectum helps minimize diffraction of the visual stimulus. (*D*) Raster plot of Ca^{2+} transients for 422 tectal neurons during periods of spontaneous activity and visual stimulus presentation (a light bar moving in a caudorostral or rostrocaudal direction [red and green arrows, respectively]).

of the larva's eye focuses the visual stimulus on the retina (Easter and Nicola 1996). The LCD screen, mounted on an *x–y–z* manipulator, was centered with respect to the larva's eye. During the imaging experiments, the larvae were not anaesthetized, as the agarose was sufficient to restrain the movement of the eyes, head, and tail.

Simultaneous Recording of Tectal Ca^{2+} Dynamics and Motor Behavior

Below the restrained larvae, a custom-made mini microscope connected to a video camera shielded with a 488-nm notch filter (in this case a confocal microscope was used), or an infrared band-pass filter (700–870 nm, optical density [OD] 8) when imaged with a two-photon laser, to avoid image saturation caused by the scattered light from the excitation laser. Transistor–transistor logic (TTL) pulses corresponding to scanning initiation and stimulus onsets were recorded in the audio track of the tail-movement videos for future timing synchronization between the visual stimulation, two-photon Ca^{2+} imaging, and motor behaviors.

Protocol

Monitoring Tectal Neuronal Activities and Motor Behavior in Zebrafish Larvae

The zebrafish larva stands out as an excellent vertebrate model. From a behavioral perspective, immediately after the larva hatches, it needs to catch prey and avoid predators. This strong evolutionary pressure leads to a rapid development of functional sensory systems (vision, in particular). By 5 d postfertilization (dpf), tectal cells show distinct visually evoked patterns of activation (Niell and Smith 2005), and the larvae are able to perform a variety of visuomotor behaviors, such as startle, optokinetic, and optomotor responses (Neuhauss 2003; Fleisch and Neuhauss 2006).

Wild-type or nacre (Lister et al. 1999) zebrafish larva, a mutant line lacking melanophores responsible for black pigmentation, of up to 15 dpf can be used. Because the nacre zebrafish larva still express the non-crest-derived retinal pigment epithelium (i.e., they exhibit a normal retina but have a transparent skin), we recommend nacre larvae for these types of experiments. Brain activity in these larvae can be monitored without the need to open the skin or use 1-phenyl-2-thiourea (PTU) to reduce pigmentation, which may affect behavior and retinal pigmentation.

The present method focuses on the use of a synthetic fluorescence dye for monitoring Ca^{2+} changes; however, neuronal activity can also be monitored using transgenic larvae expressing G-CaMP1.6. The genetically encoded G-CaMP is a high-affinity, high-signal-to-noise ratio, genetically engineered fluorescent Ca^{2+} probe, based on the conjugation of a myosin light-chain kinase fragment, a circularly permuted enhanced green fluorescent protein and calmodulin. Although this method did not permit us to monitor changes of Ca^{2+} in tectal cells, when expressing G-CaMP under the Ath5 promoter (expressed almost exclusively in the retina and predominantly in the retinal ganglion cells [RGCs]; Kay et al. 2005), we observed clear and robust Ca^{2+} changes at the RGC terminals reaching the optic tectum neuropil (the major presynaptic component at the zebrafish optic tectum).

MATERIALS

CAUTION: See Appendix 6 for proper handling of materials marked with <!>.
See the end of the chapter for recipes for reagents marked with <R>.

Reagents

Embryo medium <R>
Evans solution <R>
Evans solution <R> with or without 2% tricaine methanesulfonate <!> (MS-222, Sigma-Aldrich)
Low-melting agarose (Invitrogen)
Oregon Green BAPTA-1 AM (Invitrogen)
Pluronic F127, 20% solution in dimethylsulfoxide (DMSO) <!> (Invitrogen)
Zebrafish larvae, wild-type or nacre, 5–15 dpf

Equipment

Band-pass filter (700–870 nm, OD 8)
Capillary glass pipettes (omega dot fiber, Frederick Haer & Co.)
Custom-made mini microscope (20x)
Custom-made recording chamber (see Fig. 1B)

The dimensions of the chamber are 60 x 25 x 25 mm (length x depth x height). Its frame was made of poly(methyl methacrylate). The bottom is made of a slide and one of the long sides is made of a coverslip. The elevated stage is removable and was made with a poly(methyl methacrylate) base and a coverslip platform.

Electrode puller
Fast-infrared video camera
Heat block
High-power stereoscope (200X) preferably equipped with epifluorescence
Infrared diodes (930-nm)
Kodak 32 gelatin filter
Micromanipulator
Pneumatic pump (Picospritzer)
Small LCD display
Two-photon microscope

METHODS

Dye Preparation

1. Dissolve calcium indicator Oregon Green BAPTA-1 AM to a final concentration of 10 mM in the pluronic acid/DMSO solution.
2. Dilute the dye solution 10:1 in Evans solution.
3. Vortex the diluted solution 2 min and spin down for 2 min at 14,000 rpm.

   Because of autocleavage of the acetoxymethyl ester queue of the Ca^{2+} indicator when exposed to water, for best results the solution prepared has to be used for injection immediately.

Dye Injection

4. Pull injection pipettes using capillary glass micropipettes to obtain a tip of 2–3 µm.
5. Backfill the pipettes with the dye solution (from Step 2) and connect them to an electrically gated pneumatic pump.
6. Embed 5–15 dpf larvae, dorsal side up, in 1.2% low-melting agarose on the back side of a Petri dish cover.
7. Once the agarose is jellified, add 0.02% MS-222 in Evans solution or ~13°C Evans solution to anesthetize the larvae during the dye injections.
8. Under a high-power stereoscope (200X), bolus-inject the dye solution (two to three pressure pulses of ~30 msec duration at ~10 psi) directly into the tectal neuropil.
9. Release the larvae from the agarose using a forceps or a dissection pin.
10. Incubate the larvae in the dark, in embryo medium, for 1 h before the experiments.

    This method enables loading a very large portion of the optic tectum cells (generally ~25% of the periventricular layer, which contains 95%–98% of all tectal somata [Niell and Smith 2005; Sumbre et al. 2008]; see Fig. 1A).

Zebrafish Mounting for Imaging

11. Embed the injected larvae in 1.2% low-melting agarose and mount them dorsal-side up on the edge of a platform within the imaging chamber (see Fig. 1B,C).

    To simultaneously monitor cell activities and motor behaviors, we have used 2% low-melting agarose instead, to minimize movement artifacts.

12. After the agarose has set, fill the chamber with embryo solution and verify that the periventricular layer of the optic tectum is fluorescent.

    See *Troubleshooting*.

13. If motor activity of the tail will be monitored, use a fine dissection pin to carefully remove the agarose around the last third section of the tail, in such a way that the larva is stable under the objective but its tail is free to move (see Fig. 1B,C).

    Zebrafish larvae partially embedded in agarose show escape and swimming kinematics as observed in free-behaving larvae (Ritter et al. 2001).

Image Acquisition and Data Analysis

14. Using transmitted light or epifluorescence, locate the labeled area (periventricular layer of the optic tectum), and focus a few micrometers below the skin.
15. Start scanning while moving the plane of focus toward the ventral side of the larva until you begin to see clear labeled somata.
16. Once labeled cells are found, scan continuously, no slower than 1 Hz, and present visual stimuli (as a test it is possible to use a simple torch).

    Keep the infrared laser power low (for our system 50 mW measured before entering the scanning head).

    See *Troubleshooting.*
17. During the acquisition small drifts can occur. If this is the case, the series of images can be later registered to compensate for small drifts in the x–y plane.

TROUBLESHOOTING

Problem (Step 12): No fluorescence is seen in the tectal periventricular layer.
Solution: To properly load the periventricular-layer somata, it is important to bolus-inject the dye solution in the tectal neuropil, from where the dye is taken up by the dendrites and will diffuse to the soma. At this developmental stage (5–15 dpf), the tectal neuropil of the larva is very thin, therefore it is necessary to verify that the dye is injected within the neuropil and not underneath. It is possible to add a nonfluorescent dye (fast green) to permit visualizing the dispersion of the injected dye solution. If no fluorescence is detected in the periventricular layer and some is observed in the ventricles or other regions, the dye was injected underneath the neuropil or too high a pressure was used. If no fluorescence can be detected anywhere in the zebrafish larva, it is possible that the dye was not prepared correctly or that the skin was not penetrated and the dye was injected in the bath.

Problem (Step 16): The tectal cells are loaded with the dye but they show no changes in fluorescence when visual stimuli are presented.
Solution: (1) Focus deeper and try to find different regions that may respond to the type of stimuli presented. (2) Verify if the optic tectum has normal blood circulation compared with other brain regions. (3) If too much dye solution was injected, the high amount of DMSO can damage the injected tissue, or the dye fluorescence intensity may become insensitive to Ca^{2+} changes owing to a high concentration of the indicator in the cellular cytoplasm. Reduce injection parameters.

Problem (Step 16): The cells stop responding to visual stimuli.
Solution: Focus the objective to image a different labeled region, check that laser power is not higher than 45 mW right after the laser attenuator, and increase photomultiplier sensitivity. The infrared laser has probably heated and damaged the tissue.

DISCUSSION

The techniques described here allow simultaneous monitoring of the activity of an unprecedented number of cells (up to ~500) in an intact behaving vertebrate (Fig. 1A,D). In contrast to multielectrode arrays, two-photon Ca^{2+} imaging permits geographical and morphological identification of all the cells being monitored, even if these remain silent. Moreover, using transgenic fish lines expressing red fluorescent proteins under cell-specific promoters (e.g., VGluT [vesicular glutamate transporter] or GAD [glutamic acid decarboxylase]) will make it possible to determine if the cells monitored are inhibitory or excitatory neurons. A disadvantage of the described technique is the relatively low temporal resolution mainly owing to the necessity to scan the tissue to achieve confocality and the long dissociation constant of the Ca^{2+} dyes. Depending on the quality of labeling, the highest acquisition rate achieved was one of 4 Hz with an average rate of 1.5 Hz. Newly developed two-photon systems based on opto-acoustic deflectors and high-sensitivity photon sensors (Salomé et al. 2006; Otsu et al. 2008) may help improve the temporal resolution. Monitoring neuronal activities for long periods of time may also damage the tissue owing to photo-toxicity and the heat generated by the infrared laser.

If necessary, the acquired series of images can be later registered to compensate for small drifts in the *x–y* axis. Drifts in the *z*-axis are rare and cannot be compensated. For experiments in which the tail is free to move, moving artifacts can be expected. In our case it happened in ~30% of the experiments. The artifacts can be distinguished as a clear misalignment of sequentially scanned rows within an acquired frame.

Although it has been shown that changes in intensity of the Ca^{2+} dye fluorescence correlates with the number and rate of spikes (Niell and Smith 2005; Yaksi et al. 2006; Sumbre et al. 2008), this is not always the case (Lin et al. 2007). Overall, the described technique is best used for determining which cells are active within a large neuron population in association with sensory stimuli or motor behaviors, rather than for monitoring the precise firing pattern of single cells.

RECIPES

CAUTION: See Appendix 6 for proper handling of materials marked with <!>.
Recipes for reagents marked with <R> are included in this list.

Embryo Medium

Reagent	Final concentration
NaCl	5 mM
KCl	0.17 mM
$CaCl_2$ <!>	0.33 mM
$MgCl_2$	0.33 mM

Adjust the solution to pH 7.2.

Evans Solution

Reagent	Final concentration
NaCl	134 mM
KCl	2.9 mM
$CaCl_2$ <!>	2.1 mM
$MgCl_2$	1.2 mM
HEPES	10 mM
Glucose	10 mM

Adjust the solution to pH 7.8 and to ~290 mOsm.

REFERENCES

Arrenberg AB, Del Bene F, Baier H. 2009. Optical control of zebrafish behavior with halorhodopsin. *Proc Natl Acad Sci* **106:** 17968–17973.

Burrone J, O'Byrne M, Murthy V. 2002. Multiple forms of synaptic plasticity triggered by selective suppression of activity in individual neurons. *Nature* **420:** 414–418.

Brainard DH. 1997. The Psychophysics Toolbox. *Spat Vision* **10:** 433–436.

Brustein E, Marandi N, Kovalchuk Y, Drapeau P, Konnerth A. 2003. 'In vivo' monitoring of neuronal network activity in zebrafish by two-photon Ca^{2+} imaging. *Pflügers Arch* **446:** 766–773.

Douglass AD, Kraves S, Deisseroth K, Schier AF, Engert F. 2008. Escape behavior elicited by single, channelrhodopsin-2-evoked spikes in zebrafish somatosensory neurons. *Curr Biol* **18:** 1133–1137.

Easter SS Jr, Nicola GN. 1996. The development of vision in the zebrafish (*Danio rerio*). *Dev Biol* **180:** 646–663.

Fleisch VC, Neuhauss SC. 2006. Visual behavior in zebrafish. *Zebrafish* **3:** 191–201.

Hua JY, Smear MC, Baier H, Smith SJ. 2005. Regulation of axon growth in vivo by activity-based competition. *Nature* **434:** 1022–1026.

Kay JN, Link BA, Baier H. 2005. Staggered cell-intrinsic timing of *ath5* expression underlies the wave of ganglion cell neurogenesis in the zebrafish retina. *Development* **132:** 2573–2585.

Lin BJ, Chen TW, Schild D. 2007. Cell type–specific relationships between spiking and $[Ca^{2+}]_i$ in neurons of the *Xenopus* tadpole olfactory bulb. *J Physiol* **582:** 163–175.

Lister JA, Robertson CP, Lepage T, Johnson SL, Raible DW. 1999. Nacre encodes a zebrafish microphthalmia-related protein that regulates neural-crest-derived pigment cell fate. *Development* **126:** 3757–3767.

Neuhauss SC. 2003. Behavioral genetic approaches to visual system development and function in zebrafish. *J Neurobiol* **54:** 148–160.

Niell CM, Smith SJ. 2005. Functional imaging reveals rapid development of visual response properties in the zebrafish tectum. *Neuron* **45:** 941–951.

Otsu Y, Bormuth V, Wong J, Mathieu B, Dugue GP, Feltz A, Dieudonne S. 2008. Optical monitoring of neuronal activity at high frame rate with a digital random-access multiphoton (RAMP) microscope. *J Neurosci Methods* **173:** 259–270.

Pelli DG. 1997. The VideoToolbox software for visual psychophysics: Transforming numbers into movies. *Spat Vision* **10:** 437–442.

Ramdya P, Engert F. 2008. Emergence of binocular functional properties in a monocular neural circuit. *Nat. Neurosci.* **11:** 1083–1090.

Ritter DA, Bhatt DH, Fetcho JR. 2001. In vivo imaging of zebrafish reveals differences in the spinal networks for escape and swimming movements. *J Neurosci* **15:** 8956–8965.

Salomé R., Kremer Y, Dieudonné S, Léger JF, Krichevsky O, Wyart C, Chatenay D, Bourdieu L. 2006. Ultrafast random-access scanning in two-photon microscopy using acousto-optic deflectors. *J Neurosci Methods* **154:** 161–174.

Sumbre G, Muto A, Baier H, Poo MM. 2008. Entrained rhythmic activities of neuronal ensembles as perceptual memory of time interval. *Nature* **456:** 102–106.

Szobota S, Gorostiza P, Del Bene F, Wyart C, Fortin DL, Kolstad KD, Tulyathan O, Volgraf M, Numano R, Aaron HL, et al. 2007. Remote control of neuronal activity with a light-gated glutamate receptor. *Neuron* **54:** 535–545.

Wang S, Szobota S, Wang Y, Volgraf M, Liu Z, Sun C, Trauner D, Isacoff EY, Zhang X. 2007. All optical interface for parallel, remote, and spatiotemporal control of neuronal activity. *Nano Lett* **7:** 3859–3863.

Wyart C, Del Bene F, Warp E, Scott EK, Trauner D, Baier H, Isacoff EY. 2009. Optogenetic dissection of a behavioral module in the vertebrate spinal cord. *Nature* **461:** 407–410.

Yaksi E, Friedrich RW. 2006. Reconstruction of firing rate changes across neuronal populations by temporally deconvolved Ca^{2+} imaging. *Nat Methods* **3:** 377–383.

Zhang F, Wang LP, Brauner M, Liewald JF, Kay K, Watzke N, Wood PG, Bamberg E, Nagel G, Gottschalk A, Deisseroth K. 2007. Multimodal fast optical interrogation of neural circuitry. *Nature* **446:** 633–639.

73 Confocal Calcium Imaging of Neuronal Activity in Larval Zebrafish

Joseph R. Fetcho

Department of Neurobiology and Behavior, Cornell University, Ithaca, New York 11794

ABSTRACT

The ability to image neuronal activity with single-cell resolution is a powerful tool for investigating the neuronal basis of behavior. This protocol describes a method for using backfilling with calcium indicators to monitor neuronal activity in transparent larval zebrafish. The imaging process involves collecting a confocal time series from the cells of interest. The behavior of interest is elicited during the time series to determine whether the cells' increase in fluorescence intensity is associated with the behavior.

INTRODUCTION

Understanding the production of behavior by the nervous system requires knowing which of the many neurons in the nervous system are active during a behavior. Optical monitoring of the activity of groups of individually identified neurons in an intact vertebrate proved difficult until the advent of the ability to backfill neurons with fluorescent calcium indicators. This approach, combined with the ease of imaging in transparent larval zebrafish, led to the first in vivo imaging of neuronal activity in a vertebrate. The methods that were used to accomplish this are the focus of this chapter.

RELATED INFORMATION

This approach has been used to determine which neurons in the brain and spinal cord are active during swimming and escape behaviors in normal and mutant lines of zebrafish (Fetcho and O'Malley 1995; O'Malley et al. 1996; Ritter et al. 2001; Gahtan et al. 2002; Liu et al. 2003; McLean et al. 2007, 2008). This method is the easiest and most reliable approach to imaging activity in zebrafish and is a reliable testing ground for beginners, even if they ultimately plan to use a different method. Other, more fickle, approaches now available include using membrane-permeable calcium indicators or transgenic lines (Brustein et al. 2003; Higashijima et al. 2003; Li et al. 2005).

IMAGING SETUP

The imaging can be accomplished with any confocal microscope and even with wide-field microscopy of the spinal cord. We use a Zeiss laser-scanning confocal imaging system (LSM 510) on an inverted microscope (although an upright scope would work, too). The ideal objective is a Zeiss 40X/1.1 numerical aperture (NA) LD C-Apochromat water-immersion lens, with a total working distance of 0.79 mm. Most long-working-distance (and much less expensive) water-immersion lenses typically used for electrophysiology will work as well. In addition, 488-nm laser light is needed for Calcium Green imaging.

Protocol

Imaging Neural Activity in Zebrafish Larvae

This protocol describes the injection of calcium indicator solution into the spinal cord or muscle of transparent larval zebrafish and subsequent monitoring of neuronal activity by collecting a confocal time series from the cells of interest.

MATERIALS

CAUTION: See Appendix 6 for proper handling of materials marked with <!>.

Reagents

Agar solution (Bacto-Agar; DIFCO 214010); 1.2% or 1.4%–1.6% low-melting-point agarose (Fisher BP1360-100) (see Step 4)

Calcium Green dextran (10,000 MW; Invitrogen); 50% (w/v) dissolved in 10% Hank's solution

Hank's solution, 10% (see Westerfield 1995)

MS-222 (tricaine methanesulfonate; Tricaine-S, Western Chemical) <!>

For anesthesia (Step 4), use 0.02 g of MS-222 per 100 mL of 10% Hank's solution.

Zebrafish, 4–5-d-old larvae

Equipment

Dry block heater (VWR International)

Microelectrode glass, 1.5-mm (A-M Systems)

Micromanipulator

Microscope and analysis software: Zeiss LSM 510 and Axiovert 200M inverted microscope with the LSM 510-associated physiology software for analysis of regions of interest over time

Multiwell culture dish

Petri dishes, plastic

Petri dishes, small (3.5-cm), glass-bottomed (e.g., WPI FluoroDish FD3510)

Picospritzer device (Parker Instrumentation)

Pipette puller (e.g., Sutter Instruments model P 87 or P 97)

EXPERIMENTAL METHOD

Method for Backfilling

1. Pull the injection pipettes from 1.5-mm microelectrode glass on a pipette puller.
2. Break off the tip of the pulled pipette so that the diameter is ~10–20 µm, roughly one-fourth to one-third the width of the spinal cord in a 5-d-old larval fish.
3. Backfill the pipette with ~0.5 µL of the Calcium Green dextran solution.
4. Anesthetize the appropriate number of 4–5-d-old larval zebrafish in the Hank's solution containing MS-222. Then place the larvae in a plastic Petri dish containing a few millimeters depth of solid agar to prevent damage to the electrode and the fish during injection.

5. Pressure-inject the indicator solution into the spinal cord (or muscle) of the fish using a picospritzer device and a micromanipulator. A pressure of 10–20 psi is typically applied in brief 3-msec pulses while observing the orange/red-colored indicator entering the tissue.
6. After injection, place the fish in 10% Hank's solution in different wells of a multiwell culture dish for individual tracking.
7. Store the fish overnight at room temperature or in an incubator at 28.5°C and image the next day.

    The delay between labeling and imaging allows ample time for transport of the indicator and recovery of the fish from the injection.

Preparation of the Fish for Imaging

In our laboratory, imaging is usually performed with inverted microscopes, so the objectives come from below. The fish are held in glass-bottomed Petri dishes. The fish is embedded in agar on the surface of the coverslip.

8. Restrain the fish by embedding some or all of the larvae in agar.
    i. Coat the coverslip with a thin (about one fish thickness) layer of agar.
    ii. Place the anesthetized fish in a groove cut in the agar. Add a tiny amount of agar (kept just above its melting point in a dry block heater) to fill in the groove and cover the fish.

        For optimal imaging, the fish should be against the coverslip—it may float up when the agar is added.

    iii. Once the fish has been covered with a thin layer of agar and the agar has been allowed to set, add 10% Hank's solution on top of the agar to keep the fish moist.

        The fish is healthy if the blood cells are moving rapidly in the vessels running through the spinal cord.

Image Acquisition and Analysis

9. Using the confocal microscope, fully open the confocal aperture and increase the gain of the photomultiplier to collect and amplify as much light as possible from the specimen.

    Under these conditions, the intensity of the laser light used for imaging (15 mW argon laser 488-nm line in the Zeiss 510 at 75% power) can be attenuated to 0.1%–1% of its maximal intensity while still allowing imaging of the responses of neurons. Practice in identifying cell types is important, as it is harder to identify cells under low light levels because the details of neuronal processes are not as obvious as under brighter illumination.

10. Elicit the behavior of interest. For example, escape behavior can be elicited by a touch or an electrical shock. Acquire a series of images of the neuron(s) before and after the behavior.

    See *Troubleshooting*.

11. Use the confocal software to determine whether the cell was brighter after the behavior than before.

    This will determine the magnitude and time course of any increase in fluorescence intensity in the imaged neurons. In our experience, roughly a 10% or more increase in fluorescence ($\Delta F/F$) is observed in the soma for a single action potential.

TROUBLESHOOTING

Problem (Step 10): The cell or cells being imaged are out of focus.

Solution: To image neuronal activity, it is important to keep the preparation as still as possible. Zebrafish neurons are small (most are <10 μm in diameter), and if the fish shifts position by only a few micrometers, the cell or cells being imaged will be out of focus. The fish can use cutaneous respiration and will survive in the agar if it is kept moist. The agar does allow some movement but brings the fish back very close to its original position after the movement, so that the neurons remain in focus. Embedding the larval fish in agar without compromising its health (which usually means killing it) requires practice because the fish are very delicate.

Problem (Step 10): The cells die during imaging.

Solution: The cells must not be exposed to too much light because the indicators used for the imaging are phototoxic. The conditions useful for imaging responses are, unfortunately, not ideal for recording cell morphology. For morphology, the confocal aperture must be reduced to allow good optical sectioning of the tissue. This typically requires increased illumination of the specimen to acquire an adequate signal, which risks damaging the cells. Consequently, it is advisable to collect the physiological data first and then increase light intensity and close the confocal aperture to obtain high-quality images of morphology.

EXAMPLE OF APPLICATION

Figure 1 shows an example in which a group of hindbrain neurons in a mutant line of zebrafish was backfilled with Calcium Green to image the activity of the cells during an escape response. The approach revealed that extra hindbrain neurons, called Mauthner cells, added by a mutation in this line are activated during the behavior, revealing a potential genetic substrate for the addition and incorporation of neurons into a pre-existing network.

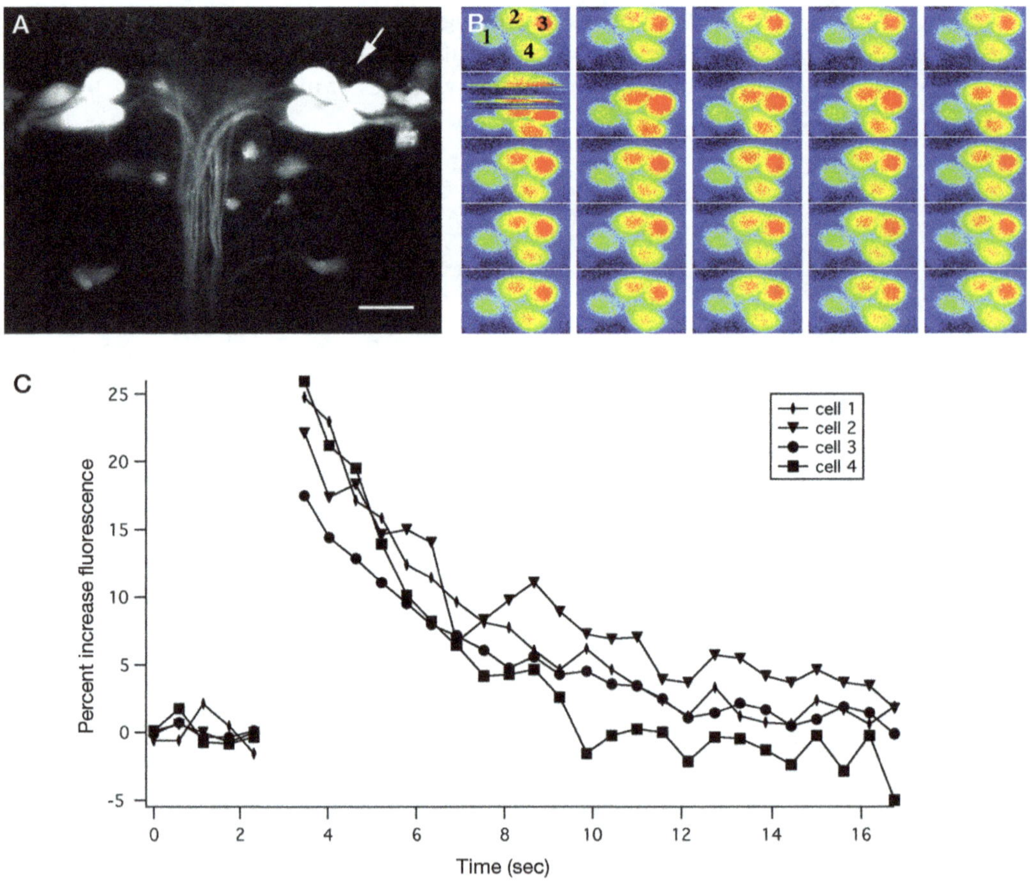

FIGURE 1. Calcium imaging of extra Mauthner cells in the mutant line *deadly seven*. (*A*) Confocal image of the hindbrain of an intact mutant fish in which the Mauthner cells were backfilled by injection of Calcium Green into caudal spinal cord. There is a cluster of Mauthner cells on both the right (arrow) and left sides of the hindbrain, where a normal fish would only have one cell. (*B*) Calcium imaging of the cells during an escape, shown in pseudocolor with red being the brightest, reveals that all of the cells increase in brightness when an escape occurs. A movement artifact from the escape is evident in the sixth panel (reading *left to right*, *top to bottom*). Frames are ordered from *left to right* with 578 msec between frames. (*C*) Quantification of the intensities in the cells labeled 1–4 in the trial in *B*. The normalized intensities ($\Delta F/F$) are plotted versus time and show that all cells responded during the escape. Scale bar, 30 μm. (Reprinted, with permission, from Liu et al. 2003, ©Society for Neuroscience.)

DISCUSSION

The key advantage of calcium imaging is that it allows easy identification of active neurons in the intact fish while it is generating behavior. This is not possible with electrophysiology, which requires paralyzed, dissected preparations. There are two major drawbacks to calcium imaging in this context. First, the injections used to label the cells may damage the circuits of interest. This can be circumvented by using other labeling approaches, such as blastomere injections or injections of AM (acetoxymethyl ester) dyes, but these typically do not label well enough to identify the neurons being imaged (Cox and Fetcho 1996; Brustein et al. 2003). The problem of damage is most effectively avoided by using genetically encoded calcium indicators, which we have applied in zebrafish (Higashijima et al. 2003). Another important problem is that, although calcium levels are related to activity, the relationship is indirect. Calcium levels can increase owing to an influx from outside the cell through the voltage-gated channels, but calcium can also be released from internal stores upon activation of a neuron. It can also enter the cell via synaptically activated channels. These different sources of calcium, along with the effects of spike timing and saturation of indicator, make it more difficult to determine the relationship between a given increase in calcium (fluorescence) and the level of activity of the cell. In vivo, the indicators are best used for determining which cells are active and for looking at patterns of activity, rather than for an exact assessment of the level of activity of individual neurons.

ACKNOWLEDGMENTS

This work was supported by National Institutes of Health grant NS-26539. Many individuals participated in the development of the techniques and in the experiments using them, with major contributions to the approaches described here coming from Drs. Don O'Malley and Katharine Liu.

REFERENCES

Brustein E, Marandi N, Kovalchuk Y, Drapeau P, Konnerth A. 2003. "In vivo" monitoring of neuronal network activity in zebrafish by two-photon Ca^{2+}imaging. *Pflugers Arch* **446:** 766–773.

Cox KJ, Fetcho JR. 1996. Labeling blastomeres with a calcium indicator: A non-invasive method of visualizing neuronal activity in zebrafish. *J Neurosci Methods* **68:** 185–191.

Fetcho JR, O'Malley DM. 1995. Visualization of active neural circuitry in the spinal-cord of intact zebrafish. *J Neurophysiol* **73:** 399–406.

Gahtan E, Sankrithi N, Campos JB, O'Malley DM. 2002. Evidence for a widespread brain stem escape network in larval zebrafish. *J Neurophysiol* **87:** 608–614.

Higashijima S, Masino MA, Mandel G, Fetcho JR. 2003. Imaging neuronal activity during zebrafish behavior with a genetically encoded calcium indicator. *J Neurophysiol* **90:** 3986–3997.

Li J, Mack JA, Souren M, Yaksi E, Higashijima S, Mione M, Fetcho JR, Friedrich RW. 2005. Early development of functional spatial maps in the zebrafish olfactory bulb. *J Neurosci* **25:** 5784–5795.

Liu KS, Gray M, Otto SJ, Fetcho JR, Beattie CE. 2003. Mutations in *deadly seven/notch1a* reveal developmental plasticity in the escape response circuit. *J Neurosci* **23:** 8159–8166.

McLean DL, Fan J, Higashijima SI, Hale ME, Fetcho JR. 2007. A topographic map of recruitment in spinal cord. *Nature* **446:** 71–75.

McLean DL, Masino MA, Koh IYY, Lindquist WB, Fetcho JR. 2008. Continuous shifts in the active set of spinal interneurons during changes in locomotor speed. *Nat Neurosci* **11:** 1419–1429.

O'Malley DM, Kao Y-H, Fetcho JR. 1996. Imaging the functional organization of zebrafish hindbrain segments during escape behaviors. *Neuron* **17:** 1145–1155.

Ritter DA, Bhatt DH, Fetcho JR. 2001. In vivo imaging of zebrafish reveals differences in the spinal networks for escape and swimming movements. *J Neurosci* **21:** 8956–8965.

Westerfield M. 1995. *The zebrafish book: A guide for the laboratory use of zebrafish (*Brachydanio rerio*)*, 3rd ed. University of Oregon Press, Eugene, OR.

74 Voltage-Sensitive Dye Imaging of Neocortical Activity

Amiram Grinvald, David B. Omer, Dahlia Sharon, Ivo Vanzetta, and Rina Hildesheim

Department of Neurobiology, The Weizmann Institute of Science, Rehovot, 76100 Israel

ABSTRACT

Neural computations underlying sensory perception, cognition, and motor control are performed by populations of neurons at different anatomical and temporal scales. Few techniques are currently available for exploring the dynamics of local and large range populations. Voltage-sensitive dye imaging (VSDI) based on organic voltage probes, reveals neural population activity in areas ranging from a few tens of micrometers to a couple of centimeters, or two areas up to ~10 cm apart. VSDI provides a submillisecond temporal resolution and a spatial resolution of ~50 μm. The dye signal emphasizes subthreshold synaptic potentials. VSDI has been applied in the mouse, rat, gerbil, ferret, tree shrew, cat, and monkey cortices to explore the lateral spread of retinotopic or somatotopic activation; the dynamic spatiotemporal pattern resulting from sensory activation, including the somatosensory, olfactory, auditory, and visual modalities; and motor preparation and the properties of spontaneously occurring population activity. In this chapter we focus on VSDI in vivo and review results obtained mostly in the visual system in our laboratory.

INTRODUCTION

The activity of highly distributed neural networks is thought to underlie sensory processing, motor coordination, and higher brain functions. These intricate networks are composed of large numbers of individual neurons, which interact through synaptic connections in complex, dynamically regulated spatiotemporal patterns. To understand network properties and functions, it is helpful to study the ensemble activity of neuronal populations, because coherent activity of many neurons is often responsible for performing the function rather than individual cells. Functionally related subnetworks of neurons are often spatially segregated, making imaging techniques ideal for monitoring population activity. Our understanding of the contribution of single neurons for generating percepts and controlling behavior can be improved within the context of the relationship between single-cell-level and population-level activity.

The remarkable performance of the mammalian brain largely arises from computations taking place in the neocortex. The neocortex is organized into cortical columns (Mountcastle 1957; Hubel and Wiesel 1962), which have lateral dimensions of a few hundred micrometers. Interactions within and between cortical columns occur on the time scale of milliseconds. To follow neuronal computations at the fundamental level of cortical columns in real time, therefore, requires a spatial resolution of ~100 µm and a temporal resolution of ~1 msec. In vivo voltage-sensitive dye imaging (VSDI) (Grinvald et al. 1984; Orbach et al. 1985) fulfills these technical requirements and should help resolve many fundamental questions.

Electrical communication in cortical networks comprises two basic signals: subthreshold potentials (reflecting synaptic input onto dendrites) and suprathreshold action potentials (forming the neuronal output). VSDI appears to relate primarily to spatiotemporal patterns of the subthreshold synaptically driven membrane potentials simply because of the relatively larger area of neocortical dendrites. Therefore, VSDI allows for monitoring of the fluctuations of the membrane potential of the population away from and toward the threshold for action potential. Single unit and multiunit recordings cannot monitor this activity, reflecting only spiking activity. The local field potential, which reflects synaptic potentials, changes its polarity depending on the activity source, and therefore provides ambiguous information about the sign of recorded activity (inhibition vs. excitation). Furthermore, its spatial resolution is far lower. Therefore, VSDI has a unique place among available techniques for measuring neural activity.

Here we describe how in vivo VSDI (Grinvald et al. 1984; Shoham et al. 1999; Grinvald and Hildesheim 2004) can be applied to anesthetized mammals as well as awake behaving monkeys. Chapter 75 provides information on cortical spatiotemporal dynamics in mice awake during behavior. We describe previous findings and technical advances, mostly from our laboratory, and how VSDI has been combined with intracortical microstimulation, single-unit recording, local-field-potential recording, and targeted injections. We conclude by discussing further technical developments to overcome current limitations. VSDI of single cells and additional types of functional imaging using other probes are discussed elsewhere in this book. Other chapters review applications of optical imaging based on intrinsic signals (see Chapters 60 and 83). This chapter begins by reviewing methodology and historical aspects of in vivo VSDI. To provide the reader with a sense for what type of questions may be resolved by this technique, we proceed by discussing results obtained in a variety of mammalian species and paradigms, from sensory cortex of anesthetized rats and cats to motor cortex of behaving monkeys performed in our laboratory.

FROM IN VITRO SINGLE-CELL RECORDINGS TO IN VIVO POPULATION IMAGING

To perform optical imaging of electrical activity, the preparation under study is first stained with a suitable voltage-sensitive dye. The dye molecules bind to the external surface of excitable membranes and act as molecular transducers that transform any changes in membrane potential into optical signals. These optical signals are observed as changes in absorption or emitted fluorescence, and they respond to membrane potential changes in microseconds. The voltage-sensitive dye signals are linearly correlated with both the membrane potential changes and the membrane area of the stained neuronal elements. These optical changes are monitored with light-imaging devices positioned in a microscope image plane. Optical signals using voltage-sensitive dyes were first recorded by Tasaki et al. (1968) in the squid giant axon, and by Cohen and his colleagues in the squid giant axon and in individual leech neurons (Salzberg et al. 1973).

Voltage-sensitive dye imaging (multiple pixels) as opposed to optical recording (single pixel rather than a full image), began with photodiode arrays in invertebrate ganglia (Grinvald et al. 1981) and was subsequently used in mammalian brain slices (Grinvald et al. 1982a) and in the isolated salamander olfactory bulb (Orbach and Cohen 1983). These initial results suggested that optical imaging could be a useful tool with which to study the mammalian brain in vivo as well. However, the initial in vivo experiments in rat visual cortex in 1982 revealed several complications that had to be overcome. One complication was the large amount of noise caused by respiratory and heartbeat pulsation. In addition, the relative opacity and packing density of the cortex limited the penetration of the excitation light and the ability of dyes to stain deep layers of the cortex. Subsequently, new, improved dyes that overcame these problems were developed (e.g., RH-414; Grinvald et al. 1982b), and an effective remedy for the heartbeat noise was found by synchronizing data acquisition with the electrocardiogram and subtracting a no-stimulus trial. These improvements facilitated in vivo imaging of several different sensory systems, including the retinotopic responses in the frog optic tectum (Grinvald et al. 1984), the whisker barrels in rat somatosensory cortex (Orbach et al. 1985), and experiments on the salamander olfactory bulb (Kauer et al. 1987; Kauer 1988; Cinelli and Kauer 1995; Cinelli et al. 1995). The development of more hydrophilic dyes improved the quality of the results obtained in cat and monkey visual cortex (e.g., RH-704 and RH-795; Grinvald et al. 1986, 1994).

The newest generation of voltage-sensitive dyes offers a 30-fold improvement in SNR over the early dyes. This was accomplished by designing dyes that are excited outside the absorption band of hemoglobin, thus minimizing pulsation and hemodynamic noise (Shoham et al. 1999). With this advance it has become possible to reveal the dynamics of cortical information processing and its underlying functional architecture at the necessary spatial and temporal resolution in both anesthetized and behaving animals. Additional advances related to the implantation of transparent artificial dura (Arieli et al. 2002) now allow chronic recordings to be taken over a long period of time. Optical imaging can be performed simultaneously with intracellular recording, extracellular recording, microstimulation, and tracer injection, thanks to the development of an electrode assembly attached to a cranial window (Arieli and Grinvald 2002).

It is important to comment on the relationship between the in vivo dye population signal and membrane potential changes. In simpler preparations, where single cells are distinctly visible, the dye signal looks just like an intracellular electrical recording (Salzberg et al. 1973, 1977; Grinvald et al. 1977, 1981, 1982b). In optical imaging from cortical tissue stained by topical application, however, the optical signal does not have single-cell resolution. Rather, it likely represents the sum of membrane potential changes in both presynaptic and postsynaptic neuronal elements, as well as a possible contribution from the depolarization of neighboring glial cells (Konnerth and Orkand 1986; Lev-Ram and Grinvald 1986). Because of the much larger area of dendrites relative to that of cell somata (~1000-fold), the voltage-sensitive dye signal in cortical tissue reflects, in the main, the postsynaptic potentials in the fine dendrites of cortical cells, rather than action potentials in cell somata. VSDI can therefore easily detect subthreshold synaptic potentials in the extensive dendritic arborization. Recent patch recordings from the rat somatosensory cortex, performed simultaneously with VSDI, indicated that the signals originate primarily from layers 2/3 and reflect dendritic postsynaptic potentials rather than somatic action potentials (Sterkin et al. 1998; Petersen et al. 2003a). Thus, dye signals are uniquely poised to supply data about subthreshold dendritic processing, providing information about aspects of neuronal processing that cannot usually be obtained from single-unit recordings or intracellular somatic recordings.

SPATIAL AND TEMPORAL RESOLUTION

Early VSDI utilized a cumbersome and low-resolution 12 x 12 "diode array camera" (Grinvald et al. 1981). Higher resolution has subsequently been achieved, due mostly to efforts by two groups in Japan: Kamino (Hirota et al. 1995) and Matsumoto (Iijima et al. 1992; Vranesic et al. 1994). Further increases in spatial resolution were achieved by Toyama and colleagues using a stroboscopic light

(Toyama and Tanifuji 1991; Tanifuji et al. 1993). Today, commercial voltage-sensitive dye cameras (charge-coupled device [CCD] or complementary metal oxide semiconductor [CMOS]), with up to 1,000,000 detectors, are available from several vendors.

Detector technology, however, is not the factor that currently limits the spatial resolution achieved with VSDI. The properties of the dyes are a major limiting factor; in particular, the SNR that can be obtained with them, and the photodynamic damage that they cause. The development of suitable voltage-sensitive dyes is a key to the successful application of optical imaging for several reasons. First, different preparations often require dyes with different properties (Ross and Reichardt 1979; Cohen and Lesher 1986; Grinvald et al. 1988). Recently, it was shown that even within the same species, different cortical areas require different dyes (e.g., a given dye provided a high-quality signal in the rat somatosensory cortex but not in the rat olfactory bulb, although the bulb was well stained [Spors and Grinvald 2002]). Second, the use of dyes is associated with difficulties that must be overcome. Under prolonged or intense illumination, dyes cause photodynamic damage. Additional difficulties include bleaching, limited depth of penetration into the cortex, and possible pharmacological side effects.

However, recent intracellular recordings in vivo have directly confirmed that stained cortical cells maintain their response properties (Sterkin et al. 1998; Grinvald et al. 1999, Fig. 23; Petersen et al. 2003a). Furthermore, long-term VSDI in awake monkeys indicated that, even after a year of imaging, monkey visual cortex continues to function normally: The animal maintained normal performance in tasks that required the cortical area be monitored. Thus, new dyes (Shoham et al. 1999) have largely alleviated the problems of pharmacological side effects and photodynamic damage. This means that extensive imaging sessions are now possible, allowing signal averaging of many repeated presentations and improved signal-to-noise ratios (SNRs), thus enhancing spatial resolution.

An additional factor that determines spatial resolution of VSDI is the structure of the cellular elements giving rise to the dye signal. Because the local dendritic trees of neurons have a diameter of ~300–500 μm, a given pixel combines activity from cells with distant somas. Because dendritic trees of neurons, tuned to different functional properties, overlap only partially, subdendrite resolution is possible. The temporal resolution of VSDI is not limited by the response time of dye molecules, because these respond within microseconds to a change in the electrical field over the membrane (Cohen et al. 1974). The temporal resolution is, however, limited by the nature of the population activity and the SNR. The power spectrum of other forms of population activity (electroencephalogram and local field potential [LFP]) suggests that most of the signal does not require a temporal resolution of > 5–10 msec. Achieving high SNRs depends largely, as explained above, on the dyes used and is somewhat more demanding in VSDI than in the sister technology of intrinsic imaging because the samples are shorter in time. This is because of the square-root relationship between the number of samples and the SNR. It is much easier to obtain a good SNR when signals are slower. For instance, the SNR in measurements of small signals is 33-fold higher when the signals have a rise time of 1 sec, as compared to a rise time of 1 msec.

Empirically, we find that current dyes and equipment allow excellent SNR at a sampling rate of 5–10 msec with each pixel looking at a 64 × 64-μm area of cortex. Finer temporal and spatial sampling does not usually provide additional data, although exceptions do exist as noted in the study of the lateral spread beyond the retinotopic border, when a sampling rate of 0.6 msec was used (Derdikman et al. 2003; Petersen et al. 2003a,b). VSDI has been said to sacrifice spatial resolution to gain better temporal resolution. The above numbers actually show, however, that no such tradeoff exists with the present state of the art, relative to intrinsic imaging. Improvements in the dyes and in the spatial resolution of fast cameras have made it possible to obtain high-resolution functional maps of orientation columns, "lighting up" in milliseconds with an SNR even better than that obtained with the slow intrinsic signals (see Fig. 2 below). Additional developments will undoubtedly introduce further improvement.

In summary, VSDI of cortical activity is a particularly attractive technique for providing new insights into the temporal aspects of mammalian brain function. Among its advantages over other methodologies are (1) direct recording of the summed membrane potential changes of neuronal populations, including fine dendritic and axonal processes; (2) ability to measure these repeatedly from the same cortical region over an extended period of time, using different experimental or stimulus conditions; (3) imaging spatiotemporal patterns of activity of neuronal populations with a submillisecond temporal resolution; and (4) selective visualization of neuronal assemblies (see below). Several related reviews have been published elsewhere (Tasaki and Warashina 1976; Waggoner and Grinvald 1977; Cohen et al. 1978; Waggoner 1979; Grinvald 1984, 1985; Cohen and Lesher 1986; De Weer and Salzberg 1986; Salzberg et al. 1986; Loew 1987; Orbach 1988; Grinvald et al. 1991; Kamino 1991; Cinelli and Kauer 1992; for more detailed reviews, see Grinvald et al. 1988, 1999; Grinvald and Hildesheim 2004: Grinvald and Petersen 2010).

DISTRIBUTED PROCESSING: THE SPREAD OF ACTIVITY FAR BEYOND THE RETINOTOPIC/SOMATOTOPIC REPRESENTATIONS

One outstanding question that has benefited from real-time optical imaging is that of the distance in the cortical surface across which activation by a sensory point stimulus spreads. As mentioned above, the voltage-sensitive dye signal in cortical tissue reflects, in the main, the postsynaptic potentials in the fine dendrites of cortical cells rather than action potentials in cell somata. VSDI is thus ideally suited to answer this question.

The frog retinotectal connections offer a system that is topographically well organized: Each spot of light on the retina activates a small region in the optic tectum. The first optical imaging study investigating the spread of activation concentrated on visualizing the topographic distribution of sensory responses in the frog (Grinvald et al. 1984). The optical signals obtained from the tectum in response to discrete visual stimuli were found to correspond well to the known retinotopic map of the tectum. However, in addition to a focus of excitation, the spatial distribution of the signals showed smaller, delayed activity (3–20 msec) covering a much larger area than would be expected on the basis of classic single-unit mapping.

In rat primary somatosensory cortex, each whisker projects to a well-defined region termed the whisker barrel, and there is a simple somatotopic organization of the different whisker barrels side by side. This preparation thus offered a convenient opportunity to explore the question of activation spread in the mammalian brain. When the tip of a whisker was gently moved, optical signals were observed in the corresponding cortical barrel field (Derdikman et al. 2003). However, a discrepancy was noted between the size of an individual barrel, as recorded optically (1300 μm) and the histologically defined barrel (300–600 μm in layer 4 of the cortex, showing neuronal somata rather than processes). The reason for this difference is probably that most of the optical signal originates from the superficial cortical layers in which neurons extend long processes to neighboring barrels (Orbach et al. 1985; Kleinfeld and Delaney 1996; Takashima et al. 2001; Derdikman et al. 2003). Glutamate antagonists blocked the spread. Thus, postsynaptic activity in the dendritic processes could account for the detected spread (Petersen et al. 2003a,b). This lateral spread is both excitatory and inhibitory, and the balance probably depends on the stimulus parameters (Fig. 1C). Net surround inhibition was also documented at late times in the surround of an activated barrel (Orbach et al. 1985; Takashima et al. 2001; Derdikman et al. 2003).

In monkey striate cortex, retinotopic imaging experiments also showed activity over a cortical area much larger than predicted on the basis of standard retinotopic measurements in layer 4, but consistent with the anatomical finding of long-range horizontal connections in visual cortex (Gilbert and Wiesel 1983). The results of these experiments were used to calculate the cortical point spread function, which reflects the extent of cortical activation by retinal point stimuli. Figure 1 illustrates the cortical point spread function in macaque primary visual cortex. To show the relationship between the observed spread and individual cortical modules, the spread function is projected on a histological section of cytochrome oxidase blobs (Fig. 1B). The stimulus used here caused spiking, activated only in neurons residing in the marked small square, which contains just four blobs. However, more than 200 blobs had access to the information carried by the signal spread, albeit at a lower amplitude. The apparent "space constant" for the spread was 1.5 mm along the cor-

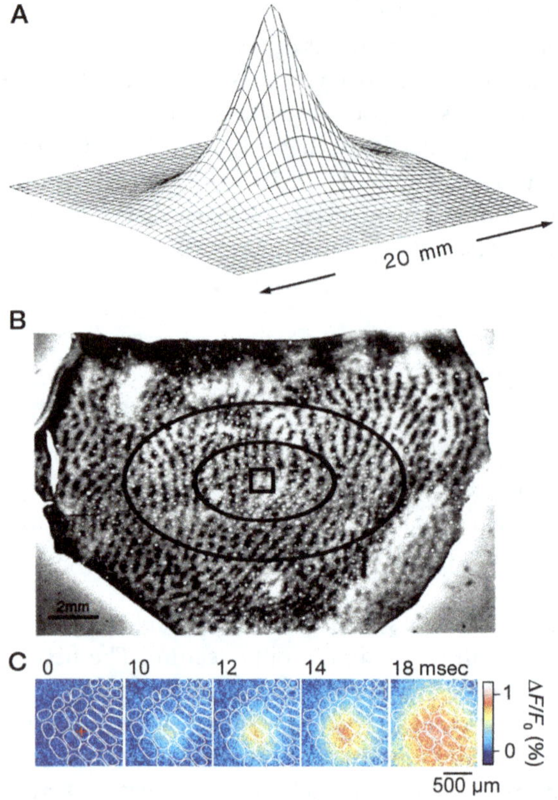

FIGURE 1. Many functional domains are activated during the processing of a small retinal image or a single whisker deflection. (*A*) Calculation of the activity spread from a small patch in layer 4 (a 1 x 1-mm square) within the upper cortical layers of a macaque. This cortical activation was produced by a retinal image of ~0.50° x 0.25° that was presented to both eyes. The space constants for the exponential activity spread measured with the dye experiments were 1.5 and 2.9 mm in the cortical axes perpendicular and parallel to the vertical meridian representation, respectively. (*B*) Direct activation in layer 4 (of the area within the square; calculated from retinotopic extracellular recordings) and the spread in layers 2/3 (elliptical contours; calculated from VSDI) are shown superimposed on a histological section showing the mosaics of cytochrome oxidase blobs, close to the border between cortical areas V1 and V2. The center ellipse shows the contour at which the amplitude of cortical activity drops to 37% of its peak. The larger ellipse shows the contour at which the spread amplitude drops to 14%. More than 10,000,000 neurons reside in the cortical area bounded by the large ellipse containing a regular mosaic of about 250 blobs. (*C*) The exact spatiotemporal pattern of the spread in the somatosensory cortex in response to a single barrel stimulation. The temporal resolution was 2 msec/frame. (*A,B*, Reprinted from Grinvald et al. 1994; *C*, data from Peterson et al. 2003a,b.)

tical axis, parallel to the ocular dominance columns, and 3 mm along the perpendicular axis. The spread velocity was 0.1–0.2 m/sec (Grinvald et al. 1994). A recent study in anesthetized cats showed that this spreading subthreshold synaptic activity is priming the cortex and the cortex then responds differently to a subsequent stimulus; it has been reported that perception of motion when only stationary stimuli were displayed could be explained by this type of activity, which is totally elusive to unit recordings (Jancke et al. 2004).

The extensive lateral spread observed beyond the retinotopic borders in layer 4, both excitatory and inhibitory, indicates that the degree of distributed processing in primary visual cortex is much larger than often conceived—certainly a nontrivial challenge for theoreticians studying cortical networks. In these early studies of the visual response using VSDI, the spatial resolution did not enable exploration of such basic attributes of the visual response as orientation selectivity. Subsequent studies, described below, addressed this issue. Much higher spatial resolution data showing similar spread in the somatosensory cortex was first reported by Derdikman et al. (2003) and Petersen et al. (2003a) (Fig. 1C) and several subsequent papers from the same group (Petersen et al. 2004; Ferezou et al. 2006, 2007; Berger et al. 2007).

The sophisticated nature of the effect of the long-range spread as revealed in the visual cortex is discussed below.

LONG-RANGE HORIZONTAL SPREAD OF ORIENTATION SELECTIVITY IS CONTROLLED BY INTRACORTICAL COOPERATIVITY

During the last three decades, the intracortical anatomical connectivity made by the horizontal connections has been extensively explored in multiple species. It has generally been concluded that horizontal axons in visual cortex bind distant columns sharing a similar orientation preference.

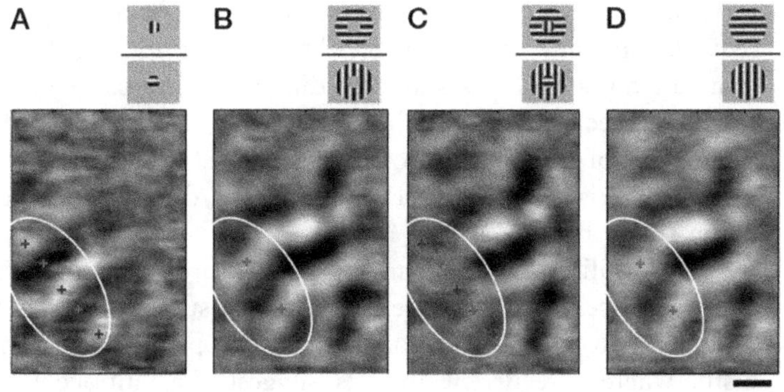

FIGURE 2. Horizontal spread and center-surround competition. (*A*) Area 18 differential map obtained by dividing the responses to horizontal vs. vertical gratings presented through a central disk aperture (stimulus diameter 4° at an eccentricity of 7.6°; see cartoons). Orientation map is confined locally. (*B*) In comparison, displaying a larger grating through an annular aperture without stimulating the central disk (stimulus inner diameter, 4°; outer diameter, 12°; same eccentricity) results in the propagation of an orientation-selective signal within the cortical representation of the central disk region. (*C*) When both stimuli compete in a composite cross-oriented configuration, the orientation map disappears within the retinotopic representation of the central disk. (*D*) Control condition with iso-oriented center-surround configurations. The white contour encircles the expected retinotopic representation of the central disk. Scale bar, 1 mm.

However, the functional selectivity of the horizontal spread has never been measured directly. One reason is that the connections made by the horizontal axons are difficult to record because they only have a subthreshold impact on their postsynaptic targets. Therefore, to unveil the functional expression of the horizontal connectivity, it is mandatory to have access to the activation of the horizontal network at the subthreshold, postsynaptic integration level. Chavane et al. (2009) reported a multiscale analysis of visually driven horizontal network activation, using direct population and intracellular measures of postsynaptic integration. Voltage-sensitive dye imaging shows that while global activation in response to a local stimulus shows long-range horizontal spread, the orientation-selective component of this response does not spread beyond the feedforward cortical imprint of the stimulus. Orientation selectivity decreases exponentially with horizontal distance. Therefore, beyond a distance of one hypercolumn, horizontal functional connectivity no longer obeys a binding rule preserving isoorientation preference with this type of stimulus. Intracellular recordings show that this loss of orientation selectivity arises from the diversity of convergence patterns of intracortical synaptic input originating from beyond the classical receptive field.

In contrast, when increasing the spatial summation evoked by the stimulus—for example, by presenting annular stimuli—orientation-selective activation spreads beyond the feedforward imprint (compare Fig. 2A,B: the orientation map). It was therefore concluded that stimulus-induced cooperativity at the network level is needed for the emergence of long-range orientation-selective spread.

Chavane et al. (2009) shows two different dynamic behaviors of the same network for two distinct stimulus configurations: a single local stimulus does not propagate orientation preference through the long-range horizontal cortical connections, whereas stimulation imposing spatial summation and temporal coherence facilitates the build up of propagating activity showing a strong orientation preference. These observations do not necessarily contradict each other. They point to the possibility that complex stimulus configurations are allowed and expected based on activation within a locally confined region in visual space, whereas an iso-oriented stimulus with a large spatial extent supports the expectation of that specific orientation throughout a yet-larger area.

DYNAMICS OF SHAPE PROCESSING AT SUBCOLUMNAR RESOLUTION

A key goal for functional brain mapping is to obtain single-condition maps: that is, to acquire the pattern of activation evoked by a single stimulus condition, without the need to perform differential imaging of two orthogonal stimuli. This is important because for resolving most of the questions requiring imaging, truly orthogonal stimuli cannot be defined, and because analyzing single-condition data reveals processes that are difficult or impossible to see with differential data (as described within this section; see also Chapter 83). With the technical advances in VSDI, we were able to obtain time series of single-condition maps millisecond by millisecond, at subcolumnar resolution. We used it to revisit fundamental questions regarding the emergence of orientation selectivity in visual cortex (Sharon and Grinvald 2002).

Two families of mechanisms have been proposed to play a role in the emergence of the highly orientation-selective responses of cortical visual neurons, a property not shared by their thalamic inputs. Feedforward-only models suggest appropriate alignment of thalamic input as the mechanism, whereas recurrent models suggest that intracortical interactions are more important. Measuring the dynamics of the response to the input by visual cortex, as enabled by VSDI, is relevant to this question because the feedforward explanation predicts that orientation selectivity should remain constant with time from stimulus onset, whereas if recurrent interactions are important, then selectivity should change as the cortical network performs its processing.

To answer the question of dynamics of orientation selectivity, we first acquire the crucial high-quality single-condition maps. A time series of the initial response is shown in Figure 3A for two orthogonal orientations. As soon as the response is observable, the two orthogonal stimuli are shown to preferentially activate complementary patches of cortex. This pattern is, of course, easily observed in the differential map, Figure 3B,C. To calculate orientation tuning curves, however, differential maps and time courses are of no use, and single-condition responses are needed (Fig. 3D).

In experiments in which six different orientations were shown, the tuning curves at each point in time could be calculated directly from the single-condition responses (Fig. 3E). The immediate impression is that of tuning curves with a constant shape but changing amplitude. Indeed, the half-width at half-height of tuning curves was steady right from response onset (Sharon and Grinvald 2002). Sustained intracortical processing, therefore, does not seem to be needed to determine orientation tuning width, at least for the majority of the population.

There are additional aspects of the response that affect orientation selectivity, not just tuning width. As shown in Figure 3C, the modulation depth of the response (difference between preferred and orthogonal) does change, decreasing after a peak at 100 msec. An intriguing phenomenon is nearly undetectable in the differential time course, but very obvious when considering the evoked single-condition responses (Fig. 3D). There is a notch in the evoked response, equivalent to a deceleration followed by acceleration in the rise time—this is termed the evoked DA (deceleration–acceleration) notch. It is likely caused by a suppressive mechanism peaking at 50–80 msec. These are the kind of dynamics that one would expect from intracortical processing. Furthermore, the evoked DA notch is more pronounced in response to the orthogonal stimulus than to the preferred stimulus (see Fig. 3D and Sharon and Grinvald 2002).

These results suggest that thalamic input may be the major determinant of orientation tuning width for most cortical neurons, but that intracortical processing is critical in amplifying the orientation-selective component of the response. They also hint that intracortical suppression contributes to this process by preventing the orthogonal response from increasing as rapidly as the response to the preferred orientation.

SELECTIVE VISUALIZATION OF NEURONAL ASSEMBLIES

It has been suggested that neurons operate in assemblies (Hebb 1949), networks of neurons that may or may not reside locally, and which communicate coherently to perform the computations required for various tasks. Thus, in VSDI, neurons at the recorded site may belong to different neuronal assemblies. To explore cortical computations, it is often interesting to look at the dynamics of individual neuronal assemblies, rather than at the activity originating from heterogeneous populations

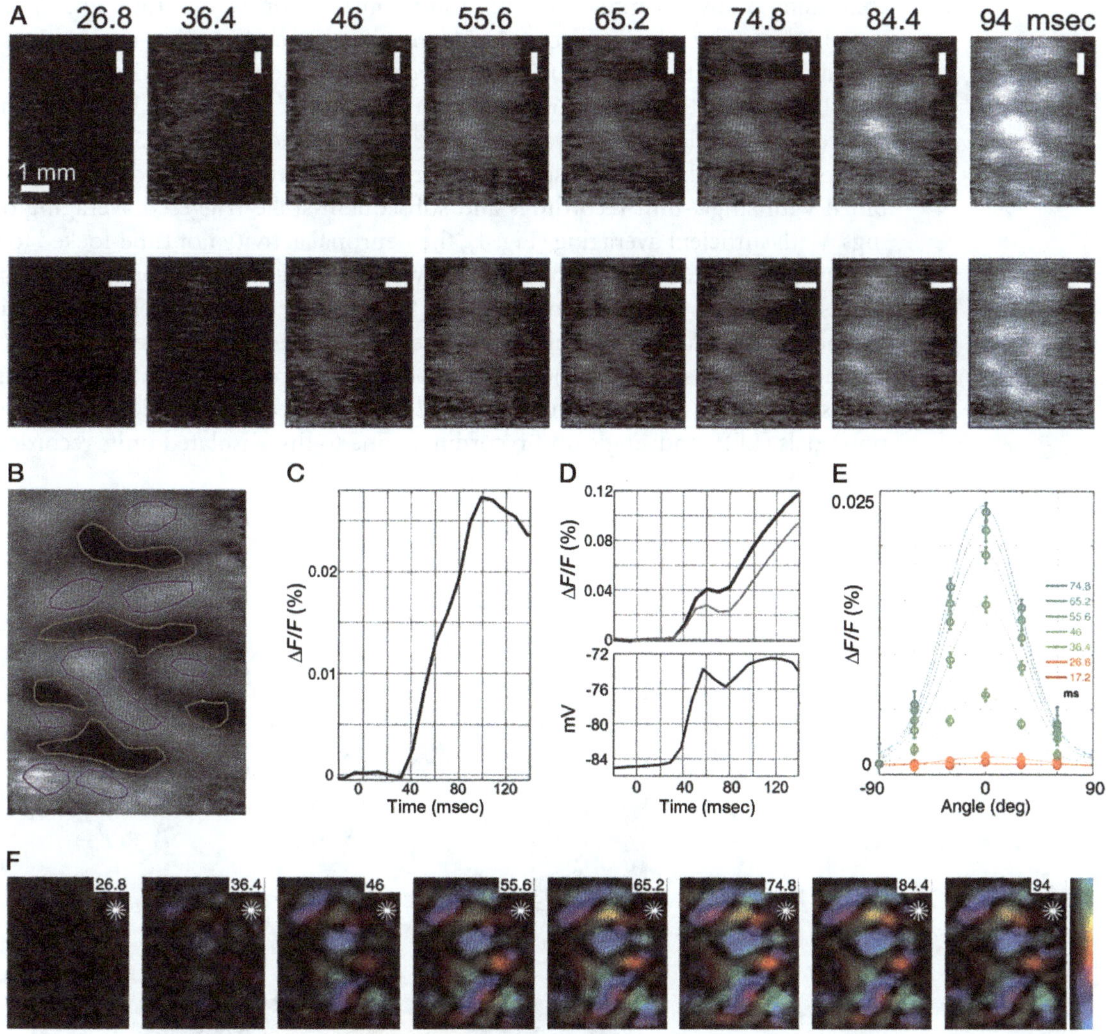

FIGURE 3. The dynamics of orientation tuning. (*A*) The two rows show the time series of single-condition orientation maps, in response to two orthogonal stimuli. (*B*) Spatial pattern revealed by differential imaging. (*C*) Temporal pattern of differential activity. The deceleration–acceleration (DA) notch is hardly visible here, thus it was overlooked in earlier studies. (*D*) (*Top*) The time course of the evoked response to preferred (black) and orthogonal (gray) orientations, calculated from the time series of single-condition maps shown in *A*. The two responses have the same onset latency (~30 msec); the response to the preferred orientation is larger from the onset. The striking feature of the evoked responses is the DA notch. *C* and *D* were calculated for pixels in the regions marked in *B*. (*Bottom*) Confirmation of the DA notch with intracellular recordings performed in our laboratory (Sterkin et al. 1998). The notch in the intracellular data appears at the same time as that in the dye data, at ~50–80 msec. After the notch, the dye response continues to grow a lot more than the intracellular response; this could be attributable to the fact that the dye measures membrane potential changes over the entire dendritic arborization, and not just in the soma. (*E*) Orientation tuning curves at different times after stimulus onset, from another experiment. The curves after response onset (green) have the same shape but different amplitudes. The baseline of the curves was shifted to facilitate comparison. (*F*) Time series of orientation preference maps in polar depiction: Color represents preferred orientation (top to bottom of color scale on right is 0°–180°), and brightness represents amplitude of differential response (*left to right* of color scale is 0%–0.05%). Preferred orientation is steady from response onset. The VSDI data in *A–D* are raw, unfiltered. (Modified from Sharon and Grinvald 2002.)

containing many assemblies. A significant contribution of real-time optical imaging has been the visualization of the dynamics of coherent neuronal assemblies (i.e., neuronal assemblies in which the activity of cells is time-locked). The firing of a single neuron is used as a time reference to selectively visualize activity that is synchronized with it (i.e., only the activity in the assembly to which the reference neuron belongs [Grinvald et al. 1989, 1991; Arieli et al. 1995]).

To isolate and study the spatiotemporal organization of neuronal assemblies, VSDI was combined with single-unit recordings and subsequent spike-triggered averaging of the optical recordings. With sufficient averaging (Fig. 4), the neuronal activity not time-locked to the reference neuron was averaged out, enabling the selective visualization of those cortical locations in which activity consistently occurred coherently with the firing of the reference neuron. The visual cortex (area 18) of anesthetized cats was stained with the dye RH-795, and either ongoing (spontaneous, in the absence of visual stimulation) or evoked activity was recorded continuously for 70 sec. We simultaneously recorded optical signals from 124 sites, together with electrical recordings of local field potentials (LFP) and single unit recordings (one to three isolated units recorded with the same elec-

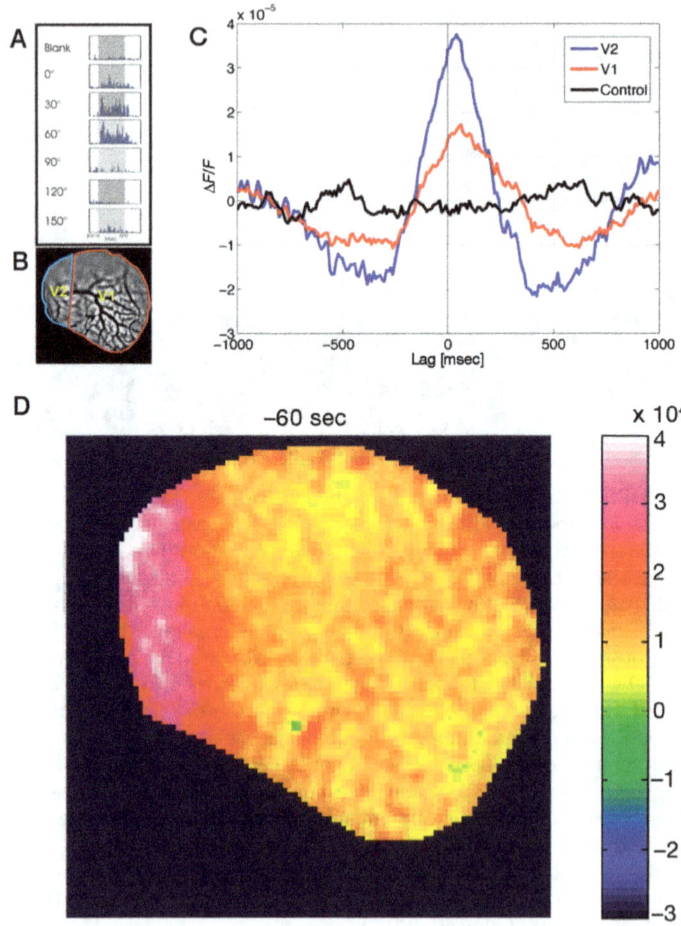

FIGURE 4. Visualization of coherent neuronal assemblies across visual areas. (*A*) The tuning properties of a single cell in V1 of primary visual cortex of the awake monkey, which was used for the spike-triggered averaging. (*B*) Cartoon of the imaged area showing the V1/V2 border that was identified by mapping the ocular dominance columns in V1. (*C*) The time course of spike-triggered average (STA) of the optical signal from V2 (blue curve) as well as the optical signal from V1 (red curve) on the time of the firing of the V1 reference neuron. Evidently, whenever this cell fired, there was activity over a large area in V1 as well as V2. In fact, in this case, whenever this cell in V1 fired there was larger coherent activity in V2 than in V1. (*D*) The average spatial pattern 60 msec before the V1 action potential occurred is shown. (Courtesy of David Omer and Amiram Grinvald; unpublished results.)

trode). The spike-triggered averaging analysis showed that the averaged optical signal at the electrode site had a peak that temporally coincided with the occurrence of a peak in the LFP. The dye signal was similar to the local field potential recorded from the same site. This indicates the expected result that many neurons next to the electrode site had coherent firing patterns. Interestingly, however, the dye signals more distant from the electrophysiological recording site were heterogeneous, indicating that VSDI provides a better spatial resolution than field potential recordings.

In 88% of the neurons recorded during spontaneous activity, a significant correlation was found between the occurrence of a spike and the optical signal recorded in a large cortical region surrounding the recording site, including cortical sites up to 6 mm away. This result indicates that spontaneous activity of single neurons is not an independent process but is time-locked to the firing or the synaptic inputs from numerous neurons, all activated in a coherent fashion, even without a sensory input. Surprisingly, it was found that the amplitude of this coherent ongoing activity, recorded optically, was often almost as large as the activity evoked by optimal visual stimulation. The amplitude of the ongoing activity, which was directly and reproducibly related to the spontaneous spikes of a single neuron, was, on average, as high as 54% of the amplitude of the visually evoked response by optimal sensory stimulation, recorded optically. Coherent activity was detected even at distant cortical sites up to 6 mm apart. On the other hand, in a few cases, it was found that the spontaneous activity of two adjacent neurons, isolated by the same electrode and sharing the same orientation preference, was correlated with two different spatiotemporal patterns of coherent activity, suggesting that adjacent neurons in the same orientation column can belong to different neuronal assemblies.

A subsequent study explored the relationship between the spatial pattern of population activity coherent with a single neuron's spikes during spontaneous and evoked activity. They were found to be very similar. Furthermore, it has been reported that the firing rate of a spontaneously active single neuron strongly depends on the instantaneous spatial pattern of ongoing population activity in a large cortical area: During spontaneous activity, whenever the instantaneous spatial population pattern correlated highly with the spatial population pattern evoked by the preferred stimulus of the recorded neuron, its firing rate increased. This was used to reconstruct the spontaneous activity of single neurons based on the global spontaneous activity (Tsodyks et al. 1999).

These results indicate that the spontaneous firing of single neurons is tightly linked to the cortical networks in which they are embedded. The idea of a neural network is a central concept in theoretical brain research, and it is finally possible to directly visualize the cortical networks and their states in action, at high spatiotemporal resolution. Exploring cortical states is likely to reveal new fundamental principles about neural strategies for cortical processing, representations of objects, memories, context, expectations, and particularly about the interplay between internal cortical representations and the sensory input in primary sensory areas.

The finding that the amplitude of spontaneous ongoing activity in neuronal assemblies is nearly as large as evoked activity suggests that it may play an important role in shaping spatiotemporal patterns evoked by sensory stimuli (Arieli et al. 1996a). Therefore, it is important to be able to study the dynamics of ongoing and evoked activity without signal averaging.

ONGOING ACTIVITY IN CORTICAL PROCESSING: IMAGING WITHOUT SIGNAL AVERAGING

Despite the large biological noise originating from respiration and heartbeat pulsations, reliable VSDI was accomplished even without signal averaging, using off-line correction procedures instead. This allowed, for the first time, exploration of the dynamics of spontaneous ongoing population activity and its interaction with evoked activity.

In the mammalian visual cortex, evoked responses to repeated presentations of the same stimulus show a large variability. It has been found that this variability results from ongoing activity, reflecting the dynamic state of the cortical network. The central finding in the initial studies of ongoing activity using VSDI without signal averaging was that, despite this large variability, the evoked responses in single trials can be predicted by taking into account the preceding ongoing activity.

The evoked response in a given single-stimulus presentation can be predicted by summing the preceding ongoing activity with the evoked response, averaged over many presentations. This prediction is valid as long as the ongoing activity pattern, which presumably continues to change during the evoked response, is still similar to the initial state (Arieli et al. 1996a).

These findings indicate that the old notions of what "noise" in brain activity represents may have to be revised. Because the ongoing activity is often very large, it would be expected to play a major role in cortical function. It may provide the neuronal substrate for the dependence of sensory information processing on context, attention, behavioral and consciousness states, memory retrieval, and other aspects of cognitive function. Preliminary experiments have shown that behavior itself is also affected by ongoing activity (Arieli et al. 1996b).

The newly improved SNR of VSDI enabled exploration of the dynamics of ongoing activity and its relation to internal representations of sensory attributes. It was found that ongoing activity in the visual cortex of the anesthetized cat is composed of dynamically switching intrinsic cortical states, many of which closely correspond to orientation maps. When such an orientation state emerged spontaneously, it spanned several cortical hypercolumns and was usually followed by a state corresponding to a proximal orientation. Otherwise, it dissolved into an unrecognized state or noisy pattern after a few tens of milliseconds (Grinvald et al. 1991; Shoham et al. 1991; Kenet et al. 1997, 2003).

The above findings raised the hypothesis that dynamically switching cortical states could express the brain's internal context, representing and influencing memory, perception, and behavior. Such studies obviously require the use of the awake behaving monkey rather than anesthetized subjects. This is indeed being explored in behaving monkeys as described by Slovin et al. (2002) and by Omer and Grinvald as described in Grinvald and Petersen (2010). They performed VSDI of ongoing cortical activity in the visual cortices of awake monkeys simultaneously with measurements of single-unit activity and the local-field potential. They found coherent activity also in the awake monkey (Fig. 4). This figure shows the tuning properties of the reference cell (Fig. 4A), the border between the two imaged areas V1 and V2 (Fig. 4B), time course of the coherent activity in area V1 (red) and V2 (blue; Fig. 4C), and the instantaneous spatial pattern of the assembly activity just before the action potential of the reference neuron (Fig. 4D). Overall, the dynamics were very different from that found in anesthetized cats and in a single epoch it was not possible to recognize any pattern that looked like the known functional architecture. Nevertheless, it was found that such patterns do appear: The pairwise correlation among pixels of known functional architecture as a function of the orientation difference between these pixels or their ocularity index had the same stereotypic shape for evoked activity and for spontaneous epochs. Similar results were also obtained in the anesthetized monkeys. However, in the anesthetized monkeys spontaneous cortical activity that resembled the functional architecture was also found just like in the anesthetized cats (D. Omer and A. Grinvald, in prep.). These results underscore the importance of carrying out experiments on awake behaving animals because the spatiotemporal patterns of activity are so drastically different.

LONG-TERM VSDI OF CORTICAL DYNAMICS IN BEHAVING MONKEYS

Chronic optical imaging based on newer voltage-sensitive dyes was developed to facilitate exploration of the spatial and temporal patterns underlying higher cognitive functions in the neocortex of behaving monkeys. The technique was used to explore cortical dynamics, with high spatial and temporal resolution, over periods of up to 1 yr from the same patch of cortex (Slovin et al. 2002). The visual cortices of trained macaques were stained one to three times a week and, immediately after each staining session, the monkey started to perform a simple behavioral task, while the primary visual cortex (V1) and the secondary area (V2) were imaged with a fast optical imaging system. Long-term, repeated VSDI from the same cortical area did not disrupt the normal cortical architecture, as confirmed repeatedly by optical imaging based on intrinsic signals. The spatial patterns of functional maps obtained by VSDI in response to full-field stimuli were essentially identical to those obtained from the same patch of cortex by imaging based on intrinsic signals. On comparing the relative amplitudes of the evoked signals and differential maps obtained using these two different imaging methods, it was found that the relative contribution of subthreshold versus spiking activity was higher in the voltage-sensitive dye signal than in the intrinsic signal. In the awake monkey, evoked activity could also be detected without signal averaging, similarly to the above-mentioned anesthetized cat experiments.

The same approach also proved useful in the exploration of the motor cortex. Electrical recording studies during Go/No-Go tasks found that primary motor cortex (M1) neurons display changes in activity which begins primarily after a Move cue, whereas dorsal premotor (PMd) neurons, known to be engaged in motor planning, display changes in activity that can begin during the Instructed Delay periods. One monkey was trained to perform a Go/No-Go task. The monkey began a trial by placing his hand over a small photodiode in front of him. This caused a small white square to appear on a computer screen. After a variable pre-cue period (3000–4000 msec), the white square was changed to red (Go mode) or green (No-Go mode). Then, after a variable Instructed Delay period (300–1500 msec), a luminance change of the square told the monkey to follow the prior instruction: Uncover (Go) or continue to cover the photodiode (No-Go). In the PMd, the VSD signal changes began shortly after the onset of the Go or No-Go instructions (i.e., at the start of the Instructed Delay periods). Relatively large signals were observed in the PMd area during both Go and No-Go. The VSD signals were much smaller in M1 during the No-Go mode. These observations suggest that the preparation to perform or withhold a motor response is associated mainly with marked subthreshold activity in the PMd (Strick et al. 2003).

This work shows a new way of recording and analyzing the dynamics of population activity in behaving monkeys, with high spatial and temporal resolution. The combination of VSDI with traditional electrical recordings can also be readily adapted to facilitate the selective visualization of coherent activity in neuronal assemblies involved in dynamic representations and processing of sensory input, as well as in the planning, control, and execution of motor output (Slovin et al. 2003; Strick et al. 2003). Another powerful combination is described below.

COMBINED VSDI AND MICROSTIMULATION IN FRONTAL AND MOTOR CORTEX

Despite extensive use of microstimulation in various brain regions to affect neuronal and behavioral responses (see, e.g., Newsome et al. 1989; Cohen and Newsome 2004), the spatiotemporal pattern of activation it evokes has not been characterized, leaving open many questions about the spread of activation, both direct, from the electric shock itself, and indirect, from synaptic activity. Microstimulation can be incorporated into VSDI studies of the frontal and motor cortex of awake monkeys.

The frontal eye field and neighboring area, 8Ar, of the primate cortex are involved in programming and execution of saccades. Electrical microstimulation in these regions elicits short-latency contralateral saccades. To determine how spatiotemporal dynamics of microstimulation-evoked activity are converted into saccade plans, the combination of VSDI and microstimulation in behaving monkeys was used. Short stimulation trains evoked a rapid and widespread wave of depolarization followed by an unexpectedly large and prolonged hyperpolarization. During this hyperpolarization, saccades are almost exclusively ipsilateral, suggesting an important role for hyperpolarization in determining saccade goal (Seidemann et al. 2002). This result shows that neural activity, with composite spatiotemporal dynamics, can be elicited by microstimulation and emphasizes the importance of further characterization of microstimulation-evoked activity for the interpretation of the behavioral effects of microstimulation.

Intracortical stimulation has also been used as a tool to discover the pattern of muscle and/or movement representation in motor areas of the cortex. VSDI was used to examine the patterns of activation evoked by different stimulation parameters in the primary motor cortex (M1). Those experiments used much finer stimulation conditions than the previous work. Single cathodal pulses in M1 (15–30 μA, 0.2-msec duration) evoked short-latency dye responses that continued for 40–50 msec after the onset of stimulation. These responses spread in an asymmetric manner from the stimulation site in M1; the width of the activated area was ~1.5–3 mm. Single pulses, >15 μA, also evoked activity at spatially separate sites within M1. Short trains of pulses in M1 (2–20 pulses, 330–500 Hz, 30–100 μA) produced activation that spread from the electrode site to adjacent premotor areas and to the contralateral hemisphere. A distinct hyperpolarization followed the activation produced by a high-current (>60 μA), short-stimulus train (2–20 pulses). Finally, long-stimulus trains (70 μA, 200 pulses, 330 Hz) evoked activation that spread over large portions of M1 and the adjacent premotor areas. The responses outlasted the stimulus train by several hundred milliseconds (Slovin et al. 2003). These results emphasize the importance of considering the spread of activation when evaluating the results of M1 stimulation.

CONCLUSIONS AND OUTLOOK

VSDI is based on the use of voltage-sensitive dyes, and the quality of results it produces is therefore limited by the properties of these dyes. The results presented here indicate that the technique has finally matured sufficiently to explore the neocortex in ways never feasible before. Furthermore, it is expected that the development of new voltage-sensitive dyes and future genetic engineering of suitable in vivo probes (see, e.g., Miyawaki et al. 1997; Siegel and Isacoff 1997; Chapters 7 and 8) will make the experiments much easier and will substantially improve the quality of the results. Of particular importance are probes that will stain only specific cell types and/or specific cellular compartments (e.g., axons, somata, dendrites). Additional developments of other technologies, using second-harmonic generation (Millard et al. 2003) and multiphoton and three-dimensional imaging, are highly desirable. Similarly, excitation at the red edge of existing VSDs has been shown to provide up to tenfold larger fractional change in the VSD signal size (Kuhn and Fromherz 2003; Kuhn et al. 2004). Because VSDI excels at revealing spatiotemporal patterns of the synaptic potential—the input to a neocortical region, but lacking sensitivity to spiking its output—it should be highly fruitful to combine it with the new approach to calcium imaging in vivo that primarily reflects the output (Stosiek et al. 2003). To increase the dimensionality of meaningful neurophysiological data obtained from the same patch of cortex, VSDI should be combined with targeted tracer injections, retrograde labeling, microstimulation, and intracellular and extracellular recordings. VSDI, combined with electrical recordings, allows investigators to obtain information about neuronal activity in coherent neuronal assemblies, within the neocortex, at subcolumnar resolution.

Although there is fast progress in the development of imaging techniques, to date no alternative imaging technique for visualizing functional organization in the living brain provides a comparable spatial and temporal resolution. It is this level of resolution that makes it possible to address questions of both *where* and *when* processing is performed, promising that this technique will contribute to the study of *how*—revealing fundamental principles of neural coding and processing strategies at the population level.

ACKNOWLEDGMENTS

We thank Yuval Toledo and Alexander Sterkin for their dedicated programming assistance, and our previous coworkers, Ron Frostig, Edmund Lieke, Doron Shoham, Daniel Glaser, Amos Arieli, Eyal Seidemann, Tal Kenet, Misha Tsodyks, Dirk Jancke, and Frederick Chavane, for their original contributions. This work was supported by grants from the Grodetsky Center, Goldsmith, Glasberg, Heineman, and Korber foundations, Ministry of Science (MOS) and Bundesministerium für Bildung, Forschung und Technologie (BMBF), the Israeli Science Foundation, and the European Commission Project 15803, Daisy.

REFERENCES

Arieli A, Grinvald A. 2002. Combined optical imaging and targeted electrophysiological manipulations in anesthetized and behaving animals. *J Neurosci Methods* **116:** 15–28.

Arieli A, Shoham D, Hildesheim R, Grinvald A. 1995. Coherent spatio-temporal pattern of ongoing activity revealed by real time optical imaging coupled with single unit recording in the cat visual cortex. *J Neurophysiol* **73:** 2072–2093.

Arieli A, Sterkin A, Grinvald A, Aertsen A. 1996a. Dynamics of ongoing activity: Explanation of the large variability in evoked cortical responses. *Science* **273**: 1868–1871.

Arieli A, Donchin O, Aertsen A, Bergman H, Gribova A, Grinvald A, Vaadia E. 1996b. The impact of ongoing cortical activity on evoked potential and behavioral responses in the awake behaving monkey. *Neuroscience* **22:** 2022. (Abstr.)

Arieli A, Grinvald A, Slovin H. 2002. Dural substitute for long-term imaging of cortical activity in behaving monkeys and its clinical implications. *J Neurosci Methods* **114:** 119–133.

Berger T, Borgdorff AJ, Crochet S, Neubauer FB, Lefort S, Fauvet B, Ferezou I, Carleton A, Luscher HR, Petersen CCH. 2007. Combined voltage and calcium epifluorescence imaging in vitro and in vivo reveals subthreshold and suprathreshold dynamics of mouse barrel cortex. *J Neurophysiol* **97:** 3751–3762.

Chavane F, Sharon D, Jancke D, Marre O, Frégnac Y, Grinvald A. 2011. Long-range horizontal spread of orientation selectivity in V1 is controlled by intracortical cooperativity. *Frontiers System Neurosci* (in press).

Cinelli AR, Kauer JS. 1992. Voltage sensitive dyes and functional-activity in the olfactory pathway. *Annu Rev Neurosci* **15:** 321–352.

Cinelli AR, Kauer JS. 1995. Salamander olfactory-bulb neuronal-activity observed by video-rate, voltage sensitive dyes imaging. 2. Spatial and temporal properties of responses evoked by electric stimulation. *J Physiol* **73:** 2033–2052.

Cinelli AR, Neff SR, Kauer JS. 1995. Salamander olfactory bulb neuronal activity observed by video rate, voltage-sensitive dye imaging. I. Characterization of the recording system. *J Neurophysiol* **73:** 2017–2032.

Cohen LB, Lesher S. 1986. Optical monitoring of membrane potential: Methods of multisite optical measurement. *Soc Gen Physiol Ser* **40:** 71–99.

Cohen MR, Newsome WT. 2004. What electrical microstimulation has revealed about the neural basis of cognition. *Curr Opin Neurobiol* **14:** 169–177.

Cohen LB, Salzberg BM, Davila HV, Ross WN, Landowne D, Waggoner AS, Wang CH. 1974. Changes in axon fluorescence during activity: Molecular probes of membrane potential. *J Membr Biol* **19:** 1–36.

Cohen LB, Slazberg BM, Grinvald A. 1978. Optical methods for monitoring neurons activity. *Annu Rev Neurosci* **1:** 171–182.

Derdikman D, Hildesheim R, Ahissar E, Arieli A, Grinvald A. 2003. Imaging spatio-temporal dynamics of surround inhibition in the barrels somatosensory cortex. *J Neurosci* **23:** 3100–3105.

De Weer P, Salzberg BM, eds. 1986. *Optical methods in cell physiology*. Society of General Physiologists Series, Vol. 40. Society of General Physiologists and Wiley-Interscience, New York.

Ferezou I, Bolea S, Petersen CC. 2006. Visualizing the cortical representation of whisker touch: Voltage-sensitive dye imaging in freely moving mice. *Neuron* **50:** 617–629.

Ferezou I, Haiss F, Gentet LJ, Aronoff R, Weber B, Petersen CC. 2007. Spatiotemporal dynamics of cortical sensorimotor integration in behaving mice. *Neuron* **56:** 907–923.

Gilbert CD, Wiesel TN. 1983. Clustered intrinsic connections in cat visual cortex. *J Neurosci* **3:** 1116–1133.

Grinvald A. 1984. Real time optical imaging of neuronal activity: From single growth cones to the intact brain. *Trends Neurosci* **7:** 143–150.

Grinvald A, Hildesheim R. 2004. VSDI: A new era in functional imaging of cortical dynamics. *Nat Rev Neurosci* **5:** 874–885.

Grinvald A, Petersen CCH. 2010. Imaging the dynamics of neocortical population activity in behaving and freely moving mammals. In *Membrane potential imaging in the nervous system: Methods and applications* (ed. Zecevic D, Canepari M), pp. 113–124. Springer Science+Business Media, New York.

Grinvald A, Salzberg BM, Cohen LB. 1977. Simultaneous recordings from several neurons in an invertebrate central nervous system. *Nature* **268:** 140–142.

Grinvald A, Cohen LB, Lesher S, Boyle MB. 1981. Simultaneous optical monitoring of activity of many neurons in invertebrate ganglia, using a 124 element "Photodiode" array. *J Neurophysiol* **45:** 829–840.

Grinvald A, Manker A, Segal M. 1982a. Visualization of the spread of electrical activity in rat hippocampal slices by voltage sensitive optical probes. *J Physiol* **333:** 269–291.

Grinvald A, Hildesheim R, Farber IC, Anglister L. 1982b. Improved fluorescent probes for the measurement of rapid changes in membrane potential. *Biophys J* **39:** 301–308.

Grinvald A, Anglister L, Freeman JA, Hildesheim R, Manker A. 1984. Real time optical imaging of naturally evoked electrical activity in the intact frog brain. *Nature* **308:** 848–850.

Grinvald A, Anglister L, Freeman JA, Hildesheim R, Manker A. 1985. Real-time optical mapping of neuronal activity: From single growth cones to the intact mammalian brain. *Annu Rev Neurosci* **8:** 263–305.

Grinvald A, Lieke E, Frostig RD, Gilbert CD, Wiesel TN. 1986. Functional architecture of cortex revealed by optical imaging of intrinsic signals. *Nature* **324:** 361–364.

Grinvald A, Frostig RD, Lieke E, Hildesheim R. 1988. Optical imaging of neuronal activity. *Physiol Rev* **68:** 1285–1366.

Grinvald A, Frostig RD, Ts'o DY, Lieke E, Arieli A, Hildesheim R. 1989. Optical imaging of activity in the visual cortex. In *Neuronal mechanisms of visual perception* (ed. Lam D, Gilbert CD), pp. 117–136. Portfolio Press, Woodlands, TX.

Grinvald A, Bonhoeffer T, Malonek D, Shoham D, Bartfeld E, Arieli A, Hildesheim R, Ratzlaff E. 1991. Optical imaging of architecture and function in the living brain. In *Memory: Organization and locus of change* (ed. Squire L), pp. 49–85. Oxford University Press, New York.

Grinvald A, Lieke EE, Frostig RD, Hildesheim R. 1994. Cortical point-spread function and long-range lateral interactions revealed by real-time optical imaging of macaque monkey primary visual cortex. *J Neurosci* **14:** 2545–2568.

Grinvald A, Shoham D, Shmuel A, Glaser ED, Vanzetta I, Shtoyerman E, Slovin H, Wijnbergen C, Hildesheim R, Sterkin A, Arieli A. 1999. In-vivo optical imaging of cortical architecture and dynamics. In *Modern techniques in neuroscience research* (ed. Windhorst U, Johansson H), pp. 893–969. Springer, New York.

Hebb D.O. 1949. The first stage of perception: Growth of the assembly. In *The organization of behavior: A neuropsychological theory*, pp. 60–78. Wiley, New York.

Hirota A, Sato K, Momosesato Y, Sakai T, Kamino K. 1995. A new simultaneous 1020 site optical recording system for monitoring neuronal activity using voltage sensitive dyes. *J Neurosci Methods* **56:** 187–194.

Hubel DH, Wiesel TN. 1962. Receptive fields, binocular interactions and functional architecture in the cat's visual cortex. *J Physiol* **160:** 106–154.

Iijima T, Matosomoto G, Kisokoro Y. 1992. Synaptic activation of rat adrenal-medula examined with a large photodiode array in combination with voltage sensitive dyes. *Neuroscience* **51:** 211–219.

Jancke D, Chavane F, Naaman S, Grinvald A. 2004. Imaging cortical correlates of a visual illusion. *Nature* **428:** 424–427.

Kamino K. 1991. Optical approaches to ontogeny of electrical activity and related functional-organization during early heart development. *Physiol Rev* **71:** 53–91.

Kauer JS. 1988. Real-time imaging of evoked activity in local circuits of the salamander olfactory bulb. *Nature* **331:** 166–168.

Kauer JS, Senseman DM, Cohen MA. 1987. Odor-elicited activity monitored simultaneously from 124 regions of the salamander olfactory bulb using a voltage-sensitive dye. *Brain Res* **25:** 255–261.

Kenet T, Arieli A, Grinvald A, Tsodyks M. 1997. Cortical population activity predicts both spontaneous and evoked single neuron firing rates. *Neurosci Lett* **48:** S27.

Kenet T, Grinvald A, Tsodyks M, Arieli A. 2003. Spontaneously occurring cortical representations of visual attributes. *Nature* **425:** 954–956.

Kleinfeld D, Delaney KR. 1996. Distributed representation of vibrissa movement in the upper layers of somatosensory cortex revealed with voltage-sensitive dyes. *J Comp Neurol* **375:** 89–108.

Konnerth A, Orkand RK. 1986. Voltage sensitive dyes measure potential changes in axons and glia of frog optic nerve. *Neurosci Lett* **66:** 49–54.

Kuhn B, Fromherz P. 2003. Anellated hemicyanine dyes in neuron membrane: Molecular stark effect and optical voltage recording. *J Phys Chem B* **107:** 7903–7913.

Kuhn B, Fromherz P, Denk W. 2004. High sensitivity of stark-shift voltage-sensing dyes by one- or two-photon excitation near the red spectral edge. *Biophys J* **87:** 631–639.

Lev-Ram R, Grinvald A. 1986. K^+ and Ca^{2+} dependent communication between myelinated axons and oligodendrocytes revealed by voltage-sensitive dyes. *Proc Natl Acad Sci* **83:** 6651–6655.

Loew LM. 1987. *Optical measurement of electrical activity.* CRC Press, Boca Raton, FL.

Millard AC, Jin L, Lewis A, Loew LM. 2003. Direct measurement of the voltage sensitivity of second-harmonic generation from a membrane dye in patch-clamped cells. *Optics Lett* **28:** 1221–1223.

Miyawaki A, Llopis J, Heim R, Caffery JM, Adams JA, Ikura M, Tsien RY. 1997. Fluorescent indicators for Ca^{2+} based on green fluorescent proteins and calmodulin. *Nature* **388:** 882–887.

Mountcastle VB. 1957. Modality and topographic properties of single neurons of cat's somatic sensory cortex. *J Neurophysiol* **20:** 408–434.

Newsome WT, Britten KH, Movshon JA. 1989. Neuronal correlates of a perceptual decision. *Nature* **341:** 52–54.

Orbach HS. 1988. Monitoring electrical activity in rat cerebral cortex. In *Spectroscopic membrane probes* (ed. Loew LM), Vol. 3, pp. 115–135. CRC Press, Boca Raton, FL.

Orbach HS, Cohen LB. 1983. Simultaneous optical monitoring of activity from many areas of the salamander olfactory bulb. A new method for studying functional organization in the vertebrate CNS. *J Neurosci* **3:** 2251–2262.

Orbach HS, Cohen LB, Grinvald A. 1985. Optical mapping of electrical activity in rat somatosensory and visual cortex. *J Neurosci* **5:** 1886–1895.

Petersen CCH, Grinvald A, Sakmann B. 2003a. Spatiotemporal dynamics of sensory responses in layer 2/3 of rat barrel cortex measured in vivo by voltage-sensitive dye imaging combined with whole-cell voltage recordings and anatomical reconstructions. *J Neurosci* **23:** 1298–1309.

Petersen CCH, Hahn T, Mehta M, Grinvald A, Sakmann B. 2003b. Interaction of sensory responses with spontaneous depolarization in layer 2/3 barrel cortex. *Proc Natl Acad Sci* **100:** 13638–13643.

Petersen CCH, Brecht M, Hahn TT, Sakmann B. 2004. Synaptic changes in layer 2/3 underlying map plasticity of developing barrel cortex. *Science* **304:** 739–742.

Ross WN, Reichardt LF. 1979. Species-specific effects on the optical signals of voltage sensitive dyes. *J Membr Biol* **48:** 343–356.

Salzberg BM, Davila HV, Cohen LB. 1973. Optical recording of impulses in individual neurons of an invertebrate central nervous system. *Nature* **246:** 508–509.

Salzberg BM, Grinvald A, Cohen LB, Davila HV, Ross WN. 1977. Optical recording of neuronal activity in an invertebrate central nervous system; simultaneous recording from several neurons. *J Neurophysiol* **40:** 1281–1291.

Salzberg BM, Obaid AL, Gainer H. 1986. Optical studies of excitation secretion at the vertebrate nerve terminal. *Soc Gen Physiol Ser* **40:** 133–164.

Seidemann E, Arieli A, Grinvald A, Slovin H. 2002. Dynamics of depolarization and hyperpolarization in the frontal cortex and saccade goal. *Science* **295:** 862–865.

Sharon D, Grinvald A. 2002. Dynamics and constancy in cortical spatiotemporal patterns of orientation processing. *Science* **295:** 512–515.

Shoham D, Ullman S, Grinvald A. 1991. Characterization of dynamic patterns of cortical activity by a small number of principal components. *Neuroscience* **17:** 1089.

Shoham D, Glaser DE, Arieli A, Kenet T, Wijnbergen C, Toledo Y, Hildesheim R, Grinvald A. 1999. Imaging cortical architecture and dynamics at high spatial and temporal resolution with new voltage-sensitive dyes. *Neuron* **24:** 791–802.

Siegel MS, Isacoff EY. 1997. A genetically encoded optical probe of membrane voltage. *Neuron* **19:** 735–741.

Slovin H, Arieli A, Hildesheim R, Grinvald A. 2002. Long-term voltage-sensitive dye imaging of cortical dynamics in the behaving monkey. *J Neurophysiol* **88:** 3421–3438.

Slovin H, Strick PL, Hildesheim R, Grinvald A. 2003. Voltage sensitive dye imaging in the motor cortex. I. Intra- and intercortical connectivity revealed by microstimulation in the awake monkey. *Soc Neuroci Abst* **554.8.**

Spors H, Grinvald A. 2002. Temporal dynamics of odor representations and coding by the mammalian olfactory bulb. *Neuron* **34:** 1–20.

Sterkin A, Lampl I, Ferster D, Grinvald A, Arieli A. 1998. Real time optical Imaging in cat visual cortex exhibits high similarity to intracellular activity. *Neurosci Lett* **51:** S41.

Stosiek C, Garaschuk O, Holthoff K, Konnerth A. 2003. In vivo two-photon calcium imaging of neuronal networks. *Proc Natl Acad Sci* **100:** 7319–7324.

Strick P, Grinvald A, Hildesheim R, Slovin H. 2003. Voltage sensitive dye imaging in the motor cortex II. Cortical correlates of Go/No-Go delayed response task. *Soc Neuroci Abst* 918.8.

Takashima I, Kajiwara R, Iijima T. 2001. Voltage-sensitive dye versus intrinsic signal optical imaging: Comparison of optically determined functional maps from rat barrel cortex. *Neuroreport* **12:** 2889–2894.

Tanifuji M, Yamanaka A, Sunaba R, Toyama K. 1993. Propagation of excitation in the visual cortex studies by the optical recording. *Jpn J Physiol* **43:** 57–59.

Tasaki I, Warashina A. 1976. Dye membrane interaction and its changes during nerve excitation. *Photochem Photobiol* **24:** 191–207.

Tasaki I, Watanabe A, Sandlin R, Carnay L. 1968. Changes in fluorescence, turbidity, and birefringence associated with nerve excitation. *Proc Natl Acad Sci* **61:** 883–888.

Toyama K, Tanifuji M. 1991. Seeing excitation propagation in visual cortical slices *Biomed Res* **12:** 145–147.

Tsodyks M, Kenet T, Grinvald A, Arieli A. 1999. Linking spontaneous activity of single cortical neurons and the underlying global functional architecture. *Science* **286:** 1943–1946.

Vranesic I, Iijima T, Ichikawa M, Matsumoto G, Knopfel T. 1994. Signal transmission in the parallel fiber Purkinje-cell system visualized by high resolution imaging. *Proc Natl Acad Sci* **91:** 13014–13017.

Waggoner AS. 1979. Dye indicators of membrane potential. *Annu Rev Biophys Bioeng* **8:** 47–63.

Waggoner AS, Grinvald A. 1977. Mechanisms of rapid optical changes of potential sensitive dyes. *Ann NY Acad Sci* **303:** 217–242.

75 Voltage-Sensitive Dye Imaging of Cortical Spatiotemporal Dynamics in Awake Behaving Mice

Carl C.H. Petersen

Laboratory of Sensory Processing, Brain Mind Institute, Faculty of Life Sciences, École Polytechnique Fédérale de Lausanne (EPFL), CH-1015 Lausanne, Switzerland

ABSTRACT

In vitro measurements show that voltage-sensitive dye (VSD) fluorescence signals faithfully report changes in membrane potential both linearly and rapidly. Some VSDs, such as RH1691, can be applied to the cortical surface in vivo to stain membranes in the upper layers of the neocortex. In vivo VSD imaging combined with whole-cell recordings in the mouse neocortex under anesthesia reveal that the VSD signals correlate closely with subthreshold membrane potential dynamics at millisecond temporal resolution and subcolumnar spatial resolution. By imaging large cortical areas, the interactions between different cortical areas and different cortical columns can be dynamically mapped. The simplicity and robustness of the VSD imaging technique have allowed it to be readily applied to the study of awake behaving mice. VSD imaging therefore now provides a technique through which we can begin to relate cortical spatiotemporal dynamics to sensory and motor behaviors.

INTRODUCTION

The cortical surface of the brain can be divided into different functional regions, which interact dynamically depending on the behavioral context. To examine such interactions in real time, we need methods for measuring cortical function with millisecond temporal resolution and subcolumnar spatial resolution. Advances in camera technology and improvements in dyes (Shoham et al. 1999; Grinvald and Hildesheim 2004) now allow routine high-resolution in vivo measurement of cortical spatiotemporal dynamics through voltage-sensitive dye (VSD) imaging (Tsodyks et al. 1999; Seidemann et al. 2002; Slovin et al. 2002; Derdikman et al. 2003; Kenet et al. 2003; Petersen et al. 2003a,b; Jancke et al. 2004; Ferezou et al. 2006; Benucci et al. 2007; Berger et al. 2007; Ferezou et al. 2007; Lippert et al. 2007; Xu et al. 2007; Han et al. 2008; Wallace and Sakmann 2008; Sigler et al. 2009).

Whole-cell recordings from supragranular neurons combined with VSD imaging of RH1691 fluorescence (Petersen et al. 2003a,b; Ferezou et al. 2006; Berger et al. 2007) in the rodent primary somatosensory barrel cortex (Petersen 2007) have provided insight into which features of cortical activity are reported by in vivo VSD imaging under anesthesia. Both spontaneous (Fig. 1A–C) and sensory-evoked (Fig. 1D–F) subthreshold membrane potentials recorded in the whole-cell configuration are highly linearly correlated with the VSD signal. Action potentials occur at low rates under these conditions and are not synchronous in nearby neurons. Action potentials recorded in individual neurons are therefore not represented in the VSD signal, which reflects population activity (Fig. 1B). On the other hand, subthreshold membrane potential changes are highly correlated in nearby neurons under these experimental conditions, and this provides the explanation for the close correlation between membrane potential in a single neuron and the population VSD signal (Petersen et al. 2003a,b). From these results obtained under anesthesia, we conclude that VSD imaging provides a useful method for imaging the ensemble membrane potential dynamics of layer-2/3 neurons. The same techniques can readily be applied to VSD imaging of awake mice as described in the protocol.

The protocol uses epifluorescence optics to collect light from a large area. Efficient collection of light from large areas can be achieved by macroscopes, which can easily be constructed from commercially available camera or video lenses (Ratzlaff and Grinvald 1991). The magnification of the tandem lens macroscope is given by the ratio of the focal lengths of the two lenses. This must be matched between the desired field of view and the size of the sensor on the camera. The other critical aspect of VSD imaging is the small fractional change in fluorescence of voltage-sensitive dyes and rapid time course of the signals. The camera should therefore be fast (operating at >100 Hz) and individual pixels should have a large well depth. To resolve a typical VSD signal with an amplitude of ~0.1%, the camera needs to collect ~1×10^6 photons, and cameras with well depths of >1×10^6 electrons are optimal.

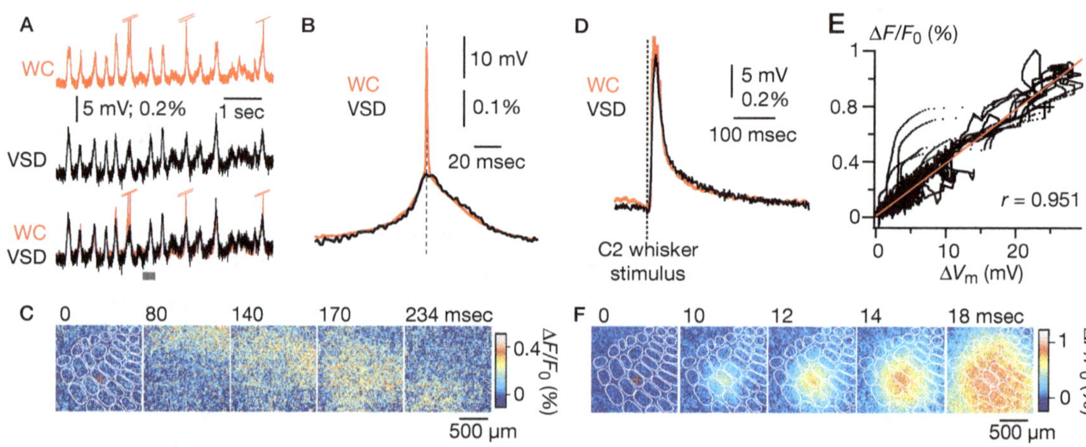

FIGURE 1. Voltage-sensitive dye signals correlate closely with membrane potential. *(A)* An example in vivo experiment with a whole-cell (WC) recording (red trace) combined with voltage-sensitive dye (VSD) imaging (black trace; fluorescence quantified in the region surrounding the recorded neuron) in a urethane-anesthetized mouse. Subthreshold spontaneous membrane potential changes are precisely correlated with changes in VSD fluorescence. However, action potentials in the recorded neuron do not correlate with large changes in the VSD signal. Action potentials are truncated (indicated by diagonal red lines). The gray bar indicates the period shown in C. *(B)* A spike-triggered average shows that the subthreshold depolarization surrounding the action potential is highly correlated in the WC recording and the VSD fluorescence trace, but that the action potential itself is not represented in the VSD signal. VSD signals therefore reflect subthreshold membrane potential changes and not action potential firing. *(C)* VSD imaging of the spontaneous activity reveals propagating waves. The red cross indicates the location of the recorded neuron. *(D)* During the same recording, the C2 whisker was deflected, evoking a cortical sensory response. The evoked depolarization follows nearly identical kinetics in the WC recording and the VSD signal. *(E)* Correlation of the membrane potential (V_m) and the VSD fluorescence changes evoked by sensory stimulation indicates a highly correlated near-linear relationship. *(F)* VSD imaging of the sensory-evoked response. Deflection of the C2 whisker evokes a response that initiates in the C2 barrel column (barrel map is superimposed in white outlines) and subsequently spreads over a large part of the barrel cortex in the next milliseconds. (Adapted, with permission, from Ferezou et al. 2006, ©Elsevier.)

Protocol

Voltage-Sensitive Dye Imaging of Cortical Spatiotemporal Dynamics in Mice

This protocol describes VSD imaging of cortical spatiotemporal dynamics in mice under anesthesia. Strategies for VSD imaging of awake mice using similar techniques are then discussed.

MATERIALS

CAUTION: See Appendix 6 for proper handling of materials marked with <!>.
See the end of the chapter for recipes for reagents marked with <R>.

Reagents

Agarose (1%), prepared in Ringer's solution <R>
Biocytin (2 mg/mL)
Cyanoacrylate glue <!>
Dental cement
Gas for anesthesia (e.g., halothane <!> or isoflurane <!>)
Hydrogen peroxide (1%) (optional; see Step 2) <!>
Intrapipette solution <R>
Paraformaldehyde (4%) <!>, prechilled to 4°C
Phosphate buffer (0.1 M), prechilled to 4°C
Ringer's solution <R>
VECTASTAIN ABC Kit (Vector Laboratories)
Voltage-sensitive dye RH1691 (Optical Imaging), dissolved to 0.1–1 mg/mL in Ringer's solution <R>

Equipment

Glass coverslip
Heating blanket for surgery
Imaging setup

This includes a fast camera with large well depth (e.g., MiCAM Ultima [SciMedia], NeuroCMOS [Red Shirt Imaging, LLC] or Imager 3001 [Optical Imaging]), macroscope (two large aperture video lenses mounted in tandem), light source (e.g., laser or halogen lamp powered by stabilized power supply), band-pass excitation filter (630 ± 10 nm), shutter, dichroic filter (650 nm), and long-pass emission filter (665 nm). Ensure that the light source is stable and that the camera has a sufficiently high signal-to-noise ratio able to resolve changes of ~0.1%. The equipment should be tested on fluorescent targets and the noise of the system measured across the relevant timescales of the desired VSD measurements.

Metal head-fixation posts for imaging head-restrained animals
Standard equipment for surgery, whole-cell recording, perfusion, and sectioning brain tissue

EXPERIMENTAL METHOD

Surgery and Staining with VSD

1. Deeply anesthetize the mouse with gas anesthesia (halothane or isoflurane). Keep the anesthetized animal on a heating blanket to maintain the rectal temperature at 37°C.

   The dose of anesthetic must be adequate for surgical tolerance before proceeding and must conform to local animal care committee guidelines. Paw withdrawal, whisker movements, and eyeblink reflexes should be largely absent. The state of the animal must be carefully monitored during the surgery and staining procedures, which take ~3 h to complete.

2. After removing or reflecting the skin covering the skull, carefully scrape the bone clean. Further cleaning of the bone can be performed, if necessary, with 1% hydrogen peroxide.
3. Apply a thin layer of cyanoacrylate glue to the surrounding region of the bone, away from where the craniotomy will be performed (this helps dental cement adhesion).
4. Glue a metal head post and recording chamber to the skull with cyanoacrylate glue and dental cement.
5. When the dental cement has hardened, immobilize the skull and minimize movement of the brain by attaching the head post firmly to a fixed holder.
6. Perform a craniotomy of the desired size by drilling within the recording chamber. Be careful not to damage the underlying brain tissue.
7. Finally, remove the dura, leaving the pia of the underlying cortex exposed.

   Alternatively, dry the dura extensively and leave it intact (Lippert et al. 2007).

8. Apply a small quantity (~250 µL) of the 0.1–1 mg/mL voltage-sensitive dye solution to the craniotomy. Seal the chamber with a glass coverslip, applying very slight pressure to prevent brain edema.
9. Leave the cortex to stain for 1–2 h. Occasional mixing of the staining solution using a micropipette is recommended during the staining period. At the end of the staining period, remove unbound dye by washing the cortex extensively with Ringer's solution.

   During this period, the dye will diffuse into the superficial layers of the neocortex. At the end of the staining period, the cortex should have a pale blue color.

10. Cover the craniotomy with 1% agarose dissolved in Ringer's solution, and place a glass coverslip on top.

    Alternatively, place coverslip so that it is in direct contact with the cortex and seal into place (this is recommended for fiber imaging).

Whole-Cell Recordings from Post Hoc Identified Neurons

11. Fill a whole-cell pipette with intrapipette solution and 2 mg/mL biocytin. Pipette resistance should be ~5 MΩ.
12. Monitor the tip resistance by applying brief voltage steps of 5 mV in the voltage-clamp mode while measuring the current flow on an oscilloscope. Apply a positive pressure of ~200 mbar to the pipette.
13. Slowly advance the electrode under the coverslip, through the agarose, and into the cortex. When the tip is close to (~50 µm) the chosen recording site, reduce the positive pressure to ~30 mbar.
14. Advance the pipette in 2-µm steps until the tip resistance suddenly increases, indicating contact with a cell membrane. Release the pressure in the pipette and apply light suction until a gigaohm seal is formed.

15. Establish a whole-cell recording configuration by rupturing the membrane in the pipette. Achieve this by applying either brief suction pulses or by slowly increasing the suction pressure.
16. After collecting data, such as the combined VSD imaging and whole-cell recording as shown in Figure 1, slowly retract the whole-cell recording pipette while monitoring whole-cell capacitance transients with 5-mV voltage steps.

    During the retraction, the excised patch configuration should be established. This ensures that the neuron remains intact and viable for later anatomical staining.
17. Perfuse the animal transcardially with cold 0.1 M phosphate buffer and then with 4% paraformaldehyde. Remove the brain from the skull and fix overnight at 4°C.
18. Section the cortex tangentially at 100 µm, and stain for biocytin using a VECTASTAIN ABC Kit (Vector Laboratories).

    The blood vessels initially imaged with green illumination now provide the link between the location of the neuronal processes visualized by the anatomical stain and the functional voltage-sensitive dye images.

Imaging during Head Restraint

For imaging awake head-restrained mice, they must first be trained to sit quietly in the experimental apparatus under mock experimental conditions. This training should occur gradually, beginning with short periods of a few minutes of head restraint. Through gradual daily training sessions, the period of head fixation can be extended up to a few hours. After this period of behavioral adaptation, the VSD staining procedure should be performed as described in Steps 1–10 of the Experimental Methods. After staining, the mouse needs several hours to recover from anesthesia, which can take place in the home cage of the mouse.

For imaging, transfer the mouse to the experimental apparatus and fix the head to a stable arm. Illuminate with green light at ~530 nm and record the blood vessel pattern on the cortical surface using the camera. Move the focal plane ~300 µm into the cortex and excite the voltage-sensitive dye with light at 630 nm. Emitted fluorescent light is long-pass filtered (665 nm), forming the voltage-sensitive dye signals; record these signals at frame rates >100 Hz. (VSD signals have small amplitudes of ~0.1% and it is therefore crucial to collect a large number of photons per pixel.)

Heartbeat-related signals and brain movements form the largest artifacts. The spatiotemporal structure of these artifacts is usually very different from VSD signals relating to cortical function. Data containing such artifacts can therefore be simply excluded from analysis or corrected through image processing procedures.

Imaging Freely Moving Animals

For VSD imaging of freely moving mice, the simplest method is to use fiber-optic image bundles that consist of well-ordered arrays of fibers. Each individual fiber can have a diameter of several micrometers, and they can be assembled into arrays covering several millimeters. The fiber can be placed in direct apposition to the cortex or it could be focused by small conventional lenses or gradient-index (GRIN) lenses. The fiber-optic image bundle should be firmly attached to the implanted recording chamber so that it is locked in position (Ferezou et al. 2006). The other end of the fiber should be placed in the focal plane of the tandem lens optics. This will allow simple and efficient transfer of excitation light to the brain and collection of the emitted fluorescence. As with imaging head-restrained animals, movement-related artifacts are the major concern.

EXAMPLE OF APPLICATION

Having found VSD imaging to be a reliable tool to investigate sensory processing in anesthetized mice, we turned our attention to examining cortical spatiotemporal dynamics in awake behaving mice. To deliver well-controlled whisker deflections in awake mice, we attached a small metal particle to the C2 whisker and evoked movement of this whisker by driving brief (1–2 msec) current pulses through an electromagnetic coil. Such a stimulus evokes a small, brief, and reproducible whisker deflection (Ferezou et al. 2006). In head-fixed mice, the evoked sensory response was imaged across a large fraction of the mouse sensorimotor cortex, which had been stained with voltage-sensitive dye RH1691 through a large craniotomy (Ferezou et al. 2007). Deflection of the C2 whisker during a period of quiet wakefulness (when the mouse was not spontaneously moving) evoked a sensory response, which at first was localized to the C2 barrel column (Fig. 2A, upper). In the following milliseconds the depolarization spread across the barrel field, presumably through local excitatory synaptic connections between neighboring cortical columns. In addition, a second localized spot of activity was evoked in the whisker motor cortex. This initially localized secondary sensory response in motor cortex also spread over the following milliseconds, and within 50 msec of the stimulus a surprisingly large cortical area had depolarized in response to a single whisker deflection. Interestingly, the mouse often began to actively whisk following such a sensorimotor response. The

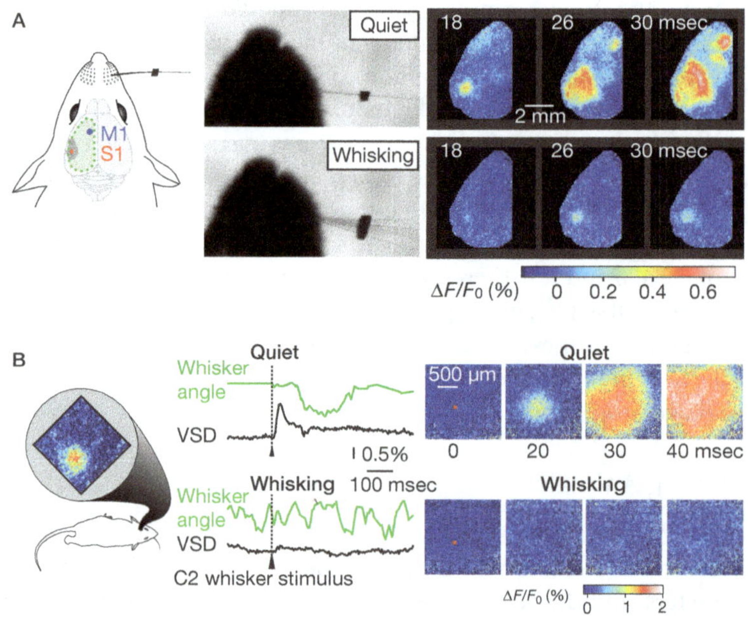

FIGURE 2. Imaging interactions between sensory and motor cortex. (*A*) A large fraction of the sensorimotor cortex, including the primary somatosensory barrel cortex (S1) and the whisker motor cortex (M1), was stained with voltage-sensitive dye (VSD). A small piece of metal was attached to the whisker, allowing brief magnetic pulses to evoke a whisker deflection and subsequent sensory processing. The whisker-related behavior was filmed simultaneously with VSD imaging, allowing direct correlation of cortical activity with behavior. The image of the mouse nose and whisker is a superposition of many frames. The *upper* image extends over a quiet prestimulus period. The *lower* image extends over a period of spontaneous whisking in the prestimulus period. Deflection of the whisker during these two different behavioral periods evokes dramatically different sensory responses. During quiet wakefulness, the response is initially localized but rapidly spreads across S1 and a secondary sensory response is observed in M1. The same stimulus applied during a period of spontaneous whisking evokes a small-amplitude sensory response that remains localized to S1. (*B*) VSD imaging of barrel cortex in freely moving mice through fiber optics provides further evidence for state-dependent sensory processing of tactile information. During a quiet period, the deflection of a whisker evoked a large-amplitude propagating sensory response. During active whisking, the same stimulus barely evoked a measurable response. (*A*, Adapted, with permission, from Ferezou et al. 2007, ©Elsevier; *B*, adapted, with permission, from Ferezou et al. 2006, ©Elsevier.)

propagating sensory response may serve as a wake-up call, bringing the mouse from an initially quiet state into one optimized for actively processing subsequent sensory stimuli. Rapid signaling of sensory information to motor cortex may be important in general for optimizing motor control. Indeed, rodents often change their pattern of whisker movements following whisker-object contact, perhaps as a way to improve the quality of sensory information.

During exploratory behavior, mice move their whiskers back and forth at high frequencies (typically ~10 Hz) scanning the environment surrounding their snouts. Mice also spontaneously whisk during head restraint and by chance some of our magnetic whisker deflections occurred during such active whisking bouts. Interestingly, the sensory-evoked responses were dramatically different during active states. The sensory responses were again initially localized to the C2 barrel column, but instead of spreading across large cortical areas, the response remained localized and had only small amplitude and a brief duration (Fig. 2A, lower). The ongoing behavior of the mouse therefore makes a dramatic difference to sensory processing.

To investigate whether similar state dependence of sensory processing occurs in freely moving mice, we used fiber-optic image bundles to visualize the membrane potential dynamics of the primary somatosensory barrel cortex (Ferezou et al. 2006). The fiber array that we used only extended across the somatosensory barrel cortex and did not include the motor cortex. Similar to the observations made in head-restrained mice, we found that a brief whisker stimulus delivered during quiet wakefulness evoked a large, spreading sensory response (Fig. 2B, upper) whereas the stimulus delivered during active whisking was strongly suppressed (Fig. 2B, lower). The propagating sensory responses evoked by a single whisker stimulation observed during anesthesia are therefore not artifacts of anesthesia, per se, but appear prominently during some behavioral states of wakefulness (Ferezou et al. 2006; Ferezou et al. 2007). That a large part of the cortex can be informed about a single whisker deflection might allow integration of sensory information from different whiskers and even different modalities, which might be crucial for associative learning and sensory perception. In future studies, it will be of great interest to image the cortical spatiotemporal dynamics of mice performing learned behaviors.

DISCUSSION

VSD imaging of awake animals is beginning to provide unique insight into the spatiotemporal dynamics of cortical processing during behavior. VSD imaging provides information on cortical dynamics with millisecond temporal resolution and subcolumnar spatial resolution. It is extremely instructive to combine VSD imaging with electrophysiological measurements. This allows the imaging signal to be calibrated to other well-established signals. Local field potential measurements are technically easy to obtain and provide an excellent correlation with the VSD signal under many different experimental conditions.

There are, of course, also many limitations to the current methodology. Perhaps most importantly, in vivo VSD imaging is currently limited to the exposed areas of the cortical surface. In the future, use of fiber-optic microendoscopes may provide a way to image from deep structures in the brain (Flusberg et al. 2005). The application of VSD to the cortical surface ensures even staining of the underlying neuronal networks, but the approach inherently lacks cell type specificity. Genetically encoded fluorescent functional reporter proteins (Miyawaki et al. 1997) would provide a possible alternative to synthesized dyes, allowing targeting to specific cell types. There has been progress in developing voltage-sensitive fluorescent proteins (Siegel and Isacoff 1997; Sakai et al. 2001; Ataka and Pieribone 2002; Dimitrov et al. 2007; Lundby et al. 2008; Akemann et al. 2010), giving hope for future prospects of cell type–specific VSD imaging.

RECIPES

CAUTION: See Appendix 6 for proper handling of materials marked with <!>.
Recipes for reagents marked with <R> are included in this list.

Intrapipette Solution

Reagent	Final concentration
Potassium gluconate <!>	135 mM
KCl	4 mM
HEPES	10 mM
Phosphocreatine	10 mM
MgATP	4 mM
Na_3GTP	0.3 mM

Adjust the pH to 7.3 with KOH <!>. Can be stored for many months at –20°C.

Ringer's Solution

Reagent	Final concentration
NaCl	135 mM
KCl	5 mM
HEPES	5 mM
$CaCl_2$ <!>	1.8 mM
$MgCl_2$	1 mM

Can be stored at 4°C for 1–2 wk.

ACKNOWLEDGMENTS

This work was funded by grants from the Human Frontiers in Science Program, Swiss National Science Foundation and SystemsX.ch.

REFERENCES

Akemann W, Mutoh H, Perron A, Rossier J, Knöpfel T. 2010. Imaging brain electric signals with genetically targeted voltage-sensitive fluorescent proteins. *Nat Methods* **7:** 643–649.

Ataka K, Pieribone VA. 2002. A genetically targetable fluorescent probe of channel gating with rapid kinetics. *Biophys J* **82:** 509–516.

Benucci A, Frazor RA, Carandini M. 2007. Standing waves and traveling waves distinguish two circuits in visual cortex. *Neuron* **55:** 103–117.

Berger T, Borgdorff AJ, Crochet S, Neubauer FB, Lefort S, Fauvet B, Ferezou I, Carleton A, Luscher HR, Petersen CCH. 2007. Combined voltage and calcium epifluorescence imaging in vitro and in vivo reveals subthreshold and suprathreshold dynamics of mouse barrel cortex. *J Neurophysiol* **97:** 3751–3762.

Derdikman D, Hildesheim R, Ahissar E, Arieli A, Grinvald A. 2003. Imaging spatiotemporal dynamics of surround inhibition in the barrels somatosensory cortex. *J Neurosci* **23:** 3100–3105.

Dimitrov D, He Y, Mutoh H, Baker BJ, Cohen L, Akemann W, Knöpfel T. 2007. Engineering and characterization of an enhanced fluorescent protein voltage sensor. *PLoS One* **2:** e440.

Ferezou I, Bolea S, Petersen CCH. 2006. Visualizing the cortical representation of whisker touch: Voltage-sensitive dye imaging in freely moving mice. *Neuron* **50:** 617–629.

Ferezou I, Haiss F, Gentet LJ, Aronoff R, Weber B, Petersen CCH. 2007. Spatiotemporal dynamics of cortical sensorimotor integration in behaving mice. *Neuron* **56:** 907–923.

Flusberg BA, Cocker ED, Piyawattanametha W, Jung JC, Cheung EL, Schnitzer MJ. 2005. Fiber-optic fluorescence imaging. *Nat Methods* **2:** 941–950.

Grinvald A, Hildesheim R. 2004. VSDI: A new era in functional imaging of cortical dynamics. *Nat Rev Neurosci* **5:** 874–885.

Han F, Caporale N, Dan Y. 2008. Reverberation of recent visual experience in spontaneous cortical waves. *Neuron* **60:** 321–327.

Jancke D, Chavane F, Naaman S, Grinvald A. 2004. Imaging cortical correlates of illusion in early visual cortex. *Nature* **428:** 423–426.

Kenet T, Bibitchkov D, Tsodyks M, Grinvald A, Arieli A. 2003. Spontaneously emerging cortical representations of visual attributes. *Nature* **425:** 954–956.

Lippert MT, Takagaki K, Xu W, Huang X, Wu JY. 2007. Methods for voltage-sensitive dye imaging of rat cortical activity with high signal-to-noise ratio. *J Neurophysiol* **98:** 502–512.

Lundby A, Mutoh H, Dimitrov D, Akemann W, Knöpfel T. 2008. Engineering of a genetically encodable fluorescent voltage sensor exploiting fast Ci-VSP voltage-sensing movements. *PLoS One* **3:** e2514.

Miyawaki A, Llopis J, Heim R, McCaffery JM, Adams JA, Ikura M, Tsien RY. 1997. Fluorescent indicators for Ca^{2+} based on green fluorescent proteins and calmodulin. *Nature* **388:** 882–887.

Petersen CCH. 2007. The functional organization of the barrel cortex. *Neuron* **56:** 339–355.

Petersen CCH, Grinvald A, Sakmann B. 2003a. Spatiotemporal dynamics of sensory responses in layer 2/3 of rat barrel cortex measured in vivo by voltage-sensitive dye imaging combined with whole-cell voltage recordings and neuron reconstructions. *J Neurosci* **23:** 1298–1309.

Petersen CCH, Hahn TTG, Mehta M, Grinvald A, Sakmann B. 2003b. Interaction of sensory responses with spontaneous depolarization in layer 2/3 barrel cortex. *Proc Natl Acad Sci* **100:** 13638–13643.

Ratzlaff EH, Grinvald A. 1991. A tandem-lens epifluorescence macroscope: Hundred-fold brightness advantage for wide-field imaging. *J Neurosci Methods* **36:** 127–137.

Sakai R, Repunte-Canonigo V, Raj CD, Knöpfel T. 2001. Design and characterization of a DNA-encoded, voltage-sensitive fluorescent protein. *Eur J Neurosci* **13:** 2314–2318.

Seidemann E, Arieli A, Grinvald A, Slovin H. 2002. Dynamics of depolarization and hyperpolarization in the frontal cortex and saccade goal. *Science* **295:** 862–865.

Shoham D, Glaser DE, Arieli A, Kenet T, Wijnbergen C, Toledo Y, Hildesheim R, Grinvald A. 1999. Imaging cortical dynamics at high spatial and temporal resolution with novel blue voltage-sensitive dyes. *Neuron* **24:** 791–802.

Siegel MS, Isacoff EY. 1997. A genetically encoded optical probe of membrane voltage. *Neuron* **19:** 735–741.

Sigler A, Mohajerani MH, Murphy TH. 2009. Imaging rapid redistribution of sensory-evoked depolarization through existing cortical pathways after targeted stroke in mice. *Proc Natl Acad Sci* **106:** 11759–11764.

Slovin H, Arieli A, Hildesheim R, Grinvald A. 2002. Long-term voltage-sensitive dye imaging reveals cortical dynamics in behaving monkeys. *J Neurophysiol* **88:** 3421–3438.

Tsodyks M, Kenet T, Grinvald A, Arieli A. 1999. The spontaneous activity of single cortical neurons depends on the underlying global functional architecture. *Science* **286:** 1943–1946.

Wallace DJ, Sakmann B. 2008. Plasticity of representational maps in somatosensory cortex observed by in vivo voltage-sensitive dye imaging. *Cereb Cortex* **18:** 1361–1373.

Xu W, Huang X, Takagaki K, Wu JY. 2007. Compression and reflection of visually evoked cortical waves. *Neuron* **55:** 119–129.

76 Two-Photon Imaging of Neural Activity in Awake Mobile Mice

Daniel Dombeck[1] and David Tank

Departments of Molecular Biology and Princeton Neuroscience Institute, Princeton University, Princeton, New Jersey 08544

ABSTRACT

A method for functional, cellular-resolution imaging of neural populations in awake and mobile mice is presented. The method is based on the use of a spherical treadmill, head restraint, and motion correction software that facilitate neural imaging in the awake brain with a fixed upright two-photon microscope. The methods have proven to be applicable to a wide range of brain regions and should help further our understanding of how neuronal population activity within the brain's microcircuitry is connected to animal behavior.

INTRODUCTION

Two-photon microscopy has become the standard in vivo method used for high-resolution imaging of the rodent brain (Kerr and Denk 2008). Because of the nonlinear relationship between excitation intensity and fluorescence excitation, the fluorescence signal is localized to the microscope objective focus, allowing for micrometer-scale resolution up to ~0.5 mm deep in brain tissue. When two-photon microscopy is used to record time-series movies of neurons labeled with calcium-sensitive dyes, it is possible to study the functional properties of large neuronal populations in vivo (Helmchen et al. 1999; Stosiek et al. 2003; Kerr et al. 2005, 2007; Ohki et al. 2005; Sato et al. 2007). Nearly all such studies have thus far been performed in anesthetized rodents; however, anesthesia is well known to alter brain function and prevents the study of most behavioral paradigms (Berg-Johnsen and Langmoen 1992; Major and Tank 2004; Greenberg et al. 2008). Here we explain the methods we have developed to perform two-photon imaging in awake, behaving mice (Dombeck et al. 2007). We describe the experimental apparatus, the surgical procedures (including head plate and cranial window design, calcium-sensitive dye loading, and methods to reduce brain motion), time-series motion-correction methods, and calcium-sensitive dye fluorescence trace signal processing. We provide example data sets showing that these methods can be used to image neuronal population activity in a wide variety of brain regions.

[1]Present address: Department of Neurobiology and Physiology, Northwestern University, Evanston, Illinois 60208.

OVERVIEW OF EXPERIMENTAL METHODS

Experimental Apparatus

The experimental apparatus is shown in Figure 1A. It consists of an upright table-mounted two-photon microscope that can be positioned such that the microscope objective is at least ~1 ft above the table surface. Our two-photon microscope was custom built using a commercial pulsed laser light source, together with an optical assembly that consists of a mixture of custom and commercial optical and electronic components and software. The optomechanical part is based on an early prototype to the movable objective microscope (MOM) now marketed by Sutter Instruments (W. Denk, pers. comm.), and is based on a modified Sutter MP285 micromanipulator, providing the ability to move the objective in *x*, *y*, and *z* above the sample as well as rotate the objective around the horizontal axis. A 40x, 0.8-NA objective (Olympus) is typically used and the illumination source is a Ti:Sapphire laser (Mira 900, Coherent, ~100-fsec pulses at 80 MHz). ScanImage is used for microscope control and image acquisition (Pologruto et al. 2003).

Below the microscope is a large (8-inch-diameter) Styrofoam ball (Floracraft) that is levitated using a thin cushion of air between the ball and custom-made casting-containing air jets (Dahmen 1980; Mason et al. 2001; Holscher et al. 2005; Stevenson et al. 2005). The casting is formed by Epoxy (BC 7062 Epoxy, BCC Products) lining the lower ~75% of the inside of a large 8.125-inch-diameter hemispherical steel ladle (McMaster-Carr). The air cushion is produced by eight symmetrical 0.25-inch-diameter holes through the casting that are pressurized with air at ~30 psi. A mouse is restrained

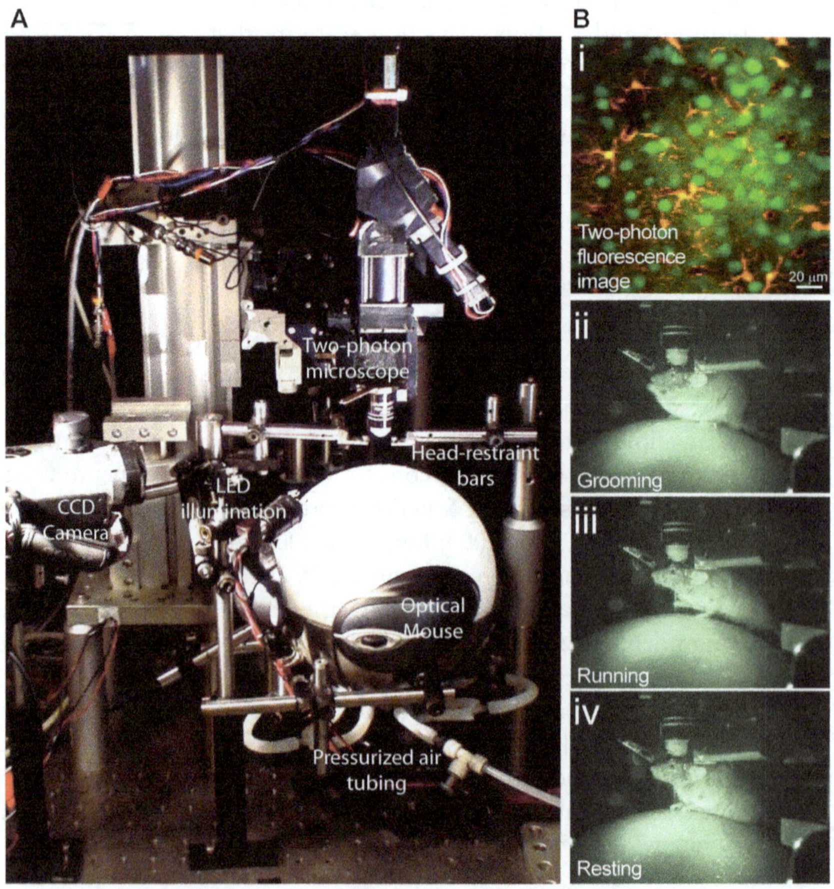

FIGURE 1. *(A)* An image of the experimental apparatus used for imaging with cellular resolution in awake behaving mice. *(B)* *(i)* A two-photon microscopy image of layer-2/3 cortical neurons and astrocytes loaded with the calcium-sensitive dye Calcium Green-1 AM (green) and SR101 (red). During two-photon imaging, mice can groom *(ii)*, run *(iii)*, and rest *(iv)* on the spherical treadmill.

via mounting bars such that its head is held motionless directly below the microscope objective and its limbs are resting on the surface of the ball. With this setup, and using appropriate methods to reduce brain motion (see below), it is possible to perform two-photon imaging (Fig. 1Bi) in the brain while the mouse grooms, runs, and rests (Fig. 1Bii–iv). To record the walking, running, and turning movements of the animal, optical computer mice are used to measure ball rotations. A Labview-based program is used to record the counts sent by the computer mice to a USB buffer. The counts to ball movement distance calibration is made by moving the ball a known distance and recording the number of counts sent to the buffer. More behavioral details can be obtained by illuminating the subject with infrared light-emitting diodes (IR LEDs) and recording the mouse movements with an IR charge-coupled device (CCD) camera; however, because the two-photon excitation light is also IR, an optical filter must be placed in front of the camera to block out the laser light but pass the LED light.

Although the spherical treadmill is useful for studying locomotion-based behaviors, it is also possible to perform imaging experiments in immobile head-restrained rodents (Ferezou et al. 2007; Greenberg et al. 2008; Kuhn et al. 2008). In fact, some experiments are likely to be more easily accomplished in awake immobilized animals in which the treadmill is not used. For example, video-based whisker tracking may be complicated by the presence of the treadmill; auditory stimuli may be masked by the noise of the treadmill produced by the pressurized air used for levitation. Although our research has focused on head-restrained imaging combined with the spherical treadmill, the surgical procedures and data analysis methods described here should apply equally well to head-restrained mobile or immobile mice.

Surgical Procedures and Dye Loading

Below is a general outline of the surgery and dye loading steps used for awake mouse two-photon imaging. A more detailed discussion follows.

1. Adapt mouse to treadmill.
2. Anesthetize mouse.
3. Cut scalp to reveal skull and mark outline of craniotomy on skull.
4. Drill outline of craniotomy and remove skull section.
5. Attach bottom piece of head plate to the skull.
6. Perform calcium-sensitive dye bolus injection.
7. Insert coverslip with attached silicone elastomer (Kwik-sil) plug.
8. Seal craniotomy with top piece of head plate.
9. Let mouse awaken and recover from surgery.
10. Head-restrain mouse on treadmill and perform imaging experiments.

All experiments should be performed in compliance with the Guide for the Care and Use of Laboratory Animals (http://www.nap.edu/readingroom/books/labrats/). Specific protocols must be approved by the Institutional Animal Care and Use Committee of the specific institution where the experiments are performed.

The design of the two-piece ~1-g titanium head plate used both to head-restrain the mouse and to hold the cranial window in place is shown in Figure 2A. Considerable effort was devoted to optimizing the custom-designed head plates and cranial windows. We found that this was crucial to reducing brain motion that would otherwise eliminate the possibility of imaging at cellular resolution.

For head plate and cranial window implantation, mice are anesthetized (typically with isoflurane or ketamine/xylazine), the scalp is cut to reveal the skull, the periosteum is rubbed off with a cotton swab, and the outline of the craniotomy position is marked. Typically the craniotomy is circular, and the diameter matches the diameter of the center hole in the head plate (either 2 or 3 mm). For some brain regions, a circular craniotomy is not possible and a half-circle craniotomy is made (e.g., over the cerebellar vermis). The outline of the craniotomy is slowly drilled using a small-diam-

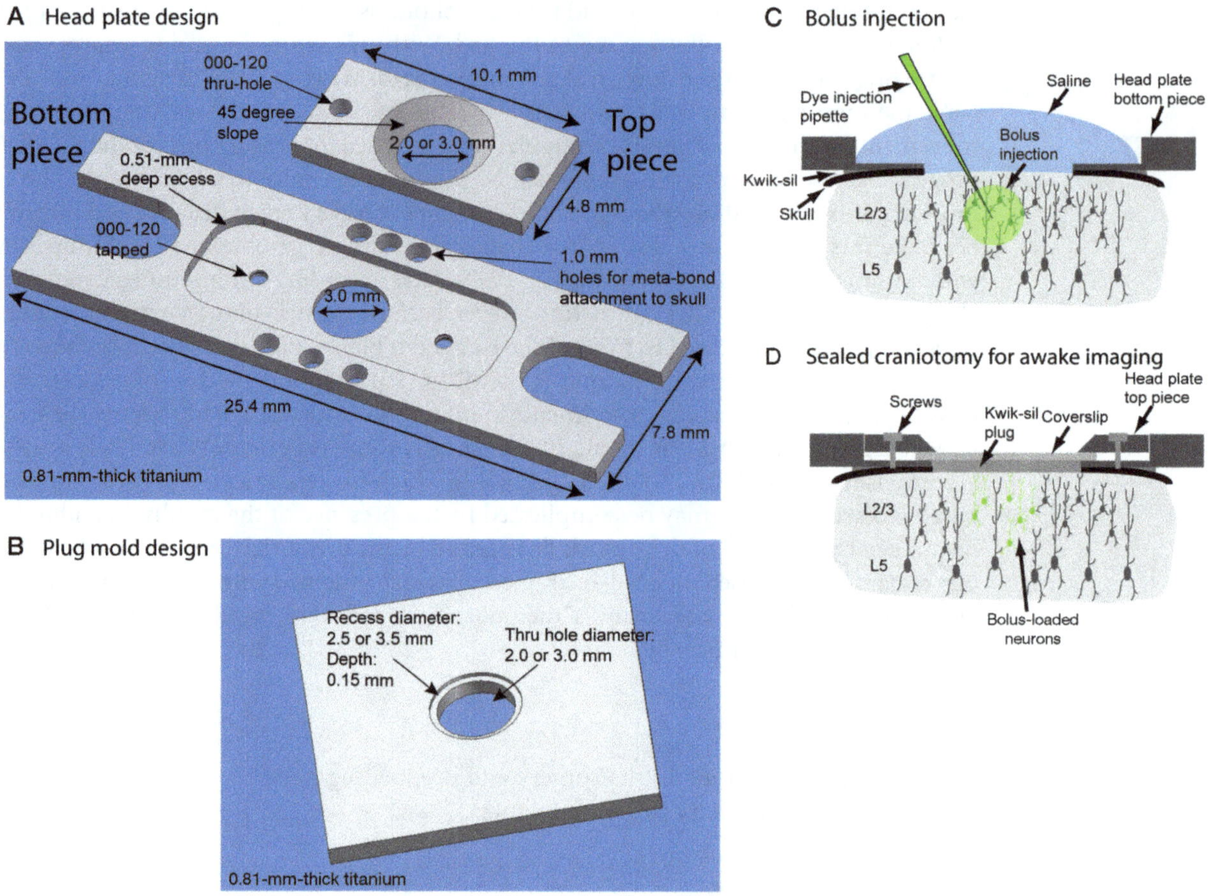

FIGURE 2. (*A*) Technical drawing of the two-piece head plate used for awake imaging. (*B*) Technical drawing of the plug mold used to form Kwik-sil plugs on circular coverslips. (*C*) Once the bottom piece of the head plate is attached (see the Surgical Procedures and Dye Loading section), bolus loading is used to load large populations of neurons with calcium-sensitive dyes. (*D*) After bolus loading, a coverslip with a Kwik-sil plug is inserted into the craniotomy and the top piece of the head plate is used to hold it in place. (Adapted from supplementary material in Dombeck et al. 2009.)

eter burr (~0.5-mm diameter) and a high-speed pneumatic dental drill. It is critical to deepen the outline of the craniotomy slowly so as to limit the amount of heating and vibration. Saline should be periodically applied to hydrate the skull in the local craniotomy region and then soaked up with a cotton swab before further drilling. Once cracks form around the craniotomy edges, the circular piece of skull is carefully removed, leaving the dura intact. A thin ring of Kwik-sil (World Precision Instruments) is then applied around the edges of the craniotomy and the bottom piece of the head plate is pushed onto the uncured Kwik-sil ring, with the center hole concentric with the craniotomy, until it sits flush against the skull. When set, the Kwik-sil has the effect of temporarily holding the head plate in place until it is permanently affixed and it forms a water- and air-tight seal between the bottom of the head plate and the skull. At this point, saline is added to the craniotomy to keep the dura from drying. Dental cement (Meta-bond, Parkell Inc.) is then used to bond the bottom piece of the head plate to the skull. To aid the bonding of the Meta-bond to the skull, the skull is scored repeatedly with the burr and is dried thoroughly.

Next, after the bottom piece of the head plate is affixed to the skull, but before the coverslip and Kwik-sil plug are implanted, dye bolus loading (Stosiek et al. 2003) is performed (Fig. 2C). Typically, neurons and astrocytes are labeled with a green calcium-sensitive dye. For offline motion correction of the acquired image time series, it is useful to simultaneously acquire fluorescence from a separate non-activity-dependent dye in a different color channel. Sulforhodamine 101 (SR101) is suitable because it labels only astrocytes (Nimmerjahn et al. 2004), it is not activity dependent, and its red emission is spectrally separable from the green calcium-sensitive dye emission. A 10-mM stock of Calcium Green-1 AM in dimethylsulfoxide (DMSO) containing 20% Pluronic F-127 (mixed fresh before each experiment) is diluted 10-fold into a solution (with pH 7.4) containing 150 mM NaCl, 2.5 mM KCl, 10 mM HEPES, and ~0.05 mM SR101. Filled pipettes (P-2000 puller, Sutter Instruments) are beveled (BV-10, Sutter) to facilitate penetration of the intact dura (postbeveling pipette resistance 4–9 MΩ). Pipettes are advanced (MP285, Sutter) through the craniotomy at 39° with respect to the brain surface, and dye injection is performed at various depths, pressures, and durations depending on the brain region. Typically, the injection is performed at or just below the neurons of interest (e.g., ~250 µm deep to label layer-2/3 neurons in the motor cortex), and longer lower-pressure injections (1.5–2.0 psi, ~8 min) are often preferred over shorter higher-pressure injections. To increase the area of labeling, multiple dye injections spaced hundreds of micrometers apart can be performed. For a successful injection, the tissue must be pristine (the surface of the dura should never be touched with any instruments and there should be no subdural bleeding).

Next, the craniotomy is sealed with a coverslip and Kwik-sil plug, and the top piece of the head plate is attached (Fig. 2D). Uniform thickness Kwik-Sil plugs (2- or 3-mm-diameter disks) were molded onto #1 thickness, 2.5- or 3.5-mm-diameter glass coverslips (custom coverslips from Erie Scientific) using a custom-made mold (Fig. 2B). The thickness of the plug was designed to apply slight pressure to the surface of the dura to reduce out-of-focal-plane (z) brain motion. After dye injection, the dura is allowed to dry until tacky, a small drop of uncured Kwik-sil is applied to the bottom of the plug, and the coverslip/plug combination is inserted into the craniotomy and onto the shelf of the bottom piece of the head plate. Because of intracranial pressure, the dural surface may appear slightly distended. The plug will press the brain surface back to a position slightly below (~100–200 µm) where it was held by the skull; however, the dural surface should never appear significantly swelled; this is an indication of trauma induced during surgery. The coverslip is then held rigidly in place with the screwed-on top piece. When cured, the small drop of Kwik-sil has the effect of bonding the plug to the dura and reducing in-focal plane (x–y) brain motion. For additional helpful suggestions on cranial window surgical procedures, see Holtmaat et al. (2009). Most previous cranial window preparations have used agarose to fill the space between the dura and coverslip. We have experimented with using agarose in our awake preparation, but have found the brain motion to be greater than what can be achieved using Kwik-sil.

Because the mice are awake, adapting them to the imaging/treadmill apparatus is an important component of the entire method. Before surgery, all mice undergo at least two ball adaptation sessions in which they move freely on the ball for ~10–15 min while the handler rotates the ball to keep the mice centered near the apex. Additional adaptation can be made possible by implanting a temporary head plate (affixed to the skull with a small amount of Meta-bond) days or weeks before the imaging experiment. The mice can then be fully adapted to head restraint on the ball before the imaging experiments take place (the temporary head plate is removed at the beginning of the craniotomy/dye injection surgery). Following surgery, the mice are head-restrained on the ball in complete darkness. Typically, for the non-head-restrained trained mice, it takes ~10–15 min for the mouse to learn to balance and then begin to walk or run; the head-restrained trained mice can balance, walk, and run immediately. Imaging typically commences at ~1 h after the mouse wakes from anesthesia.

Mice ranging in age from P35 to adults have been successfully used for awake imaging experiments. It may be possible to use younger mice; however, the inertia of the ball should provide a realistic reaction force comparable to the force needed by the mouse to start or stop free running on the ground; the younger and lighter the mice, the greater the discrepancy between the free-running and head-restrained ball rotation forces.

Data Acquisition and Processing

Time-series movies of neuronal populations loaded with calcium-sensitive dyes are typically acquired at 1 msec/line with frame rates of 8 (128 x 128 pixels), 16 (128 x 64 pixels), or 32 (128 x 32 pixels) Hz. Movie durations range anywhere from a few tens of seconds to many minutes. When determining field size, it is important to remember that although larger fields of view provide the greatest number of neurons from which to record, the decreased dwell time and number of pixels per neuron will reduce the signal to noise and the user's ability to accurately determine the borders between adjacent neurons (see the Examples of Imaging in Different Brain Regions section below for examples of typical fields of view, frame rates, and number of pixels acquired). Photodamage and photobleaching are the most significant limitations on signal to noise: Although increasing excitation laser power will increase the fluorescence emission and decrease shot noise, it will also increase bleaching and damage to the area of interest. Many tens of milliwatts of average laser power are used at the sample for in vivo imaging. Power should be kept low to avoid significant amounts of photobleaching or obvious signs of photodamage. Overt signs of photodamage that should be avoided include (1) a slow increase in fluorescence intensity throughout the calcium-sensitive dye channel during the time series and (2) astrocytic or neuronal calcium transients that are triggered by and occur only at the beginning of a time series. One way to reduce photodamage and photobleaching is to optimize the collection efficiency of the microscope; this will allow the user to reduce the illumination laser power and still collect the same amount of fluorescence emission. A two-photon microscope that is designed for maximum collection efficiency should include (1) a collection lens that is close enough and of large enough diameter to collect all of the fluorescence light emanating from the back aperture of the objective, (2) high-efficiency dichroic mirrors and optical filters (new products from companies such as Semrock and Chroma have increased efficiency compared with what was available just a few years ago), and (3) high-quantum-efficiency photomultiplier tubes (PMTs) such as the GaAsP PMTs (Zipfel et al. 2003) available from Hamamatsu (H10770PA-40, Hamamatsu).

The head plate, cranial window, and surgical procedures help to keep brain motion to a minimum while the mice maneuver on the spherical treadmill. During running, for example, there is very little out-of-plane (z) brain motion in all brain regions imaged (see the Examples of Imaging in Different Brain Regions section below): Small-scale structures typically remain in focus during the entire imaging sequence. In addition, there are small movement-related lateral (x–y) displacements of ~2–5 µm. The timescale of these brain movements is >~30 msec, meaning that time series acquired with a frame rate >~30 Hz have very few in-frame distortions. Offline motion correction of the interframe brain movements seen in these movies is relatively straightforward: The cross correlation between each frame in the movie (non-activity-dependent channel) and a frame acquired during resting can be calculated to find the maximum correlation and the appropriate distance to shift each frame to align it to the resting frame (Fig. 3Ai). An algorithm for sub-pixel alignment is available in the methods section of a recent publication (Miri et al. 2011). For time-series movies acquired with frame rates <30 Hz, in-frame distortions are often observed. These lateral shifts can be corrected using a line-by-line motion-correction algorithm (Fig. 3Aii).

One strategy for line-by-line image correction is to correct frames back to a reference image selected from the resting state using an algorithm based on a hidden Markov model (HMM) (Rabiner 1989; Dombeck et al. 2007). Each recorded line is mapped to a shifted position with respect to the reference image. The HMM model uses a maximum likelihood method to determine the most probable sequence of brain movements associated with each successive line. An intermediate strategy between whole frame and line-by-line correction was recently presented (Greenberg and Kerr 2009). The method is based on the Lucas–Kanade framework and estimates complex displacement trajectories occurring within individual image frames by gradient descent.

Once the time series are motion corrected, numerous methods exist to define the regions of interest (ROIs) within the field: manually defined (Kerr et al. 2005; Niell and Smith 2005; Dombeck et al. 2007; Flusberg et al. 2008), semiautomated (Ohki et al. 2005; Ozden et al. 2008), and automated (Brown et al. 2001; Reidl et al. 2007; Mukamel et al. 2009; Miri et al. 2011). Fluorescence versus time traces are generated for each region of interest. Slow timescale changes in the fluorescence (possibly caused by bleaching, changes in blood flow–related excitation light absorption, etc.) are removed by examining the distribution of fluorescence in a many-second duration interval (~10 sec) around each sample time point and normalizing by the 8% percentile value (Fig. 3B). For most neuron fluorescence traces, this method reliably subtracts slow changes without significantly filtering longer

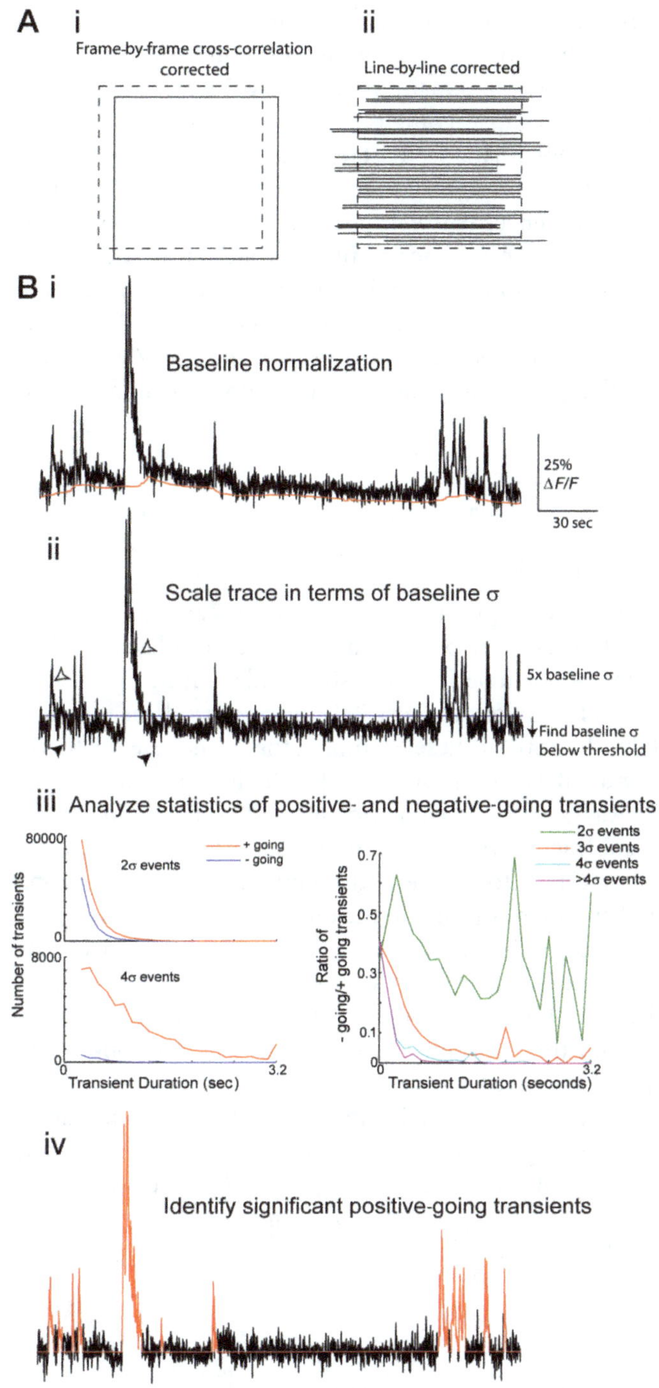

FIGURE 3. (*A*) Depending on the frame rate of the acquired time-series movie, brain-motion-induced image shifts and/or distortions can be corrected offline using frame-by-frame cross-correlation methods (*Ai*, for high-frame-rate movies) or line-by-line correction methods (*Aii*, for lower frame rate movies). (*B*) By making the assumption that calcium transients are only positive going and movements of the brain in the axial direction have an equal likelihood of causing positive- or negative-going transients, it is possible to identify significant positive-going transients that are not likely to have been caused by brain motion. (*Bi*) After regions of interest (ROIs) are selected, the fluorescence versus time trace for each ROI is first baseline normalized using a sliding window method to correct for slow drifts in the fluorescence level (red line). (*Bii*) A threshold is manually defined (blue line) such that the baseline fluorescence fluctuations (σ, below threshold) can be calculated without including calcium transients (above threshold). The trace is then scaled in units of the baseline σ, and positive- (open arrowheads in *Bii*) and negative- (filled arrowheads in *Bii*) going events of various amplitudes are identified. (*Biii*) Next, for each amplitude, the events are separated by their duration; short-duration and low-amplitude events generally have a near equal number of positive- and negative-going transients (*top left*), but longer-duration and higher-amplitude events are dominated by positive-going (calcium-related) transients (*bottom left*). The ratio of the number of negative- to positive-going events for each amplitude and duration can then be examined (*Biii, right*) to set a significance threshold, typically ~0.05. Positive-going transients with amplitude and duration defined by this threshold have a <5% chance of being caused by brain motion and a >95% chance of being a calcium transient. (*Biv*) Significant calcium transients with <5% false-positive error rate are highlighted in red.

events that have distinct onsets and offsets; however, in some traces in which much longer duration transients are present, this method may alter the longer events and hence should not be used. Instead, the baseline for these neurons may be defined by hand and then used to normalize the raw trace.

One possible concern when examining cell-activity-dependent fluorescence traces in awake behaving animals is brain-motion-induced fluorescence changes giving false signatures for cell activity. Even after in-plane motion correction, residual motion causing cells to move in and out of the focal plane may still remain. To evaluate this potential problem, it is useful to examine the ratio of positive to negative transients. Because the many tens to hundreds of neurons are distributed randomly in the *z* direction on the scale of a typical two-photon focal plane (~2–3 μm thick), out-of-plane motion is expected to cause an equal number of positive-going (cells moving into the plane and increasing the signal) and negative-going (cells moving out of the plane and decreasing the signal) false fluorescence transients. This is in stark contrast to the cell-activity-induced transient increases in calcium-sensitive dye fluorescence, in which only positive-going transients are observed. To examine the statistical differences between positive- and negative-going transients, the baseline subtracted neuron fluorescence traces are scaled in units of the baseline standard deviation (σ) (Fig. 3Bii). Histograms of positive- and negative-going fluorescence transients of varying durations and amplitudes (in units of baseline σ) are generated for all neurons in all mice in the study (Fig. 3Biii). This analysis is then used to identify significant transients with <5% false positive error rates (ratio of negative-going to positive-going transients for each amplitude and duration <0.05) (Fig. 3Biv).

DISCUSSION

Examples of Imaging in Different Brain Regions

The methods described here have proven to be useful for imaging in awake mice in a wide range of brain regions (Fig. 4). Figure 4A is an example of imaging from layer-2/3 hindlimb sensory cortex while the mouse alternated between periods of resting and running on the treadmill. Figure 4B is an example from the forelimb motor cortex layer 2/3 during periods of resting, running, and grooming (Dombeck et al. 2009). In both of these examples, the frame rate for the 128 x 128-pixel movies was ~4 Hz (Fig. 4A) and ~8 Hz (Fig. 4B); the HMM line-by-line image correction algorithm (Dombeck et al. 2007) was used and regions of interest were manually defined around neuron somata. In the baseline subtracted traces from individual neurons, a range of activity is seen, ranging from behaviorally correlated to sparse and seemingly random neuronal calcium transients. Figure 4C shows the application to imaging in the olfactory bulb. This 128 x 32-pixel time series was acquired ~150 μm below the brain surface at ~30 Hz and whole-frame cross-correlation motion correction was applied; the regions of interest around the neuronal soma (likely external tufted neurons) were manually defined. While the mouse was running on the ball, various olfactory cues were presented to the mouse and the resultant calcium transients were observed. The final example is from the molecular layer of the cerebellar vermis during mouse resting and running (Fig. 4D). The dendrites of the Purkinje neurons are loaded with a calcium-sensitive dye and calcium transients are observed during complex spiking (Sullivan et al. 2005; Ozden et al. 2008). This time series was also acquired at ~30 Hz and was motion corrected using whole-frame cross correlation. Here, regions of interest were identified using an Independent Component Analysis algorithm (Brown et al. 2001; Mukamel et al. 2007; Reidl et al. 2007).

FIGURE 4. Examples of two-photon imaging time series acquired in various brain regions in the awake behaving mouse. (*A*) (*Top*) Images of neurons and astrocytes in hindlimb sensory cortex layer 2/3 labeled with Calcium Green-1 AM (green) and SR101 (red). (*Bottom*) Fluorescence versus time traces of the neurons and neuropil regions selected in the top image; the running speed of the mouse is shown in the bottom trace. (*B*) (*Top*) Images of neurons and astrocytes in forelimb motor cortex layer 2/3 labeled with Calcium Green-1 AM (green) and SR101 (red). (*Bottom*) Fluorescence versus time traces of the neurons selected in the top image; the timing of mouse running and grooming is shown in the bottom traces. (*C*) (*Top*) Images of neurons and astrocytes in the olfactory bulb labeled with Calcium Green-1 AM (green) and SR101 (red). (*Middle*) Fluorescence versus time heat map of the neurons and neuropil region selected in the top image; the timing of the olfactory cues is shown in the bottom row. (*Bottom*) Fluorescence versus time traces of the neurons and neuropil region selected in the top image; the timing of the olfactory cues and the running speed of the mouse are shown in the bottom traces. (*D*) (*Top*) Images of the cerebellar vermis molecular layer labeled with Oregon Green BAPTA-1 AM and colored regions representing Purkinje neuron dendrite regions of interest. (*Bottom*) Fluorescence versus time traces of the dendrites selected in the top image; the timing of mouse running is shown in the bottom trace. (*A*, Adapted, with permission, from Dombeck et al. 2007, ©Elsevier; *B*, data from Dombeck et al. 2009.)

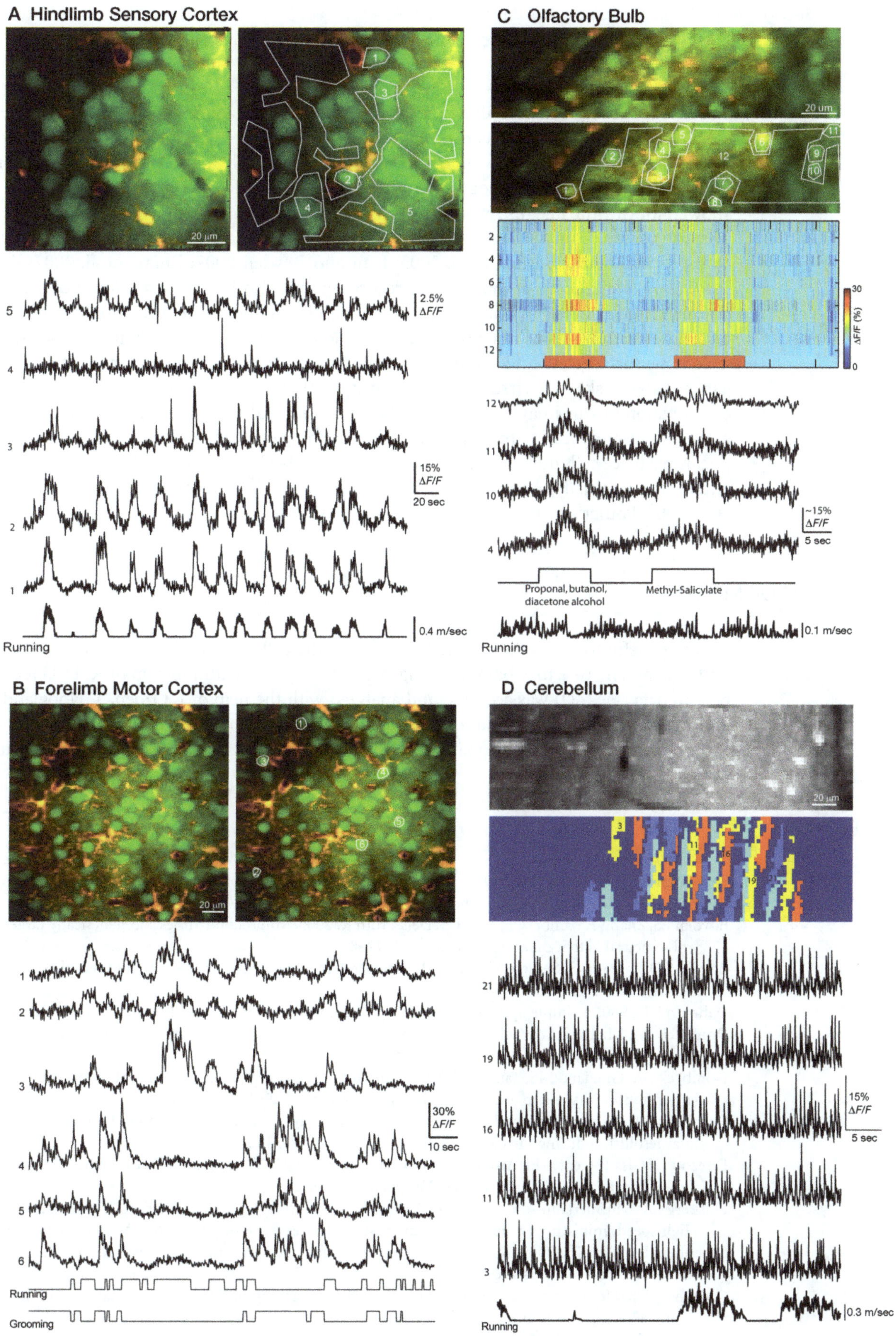

FIGURE 4. *(See facing page for legend.)*

Outlook

The methods described here allow for cellular-resolution imaging of neural activity in awake, mobile mice during behaviors such as running, resting, and grooming. The bolus-loading method used to load populations of neurons with calcium-sensitive dyes only allows for acute experiments. A recent combination of our technology with genetically encoded calcium indicators (Tian et al. 2009) allowed for chronic imaging experiments in which the activity of neurons was repeatedly monitored over the course of many weeks in awake mice. The combination of our methods with photostimulation technologies (Boyden et al. 2005; Han and Boyden 2007; Zhang et al. 2007; Rickgauer and Tank 2009) will allow not only for the recording, but also for the stimulation of activity in a large population of neurons in awake mice. Although here we have only presented methods to adapt the mice to the treadmill and then observe neural activity during spontaneous behaviors, it is also possible to perform behavioral training for tasks such as choice tasks (Komiyama et al. 2009) and navigation-based tasks in a virtual reality environment (Holscher et al. 2005; Harvey et al. 2009). The combination of our imaging technology with virtual reality and genetically encoded calcium indicators has recently lead to the capability of performing cellular resolution functional imaging of hippocampal place cells during virtual navigation (Dombeck et al. 2010). Finally, the spherical treadmill and head-restrained mouse apparatus allows for intracellular recordings in awake, mobile mice (Lee et al. 2006; Poulet and Petersen 2008; Harvey et al. 2009).

ACKNOWLEDGMENTS

We are grateful to our collaborators for aiding in the collection and analysis of some of the data presented here: The olfactory bulb data (Fig. 4C) was acquired with the help of K. Delaney, and the cerebellum data (Fig. 4D) was acquired and analyzed with the help of I. Ozden, T. Hoogland, F. Collman, and S.S. Wang. We thank C. Harvey for careful reading of this chapter. Our work has been supported by Princeton University, The National Institutes of Health (#1R01MH083686-01), and The Patterson Trust.

REFERENCES

Berg-Johnsen J, Langmoen IA. 1992. The effect of isoflurane on excitatory synaptic transmission in the rat hippocampus. *Acta Anaesthesiol Scand* **36:** 350–355.

Boyden ES, Zhang F, Bamberg E, Nagel G, Deisseroth K. 2005. Millisecond-timescale, genetically targeted optical control of neural activity. *Nat Neurosci* **8:** 1263–1268.

Brown GD, Yamada S, Sejnowski TJ. 2001. Independent component analysis at the neural cocktail party. *Trends Neurosci* **24:** 54–63.

Dahmen HJ. 1980. A simple apparatus to investigate the orientation of walking insects. *Experientia* **36:** 685–687.

Dombeck DA, Khabbaz AN, Collman F, Adelman TL, Tank DW. 2007. Imaging large-scale neural activity with cellular resolution in awake, mobile mice. *Neuron* **56:** 43–57.

Dombeck DA, Graziano MS, Tank DW. 2009. Functional clustering of neurons in motor cortex determined by cellular resolution imaging in awake behaving mice. *J Neurosci* **29:** 13751–13760.

Dombeck DA, Harvey CD, Tian L, Looger LL, Tank DW. 2010. Functional imaging of hippocampal place cells at cellular resolution during virtual navigation. *Nat Neurosci* **13:** 1433–1440.

Ferezou I, Haiss F, Gentet LJ, Aronoff R, Weber B, Petersen CC. 2007. Spatiotemporal dynamics of cortical sensorimotor integration in behaving mice. *Neuron* **56:** 907–923.

Flusberg BA, Nimmerjahn A, Cocker ED, Mukamel EA, Barretto RP, Ko TH, Burns LD, Jung JC, Schnitzer MJ. 2008. High-speed, miniaturized fluorescence microscopy in freely moving mice. *Nat Methods* **5:** 935–938.

Greenberg DS, Kerr JN. 2009. Automated correction of fast motion artifacts for two-photon imaging of awake animals. *J Neurosci Methods* **176:** 1–15.

Greenberg DS, Houweling AR, Kerr JN. 2008. Population imaging of ongoing neuronal activity in the visual cortex of awake rats. *Nat Neurosci* **11:** 749–751.

Han X, Boyden ES. 2007. Multiple-color optical activation, silencing, and desynchronization of neural activity, with single-spike temporal resolution. *PLoS One* **2:** e299.

Harvey CD, Collman F, Dombeck DA, Tank DW. 2009. Intracellular dynamics of hippocampal place cells during virtual navigation. *Nature* **461:** 941–946.

Helmchen F, Svoboda K, Denk W, Tank DW. 1999. In vivo dendritic calcium dynamics in deep-layer cortical pyramidal neurons. *Nat Neurosci* **2:** 989–996.

Holscher C, Schnee A, Dahmen H, Setia L, Mallot HA. 2005. Rats are able to navigate in virtual environments. *J Exp Biol* **208:** 561–569.

Holtmaat A, Bonhoeffer T, Chow DK, Chuckowree J, De Paola V, Hofer SB, Hubener M, Keck T, Knott G, Lee WC, et al. 2009. Long-term, high-resolution imaging in the mouse neocortex through a chronic cranial window. *Nat Protoc* **4:** 1128–1144.

Kerr JN, Denk W 2008. Imaging in vivo: Watching the brain in action. *Nat Rev Neurosci* **9:** 195–205.

Kerr JN, Greenberg D, Helmchen F. 2005. Imaging input and output of neocortical networks in vivo. *Proc Natl Acad Sci* **102:** 14063–14068.

Kerr JN, de Kock CP, Greenberg DS, Bruno RM, Sakmann B, Helmchen F. 2007. Spatial organization of neuronal population responses in layer 2/3 of rat barrel cortex. *J Neurosci* **27:** 13316–13328.

Komiyama T, Sato TR, O'Connor DH, Zhang YX, Huber D, Hooks BM, Gabitto M, Svoboda K. 2010. Learning-related fine-scale specificity imaged in motor cortex circuits of behaving mice. *Nature* **464:** 1182–1186.

Kuhn B, Denk W, Bruno RM. 2008. In vivo two-photon voltage-sensitive dye imaging reveals top-down control of cortical layers 1 and 2 during wakefulness. *Proc Natl Acad Sci* **105:** 7588–7593.

Lee AK, Manns ID, Sakmann B, Brecht M. 2006. Whole-cell recordings in freely moving rats. *Neuron* **51:** 399–407.

Major G, Tank D. 2004. Persistent neural activity: Prevalence and mechanisms. *Curr Opin Neurobiol* **14:** 675–684.

Mason AC, Oshinsky ML, Hoy RR. 2001. Hyperacute directional hearing in a microscale auditory system. *Nature* **410:** 686–690.

Miri A, Daie K, Burdine RD, Aksay E, Tank DW. 2011. Regression-based identification of behavior-encoding neurons during large scale optical imaging of neural activity at cellular resolution. *J Neurophysiol* (in press). doi: 10.1152/jn.00702.2010.

Mukamel EA, Nimmerjahn A, Schnitzer MJ. 2007. Automated sorting of intracellular calcium signals for large-scale imaging studies of neuronal and glial populations. *Soc Neurosci Abstr* Program No. 252.2.

Mukamel EA, Nimmerjahn A, Schnitzer MJ. 2009. Automated analysis of cellular signals from large-scale calcium imaging data. *Neuron* **63:** 747–760.

Niell CM, Smith SJ. 2005. Functional imaging reveals rapid development of visual response properties in the zebrafish tectum. *Neuron* **45:** 941–951.

Nimmerjahn A, Kirchhoff F, Kerr JN, Helmchen F. 2004. Sulforhodamine 101 as a specific marker of astroglia in the neocortex in vivo. *Nat Methods* **1:** 31–37.

Ohki K, Chung S, Ch'ng YH, Kara P, Reid RC. 2005. Functional imaging with cellular resolution reveals precise micro-architecture in visual cortex. *Nature* **433:** 597–603.

Ozden I, Lee HM, Sullivan MR, Wang SS. 2008. Identification and clustering of event patterns from in vivo multi-photon optical recordings of neuronal ensembles. *J Neurophysiol* **100:** 495–503.

Pologruto TA, Sabatini BL, Svoboda K. 2003. ScanImage: Flexible software for operating laser scanning microscopes. *Biomed Eng Online* **2:** 13.

Poulet JF, Petersen CC. 2008. Internal brain state regulates membrane potential synchrony in barrel cortex of behaving mice. *Nature* **454:** 881–885.

Rabiner LR. 1989. A tutorial on hidden Markov models and selected applications in speech recognition. *Proc IEEE* **77:** 257–286.

Reidl J, Starke J, Omer DB, Grinvald A, Spors H. 2007. Independent component analysis of high-resolution imaging data identifies distinct functional domains. *Neuroimage* **34:** 94–108.

Rickgauer JP, Tank DW. 2009. Two-photon excitation of channelrhodopsin-2 at saturation. *Proc Natl Acad Sci* **106:** 15025–15030.

Sato TR, Gray NW, Mainen ZF, Svoboda K. 2007. The functional microarchitecture of the mouse barrel cortex. *PLoS Biol* **5:** e189.

Stevenson PA, Dyakonova V, Rillich J, Schildberger K. 2005. Octopamine and experience-dependent modulation of aggression in crickets. *J Neurosci* **25:** 1431–1441.

Stosiek C, Garaschuk O, Holthoff K, Konnerth A. 2003. In vivo two-photon calcium imaging of neuronal networks. *Proc Natl Acad Sci* **100:** 7319–7324.

Sullivan MR, Nimmerjahn A, Sarkisov DV, Helmchen F, Wang SS. 2005. In vivo calcium imaging of circuit activity in cerebellar cortex. *J Neurophysiol* **94:** 1636–1644.

Tian L, Hires SA, Mao T, Huber D, Chiappe ME, Chalasani SH, Petreanu L, Akerboom J, McKinney SA, Schreiter ER, et al. 2009. Imaging neural activity in worms, flies and mice with improved GCaMP calcium indicators. *Nat Methods* **6:** 875–881.

Zhang F, Wang LP, Brauner M, Liewald JF, Kay K, Watzke N, Wood PG, Bamberg E, Nagel G, Gottschalk A, Deisseroth K. 2007. Multimodal fast optical interrogation of neural circuitry. *Nature* **446:** 633–639.

Zipfel WR, Williams RM, Webb WW. 2003. Nonlinear magic: Multiphoton microscopy in the biosciences. *Nat Biotechnol* **21:** 1369–1377.

77 Imaging Neuronal Population Activity in Awake and Anesthetized Rodents

David S. Greenberg, Damian J. Wallace, and Jason N.D. Kerr

Network Imaging Group, Max Planck Institute for Biological Cybernetics, 72076 Tübingen, Germany

ABSTRACT

Recent advances in in vivo two-photon imaging have extended the technique to permit detection of action potentials (APs) in populations of spatially resolved neurons in awake animals. Although experimentally demanding, this technique's potential applications include experiments to investigate perception, behavior, and other awake states. This article outlines experimental procedures for imaging neuronal populations in awake and anesthetized rodents. Details are provided on habituation to head fixation, surgery, head plate design, and dye injection. Determination of AP detection accuracy through simultaneous optical and electrophysiological recordings is also discussed. Basic problems of data analysis are considered, such as correction of signal background and baseline drift, AP detection, and motion correction. As an application of the method, the comparison of neuronal activity across arousal states is considered in detail, and some future directions are discussed.

INTRODUCTION

In addition to its excellent spatial resolution, two-photon imaging (Denk et al. 1990) works very well to record activity from cortical neurons because it is relatively noninvasive and neuronal populations can be sampled independently of their action potential firing rate. Furthermore, the same population of neurons can be revisited multiple times over the course of hours, days, weeks, and potentially months with absolute certainty about cell identity. As one of the main aims of population imaging is to accurately quantify the spiking activity simultaneously from many neurons (Kerr et al. 2005, 2007; Yaksi and Friedrich 2006; Sato et al. 2007; Greenberg et al. 2008), both single-cell and single-action-potential (AP) resolution are required. If Ca^{2+} transients are used to infer electrical spiking activity, then the relationship between the AP and the corresponding Ca^{2+} transient needs to be established. Given that the majority of population imaging is performed on layer 2/3 cortical neurons, this is especially important as spontaneous firing rates can be well below 1 Hz, and sensory stimulation can be evoked well below one spike per stimulus, on average. Simultaneous electrophysiology (cell attached) and imaging of Ca^{2+} transients can be used to infer detection fidelity as this does not perturb the intracellular contents, unlike both whole-cell and sharp-electrode recordings.

Inferring spikes from Ca^{2+} transients depends on multiple factors, such as the dye being used, the animal's health, beam properties, the speed of sampling, and the detection system. Moreover, inferring spikes is fundamentally a signal to noise problem, because the Ca^{2+} transient peak evoked from a single action potential is around 10% $\Delta F/F_0$ for organic indicators such as Oregon Green BAPTA-1 (OGB-1). Consequently, to achieve single-AP resolution it is necessary to ensure a high sampling frequency so that each transient is sampled before it decays into the noise (Kerr and Denk 2008). In addition, because ester-based loading techniques lead to dense staining in all neuronal and astrocytic processes, and this neuropil activity is highly correlated with ongoing Up and Down states (Kerr et al. 2005), it is critical to overfill the back aperture to avoid contamination by this neuropil signal in the somata signal (Kerr and Denk 2008). In this context, it is important to both understand the detection reliability for APs and also to be able to adjust imaging and collection parameters to improve fidelity when necessary.

Two-photon microscopy is ideally suited for imaging activity in neuronal populations in the headfixed awake (Greenberg et al. 2008), headfixed moving (Dombeck et al. 2007; Dombeck et al. 2009), and freely moving animal (Sawinski et al. 2009). Not surprisingly, imaging in the awake animal is more problematic because sudden animal movements interfere with the imaging process. This is an inherent problem when using raster scanning to produce an image, because the relative position of the beam in the tissue has to be known to assign the detected photons to the correct pixel positions. There are several approaches to reduce movement of the animal and "reassign" the pixels post hoc (Dombeck et al. 2007; Greenberg and Kerr 2009). Here we describe procedures that enable calcium imaging in neuronal populations in both awake and anesthetized animals and the inference of spiking activity from the measured Ca^{2+} signals. This approach can be used to establish neuronal firing rates, spike firing correlations, and responses to sensory input in the awake behaving rodent.

EXPERIMENTAL PROCEDURES

For a general description of anesthesia and surgical procedures for preparing animals for in vivo two-photon microscopy, see Chapters 49 and 51. What follows is a discussion on important points to consider when combining electrophysiology with population imaging with the goal of inferring spikes from Ca^{2+} transients.

Habituation and Training for Awake Experiments

A period of habituation of the animal to both the experimenter and the experimental setup is critical for experiments with awake animals, and is best performed gradually over a period of about a week post head plate attachment. First, familiarize the animal with being regularly handled. We routinely use three to five 5–10 min sessions over 2 d, in which the animal is allowed to move around freely on the experimenters arms and over the experimental apparatus. Restraining the animal by holding the head plate for brief periods (30–60 sec) during the later sessions is a good way of gradually introducing the animal to periods of head fixation. The next stage of the training involves sessions of head fixation in which the animal is restrained for progressively increasing periods of time, typically starting with 5 min and then increasing to 10, 15, 20, 45, and 60 min. More than one head fixation session per day is ill advised.

Head Plate and Head Plate Holders

A head plate design suitable for many different preparations in rats is shown in Figure 1Aa. Wide chamber designs with large side flanges for secure clamping are in general the most convenient option, particularly for imaging combined with electrophysiology. For experiments in awake animals, the size and weight of the head plate should be minimized. A recent design (Greenberg et al. 2008), including attachment points for chronically implanted ECoG electrode wires, is shown in Figure 1Ab. Design of the mechanism for secure clamping of the head plate is also more important

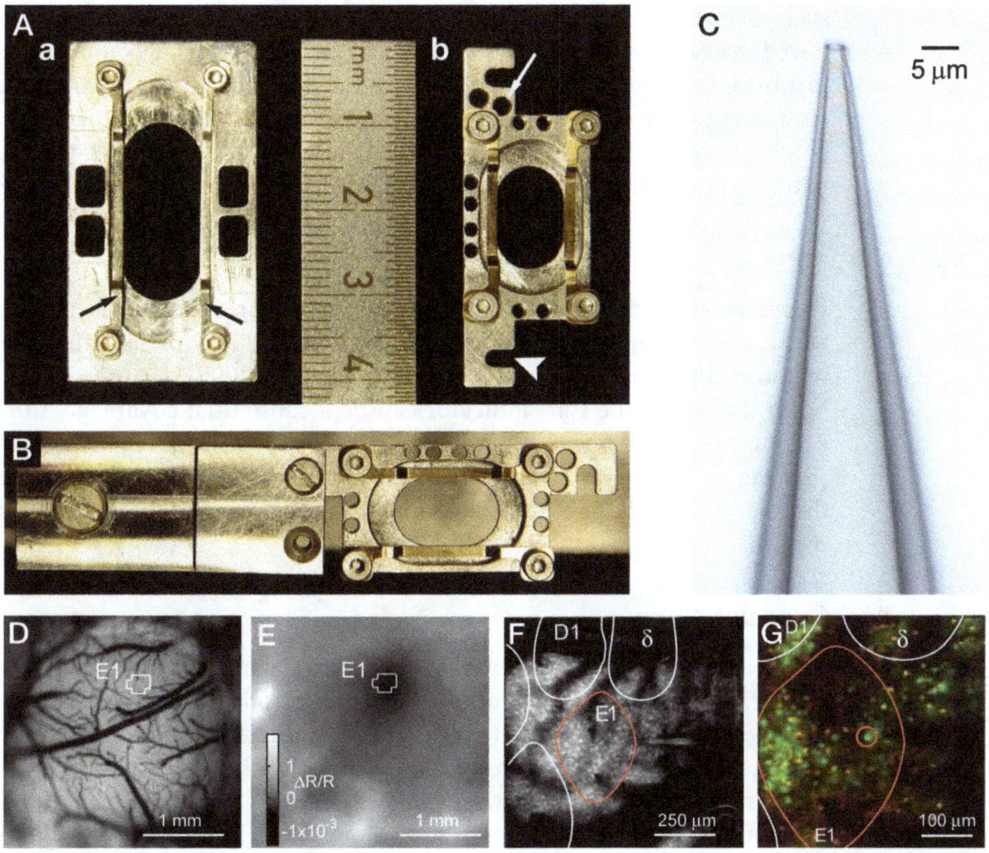

FIGURE 1. Experimental procedures. (*Aa*) Large head plate design for simultaneous two-photon imaging and electrophysiology. Spring clamps fastened with small socket head screws (indicated by the black arrows) are used to hold a coverslip in place over the craniotomy. (*Ab*) Reduced head plate design for awake experiments. The arrow indicates attachment points for chronic EEG electrode wire plugs. Arrowhead shows where the screw from the head plate holder clamps the head plate for restricting rotational motion. (*B*) Clamp design for awake recordings (note clamping screw mentioned in *A*). (*C*) Image showing indicator-loading pipette. (*D*) Craniotomy over somatosensory cortex. (*E*) Intrinsic optical signal after stimulation of E1 whisker showing the resulting "spot." The outline of the darkest pixels is overlaid in both images indicating the position of E1 barrel column. (*F*) Close-up view of the E1 area from a low-resolution two-photon image. Barrel field outlines obtained from cytochrome C staining are overlaid. Note the injection pipette tip at right. (*G*) L2/3 neurons (green) and astrocytes (yellow) stained with calcium indicator OGB-1 and sulforhodamine 101, respectively, in the E1 barrel column and surrounding tissue. A neuron within the E1 barrel border (red) was targeted with a patch electrode for cell-attached recording. (*D–G*, Reprinted from Kerr et al. 2007.)

for awake experiments. Specifically, the design should allow the animal to be restrained quickly by a single person. Designs with a single clamping point are an advantage, particularly during the habituation and training phases of the experiment, as are designs that prevent rotation of the head plate in the clamp (Fig. 1B).

Head Plate Implantation

For anesthetized preparations, dental cement provides a rigid and stable method for attaching the head plate to the skull. For good adhesion, keep fluid from the margins of the skin incision from getting between the bone and the dental cement. This is generally best achieved by carefully closing all skin incision margins with tissue glue.

For awake experiments, a more secure method for head plate attachment is required. Use of bone screws may be suitable for larger animals, but take care so that the bone and/or screw do not impinge on the pia. For smaller animals, extremely secure attachment of the head plate can be achieved using a combination of dental cement and dental bonding products usually used for tooth restorations, such as OptiBond (Kerr Corporation). We apply a layer of OptiBond onto the exposed skull, followed by a layer of Charisma (Heraeus Kulzer), a light-curing composite used for dental fillings. This forms a foundation onto which we attach the head plate using dental cement (Greenberg et al. 2008; Houweling and Brecht 2008). Postoperative analgesia is essential following these surgical procedures. Soluble analgesics, such as aspirin, can be provided in the drinking water. Alternatively or in combination, a longer-lasting analgesic, such as flunixin-meglumine can be given at the end of the surgery.

Targeting Sensory Areas

For many experiments, such as those targeting neurons in the cortical representation of a specific whisker in the somatosensory cortex (Kerr et al. 2007) or targeting the monocular or binocular regions in the primary visual cortex (Sawinski et al. 2009), accurate identification of a target area for bolus loading of the Ca^{2+} indicator can be a substantial advantage. Intrinsic optical imaging (see Chapters 60 and 83) is ideally suited for this purpose (Fig. 1E). This can be performed either after performing a craniotomy or through the thinned skull prior to opening the bone. Intrinsic imaging through thinned bone is preferable for awake experiments, because it minimizes the bone opening over the area of interest.

General Surgical Considerations

As for any in vivo preparation that is targeting recordings to the cortex, minimizing cortical insults during the surgical preparation is essential. For population imaging, particularly using sulforhodamine 101 as a counterstain (see Chapter 63), this is especially important, as proliferation and activation of glia during the surgery can dramatically affect the quality of the imaging.

It is absolutely critical to avoid contact with the cortical surface, and to remove the dura in such a way that stray edges of the dura cannot get caught by the loading or recording pipettes during pipette placement and subsequent penetration of the cortical surface. It is also imperative to minimize movement of the cortex when performing population Ca^{2+} imaging experiments. Covering the exposed area with 1%–1.3% agar and fixing a coverslip in place over the agar layer is the gold standard. Important in this procedure is to avoid excessive downward pressure on the cortex when fastening the coverslip in place, and additionally to avoid unnecessary movement of pipettes through the agar.

Preparing Dye for Injection

For calcium indicator labeling of neuronal populations, we use the multicell bolus loading technique (Stosiek et al. 2003). The main steps are provided below; for more details see Chapter 49.

1. Keep the dye desiccated at ~ –20°C.
2. Prepare a 4:1 mixture of DMSO and F-127 Pluronic. This should be made fresh each day.
3. Add the 5 µL DMSO/Pluronic mix to the dye vial, mix it by pipetting it up and down, and sonicate it for ~3 min.
4. Add 35 µL of normal rat Ringer solution and 5 µL of 40 µM Alexa Fluor 594 to the dye solution, mix it by pipetting it up and down, and mix it on a benchtop shaker for 2 min.
5. Remove bubbles by centrifugation.
6. Filter the solution through a microfilter (0.2-µm pore size).
7. Draw the solution into a glass pipette from the tip (1.5 µm internal diameter tip; Fig. 1C).
8. Backfill the pipette with additional dye solution.
9. Mount the pipette onto a headstage, and apply ~30 kPa of pressure.
10. Guide the pipette into a target area in the tissue under visual control using an *x–y–z* manipulator.

11. Check whether the pipette tip is blocked. If it is, then withdraw the pipette and prepare another one.
12. Eject the dye at low pressure (~20–30 kPa) for ~2 min.

Combining Imaging with Cell-Attached Electrophysiology

A number of sources provide equipment lists and solutions for performing electrophysiological cell-attached targeted recordings (Margrie et al. 2002; Kerr et al. 2005; Yaksi and Friedrich 2006; Kerr et al. 2007; Sato et al. 2007). There are several things to keep in mind when performing cell-attached electrophysiology.

- It is important to preserve the tissue and maintain conditions that are as close as possible to the conditions that will be used when collecting data in the absence of electrophysiology (i.e., when inferring spiking). For this purpose, we use a long tapered pipette with a tip resistance of ~4–5 MΩ, and fill it with a combination of Na-based normal rat Ringer and Alexa Fluor 594.
- Minimize lateral movements of the pipette when it is in the tissue. The best way to do this is to determine where the neurons of interest are located and, given that the angle of the pipette relative to the brain surface is constant, calculate where on the surface the pipette should enter.
- Avoid having the tip become blocked, by applying pressure through the pipette. As a general guide, apply 30 kPa when moving through the agar and entering the tissue. Reduce the pressure to ~1.2 kPa when stepping in 2-µm steps toward the target neurons. Further reduce the pressure to ~0.2–0.5 kPa when approaching the neuron of choice, and finally, neutralize or apply small negative pressure when the pipette is close to the membrane and waiting for spiking.

There are two modes for electrically recording reliable spikes: One is cell-attached mode, in which the pipette is in direct contact with the cell membrane forming a loose seal with negative pressure, and the signal-to-noise ratio is high (~10). The other is juxtasomal mode, in which the pipette tip is close to, but is not sealed against, the membrane, and the signal-to-noise ratio is ~2–3. The cell-attached mode is susceptible to mechanical damage of the neuron, which renders the data suspect, but the method has the advantage that a whole-cell configuration can be obtained, thus allowing definite identification of the target neuron. Juxtasomal recordings can last for an hour or more and do not mechanically perturb the neuron, however, identifying the structure from where the spikes originate can sometimes be problematic. Targeting the electrode to the side of the neuron, so as to not create a physical obstacle between the beam and neuron, is advised (Fig. 2A). Blind juxtasomal recordings have been successfully made in the awake animal (de Kock and Sakmann 2008), but the risk of mechanical distortions to the target neuron being imaged is increased with sudden animal movements, which will distort the calcium transients.

DATA ANALYSIS

Normalization of Fluorescence Values

Optical AP detection in vivo exploits the increased fluorescence over baseline associated with discharge of a single AP (Kerr et al. 2005, 2007; Yaksi and Friedrich 2006; Sato et al. 2007), which for OGB-1 is ~10%. To infer AP times, first obtain the normalized time-fluorescence function

$$\Delta F/F_0 = \frac{F - F_0}{F_0} \cdot (100\%),$$

where F is the measured fluorescence from a neuronal somata over time, and F_0 is the baseline fluorescence value. F is generally an average over pixel gray scale values inside a region of interest (ROI). $\Delta F/F_0$ values do not depend on the units or scaling of F, but it is critical that F does not contain an additive offset arising from hardware, software, other image channels, or autofluorescence. One

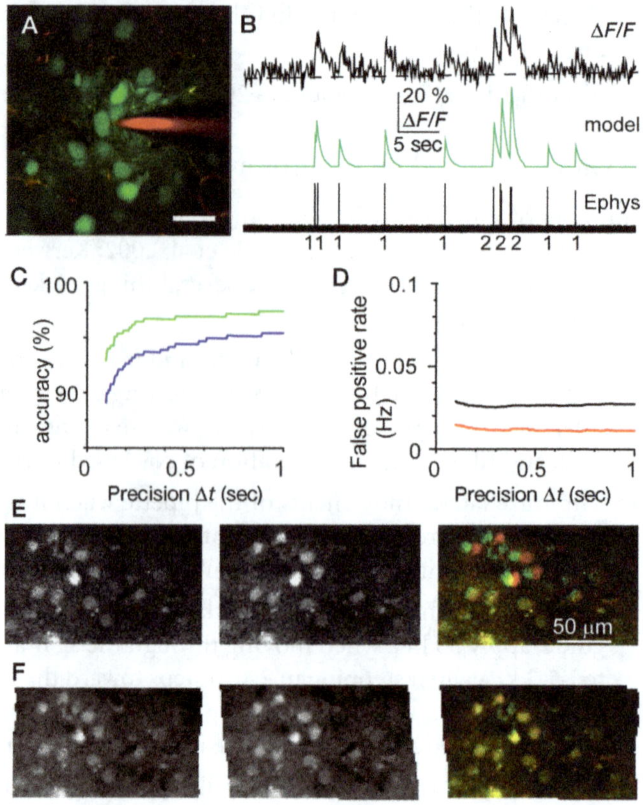

FIGURE 2. Data analysis for action potential (AP) detection and motion correction by combined optical and electrical recordings. (*A*) Image of neuronal population during cell attached recording of a single neuron; arrowhead indicates pipette (filled with Alexa 594). Scale bar, 40 µm. (*B*) Simultaneous calcium transients (upper) and cell-attached electrical recording (*lower*, number of APs indicated) obtained from the same neurons and fitted with a model (*middle*; Greenberg et al. 2008) that detected AP–evoked calcium transients and assigned the correct number of APs to each transient. (*C*) Percentage of APs detected by the algorithm (blue) with a requirement of 1:1 correspondence between electrically confirmed and inferred APs (n = 11 neurons). Electrically confirmed APs were considered detected if the algorithm inferred an AP (not already assigned to another electrically confirmed AP) with time difference $\leq \Delta t$. The green line shows the percentage of electrically confirmed APs within Δt of any inferred AP. (*D*) False positives (*right*, black line) occurred when an AP was inferred without a corresponding electrically confirmed AP (not already assigned to another inferred AP) with time difference $\leq \Delta t$. The red line shows the rate of inferred APs not within Δt of any electrically confirmed AP. (*E*) Two image frames (*left, center*) acquired immediately in sequence over periods of 100 msec each. When overlaid against each other (*right*) the image frames match at the bottom but not at the top, indicating that the brain moved ~10 µm during the acquisition of one or both frames. (*F*) Motion-corrected image frames (*left, center*), with each pixel assigned to the correct position. An overlay of the images (*right*) matches closely throughout the frame. (*A–D*, Adapted from Greenberg et al. 2008; *E,F*, adapted from Greenberg and Kerr 2009.)

common way around this is to subtract background fluorescence from within the frame, which is difficult when using bulk-labeled tissue, as all structures are labeled. Most studies, therefore, use a blood vessel for estimating background, which is valid when gray values in the blood vessel are not 0, a condition that can occur if an "offset" value is subtracted from the whole frame at the digitization stage. In addition, this approach assumes that background fluorescence is homogeneous across the frame, including the blood vessel, which is probably not the case.

Roughly speaking, F_0 is the value of F in the absence of AP activity. Because baseline fluorescence values may drift over time, especially over longer imaging times of continuous sampling, the set of F values in a moving window may be used to estimate F_0 (Greenberg et al. 2008) or to correct drift in $\Delta F/F_0$ values computed from a single choice of F_0 (Dombeck et al. 2007). Frequent discharge of APs and burst firing can bias the results of moving window techniques, especially in awake animals. This problem has been addressed by performing multiple rounds of AP detection and re-estimation of F_0 by excluding F values following a detected transient (Greenberg et al. 2008).

Estimating Noise Levels and Neuropil Contamination

To quantify noise levels without introducing a strong bias based on the rate of AP discharge, we use only the negative $\Delta F/F_0$ values, whose root-mean-square produces a result less than 8% (D.S. Greenberg and J.N. Kerr, unpubl.). In addition, the amplitude of the neuropil contamination present in a neuron's $\Delta F/F_0$ values can be determined by performing a linear regression over putatively nonspiking periods (Greenberg et al. 2008). Values of 1%–2% or less are desirable.

Correction of Movement Artifacts

In awake animals, the brain may move relative to the microscope. Such motion is more frequent during limb movement, whisking, chewing, and cleaning/grooming behaviors, but can still be >5 µm in a stationary animal (Dombeck et al. 2007; Greenberg et al. 2008). This movement can occur within an imaging frame, causing different parts of the image frame to be displaced by differing amounts and in opposite directions. So far, two approaches to motion correction for two-photon imaging of awake animals have been described. One method (Dombeck et al. 2007) is based on a hidden Markov model (HMM) and the other (Greenberg and Kerr 2009) is based on the Lucas–Kanade algorithm (LKA) (Lucas and Kanade 1981). In both methods, motion correction requires estimating the time-displacement function D describing brain motion in each image frame, and correcting each pixel's position (Fig. 2E,F). D is obtained without direct measurement of brain motion by matching image frames to a template acquired under anesthesia, or matching frames to each other. A cost function is defined indicating how well the corrected image matches the template for a given choice of D, and optionally containing additional image-independent terms that impose a prior distribution on D. D is represented using a finite set of parameters, and the parameter values are chosen to minimize the cost function. The full mathematical details for implementation of these methods can be found in the original publications (Dombeck et al. 2007; Greenberg and Kerr 2009). After motion correction, gaps in the image may appear because of fast motions, and some parts of a neuron may not be scanned at all. To avoid values of F being dependent on whether a neuron's bright or dark pixels were scanned, some normalization scheme must be used: either F must be normalized depending on which pixels are present in each frame (Greenberg and Kerr 2009) or $\Delta F/F_0$ values must be calculated for each pixel and averaged over each frame's available pixels (Dombeck et al. 2007). Initial results show motion to be greater in rats than mice, greater in older animals, and greater along the rostrocaudal axis than the mediolateral axis. Motion along scan lines is easier to correct and less likely to create gaps than motion across scan lines, so scanning on the rostrocaudal axis is advantageous.

AP Detection

Having calculated $\Delta F/F_0$ values for each neuron and each image frame, the next step is to transform these optical signals into AP counts. This is a relatively simple problem for single spikes well separated in time, such as when neurons fire infrequently with a majority of single spikes (Kerr et al. 2005), or during sensory stimulation in conditions where the majority of responses are single spikes (Kerr et al. 2007; Sato et al. 2007). The problem becomes more difficult when spiking is more complex, as is found in the awake animal (Dombeck et al. 2007, 2009; Greenberg et al. 2008; Sawinski et al. 2009) and during recordings that are continuous over many minutes. An AP detection algorithm must deal with both single isolated APs and trains of APs whose calcium transients summate and should not produce too many false-positive detections in the absence of APs (Fig. 2B). Several algorithms have been proposed, and a few have been tested by combining imaging with simultaneous targeted single cell electrophysiology, which, at this stage, is the only sure way to evaluate the accuracy of AP detection. When evaluating AP detection accuracy, it is important to evaluate both the percentage of APs that are detected (Fig. 2C) and the rate of false-positive detections (Fig. 2D), because optimizing one measure, such as detection, often comes at the expense of the other. With the approach discussed here, it was possible to detect ~95% of all spikes while keeping the false-positive rate to ~1 in every 40 sec of continuous imaging.

Once APs have been detected, computing other quantities of interest, such as firing rates, stimulus response rates, and correlations (pair-wise and higher-order), is straightforward. In general, data for which APs have been detected are amenable to the same analyses as extracellular recordings, with the exceptions that neuronal identities and spatial locations are unambiguously determined.

For populations of cortical neurons loaded with calcium indicator in vivo, the observed $\Delta F/F_0$ signals can be modeled as a linear convolution of spike counts, *s*, and a model transient, *r*. The model transient, obtained from cell-attached recordings (for review, see Kerr and Denk 2008), indicates the height and shape of a Ca^{2+} transient evoked by a single AP. The fact that a linear convolution is assumed means that the fluorescence signal from arbitrary spike patterns can be predicted by adding one model Ca^{2+} transient at the position of each AP. This means we can consider many possible spiking patterns and compare the predicted fluorescence signal from each one to the observed fluorescence data. Overall, the AP detection problem can be posed:

$$\text{Find } s \text{ such that } \Delta F/F_0 \approx s \circ r$$

or alternatively

$$\text{Find } s \text{ to minimize the error function } ||\Delta F/F_0 - s \circ r||_2^2.$$

The latter formulation, which is a least-squares minimization, is also likelihood maximization when Gaussian white noise is assumed. To be useful for AP detection in anesthetized and awake animals, an algorithm should produce an accurate estimate of the sequence of integer spike counts *s* that is not sensitive to small noise fluctuations in the optical signal $\Delta F/F_0$.

Cases in which AP firing is infrequent and contains mainly single spikes, successfully validated approaches include template matching based on cross-covariance (Kerr et al. 2005), linear regression with a moving window (Kerr et al. 2007), or regression over episodic stimulus-triggered data (Sato et al. 2007). Although effective and robust with respect to noise, these methods are limited to data sets with sparse AP firing and are not applicable to recordings in awake animals that show higher firing rates and a greater occurrence of bursts of APs (Dombeck et al. 2007, 2009; Greenberg et al. 2008; Sawinski et al. 2009).

To deal with multiple nearby AP transients, which can summate to produce complex shapes in the $\Delta F/F_0$ signals, other methods are necessary. Simply using a linear filter to directly deconvolve optical signals into spike counts is problematic, because small noise variations can introduce large oscillations into the final results. Two main approaches exist to deal with this problem: using a nonlinear filtering step to remove noise fluctuations before a linear deconvolution step, and using a nonlinear process to fit the spike counts *s* to the fluorescence data $\Delta F/F_0$.

The nonlinear prefiltering approach for nonsparse data has so far proven quite successful for AP detection in the zebrafish olfactory bulb, using explants of the intact brain and nose (Yaksi and Friedrich 2006). In this study, several rounds of an iterative nonlinear smoothing procedure were used to remove small fluorescence peaks arising from noise fluctuations. Afterward, a linear deconvolution was performed using decay constants of 3–6 sec, producing an excellent match to simultaneous electrophysiological recordings. Although this approach has yet to be tested on the shorter decay constants and higher noise levels found in mammalian cortical neurons in vivo, it may be the method of choice for neurons firing at high rates (> 10 Hz) for prolonged periods.

The second approach for nonsparse data, applying a nonlinear fitting procedure directly to the observed data, has been used to detect APs in cortical neurons in awake rats (Greenberg et al. 2008). In this study, the AP detection problem was posed as

$$\text{Find } s \text{ to minimize } ||\Delta F/F_0 - s \circ r||_2^2, \text{ such that for each image frame } i,\ s_i = 0 \text{ or } s_i \geq 1.$$

The advantage of this formulation is that it imposes realistic constraints on AP counts; for example, they cannot be negative, and half of an AP cannot occur in isolation. Because that constraint obviously applies to the true spiking activity, imposing it on the algorithm's output will in general improve the match between the two. More specifically, positive noise fluctuations in the absence of APs will not change the algorithm's output unless they are large enough to be better fit by increasing *s* from 0 to 1 at some point.

The disadvantage of this formulation is the difficulty of finding the global minimum of the least squares error. If n image frames are identified as possibly containing APs, the constraint on s gives rise to a space of possible solutions containing 2^n separate components, with each component corresponding to a different subset of image frames for which $s > 0$. Performing a separate least squares minimization on each component is not computationally feasible, but this problem can be dealt with in two ways. First, a preliminary step can reduce n by identifying image frames that cannot contain an AP transient (e.g., frames that show a fluorescence decrease compared to the previous frame). Second, the fact that an idealized AP-evoked transient is nonnegative at all time points allows, using algebraic manipulations of the observed data, some of the 2^n components to be removed from consideration. For example, it was reported that for $n = 33$ points, which gives about 8.6 billion components, only 700,000 (0.008%) of these had to be tested to find the global least squares minimum (Greenberg et al. 2008). Overall, this algorithm is more computationally intensive than other methods, but is so far the only one whose ability to detect AP activity in mammalian cortical neurons in vivo has been validated using electrophysiology, both during sparse activity and when nearby APs summated to create complex transient shapes in the $\Delta F/F_0$ signals (Fig. 2B).

Several other methods that apply nonlinear fitting steps directly to the data appear quite promising, but are still awaiting validation in vivo. These include a method based on support vector machines (Sasaki et al. 2008) and a detailed parametric model solved using particle filters (Vogelstein et al. 2009).

It should be noted that AP detection produces spike counts s_i for each imaging frame i. For imaging at 10–20 Hz, detected transients are temporally uncertain within a 50–100-msec window. In comparing detected APs to stimulation or behavior, the position of an ROI in the frame determines its timing; fluorescence measurements for neurons at the top and bottom of a single image frame are separated in time by almost a full frame duration.

EXAMPLE OF APPLICATION

Spontaneous Population Activity before and after Anesthesia

Once accurate AP detection has been confirmed and brain motions can be successfully compensated, the firing patterns obtained in awake animals can begin to yield insights into questions of biological interest. As an example, we describe a comparison of firing patterns in the same neuronal population before and after anesthesia (Greenberg et al. 2008) (Fig. 3).

In this study, infrared videography and ECoG recordings were used to separate recordings into different arousal states: awake, anesthetized, and sessile (when the animal is unanesthetized but displays no movement for extended periods, and might be asleep). The detected APs were used to calculate properties for single neurons such as firing rates (Fig. 3C) and the tendency toward burst firing, as well as population properties such as correlations (Fig. 3D). The affect of arousal state on these properties was examined, both statistically across all recorded neurons (Fig. 3D,H) and by tracking individual neurons, pairs, or subpopulations across arousal states (Fig. 3E,I). For example, although the neurons firing at the highest rates in awake animals also fired at the highest rates under anesthesia (Fig. 3D), the neuronal pairs that correlated most closely during wakefulness were not more likely to be correlated under anesthesia. Thus, under anesthesia, firing rates are mostly conserved, whereas correlations are reshaped or reset. The spatial information provided by two-photon imaging also allowed determination of the spatial distribution of arousal-mediated changes, firing rates, and correlations activity with single neuron precision (Fig. 3C,G). Finally, it had previously been reported that in acute slices, neuronal pairs with higher firing rates tended to have higher correlations (de la Rocha et al. 2007). Using two-photon imaging, it was found that this phenomenon occurs in vivo during anesthesia but not during wakefulness (Fig. 3J). Overall, the ability to track individual neurons and pairs across arousal states yielded insights that could not have been gained by recording different states in separate animals.

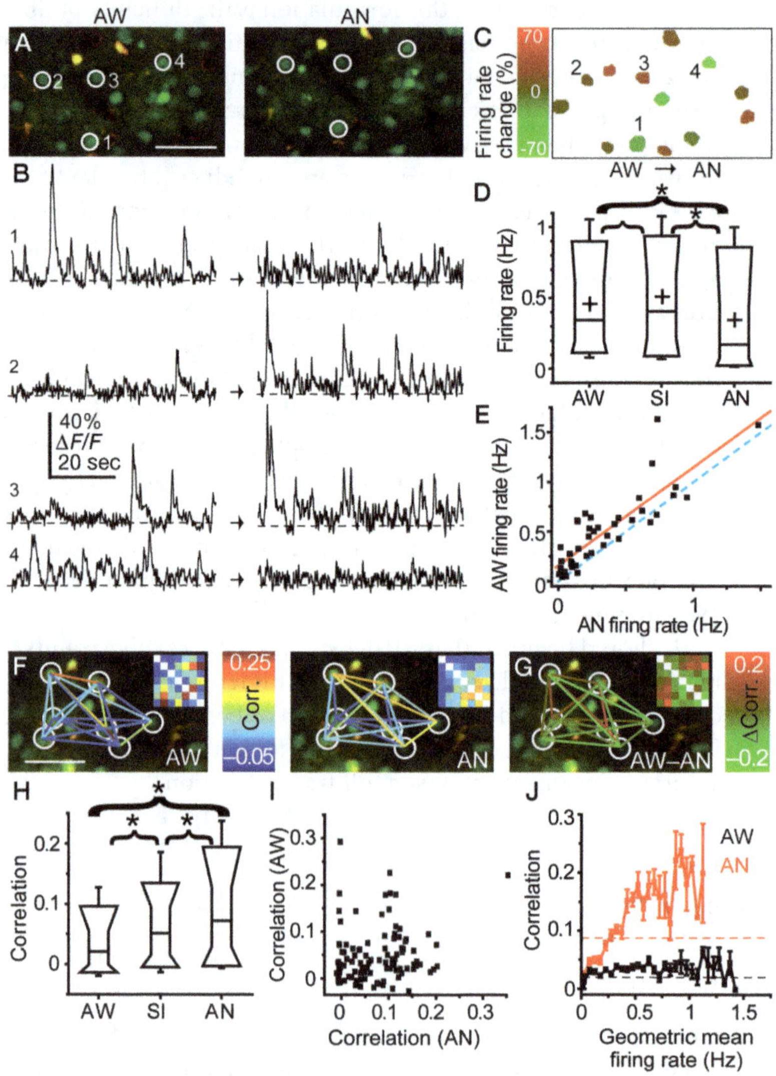

FIGURE 3. Imaging of the same neuronal population in awake and anesthetized states. (*A*) Stained L2/3 neuronal (green) and astrocyte (yellow) populations before (*left*) and after (*right*) anesthesia. Scale bar, 50 µm. (*B*) Calcium transients recorded from the same population before and after anesthesia. (*C*) Neurons pseudocolored according to percentage increase (red) or decrease (green) in firing rate on anesthesia. (*D*) Box-and-whisker plot comparing firing rates in awake, sessile, and anesthetized states. (*E*) Individual neuronal firing rates during awake and anesthetized periods. (*F*) Pair-wise correlations of spiking activity shown for awake (*left*) and anesthetized periods (*right*) as colored linkages between neurons and correlation matrices (*insets*). (*G*) Correlation difference between awake and anesthetized states. (*H*) Box-and-whisker plot comparing pair-wise correlations in awake, sessile, and anesthetized states. (*I*) Correlations in the spiking of individual neuronal pairs during awake and anesthetized periods. (*J*) Correlation between neuronal pairs during awake (black) and anesthetized periods (red), compared to pair-wise geometric mean of firing rates. Dashed lines show average correlation for AW and AN states. (Adapted from Greenberg et al. 2008.)

DISCUSSION

Future Prospects for Awake Population Imaging

The experimental and computational procedures outlined above can be used to record APs in populations of spatially resolved neurons in awake animals. So far, highly accurate detection of single APs has been confirmed only for frame scans of bulk-loaded organic calcium indicators. Consequently, the state of the art is a recording at ~10 Hz over a few hours from approximately 20 to 30 neurons simultaneously, with AP detection rates in the range of 90%–95% and false-positive rates of about 0.025 Hz. Although useful, this scenario is not close to exploiting the true potential of the technique, and we close this chapter by outlining two promising future developments that are likely to yield improvements in accuracy, temporal resolution, duration, and number of simultaneously recorded neurons.

The first important advance will be the development of new calcium indicators. Genetically encoded calcium indicators are already capable of detecting single APs in vivo (Wallace et al. 2008; Tian et al. 2009), but with lower accuracy than organic dyes, and they can suffer from uneven expression or physiological side effects. However, they promise to extend the time window in which stained neurons can be recorded from hours to months (Mank et al. 2008; Tian et al. 2009). Chronic imaging is exceptionally promising for the study of learning-related changes in population firing patterns in behaving animals. Efforts to develop new organic indicators (Lavis and Raines 2008) may improve both recording duration and the quality of fluorescence signals, thereby improving the accuracy of AP detection or allowing a reduced beam dwell time and thus more simultaneously recorded neurons. Both types of dyes may allow targeting of specific cell types.

REFERENCES

de Kock CPJ, Sakmann B. 2008. High frequency action potential bursts (≥100 Hz) in L2/3 and L5B thick tufted neurons in anaesthetized and awake rat primary somatosensory cortex. *J Physiol* **586:** 3353–3364.

de la Rocha J, Doiron B, Shea-Brown E, Josic K, Reyes A. 2007. Correlation between neural spike trains increases with firing rate. *Nature* **448:** 802–806.

Denk W, Strickler JH, Webb WW. 1990. Two-photon laser scanning fluorescence microscopy. *Science* **248:** 73–76.

Dombeck DA, Khabbaz AN, Collman F, Adelman TL, Tank DW. 2007. Imaging large-scale neural activity with cellular resolution in awake, mobile mice. *Neuron* **56:** 43–57.

Dombeck DA, Graziano MS, Tank DW. 2009. Functional clustering of neurons in motor cortex determined by cellular resolution imaging in awake behaving mice. *J Neurosci* **29:** 13751–13760.

Greenberg DS, Kerr JN. 2009. Automated correction of fast motion artifacts for two-photon imaging of awake animals. *J Neurosci Methods* **176:** 1–15.

Greenberg DS, Houweling AR, Kerr JN. 2008. Population imaging of ongoing neuronal activity in the visual cortex of awake rats. *Nat Neurosci* **11:** 749–751.

Houweling AR, Brecht M. 2008. Behavioural report of single neuron stimulation in somatosensory cortex. *Nature* **451:** 65–68.

Kerr JN, Denk W. 2008. Imaging in vivo: Watching the brain in action. *Nat Rev Neurosci* **9:** 195–205.

Kerr JN, Greenberg D, Helmchen F. 2005. Imaging input and output of neocortical networks in vivo. *Proc Natl Acad Sci* **102:** 14063–14068.

Kerr JN, de Kock CP, Greenberg DS, Bruno RM, Sakmann B, Helmchen F. 2007. Spatial organization of neuronal population responses in layer 2/3 of rat barrel cortex. *J Neurosci* **27:** 13316–13328.

Lavis LD, Raines RT. 2008. Bright ideas for chemical biology. *ACS Chem Biol* **3:** 142–155.

Lucas B, Kanade T. 1981. An iterative image registration technique with an application to stereo vision. In *Proceedings of the 7th International Joint Conference on Artificial Intelligence, Vancouver,* Vol. 2, pp. 674–679. Morgan Kaufmann, San Francisco.

Mank M, Santos AF, Direnberger S, Mrsic-Flogel TD, Hofer SB, Stein V, Hendel T, Reiff DF, Levelt C, Borst A, et al. 2008. A genetically encoded calcium indicator for chronic in vivo two-photon imaging. *Nat Methods* **5:** 805–811.

Margrie TW, Brecht M, Sakmann B. 2002. In vivo, low-resistance, whole-cell recordings from neurons in the anaesthetized and awake mammalian brain. *Pflügers Arch* **444:** 491–498.

Sasaki T, Takahashi N, Matsuki N, Ikegaya Y. 2008. Fast and accurate detection of action potentials from somatic calcium fluctuations. *J Neurophysiol* **100:** 1668–1676.

Sato TR, Gray NW, Mainen ZF, Svoboda K. 2007. The functional microarchitecture of the mouse barrel cortex. *PLoS Biol* **5:** e189.

Sawinski J, Wallace DJ, Greenberg DS, Grossmann S, Denk W, Kerr JN. 2009. Visually evoked activity in cortical cells imaged in freely moving animals. *Proc Natl Acad Sci* **106:** 19557–19562.

Stosiek C, Garaschuk O, Holthoff K, Konnerth A. 2003. In vivo two-photon calcium imaging of neuronal networks. *Proc Natl Acad Sci* **100:** 7319–7324.

Tian L, Hires SA, Mao T, Huber D, Chiappe ME, Chalasani SH, Petreanu L, Akerboom J, McKinney SA, Schreiter ER, et al. 2009. Imaging neural activity in worms, flies and mice with improved GCaMP calcium indicators. *Nat Methods* **6:** 875–881.

Vogelstein JT, Watson BO, Packer AM, Yuste R, Jedynak B, Paninski L. 2009. Spike inference from calcium imaging using sequential Monte Carlo methods. *Biophys J* **97:** 636–655.

Wallace DJ, Meyer zum Alten Borgloh S, Astori S, Yang Y, Bausen M, Kugler S, Palmer AE, Tsien RY, Sprengel R, Kerr JN, et al. 2008. Single-spike detection in vitro and in vivo with a genetic Ca^{2+} sensor. *Nat Methods* **5:** 797–804.

Yaksi E, Friedrich RW. 2006. Reconstruction of firing rate changes across neuronal populations by temporally deconvolved Ca^{2+} imaging. *Nat Methods* **3:** 377–383.

78 Miniaturization of Two-Photon Microscopy for Imaging in Freely Moving Animals

Fritjof Helmchen,[1] Winfried Denk,[2] and Jason N.D. Kerr[3]

[1]*Department of Neurophysiology, Brain Research Institute, University of Zurich, 8057 Zurich, Switzerland;*
[2]*Department of Biomedical Optics, Max Planck Institute for Medical Research, 69120 Heidelberg, Germany;*
[3]*Network Imaging Group, Max Planck Institute for Biological Cybernetics, 72076 Tübingen, Germany*

ABSTRACT

This chapter describes the development and application of miniaturized two-photon-excited fluorescence microscopes ("two-photon fiberscopes"). Two-photon fiberscopes have been developed with the aim of enabling high-resolution imaging of neural activity in freely behaving animals. They use fiber optics to deliver laser light for two-photon excitation. Their small front piece typically contains a miniature scanning mechanism and imaging optics. Two-photon fiberscopes can be made sufficiently small and lightweight to be carried by rats and mice and to allow virtually unrestricted movement within a behavioral arena. Typically mounted to the animal's skull above a cranial window, two-photon fiberscopes permit imagin g of cells down to at least 250 µm below the brain surface (e.g., in rat neocortex). In freely exploring animals, action potential–evoked calcium transients can be imaged in individual somata of visual cortex neurons bulk-labeled with a calcium indicator. Two-photon fiberscopes thus enable high-resolution optical recording of neural activity with cellular resolution during natural behaviors.

INTRODUCTION

Two-photon microscopy is highly suitable for imaging in intact tissue and in living animals owing to its reduced susceptibility to light scattering. In neuroscience, two-photon imaging has opened a new field for in vivo optical measurements of neural dynamics on the subcellular, cellular, and local population levels (for reviews, see Helmchen and Denk 2005; Kerr and Denk 2008; Grewe and Helmchen 2009; Wallace and Kerr 2010). Two-photon calcium imaging has, for example, been applied to study the functional organization of neuronal populations in primary areas of the somatosensory (Kerr et al. 2007; Sato et al. 2007), visual (Ohki et al. 2005; Mrsic-Flogel et al. 2007; Greenberg et al. 2008), and auditory cortices (Rothschild et al. 2010) as well as in the cerebellar cortex (Sullivan et al. 2005; Nimmerjahn et al. 2009).

Under anesthesia, brain state and neural dynamics are, of course, altered and it is unclear exactly how these changes affect the animal's perception. The need to work in anesthetized animals is, therefore, a major limitation and it is desirable, perhaps essential, to ultimately perform experiments in awake, behaving animals. In recent years, a number of approaches have emerged that enable optical recordings with cellular resolution in awake, behaving animals, and two main avenues have been followed.

- In the first, animals are trained to accept head fixation. An awake animal can then be placed under a standard two-photon microscope and imaging is performed as during anesthesia (Dombeck et al. 2007; Greenberg et al. 2008). The advantage of this approach is that a regular microscope can be used; the main disadvantages are that the behavior repertoire is constrained and extensive training is necessary. More details about head-fixed, awake imaging can be found in Chapters 69, 76, and 77.
- The second approach, which is the focus of this chapter, is to fiber-couple and miniaturize the microscope front piece so that it can be carried by a rat (Helmchen et al. 2001) or even a mouse (Flusberg et al. 2005b). The benefit of this approach is that animal behavior is far less restricted and it is possible to obtain high-resolution optical recordings during behavioral experiments, very much like chronic electrophysiological recordings in rodents. The difficulty with this approach lies in the special optical design. Miniaturization has to be achieved while maintaining sufficient sensitivity and resolution to measure calcium signals in individual neurons and neuronal processes in the brain.

In this chapter, we first survey the different technologies that have been explored during recent years for miniaturization of two-photon microscopy and introduce the key technological challenges. We then discuss some applications of two-photon fiberscopes, including the recent demonstration of calcium measurements with single-cell resolution in the visual cortex of freely moving rats (Sawinski et al. 2009).

IMAGING SETUP

Equipment

To build a two-photon fiberscope, several key components are the same as those required for a standard two-photon microscope. Hence, a fiberscope can be built as an addition to an existing setup. In the excitation pathway, an ultrafast laser source providing ~100-fsec laser pulses in the 700–1050-nm wavelength range is required. Typically, a tunable commercial Ti:sapphire laser system is used, although a custom-built laser may provide certain advantages (Sawinski and Denk 2007). For efficient two-photon excitation, a laser with large maximal output power is preferable because light guidance through optical fibers involves power losses, and fluorescence excitation is further reduced by pulse distortions caused by dispersion and nonlinearities (see below). Further standard components required include an intensity controller and a telescope for laser beam size adjustment. Specific to the fiberscope is a fiber coupler to stably and efficiently couple the excitation light into the single-mode optical fiber. If fluorescence is collected through a large-core fiber and detected at the fiber end remote from the front piece of the two-photon fiberscope, a standard detector system can be used. Besides these standard components, special technology is required for building a miniaturized front piece (see below). A list of useful components is given in Table 1.

Considerations for Two-Photon Fiberscope Design

First, mobilization of the microscope requires delivery of the excitation light via an optical fiber, through which it is difficult to maintain short pulse widths at high powers as is needed for efficient two-photon excitation. Second, a miniaturized scanning device is needed to scan the output beam from the optical fiber. Third, to focus the scanned beam, a small water-immersion objective suitable for imaging in the living brain is required. Fourth, fluorescence needs to be efficiently collected and detected, either directly at the fiberscope front piece using a small photodetector or with a remote detector following guidance through a large-core collection fiber (Fig. 1A–C). Finally, a simple and

TABLE 1. Components useful for designing a two-photon fiberscope

Component	Specifications	Model (vendors)
Single-mode fiber	Core size 5–6 µm; NA 0.12–0.15; cladding diameter 125 µm	F-SF, F-SPF (Newport); FS-SN-4224 (3M)
Hollow-core photonic crystal fiber	Air core 9 µm; zero-dispersion wavelength 812 nm; NA 0.12–0.15; cladding diameter 125 µm	HC-800-01 (NKT Photonics, Denmark)
Piezoelectric elements	Benders (~2 mm x 8 mm in size); tubes, etc.	(EDO; Polytec PI; Morgan Matroc; Piezo Systems; Argillon)
Tube lens	Aspheric (NA 0.15)	C280TM, C260TM (Geltech lenses; Thorlabs)
	or	
	achromat (f = 12.7 mm)	LAKN22-SFL6 (Thorlabs)
Long-pass dichroic	Cold mirror or dichroic beam splitter	M43-960 (Edmund Optics)
Short-pass dichroic	Hot mirror, near-infrared blocking filter, or dichroic beam splitter	Calflex X (Linos Photonics)
Objective	Water immersion; NA >0.8; large fluorescence collection field of view	Custom design (Throl Optical Systems) (modified from design by Syncotec; now defunct)
Collection lens	Small lens; NA matched to collection fiber	For example, C230TM (Geltech lens; Thorlabs)
Fluorescence collection fiber	Large core 0.6–1 mm; high transmission at visible wavelengths; high NA	FT-600-URT (Thorlabs); MO2-534 (Edmund Optics); plastic fiber PJU-FB1000 (Toray Deutschland)
Small photomultiplier tube (front piece detection)	High quantum efficiency at visible wavelengths; weight 4 g	R7400U (Hamamatsu)
Small DC motor	Weight 300 mg; gear ratio 1:125	Smoovy motor (Micro Precision Systems)
Position-sensitive detector (PSD)	Two-dimensional silicone-based detector with several milliliters side length	2L10SP (ON-TRAK Photonics); S5991 (Hamamatsu)

NA, numerical aperture.

flexible mounting mechanism should permit easy fixation and removal of the fiberscope onto the animal's head. In the following sections we provide some details about these various technological aspects. Further information can be found in several reviews (Helmchen 2002; Flusberg et al. 2005a) and in the original papers cited below.

Two-Photon Excitation through Optical Fibers

The standard optical fibers for delivering laser light are step-index single-mode fibers, that is, solid glass fibers with a small (~5-µm diameter) core of slightly higher refractive index. In these optical fibers, ultrafast laser pulses broaden because of material dispersion and also, at the high laser powers required for deep imaging, because of nonlinear effects (Agrawal 1995). Such broadening limits the efficiency of two-photon excitation and therefore has to be reduced to a minimum. It is possible to compensate for dispersion-induced broadening by giving the short-wavelength components a sufficient head start (a negative "prechirp") so that blue and red components arrive at the fiber end at the same time. Prechirping the laser pulses can be accomplished before coupling to the fiber with a pair of diffraction gratings (Treacy 1969) or with prism sequences (Walmsley et al. 2001; Sawinski and Denk 2007). For example, two reflection gratings (400 grooves/mm and 9.7° blaze angle; Richardson Grating Laboratory) in double-pass configuration provide around –60,000 fsec^2 at 17-cm spacing, compensating for about 1.5 m of optical fiber (Helmchen et al. 2001). The optimal spacing between gratings (or prisms) can be found by measuring the pulse width after the fiber with an autocorrelator and adjusting the spacing for minimal pulse width. Even with prechirping, however, laser pulses progressively broaden at increasing power levels because of nonlinear optical effects in the fiber core (mainly self-phase modulation; Agrawal 1995). Nonlinear pulse broadening becomes substantial starting at surprisingly low average powers (~10 mW) and is dominant at the high average power levels needed for imaging deep into tissue (Helmchen et al. 2002). As a result, relatively long laser pulses (~1 psec or longer) are used in effect at high excitation powers when standard step-index fibers are used, resulting in suboptimal two-photon excitation.

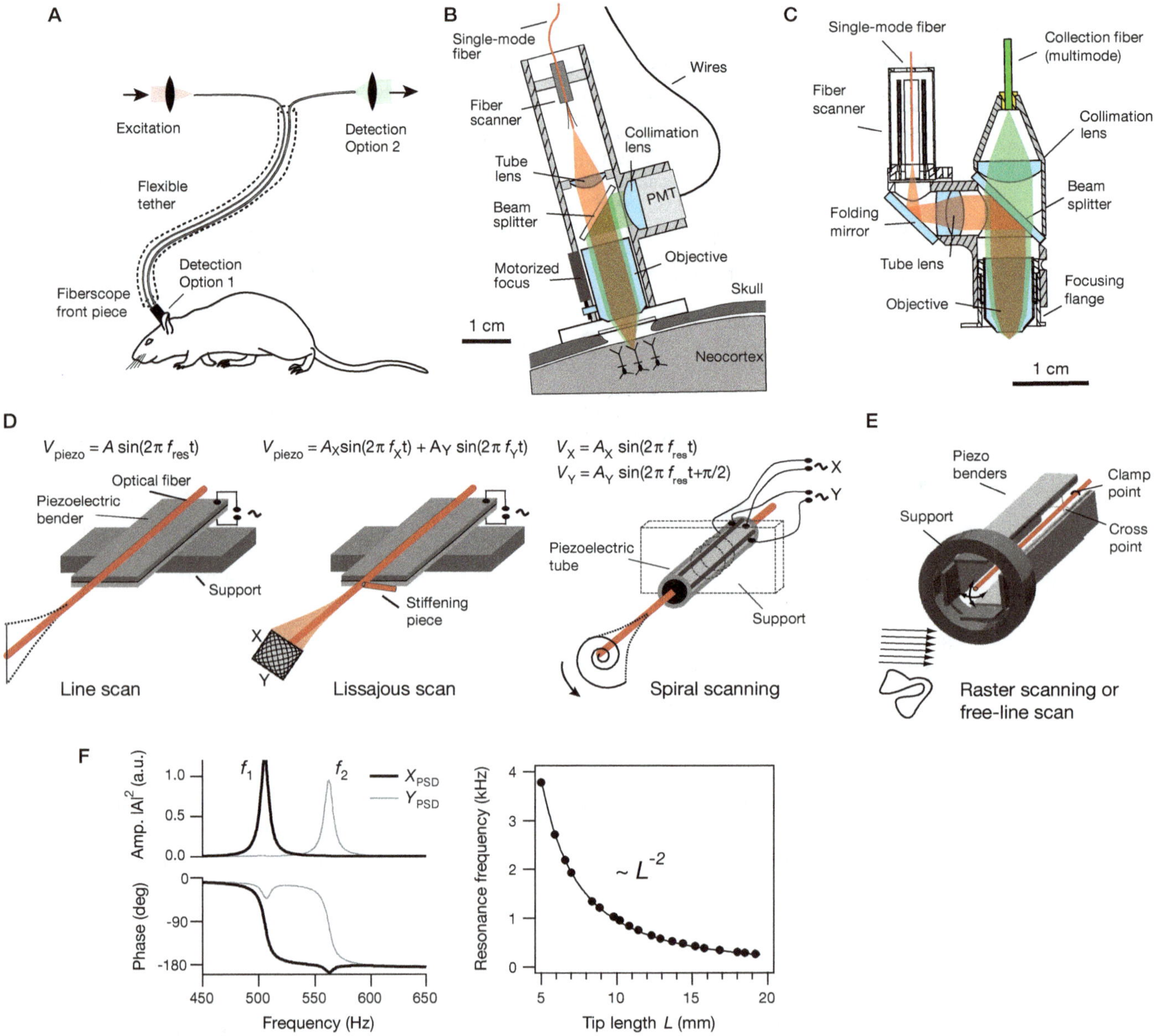

FIGURE 1. (*A*) Two-photon fiberscope setup. Excitation light is delivered through fiber optics to a miniaturized microscope front piece that is attached to the head of a freely behaving animal. Two principle options for fluorescence detection are indicated. (*B*) Example for Option 1: Original design used by Helmchen et al. (2001). Excitation light is guided through a single-mode fiber. Scanning is achieved using resonant mechanical vibration of the fiber tip. Near-infrared excitation light (red) is passed through a long-pass dichroic beam splitter and fluorescent light (green) is detected by a small photomultiplier tube (PMT) at the front piece. (*C*) Example for Option 2: Recent design used by Sawinski et al. (2009). A nonresonant fiber scanner is used and near-infrared excitation light is collimated and deflected by a short-pass dichroic beam splitter. Fluorescence is collected through a large-core multimode fiber and detected remotely. (*D*) Resonant fiber scanning. (*Left*) Resonant vibration along a line is induced by driving a single piezoelectric bender, to which the fiber is glued, near the resonance frequency (f_{res}). (*Middle*) Two-dimensional Lissajous scanning. Stiffening of the fiber end in one direction splits the resonance frequency in f_X and f_Y. Simultaneous excitation of vibrations in *x* and *y* directions creates a Lissajous pattern, more or less homogeneously covering an area. (*Right*) Spiral scanning. A piezo tube with quadrant electrodes induces spiral scanning by driving *x* and *y* directions at the same resonance frequency but with a 90° phase shift. (*E*) Nonresonant fiber scanner. This "piezolever fiber scanner" achieves fiber tip deflections by leveraging the movements of two crossed pairs of piezo benders. The fiber is glued to the tips of the cross-connected piezo pairs at the cross point close to the clamped fiber base. Driving the piezo benders with sawtooth-like waveforms (fast and slow) at frequencies below the resonance frequency creates raster scanning similar to conventional two-photon microscopes. Offset voltages to the *x* and *y* piezo benders provide the unique option to shift the scanned field laterally. Moreover, free-line scanning is possible. (*F*) Fiber resonance characteristics. (*Left*) Resonance curves measured with a position-sensitive detector (PSD) for a fiber tip stiffened in one direction. Note the distinct resonance peaks for *x* and *y* directions. (*Right*) Resonance frequency of a 125-µm diameter single-mode fiber as a function of the free length of the fiber end.

Several approaches have been devised to alleviate the problem of nonlinear pulse broadening. For example, large-core specialty fibers (Helmchen et al. 2002; Ouzounov et al. 2002) as well as spectral and temporal shaping of the input pulse (Clark et al. 2001) have been tested but are not widely used. The problem of pulse broadening can be nearly completely circumvented by using hollow-core photonic crystal fibers (Göbel et al. 2004b). In these fibers, most of the laser light travels through an air-filled core and thus encounters little dispersion or nonlinearity, minimizing pulse broadening even at high average power (Göbel et al. 2004b). No prechirping is required and the laser beam can be coupled straight into the fiber. Photonic crystal fibers have been successfully used in a number of two-photon fiberscope designs (Flusberg et al. 2005b; Engelbrecht et al. 2008; Piyawattanametha et al. 2009). A disadvantage is, however, their susceptibility to contamination of the fiber tip, probably attributable to the intricate photonic crystal structure, which poses a particular problem when the fiber end is integrated into a fiber scanner device and thus cannot be easily recleaved.

Miniaturized Fiber-Scanning Devices

A compact scanning device is essential for miniaturization. One possibility is to move the tip of the illuminating fiber, which is positioned in an optical plane conjugate to the focal plane (Giniunas et al. 1991). In this approach, the output light beam emerging from the fiber core is collimated and then refocused to a focal spot using small optics (see below), so that any movement of the fiber tip translates into a lateral movement of the focal spot. Multiple options exist for how to move the fiber tip. In particular, we need to distinguish resonant and nonresonant scanning (Fig. 1D,E). We first discuss several ways to exploit resonant vibration before describing a nonresonant fiber-scanning method.

Resonant mechanical vibration of a free-standing fiber end can be easily induced with a piezoelectric bending element ("piezo bender"). The bare fiber (with the coating stripped off) is glued to the piezo bender so that 10–20 mm of the fiber end remains free to vibrate (Fig. 1D). For a 125-µm (standard) diameter fiber, this results in a resonance frequency f_{res} ranging from 200 to 1000 Hz (Fig. 1F). In general, for transverse vibration of a cylindrical rod, f_{res} is given by the equation

$$f_{res} = \frac{s_0^2}{4\pi} \frac{R}{L^2} \sqrt{\frac{E}{\rho}},$$

where R is the radius, L is the length of the rod, ρ is the density (~2.3 g/cm^3), E is the elasticity module of the fiber (~75 GPa), and s_0 = 1.875 (Bishop and Johnson 1960). Thus, f_{res} depends inversely on the square of the length L (Fig. 1F).

Two-dimensional Lissajous scanning can be achieved by stiffening the fiber end in one direction with a short (2–4 mm) piece of bare fiber glued to the lower edge of the piezo and to the fiber at about 2–3-mm distance from the piezo (Fig. 1D). This approach results in two distinct resonance frequencies of the fiber tip, as becomes apparent by driving the piezo with a sinusoidal waveform that is swept in frequency (Fig. 1F). Because the resonance widths are typically only 5–10 Hz, one has to sweep the frequency slowly and finely enough not to miss the resonance. For accurate measurement of the resonance curve, the fiber tip can be imaged onto a two-dimensional position-sensitive detector. When the piezo bender is driven with a superposition of two sine waves at the respective resonance frequencies for the different directions, a Lissajous scan pattern is produced, which has been used in several two-photon fiberscope implementations (Helmchen et al. 2001; Flusberg et al. 2005b). With up to 100-V (peak-to-peak) drive amplitude, resonant vibrations of >1-mm peak-to-peak amplitude can be reached and the Lissajous scan pattern can be tuned so that a rectangular area is covered rather homogenously (for details on scan parameters and data acquisition, see Helmchen et al. 2001).

An alternative method of resonant fiber scanning is spiral scanning (Fig. 1D). Here, the fiber is glued inside a piezoelectric tube with quadrant electrodes that enable independent excitation of vibration in x and y directions (Seibel and Smithwick 2002; for review, see Lee et al. 2010). The advantage of spiral scanning is that two-dimensional scanning is easily achieved without stiffening of the fiber end by driving vibrations in the two orthogonal directions with a 90° phase shift (Fig. 1D). Because small piezo tubes are sufficient to drive fiber vibration, spiral scanners can be made very compact, probably smaller than is possible for Lissajous scanners. For example, in combination with a small gradient-index (GRIN) lens system, a miniature fiberscope front piece (<1 g) can be realized (Engelbrecht et al. 2008). Moreover, using a short fiber end with a correspondingly high resonance frequency of a few kHz, frame scanning at video rate is possible.

Resonant scanning lacks flexibility and, in particular, does not support lateral offsets or random access scanning. To overcome these limitations, a nonresonant, leverage-based fiber scanner has been developed (Fig. 1E) (Sawinski and Denk 2007). Here, the fiber end is deflected near its clamped base with the help of piezo benders. Because of the leverage, small deflections of the piezo benders are transduced into relatively large fiber tip deflections. Operating below the resonance frequency of the fiber tip, this design permits raster-like scanning of an area, similar to standard galvanometric scan mirrors. It also allows one to randomly access scan points within a field of view, which might be useful in future optogenetic applications. This type of scanner was used recently to measure calcium signals in individual neurons of neocortex at ~15 Hz frame rate in freely behaving rats (Sawinski et al. 2009).

Instead of deflecting the fiber tip, the beam can be scanned using a microelectromechanical system (MEMS), whereby a two-dimensional scanning mirror (Piyawattanametha et al. 2006) is inserted between the fiber end and miniature objective. The use of a coherent fiber bundle avoids scanning in the front piece at all (Göbel et al. 2004a). None of these methods, however, has become widely adopted and currently fiber tip scanning appears to provide the best results for two-photon excitation.

Small Microscope Objectives

For high-resolution in vivo imaging, a water-immersion objective with a numerical aperture (NA) of ≥0.8 is desirable. Standard high-quality microscope objectives are too large and too heavy to be useful for microscope miniaturization. Thus, the first two-photon fiberscope took advantage of a water-immersion objective, in which the core part could be removed and reassembled in a minimal holder (Helmchen et al. 2001). This type of objective is not available anymore. In the more recent fiberscope design shown in Figure 1C, a specially designed water-immersion objective with NA 0.9 (Throl Optical Systems) was used (Sawinski et al. 2009). As an alternative, GRIN lens systems have been used in two-photon fiberscopes, typically with limited NA (0.48–0.58) of the front GRIN lens (Flusberg et al. 2005b; Engelbrecht et al. 2008; Piyawattanametha et al. 2009). The recent development of thin compound lens systems, made of a GRIN lens and a plano-convex lens, now enables NA values up to 0.85 (Barretto et al. 2009). These new lens systems should be highly suitable for two-photon fiberscope designs.

Fluorescence Detection

In principle, there are two options for fluorescence collection: (1) to detect fluorescence light directly at the fiberscope head piece, and (2) to collect and guide it through an additional optical fiber and then project it onto a remote detector (Fig. 1A–C). The advantage of the first approach is its simplicity and the high detection efficiency. In the first fiberscope design (Helmchen et al. 2001), a small photomultiplier was used, but in the future a small avalanche photodiode array might provide superior quantum efficiency. The disadvantage here is that more weight is added to the animal's head. Moreover, depending on the design of the photomultiplier tube, high voltages can be present close to the animal. Collecting fluorescence through a large-core fiber suffers from some transmission losses but has the great advantage of being very flexible at the remote end of the collection fiber (Fig. 1C). Any highly sensitive detector can be used irrespective of its size and weight, and the fluorescence light can be separated into multiple spectral windows for multichannel imaging of several fluorescent dyes. As only a small coupling lens is required in the front piece, this detection scheme also helps to further reduce the size and weight of the fiberscope.

Mechanical Attachment to the Animal

Three issues arise when designing a suitable head-mounting system for the fiberscope: (1) The microscope must be easily detachable; (2) when attached, the microscope must be mechanically stable, even during high-acceleration movements (jerks) of the head; and (3) attachment must not affect the animal's behavior or viability. The second point is especially important when using a raster scan, spiral scan, or free-line scan approach, in which the presumed positions of the beam are used to reconstruct images. Although offline movement correction has been successfully applied (Dombeck et al. 2007; Greenberg et al. 2008), there is a limit as to how much movement can be corrected (Greenberg and Kerr 2009). The attachment system that has been successfully applied to the freely moving animal is based on a two-part system in which the headplate remains permanently bonded to the animal's skull and the fiberscope can be attached (Sawinski and Denk 2007; Sawinski et al. 2009). When the fiberscope is not attached, it should be replaced with a protective cover that protects the cranial window from mechanical damage, such as from the animal's paws during grooming. For the stable attachment of the headplate to the skull, we prefer using ultraviolet-cured bonding agents rather than bone screws, especially in younger animals in which the skull is thin. Another important consideration is whether it is easily possible to attach and detach the microscope to and from an awake animal while keeping handling of the animal at a minimum. This last point is important for trained animals in which excessive handling can interfere with their behavioral performance (Stüttgen and Schwarz 2008).

FLUORESCENT DYE LABELING IN VIVO

Imaging with a two-photon fiberscope depends on in vivo fluorescence labeling. Tail-vein injection of a fluorescent dye (see Chapter 84) was used for the first demonstration of two-photon fiberscope imaging in freely moving rats (Helmchen et al. 2001). For imaging neural activity, various methods of labeling neurons can be applied. Individual neurons can be filled with organic indicator dyes via intracellular recording electrodes (high-resistance sharp electrodes; Svoboda et al. 1997) or whole-cell patch pipettes (Helmchen and Waters 2002). This approach has been used for fiberscope measurements of dendritic calcium signals during anesthesia (Helmchen et al. 2001) (Fig. 2A,B). Single-cell labeling is, however, laborious and restricted to small numbers of neurons. When the activity in a larger number of neurons is to be sampled in parallel, population labeling techniques are needed. What has proven popular for fiberscope and fixed-scope experiments is the in vivo loading of populations of neocortical cells with calcium indicators by the local application of membrane-permeant dye (Stosiek et al. 2003) (see Chapter 49). Alternatively, local electroporation of calcium indicator dyes has been used (Nagayama et al. 2007) (see Chapter 50). The use of genetically encoded calcium indicators is also highly promising (Wallace et al. 2008; Tian et al. 2009; Lütcke et al. 2010) (see Chapters 8 and 57). The latter approach allows permanent labeling and should, therefore, permit repeated measurements of the same neuronal population in multiple imaging sessions over days, weeks, and perhaps even throughout an animal's life.

EXAMPLES OF APPLICATION

Using the first Lissajous-type scanning two-photon fiberscope (Helmchen et al. 2001), pyramidal neurons in neocortical layer 2/3 (loaded with calcium indicator via an intracellular electrode) could be resolved in head-restrained, anesthetized rats (Fig. 2A). Using a line-scan mode, with fiber-end vibration induced along one direction, dendritic calcium transients were resolved following intracellular current injection, providing proof-of-concept for the sensitivity of fiberscope calcium measurements (Fig. 2B). As the first awake demonstration, fluorescently labeled blood capillaries were imaged in freely moving rats (Fig. 2C), showing that stable images can be acquired in unrestrained animals, except during periods of strong head movements (Helmchen et al. 2001).

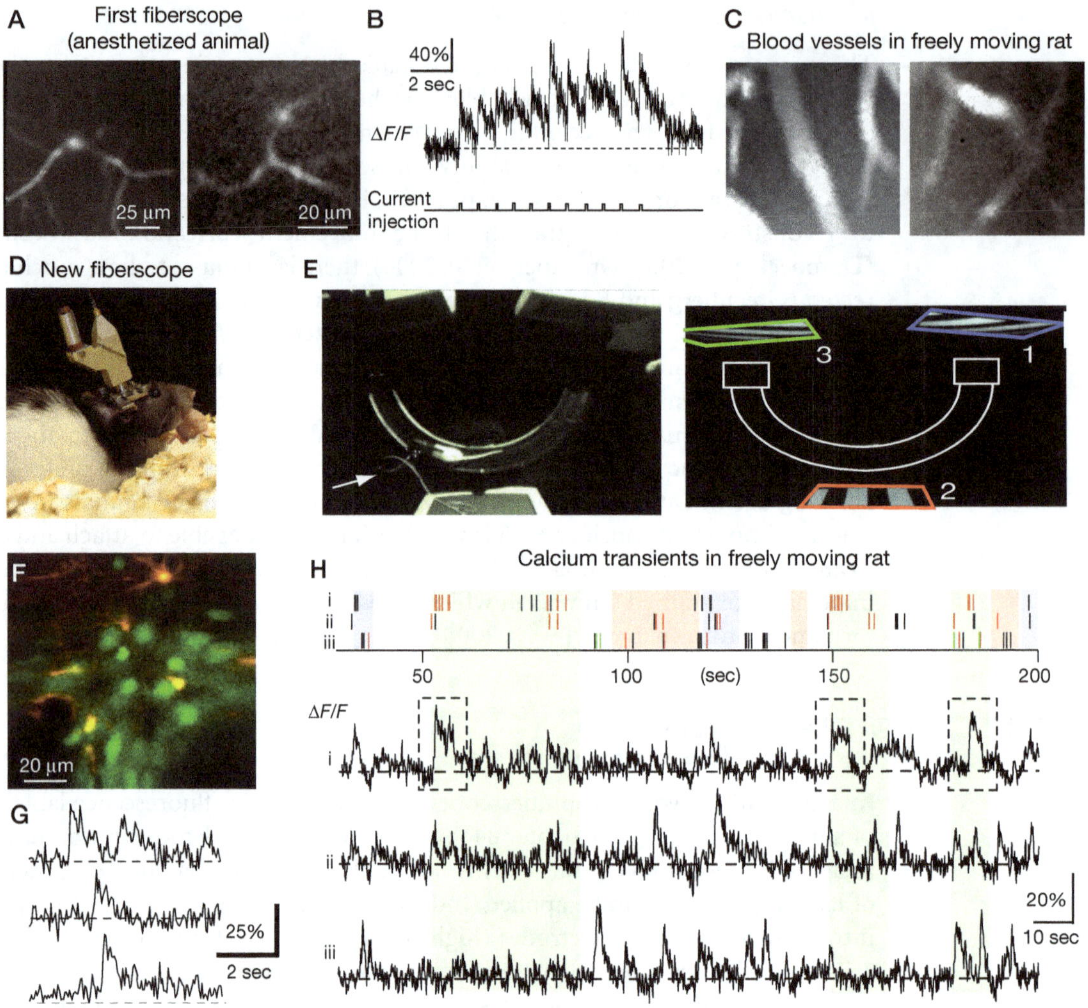

FIGURE 2. Applications of two-photon fiberscope. (*A*) In vivo images of Calcium Green-1-loaded neocortical layer 2/3 neurons acquired with the first two-photon fiberscope in a head-fixed anesthetized rat using Lissajous-type scanning (Helmchen et al. 2001). (*B*) Dendritic calcium transients evoked by somatic current injections measured with the fiberscope in line-scan mode. (*C*) Imaging in freely moving rats. Images show blood vessels stained by tail-vein injection of a fluorescent dye and were obtained in freely moving awake rats using the first two-photon fiberscope (Helmchen et al. 2001). Based on the diameter of capillaries (~5–6 μm), we estimate the fields of view to be ~85 μm and ~65 μm for the *left* and *right* image, respectively. (*D–H*) Two-photon fiberscope imaging with a piezolever fiber scanner in rat visual cortex (Sawinski et al. 2009). (*D*) Photograph showing the mounted fiberscope. (*E*) The rat was free to move on an elevated running track (*left*, top-view photograph; *right*, schematic diagram) flanked by three video monitors displaying stationary gratings at three different orientations (denoted by blue, red, and green outlines). Arrow in *left* panel indicates flexible tether containing fiber optics and electrical wires. (*F*) Fiberscope image of a neuronal population in layer 2/3 stained with the calcium-sensitive indicator Oregon-Green BAPTA-1 (green). Astrocytes were counterstained with sulforhodamine 101 (orange). (*G*) Typical action potential–evoked calcium transients recorded from neuronal somata with a fast rise, peak, and slow decay to baseline. (*H*) Examples of raster plots (*top*) and calcium transients (*bottom*) showing activity in three neurons (i–iii) during ~5 min of continuous imaging while the animal was freely exploring the elevated track (black lines correspond to single action potentials, red to doubles, and green to triples). Periods during which the animal's gaze moved across one of the three monitors are indicated by the blocks of background color (color coding shown in *E*). Note multiple large transients (dashed boxes in traces) in neuron i in response to viewing the same monitor (indicated in green) multiple times. For details on raster plot generation, see Greenberg et al. (2008).

Despite the development of various two-photon fiberscope designs during the past decade (e.g., Flusberg et al. 2005b; Engelbrecht et al. 2008; Piyawattanametha et al. 2009), the original goal of enabling functional measurements of neural activity in freely moving animals remained elusive until recently when a light, compact fiberscope (~5 g) using a leverage-based fiber scanner was combined with bulk-loading of the visual cortex with calcium-sensitive indicator (Sawinski et al. 2009). This study succeeded in recording spike-evoked calcium transients from animals freely exploring a novel environment that contained static visual stimuli (Fig. 2D–H). Intrinsic optical imaging (see Chapter 60) through the thinned skull can be used to coarsely map the cortical responses. Using the pial blood vessel pattern as a reference, this approach allows placement of the craniotomy without the need to map responses again. To decrease the stress to the animal while attaching the microscope, animals were trained over many days to tolerate repeated handling and head-fixing while awake. Once animals were trained, they were head-fixed, the calcium-sensitive indicator was injected in the area of the cortex previously mapped with intrinsic optical imaging, and the fiberscope was attached. After a calcium indicator–loaded neuronal population (approximately 10 to twenty neurons) had been located, the animal was released onto the running track (a semicircular track raised ~1.5 m above the ground), which restricted the area the animal could explore while being always fully visible for overhead and side-on cameras that were used to track the animal's position (Fig. 2E). In addition, this setup allowed the unrestricted movement of the fibers and wires attached to the fiberscope.

Full-frame (64 x 64-pixel) images were then acquired at ~10 Hz while the animal was free to explore a novel environment. This environment included three cathode ray tube monitors, each with a static grating orientated at –45°, 0°, or 45° (Fig. 2E). Using an overhead camera and four infrared light-emitting diodes attached to the fiberscope, the head of the animal could be accurately tracked in three dimensions. Movement within the two-photon frames was corrected offline and calcium transients were detected (Fig. 2G; Greenberg and Kerr 2009). Calcium transients could then be related to the head position of the animal. The relative orientation and distance to the animal's head were calculated for each screen by using video frames from the video camera, and related to the receptive field of the imaged cortical region (Fig. 2H). A subpopulation of neurons was found to become significantly more active when one specific monitor was in the receptive field, compared with other monitors (Fig. 2H). This fiberscope application shows that neuronal activity can be recorded in a freely behaving animal that is free to interact with a stimulus in a self-determined way.

DISCUSSION

During the past few years, optical fibers have been increasingly used to measure neural activity in mobile animals, mainly in connection with fluorescent calcium indicators. In the simplest case, an optical fiber provides a bulk readout of the mean calcium signal in a local population of neurons (Adelsberger et al. 2005; Murayama et al. 2007; Lütcke et al. 2010; see Chapters 60 and 83). Achieving cellular resolution with small, portable fiberscopes is much more difficult, not only because of the technical challenges associated with miniaturization but also because of strong light scattering in neural tissue. Even though calcium transients in Purkinje cell dendrites of the cerebellum could be imaged in mobile mice using single-photon imaging through a fiber bundle (Flusberg et al. 2008), single-photon approaches are highly susceptible to light scattering and are thus applicable only when the cells of interest are located near the tissue surface. In contrast, two-photon fiberscopes exploit the well-known two-photon advantages, in particular enhanced depth penetration and intrinsic optical sectioning. In principle, it should be possible to reach imaging depths of ≥400 µm with two-photon fiberscopes, similar to standard two-photon microscopes.

With the demonstration of cellular calcium imaging in freely behaving rats (Sawinski et al. 2009), the application of two-photon fiberscopes has now entered a new phase. Many potential applications and further developments can be envisaged. The expression of genetically encoded sensors in transgenic animals or following viral gene delivery (see Chapter 57) will enable long-term measurements from specific, well-identified ensembles of neurons. This method has the potential to directly relate changes in neural population dynamics to behavioral changes, such as an improved performance in a behavioral task. The fiberscope also will allow the study of social interactions as well as tasks in which the animal needs to be free to move within its environment. Although the two-photon fiberscopes discussed here first of all enable imaging studies in accessible regions such as the neocortex and the cerebellum, the use of GRIN lenses as front objectives or relay lenses (Flusberg et al. 2005b; Flusberg et al. 2008) should permit functional imaging in deeper brain regions, such as the hippocampus or the basal ganglia. Finally, another fascinating possibility is the combination of two-photon fiberscopes with optogenetic tools (see Chapters 6, 79, and 80) to eventually achieve optical control of small cell groups or even of individual neurons in freely behaving animals.

REFERENCES

Adelsberger H, Garaschuk O, Konnerth A. 2005. Cortical calcium waves in resting newborn mice. *Nat Neurosci* **8:** 988–990.

Agrawal GP. 1995. *Nonlinear fiber optics*. Academic Press, San Diego.

Barretto RP, Messerschmidt B, Schnitzer MJ. 2009. In vivo fluorescence imaging with high-resolution microlenses. *Nat Methods* **6:** 511–512.

Bishop RED, Johnson DC. 1960. *The mechanics of vibration*. Cambridge University Press, Cambridge.

Clark SW, Ilday FÖ, Wise FW. 2001. Fiber delivery of femtosecond pulses from a Ti:sapphire laser. *Opt Lett* **26:** 1320–1322.

Dombeck DA, Khabbaz AN, Collman F, Adelman TL, Tank DW. 2007. Imaging large-scale neural activity with cellular resolution in awake, mobile mice. *Neuron* **56:** 43–57.

Engelbrecht CJ, Johnston RS, Seibel EJ, Helmchen F. 2008. Ultra-compact fiber-optic two-photon microscope for functional fluorescence imaging in vivo. *Opt Express* **16:** 5556–5564.

Flusberg BA, Cocker ED, Piyawattanametha W, Jung JC, Cheung ELM, Schnitzer MJ. 2005a. Fiber-optic fluorescence imaging. *Nat Methods* **2:** 941–950.

Flusberg BA, Lung JC, Cocker ED, Anderson EP, Schnitzer MJ. 2005b. In vivo brain imaging using a portable 3.9 gram two-photon fluorescence microendoscope. *Opt Lett* **30:** 2272–2274.

Flusberg BA, Nimmerjahn A, Cocker ED, Mukamel EA, Barretto RP, Ko TH, Burns LD, Jung JC, Schnitzer MJ. 2008. High-speed, miniaturized fluorescence microscopy in freely moving mice. *Nat Methods* **5:** 935–938.

Giniunas L, Juskaitis R, Shatalin SV. 1991. Scanning fibre-optic microscope. *Electronics Lett* **27:** 724–726.

Göbel W, Kerr JND, Nimmerjahn A, Helmchen F. 2004a. Miniaturized two-photon microscope based on a flexible coherent fiber bundle and a gradient-index lens objective. *Opt Lett* **29:** 2521–2523.

Göbel W, Nimmerjahn A, Helmchen F. 2004b. Distortion-free delivery of nanojoule femtosecond pulses from a Ti:sapphire laser through a hollow-core photonic crystal fiber. *Opt Lett* **29:** 1285–1287.

Greenberg DS, Kerr JND. 2009. Automated correction of fast motion artifacts for two-photon imaging of awake animals. *J Neurosci Methods* **176:** 1–15.

Greenberg DS, Houweling AR, Kerr JND. 2008. Population imaging of ongoing neuronal activity in the visual cortex of awake rats. *Nat Neurosci* **11:** 749–751.

Grewe BF, Helmchen F. 2009. Optical probing of neuronal ensemble activity. *Curr Opin Neurobiol* **19:** 520–529.

Helmchen F. 2002. Miniaturization of fluorescence microscopes using fibre optics. *Exp Physiol* **87:** 737–745.

Helmchen F, Denk W. 2005. Deep tissue two-photon microscopy. *Nat Methods* **2:** 932–940.

Helmchen F, Waters J. 2002. Ca^{2+} imaging in the mammalian brain in vivo. *Eur J Pharmacol* **447:** 119–129.

Helmchen F, Fee MS, Tank DW, Denk W. 2001. A miniature head-mounted two-photon microscope: High-resolution brain imaging in freely moving animals. *Neuron* **31:** 903–912.

Helmchen F, Tank DW, Denk W. 2002. Enhanced two-photon excitation through optical fiber by single-mode propagation in a large core. *Appl Opt* **41:** 2930–2934.

Kerr JND, Denk W. 2008. Imaging in vivo: Watching the brain in action. *Nat Rev Neurosci* **9:** 195–205.

Kerr JND, de Kock CPJ, Greenberg DS, Bruno RM, Sakmann B, Helmchen F. 2007. Spatial organization of neuronal population responses in layer 2/3 of rat barrel cortex. *J Neurosci* **27:** 13316–13328.

Lee CM, Engelbrecht CJ, Soper TD, Helmchen F, Seibel EJ. 2010. Scanning fiber endoscopy with highly flexible, 1 mm catheterscopes for wide-field, full-color imaging. *J Biophotonics* **3:** 385–407.

Lütcke H, Murayama M, Hahn T, Margolis DJ, Astori S, Meyer zum Alten Borgloh S, Göbel W, Yang Y, Tang W, Kügler S, et al. Optical recording of neuronal activity with a genetically-encoded calcium indicator in anesthetized and freely moving mice. *Front Neural Circuits* **4:** 9.

Mrsic-Flogel TD, Hofer SB, Ohki K, Reid RC, Bonhoeffer T, Hubener M. 2007. Homeostatic regulation of eye-specific responses in visual cortex during ocular dominance plasticity. *Neuron* **54:** 961–972.

Murayama M, Perez-Garci E, Luscher HR, Larkum ME. 2007. Fiberoptic system for recording dendritic calcium signals in layer 5 neocortical pyramidal cells in freely moving rats. *J Neurophysiol* **98:** 1791–1805.

Nagayama S, Zeng S, Xiong W, Fletcher ML, Masurkar AV, Davis DJ, Pieribone VA, Chen WR. 2007. In vivo simultaneous tracing and Ca^{2+} imaging of local neuronal circuits. *Neuron* **53:** 789–803.

Nimmerjahn A, Mukamel EA, Schnitzer MJ. 2009. Motor behavior activates Bergmann glial networks. *Neuron* **62:** 400–412.

Ohki K, Chung S, Ch'ng YH, Kara P, Reid RC. 2005. Functional imaging with cellular resolution reveals precise microarchitecture in visual cortex. *Nature* **433:** 597–603.

Ouzounov DG, Moll KD, Foster MA, Zipfel WR, Webb WW, Gaeta AL. 2002. Delivery of nanojoule femtosecond pulses through large-core microstructured fibers. *Opt Lett* **27:** 1513–1515.

Piyawattanametha W, Barretto RPJ, Ko TH, Flusberg BA, Cocker ED, Ra HJ, Lee DS, Solgaard O, Schnitzer MJ. 2006. Fast-scanning two-photon fluorescence imaging based on a microelectromechanical systems two-dimensional scanning mirror. *Opt Lett* **31:** 2018–2020.

Piyawattanametha W, Cocker ED, Burns LD, Barretto RP, Jung JC, Ra H, Solgaard O, Schnitzer MJ. 2009. In vivo brain imaging using a portable 2.9 g two-photon microscope based on a microelectromechanical systems scanning mirror. *Opt Lett* **34:** 2309–2311.

Rothschild G, Nelken I, Mizrahi A. 2010. Functional organization and population dynamics in the mouse primary auditory cortex. *Nat Neurosci* **13:** 353–360.

Sato TR, Gray NW, Mainen ZF, Svoboda K. 2007. The functional microarchitecture of the mouse barrel cortex. *PLoS Biol* **5:** e189.

Sawinski J, Denk W. 2007. Miniature random-access fiber scanner for in vivo multiphoton imaging. *J Appl Phys* **102:** 034701.

Sawinski J, Wallace DJ, Greenberg DS, Grossmann S, Denk W, Kerr JND. 2009. Visually evoked activity in cortical cells imaged in freely moving animals. *Proc Natl Acad Sci* **106:** 19557–19562.

Seibel EJ, Smithwick QYJ. 2002. Unique features of optical scanning, single fiber endoscopy. *Lasers Surg Med* **30:** 177–183.

Stosiek C, Garaschuk O, Holthoff K, Konnerth A. 2003. In vivo two-photon calcium imaging of neuronal networks. *Proc Natl Acad Sci* **100:** 7319–7324.

Stüttgen MC, Schwarz C. 2008. Psychophysical and neurometric detection performance under stimulus uncertainty. *Nat Neurosci* **11:** 1091–1099.

Sullivan MR, Nimmerjahn A, Sarkisov DV, Helmchen F, Wang SS. 2005. In vivo calcium imaging of circuit activity in cerebellar cortex. *J Neurophysiol* **94:** 1636–1644.

Svoboda K, Denk W, Kleinfeld D, Tank DW. 1997. In vivo dendritic calcium dynamics in neocortical pyramidal neurons. *Nature* **385:** 161–165.

Tian L, Hires SA, Mao T, Huber D, Chiappe ME, Chalasani SH, Petreanu L, Akerboom J, McKinney SA, Schreiter ER, et al. 2009. Imaging neural activity in worms, flies and mice with improved GCaMP calcium indicators. *Nat Methods* **6:** 875–881.

Treacy EB. 1969. Optical pulse compression with diffraction gratings. *IEEE J Quantum Electron* **5:** 454–458.

Wallace DJ, Kerr JN. 2010. Chasing the cell assembly. *Curr Opin Neurobiol* **20:** 296–305.

Wallace DJ, Meyer zum Alten Borgloh S, Astori S, Yang Y, Bausen M, Kugler S, Palmer AE, Tsien RY, Sprengel R, Kerr JND, et al. 2008. Single-spike detection in vitro and in vivo with a genetic Ca^{2+} sensor. *Nat Methods* **5:** 797–804.

Walmsley I, Waxer L, Dorrer C. 2001. The role of dispersion in ultrafast optics. *Rev Sci Instrum* **72:** 1–29.

79 Optogenetics
Opsins and Optical Interfaces in Neuroscience

Antoine R. Adamantidis,[1,3] Feng Zhang,[2,3] Luis de Lecea,[1] and Karl Deisseroth[1,2]

[1]*Department of Psychiatry and Behavioral Sciences, Stanford University, Stanford, California 94305;*
[2]*Department of Bioengineering, Stanford University, Stanford, California 94305*

ABSTRACT

Optogenetics is defined as the integration of optics and genetics to control well-defined events within specified cells of living tissue. In this chapter, we focus on the basic techniques necessary for employing microbial opsins as optogenetic tools in mammalian brains. We provide a guide for the fundamentals of optogenetic application—selecting an opsin, implementing expression of opsins based on the neuroscientific experimental requirements, and adapting the corresponding optical hardware for delivery of light into mammalian brains. The diversity and structure-function properties of microbial opsin genes and our methods for applying microbial opsins to freely behaving mammals are covered in two companion chapters in this volume (see Chapters 6 and 80).

INTRODUCTION

The mammalian brain consists of many distinct types of excitatory, inhibitory, and modulatory cells. One of the central challenges of neuroscience has long been to decipher the role of distinct cell types in the organization and function of brain circuits and behavior. We have developed a technology called optogenetics, which leverages microbial opsins and molecular genetics to enable temporally precise control of genetically defined neurons in the mammalian brain in vivo using light.

Our previous reports have provided detailed characterization of the opsins described in this chapter (Boyden et al. 2005; Zhang et al. 2006; Zhang et al. 2007; Gradinaru et al. 2008; Zhang et al. 2008a; Zhao et al. 2008; Airan et al. 2009; Berndt et al. 2009; Gunaydin et al. 2010; also see Chapter 6). Likewise, the technology suitable for freely moving mammals, including optical fiber methodology, has also been described (Aravanis et al. 2007; Zhang et al. 2010). We developed and applied this technology to control specific neural populations during a variety of behavioral tests in rodents, such as sleep-to-wake transition (Adamantidis et al. 2007), conditioned place preference (Airan et al. 2009; Tsai et al. 2009; Zhang et al. 2010), and motor control (Aravanis et al. 2007), and in rodent disease models (Bi et al. 2006; Alilain et al. 2008; Abbott et al. 2009; Gradinaru et al. 2009).

[3]These authors contributed equally to this work.

OPSINS IN NEUROSCIENCE

We have engineered a number of naturally occurring and synthetic opsins to enable optical control of electrical and biochemical signaling in neurons. Although all of these opsins require retinal-based cofactors (e.g., all-*trans* retinal for the microbial opsins) to fold properly and respond to light, there are sufficient levels of retinoids present in the mammalian brain so that opsins can be functionally expressed in the intact mammalian brain without supplementation of cofactors (Zhang et al. 2006, 2007). Comparison of the various opsins suitable for use in neuroscience can be found in Table 1, and additional biophysical details are summarized in Chapter 6.

As with all of the microbial opsins, the algal light-sensitive cation-conducting channelrhodopsins (ChRs) (Nagel et al. 2002; Nagel et al. 2003; Zhang et al. 2008a) were initially characterized outside of neuroscience in nonneural systems. Two ChRs have been successfully brought to neuroscience, ChR2 from *Chlamydomonas reinhardtii* (Boyden et al. 2005; Zhang et al. 2006) and VChR1 from *Volvox carteri* (Zhang et al. 2008a). Both of these ChRs have fast kinetics so that defined trains of action potentials can be triggered using brief pulses (1–5 msec duration) of light flashes at a range of frequencies relevant to neural signaling. Although both ChR2 and VChR1 are sensitive to blue light, the activation spectrum of VChR1 is sufficiently redshifted from that of ChR2 so that neurons expressing VChR1 can be activated independently of ChR2 using orange light.

TABLE 1. Comparison table of optogenetic tools suitable for fast neural control in vivo in mammals

Opsin	Host organism	Wavelength sensitivity	Mode of control	Modulatory capabilities	Experimental systems tested
ChR2, ChR2(H134R), ChIEF, ChR1/ChR2 chimeras, ChETAs	*Chlamydomonas reinhardtii*	470 nm (maximum activation)	Depolarizing	Rapid on/off; best used for precise activation of neurons on the millisecond timescale. Can be used to evoke single spikes or defined trains of action potentials over a range of frequencies. The H134R mutation yields larger photocurrents relative to wild-type ChR2, but with slower k_{off} kinetics. ChIEF (Lin et al. 2009) and other ChR1/ChR2 chimeras (Wang et al. 2009) provide improved expression and inactivation. ChETA (Gunaydin et al. 2010) provides faster recovery from inactivation, faster deactivation kinetics, improved temporal precision at high spike rates, reduced artifactual plateau potentials and extra action potentials, better temporal stationarity, and faster frequency-following behavior up to at least 200 Hz. Two ChETA variants exist, one involving E123T and one involving E123A. The T variant is 10–15 nm redshifted by comparison with ChR2. This is advantageous in many cases for reduced scattering and deeper penetration of light, but if the experimenter is limited to a ~470 nm laser line or filter, the A variant may be preferable.	In vitro: Dissociated neuron culture (Boyden et al. 2005; Zhang et al. 2006; Zhang et al. 2007), acute brain sections from mouse and rat (Ishizuka et al. 2006; Zhang et al. 2006, 2008b; Petreanu et al. 2007, 2009; Wang et al. 2007; Zhang and Oertner 2007), and 293HEK cells (Nagel et al. 2003) In vivo: *C. elegans* (requires supplementation of ATR) (Nagel et al. 2005), *D. melanogaster* (requires supplementation of ATR) (Schroll et al. 2006; Hwang et al. 2007; Pulver et al. 2009) zebrafish (Douglass et al. 2008), chicken (Li et al. 2005), mouse (Bi et al 2006; Adamantidis et al. 2007; Aravanis et al. 2007; Arenkiel et al. 2007; Alilain et al. 2008; Huber et al. 2008; Gradinaru 2009; Tsai et al. 2009; Gunaydin et al. 2010), rat (Aravanis et al. 2007), and primate (Han et al. 2009)

(Continued)

TABLE 1. *(Continued)*

Opsin	Host organism	Wavelength sensitivity	Mode of control	Modulatory capabilities	Experimental systems tested
Step-function opsins (SFOs): ChR2(C128A), ChR2(C128S), and ChR2(C128T)	*Chlamydomonas reinhardtii*	470 nm (switching on), 542 nm (switching off for C128A and C128S mutants)	Depolarizing	Point mutants of ChR2 with slow or optically switchable deactivation. C128A and C128S mutants show the most prolonged activation and the highest light sensitivity, whereas C128T retains more temporal precision of activation. SFOs can be switched on and off with blue and green light pulses, respectively.	In vitro: Dissociated neuron culture (Berndt et al. 2009)
VChR1 and higher-conductance variants	*Volvox carteri*	535 nm (maximum activation at 589 nm, in which activation is separable from ChR2)	Depolarizing	Redshifted action spectrum relative to ChR2. Like ChR2, VChR1 can be used to drive reliable action potential firing over a range of frequencies. With 589-nm light, VChR1 can be activated independently of ChR2.	In vitro: Dissociated neuron culture (Zhang et al. 2008a)
NpHR, eNpHR3.0 variants	*Natronomonas pharaonis*	589 nm (maximum activation) 630–680 nm utility 400 nm utility if needed	Hyperpolarizing	Light-activated chloride pump. Can be used to hyperpolarize neurons with high temporal precision; capable of inhibiting single action potentials within high-frequency spike trains. Also can be used to mediate sustained inhibition of neurons over many minutes, with stable currents that recover rapidly. The most potent eNpHR variants (eNpHR3.0) achieve inhibition in the far red, and traffic to neuronal processes. Redshifted/compatible with ChR2 in same experiment.	In vivo: *C. elegans* (requires supplementation of ATR) (Zhang et al. 2007) and mouse (Gradinaru et al. 2008, 2010; Zhao et al. 2008) In vitro: Dissociated neuron culture (Han et al. 2007; Zhang et al. 2007; Gradinaru et al. 2008; Zhao et al. 2008; Gradinaru et al. 2010), and acute brain slices from mouse (Zhang et al. 2007; Zhao et al. 2008)
Bacteriorhodopsin and other regulators of H^+ conductance (e.g., eBR/Arch/Mac/AR/GtR3 and many others)	Numerous species	Broad range (most 450–550 nm)	Hyperpolarizing	Light-activated proton conductance regulators. May be useful for experiments that are not focused on simply controlling output of a cell, but in which it is important not to alter chloride gradients while allowing for juxtamembranous proton concentration changes. ~10x higher light power required to reach currents comparable to eNpHR3.0 max, and none currently as redshifted as eNpHR3.0, posing challenges for practical in vivo use and integration with ChR2, respectively.	In vivo: Mouse (Chow et al. 2010). In vitro: Dissociated neuron culture (Chow et al. 2010; Gradinaru et al. 2010)
Opto-α1AR	Synthetic protein	500 nm (maximum activation)	Biochemical	Light-activated GPCR, via the G_q pathway.	In vivo: Mouse (Airan et al. 2009) In vitro: 293HEK cell line (Airan et al. 2009)
Opto-β2AR	Synthetic protein	500 nm (maximum activation)	Biochemical	Light-activated GPCR, via the G_s pathway.	In vivo: Mouse (Airan et al. 2009) In vitro: 293HEK cell line (Airan et al. 2009)

Adapted, with permission, from Zhang et al. 2010.
ATR, all-*trans*-retinal.

Ultimately, the performance of opsins in neuroscience depends on the precision of modulation of neuronal spiking. Of note, VChR1 is slower to deactivate after "light off," with a time constant of ~120 msec in contrast to the ~12 msec of ChR2. But even the 12-msec deactivation rate of wild-type ChR2 (and its similarly slow transitions to and from the inactivated state) can be problematic, leading to occasional spike doublets, plateau potentials, nonstationary responses in long trains, and impaired frequency response of neurons under optogenetic control (Gunaydin et al. 2010). However, deactivation and inactivation rates are also tunable properties of the microbial opsins. On the one hand, the deactivation rate can be accelerated dramatically with the ChETA point mutations (E123T or E123A), which accelerate channel closing and are associated with faster inactivated-state transitions, and thereby have the effect of reducing or eliminating all of the confounds listed above and allow fast spiking up to at least 200 Hz (Gunaydin et al. 2010). On the other hand, mutations in cysteine 128 of ChR2 (Berndt et al. 2009) have been rationally designed to slow or eliminate deactivation, and allow ChRs to remain open after light has been turned off. These mutant ChRs, known as step-function opsins (SFOs), convert a brief pulse of light into a stable step in membrane potential, by virtue of deactivation that is about four orders of magnitude slower. This stable activation can nevertheless be precisely terminated with a distinct, wavelength-shifted (yellow or green) pulse of light. Additional mutants and chimeras generated by shuffling the *trans*-membrane loops of ChR1 and ChR2 have yielded improved expression and inactivation properties (Lin et al. 2009; Wang et al. 2009). Ongoing directed mutagenesis/evolution and genomic approaches will likely yield further enhancements in functionality.

Light-sensitive chloride-pumping halorhodopsins (HRs) have been known for decades to serve as light-activated single-component voltage regulators in reduced systems. We and our colleagues have found that the HR from *Natronomonas pharaonis* (NpHR) can be used to inhibit neural activity in response to redshifted light (useful activation up to 680 nm) in intact tissue and behaving animals (Zhang et al. 2007). As with ChRs, the yellow-light-sensitive NpHR can be activated within milliseconds of illumination, allowing inhibition of individual action potentials within high-frequency spike trains using brief light flashes. NpHR also displays minimal inactivation during chronic illumination, and therefore can be used to optically mimic lesions in targeted brain structures, or to test the functional significance of defined cell types. Although high levels of wild-type NpHR expression can sometimes lead to toxicity because of altered membrane trafficking and accumulation in the endoplasmic reticulum, we have engineered enhanced NpHRs (eNpHR; Gradinaru et al. 2008; Zhao et al. 2008) with appropriate targeting signals to enhance membrane trafficking and folding in mammalian neurons, which implement safe expression at high levels in vivo. At this point, the most potent optogenetic inhibition is achieved with eNpHR3.0 (Gradinaru et al. 2010), which can elicit nanoampere-scale hyperpolarizations in response to moderate-intensity (3.5 mW/mm^2) yellow light, and displays step-like currents with fast recovery and stability over behavioral timescales. Other inhibition strategies are also possible. We have found that the classical light-activated proton pump bacteriorhodopsin can also be used to hyperpolarize neurons in response to light (Gradinaru et al. 2010), and other proton conductances can also be used (Chow et al. 2010; Gradinaru et al. 2010). However, the proton conductances at present require ~10X greater light power to achieve currents comparable to eNpHR3.0, which may pose risks for operation in vivo with steady illumination.

Distinct opsins may now be selected from a broad palette to implement control of neuronal output via direct modulation of defined cellular activity (Table 1). The HR approach (eNpHR and variants) is ideal for probing the necessity of defined cellular events with high temporal precision, whereas ChR2 and VChR1 are suitable for testing sufficiency of defined spikes in circuit dynamics and behavior—as when probing the importance of spike timing in neural coding, or testing the spike-frequency dependence of behavioral efficacy in a select population of neurons or other circuit elements (Adamantidis et al. 2007; Gradinaru et al. 2009: Sohal et al. 2009; Tsai et al. 2009). When high photocurrents are needed (as in the case of axonal illumination) with less experimental emphasis on high frequency, the H134R mutant is suitable. In contrast, when a slight decrement in photocurrent magnitude is experimentally acceptable, but overall temporal performance and signal quality is at a premium, the ChETA mutations are ideal (Gunaydin et al. 2010).

Other kinds of modulation are important as well, and in many in vivo experiments it may be advisable to modulate cellular excitability with high temporal precision without driving defined spike trains. SFOs are indicated for use when it is important to modulate the basal activity level of a neuronal population in chronic fashion while minimizing light exposure, and when it is important to allow native asynchronous or sparse neural codes to propagate through the targeted cell type more effectively (Berndt et al. 2009). Finally, as another modulatory strategy, with the aim of studying the causal role of intracellular biochemical signals in circuit and animal behavior, we also have engineered a set of chimeric opsins (optoXRs) by replacing the intracellular loops of bovine rhodopsin with the signal-transducing intracellular loops of G-protein-coupled receptors (GPCRs) (Airan et al. 2009). These synthetic optoXRs undergo conformational changes when illuminated with green light and activate specific downstream biochemical cascades, including either G_s or G_q pathways. There is little detectable "dark" activity arising from these probes, in contrast to expression of microbial photoactivated adenylyl cyclases, which elevate cellular cyclic AMP (cAMP) as much as 30-fold even in the absence of light. OptoXRs can be used to recruit and control, with high spatiotemporal precision, receptor-initiated biochemical signaling pathways in genetically defined neural populations (Airan et al. 2009), and thereby optogenetically modulate cellular excitability by biochemical means in behaving mammals. Additional information on the biophysical properties of opsins is presented in Chapter 6.

GENE EXPRESSION AND TARGETING SYSTEMS

As with the genetically encoded anatomical marker green fluorescent protein (GFP), each opsin can be selectively expressed in specific subsets of neurons in the rodent brain using a variety of modern genetic targeting strategies (Adamantidis et al. 2007; Aravanis et al. 2007; Arenkiel et al. 2007; Petreanu et al. 2007; Huber et al. 2008; Kuhlman and Huang 2008; Luo et al. 2008; Airan et al. 2009; Gradinaru et al. 2009; Sohal et al. 2009; Tsai et al. 2009). High levels of opsin expression are important for achieving robust optical activation and inhibition and can be readily achieved using viral gene delivery. Both lentiviral and adeno-associated viral (AAV) vectors can be produced at high functional titer and stereotactically injected into target brain regions to mediate high-level opsin expression (Zhang et al. 2006, 2007; Sohal et al. 2009; Tsai et al. 2009). Detailed protocols describing the production of both lentiviral (Kutner et al. 2009) and AAV (Grieger et al. 2006) vectors have been published elsewhere, and protocols optimized for in vivo use in mammalian brains have been made available online at http://www.optogenetics.org and in Zhang et al. (2010). Different lentivirus and AAV serotypes may be selected for different diffusional properties (e.g., we find that AAV5 will transduce a larger brain region than AAV2 or lentivirus) and in some cases cell type specificity can also arise from specific viral vectors. We have also reported suitable promoter-based cell type specificity for several small (<4 kb) fragments, such as human synapsin I for targeting neurons, CamKIIα for targeting excitatory neurons, glial fibrillary acidic protein (GFAP) for targeting astroglia, and the ppHcrt promoter for targeting hypocretin neurons in rodents (reviewed in Zhang et al. 2010; http://www.optogenetics.org).

However, many cell-specific promoters have weak transcriptional activity and cannot drive sufficient levels of opsin expression to yield robust photocurrents or do not retain sufficient specificity when used in viral vectors. Therefore, we and others have designed Cre-recombinase-dependent AAV expression systems to enable cell-specific opsin expression using transgenic mice expressing Cre under the control of native cell-specific promoter/enhancer regions (e.g., tyrosine hydroxylase for dopaminergic neurons [Tsai et al. 2009] and parvalbumin for fast-spiking interneurons [Cardin et al. 2009; Sohal et al. 2009]). Although maintaining cell type specificity, this approach amplifies the transcriptional strength of the cell-specific promoter by allowing opsins to be expressed under the strong EF-1α promoter in Cre-expressing neurons. Given the increasing availability of a wide array of cell type–specific Cre-transgenic lines (Gong et al. 2007), researchers now have the ability to selectively probe the activity of a large and increasing number of important cell types optogenetically, in a versatile and generalizable fashion. A detailed implementation for the case of the tyrosine hydroxylase (TH)-Cre mouse driver line is presented in Chapter 80.

Cell types may also be targeted by virtue of their projections; this is particularly important for organisms in which genetic and promoter/enhancer information is not readily available. Axons may be selectively depolarized by illuminating away from cell bodies, evoking neurotransmitter release whether through spike-dependent or -independent mechanisms and thereby defining cells for control by virtue of their projections rather than genetic identity. In this setting, strong photocurrents are at a premium and are best achieved with, for example, the H134R ChR2 mutation while sacrificing some temporal precision (Gradinaru et al. 2007). Retrograde-type targeting strategies are also possible, in which axon terminals are transduced with viruses (e.g., herpesviruses or pseudotyped lentiviruses) bearing opsin genes. Finally, transsynaptic retrograde or anterograde targeting is possible with custom viruses that will transport Cre recombinase across synapses (e.g., as a wheat germ agglutinin–Cre fusion protein) to distinct circuits in which the Cre-dependent opsin-expressing viruses (Sohal et al. 2009; Tsai et al. 2009) have been separately injected (Gradinaru et al. 2010; Zhang et al. 2010). These "topological targeting" strategies powerfully complement the genetic strategies, and together make optogenetic targeting versatile and generalizable.

THE OPTICAL NEURAL INTERFACE

Selective expression of opsins in genetically defined neurons makes it possible to control a subset of neurons without affecting nearby cells and processes in the intact brain, but light must still be delivered to the target brain structure. Light scattering limits the delivery of light from the surface of the brain. Indeed, light power density for blue light will drop to ~1% of initial values at a tissue depth of 1 mm (Aravanis et al. 2007). For this reason, we have developed a fiber-optic-based optical neural interface (ONI), which allows optical access to any brain structure in freely moving mammals (Adamantidis et al. 2007; Aravanis et al. 2007; Gradinaru et al. 2007; see Fig. 1). Spatial resolution chiefly arises from the cellular genetic targeting. However, spatially resolved light delivery can be readily achieved with light-emitting diode (LED) arrays or digital micromirror-coupled devices (DMDs) or fibers, but resolution is rapidly lost in tissue because of scattering.

There are several important factors for achieving effective optical stimulation in vivo. First, it is important to choose a light source with sufficient power because ChR2 and NpHR require 1–10 mW/mm^2 of light for reliable activation (the SFOs require approximately two to three orders of magnitude less light power density). For most in vivo applications, solid-state diode lasers can be used to provide sufficient levels of light and are readily coupled to fiber optics. Both virus injection and the delivery of the multimode fiber (50–400 µm diameter) can be guided via the same stereotactically implanted cannula, to ensure registration of light illumination with opsin-expressing cells. In the following section, we provide a detailed protocol for setting up the fiber-optic-based ONI (Adamantidis et al. 2007; Aravanis et al. 2007) for in vivo optogenetic control (Zhang et al. 2010). Depending on the experiment, either bilateral or unilateral stimulation may be performed; procedures for bilateral and unilateral fibers are similar.

Protocol

Establishing a Fiber-Optic-Based Optical Neural Interface

The ONI system is constructed by modifying the small animal cannula system from PlasticsOne. The system for bilateral stimulation consists of a bilateral cannula guide that has been stereotactically implanted over the target brain region, a screw cap for securing the optical fiber to the animal's head, a fiber guard modified from the internal cannula adapter, and a bare fiber whose length is customized based on the depth of the target region (Fig. 1A,B). Stereotactic delivery of virus in the rodent brain has been described (Cetin et al. 2007: Zhang et al. 2010), as have detailed protocols for stereotactic surgery (Cetin et al. 2007). For unilateral stimulation, a single-fiber system can be constructed using unilateral cannula parts from PlasticsOne. Delivery of opsin-expressing virus and implantation of the ONI may be conducted in the same surgical session; alternatively, with a transgenic animal no opsin virus is delivered during the surgery. Typically with adeno-associated viral (AAV) or lentiviral introduction of opsins during surgery, strong expression is achieved within 2-3 wk at cell bodies, and within 8-12 wk at axonal projections. Similar procedures are useful for deep or superficial injections (even for neocortical targets, although in some cases surface LEDs or cortex-apposed fibers can be used for the most superficial cortical targets).

MATERIALS

CAUTION: See Appendix 6 for proper handling of materials marked with <!>.

Reagents

Anesthesia (1.5% isoflurane <!> *or* ketamine <!> and xylazine <!>)
Animals (adult rats or mice)
Betadine
Buprenorphine
Dental cement (C&B Metabond; Parkell Inc.)
Epoxy resin adhesive
Ethanol <!> (70%)
Hydrogen peroxide <!> (0.1%)
Ophthalmic ointment
PBS
Saline solution
Vetbond tissue adhesive
Virus preparation to be injected

The virus carries the genetically encoded opsin of interest (see Introduction).

Equipment

Arbitrary waveform generator (Agilent, cat. no. 33220A)
Bilateral cannula guide (PlasticsOne, cat. no. C232G)
Cannula guide and internal cannula

The length of the cannula guide and the internal cannula can be customized for desired target depth. The recommended length for the cannula guide is 0.5 mm above the target region and the internal cannula should have a projection length of 0.5 mm beyond the cannula guide tip. The dummy cannula should be flush with the tip of the cannula guide. For unilateral stimulation, use the C313 series of cannula parts.

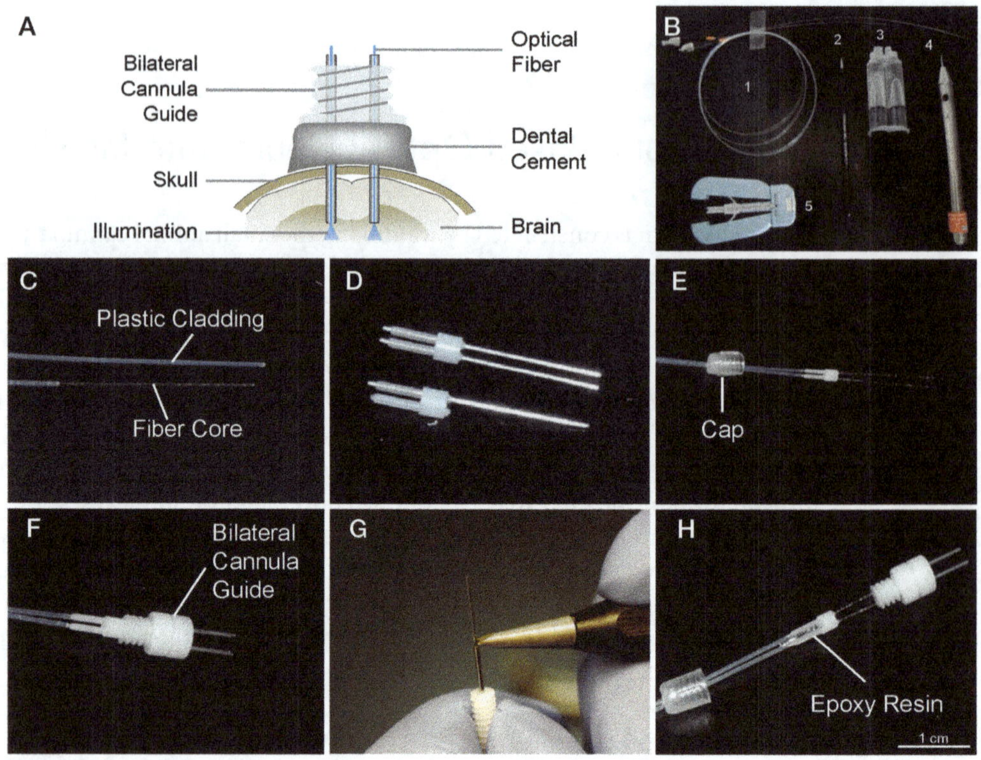

FIGURE 1. Preparation of the bilateral optical neural interface (ONI). (*A*) Diagram of the ONI system. The bilateral cannula guide is stereotactically implanted over the target brain region and secured using dental cement. Two optical fibers are inserted through the cannula guide to illuminate the target brain structure. (*B*) All the parts needed to make the ONI system: (1) optical fibers; (2) diamond scribe (see also *G*); (3) epoxy; (4) dental drill; and (5) fiber-stripping tool. (*C*) To expose the fiber, a fiber-stripping tool is used to remove the plastic cladding. (*D*) Remove the steel tubing from the internal cannula and use the plastic pedestal to fix the optical fibers. (*E*) Thread the stripped end of the optical fibers through a dust cap (modified by drilling a hole through the top) and the plastic pedestal of the internal cannula. (*F*) Insert the fiber through a bilateral cannula guide (same size as the implanted one). (*G*) Note the length of fiber projecting from the tip of the cannula guide, and use a diamond scribe to trim the fiber so that there is only 0.5 mm projecting from the tip of the cannula guide. (*H*) Use epoxy to glue the fiber to the plastic pedestal so that the fibers are securely attached. After the epoxy hardens, make sure the cap will fit over the epoxy so that the fiber can be screwed onto the implanted bilateral cannula guide.

Cannula holder (Kopf or PlasticsOne)

Diamond scribe

Diode laser (CrystaLaser, cat. no. BCL-473-050) (Specifications: 50 mW, 473 nm, TEMoo CW laser with TTL on/off, analog output intensity modulation, and PC connector at laser output)

Alternatively, blue lasers can be acquired from LaserGlow (cat. no. LRS-0473-PFM-00080-05).

Dummy cannula (PlasticsOne, cat. no. C232DC)

Fiber-stripping tool (for 300-µm fiber, use Thorlabs, cat. no. T18S31; for 200-µm fiber, use Thorlabs, cat. no. T12S25)

Fine forceps

High-precision injection pump (syringe, needle, and pump from World Precision Instruments, Sarasota, FL); see Step 5

Internal/injector cannula (PlasticsOne, cat. no. C232I)

Light power meter (Thorlabs, power sensor, cat. no. S130A; digital console, cat. no. PM100D)

Optical fiber (200-µm fiber: Thorlabs, cat. no. BFL37-200; 300-µm fiber: Thorlabs, cat. no. BFL37-300)

Fibers must have a male FC/PC connector to connect to the female FC/PC adapter on the output of the laser. The length can be specified depending on the experimental setup; we recommend that you have at least 1 m of slack fiber from the head of the animal to the first fiber fixture.

Rotating optical commutator (aluminum; custom-made; Doric Lenses, Québec, Canada)
Short-wavelength, single-mode fiber-optic splitter (Fiber Optic Network Technology, Canada)

This is used for performing bilateral stimulation using two optical fibers.

Stereotactic apparatus (Kopf Instruments, Tujunga, CA; Olympus stereomicroscope)
Surgical tools

Surgical tools must be clean and sterile; sterilize by autoclaving, heating or immersion in a disinfectant.

EXPERIMENTAL METHOD

Preparation for Surgery

Perform all surgeries under aseptic conditions. The work area should be disinfected with 70% ethanol.

1. Anesthetize the animals.
   i. Use 1.5% isoflurane (for surgeries to last longer than 1 h) or by intraperitoneal injection (use 90 mg/kg ketamine and 5 mg/kg xylazine for rats; 80 mg/kg and 15–20 mg/kg, respectively, for mice).
   ii. Use a heating pad to keep the body temperature at 37°C.
   iii. Inject saline solution subcutaneously to prevent dehydration during surgery (30 mL per kilogram of body weight or ~0.5–1.0 mL for mice and ~5.0 mL for adult rats).
   iv. Apply ophthalmic ointment to prevent eye drying.

   See *Troubleshooting*.
2. Place the animal in a stereotactic apparatus (Kopf Instruments, Tujunga, CA; Olympus stereomicroscope). The correct head position in the stereotactic apparatus allows vertical, but not lateral, movement of the head.
3. Shave and clean the head using 70% ethanol and betadine, then incise the skin (1 cm) over the target brain region. Clean the skull with cotton swabs immersed in hydrogen peroxide, and stop the reaction with PBS immediately after (using cotton swabs) and scrape skin with a scalpel blade.
4. After aligning the bregma (anterior meeting point for skull plates) and lambda (posterior meeting point for skull plates) horizontally, move to the targeted area defined by the coordinates. Make a small craniotomy (of a slightly larger diameter than the cannula guide) using a drill mounted on the stereotactic apparatus. Use fine forceps to remove the dura, to keep tissue healthy. This is crucial for cortical injections.

   See *Troubleshooting*.

Infusion of the Virus and Surgery

Infusion of the virus preparation before surgery is described in Step 5. Alternatively, viral infusion can be performed after surgery. To carry out infusion after surgery, omit Step 5 and proceed to Step 6.

5. Thaw the virus preparation on ice and microinject up to 1 µL of concentrated virus (1010 IU/mL) into the brain using a 10-µL syringe and a thin 34-gauge metal needle; control the injection volume and flow rate (1 µL at 0.1 µL/min) with a high-precision injection pump. After injection leave the needle in place for 5–10 additional minutes and then slowly withdraw.

   The volume of the virus to be infused depends on the virus titer but ideally the volume injected should be restricted to the minimum needed (no more than 0.8 µL in mice and 1 µL in rats).

   See *Troubleshooting*.
6. Place the cannula in its special holder and slowly lower the cannula guide in the exposed target area of the brain.

7. Apply a thin layer of C&B Metabond dental cement where the cement will be in contact with the skull (see Fig. 1A), including the area around and below the cannula pedestal. Allow the C&B Metabond to dry.

   If virus infusion was delivered in Step 5, proceed directly to Step 9. Otherwise carry out infusion as described in Step 8.

8. Begin the viral infusion when the dental cement is completely dry (10 min).
   i. Thaw the virus preparation on ice.
   ii. Connect the internal/injector cannula to the tubing filled with oil (which is connected to the microsyringe).
   iii. Load the virus solution and place it into the cannula guide while the animal is still anesthetized.
   iv. Deliver up to 1 μL of concentrated virus (1010 IU/mL).
   v. Allow the internal/injector cannula stand for 10–15 min after the end of the injection procedure, before removal and closure of the cannula guide with a dummy cannula (Step 9).

      Because the animal still should be under anesthesia, monitor the recovery of the animals at the end of the surgical procedure and for the following week.

   See *Troubleshooting*.

9. Withdraw the cannula holder without moving the cannula guide and secure the cannula pedestal with dental cement leaving at least two rounds for screwing the fiber later on (see Step 21 below). Glue the skin back with Vetbond tissue adhesive.
10. Deliver buprenorphine (0.03 mg/kg) subcutaneously following the surgical procedure to minimize discomfort.
11. Transfer the animal into a clean cage placed over a heating blanket until it fully recovers. Monitor the recovery of the animal carefully and thoroughly.

Optical Stimulation

12. Use a fiber-stripping tool to remove ~40 mm of plastic coating from the ends of two optical fibers to expose the bare fibers (Fig. 1).
13. Remove the two pieces of metal tubing from the plastic pedestal of an internal cannula. Thread the two optical fibers through the plastic pedestal (Fig. 1D).
14. Prepare the screw cap by drilling a hole ~2 mm in diameter centrally on the top of a dust cap to allow the optical fibers to pass through. Thread the fibers through the cap with the cap opening facing away from the FC connector (Fig. 1E).

    This screw cap will be used to secure the fibers to the cannula guide that has been implanted in the experimental animal.

15. Insert the ONI fibers through a cannula guide with same dimensions as the one implanted in the experimental animal. Do this slowly because bare fibers without plastic cladding are very brittle. Ensure that they fit tightly.

    This cannula guide serves as a template to allow accurate cutting of the fibers.

16. Use a diamond scribe to cut the fibers so that they project 0.5 mm from the tip of the cannula guide (Fig. 1G).

    It is important to use a diamond scribe because the fiber tip must be flat to ensure consistent illumination.

17. Using a stereomicroscope, check that the fiber tip is flat.
18. Apply a very thin layer of epoxy resin adhesive to the interface of coated/uncoated fiber and slide the plastic pedestal onto the fibers so it is tightly against the plastic coating on the fiber. Allow the glue to dry for 10 min (Fig. 1H).

Too much epoxy will prevent the cap from fitting over the internal cannula and screwing onto the implanted cannula guide.

19. Connect the FC end of the fibers to the PC connector on the output port of the laser. Connect the laser control box to the waveform generator.
20. Program the waveform generator with the desired stimulation protocol (i.e., pulse frequency, duration, and intensity).

    It is critical to use a power meter to measure the light output at the end of the fiber before and after behavioral testing.
21. To conduct optical stimulation in freely moving animals, insert the ONI fibers through the cannula guide implanted on the animal's head and tighten the screw cap to secure the fiber. Take extra care during fiber insertion to prevent breakage. Ensure that the screw cap is securely tightened to prevent the fiber from moving inside the cannula guide. Animal head restraint is preferred over anesthesia to insert ONI fibers into the cannula guide because this is a very brief procedure (only a few seconds for an experienced scientist). Perform scientist training on anesthetized animals.

    For most behaviors, it is essential to habituate the animals to the experimental setup by practicing fiber insertion and removal several times. The animals should be placed in the behavioral testing room for at least 3 h before behavioral testing.

    See *Troubleshooting.*
22. To begin stimulation, turn on the output on the waveform generator to trigger the laser. Fast shutters (e.g., Uniblitz) may also be used to deliver defined waveforms, which have the advantage of avoiding laser nonstationarity.
23. Verify cannula placement and opsin expression after behavioral testing by performing standard histological analyses.

    See *Troubleshooting.*

TROUBLESHOOTING

Problem (Step 1): The animal still has toe pinch.
Solution: Increase isoflurane dose or administer 10% ketamine/xylazine.

Problem (Step 4): Excessive bleeding occurs after performing the craniotomy.
Solution: Keep a soft cotton swab on the area until all bleeding stops; then wash out with sterile saline solution. Do not inject through coagulated blood; clogging of the needle will result.

Problem (Steps 5 and 8): The needle is clogged after injection.
Solution: Replace the needle with a new one and repeat the injection.

Problem (Step 21): The cap becomes loose or there is torsion of the fiber.
Solution: Use longer optical fiber (>1 m). Alternatively use a optical fiber swivel (commercially available at Doric Lenses). Note that those swivels frequently decrease light intensity (>10%); thus, always check light intensity at the tip of the optical fiber before performing experiment.

Problem (Step 21): There is excessive torsion in the fiber.
Solution: This problem can be alleviated in unilateral stimulation experiments by using fiber-optical commutators.

Problem (Step 23): There is no observable effect of light stimulation.
Solution: Ensure that the opsin is correctly expressed using immunohistochemistry and/or electrophysiology. Ensure by calculation, measurement, and c-fos assays that proper light intensity reaches the target cells in the tissue/preparation (Adamantidis et al. 2007; Aravanis et al. 2007; Gradinaru et al. 2009).

OUTLOOK

Optogenetics holds promise as a versatile tool for probing the causal role of defined neurons and circuits in freely moving mammals and other animals. There are many important areas for future investigation, ranging from further expansion of the optogenetic control spectrum to enhancing the integration of optogenetics with multiphoton microscopy and integrating complex optogenetic control with increasingly rich and precise readouts to assess circuit behavior. The common theme for ongoing development may be best summarized as "reverse engineering." The more effectively and specifically we can interrogate the input pins and output pins on the "chip" that is the neural circuit, the closer we will be able to come to inferring core computations from the relationship between complex inputs and complex outputs.

Developing improved two-photon integration will be helpful on both the input and output sides, as two-photon control of opsins will provide improved resolution and penetration within scattering tissues; also two-photon and single-photon imaging of activity in targeted cells with genetically encoded calcium indicators and voltage indicators will enrich the output stream. The output stream will be further enhanced with increasingly advanced multielectrode readouts integrated with multisource light delivery inputs; brain-computer interfaces of this novel hybrid optical/electrical character will help drive the reverse engineering efforts by virtue of greatly enriched control and quantification of the input and output information streams.

Although additional engineering advances will be continue to be made, the essential panel of opsins, targeting methodologies, and optical hardware described here form the core technology required for optogenetics. Together, these methods may allow movement toward a network engineering approach, in which complex excitable-tissue function will come to be understood in terms of system properties emerging from component dynamics.

REFERENCES

Abbott SB, Stornetta RL, Fortuna MG, Depuy SD, West GH, Harris TE, Guyenet PG. 2009. Photostimulation of retrotrapezoid nucleus phox2b-expressing neurons in vivo produces long-lasting activation of breathing in rats. *J Neurosci* **29:** 5806–5819.

Adamantidis AR, Zhang F, Aravanis AM, Deisseroth K, de Lecea L. 2007. Neural substrates of awakening probed with optogenetic control of hypocretin neurons. *Nature* **450:** 420–424.

Airan RD, Thompson KR, Fenno LE, Bernstein H, Deisseroth K. 2009. Temporally precise in vivo control of intracellular signalling. *Nature* **458:** 1025–1029.

Alilain WJ, Li X, Horn KP, Dhingra R, Dick TE, Herlitze S, Silver J. 2008. Light-induced rescue of breathing after spinal cord injury. *J Neurosci* **28:** 11862–11870.

Aravanis AM, Wang LP, Zhang F, Meltzer LA, Mogri MZ, Schneider MB, Deisseroth K. 2007. An optical neural interface: In vivo control of rodent motor cortex with integrated fiberoptic and optogenetic technology. *J Neural Eng* **4:** S143–S156.

Arenkiel BR, Peca J, Davison IG, Feliciano C, Deisseroth K, Augustine GJ, Ehlers MD, Feng G. 2007. In vivo light-induced activation of neural circuitry in transgenic mice expressing channelrhodopsin-2. *Neuron* **54:** 205–218.

Berndt A, Yizhar O, Gunaydin LA, Hegemann P, Deisseroth K. 2009. Bi-stable neural state switches. *Nat Neurosci* **12:** 229–234.

Bi A, Cui J, Ma YP, Olshevskaya E, Pu M, Dizhoor AM, Pan ZH. 2006. Ectopic expression of a microbial-type rhodopsin restores visual responses in mice with photoreceptor degeneration. *Neuron* **50:** 23–33.

Boyden ES, Zhang F, Bamberg E, Nagel G, Deisseroth K. 2005. Millisecond-timescale, genetically targeted optical control of neural activity. *Nat Neurosci* **8:** 1263–1268.

Cardin JA, Carlén M, Meletis K, Knoblich U, Zhang F, Deisseroth K, Tsai LH, Moore CI. 2009. Driving fast-spiking cells induces gamma rhythm and controls sensory responses. *Nature* **459:** 663–667.

Cetin A, Komai S, Eliava M, Seeburg PH, Osten P. 2007. Stereotaxic gene delivery in the rodent brain. *Nat Protoc* **1:** 3166–3173.

Chow BY, Han X, Dobry AS, Qian X, Chuong AS, Li M, Henninger MA, Belfort GM, Lin Y, Monahan PE, Boyden ES. 2010. High-performance genetically targetable optical neural silencing by light-driven proton pumps. *Nature* **463:** 98–102.

Douglass AD, Kraves S, Deisseroth K, Schier AF, Engert F. 2008. Escape behavior elicited by single, channelrhodopsin-2-evoked spikes in zebrafish somatosensory neurons. *Curr Biol* **18:** 1133–1137.

Gong S, Doughty M, Harbaugh CR, Cummins A, Hatten ME, Heintz N, Gerfen CR. 2007. Targeting Cre recombinase to specific neuron populations with bacterial artificial chromosome constructs. *J Neurosci* **27:** 9817–9823.

Gradinaru V, Thompson KR, Zhang F, Mogri M, Kay K, Schneider MB, Deisseroth K. 2007. Targeting and readout strategies for fast optical neural control in vitro and in vivo. *J Neurosci* **27:** 14231–14238.

Gradinaru V, Thompson KR, Deisseroth K. 2008. eNpHR: A *Natronomonas* halorhodopsin enhanced for optogenetic applications. *Brain Cell Biol* **36:** 129–139.

Gradinaru V, Mogri M, Thompson KR, Henderson JM, Deisseroth K. 2009. Optical deconstruction of Parkinsonian neural circuitry. *Science* **324:** 354–359.

Gradinaru V, Zhang F, Ramakrishnan C, Mattis J, Prakash R, Diester I, Goshen I, Thompson K, Deisseroth K. 2010. Molecular and cellular principles for diversifying and extending optogenetics. *Cell* **141:** 1–12.

Grieger JC, Choi VW, Samulski RJ. 2006. Production and characterization of adeno-associated viral vectors. *Nat Protoc* **1:** 1412–1428.

Gunaydin L, Yizhar O, Berndt A, Sohal V, Deisseroth K, Hegemann P. 2010. Ultrafast optogenetic excitation. *Nat Neuroscience* **13:** 387–392.

Han X, Boyden ES. 2007. Multiple-color optical activation, silencing, and desynchronization of neural activity, with single-spike temporal resolution. *PLoS ONE* **2:** e299.

Han X, Qian X, Bernstein JG, Zhou HH, Franzesi GT, Stern P, Bronson RT, Graybiel AM, Desimone R, Boyden ES. 2009. Millisecond-timescale optical control of neural dynamics in the nonhuman primate brain. *Neuron* **62:** 191–198.

Huber D, Petreanu L, Ghitani N, Ranade S, Hromadka T, Mainen Z, Svoboda K. 2008. Sparse optical microstimulation in barrel cortex drives learned behaviour in freely moving mice. *Nature* **451:** 61–64.

Hwang RY, Zhong L, Xu Y, Johnson T, Zhang F, Deisseroth K, Tracey WD. 2007. Nociceptive neurons protect *Drosophila* larvae from parasitoid wasps. *Curr Biol* **17:** 2105–2116.

Ishizuka T, Kakuda M, Araki R, Yawo H. 2006. Kinetic evaluation of photosensitivity in genetically engineered neurons expressing green algae light-gated channels. *Neurosci Res* **54:** 85–94.

Kuhlman SJ, Huang ZJ. 2008. High-resolution labeling and functional manipulation of specific neuron types in mouse brain by Cre-activated viral gene expression. *PLoS ONE* **3:** e2005.

Kutner RH, Zhang XY, Reiser J. 2009. Production, concentration and titration of pseudotyped HIV-1-based lentiviral vectors. *Nat Protoc* **4:** 495–505.

Li X, Gutierrez DV, Hanson MG, Han J, Mark MD, Chiel H, Hegemann P, Landmesser LT, Herlitze S. 2005. Fast noninvasive activation and inhibition of neural and network activity by vertebrate rhodopsin and green algae channelrhodopsin. *Proc Natl Acad Sci* **102:** 17816–17821.

Lin JY, Lin MZ, Steinbach P, Tsien RY. 2009. Characterization of engineered channelrhodopsin variants with improved properties and kinetics. *Biophys J* **96:** 1803–1814.

Luo L, Callaway EM, Svoboda K. 2008. Genetic dissection of neural circuits. *Neuron* **57:** 634–660.

Nagel G, Ollig D, Fuhrmann M, Kateriya S, Musti AM, Bamberg E, Hegemann P. 2002. Channelrhodopsin-1: A light-gated proton channel in green algae. *Science* **296:** 2395–2398.

Nagel G, Szellas T, Huhn W, Kateriya S, Adeishvili N, Berthold P, Ollig D, Hegemann P, Bamberg E. 2003. Channelrhodopsin-2, a directly light-gated cation-selective membrane channel. *Proc Natl Acad Sci* **100:** 13940–13945.

Nagel G, Brauner M, Liewald JF, Adeishvili N, Bamberg E, Gottschalk A. 2005. Light activation of channelrhodopsin-2 in excitable cells of *Caenorhabditis elegans* triggers rapid behavioral responses. *Curr Biol* **15:** 2279–2284.

Petreanu L, Huber D, Sobczyk A, Svoboda K. 2007. Channelrhodopsin-2-assisted circuit mapping of long-range callosal projections. *Nat Neurosci* **10:** 663–668.

Petreanu L, Mao T, Sternson SM, Svoboda K. 2009. The subcellular organization of neocortical excitatory connections. *Nature* **457:** 1142–1145.

Pulver SR, Pashkovski SL, Hornstein NJ, Garrity PA, Griffith LC. 2009. Temporal dynamics of neuronal activation by Channelrhodopsin-2 and TRPA1 determine behavioral output in *Drosophila* larvae. *J Neurophysiol* **101:** 3075–3088.

Schroll C, Riemensperger T, Bucher D, Ehmer J, Voller T, Erbguth K, Gerber B, Hendel T, Nagel G, Buchner E, et al. 2006. Light-induced activation of distinct modulatory neurons triggers appetitive or aversive learning in *Drosophila* larvae. *Curr Biol* **16:** 1741–1747.

Sohal VS, Zhang F, Yizhar O, Deisseroth K. 2009. Parvalbumin neurons and gamma rhythms enhance cortical circuit performance. *Nature* **459:** 698–702.

Tsai HC, Zhang F, Adamantidis A, Stuber GD, Bonci A, de Lecea L, Deisseroth K. 2009. Phasic firing in dopaminergic neurons is sufficient for behavioral conditioning. *Science* **324:** 1080–1084.

Wang H, Peca J, Matsuzaki M, Matsuzaki K, Noguchi J, Qiu L, Wang D, Zhang F, Boyden E, Deisseroth K, et al. 2007. High-speed mapping of synaptic connectivity using photostimulation in Channelrhodopsin-2 transgenic mice.

Proc Natl Acad Sci **104:** 8143–8148.

Wang H, Sugiyama Y, Hikima T, Sugano E, Tomita H, Takahashi T, Ishizuka T, Yawo H. 2009. Molecular determinants differentiating photocurrent properties of two channelrhodopsins from chlamydomonas. *J Biol Chem* **284:** 5685–5696.

Zhang YP, Oertner TG. 2007. Optical induction of synaptic plasticity using a light-sensitive channel. *Nat Methods* **4:** 139–141.

Zhang F, Wang LP, Boyden ES, Deisseroth K. 2006. Channelrhodopsin-2 and optical control of excitable cells. *Nat Methods* **3:** 785–792.

Zhang F, Wang LP, Brauner M, Liewald JF, Kay K, Watzke N, Wood PG, Bamberg E, Nagel G, Gottschalk A, et al. 2007. Multimodal fast optical interrogation of neural circuitry. *Nature* **446:** 633–639.

Zhang F, Prigge M, Beyriere F, Tsunoda SP, Mattis J, Yizhar O, Hegemann P, Deisseroth K. 2008a. Red-shifted optogenetic excitation: A tool for fast neural control derived from *Volvox carteri*. *Nat Neurosci* **11:** 631–633.

Zhang YP, Holbro N, Oertner TG. 2008b. Optical induction of plasticity at single synapses reveals input-specific accumulation of αCaMKII. *Proc Natl Acad Sci* **105:** 12039–12044.

Zhang F, Gradinaru V, Adamantidis AR, Durand R, Airan RD, de Lecea L, Deisseroth K. 2010. Optogenetic interrogation of neural circuits: Technology for probing mammalian brain structures. *Nat Protoc* **5:** 439–456.

Zhao S, Cunha C, Zhang F, Liu Q, Gloss B, Deisseroth K, Augustine GJ, Feng G. 2008. Improved expression of halorhodopsin for light-induced silencing of neuronal activity. *Brain Cell Biol* **36:** 141–154.

80 Optogenetics in Freely Moving Mammals

Dopamine and Reward

Feng Zhang,[1,4] Hsing-Chen Tsai,[1,4] Raag D. Airan,[1] Garret D. Stuber,[2] Antoine R. Adamantidis,[3] Luis de Lecea,[3] Antonello Bonci,[2] and Karl Deisseroth[1,3]

[1]*Department of Bioengineering, Stanford University, Stanford, California 94305;* [2]*Department of Neurology, Ernest Gallo Clinic and Research Center, University of California San Francisco, San Francisco, California 94608;* [3]*Department of Psychiatry and Behavioral Sciences, Stanford University, Stanford, California 94305*

ABSTRACT

Brain reward systems play a central role in the cognitive and hedonic behaviors of mammals. Multiple neuron types and brain regions are involved in reward processing, posing fascinating scientific questions and major experimental challenges. Using diverse approaches including genetics, electrophysiology, imaging, and behavioral analysis, a large body of research has focused on both normal functioning of the reward circuitry and on its potential significance in neuropsychiatric diseases. In this chapter, we illustrate a real-world application of optogenetics to mammalian behavior and physiology, delineating procedures and technologies for optogenetic control of individual components of the reward circuitry. Our accompanying optogenetics chapters in this volume complement the present chapter as follows: Chapter 6 provides a review of the diversity, biophysical properties, and structure–function relationships of the microbial opsins, and Chapter 79 provides a guide for the fundamental components of optogenetic technology, including selection and expression of opsins, as well as the optical hardware for delivering light into the brains of mammals. Here we describe the experimental setup and protocol for integrating optogenetic modulation of dopamine neurons with fast-scan cyclic voltammetry, conditioned place preference, and operant conditioning to assess the causal role of well-defined electrical and biochemical signals in reward-related behavior.

INTRODUCTION

The advent of microbial and engineered opsins has enabled cell type–specific, millisecond-scale modulation of electrical and biochemical neural activity patterns in heterogeneous brain tissue in vivo. Electrical excitation and inhibition of genetically defined neurons can be achieved using chan-

[4]These authors contributed equally to this work.

nelrhodopsins and halorhodopsins, respectively, whereas intracellular biochemical signaling can be optically controlled using engineered optoXRs (opsin-receptor chimeras). Previous reports have described the microbial and engineered opsins we have developed for modulating neuronal signaling in this way (Boyden et al. 2005; Zhang et al. 2006; Zhang et al. 2007; Zhang et al. 2008; Airan et al. 2009; Berndt et al. 2009). Additionally, published protocols have described in detail the procedure for preparing the optical fiber system used to activate opsin-expressing neurons in vivo (Adamantidis et al. 2007; Aravanis et al. 2007; Zhang et al. 2010). We have also previously described a set of viral vectors (Cardin et al. 2009, 2010; Sohal et al. 2009; Tsai et al. 2009; Zhang et al. 2010) suitable for achieving cell-specific opsin expression in dopaminergic neurons, subclasses of dopamine (DA) receptor neurons, or specific bundles of axonal projection; the optical conditioned place preference (CPP) paradigms described in this chapter have also been reported previously (Airan et al. 2009; Tsai et al. 2009). The same methods with minor modification can be applied to rats, monkeys, or other mammalian animal models. For detailed discussions of considerations for rodent CPP and behavioral assays in general, consult Cunningham et al. (2006), Crawley (2007), Tzschentke (2007), and Dela Cruz et al. (2009)

GENETIC STRATEGIES FOR CONTROLLING REWARD CIRCUIT COMPONENTS

The reward system spans several layers of complexity. At the macroanatomical level, brain areas including the ventral tegmental area (VTA), prefrontal cortex (PFC), nucleus accumbens (NAc), amygdala (Amy), and habenula (Hb) form an interconnected network capable of directing reward-related behaviors. Important connection pathways include the VTA to NAc dopaminergic projections and the Amy to NAc glutamatergic projections, which are formed by distinct groups of molecularly defined neurons within each reward-related brain area. Owing to the heterogeneous nature of brain tissue, in which multiple cell types and fibers of passage are intertwined, it has been difficult to selectively modulate the activity of individual circuit components using conventional electrical, chemical, or surgical modulations. Optogenetic control of neural activity overcomes this limitation by allowing select circuit components to be genetically rendered light sensitive.

Two classes of approaches (genetic and topological) may be used to target individual cell types or axonal projections, and these techniques can also be applied in combination. In the genetic approach, molecularly defined subpopulations of neurons are selectively rendered light-sensitive by delivering conditional recombinant adeno-associated virus (rAAV) vectors carrying opsin genes (Fig. 1A) into transgenic (Tg) mice expressing Cre recombinase under cell type–specific promoter/enhancer regions (e.g., tyrosine hydroxylase [TH] for dopaminergic neurons in VTA; and Drd1 for neurons expressing the dopamine receptor D1A). A wide variety of Cre driver mouse lines already exist and may be obtained from publically accessible transgenic mouse repositories such as GENSAT, Jackson Laboratory JAX repository, or EMBO's European Mouse Mutant Archive (see Chapter 12). For example, we have used the TgTH:IRES-Cre mouse to successfully express opsins in the VTA dopaminergic neurons (Fig. 1B,C). A list of optogenetic tools and currently available Cre transgenic lines suitable for reward-related studies are presented in Table 1. In the topological approach, a specific connection between two brain regions is targeted by delivering the viral vector in region A and selectively activating the input from region A to region B by directly activating the light-sensitive axon terminals in region B. With strong expression, the axon processes of the opsin-expressing neurons will also be rendered light sensitive; even light stimulation of the dopaminergic axon terminals can directly trigger DA release. This strategy for optogenetics is versatile and generalizable as more and more Cre driver lines become available. Moreover, even with the TH::Cre mouse driver line alone, different TH+ nuclei may be targeted for virus injection, as TH is also the rate-limiting step in norepinephrine (NE) synthesis, thereby allowing use of the same mice and virus to probe locus coeruleus and solitary tract nucleus contributions to circuit and animal behavior (Carter et al. 2009).

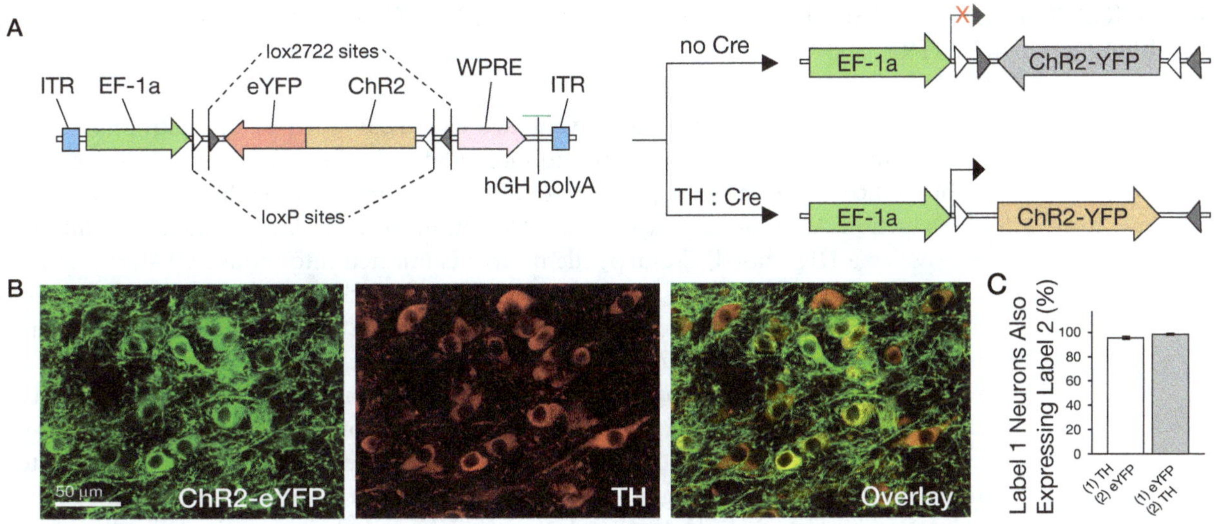

FIGURE 1. Targeted modulation of dopaminergic neurons using cell type–specific viral gene expression. (*A*) Diagram showing the double-floxed inverted opsin (DIO) adeno-associated virus (AAV) vector. The opsin gene will be conditionally inverted into a transcriptionally permissive orientation only in the presence of Cre recombinase. (*B*) Delivery of DIO ChR2-eYFP AAV vectors into the ventral tegmental area of TH::Cre transgenic mice mediates cell type–specific expression of ChR2-eYFP in DA neurons. Confocal images show that ChR2-eYFP expression (YFP: *left*, green) is restricted to TH-positive DA neurons (TH: *middle*, red); overlay shown on *right*. (*C*) Statistics of ChR2-eYFP expression in TH neurons (n = 491); error bars represent S.E.M. throughout. (Adapted from Tsai et al. 2009.)

TABLE 1. Reagents and transgenic mouse lines for targeting reward circuitry

Reagent	Purpose	Where to obtain
Conditional virus system		
Conditional ChR2-eYFP AAV vector (pAAV-EF1a-double floxed-hChR2(H134R)-eYFP-WPRE-HGHpA)	Expresses ChR2-eYFP selectively in Cre-expressing neurons	http://www.optogenetics.org or Addgene (plasmid 20298)
Conditional eNpHR-eYFP AAV vector (pAAV-double floxed-eNpHR-eYFP-WPRE-HGHpA) (eNpHR3.0 is also available; Gradinaru et al. 2010)	Expresses eNpHR-eYFP selectively in Cre-expressing neurons	http://www.optogenetics.org or Addgene (plasmid 20949)
Conditional eYFP AAV control vector (pAAV-EF1a-double floxed-eYFP-WPRE-HGHpA)	Expresses eYFP selectively in Cre-expressing neurons; used as a control vector for ChR2 and eNpHR	http://www.optogenetics.org or Addgene (plasmid 20296)
Cre transgenic mouse lines for DA neuron targeting		
Tyrosine hydroxylase (TH):IRES-Cre	Dopaminergic neurons	European Mouse Mutant Archive (EM:00254)
Dopamine transporter (DAT):IRES-Cre	Dopaminergic neurons	The Jackson Laboratory Repository (JAX Mice: 006660)
Dopamine receptor 1a(Drd1a):Cre	Drd1a neurons	GENSAT (FK150)
Dopamine receptor 2 (Drd2):Cre	Drd2 neurons	GENSAT (ER44)
Dopamine receptor 3 (Drd3):Cre	Drd3 neurons; expression restricted primarily to hippocampus (most of CA1 and dentate gyrus, less in CA2 and CA3)	GENSAT (KI198)

eYFP, enhanced yellow fluorescent protein; AAV, adeno-associated virus; DA, dopamine; GENSAT, Gene Expression Nervous System Atlas.

INTEGRATING BEHAVIORAL READOUTS WITH OPTOGENETIC MODULATION OF REWARD CIRCUITRY

Functional opsin expression may be validated with a combination of electrophysiological, immunohistochemical, and electrochemical readout methods. In the case of DA neurons, costaining with TH in brain slices of the target region is a useful method for validating the specificity and efficacy of viral transduction. As reported previously, >98% of transduced cells are indeed bona fide TH+ (DA) neurons when a DIO-based, Cre-dependent virus is injected into mouse VTA (Fig. 1; Tsai et al. 2009). Immunohistochemistry can also be used to validate both cannula targeting and efficacy of photostimulation of the targeted cell population with c-fos costaining; in this case, transcardiac perfusion is conducted >1 h after a bout of photostimulation. Next, living acute slice recordings are important with whole-cell patch clamp, targeted to opsin-expressing cells identified by XFP (fluorescent protein) fusion protein expression, to ensure that expression at a given timepoint and virus injection condition (titer, promoter, serotype) is strong enough to drive spiking in response to a given light intensity, duration, wavelength, and pulse frequency (in addition to providing additional cell type validation by virtue of identifying electrophysiological characteristics). Finally, as input resistance can be different in vitro and in vivo, additional in vivo validation is important using optrodes (optical fibers fused to tungsten recording electrodes; Gradinaru et al. 2007, 2009; Tsai et al. 2009) and voltammetry (Tsai et al. 2009; discussed in detail below) to map optimal stimulation patterns (e.g., light intensity, pulse frequency, and pulse duration). For example, it is important to test a range of light intensity and pulse durations to find parameters that achieve reliable stimulation or inhibition, while minimizing the light intensity and duration to avoid potential heating and photodamage.

When the expression conditions are well established, behavioral experiments can be explored in detail. Generally important considerations for animal behavioral studies include choice of species, strain, gender, age, housing condition, handling, and sample size (Cunningham et al. 2006; Deacon 2006; Crawley 2007; Tzschentke 2007; Dela Cruz et al. 2009). Because rodents are nocturnal, in general, tests that require the subjects to be spontaneously active and explore the apparatus for prolonged period of time are preferentially performed during the dark phase of the light cycle. Although no difference has been reported between CPP studies conducted during the light or dark phases of the light cycle, operant conditioning, a more active form of conditioning testing, is usually conducted during the dark phase. It is also important that experimenters handle each test subject before behavioral testing, so as to habituate test subjects to the testing environment and conditions.

For reward-related behavioral testing, we recommend conducting a pilot behavioral test using mice of ~8 wk of age with a cohort size of 10–12 for each testing condition. At this age, mouse skulls have matured to adult size, facilitating precise brain region targeting via stereotactic surgery, although we still recommend first testing all stereotactic coordinates in the chosen mouse strain through dye injection. As there are subtle variations between different mouse strains, it is important to make adjustments to the coordinates empirically. In addition, it is important to visually observe the test subjects after surgical procedures to check for any abnormal behavior (e.g., circling behavior, asymmetrical posture). Control cohorts should also be carefully designed for the pilot study. For example, in our previous studies (Adamantidis et al. 2007; Airan et al. 2009; Tsai et al. 2009) we used several control cohorts including control viral vector injected into TH::Cre animals (i.e., recombinant viral vector carrying only the fluorescent protein marker without the opsin gene), as well as cohorts with active viral vector injected into Cre-negative littermates, to test for surgical, viral, or light-induced side effects.

It usually takes 1 wk for the mice to recover from the surgery, and at least 3 wk for the virus to be functionally expressed. Preliminary experiments to characterize the time course of expression are recommended. During recovery, mice are usually single or pair-housed to prevent fighting-related damage to the implanted cannulae. All mice are habituated to the testing conditions (behavioral setup and handling) daily for 10 d before any behavioral assay to reduce the stress level. We recommend conducting a 3-min open-field test on the test and control cohorts using the same stimulation paradigm to determine any potential stimulation-induced side effects on stereotypic behavior, exploration, locomotor activity, and anxiety. We also recommend testing optical stimulation in the experimental cohort (Cre+ mice with active opsin injection) using light of a different wavelength that does not excite the opsin (e.g., 630-nm red light for ChR2). When applicable, positive control experiments can also be conducted using electrical stimulation, pharmacological intervention, or natural reward. It is important to also discern light-induced, nonreward-related behavioral changes from reward-related behaviors to rule out potential confounding parameters during interpretation of the behavioral results.

As noted above, on completion of a behavioral assay, all experimental subjects need to be analyzed to verify cannula location, viral expression, and any unusual brain damage or infection. Additionally, for in vivo experiments, it is also helpful to characterize the volume of activation to ensure that a sufficient number of cells are activated by the light stimulation. This can be done in two ways. First, acute brain sections of different thickness prepared from the brain region of interest can be used to determine the laser output necessary to maintain a minimum power level of ~1 mW/mm^2 throughout the target brain region (Adamantidis et al. 2007; Aravanis et al. 2007). Second, as noted above, immunohistochemical characterization of immediate early gene activation (e.g., c-fos) can also be used to determine the volume of activation (Adamantidis et al. 2007; Gradinaru et al. 2009).

Reward-related behaviors in rodents can be assessed using a number of behavioral tests, including classical and operant conditioning. Classical conditioning usually takes the form of Pavlovian conditioning or CPP, by establishing an association between two stimuli or between a behavior and a consequence. Operant conditioning, also known as instrumental conditioning, differs from classical conditioning in that the subject uses consequences to modify its behavior and environment. In operant conditioning, because the behavior immediately leads to the consequences, the parameters of this particular behavior (occurrence, rate, frequency) are monitored across repeated training sessions and used to assess each subject's reward processing. In both classical and operant conditioning, anything that increases the tracked behavior (i.e., makes it more frequent, stronger, or more likely to occur) is termed a positive reinforcer, as animals will work to strengthen the behavior that leads to these consequences. In contrast, anything that decreases the behavior (makes it less frequent, weaker, or less likely to occur) is termed a punisher or negative reinforcer, as animals will not repeat the behaviors that cause these consequences.

Conditioned Place Preference

CPP is a form of Pavlovian conditioning routinely used to study the rewarding or aversive motivational effects of objects or experiences in laboratory animals (Cunningham et al. 2006; Crawley 2007; Tzschentke 2007; Dela Cruz et al. 2009). Standard CPP protocols involve three phases: pretest (the subject explores a three-compartment chamber freely), repeated pairing (conditioning of an association between the stimulus, such as a drug of abuse and one of the two distinct chambers), and post-test (the subject is allowed to move freely). Conditioned preference is measured as the time spent in the chamber paired with stimulus before and after conditioning. We have adapted this behavioral test for use with optogenetic modulation, both in classical (Tsai et al. 2009) and modified (Airan et al. 2009) place preference conditioning.

The advantage of CPP over other behavioral assays includes its methodological simplicity (simple, relatively inexpensive equipment) and short timeline (usually ≤2 wk; sometimes one simple drug pairing exposure is sufficient to produce CPP). Our CPP apparatus consists of a rectangular cage with a left chamber measuring 27.5 x 21 cm with black walls and a black, smooth plastic floor; a center chamber measuring 15.5 x 21 cm with gray walls and a metal grill floor; and a right chamber measuring 27.5 x 21 cm with white walls and a punched metal floor (San Diego Instruments, Inc., San Diego, CA). Mouse location within the chamber during each preference test is monitored using a computerized photobeam system. The apparatus can be designed so that mice do not have a consistent marked bias for a particular chamber, and any mouse with a strong initial preference or avoidance for left or right chamber (by spending >75% or <25% of test time in either chamber) is discarded to ensure a relatively unbiased CPP (to prevent unlearned or unconditioned preferences for either chamber from confounding the CPP readouts). In this way, "ceiling" or "floor" effects are reduced during quantification.

In our previous study (Tsai et al. 2009), we used classical CPP to test the reinforcing properties of phasic DA firing in the absence of other rewards. As with drug pairing CPP, we paired each chamber with a specific optical stimulation of VTA at either phasic or tonic frequency (1 Hz tonic or 50 Hz phasic; the same number of light pulses was delivered in each condition). The conditioning protocol consisted of three phases of >4 d (Fig. 2A). On day 1, individual mice were placed in the center chamber and allowed to freely explore the entire apparatus for 15 min (pretest). Mice were conditioned for two consecutive days. In the first conditioning session (day 2), mice were confined to one of the side chambers for 30 min and paired with optical stimulation (1 Hz, 50 Hz, or 0 Hz depending on the experiment). In the second session (day 3), mice were confined to the other side chamber for 30 min with a different stimulation parameter. On day 4, as in day 1, mice were placed in the center chamber and allowed to freely explore the entire apparatus for 15 min (post-test). Both

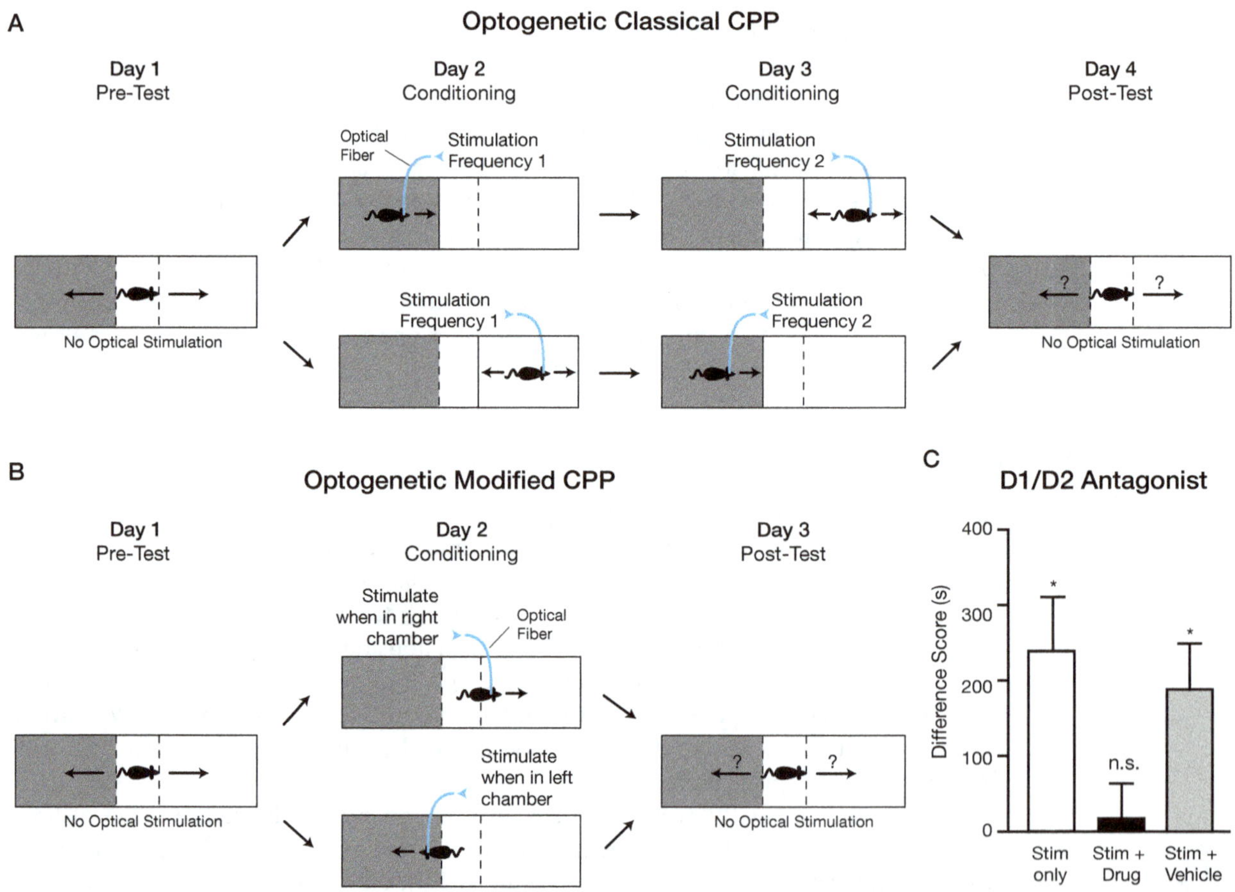

FIGURE 2. Combining optogenetic stimulation with conditioned place preference (CPP). (*A*) Conditioning protocol for classical CPP (see text for description). (*B*) Conditioning protocol for modified CPP (see text for description). (*C*) Optical stimulation-mediated CPP effect is abolished when a D1/D2 receptor antagonist cocktail (*middle*: stimulus + drug) containing SCH23390 (4 μg/μL) and raclopride (8 μg/μL) is bilaterally delivered into the NAc (0.2 μL per hemisphere) during phasic conditioning (30 min of phasic stimulation: 25 light flashes delivered at 50 Hz per min at 15 msec pulse width). The antagonist is delivered 5 min before conditioning via a bilateral cannula implanted above the NAc. The control cohort receiving vehicle administration (*right*: stimulus + vehicle) showed comparably strong conditioning effect toward the stimulation chamber as the stimulation only cohort (*left*: stimulus only). (*A*, Adapted from Tsai et al. 2009.)

post-test and pretest were conducted identically. Measurement of the reinforcing properties of VTA DA firing was achieved by quantification of the time spent in each of the compartments on day 4 (post-test) compared to day 1 (pretest), and by calculating preference score, revealing that phasic DA neuron stimulation was much more potent than tonic (even for the same number of light flashes) in driving reward learning. Notably, optogenetic stimulation in this way can be combined with pharmacological intervention to test the contribution of specific signaling pathways underlying the optically evoked behavior. For example, we have found that simultaneous administration of D1 and D2 antagonists in the downstream NAc abolished these reinforcing properties of phasic stimulation of VTA DA neurons (Fig. 2C).

In a separate study, we used a modified version of CPP (Fig. 2B) to test the involvement of intracellular biochemical signaling (via optoXRs) within NAc neurons in mediating reward (Airan et al. 2009). We targeted optical stimulation to transduced NAc as part of a three-day operant CPP assay. For in vivo optical stimulation equipment, including fiber preparation, see Chapter 79, and for complete detailed protocols, see Zhang et al. (2010). On each day of the experimental test, animals were allowed to freely explore the place preference apparatus. On day 1, animals freely explored the apparatus without optical stimulation. On day 2, whenever the animal freely entered the designated conditioned chamber, a laser-diode-coupled optical fiber registered to the transduced region delivered light pulses at 10 Hz to approximate the likely intensity of monoaminergic input during strong reward. On day 3, animals again freely explored the apparatus without optical stimulation. The time spent in the conditioned chamber was quantified, revealing potent rewarding properties of driving G-protein coupled signaling through the phosphatidylinositol pathway (Gq) in accumbens neurons.

Operant Conditioning

Operant conditioning tests the motivation of a subject to obtain substances such as nicotine and cocaine, and the behavioral reinforcing properties of such compounds (Wise 2004). During testing, subjects are conditioned to perform an operant behavior (e.g., lever press, nose poke, or wheel turn) based on the consequence (e.g., delivery of food, drug, liquid, or other rewards) (Kalivas and McFarland 2003). The operant conditioning system can be obtained from Med Associates Inc. (St. Albans, VT). Once the operant behavior is acquired and stable, the reinforcing properties of a reward stimulus can be measured using different reinforcement schedules (Kalivas and McFarland 2003). For example, in a progressive ratio schedule, a progressively increasing number of operant behaviors are required to trigger reward delivery. In addition, when the subject is conditioned in the absence of reward delivery, a previously reinforced behavior might decrease and eventually cease through a phenomenon known as extinction. Previously extinguished behavior can be reinstated using cues, stressors, or other modalities (Nair et al. 2009).

Optogenetic stimulation can be used to modulate individual reward circuit components during various phases of operant conditioning to dissect reward processing in the brain. Optical stimulation equipment can be easily integrated with standard operant conditioning instruments; for example, operant cages can be modified to accommodate the optical fiber used to deliver light into the brain of freely moving animals during the conditioning sessions. We have designed swivels and fiber commutators to automatically relieve torsion in the fiber (Gradinaru et al. 2007; Zhang et al. 2010); if needed, the experimenter can manually commutate a long fiber as well. The optical fiber can be easily placed on the test subjects before each session and removed immediately afterward. Notably, as with classical conditioning, all test subjects should be habituated to the behavioral testing setup and handling, and a sham optical fiber (i.e., optical fiber with no fiber after the cap; see Chapter 79) should be used during no-stimulation sessions to habituate the test subjects before actual optogenetic stimulation experiments. Optical control can then be used to test the sufficiency or the necessity of the distinct reward circuit component in the acquisition, extinction, and reinstatement of operant conditioning. In each case, light sources can be linked to the active lever and triggered light patterns can be delivered through the optical fiber. Alternatively, natural or artificial rewards can be combined with optical modulation to test the causal role of targeted reward circuit components in previously established addiction. Quantification of the effect of optical modulation on operant conditioning can be measured according to the number of active versus inactive lever presses during conditioning or extinction sessions.

FUNCTIONAL CIRCUIT INTERROGATION DURING OPTICAL CONTROL

Measurement of DA release in target regions such as the NAc can be performed in conjunction with optogenetic stimulation and inhibition of neural circuit elements (Tsai et al. 2009). Although optical stimulation techniques can be performed with in vivo microdialysis to quantify tonic levels of DA, microdialysis lacks the temporal resolution to resolve DA fluctuations induced by a transient burst of neuronal activation that would be optically induced via activation of ChR2 to mimic endogenous dopaminergic activity. To circumvent this, we have used fast-scan cyclic voltammetry (FSCV) to monitor fluctuations in DA signaling associated with optogenetic excitation of midbrain DA neurons. FSCV can be used to directly measure local and transient DA release in vivo (Fig. 3A,B) (Phillips et al. 2003b; Stuber et al. 2008). For example, in the anesthetized mouse, a carbon-fiber electrode can be placed in the NAc, a major DA neuron target, while using light to stimulate ChR2-expressing neurons in the VTA. By coupling FSCV with optical stimulation in vivo, it is possible to correlate electrical signaling in the form of spike frequency with chemical signaling in the form of DA concentrations. In our previous study (Tsai et al. 2009), we found that although high-frequency phasic firing in VTA DA neurons led to appreciable levels of transient DA release, low-frequency tonic firing did not lead to significant transient DA release in the NAc (Fig. 2B). Voltammetric techniques can also be used in brain slices to optically evoke DA release by selectively stimulation dopaminergic nerve terminals that are expressing opsin variants (Fig. 3C,D). Our FSCV measurement system was obtained from Scott Ng-Evans at the Electronics and Materials Engineering Shop at the University of Washington (ngevans@uw.edu) and the carbon-fiber electrodes from Cypress Systems (Chelmsford, MA; cat. no. 66-EE017).

Voltammetric measurement of DA release provides superb temporal resolution to resolve rapid changes in neurotransmitter concentration; for detailed descriptions of the methodology see Phillips et al. (2003a), Phillips et al. (2003b), and Phillips and Wightman (2004). The detection of DA at carbon-fiber microelectrodes used in voltammetric experiments relies on the electroactive properties of DA. Typically, a triangular voltage waveform (−0.4 V to +1.3 V to −0.4 V over 10 msec) is applied to the carbon-fiber electrode; the Ag/AgCl reference electrode is used to detect current caused by the oxidation and subsequent reduction of DA. The triangle waveform is typically repeated every 100 msec to resolve subsecond changes in DA concentration. The current measured at the carbon fiber as a function of applied voltage can be used to identify changes in current caused by the oxidation of DA, by comparing voltammograms seen in vivo with voltammograms obtained with standard concentrations of DA in vitro.

Optical Measurement of Dopamine Nerve Terminals in Brain Slices

Reliable DA measurements can be obtained by optically stimulating dopaminergic nerve terminals expressing opsin variants. Light capable of activating opsins can be delivered to brain slices placed in temperature-controlled recording chambers via an LED or laser diode coupled to a 50–200-µm optical fiber terminating directly above the brain slice. Alternatively, short-duration optical pulses can be introduced through the epifluorescence port of a microscope via a high-speed wavelength-switching device such as the Lambda DG-4 (Sutter Instruments), which uses a full-spectrum xenon lamp source with interchangeable excitation filters to set the wavelength. Detection of DA transients induced by this light activation of DA terminals can be detected using custom voltammetry equipment (see below) or using standard patch-clamp amplifiers such as the Axon 700B (Molecular Devices) fitted with commercially available headstages suitable for measuring the nanoampere-scale currents observed in FSCV experiments.

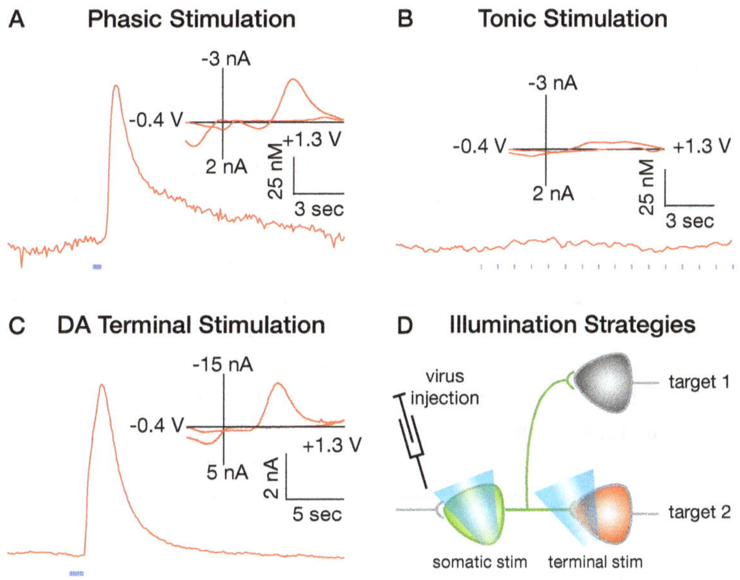

FIGURE 3. Optical strategies for targeting specific reward circuit components. (*A,B*) In vivo optical stimulation of the DA neurons in the ventral tegmental area (VTA) of the anesthetized mouse. Phasic 50-Hz light trains evoked robust transient dopamine (DA) release in the NAc, whereas tonic 1 Hz light trains did not evoke significant transient DA release, measured using fast-scan cyclic voltammetry (FSCV). Blue bars indicate the train of 20-Hz, 15-msec light flashes used for stimulation. (*C*) In vitro optical stimulation of the ChR2-eYFP-expressing DA terminals in an acute nucleus accumbens (NAc) slice (where no DA neuron cell bodies are present) leads to robust DA transient release. (*D*) In general, as in *C*, ChR2-expressing DA neurons can be either activated at the soma or at the axon terminals. The latter approach further enhances the specificity of the reward circuit component being modulated, by restricting modulation to a subset of downstream brain structures innervated by the target cell type. (*A,B*, Adapted from Tsai et al. 2009.)

Optical Measurement of Dopamine Release In Vivo

Optical stimulation of DA neurons expressing opsin variants can be performed to measure DA release in terminal fields (Tsai et al. 2009). In the anesthetized preparation, mice expressing ChR2 in TH-positive neurons of the midbrain are anesthetized and placed in a stereotaxic frame. An optical fiber coupled to a solid-state laser diode is positioned directly above midbrain DA neurons and either a commercially available or custom-made carbon-fiber microelectrode is positioned in dopaminergic terminal fields such as the dorsal or ventral striatum. The advantage of this preparation over brain slice experiments is that brain circuitry remains intact while selective perturbation of DA neuronal activity is conducted to assay release or postsynaptic effects.

For selective optical control of dopaminergic activity in freely moving mice during behavioral experiments, a few technical modifications are required. The headstage used to amplify electrochemical signals must be miniaturized to secure its placement on the mouse skull. This can be accomplished by using commercially available surface-mount operational amplifiers and associated headstage circuitry encased within electrically inert epoxy, which can be attached to the headcap on the skull by small gold pins or other miniaturized connectors. The reference electrode must be scaled down and chronically implanted in the brain hemisphere contralateral to the site of DA measurement. This is performed by cutting a small (3–4 mm) piece of silver wire, soaking it in 70% bleach for 15–30 min to generate AgCl, and then attaching the wire by epoxy to a small gold pin that can be chronically implanted. Importantly, chronic implantation of custom-made carbon-fiber electrodes is recommended instead of acute insertion of electrodes through a guide cannula (Gan et al. 2010). This approach keeps the implant size secured on the skull at a mass of <1 g (including fiber optic components), maintaining the capability of the experimental subject to explore and behave in largely unrestricted fashion. Both voltammetric data acquisition and optical stimulation can be integrated with commercially available behavioral test equipment such as operant chambers to synchronize behavioral events (e.g., lever presses) with laser stimulation pulses and DA measurements.

OUTLOOK

Together the methods, tools, and technologies we have developed and described here and in Chapters 6 and 79 allow millisecond-scale optical control and quantitative validation of activity within well-defined neurons in freely moving mammals (as illustrated in this chapter for optogenetic modulation of DA neurons). The intrinsic generalizability of these methods delivers the broad and versatile capability to assess the causal role of well-defined electrical and biochemical signals, in specific cell types, in mammalian circuit physiology.

At present more than 600 neuroscience laboratories are working with microbial opsins, fulfilling the initial promise of optogenetics. In 1979, Francis Crick suggested that the major challenge facing neuroscience was the need to control one cell type while leaving the others unaltered. He later speculated in lectures that light might be a relevant tool, but how this could be performed remained unclear. Now, with the capability to control millisecond-scale activity patterns in defined cell types deep within the brain of freely moving mammals, as shown here, it is now feasible to probe the causal role of circuit activity patterns in the brains of mammals and other animals, both in health and disease.

REFERENCES

Adamantidis AR, Zhang F, Aravanis AM, Deisseroth K, de Lecea L. 2007. Neural substrates of awakening probed with optogenetic control of hypocretin neurons. *Nature* **450:** 420–424.

Airan RD, Thompson KR, Fenno LE, Bernstein H, Deisseroth K. 2009. Temporally precise in vivo control of intracellular signalling. *Nature* **458:** 1025–1029.

Aravanis AM, Wang LP, Zhang F, Meltzer LA, Mogri MZ, Schneider MB, Deisseroth K. 2007. An optical neural interface: In vivo control of rodent motor cortex with integrated fiberoptic and optogenetic technology. *J Neural Eng* **4:** S143–S156.

Berndt A, Yizhar O, Gunaydin LA, Hegemann P, Deisseroth K. 2009. Bi-stable neural state switches. *Nat Neurosci* **12:** 229–234.

Boyden ES, Zhang F, Bamberg E, Nagel G, Deisseroth K. 2005. Millisecond-timescale, genetically targeted optical control of neural activity. *Nat Neurosci* **8:** 1263–1268.

Cardin JA, Carlen M, Meletis K, Knoblich U, Zhang F, Deisseroth K, Tsai LH, Moore CI. 2009. Driving fast-spiking cells induces gamma rhythm and controls sensory responses. *Nature* **459:** 663–667.

Cardin JA, Carlén M, Meletis K, Knoblich U, Zhang F, Deisseroth K, Tsai L-H, Moore CI. 2010. Targeted optogenetic stimulation and recording of neurons in vivo using cell-type-specific expression of Channelrhodopsin-2. *Nat Protoc* **5:** 247–254.

Carter ME, Adamantidis A, Ohtsu H, Deisseroth K, de Lecea L. 2009. Sleep homeostasis modulates hypocretin-mediated sleep-to-wake transitions. *J Neurosci.* **29:** 10939–10949.

Crawley JN. 2007. *What's wrong with my mouse: Behavioral phenotyping of transgenic and knockout mice.* Wiley-Liss, New York.

Cunningham CL, Gremel CM, Groblewski PA. 2006. Drug-induced conditioned place preference and aversion in mice. *Nat Protoc* **1:** 1662–1670.

Deacon RM. 2006. Housing, husbandry and handling of rodents for behavioral experiments. *Nat Protoc* **1:** 936–946.

Dela Cruz AM, Herin DV, Grady JJ, Cunningham KA. 2009. Novel approach to data analysis in cocaine-onditioned place preference. *Behav Pharmacol* **20:** 720–730.

Gan JO, Walton ME, Phillips PE. 2010. Dissociable cost and benefit encoding of future rewards by mesolimbic dopamine. *Nat Neurosci.* **13:** 25–27.

Gradinaru V, Thompson KR, Zhang F, Mogri M, Kay K, Schneider MB, Deisseroth K. 2007. Targeting and readout strategies for fast optical neural control in vitro and in vivo. *J Neurosci* **27:** 14231–14238.

Gradinaru V, Mogri M, Thompson KR, Henderson JM, Deisseroth K. 2009. Optical deconstruction of Parkinsonian neural circuitry. *Science* **324:** 354–359.

Gradinaru V, Zhang F, Ramakrishnan C, Mattis J, Prakash R, Diester I, Goshen I, Thompson K, Deisseroth K. 2010. Molecular and cellular principles for diversifying and extending optogenetics. *Cell* **141:** 1–12.

Kalivas PW, McFarland K. 2003. Brain circuitry and the reinstatement of cocaine-seeking behavior. *Psychopharmacology* **168:** 44–56.

Nair SG, Adams-Deutsch T, Epstein DH, Shaham Y. 2009. The neuropharmacology of relapse to food seeking: Methodology, main findings, and comparison with relapse to drug seeking. *Prog Neurobiol* **89:** 18–45.

Phillips PE, Wightman RM. 2004. Extrasynaptic dopamine and phasic neuronal activity. *Nat Neurosci* **7:** 199; author reply 199.

Phillips PE, Robinson DL, Stuber GD, Carelli RM, Wightman RM. 2003a. Real-time measurements of phasic changes in extracellular dopamine concentration in freely moving rats by fast-scan cyclic voltammetry. *Methods Mol Med* **79:** 443–464.

Phillips PE, Stuber GD, Heien ML, Wightman RM, Carelli RM. 2003b. Subsecond dopamine release promotes cocaine seeking. *Nature* **422:** 614–618.

Sohal VS, Zhang F, Yizhar O, Deisseroth K. 2009. Parvalbumin neurons and gamma rhythms enhance cortical circuit performance. *Nature* **459:** 698–702.

Stuber GD, Klanker M, de Ridder B, Bowers MS, Joosten RN, Feenstra MG, Bonci A. 2008. Reward-predictive cues enhance excitatory synaptic strength onto midbrain dopamine neurons. *Science* **321:** 1690–1692.

Tsai HC, Zhang F, Adamantidis A, Stuber GD, Bonci A, de Lecea L, Deisseroth K. 2009. Phasic firing in dopaminergic neurons is sufficient for behavioral conditioning. *Science* **324:** 1080–1084.

Tzschentke TM. 2007. Measuring reward with the conditioned place preference (CPP) paradigm: Update of the last decade. *Addict Biol* **12:** 227–462.

Wise RA. 2004. Dopamine, learning and motivation. *Nat Rev Neurosci* **5:** 483–494.

Zhang F, Wang LP, Boyden ES, Deisseroth K. 2006. Channelrhodopsin-2 and optical control of excitable cells. *Nat Methods* **3:** 785–792.

Zhang F, Wang LP, Brauner M, Liewald JF, Kay K, Watzke N, Wood PG, Bamberg E, Nagel G, Gottschalk A, et al. 2007. Multimodal fast optical interrogation of neural circuitry. *Nature* **446:** 633–639.

Zhang F, Prigge M, Beyriere F, Tsunoda SP, Mattis J, Yizhar O, Hegemann P, Deisseroth K. 2008. Red-shifted optogenetic excitation: A tool for fast neural control derived from *Volvox carteri*. *Nat Neurosci* **11:** 631–633.

Zhang F, Gradinaru V, Adamantidis AR, Durand R, Airan RD, de Lecea L, Deisseroth K. 2010. Optogenetic interrogation of neural circuits: Technology for probing mammalian brain structures. *Nat Protoc* **5:** 439–456.

81 In Vivo Calcium Recordings and Channelrhodopsin-2 Activation through an Optical Fiber

Helmuth Adelsberger, Christine Grienberger, Albrecht Stroh, and Arthur Konnerth

Institute of Neuroscience, Technical University Munich, 80802 Munich, Germany

ABSTRACT

In this chapter, we describe an approach for the fluorometric monitoring of population activity in neurons combined with the activation of optogenetic actuators in vivo.

INTRODUCTION

In many instances (e.g., when recording oscillatory wave activity from simultaneously active cells), useful information can be extracted from monitoring the activity of large neuronal populations. Therefore, we have developed a fluorescence detection system consisting of a thin multimode fiber used both for delivering the excitation light and for collecting the emitted fluorescence signals (Adelsberger et al. 2005). When combined with multicell bolus loading of calcium indicators (Stosiek et al. 2003; see also Chapter 49), this approach allows the in vivo recording of brain Ca^{2+} signals from a local cluster of coactive neurons. Such recordings may be performed in any brain region in both anesthetized and nonanesthetized animals. Importantly, the recordings can be readily combined with the stimulation of light-activated ion channels, such as channelrhodopsin-2 (ChR2) (Boyden et al. 2005), allowing the monitoring of the local calcium signals evoked by optogenetic stimulation. The feasibility of this approach was tested by performing ChR2 stimulation–calcium signal detection experiments in the mouse visual cortex in vivo.

Protocol

In Vivo Calcium Recordings and Channelrhodopsin-2 Activation through an Optical Fiber

IMAGING SETUP

A diagram of the recording and stimulation setup is shown in Figure 1A. The light for excitation of the fluorescent calcium sensitive dye, such as Oregon Green-488 BAPTA-1 (OGB-1), and activation of the channelrhodopsin ChR2 is delivered by a 10-mW solid-state laser (Sapphire, Coherent, Inc.) with a wavelength of 488 nm. The rapid control of the intensity of the laser beam is achieved by an acousto-optic modulator (AOM; 3080-125, Crystal Technology). After deflection by a dichroic mirror (Fig. 1A), the beam is focused by means of a collimator into one end of a multimode fiber (Thorlabs, Inc.) with a diameter of ≤200 µm. The emitted fluorescence light is guided back through the same fiber and focused on an avalanche photodiode (APD; S5343, Hamamatsu Photonics) with an aperture of 1 mm. The recorded fluorescence signals are digitized with a sampling frequency of 2000 Hz using a multifunction I/O device USB-6221 (National Instruments) and custom-written LabView-based software. The fluorescence signals were low-pass filtered offline with frequencies of 5–20 Hz.

MATERIALS

CAUTION: See Appendix 6 for proper handling of materials marked with <!>.

Reagents

Anesthetics appropriate for the species under investigation (see, e.g., Flecknell 2000)

For mice, we mostly use inhalation anesthesia with 0.8%–1.5% isoflurane <!> in pure O_2 <!> or intraperitoneal injection of ketamine <!>/xylazine <!> (0.1/0.01 mg/g body weight). Xylocaine <!> is used for local anesthesia.

Cyanoacrylate glue <!>

Eye protection cream (Isopto-Max, Alcon Pharma)

Mice for analysis

For combined calcium recordings and optogenetic activation, use animals expressing optogenetic actuators (e.g., ChR2) in the brain area of interest.

Reagents and equipment for multicell bolus loading of the brain tissue with calcium-sensitive dye OGB-1 AM (see Chapter 49)

Equipment

Biomonitoring system for monitoring body temperature, heartbeat, and breathing rates (e.g., AD Instruments)

Diamond Wedge scribe (Thorlabs, Inc.)

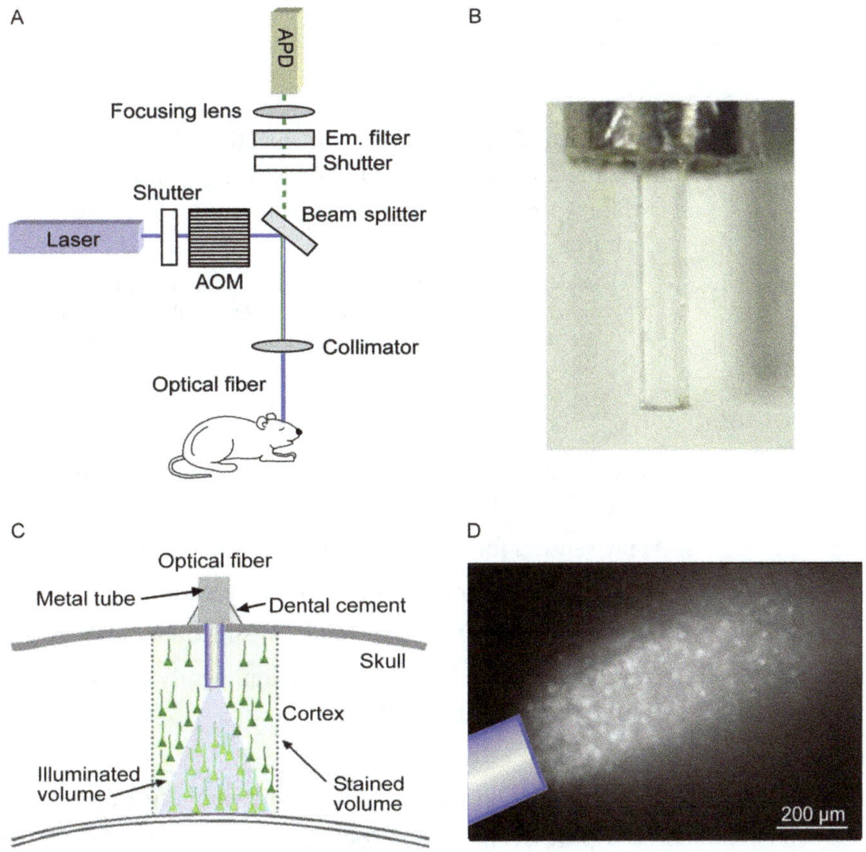

FIGURE 1. *(A)* Scheme of the optical fiber recording setup. (AOM) Acousto-optic modulator; (APD) avalanche photodiode; (Em) emission. *(B)* Image showing the tip of an optical fiber (diameter 200 µm) glued into a metal tube. *(C)* Illustration of the fiber tip inserted into the brain region with neurons labeled by the multicell bolus loading technique with calcium-sensitive dye (green). The blue cone indicates the volume in which the dye in the loaded neurons is excited by the laser light. *(D)* The tip of an optical fiber was placed perpendicular to an acute cortical slice in which the neurons were stained with Oregon Green-488 BAPTA (OGB-1) AM. The image was acquired using a charge-coupled device (CCD) camera and shows the neuron-containing fiber-illuminated cortical region.

Fiber-stripping tool (Thorlabs, Inc.)
Imaging setup, as described above
Materials for the fixation of the optical fiber in the skull
- Syntac Adhesive
- Syntac Primer
- Tetric dental cement (Ivoclar Vivadent) Total Etch (Heraeus Kulzer)
- UV lamp <!>

Optical fiber

> Typically, multimode fibers with a diameter of 200 µm and a numerical aperture of 0.48 are most commonly used. However, thinner (100-µm) or thicker (500-µm) fibers may be implanted by using the same protocol.

Stereotaxic device with ear bars
Surgical equipment (dental drill, fine scissors, fine forceps, needles)
Syringe needle (see Step 12)
Warming plate (to keep the animal's body temperature constant)

EXPERIMENTAL METHOD

Anesthesia

1. Anesthetize the animal either by injection of ketamine/xylazine intraperitoneally or by inhalation of isoflurane in pure O_2. When using isoflurane, we recommend the placement of the animal for 10 min in a closed box containing 2% isoflurane before its fixation in the stereotaxic apparatus.

   The depth of anesthesia needed for surgery is indicated by the loss of the reflexes (e.g., eyelid reflex, pinch withdrawal). At this stage, inhalation of anesthesia with isoflurane concentrations of 0.8%–1.2% is sufficient to keep the mouse at respiration rates of 80–130/min.

Preparation of the Skull

2. Place the anesthetized mouse onto a warming plate (37°C) and fix the head in a stereotaxic frame with ear bars.
3. Apply sensors for the control of the body temperature and breathing rates.
4. Apply eye protection cream onto the eyes to protect them from drying.
5. Inject 100 µL 2% Xylocaine solution subcutaneously at the site of brain surgery.
6. Ten minutes after Xylocaine injection, gently remove the skin and muscles above the desired brain region with fine scissors.
7. Dry the skull with warm air.
8. Perform the following steps to ensure stable fixation of the optic fiber:
   i. Apply a thin film of Total Etch for precisely 15 sec onto the exposed region.
   ii. Rinse the skull with water and dry with air.
   iii. Apply Syntac Primer for 15 sec and dry with air again.
   iv. Apply Syntac Adhesive for 10 sec followed by air drying.
9. Use the dental drill to thin the skull at the site of recording. Create a small opening with a diameter of ~0.5 mm by gently using fine forceps and needles.
10. Perform multicell bolus loading of the brain region of choice by using the protocol outlined in Chapter 49.

Implantation and Fixation of the Optical Fiber

11. Remove the coating from the end of the fiber with a standard fiber-stripping tool.
12. Prepare a thin metal tube with the appropriate inner diameter and a length of ~10 mm from a syringe needle.
13. Pull the metal tube over the fiber and fix it with cyanoacrylate glue.
14. Trim the protruding short end of the optic fiber (~1 cm length) with a Diamond Wedge scribe to a length corresponding to the depth of insertion into the brain (Fig. 1B). For example, trim to 200 µm when recording from a layer two-thirds into the mouse cortex.
15. Place the optic fiber transiently into a manipulator and use the site covered with the metal tube for fixation.
16. Thirty minutes after multicell bolus loading of the calcium-sensitive dye (e.g., OGB-1), insert the tip of the optic fiber gently into the brain through the skull opening.

    The optimal depth corresponds to the site with the highest fluorescence intensity.

    See *Troubleshooting*.

17. Apply Tetric dental cement around the metal tube onto the skull (Fig. 1C) and harden it with UV light for 20 sec. The fiber can now be released from the micromanipulator.

18. For the recording of calcium signals, increase the laser beam to an intensity such that it is just below the level in which bleaching is detected (typical range is 0.01–0.1 mW at the tip of the optical fiber).

    See *Troubleshooting.*

Calcium Recording and Channelrhodopsin ChR2 Activation

19. For combined calcium recordings and optogenetic activation, apply the above-described procedures to animals expressing optogenetic actuators (e.g., ChR2) in the brain area of interest (Fig. 2).

    ChR2 stimulation can be performed in combination with fluorometric calcium recordings and also without calcium recordings.

20. For combined calcium recording–ChR2 stimulation experiments, increase the excitation light intensity approximately 20 to 80 times for the period of stimulation.

    Because of the saturation of the light detected by APD, calcium recordings are not possible during ChR2 stimulation.

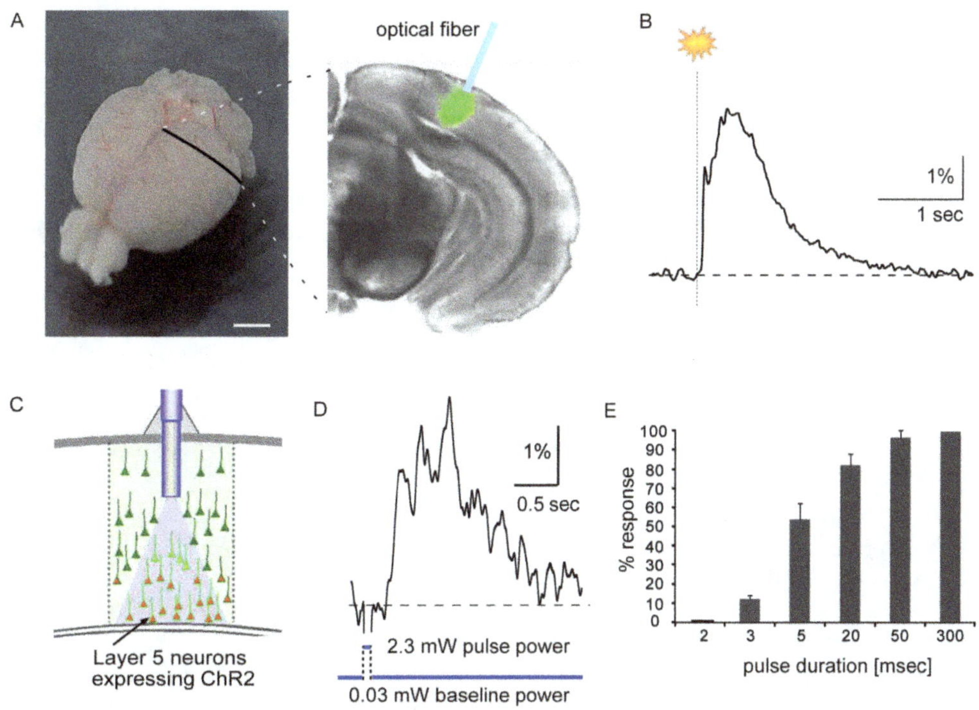

FIGURE 2. (*A*) (*Left*) Photograph of a mouse brain indicating the site of post-hoc sectioning at the level of the visual cortex. Scale bar, 2 mm. (*Right*) Section through the primary visual cortex indicating the OGB-1 AM–stained region (green in false color) and the position of the optical fiber. (*B*) Calcium transient detected in the primary visual cortex evoked by a 50-msec light flash applied to the contralateral eye in an anesthetized adult mouse. (*C*) Schematic showing the tip of an optic fiber implanted into a stained cortical region above neurons expressing ChR2 (red). (*D*) Calcium transient evoked by optogenetic stimulation of ChR2-expressing layer 5 neurons. The optical fiber was implanted at a depth of 500 µm in the visual cortex of an adult transgenic mouse. Light pulses of 488 nm wavelength and 50 msec duration, with a power of 2.3 mW, were used for stimulation. (*E*) Graph depicting, for constant excitation light intensity, the relationship between the duration of the stimulus pulse and the response probability.

TROUBLESHOOTING

Problem (Step 16): There is little or no fluorescence after implanting the optical fiber.
Solution: Aside from a failure in dye injection (see Chapter 49), this problem is mainly caused by placement of the fiber tip outside of the stained area, especially when recording from deep brain regions. To avoid this problem, take care to insert the staining pipette and the optical fiber exactly at the same position and with the same angle, by optimizing the use of the stereotaxic device. Another reason for low fluorescence levels is bleeding into the tissue after damage of a blood vessel, either with the staining pipette or with the optical fiber. Blood vessel rupture may be partially avoided by lowering the pipette and the fiber very slowly.

Problem (Step 18): The traces are too noisy.
Solution: This problem is often caused by heartbeat-mediated pulsation of a large blood vessel located close to the tip of the optical fiber. Minor changes in the depths of the fiber tip may lead to major improvements.

Problem (Step 18): There are movement artifacts.
Solution: Artifacts can occur in anesthetized animals by strong breathing excursions or, in awake animals by fast movements. Proper fixation of the fiber in the metal tube and on the skull helps to minimize movement artifacts.

DISCUSSION

The optical fiber system is used for the detection of fluorescence signals from all stained cells and the surrounding neuropile within the volume illuminated by the excitation light (Fig. 1D). For two-photon imaging recordings, the calcium signals detected by the optical fiber predominantly represent the action potential activity in the stained region (Stosiek et al. 2003; Kerr et al. 2005). From combined two-photon imaging and optical fiber experiments, we estimate that about 20 coactive neurons located near the tip are sufficient for producing a detectable calcium signal. Note that a variant of this approach has been used for the detection of calcium signals from the upper dendrites of layer 5 cortical neurons (Murayama et al. 2007; see Chapter 82).

Examples of Application

The original application of this approach was concerned with the monitoring of spontaneous calcium waves in the cortex of awake newborn mice (Adelsberger et al. 2005). An example of a population calcium signal evoked by sensory stimulation in the adult mouse is shown in Figure 2. The optical fiber was implanted in the primary visual cortex (Fig. 2A). Stimulation of the contralateral eye with a white light flash (Fig. 2B) evoked a large calcium transient lasting for ~2–3 sec. The transient occurred with a delay of ~80 msec, which corresponds closely to the range of the delays observed with electrical recordings (Porciatti et al. 1999).

For the illustration of calcium recordings combined with optogenetic stimulation, we used mice expressing ChR2 under the control of the Thy1 promoter in cortical layer 5 neurons (Arenkiel et al. 2007). Figure 2C shows the experimental arrangement of an optical fiber placed in the stained cortical region after multicell bolus loading. The calcium signals were monitored by using low light levels (0.03 mW) from the 488-nm excitation laser. For ChR2 stimulation, the excitation light intensity was increased for 50 msec to a level of 2.3 mW. The resulting calcium transients had time courses and amplitudes that were comparable to those evoked by sensory stimulation. This experimental configuration was extremely robust and allowed many trials of repeated stimulations and recordings. For example, in Fig. 2E we illustrate the result of such a long-lasting experiment, showing that for constant laser intensity, the response probability was proportional to the duration of the stimulus.

ACKNOWLEDGMENTS

Supported by the Bundesministerium für Bildung und Forschung (NGFN-2) and the Deutsche Forschungsgemeinschaft (International Graduate School 1373).

REFERENCES

Adelsberger H, Garaschuk O, Konnerth A. 2005. Cortical calcium waves in resting newborn mice. *Nat Neurosci* **8:** 988–990.

Arenkiel BR, Peca J, Davison IG, Feliciano C, Deisseroth K, Augustine GJ, Ehlers MD, Feng G. 2007. In vivo light-induced activation of neural circuitry in transgenic mice expressing channelrhodopsin-2. *Neuron* **54:** 205–218.

Boyden ES, Zhang F, Bamberg E, Nagel G, Deisseroth K. 2005. Millisecond-timescale, genetically targeted optical control of neural activity. *Nat Neurosci* **8:** 1263–1268.

Flecknell P. 2000. *Laboratory animal anesthesia*. Academic, New York.

Kerr JN, Greenberg D, Helmchen F. 2005. Imaging input and output of neocortical networks in vivo. *Proc Natl Acad Sci* **102:** 14063–14068.

Murayama M, Perez-Garci E, Lüscher H-R, Larkum ME. 2007. Fiberoptic system for recording dendritic calcium signals in layer 5 neocortical pyramidal cells in freely moving rats. *J Neurophysiol* **98:** 1791–1805.

Porciatti V, Pizzorusso T, Maffei L. 1999. The visual physiology of the wild type mouse determined with pattern VEPs. *Vision Res* **39:** 3071–3081.

Stosiek C, Garaschuk O, Holthoff K, Konnerth A. 2003. In vivo two-photon calcium imaging using multi-cell bolus loading (MCBL). *Proc Natl Acad Sci* **100:** 7319–7324.

82 Fiber-Optic Calcium Monitoring of Dendritic Activity In Vivo

Masanori Murayama[1,2] and Matthew Larkum[1]

[1]*Institute for Physiology, University of Bern, 3012, Bern, Switzerland;* [2]*Behavioral Neurophysiology Laboratory, Brain Science Institute, Riken, 2-1 Hirosawa Wako Saitama, 351-0198, Japan*

ABSTRACT

Observing dendritic activity in active animals presents great challenges and there are currently not many options available. We describe the fiber-optic-based "periscope" method for recording from populations of layer 5 neocortical pyramidal dendrites. This is an inexpensive, convenient, and robust method for recording dendritic activity during behavior. We also discuss the limitations of this approach and provide tips for how it might be adapted for different purposes and brain areas.

INTRODUCTION

Studying dendrites through imaging has a rich tradition in neuroscience. However, a great number of problems need to be overcome for dendritic recordings in vivo (Svoboda et al. 1997). Fiber-optic imaging has obvious advantages because it allows the experiments to be performed on awake, freely moving animals (Adelsberger et al. 2005). It is also relatively cheap and easy to implement and can be achieved over long periods of time, making it an ideal method for investigating various kinds of behaviors in a variety of animals. On the other hand, it is difficult to apply to dendritic recordings in vivo because dendrites are typically too small to resolve. The standard method for imaging dendrites in vivo is two-photon imaging, which is particularly suited to recording from single dendrites or small populations in the cortex (Helmchen et al. 1999; Waters et al. 2003; Kerr et al. 2005; Göbel and Helmchen 2007). Recent advances in combined two-photon/fiber-optic approaches promise to make even these recordings possible in freely moving preparations (Engelbrecht et al. 2008; Flusberg et al. 2008; Sawinksi et al. 2009; see Chapter 78).

In this chapter, we present an epifluorescence approach with some simple innovations for observing calcium changes in the apical dendrites of populations of layer 5 (L5) neocortical pyramidal neurons. This method retains all the ease of use of typical epifluorescence fiber-optic approaches, taking advantage of the distinct architecture of pyramidal neurons within the neocortical network to isolate calcium fluorescence signals from these dendrites. So far it has only proved possible with L5 pyramidal neurons (Murayama et al. 2007, 2009; Murayama and Larkum 2009a). This limitation notwithstanding, however, the method could be adapted in principle for many different purposes and should be considered especially for applications involving linking dendritic activity to behavior. The two main innovations required for the following technique are the bolus loading of L5 pyramidal cell bodies via injection directly into L5 of the cortex (Kerr et al. 2005) and the use of a prism to reflect light at 90°, to be parallel with the cortical surface. Another recently developed method combines a similar approach with an inserted prism with standard two-photon imaging (Chia and Levene 2009).

Protocol

Fiber-Optic Calcium Monitoring of Dendritic Activity In Vivo

IMAGING SETUP

Epifluorescence Microscope

The method presented here can be used with any conventional epifluorescence microscope by imaging one end of the fiber-optic cable. We built a simple, small, and inexpensive epifluorescence microscope that gave us more flexibility (Fig. 1). This consists simply of a 10× objective (Model E58-372, 0.45 NA, Edmund Optics GmbH, Germany) and a tube lens (InfiniTube standard, Infinity Photo-

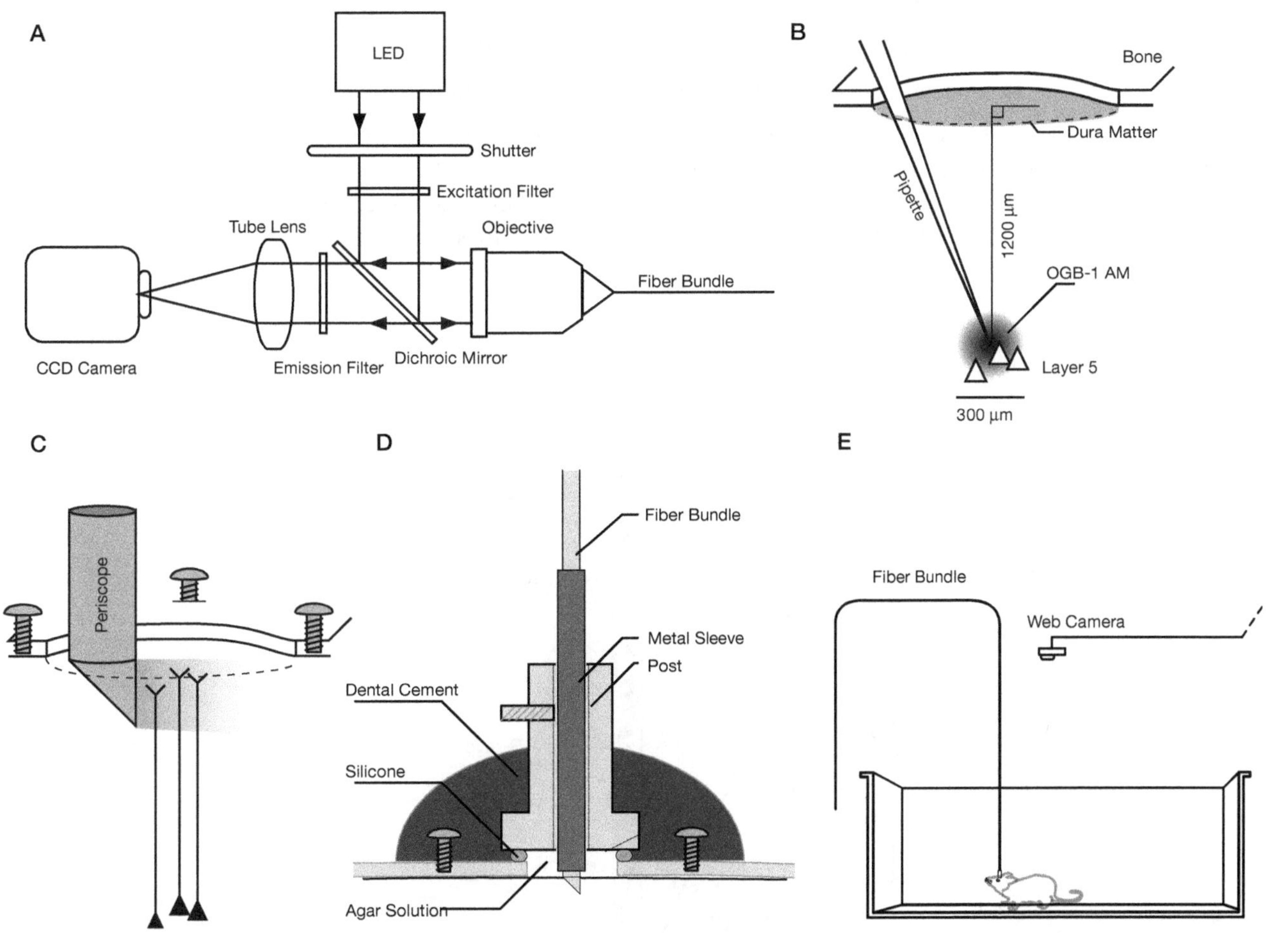

FIGURE 1. (*A*) Epifluorescence microscope for fiber-optic dendritic calcium monitoring. (*B*) Bolus loading with calcium-sensitive dye in vivo. (*C*) Periscope in neocortex. (*D*) Head mount for freely moving animals. (*E*) Recording calcium signals in freely moving animal in behavior area with simultaneous recording of behavior with a small CCD camera (e.g., a web camera, QCam for Notebooks Pro, Logitech).

Optical, Göttingen, Germany) mounted on a breadboard with standard optical components (Thorlabs, Inc.). The light source consists of a blue light-emitting diode (LED, 470 nm, 230 lm, IBF+LS30W-3W-Slim-RX, Imac Co., Ltd., Shiga, Japan). We found that the calcium fluorescent indicator Oregon Green 488 BAPTA-1 (OGB-1) acetoxymethyl (AM) gave better results than similar synthetic indicators. As a filter set we use an excitation filter (460–500 nm), a dichroic mirror (center wavelength of 507 nm), and an emission filter (510–560 nm, Chroma Technology, Rockingham, VT). Finally, a camera or imaging device is required. We use a 512 × 512 charge-coupled device (CCD) camera (MicroMax, Roper Scientific, Trenton, NJ; for details see Murayama et al. 2007). In our system, we image dendritic Ca^{2+} signals at 100-Hz sampling rate for up to 30 sec.

Fiber-Optic Construction: The "Periscope"

We refer to the fiber-optic construction as the "periscope" because it uses a similar principle to look horizontally (parallel to the cortical layers) from above the cortex (Fig. 1). The periscope contains the following key elements:

1. a fiber bundle with 17,000 elements (e.g., IGN-06/17, Sumitomo Electric Industries Tokyo, Japan)
2. a gradient index (GRIN) lens (e.g., NA. 0.5, GrinTech, Jena, Germany) (for details, see Murayama and Larkum 2009b) with a working distance of 100 µm
3. a 45° microprism angled at 90° (0.5 × 0.5 × 0.5 mm).
4. a metal sleeve for protecting the periscope from mechanical shock

These parts can be purchased already assembled by GrinTech, or purchased separately and combined by hand using UV-curing adhesive and epoxy glue (Fig. 2). Acquisition can be performed with most imaging software. In our case, analysis was made with WinView software (Princeton Instruments, Trenton, NJ) and self-written routines in Igor Pro (WaveMetrics, Portland, OR).

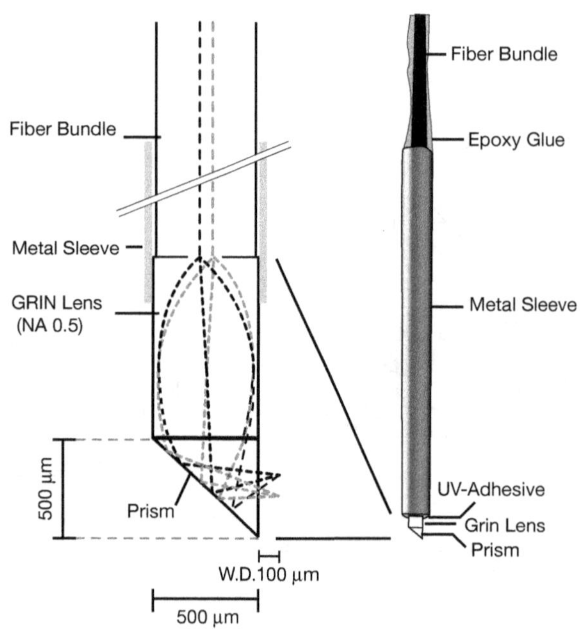

FIGURE 2. Details of periscope construction (see Imaging Setup for description). NA, numerical aperture, WD, working distance; UV, ultraviolet.

MATERIALS

CAUTION: See Appendix 6 for proper handling of materials marked with <!>.
See the end of the chapter for recipes for reagents marked with <R>.

Reagents

Where indicated, reagents are prepared using normal rat Ringer's solution (NRR); see Recipes.

Agar solution (0.5%)

This solution can be stored at room temperature for 1 yr.

Buprenorphine (0.003% [w/v] in NRR)

This solution can be stored at 4°C for 1 mo.

Dye solution <R>

Isoflurane <!>

Lidocaine <!> (2% [w/v] in NRR)

This solution can be stored at room temperature for 3 mo.

Normal rat Ringer's solution (NRR) <R>

This solution can be stored at room temperature for 3 mo.

Rats for analysis

Equipment

Anesthesia system for isoflurane (e.g., V-1 Table Top Lab Animal System, VetEquip, CA) for experiments on awake animals

Arena for freely moving animals

Binocular microscope (e.g., MZ6, Leica Microsystems, Heerbrugg, Switzerland)

Option: A longer working distance (WD) is useful for observation during experiments. We used ~297-mm WD with 50× total magnification and 4.8-mm field diameter by combination of a 0.32× objective (10422564, Leica Microsystems), 40× eyepieces (10445303, Leica Microsystems), and a 4× internal objective (changeable 0.63–4×) of MZ6. A reticule (10376119, Leica Microsystems) is also useful for measuring distance.

Body temperature controller (e.g., FHC, ME)

Bone screws (e.g., No. 19010-10, Fine Science Tools GmbH)

CCD camera (e.g., a webcam such as QuickCam Pro, Logitech, Japan) for recording animal movements

Cotton swab

Dental cement (e.g., mixture of Paladur liquid and Paladur powder, Heraeus Kulzer GmbH, Hanau, Germany)

Drill for making the craniotomy (e.g., Osada Success 40, Osada Electric Co., Tokyo, Japan)

Electric hair clipper (e.g., Moser, Wahl, Unterkirnach, Germany)

Epifluorescence microscope (see Imaging Setup and Fig. 1)

Equipment for hand-made assembly of "periscope" (see Imaging Setup)

- Epoxy glue <!> (e.g., 353NDPK, Thorlabs, Inc.)
- Mercury lamp (e.g., U-LH100HG, Olympus, Tokyo, Japan): source of UV light <!> for curing UV adhesive
- UV-curing adhesive (e.g., Norland Optical Adhesive 63, Norland Products, Cranbury, NJ) for combining optics

Gas manometer (for dye loading)

We use a custom-made gas manometer.

Green LEDs for light source during the surgery

Metal post, as a head mount, to fix the periscope on the skull (Fig. 1D)

Microglass pipette (e.g., for patch clamp)

Micromanipulator
Microneedle (#26002-10, Fine Science Tools GmbH, Heidelberg, Germany) for removing dura mater
Perfusion system (e.g., using gravity or pump system for normal in vitro experiments) for providing NRR to cortical surface
Pipette puller (e.g., Model P-87, Sutter Instruments, CA) to make microglass pipettes
Silicon lubricant (e.g., #29051-35, Fine Science Tools GmbH)
Stereotaxic device with stable manipulators (e.g., NR-5N, Narishige Group, Tokyo, Japan)
Stopwatch for timing dye loading
Surgical tools (e.g., scissors, fine forceps)
Vacuum pump (#850, Optimal, Schemel & Goetz, Offenbach am Main, Germany) for removing the NRR during surgery

EXPERIMENTAL METHOD

Preparation for Recovery Experiments

If the experimenter intends to perform experiments in awake animals, the targeted area of the brain should be exposed and prepared a number of days in advance so that the animal will not experience pain during the recording phase.

1. Prepare the surface of the rat brain by shaving and removing the scalp over the somatosensory and premotor regions under general anesthesia (isoflurane). Keep body temperature at 37°C–38°C with a body temperature controller.
2. Apply lidocaine to the wounds with a cotton swab.
3. Administer buprenorphine (an analgesic; 0.003 mg/kg) twice a day over the next 2–3 d.

   Monitor the behavior of the animal during this period to make sure that the animal is not under undue stress.

Bolus Loading

A critical aspect of the periscope method for dendritic Ca^{2+} imaging is the loading of L5 pyramidal cell dendrites, which requires relatively precise delivery of the bolus of indicator to L5.

4. Mount the rat in a stereotaxic device under anesthesia (isoflurane or preferred anesthetic).
5. Establish the region of cortex of interest.
6. With the animal under anesthesia, make a square craniotomy (~3 × ~4.4 mm) with a drill.

   After the craniotomy, locate the hind limb area of the somatosensory cortex more precisely with intrinsic imaging while stimulating the hind limb. For the hind limb area, the craniotomy is centered at 1.5 mm posterior to the bregma and 2.2 mm from the midline in the right hemisphere. Green light (e.g., from inexpensive LEDs) helps to avoid blood vessels.

7. Break tip of pipette to give a tip diameter of ~5–15 µm.
8. Fill a long-taper glass pipette with 1 µL of dye solution.
9. Prepare and position the pipette near the injection site using a micromanipulator.
10. Confirm with brief pressure that the tip is not blocked by the presence of a small droplet of dye.

    See *Troubleshooting*.

11. Clean away the droplet of dye carefully.
12. Insert pipette through dura mater to a final depth of 1200 µm.

    The total penetration distance should be trigonometrically calculated beforehand according to the angle of penetration of the pipette (see Fig. 1B).

13. Pressure-inject dye for 1 min at ~10–25 kPa into L5.

    A vital factor for success at this stage is waiting for at least 2 h before commencing the imaging experiments. Keep the brain perfused with NRR during this time. An alternative for pressure-injection is the use of a picospritzer to more closely control the injection parameters but we did not find this beneficial.

Head Mount

The correct mounting of the GRIN lens/prism/fiber-optic construction is probably the most critical step.

14. Insert three bone screws around the intended recording region for later stability.
15. Remove dura mater with a microneedle.
16. Make a "silicon barrier" by applying silicon lubricant around the craniotomy to protect it from dental cement.
17. Place the head mount piece onto the optical fiber and position it at some distance from the GRIN lens, ready for sliding down later.
18. Insert a prism 500 μm into the cortex using a mechanical manipulator.
    i. An alternative version of the fiber-optic construction that can give better signal-to-noise is to use a second, thicker fiber to collect light from above while illuminating horizontally with the first fiber attached to the inserted prism.
    ii. A second, noninvasive alternative can be constructed with an obliquely oriented fiber for illumination and a fiber to image from above.

        Care should be taken to fully submerge the illuminating fiber to prevent dispersion effects leading to unwanted illumination of the bolus of dye in L5.

        The choice of approach depends on the aim of the experiment. For questions that need to resolve the position of dendritic calcium increase in the longitudinal axis (e.g., to determine if calcium increases occur in the active calcium spike zone), the single-fiber "periscope" version is preferred. For details, see Murayama et al. (2007).
19. Apply agar solution at body temperature to the craniotomy.
20. Slide the head mount piece down on to the skull and agar solution.

    Take care that there are no gaps through which dental cement applied afterward can reach the craniotomy.
21. Cover the head mount and fixation screws with dental cement (see Fig. 1D). Wait for cement to fix completely.

Recording Calcium Signals in a Freely Moving Animal

22. Transfer the rat to the recording arena.
23. Acquire calcium transients with a CCD camera (Fig. 3).

    Care should be taken that the rat behaves normally and is not under undue stress. The fiber should not be obstructed or subjected to sudden movements. If necessary, a commutator can be used to couple the fiber to the microscope to allow rotation (Flusberg et al. 2008).

    See *Troubleshooting*.

TROUBLESHOOTING

Problem (Step 10): The pipette is blocked for injection of dye.
Solution: Use ultrasonic bath to dissolve dye and use centrifuge filtration to ensure that no particulate matter is in the solution. Clean pipette glass and avoid contact with dusty surfaces. Visually inspect solution in pipette under a microscope before use.

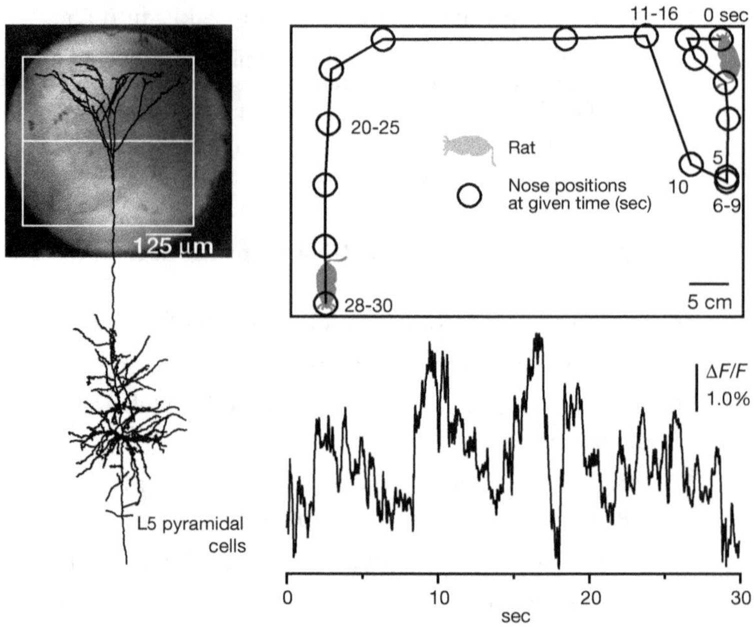

FIGURE 3. Dendritic calcium monitoring during animal behavior. (*Left*) Reconstructed pyramidal cell from in vitro experiments, demonstrating the typical regions of the dendritic tree within the imaging area. (*Upper right*) Movement of rat during 30 sec of imaging. (*Lower right*) $\Delta F/F$ at each time point indicated in *upper right*.

Problem (Step 23): There is a lack of signal or a weak signal.

Solution: Confirm the health status of the animal. Increase illumination intensity. If problem persists over several experiments, check filling of dendrites by extracting the brain, slicing the stained area, and observing under a conventional epifluorescence microscope (see Murayama et al. 2007).

Problem (Step 23): There is noise.

Solution: Noise with this system typically arises from mechanical disturbances (mostly caused by direct movement of the fiber), biological noise (i.e., fluctuations in foreground and background calcium levels), or equipment noise exacerbated by low signal. If noise is a problem, the first task is to ascertain whether it is intrinsic to the signal by blocking the possibility of calcium changes pharmacologically. We found that virtually all signal and noise could be blocked 30 min after the addition of Cd^{2+} to the cortical surface (Murayama et al. 2007). Mechanical noise, when it occurred, was mostly due to slippage of the fiber-optic cable in the mounting device. We speculate that what appears as biological noise may be mostly due to calcium increases caused by the activity of the cells themselves when backpropagating action potentials invade the dendrites (because the noise was blocked by injection of TTX into L5 and not by TTX (tetrodotoxin) in the upper layers).

DISCUSSION

Measurements of calcium changes in the dendrites with the periscope entails some compromises, and the researcher embarking on this approach might need to be ready to use parallel lines of evidence for specific questions. We envisage this method as best applied to screening for the link between dendritic calcium activity and behavior (Fig. 3). The two largest concessions that have to be made with this method are (1) signals represent combined population responses and (2) the source of calcium in the dendrites is not fully determined. With respect to the first question, another approach could be to inject or electroporate single neurons (Helmchen et al. 1999; Nagayama et al. 2007; Kitamura et al. 2008). We would anticipate from in vitro data that sufficiently large signals could be distinguished with this approach under certain circumstances (e.g., the presence of calcium spikes) that could be interpreted as originating from a single dendrite. Determining the source of calcium requires attention on a case-by-case basis. We have found that the largest signals arise from dendritic calcium spikes and smaller signals from backpropagating action potentials with little or no contribution from synaptic transmission (Murayama et al. 2007); however, this is presumably context-dependent and there may also be other sources of calcium (e.g., *N*-methyl-D-aspartate [NMDA} spikes; Larkum et al. 2009). The method could also be adapted for use in other areas of the brain. One advantage of the use of a prism is the sharp edge that it provides that can make deep insertions into the brain less damaging.

RECIPES

CAUTION: See Appendix 6 for proper handling of materials marked with <!>.
Recipes for reagents marked with <R> are included in this list.

Dye Solution

Prepare 50 µg OGB-1 AM in 5 µL of Pluronic (Pluronic F-127, –20% solution in dimethylsulfoxide <!>, Molecular Probes). Sonicate for 15 min. Dilute the solution in 28 µL of HEPES-buffered solution (125 mM NaCl, 2.5 mM KCl, 10 mM HEPES), filter (solution filter, 0.2-µM pore [e.g., #6809-4012, Semadeni, Ostermundigen, Switzerland]), and mix for another 15 min with an ultrasonic bath (e.g., Transsonic T310, Faust Laborbedarf, Schaffhausen, Switzerland). This solution can be stored at −20°C. We recommend using within 2 mo.

Normal Rat Ringer's (NRR) Solution

Substance	In MW	In mM	Grams to prepare 1 L	Grams to prepare 2 L
NaCl58.44	135	7.88	15.76	
KCl 74.56	5.4	0.40	0.80	
$MgCl_2{\cdot}6H_2O$	203.3	1	0.204	0.407
$CaCl_2$ <!>$\cdot 2H_2O$	147.02	1.8	0.265	0.529
HEPES	238.3	5	1.19	2.38

Adjust pH to 7.2 with NaOH <!>. Final osmolarity is ~290 mOsm. This solution can be stored at 4°C for 1 mo.

ACKNOWLEDGMENTS

We thank Hans-Rudolf Lüscher and Johannes J. Letzkus for their helpful comments on the manuscript and Denis Limoges and Jürg Burkhalter for their expert technical support. This work was supported by the Swiss National Science Foundation (Grant No. PP00A-102721/1) (MEL), SystemsX.ch (NEUROCHOICE) (MEL), and the Research Foundation for Opto-Science and Technology (MM).

REFERENCES

Adelsberger H, Garaschuk O, Konnerth A. 2005. Cortical calcium waves in resting newborn mice. *Nat Neurosci* **8:** 988–990.

Chia TH, Levene MJ. 2009. Microprisms for in vivo multi-layer cortical imaging. *J Neurophysiol* **102:** 1310–1314.

Engelbrecht CJ, Johnston RS, Seibel EJ, Helmchen F. 2008. Ultra-compact fiber-optic two-photon microscope for functional fluorescence imaging in vivo. *Opt Express* **16:** 5556–5564.

Flusberg BA, Nimmerjahn A, Cocker ED, Mukamel EA, Barretto RP, Ko TH, Burns LD, Jung JC, Schnitzer MJ. 2008. High-speed, miniaturized fluorescence microscopy in freely moving mice. *Nat Methods* **5:** 935–938.

Göbel W, Helmchen F. 2007. New angles on neuronal dendrites in vivo. *J Neurophysiol* **98:** 3770–3779.

Helmchen F, Svoboda K, Denk W, Tank DW. 1999. In vivo dendritic calcium dynamics in deep-layer cortical pyramidal neurons. *Nat Neurosci* **2:** 989–996.

Kerr JN, Greenberg D, Helmchen F. 2005. Imaging input and output of neocortical networks in vivo. *Proc Natl Acad Sci* **102:** 14063–14068.

Kitamura K, Judkewitz B, Kano M, Denk W, Häusser M. 2008. Targeted patch-clamp recordings and single-cell electroporation of unlabeled neurons in vivo. *Nat Methods* **5:** 61–67.

Larkum ME, Nevian T, Sandler M, Polsky A, Schiller J. 2009. Synaptic integration in tuft dendrites of layer 5 pyramidal neurons: A new unifying principle. *Science* **325:** 756–760.

Murayama M, Larkum ME. 2009a. Enhanced dendritic activity in awake rats. *Proc Natl Acad Sci* **106:** 20482–20486.

Murayama M, Larkum ME. 2009b. In vivo dendritic calcium imaging with a fiberoptic periscope system. *Nat Protoc* **4:** 1551–1559.

Murayama M, Pérez-Garci E, Lüscher HR, Larkum ME. 2007. Fiberoptic system for recording dendritic calcium signals in layer 5 neocortical pyramidal cells in freely moving rats. *J Neurophysiol* **98:** 1791–1805.

Murayama M, Pérez-Garci E, Nevian T, Bock T, Senn W, Larkum ME. 2009. Dendritic encoding of sensory stimuli controlled by deep cortical interneurons. *Nature* **457:** 1137–1141.

Nagayama S, Zeng S, Xiong W, Fletcher ML, Masurkar AV, Davis DJ, Pieribone VA, Chen WR. 2007. In vivo simultaneous tracing and Ca^{2+} imaging of local neuronal circuits. *Neuron* **53:** 789–803.

Sawinski J, Wallace DJ, Greenberg DS, Grossmann S, Denk W, Kerr JND. 2009. Visually evoked activity in cortical cells imaged in freely moving animals. *Proc Natl Acad Sci* **106:** 19557–19562.

Svoboda K, Denk W, Kleinfeld D, Tank DW. 1997. In vivo dendritic calcium dynamics in neocortical pyramidal neurons. *Nature* **385:** 161–165.

Waters J, Larkum M, Sakmann B, Helmchen F. 2003. Supralinear Ca^{2+} influx into dendritic tufts of layer 2/3 neocortical pyramidal neurons in vitro and in vivo. *J Neurosci* **23:** 8558–8567.

83 Imaging the Neocortex Functional Architecture Using Multiple Intrinsic Signals

Implications for Hemodynamic-Based Functional Imaging

Amiram Grinvald,[1] Dahlia Sharon,[1,2] David Omer,[1] and Ivo Vanzetta[1,3,4]

[1]Department of Neurobiology, The Weizmann Institute of Science, Rehovot, 76100 Israel; [2]Department of Psychology, Stanford University, Stanford, California 94305; [3]Centre National de la Recherche Scientifique, Unité Mixte de Recherche 6193, Institut de Neurosciences Cognitives de la Méditerranée, 13402 Marseille Cedex 20, France; [4]Université Aix-Marseille, 13284 Marseille Cedex 07, France

ABSTRACT

Optical imaging based on intrinsic signals has provided a new level of understanding of the principles underlying cortical development, organization, and function, providing a spatial resolution of up to 20 μm for mapping cortical columns in vivo. This chapter briefly reviews the development of this technique, the types of applications that have been pursued, and the general implications of some findings for other neuroimaging techniques based on hemodynamic responses (e.g., functional magnetic resonance imaging).

INTRODUCTION

In the mammalian neocortex, cells that perform a given function or share common functional properties are often grouped together, forming columnar clusters running from the pia mater to the white matter (Lorente de Nó 1949; Mountcastle 1957; Hubel and Wiesel 1962) (e.g., the orientation and ocular dominance columns of the visual cortex). Knowing the function performed by such "fundamental information processing units" appears essential for understanding the principles underlying the neural code and its implementation. Thus, of special importance are experimental methods that allow the visualization of the functional organization of cortical columns at the highest spatial and temporal resolution possible. To date, single-unit or multiunit extracellular recording techniques have provided the best tools for the study of the functional response properties of single cortical neurons and their synaptic interactions. Some studies have even used single-unit techniques for mapping the functional organization of larger cortical areas (Swindale et al. 1987), but because these techniques are not optimal for mapping, these studies required formidable efforts. Multielectrode techniques are more promising, and the development of nonmetallic electrode arrays compatible with functional magnetic resonance imaging (fMRI) measurements has made them suitable also for simultaneous fMRI studies. However, the size and placement of these devices still pose severe problems, and altogether the number of sampled neurons remains relatively small. Therefore, several imaging techniques that yield information about the spatial distribution of active neurons in the brain have been developed. Each of these techniques has significant advantages as well as limitations. For example, although the 2-deoxyglucose (2-DG) method permits postmortem visualization of active brain areas, or even of single cells, 2-DG is a one-shot approach: Only a single stimulus condition in a single animal can be assayed (the two-isotope 2-DG method permits the mapping of activity resulting from two stimulus conditions). Positron-emission tomography (PET) and fMRI offer spectacular three-dimensional localization of active regions in the functioning human brain, but both have low spatial resolution. Other imaging techniques have also been applied in vivo with some success, but they still suffer from a limited spatial resolution, a limited temporal resolution, or both. Among these methods are radioactive imaging of changes in blood flow, electroencephalography, magnetoencephalography, and thermal imaging.

This review summarizes, in the main, work performed in our laboratory. More general reviews describing earlier work performed on simpler preparations have been published elsewhere (Cohen 1973; Salzberg et al. 1986; Grinvald et al. 1988). More extensive reviews of optical imaging have also been published (Grinvald et al. 1988, 1999; Bonhoeffer and Grinvald 1996; Vanzetta 2006; Vanzetta and Grinvald 2008). Real-time optical imaging based on voltage-sensitive dyes in the neocortex is reviewed in Chapter 74.

OVERVIEW

Currently, the best method for imaging the functional architecture of the cortex is based on slow changes in the intrinsic optical properties of active brain tissue. This method permits visualization of active cortical regions at a spatial resolution of >50 μm and is immune to some of the problems associated with the use of extrinsic probes. The sources for these activity-dependent intrinsic signals include both changes in physical properties of the tissue itself that affect light scattering and changes in the absorption, fluorescence, or other optical properties of intrinsic molecules. The fact that, in many tissues, small intrinsic optical changes are associated with metabolic activity has been known since the pioneering experiments of Kelin and Millikan on the absorption of cytochromes (Kelin 1925) and hemoglobin (Millikan 1937). The first optical recording of neuronal activity was made more than 50 years ago by Hill and Keynes (1949), who detected light-scattering changes in active nerves. Changes in absorption or fluorescence of intrinsic chromophores were extensively investigated by Chance and colleagues (Chance et al. 1962) and by Jobsis and colleagues (Jobsis 1977; Jobsis et al. 1977; for reviews, see Cohen 1973; Mayevsky and Chance 1982). Although intrinsic optical signals are usually very small or very noisy, they are still suitable for the imaging of the functional architecture of the cortex (Grinvald et al. 1986; Frostig et al. 1990; Ts'o et al. 1990).

The basic experimental setup for optical imaging experiments is shown in Figure 1. The head of the experimental animal (anesthetized or awake) is thoroughly stabilized with respect to the detector (e.g., by a stereotaxic frame [not shown]). The brain is illuminated with flexible light guides, and digital pictures are acquired by the camera, which views the exposed cortex through a cranial window. To optimally interact with an experiment—particularly in vivo—as much information as possible is made available online. Because of the large data volume, however, a complete analysis has to be performed offline, using several different computational strategies (e.g., first-frame normalization, contrast with orthogonal conditions, principal component analysis [PCA[, independent component analysis [ICA]). (For detailed practical guidelines, see Grinvald et al. 1999.) Significant methodological advances, which may reduce the data acquisition time for some types of maps, have been reported (Kalatsky and Stryker 2003).

The initial optical imaging studies investigated the well-known structural elements of the functional architecture such as ocular dominance in the primary visual cortex and the "stripes" in V2 (Ts'o

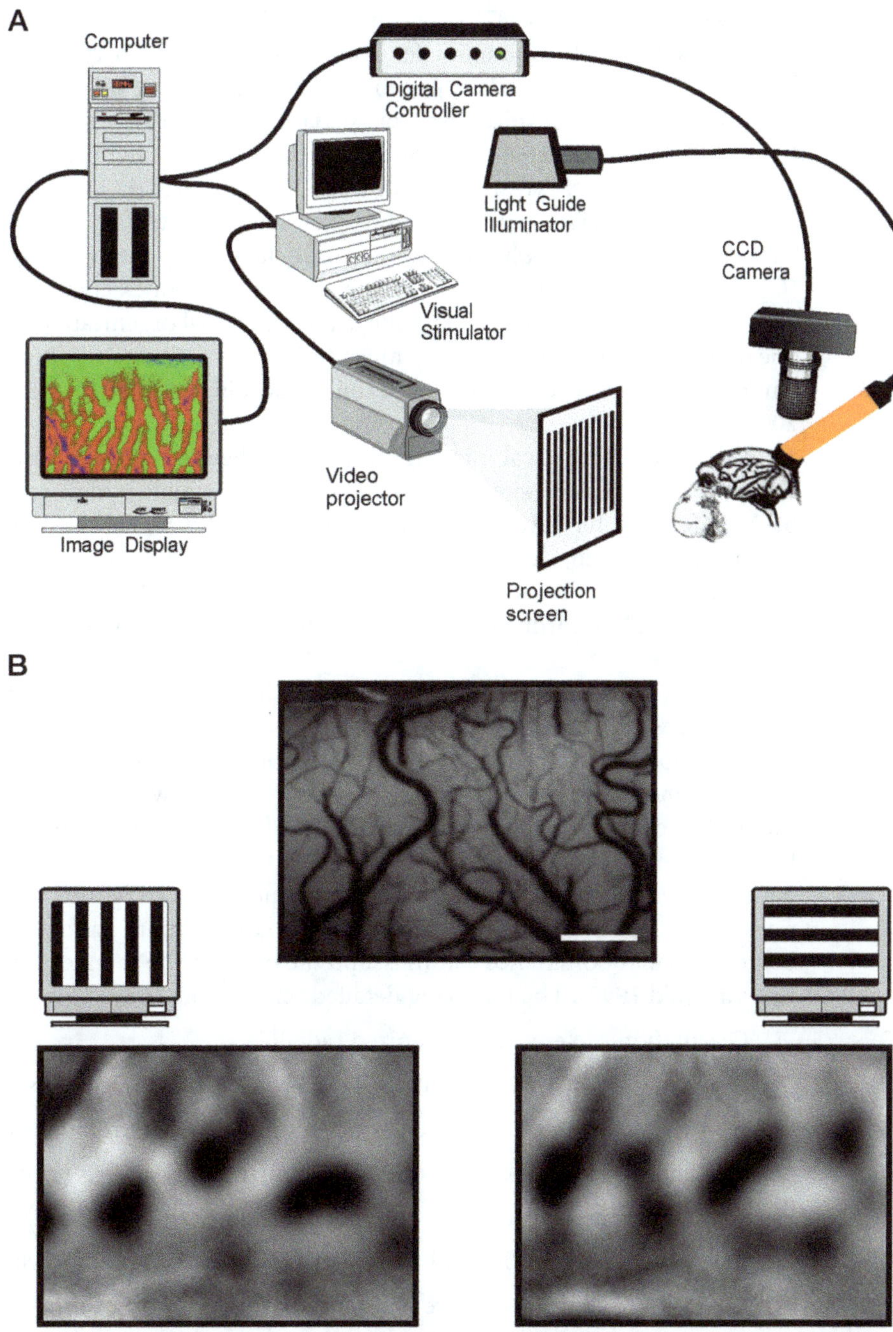

FIGURE 1. Optical imaging of functional maps in vivo. (*A*) The setup. Digital charge-coupled device (CCD) images are taken of the animal's exposed cortex, which is sealed in a cranial window. The cortex is illuminated with light of 605-nm wavelength, and a camera is used to acquire digital images of the cortex. The acquired images are digitized in a camera controller, which transfers the data to a computer that controls the entire experiment. Functional maps are subsequently analyzed and are displayed on a color monitor. To determine the quality of the maps during the imaging sessions, the data can be sent to a second computer for detailed, quasi-online analysis. (*B*) Imaging functional architecture. An activity map for one orientation is obtained straightforwardly, by dividing the image captured during presentation of this orientation by the average of the images captured during presentation of all orientations. (*Top*) The image of the cortical surface, which was illuminated with green light to emphasize the vasculature. (*Bottom*) Activity maps evoked by visual stimulation with vertical and horizontal gratings. Scale bar, 1 mm. (*A,B*, Modified from Bonhoeffer and Grinvald 1991.)

et al. 1990; Roe and Ts'o 1995), or the pinwheel-like organization of orientation preference (Bonhoeffer and Grinvald 1991, 1993; Bonhoeffer et al. 1995; Das and Gilbert 1995, 1997). Subsequently, methodological improvements made it possible to investigate more subtle features of cortical organization, such as direction-selective columns or spatial frequency columns (Malonek et al. 1994; Shmuel and Grinvald 1996; Weliky et al. 1996; Shoham et al. 1997). Similar progress has been obtained in exploring other visual areas. Ts'o and collaborators (1991, 1993) and Malach and colleagues (1994) succeeded in imaging the separate pathways in thin, thick, and pale stripes in monkey V2. It has even been possible to show functional columns in visual areas farther up the processing stream, in areas V4 and V3 (Ghose and Ts'o 1997; Kaskan et al. 2009) and MT (Malonek et al. 1994; Malach et al. 1997). Tanaka and colleagues used this method to image the functional organization in the inferotemporal (IT) area, one of the final stages of the visual pathway critical for object recognition (Wang et al. 1996). High-resolution maps were reported in the tree shrew (Bosking et al. 1997) and the ferret (see, e.g., Basole et al. 2003).

Although most optical imaging studies have been performed in the visual cortex, this is by no means the only sensory system that can be studied using this method. Indeed, it has lately proved to be a useful tool for investigating functional architecture in the somatosensory cortex of the rat (Grinvald et al. 1986; Gochin et al. 1992; Frostig et al. 1994) and the monkey (Shoham and Grinvald 2001), as well as in the auditory cortex of the guinea pig (Bakin et al. 1996) and the gerbil (Hess and Scheich 1996). Extraordinary high-resolution maps were also described in the olfactory bulb (Rubin and Katz 1999).

Here is one example of an imaging study that resolved an outstanding question that could not be resolved with alternative approaches. Hubel and Wiesel (1977) and Livingstone and Hubel (1984) have described three subsystems that are responsible for the perception of shape, depth, and color that combine in the macaque primary visual cortex in what has been referred to as the "ice cube" model. Since then, many groups have attempted to explore the nature of any geometric relationships that might exist among these three subsystems. Optical imaging has revealed how the various columnar subsystems are organized with respect to one another. Figure 2 shows the experimental data and revised ice cube model depicting, in a schematic fashion, the relationships between orientation columns, ocular dominance columns, and the blobs in the monkey's primary visual cortex (Bartfeld and Grinvald 1992). The following detailed relationships have been found.

1. Orientation preference is organized radially.
2. Orientation domains are continuous and have fuzzy boundaries.
3. Iso-orientation lines tend to cross ocular dominance borders at 90°.
4. Orientation pinwheels are centered on ocular dominance columns.
5. Blobs are centered on ocular dominance columns.
6. The centers of the blobs and the centers of the orientation pinwheels are segregated.
7. There is a regular mosaic-like organization for each type of functional domain, without an overall pattern of repeating hypercolumns, each of which contains all types of functional domains.

Similar results in the monkey have been independently reported by Blasdel (1992a,b) and by Obermayer and Blasdel (1993), using voltage-sensitive dyes, and in the cat by Hübener and colleagues (1997), using imaging of intrinsic signals.

Chronic Optical Imaging

Acute optical imaging on anesthetized animals typically lasts from 1 to 6 d. However, many important questions cannot be explored by performing acute experiments and require recording activity from behaving animals during performance of behavioral tasks. Particularly promising with regard to the feasibility of chronic optical imaging was the finding that cortical maps can be obtained through the intact dura or even through a thinned but closed skull (Frostig et al. 1990; Masino and Frostig 1996; Bosking et al. 1997; Schuett et al. 2002). These results were achieved using near-infrared light, which penetrates the tissue considerably better than does light of shorter wavelength.

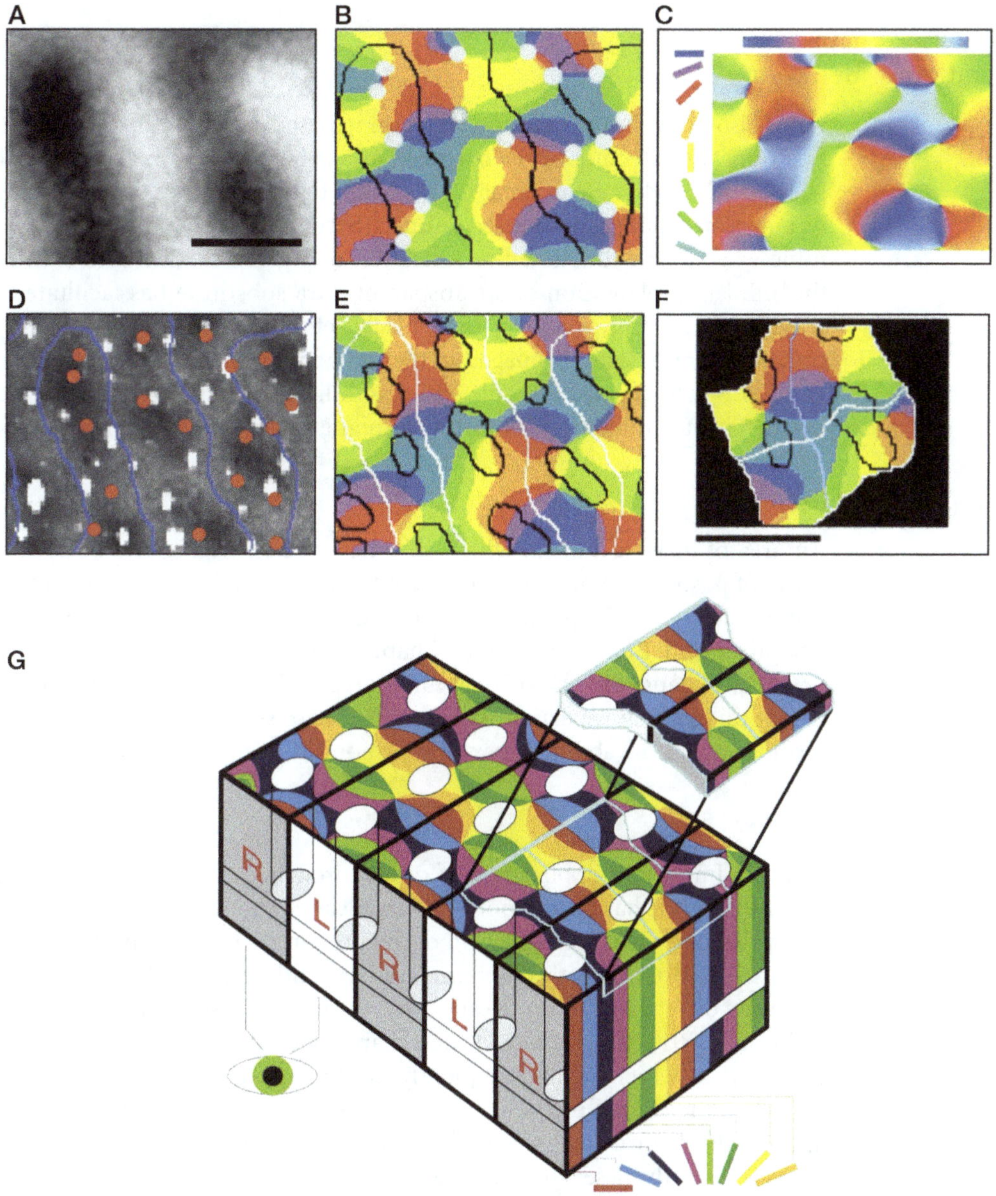

FIGURE 2. Relationship between pinwheels, ocular dominance columns, and blobs in primary visual cortex of the macaque monkey. (*A*) The optical map of ocular dominance from an $\sim$1.5 × 1-mm portion of striate cortex. The dark bands represent columns dominated by input from the right eye and the light bands by input from the left eye. Scale bar, 500 μm. (*B*) The borders of the ocular dominance columns (black lines taken from *A*) were overlaid onto the discrete "angle map" for orientation preference, obtained using eight oriented gratings (for the color code, see the bars on the *left* of *C*). The "pinwheel" centers were marked with white circles ($\sim$67% were centered on ocular dominance columns). In reality, the selectivity for orientation changes continuously: *C* shows the continuous angle map of orientation preference, obtained by interpolation from the same digital data used in *B* (the color scale on top of *C* shows the continuous color code for orientation preference). (*D*) The cytochrome oxidase (CO)-rich blobs were marked (red circles) on the histological photograph that corresponded exactly to the cortical area that was optically imaged. The blue lines mark the contour of the ocular dominance columns (*A*). (*E*) Angle map from *B* and two overlays showing the relationships between blobs, iso-orientation domains, and ocular dominance columns. (*F*) Two adjacent fundamental modules, magnified from *E*. Scale bar, 500 μm. (*G*) The revised "ice cube" model, schematically illustrating the relationships found in the above maps. Black lines mark the borders between columns of neurons that receive signals from different eyes. This segregation is partially responsible for depth perception. White ovals represent groups of neurons responsible for color perception (blobs). The pinwheels are formed by neurons involved in the perception of shape, with each color marking a column of neurons responding selectively to a particular orientation in space. Note that both the blobs and the centers of the pinwheels lie at the center of the R or L columns. The iso-orientation lines (appearing as a border between two colors) tend to cross borders of ocular dominance columns (black lines) at right angles. The top "slice" above the ice cube model depicts two adjacent fundamental modules (400 × 800 μm). Each module contains a complete set of about 60,000 neurons, processing all three features of orientation, depth, and color. This scheme is oversimplified in that clockwise and counterclockwise pinwheels are perfectly interconnected. In reality, this relationship does not exist and fundamental modules (hypercolumns) do not form an orderly mosaic.

Exploration of Behavior

It has also been shown that optical imaging based on intrinsic signals can be used to investigate the functional architecture of the cortex in awake behaving animals (Grinvald et al. 1991; Martin et al. 2002). Such studies are of great interest, first because studies of many higher brain functions cannot be performed using anesthetized animals. Second, studies of anesthetized preparations cannot determine whether the functional organization of a given cortical area is influenced by the behavior of the animal. Implantation of a transparent dura substitute has facilitated recoding from the cranial window for a long period of time (Arieli et al. 2002). Thus, ocular dominance and orientation columns were repeatedly imaged in the behaving macaque for a period of up to 1 yr (Shtoyerman et al. 2000; Slovin et al. 2002). This progress indicates that it is possible to successfully implement optical imaging techniques for the study of higher brain functions in awake behaving primates.

Developmental Studies

An area of investigation in which chronic optical imaging has been most fruitfully applied is the study of postnatal experience-dependent plasticity and development in the neocortex (see Kim and Bonhoeffer 1994; Chapman et al. 1996; Godecke and Bonhoeffer 1996; Crair et al. 1997a,b, 1998; Chapman and Bonhoeffer 1998; Chapter 60). Such studies require longitudinal experiments to determine changes in the cortical functional architecture over long periods of time. The technique of optical imaging is particularly useful in these studies, because it offers both the required spatial resolution and the ability to perform prolonged, comparative studies.

Human Clinical Studies

Another important application of optical imaging is mapping functional borders during human neurosurgery. It has been reported (MacVicar et al. 1990; Haglund et al. 1992) that optical imaging can be used to visualize activation of the human cortex in response to bipolar stimulation and during speech. More recently, Shoham and Grinvald performed optical imaging studies on humans to delineate the borders of functional areas during neurosurgery. In a preliminary report, Shoham and Grinvald (2001) described the optical mapping of the human hand representation in the somatosensory cortex, which was subsequently confirmed during the surgery with differential electroencephalogram (EEG) recording from a matrix of 16 surface electrodes. Godecke and colleagues applied the optical imaging method to map neocortical epileptic foci (I. Godecke et al., unpubl. results). In both these studies, it was found that the noise associated with the optical imaging of the human cortex was much larger than in animal experiments using the cranial window technique. Both groups also observed a large amount of activity-independent vascular noise. Despite these technical difficulties, it appears that optical imaging of functional borders in the human cerebral cortex is, in some cases, feasible (Cannestra et al. 1998). Thus, optical imaging may assist neurosurgeons in precisely locating the foci of epileptic events (Schwartz and Bonhoeffer 2001; Zhao et al. 2007; Ma et al. 2009) or the borders of functional areas close to the site of surgical procedures, as well as in obtaining high-resolution functional maps from that region.

Finally, can one hope to image human brain function noninvasively, using light, through the intact skull? The pioneering experiments by Jobsis (1977) on the cat, using transillumination with near-infrared light, and subsequent studies on human infants (Wyatt et al. 1986, 1990) have suggested that progress can be made in this direction. Furthermore, in a very innovative experiment, Chance and colleagues (1993a,b) showed that, although light reflected from the cortex is strongly attenuated by the thick skull, it can nevertheless be detected using the high sensitivity of photomultipliers. Kato and colleagues (1993) used an array of illuminator-detector pairs (optodes) to obtain low-resolution optical images of human cerebral cortex. Evoked responses in the auditory and motor cortex were detected optically, using near-infrared light, and confirmed with electroencephalography. This report and others (Hoshi and Tamura 1993; Gratton et al. 1995, 1997; Gratton 1997; Maclin et al. 2004; Tse et al. 2007) encourage the belief that it will be possible to design relatively inexpensive optical imagers for exploring cortical functional organization in human subjects, offering a spatial resolution of a few millimeters. Furthermore, a fast component of the intrinsic signal has been resolved by the Gratton group. These results suggest that this mode of optical imaging could offer the advantage of combining a spatial resolution comparable to that of PET and fMRI with a millisecond temporal resolution comparable to that of electroencephalography and magnetoencephalography (Wolf et al. 2003; Maclin et al. 2004). However, in recent experiments performed on monkeys, the fast optical signal could not be detected (Radhakrishnan et al. 2009), calling into question its usefulness for routine functional imaging. Whatever the final answer might be, the simplicity and relatively low cost of such devices for transcranial optical imaging of intrinsic signals undoubtedly justify additional efforts in this direction.

SOURCES OF INTRINSIC SIGNALS

To optimally and faithfully image functional maps in the neocortex, it is crucial to understand the mechanisms underlying the intrinsic signals and their relation to the electrical activity of neurons. More than a century ago, Roy and Sherrington (1890) postulated that "the brain possesses an intrinsic mechanism by which its vascular supply can be varied locally in correspondence with local variations of functional activity." Indeed, already in 1880, Angelo Mosso had reported changes in cerebral blood flow (CBF) produced by brain activity in patients with skull defects (Mosso 1880). More modern imaging techniques have clearly shown that there is a strong coupling between neuronal activity, local metabolic activity, and blood flow (Kety 1950; Lassen and Ingvar 1961; Sokoloff 1977; Raichle et al. 1983; Fox et al. 1986; Grinvald et al. 1986). More recently, Ogawa and colleagues (1992) and Kwong et al. (1992) have found that the intrinsic signal is useful for functional brain mapping with magnetic resonance imaging (MRI). However, only recently have the exact spatiotemporal characteristics of this coupling been revealed, through a series of experiments performed by several groups, involving optical imaging of intrinsic signals, optical imaging spectroscopy, laser Doppler recordings, phosphorescence decay time measurements, and oxygen electrode recordings.

Although the intrinsic signal has different components, which originate from different sources, functional maps obtained at different wavelengths are very similar. Therefore, all of these components can be used for functional mapping (Frostig et al. 1990). Malonek and Grinvald (1996) used optical imaging spectroscopy, a technique providing simultaneous spectral information over a large wavelength range from many cortical locations in the form of a time series of spatiospectral images, at high spatiotemporal and spectral resolution. The images obtained with imaging spectroscopy show the spectral changes at many wavelengths for each cortical point along a line (one spatial dimension versus wavelength). Using this technique, Malonek and Grinvald measured the spatial, temporal, and spectral characteristics of light reflected from the surface of the visual cortex following natural stimulation. When two-dimensional spatial information is important, fast-spinning filter wheels can be used; these, however, limit the spectral information to a few wavelength points only (Berwick et al. 2005; Nelson et al. 2005). Obtaining cortical spectra in this manner helps in (1) identifying signal sources by curve-fitting to known spectra, (2) determining the spatial precision of the signals, (3) evaluating the dynamics of the signals, and (4) exploring the dynamics in different vascular compartments. Additional technical details have been published elsewhere (Malonek and Grinvald 1996; Malonek et al. 1997; Vanzetta and Grinvald 1999, 2001; Vanzetta et al. 2004).

From the above studies, the sources of the intrinsic optical signals in reflection measurements appear to be the result of the following cellular or hemodynamic processes taking place in the tissue: activity-dependent oxygen consumption affecting hemoglobin saturation and thus its absorption spectrum; changes in blood volume affecting overall absorption as well as tissue light scattering; changes in blood flow affecting hemoglobin saturation; and activity-dependent changes in tissue light scattering. The first component of the intrinsic signal originates from activity-dependent changes in the oxygen saturation level of hemoglobin. This change in oxygenation itself contains two different components. The onset of the first component is an early one: an increase in the deoxyhemoglobin (Hbr) concentration, resulting from elevated oxygen consumption of the neurons caused by their metabolic activity. This causes a darkening of the cortex due to increased absorption by Hbr molecules. This early phase was referred to as the "initial dip" in the fMRI and related optical imaging literature. This interpretation (Grinvald et al. 1986; Frostig et al. 1990; Malonek and Grinvald 1996; Malonek et al. 1997; Vanzetta and Grinvald 2001; Vanzetta et al. 2004), however, remained controversial (Mayhew et al. 1998) for some time, as discussed below.

The second component of the intrinsic signal originates from an activity-dependent increase in cortical blood volume (CBV), increasing the amount of light absorbed by activated cortical regions in a wavelength-dependent fashion, following the absorption spectra of hemoglobin. Of great interest has been the level of spatial regulation of this component, especially because most current fMRI studies use this component of the hemodynamic response. Previous optical imaging and imaging spectroscopy work in the visual cortex of cats and monkeys (Frostig et al. 1990; Malonek and Grinvald 1996; Shtoyerman et al. 2000; Vanzetta and Grinvald 2001) have reported a much weaker colocalization of CBV changes with the electrically active columns than the increase in Hbr concentration during the initial dip. This poorer colocalization results in a lower signal-to-noise ratio (SNR) in spatial maps of activation calculated from the CBV component relative to those from the initial dip. This is especially true for maps in which the response to a single stimulus condition is normalized by the response to a blank ("single-condition" maps). For "differential" imaging, in which the response to a stimulus condition is normalized by or subtracted from the response to an "orthogonal" stimulus condition, the maps from the CBV are still somewhat poorer than from the initial dip. However, high-quality functional CBV maps have been obtained in cat and monkey visual cortex with differential imaging (Frostig et al. 1990; Vanzetta and Grinvald 2001). The differential procedure removes much of the CBV noise because it all but eliminates the contribution of the non-stimulus-specific part of the vascular response. These results prove that some of the CBV changes do colocalize with areas of increased neuronal activity; indeed, it has been shown that this local component of the CBV response is suitable for functional mapping even at the single-condition level (Vanzetta et al. 2004; Zhao et al. 2005).

The third component is delayed (Malonek et al. 1997; Jones et al. 2001): It is an activity-related increase in blood flow, causing a decrease in the Hbr concentration. This happens because the blood rushing into the activated tissue contains higher levels of oxyhemoglobin (HbO_2).

As described above, changes in blood volume, probably due to local capillary recruitment or a rapid filling of capillaries and dilation of venules in an area containing electrically active neurons, generates changes in tissue absorption due to changes in the total hemoglobin concentration. This blood-related component dominates the signal at high hemoglobin absorption wavelengths (i.e., from the ultraviolet up to ~630 nm). However, CBV changes also contribute to the optical signal via another process: Volume changes in capillaries compress the cortical tissue, giving rise to a change in light scattering that exists at all wavelengths and shows a time course typical of CBV changes (Tanner et al. 2005, 2006) rather than that of light-scattering changes that originate directly from electrical activity per se. This last significant component of the intrinsic signal arises from changes in light scattering that accompany cortical activation (Cohen et al. 1968; Tasaki et al. 1968; Grinvald et al. 1982, MacVicar and Hochman 1991). These changes are caused by ion and water movement, expansion and contraction of extracellular spaces, capillary expansion, or neurotransmitter release (for review, see Cohen 1973). Light-scattering components become a significant source of intrinsic signals above 630 nm and probably dominate the intrinsic signals in the near-infrared region above 800 nm.

The intrinsic optical signals that can be recorded from the living brain are small. In optimal cases, the change in light intensity due to neuronal activity is no more than 0.1%–6% of the total intensity of the reflected light, depending on wavelength. This means that intrinsic signals cannot be seen with the naked eye and that they have to be extracted from the cortical images with the appropriate data acquisition and analysis procedures. A major problem is that the biological noise associated with these measurements is, in many cases, larger than the signals themselves. Therefore, it is crucial to use the proper procedures to extract the small signal of interest from the raw data. Such procedures have been developed, yielding high-resolution functional maps (for details, see Grinvald et al. 1999; Reidl et al. 2007; Fekete et al. 2009; and references cited in Vanzetta et al. 2004 and Vanzetta and Grinvald 2008). To show the reliability of the data, the reproducibility of the optical maps obtained from the same area of cortex must be verified. The high degree of reproducibility that has been observed (see, e.g., Bonhoeffer and Grinvald 1993) gives confidence in the precision and reliability of cortical maps obtained in optical imaging. Note, however, that reproducibility of optical maps is a necessary, but not a sufficient, criterion. Blood vessels can give rise to signals that are reproducible but that do not correspond to neuronal activity (e.g., the patterns obtained with the macro-vascular CBV component in Vanzetta et al. 2004). Therefore, the "gold standard" for the confirmation of new findings remains, in most cases, electrophysiological verification. Combining electrophysiology with optical imaging is described below (see Combining Intrinsic Optical Imaging with Other Techniques).

THE INITIAL DIP: EVIDENCE FOR EARLY DEOXYGENATION

Initially, the very existence of the initial dip (early increase in Hbr concentration), which had been deduced from the time course of the intrinsic signal, was controversial (Malonek and Grinvald 1996; Mayhew et al. 1998; Kohl et al. 2000; Lindauer et al. 2001; Martin et al. 2002; Sheth et al. 2005). In view of the substantial implications of this dip for mechanisms underlying the neurovascular coupling, cellular metabolism, and high-resolution functional imaging, it was constructive to verify the existence of the initial dip with raw data that did not require spectroscopic models. In this line, the raw action spectra (i.e., the amount of light reflected from active cortex as a function of wavelength) obtained from an anesthetized cat and an awake monkey in our laboratory proved useful (Vanzetta and Grinvald 2001). The late action spectra clearly displayed the two HbO_2 peaks, unambiguously showing an increased concentration of HbO_2 probably caused by the delayed flow of more oxygenated blood. However, the double peak is evident only at late times of about >1 sec after the onset of electrical activity. At earlier times (e.g., ~0.6 sec), the action spectra show only a rounded shape, which cannot be explained without assuming an early increase in the concentration of Hbr. Thus, these raw action spectra suggest that there is an early increase in Hbr concentration preceding the increase in HbO_2. This is in line with the observations of Mayhew et al. (1999, 2000), Jones et al. (2001), and Malonek and coworkers (1997), who found that Hbr concentration increases before CBV changes, in this case in the rat, yet without a concomitant decrease in HbO_2. Notably, such an early, expected decrease in the concentration of HbO_2 has not been detected in most studies, and even the few positive reports (Mayhew et al. 1999; Devor et al. 2003, 2005) have been later followed by negative reports by the same groups (Mayhew et al. 2000; Jones et al. 2001; Devor et al. 2007), underscoring the necessity to carefully rule out the various pitfalls inherent in the experimental procedures, the spectroscopic models, and the stimulation paradigms.

The initial dip and its interpretation in terms of early deoxygenation have been confirmed with more direct measurement of intravascular oxygen concentration, based both on phosphorescence quenching (Vanzetta and Grinvald 1999) and on oxygen electrodes (Ances et al. 2001). It was also observed in awake subjects (Shtoyerman et al. 2000; Martin et al. 2002) and high-field blood-oxygenation-level-dependent (BOLD) fMRI (Ernst and Henning 1994; Menon et al. 1995; Hu et al. 1997; Logothetis et al. 1999; Yacoub and Hu 1999; Grinvald et al. 2000; Kim et al. 2000a,b).

Recently, direct measurements with oxygen electrodes of oxygen concentration changes in the cortical tissue itself (Thompson et al. 2003) have provided strong support for the interpretation of the origin of the initial dip in terms of increased oxygen consumption. The same conclusion was reached in two recent elegant experiments, one performed with optical imaging (Fukuda et al. 2006a) and the other with fMRI (Nagaoka et al. 2006). In both, the active vascular responses—and thus the changes in CBV—were pharmacologically blocked; yet, not only was the initial dip clearly present, it was actually enhanced, proving that the initial dip cannot be explained exclusively by CBV effects, a concern that had been raised previously (Buxton 2001).

SINGLE-CONDITION MAPPING: IMPLICATIONS FOR FUNCTIONAL IMAGING BASED ON HEMODYNAMIC RESPONSES

The ultimate goal of high-resolution functional brain mapping, at the cortical column level, is single-condition rather than differential mapping, because appropriate orthogonal stimuli are rarely available. It is important to clarify the advantages and limitations of single-condition mapping compared with differential mapping. The spatial resolution of a functional map depends on the spread of a particular imaged signal beyond the site of electrical activity, the SNR, and the instrument's spatial resolution. For example, the point spread function of the fMRI signal is large. However, two distinct but adjacent sites, activated on the cortex, can be resolved at distances much smaller than the half-width of the signal spread. This can be performed by differential imaging, provided that the SNR is adequate—that is to say, by subtracting two maps that are known to be orthogonal to each other because they are evoked by stimuli that are known to activate complementary cortical areas. Ocular dominance columns and orientation columns belong to the rare examples in which "orthogonal maps" really exist in nature. Using differential imaging, human ocular dominance columns have already been imaged with BOLD fMRI (Menon et al. 1997; Menon and Goodyear 1999; Cheng et al. 2001; Goodyear and Menon 2001) and at particularly high SNR since the availability of high-field (7-Tesla) fMRI in humans (Yacoub et al. 2007). The BOLD fMRI signal is composed of (1) a "local" component, which is stimulus-specific and probably originates from the capillaries, and (2) a component referred to as the global signal, which originates from other sources such as nonlocalized blood-volume and blood-flow changes that may not carry stimulus specificity and hemodynamic activation of venules and large draining veins far away from the electrical activation site. For this reason, when two stimuli are known to be orthogonal, in the sense that they activate neurons in complementary cortical patches, comparing the activation patterns of the two stimuli enhances the signal's local component, while eliminating many, although not all, the global components.

There are two major limitations to differential imaging, however. First, differential imaging works perfectly well only in cases in which the stimuli are orthogonal in the above sense. In the general case, orthogonal stimuli are impossible to define, particularly when it comes to domains dealing with higher cognitive functions. In such cases, subtracting responses from each other may lead to significant misinterpretation of the data. Second, by far the largest signals are from the draining veins, even veins with less than 100 μm diameter. Such large signals may spread up to 7 mm from retinotopic sites of activation (see, e.g., in monkeys; Shoham and Grinvald 2001). Importantly, the signals from such draining veins may be stimulus-specific and highly reproducible. Therefore, the use of differential imaging, high thresholds, signal averaging, and statistics does not guarantee removal of such artifacts. Single-condition mapping, in which the response to a single stimulus can be imaged directly, thus has great advantages over differential imaging and should be the ultimate goal of high-resolution functional brain imaging.

A significant advantage of optical imaging of the exposed cortical surface is that the spatial resolution of the optics itself is on the order of 1 μm, permitting unambiguous identification of blood-vessel-derived responses and distinguishing them from cortical activation per se. In addition, because reflected photons are "cheap," they can be collected in large numbers to provide an SNR that is higher than that accomplished with fMRI. This allows visualization of the hemodynamic changes in the various cortical vascular compartments, without the need for elaborate image processing or statistical methods. In other words, viewing the results of optical imaging is direct and straightforward. Kim et al. (2000a) took advantage of the optical imaging protocol and reported a striking technical advance in fMRI: improving its spatial resolution for direct and selective visualization of individual cortical columns in the anesthetized cat. The investigators used the early, negative phase of the BOLD fMRI signal (the "initial dip"; the deoxygenation phase), to obtain single-condition maps of cortical columns. They further showed that the later, positive hyperoxygenation phase of the BOLD signal, traditionally used for functional brain mapping, gives less accurate results than the early phase. These results, however, have remained controversial (Kim et al. 2000b; Logothetis 2000).

It has been suggested that perfusion-based fMRI, which is presumed to image CBF changes that are largely limited to the arterioles and capillaries, as well as water exchange with the surrounding tissue, provides single-condition maps (Duong et al. 2001). Contrast-agent-based fMRI and volume measurements also yield single-condition maps (Harrison et al. 2002, Zhao et al. 2005; Fukuda et al. 2006b). Vanzetta et al. (2004) confirmed with optical imaging that after removal of vascular signals (e.g., by PCA), "cleaned" CBV-based maps are indeed rather precise (by comparing them with the well-established "gold standard" for mapping ocular dominance columns). Therefore, a component of blood volume change is regulated at the capillary level. Although optical imaging and fMRI measure different mixtures of vascular parameters, and each method shows different sensitivity to each component (Moon et al. 2007), it appears that single-condition mapping at the fundamental columnar level is feasible with both optical imaging and fMRI.

COMBINING INTRINSIC OPTICAL IMAGING WITH OTHER TECHNIQUES

It is clear that the study of cortical organization and function can greatly benefit from a combination of optical imaging with other techniques, such as tracer injections, electrical recording, and microstimulation. Because optical imaging of functional architecture can quickly and easily provide a picture of how certain functional parameters are represented on the cortical surface, it is an ideal tool for guiding electrophysiological recordings or tracer injections. Furthermore, histological data, such as the dendritic and axonal branching of single cells, can be directly correlated to the functional organization in the very same piece of tissue, as depicted in Figure 3 (Malach et al. 1993, 1994, 1997; Kisvarday et al. 1994). Extensive work in this direction has been performed by Fitzpatrick and colleagues (Bosking et al. 1997).

In many cases, it is important to use metal or glass electrodes simultaneously with optical imaging. Furthermore, it is advantageous to be able to target the recording site to functionally identified cortical domains revealed by a previous session of optical imaging. For this purpose, it is useful to have a sealed cranial window coupled to a manipulable electrode. Arieli and Grinvald (2002) have designed such a device, which consists of a chamber with a square cover glass that is larger than the chamber diameter. The cover glass can be moved relative to the base of the cranial window. A hole in the glass covered with a rubber gasket allows the insertion of the electrode into the sealed chamber (Fig. 3A,B).

This device has proved to be very useful for targeted tracer injections, electrical confirmation of optically obtained functional maps, and targeted microelectrode recordings. Shmuel and Grinvald (1996) and Shoham et al. (1997) have used it to perform both perpendicular and nearly tangential penetrations, studying the relationships between single-neuron responses and optically imaged functional domains for orientation and direction and for spatial frequency, respectively. The same device has proved equally useful for microstimulation during optical recording experiments. Slovin et al. (2003) and Strick et al. (2003) made use of it for evaluating the extent of cortical activation in response to microstimulation in the awake monkey.

CONCLUSIONS AND OUTLOOK

Optical imaging based on intrinsic signals is a method that allows investigators to map the spatial distribution of functional domains, offering unique advantages. No alternative, noninvasive imaging technique for the visualization of functional organization in the living brain provides comparable spatial resolution. However, care must be taken to use adequate signal processing methods that enhance local hemodynamic responses and minimize macrovascular artifacts.

Following pioneering work by Chance and coworkers (Barlow et al. 1979), it has been proposed that such vascular artifacts might be minimized by imaging sensory-evoked increases in oxidative

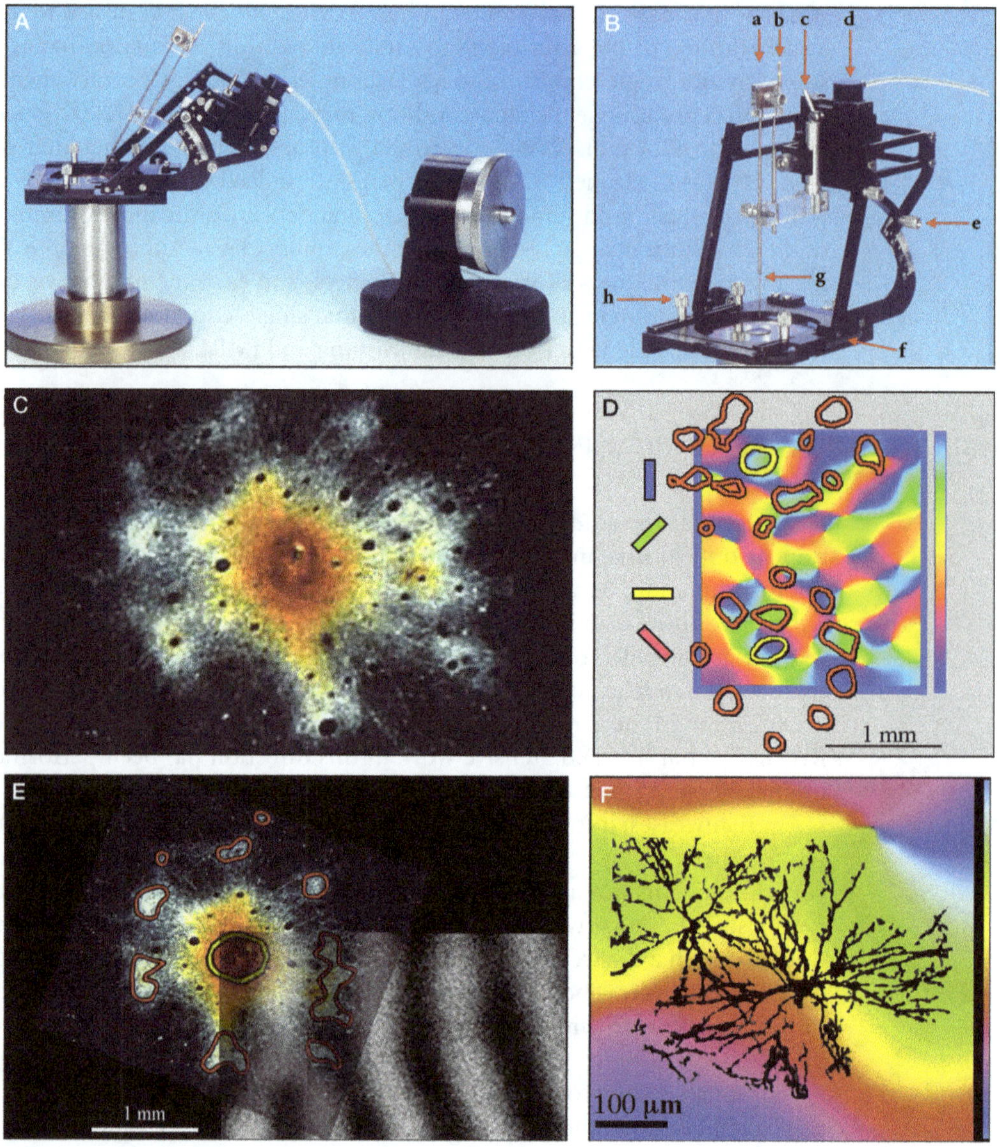

FIGURE 3. Combining optical imaging with other physiological techniques. Cranial window combined with an electrode manipulator for targeted electrical recordings or injections. (*A*) The *x–y* and axial microdrive is shown on its stand at an angle of ~60°. The hydraulic microdrive is the model MO-11N (Narishige Group), which can also be coarsely positioned with a manual manipulator. (*B*) Detail of *A*. (a) Tip of a tungsten microelectrode. (b) Pin for electrical connection with the electrode. (c) Bolt to lock coarse positioning of the manual manipulator. This enables the electrode guide to be advanced manually to penetrate the rubber gasket. Only then is the electrode pushed forward out of the protecting penetration tube, and fine movement is achieved using the hydraulic microdrive. (d) Hydraulic microdrive connected to the black tiltable holder. (e) Two screws to lock the microdrive at various angles (range of 30°–90°). (f) Sliding glass. (g) Needle, which holds and protects the microelectrode, to penetrate the rubber gasket. (h) Four screws, which lock the upper part of the microdrive to its *x–y* platform. (*C–F*) Biocytin injection targeted at a monocular site in macaque monkey area V1. (*C*) Dark-field photomicrograph of a tangential cortical section showing a biocytin injection placed in the center of an ocular dominance column. Note extensive local halo around the injection site followed by clear axonal patches farther away. (*D*) Two injections superimposed on an orientation preference continuous angle map. The two injection sites (yellow) and their associated patches (red) were superimposed on this map. (*E*) Same injection site as in *C*, but superimposed on an optically imaged map of ocular dominance columns. The injection was targeted at a contralateral eye column (coded black) at the top left corner of the optically imaged area. Note the tendency of the biocytin patches to skip over the ipsilateral eye column (coded white). (*F*) Local connections and orientation domains. Dendritic arbors of two upper-layer neurons precisely superimposed on an orientation preference angle map taken from that same patch of cortex. Note that the dendritic arbors crossed freely through cortical regions having diverse orientation preferences (Arieli and Grinvald 2002). (*C–F*, Modified from Malach et al. 1993.)

metabolism directly. In addition, the expected spatial resolution would even be better than that obtainable with hemodynamics-based optical signals. Recently, this has been performed by means of imaging the autofluorescence of flavoproteins (Shibuki et al. 2003; Reinert et al. 2004; Weber et al. 2004), which are mitochondrial respiratory pigments whose fluorescence depends on their state of oxidation. Sensory-evoked flavoprotein autofluorescence responses were recently shown not only to accurately reflect the functional architecture in an anesthetized cat primary visual cortex (Husson et al. 2007 [orientation maps]; Mallik et al. 2008 [spatial frequency maps]), but also to have quality at least as good as that of the corresponding maps obtained by intrinsic imaging of hemodynamic signals. As promising as these results are, some caution is probably warranted, notably when the technique is to be applied to other experiments, cortical areas and preparations, state of alertness, etc. Indeed, recent Monte Carlo simulations (L'Heureux et al. 2009) have shown that the autofluorescence signal depends on the structural and physiological features of the brain area considered and are sensitive to changes in blood flow and volume induced by sensory activation.

Optical imaging based on voltage-sensitive dyes (Chapter 74) does provide a comparable resolution and even offers a superior temporal resolution; however, extrinsic probes must be used. It is this level of resolution that can reveal where processing is performed—a necessary step for the understanding of the neural code at the population level. A key strength of intrinsic imaging is that the signals can be obtained in a relatively noninvasive manner over long periods of time. This is particularly important for chronic recordings, which can last up to a year. Optical imaging during chronic experiments allows the study of neocortical development in rodent (e.g., Kim and Bonhoeffer 1994; Crair et al. 1997a,b) and of plasticity (Masino and Frostig 1996; Polley et al. 1999, 2004) and higher brain functions in behaving monkeys (Slovin et al. 2002). Whereas sophisticated methods of analysis have already been introduced (see, e.g., Sirovich et al. 1996; Everson et al. 1997; Gabbay et al. 2000) along with more sophisticated data acquisition techniques (Kalatsky and Stryker 2003), additional efforts in this direction would be rewarding.

A shortcoming of optical imaging is its lack of depth resolution, particularly in contrast to fMRI, which has recently attained sublayer resolution in the rat cortex (Moon et al. 2007). Some progress in this respect has been made by using the information carried by the degree of laterality of light diffused from different depths (Hillman et al. 2007) or by using gradient-index light fibers for microendoscopy (Jung et al. 2004; Levene et al. 2004). However, at its current stage the technology is not yet appropriate for systematical optical imaging resolved for depth, owing to SNR limitations or problems linked to small field-of-view sizes. Further efforts in this direction are therefore likely to be rewarding.

Another major challenge is the application of intrinsic signal imaging to the human brain. So far, the quality of results obtained in human subjects is lower than that achieved in animal experimentation. Nevertheless, it seems likely that optical imaging will provide clinical benefits. Furthermore, during clinical use in neurological (e.g., Cannestra et al. 2001) and ophthalmic applications (Grinvald et al. 2004), the functional architecture of various sensory cortical areas can be mapped at unprecedented resolution, an order of magnitude or two better than is currently accomplished by PET or fMRI. Finally, the completely noninvasive optical imaging through the intact human skull devised by Gratton and colleagues (Gratton et al. 1995; Maclin et al. 2004) may provide an imaging tool offering both the spatial and temporal resolutions required to expand our knowledge of the principles underlying the remarkable performance of the human cerebral cortex.

ACKNOWLEDGMENTS

We thank our previous coworkers, Ron Frostig, Edmund Lieke, Charles Gilbert, Thorsten Wiesel, Dan Ts'o, Tobias Bonhoeffer, Dov Malonek, Amir Shmuel, Daniel Glaser, Amos Arieli, Evan Shtoyerman, and Doron Shoham, for their original contributions. This work was supported by grants from Minerva, Bundesministerium für Bildung und Forschung, and Ms. Enoch and the Grodetsky Center for Research of Higher Brain Functions.

REFERENCES

Ances BM, Buerk DG, Greenberg JH, Detre JA. 2001. Temporal dynamics of the partial pressure of brain tissue oxygen during functional forepaw stimulation in rats. *Neurosci Lett* **306:** 106–110.

Arieli A, Grinvald A. 2002. Combined optical imaging and targeted electrophysiological manipulations in anesthetized and behaving animals. *J Neurosci Methods* **116:** 15–28.

Arieli A, Grinvald A, Slovin H. 2002. Dural substitute for long-term imaging of cortical activity in behaving monkeys and its clinical implications. *J Neurosci Methods* **114:** 119–133.

Bakin JS, Kwon MC, Masino SA, Weinberger NM, Frostig RD. 1996. Suprathreshold auditory cortex activation visualized by intrinsic signal optical imaging. *Cereb Cortex* **6:** 120–130.

Barlow CH, Harden WR III, Harken AH, Simson MB, Haselgrove C, Chance B, O'Connor M, Austin G. 1979. Fluorescence mapping of mitochrondrial redox changes in heart and brain. *Crit Care Med* **7:** 402–406.

Bartfeld E, Grinvald A. 1992. Relationships between orientation preference pinwheels, cytochrome oxidase blobs and ocular dominance columns in primate striate cortex. *Proc Natl Acad Sci* **89:** 11905–11909.

Basole A, White LE, Fitzpatrick D. 2003. Mapping multiple features in the population response of visual cortex. *Nature* **424:** 986–990.

Berwick J, Johnston D, Jones M, Martindale J, Redgrave P, McLoughlin N, Schiessl I, Mayhew JE. 2005. Neurovascular coupling investigated with two-dimensional optical imaging spectroscopy in rat whisker barrel cortex. *Eur J Neurosci* **22:** 1655–1666.

Blasdel GG. 1992a. Differential imaging of ocular dominance and orientation selectivity in monkey striate cortex. *J Neurosci* **12:** 3115–3138.

Blasdel GG. 1992b. Orientation selectivity, preference, and continuity in monkey striate cortex. *J Neurosci* **12:** 3139–3161.

Bonhoeffer T, Grinvald A. 1991. Iso-orientation domains in cat visual cortex are arranged in pinwheel-like patterns. *Nature* **353:** 429–431.

Bonhoeffer T, Grinvald A. 1993. The layout of iso-orientation domains in area 18 of cat visual cortex: Optical imaging reveals pinwheel-like organization. *J Neurosci* **13:** 4157–4180.

Bonhoeffer T, Grinvald A. 1996. Optical imaging based on intrinsic signals: The methodology. In *Brain mapping: The methods* (ed. Toga A, Mazziotta J), pp. 893–969. Academic Press, New York.

Bonhoeffer T, Kim A, Malonek D, Shoham D, Grinvald A. 1995. The functional architecture of cat area 17. *Eur J Neurosci* **7:** 1973–1988.

Bosking WH, Zhang Y, Schofield B, Fitzpatrick D. 1997. Orientation selectivity and the arrangement of horizontal connections in tree shrew striate cortex. *J Neurosci* **17:** 2112–2127.

Buxton RB. 2001. The elusive initial dip. *Neuroimage* **13:** 953–958.

Cannestra AF, Black KL, Martin NA, Cloughesy T, Burton JS, Rubinstein E, Woods RP, Toga AW. 1998. Topographical and temporal specificity of human intraoperative optical intrinsic signals. *Neuroreport* **9:** 2557–2563.

Cannestra AF, Pouratian N, Bookheimer SY, Martin NA, Becker DP, Toga AW. 2001. Temporal spatial differences observed by functional MRI and human intraoperative optical imaging. *Cereb Cortex* **11:** 773–782.

Chance B, Cohen P, Jobsis F, Schoener B. 1962. Intracellular oxidation-reduction states in vivo. *Science* **137:** 499–508.

Chance B, Kang K, He L, Weng J, Sevick E. 1993a. Highly sensitive object location in tissue models with linear in-phase and anti-phase multi-element optical arrays in one and two dimensions. *Proc Natl Acad Sci* **90:** 3423–3427.

Chance B, Zhuang L, Unah C, Alter C, Lipton L. 1993b. Cognition-activated low-frequency modulation of light absorption in human brain. *Proc Natl Acad Sci* **90:** 3770–3774.

Chapman B, Bonhoeffer T. 1998. Overrepresentation of horizontal and vertical orientation preference in developing ferret area-17. *Proc Natl Acad Sci* **95:** 2609–2614.

Chapman B, Stryker MP, Bonhoeffer T. 1996. Development of orientation preference maps in ferret primary visual cortex. *J Neurosci* **16:** 6443–6453.

Cheng K, Waggoner RA, Tanaka K. 2001. Human ocular dominance columns as revealed by high-field functional magnetic resonance imaging. *Neuron* **32:** 359–374.

Cohen LB. 1973. Changes in neuron structure during action potential propagation and synaptic transmission. *Physiol Rev* **53:** 373–418.

Cohen LB, Keynes RD, Hille B. 1968. Light scattering and birefringence changes during nerve activity. *Nature* **218:** 438–441.

Crair MC, Ruthazer ES, Gillespie DC, Stryker MP. 1997a. Ocular dominance peaks at pinwheel center singularities of the orientation map in cat visual cortex. *J Neurophysiol* **77:** 3381–3385.

Crair MC, Ruthazer ES, Gillespie DC, Stryker MP. 1997b. Relationship between the ocular dominance and orientation maps in visual cortex of monocularly deprived cats. *Neuron* **19:** 307–318.

Crair MC, Gillespie DG, Stryker MP. 1998. The role of visual experience in the development of columns in cat visual cortex. *Science* **279:** 566–570.

Das A, Gilbert CD. 1995. Long-range horizontal connections and their role in cortical reorganization revealed by optical recording of cat primary visual cortex. *Nature* **375:** 780–784.

Das A, Gilbert CD. 1997. Distortions of visuotopic map match orientation singularities in primary visual cortex. *Nature* **387:** 594–598.

Devor A, Dunn AK, Andermann ML, Ulbert I, Boas DA, Dale AM. 2003. Coupling of total haemoglobin concentration, oxygenation, and neural activity in rat somatosensory cortex. *Neuron* **39:** 353–359.

Devor A, Ulbert I, Dunn AK, Narayanan SN, Jones SR, Andermann ML, Boas DA, Dale AM. 2005. Coupling of the cortical haemodynamic response to cortical and thalamic neuronal activity. *Proc Natl Acad Sci* **102:** 3822–3827.

Devor A, Tian P, Nishimura N, Teng IC, Hillman EM, Narayanan SN, Ulbert I, Boas DA, Kleinfeld D, Dale AM. 2007. Suppressed neuronal activity and concurrent arteriolar vasoconstriction may explain negative blood oxygenation level-dependent signal. *J Neurosci* **27:** 4452–4459.

Duong TQ, Kim D-S, Ugurbil K, Kim S-G. 2001. Localized cerebral blood flow response at submillimeter columnar resolution. *Proc Natl Acad Sci* **98:** 10904–10909.

Ernst T, Hennig J. 1994. Observation of a fast response in functional MR. *Magn Reson Med* **32:** 146–149.

Everson R, Knight BW, Sirovich L. 1997. Separating spatially distributed response to stimulation from background. I. Optical imaging. *Biol Cybern* **77:** 407–417.

Fekete T, Omer DB, Naaman S, Grinvald A. 2009. Removal of spatial biological artifacts in functional maps by local similarity minimization. *J Neurosci Methods* **178:** 31–39.

Fox PT, Mintun MA, Raichle ME, Miezin FM, Allman JM, van Essen DC. 1986. Mapping human visual cortex with positron emission tomography. *Nature* **323:** 806–809.

Frostig RD, Lieke EE, Ts'o DY, Grinvald A. 1990. Cortical functional architecture and local coupling between neuronal activity and the microcirculation revealed by in vivo high-resolution optical imaging of intrinsic signals. *Proc Natl Acad Sci* **87:** 6082–6086.

Frostig RD, Masino SA, Kwon MC. 1994. Characterization of functional whisker representation in rat barrel cortex: Optical imaging of intrinsic signals vs. single-unit recordings. *Neurosci Abstr* **20:** 566.

Fukuda M, Wang P, Moon CH, Tanifuji M, Kim SG. 2006a. Spatial specificity of the enhanced dip inherently induced by prolonged oxygen consumption in cat visual cortex: implication for columnar resolution functional MRI. *Neuroimage* **30:** 70–87.

Fukuda M, Moon CH, Wang P, Kim SG. 2006b. Mapping iso-orientation columns by contrast agent-enhanced functional magnetic resonance imaging: Reproducibility, specificity, and evaluation by optical imaging of intrinsic signal. *J Neurosci* **26:** 11821–11832.

Gabbay M, Brennan C, Kaplan E, Sirovich L. 2000. A principal components-based method for the detection of neuronal activity maps: Application to optical imaging. *Neuroimage* **11:** 313–325.

Ghose GM, Ts'o DY. 1997. Form processing modules in primate area V4. *J Neurophysiol* **77:** 2191–2196.

Gochin PM, Bedenbaugh P, Gelfand JJ, Gross CG, Gerstein GL. 1992. Intrinsic signal optical imaging in the forepaw area of rat somatosensory cortex. *Proc Natl Acad Sci* **89:** 8381–8383.

Godecke I, Bonhoeffer T. 1996. Development of identical orientation maps for two eyes without common visual experience. *Nature* **379:** 251–254.

Goodyear BG, Menon RS. 2001. Brief visual stimulation allows mapping of ocular dominance in visual cortex using fMRI. *Hum Brain Mapp* **14:** 210–217.

Gratton G. 1997. Attention and probability effects in the human occipital cortex: An optical imaging study. *Neuroreport* **8:** 1749–1753.

Gratton G, Corballis PM, Cho E, Fabiani M, Hood DC. 1995. Shades of gray matter: Noninvasive optical images of human brain responses during visual stimulation. *Psychophysiology* **32:** 505–509.

Gratton G, Fabiani M, Corballis PM, Gratton E. 1997. Noninvasive detection of fast signals from the cortex using frequency-domain optical methods. *Ann NY Acad Sci* **820:** 286–298.

Grinvald A, Manker A, Segal M. 1982. Visualization of the spread of electrical activity in rat hippocampal slices by voltage sensitive optical probes. *J Physiol* **333:** 269–291.

Grinvald A, Lieke E, Frostig RD, Gilbert CD, Wiesel TN. 1986. Functional architecture of cortex revealed by optical imaging of intrinsic signals. *Nature* **324:** 361–364.

Grinvald A, Frostig RD, Lieke EE, Hildesheim R. 1988. Optical imaging of neuronal activity. *Physiol Rev* **68:** 1285–1366.

Grinvald A, Frostig RD, Siegel RM, Bartfeld E. 1991. High resolution optical imaging of neuronal activity in awake monkey. *Proc Natl Acad Sci* **88:** 11559–11563.

Grinvald A, Shoham D, Shmuel A, Glaser ED, Vanzetta I, Shtoyerman E, Slovin H, Wijnbergen C, Hildesheim R, Sterkin A, Arieli A. 1999. In-vivo optical imaging of cortical architecture and dynamics. A. In *Modern techniques in neuroscience research* (ed. Windhorst U, Johansson H), pp. 893–969. Springer-Verlag, Berlin.

Grinvald A, Slovin H, Vanzetta I. 2000. Non-invasive visualization of cortical columns by f-MRI. *Nat Neurosci* **3:** 105–107.

Grinvald A, Bonhoeffer T, Pollack A, Aloni E, Ofri R, Nelson D. 2004. High resolution functional optical imaging:

From the neocortex to the eye. *Ophthalmol Clin North Am* **17:** 53–67.

Haglund MM, Ojemann GA, Hochman DW. 1992. Optical imaging of epileptiform and functional activity in human cerebral cortex. *Nature* **358:** 668–671.

Harrison RV, Harel N, Panesar J, Mount RJ. 2002. Blood capillary distribution correlates with hemodynamic-based functional imaging in cerebral cortex. *Cereb Cortex* **12:** 225–233.

Hess A, Scheich H. 1996. Optical and FDG mapping of frequency-specific activity in auditory cortex. *Neuroreport* **7:** 2643–2647.

Hill DK, Keynes RD. 1949. Opacity changes in stimulated nerve. *J Physiol* **108:** 278–281.

Hillman EM, Devor A, Bouchard MB, Dunn AK, Krauss GW, Skoch J, Bacskai BJ, Dale AM, Boas DA. 2007. Depth-resolved optical imaging and microscopy of vascular compartment dynamics during somatosensory stimulation. *Neuroimage* **35:** 89–104.

Hoshi Y, Tamura M. 1993. Dynamic multichannel near-infrared optical imaging of human brain activity. *J Appl Physiol* **75:** 1842–1846.

Hu X, Le TH, Ugurbil K. 1997. Evaluation of the early response in fMRI in individual subjects using short stimulus duration. *Magn Reson Med* **37:** 877–884.

Hubel DH, Wiesel TN. 1962. Receptive fields, binocular interactions and functional architecture in the cat's visual cortex. *J Physiol* **160:** 106–154.

Hubel DH, Wiesel TN. 1977. Ferrier lecture. Functional architecture of macaque monkey visual cortex. *Proc R Soc Lond B Biol Sci* **198:** 1–59.

Hübener M, Shoham D, Grinvald A, Bonhoeffer T. 1997. Spatial relationships between three columnar systems in cat area 17. *J Neurosci* **17:** 9270–9284.

Husson TR, Mallik AK, Zhang JX, Issa NP. 2007. Functional imaging of primary visual cortex using flavoprotein autofluorescence. *J Neurosci* **27:** 8665–8675.

Jobsis FF. 1977. Noninvasive, infrared monitoring of cerebral and myocardial oxygen sufficiency and circulatory parameters. *Science* **198:** 1264–1266.

Jobsis FF, Keizer JH, LaManna JC, Rosenthal M. 1977. Reflectance spectrophotometry of cytochrome aa3 in vivo. *J Appl Physiol* **43:** 858–872.

Jones M, Berwick J, Johnston D, Mayhew J. 2001. Concurrent optical imaging spectroscopy and laser-Doppler flowmetry: The relationship between blood flow, oxygenation, and volume in rodent barrel cortex. *Neuroimage* **13:** 1002–1015.

Jung JC, Mehta AD, Aksay E, Stepnoski R, Schnitzer MJ. 2004. In vivo mammalian brain imaging using one- and two-photon fluorescence microendoscopy. *J Neurophysiol* **92:** 3121–3133.

Kalatsky VA, Stryker MP. 2003. New paradigm for optical imaging: Temporally encoded maps of intrinsic signal. *Neuron* **22:** 529–545.

Kaskan PM, Lu HD, Dillenburger BC, Kaas JH, Roe AW. 2009. The organization of orientation-selective, luminance-change and binocular-preference domains in the second (V2) and third (V3) visual areas of New World owl monkeys as revealed by intrinsic signal optical imaging. *Cereb Cortex* **19:** 1394–1407.

Kato T, Kamei A, Takashima S, Ozaki T. 1993. Human visual cortical function during photic stimulation monitoring by means of near-infrared spectroscopy. *J Cereb Blood Flow Metab* **13:** 516–520.

Kelin D. 1925. On cytochrome, a respiratory pigment, common to animals, yeast, and higher plants. *Proc R Soc B* **98:** 312–339.

Kety SS. 1950. Blood flow and metabolism of the human brain in health and disease. *Trans Stud Coll Physicians Phila* **18:** 103–108.

Kim DS, Bonhoeffer T. 1994. Reverse occlusion leads to a precise restoration of orientation preference maps in visual cortex. *Nature* **370:** 370–372.

Kim D-S, Doung TQ, Kim S-G. 2000a. High-resolution mapping of iso-orientation columns by fMRI. *Nat Neurosci* **3:** 164–199.

Kim D-S, Doung TQ, Kim S-G. 2000b. Reply to "Can current fMRI techniques reveal the micro-architecture of cortex?" *Nat Neurosci* **3:** 414.

Kisvarday ZF, Kim DS, Eysel UT, Bonhoeffer T. 1994. Relationship between lateral inhibitory connections and the topography of the orientation map in cat visual cortex. *Eur J Neurosci* **6:** 1619–1632.

Kohl M, Lindauer U, Royl G, Kühl M, Gold L, Villringer A, Dirnagl U. 2000. Physical model for the spectroscopic analysis of cortical intrinsic optical signals. *Phys Med Biol* **45:** 3749–3764.

Kwong KK, Belliveau JW, Chesler DA, Goldberg IE, Weisskoff RM, Poncelet BP, Kennedy DN, Hoppel BE, Cohen MS, Turner R, et al. 1992. Dynamic magnetic resonance imaging of human brain activity during primary sensory stimulation. *Proc Natl Acad Sci* **89:** 5675–5679.

Lassen NA, Ingvar DH. 1961. The blood flow of the cerebral cortex determined by radioactive krypton. *Exp Basel* **17:** 42–43.

Levene MJ, Dombeck DA, Kasischke KA, Molloy RP, Webb WW. 2004. In vivo multiphoton microscopy of deep brain tissue. *J Neurophysiol* **91:** 1908–1912.

L'Heureux B, Gurden H, Pain F. 2009. Autofluorescence imaging of NADH and flavoproteins in the rat brain: Insights from Monte Carlo simulations. *Opt Express* **17:** 9477–9490.

Lindauer U, Royl G, Leithner C, Kuhl M, Gold L, Gethmann J, Kohl-Bareis M, Villringer A, Dirnagl U. 2001. No evidence for early decrease in blood oxygenation in rat whisker cortex in response to functional activation. *Neuroimage* **13:** 988–1001.

Livingstone MS, Hubel DH. 1984. Anatomy and physiology of a color system in the primate visual cortex. *J Neurosci* **4:** 309–356.

Logothetis N. 2000. Can current fMRI techniques reveal the micro-architecture of cortex? *Nat Neurosci* **3:** 413–414.

Logothetis NK, Guggenberger H, Peled S, Pauls J. 1999. Functional imaging of the monkey brain. *Nat Neurosci* **2:** 555–562.

Lorente de Nó R. 1949. Cerebral cortex: Architecture, intracortical connections, motor projections. In *Physiology of the nervous system* (ed. Fulton JF), pp. 288–312. Oxford University Press, New York.

Ma H, Geneslaw A, Zhao M, Suh M, Perry C, Schwartz TH. 2009. The importance of latency in the focality of perfusion and oxygenation changes associated with triggered afterdischarges in human cortex. *J Cereb Blood Flow Metab* **29:** 1003–1014.

Maclin EL, Low KA, Sable JJ, Fabiani M, Gratton G. 2004. The event-related optical signal to electrical stimulation of the median nerve. *Neuroimage* **21:** 1798–1804.

MacVicar BA, Hochman D. 1991. Imaging of synaptically evoked intrinsic optical signals in hippocampal slices. *J Neurosci* **11:** 1458–1469.

MacVicar BA, Hochman D, LeBlanc FE, Watson TW. 1990. Stimulation evoked changes in intrinsic optical signals in the human brain. *Soc Neurosci Abstr* **16:** 309.

Malach R, Amir Y, Harel M, Grinvald A. 1993. Novel aspects of columnar organization are revealed by optical imaging and in vivo targeted biocytin injections in primate striate cortex. *Proc Natl Acad Sci* **90:** 10469–10473.

Malach R, Tootell RBH, Malonek D. 1994. Relationship between orientation domains, cytochrome oxidase stripes and intrinsic horizontal connections in squirrel monkey area V2. *Cereb Cortex* **4:** 151–165.

Malach R, Schirman TD, Harel M, Tootell RBH, Malonek D. 1997. Organization of intrinsic connections in owl monkey area MT. *Cereb Cortex* **7:** 386–393.

Mallik AK, Husson TR, Zhang JX, Rosenberg A, Issa NP. 2008. The organization of spatial frequency maps measured by cortical flavoprotein autofluorescence. *Vision Res* **48:** 1545–1553.

Malonek D, Grinvald A. 1996. The imaging spectroscope reveals the interaction between electrical activity and cortical microcirculation: Implication for functional brain imaging. *Science* **272:** 551–554.

Malonek D, Tootell RBH, Grinvald A. 1994. Optical imaging reveals the functional architecture of neurons processing shape and motion in owl monkey area MT. *Proc R Soc Lond B Biol Sci* **258:** 109–119.

Malonek D, Dirnagl U, Lindauer U, Yamada K, Kanno I, Grinvald A. 1997. Vascular imprints of neuronal activity. Relationships between dynamics of cortical blood flow, oxygenation and volume changes following sensory stimulation. *Proc Natl Acad Sci* **94:** 14826–14831.

Martin C, Berwick J, Johnston D, Zheng Y, Martindale J, Port M, Redgrave P, Mayhew J. 2002. Optical imaging spectroscopy in the unanaesthetised rat. *J Neurosci Methods* **120:** 25–34.

Masino SA, Frostig RD. 1996. Quantitative long-term imaging of the functional representation of a whisker in rat barrel cortex. *Proc Natl Acad Sci* **93:** 4942–4947.

Mayevsky A, Chance B. 1982. Intracellular oxidation–reduction state measured in situ by a multichannel fiber-optic surface fluorometer. *Science* **217:** 537–540.

Mayhew J, Hu D, Zheng Y, Askew S, Hou Y, Berwick J, Coffey PJ, Brown N. 1998. An evaluation of linear model analysis techniques for processing images of microcirculation activity. *Neuroimage* **7:** 49–71.

Mayhew J, Zheng Y, Hou Y, Vuksanovic B, Berwick J, Askew S, Coffey P. 1999. Spectroscopic analysis of changes in remitted illumination: The response to increased neural activity in brain. *Neuroimage* **10:** 304–326.

Mayhew J, Johnston D, Berwick J, Jones M, Coffey P, Zheng Y. 2000. Spectroscopic analysis of neural activity in brain: Increased oxygen consumption following activation of barrel cortex. *Neuroimage* **12:** 664–675.

Menon RS, Goodyear BG. 1999. Submillimeter functional localization in human striate cortex using BOLD contrast at 4 Tesla: Implications for the vascular point-spread function. *Magn Reson Med 4:* 230–235.

Menon RS, Ogawa S, Hu X, Strupp JP, Anderson P, and Ugurbil K. 1995. BOLD based functional MRI at 4 Tesla includes a capillary bed contribution: Echo-planar imaging correlates with previous optical imaging using intrinsic signals. *Magn Reson Med* **33:** 453–459.

Menon RS, Ogawa S, Strupp JP, Ugurbil K. 1997. Ocular dominance in human V1 demonstrated by functional magnetic resonance imaging. *J Neurophysiol* **77:** 2780–2787.

Millikan G.A. 1937. Experiments on muscle hemoglobin in vivo: The instantaneous measurement of muscle metabolism. *Proc R Soc B* **123:** 218–241.

Moon CH, Fukuda M, Park SH, Kim SG. 2007. Neural interpretation of blood oxygenation level-dependent fMRI maps at submillimeter columnar resolution. *J Neurosci* **27:** 6892–6902.

Mosso A. 1880. Sulla circolazione del cervello dell'uomo. *Atti R Accad Lincei* **5:** 237–358.

Mountcastle VB. 1957. Modality and topographic properties of single neurons of cat's somatic sensory cortex. *J Neurophysiol* **20:** 408–434.

Nagaoka T, Zhao F, Wang P, Harel N, Kennan RP, Ogawa S, Kim SG. 2006. Increases in oxygen consumption without cerebral blood volume change during visual stimulation under hypotension condition. *J Cereb Blood Flow Metab* **26:** 1043–1051.

Nelson DA, Krupsky S, Pollack A, Aloni E, Belkin M, Vanzetta I, Rosner M, Grinvald A. 2005. Special report: Noninvasive multi-parameter functional optical imaging of the eye. *Ophthalmic Surg Lasers Imaging* **36:** 57–66.

Obermayer K, Blasdel GG. 1993. Geometry of orientation and ocular dominance columns in monkey striate cortex. *J Neurosci* **13:** 4114–4129.

Ogawa S, Tank DW, Menon R, Ellermann JM, Kim SG, Merkle H, Ugurbil K. 1992. Intrinsic signal changes accompanying sensory stimulation: Functional brain mapping with magnetic resonance imaging. *Proc Natl Acad Sci* **89:** 5951–5955.

Polley DB, Chen-Bee CH, Frostig RD. 1999. Two directions of plasticity in the sensory-deprived adult cortex. *Neuron* **24:** 623–637.

Polley DB, Kvasnak E, Frostig RD. 2004. Naturalistic experience transforms sensory maps in the adult cortex of caged animals. *Nature* **429:** 67–71.

Radhakrishnan H, Vanduffel W, Deng HP, Ekstrom L, Boas DA, Franceschini MA. 2009. Fast optical signal not detected in awake behaving monkeys. *Neuroimage* **45:** 410–419.

Raichle ME, Martin WRW, Herscovitz P, Minton MA, Markham JJ. 1983. Brain blood flow measured with intravenous $H_2\,^{15}O$. II. Implementation and validation. *J Nucl Med* **24:** 790–798.

Reidl J, Starke J, Omer DB, Grinvald A, Spors H. 2007. Independent component analysis of high-resolution imaging data identifies distinct functional domains. *Neuroimage* **34:** 94–108.

Reinert KC, Dunbar RL, Gao W, Chen G, Ebner TJ. 2004. Flavoprotein autofluorescence imaging of neuronal activation in the cerebellar cortex in vivo. *J Neurophysiol* **92:** 199–211.

Roe AW, Ts'o DY. 1995. Visual topography in primate V2: Multiple representation across functional stripes. *J Neurosci* **15:** 3689–3715.

Roy C, Sherrington C. 1890. On the regulation of the blood supply of the brain. *J Physiol* **11:** 85–108.

Rubin BD, Katz LC. 1999. Optical imaging of odorant representations in the mammalian olfactory bulb. *Neuron* **23:** 499–511.

Salzberg BM, Obaid AL, and Gainer H. 1986. Optical studies of excitation secretion at the vertebrate nerve terminal. *Soc Gen Physiol Ser* **40:** 133–164.

Schuett S, Bonhoeffer T, Hubener M. 2002. Mapping retinotopic structure in mouse visual cortex with optical imaging. *J Neurosci* **22:** 6549–6559.

Schwartz TH, Bonhoeffer T. 2001. In vivo optical mapping of epileptic foci and surround inhibition in ferret cerebral cortex. *Nat Med* **7:** 1063–1067.

Sheth SA, Nemoto M, Guiou MW, Walker M, Toga AW. 2005. Spatiotemporal evolution of functional haemodynamic changes and their relationship to neuronal activity. *J Cereb Blood Flow Metab* **25:** 830–841.

Shibuki K, Hishida R, Murakami H, Kudoh M, Kawaguchi T, Watanabe M, Watanabe S, Kouuchi T, Tanaka R. 2003. Dynamic imaging of somatosensorycortical activity in the rat visualized by flavoprotein autofluorescence. *J Physiol* **549:** 919–927.

Shmuel A, Grinvald A. 1996. Functional organization for direction of motion and its relationship to orientation maps in cat area 18. *J Neurosci* **16:** 6945–6964.

Shmuel A, Korman M, Sterkin A, Harel M, Ullman S, Malach R, Grinvald A. 2005. Retinotopic axis specificity and selective clustering of feedback projections from V2 to V1 in the owl monkey. *J Neurosci* **25:** 2117–2131.

Shoham D, Grinvald A. 2001. Visualization of the hand representation in the hand in macaque and human area S-I using intrinsic signal optical imaging. *J Neurosci* **21:** 6820–6835.

Shoham D, Hubener M, Grinvald A, Bonhoeffer T. 1997. Spatio-temporal frequency domains and their relationship to cytochrome oxidase staining in cat visual cortex. *Nature* **385:** 529–534.

Shtoyerman E, Arieli A, Slovin H, Vanzetta I, Grinvald A. 2000. Long term optical imaging and spectroscopy reveal mechanisms underlying the intrinsic signal and stability of cortical maps in V1 of behaving monkeys. *J Neurosci* **20:** 8111–8121.

Sirovich L, Everson RM, Kaplan E, Knight BW, O'Brien EV, Orbach D. 1996. Modeling the functional organization of the visual cortex. *Physica D* **96:** 355–366.

Slovin H, Arieli A, Hildesheim R, Grinvald A. 2002. Long-term voltage-sensitive dye imaging of cortical dynamics in the behaving monkey. *J Neurophys* **88:** 3421–3438.

Slovin H, Strick PL, Hildesheim R, Grinvald A. 2003. Voltage sensitive dye imaging in the motor cortex. I. Intra- and intercortical connectivity revealed by microstimulation in the awake monkey. *Soc Neuroci Abst* 554.8.

Sokoloff L. 1977. Relation between physiological function and energy metabolism in the central nervous system. *J Neurochem* **19:** 13–26.

Strick P, Grinvald A, Hildesheim R, Slovin H. 2003. Voltage sensitive dye imaging in the motor cortex II. Cortical correlates of Go/No-Go delayed response task. *Soc Neuroci Abst* 918.8.

Swindale NV, Matsubara JA, Cynader MS. 1987. Surface organization of orientation and direction selectivity in cat area 18. *J Neurosci* **7:** 1414–1427.

Tanner K, D'Amico E, Kaczmarowski A, Kukreti S, Malpeli J, Mantulin WW, Gratton E. 2005. Spectrally resolved neurophotonics: A case report of haemodynamics and vascular components in the mammalian brain. *J Biomed Opt* **10:** 064009.

Tanner K, Beitel E, D'Amico E, Mantulin WW, Gratton E. 2006. Effects of vasodilation on intrinsic optical signals in the mammalian brain: A phantom study. *J Biomed Opt* **1:** 064020.

Tasaki I, Watanabe A, Sandlin R, Carnay L. 1968. Changes in fluorescence, turbidity and birefringence associated with nerve excitation. *Proc Natl Acad Sci* **61:** 883–888.

Thompson JK, Peterson MR, Freeman RD. 2003. Single-neuron activity and tissue oxygenation in the cerebral cortex. *Science* **299:** 1070–1072.

Tse C-Y, Lee C-L, Sullivan J, Garnsey SM, Dell GS, Fabiani M, Gratton G. 2007. Imaging cortical dynamics of language processing with the event-related optical signal. *Proc Natl Acad Sci* **104:** 17157–17162.

Ts'o DY, Frostig RD, Lieke E, Grinvald A. 1990. Functional organization of primate visual cortex revealed by high resolution optical imaging. *Science* **249:** 417–420.

Ts'o DY, Gilbert CD, Wiesel TN. 1991. Orientation selectivity of and interactions between color and disparity subcompartments in area V2 of Macaque monkey. *Neurosci Abstr* **17:** 431.

Ts'o DY, Roe AW, Shey J. 1993. Functional connectivity within V1 and V2: Patterns and dynamics. *Neurosci Abstr* **19:** 618.

Vanzetta I. 2006. Hemodynamic responses in cortex investigated with optical imaging methods. Implications for functional brain mapping. *J Physiol* **100:** 201–211.

Vanzetta I, Grinvald A. 1999. Increased cortical oxidative metabolism due to sensory stimulation: Implications for functional brain imaging. *Science* **286:** 1555–1558.

Vanzetta I, Grinvald A. 2001. Evidence and lack of evidence for the initial dip in the anesthetized rat: Implications for human functional brain imaging. *Neuroimage* **13:** 959–967.

Vanzetta I, Grinvald A. 2008. Coupling between neuronal activity and microcirculation: Implications for functional brain imaging. *HFSP J* **2:** 79–98.

Vanzetta I, Slovin H, Omer DB, Grinvald A. 2004. Columnar resolution of blood volume and oximetry functional maps in the behaving monkey: Implications for FMRI. *Neuron* **42:** 843–854.

Wang G, Tanaka K, Tanifuji M. 1996. Optical imaging of functional organization in the monkey inferotemporal cortex. *Science* **272:** 1665–1668.

Weber B, Burger C, Wyss MT, von Schulthess GK, Scheffold F, Buck A. 2004. Optical imaging of the spatiotemporal dynamics of cerebral blood flow and oxidative metabolism in the rat barrel cortex. *Eur J Neurosci* **20:** 2664–2670.

Weliky M, Bosking WH, Fitzpatrick D. 1996. A systematic map of direction preference in primary visual cortex. *Nature* **379:** 725–728.

Wolf M, Wolf U, Choi JH, Gupta R, Safonova LP, Paunescu LA, Michalos A, Gratton E. 2003. Detection of the fast neuronal signal on the motor cortex using functional frequency domain near infrared spectroscopy. *Adv Exp Med Biol* **510:** 193–197.

Wyatt JS, Cope M, Deply DT, Wray S, Reynolds EOR. 1986. Quantitation of cerebral oxygenation and haemodynamics in sick newborn infants by near–infrared spectrophotometry. *Lancet* **2:** 1063–1066.

Wyatt JS, Cope D, Deply DT, Richardson CE, Edwards AD, Wray S, Reynolds EOR. 1990. Quantitation of cerebral blood volume in human infants by near-infrared spectroscopy. *J Appl Physiol* **68:** 1086–1091.

Yacoub E, Hu X. 1999. Detection of the early negative response in f-MRI at 1.5 Tesla. *Magn Reson Med* **41:** 1088–1092.

Yacoub E, Shmuel A, Logothetis N, Uğurbil K. 2007. Robust detection of ocular dominance columns in humans using Hahn Spin Echo BOLD functional MRI at 7 Tesla. *Neuroimage* **37:** 1161–1177.

Zhao F, Wang P, Hendrich K, Kim S-G. 2005. Spatial specificity of cerebral blood volume-weighted fMRI responses at columnar resolution. *Neuroimage* **27:** 416–424.

Zhao M, Suh M, Ma H, Perry C, Geneslaw A, Schwartz TH. 2007. Focal increases in perfusion and decreases in hemoglobin oxygenation precede seizure onset in spontaneous human epilepsy. *Epilepsia* **48:** 2059–2067.

84 Two-Photon Imaging of Blood Flow in Cortex

Jonathan D. Driscoll,[1] Andy Y. Shih,[1] Patrick J. Drew,[1] Gert Cauwenberghs,[3] and David Kleinfeld[1,2]

[1]Department of Physics, University of California at San Diego, California 92093; [2]Graduate Program in Neurosciences, University of California at San Diego, California 92093; [3]Department of Bioengineering, University of California at San Diego, La Jolla, California 92093

ABSTRACT

Cerebral blood flow plays a central role in maintaining homeostasis in the brain, and its dysfunction leads to pathological conditions such as stroke. Moreover, understanding the dynamics of blood flow is central to the interpretation of data from imaging modalities—such as intrinsic optical signaling and functional magnetic resonance imaging—that rely on changes in cerebral blood flow and oxygen level to infer changes in the underlying neural activity. Recent advances in imaging techniques have allowed detailed studies of blood flow in vivo at high spatial and temporal resolutions. We discuss techniques to accurately measure cerebral blood flow at the level of individual blood vessels using two-photon laser-scanning microscopy. By directing the scanning laser along a user-defined path, it is possible to measure red blood cell velocity and vessel diameter across multiple vessels simultaneously. The combination of these measurements permits accurate assessment of total flux with sufficient time resolution to measure fast modulations in flux, such as those caused by heartbeat, as well as slower signals caused by vasomotion and hemodynamic responses to stimulus.

INTRODUCTION

Neural activity is accompanied by both increases and decreases in local blood flow to active brain regions (Woolsey et al. 1996; Derdikman et al. 2003; Devor et al. 2008). Detection of functional changes in cerebral blood flow are the basis of wide-field brain imaging modalities, including intrinsic optical signaling (Grinvald et al. 1986; Frostig et al. 1990) and functional magnetic resonance imaging (Ogawa et al. 1990). An understanding of the fundamental mechanisms of neurovascular coupling is critical for interpreting such data and is an active area of research.

Two-photon laser scanning microscopy is an ideal method for studying blood in individual vessels, from the level of the pial vessels down to subsurface capillaries. Two-photon microscopy has been used to acquire red blood cell (RBC) velocity (Kleinfeld et al. 1998; Belayev et al. 2002; Chaigneau et al. 2003; Hirase et al. 2004; Chuquet et al. 2007; Zhang and Murphy 2007) or lumen diameter (Devor et al. 2007; Devor et al. 2008; Shih et al. 2009) from a single vessel in a given imaging trial. However, the oxygen- and nutrient-carrying capacity of a vessel is proportional to its flux, which is a function of both RBC velocity and vessel diameter. Since these two parameters can change independently, both must be measured simultaneously to accurately access total blood flow (Kontos 1989; Shih et al. 2009).

Here, we review general techniques for animal preparation and measurement of blood flow with two-photon microscopy. We incorporate extensions to existing methods to accurately acquire flux data simultaneously across multiple vessels in a single trial. Central to these measurements is the ability to generate scan paths that smoothly connect user-defined lines of interest while maintaining high accuracy of the scan path.

Protocol

Cranial Window Preparation in the Rat for Two-Photon Imaging of Blood Flow in the Cortex

A craniotomy is made above the cortical area of interest; see Mostany and Portera-Caillau (2008) for a demonstration of the original procedure (Kleinfeld et al 1998). In our example data, this corresponds to the somatosensory area of parietal cortex in rat. A custom-made head frame is cemented to the skull to provide a means to secure the animal under the microscope (Fig. 1). As a means to protect the brain and prevent motion artifacts, the exposed region of cortex is covered with agarose prepared in an artificial cerebrospinal fluid and sealed with a microscope coverslip. Note that an alternative technique for mice makes use of a thinned and reinforced skull to avoid exposing the brain at the price of restricted optical access (Drew et al. 2010b). All procedures must be performed in accordance with the relevant animal care guidelines.

MATERIALS

CAUTION: See Appendix 6 for proper handling of materials marked with <!>.
See the end of the chapter for recipes for reagents marked with <R>.

Reagents

Agarose, low melting point (Type III-A, Sigma-Aldrich)
Dextrose, 5% (w/v) in saline
Fluoroscein-dextran (2 MDa, Sigma-Aldrich) or Texas Red-dextran (70 kDa, Invitrogen), 5% (w/v) in saline
Freeze the dextran-conjugated dye solution(s) in aliquots at –20°C for later use.
Heparin <!>, 20 U/mL in saline
Isoflurane anesthesia <!> (for survival studies)
Lidocaine solution <!>, 2% (v/v)
Modified artificial cerebral spinal fluid (mACSF), free of carbonate and phosphate <R>
Ophthalmic ointment
Rose Bengal, 1% (w/v) in saline (filtered before use)
Freeze the 1% Rose Bengal solution in aliquots at –20°C for later use.
SurgiFoam (Edgepark)
Urethane <!> or α-cholarlose anesthesia (for terminal studies)
VetBond (3M)

Equipment

Blood gas monitor (RapidLab 248; Bayer)
Blood pressure monitor (BP1, World Precision Instruments for intra-arterial; XBP1000, Kent Scientific for tail cuff measurements)
Catheter (Surflo; Terumo)
Cover Glass, no. 1 thickness
Dental acrylic (Grip Cement; Dentsply)

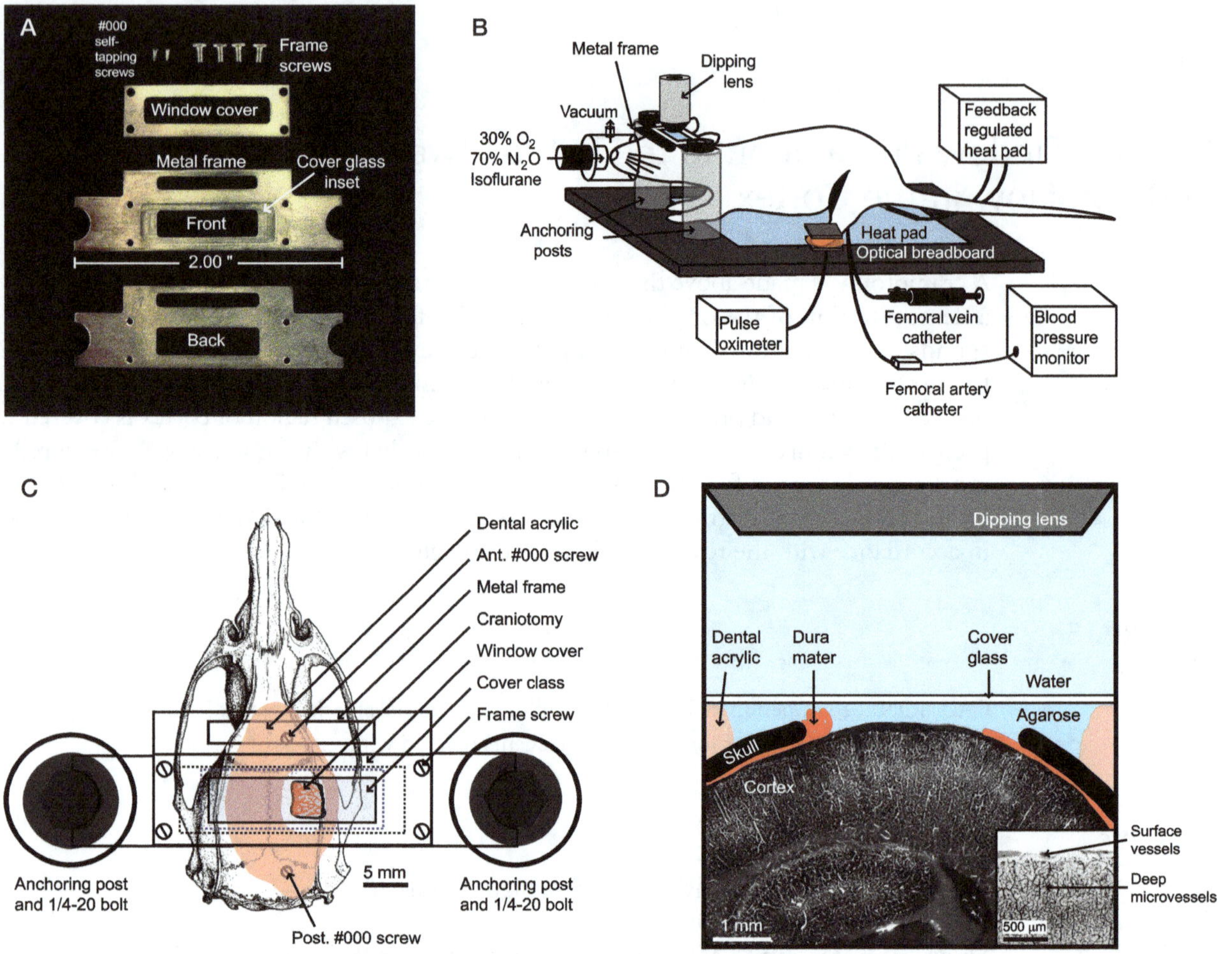

FIGURE 1. Setup for in vivo imaging of blood flow through a cranial window. (*A*) To immobilize the head of the animal during imaging, we designed a metal frame that could be cemented to the skull, and then anchored to an optical setup. The frame is constructed of type 410 stainless steel with dimensions of 2.00 inches long, 0.61 inches wide, and 0.029 inches thick, and can be secured between two posts in a standard optical breadboard. An inset region, 0.015 inches deep, borders the frame window to hold a cover glass over the craniotomy. The window cover, 0.015 inches thick of type 301 stainless steel, is then secured to the frame with four screws, sandwiching a no. 1 cover glass in place. (*B*) A typical experimental setup. Note: The metal frame attached to the skull is immobilized between two anchoring posts inserted into an optical breadboard. (*C*) Diagram of cranial window preparation for rat. (*D*) A cross-sectional view of the cranial window. (*Inset*) An inverted coronal view of surface vessels and deep microvessels that are targeted for occlusion.

Dental drill, air-powered (Silent Air)
Drill burrs, 0.5- and 0.25-mm tip sizes (Henry Schein)
Forceps, extra-sharp (Dumont no. 55; Fine Science Tools)
Glass cutter
Head-frame, custom-made (Fig. 1A)
Heat pad, feedback-regulated (Harvard for rats; FHC for mice)
Hemostats
Isoflurane vaporizer (IsoTec)
Kimwipes
Optical breadboard with head holder (Fig. 1B) and devices for physiological support and monitoring
PE50 tubing, for femoral artery/vein catheters (Intramedic)
Periosteal elevator (Roboz)

Pulse oximeter (Nonin for rats; Starr Life Sciences for mice)
Scalpel blades
Screwdriver, miniature
Screws, self-tapping (#000-3/32; Small Parts Inc.)
Stereotaxic frame (Kopf)
Syringe needle, 26-gauge
Two-photon laser-scanning microscope

Our TPLSM is a custom-design optimized for in vivo studies (Nguyen et al. 2006, 2009; Tsai and Kleinfeld 2009).

EXPERIMENTAL METHOD

Anesthetizing the Rat

1. Anesthetize the rat.

   Common anesthetic choices include (1) urethane delivered intraperitoneally, that is, 1000 mg/kg body weight initial dose with 100 mg/kg supplements as required (Kleinfeld and Denk 2000), or (2) initial isoflurane for surgery followed by a transition to intravenous delivery of α-chloralose for imaging, that is, an initial bolus at 50 mg/kg body weight for induction and continuous delivery of 40 mg/kg for maintenance (Devor et al. 2007). Both urethane and α-chloralose should be prepared fresh on the day of the experiment. Heating to ~60°C and agitation is necessary to dissolve α-chloralose. These anesthetics are not suitable for survival experiments.

2. Check for lack of toe pinch reflex to ensure an adequate level of anesthesia.
3. Secure the rat in a stereotaxic frame.
4. Apply ophthalmic ointment to the eyes to keep them moist.
5. Inject 0.1 mL of 2% (v/v) lidocaine subcutaneously into the scalp as a local anesthetic before incision (in Step 11).

Monitoring Requirements

The following procedures need to be performed throughout the surgery and during imaging.

6. Monitor heart and breathing rates with a pulse oximeter. They should remain within a normal range: 300–400 beats/min and 60–120 events/min.
7. Maintain body temperature at 37°C using a feedback-regulated rectal probe and heat pad.
8. Inject 5% (w/v) dextrose in saline intraperitoneally at 3 mL/kg body weight every 2 h to maintain body fluids and energy requirements.
9. Install a femoral artery catheter to collect blood samples and a femoral vein catheter to administer drugs. Monitored blood gases every 2 h.
10. Monitor arterial blood pressure continuously from the femoral arterial line. Alternatively, the tail cuff method can be used to noninvasively measure blood pressure at intermittent time points.

Cranial Window Surgery

11. Shave the surgical area and make a 4–5-cm incision down the midline of the scalp. Use a periosteal elevator to remove the thin periosteum from the surface of the skull.
12. Demarcate the location of the cranial window. We typically consider measurements over primary somatosensory cortex, which is the part of parietal cortex that nominally lies between –1 and –5 mm relative to the Bregma point and between 1 and 7 mm from the midline on the mediolateral axis for rats (Paxinos and Watson 1986).

13. Attach a custom metal frame to the skull with dental acrylic. The frame holds the head of the animal rigidly to the optical apparatus (Figs. 1A,B).
    i. Clean the soft tissue from the contact regions of the bone.
    ii. Apply a thin layer of VetBond.
    iii. Introduce small self-tapping screws into the anterior and posterior aspects of the skull. (Note that one of the screws passes through an opening in the frame.) Mechanically link the screws to the frame with dental cement (Fig. 1C).
    iv. Secure the threads of the screws to the bone with a small dab of VetBond.

        Note that the temporalis muscle may need to be retracted in some cases, and the temporal ridge may need to be flattened with a dental drill. This is to ensure that the metal head frame is cemented tangential to the cortical surface when imaging the lateral aspects of the barrel cortex.

14. Perform a craniotomy above the brain region of interest using a high-speed drill.
    i. Thin the skull throughout the entire window to approximately one-quarter of its original thickness using a 0.5-mm drill burr until the underlying pial vasculature becomes visible following application of mACSF. During drilling, flush the window regularly with mACSF to reduce heat buildup and to remove blood and bone shavings.
    ii. Carefully thin the edges of the window with a 0.25-mm burr until the bone begins to craze.
    iii. Use forceps to gently separate the bone flap from the skull without protruding too deeply.
    iv. Use two forceps to grasp adjacent corners of the loosened bone flap, and slowly peel it away from the underlying dura mater.
15. Reflect the dura to the edges of the window (Fig. 1D).
    i. Make a small incision using the cutting edge of a 26-gauge syringe needle. Bend the needle to an obtuse angle with hemostats to ensure that the cutting edge approaches the dura at a suitable angle.
    ii. Use two sharp no. 55 forceps to gently lift the dura away from the cortical surface, starting at the incision site, and tearing in small increments. Whenever possible, tear around large dural vessels to avoid bleeding.

        Limit any dural bleeding with small pieces of Surgifoam soaked in mACSF, and Kimwipes twisted to a fine point with the fingers. Keep the cranial window moist with a piece of moist Surgifoam.

    iii. Retract the dural flaps to the side of the window, and flush the cortical surface with mACSF.

        It is crucial to avoid any damage to pial vessels. Hemorrhaging will alter cerebral blood flow, accelerate edema, and severely degrade imaging quality.

16. Fill the interior of the chamber with 1.5% (w/v) low-melting-point agarose dissolved in mACSF (Fig. 1D) (Kleinfeld and Delaney 1996).

    Dissolve the agarose by heating it in a microwave. The temperature of the agarose must not exceed 37°C when it is applied to the brain.

17. Immediately seal the chamber using a cover glass as a window (Fig. 1D). Resealing the craniotomy is crucial to protect the cortex and suppress motion from cranial pressure fluctuations due to heartbeat and breathing. One edge of the window can remain uncovered to allow insertion of electrodes or micropipettes.
18. Suture the skin together around the frame, and trail agarose around the cover glass to hold water for the dipping lens.
19. Stabilize the animal on an optical breadboard for imaging, using the frame as a head support.

    Our separate plate can be transported between surgical and imaging suites with the animal and all physiological monitoring devices assembled as one unit (Fig. 1B).

EXAMPLE APPLICATION

A cranial window was generated over the rat somatosensory cortex (Fig. 2). Intrinsic optical imaging was used to determine the locations of the hindlimb and forelimb cortical representation (Drew and Feldman 2009). The cerebral vasculature of the hindlimb somatosensory region was mapped using a TPLSM image stack obtained with a low magnification 4x objective (Fig. 2A). A single penetrating arteriole and neighboring ascending venule were selected for measurement under a 40x objective (Fig. 2B).

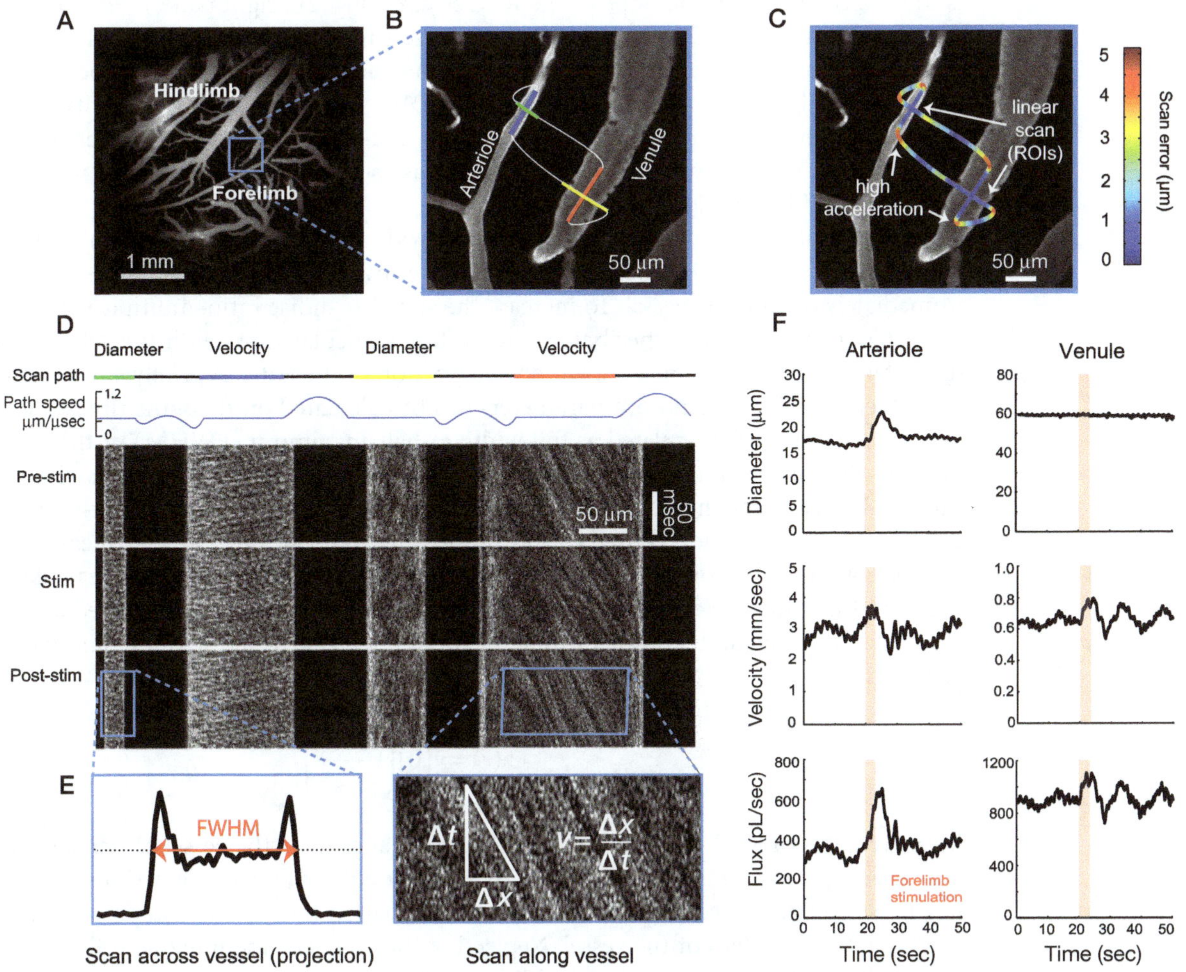

FIGURE 2. Simultaneous measurement of diameter and velocity in multiple vessels using spatially optimized line scans. *(A)* Image of fluorescently stained vessels in somatosensory cortex of a Sprague Dawley rat. The forelimb and hindlimb representations across cortex were mapped using intrinsic optical imaging. *(B)* Image of a surface arteriole and venule, with scan pattern superimposed. Portions of the scan path along the length are used to calculate red blood cell (RBC) velocity, while portions across the diameter of the vessels are used to calculate diameter. Scans were acquired at a rate of 735 lines per second. *(C)* Scan path, colored to show the error between the desired scan path and the actual path the mirrors traversed. The error along linear portions of the image is ~1 μm, and increases when the mirrors undergo rapid acceleration. The error between successive scans of the same path is <0.15 μm, several times lower than the point-spread function of a TPLSM. *(D)* Mirror speed as a function of time. Note that portions used to acquire diameter and velocity data are constant speed *(top)*. The line scans generated from the path can be stacked sequentially as a function of time to produce a raw cascade image *(bottom)*. *(E)* Vessel diameter is calculated as the full width at half-maximum of a time average of several scans across the width of a vessel *(left)*. RBC velocity calculated from the angle of the RBC streaks. *(F)* Data traces of diameter, velocity, and flux for the arteriole and venule, processed to remove heart rate and smoothed with a running window. Both vessels show an increase in flux in response to forelimb stimulation. In the arteriole, this flux increase is due to simultaneous increase of lumen diameter and RBC velocity. In contrast, flux increase in the venule is due only to an increase in RBC velocity, as diameter is unchanged by stimulation.

A scan path was created to traverse along the length of the lumen center, and across the lumen width for each vessel. The accuracy of the scan path can be verified by comparing the value of the mirror position encoder with the control voltage. Scan accuracy in user-selected linear scan paths is ~1 μm, whereas intervening segments that connect the user-defined regions and are not used in data analysis have a higher error, up to ~5 μm (Fig. 2C). The resulting line scan is a space-time image (Fig. 2D).

Portions of the scan path along the vessel length appear as streaks within the line-scan image. These streaks represent nonfluorescent RBCs that move through a fluorescent background. The x-axis represents the distance traveled by the RBCs, and the y-axis is the time. The centerline velocity is then calculated from the slope of the RBC streaks (Fig. 2E, right panel). Previously, singular value decomposition was used to find the slope of these streaks and calculate the RBC velocity (Kleinfeld et al. 1998; Schaffer et al. 2006). A faster method that can calculate the velocity in near real time makes use of the Radon transform (Drew et al. 2010a). In either case, the velocity is taken using windowed portions of the line scans, typically 25 msec of data for a Nyquist frequency of 20 Hz (see Fig. 1 in Chapter 7). This is sufficient to capture the fastest biological signal, the heart rate, without aliasing the second harmonic of the rate.

Portions of the scan path across the vessel width capture the diameter of the vascular lumen, because the fluorescently labeled blood plasma provides high contrast with the unlabeled region immediately outside the vessel. To increase the signal-to-noise ratio, multiple scans across the vessel diameter are averaged together before the profile is calculated. As with the velocity calculation, the diameter transform is made on a windowed portion of the data. Typically, the same window size as for velocity is used so that both parameters can be calculated on the same timescale.

Vessel diameter is calculated as full width at half-maximum (FWHM) of the vessel profile (Fig. 2E, left). Because the intensity profile tends to decrease in the center of the vessel, where a large volume of RBCs exclude the fluorescently labeled plasma, the two outermost half-maximum points are used. Linear interpolation is used to add subpixel accuracy to the diameter measurement.

We now consider combining the velocity and diameter measurements to calculate flux. In laminar flow, the velocity $v(r)$ of the fluid through the pipe with radius R measured at a distance r from the center has a quadratic profile, with a maximal velocity $v(0)$ in the center and tapering to 0 at the edges, that is,

$$v(r) = v(0)\left(1 - \frac{r^2}{R^2}\right). \tag{1}$$

Measurements of RBC velocity as a function of distance from the vessel center show that this model is a close approximation (Schaffer et al. 2006), although systematic deviations result from the non-negligible size of the red blood cells. This tends to flatten the profile, and prevents the velocity from reaching 0 at the edges of the vessel. Neglecting these effects, the integrated flux through the vessel is given by (Helmchen and Kleinfeld 2008)

$$F = (1/2)\ v(0)\ \pi R^2. \tag{2}$$

In our example, both vessel diameter and RBC velocity in the arteriole respond to somatotopic stimulation. The flux through the arteriole increases by ~100% over baseline, compared with 35% and 40% for diameter and velocity measurements alone, respectively (Fig. 2F). Since the scan path runs at 733 Hz in this example, the diameter and RBC velocity traces are collected nearly simultaneously from both the penetrating arteriole and ascending venule.

GENERATION OF SPATIALLY OPTIMIZED LINE SCANS

The microscope laser is directed by a pair of fast x–y scan mirrors (Cambridge Technology, 6210H galvometer optical scanner with MicroMax 673xx dual-axis servo driver). The mirror controller employs a closed loop position feedback system to accurately map control voltage to mirror position. The position accurately tracks the control voltage, with ~90 μsec delay.

To create an arbitrary scanning path, several full-frame images of the region containing the desired vessels are first acquired and averaged together to increase the signal-to-noise ratio. This image is loaded by custom software written in MATLAB (MathWorks), which allows users to interactively create lines of interest (i.e., scans across or along vessels) on the full frame image.

The mirror voltages used to acquire the original full frame image are known, so that positions on the image can be mapped one-to-one to mirror voltages. Linear portions of the scan path, such as those used to track along a vessel to measure RBC speed and those that span the vessel to measure diameter, are given by

$$\mathbf{P} = \mathbf{P}_0 + \mathbf{V}_{\text{linear}}\, t, \tag{3}$$

where **P** and **V** are two-dimensional vectors that contain x and y coordinates.

Linear portions of the scan path, scanned at constant velocity, must be connected by a function that creates a physically realistic path for the mirrors to follow (Lillis et al. 2008). Although there are an arbitrary number of such functions, the simplest is a third-order polynomial given by

$$\mathbf{P}_{\text{spline}} = \mathbf{P}_i + \mathbf{V}_i t + \mathbf{C}t^2 + \mathbf{D}t^3. \tag{4}$$

For computational convenience, the connecting spline is taken to start at $t = 0$, and end at $t = \tau$. The initial position $\mathbf{P}_i$ and velocity $\mathbf{V}_i$ of the spline are set to match the position and velocity of the end of the linear region preceding the spline.

The additional spline parameters depend on the length (in time) of the spline, τ, and the final position and velocity of the spline, and are given by

$$\mathbf{C} = \frac{3\mathbf{P}_f}{\tau^2} - \frac{3\mathbf{P}_i}{\tau^2} - \frac{2\mathbf{V}_i}{\tau} - \frac{\mathbf{V}_f}{\tau} \tag{5}$$

and

$$\mathbf{D} = \frac{\mathbf{V}_f}{3\tau^2} - \frac{\mathbf{V}_i}{3\tau^2} - \frac{2\mathbf{C}}{3\tau}\,. \tag{6}$$

The value of τ used should be the smallest positive real value that does not subject the mirrors to an acceleration larger than a user-defined maximal value, m, typically 100 V/msec^2. Because the acceleration of the spline is a linear function of time, the extreme values of acceleration occur at the beginning and the end of the spline.

Candidate values for the shortest possible spline length are found by setting the mirror acceleration to $\pm m$ at the beginning and end of spline, and finding all positive real values for τ:

$$0 = \pm m\,\tau^2 + (2V_{i,x} + 4V_{f,x})\tau + (6P_{i,x} - 6P_{f,x}), \tag{7}$$

$$0 = \pm m\,\tau^2 + (2V_{i,y} + 4V_{f,y})\tau + (6P_{i,y} - 6P_{f,y}), \tag{8}$$

$$0 = \pm m\,\tau^2 + (4V_{i,x} + 2V_{f,x})\tau + (6P_{i,x} - 6P_{f,x}), \tag{9}$$

$$0 = \pm m\,\tau^2 + (4V_{i,y} + 2V_{f,y})\tau + (6P_{i,y} - 6P_{f,y}). \tag{10}$$

The actual time used is the smallest value that keeps the acceleration within limits at the beginning and end of the spline; that is, all of $|2C_x| < m$, $|2C_y| < m$, $|2C_x + 6D_x\tau| < m$, and $|2C_y + 6D_x\tau| < m$ are true.

SUMMARY

Two-photon laser scanning microscopy offers a means to obtain high-resolution images of RBC velocity and vessel diameter in vivo. These measurements can be combined to calculate the flux for a given vessel, which is a more accurate metric of the oxygen- and nutrient-carrying capability of blood.

Velocity and diameter measurements can change independently (Fig. 2F) and thus they must be measured nearly simultaneously to accurately access flux. This can be achieved with the spatially optimized line-scan algorithm described above. This technique can be extended to image other types of fluorescent signals, for instance, neural activity as indicated by calcium-sensitive dyes. In principle, this technique can be readily extended to scan in three dimensions as well. However, the speed of current mechanical *z*-axis scanners is currently much slower than what can be achieved when scanning in the *x–y* plane alone (Göbel and Helmchen 2007; Göbel et al. 2007).

RECIPE

CAUTION: See Appendix 6 for proper handling of materials marked with <!>.
Recipes for reagents marked with <R> are included in this list.

Modified Artificial Cerebral Spinal Fluid (mACSF), Free of Carbonate and Phosphate

NaCl	125 mM
Glucose	10 mM
HEPES	10 mM
$CaCl_2$ <!>	3.1 mM
$MgCl_2$	1.3 mM

Adjust the pH to 7.4 (from Kleinfeld and Delaney 1996).

ACKNOWLEDGMENTS

We thank Quoc-Thang Nguyen for useful discussions on MPScope scan software. This work was funded by the National Institutes of Health (grants EB003832, NS059832, and RR021907 to DK and AG029681 to GC), the National Science Foundation (grant DBI 0455027), and a postdoctoral fellowship from the American Heart Association to AYS.

REFERENCES

Belayev L, Pinard E, Nallet H, Seylaz J, Liu Y, Riyamongkol P, Zhao W, Busto R, Ginsberg MD. 2002. Albumin therapy of transient focal cerebral ischemia: *In vivo* analysis of dynamic microvascular responses. *Stroke* **33:** 1077–1084.

Chaigneau E, Oheim M, Audinat E, Charpak S. 2003. Two-photon imaging of capillary blood flow in olfactory bulb glomeruli. *Proc Natl Acad Sci* **100:** 13081–13086.

Chuquet J, Hollender L, Nimchinsky EA. 2007. High-resolution in vivo imaging of the neurovascular unit during spreading depression. *J Neurosci* **27:** 4036–4044.

Derdikman D, Hildesheim R, Ahissar E, Arieli A, Grinvald A. 2003. Imaging spatiotemporal dynamics of surround inhibition in the barrels somatosensory cortex. *J Neurosci* **23:** 3100–3105.

Devor A, Tian P, Nishimura N, Teng IC, Hillman EM, Narayanan SN, Ulbert I, Boas DA, Kleinfeld D, Dale AM. 2007. Suppressed neuronal activity and concurrent arteriolar vasoconstriction may explain negative blood oxygenation level–dependent signaling. *J Neurosci* **27:** 4452–4459.

Devor A, Hillman EM, Tian P, Waeber C, Teng IC, Ruvinskaya L, Shalinsky MH, Zhu H, Haslinger RH, Narayanan SN, et al. 2008. Stimulus-induced changes in blood flow and 2-deoxyglucose uptake dissociate in ipsilateral somatosensory cortex. *J Neurosci* **28:** 14347–14357.

Drew PJ, Feldman DE. 2009. Intrinsic signal imaging of deprivation-induced contraction of whisker representations in rat somatosensory cortex. *Cereb Cortex* **19:** 331–348.

Drew PJ, Blinder P, Cauwenberghs G, Shih AY, Kleinfeld D. 2010a. Rapid determination of particle velocity from space-time images using the Radon transform. *J Comput Neurosci* **29:** 5–11.

Drew PJ, Shih AY, Driscoll JD, Knutsen PM, Blinder P, Davalos D, Akassoglou K, Tsai PS, Kleinfeld D. 2010b. Chronic optical access through a polished and reinforced thinned skull. *Nature Meth* **7:** 981–984.

Frostig RD, Lieke EE, Ts'o DY, Grinvald A. 1990. Cortical functional architecture and local coupling between neuronal activity and the microcirculation revealed by *in vivo* high-resolution optical imaging of intrinsic signals. *Proc Natl Acad Sci* **87:** 6082–6086.

Göbel W, Helmchen F. 2007. New angles on neuronal dendrites in vivo. *J Neurophysiol* **98:** 377–379.

Göbel W, Kampa BM, Helmchen F. 2007. Imaging cellular network dynamics in three dimensions using fast 3D laser scanning. *Nat Methods* **4:** 73–79.

Grinvald A, Lieke EE, Frostig RD, Gilbert CD, Wiesel TN. 1986. Functional architecture of cortex revealed by optical imaging of intrinsic signals. *Nature* **324:** 361–364.

Helmchen F, Kleinfeld D. 2008. In vivo measurements of blood flow and glial cell function with two-photon laser scanning microscopy. *Methods Enzymol* **444:** 231–254.

Hirase H, Creso J, Buzsaki G. 2004. Capillary level imaging of local cerebral blood flow in bicuculline-induced epileptic foci. *Neuroscience* **128:** 209–216.

Kleinfeld D, Delaney KR. 1996. Distributed representation of vibrissa movement in the upper layers of somatosensory cortex revealed with voltage sensitive dyes. *J Comp Neurol* **375:** 89–108.

Kleinfeld D, Denk W. 2000. Two-photon imaging of neocortical microcirculation. in *Imaging neurons: A laboratory manual* (ed. Yuste R, et al.), pp. 23.1–23.15. Cold Spring Harbor Laboratory Press, Cold Spring Harbor, NY.

Kleinfeld D, Mitra PP, Helmchen F, Denk W. 1998. Fluctuations and stimulus-induced changes in blood flow observed in individual capillaries in layers 2 through 4 of rat neocortex. *Proc Natl Acad Sci* **95:** 15741–15746.

Kontos HA. 1989. Validity of cerebral arterial blood flow calculations from velocity measurements. *Stroke* **20:** 1–3.

Lillis KP, Eng A, White JA, Mertz J. 2008. Two-photon imaging of spatially extended neuronal network dynamics with high temporal resolution. *J Neurosci Methods* **172:** 178–184.

Mostany R, Portera-Cailliau C. 2008. A craniotomy surgery procedure for chronic brain imaging. *J Vis Exp* **4:** 1128–1144.

Nguyen Q-T, Tsai PS, Kleinfeld D. 2006. MPScope: A versatile software suite for multiphoton microscopy. *J Neurosci Methods* **156:** 351–359.

Nguyen Q-T, Dolnick EM, Driscoll J, Kleinfeld D. 2009. MPScope 2.0: A computer system for two-photon laser scanning microscopy with concurrent plasma-mediated ablation and electrophysiology. In *Methods for in vivo optical imaging,* 2nd ed. (ed. Frostig RD), pp. 117–142. CRC Press, Boca Raton, FL.

Ogawa S, Lee TM, Kay AR, Tank DW. 1990. Brain magnetic resonance imaging with contrast dependent on blood oxygenation. *Proc Natl Acad Sci* **87:** 9868–9872.

Paxinos G, Watson C. 1986. *The rat brain in stereotaxic coordinates.* Academic, San Diego.

Schaffer CB, Friedman B, Nishimura N, Schroeder LF, Tsai PS, Ebner FF, Lyden PD, Kleinfeld D. 2006. Two-photon imaging of cortical surface microvessels reveals a robust redistribution in blood flow after vascular occlusion. *Pub Libr Sci Biol* **4:** 258–270.

Shih AY, Friedman B, Drew PJ, Tsai PS, Lyden PD, Kleinfeld D. 2009. Active dilation of penetrating arterioles restores red blood cell flux to penumbral neocortex after focal stroke. *J Cereb Blood Flow Metab* **29:** 738–751.

Tsai PS, Kleinfeld D. 2009. In vivo two-photon laser scanning microscopy with concurrent plasma-mediated ablation: Principles and hardware realization. In *Methods for in vivo optical imaging,* 2nd ed. (ed. Frostig RD), pp. 59–115. CRC Press, Boca Raton, FL.

Woolsey TA, Rovainen CM, Cox SB, Henger MH, Liange GE, Liu D, Moskalenko YE, Sui J, Wei L. 1996. Neuronal units linked to microvascular modules in cerebral cortex: Response elements for imaging the brain. *Cereb Cortex* **6:** 647–660.

Zhang S, Murphy TH. 2007. Imaging the impact of cortical microcirculation on synaptic structure and sensory-evoked hemodynamic responses in vivo. *Pub Libr Sci Biol* **5:** e119.

85 Optically Induced Occlusion of Single Blood Vessels in Neocortex

Andy Y. Shih,[1] Nozomi Nishimura,[2] John Nguyen,[2] Beth Friedman,[3] Patrick D. Lyden,[4] Chris B. Schaffer,[2] and David Kleinfeld[1]

[1]*Department of Physics, University of California at San Diego, La Jolla, California 92093;* [2]*Department of Biomedical Engineering, Cornell University, Ithaca, New York 14853;* [3]*Department of Pharmacology, University of California at San Diego, La Jolla, California 92093;* [4]*Department of Neurology, Cedars-Sinai Medical Center, Los Angeles, California 90048*

ABSTRACT

The ability to form targeted vascular occlusions in small vessels of the brain is an important technique for studying the microscopic basis of cerebral ischemia. We describe two complementary methods that enable targeted occlusion of any single blood vessel within the upper 500 µm of rodent neocortex. Our goal is to generate highly localized regions of ischemia by blocking penetrating arterioles and ascending venules, which are bottlenecks of flow in the cortical angioarchitecture. One method makes use of linear optical absorption by a photosensitizer, transiently circulated in the blood stream, to induce a clot in a surface or near-surface segment of a vessel. The second method makes use of nonlinear optical interactions, without the need to introduce an exogenous absorber, to induce clots in subsurface segments of penetrating vessels, as well as subsurface microvessels and capillaries. We provide details on concurrent high-resolution in vivo imaging and auxiliary laser setups, occlusion protocols, and post hoc histological procedures.

INTRODUCTION

Advances in imaging techniques, such as two-photon laser scanning microscopy, combined with various methods to label and record cellular activity in vivo, have greatly improved both spatial and temporal resolution for studying pathological processes in real time (Misgeld and Kerschensteiner 2006). These methods are important for the study of pathological mechanisms that underlie ischemic injury, including abnormal changes in cell signaling and structure, vascular dysfunction, and inflammation.

We consider two complementary approaches to selectively occlude single vessels anywhere within the upper 500 µm of cerebral cortex. One method, termed "photothrombosis," uses linear optical absorption by a photosensitizer, transiently circulating in the blood stream, to initiate a clotting cascade in vessels on the cortical surface (Watson et al. 1985; Schaffer et al. 2006; Nishimura et al. 2007; Sigler et al. 2008). The second method, termed "plasma-mediated ablation," makes use of nonlinear optical interactions, without the need to introduce an exogenous absorber, to induce clots in subsurface vessels and capillaries (Nishimura et al. 2006, 2010; Nguyen et al. 2007). Both techniques complement an existing variety of pre-clinical stroke models that generate larger areas of ischemia (Ginsberg and Busto 1989). The advantages of this targeted approach include the ability to (1) select a specific class of cortical vessel (i.e., capillary vs. surface arteriole or venule); (2) select a specific location for ischemia within the imaging window (i.e., a cortical column); (3) make pre- and post-measurements of hemodynamic and neuronal state variables; and (4) initiate the ischemic event in situ under the microscope.

Here, our goal is to exploit the physical localization of highly focused laser light to form occlusions within "bottleneck" vessels of the cortical angioarchitecture. Occlusions of either penetrating arterioles (Nishimura et al. 2007) or ascending venules (Nguyen et al. 2007) have a large impact on cortical blood flow. Furthermore, the columnar regions of ischemia generated by such occlusion are capable of developing into cortical micro-infarctions over several days.

Protocol

Optically Induced Occlusion of Single Blood Vessels in Neocortex

IMAGING SETUP

We integrate additional lasers into a custom-built two-photon imaging system for photothrombosis and plasma-mediated ablation (Fig. 1A; Majewska et al. 2000; Tsai and Kleinfeld 2009). Additionally, Sigler et al. (2008) have described methods to convert an Olympus BX series microscope for photothrombosis of single vessels. These and other schemes that combine imaging and ablation beams before the scan mirrors may be adapted for a commercial instrument, the Ultima Multiphoton Microscopy System with an optional high-speed two-photon optics set and external detector (Prairie Technologies).

Photothrombosis

Targeted photothrombosis can be used to occlude individual arterioles and venules that lie on or near the cortical surface. A 50-mW green laser light (NT56-487, Edmund Optics) is introduced into the imaging beam path with dichroic mirror 1 (625 DRLP, Chroma) (Fig. 1A). The focal point of the beam is fixed, centered along the optical axis, and controlled by a shutter (LS3Z2 Uniblitz and VMM-D1 driver, Vincent). The beam is adjusted to pass through a 3.5-mm-diameter clearing that was etched in the dielectric coating of dichroic mirror 2 (700 DCXRU, Chroma) (Fig. 1A). This allows transmission of the green laser while still reflecting >90% of emitted light from the sample toward the photomultiplier tubes (PMTs). As the interval of irradiation is typically long (i.e., 50–100 sec), irradiation is periodically interrupted for brief epochs of imaging, typically 1 frame (0.2 sec per frame) every 1 sec for an 80% duty cycle. This permits formation of the clot to be observed in real time. Alternatively, the PMT assembly can be temporarily moved out of the path to allow the green laser light to pass. However, observation of the formation of the clot is recommended to avoid excessive irradiation and unintended photothrombosis in neighboring vessels.

The green laser is brought to the same focal plane as the imaging beam of the two-photon microscope using an appropriate telescope and centered in the imaging field. When using a 40x, 0.8-NA dipping objective with an underfilled back aperture, the final green laser spot size is ~3 μm in diameter. In all cases, the intensity of the laser light at the focus should initially be adjusted to ~1 mW using a tunable neutral density filter and may be increased as needed. The motorized/encoding *x–y* stage of the microscope is used to reposition the animal to change the location of photoactivation (Fig. 1A). Alternatively, additional lasers may be introduced through a second set of scanners so that the position of irradiation may be shifted electromechanically (i.e., the Ultima with an optional second set of scanners).

Plasma-Mediated Ablation

Plasma-mediated ablation makes use of the dissociation of matter by high fluence, 100–300-fsec pulses of near-infrared light (Schaffer et al. 2001; Tsai et al. 2004, 2009). These pulses ionize the material, such as blood plasma or vascular endothelium, within a femtoliter-sized focal volume of the incident laser pulse. This method allows occlusions of microvessels and capillaries deep below the surface of the cortex, typically down to 500 μm, but in principle down to 1000 μm below the pia, without disruption of the neighboring tissue. Plasma-mediated ablation is also well-suited for inducing clots in penetrating arterioles and ascending venules, by targeting subsurface segments of the vascular tree. However, this method is not suitable for targeting arterioles that lie on the cortical surface, which tend to hemorrhage and obscure subsequent imaging.

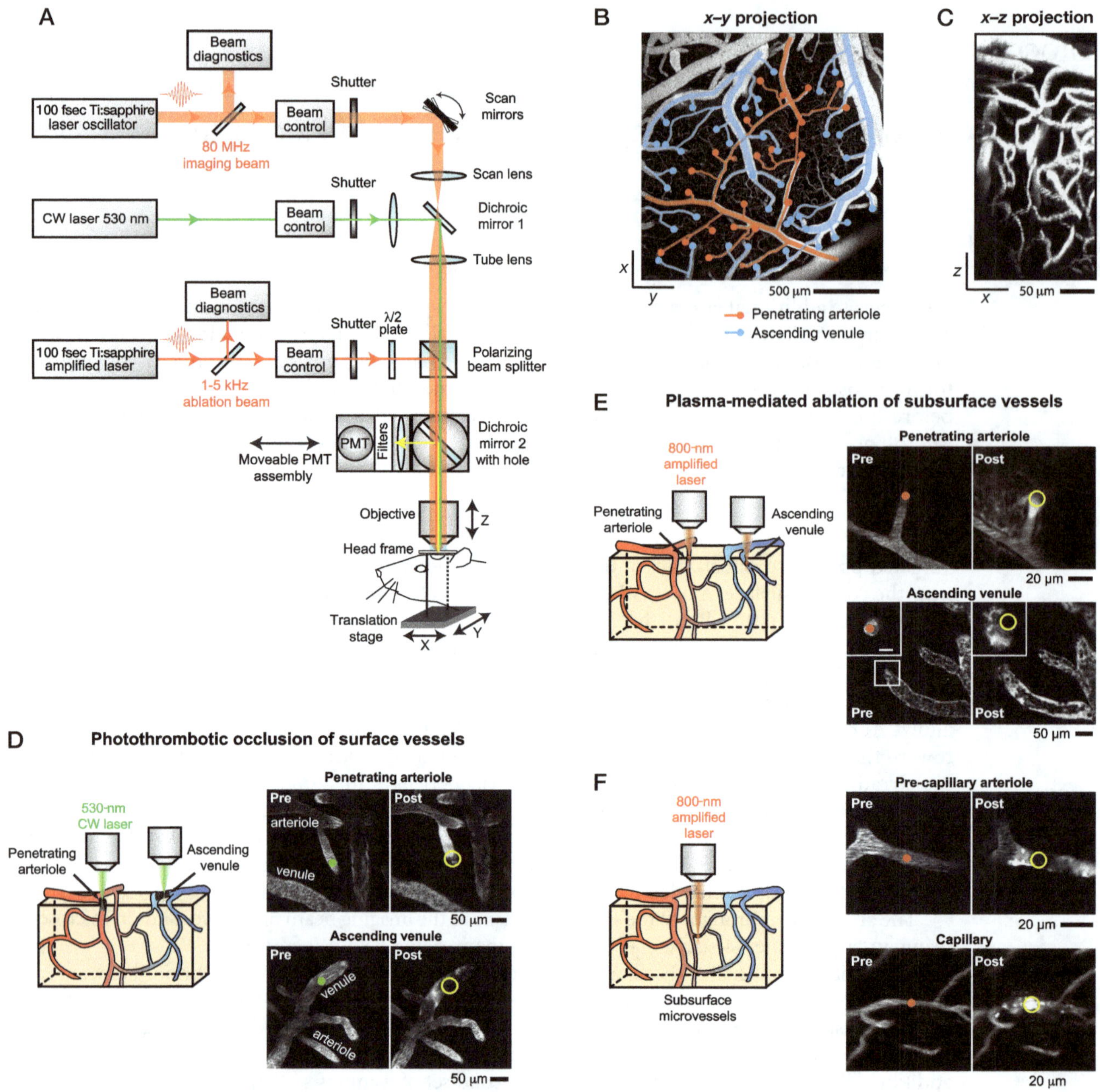

FIGURE 1. Forming intravascular occlusions in single cortical vessels using photothrombosis or plasma-mediated ablation. (*A*) Schematic of a two-photon laser scanning microscope (TPLSM) system modified to include a continuous-wave green-light laser for photothrombosis and an amplified pulsed laser for plasma-mediated ablation. (CW) Continuous wave; (PMT) photomultiplier tube. The beam control module refers to a device for control of laser intensity, a pair of lenses for beam shaping, and a prism pair for pulse dispersion compensation, if required. (*B*) *x-y* maximal projection of a TPLSM image stack showing pial vessels and fine microvessels through a rat cranial window. A manual tracing of arteriole (red) and venule (blue) networks is overlaid on the image. The locations of penetrating arterioles and ascending venules are marked with circles. (*C*) *x-z* maximal projection of TPLSM image stack showing a single penetrating arteriole, typical of those selected for targeted occlusion. (*D*) (*Left*) Schematic of vessel types that can be targeted by focused photothrombosis (i.e., penetrating arterioles and ascending venules). (*Right*) Examples of pre- and post-occlusion images from rat. (*E–F*) Schematic of vessel types that can be targeted by plasma-mediated ablation (i.e., penetrating arterioles, ascending venules, or deep microvessels). Pre- and post-occlusion examples for each vessel class are shown. The *inset* within the ascending venule example shows a subsurface imaging plane 100 µm deep where the clot was formed. The capillary image in *F* was obtained from an Alzheimer's mouse model. All other example images are from adult rat.

An amplified 100-fsec light source is introduced into the beam path through a polarizing beam splitter (Fig. 1A). As with the case of green laser light, it may also be introduced in other locations along the path of the imaging beam. This beam is focused to the same plane as the imaging laser and aligned such that photodisruption occurs at the center of the imaged field. The energy per pulse of the photodisruption beam is coarsely varied with a tunable neutral density filter and finely varied with a $\lambda/2$ plate followed by a polarizing beam splitter. The number of pulses is controlled by a fast shutter (LS3Z2 Uniblitz and VMM-D1 driver).

The diameter of the amplified beam is adjusted with an appropriate telescope to overfill the back aperture of the objective. This is necessary to obtain the smallest possible spot size for the amplified beam, which is on the order of a femtoliter. The energy per pulse is initially set to 30 nJ at the focus of a 40×, 0.8-NA dipping objective, which is near the threshold fluence for damage of ~1 J/cm^2. For a 5-kHz, 100-fec amplified laser, this corresponds to an average power of 0.15 mW.

MATERIALS

Reagents

Adult rats or mice
Fluorescein-dextran (2 MDa; Sigma-Aldrich) or Texas Red-dextran (70 kDa; Invitrogen), 5% (w/v) in saline
Fluorescein-dextran-agarose mixture (see Step 17)
Rose Bengal, 1% (w/v) in saline (filtered before use)
Sucrose, 30% (w/v)

Equipment

Amplified Ti:sapphire 100-fsec laser source (Libra, Coherent Inc.)
Brain matrix (RSMAC-2, Plastics One Inc.)
Continuous-wave 530-nm laser source (NT54-551, Edmund Optics)
Fine tungsten electrodes
Glass slides
Two-photon laser-scanning microscope (Tsai and Kleinfeld 2009) and associated software (Nguyen et al. 2006, 2009)

EXPERIMENTAL METHOD

1. Prepare a cranial window suitable for imaging in the rat.

   Detailed procedures for animal preparation and generation of a cranial imaging window in the rat are provided in Chapter 84 and previous publications (Kleinfeld and Delaney 1996; Kleinfeld and Denk 2000; Kleinfeld et al. 2008). The approaches described here can also be applied to mice. Details of mouse cranial window preparations can be found in Chapters 31 and 32 and previous publications (Xu et al. 2007; Holtmaat et al. 2009; Drew et al. 2010).

   Details for performing both photothrombosis and plasma-mediated occlusion of individual vessels are provided below. Choose the method based on the location of the vessels being studied and the study objectives. The specific arrangements of the optical equipment are discussed above (see Imaging Setup).

Photothrombotic Occlusion

2. Inject a 1.0 mL/kg bolus of 5% (w/v) fluorescein-dextran solution in saline into either a femoral vein catheter or directly into a tail vein.

   Label the blood plasma with the 2-MDa fluorescein-dextran providing contrast between the red blood cells (RBCs) and the plasma (Kleinfeld et al. 1998). This high-molecular-weight dextran provides stable labeling, typically for at least 3 h. Supplement the dye at 0.3 mL/kg, as needed during imaging.

3. Obtain a three-dimensional stack of images of the cerebral vessels throughout the cranial window first at low magnification, typically a 4×, 0.28-NA objective (UB950, Olympus) or a 10×, 0.3-NA objective (W Achroplan, Zeiss), to use as a navigational map (Fig. 1B).

   Higher-resolution imaging of local vascular architecture and vasodynamics, such as RBC speed and vessel diameter (Shih et al. 2009), are typically obtained with a 40×, 0.8-NA objective (UM568, Olympus).

4. Select a penetrating arteriole or ascending venule to target for occlusion (Fig. 1C). Targeting vessels that have a segment coursing parallel to the cortical surface allows for easy verification of clot formation.

   Note that cortical arterioles appear smoother and straighter than venules and have fewer branching points. The location of penetrating arterioles and ascending venules can be identified by manually tracing through the pial arterial and venous networks (Fig. 1B). Arterioles and venules can also be differentiated based on direction of flow (i.e., into or out of the cortex).

5. Inject a 1.0 mL/kg bolus of 1% (w/v) Rose Bengal solution in saline through the femoral vein catheter or tail vein.

   Rose Bengal should be injected immediately before forming the occlusion, since it is cleared from the blood stream within minutes (Zhang et al. 2005).

6. Initiate photothrombosis by aiming the focal point of the green laser within the lumen of the target vessel, near the vessel wall (Fig. 1D). This will nucleate the clotting cascade and create an attachment point for the growing clot. If a clot does not occur within 5–10 sec of irradiation, inject a 0.5 mL/kg supplement of Rose Bengal solution.
7. Traverse the focus of the green laser across the diameter of the blood vessel to facilitate thrombus formation. Occlusion may require 10–100 sec of irradiation. Larger vessels with relatively fast RBC flow require more time to fully occlude; the exact time must be determined empirically. A successful occlusion appears densely packed with immobile nonfluorescent RBCs often surrounded by a brighter region of stagnant fluorescent plasma (Fig. 1D).
8. Collect post-occlusion measurements, as desired.

Plasma-Mediated Occlusion

9. Collect baseline vascular image stacks. To resolve deep microvessels, this should be done at high resolution using a 40× objective.
10. Select a subsurface segment of vasculature for targeted occlusion. This could be penetrating arterioles, ascending venules, or subsurface microvessels.

To occlude penetrating arterioles and ascending venules

11. Place the focus of the amplified beam about halfway between the brain surface and the first capillary branch off of the vessel (Fig. 1E). The laser focus should be located within the vessel lumen a few micrometers from the vessel wall.
12. Irradiate with one pulse using an energy that is below the expected damage threshold. This starting energy is depth dependent (i.e., ~50 nJ at 20 µm depth and 150 nJ at 150 µm depth).
13. Increase the number of pulses by factors of 10 with the same energy, often trying each pulse number a few times while watching for signs of laser damage. The first visual indication of photodisruption is extravasation of fluorescein-dextran outside the vessel lumen (Fig. 1E). If extravasation fails to occur, increase the pulse energy by ~25% and repeat the procedure described above.
14. Once extravasation occurs, repeat the irradiation with the threshold energy and number of pulses, while targeting multiple nearby areas along the inner wall of the vessel segment.
15. Continue to irradiate the targeted segment until there is a complete cessation of RBC movement in the target vessel. To verify successful occlusion, return to the cortical surface and ensure that flow has ceased in the main branch of the arteriole or venule.

Increasing the energy too high results in the rupture of the vessel and a large hemorrhage (Nishimura et al. 2006).

To occlude subsurface microvessels

16. Aim the focal point of the amplified beam into the vessel lumen, a few micrometers away from the vessel wall (Fig. 1F). Use a similar graded approach to irradiation, as described above. It is critical to identify a segment of the targeted vessel where RBC movement can be readily visualized to confirm successful occlusion.

Histology

It is often desirable to examine the consequences of single vessel occlusion using post hoc histology. A number of strategies can be used to relocate the occluded vessel and the corresponding cyst in perfused tissue (Fig. 2A–C).

A fluorescent vascular casting method can be used to visualize the pial vasculature during histological processing.

17. Inject a fluorescein-dextran-agarose mixture transcardially, as described in detail previously (Tsai et al. 2005; Shih et al. 2009).
18. Because this dye remains in situ after tissue fixation, use it to map the exact location of the targeted vessel, based on unique features in the pial vascular topology that are visible through the imaging window.

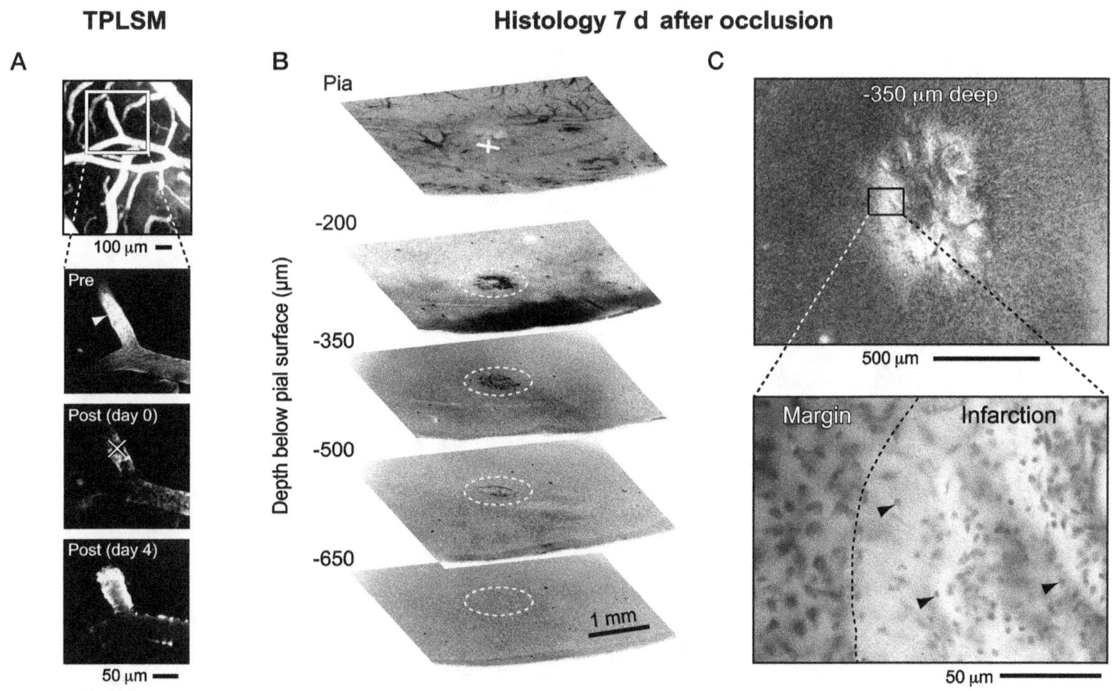

FIGURE 2. Photothrombotic occlusion of penetrating arterioles using Rose Bengal. (*A*) *x-y* maximal projected two-photon laser scanning microscope (TPLSM) image showing a single penetrating arteriole to be occluded by photothrombosis. (*Inset* images) The occluded vessel before and after occlusion. Note the extravasation of fluoroscein-dextran dye from the vessel 4 d after stroke, which indicates damage to the vascular wall. (*B*) Examination of the cyst volume 7 d after occlusion in Cresyl Violet–stained tissue sections. The animal was perfused with a vascular cast, and the cortex was flattened. The occluded vessel, marked by an x, was located in the fixed tissue by using a fluorescent vascular cast as a map. The fluorescent vessels are visible at the pial surface. Electrolytic fiducials (not shown) were made through the thickness of the cortex at the time of perfusion, and fixation aided tissue re-alignment after serial sectioning. The image intensities were inverted for improved visualization. (*C*) Magnified view of Cresyl Violet–stained brain sections. Note the shrunken pynotic cells that are widespread within the core of the micro-infarction (black arrowheads).

19. Excise the region of interest in a coronal orientation using a brain matrix.
20. Cryoprotect the blocked tissue in 30% sucrose overnight and further cryosection it for immunohistology.

The micro-infarction can also be viewed in a flattened cortex preparation that is cryosectioned tangentially to the cortical surface.

21. The micro-infarction can also be viewed in a flattened cortex preparation after fluorescent casting of the vasculature (Fig. 2B). Excise the cortex, sandwich it between two glass slides separated by a distance of 2.5 mm, and post-fix it overnight.
22. Locate the target vessel and then generate fiducial marks in its vicinity by applying current through fine tungsten electrodes that are advanced through the thickness of the cortex. This creates burn marks in each section that are used for alignment of serial tissue sections after cryosectioning tangentially to the cortical surface.

EXAMPLE APPLICATION

The techniques described here are appropriate for chronic as well as acute studies. An example longitudinal study shows that occlusion of even a single penetrating arteriole leads to discrete regions of cortical ischemia and tissue damage (Fig. 2B,C; see Blinder et al. 2010). The accuracy in this technique allows one to investigate the functional consequences of ischemia on cells neighboring the targeted vessel, by, for example, using mice with genetically labeled cell populations (Feng et al. 2000) or after loading tissue with organic Ca^{2+} dyes (Stosiek et al. 2003; Ding et al. 2009). Furthermore, these occlusions provide a means to model and study the consequences of vascular disease and small-scale strokes that occur in the aging human cortex (Suter et al. 2002; Kovari et al. 2004). Although the examples in this chapter are from the rat, it is anticipated that future studies will take advantage of various genetically engineered mouse lines, which appear to be particularly amenable to longitudinal imaging after photothrombosis (Brown et al. 2007; Drew et al. 2010).

LIMITATIONS

Two caveats of photothrombosis are noteworthy. First, small thrombi formed during the period of illumination may travel downstream rather than stick at the site of irradiation. It is difficult to track the location and stability of these unintentional thrombi, and thus their impact on blood flow is unknown. Second, even well-formed clots tend to dislodge or disintegrate over the long term, particularly in fast-flowing arterioles. Although most clots are stable for at least 5–6 h, over the course of a typical experiment, the exact timing of natural de-occlusion can vary greatly from vessel to vessel. Thus, the effects of reperfusion injury would be difficult to assess in this model. However, future approaches may be able to improve control over re-canalization by irradiating the clot with ultraviolet light (Watson et al. 2002). Lastly, one caveat of the plasma-mediated ablation method is that excessive damage to the vessel wall is likely to cause hemorrhaging.

ACKNOWLEDGMENTS

We thank Jonathan D. Driscoll for critical reading of the manuscript. This work was funded by the National Institutes of Health (grants EB003832, MH085499, and NS059832 to DK and NS43300 and NS052565 to PDL), the Ellison Medical Foundation (grant AG-NS-0330-06 to CBS), the American Heart Association (grant 0735644T to CBS), the VA Medical Research Foundation (grant to PDL), and a postdoctoral fellowship from the American Heart Association to AYS.

REFERENCES

Blinder P, Shih AY, Rafie C, Kleinfeld D. 2010. Topological basis for the robust distribution of blood to rodent neocortex. *Proc Natl Acad Sci* **107:** 12670–12675.

Brown CE, Li P, Boyd JD, Delaney KR, Murphy TH. 2007. Extensive turnover of dendritic spines and vascular remodeling in cortical tissues recovering from stroke. *J Neurosci* **27:** 4101–4109.

Ding S, Wang T, Cui W, Haydon PG. 2009. Photothrombosis ischemia stimulates a sustained astrocytic Ca^{2+} signaling in vivo. *Glia* **57:** 767–776.

Drew PJ, Shih AY, Driscoll JD, Knutsen PM, Blinder P, Davalos D, Akassoglou K, Tsai P, Kleinfeld D. 2010. Chronic optical access through a polished and reinforced thinned skull. *Nature Methods* **7:** 981–984.

Feng G, Mellor RH, Bernstein M, Keller-Peck C, Nguyen QT, Wallace M, Nerbonne JM, Lichtman JW, Sanes JR. 2000. Imaging neuronal subsets in transgenic mice expressing multiple spectral variants of GFP. *Neuron* **28:** 41–51.

Ginsberg MD, Busto R. 1989. Rodent models of cerebral ischemia. *Stroke* **20:** 1627–1642.

Holtmaat A, Bonhoeffer T, Chow DK, Chuckowree J, De Paola V, Hofer SB, Hübener M, Keck T, Knott G, Lee WC, et al. 2009. Long-term, high-resolution imaging in the mouse neocortex through a chronic cranial window. *Nat Protoc* **4:** 1128–1144.

Kleinfeld D, Delaney KR. 1996. Distributed representation of vibrissa movement in the upper layers of somatosensory cortex revealed with voltage sensitive dyes. *J Comp Neurol* **375:** 89–108.

Kleinfeld D, Denk W. 2000. Two-photon imaging of neocortical microcirculation. In *Imaging neurons: A laboratory manual* (ed. Yuste R, et al.), pp. 23.1–23.15. Cold Spring Harbor Laboratory Press, Cold Spring Harbor, NY.

Kleinfeld D, Mitra PP, Helmchen F, Denk W. 1998. Fluctuations and stimulus-induced changes in blood flow observed in individual capillaries in layers 2 through 4 of rat neocortex. *Proc Natl Acad Sci* **95:** 15741–15746.

Kleinfeld D, Friedman B, Lyden PD, Shih AY. 2008. Targeted occlusion to surface and deep vessels in neocortex via linear and nonlinear optical absorption. In *Animal models of acute neurological injuries* (ed. Chen J, et al.), pp. 169–185. Humana Press, Totowa, NJ.

Kovari E, Gold G, Herrmann FR, Canuto A, Hof PR, Michel JP, Bouras C, Giannakopoulos P. 2004. Cortical microinfarcts and demyelination significantly affect cognition in brain aging. *Stroke* **35:** 410–414.

Majewska A, Yiu G, Yuste R. 2000. A custom-made two-photon microscope and deconvolution system. *Pflugers Arch* **441:** 398–408.

Misgeld T, Kerschensteiner M. 2006. In vivo imaging of the diseased nervous system. *Nat Rev Neurosci* **7:** 449–463.

Nguyen Q-T, Tsai PS, Kleinfeld D. 2006. MPScope: A versatile software suite for multiphoton microscopy. *J Neurosci Methods* **156:** 351–359.

Nguyen J, Nishimura N, Iadecola C, Schaffer CB. 2007. Single venule occlusions induced by photodisruption using femtosecond laser pulses cause decreased blood flow in rat cortex. In *Society for neuroscience.* Society for Neuroscience, San Diego.

Nguyen Q-T, Dolnick EM, Driscoll J, Kleinfeld D. 2009. MPScope 2.0: A computer system for two-photon laser scanning microscopy with concurrent plasma-mediated ablation and electrophysiology. In *Methods for in vivo optical imaging,* 2nd ed. (ed. Frostig RD), pp. 117–142. CRC Press, Boca Raton, FL.

Nishimura N, Schaffer CB, Friedman B, Tsai PS, Lyden PD, Kleinfeld D. 2006. Targeted insult to individual subsurface cortical blood vessels using ultrashort laser pulses: Three models of stroke. *Nat Methods* **3:** 99–108.

Nishimura N, Schaffer CB, Friedman B, Lyden PD, Kleinfeld D. 2007. Penetrating arterioles are a bottleneck in the perfusion of neocortex. *Proc Natl Acad Sci* **104:** 365–370.

Nishimura N, Rosidi NL, Iadecola C, Schaffer CB. 2010. Limitations of collateral flow after occlusion of a single cortical penetrating arteriole. *J Cereb Blood Flow Metab* **30:** 1914–1927.

Schaffer CB, Brodeur A, Mazur E. 2001. Laser-induced breakdown and damage in bulk transparent materials induced by tightly-focused femtosecond laser pulses. *Meas Sci Technol* **12:** 1784–1794.

Schaffer CB, Friedman B, Nishimura N, Schroeder LF, Tsai PS, Ebner FF, Lyden PD, Kleinfeld D. 2006. Two-photon imaging of cortical surface microvessels reveals a robust redistribution in blood flow after vascular occlusion. *PLoS Biol* **4:** e22. doi: 10.1371/journal.pbio.0040022.

Shih AY, Friedman B, Drew PJ, Tsai PS, Lyden PD, Kleinfeld D. 2009. Active dilation of penetrating arterioles restores red blood cell flux to penumbral neocortex after focal stroke. *J Cereb Blood Flow Metab* **29:** 738–751.

Sigler A, Goroshkov A, Murphy TH. 2008. Hardware and methodology for targeting single brain arterioles for photothrombotic stroke on an upright microscope. *J Neurosci Methods* **170:** 35–44.

Stosiek C, Garaschuk O, Holthoff K, Konnerth A. 2003. In vivo two-photon calcium imaging of neuronal networks. *Proc Natl Acad Sci* **100:** 7319–7324.

Suter OC, Sunthorn T, Kraftsik R, Straubel J, Darekar P, Khalili K, Miklossy J. 2002. Cerebral hypoperfusion generates cortical watershed microinfarcts in Alzheimer disease. *Stroke* **33:** 1986–1992.

Tsai PS, Kleinfeld D. 2009. In vivo two-photon laser scanning microscopy with concurrent plasma-mediated ablation: Principles and hardware realization. In *Methods for in vivo optical imaging,* 2nd ed. (ed. Frostig RD), pp. 59–115. CRC Press, Boca Raton, FL.

Tsai PS, Friedman B, Squier JA, Kleinfeld D. 2004. Ultrashort pulsed laser light: A cool tool for ultraprecise cutting of tissue and cells. *Optics & Photonic News* **14:** 24–29.

Tsai PS, Friedman B, Schaffer CB, Squier JA, Kleinfeld D. 2005. All-optical, in situ histology of neuronal tissue with femtosecond laser pulses. In *Imaging in neuroscience and development: A laboratory manual* (ed. Yuste R, Konnerth A), pp. 815–826. Cold Spring Harbor Laboratory Press, Cold Spring Harbor, NY.

Tsai PS, Blinder P, Migliori BJ, Neev J, Jin Y, Squier JA, Kleinfeld D. 2009. Plasma-mediated ablation: An optical tool for submicrometer surgery on neuronal and vascular systems. *Curr Opin Biotechnol* **20:** 90–99.

Watson BD, Dietrich WD, Busto R, Wachtel MS, Ginsberg MD. 1985. Induction of reproducible brain infarction by photochemically initiated thrombosis. *Ann Neurol* **17:** 497–504.

Watson BD, Prado R, Veloso A, Brunschwig JP, Dietrich WD. 2002. Cerebral blood flow restoration and reperfusion injury after ultraviolet laser–facilitated middle cerebral artery recanalization in rat thrombotic stroke. *Stroke* **33:** 428–434.

Xu HT, Pan F, Yang G, Gan WB. 2007. Choice of cranial window type for in vivo imaging affects dendritic spine turnover in the cortex. *Nat Neurosci* **10:** 549–551.

Zhang S, Boyd J, Delaney KR, Murphy TH. 2005. Rapid reversible changes in dendritic spine structure in vivo gated by the degree of ischemia. *J Neurosci* **25:** 5333–5338.

86 Two-Photon Imaging of Neuronal Structural Plasticity in Mice during and after Ischemia

Timothy H. Murphy

Kinsmen Laboratory, Department of Psychiatry, University of British Columbia, Vancouver, British Columbia V6T 1Z3, Canada

ABSTRACT

To gain insight into the dendritic structural changes that occur within minutes of ischemia, as well as circuit-level plasticity that occurs over days to weeks, it is necessary to devise stroke models for anesthetized transgenic mice that are compatible with in vivo imaging methods. In particular, we are interested in examining events during the first few minutes of ischemia that are associated with a wave of ischemic depolarization. Given the short latency to ischemic depolarization (~2 min), we have found it necessary to develop ischemia models that can be initiated while the animal is within the two-photon microscope for in vivo imaging. We describe both a global ischemia model and a targeted photothrombotic ischemia model in green fluorescent protein– and yellow fluorescent protein–expressing transgenic mice in which in vivo imaging of dendritic structure can be performed. The advantage of the global model is the speed with which ischemia can be induced and reversed, whereas the focal model lacks precise temporal control but can be targeted to a select region.

INTRODUCTION

Here, we describe both a global ischemia model and a targeted photothrombotic ischemia model in anesthetized green fluorescent protein (GFP)- and yellow fluorescent protein (YFP)-expressing transgenic mice (Feng et al. 2000), where fluorescent layer V cortical neuron dendritic arbors can be viewed using in vivo two-photon imaging. The global ischemia model uses C57BL/6 mice that have poorly functioning posterior communicating arteries, permitting forebrain global ischemia to be reversibly induced by tensioning sutures looped around the common carotid arteries (Yang et al. 1997). In the targeted stroke model, we use green laser light (Sigler et al. 2008) to photoactivate the dye Rose Bengal (Watson et al. 1985) and initiate clotting cascades within pre-selected arterioles (photothrombosis). The protocol outlines steps to perform these procedures in combination with two-photon imaging.

Protocol

Two-Photon Imaging of Neuronal Structural Plasticity in Mice during and after Ischemia

IMAGING SETUP

We have modified commercial upright microscopes (Olympus BX50 or BX51) to allow in vivo two-photon imaging. Fluorescence of YFP, GFP, and Texas Red is two-photon excited by 100-fsec pulses at 900-nm wavelength with a Coherent Mira 900 Ti:sapphire laser that is pumped by a 5-W Verdi laser. We take fluorescence images with Olympus IR-LUMPlanFl water-immersion objectives (40×/0.8 NA or 60×/0.9 NA) and record with custom software (Igor, Wavemetrics). The mouse is stabilized by a cranial support frame (Kleinfeld and Denk 2000). For imaging of epifluorescence and reflected light, the microscope is equipped with a charge-coupled device (CCD) video camera (Dalsa 1M60 Pantera) for mapping vasculature and for laser speckle imaging (Briers and Webster 1996). For photoactivating Rose Bengal and monitoring photothrombosis, we use a removable 532-nm custom-made laser light source, ≈10 mW (Beta Electronics), which is inserted into the filter slot of the microscope (Sigler et al. 2008). We record video images of surface blood flow and laser speckle with a PIXCI E4DB frame grabber driven by XCAP-standard version 2.2 imaging software (both EPIX), and analyze them with ImageJ (NIH, http://rsbweb.nih.gov/ij).

MATERIALS

CAUTION: See Appendix 6 for proper handling of materials marked with <!>.

Reagents

Agarose (gel point 36°C ± 1.5°C; Type III-A, Sigma-Aldrich A9793)
Anesthesia <!> (see Step 1)
Brain buffer (135 mM NaCl, 5.4 mM KCl, 1 mM $MgCl_2$, 5 mM HEPES, pH to 7.4)
Dental curing crown acrylic resin (Jet Tooth Shade Powder and Jet Liquid, Lang Dental)
Glass coverslips no. 1, cut to fit craniotomy frame
Instant adhesive (Krazy Glue) <!>
Lidocaine <!>, 2% (Ayerst)
Rose Bengal, Na salt (diluted to 10 mg/mL in ACSF; Sigma-Aldrich R3877)
Suture (5-0 silk, Ethicon)
Syringes, 0.5 mL with 28½-gauge needle, for intraperitoneal injection (Insulin Syringe U-100, Becton-Dickinson)
Syringes, 1 mL with 26½-gauge needle, for intravenous injection into the tail vein
Texas Red–labeled dextran, 70 kDa (Invitrogen D-1830), for determination of blood flow using two-photon imaging
Transgenic mice

For studying dendrite morphology, we use 2–5-mo-old YFP-H (Jackson Labs) and GFP-M transgenic C57BL/6 mice (Feng et al. 2000). For the global ischemia model, it is critical that C57BL/6 mice be used because they have poorly functioning posterior communicating arteries and will be subject to global ischemia with only occlusion of both common carotid arteries (Yang et al. 1997).

Equipment

Dental drill (NSK EX-203 [22,000 rpm] or Midwest Stylus 540S)
External flexible fiber-optic illuminator (Edmund Optics), equipped with a blue band-pass filter, 395–480-nm (Hoya B-440 or Edmund Optics NT46-541)
Heating pad, feedback-regulated (FHC EI-08B)
Laser, 784-nm, 32-mW (StockerYale SNF-XXX-785S-35), with an anamorphic beam expander (Edmund Optics T47274) mounted on a micromanipulator, for laser speckle imaging
Surgical tools for craniotomy and dissecting the carotid artery
Test slide with orange or red fluorescence (Chroma) (optional for laser adjustment; see Step 21)
Two-photon microscope and associated optics (see Imaging Setup)

EXPERIMENTAL METHOD

The experimental protocols described here have been approved by the University of British Columbia animal care committee and are consistent with Canadian Council on Animal Care and Use guidelines.

Global Ischemia Model: Common Carotid Artery Occlusion

The advantage of this model is that ischemia can be induced within seconds of tensioning sutures looped around the carotids, and full reperfusion can be induced within seconds of reducing the tension. Global ischemia in this model affects blood flow within a large region of both forebrain hemispheres. Figure 1 demonstrates its application for studying the effects of stroke and reperfusion on dendrite structure (Murphy et al. 2008).

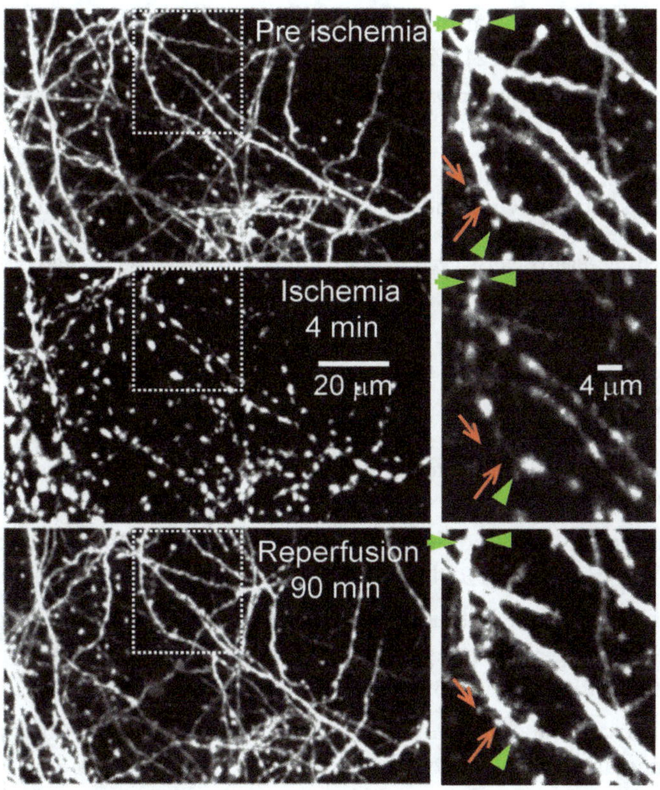

FIGURE 1. Two-photon imaging of spiny dendritic structure loss during bilateral common carotid artery (CCA) occlusion. Images of layer 5 neuron tuft dendrites (maximal intensity projection) before and 4 min after ischemia (with "blebbed" dendrites), and after reperfusion. The white-boxed area in the *top left* corresponds to a region shown in the close-up images showing dendritic spines present throughout the ischemia/reperfusion (green arrows) and transiently lost spines (red arrows). (Reprinted, with permission, from Murphy et al. 2008, ©2008 Society for Neuroscience.)

1. Anesthetize the mouse as follows:

   For acute experiments:

   Intraperitoneally inject 1200 mg urethane/kg body weight (Zhang and Murphy 2007) to provide long-lasting anesthesia (Kleinfeld and Denk 2000).

   For chronic imaging experiments:

   Anesthetize with inhalation of 1% isoflurane.

   When isoflurane is unavailable, ketamine/xylazine can be intraperitoneally administered (100 mg/kg ketamine, 10 mg/kg xylazine) to maintain anesthesia for an imaging session of ~1–2 h.

2. Confirm a suitable level of anesthesia by the lack of a toe pinch reflex.
3. Maintain body temperature at 37 ± 0.5°C using a heating pad and feedback regulation from a rectal temperature probe, and maintain hydration by intraperitoneal or subcutaneous injection of either 200–300 µL of brain buffer with 20 mM glucose at 1–2-h intervals.
4. Prepare the neck area for surgery to expose the carotid arteries.
5. Place the anesthetized mouse on its back with its paws secured to a backing plate by masking tape. Place a 1-mL syringe under its neck to better expose the trachea.
6. Inject 20 µL of 2% lidocaine subcutaneously into the neck region.
7. Make an incision over the thyroid gland. Using blunt dissecting tools, carefully dissect the thyroid glands and their associated musculature and connective tissue away from the trachea. Separate the common carotid arteries from the vagus nerves and muscles using blunt forceps.
8. Loop a suture around each common carotid artery.

   Do not confuse the jugular veins with the arteries (veins are more superficial, larger, and flaccid).

9. Moisten the area around the common carotid arteries and looped sutures with saline. Close the incision with two stitches, and pull the suture ends gently away from the incision through a space in the skin.
10. To set up for imaging, turn the animal upright, attach an imaging frame, and perform a craniotomy as described (Chapter 32) (Zhang et al. 2005; Mostany and Portera-Cailliau 2008).

    Currently there are two major preparations for two-photon microscopy of the superficial cortex: the craniotomy and the thinned skull. For the craniotomy preparation, a section of bone is removed and replaced by a glass coverslip, permitting imaging through the coverslip (Trachtenberg et al. 2002). With the thinned skull preparation, the surface skin is opened and the overlying bone is thinned (Grutzendler et al. 2002), allowing imaging through the thinned bone. For all of our experiments the craniotomy preparation was used and the cortex was covered with 1.3% "low-melting" agarose dissolved in brain buffer and sealed with a glass coverslip as in Zhang and Murphy (2007).

11. To standardize imaging areas between experiments, perform functional mapping with intrinsic optical signal imaging or use stereotaxic coordinates to ensure that stroke is induced within the forelimb sensory area.

    The presence of sensory responses also confirms that the cortex is healthy after the craniotomy. It is possible that regions more posterior to the forelimb and hindlimb sensory cortex are less sensitive to this model.

12. To permit determination of blood flow velocity (Murphy et al. 2008), inject into the tail vein a 0.1-mL bolus of 5 mg/mL Texas Red dextran (70 kDa) in brain buffer or H_2O.
13. Collect baseline images before inducing ischemia. To ensure that the best optical conditions are present for spine imaging, most images of dendrites are taken within the first 50 µm of cortex, where the dendrites of layer 5 neurons project laterally up to layer 1 (Fig. 1).
14. To induce ischemia, carefully pull on both sutures (around each carotid) by reaching inside the curtain (which shields the two-photon detectors from ambient light) and gently tensioning the sutures. Confirm that blood flow is blocked by using video microscopy to monitor surface blood flow. Use a green or blue filter (to increase contrast) with reflected light (wide-field) microscopy.

See *Troubleshooting.*

Two-photon imaging could be used to monitor ischemia induction, but it may be difficult to tension the sutures by hand in the darkness, which is necessary to obtain good-quality images. In addition, the refresh rate of two-photon microscopy may be too slow to monitor surface blood flow over large areas while interactively tensioning the sutures. Occlusion can be confirmed by in vivo two-photon imaging of blood flow as previously described (Murphy et al. 2008) or optional laser speckle imaging (Fig. 2A).

15. Once blood flow has greatly slowed, secure the tensioned sutures with masking tape.
16. Change light paths from video microscopy of reflected light to two-photon imaging and begin imaging for ~5–7 min (see note below) (see Fig. 1 for examples). For in vivo time-lapse imaging of dendritic structure during ischemia, record multiple z-stacks with step sizes of 1 µm at intervals of ≈1 min. Steps of 1 µm are sufficient to resolve spines given the relatively poor axial resolution of two-photon imaging in vivo. To determine stroke-related morphological changes in dendrites, take images within 100 µm of the pial surface and in layer I, and quantify structure within the first 50 µm where damage is more pronounced.

    To avoid potentially irreversible depression of the heart rate, limit the ischemic period to 6–8 min (although some mice can tolerate longer ischemia).

    See *Troubleshooting.*
17. Induce reperfusion by removing the tension on the sutures. Because it may be necessary to illuminate the area around the sutures (with a flashlight) to remove the masking tape, two-photon

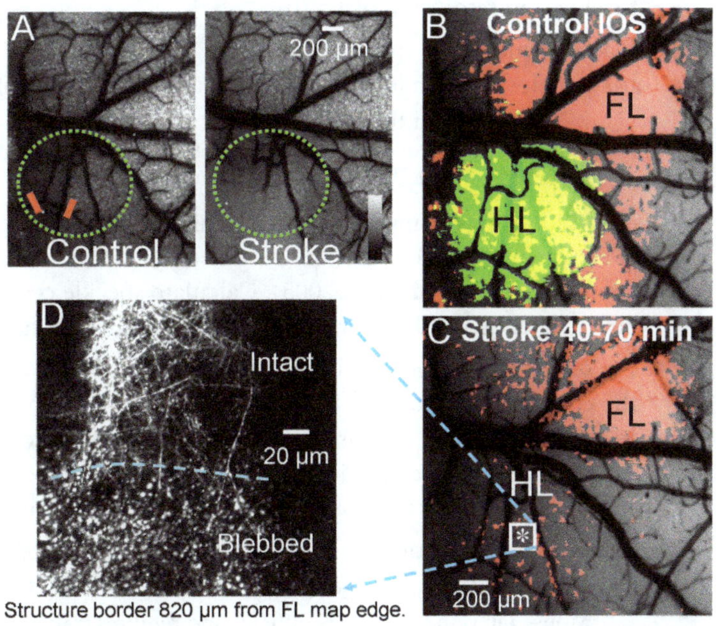

FIGURE 2. Determination of the areas of sensomotor function loss and dendrite damage after targeted ischemic stroke. (*A*) Laser speckle surface blood flow image (dark areas indicate flow) showing the approximate location of the hindlimb territory (green circle) and two sites where photoactivation of Rose Bengal (RB) was performed using a fluorescence arc lamp (red lines). (*Right*) Laser speckle image 20 min after stroke induction shows a loss of surface blood flow to about half of the hindlimb territory. (*B*) Quantification of sensomotor activity after stroke with intrinsic optical imaging (IOS) (Harrison et al. 2009) after contralateral limb stimulation, in the forelimb (FL) and hindlimb (HL) area of the somatosensory cortex. IOS maps for hindlimb and forelimb regions in the animal before stroke induction, in overlay of the vasculature map. (*C*) Overlay of IOS maps as in *B*, but after stroke, with the vasculature map recorded before stroke. The hindlimb response was completely lost (no significant response was observed so a color map could not be created) 40–70 min after photothrombotic stroke, while a response in the forelimb area was still present. (*D*) The white box and enlarged image to the *left* indicate the area where two-photon imaging established the border of dendritic damage. (Reprinted from Zhang and Murphy 2007.)

imaging may need to be stopped briefly (or at least the detectors turned off). During prolonged reperfusion, record stacks less frequently to avoid photodamage.

18. Continue imaging for up to 6 h.

Laser-Targeted Focal Photothrombotic Stroke Model

Laser-mediated photothrombosis permits targeting identified cortical surface arteries for occlusion, providing a means of disrupting flow within defined cortical domains (Schaffer et al. 2006). “Photothrombosis” refers to activation of a photosensitizer (Rose Bengal) to induce blood clotting cascades (Watson et al. 1985). We have previously described in detail the hardware for photothrombosis induction in cortical blood vessels (Sigler et al. 2008). Arterioles that feed the subsurface cortex usually run along the pial surface before penetrating into the cortex (Nishimura et al. 2007). This allows targeting of upstream arterioles that feed large regions of cortex from the surface. The following steps are optimized for acute imaging following photothrombotic induction of stroke. For modifications to the procedure when performing chronic imaging of stroke-related changes, see Discussion.

19. Anesthetize the mouse as described in Steps 1–3.

20. Perform a craniotomy as described in Steps 10–12.

21. Mark and optimize the position of the collimated 532-nm photoactivation laser that is directed into the microscope’s epifluorescence excitation light path (Sigler et al. 2008). Optimize the position (within the microscope field) of the photoactivation spot by adjusting the mirrors within the photoactivation device using an orange or red fluorescent polymer test slide. Use a 40×, 0.8-NA objective and the optics described previously (Sigler et al. 2008) to achieve photoactivation resolution of >5 μm.

22. Place the mouse on the microscope stage and visually select target arterioles (typically <60 μm in diameter) using a low-power 4× objective (as shown in Fig. 3C,D).

    Include a filter in the optical pathway that blocks light >480 nm. This will be useful once Rose Bengal is injected into the mouse (Step 25), because much of the light >480 nm would be absorbed by Rose Bengal and could cause spurious blood clotting.

23. Map blood flow using laser speckle microscopy as described previously (Briers and Webster 1996; Zhang and Murphy 2007; Sigler et al. 2008). Collect 50 images at 10 Hz using 10-msec exposure time (Yuan et al. 2005). Calculate speckle contrast as described (Zhang and Murphy 2007; Sigler et al. 2008). Areas with reduced variance have higher blood flow (Fig. 2A).

24. For selection of targeting sites for photothrombosis, switch back to the 40× objective and mark the coordinates of one or two arterioles 10–60 μm in diameter that are supplying the region selected for stroke. Because surface blood vessels often possess redundant flow paths, target multiple sites in one vessel (Schaffer et al. 2006). A motorized stage that allows for the storage of several positions along an arteriole selected for photothrombosis is desirable.

25. To induce photothrombosis, inject 30 mg of 10 mg/mL Rose Bengal per kilogram of mouse into the tail vein using a 26½-gauge needle after warming the tail with a heat lamp.

    Mouse tail vein injection is difficult and requires training. To avoid eye injury, make sure that no significant amount of green laser light can leak into the eyepieces before viewing the photoactivation process (Sigler et al. 2008). Dim the room lights after the Rose Bengal injection to prevent spurious photoactivation.

26. Begin photoactivation of arterioles with the green laser (use 0.7–3 mW of power measured at the objective back aperture). Monitor the process by examining red epifluorescence generated from Rose Bengal photoactivation (Fig. 3). A fluorescence arc lamp can be used in place of the laser (Fig. 2).

    See *Troubleshooting*.

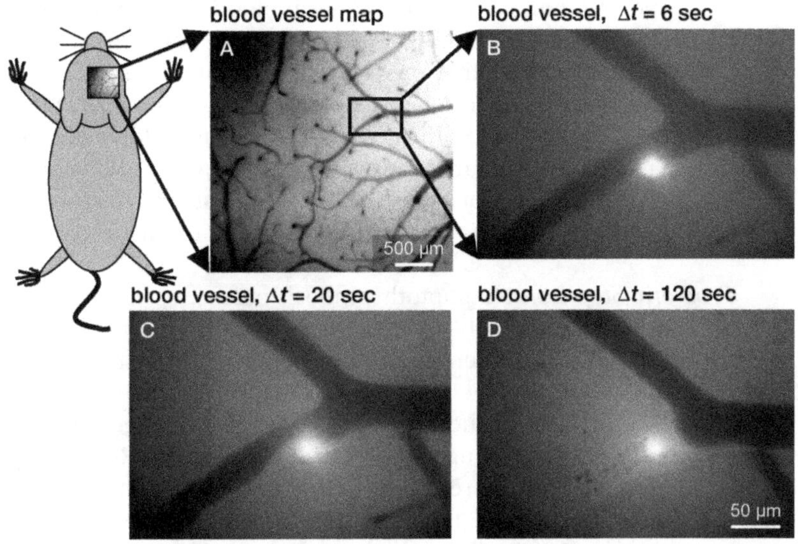

FIGURE 3. Targeting individual brain arterioles for photoactivation of Rose Bengal. (*A*) Schematic showing experimental subject (C57BL/6 mouse) and brain vasculature with regions for photoactivation indicated by arrows. (*B–D*) Photoactivation spot visualized by Rose Bengal fluorescence following excitation with 532-nm laser light in the arteriole segment shown in *A*. Blood flow is blocked within 2 min of photoactivation. See also Movie 86.1. (Reprinted, with permission, from Sigler et al. 2008, ©Elsevier.)

It is also possible to view the process using co-injection of Texas Red (or fluorescein)–labeled dextran (Schaffer et al. 2006; Zhang and Murphy 2007; Ding et al. 2009).

27. For the best results, rotate between two or three targeted sites, dwelling no more than 1–2 min at each site. The clotting procedure should be completed within 10 min after injection, since the mouse will clear Rose Bengal with a decay time $\tau = 3$ min (Zhang and Murphy 2007).

    See *Troubleshooting.*

    The Rose Bengal clearance appears to be slower after intraperitoneal injection, although this was not measured.

28. Confirm that blood flow in the arterioles is blocked by monitoring with either laser speckle or two-photon imaging, as was done before Rose Bengal injection (Fig. 2A).

TROUBLESHOOTING

Problem (Step 14): A carotid artery ruptures while trying to stop surface blood flow.
Solution: In rare cases, the mouse has a well-developed posterior communicating artery or is not a C57BL/6 mouse and sutures are over-tensioned while attempting to block blood flow leading to vessel rupture. In such instances, replace the subject. More commonly, adjusting suture tension until surface flow is nearly stopped alleviates this problem. Note that in most animals flow does not stop completely in large vessels but can be seen as a turbulent trickle by video microscopy. Flow should appear to stop in most capillaries. If the protocol steps are followed, two-photon imaging will indicate a reduction in capillary red blood cell velocity to <10% of pre-stroke levels (Murphy et al. 2008).

Problem (Step 16): Ischemic depolarization is observed, but minimal local structural change in dendrites is apparent.
Solution: It is important to know that although dendritic structural changes are triggered by ischemic depolarization, damage to dendrites may only occur in regions where both ischemic

depolarization is initiated and local blood flow (and oxygenation) is at a very low level (<10% of baseline). Therefore, it is important to confirm ischemia using local blood flow measurements. To ensure that there are not local unblocked routes of blood flow just outside of the imaging field, try to image dendrites that are in areas where relatively few major surface artery branches are present. Images should be acquired in areas toward the center of the forelimb map that is relatively more sensitive to bilateral CCA occlusion than posterior medial areas. Because previous trauma or ischemia can affect stroke outcome, do not use animals that have had hypoxic episodes (O_2 saturation <90%) or extensive cortical damage during the craniotomy.

Problem (Step 26): No photothrombosis occurs.

Solution: This is probably due to insufficient injection of Rose Bengal. Confirm this possibility by examining Rose Bengal epifluorescence during the clotting process using the eyepieces or a video camera. If the injection was insufficient, little or no red Rose Bengal epifluorescence will be observed with the stroke-targeting laser focused on an artery (using a Texas Red–style filter set). By not injecting Texas Red dextran before Rose Bengal, it is easier to confirm that Rose Bengal injections are successful. If necessary, re-inject the animal again through the other side of the tail, because two veins are present. If the second vein fails, perform an intraperitoneal injection (100 mg/kg). Disadvantages of an intraperitoneal injection are slower pharmacokinetics and (likely) lower peak concentration in vivo following injection.

Problem (Step 27): Widespread clotting is evident despite only local photoactivation of Rose Bengal.

Solution: It is possible to observe nonspecific light-induced clotting after injection of Rose Bengal from non-laser light sources. Use a surgery light with a blue safety filter. Dim fluorescent room lights. Do not examine the preparation using epifluorescence or two-photon imaging when the concentration of Rose Bengal levels may be high (i.e., initial 5–10 min after injection).

DISCUSSION

Expected Results from the Global Ischemia Model

With the global ischemia model, one should observe blood flow slowing within seconds of tensioning the sutures around each carotid. Following ischemic induction, after ~2 min, ischemic depolarization should be observed if a DC recording of EEG activity is performed (Murphy et al. 2008). Two-photon images of dendritic structure should show a swelling and beading during ischemic depolarization (Fig. 1). Structural changes will be most prominent within the first 70 µm of cortical layer 1. Astrocyte structure should also be affected during global ischemia (Risher et al. 2009). If reperfusion is successfully induced after ~7 min of occlusion, restoration of blood flow should be observed by two-photon imaging or surface video analysis. Following the restoration of blood flow, a recovery of swollen dendritic structure will occur over tens of minutes. Coincident with this recovery of dendritic structure will be a partial recovery of spontaneous EEG activity. Following reperfusion if blood flow levels fully recover, the recovered dendritic structure should be relatively intact over at least the next 4 h, and fine analysis of dendrites will reveal some sustained spine loss after a single ischemic episode (Murphy et al. 2008).

Expected Results from the Focal Ischemia Model

Following the initiation of Rose Bengal photoactivation during targeted focal ischemia, several results are expected. Immediately upon photoactivation, it should be possible to see fluorescent Rose Bengal and the production of localized clots and turbulent flow within the photoactivation site (Fig. 3). Following the induction of local clotting at several points, it should be possible to observe a maintained blockade of blood flow within several segments of the surface artery of interest (Fig. 2A). This reduction in flow should be relatively stable, and the best strokes are those where blood flow truly goes to 0 and the clots become stable. After formation of stable clots, it should be possible to observe beaded dendrites within the vicinity (Figs. 1 and 2). One should also observe a transition zone between damaged and intact dendrites near areas where blood flow is apparent in unclotted regions (Zhang and Murphy 2007). Dendritic damage should be relatively stable over time unless the preparation undergoes spontaneous reperfusion (Zhang et al. 2005), and then it is possible to observe a reversal of damaged structure as shown in Figure 1A for the global ischemia model.

Imaging Long-Term Changes Caused by Induced Stroke

When inducing stroke that will then be chronicled by repeated imaging of dendrite structure over weeks, a larger (diameter ≈ 1 mm) region of interest, such as the entire forelimb sensory area, can be targeted with a wider collimated photoactivation laser beam (1–1.5 mm in diameter), as in Brown et al. (2007). This larger beam strikes brain directly (without use of an objective) through the thinned skull using powers in the range of 15 mW and photoactivation times of 10–15 min. To restrict the photoactivation area, only the skull over the targeted region of cortex is thinned, and adjacent areas are masked with the ink from a black lab marker pen. In this application, the procedure can be further simplified using intraperitoneal injection of a 1% Rose Bengal solution made in phosphate-buffered saline (100 mg/kg). If a laser is not available, a fluorescence arc lamp and 10x, 0.3-NA objective with a green excitation filter is another suitable light source (Zhang and Murphy 2007). In this case, the size of the photoactivation area can be set with the field aperture. If Rose Bengal photoactivation is performed through a glass coverslip, rather than thinned bone, the transfer of light will be much more efficient so the laser or lamp power should be reduced to <3 mW to prevent excessive clot formation, which could have systemic effects.

ACKNOWLEDGMENTS

This work was supported by operating grants to THM and the In Vivo Imaging Centre from the Canadian Institutes of Health Research and a Grant in Aid from the Heart and Stroke Foundation of British Columbia and the Yukon. We thank Albrecht Sigler for assistance with the production of this manuscript. THM is a recipient of a Vascular Cognitive Impairment team grant from the Canadian Stroke Network.

REFERENCES

Briers JD, Webster S. 1996. Laser speckle contrast analysis (LASCA): A nonscanning, full-field technique for monitoring capillary blood flow. *J Biomed Opt* **1:** 174–179.

Brown CE, Li P, Boyd JD, Delaney KR, Murphy TH. 2007. Extensive turnover of dendritic spines and vascular remodeling in cortical tissues recovering from stroke. *J Neurosci* **27:** 4101–4109.

Ding S, Wang T, Cui W, Haydon PG. 2009. Photothrombosis ischemia stimulates a sustained astrocytic Ca^{2+} signaling in vivo. *Glia* **57:** 767–776.

Feng G, Mellor RH, Bernstein M, Keller-Peck C, Nguyen QT, Wallace M, Nerbonne JM, Lichtman JW, Sanes JR. 2000. Imaging neuronal subsets in transgenic mice expressing multiple spectral variants of GFP. *Neuron* **28:** 41–51.

Grutzendler J, Kasthuri N, Gan WB. 2002. Long-term dendritic spine stability in the adult cortex. *Nature* **420:** 812–816.

Harrison TC, Sigler A, Murphy TH. 2009. Simple and cost-effective hardware and software for functional brain mapping using intrinsic optical signal imaging. *J Neurosci Methods* **182:** 211–218.

Kleinfeld D, Denk W. 2000. Two-photon imaging of cortical microcirculaton. In *Imaging neurons: A laboratory manual* (ed. Yuste R, et al.), pp. 23.1–23.15. Cold Spring Harbor Laboratory Press, Cold Spring Harbor, NY.

Mostany R, Portera-Cailliau C. 2008. A craniotomy surgery procedure for chronic brain imaging. *J Vis Exp* **12:** 680. doi: 10.3791/680.

Murphy TH, Li P, Betts K, Liu R. 2008. Two-photon imaging of stroke onset in vivo reveals that NMDA-receptor independent ischemic depolarization is the major cause of rapid reversible damage to dendrites and spines. *J Neurosci* **28:** 1756–1772.

Nishimura N, Schaffer CB, Friedman B, Lyden PD, Kleinfeld D. 2007. Penetrating arterioles are a bottleneck in the perfusion of neocortex. *Proc Natl Acad Sci* **104:** 365–370.

Risher WC, Andrew RD, Kirov SA. 2009. Real-time passive volume responses of astrocytes to acute osmotic and ischemic stress in cortical slices and in vivo revealed by two-photon microscopy. *Glia* **57:** 207–221.

Schaffer CB, Friedman B, Nishimura N, Schroeder LF, Tsai PS, Ebner FF, Lyden PD, Kleinfeld D. 2006. Two-photon imaging of cortical surface microvessels reveals a robust redistribution in blood flow after vascular occlusion. *PLoS Biol* **4:** e22. doi: 10.1371/journal.pbio.0040022.

Sigler A, Goroshkov A, Murphy TH. 2008. Hardware and methodology for targeting single brain arterioles for photothrombotic stroke on an upright microscope. *J Neurosci Methods* **170:** 35–44.

Trachtenberg JT, Chen BE, Knott GW, Feng G, Sanes JR, Welker E, Svoboda K. 2002. Long-term in vivo imaging of experience-dependent synaptic plasticity in adult cortex. *Nature* **420:** 788–794.

Watson BD, Dietrich WD, Busto R, Wachtel MS, Ginsberg MD. 1985. Induction of reproducible brain infarction by photochemically initiated thrombosis. *Ann Neurol* **17:** 497–504.

Yang G, Kitagawa K, Matsushita K, Mabuchi T, Yagita Y, Yanagihara T, Matsumoto M. 1997. C57BL/6 strain is most susceptible to cerebral ischemia following bilateral common carotid occlusion among seven mouse strains: Selective neuronal death in the murine transient forebrain ischemia. *Brain Res* **752:** 209–218.

Yuan S, Devor A, Boas DA, Dunn AK. 2005. Determination of optimal exposure time for imaging of blood flow changes with laser speckle contrast imaging. *Appl Opt* **44:** 1823–1830.

Zhang S, Murphy TH. 2007. Imaging the impact of cortical microcirculation on synaptic structure and sensory-evoked hemodynamic responses in vivo. *PLoS Biol* **5:** e119. doi: 10.1371/journal.pbio.0050119.

Zhang S, Boyd J, Delaney K, Murphy TH. 2005. Rapid reversible changes in dendritic spine structure in vivo gated by the degree of ischemia. *J Neurosci* **25:** 5333–5338.

MOVIE LEGEND

Movies are freely available online at www.cshprotocols.org/imaging.

MOVIE 86.1. Targeting individual brain arterioles for photoactivation of Rose Bengal. The movie has been recorded as epifluorescence with a charge-coupled device (CCD) camera on a microscope equipped with a 40×, 0.8-NA water-immersion lens over a time period of 120 sec and shows the blood vessel in sextuple time (movie plays in 20 sec) during targeting with 532-nm laser light. While the laser beam position is fixed (bright spot in the middle of the image), the specimen is being moved to target multiple sites within the blood vessel. Approximately 80 sec after start of photoactivation (≈14 sec in the movie), the blood flow appears to be blocked. See Figure 3 for single frames of this movie and further description. A small amount of background light was added to permit viewing of areas not subjected to photoactivation; this procedure was only used for creation of this movie and was never employed during an actual experiment. (Reprinted, with permission, from Sigler et al. 2008, ©Elsevier.)

87 Two-Photon Imaging of Microglia in the Mouse Cortex In Vivo

Axel Nimmerjahn[1]

Departments of Biology and Applied Physics, James H. Clark Center for Biomedical Engineering and Sciences, Stanford University, Stanford, California 94305

ABSTRACT

Microglia are the primary immune effector cells of the brain parenchyma. They are distributed throughout the brain at various densities. Two-photon fluorescence microscopy, together with expression of fluorescent proteins in microglia, has enabled study of these fascinating cells in vivo. Imaging studies have shown, for example, that microglia continually survey their cellular environment and immediately respond to injury. However, we still know very little about their roles in various parts of the developing and adult brain or their diverse effector functions in aging and different disease states. These protocols describe experimental procedures for minimally invasive short- and long-term two-photon imaging of microglial cells in cortical regions of the intact mouse brain. Detailed information on surgical preparation and fluorescence imaging is provided. From the beginning to the end of surgery, the procedure takes ~2 h. Technical pitfalls, limitations, and alternative approaches are discussed.

INTRODUCTION

Microglia form the brain's first line of defense against injury and infection. They are highly sensitive to even slight changes in their cellular environment. Disturbances of central nervous system homeostasis lead to rapid changes in microglia phenotype (Raivich et al. 1999; Ransohoff and Perry 2009). Studying normal microglia physiology in vitro has therefore been challenging given the disturbances that cell isolation and tissue slicing introduce.

Various techniques, including bioluminescence imaging, magnetic resonance imaging, positron emission tomography, and two-photon microscopy, now enable minimally invasive study of microglia in the intact brain (Wiart et al. 2006; Lalancette-Hebert et al. 2009; Ransohoff and Perry

[1]Present address: Salk Institute for Biological Studies, La Jolla, California 92037.

2009). Two-photon microscopy is presently the only technique that provides submicrometer spatial and subsecond temporal resolution for the cellular level study of microglia physiology in vivo (Garcia-Alloza and Bacskai 2004). Owing to its superior resolution, pathological changes can be detected and followed at much earlier stages than with other minimally invasive techniques (Garcia-Alloza and Bacskai 2004). However, given its restricted depth penetration, two-photon imaging is limited to several hundred micrometers below the cortical surface (Helmchen and Denk 2005). Nevertheless, the ability to image with single-cell resolution in the intact brain is a powerful tool for studying the role of microglia in health and disease.

Two-photon imaging has yielded intriguing findings about microglia. In the healthy adult brain, highly branched microglial processes continuously palpate their microenvironment, a trait likely serving an immune surveillance function (Nimmerjahn et al. 2005). After local injury, microglia processes rapidly (within minutes) extend toward the site of injury, shielding and scavenging damaged tissue (Davalos et al. 2005; Nimmerjahn et al. 2005; Kim and Dustin 2006). Chemical cues such as ATP and its breakdown products released by affected tissue and detected by microglia receptors guide process extension (Davalos et al. 2005; Haynes et al. 2006). Morphological transformation proceeds gradually through several phenotypically distinct phases. Transformation can lead to an amoeboid morphology and cell body migration over the course of days (Kim and Dustin 2006). However, not all brain injuries trigger a rapid and directed response, as in the case of traumatic brain injury. Focal cerebral ischemia, for example, leads to widespread microglial activation without unidirectional process movement (Wake et al. 2009). Microglial response to tissue injury therefore depends on the type and severity of tissue damage. Using mouse models of disease, two-photon imaging allows disease progression and microglial response to treatment to be assessed in vivo (Koenigsknecht-Talboo et al. 2008). In summary, two-photon imaging has yielded valuable insight into some of the cellular and molecular mechanisms underlying microglia physiology.

These protocols describe procedures for stable, minimally invasive short- and long-term two-photon imaging of microglial cells in the cortex of anesthetized mice. These techniques can be used to develop new studies of microglial cell function in various parts of the intact healthy and diseased brain.

Protocol A

Surgical Implantation of a Head Plate in Mice in Preparation for In Vivo Two-Photon Imaging

MATERIALS

CAUTION: See Appendix 6 for proper handling of materials marked with <!>.

Reagents

Analgesic (e.g., buprenorphine)

Cx3cr1-eGFP (enhanced green fluorescent protein) knockin mice or Iba1-eGFP transgenic mice

Cx3cr1-eGFP knockin mice are available from The Jackson Labotatory (stock # 005582). In this mouse line, the *Cx3cr1* gene encoding the fractalkine receptor is replaced by the reporter gene eGFP, resulting in labeling of microglia in the central nervous system as well as monocytes, subsets of natural killer (NK), T, and dendritic cells in the circulation. Heterozygous mice retain fractalkine receptor function; see Jung et al. (2000) for details. For details on Iba1-eGFP transgenic mice see Hirasawa et al. (2005).

Dental acrylic (e.g., item H00335, Coltène Whaledent)

Eye ointment (e.g., Puralube Vet Ointment, PharmaDerm)

Hair removal cream (e.g., Nair, Church & Dwight Co., Inc.)

Local anesthetic (e.g., 1% lidocaine <!>)

Physiologic saline/lactated Ringer's solution (e.g., from Henry Schein Medical)

Skin disinfectant (e.g., Techni-Care, Care-Tech Laboratories, Inc.)

Surface disinfectant (e.g., Virkon-S, DuPont)

Tissue adhesive (e.g., Vetbond, 3M)

Equipment

Aseptic instruments/surgical tools (e.g., from Fine Science Tools)

Aseptic working field

Balance (e.g., Mettler Toledo PG503-S), for weighing animals (particularly important if injectable instead of gaseous anesthetics are used)

Flexible-arm magnetic-base holder (e.g., item 20755A23, McMaster-Carr)

Glass bead sterilizer (e.g., model BS-500, Dent-EQ)

Hair clipper (e.g., Wahl Mini Arco)

Head holder, custom-made 6.4 x 76 x 6.4-mm (W x L x H) stainless steel (Fig. 1)

Head plate assembly, custom-made, including an anodized 9 x 15 x 1-mm (W x L x H) aluminum head plate and a C-shaped 7.5-mm-diameter phosphor bronze metal ring (Fig. 1)

Laboratory animal anesthesia system (e.g., VetEquip Inc. item 901806), anesthetic gas (e.g., isoflurane <!> from Southmedic, Inc.), carrier gas tank (e.g., medigrade oxygen from Praxair); recommended: waste anesthetic gas system (e.g., VetEquip Inc. item 933101)

Pressurized air container for removal of skull/hair shavings (e.g., UltraJet Air Duster System, Chemtronics)

Stereomicroscope (e.g., Leica MZ12.5)

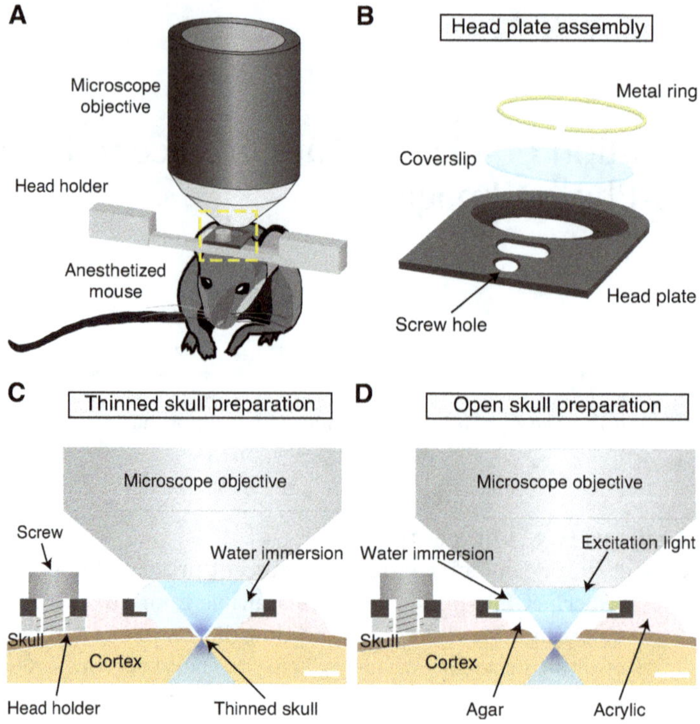

FIGURE 1. Experimental procedures for two-photon fluorescence imaging in the cortex of anesthetized mice. (*A*) Schematic view of experimental setup. Stable fluorescence imaging is enabled by restraining the animal's head via a head plate and head holder. A blow-up of the head plate (yellow dashed box) is shown in *B*. (*B*) Firm and repeatable head restraint is facilitated by implanting a lightweight head plate. Coverslip and metal ring are used to stabilize and seal open skull preparations. (*C,D*) Lateral view showing how skull, head plate, and head holder interface with each other. Scale bar, 2 mm. (*C*) Schematic view of a thinned skull preparation. A restricted area of the skull (typically <350 µm in diameter) is reduced to an ~25–35-µm-thin layer. Cells and their processes can be visualized over short and long timescales up to a depth of ~250 µm. (*D*) Open skull preparation. To reduce heartbeat- and breathing-induced motion artifacts as well as optical aberrations, the exposed brain is covered with agarose gel and a coverslip. This preparation is particularly useful for short-term investigation of deep cortical regions and large contiguous brain areas.

Stereotaxic apparatus/Dissecting setup (custom-made or from David Kopf Instruments)
Surgical eye spears (e.g., item 1556455, Henry Schein Medical)
Temperature regulation system (e.g., item 40-90-8, 40-90-5, 40-90-2-07, FHC Inc.)
Waste liquid suction line, custom-made

EXPERIMENTAL METHOD

All procedures described in this and the following protocols were developed for two-photon imaging of microglial cells in the cortex of anesthetized mice (Nimmerjahn et al. 2005). Study of microglial cells in other central nervous system regions or model organisms requires different procedures (Davalos et al. 2008; Eter et al. 2008; Peri and Nüsslein-Volhard 2008). All procedures described in this and the following protocols were approved by the Max Planck Society's and Stanford's Institutional Animal Care and Use Committee (IACUC). Some steps involve handling of animals and use of drugs. Exposure to laboratory animal allergens can cause allergies (Figler 2004). Improper use of drugs may constitute a health hazard. Wearing appropriate personal protective equipment and carefully reviewing material safety data sheets is therefore recommended. All surgical procedures should be performed uninterrupted and under conditions as sterile as possible. Use proper disinfectants to thoroughly clean surgical tools and equipment. This surgical procedure takes ~1.5 h.

1. Weigh the animal. Fluid and drug amounts to be administered during the experiment are based on body mass. To prevent dehydration of the animal, inject ~0.25–0.5 mL of warm fluids (physiologic saline or lactated Ringer's solution) per 15–25 g body mass/h.
2. Transfer the animal to an anesthesia induction chamber. Anesthetize the animal with gaseous isoflurane (3% in 1 L/min O_2).

   Gaseous isoflurane is particularly useful for juvenile mice and long-term investigations because it allows control of anesthesia level and duration without the need for repeated anesthetic injections.
3. Place the animal in a stereotaxic apparatus that is equipped with a nose cone for maintaining isoflurane anesthesia (typically 1%–2% in 1 L/min O_2 during surgery) and allows for head rotation and tilt. Maintain animal's body temperature (37°C) using a rectal probe and heating pad. Fix the animal's head with ear bars.
4. Protect animal's eyes with ophthalmic ointment.
5. Shave the fur from the top of the head (from the back of the neck up to the eyes). Remove the remaining fur using hair removal cream.
6. Disinfect the shaved scalp (using, e.g., Techni-Care surgical scrub).
7. Inject a local anesthetic (e.g., 1% lidocaine) subcutaneously into the scalp.
8. Begin surgery only after the animal is nonresponsive, as assessed by absence of paw withdrawal to a toe pinch. Monitor breathing rate, mucous membrane color, and withdrawal reflex of the animal periodically and adjust anesthetic level as necessary.
9. Using aseptic instruments, make an incision along the midline of the scalp from the level of the eyes to the ears. Remove the skin on both sides of the incision down to the ridge of the temporal bone where the jaw and neck muscles attach. Clean the surface of the skull (i.e., clear all periosteum and resolve any bleeding at the skull surface).
10. Carefully retract cervical muscle as necessary for imaging or head plate attachment.
11. Mark the position for the center of the optical window above the brain area of interest (e.g., 0.5 mm posterior and 3.5 mm lateral of bregma for S1 barrel field in an adult C57BL/6J mouse; see Paxinos and Franklin 2003).
12. Clean scalp edges and attach to skull using tissue adhesive.
13. Thoroughly dry skull surface (e.g., using surgical eye spears). This is critical for firm head plate attachment.
14. Connect head plate to head holder with an appropriate screw (Fig. 1). Using a flexible-arm magnetic-base holder to hold the head holder by one of its distal ends, center the head plate over the marked position. Lower the head plate until its lower surface is level with the skull surface. Tilt head plate until it is parallel to the underlying brain surface at the marked position.

    Note that in certain regions, brain and overlying skull surface may not be parallel.
15. Attach head plate to the skull using dental acrylic while keeping the area over which the optical window will be located uncovered. Dental acrylic should cover all wound margins but should not extend beyond the upper head plate surface (to facilitate head alignment and avoid interference with short working distance objectives) (Fig. 1C,D).
16. After the dental acrylic has set (~10 min; test firmness with scalpel), detach the head plate from head holder. Remove ear bars. Attach head holder to stereotaxic apparatus. Connect head plate to head holder. Fix animal's head in a position suitable for optical window preparation and imaging (Fig. 1).

Protocol B

Optical Window Preparation for Two-Photon Imaging in Mice

Two methods are presented in this protocol. The thinned skull method should be used whenever possible (see Discussion). Skull thinning enables transcranial two-photon imaging while minimizing external influences that might disturb normal microglia physiology and brain homeostasis. The sealed craniotomy preparation is useful for short-term investigation of microglia (see Discussion). See Holtmaat et al. (2009) for suggested methods regarding long-term imaging using an open skull preparation.

MATERIALS

See the end of the chapter for recipes for reagents marked with <R>.

Reagents

Agarose, type III-A (Sigma-Aldrich)
Artificial cerebral spinal fluid (ACSF; custom-made or, e.g., from Harvard Apparatus) <R>
Cx3cr1-eGFP knockin mice or Iba1-eGFP transgenic mice (head plate implanted; see Protocol A)
Dental acrylic (e.g., item H00335, Coltène Whaledent)
Eye ointment (e.g., Puralube Vet Ointment, PharmaDerm)
Physiologic saline/lactated Ringer's solution (e.g., from Henry Schein Medical)
Skin disinfectant (e.g., Techni-Care, Care-Tech Laboratories, Inc.)
Surface disinfectant (e.g., Virkon-S, DuPont)

Equipment

Aseptic instruments/surgical tools (e.g., from Fine Science Tools)
Aseptic working field
Balance (e.g., Mettler Toledo PG503-S), for weighing animals (particularly important if injectable instead of gaseous anesthetics are used)
Cold light source (e.g., model I-150, CUDA Products Corp.)
Coverslips, #1 thickness, 7-mm-diameter, round (e.g., from ThermoFisher Scientific)
Cyanoacrylate adhesive, clear
Epifluorescence microscope
Flexible-arm magnetic-base holder (e.g., item 20755A23, McMaster-Carr)
Gel foam (e.g., from Henry Schein Medical)
Glass bead sterilizer (e.g., model BS-500, Dent-Eq)
Head holder, custom-made, 6.4 x 76 x 6.4-mm (W x L x H), stainless steel (Fig. 1)
Heat block or water bath
Laboratory animal anesthesia system (e.g., VetEquip Inc. item 901806), anesthetic gas (e.g., isoflurane <!> from Southmedic, Inc.), carrier gas tank (e.g., medigrade oxygen from Praxair); recommended: waste anesthetic gas system (e.g., VetEquip Inc. item 933101)
Microdrill (e.g., Osada EXL-M40)
Microsurgical blades (e.g., Surgistar item 8610)
Microwave (for agarose gel preparation)

Permanent marker
Pressurized air container for removal of skull/hair shavings (e.g., UltraJet Air Duster System, Chemtronics)
Stereomicroscope (e.g., Leica MZ12.5)
Stereotaxic apparatus/Dissecting setup (custom-made or from David Kopf Instruments)
Surgical eye spears (e.g., item 1556455, Henry Schein Medical)
Temperature regulation system (e.g., item 40-90-8, 40-90-5, 40-90-2-07, FHC Inc.)
Waste liquid suction line, custom-made

EXPERIMENTAL METHOD

Method 1: Thinned Skull

Skull thinning proceeds in two steps. First, the outer dense layer of compact bone and the majority of trabecular bone are removed by drilling. In the second step, the remaining bone is thinned carefully by scraping. The mature mouse skull is a flat bone typically consisting of three layers—two parallel layers of lamellated compact bone sandwiching a thin layer of trabecular bone also called cancellous or spongy bone because of its open cell porous network. The trabecular bone contains most of the bone arteries and veins. The thickness of each layer (typically <250 μm) depends on mouse strain, age, and skull region. This procedure takes <30 min to complete.

1. Apply ACSF immersion to the cleaned surface of the skull. Under a stereomicroscope and bright cold light illumination, large blood vessels in the skull and on the brain surface should become apparent.

   Some skull regions may require slight thinning (cf. Step 3 in Method 2) to visualize surface blood vessels.

2. Using a microdrill (20,000–40,000 rpm) equipped with a new drill bit (0.5 mm diameter), begin thinning the bone in an ~1.0 mm diameter region surrounding the marked target point and devoid of large blood vessels. To maintain bone stability, create a conical excavation. Drill under ample ACSF immersion and pause intermittently to allow heat to dissipate. Exchange the ACSF frequently to remove bone shavings. To reduce tremor/vibrations during drilling support your forearms and wrists. Avoid exerting pressure onto the skull. Immediately resolve any bleeding (e.g., using wet gel foam; small blood vessels in the bone often stop bleeding spontaneously).

3. Continue thinning until fine blood vessels at the brain surface can be seen through the thinned skull under ACSF immersion. Near the excavation center small cavities within the bone (characteristic of trabecular bone matter) should be visible under the stereomicroscope at high magnification.

4. Create a uniformly thin area in the excavation center by scraping very gently with a microsurgical blade. To reduce the risk of causing bone cracks, restrict scraping to a closely confined central region (~350 μm in diameter). Avoid pushing downward onto the skull. Scraping can be performed without ACSF immersion. However, ACSF immersion should be applied frequently to dissipate heat, remove bone shavings, and assess thinning progress.

5. Continue thinning until fluorescently labeled cells and their processes can be seen through the thinned skull under an epifluorescence microscope. Keep overall exposure to and power of fluorescence excitation light to a minimum.

   Skull thickness can be estimated using two-photon microscopy (cf. Step 6, Protocol C).

6. Use a permanent marker to label areas surrounding the excavation. These marks will facilitate locating the thinned skull area under a high-magnification objective.

7. If long-term imaging is intended, take a high-resolution image of the blood vessel pattern underlying the thinned skull area (e.g., using a stereomicroscope-mounted digital or charge-coupled device [CCD] camera). To improve the stability of the thinned area of the skull and

prevent bone regrowth, a thin layer of clear cyanoacrylate adhesive (refractive index typically ~1.4–1.5) can be applied with or without coverslip. This approach may also be used to prepare larger (i.e., up to several millimeters in diameter) thinned skull areas for long-term imaging (Masino and Frostig 1996; Drew et al. 2010).

Method 2: Sealed Craniotomy

This procedure takes <30 min to complete.

1. Prepare 1.5%–2% agar in ACSF for optical window creation. Keep the agar on a heat block or in a water bath set to >37°C while not in use.
2. Apply ACSF immersion to cleaned surface of the skull. Under a stereomicroscope and bright cold light illumination, large blood vessels in the skull and on the brain surface should become apparent. Some skull regions may require slight thinning (see next step) to visualize surface blood vessels.
3. If slight thinning is required, use a microdrill (20,000–40,000 rpm) equipped with a new drill bit (0.5-mm-diameter) to remove the top layer of the skull in an ~1.5 mm diameter region surrounding the marked target point. Drill under ACSF immersion to dissipate heat. Exchange ACSF immersion frequently to remove bone shavings. Immediately resolve any bleeding from bone vessels (e.g., using wet gel foam or drilling toward the bleeding site to remove the bleeding vessel; small bone vessels often stop bleeding spontaneously). Stop thinning after large blood vessels at the brain surface can be seen through the thinned skull under ACSF immersion.
4. Drill carefully around the edges of the desired target area. Ideally, this area should be devoid of large blood vessels and bone sutures. Drilling across these structures can cause substantial bleeding. Continue drilling until fine blood vessels at the brain surface can be seen clearly through the thinned skull. Avoid exerting pressure onto the skull.
5. Using aseptic forceps, lift away the center bone in one piece. Do not remove the dura mater. Immediately resolve any bleeding (e.g., using wet gel foam).

   See *Troubleshooting*.
6. Create an optical window for imaging. To do so, first resolve any remaining surface bleeding. Hemoglobin shows large two-photon absorptivity and may interfere with imaging (Clay et al. 2007). Remove excess ACSF from the surface of the skull (using surgical eye spears or other tool) while keeping a thin layer of the liquid on top of the brain. Apply agar while holding a clean coverslip of desired size in place. To avoid heat-related damage to the brain, let agar cool until it has nearly solidified (close to physiological temperature) before applying. Let agar harden (~1 min). Then, remove excess gel around and on top of the coverslip. Fix coverslip (e.g., using a small metal ring that fits snugly within the head plate; Fig. 1). Clean exposed coverslip surface.
7. Take a high-resolution image of the surface blood vessel pattern (e.g., using a stereomicroscope mounted digital or CCD camera).
8. If access to the brain is not required, coverslip edges can be sealed with a thin layer of dental acrylic.

TROUBLESHOOTING

Problem (Step 5 of Method 2): You have trouble removing the bone with forceps during open skull preparation.

Solution: Thin the edges of the target area further. Alternatively, in short-term experiments, use a 30-gauge needle to perforate the edges superficially. The needle should penetrate the thinned skull easily without much force. To minimize damage to the underlying cortex, hold the needle at a shallow angle and point its tip away from the area to be imaged. After it is perforated, use aseptic forceps to carefully lift away the bone in one piece. Restrict imaging to areas far from skull edges.

Protocol C

Two-Photon Imaging of Microglia in the Mouse Cortex In Vivo

IMAGING SETUP

Equipment

- Two-photon microscope (custom-made) equipped with an ultrafast Ti-sapphire laser, a 10-W all-solid-state pump laser (Spectra-Physics), a motorized micromanipulator-controlled (Sutter Instruments) *x–y* table (New England Affiliated Technologies, Inc.), a motorized objective focusing nose piece (e.g., using a Vexta stepping motor), and one or more fluorescence detection channels (filter sets depend on application; for excitation/emission spectra of common fluorescent proteins, see Spiess et al. 2005)
- Electro-optic laser power controller (e.g., model 302RM and 350-80-LA, Conoptics)
- Beam shutter (e.g., item SH05, Thorlabs, Inc.)
- Galvanometer-based optical scanner (e.g., item 6215H, Cambridge Technology, Inc.)
- High-numerical-aperture (NA) water-immersion objective (e.g., Zeiss Achroplan 40x 0.80 NA)
- Variable-gain, low-noise current amplifiers (e.g., DLPCA-200, Femto)
- High-speed digital signal integrator, custom-made (as per Nguyen et al. 2006)
- Data acquisition hardware (e.g., items NI PCI-6110, NI BNC-2110, NI SHC68-68-EP, National Instruments)
- Data acquisition software (e.g., MPScope; see Nguyen et al. 2006)
- Optical power meter (e.g., item FM203, Thorlabs, Inc.)
- Cold light source (e.g., KL 1500, Schott)
- Digital or CCD camera (e.g., from Andor Technology)
- Infrared viewer (e.g., FIND-R-SCOPE, FJW Optical Systems, Inc.) or viewing card (Thorlabs, Inc.)

Imaging Parameters

- Excitation wavelength: ~950 nm (this wavelength is optimal for eGFP excitation; see Spiess et al. 2005)
- Excitation intensity at sample surface: typically, <15 mW (depending on details of the optical setup, animal age, surgical preparation, imaging depth, etc.)
- Images/frame rate: 512 x 512 pixels (4.07 Hz), 256 x 256 pixels (10.17 Hz)
- Average: typically one to two frames
- Recording duration: minutes to months (depending on biological process investigated)
- Sampling interval: seconds to days (depending on biological process investigated and volume sampled)

Image Processing

- ImageJ (e.g., for lateral image alignment, projection images)
- Custom-written software (e.g., using IGOR Pro from WaveMetrics, Inc.)
- Commercial software packages (e.g., Amira or Imaris from Visage Imaging and Bitplane, respectively)

MATERIALS

See the end of the chapter for recipes for reagents marked with <R>.

Reagents

Artificial cerebral spinal fluid (ACSF; custom-made or from Harvard Apparatus) <R>
Cx3cr1-eGFP knockin mice or Iba1-eGFP transgenic mice (head plate implanted, prepared with either thinned skull or optical window; see Protocols A and B)
Eye ointment (e.g., Puralube Vet Ointment, PharmaDerm)
Skin disinfectant (e.g., Techni-Care, Care-Tech Laboratories, Inc.)
Surface disinfectant (e.g., Virkon-S, DuPont)
Transparent dressing (e.g., Bioclusive, Johnson & Johnson)

Equipment

Cold light source (CUDA Products Corp., model I-150)
Flexible-arm magnetic-base holder (McMaster-Carr, item 20755A23)
Glass bead sterilizer (Dent-EQ, model BS-500)
Head holder, custom-made, 6.4 x 76 x 6.4-mm (W x L x H), stainless steel (Fig. 1)
Laboratory animal anesthesia system (VetEquip Inc., item 901806), anesthetic gas (isoflurane <!> from Southmedic, Inc.), carrier gas tank (medigrade oxygen from Praxair); recommended: waste anesthetic gas system (VetEquip Inc., item 933101)
Optical power meter (Thorlabs, item FM203)
Stereomicroscope (Leica MZ12.5)
Stereotaxic apparatus/dissecting setup (custom-made or from David Kopf Instruments)
Surgical eye spears (Henry Schein Medical, item 1556455)
Temperature regulation system (FHC Inc., item 40-90-8, 40-90-5, 40-90-2-07)
Two-photon microscope (see Imaging Setup)

EXPERIMENTAL METHOD

The initial imaging session typically takes ~1.5–2 h but up to 5 h depending on the number of recording sites and biological processes investigated. Subsequent imaging sessions last ~90 min or less.

Two-Photon Imaging

1. Tune laser to desired wavelength (e.g., ~950 nm for eGFP excitation)
2. Make sure the back aperture of the objective is as evenly illuminated as possible. Use an infrared viewer or infrared viewing card to visualize the laser beam. Adjust microscope mirrors as necessary. Wear appropriate protective eyewear and use minimal laser power during this process.
3. Using an optical power meter, measure the average laser power exiting the microscope objective for different beam deflector settings. Excessive tissue illumination can cause microglia activation. During imaging keep average laser power levels well below the previously determined photodamage threshold (see Discussion).

4. Transfer animal to microscope. Using adjustments on the stereotaxic apparatus, tilt the animal's head such that the head plate's upper surface is perpendicular to the optical axis. If imaging at later time points is required, mark stereotaxic position and settings. Alternatively, position a laser pointer inside the imaging setup such that its beam reflects off a reflective particle/area on the head plate into a target area. Keep laser pointer and target area fixed for precise head alignment during subsequent imaging sessions.
5. Using the microscope's ocular (or camera) and cold light illumination, locate the brain surface by translating the microscope stage or objective. Switch to fluorescence mode.
6. Start imaging using minimal excitation intensity. If transcranial imaging is performed, assess skull thickness by measuring the distance at which autofluorescence of the skull is observed. For a skull thickness of ~25–35 µm fluorescently labeled cells and their fine processes should be resolved up to a depth of ~200 µm below the pia (Fig. 2).
7. Identify an area of interest. Take a low-magnification but high-resolution fluorescence image stack for later reference.
8. Switch back to cold light illumination. Inspect corresponding surface blood vessel patterns. Mark the recording site location on the high-resolution surface blood vessel map recorded earlier (Protocol B, Step 7). Set micromanipulator coordinates to 0.

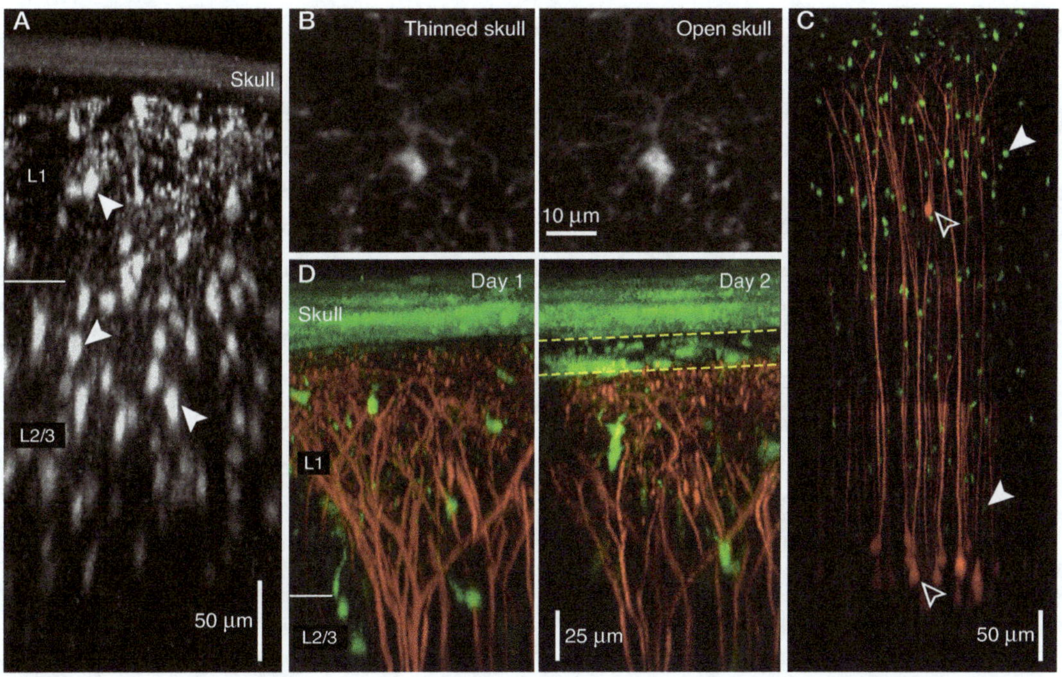

FIGURE 2. Selected aspects of thinned and open skull preparations. (*A*) Example image showing a maximum-intensity side projection of enhanced green fluorescent protein (eGFP)-positive cells in the neocortex of a heterozygous Cx3cr1-eGFP mouse. Images were acquired through the thinned, autofluorescent skull (visible on *top*). Note that microglia somata (closed arrowheads) and their fine processes can be resolved clearly in layer 1 (L1) and parts of layer 2/3 (L2/3). Image quality degrades rapidly with depth. (*B*) (*Left*) Fluorescence images showing a transcranially imaged microglial cell in L1. (*Right*) The same microglial cell after skull removal. Fine processes can be resolved in both cases. (*C*) Example image showing a maximum-intensity side projection of enhanced yellow fluorescent protein (eYFP)-expressing neurons (red, open arrowheads) and eGFP-positive microglia (green, closed arrowheads) in the neocortex of a Thy1-eYFP x Cx3cr1-eGFP mouse implanted with an open skull window. Note increased depth penetration and spatial resolution compared to *A*. (*D*) (*Left*) Side projection showing neuronal dendrites (red) and microglial cells (green) in the neocortex of a Thy1-eYFP x Cx3cr1-eGFP mouse immediately after thinned skull preparation. (*Right*) The same tissue volume imaged 1 d later without skull rethinning. Improper skull thinning can lead to inflammation in regions near the skull surface (yellow dashed lines). (*A,B*, Modified, from supplementary online material, from Nimmerjahn et al. 2005.)

9. Switch back to fluorescence mode. Acquire imaging data as necessary. Record micromanipulator coordinates relative to the brain surface and other microscope settings. For additional recording sites, repeat Steps 7–9. During prolonged imaging sessions, monitor animal periodically and treat as necessary (e.g., inject warm fluids or adjust anesthetic level).

   See *Troubleshooting.*
10. At the end of a thinned skull imaging session, dry and disinfect the skull surface (e.g., using surgical eye spears and Techni-Care, respectively) and cover the head plate surface (e.g., using a transparent dressing). Detach head plate from head holder and return animal to its cage (keep on heating pad and monitor until awakening).
11. Clean the microscope's objective and record transmitted average laser power for all beam deflector settings used during the experiment.
12. Thoroughly disinfect all areas and instruments (e.g., with a glass bead sterilizer to autoclave tips of surgical instruments).

Two-Photon Imaging at Later Time Points

13. Transfer the animal to an anesthesia induction chamber. Anesthetize the animal with isoflurane (cf. Protocol A, Step 2).
14. Place the animal in the stereotaxic apparatus used previously. Fix head plate to head holder. Maintain animal's body temperature using a rectal probe and heating pad. Assess animal condition periodically and treat as necessary.

    See *Troubleshooting.*
15. Protect animal's eyes with ophthalmic ointment.
16. Remove head plate cover (Step 10). Inspect optical window condition under a stereomicroscope. Clean/rinse skull and optical window surface.
17. Transfer animal to microscope. Adjust head position as necessary (cf. Step 4).
18. Using the reference image(s)/stack(s)/coordinates recorded during the first imaging session (see Step 7 in Protocol B, and Steps 7 and 9 in this protocol), locate previous recording sites. If transcranial imaging is performed skull rethinning may be necessary (see Discussion).

    See *Troubleshooting.*
19. Acquire imaging data as necessary (Fig. 3; Movies 87.2–87.4). If possible, use the same microscope settings as in previous session(s).

    See *Troubleshooting.*
20. At the end of the imaging session, dry and disinfect skull/optical window surface and cover head plate surface (cf. Step 10).
21. Detach head plate from head holder and return animal to its cage.
22. Clean the microscope's objective and record transmitted average laser power for all beam deflector settings used during the experiment.
23. Thoroughly disinfect all areas and instruments.

TROUBLESHOOTING

Problem (Step 9): Motion artifacts occur during imaging, particularly at high magnification.
Solution: In anesthetized animals, motion artifacts during imaging can arise from heartbeat or breathing, leading to micrometer-scale displacements of the brain relative to the anchored skull. To minimize motion artifacts in an open skull preparation, craniotomy size should be kept small

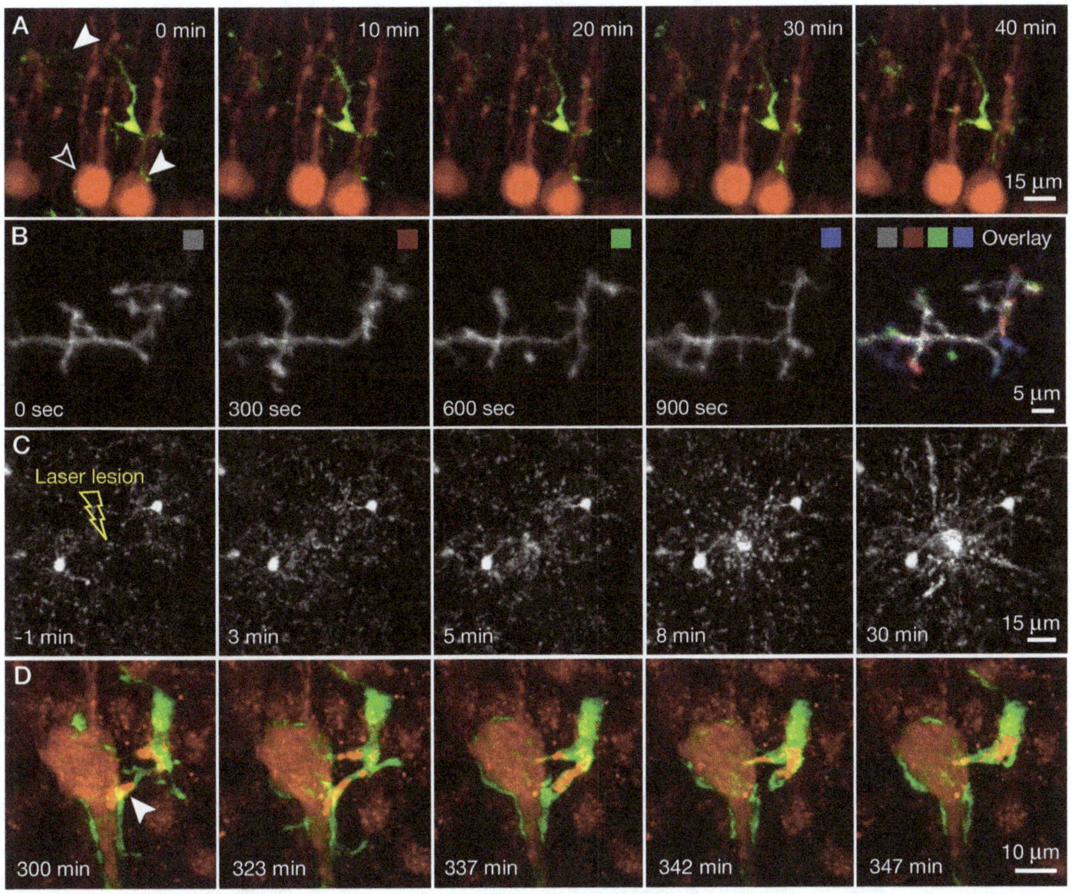

FIGURE 3. Microglia dynamics in the healthy and injured adult brain. (*A*) Overlay images from a time-lapse recording showing eGFP-expressing microglia (green) and eYFP-expressing Purkinje cells (red, open arrowhead) in the cerebellar cortex of an anesthetized Thy1-eYFP x Cx3cr1-eGFP mouse. Images were recorded near a bend in the cerebellar folium, which allowed the molecular and Purkinje cell layers to be visualized within a single image. Note that microglial processes (closed arrowheads) are highly motile. Neurons appear structurally stable during the same time interval. (*B*) Image frames showing a microglial process at high magnification. Fine protrusions continually palpate the cellular environment. The image on the far right is an overlay of the four images on the left pseudo-colored in gray, red, green, and blue. (*C*) Microglia response to laser lesion. Within minutes, microglia extend their processes toward the lesion site where they shield injured tissue. Time before and after laser lesion is indicated. (*D*) Phagocytosis of cellular debris by microglia. Overlay images showing a mechanically injured Purkinje cell (red) braided by microglia processes. Time after mechanical injury is indicated. Sulforhodamine 101 was used to stain Bergmann glia and injured cell bodies. (*C*, Images are based on data from supplementary online material from Nimmerjahn et al. 2005.)

(<1.5 mm). To further dampen brain motion, increase agar viscosity. Exerting increased pressure onto the brain via the coverslip is not recommended because of the risk of causing tissue damage. In time-lapse recordings, image stability may be improved by triggering image acquisition from the animal's heartbeat or breath. Changing the animal's body position or anesthetic level may help to diminish breathing-induced motion artifacts. Higher scan rates (at lower resolution) may help to reduce within-frame image distortions. Lateral motion artifacts remaining in and between recorded images can be largely corrected offline by image registration algorithms. However, correction of axial motion artifacts remains challenging. See Dombeck et al. (2007), Greenberg and Kerr (2009), and Nimmerjahn et al. (2009) for example algorithms.

Problem (Step 9): Microglia become activated during two-photon imaging.

Solution: Use as little laser light exposure as possible. If standard microscope settings are insufficient to achieve adequate signal-to-noise ratio, fluorescence excitation or detection efficiency can be boosted using, for example, high-NA objectives in combination with large-aperture detection optics (Oheim et al. 2001), dispersion precompensation (Tang et al. 2006), supplementary fiber-optic light collection (Engelbrecht et al. 2009), passive pulse splitting (Ji et al. 2008), bunched pulse excitation (Donnert et al. 2009), or adaptive wave-front correction (Rueckel et al. 2006). In addition, some improvement in signal-to-noise ratio may be achieved offline using computational methods (e.g., blind image deconvolution).

Problem (Step 14): The head plate detaches from the skull during head fixation.

Solution: This is a highly unlikely scenario that may be caused by improper head plate implantation or mouse handling during head fixation. To improve the adhesion of dental cement to the skull in Step 15 of Protocol A, a thin layer of cyanoacrylate adhesive (e.g., Vetbond) can be applied to the skull surface in Step 13 of Protocol A, keeping the area over which the optical window will be performed uncovered. Use great care when restraining the animal's head. Avoid shear stress.

Problem (Step 18): Compared to previous imaging sessions, microglia underneath the thinned skull optical window appear activated.

Solution: Microglia activation may be caused by improper surgical preparation (Fig. 2D). Using postmortem histology, assess the extent of glial activation in relation to optical window position. In future preparations, thin less, avoid pushing the skull toward the cortical surface, or allow sufficient heat dissipation.

Problem (Step 18): Using the same settings as in previous imaging sessions, transcranially acquired fluorescence images appear dimmer or more distorted.

Solution: Assuming this is not caused by light collection problems, for example, from low water immersion or bubbles in water immersion, reduced fluorescence levels/image distortions may be caused by skull regrowth. Rethinning may improve optical conditions (see Discussion). Low fluorescence levels may be compensated by increasing excitation intensity, although at the increased risk of causing photodamage.

Problem (Step 19): The imaged volume appears rotated compared to previous imaging sessions (e.g., because of imprecise head alignment).

Solution: Compensate image rotation online as much as possible (cf. Step 5). Remaining image rotations can be largely corrected offline (using ImageJ or commercial software packages such as Amira).

DISCUSSION

Two-photon imaging of cellular processes in the brain of live mice involves creation of an optical window. This window can either be a thinned skull region or a sealed craniotomy. Both thinned skull (Davalos et al. 2005; Nimmerjahn et al. 2005; Kim and Dustin 2006; Wake et al. 2009) and open skull preparations (Davalos et al. 2005; Nimmerjahn et al. 2005; Haynes et al. 2006; Kim and Dustin 2006; Bolmont et al. 2008; Koenigsknecht-Talboo et al. 2008; Mandrekar et al. 2009; Wake et al. 2009) have been used in the study of microglia physiology. Microglia in various parts of the mouse neocortex have been studied, but most of these studies have been restricted to periods of a few hours. Few studies have tracked microglial cells for extended periods (i.e., >24 h) (Kim and Dustin 2006; Bolmont et al. 2008). All studies so far have been conducted under various types of anesthesia.

The choice of optical window type depends on experimental requirements. The preferred method for short- and long-term imaging experiments is the thinned skull preparation because, when performed correctly, it is the least invasive to parenchymal tissue. However, some applications may require use of an open skull window. Skull removal provides direct access to the brain parenchyma allowing investigation of molecular mechanisms underlying microglia function by means of pharmacology (Davalos et al. 2005; Nimmerjahn et al. 2005; Haynes et al. 2006). Other considerations in choosing optical window type are desired imaging depth, spatial resolution, and cortical area to be monitored. Compared to an open skull preparation, the thinned skull window provides somewhat lower spatial resolution and depth penetration into tissue (Helm et al. 2009). The extent to which these parameters are affected depends on the thickness and texture of the bone remaining after the thinning procedure. Most texture and thickness irregularities can be eliminated by careful scraping with dedicated surgical blades. Remaining irregularities in skull thickness can lead to image distortions (Helm et al. 2009), some of which may be diminished, for example, using adaptive wave-front correction (Rueckel et al. 2006). However, because the risk of causing bone cracks increases with skull area to be thinned, imaging of large contiguous cortical areas (several millimeters in diameter) has been challenging. High-resolution two-photon imaging of extensive cortical areas or micrometer-scale structures deep inside the cortex (>250 µm) may therefore require the use of an open skull preparation (Holtmaat et al. 2009).

Experimenters using an open skull preparation should bear in mind that skull removal may lead to mechanical injuries to the cortical surface or immediate disturbances in local blood perfusion, blood–brain barrier permeability, and brain homeostasis (Navari et al. 1978; Kawamura et al. 1990). Hence, imaging of microglial cells close to the cortical surface should be avoided. External influences such as temperature gradients between cortical tissue and the water-immersion objective should be minimized by using, for example, an objective heater system. Long-term studies should take into account that opening the cranial cavity can lead to tissue inflammation involving glial activation that may persist for up to several weeks (Bacskai et al. 2001; Xu et al. 2007). Treatment with anti-inflammatory drugs (dexamethasone, carprofen) may reduce tissue reactions (Holtmaat et al. 2009) but may not prevent the transient increase in optical window opacity often seen within the first 1–2 weeks after craniotomy (Xu et al. 2007). Hence, if use of an open skull preparation is inevitable, control experiments should be performed to ensure that observed cellular processes are nonpathological.

Similar considerations apply to thinned skull preparations. Although parenchymal cells have been imaged successfully over days without apparent signs of inflammation, cellular structures in the meninges (which lie outside the blood–brain barrier) can be affected by the manual thinning procedure (Kim and Dustin 2006). Thinning approaches that exert minimal pressure onto the skull are desirable, particularly for studies in juvenile mice (postnatal day 12 or younger) whose skull is soft and flexible, and for long-term imaging experiments in which efficient skull rethinning is difficult and typically limited to a few repeats (Holtmaat et al. 2009). Alternative approaches to manual thinning may involve laser ablation (Walter et al. 1999). No matter which thinning approach is used, reducing the bone to a thin layer eliminates bone blood vessels and marrow. This can lead to local bone inflammation or degeneration. Repeated head plate detachment and suturing of the animal's skin during long-term experiments (Christie et al. 2001; Grutzendler et al. 2002) may compromise tissue over time. Chronic implantation of a lightweight head plate (Fig. 1) provides an alternative solution. Head plate implantation allows easy and reproducible head fixation in a manner less stressful to the animal. However, as the skin above the optical window is removed, which normally provides a protective barrier, great care must be taken to avoid infection of or mechanical damage to the optical window.

Another critical aspect in microglia imaging is the potential effect of two-photon imaging on tissue physiology itself. Excessive illumination can cause pathological tissue changes. Photodamage threshold depends on various parameters, including pulse duration, exposure time, and excitation intensity (Koester et al. 1999; Hopt and Neher 2001) and therefore needs to be determined experimentally. This can be achieved, for example, by exposing stained cortical tissue to various illumination conditions and then quantifying any changes in microglia morphology over the duration of light exposure or using immunohistochemistry after imaging. Generally, fluorescence imaging should be conducted using as little light exposure as possible.

Data analysis can be performed using public domain image processing software (e.g., ImageJ) or commercial software packages (Imaris from Bitplane). To study the interaction of microglia with other cellular elements in the brain, a variety of fluorescent markers and minimally invasive labeling approaches is available (Nimmerjahn et al. 2008). Tail vein injection of dextran-conjugated fluorescent dyes can be used to visualize blood vessels and blood–brain barrier leaks in the brain (Nimmerjahn et al. 2005; Bolmont et al. 2008; Koenigsknecht-Talboo et al. 2008). Co-staining of parenchymal cells other than microglia can be achieved by crossbreeding different transgenic mouse lines (Kim and Dustin 2006; Wake et al. 2009) (Figs. 2 and 3; Movies 87.1 and 87.2). In mouse models of Alzheimer's disease, amyloid can be stained by intraperitoneal injection of blood–brain barrier-permeable methoxy-X04 (Bolmont et al. 2008; Koenigsknecht-Talboo et al. 2008). In the future, expression of genetically encoded calcium indicators in microglia may allow study of their calcium signaling.

In summary, two-photon imaging in the brain of transgenic mice is a powerful tool in the study of relationships between microglia dynamics, phenotype, state of tissue health, and associated molecular mechanisms.

RECIPE

CAUTION: See Appendix 6 for proper handling of materials marked with <!>.
Recipes for reagents marked with <R> are included in this list.

ACSF

NaCl	125 mM
KCl	5 mM
D-glucose	10 mM
HEPES	10 mM
$MgSO_4$ <!>	2 mM
$CaCl_2$ <!>	2 mM

Adjust pH to ~7.3 with NaOH <!>. Osmolarity should be ~300.

ACKNOWLEDGMENTS

The author thanks Richard Luong, Alessio Attardo, Jay Jiang, and Jill Helms for helpful discussions; Mark Schnitzer and Ben Barres for generous support; and Frank Kirchhoff for transgenic mice. This work was supported by postdoctoral fellowships to AN from the Alexander von Humboldt-Foundation and the International Human Frontier Science Program Organization.

REFERENCES

Bacskai BJ, Kajdasz ST, Christie RH, Carter C, Games D, Seubert P, Schenk D, Hyman BT. 2001. Imaging of amyloid-β deposits in brains of living mice permits direct observation of clearance of plaques with immunotherapy. *Nat Med* **7:** 369–372.

Bolmont T, Haiss F, Eicke D, Radde R, Mathis CA, Klunk WE, Kohsaka S, Jucker M, Calhoun, ME. 2008. Dynamics of the microglial/amyloid interaction indicate a role in plaque maintenance. *J Neurosci* **28:** 4283–4292.

Christie RH, Bacskai BJ, Zipfel WR, Williams RM, Kajdasz ST, Webb WW, Hyman BT. 2001. Growth arrest of individual senile plaques in a model of Alzheimer's disease observed by in vivo multiphoton microscopy. *J Neurosci* **21:** 858–864.

Clay GO, Schaffer CB, Kleinfeld D. 2007. Large two-photon absorptivity of hemoglobin in the infrared range of 780–880 nm. *J Chem Phys* **126:** 025102.

Davalos D, Grutzendler J, Yang G, Kim JV, Zuo Y, Jung S, Littman DR, Dustin ML, Gan WB. 2005. ATP mediates rapid microglial response to local brain injury in vivo. *Nat Neurosci* **8:** 752–758.

Davalos D, Lee JK, Smith WB, Brinkman B, Ellisman MH, Zheng B, Akassoglou K. 2008. Stable in vivo imaging of densely populated glia, axons, and blood vessels in the mouse spinal cord using two-photon microscopy. *J Neurosci Methods* **169:** 1–7.

Dombeck DA, Khabbaz AN, Collman F, Adelman TL, Tank DW. 2007. Imaging large-scale neural activity with cellular resolution in awake, mobile mice. *Neuron* **56:** 43–57.

Donnert G, Eggeling C, Hell SW. 2009. Triplet-relaxation microscopy with bunched pulsed excitation. *Photochem Photobiol Sci* **8:** 481–485.

Drew PJ, Shih AY, Driscoll JD, Knutsen PM, Blinder P, Davalos D, Akassoglou K, Tsai PS, Kleinfeld D. 2010. Chronic optical access through a polished and reinforced thinned skull. *Nat Methods* **7:** 981–984.

Engelbrecht CJ, Göbel W, Helmchen F. 2009. Enhanced fluorescence signal in nonlinear microscopy through supplementary fiber-optic light collection. *Opt Express* **17:** 6421–6435.

Eter N, Engel DR, Meyer L, Helb HM, Roth F, Maurer J, Holz FG, Kurts C. 2008. In vivo visualization of dendritic cells, macrophages, and microglial cells responding to laser-induced damage in the fundus of the eye. *Invest Ophthalmol Vis Sci* **49:** 3649–3658.

Figler N. 2004. Laboratory animal allergies: Overview of causation and prevention. *Lab Anim* **33:** 25–27.

Garcia-Alloza M, Bacskai BJ. 2004. Techniques for brain imaging in vivo. *Neuromolecular Med* **6:** 65–78.

Greenberg DS, Kerr JN. 2009. Automated correction of fast motion artifacts for two-photon imaging of awake animals. *J Neurosci Methods* **176:** 1–15.

Grutzendler J, Kasthuri N, Gan WB. 2002. Long-term dendritic spine stability in the adult cortex. *Nature* **420:** 812–816.

Haynes SE, Hollopeter G, Yang G, Kurpius D, Dailey ME, Gan WB, Julius D. 2006. The $P2Y_{12}$ receptor regulates microglial activation by extracellular nucleotides. *Nat Neurosci* **9:** 1512–1519.

Helm PJ, Ottersen OP, Nase G. 2009. Analysis of optical properties of the mouse cranium—Implications for in vivo multiphoton laser scanning microscopy. *J Neurosci Methods* **178:** 316–322.

Helmchen F, Denk W. 2005. Deep tissue two-photon microscopy. *Nat Methods* **2:** 932–940.

Hirasawa T, Ohsawa K, Imai Y, Ondo Y, Akazawa C, Uchino S, Kohsaka S. 2005. Visualization of microglia in living tissues using Iba1-EGFP transgenic mice. *J Neurosci Res* **81:** 357–362.

Holtmaat A, Bonhoeffer T, Chow DK, Chuckowree J, De Paola V, Hofer SB, Huebener M, Keck T, Knott G, Lee W-CA, et al. 2009. Long-term, high-resolution imaging in the mouse neocortex through a chronic cranial window. *Nat Protoc* **4:** 1128–1144.

Hopt A, Neher E. 2001. Highly nonlinear photodamage in two-photon fluorescence microscopy. *Biophys J* **80:** 2029–2036.

Ji N, Magee JC, Betzig E. 2008. High-speed, low-photodamage nonlinear imaging using passive pulse splitters. *Nat Methods* **5:** 197–202.

Jung S, Aliberti J, Graemmel P, Sunshine MJ, Kreutzberg GW, Sher A, Littman DR. 2000. Analysis of fractalkine receptor CX_3CR1 function by targeted deletion and green fluorescent protein reporter gene insertion. *Mol Cell Biol* **20:** 4106–4114.

Kawamura S, Schuerer L, Goetz A, Kempski O, Schmucker B, Baethmann A. 1990. An improved closed cranial window technique for investigation of blood–brain barrier function and cerebral vasomotor control in the rat. *Int J Microcirc Clin Exp* **9:** 369–383.

Kim JV, Dustin ML. 2006. Innate response to focal necrotic injury inside the blood–brain barrier. *J Immunol* **177:** 5269–5277.

Koenigsknecht-Talboo J, Meyer-Luehmann M, Parsadanian M, Garcia-Alloza M, Finn MB, Hyman BT, Bacskai BJ, Holtzman DM. 2008. Rapid microglial response around amyloid pathology after systemic anti-Aβ antibody administration in PDAPP mice. *J Neurosci* **28:** 14156–14164.

Koester HJ, Baur D, Uhl R, Hell SW. 1999. Ca^{2+} fluorescence imaging with pico- and femtosecond two-photon excitation: Signal and photodamage. *Biophys J* **77:** 2226–2236.

Lalancette-Hebert M, Phaneuf D, Soucy G, Weng YC, Kriz J. 2009. Live imaging of Toll-like receptor 2 response in cerebral ischaemia reveals a role of olfactory bulb microglia as modulators of inflammation. *Brain* **132:** 940–954.

Mandrekar S, Jiang QG, Lee CYD, Koenigsknecht-Talboo J, Holtzman DM, Landreth GE. 2009. Microglia mediate the clearance of soluble Aβ through fluid phase macropinocytosis. *J Neurosci* **29:** 4252–4262.

Masino SA, Frostig RD. 1996. Quantitative long-term imaging of the functional representation of a whisker in rat barrel cortex. *Proc Natl Acad Sci* **93:** 4942–4947.

Navari RM, Wei EP, Kontos HA, Patterson JL. 1978. Comparison of the open skull and cranial window preparations in the study of the cerebral microcirculation. *Microvasc Res* **16:** 304–315.

Nguyen QT, Tsai PS, Kleinfeld D. 2006. MPScope: A versatile software suite for multiphoton microscopy. *J Neurosci Methods* **156:** 351–359.

Nimmerjahn A, Kirchhoff F, Helmchen F. 2005. Resting microglial cells are highly dynamic surveillants of brain parenchyma in vivo. *Science* **308:** 1314–1318.

Nimmerjahn A, Theer P, Helmchen F. 2008. Two-photon laser scanning microscopy. In *Ultrashort laser pulses in biology and medicine* (ed. Braun M, Gilch P, Zinth W), pp. 29–51. Springer-Verlag, Heidelberg.

Nimmerjahn A, Mukamel EA, Schnitzer MJ. 2009. Motor behavior activates Bergmann glial networks. *Neuron* **62:** 400–412.

Oheim M, Beaurepaire E, Chaigneau E, Mertz J, Charpak S. 2001. Two-photon microscopy in brain tissue: Parameters influencing the imaging depth. *J Neurosci Methods* **111:** 29–37.

Paxinos G, Franklin KBJ. 2003. *The mouse brain in stereotaxic coordinates*. Academic, New York.

Peri F, Nüsslein-Volhard C. 2008. Live imaging of neuronal degradation by microglia reveals a role for v0-ATPase a1 in phagosomal fusion in vivo. *Cell* **133:** 916–927.

Raivich G, Bohatschek M, Kloss CUA, Werner A, Jones LL, Kreutzberg GW. 1999. Neuroglial activation repertoire in the injured brain: Graded response, molecular mechanisms, and cues to physiological function. *Brain Res Rev* **30:** 77–105.

Ransohoff RM, Perry VH. 2009. Microglial physiology: Unique stimuli, specialized responses. *Annu Rev Immunol* **27:** 119–145.

Rueckel M, Mack-Bucher JA, Denk W. 2006. Adaptive wavefront correction in two-photon microscopy using coherence-gated wavefront sensing. *Proc Natl Acad Sci* **103:** 17137–17142.

Spiess E, Bestvater F, Heckel-Pompey A, Toth K, Hacker M, Stobrawa G, Feurer T, Wotzlaw C, Berchner-Pfannschmidt U, Porwol T, Acker H. 2005. Two-photon excitation and emission spectra of the green fluorescent protein variants ECFP, EGFP, and EYFP. *J Microsc* **217:** 200–204.

Tang S, Krasieva TB, Chen Z, Tempea G, Tromberg BJ. 2006. Effect of pulse duration on two-photon excited fluorescence and second harmonic generation in nonlinear optical microscopy. *J Biomed Opt* **11:** 020501.

Wake H, Moorhouse AJ, Jinno S, Kohsaka S, Nabekura J. 2009. Resting microglia directly monitor the functional state of synapses in vivo and determine the fate of ischemic terminals. *J Neurosci* **29:** 3974–3980.

Walter ML, Domes ME, Diller RA, Sproedt J, Joosten UH. 1999. Photoablation of bone by excimer laser radiation. *Lasers Surg Med* **25:** 153–158.

Wiart M, Davoust N, Pialat J-B, Desestret V, Moucharrafie S, Cho T-H, Mutin M, Langlois J-B, Beuf O, Honnorat J, et al. 2006. MRI monitoring of neuroinflammation in mouse focal ischemia. *Stroke* **38:** 131–137.

Xu HT, Pan F, Yang G, Gan WB. 2007. Choice of cranial window type for in vivo imaging affects dendritic spine turnover in the cortex. *Nat Neurosci* **10:** 549–551.

MOVIE LEGENDS

Movies are freely available online at www.cshprotocols.org/imaging.

MOVIE 87.1. Three-dimensional distribution of enhanced yellow fluorescent protein (eYFP)-expressing neurons (yellow) and enhanced green fluorescent (GFP)-positive microglia (green) in the neocortex of a Thy1-eYFP x Cx3cr1-eGFP mouse implanted with an open skull window. The images are maximum-intensity side projections from a stack of fluorescence images. Individual focal planes were recorded in 2.5-µm steps starting from 800 µm depth below the pia mater to the cortical surface.

MOVIE 87.2. A time-lapse recording showing enhanced green fluorescent protein (eGFP)-expressing microglia (green) and enhanced yellow fluorescent protein (eYFP)-expressing Purkinje cells (red) in the cerebellar cortex of an anesthetized adult Thy1-eYFP x Cx3cr1-eGFP mouse. Images were recorded near a bend in the cerebellar folium, which allowed the molecular and Purkinje cell layers to be visualized within a single image. Microglial processes are highly motile. Neurons appear structurally stable during the same time interval. Elapsed time in minutes is shown in the *upper right* corner. Scale bar, 15 µm.

MOVIE 87.3. A video clip taken in the cerebellar cortex of an anesthetized mouse expressing enhanced green fluorescent protein (eGFP) in microglia. The video shows a microglial process at high magnification. Fine protrusions continually palpate the cellular environment. Elapsed time in minutes is indicated in the *upper right* corner. Scale bar, 5 µm.

MOVIE 87.4. This exemplary video shows phagocytosis of cellular debris by microglia. Fluorescence data were recorded in the cerebellar cortex of an adult Cx3cr1-eGFP mouse expressing enhanced green fluorescent protein (eGFP) in microglia (*left*) following mechanical injury to the cortex and topical application of the red fluorescent dye sulforhodamine 101 (SR101) (*center*). The *right* panel shows an overlay of the eGFP and SR101 signals. The bright red cell is an injured Purkinje neuron that has accumulated SR101. Time after mechanical injury is indicated in the *upper right* corner. Scale bar, 10 µm.

88 Two-Photon Imaging of Immune Cells in Neural Tissue

Raluca Niesner,[1,2] Volker Siffrin,[1,3] and Frauke Zipp[1,3]

[1]*Max-Delbrück Center for Molecular Medicine Berlin-Buch, 13125 Berlin, Germany;* [2]*German Rheumatism Research Center, 10115 Berlin, Germany;* [3]*Neurology Department, University Medical Center Mainz, Johannes Gutenberg University Mainz, 55131 Mainz, Germany*

ABSTRACT

To gain deeper insight into the pathological phenomena that occur during chronic neuroinflammation of the central nervous system, we use two-photon imaging to view acute murine hippocampal slices cocultured with different subpopulations of immune cells. Two-photon imaging is also used to view in vivo the brain stem of transgenic mice affected by experimental autoimmune encephalomyelitis. Preparation of cocultures of hippocampal slices with immune cells and of the brain stem in living mice, as well as new developments in two-photon microscopy that allow for greater flexibility in the choice of fluorophores, are described in this chapter.

INTRODUCTION

To develop new therapeutic strategies for many central nervous system (CNS) diseases, it is essential to observe the motility and function of immune cells within neural tissue. Two-photon laser-scanning microscopy is an outstanding technique for imaging these phenomena under in vivo–like conditions. We use it to investigate cocultures of acute or organotypic brain slices with encephalitogenic immune cells, and to look intravitally at the brain stem in anesthetized mice. This chapter describes the preparation of cocultures of acute hippocampal slices with antigen-specific T helper 17 (Th17) cells migrating into the parenchyma, and the preparation of anesthetized mice for imaging the brain stem. It also provides technical aspects of dual-color, two-photon laser-scanning microcopy that is used to image these samples. The protocols described here are used in our group to investigate murine models of multiple sclerosis (i.e., experimental autoimmune encephalomyelitis [EAE]) on the cellular and molecular levels. We perform intravital imaging in the brain stem of EAE mice because the disease with all of its pathologic features (inflammation, demyelination, oligodendrocyte and upper motor neuron injury, axon pathology, microglia activation, and astrocytosis) can be best visualized in this area.

Protocol

Two-Photon Imaging of Immune Cells in Murine Neural Tissue

IMAGING SETUP

We use a specialized two-photon laser-scanning microscope based on a commercial scan head (TriMScope, LaVision BioTec) for imaging both cocultures of acute hippocampal slices with immune cells and anesthetized transgenic mice. The setup (Fig. 1) allows for dual near-infrared (NIR; 690–1080 nm) and infrared (IR; 1050–1600 nm) excitation. Pulsed NIR radiation is generated by an automatically tunable Ti:sapphire laser (Mai Tai HP, Spectra-Physics), 10% of which is coupled into the scan head via routing mirrors. Inside the scan head, light first passes through a motorized attenuator and a variable three-lens telescope that is used to adapt the beam diameter to the objective lens entrance aperture for maximum penetration depth in tissue (i.e., underfilling the back aperture of the objective lens). Next, it passes a prechirp prism-based arrangement that enables a pulse width of ~120 fsec at the sample. The other 90% of the Ti:sapphire-laser power is coupled into a synchronously pumped optical parametric oscillator (OPO, APE, Germany). The OPO used in our setup can be pumped at different wavelengths of Ti:sapphire (780–900 nm) while being tuned within the output wavelength range 1050–1600 nm (Rimke et al. 2009). The generated OPO beam first passes through spectral filters that extract the IR radiation and then through a motorized attenuator before entering the scan head and overlapping with the Ti:sapphire beam. Overlapping of the Ti:sapphire and OPO beams occurs by means of a dichroic mirror. The pulse width of the OPO beam is ~140 fsec in the sample (Rimke et al. 2009). The colocalized (Ti:sapphire and OPO) beams are coupled into an upright microscope (BX-51WI, Olympus) through the scan lens/tube lens combination. A dichroic mirror (DM) reflects the excitation beams toward the objective lens (20x, numerical aperture [NA] 0.95, Olympus), which focuses them onto the sample at the same spot. Laser beam scanning is performed with a conventional pair of galvanometer scanners that permit free selection of the position and size of the field of view. In combination with the motorized microscope *z*-drive, three-dimensional objects can be visualized. Fluorescence is collected by the same objective lens and directed to a spectrally resolving detection unit containing dichroic mirrors, interference filters, and up to three non-descanned photomultiplier tubes (H7422-40, Hamamatsu). To increase the penetration depth within tissue, the detection unit is placed in the vicinity of the sample, so that a maximum detection efficiency of the scattered fluorescence photons is achieved. Applied average laser power (2–10 mW) measured at the sample ensures negligible photobleaching of the fluorophores and photodamage of tissue.

MATERIALS

CAUTION: See Appendix 6 for proper handling of materials marked with <!>.
See the end of the chapter for recipes for reagents marked with <R>.

Reagents

Agarose, sterile 0.5% (see Step 16)
Antibodies and cytokines for stimulating T cells (see Step 6)
Antigen-presenting cells (APCs)
Artificial cerebrospinal fluid (ACSF) <R>

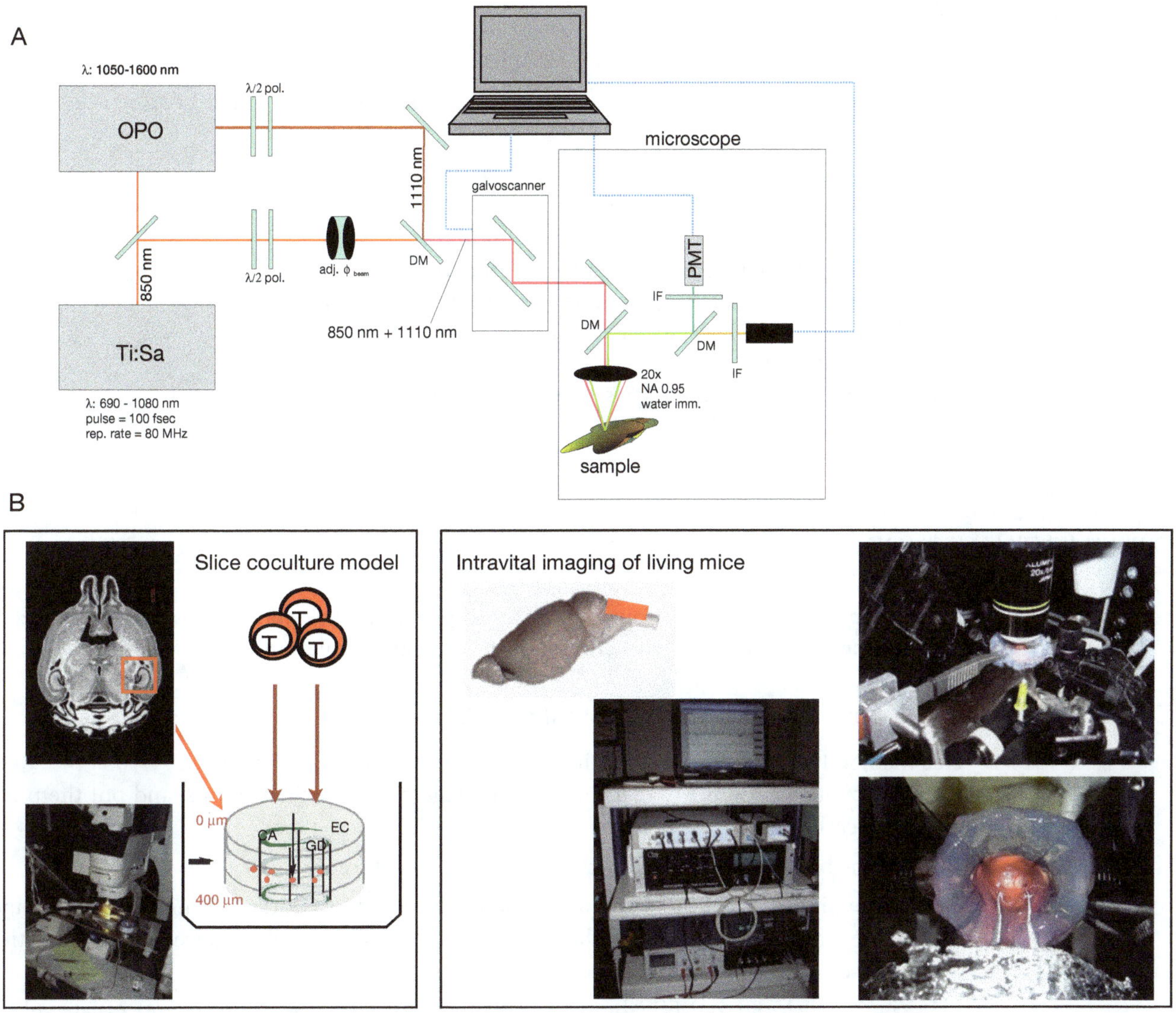

FIGURE 1. (*A*) Dual-color two-photon microscopy imaging setup. Axial brain slice in which the hippocampus is highlighted (CA, cornu ammonis; EC, entorhinal cortex; GD, dentate gyrus; *upper* photo). Slices are set in a heated chamber that is continually perfused with warmed and hyperoxygenated (95% oxygen, 5% carbogen) artificial cerebrospinal fluid (ACSF). Immune T helper cells are pipetted on the acute slices in the slice chamber. TPLSM is performed by using an upright microscope with a 20x long-distance objective (NA 0.95), which is immersed in the slice chamber. (*B*) Preparation of cocultures of acute hippocampal slices with immune cells (*left* box) and of the brain stem of anesthetized mice (*right* box) for intravital imaging. The area of interest is the lower brain stem (oblongate medulla) and cerebellum (*upper left*), which is accessible between skull bone and atlas (first cervical vertebra) if the neck is bent of a hanging mouse (photos on the *left*). Fixation is done by ear holders and pincers at the posterior tubercle of the atlas. Continuous monitoring of the animal ensures long imaging periods (*lower left*: setup for CO_2, temperature, ventilation, ECG control, and registration).

Carbogen (95% O_2 <!> and 5% CO_2 <!>) for the acute hippocampal slices
C57BL/6 mice, either pups P10–P12) or adults (6–10 wk), source of neural tissue
$CD4^+$ $CD62L^+$ T cell isolation kit II (Miltenyi Biotec; item 130-093-227)
CD90-depleted, irradiated spleen cells (30 Gy) from naïve syngenic C57BL/6 mice
hrIL-2 (Chiron Therapeutics)
Isoflurane <!>; O_2 and N_2O for anesthesia
MOG peptide (myelin oligodendrocyte glycoprotein peptide)

NaCl solution, 0.9%
Ringer solution
RPMI 1640 cell culture medium <R>
2d2 FP (fluorescent protein) T cell receptor (TCR) transgenic mice (source of spleen-derived T cells)

Equipment

Heating cover and rectal thermometer
Imaging analysis software (Volocity, Improvision, UK)
Microcapnograph (CI-240; Columbus Instruments)
Respirator (advanced safety respirator; Harvard Apparatus, Hugo Sachs)
Slice chamber set at 37°C (Luigs & Neumann)
Sterile surgical instruments
Two-photon laser-scanning microscope (LaVision BioTec; see Imaging Setup)
Vibratome <!> (VT 1200 S; Leica)

EXPERIMENTAL METHOD

Animal experiments must be approved by the appropriate state committees for animal welfare and be performed in accordance with current guidelines and regulations.

Cocultures of Acute Hippocampal Slices with Immune Cells

This method has been used by Nitsch et al. (2004) and Siffrin et al. (2009).

1. Decapitate C57BL/6 P10–P12 pups, immediately remove their brains, and put them into aerated (carbogen) ACSF at 4°C.
2. Cut the brain into 400-µm-thick slices with a vibratome.
3. Isolate hippocampal slices and allow them to recover for at least 45 min at room temperature before transferring them to a slice chamber set at 37°C, in which slices must be continuously perfused with 37°C carbogen-aerated ACSF.
4. Kill 6–10-wk-old *2d2 FP* (fluorescent protein) TCR transgenic mice, and isolate spleen cells.
5. Sort spleen cells to isolate naïve T cells (specifically, $CD4^+$ $CD62L^+$ T cells) using T cell isolation kit.
6. Stimulate with 12.5 µg/mL MOG peptide presented by antigen presenting cells (APCs). APCs were prepared by MACS-CD90-depletion of mouse spleen cells and irradiated (30 Gy) before coculture. Stimulation was performed in the presence of 20 ng/mL IL-6, 3 ng/mL TGFβ, 20 ng/mL IL-23 (Sf 21), and 5 µg/mL anti-IL-4 (11B11). Also include 5 µg/mL anti-IL-12 (C17.8) and anti-IFN-γ (AN18.17.24) for differentiation toward the Th17 phenotype.
7. Keep cells in RPMI 1640 cell culture medium supplemented with 50–100 IU hrIL-2.
8. For restimulation, T cells were harvested, washed, and cocultured with APC (see Step 6) in the presence of MOG peptide and cytokines as described. Allow these cultures to grow for 4–7 d before using the T cells in brain slice coculture experiments.

   Using 4–7-d–old T cells will allow the activation status of various FPs T cell populations to be compared in the brain slices.

9. Stop the carbogen-aerated ACSF perfusion of the hippocampal slices. Prepare a concentrated suspension of T cells (from Step 8) in 10 µL. Pipette the T cells onto a hippocampal slice and allow the T cells to invade the slice for ~1 h before imaging.

10. Begin image acquisition, which is typically performed for ~2–3 h per slice. Jump to Step 18 for details on image acquisition.

Preparation of the Brain Stem of Anesthetized Mice for Intravital Imaging

This method was used by Siffrin et al. (2009, 2010).

11. Anesthetize the mouse using 1.5% isoflurane in oxygen/nitrous oxide (2:1) dispensed through a face mask. Experimental animals can be any transgenic for fluorescence expression in the CNS or immune cells. We mainly used eGFP-transgenic animals that express eGFP under the Thy1 promoter, which is exclusively expressed in neurons within the CNS. Immune cells were grafted by bone marrow transplant after irradiation of the mice and expressed tdRFP by a ubiquitous promoter.

    Maintain body temperature at 37°C during surgery and microscopy by using a heating cover, and monitor temperature with a rectal thermometer. Control anesthesia depth by continuous CO_2 measurements of exhaled gas recorded with a microcapnograph.

12. Perform a tracheotomy on the mouse. Connect the mouse to a respirator and begin continuous respiration. Typically, a 20-g mouse is respirated at a rate of 140 min^{-1} with a volume of 0.25 cm^3.

13. Transfer the anaesthetized animal to an appropriate (in our case, custom-built) operation and microscopy table, as shown in Figure 1 and secure it in a suspended position (adapted protocols for cortical imaging).

14. Expose the brain stem by carefully removing musculature above the dorsal neck area and removing dura mater between the first cervical vertebra and occipital skull bone.

15. Incline the head for access to deeper brain stem regions and superfuse the brain stem with isotonic Ringer solution.

16. Place a sterile agarose patch (0.5% in 0.9% NaCl solution) on the now exposed brain surface to reduce heartbeat and breathing artifacts (Fig. 1).

17. To better control the anesthesia depth of the mouse, apply an Einthoven three-lead electrocardiogram (ECG) throughout the imaging period.

Image Acquisition and Analysis

18. Three-dimensional fluorescence image stacks are typically collected within a scan field of 50 x 50 µm to 500 x 500 µm, normally at 512 x 512-pixel resolution and a *z*-plane distance of 2 µm at a frame rate of 50–1200 Hz. Depending on the phenomenon under study, such three-dimensional (3D) stacks of 50–100 µm thickness are recorded every 15–60 sec over a period of 2–3 h in the acute brain slices and for >7 h in the brain stem of anesthetized mice.

    In acute brain slices, imaging must be performed in 50-µm depth or deeper (typically up to 200 µm), so that tissue injuries at the surface of the slice do not lead to errors in data interpretation.

    See *Troubleshooting.*

19. To avoid the rather strong breathing artifacts in the brain stem of anesthetized mice, use the signal of the ECG as an external trigger for the image acquisition software, which controls the hardware of the microscope setup. Thus, each fluorescence stack will be recorded at the same respiration phase of the mouse and also in the same tissue region.

20. Analyze 4D data with a corresponding evaluation program (e.g., Volocity, PerkinElmer). The analysis provides insight into cell motility patterns, interactions between different cell populations by colocalization measurements, and cell functionality (e.g., by means of FRET [Förster resonance energy transfer]).

    Quantitative evaluation of the fluorescence signal (independent of the investigated sample) must take into account its decrease related to the increasing imaging depth in tissue and to photobleaching of the fluorophores.

TROUBLESHOOTING

Problem (Step 18): Despite external triggering of the microscope with the ECG signal of the mouse, the region of interest is not stable axially (in the *z* direction).

Solution: If the curve of CO_2 percentage of exhaled gas is nonhomogeneous, either the anesthesia is insufficient (CO_2 percentage is too high, >2%) or the mouse is not optimally ventilated (CO_2 percentage is low, ≤1%). In the first case, increase the dosage of isoflurane on the respiratory machine to reduce CO_2 percentage to <2%. In the second case, increase either the respiration volume or the respiration rate, so that a higher CO_2 percentage and a higher CO_2 curve homogeneity are attained. To perform intravital experiments it is essential to perform the tracheotomy (Step 12) with precision, because imprecise positioning of the tube in the trachea cannot be corrected during imaging.

Problem (Step 18): Immediately after transfer from the preparation laboratory to the microscope laboratory, the region of interest continuously drifts along the *z*-axis (axially).

Solution: Usually this "*z* drift" is caused by temperature changes within the mouse as it acclimates to the cooler conditions within the microscope laboratory. Wait 10–15 min for the mouse to acclimate to the lower temperature and increase the heat output of the cover. In addition, heating the microscope stage (underneath the mouse) makes it easier for the mouse to acclimate.

DISCUSSION

Two-photon laser-scanning microscopy of mice prepared as described in this chapter allows for numerous studies of the motility and interactions of immune cell subpopulations in the CNS, as well as interactions between immune cells and cells of the CNS (Fig. 2, Movie 88.1). Imaging cocultures of acute brain slices with immune cells allows for near-real conditions, whereas the brain stem preparation allows for real-time intravital imaging. Two-photon microscopy is superior to confocal and wide-field microscopy for thick-tissue dynamic imaging and especially in intravital imaging for two reasons. First, the use of long wavelength (near infrared) excitation (i.e., ballistic photons excitation) permits imaging at increased penetration depths in tissue, and second, the intrinsic axial resolution, which is owing to nonlinear excitation, ensures reduced photobleaching and photodamage outside of the focal plane.

Furthermore, the rather new development of dual-color two-photon microscopy (i.e., simultaneous near-infrared and infrared excitation) is a dramatic improvement over classical two-photon microscopy owing to its broader excitation wavelength range and to the possibility of infrared (i.e., long) wavelength excitation. The technique allows for simultaneous imaging of fluorophores with extremely different spectral characteristics (e.g., enhanced green fluorescent protein [eGFP] and tdRFP [Fig. 2, Movie 88.1]), and is able to counteract the deterioration of spatial resolution owing to increasing imaging depth and to increase penetration depth because there is a much lower scattering of photons with wavelengths >1000 nm.

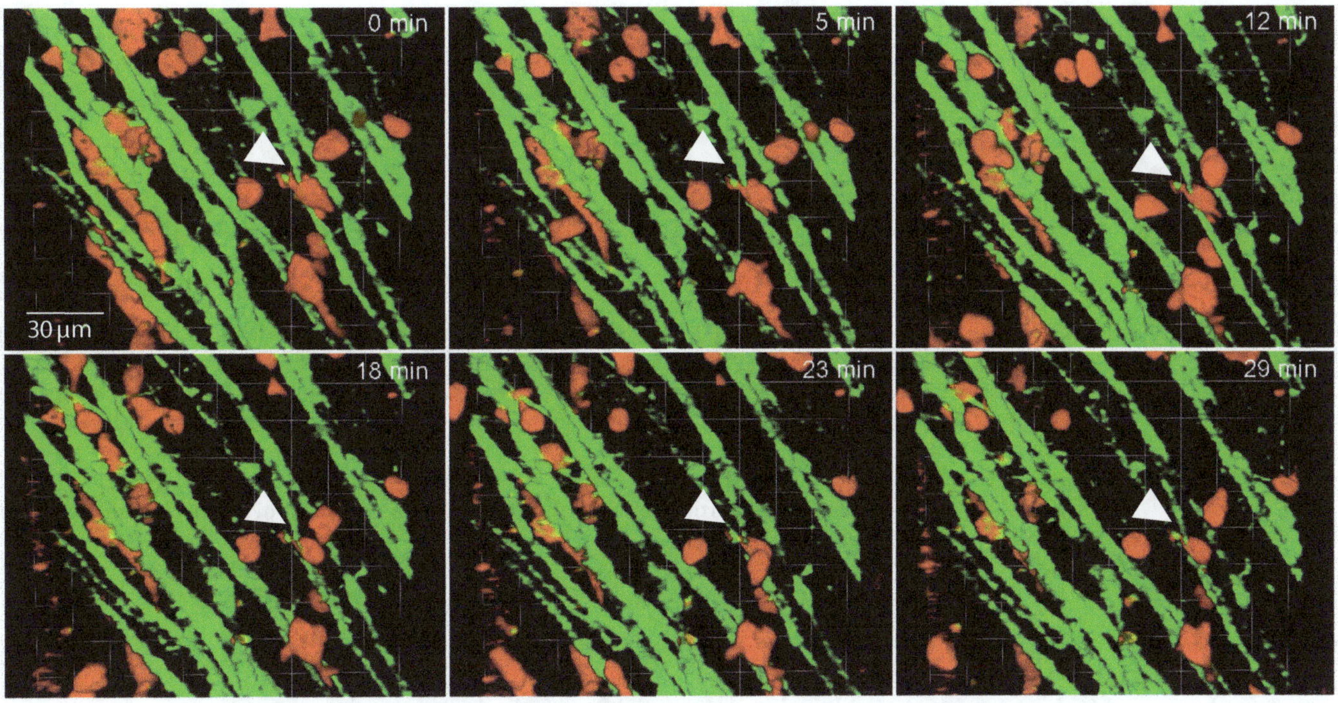

FIGURE 2. Immune cells expressing tdRFP (red) are infiltrating the brain stem parenchyma of a mouse showing signs of experimental autoimmune encephalomyelitis (EAE) and are attacking neural processes, mainly axons, expressing eGFP (green). The fluorescence images were captured after simultaneous excitation at 850 nm (2 mW of Ti:sapphire) and at 1110 nm (4 mW of OPO). Hence, both the eGFP-expressing neurons and the tdRFP-expressing immune cells could be imaged. This allows for a detailed study of interactions between immune cells and neural processes and of typical motility patterns of immune cells in the central nervous system. For instance, the white arrowhead in the displayed images indicates the dynamic contact between an immune cell (red) and an axon (green).

The experimental methods described in this chapter are particularly useful for the investigation of pathological mechanisms in the CNS. Thus, whereas the success of experiments depends on careful preparation of cells, tissues, or organisms, and of thoughtful image acquisition and analysis, it is equally important to develop transgenic mice that show desirable cell-specific expression of fluorescent proteins within appropriate disease models.

RECIPES

CAUTION: See Appendix 6 for proper handling of materials marked with <!>.
Recipes for reagents marked with <R> are included in this list.

Artificial Cerebrospinal Fluid (ACSF)

NaCl	124 mM
NaH_2PO_4	1.25 mM
$NaHCO_3$	26 mM
KCl	3 mM
$CaCl_2$ <!>	1.6 mM
$MgSO_4$ <!>	1.8 mM
Glucose	10 mM

Adjust to pH 7.35.

RPMI 1640 Cell Culture Medium

Supplement RPMI 1640 medium with 2 mM L-glutamine, 100 U/mL penicillin <!>, 100 µg/ mL penicillin/streptomycin <!>, and 10% fetal calf serum.

REFERENCES

Nitsch R, Pohl EE, Smorodchenko A, Infante-Duarte C, Aktas O, Zipp F. 2004. Direct impact of T cells on neurons revealed by two-photon microscopy in living brain tissue. *J Neurosci* **24:** 2458–2464.

Rimke I, Siffrin V, Niesner R, Leuenberger T, Zipp F, Büttner E, Stilbenz G. 2009. A new light source for multimodal multiphoton microscopy including CARS. *Proc SPIE* **7183:** 718314.

Siffrin V, Brandt AU, Radbruch H, Herz J, Boldakowa N, Leuenberger T, Werr J, Hahner A, Schulze-Tophoff U, Nitsch R, Zipp F. 2009. Differential immune cell dynamics in the CNS cause CD^{4+} T cell compartmentalization. *Brain* **132:** 1247–1258.

Siffrin V, Radbruch H, Glumm R, Niesner R, Paterka M, Herz J, Leuenberger T, Lehmann SM, Luenstedt S, Rinnenthal JL, et al. 2010. In vivo imaging of partially reversible th17 cell–induced neuronal dysfunction in the course of encephalomyelitis. *Immunity* **33:** 424–436.

MOVIE LEGEND

Movies are freely available online at www.cshprotocols.org/imaging.

MOVIE 88.1. The dynamic interaction between immune cells expressing tdRFP (red) and neural processes expressing eGFP (green) in an experimental autoimmune encephalomyelitis (EAE)-affected mouse. The visualization was performed by dual near-infrared/infrared (NIR/IR) excitation two-photon microscopy in the brain stem of anesthetized mice.

89 Two-Photon Imaging of Structure and Function in Alzheimer's Disease

Michal Arbel-Ornath, Monica Garcia-Alloza, Kishore V. Kuchibhotla, Tara Spires-Jones, and Brian J. Bacskai

MassGeneral Institute for Neurodegenerative Diseases, Massachusetts General Hospital, Charlestown, Massachusetts 02129

ABSTRACT

Alzheimer's disease is a progressive neurodegenerative disease diagnosed conclusively in postmortem tissue by the presence of amyloid plaques, neurofibrillary tangles, and severe neuronal loss. The development of specific, sensitive, pre-morbid biomarkers is of the highest priority. Clinical approaches for imaging amyloid plaques using magnetic resonance imaging or positron emission tomography have received considerable attention, and the PET approaches have met with a great deal of recent success. The generation of transgenic mouse models that develop the amyloid or tangle pathology (or both) have led to significant insights into the pathophysiology of Alzheimer's disease. Two-photon microscopy approaches have allowed direct visualization of these pathologies in the intact, living brain and have permitted longitudinal imaging of both the pathologies as well as co-occurring structural and functional readouts in specific cell populations in the brain. This has allowed elucidation of the spatial and temporal sequence of cellular and molecular events, analysis of therapeutic approaches, and validation of clinically relevant molecular imaging probes. This chapter reviews the recent advances in Alzheimer's disease research using advanced microscopy approaches, transgenic mouse models, and fluorescent imaging reagents.

INTRODUCTION

Alzheimer's disease (AD) is the most prevalent neurodegenerative disease and the most common form of dementia in the elderly. Macroscopic postmortem examination of AD brain tissue reveals gross cortical atrophy with thinning of the medial temporal lobe. Microscopically, there is a widespread neuronal and synaptic loss accompanied by reactive gliosis and the pathological hallmarks of the disease: extracellular amyloid deposits and intracellular neurofibrillary tangles (Hyman and Trojanowski 1997). Although the former are mainly composed of aggregated amyloid β (Aβ) peptides, the proteolytic products of amyloid precursor protein (APP), the latter are intraneuronal inclusions mainly composed of abnormally hyperphosphorylated tau protein (Buée et al. 2000).

To date, the diagnosis of AD is based on neuropsychological examinations that enable detection at mild to late stages of the disease. Structural neuroimaging techniques using magnetic resonance imaging (MRI) are commonly used in the clinical evaluation of AD patients, and longitudinal imaging of hippocampal atrophy remains a sensitive and predictive indicator of AD. With recent advances in molecular imaging techniques based on positron emission tomography (PET) or single-photon emission tomography (SPECT) that allow quantitative measurement of amyloid pathology in the living brain, there is the hope that the natural history of Alzheimer's disease will be revealed for the first time in patient populations. In many cases, these clinical imaging approaches are first developed in animal models of the disease, and validation with high-resolution optical approaches for amyloid imaging has accelerated progress. Because of the subcellular spatial resolution of two-photon microscopy, individual amyloid plaques and identified cells in the living brain can be analyzed, highlighting the great potential that MPM holds for increasing our understanding of mechanisms that are involved in the pathogenesis of AD, for developing diagnostic tools, and for testing putative drugs.

Many efforts are directed at engineering lipophilic fluorescent Aβ binding ligands that cross the blood–brain barrier (BBB) and label plaques in vivo, for imaging purposes. Methoxy-XO_4, a Congo red derivative, and Pittsburg compound B (PiB), a thioflavin T derivative, can both cross the BBB and decorate plaques (Klunk et al. 2002; Mathis et al. 2002). Constructed for PET imaging, PiB is radiolabeled with ^{11}C, but it also retains the fluorescent properties of its parent molecule. Recent human imaging studies with PiB showed very promising results, as the compound helped discriminate AD from frontotemporal lobar degeneration (Rabinovici et al. 2007). Importantly, two in vivo correlation studies have shown excellent correspondence between the PiB in vivo imaging and postmortem measurements of Aβ (Bacskai et al. 2007; Ikonomovic et al. 2008). To date, PiB is used for clinical imaging in more then 30 sites worldwide.

NEAR-INFRARED IMAGING OF AD

Near-infrared fluorescence (NIRF) tomography is a relatively new approach for noninvasive amyloid imaging. This technique relies on the deep transmission of near-infrared light through the scattering tissue and potentially allows optical imaging with PET-like resolution in intact transgenic mice and humans. The spatial resolution of near-infrared imaging varies with depth but can approach 1 mm at up to several centimeters into the brain tissue. The technique requires targeted molecular imaging probes that are fluorescent and have excitation and emission spectra within the range of 650–950 nm. The advantage of using optical probes is that changes in fluorescence quantum yield, lifetime, and emission spectra upon Aβ binding would allow quantitative imaging of bound versus unbound probe and thus increase the sensitivity of amyloid detection (Raymond et al. 2008). Integration of NIRF probes for amyloid imaging has significant advantages over the more conventional PET and SPECT, because this technique does not require radioactive tracers and the instrumentation involved is less expensive.

TWO-PHOTON IMAGING OF AD

Two-photon microscopy is based on the nearly simultaneous arrival (within ~0.5 fsec) of two or more photons at a fluorescent molecule that then combine their energies to promote the molecule to an excited state. The excited fluorophore can then undergo fluorescence emission (Denk et al. 1990; for an elaborated review, see Denk 2011). Two-photon microscopy presents some advantages over other in vivo imaging approaches described above, reaching ~1 μm spatial resolution, which allows for visualization of cellular and subcellular structures. Two-photon microscopy also has some distinct advantages over its single-photon counterpart, confocal microscopy. First, the use of near-infrared light as the excitation source reduces absorption and scattering within the tissue, minimizing phototoxicity and allowing higher depth penetration. Second, two-photon absorption is spatially confined to the point of focus of the microscope objective, reducing out-of-focus fluorescence and,

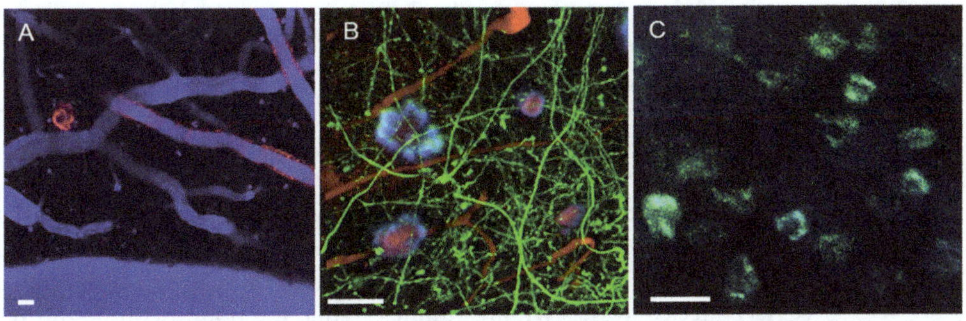

FIGURE 1. Live imaging of the pathological hallmarks of Alzheimer's disease. (*A*) In vivo imaging of amyloid deposits in the parenchyma and blood vessels of aged APP/PS1 transgenic mice. Methoxy-XO_4 (red) and Texas red (blue, injected via tail vein) were used to label amyloid deposits and brain vasculature, respectively. (*B*) Simultaneous imaging of neuronal processes (green) and amyloid deposits (blue) along with a Texas red dextran angiogram (red, injected via tail vein). Tg2576 APP transgenic mice were injected with adeno-associated virus encoding for GFP in layer 5 of the cortex. GFP fills the neuron soma and neurites (green) both nearby and far from plaques (co-labeled with methoxy-XO_4 and 3D6 antibody in blue and red, respectively). (*C*) Live imaging of tangle-bearing cortical neurons in tg4510 tau transgenic mouse. Tangles were labeled with thioflavin-S applied topically to the brain surface before the imaging session. Scale bars, 20 µm.

in turn, reducing phototoxicity. Furthermore, different fluorophores with distinct emission spectra can all be excited with a single excitation laser, allowing simultaneous fluorescence multiplexing.

Despite its superior resolution compared with other imaging methodologies, two-photon microscopy requires removal or thinning of a portion of the skull for imaging, which limits its clinical applications. The depth penetration of two-photon microscopy is also limited to <1 mm, so only superficial volumes of brain can be imaged. However, the advantages presented by two-photon microscopy for imaging deep within an intact live tissue at subcellular resolution have prompted its use in various fields of biological research, allowing the study of cellular metabolic state, neuronal circuits, neurites, and spine morphologies, as well as pathological abnormalities that are affected by neurodegenerative processes.

To date, a wide range of transgenic mouse models exist that mimic some aspects of AD pathogenesis including Aβ plaques, neurofibrillary tangles (NFTs) (or both), neuronal and synaptic loss, and behavioral deficits (for review, see Morrissette et al. 2009). Although these mice mainly express mutant genes that are implicated in the genetic forms of AD (including APP, tau, presenilin 1 and 2 [PS1 and PS2], and different APOE isoforms), they are extensively used in AD research and have been highly instrumental in studying disease progression and in screening for putative disease-modifying drugs.

The combination of in vivo two-photon microscopy with the available mouse models provides a powerful tool to follow both functional and structural changes in the central nervous system during disease progression in living mice (Fig. 1). As such, it can be exploited to investigate mechanisms that contribute to the neuronal and synaptic loss in AD, as well as to screen for new imaging agents that can then be integrated for use in PET, SPECT, or MRI methodologies. Importantly, in vivo imaging of the disease-related pathologies is particularly valuable for monitoring the effect of putative drugs. Here we summarize several examples of structural and functional live imaging in AD models using two-photon microscopy.

Structural In Vivo Imaging

Aβ Plaques

The high spatial resolution and high signal-to-noise ratio of two-photon microscopy has enabled the detection of individual Aβ plaques in the brains of transgenic mouse models of AD. With a cranial window, longitudinal imaging of the same plaques can be performed, and we have done this at intervals up to 6 months in the living mouse. The first demonstration of plaque imaging in vivo exploited the amyloid binding dye thioflavin-S, a nontoxic fluorophore that binds highly ordered β-sheet structures and is known to decorate the dense core of amyloid plaques in brain sections from AD patients and mouse models. As thioflavin-S cannot cross the BBB, it was applied topically to the surface of the brain of living APP transgenic mice to label Aβ plaques (Christie et al. 2001). Aβ plaques deposited at the parenchyma and blood vessels were detected up to 400 μm deep in the cortex. The development of BBB-permeable probes, such as methoxy-XO_4 and PiB, allowed live imaging of plaques after a single peripheral administration of the probe (Klunk et al. 2002).

Using two-photon microscopy, we were able to image PiB binding to Aβ deposits at subcellular resolution in the brains of living transgenic mice following peripheral application (Bacskai et al. 2003). PiB crosses the BBB within seconds, and labeled plaques and cerebral amyloid angiopathy (CAA) affected vessels within minutes. While the bound compound remained associated with the deposits for several days, the unbound PiB was cleared quickly, meeting one of the criteria for a good amyloid imaging agent.

The ability to perform live imaging at subcellular resolution and the availability of different transgenic mouse strains make it possible to explore the temporal relation between plaque formation and changes in brain microarchitecture. Although a rare event, plaque formation occurs rapidly over the course of 24 h and recruits microglia cells to that site. Longitudinal imaging of these new plaques showed that they do not change in size after the first 24 h and remain stable for up to 5 mo, regardless of their initial size (Christie et al. 2001; Meyer-Luehmann et al. 2008). Using either double-transgenic mice that express both mutated APP and YFP in neurons or APP mice previously injected with adeno-associated virus (AAV) encoding for GFP, it was possible to follow neuronal soma, neurites, and spines. Neuritic curvature around the generated plaques is increased gradually to a maximal extent over the first week. Dendritic spine density is decreased in the area of plaques (Tsai et al. 2004; Spires et al. 2005) as spine elimination is increased in these areas (Spires-Jones et al. 2007).

As the region of interest in the tissue can be imaged before and after treatments in living mice, imaging amyloid plaques can serve as a powerful tool to evaluate the efficacy of therapeutic approaches designed to interfere with the amyloid cascade. As the most advanced approach currently in clinical trails, the effect of Aβ immunotherapy was explored. One dose of Aβ antibody, applied to the brain surface, cleared dense core Aβ deposits from brain parenchyma over the course of 3 d (Bacskai et al. 2001), led to recovery of neurite trajectory within days (Lombardo et al. 2003), and increased spine formation far from plaques within 1 h (Spires-Jones et al. 2009). Single exposure of the antibody resulted in temporary mild regression of CAA in leptomeningeal vessels, whereas continuous antibody infusion for 14 d showed a more robust and sustained effect. Unlike the plaques that remained clear after a single antibody dose, the effect of the antibody on CAA was short-lived (Prada et al. 2007).

We have also investigated how interference in earlier steps in the amyloid cascade, namely, Aβ production, will affect Aβ pathology and neuritic curvature (Garcia-Alloza et al. 2009a). To that end, we have explored the effect of a short treatment with the orally active γ-secretase inhibitor LY-411575 on Aβ pathology in APP/PS1 mice. Despite the reduced plasma and brain Aβ levels in the treated regimen, neither the size of existing plaques nor the curvature and dystrophies of neurites in the vicinity of these plaques were changed throughout the treatment. These data strengthen the close association between neuritic abnormalities and the physical presence of plaques and thus suggest that such a therapeutic approach may require a combination treatment aimed at both suppressing Aβ production and enhancing its clearance.

As both oxidative stress and inflammation are implicated in AD, we tested the effect of curcumin—which possess antioxidant, anti-inflammatory, and anti-fibrillogenic properties—on Aβ-related pathology using two-photon microscopy. Longitudinal imaging of individual plaques revealed that systemic treatment with curcumin clears and reduces existing plaques with a limited reversal of abnormal curvature and dystrophy size (Garcia-Alloza et al. 2007).

Neurofibrillary Tangles

NFTs are intraneuronal aggregates of hyperphosphorylated tau protein that are small but can be stained with β-sheet-binding small molecules like thioflavin-S. We have recently developed imaging methodology for NFTs based on topical application of thioflavin-S to the brain surface of mice expressing human mutated tau (Spires-Jones et al. 2008). As with Aβ plaques, developing reliable in vivo imaging of NFTs may be useful in exploring the relationship between NFTs and neurodegeneration, the temporal relations between plaques and NFTs (using double-transgenic mice), and in evaluating treatments aimed at interfering with NFT pathology in AD and other tauopathies.

Functional In Vivo Imaging

Imaging the brain with cellular resolution provides essential anatomical-level information, as discussed above. It can also elucidate cellular functions and thus can serve as a powerful tool for studying neuronal activity and signaling pathways. Currently, a plethora of molecular physiological indicators are available for measuring the concentration of H^+, Zn^{2+} (Thompson et al. 2000), Cl^- (Jayaraman et al. 2000), and Ca^{2+} (Baird et al. 1999; Miyawaki et al. 1999), as well as for monitoring oxidative metabolism and protease activity (Garcia-Alloza et al. 2010) among others. Use of genetically encoded indicators based on GFP or spectral variants also provides the opportunity to incorporate biologically active sequences, such as an organelle-specific retention signal, into the fluorescent probe (Miyawaki et al. 1997, 1999; see Chapter 8). Furthermore, low-molecular-weight cell-permeable fluorescent indicators also serve as powerful research tools for studying identified cell populations, because the indicators can be loaded into live cells within the intact brain (Stosiek et al. 2003; Kuchibhotla et al. 2009).

Calcium Imaging

Calcium is an intracellular messenger involved in many cellular functions, including secretion, excitability, and neuronal stability. Since calcium fluxes are tightly linked to neuronal electrical activity, recordings of intracellular calcium levels are used to indirectly monitor the electrical activity of an individual neuron.

Currently, calcium imaging is pursued using synthetic molecules that show changes either in quantum yield, lifetime, or emission spectrum upon calcium binding (Tsien 1989) and with fluorescent proteins fused to calcium-sensitive moieties (Miyawaki et al. 1999).

Accumulating evidence suggests that calcium dyshomeostasis plays a pivotal role in neurodegeneration (for reviews, see Holscher 1998; Bezprozvanny and Mattson 2008). We have recently developed in vivo calcium imaging methodologies that can be used to accurately measure changes in resting calcium levels in both neuronal processes (Kuchibhotla et al. 2008) and astrocytes (Kuchibhotla et al. 2009) based on two-photon microscopy. APP transgenic mice were stereotactically injected with AAV encoding the YC3.6 ratiometric calcium indicator (Fig. 2A,B). We calibrated this indicator to allow quantitative imaging of intracellular calcium in vivo. In transgenic mice with plaques, a large fraction of neurites exhibited significantly elevated calcium levels compared with age-matched controls or younger transgenic animals (~20% of the processes in APP transgenic mice compared with only 2% in the wild-type littermate). A greater proportion of neurites with calcium overload was detected near the senile plaques, but calcium overload was also detected in areas far from plaques. We also observed a correlation between calcium concentration and morphology of the neurites. At moderately elevated calcium levels in the neurites (200–400 nM), the calcium in spines was also elevated, demonstrating that spino-dendritic compartmentalization was lost. At calcium concentrations of >700 nM, neurites showed altered morphologies that included a beaded appearance. Together these results demonstrated alterations in both structure and function in individual neurites in the living brain as a result of the accumulation of amyloid plaques.

In a different study, injection of the calcium indicator Oregon Green BAPTA-1 (OGB-1) AM into the cortex layers 2/3 via a multicell bolus loading technique made it possible to follow the function of individual neurons (Busche et al. 2008). This study revealed a decrease in activity of ~33% of the neurons in layers 2/3, and an increase in the activity of 21% of the neurons in the same area, as measured by spontaneous calcium bursts. Notably, the hyperactive neurons were found directly in the

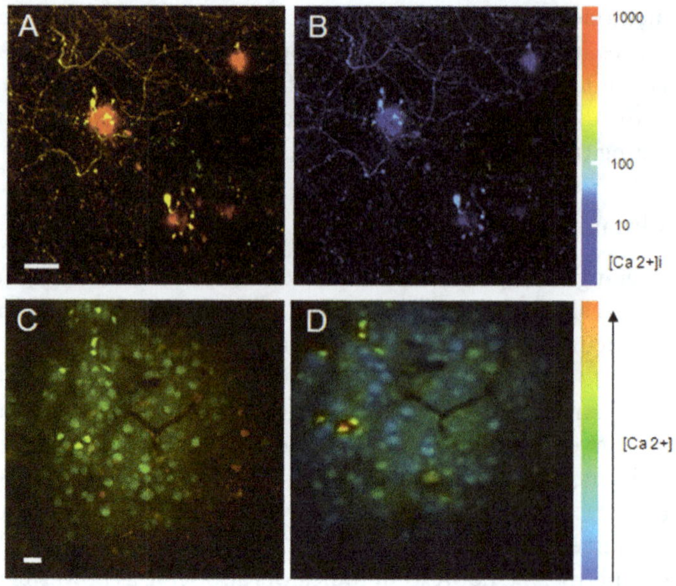

FIGURE 2. Live calcium imaging. (*A,B*) In vivo two-photon imaging deep in the neocortex of adult B6C3 APP/PS1 transgenic mice previously injected with YC3.6 AAV. YC3.6-filled dendrites and axons are presented (*A*, yellow) nearby and far from amyloid deposits (red). The CFP and YFP intensities are colored in red and green, respectively, giving rise to a yellow-to-green signal gradient according to resting calcium levels. YFP/CFP ratios were calculated and converted to calcium levels, which are shown in the pseudocolored image presented in nanomolar scale (*B*). (*C,D*) Two-photon fluorescent lifetime imaging microscopy (FLIM) to measure resting calcium levels in neurons and astrocytes of live APP:PS1 mice. Multicell bolus loading of neurons and astrocytes with Oregon Green BAPTA-1 AM (green) and sulforhodamine 101 (red) for detection of resting calcium levels and astrocytes, respectively. (*C*) Fluorescence emission was sent to either an intensity-based photomultiplier tube (PMT) module or a 16-channel multispectral detector and time-correlated single-photon counting (TCSPC) acquisition card (Becker & Hickl). Fluorescence decay curves were fit with a calcium-bound lifetime (2400 psec) and an unbound calcium lifetime (570 psec) for each pixel. (*D*) The pixel data were averaged to obtain single-cell calcium levels, depicted with a calibrated color bar. Scale bars, 20 μm.

vicinity of amyloid plaques (similar to our observations), emphasizing the notion that amyloid plaques are a focal source of toxicity.

Combining the calcium indicator OGB-1 AM with lifetime imaging microscopy, we have quantitatively imaged astrocytic calcium in AD transgenic mice (Kuchibhotla et al. 2009). As the bolus loading is not specific to the different cell types, astrocytes were distinguished from neurons using sulforhodamine 101 dye (Fig. 2; see Chapter 63). Differing from our observations in neurons, resting calcium levels in astrocytes were globally higher in transgenic mice compared with their age-matched littermates, independent of proximity to plaques. Time-lapse imaging revealed that calcium transients in astrocytes were more frequent, synchronously coordinated across long distances, and uncoupled from neuronal activity. We have also detected rare events of intercellular calcium waves in mice with plaques, originating near plaques and spreading radially over 200 μm.

Oxidative Stress

Oxidative stress has been strongly implicated in neurodegeneration and considered as a plausible mechanism that underlies the damage to the cerebral tissue in AD. The brain has a high lipid content and poor antioxidant defenses. Using two-photon microscopy together with the fluorogenic oxidation reporter Amplex Red, we have recently established an in vivo imaging methodology to quantitatively evaluate oxidative stress (Garcia-Alloza et al. 2006). Imaging Aβ deposits and oxidative stress in animal models that develop plaques and CAA revealed that these deposits are sources of oxidative stress. Inhibiting the generation of reactive oxygen species with antioxidants reduced the oxidative stress but had no effect on the size of senile plaques or the progression of CAA over time.

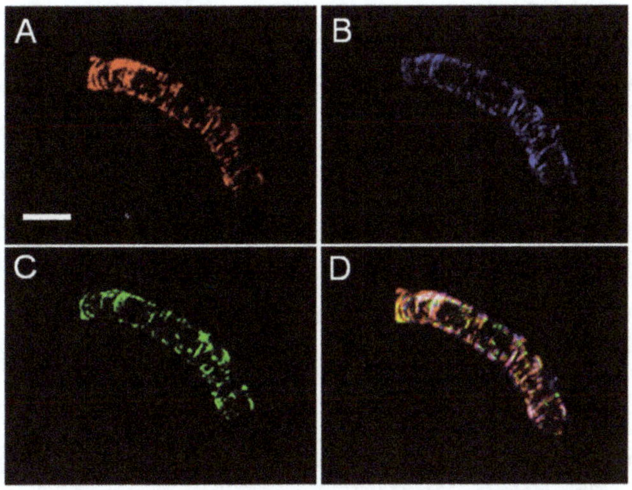

FIGURE 3. Imaging oxidative stress and matrix metalloproteinase activity in cerebral amyloid angiopathy (CAA)-affected vessels. Representative images of CAA, oxidative stress, and matrix metalloproteinase (MMP) activation in the same vessel segment from a living B6C3 APP/PS1 mouse. (*A*) Thioflavin S was used to visualize amyloid deposits, (*B*) Amplex Red was used to detect oxidative stress, and (*C*) fluorescein conjugate DQ gelatin enabled imaging of MMP activity. (*D*) All three measures spatially coincide as shown after thresholding, segmenting, and aligning the image from each marker. Scale bar, 100 µm.

Enzymatic Activities

Altered enzymatic activity is known to be implicated in AD pathogenesis and in cell death, including proteases such as APP-cleaving enzymes and Aβ-degrading enzymes in the case of AD and caspases in the case of apoptosis. Numerous fluorescent assays were developed to monitor caspase activity and thus apoptotic cells (Hanson and Hanson 2008). Fluorescent caspase indicators were recently used in combination with NFT imaging to monitor apoptosis in vivo and to determine the relationship between these two phenomena in the pathogenesis of AD (Spires-Jones et al. 2008). Caspase activation was detected in a small subset of neurons, the vast majority of which contained tangles. Surprisingly, this activation did not lead to cell death as these caspase-positive neurons remained morphologically normal for up to 4 h of continuous imaging.

Matrix metalloproteinases (MMPs) have recently been shown to be activated by CAA, playing a role in degrading the vessel basement membrane (Asahina et al. 2001). We have recently used the fluorogenic DQ gelatin as an MMP substrate for in vivo imaging in transgenic mice that develop parenchymal and vascular Aβ deposits to investigate the relationship between CAA, MMP, and oxidative stress (Fig. 3). The results of this study revealed strong associations between MMP activation, oxidative stress, and CAA deposition in leptomeningeal vessels. MMP inhibitors had no effect on CAA progression; however, they were able to reduce Aβ-induced MMP activation as well as oxidative stress in the affected vessels (Garcia-Alloza et al. 2009b).

CONCLUSION

Throughout this chapter we have described imaging methodologies based on MPM in mouse models of AD. We have used a variety of structural and functional indicators to address questions in specific cell types in the living brain. The combination of MPM with the available transgenic mouse models has and will surely continue to increase our understanding of functional and structural pathological processes that underlie neurodegeneration. Because of the capacity for longitudinal imaging, MPM serves as an excellent tool to study disease progression and screen potential disease-modifying drugs.

REFERENCES

Asahina M, Yoshiyama Y, Hattori T. 2001. Expression of matrix metalloproteinase-9 and urinary-type plasminogen activator in Alzheimer's disease brain. *Clin Neuropathol* **20:** 60–63.

Bacskai BJ, Kajdasz ST, Christie RH, Carter C, Games D, Seubert P, Schenk D, Hyman, BT. 2001. Imaging of amyloid-β deposits in brains of living mice permits direct observation of clearance of plaques with immunotherapy. *Nat Med* **7:** 369–372.

Bacskai BJ, Hickey GA, Skoch J, Kajdasz ST, Wang Y, Huang GF, Mathis CA, Klunk WE, Hyman BT. 2003. Four-dimensional multiphoton imaging of brain entry, amyloid binding, and clearance of an amyloid-β ligand in transgenic mice. *Proc Natl Acad Sci* **100:** 12462–12467.

Bacskai BJ, Frosch MP, Freeman SH, Raymond SB, Augustinack JC, Johnson KA, Irizarry MC, Klunk WE, Mathis CA, Dekosky ST, et al. 2007. Molecular imaging with Pittsburgh Compound B confirmed at autopsy: A case report. *Arch Neurol* **64:** 431–434.

Baird GS, Zacharias DA, Tsien RY. 1999. Circular permutation and receptor insertion within green fluorescent proteins. *Proc Natl Acad Sci* **96:** 11241–11246.

Bezprozvanny I, Mattson MP. 2008. Neuronal calcium mishandling and the pathogenesis of Alzheimer's disease. *Trends Neurosci* **31:** 454–463.

Buée L, Bussiere T, Buée-Scherrer V, Delacourte A, Hof PR. 2000. Tau protein isoforms, phosphorylation and role in neurodegenerative disorders. *Brain Res Brain Res Rev* **33:** 95–130.

Busche MA, Eichhoff G, Adelsberger H, Abramowski D, Wiederhold KH, Haass C, Staufenbiel M, Konnerth A, Garaschuk O. 2008. Clusters of hyperactive neurons near amyloid plaques in a mouse model of Alzheimer's disease. *Science* **321:** 1686–1689.

Christie RH, Bacskai BJ, Zipfel WR, Williams RM, Kajdasz ST, Webb WW, Hyman BT. 2001. Growth arrest of individual senile plaques in a model of Alzheimer's disease observed by in vivo multiphoton microscopy. *J Neurosci* **21:** 858–864.

Denk W. 2011. Introduction to multiphoton-excitation fluorescence microscopy. In *Imaging: A laboratory manual* (ed. Yuste R), pp. 105–110. Cold Spring Harbor Laboratory Press, Cold Spring Harbor, NY.

Denk W, Strickler JH, Webb WW. 1990. Two-photon laser scanning fluorescence microscopy. *Science* **248:** 73–76.

Garcia-Alloza M, Dodwell SA, Meyer-Luehmann M, Hyman BT, Bacskai BJ. 2006. Plaque-derived oxidative stress mediates distorted neurite trajectories in the Alzheimer mouse model. *J Neuropathol Exp Neurol* **65:** 1082–1089.

Garcia-Alloza M, Borrelli LA, Rozkalne A, Hyman BT, Bacskai BJ. 2007. Curcumin labels amyloid pathology in vivo, disrupts existing plaques, and partially restores distorted neurites in an Alzheimer mouse model. *J Neurochem* **102:** 1095–1104.

Garcia-Alloza M, Subramanian M, Thyssen D, Borrelli LA, Fauq A, Das P, Golde TE, Hyman BT, Bacskai BJ. 2009a. Existing plaques and neuritic abnormalities in APP:PS1 mice are not affected by administration of the gamma-secretase inhibitor LY-411575. *Mol Neurodegener* **4:** 19. doi: 10.1186/1750-1326-4-19.

Garcia-Alloza M, Prada C, Lattarulo C, Fine S, Borrelli LA, Betensky R, Greenberg SM, Frosch MP, Bacskai BJ. 2009b. Matrix metalloproteinase inhibition reduces oxidative stress associated with cerebral amyloid angiopathy in vivo in transgenic mice. *J Neurochem* **109:** 1636–1647.

Garcia-Alloza M, Borrelli LA, Hyman BT, Bacskai BJ. 2010. Antioxidants have a rapid and long-lasting effect on neuritic abnormalities in APP:PS1 mice. *Neurobiol Aging* **31:** 2058–2068. doi: 10.1016/j.neurobiolaging.2008.11.006.

Hanson GT, Hanson BJ. 2008. Fluorescent probes for cellular assays. *Comb Chem High Throughput Screen* **11:** 505–513.

Holscher C. 1998. Possible causes of Alzheimer's disease: Amyloid fragments, free radicals, and calcium homeostasis. *Neurobiol Dis* **5:** 129–141.

Hyman BT, Trojanowski JQ. 1997. Consensus recommendations for the postmortem diagnosis of Alzheimer disease from the National Institute on Aging and the Reagan Institute Working Group on diagnostic criteria for the neuropathological assessment of Alzheimer disease. *J Neuropathol Exp Neurol* **56:** 1095–1097.

Ikonomovic MD, Klunk WE, Abrahamson EE, Mathis CA, Price JC, Tsopelas ND, Lopresti BJ, Ziolko S, Bi W, Paljug WR, et al. 2008. Post-mortem correlates of in vivo PiB-PET amyloid imaging in a typical case of Alzheimer's disease. *Brain* **131:** 1630–1645.

Jayaraman S, Haggie P, Wachter RM, Remington SJ, Verkman AS. 2000. Mechanism and cellular applications of a green fluorescent protein-based halide sensor. *J Biol Chem* **275:** 6047–6050.

Klunk WE, Bacskai BJ, Mathis CA, Kajdasz ST, McLellan ME, Frosch MP, Debnath ML, Holt DP, Wang Y, Hyman BT. 2002. Imaging Aβ plaques in living transgenic mice with multiphoton microscopy and methoxy-X04, a systemically administered Congo red derivative. *J Neuropathol Exp Neurol* **61:** 797–805.

Kuchibhotla KV, Goldman ST, Lattarulo CR, Wu HY, Hyman BT, Bacskai BJ. 2008. Aβ plaques lead to aberrant regulation of calcium homeostasis in vivo resulting in structural and functional disruption of neuronal networks. *Neuron* **59:** 214–225.

Kuchibhotla KV, Lattarulo CR, Hyman BT, Bacskai BJ. 2009. Synchronous hyperactivity and intercellular calcium waves in astrocytes in Alzheimer mice. *Science* **323:** 1211–1215.

Lombardo JA, Stern EA, McLellan ME, Kajdasz ST, Hickey GA, Bacskai BJ, Hyman BT. 2003. Amyloid-β antibody treatment leads to rapid normalization of plaque-induced neuritic alterations. *J Neurosci* **23:** 10879–10883.

Mathis CA, Bacskai BJ, Kajdasz ST, McLellan ME, Frosch MP, Hyman BT, Holt DP, Wang Y, Huang GF, Debnath ML, et al. 2002. A lipophilic thioflavin-T derivative for positron emission tomography (PET) imaging of amyloid in brain. *Bioorg Med Chem Lett* **12:** 295–298.

Meyer-Luehmann M, Spires-Jones TL, Prada C, Garcia-Alloza M, de Calignon A, Rozkalne A, Koenigsknecht-Talboo J, Holtzman DM, Bacskai BJ, Hyman BT. 2008. Rapid appearance and local toxicity of amyloid-β plaques in a mouse model of Alzheimer's disease. *Nature* **451:** 720–724.

Miyawaki A, Llopis J, Heim R, McCaffery JM, Adams JA, Ikura M, Tsien RY. 1997. Fluorescent indicators for Ca^{2+} based on green fluorescent proteins and calmodulin. *Nature* **388:** 882–887.

Miyawaki A, Griesbeck O, Heim R, Tsien RY. 1999. Dynamic and quantitative Ca^{2+} measurements using improved cameleons. *Proc Natl Acad Sci* **96:** 2135–2140.

Morrissette DA, Parachikova A, Green KN, LaFerla FM. 2009. Relevance of transgenic mouse models to human Alzheimer disease. *J Biol Chem* **284:** 6033–6037.

Prada CM, Garcia-Alloza M, Betensky RA, Zhang-Nunes SX, Greenberg SM, Bacskai BJ, Frosch MP. 2007. Antibody-mediated clearance of amyloid-β peptide from cerebral amyloid angiopathy revealed by quantitative in vivo imaging. *J Neurosci* **27:** 1973–1980.

Rabinovici GD, Furst AJ, O'Neil JP, Racine CA, Mormino EC, Baker SL, Chetty S, Patel P, Pagliaro TA, Klunk WE, et al. 2007. ^{11}C-PIB PET imaging in Alzheimer disease and frontotemporal lobar degeneration. *Neurology* **68:** 1205–1212.

Raymond SB, Skoch J, Hills ID, Nesterov EE, Swager TM, Bacskai BJ. 2008. Smart optical probes for near-infrared fluorescence imaging of Alzheimer's disease pathology. *Eur J Nucl Med Mol Imaging* **35:** S93–S98.

Spires TL, Meyer-Luehmann M, Stern EA, McLean PJ, Skoch J, Nguyen PT, Bacskai BJ, Hyman BT. 2005. Dendritic spine abnormalities in amyloid precursor protein transgenic mice demonstrated by gene transfer and intravital multiphoton microscopy. *J Neurosci* **25:** 7278–7287.

Spires-Jones TL, Meyer-Luehmann M, Osetek JD, Jones PB, Stern EA, Bacskai BJ, Hyman BT. 2007. Impaired spine stability underlies plaque-related spine loss in an Alzheimer's disease mouse model. *Am J Pathol* **171:** 1304–1311.

Spires-Jones TL, de Calignon A, Matsui T, Zehr C, Pitstick R, Wu HY, Osetek JD, Jones PB, Bacskai BJ, Feany MB, et al. 2008. In vivo imaging reveals dissociation between caspase activation and acute neuronal death in tangle-bearing neurons. *J Neurosci* **28:** 862–867.

Spires-Jones TL, Mielke ML, Rozkalne A, Meyer-Luehmann M, de Calignon A, Bacskai, BJ, Schenk D, Hyman BT. 2009. Passive immunotherapy rapidly increases structural plasticity in a mouse model of Alzheimer disease. *Neurobiol Dis* **33:** 213–220.

Stosiek C, Garaschuk O, Holthoff K, Konnerth A. 2003. In vivo two-photon calcium imaging of neuronal networks. *Proc Natl Acad Sci* **100:** 7319–7324.

Thompson RB, Whetsell WO Jr, Maliwal BP, Fierke CA, Frederickson CJ. 2000. Fluorescence microscopy of stimulated Zn(II) release from organotypic cultures of mammalian hippocampus using a carbonic anhydrase-based biosensor system. *J Neurosci Methods* **96:** 35–45.

Tsai J, Grutzendler J, Duff K, Gan WB. 2004. Fibrillar amyloid deposition leads to local synaptic abnormalities and breakage of neuronal branches. *Nat Neurosci* **7:** 1181–1183.

Tsien RY. 1989. Fluorescent probes of cell signaling. *Annu Rev Neurosci* **12:** 227–253.

90 Two-Photon Imaging of Neural Networks in a Mouse Model of Alzheimer's Disease

Gerhard Eichhoff and Olga Garaschuk

University of Tübingen, Institute of Physiology II, 72074 Tübingen, Germany

ABSTRACT

In humans, Alzheimer's disease develops over many years. It comprises a chain of subtle yet irreversible alterations in brain function finally leading to impairment of memory and cognition. Presymptomatic and thus invisible in humans, these alterations can be studied in the animal models of Alzheimer's disease. This, however, requires in vivo monitoring of brain function in aged living animals at both a single-cell and network level. Here we describe an approach for addressing this issue. Cell type–specific labeling of neurons and glia is combined with in vivo two-photon calcium imaging and fluorescent labeling of amyloid plaques to study functional properties of cortical circuits in a mouse model of Alzheimer's disease.

INTRODUCTION

With more than 15 million affected individuals worldwide, Alzheimer's disease (AD) is the most prevalent and costly neurodegenerative disorder. It causes progressive deterioration of mental capabilities and cognitive and functional impairments accompanied by severe memory loss. The disease is clearly age-dependent, with advancing age being the number one risk factor for developing AD. The probability of being diagnosed with AD nearly doubles every 5 years after the age of 65. Around 95% of all AD cases are sporadic and only ~5% are due to autosomal dominant (familial) mutations, mostly in three AD-related genes: amyloid precursor protein (APP) and presenilins 1 and 2 (Waring and Rosenberg 2008). Mouse models of the disease expressing AD-related proteins with familial mutations reproduce several pathological hallmarks of AD including (i) accumulation of amyloid β-containing plaques, (ii) intraneuronal aggregation of hyperphosphorylated protein tau, (iii) inflammatory response present in AD, and (iv) learning and memory deficits (Morrissette et al. 2009). The models, however, do not recapitulate the abundant neuronal loss seen in humans. Nonetheless, these mouse models offer a unique opportunity to learn more about synaptic and cellular mechanisms underlying the disease (both in their essence and in their temporal sequence) through in vivo analyses of brain function. Tools developed over the last several decades can be used to selectively mark and to visualize in vivo many important elements of the diseased brain parenchyma, such as: amyloid plaques, individual neurons, and glial and microglial cells. In this chapter we describe how to combine these tools and to use them in vivo for high-resolution functional analysis of cortical function.

Protocol

Two-Photon Imaging of Neural Networks in a Mouse Model of Alzheimer's Disease

IMAGING SETUP

Any commercially available two-photon imaging system can be used. Such systems are available from several providers (e.g., Carl Zeiss, Olympus, Leica Microsystems, and Prairie Technologies). We currently use a custom-built setup based on a mode-locked Ti:sapphire laser with automated dispersion compensation (Mai Tai HP DeepSee, Spectra-Physics) and a laser-scanning system (Olympus Fluoview) coupled to an upright microscope (BX51WI, Olympus) and equipped with a 40×, 0.80-NA water-immersion objective (CFI Apo 40×, Nikon). A custom-built system such as this can be assembled following instructions found in Majewska et al. (2000) and Nikolenko and Yuste (2005).

For in vivo calcium imaging we use the calcium indicator dye Oregon Green 488 BAPTA-1 (OGB-1), which can be excited at any wavelength between 800 and 930 nm. The emitted fluorescence of the dye is then collected between 400 and 720 nm. For simultaneous visualization of OGB-1 and an amyloid plaque marker Thioflavin-S, the dyes are excited at 800 nm and the emitted light is separated at 515 nm using a beam splitter. The collected images are background-corrected and analyzed off-line with the ImageJ program (http://rsb.info.nih.gov/ij/) and a LabView-based software package (National Instruments).

MATERIALS

CAUTION: See Appendix 6 for proper handling of materials marked with <!>.
See the end of the chapter for recipes for reagents marked with <R>.

Reagents

Agarose, low-melting-point
Cyanoacrylate glue <!>
Eye ointment (e.g., Bepanthen)
Isoflurane <!>, applied through a vaporizer, vaporized in pure oxygen <!>
Lidocaine <!> or equivalent local anesthetic agent
Membrane-permeable calcium indicator dye (e.g., Oregon Green 488 BAPTA-1 AM [OGB-1]; Molecular Probes)
Pluronic F-127, 20% (w/v) in DMSO <!> (Sigma-Aldrich)
Standard external saline <R>
Standard pipette solution <R>
Sulforhodamine 101 (SR101; Sigma-Aldrich)
Thioflavin-S, 0.001% in standard pipette solution (Sigma-Aldrich)

Equipment

Anesthesia unit including a chamber for preanesthetic medication, a flow meter, and a vaporizer (latter items are for volatile anesthetic agents only)

Consult the literature (e.g., Flecknell 2000) for the best choice of anesthesia for your species.

Animal monitoring system for monitoring respiration and pulse rate, body temperature, and blood pressure (e.g., AD Instruments)
Centrifugal filters (Ultrafree-MC; Millipore)
Dental hand drill (e.g., NSK-Nakanishi), drill bits with 0.5-mm diameter
Glass capillaries (e.g., Hilgenberg GmbH)
Imaging setup (see above)
Micromanipulator (e.g., Luigs & Neumann GmbH)
Needles, 30-gauge, attached to a micromanipulator
Pasteur pipettes, plastic
Patch-clamp amplifier (e.g., HEKA)
Pipette puller (e.g., PP830; Narishige)
Pressure application system (e.g., npi electronic GmbH or Toohey Company)
Recording chamber with central access opening: custom-made from a standard tissue-culture dish (35-mm-diameter; for details, see Garaschuk et al. 2006])
Stereo dissecting microscope equipped with variable magnification lens (~1X–4X)
Stereotaxic apparatus (e.g., TSE-Systems)
Surgical equipment
Warming blanket (e.g., TSE-Systems)

EXPERIMENTAL METHOD

Surgical Procedure

1. Anesthetize the animal. Test the toe pinch withdrawal reflex to ascertain that the surgical level of anesthesia has been reached.

   Note that the required amount of anesthetic can vary with age and/or the individual state of the animal. For easily adjusting the amount of anesthetic, use volatile anesthetic isoflurane. To induce anesthesia, we use 2.5% isoflurane (in the anesthesia induction chamber), and for surgery, 1.5% isoflurane.

2. Place the animal on a warming blanket preheated to 38°C and monitor the animal's temperature and breathing rate. To prevent the animal's eyes from drying out, apply eye ointment.
3. Inject a local anesthetic under the skin above the animal's skull, wait for the anesthetic to work and carefully remove the skin. Start with a horizontal cut at the base, followed by two oblique cuts toward the eyelids of the mouse converging at the midline. Gently scrape any resting tissue on top of the animal's skull with a surgical blade and dry the bone completely using compressed air. Use a stereotaxic device to identify the location of the brain area to be targeted for recordings.
4. Apply cyanoacryl glue around the defined recording area and glue a recording chamber to the dried skull. Immediately remove any glue covering the recording area. Allow the glue to dry (wait 10–20 min) and then fix the recording chamber in an appropriate holder under a dissecting microscope.
5. Use a dental drill with appropriate drill bits to evenly thin the skull under the opening of the recording chamber. Use a high drilling speed (50,000 rev/min) and avoid pressing onto the skull while drilling.

   The skull of older mice often contains many small blood vessels. It is not necessary to avoid them, just continue drilling evenly. Regularly use compressed air to remove the bone dust and a few drops of standard external saline to absorb the heat produced by drilling. Stop drilling when the brain vessels located under the dura become visible. The skull is sufficiently thinned if you (i) can clearly see the pattern of blood vessels located under the dura (apply standard external saline on top of the skull to improve skull transparency), and (ii) see the skull bending when it is touched with thin tweezers.

6. Transfer the animal to the experimental setup, and connect it to a monitoring system for continuous monitoring of respiration and pulse rate, body temperature, and blood pressure.
7. Perfuse the recording chamber with 37°C standard external saline.
8. Perform a craniotomy making a rectangular hole that is 0.5–1 mm on each side. For good optical control of your manipulations, perform the craniotomy under a 10x objective.
   i. Keep the chamber filled with standard external saline.

      Covering the skull with the fluid increases tissue transparency and keeps the dura/cortex from drying out.

   ii. Choose an area devoid of large blood vessels. Open the skull with a 30-gauge needle attached to either a syringe and guided by your steady hand or attached to a hand-driven micromanipulator. Form a rectangle by making four straight, slightly overlapping cuts through the bone only; do not cut the dura. Ensure that the edges of the rectangle do not adhere to the rest of the skull.
   iii. Lift and remove the cut rectangle of bone. It should come out easily. Do not remove the dura.

      Leaving the dura in place protects and stabilizes the underlying tissue.

   See *Troubleshooting.*
9. Depending on your experimental strategy, either proceed with Step 10 if staining plaques, or skip to Step 16 for staining neurons and glial cells.

Plaque-Staining Procedure

Stain plaques in the area of interest using Thioflavin-S, a selective marker.

10. Dissolve Thioflavin-S in DMSO to 1%. Dilute this stock solution with standard pipette solution to a final Thioflavin-S concentration of 0.001%. Filter the staining solution using an Ultrafree-MC centrifugal filter.
11. Pull glass micropipettes so that they have a resistance of 5–6 MΩ when filled with the staining solution.
12. Fill a micropipette with the Thioflavin-S staining solution. Use a 4x objective and an LN-Mini manipulator to place the dye-containing pipette above the skull opening. Switch to a 40x long-working-distance objective and maneuver the pipette directly onto the cortical surface.

    It is preferable to insert the pipette at the steepest angle allowed by the geometry of the system.

13. Insert the pipette into the cortical tissue by moving it along the pipette axis until it is 200 µm under the cortical surface. Inject the Thioflavin-S for 30 sec at 55 kPa.
14. Repeat the application every 50 µm while withdrawing the pipette until a depth of 50 µm under the dura is reached (Fig. 1B).

    Thioflavin-S does not diffuse as well as the calcium indicator dyes, so only a small amount should be administered per injection site. Insert the staining pipette in up to four different locations to stain all plaques under the craniotomy.

15. Wait for up to 25 min for the unbound Thioflavin-S to wash out before assessing the quality of plaque staining by means of two-photon imaging (Fig. 1).

Staining Neurons and Glia with Calcium Indicator Dyes

To stain neurons and glia in mouse models of AD with calcium indicator dyes, we modified a multicell bolus loading technique (MCBL) (Stosiek et al. 2003; see also Chapter 49).

16. Dissolve OGB-1 AM in a solution containing 20% (w/v) Pluronic F-127 in DMSO to yield 10 mM OGB-1 AM. Dilute this solution with standard pipette solution to yield a final dye concentration of 0.5 mM. Filter the staining solution through an Ultrafree-MC centrifugal filter directly before application.

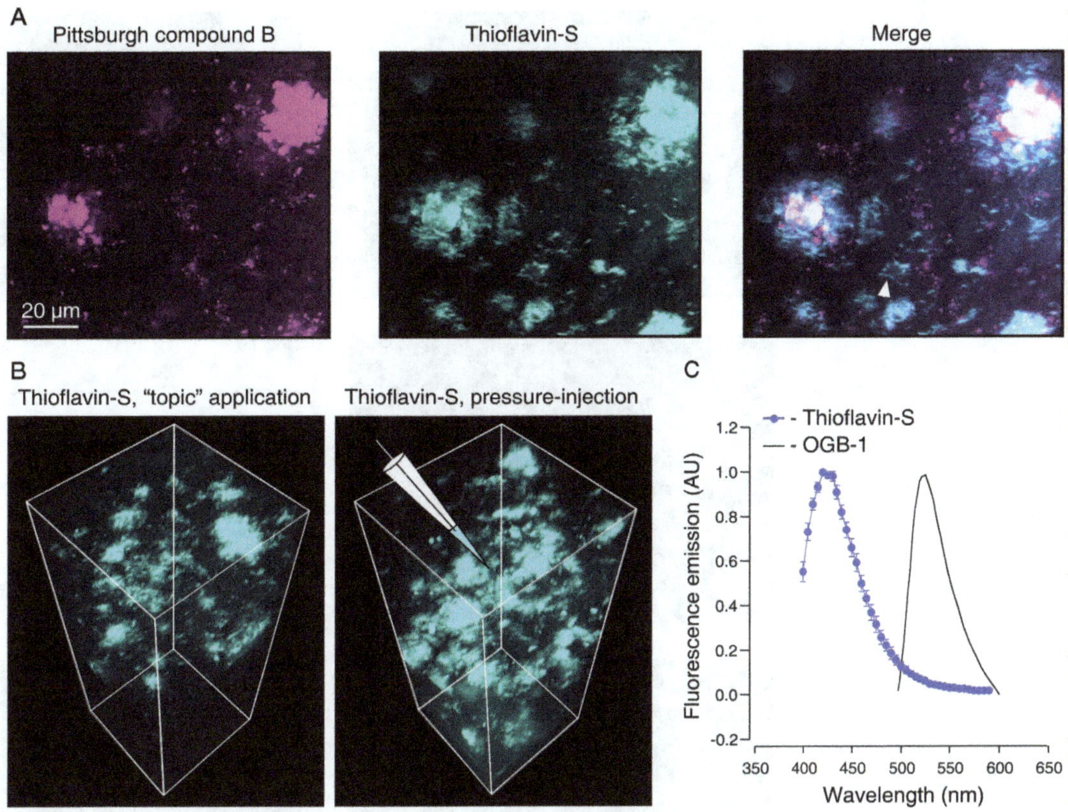

FIGURE 1. Labeling of amyloid plaques. (*A*) Amyloid plaques in the cortex of an APP23 x PS45 mouse (Busche et al. 2008) stained by pressure injection of both Pittsburgh compound B (*left*) and Thioflavin-S (*middle*). The image to the *right* is an overlay of the two images. The arrowhead points to fiber-like structures, which have selectively bound Thioflavin-S. (*B*) (*Left*) Amyloid plaques stained by topical application of 0.005% Thioflavin-S for 20 min. (*Right*) Plaques are stained by pressure injection of Thioflavin-S as described in Steps 13 and 14. Both three-dimensional reconstructions show a 60 × 60 × 260-µm volume of the mouse cortex. (*C*) Emission spectra of Thioflavin-S (two-photon excitation) and OGB-1 (one-photon excitation, taken from Molecular Probes Catalog). (*A*,*B*, Reproduced, with permission from Springer Science+Business Media, from Eichhoff et al. 2008.)

17. Pull glass micropipettes so that they have a resistance of 5–6 MΩ when filled with the staining solution.
18. Fill a micropipette with the OGB-1 AM solution. Use a 4x objective and an LN-Mini manipulator to place the dye-containing pipette above the skull opening. Switch to a 40x long-working-distance objective and maneuver the pipette directly onto the cortical surface.
19. To obtain good staining of cortical layer 2/3 neurons in AD mice, use up to four staining locations per 0.5–1 mm^2. Place the pipette directly on top of the dura at the steepest possible angle and advance it slowly to a depth of ~250–300 µm. Inject the dye for 1–2 min at 55–60 kPa.
20. Wait at least 60 min for diffusion, deesterification, and wash out of the extracellular dye before imaging. Check the quality of the obtained staining.

    See *Troubleshooting*.

Calcium Imaging of Neuronal Function

21. If plaques were previously stained, then split Thioflavin-S and OGB-1 fluorescence using a 500-nm beam splitter (Figs. 1C and 2A).

    For technical reasons, we use a 515-nm beam splitter in our setup.

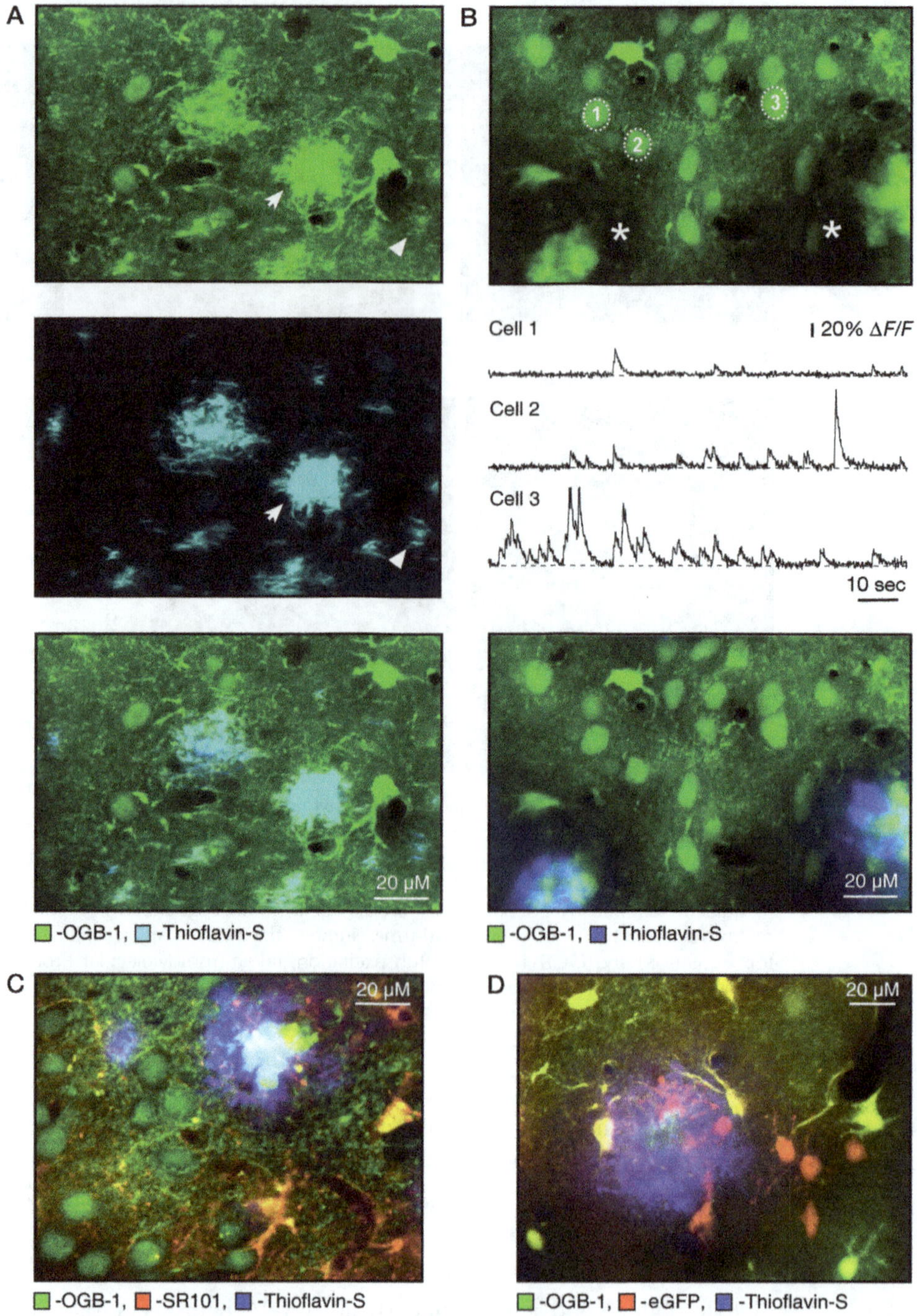

FIGURE 2. In vivo imaging of different cell types in a mouse model of Alzheimer's disease. (*A*) Images of cortical layer 1 stained with pressure application of OGB-1 (*top*) and Thioflavin-S (*middle*). The *bottom* panel shows an overlay of the two images. The dyes are excited at 800 nm and the emission light is split at 515 nm. Note that large plaques are visible even in the OGB-1 channel (arrow), small fiber-like structures, however, can be recognized only when stained with Thioflavin-S (arrowhead). (*B*) Intracellular calcium transients (*middle*) recorded from cells marked with corresponding numbers in the *top* panel. In this preparation, stained with OGB-1 only, amyloid plaques appear as bright spherical areas with surrounding dark regions (asterisks). Subsequent staining of the same area with Thioflavin-S (*bottom*) confirms the deduced location of the plaques. (*C*) Microphotograph of the cortical layer 2/3 stained with OGB-1 (green), sulforhodamine 101 (red), and Thioflavin-S (blue). Neurons, astrocytes, and plaques are visualized by consecutive splitting of the emission light at 515 and 570 nm. (*D*) An image of cortical layer 1 in a triple mutant mouse made by crossing APP23 x PS45 mice with CX_3CR1 mice expressing eGFP-labeled microglia (Jung et al. 2000). OGB-1 staining is shown in green, Thioflavin-S in blue, and eGFP in red. OGB-1 and Thioflavin-S were excited at 800 nm and their fluorescence was separated by a 515-nm beam splitter. eGFP was excited at 930 nm and the emitted light was sampled between 475 and 515 nm. (*B*, Reproduced, with permission from AAAS, from Busche et al. 2008.)

22. To assure that your preparation is vital, monitor spontaneous neuronal activity as described in Busche et al. (2008).

    See *Troubleshooting.*

23. Good-quality recordings of spontaneous neuronal activity (Fig. 2B) can be routinely obtained for up to 5 h. To resolve the time course of neuronal calcium transients, use a recording speed of at least 10 Hz.

24. If plaques are not yet stained with Thioflavin-S, deduce their location based on OGB-1 staining (Fig. 2A,B) as follows. In the brain tissue labeled with OGB-1, only plaques appear as large bright sphere-shaped areas (arrow in Fig. 2A, upper image) often surrounded by dark areas devoid of labeled cells (asterisks in Fig. 2B). Thus, it is also possible to record neuronal or glial activity first and then stain the preparation with Thioflavin-S. There are two possible advantages of this scenario: (1) recordings can begin earlier and are thus made during the time when the animal is in the best shape, and (2) any possible Thioflavin-S-mediated modification of cellular activity is excluded. If one chooses this approach, we recommend a careful three-dimensional reconstruction of the imaged cells/area at the end of the recording session. The reconstructed volume should contain characteristic landscape marks, like blood vessel pattern and bright cells with processes. This enables easy identification of the imaged area after plaque staining (Fig. 2B, bottom).

Multicolor Imaging

25. Using cell type–specific markers and appropriate optics, visualize and monitor different cells types within the cortical network (Fig. 2C,D). Astrocytes, for example, can be labeled with sulforhodamine 101 (SR101), a specific astroglial marker (Nimmerjahn et al. 2004). SR101 can be injected simultaneously with the calcium indicator dye using a protocol described in Garaschuk et al. (2006). Because of its red-shifted fluorescence spectrum, SR101 can be easily separated from OGB-1 and Thioflavin-S by using, for example, a 570-nm beam splitter (Fig. 2C).

26. Visualize microglial cells in vivo using transgenic mice in which this cell type is labeled with enhanced green fluorescent protein (eGFP) (Jung et al. 2000; Hirasawa et al. 2005). Using multicolor two-photon imaging, microglial cells can be visualized in transgenic mice containing both eGFP-labeled microglia and amyloid plaques (Fig. 2D). To separate the eGFP fluorescence, we use excitation splitting, making use of the fact that eGFP is very efficiently excited at 930 nm but not at 800 nm (Sohya et al. 2007).

TROUBLESHOOTING

Problem (Step 8): Bleeding is seen during the craniotomy.

Solutions: Take care to cut the skull only and not to touch or pull or damage the underlying blood vessels. Because blood vessels are usually located under the dura, take care not to cut the dura. If you pull blood vessels when cutting, stop immediately and reposition the cutting blade. The skull should be sufficiently thinned beforehand to minimize the risk of blood vessel damage during the craniotomy. If bleeding occurs, try to stop bleeding by repetitively rinsing the bleeding spot with standard external saline applied with a plastic pasteur pipette. This will also prevent red blood cells from adhering to the dura and thus reducing the optical transparency of the preparation. Should blood circulation in a large blood vessel be disrupted as a result of damage during the craniotomy, stop and attempt another craniotomy in a sufficiently remote location.

Problem (Step 20): Poor or blurry staining is seen.

Solutions: Make sure that no large blood vessels were damaged during the craniotomy, that the surface of the brain is not covered by erythrocytes, and that you are not imaging below a large blood vessel. Trying to dye-label a damaged brain area results in so-called "salt-and-pepper" staining (see Fig. 3 in Garaschuk et al. 2006).

If the preparation looks healthy but unstained, consider the following: (1) The dye-injection pipette may be clogged. Monitor pipette resistance during the staining procedure. (2) Application of the dye is too superficial. Monitor pipette resistance to estimate when the pipette is touching the dura. Especially when labeling neurons in adult and aged tissue, it is sometimes necessary to wait longer (e.g., for an additional 30 min) for the staining to develop. In general, the more plaques the animal has, the more challenging it is to obtain high-quality labeling of layer 2/3 neurons. Therefore, first familiarize yourself with the technique in control littermates and then in animals that have not yet reached their maximal plaque load.

Problem (Step 22): There is a lack of spontaneous neuronal activity despite good labeling with the indicator dye.

Solution: Check the condition of the experimental animal. Breathing rate should not fall below 85 breaths/min and the body temperature should stay above 37°C. These criteria should be applied throughout the experiment (e.g., surgery, staining, and recording). Surgical and staining procedures should not exceed 2 h before the commencement of recording.

Problem (Step 22): Unstable recording conditions and/or movement artifacts are seen.

Solutions: Movement artifacts observed during high-resolution in vivo imaging can be subdivided into two classes: (1) a slow drift of the image plane, and (2) fast movement artifacts caused either by breathing or the heartbeat of the animal. To prevent slow drifts it is advisable to keep constant both body temperature and the temperature of the superfusing standard external saline. Breathing artifacts can be reduced by (i) stabilizing the animal's breathing rate above 85 breaths/min, and (ii) assuring that the recording chamber is tightly attached to the skull and tightly fixed in the experimental setup. Heartbeat pulsations can cause artifacts at a frequency of ~400 beats/min. These are very prominent in the vicinity of large blood vessels and are rather difficult to avoid. Therefore we prefer to record from brain regions devoid of large blood vessels. In general, however, movement artifacts are less of a problem when imaging cortical areas in adult/aged animals compared to young/juvenile ones. To control movement artifacts, it is often sufficient simply to leave the dura intact. As in young animals, movement artifacts can be further minimized by covering the skull with 2% low-melting-point agarose in standard external saline and a glass coverslip (Svoboda et al. 1999).

DISCUSSION

In Vivo Labeling of Amyloid Plaques

Several fluorescent compounds can be used to label amyloid plaques in vivo. These include Thioflavin-S (or Thioflavin-T; Bacskai et al. 2001), Pittsburgh compound B (PiB; Bacskai et al. 2003), and methoxy-X04 (Klunk et al. 2002), as well as its parent substances Congo Red and Chrysamine-G (Nesterov et al. 2005). These compounds bind to fibrillar β-sheet amyloid deposits (Klunk et al. 2002; Rak et al. 2007) and thus recognize dense core plaques but not diffuse ones. However, when applied in the same concentration to the same plaques, PiB labeled only the very core of the plaque (note that similar images were obtained by Klunk et al. [2002] when using methoxy-X04), whereas Thioflavin-S also labeled surrounding fibril-like structures (Fig. 1; Eichhoff et al. 2008). There are several possible explanations for this finding: Thioflavin-S may bind in vivo with higher affinity, have better fluorescence quantum yield or be capable of recognizing structures with lower β-sheet content. Surprisingly, the core of the plaque is also reasonably well stained with OGB-1 (Fig. 2) and with other fluorophores, such as calcium-insensitive dye Alexa Fluor 594 (not shown). Although the precise mechanism of this binding remains unclear, such binding allows for convenient and early identification of plaques in the preparations stained (for example) with OGB-1 only. It has to be stressed, however, that small fibril-like, Thioflavin-positive structures (arrowheads in Figs. 1 and 2) cannot be identified in the OGB-1 labeled tissue.

When deciding which dye to use, it is necessary to also consider the method for administering the dye. PiB and methoxy-X04 can cross the blood–brain barrier and therefore can be applied intravenously or, in the case of methoxy-X04, even intraperitoneally. In contrast, Thioflavin-S must be applied directly to the area of interest either topically (on the top of the cortex) or by pressure-injection. Thus, the choice of dye critically depends on experimental design. Experiments aiming at monitoring plaques over prolonged periods of time will preferably use methoxy-X04 or PiB for plaque staining. However, bear in mind that intravenous/intraperitoneal application of any drug or dye results in its delayed delivery to the brain area of interest. The kinetics of dye delivery as well as the concentration of the dye remain unknown and may vary from animal to animal and from brain area to brain area depending, for example, on inhomogeneities of blood circulation. Especially in aged mouse models of AD with substantial deposition of cerebrovascular and parenchymal amyloid, differences in blood flow may substantially influence the labeling pattern and quality (see Zou et al. 2008 for additional concerns regarding intraperitoneal dye administration). Therefore, we prefer to use Thioflavin-S for combined imaging of neurons and amyloid plaques. Because a craniotomy is a necessity in such experiments, labeling plaques with Thioflavin-S requires little additional effort. Because Thioflavin-S shows limited diffusion (Fig. 1B), we prefer to target dye delivery directly to the area of interest, as described in Steps 10–15 (see above).

Using MCBL to Label Adult and Aged Brain Tissue

Originally MCBL was developed to stain juvenile (Stosiek et al. 2003), and newborn (Adelsberger et al. 2005) tissue. As shown in Figures 2 and 3, MCBL can also be used to label adult and aged cortical tissue. To our knowledge this is the only technique for intravital labeling of aged brain tissue with small molecule calcium indicators. However, image contrast is decreased in the aged compared to juvenile tissue (Fig. 3). This reduction of labeling quality is most probably caused by (i) less effective diffusion of the dye within the tissue, (ii) reduced activity of intracellular esterases, and (iii) impeded wash out of the dye from the extracellular space. These factors make the use of MCBL in adult/aged mice more challenging. The following strategy was used to enable MCBL-based labeling of aged tissue: the amount of the dye within the staining pipette and the duration of the dye injection were reduced (to reduce the amount of the dye delivered at once), and the number of injections was increased with injection spots homogeneously distributed all over the area of interest. In addition, longer waiting periods were used to allow better dye wash out and deesterification.

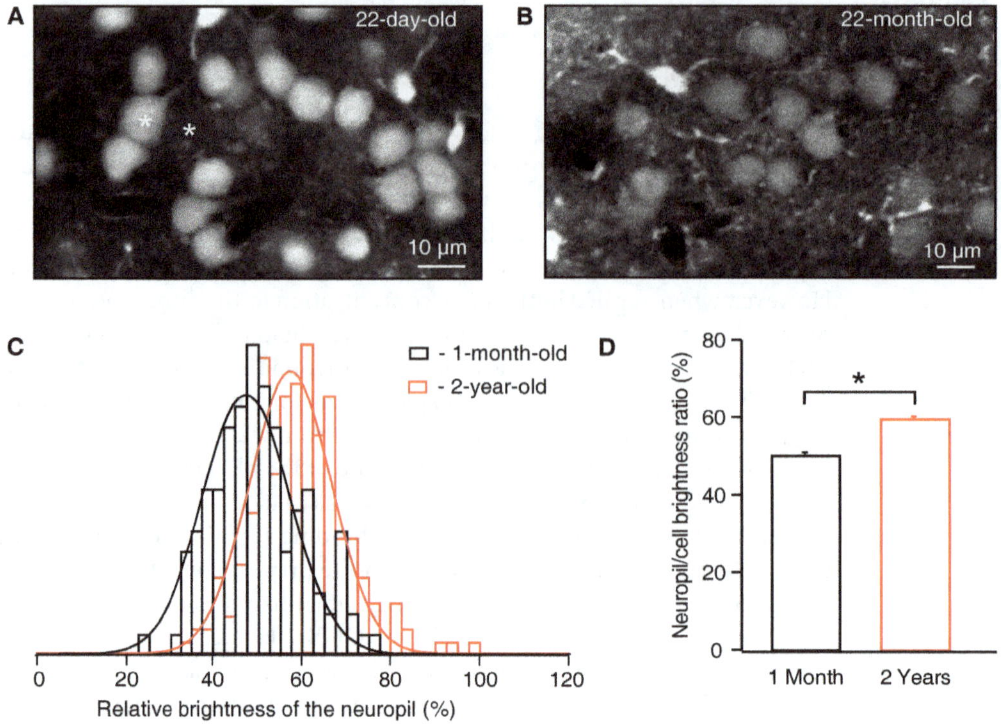

FIGURE 3. Quality of neuronal staining in juvenile and aged mice. (*A*,*B*) Images of cortical layer 2/3 cells stained with OGB-1 in a juvenile (*A*) and an aged (*B*) mouse. (*C*,*D*) Histograms (*C*) and a corresponding bar graph (*D*) show the distributions and the mean values of the neuropil/cell brightness ratio in the juvenile (black) and aged (red) mouse cortex. The brightness of the neuropil was measured by using a region of interest identical to the one used to analyze the corresponding cell placed in the immediate neighborhood of the cell of interest (see asterisks in *A*). (*B*–*D*, Reproduced, with permission from Springer Science+Business Media, from Eichhoff et al. 2008.)

Another factor to consider when imaging adult/aged tissue is the "aging pigment" lipofuscin. Lipofuscin is the product of the breakdown and oxidation of unsaturated fatty acids (Terman and Brunk 2004). It accumulates in lysosomes throughout the entire life. However, the speed of its intraneuronal accumulation is increased by neurodegenerative diseases, such as Alzheimer's and Parkinson's disease (Brunk and Terman 2002; Meredith et al. 2002). Lipofuscin is strongly fluorescent and has a broad emission spectrum ranging from 450–700 nm (Bindewald-Wittich et al. 2006; Eichhoff et al. 2008). Thus, lipofuscin interferes with many commonly used indicator dyes. When using OGB-1, however, it is possible to make use of the long-wavelength part of the lipofuscin emission spectrum to color-code and thus to visualize lipofuscin granules (see Fig. 4 in Eichhoff et al. 2008). It has to be noted, however, that the largest portion of photons emitted by lipofuscin has the same spectral properties as OGB-emitted photons (Fig. 4 in Eichhoff et al. 2008). Therefore, to avoid interference between lipofuscin and OGB-1 fluorescence, we do not study neurons containing large lipofuscin granules.

With all of these precautions in mind, the combination of MCBL, two-photon microscopy, and multicolor imaging provides a versatile technique for monitoring in vivo activity of many different elements of the cortical network in the aging and diseased brain.

RECIPES

CAUTION: See Appendix 6 for proper handling of materials marked with <!>.
Recipes for reagents marked with <R> are included in this list.

Standard External Saline

NaCl	125 mM
KCl	4.5 mM
$NaHCO_3$	26 mM
NaH_2PO_4	1.25 mM
$CaCl_2$ <!>	2 mM
$MgCl_2$	1 mM
Glucose	20 mM

The pH should be 7.4 when the solution is bubbled with 95% O_2 <!> and 5% CO_2 <!>.

Standard Pipette Solution

NaCl	150 mM
KCl	2.5 mM
HEPES (pH 7.4)	10 mM

ACKNOWLEDGMENTS

This work was supported by grants of the Deutsche Forschungsgemeinschaft (SFB 596, GA 654/1-1) to OG.

REFERENCES

Adelsberger H, Garaschuk O, Konnerth A. 2005. Cortical calcium waves in resting newborn mice. *Nat Neurosci* **8:** 988–990.

Bacskai BJ, Kajdasz ST, Christie RH, Carter C, Games D, Seubert P, Schenk D, Hyman BT. 2001. Imaging of amyloid-β deposits in brains of living mice permits direct observation of clearance of plaques with immunotherapy. *Nat Med* **7:** 369–372.

Bacskai BJ, Hickey GA, Skoch J, Kajdasz ST, Wang Y, Huang GF, Mathis CA, Klunk WE, Hyman BT. 2003. Four-dimensional multiphoton imaging of brain entry, amyloid binding, and clearance of an amyloid-β ligand in transgenic mice. *Proc Natl Acad Sci* **100:** 12462–12467.

Bindewald-Wittich A, Han M, Schmitz-Valckenberg S, Snyder SR, Giese G, Bille JF, Holz FG. 2006. Two-photon-excited fluorescence imaging of human RPE cells with a femtosecond Ti:Sapphire laser. *Invest Ophthalmol Vis Sci* **47:** 4553–4557.

Brunk UT, Terman A. 2002. Lipofuscin: Mechanisms of age-related accumulation and influence on cell function. *Free Radic Biol Med* **33:** 611–619.

Busche MA, Eichhoff G, Adelsberger H, Abramowski D, Wiederhold KH, Haass C, Staufenbiel M, Konnerth A, Garaschuk O. 2008. Clusters of hyperactive neurons near amyloid plaques in a mouse model of Alzheimer's disease. *Science* **321:** 1686–1689.

Eichhoff G, Busche MA, Garaschuk O. 2008. In vivo calcium imaging of the aging and diseased brain. *Eur J Nucl Med Mol Imaging* (suppl 1) **35:** S99–S106.

Flecknell P. 2000. *Laboratory animal anaesthesia*. Academic, San Diego.

Garaschuk O, Milos RI, Konnerth A. 2006. Targeted bulk-loading of fluorescent indicators for two-photon brain imaging in vivo. *Nat Protoc* **1:** 380–386.

Hirasawa T, Ohsawa K, Imai Y, Ondo Y, Akazawa C, Uchino S, Kohsaka S. 2005. Visualization of microglia in living tissues using Iba1-eGFP transgenic mice. *J Neurosci Res* **81:** 357–362.

Jung S, Aliberti J, Graemmel P, Sunshine MJ, Kreutzberg GW, Sher A, Littman DR. 2000. Analysis of fractalkine receptor CX(3)CR1 function by targeted deletion and green fluorescent protein reporter gene insertion. *Mol Cell Biol* **20:** 4106–4114.

Klunk WE, Bacskai BJ, Mathis CA, Kajdasz ST, McLellan ME, Frosch MP, Debnath ML, Holt DP, Wang Y, Hyman BT. 2002. Imaging Aβ plaques in living transgenic mice with multiphoton microscopy and methoxy-X04, a systemically administered Congo red derivative. *J Neuropathol Exp Neurol* **61:** 797–805.

Majewska A, Yiu G, Yuste R. 2000. A custom-made two-photon microscope and deconvolution system. *Pflügers Arch* **441:** 398–408.

Meredith GE, Totterdell S, Petroske E, Santa Cruz K, Callison RC Jr, Lau YS. 2002. Lysosomal malfunction accompanies alpha-synuclein aggregation in a progressive mouse model of Parkinson's disease. *Brain Res* **956:** 156–165.

Morrissette DA, Parachikova A, Green KN, LaFerla FM. 2009. Relevance of transgenic mouse models to human Alzheimer disease. *J Biol Chem* **284:** 6033–6037.

Nesterov EE, Skoch J, Hyman BT, Klunk WE, Bacskai BJ, Swager TM. 2005. In vivo optical imaging of amyloid aggregates in brain: Design of fluorescent markers. *Angew Chem Int Ed Engl* **44:** 5452–5456.

Nikolenko V, Yuste R. 2005. How to build a two-photon microscope using a confocal scan head. In *Imaging in neuroscience and development: A laboratory manual* (ed. Yuste R, Konnerth A), pp. 75–78. Cold Spring Harbor Laboratory Press, Cold Spring Harbor, NY.

Nimmerjahn A, Kirchhoff F, Kerr JND, Helmchen F. 2004. Sulforhodamine 101 as a specific marker of astroglia in the neocortex in vivo. *Nat Methods* **1:** 31–37.

Rak M, Del Bigio MR, Mai S, Westaway D, Gough K. 2007. Dense-core and diffuse Aβ plaques in TgCRND8 mice studied with synchrotron FTIR microspectroscopy. *Biopolymers* **87:** 207–217.

Sohya K, Kameyama K, Yanagawa Y, Obata K, Tsumoto T. 2007. GABAergic neurons are less selective to stimulus orientation than excitatory neurons in layer II/III of visual cortex, as revealed by in vivo functional Ca^{2+} imaging in transgenic mice. *J Neurosci* **27:** 2145–2149.

Stosiek C, Garaschuk O, Holthoff K, Konnerth A. 2003. In vivo two-photon calcium imaging of neuronal networks. *Proc Natl Acad Sci* **100:** 7319–7324.

Svoboda K, Helmchen F, Denk W, Tank DW. 1999. Spread of dendritic excitation in layer 2/3 pyramidal neurons in rat barrel cortex in vivo. *Nat Neurosci* **2:** 65–73.

Terman A, Brunk UT. 2004. Lipofuscin. *Int J Biochem Cell Biol* **36:** 1400–1404.

Waring SC, Rosenberg RN. 2008. Genome-wide association studies in Alzheimer disease. *Arch Neurol* **65:** 329–334.

Zou K, Maeda T, Michikawa M, Komano H. 2008. New amyloid plaques or a game of hide-and-seek? *Int J Biol Sci* **4:** 200–201.

91 Imaging Tumors in the Brain

Kelley S. Madden,[1] Martha L. Zettel,[2] Ania K. Majewska,[2] and Edward B. Brown[1,2]

[1]*Department of Biomedical Engineering, University of Rochester Medical Center, Rochester, New York 14627;*
[2]*Department of Neurobiology and Anatomy, University of Rochester Medical Center, Rochester, New York 14642*

ABSTRACT

The brain constitutes a unique environment for tumor growth. Despite progress in treating primary tumors in the periphery, tumors that initiate in the central nervous system (CNS) or that preferentially migrate from peripheral primary or secondary metastatic sites into the brain remain refractory to treatment. When brain tumor cell lines are endogenously labeled with a fluorescent protein and injected into an optically accessible cranial window, advanced imaging techniques, such as two-photon laser scanning microscopy, can be used to investigate aspects of tumor pathogenesis in real time in a living animal. The first protocol in this chapter describes the use of two-photon microscopy to assess the development of glioma, using the CNS-1 glioma cell line transfected with a yellow fluorescent protein variant, called Venus. To study metastasis to the brain, we employ a breast cancer cell line that selectively migrates to the brain (MB-231BR). The second protocol describes intracardiac injection of Venus-transfected MB-231BR and subsequent imaging of metastatic foci in the brain by standard one-photon imaging techniques. The advantages and disadvantages of both tumor delivery systems and the subsequent imaging techniques are discussed. These models are useful for understanding the molecular pathways involved in brain tumor pathogenesis and for identifying molecular targets for potential therapeutic agents.

INTRODUCTION

Despite progress in treating primary tumors in the periphery, tumors that initiate in the central nervous system (CNS) or that preferentially migrate from peripheral primary or secondary metastatic sites into the brain remain refractory to treatment (King et al. 2005; Lin et al. 2004; Weil et al. 2005). Neurons and the supporting glial cells of the brain create a distinct microenvironment for primary brain tumors, and metastastic tumor cells must adapt to this unique environment (Weil et al. 2005). Furthermore, to enter the brain, tumor cells and therapeutics must cross the blood–brain barrier. Animal models that recapitulate all aspects of brain tumor pathogenesis are critical to improve our understanding of CNS tumors and to develop more effective therapies that inhibit their development.

Glioblastoma is a highly invasive and aggressive brain tumor that is very difficult to treat (King et al. 2005). The rat CNS-1 cell line is a widely used, well-characterized model of infiltrative glioma. Here we describe how CNS-1 cells transfected with Venus, a variant of yellow fluorescent protein, are grown in cranial windows to study tumor growth and spread using two-photon microscopy.

Brain metastases from primary or secondary breast tumors are difficult to model in the mouse. When metastatic breast cancer cell lines are injected directly into the arterial circulation, only a small fraction of cells enter the brain to form metastatic foci. Therefore, several laboratories have used multiple cycles of in vivo passaging and isolation of cells from the brain to select for variants that cross the blood–brain barrier and produce consistent, quantifiable brain metastasis (Yoneda et al. 2001; Kim et al. 2004; Stark et al. 2007). To facilitate tumor visualization, we have transfected one of these brain-homing cell lines (MB-231BR from Yoneda and colleagues) with Venus. Here we describe the intracardiac injection required to elicit brain metastasis and the quantification of metastatic foci using fluorescent image analysis. The advantages and disadvantages of both tumor models and imaging techniques will be discussed.

Protocol A

Brain Tumor Imaging: Live Imaging of Glioma by Two-Photon Microscopy

We have used CNS-1, a rat glioma cell line, to study tumors that initiate within the brain parenchyma. The CNS-1 cells were stably transfected with the yellow fluorescent protein (YFP) variant Venus to enable standard one-photon and two-photon imaging. We describe how to prepare a cranial window and how to inject the tumor cells into the cerebral cortex at a depth suitable for two-photon imaging. Imaging can begin 24 h after implantation of the cells. Two-photon imaging employs a long excitation wavelength (920 nm); the emission spectrum is the same for the fluorophore as for standard one-photon imaging (emission maximum = 528 nm). Two-photon microscopy permits tissue imaging to depths of 500 µm with three-dimensional (3D) high resolution and minimal photodamage to surrounding tissues. Multiple imaging sessions can be conducted over weeks in the same animal, depending on how long the cranial window remains clear and how quickly the tumor grows. Using this technique, the kinetics of tumor growth and invasion into the surrounding brain parenchyma can be measured in the same animal. This model can be used for determining the molecular and cellular players in brain tumor growth and invasion and for testing potential drug therapies to prevent brain metastasis.

IMAGING SETUP

For two-photon microscopy, a mode-locked, tunable Ti:sapphire laser (MaiTai; Spectra Physics) provides 920-nm excitation light. Beam scanning is accomplished by a Fluoview 3 scanner (Olympus) via an XLUMPlan Fl 20x/0.95 NA objective water-immersion lens. Fluorescence passes through a 690-nm short-pass dichroic mirror and a 550DF150 emission filter (Chroma) and is detected by external (i.e., non-descanned) HC125-02 photomultiplier tubes (Hamamatsu). Images were scanned at the slow setting with 1024 x 1024-pixel images in the *x*-, *y*-, and *z*-directions with a step size 3–5 µm in the *z*-direction.

Alternatively, to produce epifluorescence images of excised brain sections, an Olympus BX51 epifluorescence microscope, equipped with filters for yellow fluorescent protein (YFP) (emission maximum = 528 nm) and DAPI (emission maximum = 461 nm), a Spot Pursuit camera, and SPOT Advanced software, was used. ImageJ software was used for analysis of fluorescence images.

MATERIALS

CAUTION: See Appendix 6 for proper handling of materials marked with <!>.

Reagents

Avertin <!>
Buprenorphine
CNS-1 cells
Cyanoacrylate glue <!>
Immunocompromised nude (nu/nu) mice
Paraformaldehyde <!>, 4%

Phosphate-buffered saline (pH 7.4)
Prolong Gold Antifade Reagent with DAPI (Invitrogen item P-36931)
Saline, sterile
Sucrose, 10% and 30%
Tissue culture reagents (Invitrogen; keep all solutions sterile)
  Dulbecco's phosphate-buffered saline (PBS) containing 0.5% FBS (both at room temperature and ice cold)
  Fetal calf serum (FCS)
  RPMI medium containing 10% FBS
  RPMI medium containing 0.1 mg/mL Hygromycin B <!> and 10% fetal bovine serum (FBS)
  Trypsin <!>, 0.25% with EDTA <!>

Equipment

Cell counter
Centrifuge
Cover glass, 7-mm-diameter (Fisher)
Dental cement
Drill, high-speed, with a sterile burr tip
Epifluorescence microscope (Olympus) equipped with SPOT Pursuit digital camera and attached computer equipped with the SPOT Advanced software package
Gelfoam
Hamilton syringe (low-volume) and 31-gauge needle
Heat pads
ImageJ software (download from: http://rsb.info.nih.gov/ij/download.html)
Incubator set at 37°C
Malis dissector
Microslides and coverslips
Microtome
Stainless steel bar with tapped screw holes (see Step 21)
Stereotactic frame with earbars and bite bars
Surgical instruments, sterile

EXPERIMENTAL METHOD

Preparation of Cells for Injection

1. Maintain CNS-1 cells in RPMI medium containing 0.1 mg/mL Hygromycin B and 10% fetal bovine serum.
2. When cells reach 70%–80% confluence, trypsinize them with 0.25% trypsin/EDTA for 3 min at 37°C.
3. Stop the trypsin reaction by adding RPMI containing 10% FBS.
4. Centrifuge the cells at 200*g* for 5 min at 4°C.
5. Resuspend the cells in 5 mL of Dulbecco's PBS containing 0.5% FBS.
6. Centrifuge the cells at 200*g* for 5 min at 4°C.
7. Resuspend in 1 mL of ice-cold Dulbecco's PBS containing 0.5% FCS.
8. Count the cells and adjust volume to yield a concentration of 10^7 cells/mL.
9. Place cells on ice until injection.

Preparation of a Cranial Window

The duration of the entire surgical procedure (Steps 10–21), including anesthesia and glue-drying time, is ~1 h.

10. Pretreat immunocompromised nude (nu/nu) mice (2–6 months of age) with 0.5 mg/kg buprenorphine subcutaneously, then anesthetize with 200 mg/kg Avertin (intraperitoneally) and place on a heat pad; stabilize the head in a stereotactic frame using a bite bar and earbars.
11. Make a longitudinal incision of the skin between the occiput and the forehead.
12. Remove a disk of skin from the top of the skull, and scrape off the underlying periosteum to the temporal crests.
13. Draw a 6-mm circle over the frontal and parietal regions of the skull.
14. Using a sterile high-speed drill with a burr tip, make a groove on the margin of the drawn circle.
15. Utilize repetitive gentle drilling to deepen the groove until the bone flap becomes loose.
16. Using a Malis dissector, separate the bone flap from the underlying dura mater and remove the bone. Place Gelfoam on the cut edge if needed.
17. Rinse the dura mater continuously with sterile saline.

    Continuous rinsing minimizes window dehydration and heat buildup that can damage the preparation.

Tumor Cell Injection into the Brain

18. Using a special low-volume Hamilton syringe, aseptically inject a 1-µL volume containing 1×10^4 cells into the center of the craniotomy.

    For this injection, the needle is placed 1.5 mm below the surface and then retracted back to 1.0 mm to create a small space for the cells. Over a period of 5 min, the cells are slowly injected into the frontal cortex of the brain as the needle is gradually retracted.

19. Dip the 7-mm round cover glass into sterile saline, place over the craniotomy, and secure to the surrounding bone with cyanoacrylate glue.
20. Build up a shallow well with glue around the edge to hold water for imaging using the water-immersion lens (see Fig. 1).
21. Mount a stainless steel bar with tapped screw holes onto the front of the skull using dental cement for head stabilization during imaging.
22. Maintain the animals until ready for imaging.

    See *Troubleshooting.*

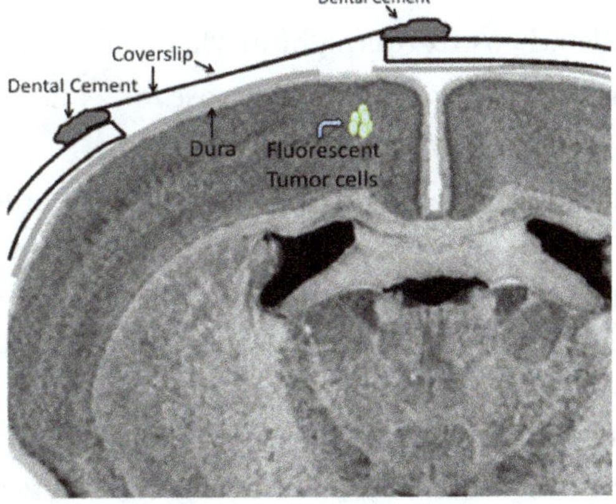

FIGURE 1. Chronic cranial window for imaging brain tumors. The diagram shows a cranial window in cross section. A craniotomy is performed and the dura (gray) is broken in a small peripheral area in order to inject brain cancer cells (green) into the cerebral cortex. The injection pipette is withdrawn, the craniotomy is filled with sterile saline, and a small coverslip is secured in place using dental cement to allow chronic optical access to the brain.

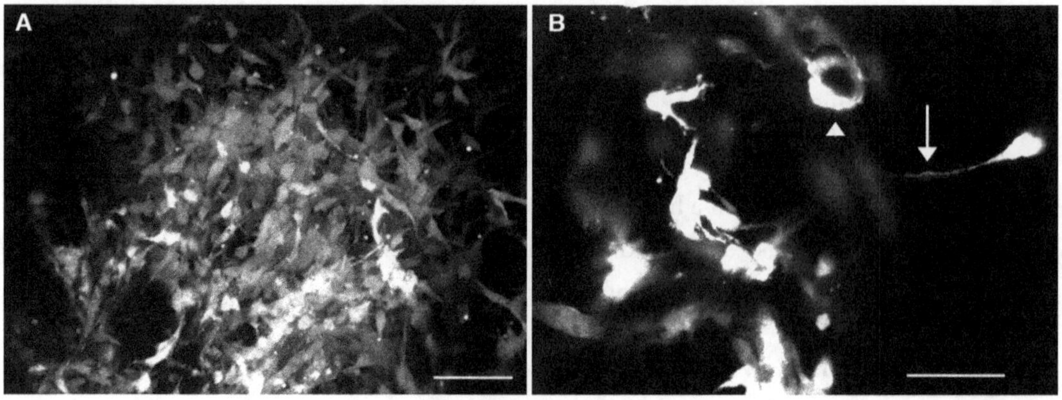

FIGURE 2. Imaging gliomas in vivo. (*A*) Two-photon image of CNS-1 Venus-expressing cells forming a tumor in the brain imaged under a cranial window. The image was taken 5 d after intracranial cell injection and window implantation and is 800 µm across. (*B*) Higher-resolution image showing individual CNS-1 cells in the brain. Notice that some cells have many small processes that invade the brain (arrow), whereas other cells wrap around blood vessels and have a doughnut-like shape (arrowhead). Scale bars, 100 µm (*A*); 20 µm (*B*).

Two-Photon Imaging

Imaging can begin as early as 1 d after injection of the CNS-1 cells.

23. Anesthetize the mouse with 200 mg/kg Avertin intraperitoneally. Place the anesthetized mouse on a heat pad.
24. To minimize motion artifacts, stabilize the head by attaching the head bar to a frame.
25. Use the imaging parameters for two-photon imaging listed in Imaging Setup. Take low-magnitude (1× digital zoom) images to reveal the organization of the tumor mass. Use the patterns of blood vessels as landmarks for alignment purposes between imaging sessions. Take higher resolution (5×–10× digital zoom) images to reveal the morphology and dynamics of individual cells within the tumor and at the tumor margins. See Figure 2 for maximum intensity projections of two-photon images of CNS-1 Venus cells.

    Imaging sessions can be repeated periodically until the window becomes cloudy or the tumor becomes too large.

Postimaging Analysis

26. After the last imaging session, perfuse the mouse intracardially with PBS (pH 7.4) followed by 4% paraformaldehyde for fixation.
27. Remove the brain from the skull and postfix overnight in 4% paraformaldehyde.
28. Cryoprotect in 10% sucrose for 24 h, followed by 30% sucrose for 24 h.
29. Cut 50-µm frozen sections from the area of the injection.
30. Mount the sections on microslides, counterstain with Prolong Gold Antifade with DAPI, and coverslip.
31. Examine sections under epifluorescence with appropriate filters for Venus and DAPI (see Imaging Setup). Take photomicrographs of the tumor with a SPOT Pursuit digital camera in conjunction with the SPOT Advanced software package.
32. Analyze the size and spread of the tumor and the number of fluorescent foci per section using NIH Image J software and compare with the results obtained with two-photon microscopy.

TROUBLESHOOTING

Problem (Step 22): Rapid tumor growth is observed.
Solution: As CNS-1 tumors enlarge, brain edema and neurological damage may lead to neurological impairments such as seizures. Animals should be closely monitored. Kill animals that display neurological impairment. Inject fewer cells if impairment occurs too rapidly.

Problem (Step 22): A cloudy cranial window results in poor imaging.
Solution: Inflammation and bleeding may result in a cloudy window that is unsuitable for imaging. Maintain sterility and minimize damage to the brain and brain dura during preparation of the cranial window. Observe the window daily. Cloudy windows may clear up on their own in a day or two. If the damage is too severe, the window will not clear up, and the animal should be killed.

DISCUSSION

Glioblastoma is an invasive brain tumor with a high mortality rate, even with treatment (King et al. 2005). The rat CNS-1 cell line is a widely used, well-characterized model of infiltrative glioma grown in nude mice. CNS-1 cells are implanted in nude mice because these mice are deficient in T lymphocytes and unable to mount an acquired response to the tumor, although their innate response is normal. Cranial windows provide long-term optical accessibility to the brain for high-resolution two-photon microscopy, creating a 3D view of the region in which CNS-1 cells are growing and spreading. With clear, intact cranial windows, tumors can be imaged many times over weeks in the same animal until the tumor grows large enough to induce neurological impairment. The injection technique controls the placement of tumor cells, reducing animal-to-animal variability due to microenvironmental differences. By incorporating other fluorophores, tumor cell interactions with neurons, glial cells, and blood vessels can be studied and manipulated. This model is not useful for understanding the process of brain metastasis, which requires tumor cells to cross the blood–brain barrier. Also, although two-photon microscopy permits imaging to a depth of 500 μm (deeper than standard one-photon imaging), many tumors grow in regions of the brain that are not optically accessible, and therefore are not amenable to the high-resolution advantages of two-photon microscopy. With this model, experiments manipulating the surrounding environment and the tumor cells will elicit a greater understanding of the molecular pathways critical to glioma growth and spread, leading to more effective therapeutics directed against this aggressive brain tumor.

Protocol B

Brain Tumor Imaging: Imaging Brain Metastasis Using a Brain-Metastasizing Breast Adenocarcinoma

To study the molecular and cellular mechanisms of brain metastasis, we have transfected MB-231BR, a brain-homing derivative of a human breast adenocarcinoma line MDA-MB-231, with the yellow fluorescent protein (YFP) variant Venus. MB-231BR selectively enters the brain after intracardiac injection into the arterial circulation, resulting in accumulation of fluorescent foci of cells in the brain that can be viewed by standard fluorescence imaging procedures. We describe how to perform the intracardiac injection and the parameters used to quantify brain metastasis in brain sections by standard one-photon fluorescence imaging. The disadvantage of this model is that the kinetics of growth over time cannot be determined in the same animal. In addition, the injection technique does not permit precise placement of tumor cells within the brain. This model is useful for determining the molecular determinants of brain tumor metastasis.

IMAGING SETUP

- An Olympus BX51 epifluorescence microscope equipped with filters for YFP (emission maximum = 528 nm) and DAPI (emission maximum = 461 nm)
- A Spot Pursuit camera
- SPOT Advanced software
- ImageJ software for producing and analyzing the fluorescent images

MATERIALS

CAUTION: See Appendix 6 for proper handling of materials marked with <!>.

Reagents

Avertin (200 mg/kg) <!>
Buprenorphine (0.5 mg/kg)
Gentamycin (0.8 mg/mL)
High-glucose Dulbecco's modified Eagle's medium (DMEM)
Immunocompromised nude (nu/nu) mice
MB-231BR cells
Paraformaldehyde <!>, 4%
Prolong Gold Antifade Reagent with DAPI (Invitrogen item P-36931)
Sucrose, 10% and 30%
Tissue culture reagents (Invitrogen; Keep all solutions sterile.)
Trypsin <!>, 0.25% with EDTA
　DMEM containing 10% FBS
　Dulbecco's PBS containing 0.5% FBS, ice cold

Equipment

Cell counter
Epifluorescent microscope (Olympus; see Imaging Setup)
Heat pads
ImageJ software (download from http://rsb.info.nih.gov/ij/download.html)
Incubator set at 37°C
Micromanipulator
Microslides and coverslips
Microtome
Syringe, 1-mL and 26 3/8-gauge needle

EXPERIMENTAL METHOD

Preparation of Cells for Injection

1. Maintain MB-231BR cells transfected with Venus in high glucose DMEM medium supplemented with 10% FBS and 0.8 mg/mL gentamycin.
2. When cells reach 70%–80% confluence, trypsinize them with 0.25% trypsin/EDTA for 3 min at 37°C.
3. Stop the trypsin reaction by adding DMEM containing 10% FBS.
4. Centrifuge the cells two times in ice-cold Dulbecco's PBS containing 0.5% FBS.
5. Count the cells and resuspend them in ice-cold Dulbecco's PBS containing 0.5% FBS at a concentration of 2×10^6 cells/mL.
6. Place cells on ice until injection.

Tumor Cell Injection by Intracardiac Injection

7. Pretreat immunocompromised nude (nu/nu) mice with 0.5 mg/kg buprenorphine, subcutaneously. Anesthetize with 200 mg/kg Avertin (intraperitoneally) and place the mice on a heat pad.
8. Place a 1-mL syringe equipped with a 26-gauge hypodermic needle in a micromanipulator to ensure stability throughout the injection.
9. Determine the position of the heart by measuring the distance from the top of the rib cage to the bottom and dividing the distance by 0.5.
10. Insert the syringe at this halfway point just to the left of the sternum, and inject 2×10^5 cells in 100 µL into the left ventricle.

    Correct placement of the needle into the left ventricle is indicated by the back pulsing of blood into the syringe throughout the injection.
11. Mice should be observed every 3 d for behavioral abnormalities indicating excessive growth of brain metastasis.

    In our experience, no behavioral abnormalities are observed for up to 6 wk after tumor injection.

Analysis of Tumor Metastasis by Fluorescence Imaging

12. After 3–6 wk postinjection, perfuse mice with 4% paraformaldehyde.
13. Remove the brains, postfix in 4% paraformaldehyde for 24 h, and cryoprotect for 24 h each in 10% sucrose, followed by 30% sucrose.

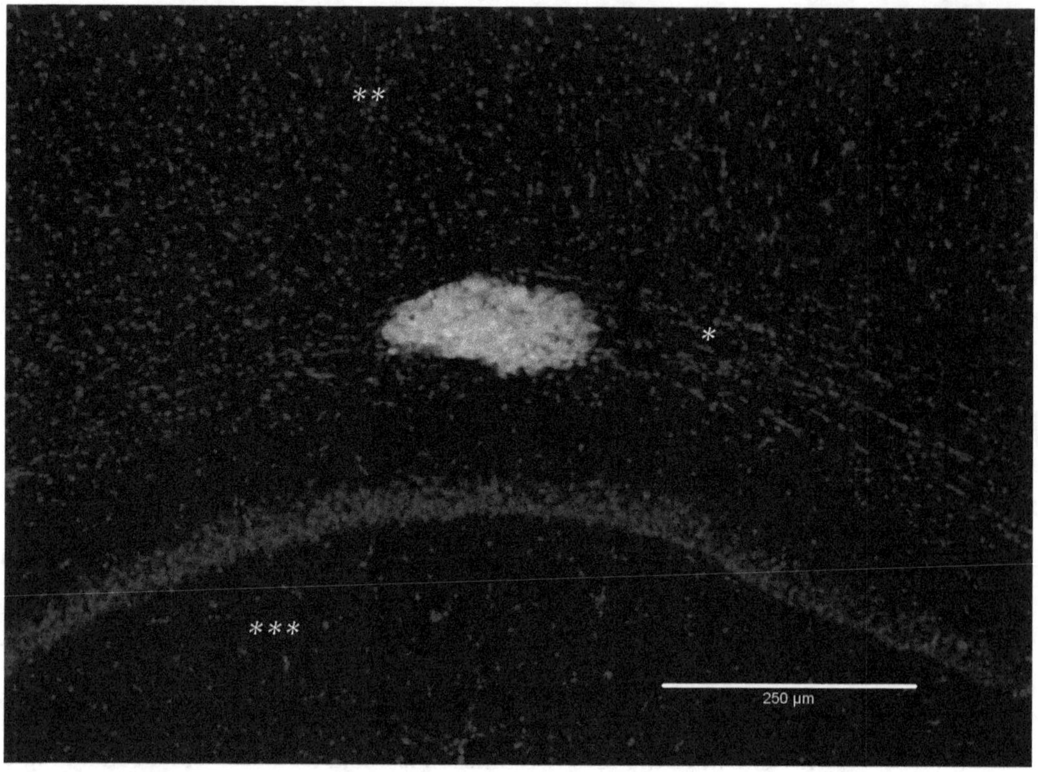

FIGURE 3. Breast tumor metastasis imaged in mouse brain. MB-231BR-Venus cells (white), injected intracardially, are visualized using epifluorescence against a background of DAPI nuclear counter stain (gray). Tumor is located in the external capsule fiber tract (*), which separates the cortex (**) from the hippocampus (***). Scale bar, 250 µm.

14. Prepare 50-µm frozen brain sections.
15. Mount every fourth section onto a microslide, counterstain with Prolong Gold Antifade with DAPI, and coverslip.
16. Examine the sections under epifluorescence with appropriate filters for Venus and DAPI. (See Imaging Setup above.)

    See *Troubleshooting*.
17. Use ImageJ or other imaging software to quantify the number of fluorescent foci in each section and to measure the size of each metastatic foci (see Fig. 3). DAPI fluorescence can be used to localize the tumor to white versus gray matter.

TROUBLESHOOTING

Problem (Step 16): No fluorescent metastatic foci are detectable.
Solution: There may be a number of reasons for no detectable metastasis. Incorrect needle placement during intracardiac injection would preclude cells from entering the arterial circulation. Make sure to observe back pulsing of blood during the injection. The optimal metastatic time point following injection can be variable. Pilot experiments should be conducted to determine the optimal time point(s) to kill the mice following injection of the tumor cells.

DISCUSSION

Here we describe a method for studying brain metastasis using a brain-homing variant of the breast adenocarcinoma MB-231. Two features of this protocol allow for consistent, measurable brain metastasis: (1) use of cells that selectively migrate to the brain and (2) intracardiac injection to deliver cells directly to the brain via the arterial circulation, bypassing the liver and lungs.

The disadvantage of this technique is that precise control of the brain region(s) to which the cells metastasize is not possible, rendering it problematic to use in conjunction with chronic cranial windows. Furthermore, the cells generally lodge deep within the brain parenchyma, where in vivo imaging with two-photon microscopy is not possible.

There are several advantages of this technique. Multistep, time-consuming staining procedures are not required, and standard imaging software can be used for rapid, subjective analysis of the experimental results.

Most importantly, this model is useful for studying the molecular pathways involved in tumor cells crossing the blood–brain barrier and unique aspects of the brain parenchyma that contribute to the size and infiltration of the metastastic foci. For example, brain-homing variants of breast cancer cell lines have been used to identify key molecular differences in brain metastasis (Palmieri et al. 2007; Stark et al. 2007) and to assess the therapeutic efficacy of drugs designed to inhibit brain metastasis in a preclinical model of metastatic cancer (Gril et al. 2008). With this model, we hope to achieve a greater understanding of the molecular pathways that facilitate tumor entry into the brain.

ACKNOWLEDGMENTS

This work was supported by a Department of Defense Era of Hope Scholar Award (to EBB), a Synergistic Idea Award (to AKM and EBB), a Burroughs Wellcome Fund Career Award in the Biomedical Sciences (to AKM), and a Pew Scholar in the Biomedical Sciences Award (to EBB).

REFERENCES

Gril B, Palmieri D, Bronder JL, Herring JM, Vega-Valle E, Feigenbaum L, Liewehr DJ, Steinberg SM, Merino MJ, Rubin SD, Steeg PS. 2008. Effect of lapatinib on the outgrowth of metastatic breast cancer cells to the brain. *J Natl Cancer Inst* **100:** 1092–1103.

Kim LS, Huang S, Lu W, Lev DC, Price JE. 2004. Vascular endothelial growth factor expression promotes the growth of breast cancer brain metastases in nude mice. *Clin Exp Metastasis* **21:** 107–118.

King GD, Curtin JF, Candolfi M, Kroeger K, Lowenstein PR, Castro MG. 2005. Gene therapy and targeted toxins for glioma. *Curr Gene Ther* **5:** 535–557.

Lin NU, Bellon JR, Winer EP. 2004. CNS metastases in breast cancer. *J Clin Oncol* **22:** 3608–3617.

Palmieri D, Bronder JL, Herring JM, Yoneda T, Weil RJ, Stark AM, Kurek R, Vega-Valle E, Feigenbaum L, Halverson D, et al. 2007. Her-2 overexpression increases the metastatic outgrowth of breast cancer cells in the brain. *Cancer Res* **67:** 4190–4198.

Stark AM, Anuszkiewicz B, Mentlein R, Yoneda T, Mehdorn HM, Held-Feindt J. 2007. Differential expression of matrix metalloproteinases in brain- and bone-seeking clones of metastatic MDA-MB-231 breast cancer cells. *J Neurooncol* **81:** 39–48.

Weil RJ, Palmieri DC, Bronder JL, Stark AM, Steeg PS. 2005. Breast cancer metastasis to the central nervous system. *Am J Pathol* **167:** 913–920.

Yoneda T, Williams PJ, Hiraga T, Niewolna M, Nishimura R. 2001. A bone-seeking clone exhibits different biological properties from the MDA-MB-231 parental human breast cancer cells and a brain-seeking clone in vivo and in vitro. *J Bone Miner Res* **16:** 1486–1495.

92 Fluorescence Molecular Tomography of Brain Tumors in Mice

Nikolaos C. Deliolanis and Vasilis Ntziachristos

Institute for Biological and Medical Imaging, Helmholtz Zentrum München and Technische Universität München, 85764 Neuherberg, Germany

ABSTRACT

In this chapter we present a procedure to tomographically reconstruct the biodistribution of fluorescence in small animals. The imaging system and equipment are described, the step-by-step image acquisition and preliminary image processing methods are presented, and the tomographic reconstruction procedure is outlined. Finally, the method is showcased by imaging the fluorescence activity of a brain tumor of a glioblastoma mouse model.

INTRODUCTION

Fluorescence molecular tomography of tissues is a method that three-dimensionally resolves fluorescence biodistribution in vivo, with applications in small animal research and pre-clinical diagnostics. The method typically combines theoretical models of photon propagation in diffusive media and boundary measurements of light propagating through tissue using multiple source–detector pairs using a mathematical inversion scheme. The tomographic performance of the new generation of free-space charge-coupled device (CCD)-based 360° projection tomographic systems is increased over the prototypes that utilized a relatively small number of source–detector pairs, by bringing fiber pairs into contact with the tissue. This new generation approach can result in a high-quality data set, as it enables high spatial sampling of photon fields propagating through tissues at any preferred combination of imaging angles.

This imaging protocol is adapted from methods presented in Deliolanis et al. (2007, 2009). The animal model details and procedures can be found in Deliolanis et al. (2008) and Wurdinger et al. (2008). Information specific to the fluorescent probes can be found in Weissleder et al. (1999) and Bremer et al. (2001). The protocol herein refers to tomographic imaging of tumors either targeted with fluorescent probes or labeled with red-shifted fluorescent proteins, such as mCherry, mRaspberry, or Katushka (Shaner et al. 2004; Shcherbo et al. 2007).

Protocol

Fluorescence Molecular Tomography of Brain Tumors in Mice

There are many alternative imaging geometries in optical tomographic experimental systems, but in general, all imaging setups consist of four subsystems: illumination, animal mount, imaging, and automation and data acquisition (i.e., electronics and computer). Generally, optical tomographic systems are either fiber-based or CCD-based systems. Fiber-based systems deliver and collect the light using fibers that physically touch the tissue surface. In CCD-based setups, light is focused in small spots on the tissue surface, and the propagating photon field at the surface is imaged with the use of a lens and a CCD camera. In this protocol we refer to CCD-based systems that work in trans-illumination (i.e., illumination and detection occur on opposite sides of the subject), while the mouse is rotated through 360° to allow photon acquisition from multiple projections. A schematic of the system is depicted in Figure 1.

IMAGING SETUP

Illumination

- Continuous wave laser (wavelength depends on fluorochrome excitation peak; see Table 1)
- Multimode optical fiber (50-µm core/125-µm cladding)
- Focusing optics (one fiber collimation lens and one focusing 50-mm lens)
- An *x–y* translation system consisting of two motorized micropositioning translation stages mounted in orthogonal directions, which is used to translate the focusing optics

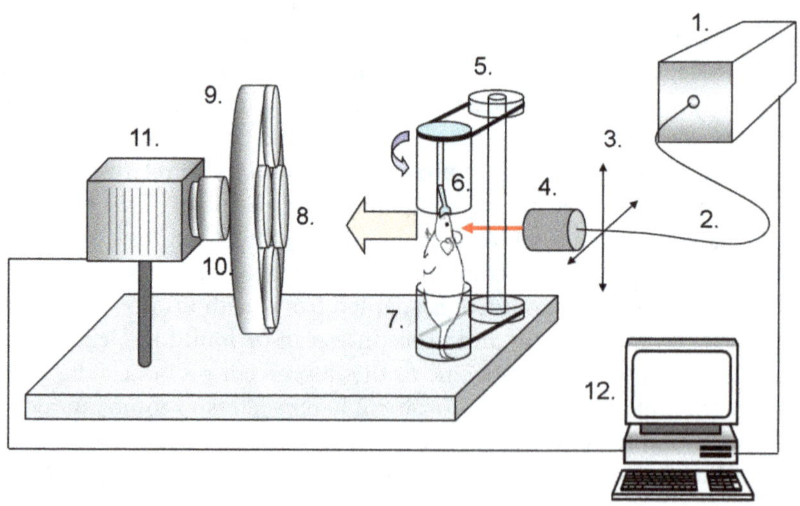

FIGURE 1. Schematic of the fluorescence tomography system. (1) Continuous-wave laser, (2) multimode optical fiber, (3) *x–y* translation system, (4) focusing optics, (5) motorized 360° rotation state, (6) snout clamp, (7) mechanical support, (8) two interference band-pass filters, (9) filter wheel, (10) photographic macrolens, (11) CCD camera, (12) computer.

TABLE 1. Excitation laser lines and filter specifications for 680 and 750 fluorescent probes and red fluorescent proteins (FPs)

	Excitation laser line (nm)	Excitation filter (nm)	Emission filter (nm)
680 probes	680	675–685	705–715
750 probes	750	745–755	795–805
Red FPs	593	588–598	615–625

Mounting

- Motorized 360° rotation stage
- Snout clamp (Harvard Apparatus) attached to the rotation stage to immobilize the head
- Mechanical support for the body also attached to the rotation stage

Imaging

- Two interference band-pass filters (10 nm at FWHM [full width at half-maximum]), one for the peak excitation and one for the peak emission wavelength. Emission filter should have a 6 OD at the excitation wavelength. The excitation wavelengths, as well as the excitation and emission band-pass filters used, depend on the fluorescent agent or protein used (see Table 1).
- Filter wheel
- Photographic macrolens (focal length ~60 mm, working distance <50 mm, f/2.8)
- High-sensitivity, low-noise, cooled CCD camera (preferably 16 bit, 1024 × 1024 pixels). The imaging resolution with this configuration is ~100 μm.

Automation and Data Acquisition

- Computer
- Motorized-rotation controllers
- CCD digital acquisition card
- Automation software

MATERIALS

CAUTION: See Appendix 6 for proper handling of materials marked with <!>.

Reagents

Fluorescent probes (e.g., ProSense, MMPSense, AngioSense, IntegriSense; available at 680 and 750 nm from VisEn Medical)

For most fluorescent probes, slowly inject 2 nmol in a volume of 150 μL via the tail vein with a 27-gauge syringe, and image the mouse within 24 h. Probe injection instructions and imaging time windows may depend on the probe, as specified in the manufacturer's documentation.

Isoflurane <!>, 1.5% in a gas mixture

Mice previously injected with fluorescent markers (see above)

Equipment

Diffuser plate or white paper, >100 g/m^2

Laser and optical system (see Imaging Setup)

Medical tape

Object for calibrating images (see Step 10)
Rubber mask for delivering anesthesia
Snout clamp and rotation stage (see Imaging Setup)

EXPERIMENTAL METHOD

1. Anesthetize the mouse with a gas vapor mixture of 1.5% isoflurane at a flow rate of 1 L/min.

   Use isoflurane because it can safely sustain a mouse under anesthesia for >30 min. On the other hand, prolonging anesthesia with injected anesthetic for >30 min may require a higher dose (which may be fatal) or reboosting doses (which is not practical).

2. Mount the mouse to the snout clamp that is attached to the rotation stage. Supply anesthesia directly into the respiratory opening with the help of a rubber mask. Tape the ears to the shoulders of the mouse with small pieces of medical tape so that they do not impede image acquisition.
3. Illuminate a single spot on the mouse's head in the area of the brain with the focused laser beam (Fig. 2A).
4. Acquire an image of the excitation photon field propagating through the head with the CCD camera using the excitation filter (Fig. 2B). For a spot intensity of 30 mW, a typical exposure time is 0.1 sec.
5. Acquire a second image of the fluorescence emanating from the brain with the fluorescence filter (Fig. 2C). Typical exposure time is 0.2–2 sec.
6. Translate the illuminating spot to a new position on the head with the use of the *x–y* translation stages. Repeat Steps 4 and 5. In total, scan the head surface above the brain area with the focused beam at 20 to 40 different positions, repeating Steps 4 and 5 for every beam position.
7. Rotate the mouse to a new angular projection. Repeat Steps 4–6 for each projection; 18 to 36 projections are sufficient.

   See Image Acquisition Considerations below.

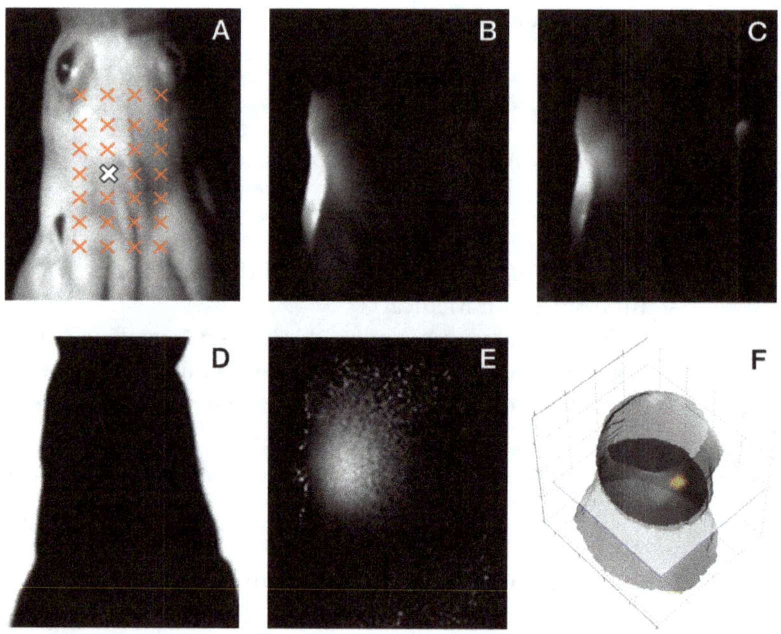

FIGURE 2. Imaging of a brain tumor. (*A*) Focused spot positions; (*B*) excitation image; (*C*) fluorescence image; (*D*) head silhouette; (*E*) normalized image; and (*F*) three-dimensional reconstruction.

8. Acquire a silhouette image of the head against a bright background (Fig. 2D). Repeat the silhouette image acquisition for all angular projections.
9. After completing image acquisition, remove the mouse from the rotation stage, discard the tape around the ears, and place the animal on a lukewarm heating pad to help it recover.
10. Place an object of a known size in the field of view, and acquire an image with the same lens setting that will be used to calibrate the images.
11. Insert a strong diffuser plate (or a piece of thick white paper, i.e., >100 g/m^2) approximately at the original mouse position, and take images of the diffusing spots for all the focused spot positions. These images will serve as identifiers of the source positions.
12. Acquire a camera background noise picture, by turning off any light source and using the same exposure time as in Step 5.

Initial Image Processing

The following procedure will result in normalized fluorescence two-dimensional (2D) images that are very useful for a fast and qualitative estimation of the brain tumor. This image processing can be performed with standard imaging programs (e.g., ImageJ), but since the bulk of the data is rather large, writing a processing routine in a programming language is more suitable.

13. Measure the mean pixel value from the background picture acquired at Step 12, and subtract this mean value from every other picture acquired.
14. Apply a median filter (usually 3 x 3) to every image to smooth out noise.
15. Threshold the image by setting to 0 the pixel values that are below a predetermined minimum value (e.g., <5).
16. Divide the fluorescence with the excitation image for each source position and projection. Then set all infinity values and undetermined values to 0. The ratio of the images normalizes the fluorescence for tissue and illumination inhomogeneities. In Figure 2B the normalized fluorescence is located at the right hemisphere, where the tumor was originally implanted.

    See *Troubleshooting*.

Image Acquisition Considerations

A. The image acquisition sequence can differ from that described in Steps 4–7 and can be implemented with a different order between filters—beam positions—angular projections. The optimum order depends on the idle time for moving between different filters, beam positions, and projections. The most time-consuming movements should be executed in an outer nested loop to minimize total acquisition time. A singular value decomposition analysis on the optimum number of scanning positions and angular positions is given in Lasser and Ntziachristos (2007).

B. The scanning pattern of the laser beam is a rectangular array, and the focused spot should never be positioned beyond the boundaries of the head or it will be directly imaged on the CCD surface and cause permanent damage.

C. Because of the varying thickness of tissue through which the light propagates (the cross section of the head is approximately round) and the inhomogeneity of tissue optical properties over different areas of the head, the photon field is differentially attenuated. Consequently, the dynamic range of the intensity of the photon fields emanating from different areas of the head can easily exceed the dynamic range of the CCD camera. In particular, light is "wave-guided" under the subcutaneous fat, while it is strongly attenuated when propagating through the center of the head (see Fig. 2B,C). Since tumor fluorescence is emanating from the center of the head, a convenient compromise is to set an exposure time such that the pixels of the image corresponding to the side boundaries of the head are slightly saturated. Additionally, the scanning pattern can be limited to 2–3 mm inside the side boundaries.

Tomographic Reconstruction

Tomographic reconstruction of the fluorescence is based on the normalized Born approximation used to predict the forward field of fluorescent photon propagation in diffuse media (Ntziachristos and Weissleder 2001). The procedure in noncontact CCD-based systems is described in Deliolanis et al. (2007, 2009). The normalized experimental data are extracted from the ratio of images as calculated in Step 16.

The image pixels are binned in larger clusters (usually 4 x 4) to form virtual detectors that are assumed to be located in the surface of the head. For the calculation of the forward model inside a volume with arbitrary shape, like the head, it is necessary to accurately determine its surface. The algorithm for surface extraction from silhouettes that are acquired as in Step 8 has been described in detail (Potmesil 1987; Lasser et al. 2008). The normalized fluorescence intensity at the surface of the head is calculated for each source–detector pair and a weight. The mouse body is discretized in a rectangular mesh of $M = 24 \times 14 \times 22$ nodes, and the normalized intensity for each source–detector pair is related to the unknown fluorescence value in each of the nodes via the normalized Born approximation. The unknown fluorescence values are then calculated by inverting the resulting weight matrix with an algebraic reconstruction technique (ART) (Kak and Slaney 1999). As an example, the three-dimensional (3D) surface rendering and the tomographic reconstruction of an mCherry-labeled glioma tumor are presented in Figure 2F.

Although the original image resolution is ~100 µm, the reconstructed image resolution is between 0.5 and 1 mm, because of strong light scattering by tissue. The sensitivity of this method is sufficient to image the signal from fluorescent probes accumulating in tissue areas at ~100 nM. The limiting factor in the sensitivity of this method is the background autofluorescence that is intrinsic in tissue, not the electronic noise.

TROUBLESHOOTING

Problem (Step 16): The normalized images appear noisy, and tumor signals do not appear in any projection.

Solution: Repeat the initial image processing procedure with increased median filter size (Step 14) and higher threshold values (Step 15).

CONCLUSION

Fluorescence molecular tomography in free-space 360° projection imaging geometry is a powerful tool for imaging the 3D distribution of fluorescence in small animal tumor models in vivo. The advantages of this particular geometrical implementation are twofold. First, it is a more natural and less stressful procedure for the mouse because there are no matching fluids and no compression, also given the ability to image the head. Second, it enables the high spatial frequency acquisition of data over multiple projections. The general aspects of this imaging and data processing protocol apply also to the future generations of hybrid FMT X-ray CT tomographic systems, in which the mouse is immobile, and the illumination and imaging systems are mounted on a gantry rotating around the mouse.

REFERENCES

Bremer C, Tung CH, Weissleder R. 2001. In vivo molecular target assessment of matrix metalloproteinase inhibition. *Nat Med* **7:** 743–748.

Deliolanis NC, Lasser T, Hyde D, Soubret A, Ripoll J, Ntziachristos V. 2007. Free-space fluorescence molecular tomography utilizing 360° geometry projections. *Opt Lett* **32:** 382–384.

Deliolanis NC, Kasmieh R, Wurdinger T, Tannous BA, Shah K, Ntziachristos V. 2008. Performance of the red-shifted fluorescent proteins in deep-tissue molecular imaging applications. *J Biomed Opt* **13:** 044008. doi: 10.1117/1.2967184.

Deliolanis NC, Dunham J, Wurdinger T, Figueiredo JL, Bakhos T, Ntziachristos V. 2009. In-vivo imaging of murine tumors using complete-angle projection fluorescence molecular tomography. *J Biomed Opt* **14:** 030509. doi: 10.1117/1.3149854.

Kak AC, Slaney M. 1999. *Principles of computerized tomographic imaging*. IEEE Press, New York.

Lasser T, Ntziachristos V. 2007. Optimization of 360° projection fluorescence molecular tomography. *Med Image Anal* **11:** 389–399.

Lasser T, Soubret A, Ripoll J, Ntziachristos V. 2008. Surface reconstruction for free-space 360° fluorescence molecular tomography and the effects of animal motion. *IEEE Trans Med Imaging* **27:** 188–194.

Ntziachristos V, Weissleder R. 2001. Experimental three-dimensional fluorescence reconstruction of diffuse media by use of a normalized Born approximation. *Opt Lett* **26:** 893–895.

Potmesil M. 1987. Generating octree models of 3D objects from their silhouettes in a sequence of images. *Comput Vis Graph Image Process* **40:** 1–29.

Shaner NC, Campbell RE, Steinbach PA, Giepmans BNG, Palmer AE, Tsien RY. 2004. Improved monomeric red, orange and yellow fluorescent proteins derived from *Discosoma* sp red fluorescent protein. *Nat Biotechnol* **22:** 1567–1572.

Shcherbo D, Merzlyak EM, Chepurnykh TV, Fradkov AF, Ermakova GV, Solovieva EA, Lukyanov KA, Bogdanova EA, Zaraisky AG, Lukyanov S, Chudakov DM. 2007. Bright far-red fluorescent protein for whole-body imaging. *Nat Methods* **4:** 741–746.

Weissleder R, Tung CH, Mahmood U, Bogdanov A. 1999. In vivo imaging of tumors with protease-activated near-infrared fluorescent probes. *Nat Biotechnol* **17:** 375–378.

Wurdinger T, Badr C, Pike L, de Kleine R, Weissleder R, Breakefield XO, Tannous BA. 2008. A secreted luciferase for ex vivo monitoring of in vivo processes. *Nat Methods* **5:** 171–173.

Appendix 1

Electromagnetic Spectrum

Marilu Hoeppner

The Electromagnetic Spectrum

The different forms of electromagnetic radiation include radio waves, infrared (IR) rays, visible light, ultraviolet (UV) light, X-rays, and γ rays. All electromagnetic rays exhibit the properties of waves, each having a characteristic wavelength (λ) and frequency (ν). Electromagnetic waves travel with the speed of light c, which in vacuum is ~3 x 10^8 m/sec. The relationship between the wavelength, the frequency, and the speed of light is given by $c = \lambda\nu$. Thus, it is possible to refer to electromagnetic radiation in terms of wavelength or frequency. Electromagnetic radiation may also be considered in terms of quanta or photons. The relationship between the energy of the electromagnetic wave E and the frequency is given by $E = h\nu$, where h is Planck's constant (h = 6.63 x 10^{-34} J sec). The higher the frequency of the electromagnetic wave (i.e., the shorter the wavelength), the greater the amount of energy carried by a photon.

Abbreviations

Photon energy: eV, electron volt (1 eV = 1.60219 x 10^{-19} joule); keV, kiloelectron volt (10^3 eV); MeV, megaelectron volt (10^6 eV).

Frequency: Hz, hertz (cycles per second); kHz, kilohertz (10^3 Hz); MHz, megahertz (10^6 Hz); GHz, gigahertz (10^9 Hz); THz, terahertz (10^{12} Hz); PHz, petahertz (10^{15} Hz); EHz, exahertz (10^{18} Hz).

Wavelength: m, meter; km, kilometer (10^3 m); Mm, megameter (10^6 m); mm, millimeter (10^{-3} m); µm, micrometer (10^{-6} m); nm, nanometer (10^{-9} m); pm, picometer (10^{-12} m); fm, femtometer (10^{-15} m).

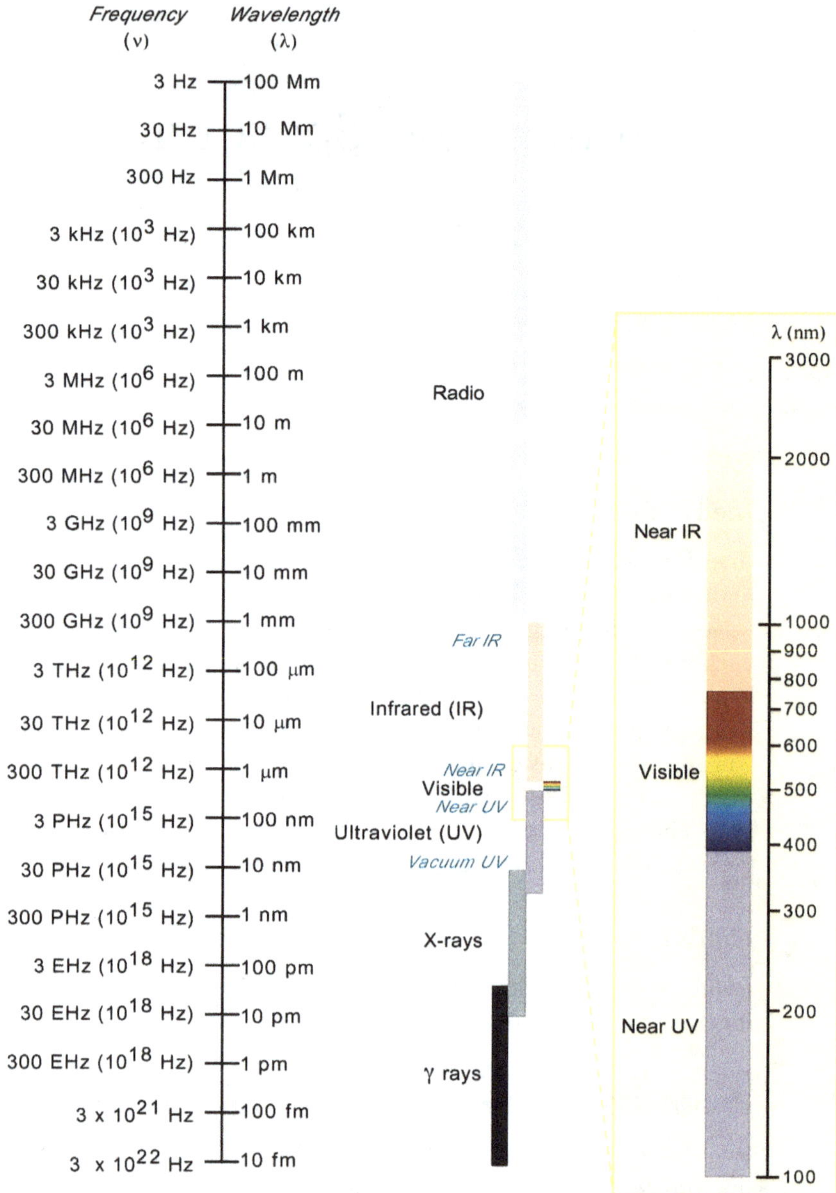

FIGURE 1. The electromagnetic spectrum. (Provided by Marilu Hoeppner.)

Appendix 2 Fluorescence Microscopy Filters and Excitation/Emission Spectra

FILTERS USED IN FLUORESCENCE MICROSCOPY

The primary filtering element in the epifluorescence microscope is the set of three filters housed in the fluorescence **filter cube**, also called the **filter block**: the **excitation filter**, the **emission filter**, and the **dichroic beam splitter**. A typical filter cube is illustrated schematically in Figure 1.

- The **excitation filter** (also called the **exciter**) is a color filter that transmits only those wavelengths of the illumination light that efficiently excite a specific dye. Common filter blocks are named after the type of excitation filter: UV or U (ultraviolet) for exciting DAPI, Indo-1, etc.; B (blue) for exciting FITC; and G (green) for exciting TRITC, Texas Red, etc. Although short-pass filter designs were used in the past, band-pass filter designs are now used.
- The **emission filter** (also called the **barrier filter** or **emitter**) is a color filter that attenuates all of the light transmitted by the excitation filter and very efficiently transmits any fluorescence emitted by the specimen. This light is always of longer wavelength (more to the red) than the excitation color.

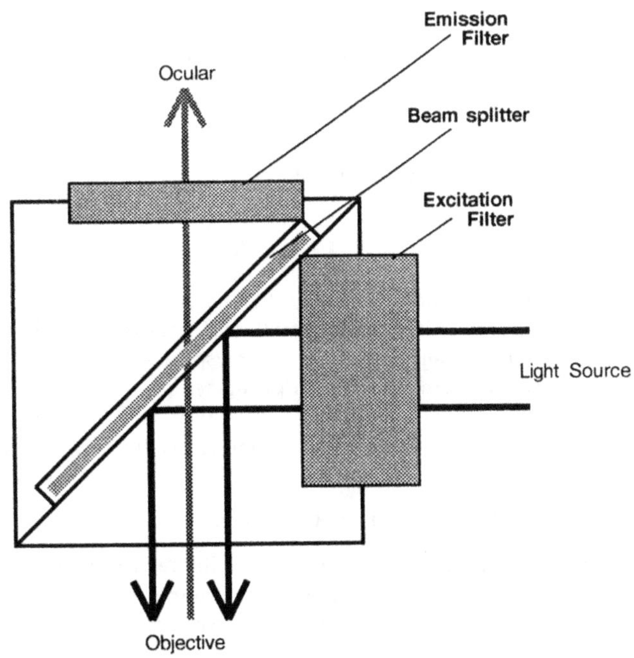

FIGURE 1. Schematic of a fluorescence filter cube.

These can be either band-pass filters or long-pass filters. Common barrier filter colors are blue or pale-yellow in the U-block; green or deep-yellow in the B-block; and orange or red in the G-block.

- The **dichroic beam splitter** (also called the **dichroic mirror** or **dichromatic beam splitter**) is a thin piece of specially coated glass (the substrate) set at an angle 45° to the optical path of the microscope. This coating has the unique ability to reflect one color (the excitation light) but transmit another color (the emitted fluorescence). Current dichroic beam splitters achieve this with great efficiency—that is, with >90% reflectivity of the excitation along with ~90% transmission of the emission. This is a great improvement over the traditional gray half-silvered mirror, which reflects only 50% and transmits only 50%, giving only ~25% efficiency.

Most microscopes have a slider or turret that can hold from two to four individual filter cubes. It must be noted that the filters in each cube are a matched set, and mixing filters and beam splitters should be avoided unless the complete spectral characteristics of each filter component are known.

Other optical filters can also be found in fluorescence microscopes.

- A heat filter (also called a hot mirror) is incorporated into the illuminator collector optics of most but not all microscopes. It attenuates infrared light (typically wavelengths >800 nm) but transmits most of the visible light.
- Neutral-density filters, usually housed in a filter slider or filter wheel between the collector and the aperture diaphragm, are used to control the intensity of illumination.
- Filters used for techniques other than fluorescence, such as color filters for transmitted-light microscopy and linear polarizing filters for polarized-light microscopy, are sometimes installed. (Information courtesy of Chroma Technology Corp.)

REFERENCES

Bieber F. 1994. Microscope and image analysis. In *Current protocols in human genetics* (ed. Dracopoli NC, et al.), p. 4.4.5. John Wiley, New York.

Spector DL, Goldman RD, Leinwand LA, eds. 1998. Microscopy: Lenses, filters, and emission/excitation spectra. In *Cells: A laboratory manual*, Vol. 3. *Subcellular localization of genes and their products*, pp. A3.1–A3.10. Cold Spring Harbor Laboratory Press, Cold Spring Harbor, NY.

WWW RESOURCES: FILTER SPECIFICATIONS

Filter Sets

http://chroma.com/products/catalog/
http://www.olympusmicro.com/primer/java/fluorescence/matchingfilters/index.html
https://www.micro-shop.zeiss.com/us/us_en/spektral.php?f=fi
http://www.leica-microsystems.com/products/light-microscopes/accessories/filtercubes-and-fluochromes/find-your-filtercubes/
http://www.microscopyu.com/articles/fluorescence/filtercubes/filterindex.html

Useful Tools for Matching Spectra with Filter Blocks

http://www.olympusmicro.com/primer/java/fluorescence/matchingfilters/index.html
http://www.invitrogen.com/site/us/en/home/support/Research-Tools/Fluorescence-SpectraViewer.html
https://www.micro-shop.zeiss.com/us/us_en/spektral.php?f=fa

Two-Photon Action Cross Sections

http://www.drbio.cornell.edu/cross_sections.html

TABLE 1. Excitation (Exc) and emission (Em) maxima for typical fluorochromes

	Exc (nm)	Em (nm)		Exc (nm)	Em (nm)
4-methylumbelliferone	360	450	DiI	557	565
6-amino-quinoline	360	443	DiO	551	569
7-amino-actinomycin D	555	655	DsRed	558	583
Acid Fuchsin	540	630	DTAF	495	528
Acridine orange + DNA	502	526	dTomato	554	581
Acridine orange + RNA	460	650	Ethidium bromide + DNA	510	595
Acridine yellow	470	550	Euchrysin	430	540
Acriflavin-Feulgen	480	550–600	Evans Blue	550	611
Alexa Fluor 350	346	442	eYFP	514	527
Alexa Fluor 405	401	421	Feulgen	480	560
Alexa Fluor 430	433	541	FITC	495	517
Alexa Fluor 488	496	519	Fluo-3	506	526
Alexa Fluor 532	532	553	Fluo-4	494	516
Alexa Fluor 546	556	573	Fluorescein	494	521
Alexa Fluor 555	555	565	Fluoro Gold	350–395	530–600
Alexa Fluor 568	578	603	FluoSpheres (blue)	360	415
Alexa Fluor 594	590	617	FluoSpheres (crimson)	625	645
Alexa Fluor 610	612	628	FluoSpheres (dark red)	650	690
Alexa Fluor 633	632	647	FluoSpheres (red)	580	605
Alexa Fluor 635	633	647	FluoSpheres (yellow-green)	490	515
Alexa Fluor 647	650	665	FM1-43	479	598
Alexa Fluor 660	663	690	FM4-64	508	751
Alexa Fluor 680	679	702	Fura Red	420/480	640/665
Alexa Fluor 700	702	723	Fura-2	340/380	500/530
Alexa Fluor 750	749	775	GFP (enhanced)	489	509
Alexa Fluor 790	784	814	Hoechst 33258, 33342	352	461
Alizarin complex	530–560	580	Indo-1	328/352	403/520
Allphycocyanin	630	660	Lissamine Rhodamine B	575	595
AMCA	345	425	Lucifer Yellow	428	540
Amino-methylcumarin	354	441	Magdala Red	524	600
Auramine	460	550	Mag-indo-1	331	415
BCECF	430/480	520	mBanana	540	553
Berberine sulfate	430	550	mCFP	433	475
BFP (Blue Fluorescent Protein)	382	448	mCherry	587	610
BOBO-1	462	481	Merocyanine	555	578
BOBO-3	570	602	mHoneydew	487/504	537/562
BODIPY	503	512	Mithramycin	420	575
BODIPY 581/591 phalloidin	584	592	mOrange	548	562
Calcein	495	500–550	mPlum	590	649
Calcein Blue	375	420–450	mRaspberry	598	625
Calcium Crimson	590	615	mRFP1	584	607
Calcium Green	505	535	mStrawberry	574	596
Calcium Orange	549	576	mTangerine	568	585
Cascade Blue	376/399	423	NBD–amine	460–485	534–542
Catecholamine	410	470	NBD–chloride	480	510–545
Cerulean	433	475	Nile Red	559	637
Chromomycin	436–460	470	Olivomycin	350–480	470–630
Coriphosphine	460	575	Oregon Green 488	498	526
Coumarin maleimide	385	471	Pararosaniline-Feulgen	560	625
Cyanine Cy3.5	578	691	PBFI	340/380	420
Cyanine Cy2	489	505	Phosphine	465	565
Cyanine Cy3	549	562	Phycoerythrin	565	575
Cyanine Cy5	640	664	POPO-1	434	456
Cyanine Cy5.5	673	692	POPO-3	534	570
Dansyl chloride	380	475	Primuline	410	550
DAPI + DNA	359	461	Propidium iodide	536	617
DiBAC4	439	516	Pyrene	343	380–400
DIDS thiocyanatostilbene	342	418	Pyronine	410	540
Di-2 ANEPEQ (JPW 1114)	517	721	Quin-2	340/365	490
Di-8 ANEPPS	469	630	Quinacrine mustard	440	510

(Continued)

TABLE 1. *(Continued)*

	Exc (nm)	Em (nm)		Exc (nm)	Em (nm)
Resorufin	571	585	Tetracycline	390	560
Rhodamine	551	573	Tetramethylrhodamine	540	566
Rhodamine 123	507	529	Texas red	595	620
Rhodol green	500	525	Thiazine red	510	580
R-phycoerythrin	565	578	Thioflavin 5	430	550
SBFI	340	555	TOTO-1 + DNA	509	533
SITA	365	460	TOTO-3 + DNA	642	661
SNARF	563	639	TRITC	557	580
Stilbene	335	440	XRITC	582	601
Sulfaflavin	380–470	470–580	Xylenol orange	377	610
Sulforhodamine 101	579	594	YOYO-1	491	509
tdTomato	554	581	YOYO-3	612	631

Provided by Adam Packer, Columbia University.

Appendix 3 Safe Operation of a Fluorescence Microscope

George McNamara

Fluorescence microscopes are used by thousands of researchers every day with few publicized reports of major accidents. This may be a matter of luck! Here we highlight several safety issues.

The standard illuminator for fluorescence microscopy is the high-pressure mercury arc lamp (sometimes called an Hg bulb or Hg burner), which can explode, spraying shards of glass. The glass can ruin the mirror and collector lens and injure people in the vicinity. Explosions typically occur either during startup or when someone is trying to change the bulb position by adjusting one or more controls on the lamphouse. Safety recommendations include the following.

1. New users should be trained by an experienced user in the laboratory or by the microscope sales representative.
2. Never operate an arc lamp without its housing.
3. If the lamphouse does not incorporate an explosion trap, then replace it with one that does.

A famous cartoon in many laser laboratories shows a scientist in a lab coat with an eye patch and the caption, "Do not look at laser with remaining eye." Commercial laser microscope systems should have one or more safety interlocks to prevent laser light from illuminating the specimen if the condenser arm is raised back. This is especially a concern for total internal reflection fluorescence (TIRF) microscopes, whose safety interlocks (if ever present) can be defeated by the user to enable shining the light up to the ceiling for inverted systems or down to the microscope stand on upright systems (if the condenser has been removed). Illuminating the ceiling may facilitate visualizing that the TIRF critical angle has been exceeded (the illumination disappears because it is trapped in the cover glass), but it is a major safety issue. Standard confocal and TIRF microscope lasers range in power from a modest 1 mW to a fairly powerful 100 mW up to >3 W for the average power of a multiphoton laser. Almost all multiphoton lasers emit ~100 femtosecond (100×10^{-15} sec) pulses at an 80-MHz pulse frequency (80×10^{6} times per second). The total pulse time is thus 8×10^{-6} sec, and the peak power is the product of the inverse of this number times the average power. For 1 W, the peak power is 125,000 W, which is significantly greater than the threshold required for permanent eye damage. When operating any high-power laser (i.e., >10 mW for a CW laser and for ultrafast lasers of any average power), observe the following.

1. Do not remove the safety interlocks.
2. Do not insert reflective objects into the laser light path during operation—especially metal or glass at the output of the objective lens.
3. Operate the laser behind blackout curtains.
4. Do not design the optical light path such that anyone's eyes are going to be at laser output height.

This latter caution is especially important for multiphoton laser systems in which the laser beam is guided around the optical table in an open-path design. Redesign these systems so that they are enclosed.

Most modern arc lamps have the high-voltage starter unit integrated into the lamphouse. With many older arc lamp systems, however, the starter unit is in the power supply. Starting an older unit requires sending an electromagnetic pulse (EMP) down the cable from the power supply to the lamphouse. This EMP can cross over from its cable to other electrical cables, especially if they are aligned in parallel to the arc lamp cable. The EMP can destroy electronics attached to the other cables, such as the computer, filter wheel and its controller, or other peripherals attached to your instrument. Therefore, do not bundle the arc lamp cable with other cables. In addition, arc lamps should not be attached to power line conditioners or to standard uninterruptible power supplies. The arc lamp power supply should not have an automatic restart, because power outages (especially multiple cycles) can trigger a lamp explosion. There should be no cables, paper notes, or walls near an arc lamp and there should be free air circulation under and above the lamp. The use of a liquid light guide or fiber optic is a good investment (see below).

The newest generation of fluorescence illuminators may be safer, as well as having longer lasting bulbs and (hopefully) much more stable light output. Interesting designs include LED illuminators such as the Colibri (Zeiss) or precisExcite (CoolLED), and metal halide lamps such as the X-Cite 120 (EXFO), PhotoFluor II (Chroma Technology or 89North.com), or Lambda XL plasma lamp (Sutter Instruments). Xenon arc lamps, such as the DG-4 or Lambda LS300 (Sutter Instruments), may be preferable because they do not use high pressure, may have more favorable spectral output, and may have longer bulb lifetimes than do mercury arc lamps. All lamps can—and ideally should—be attached to the microscope by a liquid light guide or fiber-optic scrambler, which scrambles the lamp image generating much more uniform illumination than is provided by a standard arc lamp attached directly to the microscope. Protect the light guide or fiber-optic scrambler from heat by including IR (infrared)-cutoff filters in the lamphouse, and protect it from users, because a kink could ruin a $1000 light guide. Each liquid light guide should be regarded as a consumable, because even if well maintained, it will become dimmer over time (1–3 yr).

Many laboratories will keep their standard mercury arc lamp fluorescence illuminators for a long time. The standard 100-W Hg arc lamp bulb (~$125/bulb) is rated for 200–300 h. In our laboratory, the first user of the day turns on the lamp and the last user or last person to leave the laboratory turns off the arc lamp at the end of the workday. When using our Zeiss FluoArc lamphouse/power supplies, we turn the power down to 20% between sessions. When replacing a bulb, wear gloves to prevent finger oils from contaminating the glass envelope, and gently rub the glass envelope with an ethanol-moistened Kimwipe to remove any chemical residues picked up during manufacture or in storage (tip from J. Zhang, Leica Microsystems, via a Leica customer). This cleaning and handling procedure can triple the useful life of the Hg bulb.

Appendix 4 Microscope Objective Lenses

Adapted from Inoué 1986 by A. Packer (Columbia University)

CARE AND CLEANING OF OPTICAL EQUIPMENT

Purpose of Cleanliness

Microscopes provide optimal images only if lens surfaces are kept clean. Cleanliness requirements are most stringent for polarizing and interference microscopes, but a fingerprint on any lens of a microscope can spoil the image. Similarly, it is important to use clean, unscratched, high-quality slides and coverslips and to keep fingerprints and all other dirt from their surfaces.

How to Clean an Optical Surface

1. Use only high-quality lens paper.
2. Do not touch a lens surface except as a last resort. Dust can often be removed from the surface of a lens by using a small bulb to puff air lightly over the surface. Do not try to blow off the surface with your mouth as you are likely to deposit small droplets of spit. Never rub a lens with lens paper or other items such as Q-tips, commercial facial tissue, or Kimwipes. Small dust particles are always present in the air and, if rubbed into a microscope lens, these dust particles can cause microscopic scratches in the lenses and the image will be degraded. Do not clean lenses with commercial facial tissue or bathroom tissue, which contain diatom frustules (glass) as filler. Lens coatings must be treated with respect.
3. If deposits have dried onto the surface of the lens, first attempt to clean the lens with distilled water by placing a drop on the optical surface and sliding a pristine piece of high-quality lens paper across the surface. If repeated attempts of this procedure do not produce a clean optical surface, then try alcohol as an alternative solvent. Be particularly cautious with filters as they are coated with antireflective coatings—read the cleaning instructions that come with the filter (e.g., http://chroma.com/support/filter-support).
4. Inspect the lens using an inverted ocular from the microscope as a magnifier. Observe the objective lens in the reflected light from a lamp (generally in the ceiling). To observe microscopic imperfections and dirt, place the lens to be inspected and cleaned under a dissecting microscope and focus on the lens. The lens will need to be illuminated from above.

Keeping Lenses Clean

1. Only use oil on lenses designed to be used with oil, as it can be time-consuming to clean a lens that has accidentally been immersed in oil. Clean oil off a lens after each session.

2. Keep objectives that are not in use in the screw-top containers in which they came. Always inspect lens surfaces with an inverted ocular before putting them away.
3. Some microscopists find it useful not to clean the lens between slides if they are continually using an oil-immersion lens with a series of slides. Instead, wipe off excess oil (especially important in inverted microscopes) with lens paper and place new slide with a bit of new oil onto the old oil, being careful to avoid air bubbles.
4. If using seawater, be especially careful to remove it from all metallic and lens surfaces as it will corrode metals and etch away the antireflective coating of some lenses.
5. Keep condensers, compensators, etc., in boxes, plastic "baggies," or Petri dishes until used.
6. Prohibit all smoking in microscopy laboratories because tobacco tar accumulates on lens surfaces and reduces their efficiency.

Personal Safety

See Appendix 3 for additional safety information.

1. Never open an ether bottle if there is a flame in the room.
2. Observe the no smoking regulation.
3. Avoid damage to the retinas of your eyes by protecting them from unfiltered arc sources (e.g., mercury, xenon), strobe lamps, and especially lasers.
4. If a high-pressure mercury arc lamp should explode or even stop working, *evacuate the room without delay.* If possible, open a window as you exit and then notify your institution's safety office.

REFERENCES

Inoué S. 1986. *Video microscopy.* Plenum Press, New York.

Spector DL, Goldman RD, Leinwand LA, eds. 1998. Microscopy: Lenses, filters, and emission/excitation spectra. In *Cells: A laboratory manual,* Vol. 3. *Subcellular localization of genes and their products,* pp. A3.1–A3.10. Cold Spring Harbor Laboratory Press, Cold Spring Harbor, NY.

WWW RESOURCES: OBJECTIVE LENSES

http://www.olympusamerica.com/seg_section/uis2/seg_uis2.asp The Olympus website is a searchable database of their complete line of objectives. Suitable imaging modalities are suggested for each objective, and a resource table and a transmittance curve are provided for each objective. A brochure in pdf format with a comprehensive resource table can be accessed from the website.

https://www.micro-shop.zeiss.com/us/us_en/objektive.php?cp_sid= The Carl Zeiss website is a searchable database of their full line of objectives. Search returns are flexible, allowing comparisons among user-selected objectives and providing detailed resource information including transmittance curves.

http://www.leica-microsystems.com/products/light-microscopes/accessories/objectives/ The Leica Microsystems website is a searchable database of their complete line of objectives. Detailed resource information is provided for each objective.

http://www.nikoninstruments.com/Products/Optics-Objectives The Nikon Instruments website provides information on their complete line of objectives, suggested uses, a selection guide, and a brochure in pdf format that contains a comprehensive resource table.

Appendix 5 Glossary of Imaging Terms

Absorption: The transfer of energy from light to the material it is passing through, generally caused by its conversion to other energy forms (typically heat). At the molecular level, absorption is the process by which a quantum of energy is transferred from the radiation field to the bound electrons of the molecule. It is the cause of the attenuation of the light beam. In general, absorption is a resonant process, in which the frequency of the light must match a difference frequency of oscillation between the initial and final quantum states of the molecule. In the radiation field, each energy quantum is carried by a photon, one of which is annihilated for every absorption event. *See also* Attenuation, Extinction coefficient, Multiphoton absorption, Photon.

Absorption spectrum: A graph showing the frequency dependence of the efficiency or rate of absorption of light in a material, on a relative or per-molecule basis. Because photon energy is proportional to frequency (given by Planck's formula; $E = h\nu$), and frequency is inversely proportional to wavelength ($\nu = c/\lambda$, in vacuo), the graph may be plotted versus frequency (Hz), energy (eV), wavelength (nm), or wave number ($1/\lambda$). In biological applications, by far the most common plot is extinction coefficient versus wavelength. *See also* Emission spectrum, Excitation spectrum.

Access resistance (or series resistance): The electrical resistance (usually in megaohms) of the connection between a measurement device and a biological cell. Typically, the device is a micropipette electrode.

Achromatic: Free of color; absence of primary chromatic aberration. In a simple lens, or singlet, the focal length and magnification vary slightly with wavelength, mainly because of slight variation in refractive index with wavelength. This leads to color-dependent sharpness of focus or color-dependent magnification in the image, known as longitudinal and lateral chromatic aberration, respectively. An Achromat is a multi-element lens that is designed to have a null point in chromatic aberration in the mid-visible spectrum and improved performance across the spectrum. An Apochromat is more broadly corrected, with significantly improved performance over an achromat.

Acousto-optic deflector (AOD): An electro-optic device used as a laser-scanning device. An AOD is composed of a birefringent crystal, to which is coupled a piezoelectric transducer. Driving the transducer with a radio frequency electronic signal produces an acoustic wave in the crystal that causes it to act as a diffraction grating. The diffraction angle is dependent on the frequency of the acoustic wave, which can be changed within a few microseconds depending on the size of the AOD aperture. *See* Chapter 53, this volume.

Acousto-optic tunable filter (AOTF): An electro-optic device used as a tunable band-pass filter. An AOTF is composed of a birefringent crystal to which is coupled one or more piezoelectric transducers. Driving the transducers with a radio frequency electronic signal produces an acoustic wave train in the crystal that causes it to act as a diffraction grating with a band-pass characteristic. The center wavelength of the filter can be shifted (on the microsecond timescale) over much of the visible spectrum by variation of the radio frequency in the 40–68-MHz range. The bandwidth of the filter can be adjusted electronically in advanced designs.

Afocal: An optical system in which two lenses are positioned on a common axis with a separation equal to the sum of the two focal lengths. In this configuration, entering parallel rays are brought to a focus by the first lens, and then recollimated by the second. The combination therefore has no finite focal length, in the usual sense. However, for an object located in the front focal plane of one lens, a real image is formed in the back focal plane of the other.

Airy disk: In an aberration-free circularly symmetric optical system, the central bright peak in the in-focus image of a point source of light. The central zone of the Airy pattern.

Amplifier: An electronic or optical device that actively enlarges or strengthens a signal without significant distortion. Power amplifiers drive output devices such as motors or loudspeakers; isolation amplifiers buffer a weak signal source against drainage of power by a measuring device. The operational amplifier (Op-Amp) is an integrated differential amplifier with high gain and high input impedance. Op-Amps used with simple feedback networks form the basis for many instrumentation systems such as voltage and current clamps.

Amplitude: The strength of a wave field at a particular location and instant. For light, which is an electromagnetic wave, amplitude usually refers to the electric field strength of the wave. Because this is a vector quantity, the amplitude can be resolved into directional, or polarized, components. In general, the intensity of light is proportional to the (time-averaged) square of the amplitude.

Analog: A device or record in which a signal is represented by a proportional voltage, current, brightness, magnetization, or other physical variable. A continuous (nondigital) representation.

Analog-to-digital (A/D) converter: A device that converts an analog signal, that is, a signal in the form of a continuous variable voltage or current, to a digital signal, in the form of bits.

Antireflection coating: A thin layer of material applied to a lens surface to reduce reflected light. Ideally, the index of refraction of the film should be equal to the geometric mean of the refractive indices of the materials on either side of the coating. The ideal thickness for a single layer coating is one-quarter of the wavelength at which reflectance is to be minimized. For the most common case (glass–air interface), magnesium fluoride (MgF_2) is used.

Aperture: A measure of the angular range over which an optical system can accept entering rays. In photography, aperture is specified by f-number (f/#). In microscopy, it is specified by numerical aperture (NA). For an objective lens, NA = $n \sin \theta_{max}$, where n is the refractive index of the specimen and immersion medium, and θ_{max} is the largest angle at which light may enter the objective. The illuminating numerical aperture (INA) is defined similarly, for rays exiting the condenser into the specimen.

Aperture stop: An iris or circular mask at a location in an optical system where the diameter of the iris sets the maximum angle at which light may enter or leave the objective. *See also* Field stop.

Aplanatic: A lens system that is free from spherical aberration and coma.

Arc lamp: A light source in which emission occurs from an ionized plasma in an electric discharge between two electrodes. The electrodes are contained in a glass or fused quartz bulb containing a specific gas or vapor. For fluorescence microscopy, intense mercury-vapor and xenon lamps are most common. The 100-W mercury direct-current short arc is the brightest available nonlaser light source that is commonly available for a microscope.

Attenuation: The gradual loss of intensity of a light beam as it passes through a material. Whereas absorption relates to the process of transferring energy from the light to the material, attenuation is the resultant effect on the light beam itself. In tomographic techniques the beam attenuation at multiple orientations through the specimen is the key measurement allowing reconstruction of the sample. *See also* Absorption.

Axial resolution: The sharpness with which an optical system can discriminate between features that are superimposed in an image (i.e., that differ only in best focus setting). Axial resolution is related to depth of field and to the degree to which out-of-focus features are blurred or filtered out of the image. Strictly, axial resolution is defined by the shape of the optical transfer function.

Back focal plane (of an objective): The distance from the last optical element (of an objective) to the rear focal point.

Band-pass filter: A filter with high transmissivity in a particular wavelength interval and high blocking power outside of this range.

Beam: (1) A bundle of light rays that may be parallel, converging, or diverging. (2) A concentrated, unidirectional stream of particles. (3) A concentrated, unidirectional flow of electromagnetic waves.

Beam expander: A system of optical components designed to increase the diameter of a laser beam. Usually an afocal system composed of a low-power microscope objective used in reverse plus a conventional achromatic collimating lens.

Beam splitter: An optical element for dividing a beam into two or more separate beams. Usually, a beam splitter is a semitransparent plate on which an incident beam is partially reflected and partially transmitted. Beam splitters may be polarization or wavelength selective. *See also* Dichroism.

Binning: Detector readout and digitization process in which pixels are grouped prior to measuring signal level. For example, a 512 x 512-pixel camera array may be read out and digitized as a 256 x 256-pixel image by measuring the total accumulated charge in contiguous 2 x 2-pixel squares. Binning increases readout speed and (under some circumstances) signal-to-noise ratio at a cost of reduced resolution.

Bioluminescence: Light emission from certain living organisms caused by the enzyme-catalyzed oxidation of an organic substrate. In general, the enzyme is known as a luciferase, and the substrate is a luciferin. The catalyzed reaction between oxygen and luciferin leaves the oxidized molecule in an electronic excited state from which emission of a photon occurs.

Bit: A contraction of binary digit; the fundamental unit of digital computing. A bit is either 1 or 0, expressing the binary state of on or off, true or false, etc.

Blaze: A method of cutting the parallel grooves in a diffraction grating such that the spectrally dispersed light is concentrated in one diffraction order. In a blazed grating, the individual grooves are formed with flat smooth faces inclined to the surface by an angle known as the groove angle.

Birefringence: The optical property resulting from anisotropy of the refractive index in a material. In a birefringent medium, the effective refractive index depends on the direction of travel of a light wave and on its polarization. Many biological materials, crystals, liquid crystals, and synthetic polymers are birefringent (e.g., mitotic spindle, collagen fibrils, calcite, quartz, asbestos, cellophane, molded polystyrene). A thin flake of birefringent material placed between crossed polarizers will show brightness and possibly color variation as the flake is rotated relative to the polarizer axes. In crystals, birefringence gives rise to double refraction, in which a beam of incident light is separated into two beams of orthogonal polarization.

Bright field: Standard mode of transmitted-light microscopy in which the specimen is evenly back-illuminated through a condenser. If the specimen is pigmented or stained, or otherwise attenuates light, it appears in color or dark against a light background. In comparison, phase-contrast optics is used to view specimens where there is little or no attenuation of transmitted light.

Broadband: Indicating a capability to deal with a relatively wide spectral bandwidth.

Calcium flux: The net rate of calcium ion (Ca^{2+}) permeation of a membrane. Normally, this occurs through voltage-gated or receptor-type calcium channels.

Caged compound: A bioactive molecule or tracer that has been rendered inactive by chemical coupling to a light-sensitive residue. Excitation of the residue (usually by exposure to ultraviolet light) results in a photochemical reaction that regenerates the active form, in situ. Calcium ions, neurotransmitters, ATP, inositol phosphates, fluorescent dyes, and transcription factors have been caged and used in biological experiments.

Capacitance. *See* Membrane capacitance.

CARS: *See* Coherent anti-Stokes Raman scattering.

CCD: *See* Charge-coupled device.

Centration: The precision with which two rotationally symmetric optical elements are aligned on a common axis.

Charge-coupled device (CCD): A light-sensitive semiconductor detector array in the form of an integrated circuit that forms the basis for most modern video cameras and precision electronic imagers. In a CCD, each light-sensitive element is a capacitor, which stores a displaced electron for each incident photon that is absorbed within the confines of that element. The stored charge pattern in a CCD array is a pixelated linear representation of the image irradiance. Under control of a clock signal, the rows of the array can be read out in order through a charge-sensitive amplifier followed by an A/D converter.

Chromatic aberration: The lens aberration resulting from the normal increase in refractive index of all common lens materials toward the blue end of the spectrum. Change in best focus setting with color is known as longitudinal (axial) chromatic aberration. The change in image size with color is known as lateral (transverse) chromatic aberration, lateral color, or chromatic difference of magnification. Chromatic aberration gives rise to colored fringes or halos around points and edges in the image. *See also* Achromatic.

Coherence: The time-averaged degree of similarity in amplitude variation and relative phase between two locations in a wave field, or at a single location for a fixed time difference. In a highly coherent light source, such as a laser, all of the waves exit in phase. Temporal coherence, also known as longitudinal coherence, is a measure of the monochromaticity of the light source (i.e., the regularity of the wave train); white light has low time coherence, whereas laser sources usually have high time coherence. Spatial coherence, also known as transverse coherence, is a measure of the regularity in wave-front shape and is related to the effective size of the light source as seen from the illuminated field; single point sources necessarily produce spherical waves that have perfect spatial coherence (even if not monochromatic). A classic light source, such as a lamp, is composed of a large number of independent (incoherent) point sources spread out over a finite surface or volume. Therefore, waves exiting the lamp are highly irregular and variable. At a great distance from the lamp, however, the regularity becomes significant. Degree of coherence is usually measured by use of an interferometer. Highly coherent light will form interference fringes of high contrast. Most lasers produce light that is both temporally and spatially coherent to a high degree.

Coherent anti-Stokes Raman scattering (CARS): *See* Chapter 48, Book 1, for a complete description.

Collimation: (1) The process of aligning the optical axes of optical systems to the reference mechanical axes or surfaces of an instrument. (2) The adjustment of two or more optical axes with respect to each other. (3) The process by which a divergent beam of radiation or particles is converted into a parallel beam. Collimation of an incoherent light source such as a lamp involves focusing an image of the lamp onto a small iris or pinhole placed at the focus of a lens. Spherical waves emanating from the other side of the pinhole are converted into plane waves (collimated rays) by the lens. Effectively, a plane-wave component of the lamp emission is selected by blocking out much of the full lamp output. The collimated light has increased transverse coherence.

Command voltage: The input signal to a voltage- or current-clamp amplifier system.

Condenser: A single positive lens or group of lenses used in a projection system to collect light from a source and cause it to evenly illuminate the object to be projected.

Confocal optics: A microscope optical system in which the condenser and objective lenses both focus onto one single point in the specimen, and the image of that point is focused onto a pinhole followed by a detector. An image is generated by scanning the beam over the object. The pinhole mask effectively blocks waves that originate at points other than the scan point (i.e., from out-of-focus or offset points in the object). Therefore, confocal optics improve both transverse and axial resolution. *See also* Descanning.

Conjugate plane: The idealized plane in which the image of a plane object is formed, or the idealized plane in which the object is located for a given image plane. Conjugate distances (object distance, image distance) are measured from each conjugate plane to the corresponding principal point of the lens system. For a simple thin lens, the conjugate distances are measured from the lens to the object and image points (s and s') and are related by $1/f = 1/s + 1/s'$, where f is the focal length.

Continuous-wave (cw) laser: A laser that emits radiation continuously rather than in short bursts, as in a pulsed laser.

Convolution: A linear process in which a variable input signal is represented as a sequence of scaled impulses. Each scaled impulse gives rise to a proportionally weighted and delayed impulse response. The system output is the summation or integral of the sequence of incrementally delayed impulse responses. A convolution integral or discrete convolution is the input–output relation for a linear system that is shift invariant. In this case, the impulse response, also known as the point-spread function (PSF), carries all of the information on the transformation caused by the system on the input signal. In fluorescence microscopy, the image is modeled as a convolution of the true object with the PSF of the microscope.

Dark-field: A microscope optical system in which the sample is illuminated by oblique light incident over a limited range of angles larger than the maximum acceptance angle of the objective. In this case only scattered light is seen, so that the features of the specimen appear bright against a dark background.

Deblur: Removal of the out-of-focus component in an image, or restoration of movement-blurred features.

Deconvolution: Computational procedure for estimating the input signal to a linear system when the output signal is the experimental data. The most common case in microscopy is computation of optical sections of a fluorescently labeled object from image data that contains out-of-focus features superimposed on in-focus features. Using the point-spread function (PSF) of the microscope as a known factor, deconvolution methods can significantly attenuate blurred features and sharpen in-focus features in each image plane of a serial-focus stack. In blind deconvolution, optical sections of both the object and the a priori unknown PSF are jointly estimated.

Depth of field: The range in the object field throughout which an optical system produces a sharp image. In photography, depth of field increases with f-number (f/#). In microscopy, depth of field decreases sharply with numerical aperture (NA). A shallow depth of field is essential for 3D imaging by serial-focus image acquisition.

Depth of focus: The range in image distance that corresponds to the range of object distance covered by the depth of field. In an image-forming instrument, the detector (film, video pickup tube, CCD array) must be located at the plane where the geometric image is formed. The depth of focus is the tolerance on that location.

Descanning: The process by which the time-dependent deflection of a light beam in a scanning optical system is compensated or nulled out in the back-reflected or returning light path. In many confocal scanning fluorescence microscopes, deflection of the laser beam is controlled by galvanometrically driven mirrors. Fluorescence captured by the objective travels the reverse path in the microsope, so that the mirrors exactly cancel the deflection. The emergent light can therefore be brought to a stationary focus on a detector.

Dichroism: (1) In reference to anisotropic materials, the selective absorption of light rays that are polarized in one particular plane relative to the crystalline axes or principal axes. (2) In reference to isotropic materials, the selective reflection and transmission of light as a function of wavelength, regardless of its polarization. The color of these materials varies with thickness. (3) An optical interference filter (dichroic reflector) that transmits one or more bands of wavelengths while reflecting the complementary bands.

Diffraction: The physical phenomena observed when any wave spreads out in space because of the apparent bending of that wave around small obstacles. Diffraction can also be visualized as the accumulation of the infinitesimal interference effects of a wave with itself.

Diffraction-limited: The property of an optical system whereby only the effects of diffraction determine the quality of the image it produces.

Digital: Denoting the use of binary notation; that is, the representation of data numerically by bits (1 or 0).

Digitize: To put in a digital representation. *See* Digital.

Drift: A gradual, uncontrolled change in the setting of an instrument or output of a circuit over time.

Dynamic range: The number of distinguishable values of a signal that may be measured by use of a particular detection device. In general, dynamic range is the number obtained by dividing the full output range of the device by the root-mean-square (rms) noise level. For example, an instrumentation amplifier with a full range of 10 V and an intrinsic noise level of 10 μV rms has a dynamic range of one million. Dynamic range is often expressed logarithmically, in bits or decibels.

Electron multiplier: A device containing a material that, when bombarded with a single electron, emits more than one electron. Many stages of such material are often combined in the presence of an electric field, which causes the emitted electrons to accelerate into subsequent stages, leading to an overall exponential multiplicative gain in the total number of electrons.

Electroporation: A method to introduce substances (e.g., fluorescent dyes or DNA plasmids) into cells through permeabilization of the cells' plasma membrane with an externally applied electric field. Electroporation can be performed on large cell populations (e.g., in cell cultures) using large electrodes or on single cells by applying voltage pulses through a glass pipette targeted to individual cells. *See* Chapter 50, this volume.

Emission: Conversion of the internal energy of an excited atom or molecule into a radiated photon. In general, the energy (color) of the radiation will be determined by the difference in energy between the initial and final quantum states. Emission can be spontaneous, as in fluorescence, or can be stimulated, as in a laser, by driving the excited atom or molecule at the resonant frequency for the transition. Absorption, the reciprocal process, is necessarily stimulated.

Emission spectrum: The wavelength composition of light emitted in the radiative de-excitation of atoms or molecules. For a fluorescent or phosphorescent dye, usually the graph showing the spectral composition of the emitted light versus wavelength. The absorption spectrum (and the closely related excitation spectrum) rise on the short-wavelength side of the emission spectrum and overlap it to some extent. Generally, the emission spectrum is insensitive to the exact choice of excitation wavelength, as long as that wavelength is not in the overlap region. *See also* Absorption, Excitation spectrum.

Evanescent wave: Short-reaching electromagnetic wave at the boundary where light undergoes total internal reflection. Evanescent waves show an exponential decrease with distance from the boundary and penetrate to only ~100 nm into the medium. They are used for thin-sheet illumination (e.g., to image cellular processes occurring at the plasma membrane). *See* Chapter 19, this volume.

Excitation: (1) The process by which an atom or molecule acquires energy sufficient to raise it to a quantum state higher than its ground state. (2) More specifically with respect to lasers, the process by which the material in the laser cavity is energized by light or other means, so that atoms are converted to a semi-stable state, initiating the lasing process.

Excitation spectrum: The graph showing the brightness of fluorescence from a dye versus the excitation wavelength. Usually, the excitation spectrum is determined at the emission wavelength at which the dye is most fluorescent. The excitation spectrum clearly depends on the absorption spectrum but also includes the efficiency with which the absorbed energy is recovered as emission. This factor is the **quantum yield** of fluorescence. *See also* Absorption, Emission spectrum.

Extinction coefficient: The wavelength-dependent constant of proportionality between the absorbance and the molar concentration of a dye. The molar extinction coefficient (ε) is defined in Beer's Law, the relation between attenuation (I/I_0), concentration (c), absorbance (A), and path length (L): $A = \log I_0/I = \varepsilon cL$.

F-number (f/#): The focal length of a lens divided by the entrance pupil, or diameter, of the lens.

Ferroelectric: Materials that can be electrostatically polarized by an external electric field, and that retain the direction and degree of polarization. A polarized ferroelectric material is known as an electret (analogous to magnet), which has a positive pole, a negative pole, and establishes an electrostatic field in its vicinity.

Fiber optics: Optical fibers made of glass that can be flexibly used for guiding light. Typical optical fibers are 125–1000 μm in diameter and consist of a glass core surrounded by a glass cladding of slightly lower refractive index. Light is confined to the core by total internal reflection at the core–cladding interface if light rays enter the core below a certain angle. So-called "single-mode" fibers have small core diameters (4–8 μm) and are suitable for guiding laser beams while preserving the ability to recollimate the laser beam. *See* Chapters 78 and 82, this volume.

Field of view: The maximum area that can be seen through a lens or an optical instrument.

Field stop: An iris located in an optical system such that its open diameter delimits the field of view. Generally, the field stop is located between the light source and the condenser, so that its image is formed in the specimen and therefore sets the diameter of the illuminated field. It is also possible to put a field stop between the objective and the detector, so that the iris directly masks the view. *See also* Aperture stop.

Flashlamp: A type of arc lamp that converts stored electrical energy into light by means of a sudden electrical discharge or spark. The discharge is generally between two electrodes in a gas or vapor contained in a glass or fused quartz tube. Depending on the needed spectral output, the lamp may be filled with air, xenon, mercury vapor, other gases, or gaseous mixtures.

Fluorescence: One of several processes in which there is emission of light by a molecule subsequent to the excitation of that molecule by absorption of light. Because of internal relaxation processes in which some energy is lost as heat, the emitted photon is of a lower energy (longer wavelength) than the initially absorbed photon. Fluorescence is characterized by conservation of the electronic spin state of the molecule (usually singlet/singlet), and therefore by strongly allowed transitions. The excited-state lifetime for fluorescence is generally in the nanosecond range. This is in contrast to phosphorescence, characterized by a spin change between excited state and ground state (usually triplet/singlet), weakly allowed transitions, and excited-state lifetimes ranging from microseconds to many minutes.

Focal length: For a simple thin lens, the focal length is the distance beyond the lens at which incoming collimated light is converged to a focused point. For diverging lenses the focal length is negative, with the virtual focus before the lens. In a lens system, the front focal length and the back focal length are measured from mathematically defined principal points on the axis of the lens group. In general, the front and back focal lengths will be in the ratio of the refractive indices of the media on either side of the lens; that is, for a lens group in air, the two focal lengths will be equal. For an immersion objective, the front focal length will exceed the back by a ratio equal to the immersion index.

Focal plane: (1) An idealized plane (including the focal point) normal to the axis of a lens or mirror. (2) It is most commonly used in microscopy to describe an idealized plane located at the position in an optical instrument at which the sharpest image is formed.

Frame: (1) To center an object in the field of a camera. (2) A single image from among the sequential images on motion-picture film. (3) In raster-scanned television, a complete 525-line image. In conventional video cameras, a frame is composed of two sequential interlaced 262.5-line fields. (4) The complete readout of a single image from a digital camera. *See also* Video.

Frame grabber: Electronic image acquisition device, usually in the form of a circuit card for a computer, that samples, digitizes, and stores a video camera frame in computer memory.

Full-well capacity: The number of electrons that each pixel element in a charge-coupled device (CCD) can hold without overflowing and causing image blooming. Generally, this will be in the range 10^4–10^5, depending on the size of the CCD elements.

Gain: Also known as amplification. (1) The numerical factor by which an amplifier or a concentrator (such as an antenna) increases signal strength. (2) In optics, a material that exhibits amplification rather than absorption of electromagnetic waves. This type of behavior provides the basis for laser operation. (3) In a light detector, the number of electrons or holes counted per detected photon; in a CCD, the gain is 1. In a photomultiplier, the gain may range from 10^2 to 10^6.

Gray scale: In image processing, the number of usable gray levels. In an 8-bit system, the gray scale runs from 0 to 255. The range is normally defined as running from black to white.

Group velocity dispersion: The time-dependent alteration of the amplitude and width of an electromagnetic wave pulse propagating through a material because of differences in the wave speed of the monochromatic components of the pulse. Perfectly monochromatic light is ideally a wave train of infinite duration; however, any finite pulse contains a band of monochromatic components (roughly, pulsewidth and bandwidth vary inversely) having specific relative phase. For femtosecond pulses, the bandwidth is large enough that dispersion in the refractive index of lenses, or even in the air path of the beam, may have significant effect on pulse shape and peak intensity.

Harmonic generation: In optics, the conversion of electromagnetic radiation at a given frequency into radiation at an integer multiple of the frequency. This can occur when an intense coherent light beam passes through a material having a nonlinear polarizability. The harmonic appears as a beam collinear with and coherent with the fundamental but can be separated by use of a dichroic filter, prism, or grating. The most common example of harmonic generation is frequency doubling; for example, the 1064-nm near-infrared output from a YAG laser can be frequency-doubled to 532-nm green light.

Holographic grating: A diffraction grating made by recording a holographic interference pattern in a photosensitive material rather than by the mechanical ruling of grooves. The medium is then processed so that the hologram is encoded as a refractive index pattern. Holographic gratings are phase gratings (i.e., block no light), and therefore have very high optical efficiency.

Image enhancement: A process by which an image is scaled or manipulated to increase the amount of information perceivable by the human eye. The simplest image enhancement operations are brightness and contrast adjustment.

Image intensifier: An electro-optic device that is essentially a vacuum-tube photomultiplier that preserves spatial information. When an intensifier is placed in the image plane of an optical instrument, each detected incoming photon causes ejection of an electron from the photocathode surface in the tube. Ejected electrons are accelerated in a high electric field within the tube and collide with a shielded phosphor screen at the output end. In the collision, a burst of photons is released, resulting in a bright secondary image that can be focused by conventional means onto a television or CCD camera. The field in the intensifier is shaped so that the electron trajectories preserve image coordinates. Modern intensifiers utilize microchannel plate (MCP) technology in which several stages of secondary electron multiplication precede the output phosphor screen. *See also* Photomultiplier tube.

Image plane: An idealized plane, perpendicular to the axis of a lens, in which an image is formed. A real image formed by a positive lens would be visible upon a screen located in this plane.

Incoherent: In optics, the term denoting the lack of a fixed phase relationship between two or more waves. If two incoherent waves are superimposed, interference effects are not stable, and no intereference pattern is observed. *See also* Coherence.

Intrinsic signal: A change in intrinsic optical tissue properties, such as transmittance or reflectance, that report neural activation without the addition of external indicators. Intrinsic optical signals are small but can be picked up with a sensitive camera. *See* Chapters 60 and 83, this volume.

Iris diaphragm: Traditionally, the iris located in the back focal plane of the condenser. The open diameter of the iris diaphragm sets the illuminating numerical aperture (INA) of the condenser. *See also* Aperture.

Isosbestic wavelength: In spectroscopy, the occurrence of a wavelength where a family of spectral curves all intersect. Isosbestic points usually signify the proportional interconversion of two different molecular species, where the two extreme spectra in the family represent the two species. The most well-known example is the isosbestic point at 360 nm in the family of Fura-2 excitation spectra over the nanomolar–micromolar range in Ca^{2+} concentration. At this wavelength, the excitation of fluorescence is essentially independent of calcium concentration.

Köhler illumination: A two-stage illuminating system for a microscope, in which the light source is imaged in the back focal plane of the substage condenser by a collector lens, the substage condenser in turn uniformly illuminating the object. In the Köhler setup, an iris located in the back focal plane of the condenser controls the maximum angle at which light exits the condenser into the specimen. This iris is the aperture stop. A second iris, usually located closer to the collector lens, is imaged on the specimen by the condenser. This iris is the field stop.

Lag: A term applied to an electric charge image in a video camera tube that persists for a period of a few frames after the initial light exposure. Lag is particularly noticeable in high-gain video cameras operating at low light level.

Laser: An acronym for light amplification by stimulated emission of radiation. The variations in laser type have become numerous, but the basic device consists of an optical gain medium within an optical cavity composed of two or more reflectors or gratings. The gain medium may be a gas or vapor of excited atoms or ions, excited dye molecules in a solvent, excited atoms or molecules in a crystal or glassy solid, or excited charge carriers in a semiconductor. The excitation process serves to create an inverted population, in which more atoms are poised to emit a photon than are able to absorb; that is, emission can exceed absorption. The cavity serves to establish electromagnetic traveling or standing wave modes in the gain medium, thereby causing in-phase stimulated emission at specific wavelengths from the inverted population. The amplified light partially or completely escapes the cavity as an intense (usually monochromatic) coherent beam.

Liquid-crystal tunable filter (LCTF): An active optical filter composed of a variable liquid crystal retarder between polarizing filters. The retarder is composed of a liquid-crystalline medium sandwiched between two flat windows that support transparent electrode films. The birefringence of the liquid crystal can be adjusted by applying a voltage across the electrodes. By itself, the device can be used as a variable retarder. When inserted between polarizing filters, wavelength-dependent transmission results. To improve band-pass shape and extinction, several such elements are used in series in an LCTF, along with computer control of the individual stages. *See also* Retardation.

Magnification: The ratio of the size of the image of an object to the size of the object defines the transverse magnification. For a simple lens, transverse magnification varies with object distance. In a microscope, where the camera position is fixed relative to the optics and focus is set by moving the stage axially relative to the instrument, the transverse magnification is constant. Angular magnification is the ratio of the apparent angular size of the image observed through an optical device to that of the object viewed by the unaided eye. Axial magnification is defined in terms of focus setting as the ratio of the axial distance moved by a geometric image point (in image space) to the actual stage focus increment. Generally, axial magnification is proportional to the square of the transverse magnification.

Membrane capacitance: The charge-storage capability of a cell membrane due to its basic structure—a thin, highly insulating lipid bilayer separating two conductive ionic aqueous media, the cytoplasm and the extracellular fluid. When unbalanced charge (usually ions or, in some cases, electrons) is moved across a membrane, the excess ions of one polarity are attracted to ions of the opposite charge remaining on the other side of the membrane. This results in the unbalanced charge appearing in a pair of subnanometer aqueous layers, one associated with each face of the membrane. The energy of the separated charges is stored as an electric field within the bilayer, the constant of proportionality between charge and membrane voltage being the capacitance. For biological membranes, the capacitance is generally in the range 0.5–1.0 $\mu F/cm^2$.

Mode-locked laser: A laser that functions by modulating the energy content of each electromagnetic mode in its optical cavity to give rise selectively to energy bursts of high peak power and short duration. The pulse rate of a mode-locked laser is the round-trip time for a light pulse in the cavity.

Monochromator: Any device used for isolating portions of the spectrum of a light source. Monochromators generally contain a rotatable prism or diffraction grating, along with input and output masks (or slits) to limit the angular range of rays that enter or exit. The entering light is dispersed in angle by the prism or grating, and the exiting wavelength band is selected by rotating the dispersive element relative to the slits. For high spectral purity, two monochromators can be used in tandem in an additive-dispersion configuration.

Multiphoton absorption: The nonlinear optical process in which a molecular electronic excited state is produced by the simultaneous absorption of two or more photons. This differs from sequential absorption in that no intervening stationary quantum state exists as an intermediate. Therefore, extremely high irradiation intensities are required for significant multiphoton absorption rates to occur. Also, the absorption rate is proportional to the square of the intensity for two-photon excitation (the cube of the intensity for three-photon processes, etc.), in contrast to conventional one-photon excitation. In the case of fluorescent dyes, the summed energy of the absorbed photons approximately equals the energy of the single photon that would be required to cause the transition from ground state to the first excited singlet state. The most well-known example of multiphoton absorption is two-photon excitation fluorescence microscopy, in which pairs of near-infrared photons excite ultraviolet/visible-band dyes. This method combines the high intensity of a femtosecond-pulsed laser system with the sharp focusing of a microscope objective to achieve the excitation rates needed to produce photons in sufficient number for practical imaging.

NA: *See* Numerical aperture.

Neutral-density filter: Also known as a gray filter. A light filter that decreases the intensity of light without altering the relative spectral distribution of the energy.

Noise: The unwanted and unpredictable fluctuations that distort a received signal and hence tend to obscure the signal of interest. In microscopy, there is a fundamental statistical limit on the noise in a measurement of the number of photons in a given location. This noise is equal to the square root of the number of photons counted, which is confusingly referred to as shot noise, photon noise, counting noise, and even photon counting noise. In electronic equipment, noise is also generated by thermal processes. This noise is known as Johnson or thermal noise and can be reduced by lowering the temperature of the components performing the measurement. In addition, fluctuation in the illumination source such as the flicker or drift that can be produced by arc lamps will be reflected in the recorded signal as noise. Finally, read or readout noise in an electronic camera refers to the noise inherent in converting photons into electrons and the subsequent analog-to-digital conversion. Steps that can be taken to limit the different sources of noise include shielding, signal averaging or integration, cooling of measurement devices, tuned or phase-locked detection, and proper staging of amplification.

Numerical aperture (NA): A fundamental numerical quantity that is a measure of the range of angles over which a lens system can accept entering light rays. The transverse resolution of a microscope improves directly with NA, and axial resolution approximately with the square of NA. *See also* Aperture.

Objective: The optical element that receives light from the object and forms the first or primary image in telescopes and microscopes. In cameras, the image produced by the objective is the final image. In telescopes and microscopes, when used visually, the image formed by the objective is magnified by an eyepiece.

Observation volume: In a confocal or multiphoton microscope, the specimen volume from which emitted light is detected. For a high-NA diffraction-limited system, this volume can be as small as 0.044 μm^3 (0.044 fL).

Optical sectioning: The property of certain optical systems to attenuate or remove contributions to the image originating from out-of-focus parts of the specimen. In this case, the image contains only in-focus features and therefore represents a planar section of the object. Optical sectioning depends on the shallow depth of field in high-aperture systems. Confocal optical systems contain a spatial filter that blocks light waves that do not originate in the immediate neighborhood of the scan point. In multiphoton microscopes, nonlinear interaction of the focused scan beam with the fluorescent label brings about a similar spatial weighting. In conventional microscopes that form an image directly, quantitative optical sectioning is approached through computational deblurring.

Orthogonal: In comparing two vectors or functions, the property that the projection of one function on the other is 0. Mathematically, orthogonal objects have an inner product equal to 0. In the simplest case, two vectors are orthogonal in two or three dimensions when they are perpendicular. In the case of functions, sine and cosine are orthogonal, as are any two distinct harmonics of a sine or cosine wave. In quantum mechanics, the stationary state wave functions for an atomic or molecular system form a normalized orthogonal basis set.

Paraboloidal mirror: A concave mirror that has the form of a paraboloid of revolution. The paraboloidal mirror may have only a portion of a paraboloidal surface through which the axis does not pass, and is known as an off-axis paraboloidal mirror. All axial, parallel light rays are focused at the focal point of the paraboloid without spherical aberration, and conversely, all light rays emitted from an axial source at the focal point are reflected as a bundle of parallel rays without spherical aberration.

Paraxial optics: Ray diagrams and optical formulas derived by considering only rays that propagate close to the axis of an optical system and do not make a significant angle with respect to the axis. In this limit, the Law of Refraction can be linearized, greatly simplifying the derivation of conjugate distance formulas and other lens equations.

Phosphorescence: Emission of a photon of light by an atom or molecule when the transition is accompanied by a change in electron spin pairing. In biological imaging and spectroscopy, phosphorescence usually arises in the transition from an excited triplet state (unpaired spins) to a lower-energy singlet state (paired spins) of a molecule. Because singlet–triplet transitions are symmetry-forbidden in isolated systems of low atomic number, phosphorescence is a slow process relative to fluorescence or quenching. Molecular oxygen is a very effective quencher of triplet states, therefore phosphorescence spectroscopy or imaging is often most useful in systems where the oxygen partial pressure can be reduced. *See also* Fluorescence.

Photobleaching: The photophysical or photochemical conversion of a dye to a colorless form. In microscopy, photobleaching usually refers to the irreversible loss of fluorescence in dyes that are intensely irradiated. In some cases, photobleaching is reversible. Molecular oxygen is a common mediator of photochemical reactions leading to photobleaching, becoming activated upon quenching dye triplet states. Excited-state molecular oxygen can react not only with dye molecules in the specimen but also with other biochemicals and macromolecules in a cell. This type of activity can lead to phototoxicity.

Photodiode: (1) A two-electrode, light-sensitive junction formed in a semiconductor material in which the reverse current varies with illumination. Photodiodes are used for the detection of light and for the conversion of radiant energy to electrical power. (2) Originally, a vacuum tube diode in which the cathode was coated with a metal film such as sodium so as to make the cathode photo-emissive.

Photoelectric effect: The excitation of a bound electron in a material by absorption of a photon. In the classic photoelectric effect, one electron is ejected from a metal film into vacuum for each photon absorbed. The energy of the photon must exceed the potential energy needed to remove the electron from the metal, with the excess energy appearing as kinetic energy of the electron. The photoelectric effect is one of the primary pieces of experimental evidence for the quantum nature of the electromagnetic field.

Photomultiplier tube (PMT): A vacuum tube light detector consisting of a glass envelope with an internal photocathode followed by a series of dynodes (electrodes) and an anode. In operation, the dynodes and anode are held at sequentially more-positive voltages than the cathode. The cathode emits electrons via the photoelectric effect when exposed to light. The electrons are accelerated by the electrostatic field to collide with the first dynode where secondary electrons are ejected. The secondary group is accelerated to the next dynode, where there is another multiplicative gain in released charge. The dynode chain acts as an electron multiplier, providing a high charge gain at the anode output. The unit is essentially a photoelectric cell followed by a high-gain electron multiplier. PMTs can be used with an analog electrometer or with a discriminator and counter circuit.

Photon: The massless quantum particle of the electromagnetic field. The energy of a photon is given by Planck's formula, $E = h\nu$, where h is Planck's constant (6.626×10^{-34} J-sec) and ν is the frequency of the radiation in hertz. Photons propagate in free space at the speed of light ($c = 3.00 \times 10^{8}$ m/sec) and carry momentum, angular momentum, and a polarization. When an electromagnetic wave field gives up energy to an atomic or molecular system, the exchange occurs via the absorption of photons.

Pixel: Contraction of "picture element." A small element of a scene, often the smallest resolvable area, in which an average brightness value is determined and used to represent that portion of the scene. Pixels are usually arranged in a square array to form a complete image.

Planar illumination: Use of a thin light sheet for fluorescence excitation to obtain optical sectioning. Imaging is performed perpendicular to the light sheet. *See* Chapter 56, this volume.

PMT: *See* Photomultiplier tube.

Point-spread function (PSF): In an optical system, the energy intensity distribution about the image of a point source. This function defines the resolution of the optical system as convolution of the PSF with a specimen mathematically defines the resulting image.

Polarized light: Light in which the electric field oscillates in a particular direction transverse to the direction of travel (linear polarization), or in which the field components rotate in a plane transverse to the direction of travel (elliptical or circular polarization). In an isotropic medium, light propagates as a transverse wave in which the electric field and the magnetic field.

PSF: *See* Point-spread function.

QE: *See* Quantum efficiency.

Quantum efficiency (QE): (1) With respect to a source of radiant flux, the ratio of the number of quanta of radiant energy (photons) emitted per second to the number of electrons flowing per second. (2) For a detector, the ratio of induced current to incident flux. Often measured in electrons per photon (dimensionless) or amperes per watt.

Quantum yield (QY): The fractional probability of a photophysical or photochemical event occurring subsequent to the absorption of a photon. In photosynthesis, the quantum yield is the probability that the energy from an absorbed photon gets coupled to the reaction center rather than lost as heat. For a fluorescent dye, the QY is the probability that a photon will be emitted for each photon absorbed. For different dyes and conditions, QY ranges from <0.1 to >0.95.

QY: *See* Quantum yield.

Raster: The pattern of lines traced by rectilinear scanning in display systems and scanning microscopes.

Ratiometric indicator: Fluorescent indicator dyes that undergo changes in the excitation or emission spectrum that can be utilized to approximately correct for unknown indicator concentration and the optical effect of irregularities in the shape of the specimen. The procedure, known as ratio imaging, involves normalization of one fluorescence image by a second image of the same object made with altered excitation or emission wavelength. In this way, the physiological response of the indicator can be estimated without inaccuracies due to concentration variations or cell shape–dependent brightness variations. The exact procedure used is dependent on the type of spectral changes that occur in the dye with changes in the physiologically relevant variable. The most well-known ratiometric indicator is the Ca^{2+}-sensitive dye Fura-2, which shows an altered excitation spectrum. In ratiometric Fura imaging, a single blue-green emission filter is used along with an alternate pair of ultraviolet excitation filters, usually 340 nm and 380 nm. The ratio image is computed digitally as the background-corrected 340-nm-excited image divided by the background-corrected 380-nm-excited image.

Refractive index: A fundamental unitless number that is the ratio of the speed of monochromatic light in a vacuum to the speed (phase velocity) in a particular material or medium: $n = c/v$. Ray paths are redirected at a refractive boundary according to the Law of Refraction; $n \sin \theta = n' \sin \theta'$, where n and n' are the refractive indices of the materials on either side of the boundary, and the angles are measured from the local normal to the boundary. In a material, the vacuum wavelength of the light is altered from λ to λ/n. In a more general sense, the refractive index is a complex number related to the phase of the optical response of a material to the light wave. The conventional index is the real part of the complex index, and the absorption (or stimulated emission) coefficient is the imaginary part.

Retardation: The phase change in a light wave, relative to a reference beam, due to propagation through a region of different refractive index. In phase-contrast microscopy, light propagating through an object of increased index will lag in phase relative to light that bypasses the structure. In a birefringent filter, two orthogonal linearly polarized light waves will propagate at different speeds, causing one to become retarded relative to the other. This causes the polarization to shift between linear, elliptical, and circular, depending on the thickness of the filter, the degree of birefringence, and the wavelength of the light. A retarder or wave plate is a birefringent flat optic, usually of mica, quartz, or calcite. The thickness of the plate is usually finished so that the retardance for one principal polarization relative to the other is either one-quarter or one-half wave. A properly oriented quarter-wave plate will convert linearly polarized light to circular, and vice versa. A half-wave plate will rotate linear polarization by 90° or will interconvert right-hand and left-hand circular polarizations. In a liquid-crystal retarder the birefringence can be controlled electronically. *See also* Liquid-crystal tunable filter.

Scattering (electromagnetic): A fundamental process that can be pictured as the result of a plane wave beam impinging on an atom, molecule, or small particle. The oscillating electromagnetic field polarizes the particle and causes it to act as a source of spherical waves. The scattered waves carry energy off in other directions and cause attenuation of the beam. When a beam of light propagates through a medium containing many such particles, the scatter makes the path of the beam visible. Light scattering is seen whenever there is refractive index heterogeneity within an otherwise transparent material, or on a surface. Milk appears white (in white light) because the numerous suspended lipoprotein particles backscatter essentially all the incident light. The sky appears blue because of the wavelength-dependent scatter of sunlight by air molecules. Refraction is the net result of scatter when the density of scatterers is both high and uniform on a scale comparable to the wavelength. A refracted wave can be pictured as the coherent superposition of the incident wave field and the scattered field.

Signal-to-noise ratio (SNR): The ratio of the power in a desired signal to the undesirable noise present in the absence of a signal.

SLM: *See* Spatial light modulator.

SNR: *See* Signal-to-noise ratio.

Spatial light modulator (SLM): *See* Chapter 56, Book 1.

Superresolution: Any of a number of methods that produce an image with a resolution below the diffraction limit. Also referred to as subdiffraction microscopy. *See also* Chapters 31–37, Book 1.

Stimulated emission: Interaction of an atom or molecule in an excited state with an electromagnetic wave of a certain frequency, which causes the atom or molecule to drop to a lower energy level while transferring its energy to that field.

Stimulated emission depletion (STED): De-excitation of an excited fluorescent dye by using stimulated emission to deplete the excited state and prevent spontaneous fluorescence emission, effectively turning off the fluorescent molecule.

Total internal reflection: An optical phenomenon at the boundary of two media with different refractive indices. If light rays hit the boundary from the side of the higher refractive index medium at an angle larger than a particular critical angle with respect to the normal to the surface, then no light can pass through and all of the light is reflected. An evanescent, sharply decaying electromagnetic field reaches into the lower refractive medium, though. *See* Evanescent wave.

Transmission: The process by which light (or electromagnetic radiation in general) passes through a material. In tomography the transmission mode contrasts with the emission mode. In the former case, an external source of radiation is employed to create beams that are transmitted through the sample and whose attenuation is measured by detectors on the other side of the sample (e.g., computed tomography). In the latter case, radiation is stimulated within the sample and is therefore emitted outward toward the detectors (e.g., positron emission tomography).

Video: The encoded picture signal output of a video camera, consisting of a sequential horizontal scan line analog signal interleaved with horizontal synchronization pulses and terminated by a vertical sync pulse sequence. In the U.S. system, a video picture consists of 525 horizontal scan lines, presented as 60 262.5-line interlaced fields/sec (30 frames/sec). In the European standard, the frame consists of 625 lines presented as 50 312.5-line interlaced fields/sec (25 frames/sec). The 1-V peak-to-peak (p-p) signal reserves 0.7 V for the monochrome image brightness signal and 0.3 V for the negative-going synchronization pulses. Video digitization is normally set at 8 bits, which can be increased by frame averaging.

Wavelength: Electromagnetic energy such as light is transmitted in the form of a periodic or quasiperiodic wave. In monochromatic light, the wave can be pictured as a sinusoidally varying electric field and magnetic field. The wavelength is the spatial dimension of one cycle of this wave field; wavelength is inversely proportional to frequency. The visible band of wavelengths runs from 390 nm (violet) to 750 nm (far-red).

Wave number: The frequency of a wave divided by its velocity of propagation; the reciprocal of the wavelength.

Appendix 6 Cautions

GENERAL CAUTIONS

Please note that the Cautions Appendix in this manual is not exhaustive. Readers should always consult individual manufacturers and other resources for current and specific product information. Chemicals and other materials discussed in text sections are not identified by the icon <!> used to indicate hazardous materials in the protocols. However, without special handling, they may be hazardous to the user. Please consult your local safety office or the manufacturer's safety guidelines for further information.

The following general cautions should always be observed.

- **Before beginning the procedure,** become completely familiar with the properties of substances to be used.
- **The absence of a warning** does not necessarily mean that the material is safe, because information may not always be complete or available.
- **If exposed** to toxic substances, contact your local safety office immediately for instructions.
- **Use proper disposal procedures** for all chemical, biological, and radioactive waste.
- **For specific guidelines on appropriate gloves to use,** consult your local safety office.
- **Handle concentrated acids and bases** with great care. Wear goggles and appropriate gloves. A face shield should be worn when handling large quantities.

  Do not mix strong acids with organic solvents because they may react. Sulfuric acid and nitric acid especially may react highly exothermically and cause fires and explosions.

  Do not mix strong bases with halogenated solvent because they may form reactive carbenes which can lead to explosions.
- **Handle and store pressurized gas containers** with caution because they may contain flammable, toxic, or corrosive gases; asphyxiants; or oxidizers. For proper procedures, consult the Material Safety Data Sheet that must be provided by your vendor.
- **Never pipette** solutions using mouth suction. This method is not sterile and can be dangerous. Always use a pipette aid or bulb.
- **Keep halogenated and nonhalogenated** solvents separately (e.g., mixing chloroform and acetone can cause unexpected reactions in the presence of bases). Halogenated solvents are organic solvents such as chloroform, dichloromethane, trichlorotrifluoroethane, and dichloroethane. Nonhalogenated solvents include pentane, heptane, ethanol, methanol, benzene, toluene, *N,N*-dimethylformamide (DMF), dimethyl sulfoxide (DMSO), and acetonitrile.
- **Laser radiation,** visible or invisible, can cause severe damage to the eyes and skin. Take proper precautions to prevent exposure to direct and reflected beams. Always follow the manufacturer's safety guidelines and consult your local safety office. See caution below for more detailed information.
- **Flash lamps,** because of their light intensity, can be harmful to the eyes. They also may explode on occasion. Wear appropriate eye protection and follow the manufacturer's guidelines.
- **Photographic fixatives, developers, and photoresists** also contain chemicals that can be harmful. Handle them with care and follow the manufacturer's directions.

- **Power supplies and electrophoresis equipment** pose serious fire hazard and electrical shock hazards if not used properly.
- **Microwave ovens and autoclaves** in the laboratory require certain precautions. Accidents have occurred involving their use (e.g., when melting agar or Bacto Agar stored in bottles or when sterilizing). If the screw top is not completely removed and there is inadequate space for the steam to vent, the bottles can explode and cause severe injury when the containers are removed from the microwave or autoclave. Always completely remove bottle caps before microwaving or autoclaving. An alternative method for routine agarose gels that do not require sterile agar is to weigh out the agar and place the solution in a flask.
- **Ultrasonicators** use high-frequency sound waves (16–100 kHz) for cell disruption and other purposes. This "ultrasound," conducted through air, does not pose a direct hazard to humans, but the associated high volumes of audible sound can cause a variety of effects, including headache, nausea, and tinnitus. Direct contact of the body with high-intensity ultrasound (not medical imaging equipment) should be avoided. Use appropriate ear protection and display signs on the door(s) of laboratories where the units are used.
- **Use extreme caution when handling cutting devices,** such as microtome blades, scalpels, razor blades, or needles. Microtome blades are extremely sharp! Use care when sectioning. If unfamiliar with their use, have an experienced user demonstrate proper procedures. For proper disposal, use the "sharps" disposal container in your lab. Discard used needles *unshielded*, with the syringe still attached. This prevents injuries and possible infections when manipulating used needles because many accidents occur while trying to replace the needle shield. Injuries may also be caused by broken pasteur pipettes, coverslips, or slides.
- **Procedures for the humane treatment of animals** must be observed at all times. Consult your local animal facility for guidelines. Animals, such as rats, are known to induce allergies that can increase in intensity with repeated exposure. Always wear a lab coat and gloves when handling these animals. If allergies to dander or saliva are known, wear a mask.

GENERAL PROPERTIES OF COMMON CHEMICALS

The hazardous materials list can be summarized in the following categories.

- Inorganic acids, such as hydrochloric, sulfuric, nitric, or phosphoric, are colorless liquids with stinging vapors. Avoid spills on skin or clothing. Spills should be diluted with large amounts of water. The concentrated forms of these acids can destroy paper, textiles, and skin and cause serious injury to the eyes.
- Inorganic bases, such as sodium hydroxide, are white solids that dissolve in water and under heat development. Concentrated solutions will slowly dissolve skin and even fingernails.
- Salts of heavy metals are usually colored, powdered solids that dissolve in water. Many of them are potent enzyme inhibitors and therefore toxic to humans and the environment (e.g., fish and algae).
- Most organic solvents are flammable volatile liquids. Avoid breathing the vapors, which can cause nausea or dizziness. Also avoid skin contact.
- Other organic compounds including organosulphur compounds, such as mercaptoethanol or organic amines, can have very unpleasant odors. Others are highly reactive and should be handled with appropriate care.
- If improperly handled, dyes and their solutions can stain not only your sample, but also your skin and clothing. Some are also mutagenic (e.g., ethidium bromide), carcinogenic, and toxic.
- Nearly all names ending with "ase" (e.g., catalase, β-glucuronidase, or zymolyase) refer to enzymes. There are also other enzymes with nonsystematic names such as pepsin. Many of them are provided by manufacturers in preparations containing buffering substances, etc. Be aware of the individual properties of materials contained in these substances.
- Toxic compounds are often used to manipulate cells. They can be dangerous and should be handled appropriately.
- Be aware that several of the compounds listed have not been thoroughly studied with respect to their toxicological properties. Handle each chemical with appropriate respect. Although the toxic effects of a compound can be quantified (e.g., LD_{50} values), this is not possible for carcinogens or mutagens where one single exposure can

have an effect. Also realize that dangers related to a given compound may also depend on its physical state (fine powder vs. large crystals/diethylether vs. glycerol/dry ice vs. carbon dioxide under pressure in a gas bomb). Anticipate under which circumstances during an experiment exposure is most likely to occur and how best to protect yourself and your environment.

HAZARDOUS MATERIALS

Note: In general, proprietary materials are not listed here. Kits and other commercial items as well as most anesthetics, sedatives, dyes, fixatives, embedding media, stains, herbicides, and fungicides are also not included. Anesthetics and antibiotics also require special care. Follow the manufacturer's safety guidelines that accompany these products.

Acetic acid (concentrated) must be handled with great care. It may be harmful by inhalation, ingestion, or skin absorption. Wear appropriate gloves and goggles. Use in a chemical fume hood.

Acetic acid (glacial) is highly corrosive and must be handled with great care. It may be a carcinogen. Liquid and mist cause severe burns to all body tissues. It may be harmful by inhalation, ingestion, or skin absorption. Wear appropriate gloves and goggles and use in a chemical fume hood. Keep away from heat, sparks, and open flame.

Acetone causes eye and skin irritation and is irritating to mucous membranes and upper respiratory tract. Do not breathe the vapors. It is also extremely flammable. Wear appropriate gloves and safety glasses. Keep away from heat, sparks, and open flame.

Acrylamide (unpolymerized) is a potent neurotoxin and is absorbed through the skin (the effects are cumulative). Avoid breathing the dust. Wear appropriate gloves and a face mask when weighing powdered acrylamide and methylene-bisacrylamide. Use in a chemical fume hood. Polyacrylamide is considered to be nontoxic, but it should be handled with care because it might contain small quantities of unpolymerized acrylamide.

α-bungarotoxin is a potent neurotoxin and is harmful by inhalation, ingestion, or skin absorption. Wear appropriate gloves and safety glasses and always use in a chemical fume hood. Do not breathe the dust.

α-chloralose is toxic if swallowed. It is harmful by inhalation, ingestion, or skin absorption. Wear appropriate gloves and safety glasses. Do not breathe the dust.

Ammonium chloride, NH_4Cl, may be harmful by inhalation, ingestion, or skin absorption. Wear appropriate gloves and safety glasses and use in a chemical fume hood.

Ammonium persulfate, $(NH_4)_2S2O_8$, is extremely destructive to tissue of the mucous membranes and upper respiratory tract, eyes, and skin. Inhalation may be fatal. Wear appropriate gloves, safety glasses, and protective clothing, and use only in a chemical fume hood. Wash thoroughly after handling.

Ampicillin may be harmful by inhalation, ingestion, or skin absorption. Wear appropriate gloves and safety glasses and use in a chemical fume hood.

Anesthetics, follow manufacturer's safety guidelines

Atropine may be fatal if inhaled or ingested. It is also harmful by skin absorption. Wear appropriate gloves and safety glasses. Do not breathe the dust.

Avertin, *see* **2,2,2-tribromoethanol**

Azide, *see* **Sodium azide**

BAPTA, *see* **1,2-*bis*(2-aminophenoxy)-ethane-*N,N,N′,N′*-tetraacetic acid**

β-mercaptoethanol (2-mercaptoethanol, α-mercaptoethanol), $HOCH_2CH_2SH$, may be fatal if inhaled or absorbed through the skin and is harmful if ingested. High concentrations are extremely destructive to the mucous membranes, upper respiratory tract, skin, and eyes. β-mercaptoethanol has a very foul odor. Wear appropriate gloves and safety glasses and always use in a chemical fume hood.

1,2-*bis*(2-aminophenoxy)-ethane-*N,N,N′,N′*-tetraacetic acid (BAPTA), may be harmful by inhalation, ingestion, or skin absorption. It is irritating to mucous membranes, upper respiratory tract, the skin, and the eyes. Wear appropriate gloves and safety glasses and use in a chemical fume hood.

Bromophenol blue may be harmful by inhalation, ingestion, or skin absorption. Wear appropriate gloves and safety glasses and use in a chemical fume hood.

$CaCl_2$, *see* **Calcium chloride**

Cacodylate contains arsenic, is highly toxic, and may be fatal if inhaled, ingested, or absorbed through the skin.

Wear appropriate gloves and safety glasses and use in a chemical fume hood. *See also* **Sodium cacodylate.**

Calcium chloride, $CaCl_2$, is hygroscopic and may cause cardiac disturbances. It may be harmful by inhalation, ingestion, or skin absorption. Do not breathe the dust. Wear appropriate gloves and safety goggles.

Carbogen, *see* **Carbon dioxide**

Carbon dioxide, CO_2, in all forms may be fatal by inhalation, ingestion, or skin absorption. In high concentrations, it can paralyze the respiratory center and cause suffocation. Use only in well-ventilated areas. In the form of dry ice, contact with carbon dioxide can also cause frostbite. Do not place large quantities of dry ice in enclosed areas such as cold rooms. Wear appropriate gloves and safety goggles.

Carprofen is highly toxic if swallowed. It may be harmful by inhalation, ingestion, or skin absorption. Wear appropriate gloves and safety glasses. Do not breathe the dust.

Cesium chloride, CsCl, may be harmful by inhalation, ingestion, or skin absorption. Wear appropriate gloves and safety glasses.

Cesium hydroxide, CsOH, may be harmful by inhalation, ingestion, or skin absorption. It is extremely destructive to the mucous membranes and upper respiratory tract; inhalation may be fatal. Do not breathe the dust. Wear appropriate gloves and safety glasses and always use in a chemical fume hood.

$C_6H_5CH_3$, *see* **Toluene**

CO_2, *see* **Carbon dioxide**

Coomassie brilliant blue (CBR) may be harmful by inhalation, ingestion, or skin absorption. Wear appropriate gloves and safety glasses.

Cryostat blades, *see* **Microtome blades**

CsCl, *see* **Cesium chloride**

CsOH, *see* **Cesium hydroxide**

Cyanoacrylate adhesives are harmful by inhalation, ingestion, and skin absorption. Immediate bonding of tissues can occur. Do not pull apart. Inhalation may cause lightheadedness. Wear appropriate gloves and safety glasses. Follow manufacturer's safety guidelines.

DAB, *see* **3,3′-diaminobenzidine**

DAPI, *see* **4,6-diamidine-2-phenylindole dihydrochloride**

DCM, *see* **Dichloromethane**

Dexamethasone may be harmful by inhalation, ingestion, or skin absorption. Overexposure may cause reproductive disorder(s) and possible risks of harm to the unborn child. Wear appropriate gloves and safety glasses. Do not breathe the dust.

4,6-diamidine-2 phenylindole dihydrochloride (DAPI) is a possible carcinogen. It may be harmful by inhalation, ingestion, or skin absorption. It may also cause irritation. Avoid breathing the dust and vapors. Wear appropriate gloves and safety glasses and use in a chemical fume hood.

3,3′-diaminobenzidine (DAB) is a carcinogen. Handle with extreme care. Avoid breathing vapors. Wear appropriate gloves and safety glasses and use in a chemical fume hood.

1,3-diamino-2-hydroxypropanol-*N,N,N′,N′*-tetraacetic acid (DPTA) is an irritant and may be harmful by inhalation, ingestion, or skin absorption. Wear appropriate gloves and safety glasses. Do not breathe the dust.

Dichloromethane (DCM), CH_2Cl_2 (also known as **Methylene chloride**) is toxic if inhaled, ingested, or absorbed through the skin. It is also an irritant and is suspected to be a carcinogen. Wear appropriate gloves and safety goggles and use in a chemical fume hood. Do not breathe the vapors.

Dimethylsulfoxide (DMSO) may be harmful by inhalation or skin absorption. It easily penetrates the skin and anything dissolved or mixed with it will be absorbed. Wear appropriate gloves and safety glasses and use in a chemical fume hood. DMSO is also combustible. Store in a tightly closed container. Keep away from heat, sparks, and open flame.

DMSO, *see* **Dimethylsulfoxide**

Doxycycline is an irritant and may be harmful by inhalation, ingestion, or skin absorption. Wear appropriate gloves and safety glasses. Do not breathe the dust. It must never be used by pregnant or breast-feeding women as it may cause serious health effects in both nursing and unborn babies.

DPTA, *see* **1,3-diamino-2-hydroxypropanol-*N,N,N′,N′*-tetraacetic acid**

Durcupan resin may be harmful by inhalation, ingestion, or skin absorption. Avoid breathing the vapors and dusts. Wear appropriate gloves and safety goggles and always use in a chemical fume hood. Sensitivity to these chemicals may develop with repeated contact. Keep away from heat, sparks, and open flame.

Dyes, *follow* **manufacturer's safety guidelines**

EDTA, *see* **Ethylenediamenetetraacetic acid**

Epoxy and acrylic resins, *see* **Resins**

Ethanol (EtOH), CH_3CH_2OH, is highly flammable and may be harmful by inhalation, ingestion, or skin absorption. Wear appropriate gloves and safety glasses. Keep away from heat, sparks, and open flame.

Ethidium bromide is a powerful mutagen, is toxic, and may cause irreversible effects. Consult the local institutional safety officer for specific handling and disposal procedures. Avoid breathing the dust. Wear appropriate gloves when working with solutions that contain this dye.

Ethylenediaminetetraacetic acid (EDTA) may be harmful by inhalation, ingestion, or skin absorption. Wear appropriate gloves and safety glasses. Severe over-exposure can result in death.

EtOH, *see* **Ethanol**

Fast Green is a carcinogen and may be harmful by inhalation, ingestion, or skin absorption. Wear appropriate gloves and safety glasses and always use in a chemical fume hood.

Fentanyl is highly toxic and may be fatal if inhaled or absorbed through the skin. It is harmful by inhalation, ingestion, or skin absorption. Wear appropriate gloves and safety goggles and use respiratory protection. Keep away form heat, sparks, and open flame. Do not breathe the dust.

FuGENE 6 transfection reagent is highly flammable and may be harmful by inhalation, ingestion, or skin absorption. Wear appropriate gloves and safety glasses. The vapor forms explosive mixture with air. Keep away from heat, sparks, and open flame.

Glacial acetic acid, *see* **Acetic acid (glacial)**

Glutaraldehyde is toxic. It is readily absorbed through the skin and is irritating or destructive to the skin, eyes, mucous membranes, and upper respiratory tract. Wear appropriate gloves and safety glasses and always use in a chemical fume hood.

Halothane is an anesthetic and may be harmful by inhalation or skin absorption. It is also highly volatile. Wear appropriate gloves and safety glasses and work in a well-ventilated area.

HCl, *see* **Hydrochloric acid**

Heparin is an irritant and may act as anticoagulant subcutaneously or intravenously. It may be harmful by inhalation, ingestion, or skin absorption. Wear appropriate gloves and safety glasses.

H_2O_2, *see* **Hydrogen peroxide**

Hydrochloric acid, HCl, is volatile and may be fatal if inhaled, ingested, or absorbed through the skin. It is extremely destructive to mucous membranes, upper respiratory tract, eyes, and skin. Wear appropriate gloves and safety glasses and use with great care in a chemical fume hood. Wear goggles when handling large quantities.

Hydrogen peroxide, H_2O_2, is corrosive, toxic, and extremely damaging to the skin. It may be harmful by inhalation, ingestion, and skin absorption. Wear appropriate gloves and safety glasses and use only in a chemical fume hood.

6-hydroxy-2,5,7,8-tetramethylchromane-2-carboxylic acid (Trolox) is an irritant and may be harmful by inhalation, ingestion, or skin absorption. Wear appropriate gloves and safety glasses. No not breathe the dust.

Hygromycin B is highly toxic and may be fatal if inhaled, ingested, or absorbed through the skin. Wear appropriate gloves and safety goggles and use only in a chemical fume hood. Do not breathe the dust.

Iodine crystals are volatile at room temperature and form a purple vapor that is intensely irritating to the eyes and mucous membranes. Always wear appropriate gloves and safety glasses when handling iodine and use in a chemical fume hood.

Isoflurane (Isofluorane) is an irritant and may be harmful by inhalation, ingestion, or skin absorption. Chronic exposure may be harmful. Wear appropriate gloves and safety glasses.

Isopentane (2-methylbutane) is extremely flammable. Keep away from heat, sparks, and open flame. It may be harmful by inhalation, ingestion, or skin absorption. Wear appropriate gloves and safety glasses.

Isopropanol is flammable and irritating. It may be harmful by inhalation, ingestion, or skin absorption. Wear appropriate gloves and safety glasses. Do not breathe the vapor. Keep away from heat, sparks, and open flame.

KCl, *see* **Potassium chloride**

Ketamine and **Ketamine hydrochloride** are toxic and may be harmful by inhalation, ingestion, or skin absorption. Wear appropriate gloves and safety glasses.

K-gluconate, *see* **Potassium gluconate**

KOH, *see* **Potassium hydroxide**

Laser radiation, both visible and invisible, can be seriously harmful to the eyes and skin and may generate airborne contaminants, depending on the class of laser used. High-power lasers cause permanent eye damage and can burn exposed skin, ignite flammable materials, and activate toxic chemicals that release hazardous by-products. Avoid eye or skin exposure to direct or scattered radiation. Do not stare at the laser and do not point the laser at anyone. Wear appropriate eye protection and use suitable shields that are designed to offer protection for the specific type of wavelength, mode of operation (continuous wave or pulsed), and power output (watts) of the laser being used. Avoid wearing jewelry or other objects that may reflect or scatter the beam. Some non-beam hazards include electrocution, fire, and asphyxiation. Entry to the area in which the laser is being used must be controlled and posted with warning signs that indicate when the laser is in use. Always follow suggested safety guidelines that accompany the equipment and contact your local safety office for further information.

Ion lasers present a hazard because of high-voltage high-current power supplies. Always follow manufacturer's suggested safety guidelines.

Ultraviolet lasers present a hazard because of invisible beam, high-energy radiation. Always use beam traps, scattered light shields, and fluorescent beamfinder cards.

Blue-green lasers present a hazard because of photothermal coagulation. Blue and green wavelengths are readily absorbed by blood hemoglobin.

Lead citrate is a potential cancer hazard and may cause possible death. There is a danger of cumulative effects. It is harmful by inhalation, ingestion, or skin absorption. Wear appropriate gloves and safety goggles and always use in a chemical fume hood. Do not breathe the dust.

Lead nitrate, $Pb(NO_3)_2$, and **all other lead salts** may be harmful by inhalation, ingestion, or skin absorption. Wear appropriate gloves and safety glasses and always use in a chemical fume hood.

Lidocaine is toxic if swallowed. It may be harmful by inhalation, ingestion, or skin absorption. Wear appropriate gloves and safety glasses. Do not breathe the dust.

Liquid nitrogen (LN_2) can cause severe damage because of extreme temperature. Handle frozen samples with extreme caution. Do not breathe the vapors. Seepage of liquid nitrogen into frozen vials can result in an exploding tube upon removal from liquid nitrogen. Use vials with O-rings when possible. Wear cryo-mitts and a face mask. Do not allow the liquid nitrogen to spill onto your clothes. Do not breathe the vapors.

LN_2, *see* **Liquid nitrogen**

Magnesium chloride, $MgCl_2$, may be harmful by inhalation, ingestion, or skin absorption. Wear appropriate gloves and safety glasses and use in a chemical fume hood.

Magnesium sulfate, $MgSO_4$, presents chronic health hazards and affects the central nervous system and the gastrointestinal tract. It may be harmful by inhalation, ingestion, or skin absorption. Wear appropriate gloves and safety glasses and use in a chemical fume hood.

Medetomidine may be harmful by inhalation, ingestion, or skin absorption. Wear appropriate gloves and safety goggles. Follow manufacturer's directions.

MeOH or **H_3COH,** *see* **Methanol**

Methanol, MeOH or **H_3COH,** is toxic, can cause blindness, and is highly flammable. It may be harmful by inhalation, ingestion, or skin absorption. Adequate ventilation is necessary to limit exposure to vapors. Avoid inhaling these vapors. Wear appropriate gloves and safety goggles and use only in a chemical fume hood.

2-methylbutane, *see* **Isopentane**

Methylene chloride, *see* **Dichloromethane**

$MgCl_2$, *see* **Magnesium chloride**

$MgSO_4$, *see* **Magnesium sulfate**

Microtome blades are extremely sharp! Use care when sectioning. If unfamiliar with the use of a microtome, have someone demonstrate its use.

MQAE, *see* ***N*-(ethoxycarbonylmethyl)-6-methoxyquinolinium bromide**

NaN_3, *see* **Sodium azide**

NaOH, *see* **Sodium hydroxide**

Nembutol, *see* **Pentobarbital**

***N*-(ethoxycarbonylmethyl)-6-methoxyquinolinium bromide** (MQAE) is potentially harmful by inhalation, ingestion, or skin absorption. Wear appropriate gloves and safety glasses.

NH_4Cl, *see* **Ammonium chloride**

$(NH_4)_2S_2O_8$, *see* **Ammonium persulfate**

Nigericin is highly toxic and may be fatal by inhalation, ingestion, or skin absorption. It is irritating to the eyes, skin, mucous membranes, and upper respiratory tract. Wear appropriate gloves and safety goggles and use in a chemical fume hood. Do not breathe the dust.

Nitrous oxide is potentially fatal if inhaled. It may be harmful by inhalation or skin absorption. Wear appropriate gloves and safety glasses. Do not expose containers to heat.

Norepinephrine is highly toxic and a reproductive hazard. It is fatal by inhalation, ingestion, or skin absorption. Wear appropriate gloves, safety goggles, and respiratory protection. Do not breathe the dust, mist, or vapors.

O_2, *see* **Oxygen**

OCT is composed of polyvinyl alcohol, polyethylene glycol, and dimethyl benzyl ammonium chloride. Follow the manufacturer's guidelines for handling OCT.

OsO_4, *see* **Osmium tetroxide**

Osmium tetroxide (osmic acid), OsO_4, is highly toxic if inhaled, ingested, or absorbed through the skin. Vapors can react with corneal tissues and cause blindness. There is a possible risk of irreversible effects. Wear appropriate gloves and safety goggles and always use in a chemical fume hood. Do not breathe the vapors.

Oxygen, O_2, presents a fire and explosion hazard. The gas is heavier than air and is a strong oxidant. Vapors may cause dizziness or asphyxiation without warning. Keep away from heat, sparks, and open flame.

Pancuronium bromide is toxic if swallowed. It is harmful by inhalation, ingestion, or skin absorption. Wear appropriate gloves and safety goggles. Do not breathe the dust.

Paraformaldehyde is highly toxic and may be fatal. It may be a carcinogen. It is readily absorbed through the skin and is extremely destructive to the skin, eyes, mucous membranes, and upper respiratory tract. Avoid breathing the dust or vapor. Wear appropriate gloves and safety glasses and use in a chemical fume hood. Keep away from heat, sparks, and open flame.

$Pb(NO_3)_2$, *see* **Lead nitrate**

Penicillin G (Procaine Salt) may cause allergic respiratory and skin reactions and may be harmful by inhalation, ingestion, or skin absorption. Wear appropriate gloves. Do not breathe the dust.

Pentobarbital (Nembutal) is toxic and may be harmful by inhalation, ingestion, or skin absorption. It can induce respiratory depression and sedation and presents a risk to the unborn child. Do not breathe the dust. Wear appropriate gloves and safety glasses and use in a chemical fume hood.

Permount, *see* **Toluene**

PFA, *see* **Paraformaldehyde**

Phenylephrine hydrochloride is toxic and may be harmful by inhalation, ingestion, or skin absorption. Wear appropriate gloves and safety glasses and use in a chemical fume hood. Do not breathe the dust, vapors, or mist.

Picric acid powder (Trinitrophenol) is caustic and potentially explosive if it is dissolved and then allowed to dry out. Care must be taken to ensure that stored solutions do not dry out. Handle all concentrated acids with great care. It is also highly toxic and may be harmful by inhalation, ingestion, or skin absorption. Wear appropriate gloves and safety goggles and use in a chemical fume hood.

Polyacrylamide is considered to be nontoxic, but it should be treated with care because it may contain small quantities of unpolymerized material (*see* **Acrylamide**).

Potassium chloride, KCl, may be harmful by inhalation, ingestion, or skin absorption. Wear appropriate gloves and safety glasses.

Potassium gluconate may be harmful by inhalation, ingestion, or skin absorption. Wear appropriate gloves and safety glasses.

Potassium hydroxide, KOH, and **KOH/methanol** are highly toxic and may be fatal if swallowed. They may be harmful by inhalation, ingestion, or skin absorption. Solutions are corrosive and can cause severe burns. They should be handled with great care. Wear appropriate gloves and safety goggles.

Propylene oxide is highly flammable, toxic, and may be carcinogenic. High concentrations are extremely destructive to the mucous membranes and upper respiratory tract. It may be harmful by inhalation, ingestion, or skin absorption. Wear appropriate gloves and safety glasses and use only in a chemical fume hood. Keep away from heat, sparks, and open flame.

Resins are flammable and are suspected carcinogens. The unpolymerized components and dusts may cause toxic reactions, including contact allergies with long-term exposure. Avoid breathing the vapors and dusts. Wear appropriate gloves and safety goggles and always use in a chemical fume hood. Sensitivity to these chemicals may develop with repeated contact. Keep away from heat, sparks, and open flame.

R101, *see* **Rhodamine 101**

Rhodamine 101 should be handled as a possible carcinogen. Wear safety glasses and gloves and work with effective ventilation.

SDS, *see* **Sodium dodecyl sulfate**

Sodium azide, NaN_3, is highly poisonous. It blocks the cytochrome electron transport system. Solutions containing sodium azide should be clearly marked. It may be harmful by inhalation, ingestion, or skin absorption. Wear appropriate gloves and safety goggles and handle it with great care. Sodium azide is an oxidizing agent and should not be stored near flammable chemicals.

Sodium cacodylate may be carcinogenic and contains arsenic. It is highly toxic and may be fatal by inhalation, ingestion, or skin absorption. It also may cause harm to the unborn child. Effects of contact or inhalation may be delayed. Do not breathe the dust. Wear appropriate gloves and safety goggles and use only in a chemical fume hood. *See also* **Cacodylate.**

Sodium deoxycholate is irritating to mucous membranes and the respiratory tract and may be harmful by inhalation, ingestion, or skin absorption. Wear appropriate gloves and safety glasses when handling the powder. Do not breathe the dust.

Sodium dodecyl sulfate (**SDS**) is toxic, an irritant, and poses a risk of severe damage to the eyes. It may be harmful by inhalation, ingestion, or skin absorption. Wear appropriate gloves and safety goggles. Do not breathe the dust.

Sodium hydroxide, NaOH, and **solutions containing NaOH** are highly toxic and caustic and should be handled with great care. Wear appropriate gloves and a face mask. All other concentrated bases should be handled in a similar manner.

Sodium pentobarbital is toxic and may be harmful by inhalation, ingestion, or skin absorption. It can induce respiratory depression and sedation and presents a risk to the unborn child. Do not breathe the dust. Wear appropriate gloves and safety glasses and use in a chemical fume hood.

Sodium pentobarbitone may be harmful by inhalation, ingestion, or skin absorption. Wear appropriate gloves and safety glasses.

Streptomycin is toxic and a suspected carcinogen and mutagen. It may cause allergic reactions. It may be harmful by inhalation, ingestion, or skin absorption. Wear appropriate gloves and safety glasses.

TAE buffer contains **Tris** and **Glacial acetic acid.**

Tamoxifen may damage fertility or the unborn child. It may be harmful by inhalation, ingestion, or skin absorption. Wear appropriate gloves and safety goggles. Do not breathe the dust.

TEA-CL, *see* **Tetraethylammonium chloride**

TEMED, *see* ***N,N,N′,N′*-tetramethylethylenediamine**

Tetraethylammonium chloride (**TEA-CL**) may cause allergic skin reaction and may be harmful by inhalation, ingestion, or skin absorption. It is irritating to the mucous membranes and upper respiratory tract. Do not breathe the dust. Wear appropriate gloves and safety glasses.

***N,N,N′,N′*-tetramethylethylenediamine** (**TEMED**) is highly caustic to the eyes and mucous membranes and may be harmful by inhalation, ingestion, or skin absorption. Wear appropriate gloves and tightly sealed safety goggles.

Toluene, $C_6H_5CH3_6$, vapors are irritating to the eyes, skin, mucous membranes, and upper respiratory tract. Toluene can exert harmful effects by inhalation, ingestion, or skin absorption. Do not inhale the vapors. Wear appropriate gloves and safety glasses and use in a chemical fume hood. Toluene is extremely flammable. Keep away from heat, sparks, and open flame.

2,2,2-tribromoethanol (**Avertin**) may be harmful by inhalation, ingestion, or skin absorption. The vapor is also irritating to the eyes, mucous membranes, and upper respiratory tract. Wear appropriate gloves and safety glasses.

Tributyltin is toxic and harmful by inhalation, ingestion, or skin absorption. It may cause serious damage to the eyes and is extremely destructive to the respiratory system, eyes, and skin. It may cause burns. Wear appropriate gloves and safety goggles. Do not breathe the dust.

Tricaine, *see* **Tricaine methanesulfonate**

Tricaine methanesulfonate is an irritant and may be harmful by inhalation, ingestion, or skin absorption. Wear appropriate gloves and safety glasses.

Trinitrophenol, *see* **Picric acid**

Tris is an irritant and may be harmful by inhalation, ingestion, or skin absorption. Wear appropriate gloves and safety glasses.

Triton X-100 causes severe eye irritation and burns. It may be harmful by inhalation, ingestion, or skin absorption. Wear appropriate gloves and safety goggles. Do not breathe the vapor.

Trolox, *see* **6-hydroxy-2,5,7,8-tetramethylchromane-2-carboxylic acid**

Trypsin may cause an allergic respiratory reaction. It may be harmful by inhalation, ingestion, or skin absorption. Do not breathe the dust. Wear appropriate gloves and safety goggles. Use with adequate ventilation.

Uranyl acetate is toxic if inhaled, ingested, or absorbed through the skin. Wear appropriate gloves and safety glasses and use in a chemical fume hood.

Urethane is a mutagen and suspected carcinogen. It is also highly toxic and is readily absorbed through the skin. It is harmful by inhalation, ingestion, or skin absorption. Wear appropriate gloves and safety glasses. Do not breathe the dust and use only in a chemical fume hood.

UV light and/or **UV radiation** is dangerous and can damage the retina. Never look at an unshielded UV light source with naked eyes. Examples of UV light sources that are common in the laboratory include hand-held lamps and transilluminators. View only through a filter or safety glasses that absorb harmful wavelengths. UV radiation is also mutagenic and carcinogenic. To minimize exposure, make sure that the UV light source is adequately shielded. Wear protective appropriate gloves when holding materials under the UV light source.

Vecuronium bromide is toxic if swallowed. It is harmful by inhalation, ingestion, or skin absorption. Wear appropriate gloves and safety glasses. Do not breathe the dust.

Xylazine and **Xylazine hydrochloride** are toxic and may be harmful by inhalation, ingestion, or skin absorption. Wear appropriate gloves and safety glasses.

Xylocaine may be harmful by inhalation, ingestion, or skin absorption. Wear appropriate gloves and safety glasses.

Index

Page references followed by f denote figures; page references followed by t denote tables.

B

E

F

G

H

I

J

K

L

N

T

U

V

W

X

Y

Z

www.ingramcontent.com/pod-product-compliance
Lightning Source LLC
LaVergne TN
LVHW080100011026
849160LV00013B/996
9780879699383